George Bebis   Richard Boyle
Bahram Parvin   Darko Koracin
Yoshinori Kuno   Junxia...
Renato Pajarola   Peter I...
André Hinkenjann   Mig...
Cláudio T. Silva   Daniel Coming (Eds.)

Advances in Visual Computing

5th International Symposium, ISVC 2009
Las Vegas, NV, USA, November 30 – December 2, 2009
Proceedings, Part I

 Springer

Volume Editors

George Bebis, E-mail: bebis@cse.unr.edu

Richard Boyle, E-mail: richard.boyle@nasa.gov

Bahram Parvin, E-mail: parvin@hpcrd.lbl.gov

Darko Koracin, E-mail: darko@dri.edu

Yoshinori Kuno, E-mail: kuno@cv.ics.saitama-u.ac.jp

Junxian Wang, E-mail: junxianw@microsoft.com

Renato Pajarola, E-mail: pajarola@ifi.uzh.ch

Peter Lindstrom, E-mail: pl@llnl.gov

André Hinkenjann, E-mail: andre.hinkenjann@fh-bonn-rhein-sieg.de

Miguel L. Encarnação, E-mail: lme@computer.org

Cláudio T. Silva, E-mail: csilva@cs.utah.edu

Daniel Coming, E-mail: daniel.coming@dri.edu

Library of Congress Control Number: 2009938706

CR Subject Classification (1998): I.3, H.5.2, I.4, I.5, I.2.10, J.3, F.2.2, I.3.5

LNCS Sublibrary: SL 6 – Image Processing, Computer Vision, Pattern Recognition, and Graphics

ISSN	0302-9743
ISBN-10	3-642-10330-8 Springer Berlin Heidelberg New York
ISBN-13	978-3-642-10330-8 Springer Berlin Heidelberg New York

springer.com

© Springer-Verlag Berlin Heidelberg 2009
Printed in Germany

Typesetting: Camera-ready by author, data conversion by Scientific Publishing Services, Chennai, India
Printed on acid-free paper   SPIN: 12793846   06/3180   5 4 3 2 1 0

Preface

It is with great pleasure that we present the proceedings of the 5th International Symposium on Visual Computing (ISVC 2009), which was held in Las Vegas, Nevada. ISVC offers a common umbrella for the four main areas of visual computing including vision, graphics, visualization, and virtual reality. The goal is to provide a forum for researchers, scientists, engineers, and practitioners throughout the world to present their latest research findings, ideas, developments, and applications in the broader area of visual computing.

This year, the program consisted of 16 oral sessions, one poster session, 7 special tracks, and 6 keynote presentations. Also, this year ISVC hosted the Third Semantic Robot Vision Challenge. The response to the call for papers was very good; we received over 320 submissions for the main symposium from which we accepted 97 papers for oral presentation and 63 papers for poster presentation. Special track papers were solicited separately through the Organizing and Program Committees of each track. A total of 40 papers were accepted for oral presentation and 15 papers for poster presentation in the special tracks.

All papers were reviewed with an emphasis on potential to contribute to the state of the art in the field. Selection criteria included accuracy and originality of ideas, clarity and significance of results, and presentation quality. The review process was quite rigorous, involving two to three independent blind reviews followed by several days of discussion. During the discussion period we tried to correct anomalies and errors that might have existed in the initial reviews. Despite our efforts, we recognize that some papers worthy of inclusion may have not been included in the program. We offer our sincere apologies to authors whose contributions might have been overlooked.

We wish to thank everybody who submitted their work to ISVC 2009 for review. It was because of their contributions that we succeeded in having a technical program of high scientific quality. In particular, we would like to thank the ISVC 2009 Area Chairs, the organizing institutions (UNR, DRI, LBNL, and NASA Ames), the industrial sponsors (Intel, DigitalPersona, Equinox, Ford, Hewlett Packard, Mitsubishi Electric Research Labs, iCore, Toyota, Delphi, General Electric, Microsoft MSDN, and Volt), the International Program Committee, the special track organizers and their Program Committees, the keynote speakers, the reviewers, and especially the authors that contributed their work to the symposium. In particular, we would like to thank *Mitsubishi Electric Research Labs*, *Volt*, *Microsoft MSDN*, and *iCore*, for kindly sponsoring several "best paper awards" this year.

We sincerely hope that ISVC 2009 offered opportunities for professional growth and that you enjoy these proceedings.

September 2009   ISVC09 Steering Committee and Area Chairs

Organization

ISVC 2009 Steering Committee

Bebis George	University of Nevada, Reno, USA
Boyle Richard	NASA Ames Research Center, USA
Parvin Bahram	Lawrence Berkeley National Laboratory, USA
Koracin Darko	Desert Research Institute, USA

ISVC 2009 Area Chairs

Computer Vision

Kuno Yoshinori	Saitama University, Japan
Wang Junxian	Microsoft, USA

Computer Graphics

Pajarola Renato	University of Zurich, Switzerland
Lindstrom Peter	Lawrence Livermore National Laboratory, USA

Virtual Reality

Hinkenjann Andre	Bonn-Rhein-Sieg University of Applied Sciences, Germany
Encarnacao L. Miguel	Humana Inc., USA

Visualization

Silva Claudio	University of Utah, USA
Coming Daniel	Desert Research Institute, USA

Publicity

Erol Ali	Ocali Information Technology, Turkey

Local Arrangements

Mauer Georg	University of Nevada, Las Vegas, USA

Special Tracks

Porikli Fatih	Mitsubishi Electric Research Labs, USA

ISVC 2009 Keynote Speakers

Perona Pietro	California Institute of Technology, USA
Kumar Rakesh (Teddy)	USarnoff Corporartion, USA
Davis Larry	University of Maryland, USA
Terzopoulos Demetri	University of California at Los Angeles, USA
Ju Tao	Washington University, USA
Navab Nassir	Technical University of Munich, Germany

ISVC 2009 International Program Committee

(Area 1) Computer Vision

Abidi Besma	University of Tennessee, USA
Aggarwal J.K.	University of Texas, Austin, USA
Agouris Peggy	George Mason University, USA
Argyros Antonis	University of Crete, Greece
Asari Vijayan	Old Dominion University, USA
Basu Anup	University of Alberta, Canada
Bekris Kostas	University of Nevada at Reno, USA
Belyaev Alexander	Max-Planck-Institut für Informatik, Germany
Bhatia Sanjiv	University of Missouri-St. Louis, USA
Bimber Oliver	Weimar University, Germany
Bioucas Jose	Instituto Superior Tecnico, Lisbon, Portugal
Birchfield Stan	Clemson University, USA
Bischof Horst	Graz University of Technology, Austria
Goh Wooi-Boon	Nanyang Technological University, Singapore
Bourbakis Nikolaos	Wright State University, USA
Brimkov Valentin	State University of New York, USA
Cavallaro Andrea	Queen Mary, University of London, UK
Chellappa Rama	University of Maryland, USA
Cheng Hui	Sarnoff Corporation, USA
Chung, Chi-Kit Ronald	The Chinese University of Hong Kong, Hong Kong
Darbon Jerome	UCLA, USA
Davis James W.	Ohio State University, USA
Debrunner Christian	Colorado School of Mines, USA
Duan Ye	University of Missouri-Columbia, USA
El-Gammal Ahmed	University of New Jersey, USA
Eng How Lung	Institute for Infocomm Research, Singapore
Erol Ali	Ocali Information Technology, Turkey
Fan Guoliang	Oklahoma State University, USA
Ferri Francesc	Universitat de Valencia, Spain
Filev, Dimitar	Ford Motor Company, USA
Foresti GianLuca	University of Udine, Italy

Fukui Kazuhiro	The University of Tsukuba, Japan
Galata Aphrodite	The University of Manchester, UK
Georgescu Bogdan	Siemens, USA
Gleason, Shaun	Oak Ridge National Laboratory, USA
Guerra-Filho Gutemberg	University of Texas Arlington, USA
Guevara, Angel Miguel	University of Porto, Portugal
Guerra-Filho Gutemberg	University of Texas Arlington, USA
Hammoud Riad	Delphi Corporation, USA
Harville Michael	Hewlett Packard Labs, USA
He Xiangjian	University of Technology, Australia
Heikkilä Janne	University of Oulu, Finland
Heyden Anders	Lund University, Sweden
Hou Zujun	Institute for Infocomm Research, Singapore
Imiya Atsushi	Chiba University, Japan
Kamberov George	Stevens Institute of Technology, USA
Kampel Martin	Vienna University of Technology, Austria
Kamberova Gerda	Hofstra University, USA
Kakadiaris Ioannis	University of Houston, USA
Kettebekov Sanzhar	Keane inc., USA
Kim Tae-Kyun	University of Cambridge, UK
Kimia Benjamin	Brown University, USA
Kisacanin Branislav	Texas Instruments, USA
Klette Reinhard	Auckland University, New Zeland
Kokkinos Iasonas	Ecole Centrale Paris, France
Kollias Stefanos	National Technical University of Athens, Greece
Komodakis Nikos	Ecole Centrale de Paris, France
Kozintsev, Igor	Intel, USA
Lee D.J.	Brigham Young University, USA
Li Fei-Fei	Princeton University, USA
Lee Seong-Whan	Korea University, Korea
Leung Valerie	Kingston University, UK
Li Wenjing	STI Medical Systems, USA
Liu Jianzhuang	The Chinese University of Hong Kong, Hong Kong
Little Jim	University of British Columbia, Canada
Ma Yunqian	Honyewell Labs, USA
Maeder Anthony	CSIRO ICT Centre, Australia
Makris Dimitrios	Kingston University, UK
Maltoni Davide	University of Bologna, Italy
Maybank Steve	Birkbeck College, UK
McGraw Tim	West Virginia University, USA
Medioni Gerard	University of Southern California, USA
Melenchón Javier	Universitat Oberta de Catalunya, Spain
Metaxas Dimitris	Rutgers University, USA
Miller Ron	Ford Motor Company, USA

Mirmehdi Majid	Bristol University, UK
Monekosso Dorothy	Kingston University, UK
Mueller Klaus	SUNY Stony Brook, USA
Mulligan Jeff	NASA Ames Research Center, USA
Murray Don	Point Grey Research, Canada
Nachtegael Mike	Ghent University, Belgium
Nait-Charif Hammadi	Bournemouth University, UK
Nefian Ara	NASA Ames Research Center, USA
Nicolescu Mircea	University of Nevada, Reno, USA
Nixon Mark	University of Southampton, UK
Nolle Lars	The Nottingham Trent University, UK
Ntalianis Klimis	National Technical University of Athens, Greece
Papadourakis George	Technological Education Institute, Greece
Papanikolopoulos Nikolaos	University of Minnesota, USA
Pati Peeta Basa	First Indian Corp., India
Patras Ioannis	Queen Mary University, London, UK
Petrakis Euripides	Technical University of Crete, Greece
Peyronnet Sylvain	LRDE/EPITA, France
Pinhanez Claudio	IBM Research, Brazil
Piccardi Massimo	University of Technology, Australia
Pietikäinen Matti	LRDE/University of Oulu, Finland
Porikli Fatih	Mitsubishi Electric Research Labs, USA
Prabhakar Salil	DigitalPersona Inc., USA
Prati Andrea	University of Modena and Reggio Emilia, Italy
Prokhorov Danil	Toyota Research Institute, USA
Qian Gang	Arizona State University, USA
Raftopoulos Kostas	National Technical University of Athens, Greece
Reed Michael	Blue Sky Studios, USA
Regazzoni Carlo	University of Genoa, Italy
Remagnino Paolo	Kingston University, UK
Ribeiro Eraldo	Florida Institute of Technology, USA
Robles-Kelly Antonio	National ICT Australia (NICTA), Australia
Ross Arun	West Virginia University, USA
Salgian Andrea	The College of New Jersey, USA
Samal Ashok	University of Nebraska, USA
Sato Yoichi	The University of Tokyo, Japan
Samir Tamer	Ingersoll Rand Security Technologies, USA
Sarti Augusto	DEI, Politecnico di Milano, Italy
Schaefer Gerald	Aston University, UK
Scalzo Fabien	University of Rochester, USA
Shah Mubarak	University of Central Florida, USA
Shi Pengcheng	The Hong Kong University of Science and Technology, Hong Kong

Shimada Nobutaka   Ritsumeikan University, Japan
Singh Meghna   University of Alberta, Canada
Singh Rahul   San Francisco State University, USA
Skurikhin Alexei   Los Alamos National Laboratory, USA
Souvenir, Richard   University of North Carolina - Charlotte, USA
Su Chung-Yen   National Taiwan Normal University, Taiwan
Sugihara Kokichi   University of Tokyo, Japan
Sun Zehang   eTreppid Technologies, USA
Syeda-Mahmood Tanveer   IBM Almaden, USA
Tan Kar Han   Hewlett Packard, USA
Tan Tieniu   Chinese Academy of Sciences, China
Tavares, Joao   Universidade do Porto, Portugal
Teoh Eam Khwang   Nanyang Technological University, Singapore
Thiran Jean-Philippe   Swiss Federal Institute of Technology
                           Lausanne (EPFL), Switzerland
Trucco Emanuele   University of Dundee, UK
Tsechpenakis Gabriel   University of Miami, USA
Tubaro Stefano   DEI, Politecnico di Milano, Italy
Uhl Andreas   Salzburg University, Austria
Velastin Sergio   Kingston University London, UK
Verri Alessandro   Università di Genova, Italy
Wang Song   University of South Carolina, USA
Wang Yunhong   Beihang University, China
Webster Michael   University of Nevada, Reno, USA
Wolff Larry   Equinox Corporation, USA
Wong Kenneth   The University of Hong Kong, Hong Kong
Xiang Tao   Queen Mary, University of London, UK
Xu Meihe   University of California at Los Angeles, USA
Yang Ruigang   University of Kentucky, USA
Yi Lijun   SUNY at Binghampton, USA
Yu Ting   GE Global Research, USA
Yuan Chunrong   University of Tübingen, Germany
Zhang Yan   Delphi Corporation, USA
Zhang Yongmian   eTreppid Technologies, USA

(Area 2) Computer Graphics

Abram Greg   IBM T.J. Watson Reseach Center, USA
Agu Emmanuel   Worcester Polytechnic Institute, USA
Andres Eric   Laboratory XLIM-SIC, University of Poitiers,
                           France
Artusi Alessandro   Warwick University, UK
Baciu George   Hong Kong PolyU, Hong Kong
Barneva Reneta   State University of New York, USA
Bartoli Vilanova Anna   Eindhoven University of Technology,
                           The Netherlands

Belyaev Alexander	Max-Planck-Institut für Informatik, Germany
Benes Bedrich	Purdue University, USA
Bilalis Nicholas	Technical University of Crete, Greece
Bimber Oliver	Weimar University, Germany
Bohez Erik	Asian Institute of Technology, Thailand
Bouatouch Kadi	University of Rennes I, IRISA, France
Brimkov Valentin	State University of New York, USA
Brown Ross	Queensland University of Technology, Australia
Callahan Steven	University of Utah, USA
Chen Min	University of Wales Swansea, UK
Cheng Irene	University of Alberta, Canada
Chiang Yi-Jen	Polytechnic Institute of New York University, USA
Choi Min	University of Colorado at Denver, USA
Comba Joao	Univ. Fed. do Rio Grande do Sul, Brazil
Cremer Jim	University of Iowa, USA
Damiand Guillaume	SIC Laboratory, France
Debattista Kurt	University of Warwick, UK
Deng Zhigang	University of Houston, USA
Dick Christian	Technical University of Munich, Germany
DiVerdi Stephen	Adobe, USA
Dingliana John	Trinity College, Ireland
El-Sana Jihad	Ben Gurion University of The Negev, Israel
Entezari Alireza	University of Florida, USA
Fiorio Christophe	Université Montpellier 2, LIRMM, France
Floriani Leila De	University of Genova, Italy
Gaither Kelly	University of Texas at Austin, USA
Gotz David	IBM, USA
Gooch Amy	University of Victoria, Canada
Gu David	State University of New York at Stony Brook, USA
Guerra-Filho Gutemberg	University of Texas Arlington, USA
Hadwiger Markus	VRVis Research Center, Austria
Haller Michael	Upper Austria University of Applied Sciences, Austria
Hamza-Lup Felix	Armstrong Atlantic State University, USA
Han JungHyun	Korea University, Korea
Hao Xuejun	Columbia University and NYSPI, USA
Hernandez Jose Tiberio	Universidad de los Andes, Colombia
Huang Zhiyong	Institute for Infocomm Research, Singapore
Joaquim Jorge	Instituto Superior Técnico, Portugal
Ju Tao	Washington University, USA

Sapidis Nickolas	University of the Aegean, Greece
Sarfraz Muhammad	Kuwait University, Kuwait
Scateni Riccardo	University of Cagliari, Italy
Schaefer Scott	Texas A&M University, USA
Sequin Carlo	University of California-Berkeley, USA
Shead Tinothy	Sandia National Laboratories, USA
Sorkine Olga	New York University, USA
Sourin Alexei	Nanyang Technological University, Singapore
Stamminger Marc	REVES/INRIA, France
Su Wen-Poh	Griffith University, Australia
Staadt Oliver	University of Rostock, Germany
Tan Kar Han	Hewlett Packard, USA
Teschner Matthias	University of Freiburg, Germany
Umlauf Georg	University of Kaiserslautern, Germany
Wald Ingo	University of Utah, USA
Weyrich Tim	University College London, UK
Wimmer Michael	Technical University of Vienna, Austria
Wylie Brian	Sandia National Laboratory, USA
Wyman Chris	University of Iowa, USA
Yang Ruigang	University of Kentucky, USA
Ye Duan	University of Missouri-Columbia, USA
Yi Beifang	Salem State College, USA
Yin Lijun	Binghamton University, USA
Yoo Terry	National Institutes of Health, USA
Yuan Xiaoru	Peking University, China
Zhang Eugene	Oregon State University, USA
Zordan Victor	University of California at Riverside, USA

(Area 3) Virtual Reality

Alcañiz Mariano	Technical University of Valencia, Spain
Arns Laura	Purdue University, USA
Behringer Reinhold	Leeds Metropolitan University, UK
Benes Bedrich	Purdue University, USA
Bilalis Nicholas	Technical University of Crete, Greece
Blach Roland	Fraunhofer Institute for Industrial Engineering, Germany
Blom Kristopher	University of Hamburg, Germany
Borst Christoph	University of Louisiana at Lafayette, USA
Brady Rachael	Duke University, USA
Brega Jose Remo Ferreira	Universidade Estadual Paulista, Brazil
Brown Ross	Queensland University of Technology, Australia
Bruce Thomas	The University of South Australia, Australia
Bues Matthias	Fraunhofer IAO in Stuttgart, Germany
Chen Jian	Brown University, USA

Cheng Irene	University of Alberta, Canada
Coquillart Sabine	INRIA, France
Craig Alan	NCSA University of Illinois at Urbana-Champaign, USA
Crawfis Roger	Ohio State University, USA
Cremer Jim	University of Iowa, USA
Figueroa Pablo	Universidad de los Andes, Colombia
Fox Jesse	Stanford University, USA
Friedman Doron	IDC, Israel
Froehlich Bernd	Weimar University, Germany
Gregory Michelle	Pacific Northwest National Lab, USA
Gupta Satyandra K.	University of Maryland, USA
Hachet Martin	INRIA, France
Haller Michael	FH Hagenberg, Austria
Hamza-Lup Felix	Armstrong Atlantic State University, USA
Harders Matthias	ETH Zürich, Switzerland
Hollerer Tobias	University of California at Santa Barbara, USA
Julier Simon J.	University College London, UK
Klinger Evelyne	Arts et Metiers ParisTech, France
Klinker Gudrun	Technische Universität München, Germany
Klosowski James	IBM T.J. Watson Research Center, USA
Kozintsev, Igor	Intel, USA
Kuhlen Torsten	RWTH Aachen University, Germany
Liere Robert van	CWI, The Netherlands
Lok Benjamin	University of Florida, USA
Luo Gang	Harvard Medical School, USA
Majumder Aditi	University of California, Irvine, USA
Malzbender Tom	Hewlett Packard Labs, USA
Mantler Stephan	Technical University of Vienna, Austria
Meyer Joerg	University of California, Irvine, USA
Molineros Jose	Teledyne Scientific and Imaging, USA
Moorhead Robert	Mississippi State University, USA
Muller Stefan	University of Koblenz, Germany
Paelke Volker	Leibniz Universität Hannover, Germany
Papka Michael	Argonne National Laboratory, USA
Peli Eli	Harvard University, USA
Pettifer Steve	The University of Manchester, UK
Pugmire Dave	Los Alamos National Lab, USA
Qian Gang	Arizona State University, USA
Raffin Bruno	INRIA, France
Redon Stephane	INRIA, France
Reiners Dirk	University of Louisiana, USA
Richir Simon	Arts et Metiers ParisTech, France
Rodello Ildeberto	University of Sao Paulo, Brazil

Santhanam Anand   MD Anderson Cancer Center Orlando, USA
Sapidis Nickolas   University of the Aegean, Greece
Schmalstieg Dieter   Graz University of Technology, Austria
Schulze, Jurgen   University of California - San Diego, USA
Slavik Pavel   Czech Technical University in Prague,
                            Czech Republic
Sourin Alexei   Nanyang Technological University, Singapore
Stamminger Marc   REVES/INRIA, France
Srikanth Manohar   Indian Institute of Science, India
Staadt Oliver   University of Rostock, Germany
Stefani Oliver   COAT-Basel, Switzerland
Thalmann Daniel   EPFL VRlab, Switzerland
Varsamidis Thomas   Bangor University, UK
Vercher Jean-Louis   Université de la Méditerranée, France
Wald Ingo   University of Utah, USA
Yu Ka Chun   Denver Museum of Nature and Science, USA
Yuan Chunrong   University of Tübingen, Germany
Zachmann Gabriel   Clausthal University, Germany
Zara Jiri   Czech Technical University in Prague, Czech
Zyda Michael   University of Southern California, USA

(Area 4) Visualization

Andrienko Gennady   Fraunhofer Institute IAIS, Germany
Apperley Mark   University of Waikato, New Zealand
Avila Lisa   Kitware, USA
Balázs Csébfalvi   Budapest University of Technology and
                            Economics, Hungary
Bartoli Anna Vilanova   Eindhoven University of Technology,
                            The Netherlands
Brady Rachael   Duke University, USA
Benes Bedrich   Purdue University, USA
Bilalis Nicholas   Technical University of Crete, Greece
Bonneau Georges-Pierre   Grenoble University, France
Brown Ross   Queensland University of Technology,
                            Australia
Bühler Katja   VRVIS, Austria
Callahan Steven   University of Utah, USA
Chen Jian   Brown University, USA
Chen Min   University of Wales Swansea, UK
Cheng Irene   University of Alberta, Canada
Chiang Yi-Jen   Polytechnic Institute of New York University,
                            USA
Chourasia Amit   University of California - San Diego, USA
Dana Kristin   Rutgers University, USA
Dick Christian   Technical University of Munich, Germany

DiVerdi Stephen	Adobe, USA
Doleisch Helmut	VRVis Research Center, Austria
Duan Ye	University of Missouri-Columbia, USA
Dwyer Tim	Monash University, Australia
Ebert David	Purdue University, USA
Entezari Alireza	University of Florida, USA
Ertl Thomas	University of Stuttgart, Germany
Floriani Leila De	University of Maryland, USA
Fujishiro Issei	Keio University, Japan
Gotz David	IBM, USA
Grinstein Georges	University of Massachusetts Lowell, USA
Goebel Randy	University of Alberta, Canada
Gregory Michelle	Pacific Northwest National Lab, USA
Hadwiger Helmut Markus	VRVis Research Center, Austria
Hagen Hans	Technical University of Kaiserslautern, Germany
Hamza-Lup Felix	Armstrong Atlantic State University, USA
Heer Jeffrey	Armstrong University of California at Berkeley, USA
Hege Hans-Christian	Zuse Institute Berlin, Germany
Hochheiser Harry	Towson University, USA
Hollerer Tobias	University of California at Santa Barbara, USA
Hong Lichan	Palo Alto Research Center, USA
Hotz Ingrid	Zuse Institute Berlin, Germany
Joshi Alark	Yale University, USA
Julier Simon J.	University College London, UK
Kao David	NASA Ames Research Center, USA
Kohlhammer Jörn	Fraunhofer Institut, Germany
Kosara Robert	University of North Carolina at Charlotte, USA
Laramee Robert	Swansea University, UK
Lee Chang Ha	Chung-Ang University, Korea
Lewis Bob	Washington State University, USA
Liere Robert van	CWI, The Netherlands
Lim Ik Soo	Bangor University, UK
Linsen Lars	Jacobs University, Germany
Liu Zhanping	Kitware, Inc., USA
Ma Kwan-Liu	University of California-Davis, USA
Maeder Anthony	CSIRO ICT Centre, Australia
Majumder Aditi	University of California, Irvine, USA
Malpica Jose	Alcala University, Spain
Masutani Yoshitaka	The University of Tokyo Hospital, Japan
Matkovic Kresimir	VRVis Forschungs-GmbH, Austria
McGraw Tim	West Virginia University, USA

Melançon Guy	CNRS UMR 5800 LaBRI and INRIA
	Bordeaux Sud-Ouest, France
Meyer Joerg	University of California, Irvine, USA
Miksch Silvia	Vienna University of Technology, Austria
Monroe Laura	Los Alamos National Labs, USA
Moorhead Robert	Mississippi State University, USA
Morie Jacki	University of Southern California, USA
Mueller Klaus	SUNY Stony Brook, USA
Museth Ken	Linköping University, Sweden
Paelke Volker	Leibniz Universität Hannover, Germany
Papka Michael	Argonne National Laboratory, USA
Pettifer Steve	The University of Manchester, UK
Pugmire Dave	Los Alamos National Lab, USA
Rabin Robert	University of Wisconsin at Madison, USA
Raffin Bruno	Inria, France
Razdan Anshuman	Arizona State University, USA
Rhyne Theresa-Marie	North Carolina State University, USA
Santhanam Anand	MD Anderson Cancer Center Orlando, USA
Scheuermann Gerik	University of Leipzig, Germany
Shead Tinothy	Sandia National Laboratories, USA
Shen Han-Wei	Ohio State University, USA
Sips Mike	Stanford University, USA
Slavik Pavel	Czech Technical University in Prague,
	Czech Republic
Sourin Alexei	Nanyang Technological University, Singapore
Theisel Holger	University of Magdeburg, Germany
Thiele Olaf	University of Mannheim, Germany
Toledo de Rodrigo	Petrobras PUC-RIO, Brazil
Tricoche Xavier	Purdue University, USA
Umlauf Georg	University of Kaiserslautern, Germany
Viegas Fernanda	IBM, USA
Viola Ivan	University of Bergen, Norway
Wald Ingo	University of Utah, USA
Wan Ming	Boeing Phantom Works, USA
Weinkauf Tino	Courant Institute, New York University, USA
Weiskopf Daniel	University of Stuttgart, Germany
Wischgoll Thomas	Wright State University, USA
Wylie Brian	Sandia National Laboratory, USA
Yeasin Mohammed	Memphis University, USA
Yuan Xiaoru	Peking University, China
Zachmann Gabriel	Clausthal University, Germany
Zhang Eugene	Oregon State University, USA
Zhukov Leonid	Caltech, USA

ISVC 2009 Special Tracks

1. 3D Mapping, Modeling and Surface Reconstruction

Organizers

Nefian Ara	Carnegie Mellon University/NASA Ames Research Center, USA
Broxton Michael	Carnegie Mellon University/NASA Ames Research Center, USA
Huertas Andres	NASA Jet Propulsion Lab, USA

Program Committee

Hancher Matthew	NASA Ames Research Center, USA
Edwards Laurence	NASA Ames Research Center, USA
Bradski Garry	Willow Garage, USA
Zakhor Avideh	University of California at Berkeley, USA
Cavallaro Andrea	University Queen Mary, London, UK
Bouguet Jean-Yves	Google, USA

2. Object Recognition

Organizers

Andrea Salgian	The College of New Jersey, USA
Fabien Scalzo	University of Rochester, USA

Program Committee

Bergevin Robert	University of Laval, Canada
Leibe Bastina	ETH Zurich, Switzerland
Lepetit Vincet	EPFL, Switzerland
Matei Bogdan	Sarnoff Corporation, USA
Maree Raphael	Universite de Liege, Belgium
Nelson Randal	University of Rochester, USA
Qi Guo-Jun	University of Science and Technology of China, China
Sebe Nicu	University of Amsterdam, The Netherlands
Tuytelaars Tinne	Katholieke Universiteit Leuven, Belgium
Vedaldi Andrea	Oxford University, UK
Vidal-Naquet Michel	RIKEN Brain Science Institute, Japan

3. Deformable Models: Theory and Applications

Organizers

Terzopoulos Demetri	University of California, Los Angeles, USA
Tsechpenakis Gavriil	University of Miami, USA
Huang Xiaolei	Lehigh University, USA

Discussion Panel

Metaxas Dimitris (Chair)   Rutgers University, USA

Program Committee

Angelini Elsa	Ecole Nationale Supérieure de Télécommunications, France
Breen David	Drexel University, USA
Chen Ting	Rutgers University, USA
Chen Yunmei	University of Florida, USA
Delingette Herve	INRIA, France
Delmas Patrice	University of Auckland, New Zealand
El-Baz Ayman	University of Louisville, USA
Farag Aly	University of Louisville, USA
Kimia Benjamin	Brown University, USA
Kambhamettu Chandra	University of Delaware, USA
Magnenat-Thalmann Nadia	University of Geneva, Switzerland
McInerney Tim	Ryerson University, Canada
Metaxas Dimitris	Rutgers University, USA
Palaniappan Kammappan	University of Missouri, USA
Paragios Nikos	Ecole Centrale de Paris, France
Qin Hong	Stony Brook University, USA
Salzmann Mathieu	UC Berkeley, USA
Sifakis Eftychios	University of California at Los Angeles, USA
Skrinjar Oskar	Georgia Tech, USA
Szekely Gabor	ETH Zurich, Switzerland
Teran Joseph	University of California at Los Angeles, USA
Thalmann Dabiel	EPFL, Switzerland

4. Visualization-Enhanced Data Analysis for Health Applications

Organizers

Cheng Irene	University of Alberta, Canada
Maeder Anthony	University of Western Sydney, Australia

Program Committee

Bischof Walter	University of Alberta, Canada
Boulanger Pierre	University of Alberta, Canada
Brown Ross	Queensland University of Technology, Australia
Dowling Jason	CSIRO, Australia
Figueroa Pablo	Universidad de los Andes, Colombia
Liyanage Liwan	University of Western Sydney, Australia

Malzbender Tom	HP Labs, USA
Mandal Mrinal	University of Alberta, Canada
Miller Steven	University of British Columbia, Canada
Nguyen Quang Vinh	University of Western Sydney, Australia
Shi Hao	Victoria University, Australia
Shi Jiambo	University of Pennsylvania, USA
Silva Caludio	University of Utah, USA
Simoff Simeon	University of Western Sydney, Australia
Yin Lijun	University of Utah, USA
Zabulis Xenophon	Institute of Computer Science-FORTH, Greece
Zanuttigh Pietro	University of Padova, Italy

5. Computational Bioimaging

Organizers

Tavares João Manuel R.S.	University of Porto, Portugal
Jorge Renato Natal	University of Porto, Portugal
Cunha Alexandre	Caltech, USA

Program Committee

Santis De Alberto	Università degli Studi di Roma "La Sapienza", Italy
Falcao Alexandre Xavier	University of Campinas, Brazil
Reis Ana Mafalda	Instituto de Ciências Biomédicas Abel Salazar, Portugal
Barrutia Arrate Muñoz	University of Navarra, Spain
Calco Begoña	University of Zaragoza, Spain
Kotropoulos Constantine	Aristotle University of Thessaloniki, Greece
Iacoviello Daniela	Università degli Studi di Roma "La Sapienza", Italy
Rodrigues Denilson Laudares	PUC Minas, Brazil
Shen Dinggang	University of Pennsylvania, USA
Ziou Djemel	University of Sherbrooke, Canada
Pires Eduardo Borges	Instituto Superior Técnico, Portugal
Sgallari Fiorella	University of Bologna, Italy
Perales Francisco	Balearic Islands University, Spain
Rohde Gustavo	Carnegie Mellon University, USA
Peng Hanchuan	Howard Hughes Medical Institute, USA
Rodrigues Helder	Instituto Superior Técnico, Portugal
Pistori Hemerson	Dom Bosco Catholic University, Brazil
Zhou Huiyu	Brunel University, UK

Yanovsky Igor	Jet Propulsion Laboratory, USA
Corso Jason	SUNY at Buffalo, USA
Maldonado Javier Melenchón	Open University of Catalonia, Spain
Barbosa Jorge M.G.	University of Porto, Portugal
Marques Jorge	Instituto Superior Técnico, Portugal
Aznar Jose M. García	University of Zaragoza, Spain
Tohka Jussi	Tampere University of Technology, Finland
Vese Luminita	University of California at Los Angeles, USA
Reis Luís Paulo	University of Porto, Portugal
El-Sakka Mahmoud	The University of Western Ontario London, Canada
Hidalgo Manuel González	Balearic Islands University, Spain
Kunkel Maria Elizete	Universität Ulm, Germany
Gurcan Metin N.	Ohio State University, USA
Liebling Michael	University of California at Santa Barbara, USA
Dubois Patrick	Institut de Technologie Médicale, France
Jorge Renato M.N.	University of Porto, Portugal
Barneva Reneta	State University of New York, USA
Bellotti Roberto	University of Bari, Italy
Tangaro Sabina	University of Bari, Italy
Newsam Shawn	University of California at Merced, USA
Silva Susana Branco	University of Lisbon, Portugal
Pataky Todd	University of Liverpool, UK
Brimkov Valentin	State University of New York, USA
Zhan Yongjie	Carnegie Mellon University, USA

6. Visual Computing for Robotics

Organizers

Chausse Frederic	Clermont Universite, France

Program Committee

Didier Aubert Didier	LIVIC, France
Thierry Chateu Thierry	Clermont Université, France
Chapuis Roland	Clermont Université, France
Hautiere Nicolas	LCPC/LEPSIS, France
Royer Eric	Clermont Université, France
Bekris Kostas	University of Nevada, Reno, USA

7. Optimization for Vision, Graphics and Medical Imaging: Theory and Applications

Organizers

Komodakis Nikos   University of Crete, Greece
Langs Georg   University of Vienna, Austria

Program Committee

Paragios Nikos   Ecole Centrale de Paris/INRIA Saclay
                         Ile-de-France, France
Bischof Horst   Graz University of Technology, Austria
Cremers Daniel   University of Bonn, Germany
Grady Leo   Siemens Corporate Research, USA
Navab Nassir   Technical University of Munich, Germany
Samaras Dimitris   Stony Brook University, USA
Lempitsky Victor   Microsoft Research Cambridge, UK
Tziritas Georgios   University of Crete, Greece
Pock Thomas   Graz University of Technology, Austria
Micusik Branislav   Austrian Research Centers GmbH - ARC,
                         Austria
Glocker Ben   Technical University of Munich, Germany

8. Semantic Robot Vision Challenge

Organizers

Rybski Paul E.   Carnegie Mellon University, USA
DeMenthon Daniel   Johns Hopkins University, USA
Fermuller Cornelia   University of Maryland, USA
Fazli Pooyan   University of British Columbia, Canada
Mishra Ajay   National University of Singapore, Singapore
Lopes Luis   Universidade de Aveiro, Portugal
Roehrbein Florian   Universität Bremen, Germany
Gustafson David   Kansas State University, USA
Nicolescu Mircea   University of Nevada at Reno, USA

Additional Reviewers

Vo Huy   University of Utah, USA
Streib Kevin   Ohio State University, USA
Sankaranarayanan
  Karthik   Ohio State University, USA
Guerrero Paul   Vienna University of Technology, Austria
Brimkov Borris   University at Buffalo, USA
Kensler Andrew   University of Utah, USA

Organizing Institutions and Sponsors

Table of Contents – Part I

Visualization I

ST: Visual Computing for Robotics

Feature Extraction and Matching

Medical Imaging

Motion

Virtual Reality I

ST: Computational Bioimaging

Computer Graphics II

ST: 3D Mapping, Modeling and Surface Reconstruction

Face Processing

Reconstruction I

ST: Deformable Models: Theory and Applications

ST: Visualization Enhanced Data Analysis for Health Applications

Virtual Reality II

ST: Optimization for Vision, Graphics and Medical Imaging: Theory and Applications

Table of Contents – Part II

Detection and Tracking

Reconstruction II

Applications

Video Analysis and Event Recognition

Poster Session

Which Shape Representation Is the Best for Real-Time Hand Interface System?

Serkan Genç[1] and Volkan Atalay[2]

[1] Computer Technology and Information Systems, Bilkent University, Ankara, Turkey
sgenc@bilkent.edu.tr
[2] Department of Computer Engineering, METU, Ankara, Turkey
volkan@ceng.metu.edu.tr

Abstract. Hand is a very convenient interface for immersive human-computer interaction. Users can give commands to a computer by hand signs (hand postures, hand shapes) or hand movements (hand gestures). Such a hand interface system can be realized by using cameras as input devices, and software for analyzing the images. In this hand interface system, commands are recognized by analyzing the hand shapes and its trajectories in the images. Therefore, success of the recognition of hand shape is vital and depends on the discriminative power of the hand shape representation. There are many shape representation techniques in the literature. However, none of them are working properly for all shapes. While a representation leads to a good result for a set of shapes, it may fail in another one. Therefore, our aim is to find the most appropriate shape representation technique for hand shapes to be used in hand interfaces. Our candidate representations are Fourier Descriptors, Hu Moment Invariant, Shape Descriptors and Orientation Histogram. Based on widely-used hand shapes for an interface, we compared the representations in terms of their discriminative power and speed.

Keywords: Shape representation, hand recognition, hand interface.

1 Introduction

Hands play very important role in inter-human communication and we use our hands for pointing, giving commands and expressing our feelings. Therefore, it is reasonable to mimic this interaction in human-computer interaction. In this way, we can make computer usage natural and easier. Although several electro-mechanical and magnetic sensing devices such as gloves are now available to use with hands in human computer interaction, they are expensive and uncomfortable to wear for long times, and require considerable setup process. Due to these disadvantages, vision based systems are proposed to provide immersive human computer interaction. Vision systems are basically composed of one or more cameras as input devices, and processing capabilities for captured images. Such a system is so natural that a user may not be aware of interacting with a computer system. However, there is no unique vision based hand interface system that can be used in all types of applications. There are several reasons for this. First, there is no computer vision algorithm which reconstructs a hand from an image. This is because a hand has a very complex model with 27 degrees of

G. Bebis et al. (Eds.): ISVC 2009, Part I, LNCS 5875, pp. 1–11, 2009.

freedom (DOF) [1]. Modeling the kinematic structure and dynamics are still open problems, and need further research [2]. Second, even if there was an algorithm which finds all 27 parameters of a hand, it would be very complex, and it may not be appropriate for real-time applications. Third, it is unnecessary to use complex algorithms for a simple hand interface, since it consumes considerable or even all available computing power of the system. Appearance-based techniques that analyze the image without using any 3-dimensional hand model work faster than 3-dimensional model based techniques and they are more appropriate for real time hand interface applications [2],[3].

This study mainly focuses on appearance-based methods for static hand posture systems. However, the shape representations presented in this study can be incorporated as feature vectors to standard spatio-temporal pattern matching methods such as Hidden Markov Models (HMM) [18] or Dynamic Time Warping (DTW) [22] to recognize dynamic hand movements or hand gestures.

Our initial motivation was to develop an application which was controlled by hand. In this application, the setting was composed of a camera located on top of the desk, and the user gave commands by hand. Although capturing from above limits possible hand shapes, this is very frequent setup for hand interface systems. For example, the system described by Quek *et al.* controls the behavior of a mouse by a hand on keyboard [4]. ITouch uses hand gestures appearing on the monitor similar to touch screen monitors [5]. Licsar and Sziranyi present another example that enables a user to manage presentation slides by hands [6]. Freeman *et al.* let the users play games by their hands [7]. Nevertheless, there is no study comparing techniques employed in such a setup. The aim of our paper is to assess various representations for hand shape recognition system having a setup where a camera is located above the desk and is looking downward to acquire the upper surface of the hand.

Usually, a hand interface system is composed of several stages: image acquisition, segmentation, representation, and recognition. Among those stages, segmentation is the main bottleneck in developing general usage HCI applications. Although there are many algorithms attempting to solve segmentation such as skin color modeling [24], Gauss Mixture Model for Background Subtraction [23] and Neural Network methods, all of them impose constraints on working environments such as illumination condition, stationary camera, static background, uniform background, etc. When the restriction is slightly violated, a clean segmentation is not possible, and the subsequent stages fail. Remedy to this problem is to use more complex representations or algorithms to compensate the deficiency of segmentation. However, there is also a limit on compensation. As a result, for the time being, even using the state-of-the-art segmentation algorithms for color images, a robust HCI application is not possible. However, there is a good news recently on segmentation with a new hardware technology, which is called Time-of-Flight (ToF) depth camera [19]. It captures depth information for each pixel in the scene and the basic principle is to measure the distances for each pixel using the round-trip delay of light, which is similar to radar systems. This camera is not affected by illumination changes at all. With this technology, clean segmentation for HCI applications is possible using depth keying technique [20]. Microsoft's Natal Project uses this technology to solve segmentation problem and enable users to interact with their body to control games [21]. Therefore, shape representations from clean segmentation can be used with this technology, and we used clean segmented hand objects in our study. We believe that future HCI applications will be using ToF camera to solve segmentation problem.

The next stage after segmentation is representation, hand pixels are transformed into a meaningful representation which is useful at recognition stage. Representation is very important for recognition since unsuccessful representation gives unsatisfactory results even with state-of-the-art classifiers. On the other hand, good representation always results in an acceptable result with an average classifier [8].

This paper compares four representation techniques which can be used in shape recognition systems. In the selection of these representations, the following criterions are used: discriminative power, speed and invariance to scale, translation and rotation. Selected representations are Fourier descriptors, Hu moment invariants, shape descriptors, and orientation histogram. In Section 2, these selected representations are explained in detail. To assess the representations, bootstrapping is used to measure the quality of representations while decision tree is used as the classifier. Section 3 gives all the details concerning tests. In Section 4, we comment on the results in terms of discriminative power and real-time issues. Finally, we conclude the paper.

2   Shape Representations

Recognizing commands given by hand depends on the success of shape recognition, and thus, it is closely related to shape representation. Therefore, it is vital to select the appropriate shape representation for hand interfaces. Unfortunately, there is no unique representation that works for all sets of shapes. This is the motivation that leads us to compare and assess popular hand shape representations. Techniques for shape representation can be mainly categorized as contour-based and region based [9]. Contour based techniques use the boundary of the shape while region based techniques employ all the pixels belonging to a shape. Each category is divided into two subcategories: structural and global. Structural methods describe the shape as a combination of segments called primitives in a structural way such as a tree or graph. However, global methods consider the shape as a whole. Although there are many shape representation techniques in these categories, only some of them are eligible to be used in hand interfaces. We took into account certain criterions for selection. The first criterion is the computational complexity of finding the similarity of two shapes, i.e., matching. Contour based and region based structural methods such as polygon approximation, curve decomposition, convex hull decomposition, medial axis require graph matching algorithm for similarity, thus they are computationally complex [9]. Zhang *et al.* show that Fourier Descriptor (FD) which is a contour based global method performs better than Curvature Scale Space (CSS) which is also a contour based global method, in terms of matching and calculating representations [10]. Another contour-based global method, Wavelet Descriptor requires shift matching for similarity, which is costly compared to FD. Therefore, we have selected FD as a candidate. Freeman *et al.* use Orientation Histogram which is a region-based global method, for several applications controlled by hand [7],[11]. Since there is no comparison of Orientation Histogram with others, and authors promote it in terms of both speed and recognition performance, we have also chosen it. Peura *et al.* claim that practical application does not need too sophisticated methods, and they use the combination of simple shape descriptors for shape recognition [12]. Since Shape Descriptors are semantically simple, fast and powerful, we have chosen Shape Descriptors (SD) as well. Each Shape Descriptor is either a contour or a region based global method. Flusser asserts that moment-invariants such as Hu Moment Invariants are important [13] since they are fast to compute, easy to implement and invariant to rotation, scale and translation. Therefore, Hu Moment Invariants are also selected for hand interface.

In conclusion, we have opted for four shape representation techniques; Shape Descriptors, Fourier Descriptors, Hu Moment Invariants and Orientation Histogram to assess their discrimination power and speed on a hand shape data set. In the rest of this section, we describe each selected method.

2.1 Shape Descriptors

Shape Descriptor is a quantity which describes a property of a shape. Area, perimeter, compactness, rectangularity are examples of shape descriptors. Although a single descriptor may not be powerful enough for discrimination, a set of them can be used for shape representation [12]. In this study, five shape descriptors; compactness, ratio of principal axes, elliptical ratio, convexity and rectangularity are chosen because they are invariant to scale, translation and rotation, and easy to compute. Also they are reported as successful descriptors in [9],[12].

2.1.1 Compactness
A common compactness measure, called the circularity ratio, is the ratio of the area of the shape to the area of a circle (the most compact shape) having the same perimeter. Assuming P is the perimeter and A is the area of a hand shape, circularity ratio is defined as follows.

$$\Psi = \frac{4 \pi A}{P^2}$$

For a circle, circularity ratio is 1, for a square, it is $\frac{\pi}{4}$, and for an infinite long and narrow shape, it is 0.

2.1.2 Ratio of Principal Axes
Principal axes of a 2-dimensional object are two axes that cross each other orthogonally in the centroid of the object and the cross-correlation of boundary points on the object in this coordinate system is zero [12]. Ratio of principal axes, ρ provides the information about the elongation of a shape. For a hand shape boundary B which is an ordered list of boundary points, ρ can be determined by calculating covariance matrix Σ, of a boundary B, and then finding the ratio of Σ's eigenvalues; λ_1 and λ_2. Eigenvectors e_1, e_2 of Σ are orthogonal and cross-correlation of points in B with e_1 and e_2 is zero since Σ is a diagonal matrix. The values of λ equal to the length of the principal axes. However, to find the ratio of λ_1 and λ_2 or principal axes, there is no need to explicitly compute eigenvalues, and ρ can be calculated as follows [12]:

$$\Sigma = \begin{vmatrix} \Sigma_{xx} & \Sigma_{xy} \\ \Sigma_{yx} & \Sigma_{yy} \end{vmatrix}, \text{ where } \Sigma_{ij} \text{ represents covariance of } i \text{ and } j.$$

$$\rho = \frac{\Sigma_{yy} + \Sigma_{xx} - \sqrt{(\Sigma_{yy} + \Sigma_{xx})^2 - 4(\Sigma_{yy}\Sigma_{xx} - \Sigma_{xy}{}^2)}}{\Sigma_{yy} + \Sigma_{xx} + \sqrt{(\Sigma_{yy} + \Sigma_{xx})^2 - 4(\Sigma_{yy}\Sigma_{xx} - \Sigma_{xy}{}^2)}}$$

2.1.3 Elliptical Ratio

Elliptical Ratio, ε is the ratio of minor axis, b to major axis, a of an ellipse which is fitted to boundary points.

$$\varepsilon = \frac{b}{a}$$

Ellipse fitting is performed using a least-square fitness function. In the implementation, ellipse fitting algorithm proposed by Fitzgibbon *et al.* [14] and provided by OpenCV is used [17].

2.1.4 Convexity

Convexity is the ratio of perimeter of the convex-hull, $\pi_{convexHull}$ to the perimeter of the shape boundary, π_{Shape}, where convex hull is the minimum convex polygon covering the shape.

$$\varsigma = \frac{\pi_{convexHull}}{\pi_{Shape}}$$

2.1.5 Rectangularity

Rectangularity measures the similarity of a shape to a rectangle. This can be calculated by the ratio of the area of the hand shape, $A_{handShape}$ to the minimum bounding box of hand shape, A_{MBB}. Minimum bounding box (MBB) is the smallest rectangle covering the shape.

$$\Gamma = \frac{A_{handShape}}{A_{MBB}}$$

For a rectangle, rectangularity is 1; for a circle, it is $\frac{\pi}{4}$.

2.2 Fourier Descriptors

Fourier Descriptors (FDs) represent the spectral properties of a shape boundary. Low frequency components of FDs correspond to overall shape properties; while high frequency components describe the fine details of the shape.

FDs are calculated using Fourier Transform of shape boundary points, (x_k, y_k), $k=0,...,\ N-1$ where N is the number of points in the boundary. Boundary can be represented by an ordered list of complex coordinates called complex coordinate signature, as $p_k = x_k + i\ y_k$, $k=0,...,N-1$ or a boundary can be represented by an ordered list of distances, r_k, of each boundary point (x_k, y_k) to centroid of the shape (x_c, y_c) called centroid distance signature.

Zhang and Lu compared the effect of four 1-dimensional boundary signatures for FDs; these signatures are complex coordinates, centroid distances, curvature signature and cumulative angular function [15]. The authors concluded that FDs derived from centroid distance signature is significantly better than the others. Therefore, we use centroid signature of the boundary. To calculate FDs of a boundary, the following steps are pursued.

1. Calculate the centroid of the hand shape boundary

$$x_c = \frac{1}{N}\sum_{k=0}^{N-1} x_k \ \ and \ y_c = \frac{1}{N}\sum_{k=0}^{N-1} y_k$$

2. Convert each boundary point (x_k,y_k) to centroid distance r_k,

$$r_k = [(x_k - x_c)^2 + (y_k - y_c)^2], k = 0, ..., N-1$$

3. Use Fourier Transform to obtain FDs.

$$FD_f = \frac{1}{N}\sum_{k=0}^{N-1} r_k \cdot e^{-j2\pi fk/N}$$

$FD_f, f=0,...,N-1$ are Fourier coefficients.

4. Calculated FDs are translation invariant since centroid distance is relative to the centroid. In Fourier Transform, rotation in spatial domain means phase-shift in frequency domain so using magnitude values of coefficients make FDs rotationally invariant. Scale invariance is achieved by dividing FDs by FD_0. Since each r_k is real valued, first half of FDs are the same with second half. Therefore, half of the FDs are enough to represent shape. As a result, a hand shape boundary is represented as follows [15]:

$$FD = \left[\frac{|FD_1|}{|FD_0|}, \frac{|FD_2|}{|FD_0|}, ..., \frac{|FD_{N/2}|}{|FD_0|}\right]$$

2.3 Hu Moment Invariants, Φ

Hu derived 7 moments which are invariant to translation, rotation and scaling [13],[16]. This is why Hu moments are so popular and many applications use them in shape recognition systems. Each moment shows a statistical property of the shape. Hu Moment Invariants can be calculated as follows. Note that μ shows the 2nd and 3rd order central and normalized moments.

$\phi_1 = \mu_{20} + \mu_{02}$

$\phi_2 = (\mu_{20} - \mu_{02})^2 + 4\mu_{11}^2$

$\phi_3 = (\mu_{30} - 3\mu_{12})^2 + (3\mu_{21} - \mu_{03})^2$

$\phi_4 = (\mu_{30} + \mu_{12})^2 + (\mu_{21} + \mu_{03})^2$

$\phi_5 = (\mu_{30} - 3\mu_{12})(\mu_{30} + \mu_{12})((\mu_{30} + \mu_{12})^2 - 3(\mu_{21} + \mu_{03})^2) + (3\mu_{21} - \mu_{03})(\mu_{21} + \mu_{03})(3(\mu_{30} + \mu_{12})^2 - (\mu_{21} + \mu_{03})^2)$

$\phi_6 = (\mu_{20} - \mu_{02})((\mu_{30} + \mu_{12})^2 - (\mu_{21} + \mu_{03})^2) + 4\mu_{11}(\mu_{30} + \mu_{12})(\mu_{21} + \mu_{03})$

$\phi_7 = (3\mu_{21} - \mu_{03})(\mu_{30} + \mu_{12})((\mu_{30} + \mu_{12})^2 - 3(\mu_{21} + \mu_{03})^2) - (\mu_{30} - 3\mu_{12})(\mu_{21} + \mu_{03})(3(\mu_{30} + \mu_{12})^2 - (\mu_{21} + \mu_{03})^2)$

The main problem of Hu moments in classification are the large numerical variances in the values of moment invariants. Therefore, the use of Euclidian distance to compute similarity is not possible. In our implementation, decision tree is used as the classifier.

2.4 Orientation Histogram

Orientation Histogram (OH) is a histogram of local orientations of pixels in the image [11]. Freeman *et al.* applied the idea of OH to create fast and simple hand interfaces [7]. The basic idea of OH is that hand pixels may vary in illumination, and pixel-by-pixel difference leads to huge error in total. Instead of using pixels themselves for comparison, their orientations are used to overcome the illumination problem. To make it translation invariant, orientations are collected in a histogram with 36 bins where each bin represents 10 degree. Scale and rotation invariance are not pointed in [11]. However, our implementation normalized the magnitude of the histogram to overcome scaling problem, and updated by shifting all orientations relative to peak one to make it rotationally invariant. Instead of using the whole image, its dimension is reduced to about 100 by 80 pixels for faster computation [11]. The problems of the method are also reported as two similar shapes can produce very different histograms, and hand shape must not be a small part of the image. Thinking each normalized histogram as a vector in 36 dimensional space, we classified them using a decision tree algorithm similar to other three methods.

3 Experimental Results

To evaluate the performance of 4 shape representations, we collected 10160 samples of widely-used 15 hand shapes from 5 different people. There is approximately the same number of samples for each person and hand commands. A sample set of collected 15 hand shapes are shown in Fig. 1. The evaluation is based mainly on discrimination power and speed. Furthermore, we have also investigated the performance with respect to the number of people and samples in the training set.

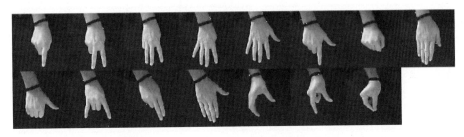

Fig. 1. Hand commands used in our experiments

We have first divided the samples into two sets: training set and test set. Training set is used to train a decision tree for each representation, and samples in the test set are classified by the corresponding decision tree. Hit ratio, which is the percentage of correctly classified samples in the test set, is used as the measurement for discrimination power of each representation. The division of training and test set is based on two parameters: number of people in the training set, and the percentage of the samples of each person selected for training.

We first assessed the effect of training set size when the system is trained with samples only from one person. We selected randomly one person among 5 people, and used 10% of samples from each hand command of the selected person in training. All the remaining samples, that is, remaining 90% samples of the same person and all samples from other 4 people, were employed in test set. This test is repeated 5 times and the average of the hit ratios is used as the measurement of discrimination power for each representation. This procedure is repeated with 30% and 50% of samples in training for one person, and Fig. 2.a shows the results graphically. We repeated the above procedure for 2, 3, 4, and 5 people, and only the results for 5 people is depicted in Fig. 2.b. Results for all number of people with all number of training set size can be found in Table 1.

Table 1. Hit ratios for all parameters used in the experiment

Parameters		Hit Ratios (%)			
Num. of people	Training Size (%)	SD	FD	HU	OH
1	10	55.63	65.08	62.1	38.15
	30	64.99	77.5	70.97	45.87
	50	62.2	76.45	70.16	47.24
2	10	75.47	86.09	85.19	49.53
	30	75.54	88.11	86.48	55.82
	50	75.97	87.9	85.6	57.12
3	10	81.05	89.29	89.81	53.23
	30	83.44	92.98	91.8	63.33
	50	82.73	93	91.1	65.48
4	10	83.89	93.56	93.86	58.56
	30	87.05	94.12	94.82	67.38
	50	87.07	94.05	94.27	69.65
5	10	87.29	96.84	95.73	63.45
	30	91.18	98.64	98.13	70.17
	50	92.35	99.61	98.58	75.43

Fig. 2.a and Fig. 2.b show the hit ratios of representations when 10%, 30% and 50% of samples from the people used in training. It is observed that the performances of SD, FD and HU are not influenced considerably by increasing the number of samples from the same person. As a result, SD, FD and HU representations produce low variances for the representations of similar hand shapes from the same person. This is a desired property since a few training samples from a person are adequate to train the system for that person.

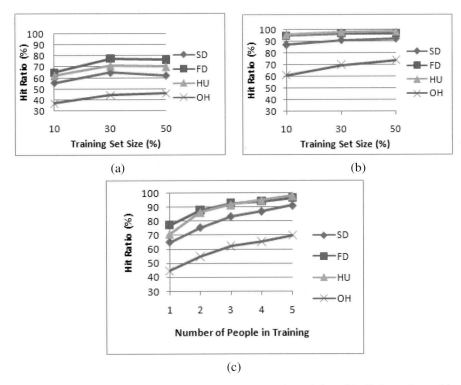

Fig. 2. Hit ratios for (a) randomly selected one person for training, (b) all 5 people used in training, (c) 30% of samples of selected number of people for training

Fig. 2.c indicates the influence of number of people in training on hit ratio. It is apparently observed that hit ratios of all representations improve drastically when the number of people participating in training increases. Thus, it is reasonable to use many people to train a hand interface system which uses one of four shape representations described in this paper.

Our aim is to find out the best of four representations in terms of discrimination power and speed. According to Fig. 2, FD and HU outperform SD and OH, and the results of FD and HU are very close to each other in terms of hit ratio. FD is slightly better than HU.

Real-time is an indispensable requirement for a hand interface system. Noticeable delay prevents user from immersive usage of the system. Therefore, we analyzed the running time performance of representations. To measure running time performances of each representation, we computed the total elapsed time of each representation for all samples. Fig. 3 shows the results of total elapsed time of 10160 samples for each representation. According to the results, HU is the fastest method among these four representations. Remark that FD is also a relatively fast method. On the average, calculation of a HU or FD representation of a segmented image is less than a millisecond (dividing total application time by total samples according to Fig. 3) with a Pentium IV – 3GHz computer with 1GB RAM of memory.

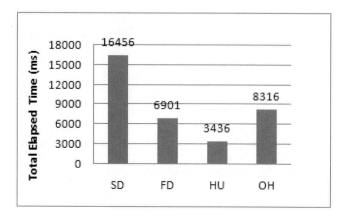

Fig. 3. Running times of 4 shape representations for 10160 samples

In conclusion, both HU and FD provide acceptable results in terms of discrimination power and speed, and they can be used for computer systems which employ static hand shapes as commands.

4   Conclusion

The main component of a hand interface is the part that recognizes the given hand command. Shape is an important property of a hand, and it can be used for representation of a hand. In this paper, four representation techniques; Shape Descriptors (SD), Fourier Descriptors(FD), Hu Moment Invariants (HU) and Orientation Histogram (OH) are selected, and compared in terms of their discrimination power and speed. When forming the test environment, a widely-used hand interface setup is chosen. In this setup, a camera is located above the desk in such a way that its viewing direction is to the top of the desk. In the experiments, there are totally 10160 samples of 15 different hand commands from 5 people. Those samples are divided into training set and test set. Each sample in training set is converted to four representations, and each representation is used to train a decision tree classifier for recognition. Furthermore, running times of calculating the representations are accumulated and used as the measurement of speed for each representation. According to test results, HU and FD outperform SD and OH in terms of both discrimination power and speed. Therefore, HU and FD are reasonable to use in hand shape recognition systems as posture recognizer or as a feature vector to a spatio-temporal pattern recognizers such as HMM.

References

1. La Viola, J.J.: A Survey of Hand Posture and Gesture Recognition Techniques and Technology. Technical Report CS-99-11, Department of Computer Science, Brown University (1999)
2. Erol, A., Bebis, G., Nicolescu, M., Boyle, R.D., Twombly, X.: A Review on Vision Based Full DOF Hand Motion Estimation. In: IEEE Workshop on Vision for Human-Computer Interaction (in conjunction with CVPR 2005), San Diego, CA, June 21 (2005)

3. Wu, Y., Huang, T.S.: Hand Modeling, Analysis and Recognition. IEEE Signal Processing Magazine, 51–58 (May 2001)
4. Quek, F.K.H., Mysliwiec, Y., Zhao, M.: FingerMouse: a Freehand Pointing Interface. In: Proc. of Int. Workshop on Automatic Face and Gesture Recognition, pp. 372–377 (1995)
5. Genc, S., Atalay, V.: ITouch: Vision-based Intelligent Touch Screen in a Distributed Environment. In: Int. Conf. on Multimodal Interfaces (ICMI), Doctoral Spotlight (October 2005)
6. Licsar, A., Sziranyi, T.: Dynamic Training of Hand Gesture Recognition System. In: Proc. Intl. Conf. on Pattern Recognition, ICPR 2004, vol. 4, pp. 971–974 (2004)
7. Freeman, W.T., Anderson, D., Beardsley, P., Dodge, C., Kage, H., Kyuma, K., Miyake, Y., Roth, M., Tanaka, K., Weissman, C., Yerazunis, W.: Computer Vision for Interactive Computer Graphics. IEEE Computer Graphics and Applications 18(3), 42–53 (1998)
8. Duda, R.O., Hart, P.E., Stork, D.G.: Pattern Classification, 2nd edn. Wiley-Interscience, Hoboken (2000)
9. Zhang, D., Lu, G.: Review of shape representation and description techniques. Pattern Recognition 37(1), 1–19 (2004)
10. Zhang, D., Lu, G.: Content-Based Shape Retrieval Using Different Shape Descriptors: A Comparative Study. In: Proc. of IEEE Conference on Multimedia and Expo., Tokyo, August 2001, pp. 317–320 (2001)
11. Freeman, W.T., Roth, M.: Orientation Histograms for Hand Gesture Recognition. In: Intl. Workshop on Automatic Face and Gesture Recognition (1995)
12. Peura, M., Livarinen, J.: Efficiency of Simple Shape Descriptors. In: Aspects of Visual Form, pp. 443–451. World Scientific, Singapore (1997)
13. Flusser, J.: Moment Invariants in Image Analysis. In: Proceedings of World Academy of Science, Engineering and Technology, February 2006, vol. 11 (2006) ISSN 1307-6884
14. Fitzgibbon, A., Pilu, M., Fisher, R.B.: Direct Least Square Fitting of Ellipses. Pattern Analysis and Machine Intelligence 21(5) (May 1999)
15. Zhang, D., Lu, G.: A Comparative Study on Shape Retrieval Using Fourier Descriptors with Different Shape Signatures. In: Proc. of International Conference on Intelligent Multimedia and Distance Education, Fargo, ND, USA, pp. 1–9 (2001)
16. Hu, M.K.: Visual Pattern Recognition by Moment Invariants. IRE Transactions on Information Theory IT-8, 179–187 (1962)
17. Gary, K.A.: Learning OpenCV. O'Reilly, Sebastopol (2008) (first print)
18. Chen, F.S., Fu, C.M., Huang, C.L.: Hand Gesture Recognition Using a Real-time Tracking Method and Hidden Markov Models. Image and Vision Computing 21, 745–758 (2003)
19. Kolb, A., Barth, E., Koch, R.: ToF-sensors: New dimensions for realism and interactivity. In: IEEE Conf. on Computer Vision and Pattern Recognition (CVPR), Workshop on ToF Camera based Computer Vision (TOF-CV), pp. 1–6 (2008), doi:10.1109/CVPRW
20. Gvili, R., Kaplan, A., Ofek, E., Yahav, G.: Depth Keying,
http://www.3dvsystems.com/technology/DepthKey.pdf
21. Microsoft Natal Project: http://www.xbox.com/en-US/live/projectnatal
22. Nakagawa, S., Nakanishi, H.: Speaker-Independent English Consonant and Japanese word recognition by a Stochastic Dynamic Time Warping Method. Journal of Institution of Electronics and Telecommunication Engineers 34(1), 87–95 (1988)
23. Stauffer, C., Grimson, W.E.L.: Adaptive Background Mixture Models for Real-Time Tracking. In: IEEE Computer Society Conference on Computer Vision and Pattern Recognition (Cat. No PR00149), pp. 246–252 (1999)
24. Jones, M.J., Rehg, J.M.: Statistical Color Models with Application to Skin Detection. Int. J. of Computer Vision 46(1), 81–96 (2002)

Multi-target and Multi-camera Object Detection with Monte-Carlo Sampling

Giorgio Panin, Sebastian Klose, and Alois Knoll

Technische Universität München, Fakultät für Informatik
Boltzmannstrasse 3, 85748 Garching bei München, Germany
{panin,kloses,knoll}@in.tum.de

Abstract. In this paper, we propose a general-purpose methodology for detecting multiple objects with known visual models from multiple views. The proposed method is based Monte-Carlo sampling and weighted mean-shift clustering, and can make use of any model-based likelihood (color, edges, etc.), with an arbitrary camera setup. In particular, we propose an algorithm for automatic computation of the feasible state-space volume, where the particle set is uniformly initialized. We demonstrate the effectiveness of the method through simulated and real-world application examples.

1 Introduction

Object detection is a crucial problem in computer vision and tracking applications. It involves a global search over the feasible state-space, and requires to cope with an unknown number of targets, possible mutual occlusions, as well as false measurements, arising from background clutter.

Using multiple cameras can greatly improve the detection results in terms of precision and robustness, since the joint likelihood will be much more focused on real targets, and mutual occlusions from a given view will be solved by the others. Moreover, multiple cameras constrain the state-space of visible objects to a smaller volume, where a target appears in all visual fields. This reduces the search space of a great amount, and therefore facilitates the detection process.

For this purpose, a typical *bottom-up* approach usually consists of sampling image features (e.g. segmenting color *blobs*) and matching them between cameras, in order to perform a 3D triangulation and object localization: however, this approach requires to explore all possible combinations of data that can be associated to similar targets, possibly in presence of missing detections and false alarms, as well as partial occlusions, which can make the problem of an intractable complexity.

In a *top-down* approach, instead, a detection task can be seen as a global optimization of a multi-modal *likelihood* function in state-space, which presents strong local maxima around each target (detected by the optmization method), as well as smaller peaks around false measurements. This optimization problem involves generating and testing a number of state-space *hypotheses*, by projecting

G. Bebis et al. (Eds.): ISVC 2009, Part I, LNCS 5875, pp. 12–21, 2009.

the relevant model features on each camera view, and comparing them locally with the image measurements.

When two targets are too close with respect to the covariance of the measurement noise, the related peaks merge to some extent, and are not anymore distinguishable by the search method. Therefore, the measurement covariance by itself sets a limit to the state-space *resolution* of the detector.

Evolutionary and Genetic Algorithms are well-known in the literature, in order to cope with such multi-modal optimization problems [9]; however, their computational complexity limits the application field, particularly when a real-time (or near real-time) performance is required, such as object tracking.

In order to approach the problem from a general point of view, not restricted to a particular form of the likelihood, or a given camera configuration, we choose instead a Monte-Carlo based strategy, followed by unsupervised clustering of state hypotheses, according to the respective likelihoods. This approach has the advantage of neither requiring any prior assumption about the number of targets, nor about the form of the likelihood, provided that a significant local maximum is present around each target state.

The paper is organized as follows: Section 2 describes the general clustering strategy, based on kernel representation and weighted mean-shift; Section 3 the introduces the uniform sampling strategy for multiple camera views, on the joint viewing volume; Section 4 provides simulated and experimental results, and Section 5 concludes the work with proposed future developments.

2   The Particle-Based Detector

In order to detect targets, we basically look for local maxima (or *modes*) of a given likelihood, provided by any visual property of each target, and a suitable matching strategy between model and image features. In general, this function can integrate multiple visual cues, as well as data from multiple cameras. Such a general formulation, together with an arbitrary number and relative location of targets, makes the estimation problem of a complex and nonlinear nature.

Therefore, we approach the problem by means of a general and flexible method, such as Monte-Carlo sampling. In particular, we represent our likelihood through a discrete *particle set* (s^i, w^i), where s_i are state hypotheses, weighted by their likelihood w_i. This representation is well-known in a tracking framework [7], and can cope with nonlinear and multi-modal distributions.

In absence of any prior information about the possible location of targets, the particle set is initialized with uniform distribution, covering the feasible state-space volume where targets can be viewed by the multi-camera setup (Sec. 3).

Each peak of the likelihood will provide a cluster of high-weighted particles around it, and therefore a weighted *state-space clustering* algorithm can be run, in order to identify them. However, if the likelihood peaks are too large and partially overlapping, the clusters will overlap as well, and the algorithm will fail to separate them properly.

In a computer vision application, the width of the likelihood modes depends on the modality used (edges, color, etc.), and on the related covariance. This parameter can be externally set (or internally computed), and it reflects the uncertainty of the measurement process in feature- or state-space: a high-resolution measurement has low covariance, with narrow peaks well-located around the targets, but also many local maxima in the neighborhood; on the other hand, a low-resolution measurement will have a higher covariance, larger and less localized peaks.

In order to identify the modes, we need a smooth representation that can be locally optimized, such as a *kernel-based* representation [2][4]. More in detail, if x is a one dimension variable, a weighted kernel density is represented by

$$p\left(x|\,\theta\right) = \frac{1}{N} \sum_{i=1}^{N} \frac{w_i}{h} k\left(\frac{x - x_i}{h}\right) \tag{1}$$

In this formula, k is the kernel, which has a maximum value in $x = 0$, and quickly decays in a neighborhood of the origin; h is called *bandwidth*, and regulates the width of the kernel around each point x_i. The number of *modes* N is also a parameter of this distribution, overall represented by the set of values

$$\theta = (N, h, x_1, ..., x_N, w_1, ..., w_N) \tag{2}$$

A typical choice for the kernel is the Gaussian distribution

$$k\left(x\right) = \frac{1}{\sqrt{2\pi}} e^{-\frac{x^2}{2}} \tag{3}$$

for which $h = \sigma$ is the standard deviation, so that (1) represents a Mixture of Gaussians. In a multi-dimensional space, the kernel representation generalizes to

$$p\left(\mathbf{x}|\,\theta\right) = \frac{1}{N} \sum_{i=1}^{N} \frac{w_i}{\det H} \mathcal{K}\left(H^{-1}\left(\mathbf{x} - \mathbf{x}_i\right)\right) \tag{4}$$

where the multi-variate kernel $\mathcal{K}$ is obtained as the product of univariate ones

$$\mathcal{K}\left(\mathbf{x}\right) = \prod_{d=1}^{D} k\left(x_d\right) \tag{5}$$

with D the space dimension. In the Gaussian case, $\Sigma = HH^T$ is the *covariance matrix* of the multi-variate kernel.

Concerning the clustering method, in order to keep the most general setting, we make use of unsupervised clustering, through the weighted Mean-Shift algorithm [2]. Mean-shift is a kernel-based, non-parametric and unsupervised clustering method, that finds local maxima of the kernel density by gradient ascent, starting from each sample point, and assigns to the same cluster all paths that converge to the same peak; therefore, it simultaneously finds the number and location of modes, and assigns the sample points to each cluster as well.

By restricting the attention to isotropic kernels ($H = hI$), the density can be locally optimized by computing the weighted mean-shift vector

$$\mathbf{m}_h(\mathbf{x}) = \left[\frac{\sum_{i=1}^{n} w_i g\left(\left\| \frac{\mathbf{x} - \mathbf{x}_i}{h} \right\|^2 \right) \mathbf{x}_i}{\sum_{i=1}^{n} w_i g\left(\left\| \frac{\mathbf{x} - \mathbf{x}_i}{h} \right\|^2 \right)} - \mathbf{x} \right] \tag{6}$$

where $g = -k'$ its first derivative of the kernel, and afterwards updating the position $\mathbf{x} \rightarrow \mathbf{x} + \mathbf{m}_h$. The iteration is stopped when the update vector becomes smaller than a given threshold: $\|\mathbf{m}_h\| < \epsilon$.

Chosing the correct bandwidth h can be critical, in order to ensure that the correct number of particle clusters will be found. For our purposes, we simply choose h proportional to the minimum distance between detectable targets in state-space (resolution of the detector), which is of course application-dependent: for example, if the detected targets are small objects, the minimum distance will be smaller than for people detection.

During mean-shift clustering, it still may happen that small, spurious clusters of a few sample points are detected. These clusters are stationary points (where the mean-shift gradient is zero) but usually located on non-maxima, such as saddle points. Therefore, they are removed by a simple procedure: if a cluster center, located on a local maximum, is perturbed by a small amount, and the mean-shift algorithm is run again from this location, then it will converge again to the same point. Otherwise, the cluster center must be located on a saddle point.

3 Redundant Multi-camera Setup: Sampling from the View-Volume Intersection

In a multi-camera context, we need first to initialize the particle set with a uniform distribution in 3D space. This requires defining the *sampling volume* for this distribution, in particular concerning the positional degrees of freedom (x, y, z translation).

In general, we consider here *redundant* multi-camera settings (Fig. 1), as opposed to *complimentary* ones. In a redundant configuration, the fields of view overlap to a large extent, so that the object can simultaneously be seen from all cameras, at any pose. This has the advantage of a more informative measurement set, which allows 3D tracking of complex objects. By contrast, a complimentary setup consists of almost non-overlapping camera views, where the object to be tracked can be completely seen only by one camera at a time.

In particular, when dealing with a redundant configuration, we need to sample hypotheses uniformly from the subset of state-space configurations that are visible from all cameras. This requires computing the joint *viewing volume* of m cameras. For this purpose, each camera provides 6 *clipping planes*, which overall define a truncated pyramid: 4 lateral planes defined by the 4 image sides, and the focal length, while the two frontal planes define the minimum and maximum depth of detectable objects.

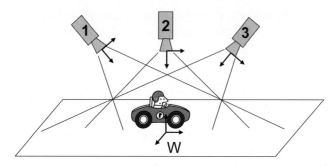

Fig. 1. Redundant, multi-camera configuration for object tracking

These planes can be expressed (in camera-centered or world-centered coordinates) by means of 3 points in space. For example, the left clipping plane contains the upper-left and lower-left corners of the image, plus the camera center. In camera coordinates, then we have

$$^c\mathbf{x}_{u,l} = (-\frac{r_x}{2}, -\frac{r_y}{2}, f) \tag{7}$$

$$^c\mathbf{x}_{l,l} = (-\frac{r_x}{2}, \frac{r_y}{2}, f) \tag{8}$$

$$^c\mathbf{c} = (0,0,0)$$

These points can be transformed to world coordinates, by applying the respective camera transformation matrix $T_{W,c}$

$$^W\mathbf{x} = T_{W,c} \cdot {}^c\mathbf{x} \tag{9}$$

Therefore, the left clipping plane of camera c, $\pi_{c,1}$ is given in homogeneous coordinates by the null-space of the (3×4) matrix [6] (dropping the reference frame W)

$$\pi = null\left(\begin{bmatrix}\mathbf{x}_{u,l}^T \\ \mathbf{x}_{l,l}^T \\ \mathbf{c}^T\end{bmatrix}\right) \tag{10}$$

so that all visible points from camera c must lie in the half-space defined by

$$\pi_{c,1}^T\mathbf{x} \le 0 \tag{11}$$

where the sign of π can be chosen, for example, in order to make sure that the image center $(0,0,f)$ (expressed in world coordinates) is contained in the half-space. The same procedure can be applied to the other clipping planes in a similar way.

If we denote by $\pi_{c,i}, i = 1, ..., 6$ the world-related planes of camera c, its viewing polyhedron is defined by the homogeneous inequalities

$$A_c\mathbf{x} \le \mathbf{0}; \ A_c \equiv \begin{bmatrix}\pi_{c,1}^T \\ ... \\ \pi_{c,6}^T\end{bmatrix} \tag{12}$$

and finally, the overall intersection is given by the convex polyhedron, defined by

$$Ax \leq 0; \ A \equiv \begin{bmatrix} A_1 \\ \dots \\ A_C \end{bmatrix} \quad (13)$$

This equation could be used in principle to directly select a uniform sample of visible points inside it. For this purpose, most popular methods in the literature refer to Markov-Chain Monte-Carlo (MCMC) strategies, starting from the well-known work [5]. However, due to its computational complexity and the presence of several parameters in the algorithm, in the present work we propose a simpler approach, that consists in uniformly sampling from the 3D *bounding box* of the polyhedron, and discarding all samples which do not satisfy (13). This will produce uniformly distributed points, at the price of discarding many samples, and therefore requiring a longer (and less predictable) time before reaching the desired number of valid points.

In order to compute the bounding box of the polyhedron, we also need to explicitly compute its vertices in 3D space, from the implicit formula (13). This is known as the *vertex enumeration* problem, and can be solved via the *primal-dual* method of [1].

A final note concerns the choice of the two main parameters for our algorithm (namely, the kernel size and the number of hypotheses), for which we employ the following criterion:

- The kernel width h determines the resolution of our detector, since two likelihood peaks closer than h lead to a single, detected mode in the mean-shift optimization.
- The number of hypotheses n determines the spatial density of the sample, which depends on the kernel size h: we need to make sure that at least one sample point falls into any sphere of radius h, in order for the mean-shift algorithm to work and not getting stuck into zero-density regions. Therefore, if $V_S(h)$ is the volume of a sphere of radius h, and V_B is the volume of the bounding box for sampling (which is larger than the polyhedron volume), we can choose $n = V_B/V_S$.

4   Applications and Experimental Results

In Fig. 2, we show an example result of the sampling procedure, applied to a 3-camera configuration. The three viewing volumes (indicated with different colors) intersect in the central polyhedron, which is filled by uniformly distributed points. Its bounding box is also shown in black.

As a first experiment, we run the proposed system on a simulated scenario: a set of randomly chosen targets provide "virtual measurements", by generating for each target o a measurement z_o around the true state $\bar{s}_o$, plus Gaussian noise v_o

$$z_o = \bar{s}_o + v_o \quad (14)$$

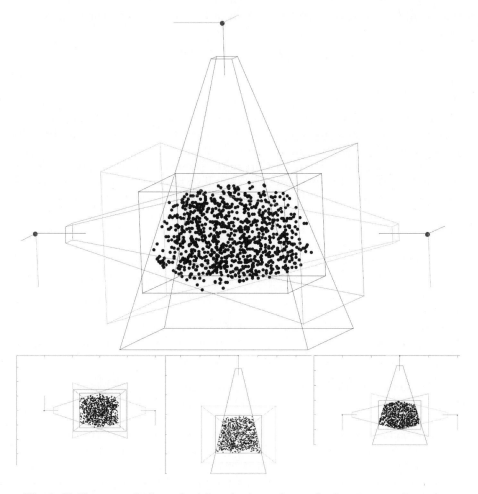

Fig. 2. Uniform sample from the joint viewing volume of a 3-camera configuration

with the same covariance matrix V for all targets. This model corresponds to a Mixture of Gaussians likelihood

$$P\left(z\mid s\right) \propto \sum_{o} exp(-\frac{1}{2}\left(z_o - s\right)^T V^{-1}\left(z_o - s\right)) \qquad (15)$$

for any state hypothesis s within the joint volume. The state here is represented by 3D position, $s = (x, y, z)$.

Four targets are selected at random within the viewing volume. In this example, all targets are separated in space by more than $100mm$, and the kernel size is $h = 30$, so that the detector has no difficulties in distinguishing them. A set of $n = 2000$ sample points is drawn within the volume, and their likelihood values are computed. After performing mean-shift clustering, the detected modes are shown with different colors on the right side of Fig. 3.

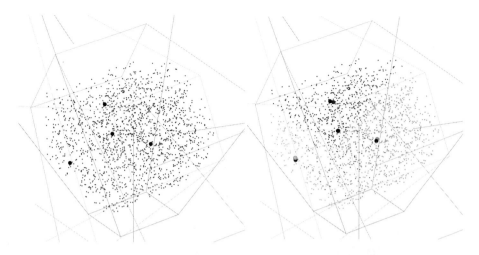

Fig. 3. Result of the simulated experiment, with 4 targets and a Gaussian likelihood on each target. Left: true targets (black dots) and the uniform Monte-Carlo set. Right: result of weighted mean-shift clustering, and respective cluster centers (the red cluster center is not visible, because almost coincident with the true target position).

As it can be seen, the detected modes are in the correct number, and their location is in a good accordance with the real target positions.

Subsequently, we tested the system on real camera images. As image likelihood, we compute the Bhattacharyya distance between color histograms, often used for object tracking [4][8]

$$B\left(q, p\left(s\right)\right) = \left[1 - \sum_b \sqrt{q_b p_b}\right]^{\frac{1}{2}} \tag{16}$$

where q is a reference histogram, collected from an image of the object, and $p(s)$ is the observed histogram, from image pixels underlying the projected object area, under pose hypothesis s (Fig. 4, left side). The sum is performed over $(D \times D)$ histogram bins ($D = 10$ is a common choice in Hue-Saturation space).

On a multi-camera setup, by assuming independence of the camera measurements, the corresponding likelihood is

$$P\left(z|s\right) \propto \prod_c \exp\left(-\frac{B^2\left(q, p_c\left(s\right)\right)}{2\sigma^2}\right) \tag{17}$$

where $p_c(s)$ is the image histogram at pose s, projected on camera c, and σ^2 is the measurement noise covariance (the same for all cameras and targets).

In Fig. 4, we can see the detection result for 4 real targets. The object model is given by a yellow sphere of radius $65mm$, and the reference histogram is collected from a single image of the object. On the left side of the picture, we

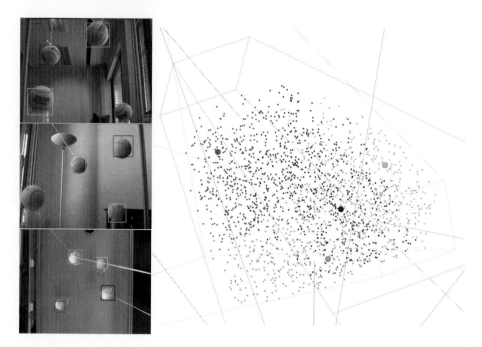

Fig. 4. Left: Camera images, with re-projection of the detected targets. Right: detected targets in 3D space, and particle clusters after mean-shift optimization.

can see the 3 camera images with superimposed projections of the estimated target locations; on the right side, the 3D positions of the detected targets in the common viewing volume, together with the particle clusters after mean-shift optimization, are shown.

5   Conclusion and Future Work

We presented a Monte-Carlo methodology for generic multi-camera, multi-target detection. The proposed method can be applied to a variety of likelihood functions in computer vision, as well as to generic, calibrated camera setups.

One limitation of the proposed system is the number of targets that can simultaneously be detected, still limited to a few units: a maximum of 7-8 targets have successfully been detected with the simulated experiment of Sec. 4 which, as explained at the end of Sec. 3, depends on the spatial resolution desired (i.e. the kernel bandwidth h).

A possible improvement of the system may use an adaptive version of mean-shift algorithm [3], where the bandwidth parameter is selected and modified according to the data points, in order to give the best clustering results.

Acknowledgements

This work is partly supported by the German Research Council (DFG), under the excellence initiative cluster *CoTeSys - Cognition for Technical Systems*[1] within the project *ITrackU (Image-based Tracking and Understanding)*.

References

1. Bremner, D., Fukuda, K., Marzetta, A.: Primal-dual methods for vertex and facet enumeration (preliminary version). In: SCG 1997: Proceedings of the thirteenth annual symposium on Computational geometry, pp. 49–56. ACM, New York (1997)
2. Comaniciu, D., Meer, P.: Mean shift: A robust approach toward feature space analysis. IEEE Trans. Pattern Anal. Mach. Intell. 24(5), 603–619 (2002)
3. Comaniciu, D., Ramesh, V., Meer, P.: The variable bandwidth mean shift and data-driven scale selection. In: IEEE International Conference on Computer Vision, vol. 1, p. 438 (2001)
4. Comaniciu, D., Ramesh, V., Meer, P.: Kernel-based object tracking. IEEE Trans. Pattern Anal. Mach. Intell. 25(5), 564–575 (2003)
5. Dyer, M., Frieze, A., Kannan, R.: A random polynomial-time algorithm for approximating the volume of convex bodies. J. ACM 38(1), 1–17 (1991)
6. Hartley, R.I., Zisserman, A.: Multiple View Geometry in Computer Vision, 2nd edn. Cambridge University Press, Cambridge (2004)
7. Isard, M., Blake, A.: Condensation – conditional density propagation for visual tracking. International Journal of Computer Vision (IJCV) 29(1), 5–28 (1998)
8. Pérez, P., Hue, C., Vermaak, J., Gangnet, M.: Color-based probabilistic tracking. In: Heyden, A., Sparr, G., Nielsen, M., Johansen, P. (eds.) ECCV 2002. LNCS, vol. 2350, pp. 661–675. Springer, Heidelberg (2002)
9. Singh, G., Deb, K.: Comparison of multi-modal optimization algorithms based on evolutionary algorithms. In: GECCO 2006: Proceedings of the 8th annual conference on Genetic and evolutionary computation, pp. 1305–1312. ACM, New York (2006)

[1] http://www.cotesys.org

Spatial Configuration of Local Shape Features for Discriminative Object Detection

Lech Szumilas[1] and Horst Wildenauer[2]

[1] Industrial Research Institute for Automation and Measurements
Al. Jerozolimskie 202, 02-486 Warsaw, Poland
lech.szumilas@gmail.com
http://www.piap.pl
[2] Vienna University of Technology
Favoritenstraße 9/1832, A-1040 Wien, Austria
horst.wildenauer@gmail.com
http://www.acin.tuwien.at.at

Abstract. This paper proposes a discriminative object class detection and recognition based on spatial configuration of local shape features. We show how simple, redundant edge based features overcome the problem of edge fragmentation while the efficient use of geometrically related feature pairs allows us to construct a robust object shape matcher, invariant to translation, scale and rotation. These prerequisites are used for weakly supervised learning of object models as well as object class detection. The object models employing pairwise combination of redundant shape features exhibit remarkably accurate localization of similar objects even in the presence of clutter and moderate view point changes which is further exploited for model building, object detection and recognition.

1 Introduction

The study of shape for object description and recognition has a long research tradition, dating back into the early days of the computer vision field. Several recent works have explored the idea of coupling local, contour-based features together with their geometric relations as effective means of discriminating object categories using shape. Promising results in the context of object class recognition and object localization have been achieved, solely operating on boundary-based representations. In [12,10] codebooks of class discriminative shape features, drawn from a corpus of training images, are augmented with geometric relations encoded in pointers to an object instances centroid. A similar representation was suggested by Ferrari et al. [5], however building on a more generic alphabet of shape features, derived from groups of adjacent contour segments. In contrast, [9,3] model global shape by means of ensembles of pairwise relations between local contour features.

In our work we exploit pairwise relationships between local shape fragments to construct a robust shape matching technique that is invariant to translation, scale and rotation. This method allows us to localize objects in the scene using the model based on spatial arrangement of local shape fragments discussed in Section 3. The use of this technique for model extraction as well as object detection and classification is investigated in Sections 4 and 5.

G. Bebis et al. (Eds.): ISVC 2009, Part I, LNCS 5875, pp. 22–33, 2009.
© Springer-Verlag Berlin Heidelberg 2009

2   The Shape Based Object Model

In this work, an object's shape is represented by an overcomplete set of weak local features together with their spatial relations: Local, edge-based features are embedded in a global geometric shape model, organized as a star-like configuration around an object's centroid. The aspects of this type of representation has been extensively investigated in the context of part-based object class recognition [11,4], showing the advantage of considerably reduced computational burden during training and testing compared to fully connected constellation models. Star-like representations also found successful application in contour-based object detection methods: Opelt et al. [10] and Shotton et al. [12] concentrated on the use of class-discriminative codebooks of boundary fragments, while Ferrari et al. [5] employ a generic alphabet of shape features, derived from groups of adjacent contour segments.

To further increase the explanatory power of the rather weak features, we group them into pairs and exploit their pairwise constraints and their geometric relationship to the centroid to arrive at an efficient matching process which is invariant to changes in translation, scale, and orientation while still being able to handle moderate shape deformations. Ensembles of pairwise relations between edge-based features have been previously used in the same context by [9]. However, the proposed encoding only allowed for translational invariance during the matching process.

3   Features and Matching

Our choice of shape features is based on two criteria: Achieving invariance to translation, scaling and rotation and minimizing the sensitivity w.r.t. edge fragmentation. We start with edges obtained by Canny's edge detector and then coarsely segment each edge into a chain of straight segments by splitting at high curvature points (see Figure 1). Similar in spirit to [5], these edge fragments are used to construct a basic feature that represent local contours in the form of a pair of adjacent segments and the key point at the segment intersection. Segment adjacency is defined in terms of overall distance from the key point to the associated segment boundaries (referred later as "relaxed segment adjacency") – thus allowing pairing of segments that correspond to different edges or non-consecutive segments along the same edge. Although this is a very weak feature that does not allow for reliable scale and orientation estimation, the negative effect of edge fragmentation is compensated by relaxed segment adjacency and the introduced redundancy.

In order to create a feature that is fully invariant to similarity transformation and increase its discriminative power we pair individual key points as shown in Figure 1 (lower-right corner). The key point pair is described by two sets of parameters:

- matching features $f_{ij} = [\beta_{ij1}, \beta_{ij2}, \beta_{ij1}, \beta_{ij2}] \in \mathbb{F}^4$ where $\mathbb{F}$ represents real numbers in the range $-\pi..\pi$ that describe segment angles relative to the vector connecting key points i and j. Note that f_{ij} is invariant to similarity transformations.
- geometrical relationships used for estimation of relative scale, orientation and object centroid location during feature matching $g_{ij} = [d_{ij}, \alpha_{ij}, \Delta x_{ij}, \Delta y_{ij}, \Delta x_{ic}, \Delta y_{ic}]$.

The Δx_{ic} and Δy_{ic} represent spatial relation between $i - th$ key point and the object centroid c. Object centroid is either known (model) or estimated during object detection.

Note that such defined feature pair is an ordered set of key points i and j.

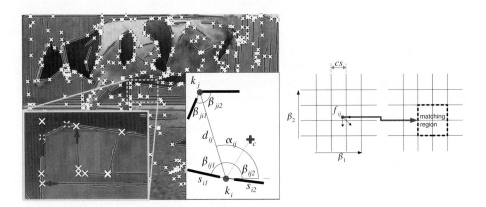

Fig. 1. Left: Example of local shape features. Green lines and white markers depict edge segments and the associated key points respectively. The lower-left corner shows an example of key point association to segments belonging to different edges or non-consecutive segments along the same edge. The lower-right corner shows an example of key point pair. Right: Example of feature discretization (2D case for clarity) that allows for matching of feature sub-sets instead of exostive correspondence estimation (see Section 3.1).

Geometrical relations between local image features have been previously used to disambiguate feature correspondences in object recognition, see Section 2. Here we extend the use of feature pairs (referred as features from now on) to obtain feature matching, registration and object detection that is invariant to translation, scale and orientation.

The problem of feature matching and subsequently fitting the object model m to the feature set from target image t is defined as a three stage process:

1. Feature matching that estimates feature similarity, relative scale and orientation between the model and the target sets as well as the centroid position in the target image. Feature matching also produces soft correspondences between model and target features, where each feature in the model set correspond to k most similar target features. We have chosen $k = 20$ which gives a good balance between accuracy and efficiency of the model fitting.
2. Estimation of potential centroid locations in the target image with Hough-style voting.
3. Iterative model fitting around detected centroids combined with feature correspondence pruning. The model fitting establishes relative scale and orientation between uniquely corresponding features that minimizes global fitting error.

3.1   Feature Matching

The feature matching between model and target sets involves estimation of similarity between key point pairs in the compared features, estimation of relative scale, orientation as well as centroid location for the target features. The dissimilarity between feature p, corresponding to key point pairs ij, from the model set m and feature q, corresponding to key point pairs $i'j'$, from the target set t is obtained by comparing $f_{m,p}$ and $f_{t,q}$:

$$\epsilon_f(p, q) = (|f_{m,p} - f_{t,q}| \mod \pi) \tag{1}$$

The relative scale and orientation are given by $\zeta_{p,q} = d_p/d_q$ and $\omega_{p,q} = ((\alpha_p - \alpha_q) \mod \pi)$ respectively. Note that because of using key point pairs the relative scale and orientation are non-ambiguous. The estimation of centroid location in the target image is given by:

$$
\begin{aligned}
x_{ct}(p, q) &= \Big(\Delta x_{ic}(\Delta x_p \Delta x_q + \Delta y_p \Delta y_q) + \\
&\quad \Delta y_{ic}(\Delta y_p \Delta x_q - \Delta x_p \Delta y_q) \Big) / d_p^2 + x_{i'} \\
y_{ct}(p, q) &= \Big(\Delta x_{ic}(\Delta x_p \Delta y_q - \Delta y_p \Delta x_q) + \\
&\quad \Delta y_{ic}(\Delta x_p \Delta x_q + \Delta y_p \Delta y_q) \Big) / d_p^2 + y_{i'}
\end{aligned}
\tag{2}
$$

where $x_{i'}$ and $y_{i'}$ correspond to the position of the first key point in the feature q.

The negative aspect of using feature pairs is higher computational complexity. In a typical case feature matching would compare all possible combinations of features in two feature sets – if the model and target sets contain $K = 1000$ key points each[1] the matching procedure has to compare $(K^2 - K) \times (K^2 - K) \approx 10^{12}$ pair combinations which corresponds to quadratic complexity and leads to prohibitively high execution times. However, due to the simplicity of the feature descriptor we can partition features into a 4D array $\mathbf{F}$ representing a discrete space of $\mathbb{F}^4$. Each cell of the array $\mathbf{F}$ contains a sub-set of features corresponding to the cell span in $\mathbb{F}^4$, thus the matching of features in a single cell of the array $\mathbf{F}_m$ (model features) is confined only to the same cell in the array $\mathbf{F}_t$ (target features) and adjacent cells as shown in Figure 1. The cell span cs defines the similarity threshold at which relative segment angles are no longer compared. We have chosen a conservative value of $cs = 30°$ which allows maximum angle difference between two segments of $45°$ and produces 12^4 cells in the array. The efficiency benefit of this solution depends on the particular distribution of features in $\mathbf{F}_{m,t}$ spaces e.g. when features are distributed uniformly the speed up factor equals to $\frac{1}{3}12^4$. Typical matching times range from below a second ($K = 200$) to about 30 seconds ($K = 1000$ - complex scenes) on a 3GHz multi-core processor (which can be further improved by using GPU)[2].

The feature matching produces a set of soft correspondences, allowing association between a single model feature and k target features, which is a necessary measure to account for feature ambiguity and presence of multiple similar objects in the scene. The feature correspondences produce a centroid estimates according to (2) which are accumulated to generate hypotheses about the object location in the analyzed scene using a

---

[1] The images in the tested databases produce between 200 and 1000 key points.

[2] For simplicity we assumed that model and target have identical number of key points.

Fig. 2. The top row shows locations of estimated centroids in the target image (right image) given the model obtained from the features enclosed by the bounding box (left image). Despite the noisy model that contains also elements of the background, significant edge fragmentation and the differences in appearance (textures resulting in additional clutter) the strongest voting maximum is closely aligned with the true location of the object centroid. The bottom row shows accumulator arrays and voting maxima tracking across different resolutions. Images are best viewed in color.

Fig. 3. Example of the object localisation by matching and fitting a noisy model (contents of the bounding box in the left image). The second image shows the features extracted from the target image and the estimated location of the object centroid (strongest voting maximum). The right figure shows the alignment of the model (green, thick lines) with the uniquely corresponding features in the target image (red, thin lines). Note the amount of clutter in the target image.

Hough-like voting scheme [2]. The spread of the accumulated centroid votes depends on factors such as the amount of clutter present, overall shape similarity between model and target objects, and the relative scale at which features have been extracted [7]. Since this cannot be established a-priori, we adopt a simple multi-resolution refinement step, searching for voting maxima which are stable across different levels of granularity of the accumulator array as shown in Figure 2.

3.2 Model Fitting

Depending on the allowable degrees of freedom (e.g., rigid and non-rigid deformations), finding correspondences between model and image features often poses a costly combinatorial problem which gets quickly out of hands for more than a rather moderate

number of features involved. Typically, efficient strategies for searching less then optimal matching solutions are adopted to make the problem more tractable. Among those,
approaches based on Integer Quadratic Programming [3], graph cuts [13], and spectral
matching [8] have been shown to give excellent results in the context of object class
recognition.

However, despite their efficiency, the number of features that can be coped with is
limited to few hundred. Since our approach operates on a large number of feature pairs,
the amount of initial pair correspondences to be optimized requires a more efficient approach. E.g. for $K = 1000$ model key points and $k = 20$ correspondences per key point
pair we obtain up to $(K^2 - K) \times k \approx 20 \times 10^6$ soft correspondences, unfortunately
ruling out the use of the aforementioned methods. Therefore, we adopted a more practically usable and efficient procedure based on iterations of coarse model alignment and
feature pruning.

Specifically, an initial model position is obtained from the estimated centroid, while
relative scale and orientation are estimated from the soft feature correspondences that
casted votes for the centroid[3]. Since the initial estimation of position, scale and orientation cannot be expected to be accurate, it is optimized in an iterative process that
combines model fitting and soft-correspondence pruning until unique correspondences
are found. Here, due to the centroidal alignment, only a moderately sized sub-set of soft
correspondences voting on the centroid has to be processed in subsequent iterations of
the fitting procedure.

The following simplified fitting procedure is repeated for every centroid:

1. Obtain a list of soft correspondences that casted votes for the centroid (the list is
   produced during voting for each maximum in the voting accumulator at lowest
   resolution) $\mathbf{C} = [(p_1, q_1), (p_2, q_2), ..., (p_M, q_M)]$, where (p_m, q_m) are indexes of
   corresponding pairs in the model and target sets respectively. The correspondences
   are weighted (w_m) inversely proportional to the distance between their vote and the
   position of maximum in the voting accumulator.
2. Estimate scale $\bar{\varsigma}$ and orientation $\bar{\omega}$ of model relative to the corresponding target
   features:

$$\bar{\varsigma} = \frac{1}{\sum_M w_m} \exp \left(\sum_M w_m \log \left(\frac{d(q_m)}{d(p_m)} \right) \right) \tag{3}$$

where $d(p_m)$ and $d(q_m)$ are spatial distances between key point pairs p and q respectively.

$$\bar{\omega} = \frac{1}{\sum_M w_m} \sum_M w_m \left((\alpha_q - \alpha_p) \mod \pi \right) \tag{4}$$

3. Transform the model: scale by the factor $\bar{\varsigma}$, rotate by $\bar{\omega}$ and translate to the target
   centroid.
4. Estimate a similarity score $s_{p,q}$ for corresponding features that is a combination of
   spatial misalignment $\epsilon_s(p, q)$ and feature similarity $\epsilon_f(p, q)$:

$$s(p, q) = \exp \left(-\frac{(\epsilon_s(p, q) + \bar{\varsigma}\sigma_s\epsilon_f(p, q))^2}{2(\bar{\varsigma}\sigma_s)^2} \right) \tag{5}$$

---

[3] The fitting is repeated for each centroid detected in the target image.

where $\epsilon_s(p, q)$ is an Euclidean distance between transformed model key points and target key points in the corresponding features and σ_s is a parameter which binds spatial and angular alignment errors. Proposed measure produces similarity score 1 for perfectly aligned features ($\epsilon_s(p, q) = 0$ and $\epsilon_f(p, q) = 0$) and approaches 0 when $\epsilon_s(p, q) >> \sigma_s$ or $\epsilon_f(p, q) \rightarrow 2\pi$. All results presented in this paper were achieved with σ_s set to 0.1 of the maximum model extent (bounding box) although our experiments has shown that range between 0.05 and 0.2 produces almost identical fitting results.

5. Find all model features p_r that correspond to more than one target feature and for each feature p_r discard a correspondence that produced minimum similarity score.
6. Return to step 2 if any of the model features correspond to multiple target features.

Examples of feature matching and model fitting are shown in Figure 4.

4   Model Extraction

Our primary concern is the construction of object class model that contains a sufficient number of discriminative and repeatable features to maximize accuracy of object detection and classification.

In [10] and [12] the initial set of training features is reduced using a simple clustering technique and the discriminative features are selected by a training stage based on AdaBoost. Our approach follows this scheme. However, instead of initial feature reduction we produce a set of "sub-models" that represent groups of geometrically similar object instances in the training data set. The aim of sub-models is to capture a distinctive shape variations within the whole training set in terms of overall shape similarity and centroid localisation accuracy (see Figure 5). Such partitioning allows as to a) build more specific object models that increase fitting accuracy, b) minimize matching complexity and c) obtain more accurate feature alignment than it is possible with ordinary clustering approach.

The extraction of sub-models is a pre-processing step before the learning discriminative model, meant as a coarse data partitioning. The purpose of sub-models is to obtain a compact feature set from similar object instances and ensure that each sub-model preserves geometrical characteristics of represented shape. The sub-model extraction procedure consist of object instance grouping and feature compacting as follows:

1. Grouping starts with matching object instances in the training set, giving an estimate of global shape similarity and centroid estimation accuracy for every matched pair. The global shape similarity between an instance a and b is an average of feature similarities (5) obtained from model fitting $S_{a,b} = (\sum_M s(p_m, q_m))/M$ (instance a is the model and instance b is the target). Note that these estimates are asymmetric in general ($S_{a,b} \neq S_{b,a}$) due to different number of features in both instances and potential presence of non-repeatable background inside the bounding boxes. For that reason a symmetric similarity between two instances is defined as $\widehat{S}_{a,b} = S_{a,b} + S_{b,a}$.
2. Object instances are grouped using a hierarchical clustering on the global shape similarity $\widehat{S}$ with an additional constraint on maximum allowed centroid error $e_c(a, b)$

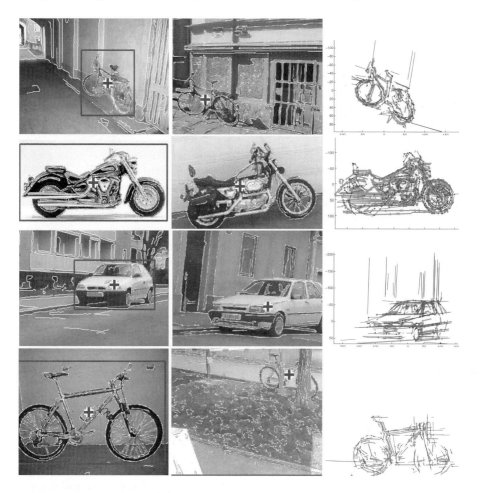

Fig. 4. Examples of object localisation (using noisy models) in the presence of scale, orientation, view point change and occlusion. The first column contains images of the model (enclosed by the bounding box) that are matched to the target images in the second column. Estimated centroids are shown in the target images while the model fitting is visualized in the third column.

and scale estimation error $e_s(a, b)$. The centroid localisation error is an Euclidean distance between detected and true positions of the centroid relative to the bounding box of instance b while the scale estimation error compares $\bar{\varsigma}$ (3) to the relative scale of bounding boxes in instances a and b. These constraints ensure that object instances with high centroid and scale estimation errors will not be grouped together. We have used conservative error thresholds $e_c < 2\sigma_s$ and $0.75 < e_s < 1.3$ (relative scale) for all evaluated image databases. The centroid and scale accuracy constraints typically result in 8-12 groups (see Figure 5 as an example).

3. In the final step features are compacted within each group of object instances. Corresponding key points from different object instances (exhibiting both feature similarity and global spatial alignment) form cliques that are averaged into a single

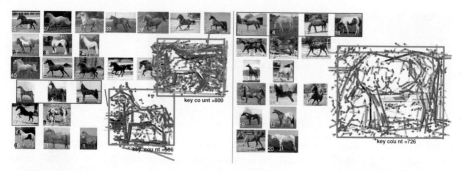

Fig. 5. Example of object instance grouping in the training data set based on overall shape similarity, centroid detection and scale estimation accuracy. Each row contains a group of similar object instances (note the figure is split into two columns). The final result depends on the training data, intra-class variability and the bounding box background variability. This example shows that it is possible to obtain meaningful groups of real objects with a low number of outliers. Resulting sub-models (only a subset is shown for clarity) display a reduced set of features, each visualized with a gray intensity corresponding to the associated strength.

model key point. A key point clique can be viewed as a connected graph with key point based nodes and correspondence based edges. The strength of the resulting key point is a sum of similarity scores (5) from all correspondences between merged key points and is used as a weight during casting centroid votes. Examples of sub-models produced by key point merging are shown in Figure 5.

The outline of our final feature selection and classifier learning is as follows. We combine sub-models into a global set of a spatially related features which will be pruned during the learning process. The sub-models are matched to the validation images to obtain a set of positive and negative training examples in terms of similarity and alignment of individual features. The role of sub-models is to localise and estimate pose of similar objects or shape structures in the validation images. The positive and negative examples however contain similarity scores (5) of every feature in the global model set that are transformed according to previously estimated pose and centroid location. Positive examples are obtained whenever one or more sub-models locates the same type of object in the validation image while negative examples are drawn from other object types and the background. The final object classifier and feature selection are produced by applying the Gentle-Boost learner to the set of positive and negative examples.

The Gentle-Boost classifier has a typical form of linear combination of weak classifiers:

$$H(d) = \sum_M a_m \left(s(p_{m,d}, q_{m,d}) > \theta_m \right) + b_m \tag{6}$$

where a_m, b_m and θ_m are learned parameters, d indicates a particular centroid/pose detection (more than one possible per image) while $p_{m,d}$ and $q_{m,d}$ are corresponding model and target features (model features are transformed). Depending on the training data sets the typical number of discriminative features selected varied between 300 and 400. The features that were dropped during classifier training are also removed from sub-models.

The extraction of training examples plays a critical role in obtaining a robust classifier. These examples must account for inaccuracy in centroid and pose estimation that is caused by the intra-class shape variability or change of view related to projective transformation. The examples produced by matching of sub-models to the validation images must be therefore artificially expanded by injecting potential errors into centroid and pose estimation in a similar manner as in [6,12]. These additional examples are produced by computing alignment of features and thus similarity measures for displaced centroid positions and slight scale variations. This procedure produces not only an additional positive examples (around the true centroid position) but also negative examples when the model is shifted toward the boundary of the bounding box in the validation image.

The process of feature selection and classifier training is repeated for every database separately. The negative examples used for boosting are obtained from the training images of the trained class (background outside of the bounding box) and training sets of other classes. This is done to obtain an object class detector that is not only able to discriminate object of particular type from a typical background but also to discriminate it from other object classes.

5   Evaluation

We test our approach on five databases listed in Table 1 that has been previously used for evaluation of other shape based detectors. We select a relatively small number of images ($< 10\%$) for sub-model extraction and another set of images to serve as validation data. The overall training data set do not exceed 25% of the whole database in each case. Tests were conducted on the combined set of test images drawn together from all databases.

To evaluate our approach we measure object detection and image classification accuracy for each object class separately. By object detection we understand localisation and classification of object instances as follows. We use sub-models to produce hypotheses d on object location (centroid) as described in Section 3.1 and 3.2. Next, the classification score $H(d)$ (6) is computed for each hypothesis. We use a simple non-maxima suppression on $|H(d)|$ to locally eliminate "weak" detections in overlapping regions (within 50% of the bounding box area). Remaining hypotheses d are classified as an "object" $H(d) > \Theta$ or "background" $H(d) \leqslant \Theta$, where Θ is a global confidence threshold regulating trade off between true and false positives. Resulting classification is compared against the ground truth to produce statistics on the number of true and false positives as a function of threshold Θ. A particular detection is associated with the object class if the area overlap between the detected (scaled) bounding box and the annotated bounding box is greater than 50% (assuming it contains the same object type) [1]. For the image classification results, the detection exhibiting the strongest classification confidence $\max_d |H(d)|$ is used to decide whether an instance of the object class is present in the image or not.

Table 1 provides the image classification and object detection accuracy along with the receiver operating characteristic (ROC) curve for the image classification case. The result of our evaluation gives an indication of how well the particular object class is discriminated against the background and *other object classes*. This is in our opinion

Table 1. Classification and detection performance on 5 image databases. The second column shows the true positive ratio for image classification at equal error rate (EER). Third column shows the area under ROC curve (ROC-AUC). The fourth column represents the area under Precision-Recall curve (PR-AUC) for the detection of object instances. Right: The ROC curve represents image classification accuracy for each of the tested databases, showing a trade off between true positives and false positives as the global confidence threshold is varied.

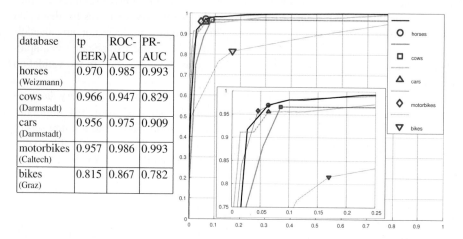

database	tp (EER)	ROC-AUC	PR-AUC
horses (Weizmann)	0.970	0.985	0.993
cows (Darmstadt)	0.966	0.947	0.829
cars (Darmstadt)	0.956	0.975	0.909
motorbikes (Caltech)	0.957	0.986	0.993
bikes (Graz)	0.815	0.867	0.782

a more realistic and challenging test scenario than the typical object detection against background only [12,10].

We benchmark our method against state-of-the-art approach from Shotton et al. [12]. Our approach achieves particularly good performance on the database of horses (PR-AUC of our method 0.993 vs. 0.968/0.785 in [12]) and bicycles (our 0.782 vs. 0.6959 [12]) considering that they were not split into side/front views as in [12]. Detection accuracy of motorbikes is almost identical in the two methods. Detection accuracy of Cows is worse than in [12], however the problem is primarily related to cows being confused with horses (not done in [12]) as well as imbalance in the number of test images (5:1) between these two databases.

6   Conclusions

We have presented a novel shape matcher and its application to discriminative object recognition. The shape matcher efficiently utilizes pairs of local shape fragments for robust model localisation. Although feature pairs have been previously exploited for matching and object recognition, we extend their use to provide invariance to rotation and scale effortlessly. Reported results show that our approach tolerates moderate view point changes, clutter and partial object occlusion (see Figures 4). Evaluation of object detection accuracy proves that the method is capable of outperforming state-of-the art detectors on challenging databases, containing multiple views of the same object class.

Our analysis of the method properties indicates that the combination of redundant features and the use of feature pairs plays a crucial role in object localisation and pose estimation while the use of sub-models (Section 4) significantly improves object detection accuracy. The use of multiple object models per class, feature sharing between these models and verification of different model extraction approaches is a primary focus of future work.

Acknowledgements

The research has received funding from the EC grant FP6-2006-IST-6-045350 (robots@ home).

References

1. http://www.pascal-network.org/challenges/VOC
2. Ballard, D.H.: Generalizing the hough transform to detect arbitrary shapes. Pattern Recognition 2(13) (1981)
3. Berg, A.C., Berg, T.L., Malik, K.: Shape matching and object recognition using low distortion correspondences. In: CVPR, pp. 26–33 (2005)
4. Crandall, D., Felzenszwalb, P., Huttenlocher, D.: Spatial priors for part-based recognition using statistical models. In: CVPR, pp. 10–17 (2005)
5. Ferrari, V., Fevrier, L., Jurie, F., Schmid, C.: Groups of adjacent contour segments for object detection. T-PAMI 30(1) (2008)
6. Laptev, I.: Improving object detection with boosted histograms. In: Image and Vision Computing (2008)
7. Leibe, B., Leonardis, A., Schiele, B.: Robust object detection with interleaved categorization and segmentation. IJCV 77(1-3), 259–289 (2008)
8. Leordeanu, M., Hebert, M.: A spectral technique for correspondence problems using pairwise constraints. In: ICCV, vol. 2, pp. 1482–1489 (2005)
9. Leordeanu, M., Hebert, M.: Beyond local appearance: Category recognition from pairwise interactions of simple features. In: CVPR, pp. 1–8 (2007)
10. Opelt, A., Pinz, A., Zisserman, A.: A boundary-fragment-model for object detection. In: Leonardis, A., Bischof, H., Pinz, A. (eds.) ECCV 2006. LNCS, vol. 3952, pp. 575–588. Springer, Heidelberg (2006)
11. Fergus, P.P.R., Zisserman, A.: A sparse object category model for efficient learning and exhaustive recognition. In: CVPR, pp. 380–387 (2005)
12. Shotton, J., Blake, A., Cipolla, R.: Multiscale categorical object recognition using contour fragments. T-PAMI 30(7), 1270–1281 (2008)
13. Win, J., Jojic, N.: Locus: learning object classes with unsupervised segmentation. In: ICCV, pp. 756–763 (2005)

A *Bag of Features* Approach for 3D Shape Retrieval

Janis Fehr[1,2], Alexander Streicher[2], and Hans Burkhardt[2]

[1] HCI, University Heidelberg, Germany
[2] LMB, University Freiburg, Germany
janis.fehr@iwr.uni-heidelberg.de

Abstract. In this paper, we present an adaptation the *Bag of Features* (BoF) concept to 3D shape retrieval problems. The BoF approach has recently become one of the most popular methods in 2D image retrieval. We extent this approach from 2D images to 3D shapes. Following the BoF outline, we address the necessary modifications for the 3D extension and present novel solutions for the parameterization of 3D patches, a 3D rotation invariant similarity measure for these patches and a method for the codebook generation. We experimentally evaluate the performance of our methods on the *Princeton Shape Benchmark*.

1   Introduction

The retrieval of digitized 3D objects is a rising topic. Similar to 2D image retrieval, which recently has become a very popular research topic, the constantly growing size of available 3D data triggers the need for effective search methods. There have been several practically important applications to 3D object retrieval, such as retrieving 3D protein structures from very large databases in bio-informatics and computational chemistry [1] or the retrieval of 3D objects from depth images (laser range scans) in robot navigation [2].

We apply our methods to a more academic problem setting given by the Princeton Shape Benchmark (PSB) [3], which has become a standard benchmark for 3D shape retrieval.

1.1   Related Work

We limit our brief review of the related work to methods which have been applied to the Princeton Shape Benchmark and thus can be compared to our results later on.

The **Spherical Extent Function (EXT)** [4] projects the distance of the object center to each point of the object surface onto the enclosing outer sphere. The resulting spherical distance map is then expanded in Spherical Harmonics from which the $\mathcal{SH}_{abs}$ (4) feature is extracted. The **Spherical Harmonic Descriptor (SHD)** [5] is very similar to EXT, it also computes $\mathcal{SH}_{abs}$ over several radii, but organizes the results in a 2D histogram. The **Light Field Descriptor**

G. Bebis et al. (Eds.): ISVC 2009, Part I, LNCS 5875, pp. 34–43, 2009.

(LFD) [6] uses multiple 2D views of 3D shapes. Rotation invariance is achieved by a collection of 100 2D views per object, which are rendered orthogonal to the outer enclosing sphere of the object. Then a set of 2D features (mostly geometric and Zernike moments) is computed for each 2D view. Currently, LFD is the best performing approach on the PSB.

All of these methods have in common that they try to model object shapes at a global level which has the disadvantage that the assumption that objects of the same class are sharing the same base shape is not always adequate - especially when one considers more semantic groupings with high intra-class variance as presented by the PSB. In 2D image retrieval, these problems have been approached quite successfully by BoF methods (see section 2). Hence, there have been several previous attempts to introduce a BoF approach for 3D shape retrieval, like [7], using Spin Images as local 3D patch descriptors. However, the results of these attempts were rather poor (see experiments), which we suspect to

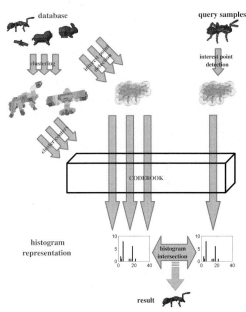

Fig. 1. Schematic overview of the "Bag of Features" concept

be mostly due to the limited discrimination power of spin images. We try to overcome these problems by the use of other local 3D patch descriptors.

2   3D Shape Retrieval with Local Patches

2.1   The *Bag of Features* Concept

One "state of the art" approach in modern 2D image retrieval is commonly known under the name *Bag of Features* (BoF) [8][9]. The method of BoF is largely inspired by the *Bag of Words* [10] concept which has been used in text retrieval for quite some time.

Even though there are countless variations of retrieval algorithms emerging under the label *Bag of Features* [8][9] and it is hard to capture *the* actual BoF algorithm, there is a common concept which is shared by all of theses methods.

Local Features: The central aspect of the BoF concept is to move away from a global image description and to represent images as a collection of local properties. These local properties are derived in form of (invariant) **image features**, e.g. the very popular SIFT features [11], which are computed on small sub-images called **patches**. The patches, are simply small rectangular regions which are extracted around **interest points** (see section 4).

Codebook Representation: The second main aspect of the "Bag of Features" concept is the way images are represented as collections of local features and how two or more of these representations can be compared. The basic idea here is to use a class independent **clustering** over the feature representations of all patches (from all database samples). The representatives of the resulting clusters are then used as entries of a unified (class independent) **codebook** [12]. Each patch is then mapped against the entries of this codebook, such that an image is represented as the **histogram** over the best codebook matches of its patches. The similarity of images can then be obtained by comparing the BoF histograms, e.g. by histogram intersection.

Figure 1 gives a schematic overview of the computation steps in the codebook generation and retrieval stage of the 'Bag of Features" concept.

3   Mathematical Foundations

A key aspect of our 3D extension of the BoF concept is the idea to parameterize the 3D patches as spheres and to provide a fast rotation invariant similarity measure for these spherical patches. The spherical representation is a natural choice for local 3D patches which allows us to rely on the well established mathematical foundations of the *angular momentum theory* [13] to perform all necessary computation in the harmonic domain.

Spherical Harmonics (SH): [13] form an orthonormal base on the 2-sphere. Analogical to the Fourier Transform, any given real valued and continuous signal f on a sphere with its parameterization over the angles Θ, Φ (latitude and longitude of the sphere) can be represented by an expansion in its harmonic coefficients:

$$f(\Theta, \Phi) = \sum_{l=0}^{\infty} \sum_{m=-l}^{m=l} \hat{f}_m^l \, \overline{Y_m^l}(\Theta, \Phi), \tag{1}$$

where l denotes the band of expansion, m the order for the l-th band, $\hat{f}_m^l$ the harmonic coefficients and $\overline{Y_m^l}$ the complex conjugate of the harmonic base functions Y_m^l that are computed as:

$$Y_m^l(\Theta, \Phi) = \sqrt{\frac{2l+1}{4\pi} \frac{(l-m)!}{(l+m)!}} P_m^l(\cos \Theta) e^{im\Phi}, \tag{2}$$

where P_m^l is the associated Legendre polynomial.

Rotations in SH: Rotations $\mathcal{R}(\varphi, \theta, \psi)f$ in the Euclidean space find their equivalent representation in the harmonic domain in terms of the so called Wigner

D-Matrices [13], which form an irreducible representation of the rotation group $\mathcal{SO}(3)$. For each band l, $D^l(\varphi, \theta, \psi)$ (or abbreviated $D^l(\mathcal{R})$) defines a band-wise rotation in the $\mathcal{SH}$ coefficients, using the Euler notation in zyz-convention with $\varphi, \psi \in [0, 2\pi[$ and $\theta \in [0, \pi[$ to parameterize the rotations $\mathcal{R} \in \mathcal{SO}(3)$. Hence, a rotation in the Euclidean space can be estimated in the harmonic domain (with a maximum expansion band b_{max}), by

$$\mathcal{R}f \approx \sum_{l=0}^{b_{max}} \sum_{m=-l}^{l} \sum_{n=-l}^{l} D_{mn}^l(\mathcal{R}) \hat{f}_m^l Y_m^l. \tag{3}$$

The $\mathcal{SH}_{abs}$ Feature: A $\mathcal{SH}$ representation of spherical signals raises the demand for a (rotational invariant) similarity measure between two or more signals. A popular choice [5] is to use the band-wise absolute values of the harmonic power-spectrum, which we refer to as $\mathcal{SH}_{abs}$ feature:

$$\mathcal{SH}_{abs}(\hat{f}^l) := \sum_{m=-l}^{m=l} \|\hat{f}_m^l\|. \tag{4}$$

The main drawback of the $\mathcal{SH}_{abs}$ features is that it obtains its rotation invariance by neglecting the phase information. Hence, $\mathcal{SH}_{abs}$ is an incomplete feature which suffers from its ambiguities.

Fast Correlation in $\mathcal{SH}$: We follow a different approach to obtain a rotation invariant similarity measure between harmonic expansions: the full correlation over all possible rotation angles: The full correlation $f\#g : \mathcal{SO}(3) \to \mathbb{R}$ of two signals f and g under the rotation $\mathcal{R} \in \mathcal{SO}(3)$ on a 2-sphere is given as:

$$(f\#g)(\mathcal{R}) := \int_{S^2} f(\mathcal{R}g) \quad d\phi d\theta d\psi. \tag{5}$$

Fehr et. all. [14] proposed a method for a fast computation of (5) in the harmonic domain by use the *Convolution Theorem*. Starting from the substitution of f and g in (5) by their SH expansions

$$(f\#g)(\mathcal{R}) = \sum_{lmn} \overline{D_{mn}^l(\mathcal{R})} \hat{f}_m^l \overline{\hat{g}_n^l}, \tag{6}$$

their method provides the correlation value for each possible ration in a discrete 3D matrix $C^\#$ which represents the angular space over the rotation angles (ϕ, θ, ψ):

$$C^\# = \mathcal{F}^{-1}(\widehat{C^\#}), \tag{7}$$

with

$$\widehat{C^\#}(m, h, m') = \sum_{l=0}^{b_{max}} d_{mh}^l(\pi/2) d_{hm'}^l(\pi/2) \hat{f}_m^l \overline{\hat{g}_{m'}^l} \tag{8}$$

and $m, h, m' \in \{-l, \ldots, l\}$. The rotation invariant correlation maximum is then simply the maximum value in $C^\#$. Please refer to [14] details on (7) and proofs.

Normalized Cross-Correlation: We follow an approach which is widely known from the normalized cross-correlation of 2D images: First, we subtract the mean from both functions prior to the correlation and then divide the results by the variances:

$$(f\#g)_{norm}(\mathcal{R}) := \int_{\mathcal{S}^2} \frac{\big(f - \mathbf{E}(f)\big)\big(\mathcal{R}(g - \mathbf{E}(g))\big)}{\sigma_f \sigma_g} \quad \sin\Theta d\Phi d\Theta. \qquad (9)$$

Analogous to Fourier transform, we obtain the expectation values $\mathbf{E}(f)$ and $\mathbf{E}(g)$ directly from the 0th $\mathcal{SH}$ coefficient. The variances σ_f and σ_g can be estimated from the band-wise energies:

$$\sigma_f \approx \sqrt{\sum_l |\widehat{f}_l|^2}. \qquad (10)$$

Discrete Spherical Harmonic Expansions: For practical applications, we need a discrete version of the Spherical Harmonic transform, i.e. we need to obtain the frequency decomposition of 3D signals at discrete positions $\mathbf{x} \in X$ on discrete spherical surfaces $\mathcal{S}$ of radius r:

$$\mathcal{S}[r](\mathbf{x}) := \{\mathbf{x}' \in \mathbb{R}^3 | \quad \|\mathbf{x} - \mathbf{x}'\|_2 = r\}. \qquad (11)$$

To obtain the discrete Vectorial Harmonic transformation $\mathcal{SH}\big(\mathcal{S}[r](\mathbf{x})\big)$, we pre-compute discrete approximations $\tilde{Y}_m^l[r]$ of the orthonormal harmonic base functions as:

$$\mathcal{SH}\big(X|_{\mathcal{S}[r](\mathbf{x})}\big)_m^l := \sum_{\mathbf{x_i} \in \mathcal{S}[r](\mathbf{x})} X(\mathbf{x_i})\tilde{Y}_m^l[r](\mathbf{x_i}). \qquad (12)$$

In order to compute the harmonic transformation of the neighborhoods around each voxel of X, we perform a fast convolution of the pre-computed based functions with the discrete input data:

$$\mathcal{SH}[r](X)_m^l = X * \tilde{Y}_m^l[r]. \qquad (13)$$

4   Algorithm

Our approach directly follows the *Bag of Features* scheme (see figure 1). With exception of an additional preprocessing, we simply walk through the BoF pipeline step by step and replace 2D specific algorithms with our own 3D methods.

Preprocessing: Prior to the actual BoF pipeline, we apply a series of pre-processing steps to the objects in the PSB database: primarily, we have to render the objects from triangular mesh format to a volume representation where the voxels inside the object are set to 1 and the voxels outside to 0. We use this rendering step to align the object in the geometric center of the volume and to normalize the object size to a fixed height of the object bounding box. Thus, we obtain translation and scale invariant volume representations of the models.

Sampling Points: The next step, and first in the actual BoF pipeline, is to determine the location of the local patches. In the original 2D setting, where the objects of interest are located in more or less cluttered scenes, the detection of interest points is a crucial step: important parts of the target objects should not be missed, while the overall number of interest points directly affects the computational complexity, so that one tries to avoid large numbers of false positive points. For our 3D case, the selection of the interest points is by far less crucial since we have already segmented objects. Hence, we simply apply a simple equidistant sampling on the object surface.

Extracting Local Patches: The next step is to extract the local patches $p(\mathbf{x})$ at the location of the sampling points. In contrast to the original 2D case, were the patches are rectangular areas, we extract spherical patches which are centered in the respective sampling points.

Given the volume rendering of a model X and sampling points at positions $\mathbf{x}$, the associated patches are then represented by a series of m concentric spherical neighborhoods $X|_{\mathcal{S}[r_i](\mathbf{x})}$ (12) at different radii $r_i \in \{r_1, \ldots, r_n\}$. We then expand the local patches radius by radius in Spherical Harmonics. Hence, we define a patch $p(\mathbf{x})$ as collection of radius-wise harmonic expansions up to some upper band $l = b_{max}$ around $\mathbf{x}$:

$$p(\mathbf{x}) := \left\{ \mathcal{SH}\big(X|_{\mathcal{S}[r_1](\mathbf{x})}\big), \ldots, \mathcal{SH}\big(X|_{\mathcal{S}[r_n](\mathbf{x})}\big) \right\}. \tag{14}$$

Figure 3a and 3b illustrate the patch extraction. The motivation to use spherical instead of rectangular patches is obvious considering that we need to obtain full rotation invariance, which often times can be neglected in the case of 2D image retrieval.

Generating the Codebook: While the preprocessing and patch extraction has to be done for all reference and query objects, we now turn to the off-line procedure which is only performed on the initial database. The off-line stage has two different stages: first, we have to generate a problem specific but class independent

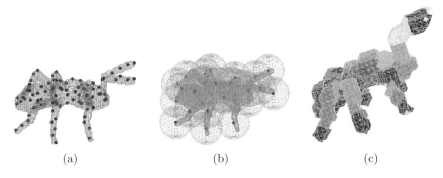

(a)   (b)   (c)

Fig. 3. (a) Extracting spherical patches at the sampling points. (b) extraction of spherical neighborhoods (patches) around the sampling points. (c) Example clustering results where patches were extracted at all object voxels.

Fig. 4. Sample generalized codebook entry: the figure illustrates the location (blue circle) of a sample codebook entry on several different objects

codebook of local patches, which is done via clustering, and then, we have to represent the database samples in terms of histograms over the codebook.

After the extraction of the patches, we use a simple radius-wise k-means clustering [15] to obtain k patch clusters for each radius of the patch parameterization independently. The key for the clustering is the choice of the similarity function d: we apply the normalized correlation (10) for the Spherical Harmonic domain to measure the rotation invariant similarity of two patches:

$$d\big(p(\mathbf{x_i}), p(\mathbf{x_j})\big) := p(\mathbf{x_i}) \# p(\mathbf{x_j}). \tag{15}$$

In order to reduce the computational complexity, we do not apply the clustering on all patches from all database samples. Our experiments showed (see 5), that it is sufficient to use a small random subset of 10% of the database to generate a stable set of clusters.

It should be noted, that the class label of the database samples is completely neglected during the clustering since our goal is to obtain a general, class independent representation of local patches in the later codebook. Figure 3c shows example results of the clustering. The final step towards the generation of the codebook is based on the previous clustering. We simply use the cluster centers as representatives in the codebook. Since we perform the clustering for each radius of the patch parameterization independently, we obtain separate codebooks for each radius.

4.1   Retrieval by Feature Histograms

After we learned the codebook based on a small subsection of the database, we can pursuit the BoF approach without further changes. As in the original *Bag of Features* concept, all off the database samples are represented as histograms over the codebook entries. We simply use our fast normalized cross-correlation (9) to match all patches of an object with the codebook and rise the count of the histogram bin associated with the best match. Figure 1 illustrates a example codebook histogram representation of an object.

Features	PSB *base* level
LFD	$65.7\%^{\dagger}(61.9\%)^{*}$
BoF with $\mathcal{SH}_{corr}$	**62.4%**
SHD	$55.6\%^{\dagger}(52.3\%)^{*}$
EXT	$54.9\%^{\dagger}$
BoF with $\mathcal{SH}_{abs}$	54.5%
BoF with Spin Images	33.5%

Fig. 5. Results for the 3D shape retrieval on the PSB. Results taken from the literature are marked with †, results from our own implementations are marked with ∗. Unfortunately, we were not able to exactly reproduce the results given in the literature. This could be caused by a different initial rendering, which is not discussed in the given literature.

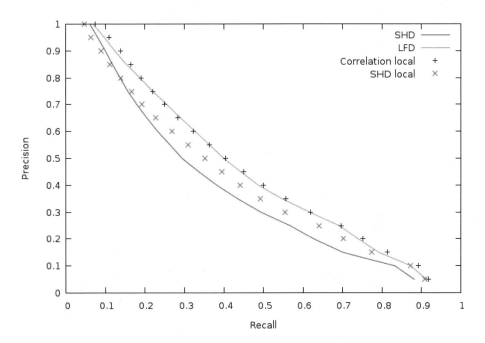

Fig. 6. Precision recall graph for our approach on the PSB *base* test set. The results of our implementation of LFD and SHD reference methods are plotted as lines. We also compare the BoF results with our $\mathcal{SH}_{corr}$ features compared to the use of the $\mathcal{SH}_{abs}$ features as patch descriptor.

Retrieval Query: Given an query object, we perform the preprocessing and patch extraction steps and then compute its codebook histogram representation just as we do it for the database. We then use a normalized histogram intersection as similarity measure to find the best matches in the database.

5   Experimental Evaluation

We evaluate our proposed approach on the standard PSB experimental setup, as described in [3]. We use the *base* scheme, where the 1814 shapes of the PSB are split into equally large database and query sets.

General Experimental Setup: We use a rendering normalized to the size of 64 voxels on the longest edge of the bounding box (see 4). The codebook is built form a random selection of 10% of the training set (more samples simply increase the computation time without notable effect on the later recognition rate). 8 different radii with $r_i \in \{3, 4, 5, 6, 7, 8, 10, 12\}$ are used to compute the codebook, where b_{max} of the harmonic expansion is increased according with the radius (from 3 to 7). The codebook size k is set to 150 bins per radius.

Given these fixed parameters, we obtained the following results for our approach on the PSB *base* test set: table 5 shows the $k = 1$ nearest neighbor results of our method compared to the results known from literature. Figure 6 shows the precision-recall plots provided by the standardized PSB evaluation. In order to emphasize the use of our $\mathcal{SH}_{corr}$ feature, we additionally implemented a BoF approach where we used the $\mathcal{SH}_{abs}$ feature as patch descriptor.

6   Conclusions and Outlook

The experiments showed that our approach achieves competitive results on the difficult PSB. The main drawback of our method is that we cannot be sure if the given results are actually representing the global optimum of what can be achieved with our method or if we are stuck in a local maximum of the parameter space. Due to the large number of parameters, we face the problem that the maximum search in the parameter space turns out to be quite tedious. A possible further extension of our histogram approach could be to localize the patch positions. Similar to the 2D approach in [16], we could increase the discrimination power by replacing the global histogram with a localized patch histogram.

References

1. Reisert, M., Burkhardt, H.: Irreducible group representation for 3d shape description. In: Franke, K., Müller, K.-R., Nickolay, B., Schäfer, R. (eds.) DAGM 2006. LNCS, vol. 4174, pp. 132–141. Springer, Heidelberg (2006)
2. Wolf, J., Burgard, W., Burkhardt, H.: Robust vision-based localization by combining an image retrieval system with Monte Carlo localization. IEEE Transactions on Robotics 21(2) (2005)
3. Shilane, P., Min, P., Kazhdan, M., Funkhouser, T.: The princeton shape benchmark. In: Shape Modeling International, Genova, Italy (2004)
4. Saupe, D., Vranic, D.V.: 3d model retrieval with spherical harmonics and moments. In: Radig, B., Florczyk, S. (eds.) DAGM 2001. LNCS, vol. 2191, pp. 392–397. Springer, Heidelberg (2001)

5. Kazhdan, M., Funkhouser, T., Rusinkiewicz, S.: Rotation invariant spherical harmonic representation of 3d shape descriptors. In: Symposium on Geometry Processing (2003)
6. Chen, D.Y., Ouhyoung, M., Tian, X.P., Shen, Y.T.: On visual similarity based 3d model retrieval. In: Computer Graphics Forum, pp. 223–232 (2003)
7. Li, X., Godil, A., Wagan, A.: Spatially enhanced bags of words for 3d shape retrieval. In: Bebis, G., Boyle, R., Parvin, B., Koracin, D., Remagnino, P., Porikli, F., Peters, J., Klosowski, J., Arns, L., Chun, Y.K., Rhyne, T.-M., Monroe, L. (eds.) ISVC 2008, Part I. LNCS, vol. 5358, pp. 349–358. Springer, Heidelberg (2008)
8. Mikolajczyk, K., Leibe, B., Schiele, B.: Local features for object class recognition. In: ICCV 2005: Proceedings of the Tenth IEEE International Conference on Computer Vision, pp. 1792–1799. IEEE Computer Society, Los Alamitos (2005)
9. Sivic, J., Russell, B.C., Efros, A.A., Zisserman, A., Freeman, W.T.: Discovering objects and their location in images. In: Tenth IEEE International Conference on Computer Vision, 2005. ICCV 2005, vol. 1, pp. 370–377 (2005)
10. Blei, D., Ng, A., Jordan, M.: Latent dirichlet allocation. J. Mach. Learn. Res. 3, 993–1022 (2003)
11. Lowe, D.: Distinctive image features from scale-invariant keypoints. International Journal of Computer Vision 60, 91–110 (2004)
12. Jurie, F., Triggs, B.: Creating efficient codebooks for visual recognition. In: ICCV 2005: Proceedings of the Tenth IEEE International Conference on Computer Vision (ICCV 2005), vol. 1, pp. 604–610. IEEE Computer Society, Los Alamitos (2005)
13. Brink, D.M., Satchler, G.R.: Angular Momentum, 2nd edn. Clarendon Press, Oxford (1968)
14. Fehr, J., Reisert, M., Burkhardt, H.: Fast and accurate rotation estimation on the 2-sphere without correspondences. In: Forsyth, D., Torr, P., Zisserman, A. (eds.) ECCV 2008, Part II. LNCS, vol. 5303, pp. 239–251. Springer, Heidelberg (2008)
15. Bishop, C.M.: Pattern Recognition and Machine Learning (Information Science and Statistics). Springer, Heidelberg (2006)
16. Leibe, B., Schiele, B.: Interleaved object categorization and segmentation. In: British Machine Vision Conference (BMVC 2003), Norwich, UK, September 9-11 (2003)

Efficient Object Pixel-Level Categorization Using Bag of Features

David Aldavert[1], Arnau Ramisa[2], Ricardo Toledo[1],
and Ramon Lopez de Mantaras[2]

[1] Computer Vision Center (CVC)
Dept. Ciències de la Computació
Universitat Autònoma de Barcelona (UAB), 08193, Bellaterra, Spain
{aldavert,ricard}@cvc.uab.cat
[2] Artificial Intelligence Research Institute (IIIA-CSIC)
Campus de la UAB, 08193, Bellaterra, Spain
{aramisa,mantaras}@iiia.csic.es

Abstract. In this paper we present a pixel-level object categorization method suitable to be applied under real-time constraints. Since pixels are categorized using a bag of features scheme, the major bottleneck of such an approach would be the feature pooling in local histograms of visual words. Therefore, we propose to bypass this time-consuming step and directly obtain the score from a linear Support Vector Machine classifier. This is achieved by creating an integral image of the components of the SVM which can readily obtain the classification score for any image sub-window with only 10 additions and 2 products, regardless of its size. Besides, we evaluated the performance of two efficient feature quantization methods: the Hierarchical K-Means and the Extremely Randomized Forest. All experiments have been done in the Graz02 database, showing comparable, or even better results to related work with a lower computational cost.

1 Introduction

A method for robustly localizing objects is of major importance towards creating smart image retrieval tools able to search in digital image collections. In the last years, object recognition in images has seen impressive advances thanks to the development of robust image descriptors [1] and simple yet powerful representation method such as the *bag of features* [2,3,4,5]. Furthermore, the ever-growing collections of images available on the Internet make computational efficiency an imperative for methods that aim to be used in such a scenario.

In this work we propose a new method for fast pixel-wise categorization based on the bag of features object representation. Given that the method will have to be applied at every pixel of the image, it is essential to optimize it to perform in the least possible time. Although different bag of features approaches have been proposed, all of them consist on four basic steps. Namely, feature extraction from the image, feature quantization into visual words, accumulation of visual word into histograms and classification of the resulting histogram.

G. Bebis et al. (Eds.): ISVC 2009, Part I, LNCS 5875, pp. 44–54, 2009.

In order to accelerate the quantization step, Nister and Stewenius [3] proposed to use a Hierarchical K-Means vocabulary tree (HKM) created by recursively applying k-means to the clusters from the previous level. With this technique they were able to improve the recognition results by using larger dictionaries in a reasonable time. As an alternative to the HKM, Moosmann et al. [6] and Shotton et al. [7] proposed to use Extremely Randomized Forests (ERF) of K-D trees to improve the classification accuracy. In the approach proposed by Moosemann et al. random features are selected at high saliency areas and used in a bag of features scheme to classify the whole image. Besides, the probability of each object class for each individual feature is determined and used in an object probability map that iteratively decides the geometric parameters of new random features.

The accumulation step is an even more critical bottleneck for an object localization using bag of features. Since no geometrical information is used, many sub-windows of the image have to be evaluated. Some authors addressed this problem by reducing the number of evaluated windows, either by using a pre-processing step [8] or by searching the best sub-window as in an optimization problem [9]. This is done by defining an upper bound on the SVM classification, and using branch and bound to discard uninteresting areas. Although not used here, the strategy proposed by Lampert et al. is also applicable in our method.

Other authors have focused on accelerating the accumulation step. In the approach by Fulkerson et al. [4], the authors speed up the accumulation step using integral images in a sliding windows based analysis of the image. For this speed-up measure to be effective, it is important to use small dictionaries. However, various works [10,3] show that large dictionaries typically obtain better classification results. Therefore, in order to compress the dictionary without losing classification accuracy, they propose to use Agglomerative Information Bottleneck (AIB) to create a coarse-to-fine-to-coarse architecture that is optimized for discrimination of object versus non-object. This approach shares some similitudes with the one presented here. However, In contrast to Fulkerson et al. , we propose to bypass the descriptor accumulation step, and make every computed feature vote directly with its classifier score in an integral image to reduce the computational cost of classifying an image sub-window to only 10 additions and 2 multiplications, regardless of its size.

The rest of the paper is organized as follows: In Section 2 the proposed methodology for object classification and localization in images is described. Then, in Section 3, our proposed method is evaluated with the Graz02 dataset [11] and results are presented and discussed. Finally, in Section 4 the contributions and conclusions of this work, as well as future research directions, are summarized.

2   Efficient Pixel-Level Categorization

Our method[1] uses an efficient categorization algorithm to assign a category label to each pixel of an image: First, region descriptors are densely sampled from the

---

[1] Additional information available at http://www.cvc.uab.cat/~aldavert/plor/

a)   b)   c)

Fig. 1. Examples of pixel-level categorization results obtained with our method for a) bikes, b) cars and c) person in the Graz02 database

image and quantized into visual words using a codebook. Then, a sliding window scheme is used to assign a category label to each pixel of the image. Visual words within a window are accumulated in a histogram, which is later classified using a linear support vector machine. In Fig. 1 some pixel-level categorization results obtained using the proposed method are shown. Categorizing all the pixels from an image with this brute-force approach in a reasonable time requires each of the previous steps to be executed in a very efficient way.

2.1 Dense Features

As previously mentioned, we use densely sampled image descriptors as input data. Dense sampling has several advantages when compared to keypoint-based approaches, such as extracting more information from the underlying image, and avoiding the time-consuming keypoint detection step [10]. Furthermore, if robust descriptors can be computed in an efficient way, it can even become faster than the keypoint-based alternative despite the larger number of descriptors computed.

With this in mind, we have decided to use the Integral Histograms of Oriented Gradients (IHOG) descriptor [12]. The IHOG is an approximation to the Histograms of Oriented Gradients (HOG) descriptor [13], which speeds up the descriptor extraction using integral images. First, each pixel votes according to its gradient orientation, weighted by its gradient magnitude, in a histogram of N orientation bins. Then, an integral image is generated for each one of the N orientation bins. Using these integral images, to compute an IHOG descriptor with N orientation bins and $P \times P$ position bins (i.e. a $N \times P \times P$ dimensions descriptor) we need just $N \times (P-1)^2$ memory accesses and $N \times P^2$ additions, regardless of the feature region size in pixels.

Unlike the HOG descriptor, the IHOG descriptor is incompatible with the Gaussian mask and the tri-linear interpolation to weight the contribution of the gradient module in the spatial bins of the descriptor used in the HOG. Another difference is that the IHOG descriptor uses L1 normalization instead the L2 normalization. Nevertheless, despite all these simplifications, the performance of the IHOG descriptor is only slightly worse than that of the HOG descriptor [12]. Moreover, neither HOG nor IHOG descriptors are rotation invariant. However, according to Zhang et. al. [14], the use of rotation invariant descriptors has a negative effect in the performance of bag of features approaches.

2.2   Codebook Generation

Once all descriptors have been computed from the image, it is necessary to quantize them into visual words using a codebook. The computational cost of quantizing a D-dimensional descriptor using linear codebook of V visual words is $O(DV)$. From the various alternatives that have been proposed to reduce this computational cost, in this work we have evaluated two: the Hierarchical K-Means (HKM) and the Extremely Randomized Forest (ERF).

The HKM defines a hierarchical quantization of the feature space. Instead of k being the final number of visual words of the codebook, it determines the branch factor (number of children of each node) of a tree. Given a set of training descriptors, an HKM is generated as follows: First, the k-means algorithm is used to split the training data into k groups. Then, this clustering process is recursively applied to the groups from the previous level until a maximum depth is reached. This recursive method creates a vocabulary tree (i.e. codebook) with a reduced computational cost both in the training and descriptor quantization phases. The computational complexity of quantizing a D-dimensional descriptor using a HKM with V visual words is $O(Dk \log_k V)$. In the original implementation of the HKM, all nodes of the tree are used as visual words to alleviate misclassification problems in the superior levels of the tree and the contribution of each node of the histogram is weighted using a TF-IDF scheme [3]. However, the use of these two refinements have a modest impact in the performance of the HKM. Therefore, we have removed them from our implementation.

The ERF [6] uses a combination of several random K-D trees in order to quantize the feature space. Given a set of labeled training descriptors (i.e. descriptors with a category label associated), the K-D trees of the ERF are built recursively in a top-down manner as follows: Every node of the K-D trees splits the training descriptors from the previous level in two disjoint sets with a boolean test in a random descriptor vector position. The boolean test consists in dividing the descriptors in two groups according to a random threshold θ_t applied at descriptor vector dimension D_t, also chosen randomly. For each node, the random boolean test is scored using the Shannon entropy until a minimum value S_{min} is attained or a maximum number of trials T_{max} has been reached. Then, the selected random boolean test is the one that has a highest score. Parameter S_{min} can be used to select the randomness of the obtained K-D trees. For instance $S_{min} = 1$ creates a highly discriminant tree while $S_{min} = 0$ creates a completely random tree. The main advantage of the random K-D tree compared to other quantization methods is its low computational cost. Quantizing a D-dimensional descriptor vector using a random K-D tree with V visual words is $O(\log_2 V)$. Since a random K-D tree usually has less discriminative power than other clustering methods, like the k-means or the HKM, several K-D trees are combined together to obtain a more discriminative codebook. Finally, the resulting histogram of the ERF is created by concatenating the histograms generated by each K-D tree of the forest.

2.3   Integral Linear Classifiers

The "integral image" representation has been first introduced by Viola and Jones to quickly extract Haar-wavelet type features [15]. Since then, integral images have been applied to many different tasks like invariant feature extraction [16], local region descriptors [12], to compute histograms over arbitrary rectangular image regions [17] or to compute bag of feature histograms [4]. Inspired by these previous works, we propose the use of an integral image to quickly calculate the output score of the linear classifier which is applied to bag of features histograms.

To categorize a V dimensional histogram of visual words, we use a linear classifier with weight vector $\boldsymbol{W}$ and bias b. Then, the output score of the linear classifier is:

$$\frac{1}{\|\boldsymbol{X}\|} \sum_{i=0}^{V} x_i w_i + b > 0 \tag{1}$$

where x_i is the frequency of the i-th visual word of the codebook, $\|\boldsymbol{X}\|$ is the norm of histogram $\boldsymbol{X}$ and w_i is the i-th component of the linear classifier weight vector $\boldsymbol{W}$. If all components of $\boldsymbol{W}$ are positive, then, the sum of the previous

a)   b)

Fig. 2. Image containing the components of a linear classifier for bikes b) obtained from extracting dense features every four pixels in the original image a)

equation can be calculated using an integral image. Therefore, we define the classifier weight vector $\tilde{W}$ components as:

$$\tilde{w}_i = w_i - W_m \tag{2}$$

where W_m is the w_i component with the lowest value. Then, replacing W by $\tilde{W}$ in Eq. 1 the output score of the linear classifier is:

$$\frac{1}{\|X\|} \sum_{i=0}^{V} x_i \tilde{w}_i + \frac{W_m}{\|X\|} \sum_{i=0}^{V} x_i + b > 0 \tag{3}$$

We normalize the histogram X using L1 norm (i.e. the amount of visual words that casted a vote in the histogram) since it is fast to compute using an integral image. Then, Eq. 3 becomes:

$$\frac{1}{N} \sum_{i=0}^{V} x_i \tilde{w}_i + W_m + b > 0 \tag{4}$$

where N is the L1 normalization of histogram $\|X\|$. Once all $\tilde{W}$ components are positive, the integral image can be used to calculate the sum in Eq. 4. For each linear classifier c, let $L_c(x, y)$ be the sum of components $\tilde{w}_i^c$ corresponding to the visual words at pixel (x, y). In Fig. 2 an example of L_c image for the *bikes* classifier is shown. Then, each image L_c is transformed into an integral image I_c, so that, the sum of Eq. 4 of a rectangular image region R can be calculated using the integral image I_c:

$$H_R = I_c(x_u, y_u) + I_c(x_b, y_b) - I_c(x_u, y_b) - I_c(x_b, y_u) \tag{5}$$

where (x_u, y_u) and (x_b, y_b) are respectively the upper left and bottom right corner coordinates of region R. Then, the output score of a linear classifier applied to any rectangular image region can be calculated as follows:

$$\frac{1}{N} H_R + W_m + b > 0 \tag{6}$$

Using integral images, the computational complexity of classifying any rectangular image region is reduced to 8 memory access, 10 additions and 2 products, independently of the size of rectangular region.

3   Experiments

We have evaluated the performance of our pixel-level categorization method on the Graz02 database [11]. The Graz02 database is a challenging database consisting on three categories (bikes, cars and people) where objects have an extreme variability in pose, orientation, lighting and different degrees of occlusion. The Graz02 annotation only provides a pixel segmentation mask for each image, so that, it is impossible to know how many object instances are present in the image. In consequence, to evaluate the performance of our pixel-level categorization method we use the pixel-based precision-recall curves as in [18]. Active pixels of the ground truth segmentation mask scorrectly categorized as object are counted as true positives, and as false negatives otherwise. Also, incorrectly classified background pixels of the ground truth segmentation mask are counted as false positives. Finally, we have taken the odd images as train and the even as test as in [18,4]. However, due to the high variation we observed in the results depending on the train/test sets, we decided to also use random selection to split half of the images for train and half for test. The final result of a test when using the random sampling is the mean of a 1,000-repetitions experiment to ensure statistical invariance of the selected train/test sets.

3.1   Parameter Setting

The results were obtained using the same parameters in each experiment. The IHOG descriptors have been densely sampled each four pixels. Descriptors that have a low gradient magnitude before normalization are discarded as in [4]. Each IHOG descriptor is extracted from a 40×40 pixels patch and it has 8 orientation bins and 4 positional bins (i.e. a 32 dimensional descriptor). Therefore, as Graz02 images have a regular size of 640×480, a maximum of 16,500 descriptors are extracted per image. Then, bag of features histograms are computed accumulating the visual words that are inside a region of 80×80 pixels. Later, those histograms are categorized using a support vector machine. The SVM has been trained using logistic regression (LR-SVM)[19] with the LIBLINEAR software package [20]. Finally, the shown times results have been obtained using laptop with an Intel T7700 Core Duo CPU and 2Gb of RAM.

3.2   Parameters of the ERF

The performance of the ERF depends on the K-D tree randomness parameter and the amount of trees in the forest. Therefore, we wanted to evaluate

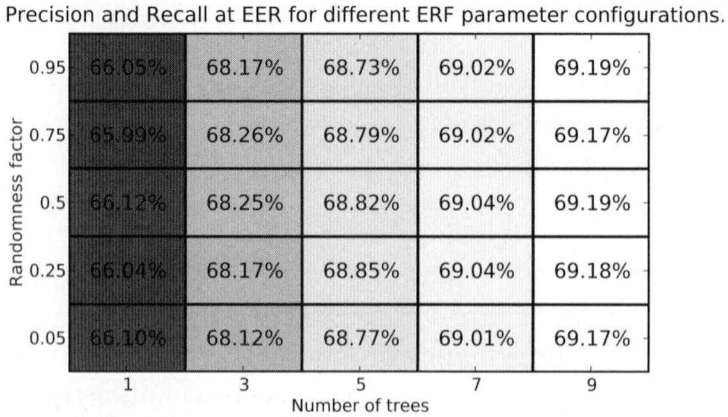

Precision and Recall at EER for different ERF parameter configurations.

Randomness factor	Number of trees				
	1	3	5	7	9
0.95	66.05%	68.17%	68.73%	69.02%	69.19%
0.75	65.99%	68.26%	68.79%	69.02%	69.17%
0.5	66.12%	68.25%	68.82%	69.04%	69.19%
0.25	66.04%	68.17%	68.85%	69.04%	69.18%
0.05	66.10%	68.12%	68.77%	69.01%	69.17%

Fig. 3. Precision-recall at EER comparison for the ERF using different randomness factor and number of trees

which combination of those parameters gives better results for our categorization method. In Fig. 3 the mean precision-recall values at Equal Error Rate (EER) obtained for the different parameter combinations are shown. The results shows that the performance for the ERF largely depends on the amount of trees, while the randomness factor has little, if any, effect in the performance. For the remaining experiments, we have selected a randomness factor of 0.05 (i.e. a completely random forest) and 5 trees, which are a good compromise between performance and computational cost.

3.3   Comparison between HKM and ERF

To compare the performance of the HKM and the ERF, dense features have been computed for all the 450 training images, resulting in about 6,000,000

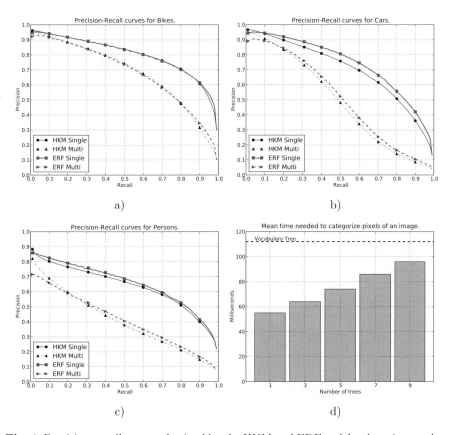

Fig. 4. Precision-recall curves obtained by the HKM and ERF codebooks using random sampling for the a) bikes, b) cars and c) persons categories of the Graz02 database. For each category, both methods have been evaluated using only the category images (Single) and using all testing images (Multi). In d) the mean time spent evaluating an image for the ERF and the HKM is shown.

Table 1. Comparison of the precision-recall values obtained at equal error rate on the Graz02 database both using odd/even and random train/test sampling. The categorization methods have been evaluated using only images that contain objects from the searched category in the "Single" test and using all the images in the "Multi" test.

Sampling	Test	Method	Bikes	Cars	Persons
Even/Pair	Single	HKM	**73.77% ± 0.20%**	63.90% ± 0.17%	61.80% ± 0.40%
		ERF	73.63% ± 0.32%	**65.68% ± 0.60%**	**62.59% ± 0.30%**
	Multi	HKM	**63.42% ± 0.24%**	47.09% ± 0.45%	44.36% ± 0.31%
		ERF	63.27% ± 0.32%	**47.62% ± 1.45%**	**46.34% ± 0.43%**
Random	Single	HKM	**74.33% ± 1.21%**	65.70% ± 1.45%	62.13% ± 1.30%
		ERF	74.17% ± 1.22%	**68.39% ± 1.44%**	**63.27% ± 1.38%**
	Multi	HKM	**64.55% ± 1.26%**	49.60% ± 1.61%	42.78% ± 1.86%
		ERF	63.98% ± 1.28%	**51.30% ± 2.03%**	**44.30% ± 2.12%**

training features. For the HKM we have selected a branch factor of 10 as in [4] to generate a codebook with 200,000 visual words in average. For the ERF, using the parameters selected in the previous section, we have obtained a codebook of 150,000 visual words in average. We have done two different tests: the "Single" test only uses the images containing objects from the tested category (e.g. using the bike classifier only for the images where a bicycle can be actually found), and the "Multi" test uses all test images (e.g. using the bike classifier for all 450 test images). As can be seen in Table 1 the precision-recall values obtained at EER show that the ERF performs slightly better than the HKM, both in the "Single" and the "Multi" tests. In the "Multi" test we can see that, when 300 images not containing objects are added to the test set, precision decreases a reasonable 32%. Finally, Fig. 4 shows the precision-recall curves for the different categories of the Graz02 database. As can be seen, the ERF has a slightly better performance than the HKM.

3.4   Time Cost Evaluation

Finally, regarding the computational cost of the categorization approach, the average time needed to construct the HKM is of 5 minutes, while that of the ERF depends on the randomness factor and the number of trees used, ranging from 100 milliseconds for a completely random ERF with a single tree, to 12 minutes for a highly discriminative ERF with 9 trees. The cost of training a linear classifier using the LR-SVM is of about 2 minutes for the histograms generated with HKM codebook, and from 2 to 5 minutes for those generated with the ERF (it depends on the amount of trees of the ERF). In Fig. 4.d) we can see the average time needed to categorize all image pixels using a HKM and ERF codebooks. Although being a bit slower in the training phase, the ERF is faster than the HKM in the categorization phase, where speed is truly essential. Using a ERF with 5 K-D trees, the whole scheme needs about 72.6 milliseconds to categorize an image, so that, we can process about 13 images per second.

4   Conclusions

In this paper we have presented an efficient method to assign a category label to each pixel of an image. Our main contribution is the introduction of integral linear classifier, which is used to bypass the accumulation step and directly obtain the classification score for an arbitrary sub-window of the image. Besides, we have compared the performance of the Hierarchical K-Means (HKM) and Extremely Randomized Forest (ERF). The obtained results show that the ERF performs slightly better than the HKM and with a lower computational cost. We have shown that the proposed method with the ERF feature quantization approach is suitable for real-time applications. In future work, we plan to improve this method and use it in an online learning scheme with a mobile robot.

Acknowledgements

This work was supported by TIN 2006-15308-C02-02 project grant of the Ministry of Education of Spain, the CSD2007-00018 and the MIPRCV Consolider Ingenio 2010, the European Social Fund, the project grant 200450E550, FEDER funds, and the grant 2009-SGR-1434.

References

1. Lowe, D.: Distinctive image features from scale-invariant keypoints. Int. Journal of Computer Vision 60, 91–110 (2004)
2. Csurka, G., Bray, C., Dance, C., Fan, L.: Visual categorization with bags of keypoints. In: Workshop on Stat. Learning in Computer Vision, ECCV, pp. 1–22 (2004)
3. Nister, D., Stewenius, H.: Scalable recognition with a vocabulary tree. In: Proc. of Computer Vision and Pattern Recognition, pp. 2161–2168 (2006)
4. Fulkerson, B., Vedaldi, A., Soatto, S.: Localizing objects with smart dictionaries. In: Proc. of European Conference on Computer Vision, pp. 179–192 (2008)
5. Sastre, R., Tuytelaars, T., Bascon, S.: Class representative visual words for category-level object recognition. In: IbPRIA 2009: Proceedings of the 4th Iberian Conference on Pattern Recognition and Image Analysis, pp. 184–191 (2009)
6. Moosmann, F., Nowak, E., Jurie, F.: Randomized clustering forests for image classification. IEEE Trans. on Pat. Anal. and Machine Intel. 30, 1632–1646 (2008)
7. Shotton, J., Johnson, M., Cipolla, R., Center, T., Kawasaki, J.: Semantic texton forests for image categorization and segmentation. In: Proc. of Computer Vision and Pattern Recognition, pp. 1–8 (2008)
8. Ramisa, A.: Localization and Object Recognition for Mobile Robots. PhD thesis, Universitat Autonoma de Barcelona (2009)
9. Lampert, C.H., Blaschko, M.B., Hofmann, T.: Beyond sliding windows: Object localization by efficient subwindow search. In: Proc. of Computer Vision and Pattern Recognition, pp. 1–8 (2008)
10. Nowak, E., Jurie, F., Triggs, B.: Sampling strategies for bag-of-features image classification. In: Proc. of European Conference on Computer Vision, pp. 490–503 (2006)

11. Opelt, A., Pinz, A., Fussenegger, M., Auer, P.: Generic object recognition with boosting. IEEE Trans. on Pat. Anal. and Machine Intel. 28(3), 416–431 (2006)
12. Zhu, Q., Yeh, M.C., Cheng, K.T., Avidan, S.: Fast human detection using a cascade of histograms of oriented gradients. In: Proc. of Computer Vision and Pattern Recognition, pp. 1491–1498 (2006)
13. Dalal, N., Triggs, B.: Histograms of oriented gradients for human detection. In: Proc. of Computer Vision and Pattern Recognition, pp. 886–893 (2005)
14. Zhang, J., Marszalek, M., Lazebnik, S., Schmid, C.: Local features and kernels for classification of texture and object categories: A comprehensive study. Int. Journal of Computer Vision 73, 213–238 (2007)
15. Viola, P., Jones, M.: Rapid object detection using a boosted cascade of simple features. In: Proc. of Computer Vision and Pattern Recognition, pp. 511–518 (2001)
16. Bay, H., Ess, A., Tuytelaars, T., Gool, L.V.: Surf: Speeded up robust features. Computer Vision and Image Understanding (CVIU) 110, 346–359 (2008)
17. Porikli, F.: Integral histogram: A fast way to extract histograms in cartesian spaces. In: Proc. of Computer Vision and Pattern Recognition, pp. 829–836 (2005)
18. Marszalek, M., Schmid, C.: Accurate object localization with shape masks. In: Proc. of Computer Vision and Pattern Recognition, pp. 1–8 (2007)
19. Lin, C.-J., Weng, R.C., Keerthi, S.S.: Trust region newton methods for large-scale logistic regression. In: Int. Conf. on Machine Learning, pp. 561–568 (2007)
20. Fan, R.-E., Chang, K.-W., Hsieh, C.-J., Wang, X.-R., Lin, C.-J.: Liblinear: A library for large linear classification. J. Mach. Learn. Res. 9, 1871–1874 (2008)

Relighting Forest Ecosystems

Jay E. Steele and Robert Geist

Clemson University
jesteel@cs.clemson.edu,
geist@cs.clemson.edu

Abstract. Real-time cinematic relighting of large, forest ecosystems remains a challenging problem, in that important global illumination effects, such as leaf transparency and inter-object light scattering, are difficult to capture, given tight timing constraints and scenes that typically contain hundreds of millions of primitives. A solution that is based on a lattice-Boltzmann method is suggested. Reflectance, transmittance, and absorptance parameters are taken from measurements of real plants and integrated into a parameterized, dynamic global illumination model. When the model is combined with fast shadow rays, traced on a GPU, near real-time cinematic relighting is achievable for forest scenes containing hundreds of millions of polygons.

1 Introduction

Rendering large-scale, forest ecosystems is a topic of wide interest, with applications ranging from feature film production to environmental monitoring and ecosystem management. Important global illumination effects, such as leaf transparency and inter-object scattering, are difficult to capture, given tight timing constraints and models that potentially contain hundreds of millions of primitives.

Early approaches to rendering such systems relied on conventional rasterization using billboard clouds [1]. More recently, focus has shifted to geometry-based techniques that rely on ray-tracing. Ray-tracing generally provides superior visual results, and new acceleration structures combined with low-cost, highly parallel execution environments can make ray-tracing competitive in execution time. In this vein, Geist et al. [2] extended the approach of Dietrich et al. [3] to include diffuse leaf transparency and inter-object scattering while maintaining near real-time rendering for scenes that comprised hundreds of millions of polygons. They used a lattice-Boltzmann (LB) photon transport model [4] to solve for global illumination in a pre-processing step and a highly parallel ray tracer, built on NVIDIA's Compute Unified Device Architecture (CUDA) [5], to achieve acceptable execution time.

Nevertheless, as noted in [2], there remain drawbacks to this work. First, although the LB lighting solution captured the desired global effects, it was constrained by the use of geometric instancing. The scenes of [2] each show approximately 1,000 trees, but there are only 5 distinct tree geometries. The LB transport model was solved only once per geometry. As a result, the common

G. Bebis et al. (Eds.): ISVC 2009, Part I, LNCS 5875, pp. 55–66, 2009.

method of achieving desired forest scene variability, random instance rotation, yields an incorrect lighting solution for any rotated tree. Second, relighting a scene (for example, in response to the movement of the sun) required re-computing the full LB lighting solution for each tree model based on the current sun direction, which, for near real-time frame rates and complex scenes, is not feasible with today's hardware. Finally, the LB lighting solution for each tree model was computed assuming there was no external occlusion of indirect illumination, which is not valid for large forest scenes.

The principal goal of this effort is to extend the approach of [2] to obtain a compact, parameterized lighting solution that is accurate for any light source position and intensity. This parameterized solution allows both arbitrary rotation and translation of geometric instances as well as dynamic global illumination, i.e., real-time movement of the sun.

2   Background

2.1   Related Work

Dorsy et al. [6] achieved beautiful results in rendering small collections of plant leaves using carefully constructed bidirectional reflectance and transmittance functions that were based on measurements of real plants. Their method is computationally intensive, with large memory requirements, and so as yet unsuitable for real-time rendering of large-scale, high-density ecosystems.

Reeves et al. [7] represented trees as a particle system. A probabilistic shading model shaded each particle based on the particle's position and orientation. Hegeman et al. [8] ignore physical accuracy in a technique that attempts to achieve visual plausibility and fast computation through approximating ambient occlusion. Trees are approximated by bounding spheres containing randomly distributed small occluders. Fragments are shaded based on the average probability of the fragment being occluded given its position within the bounding sphere. Though simple to compute, this method, as mentioned by the authors, only considers local information, and results can differ widely from more physically accurate approaches. Luft et al. [9] were able to capture ambient occlusion in rendering foliage through the addition of a preprocessing step in which overall tree geometry was simplified to an implicit surface, i.e., a density field, using Marching Cubes [10]. The ambient coefficient in a standard, local illumination model was then modified by a transfer function that was exponentially decreasing in the field density. They also realigned leaf normal vectors to match the implicit surface in order to reduce lighting noise.

The overall technique presented in this study is similar, in spirit, to both precomputed radiance transfer of Sloan et al. [11] and photon mapping of Jensen et al. [12], in that a preprocessing step is used to compute and store lighting information within the scene itself. Comparatively, the LB lighting preprocessing step is very fast. Hašan et al. [13] introduced the direct-to-indirect transfer for cinematic relighting, which requires a long preprocessing step and does not provide full anisotropic scattering.

2.2   Lattice-Boltzmann Lighting

Our goal, cinematic relighting, requires fast global illumination of forest ecosystems. Geist et al. [4] describe a lighting technique that was adapted in [2] to provide static forest illumination. The original lighting technique offered a new solution to the standard volume radiative transfer equation for modeling light in a participating medium:

$$(\boldsymbol{\omega} \cdot \nabla + \sigma_t) L(\boldsymbol{x}, \boldsymbol{\omega}) = \sigma_s \int p(\boldsymbol{\omega}, \boldsymbol{\omega}') L(\boldsymbol{x}, \boldsymbol{\omega}') d\boldsymbol{\omega}' + Q(\boldsymbol{x}, \boldsymbol{\omega}) \qquad (1)$$

where L denotes radiance, $\boldsymbol{\omega}$ is spherical direction, $p(\boldsymbol{\omega}, \boldsymbol{\omega}')$ is the phase function, σ_s is the scattering coefficient of the medium, σ_a is the absorption coefficient of the medium, $\sigma_t = \sigma_s + \sigma_a$, and $Q(\boldsymbol{x}, \boldsymbol{\omega})$ is the emissive field in the volume [14]. The solution, which is applicable to simulating photon transport through participating media such as clouds, smoke, or haze, was based on a lattice-Boltzmann (LB) method. At each iteration, photons scatter from a lattice site to its neighboring sites. Neighboring sites are defined by a standard approach, due to d'Humières, Lallemand, and Frisch [15], that uses 24 points, equidistant from the origin in 4D space, projected onto 3D. The points are:

$$(\pm 1, 0, 0, \pm 1) \; (0, \pm 1, \pm 1, 0) \; (0, \pm 1, 0, \pm 1)$$
$$(\pm 1, 0, \pm 1, 0) \; (0, 0, \pm 1, \pm 1) \; (\pm 1, \pm 1, 0, 0)$$

and projection is truncation of the fourth component, which yields 18 directions. Axial directions then receive double weights, thus ensuring isotropic flow. A final direction, a rest direction that points from a lattice site to itself, is added to facilitate the representation of several phenomena, including energy absorption and energy transmission. This gives 19 directions, $\boldsymbol{c}_m$, $m \in \{0, 1, ..., 18\}$. Each lattice site contains a per-node photon density, $f_m(\boldsymbol{r}, t)$, which is the density arriving at lattice site $\boldsymbol{r} \in \Re^3$ at time t in direction $\boldsymbol{c}_m$. Given a lattice spacing λ and a time step τ, the simulation amounts to a synchronous update:

$$f_m(\boldsymbol{r} + \lambda \boldsymbol{c}_m, t + \tau) - f_m(\boldsymbol{r}, t) = \Omega_m \cdot f(\boldsymbol{r}, t) \qquad (2)$$

where Ω_m denotes row m of a 19×19 matrix, Ω, that describes scattering, absorption, and wavelength shift at each site.

For any LB method, the choice of Ω is not unique. In [4], for the case of isotropic scattering, Ω was chosen as follows:
For row 0:

$$\Omega_{0j} = \begin{cases} -1 & j = 0 \\ \sigma_a & j > 0 \end{cases} \qquad (3)$$

For the axial rows, $i = 1, ..., 6$:

$$\Omega_{ij} = \begin{cases} 1/12 & j = 0 \\ \sigma_s/12 & j > 0, j \neq i \\ -\sigma_t + \sigma_s/12, & j = i \end{cases} \qquad (4)$$

For the non-axial rows, $i = 7, ..., 18$:

$$\Omega_{ij} = \begin{cases} 1/24 & j = 0 \\ \sigma_s/24 & j > 0, j \neq i \\ -\sigma_t + \sigma_s/24, & j = i \end{cases} \tag{5}$$

Entry i, j controls scattering from direction c_j into direction c_i, and directional density f_0 holds the absorption/emission component. Note that the axial rows, $i = 1, ..., 6$, are weighted twice as much as the non-axial rows, $i = 7, ..., 18$. The entries of Ω were then multiplied by the density of the medium at each lattice site, so that a zero density yielded a pass-through in (2), and a density of 1 yielded a full scattering.

In terms of total site photon density, $\rho = \sum_{i=0}^{18} f_i(\boldsymbol{r}, t)$, the synchronous update (2) was shown, in the limit $(\lambda, \tau \to 0)$, to yield a diffusion equation

$$\frac{\partial \rho}{\partial t} = \left(\frac{\lambda^2}{\tau} \right) \left[\frac{(2/\sigma_t) - 1}{4(1 + \sigma_a)} \right] \nabla_r^2 \rho \tag{6}$$

Previous approaches ([16,17]) have demonstrated that multiple photon scattering events invariably lead to diffusion processes.

In [2], this model was extended to provide global illumination effects for forests scenes. Wavelength-dependent, anisotropic scattering was achieved through modifying Ω. Each σ_s that appears in entry $\Omega_{i,j}$ was multiplied by a normalized phase function:

$$pn_{i,j}(g) = \frac{p_{i,j}(g)}{\left(\sum_{i=1}^{6} 2p_{i,j}(g) + \sum_{i=7}^{18} p_{i,j}(g) \right)/24} \tag{7}$$

where $p_{i,j}(g)$ is a discrete version of the Henyey-Greenstein phase function [18],

$$p_{i,j}(g) = \frac{1 - g^2}{(1 - 2g\boldsymbol{n}_i \cdot \boldsymbol{n}_j + g^2)^{3/2}} \tag{8}$$

Here $\boldsymbol{n}_i$ is the normalized direction, c_i. Parameter $g \in [-1, 1]$ controls scattering direction. Value $g > 0$ provides forward scattering, $g < 0$ provides backward scattering, and $g = 0$ yields isotropic scattering. For each primary wavelength, g was computed from leaf characteristics (transmission and reflectance) of real plants measured by Knapp and Carter [19], to provide accurate scattering.

3   Cinematic Relighting

Our goal is to allow cinematic relighting, that is real-time rendering, of complex forest scenes with dynamic, user-controllable light sources, while capturing global illumination effects such as scattering, transmission, and absorption. Our target scenes consist of one infinite point light (the sun) and ambient light due to reflection and scattering of the sky, ground, and forest. (Supporting multiple infinite point lights is a trivial extension.) Due to the large memory requirements of plant models, individual plants in a forest are rotated and translated instances of a small set of shared plant models. Thus, a typical scene may contain hundreds of individual plants, but less than ten unique plant models.

First, we detail a new technique that utilizes the previously described static LB lighting model to allow for dynamic, global illumination, based on the current, run-time sun position and intensity. Then, we introduce a coarse-grained illumination model that accounts for global scene effects, such as plant-plant occlusion, that the fine-grained illumination model does not account for by itself.

3.1 Dynamic Lighting

In the technique of [2], plant instances that reference the same plant model, but have different orientations relative to a given sun direction, require individual lighting solutions, as a lighting solution is only valid for one orientation. Storing a solution per plant instance would quickly exhaust the available memory of today's hardware. Computing a solution per frame for each plant instance would exceed the real-time capability of today's hardware. We now describe a technique that removes these restrictions, thus enabling dynamic global illumination for scenes containing many plant models while allowing both translated and rotated instances. The approach is conceptually simple. A set of base lighting solutions is precomputed per plant model, not per instance. These base lighting solutions are combined, at run-time, based on each instance's orientation relative to the current sun direction, thus allowing dynamic lighting updates to complex scenes containing thousands of plants in real-time.

First, for each plant model (not instance) we precompute 19 base lighting solutions $\{B_j | j = 0, 1, ..., 18\}$ by using (2) with boundary conditions based on direction index j. For solution B_j with $j > 0$, all boundary nodes have fixed densities $f_i(\boldsymbol{r}, t) = \delta_{ij}$ (Kronecker delta), all i. For solution B_0, the ambient solution, all boundary nodes have $f_i(\boldsymbol{r}, t) = 1$, all i.

At run-time, we employ a shader that computes the lighting for each fragment by combining multiple, weighted base solutions based on the sun direction, transformed into each instance's local coordinate space. The selection of the base solutions and weights is geometrically straightforward. Consider a convex polyhedron, consisting of 32 faces (triangles), formed by adjacent triples of unit length directions, $\boldsymbol{c}_i / \|\boldsymbol{c}_i\|, i = 1, ..., 18$. A ray from the origin in the sun direction will intersect one triangle. If the vertices of that triangle have associated directions $\boldsymbol{c}_{i_0}$, $\boldsymbol{c}_{i_1}$, and $\boldsymbol{c}_{i_2}$, then we use base solutions B_{i_0}, B_{i_1}, and B_{i_2}, with weights determined by the barycentric coordinates of the intersection point. The weights sum to one, thus conserving total energy.

At run-time, for each fragment, we sample the three base solutions for that fragment's instance and linearly combine them using the weights. Ambient light is added by sampling from the ambient base solution B_0, which has scaling controllable by the user. As described, dynamic lighting requires at most 32 ray-triangle intersections per plant instance and four 3D texture lookups per fragment.

Fig. 1. Final LB lighting solution and three base LB lighting solutions

Figure 1 illustrates the dynamic lighting technique. The left image visualizes the final LB lighting, which is computed by combining the three base LB lighting solutions that appear in the remaining three images. (Ambient light is also used, but not shown.) The base LB lighting solutions are selected and weights are computed based on the plant instance's orientation relative to the current sun position.

3.2   Hierarchical Lighting Model

Our LB lighting method, as described to this point, computes indirect illumination effects under the assumption that global illumination of the plant is not occluded by other plants. (Occlusion of direct illumination is handled by shadow rays, as described in the next section.) This can lead to too much light energy at the boundary nodes, resulting in too much indirect illumination for occluded plants. We now describe an extension that accounts for such inter-object (plant-plant) occlusions.

We combine our fine-grained, local LB lighting solution with a coarse-grained, global LB lighting solution. A coarse-grained, global LB lighting grid can be imposed upon an instanced forest system. Each node in the coarse-grained grid has a density factor estimated from the tree instances that intersect the node. Solution of this coarse-grained grid over the entire forest, using iterations of (2), simply provides scale factors that weight the indirect illumination of a fragment computed from the fine-grained, local LB lighting solution.

As with the fine-grained, local LB solution, a coarse-grained, global LB solution can be precomputed for a scene. The same technique described in Section 3.1 can be applied to the coarse-grained, global LB lighting solution to allow dynamic updates in real-time as the sun traverses the sky.

4   Implementation

The preprocessing steps of sections 3.1 and 3.2 are implemented in CUDA. We precompute the fine-grained LB lighting separately for the three primary wavelengths for each lattice direction. We only precompute lighting intensity (luminance) for the coarse-grained LB lighting.

Visible fragments are generated through ray-tracing with CUDA, though rasterization is a viable alternative. A perspective grid (see Hunt and Mark [20]) is used to accelerate primary ray/instance intersection testing. For the scenes provided here, we compute local diffuse lighting of the form $k_d(\boldsymbol{n}\cdot\boldsymbol{l})$, where $\boldsymbol{l}$ is sun direction, $\boldsymbol{n}$ is the surface normal, and k_d is the combined sun color and sampled texture color. Specular lighting is of the form $k_s(\boldsymbol{v}\cdot\boldsymbol{r})^s$, where $\boldsymbol{v}$ is the viewer position vector, $\boldsymbol{r}$ is the sun reflection vector, s controls highlight dissipation, and k_s is the combined sun color and leaf/wood/water specular color. Our LB lighting, such as that shown in the left image of Figure 1, is modulated by texture color and added to the local, direct illumination, to produce final fragment color.

4.1   Shadow Rays

To accelerate shadow rays (the dynamic, global illumination technique does not provide high frequency shadows), we employ a sun-aligned grid, which is similar in spirit to the perspective grids of Hunt and Mark [20] and is a two-dimensional data structure, consisting of multiple tiles arranged as a grid on a plane whose normal is parallel to the sun direction. Each tile contains a list of those model instances that intersect that tile in sun-space, which is an orthographic projection of the scene translated so that the sun is located at infinity on the z-axis and the direction of sunlight is the negative z-axis. For each plant instance in our scene, the associated model's axis-aligned bounding box (AABB) is projected into world-space, followed by a projection into sun-space. The sun-aligned grid is recomputed as the sun moves.

Once a shadow ray intersects an instance's AABB, we transform the shadow ray to model space and test for occlusion with the model's geometry by traversing a uniform grid. We found that traversing uniform grids, as described by Lagae and Dutré [21], offered better performance than multiple kd-tree traversal algorithms when traversing aggregate models such as trees.

4.2   Compression

The LB lighting model produces volumetric lighting data, which quickly consumes large amounts of memory per plant model. Each node value requires 3 floats (1 for each primary wavelength). For each plant model, 19 base solutions on a 128^3 lattice requires $(2M \times 19 \times 3 \times 4)$ 456MB of LB lighting data. Our implementation supports compression of this data through the use of Haar wavelets. Several authors have demonstrated that Haar wavelets can significantly compress volumetric data while supporting fast, random access to individual voxels. As noted by Westermann [22], other wavelet transforms may provide better compression rates, but, the simplicity of Haar wavelets results in faster reconstruction times, which is vital to our application.

To avoid visible discontinuities that result from nearest-neighbor sampling, we use linear filtering when sampling from the LB lighting. Thus, we need fast, random access to each 2^3 block of texels. We compress sub-blocks of size 3^3, but include the forward neighbors of all nodes in a sub-block. Our final sub-block size is thus 4^3, for which 37 of 64 values stored are redundant. Although this reduces our compression rate, the redundancy significantly improves rendering times by reducing the amount of work required to access those forward neighbors of a texel needed when applying linear filtering. We compress by decimating coefficients below a user supplied threshold. As in the work of Bajaj et al. [23], we do not store data that will not be accessed.

5   Results

All results are reported for lattices of size 128^3. Execution time for the dynamic LB lighting preprocessing step, described in Section 3.1, (computing 19 base solutions, three wavelengths each, on a GTX 280) averages 3 minutes 20 seconds for multiple plants, which includes computation of σ_s and σ_a from the model data. Each wavelength for each base solution averages 2.926 seconds. The number of iterations required to achieve convergence to steady-state is approximately twice the longest edge dimension of the grid.

The effect of applying the hierarchical method described in Section 3.2 is shown in Figure 2. In this scene, the sun is located towards the upper right of the image. The two smaller tree instances both reference the same tree model and thus have the same precomputed LB lighting solution, although each accounts

Table 1. Execution times for relighting, Tesla S1070

Catalapa			Canopy		
Number of GPUs	1 ray/pixel	4 rays/pixel	Number of GPUs	1 ray/pixel	4 rays/pixel
1	0.072 s	0.258 s	1	0.259 s	0.858 s
2	0.040 s	0.150 s	2	0.132 s	0.432 s
4	0.023 s	0.096 s	4	0.073 s	0.247 s

Local LB only   Local and global LB

Fig. 2. Rendering comparison: Local LB vs. local and global LB

Fig. 3. Relighting scenes. Top: Catalapa. Bottom: Canopy.

Fig. 4. River scene

for instance translation and rotation. The image on the left combines direct illumination with the fine-grained indirect illumination from LB lighting, while the image on the right combines direct illumination with the fine-grained indirect illumination from LB lighting and the coarse-grained indirect illumination. Note that in the image on the left, the tree on the left receives too much light for the position of the sun, while in the image on the right, the same tree is appropriately darkened.

Figure 3 shows multiple frames captured from two scenes as a user moves the sun from left to right. The top scene, consisting of a single Catalpa tree, contains 316,727 triangles. The bottom scene, a canopy view of multiple plants, contains 109 million triangles. Execution times for relighting each scene in Figure 3 at a resolution of 512×512 pixels with 1 ray per pixel and 4 rays per pixel are shown in Table 1, for one, two, and all four GPUs of a Tesla S1070. Relighting time includes the time required to compute indirect illumination by sampling and combining four base solutions (including decompressing) and the time to compute direct shadows with shadow rays. Table 1 shows the average execution time for 24 sun positions, ranging 90 degrees about the zenith.

Without compression, the precomputed LB lighting solution for the canopy scene of Figure 3 would require 2.28 GB (456 MB per model, 5 models). With compression, the total LB lighting for the scene is 366 MB. The root mean square error between renders of the canopy scene with compression and without compression is 0.00843. Finally, a high fidelity scene rendered with these techniques is shown in Figure 4.

6   Conclusion

Incorporating global illumination effects, such as leaf transparency and inter-object reflection, in real-time rendering of large, forest ecosystems is a challenging problem. One approach to achieving such effects, suggested herein, is through extending a lattice-Boltzmann lighting model that approximates indirect illumination to allow for dynamic sampling at run-time.

There are several drawbacks to the original LB lighting technique [2,4] that have been addressed in this work. First, we use a preprocessing step that allows dynamic, run-time updates of positions of lights at an infinite distance, such as the sun. This also allows plant instances that reference the same plant model to share the same LB lighting solution, regardless of each instance's orientation. Second, we employ a two-stage hierarchy to capture occlusion of global, indirect illumination. Third, we have shown that compressing the volumetric, LB lighting data using Haar wavelets is viable, in that relatively high compression rates can be achieved while maintaining final image quality and rendering performance.

Cinematic relighting is achieved by combining local, direct illumination at any visible fragment point with an indirect value obtained by interpolating values from a set of preprocessed LB lighting solutions. Shadowing, a vital component of any direct illumination technique, is computed by shadow rays that traverse a sun-aligned grid. Near real-time performance is obtained by mapping the ray-tracing engine, as well as the LB lighting model, to NVIDIA's Compute Unified Device Architecture and then distributing across multiple GPUs.

Future work will focus on speeding visibility computations necessary for cinematic relighting. It is worth investigating the techniques of Lacewell et al. [24], where occlusion is prefiltered, and how it may be extended or incorporated into our current technique.

Acknowledgements

This work was supported in part by an NVIDIA Fellowship grant, an NVIDIA equipment donation through the Professor Partnership Program, and by the U.S. National Science Foundation under Award 0722313.

References

1. Behrendt, S., Colditz, C., Franzke, O., Kopf, J., Deussen, O.: Realistic real-time rendering of landscapes using billboard clouds. Computer Graphics Forum 24, 507–516 (2005)
2. Geist, R., Steele, J.: A lighting model for fast rendering of forest ecosystems. In: IEEE/EG Symposium on Interactive Ray Tracing 2008, IEEE/EG, pp. 99–106, back cover (2008)
3. Dietrich, A., Colditz, C., Deussen, O., Slusallek, P.: Realistic and Interactive Visualization of High-Density Plant Ecosystems. In: Eurographics Workshop on Natural Phenomena, Dublin, Ireland, pp. 73–81 (2005)
4. Geist, R., Rasche, K., Westall, J., Schalkoff, R.: Lattice-boltzmann lighting. In: Rendering Techniques 2004 (Proc. Eurographics Symposium on Rendering), Norrköping, Sweden, pp. 355–362, 423 (2004)
5. NVIDIA Corp.: Nvidia cuda programming guide, version 2.1 (2008), http://www.nvidia.com/object/cuda_get.html
6. Wang, L., Wang, W., Dorsey, J., Yang, X., Guo, B., Shum, H.-Y.: Real-time rendering of plant leaves. ACM Trans. Graph. 24(3), 712–719 (2005)
7. Reeves, W.T., Blau, R.: Approximate and probabilistic algorithms for shading and rendering structured particle systems. In: SIGGRAPH 1985: Proceedings of the 12th annual conference on Computer graphics and interactive techniques, pp. 313–322. ACM, New York (1985)
8. Hegeman, K., Premože, S., Ashikhmin, M., Drettakis, G.: Approximate ambient occlusion for trees. In: I3D 2006: Proceedings of the 2006 symposium on Interactive 3D graphics and games, pp. 87–92. ACM, New York (2006)
9. Luft, T., Balzer, M., Deussen, O.: Expressive illumination of foliage based on implicit surfaces. In: Natural Phenomena 2007 (Proc. of the Eurographics Workshop on Natural Phenomena), Prague, Czech Republic, pp. 71–78 (2007)
10. Lorensen, W., Cline, H.: Marching cubes: A high resolution 3d surface construction algorithm. In: Proc. SIGGRAPH 1987, pp. 163–169 (1987)
11. Sloan, P.P., Kautz, J., Snyder, J.: Precomputed radiance transfer for real-time rendering in dynamic, low-frequency lighting environments. In: SIGGRAPH 2002: Proceedings of the 29th annual conference on Computer graphics and interactive techniques, pp. 527–536 (2002)
12. Jensen, H.W.: Realistic Image Synthesis Using Photon Mapping. A.K. Peters, Natick (2001)
13. Hašan, M., Pellacini, F., Bala, K.: Direct-to-indirect transfer for cinematic relighting. In: SIGGRAPH 2006: ACM SIGGRAPH 2006 Papers, pp. 1089–1097. ACM, New York (2006)
14. Arvo, J.: Transfer equations in global illumination. In: Global Illumination, SIGGRAPH 1993 Course Notes (1993)
15. d'Humières, D., Lallemand, P., Frisch, U.: Lattice gas models for 3d hydrodynamics. Europhysics Letters 2, 291–297 (1986)

16. Jensen, H., Marschner, S., Levoy, M., Hanrahan, P.: A practical model for subsurface light transport. In: Proceedings of SIGGRAPH 2001, pp. 511–518 (2001)
17. Stam, J.: Multiple scattering as a diffusion process. In: Proc. 6th Eurographics Workshop on Rendering, Dublin, Ireland, pp. 51–58 (1995)
18. Henyey, G., Greenstein, J.: Diffuse radiation in the galaxy. Astrophysical Journal 88, 70–73 (1940)
19. Knapp, A., Carter, G.: Variability in leaf optical properties among 26 species from a broad range of habitats. American Journal of Botany 85, 940–946 (1998)
20. Hunt, W., Mark, W.R.: Ray-specialized acceleration structures for ray tracing. In: IEEE/EG Symposium on Interactive Ray Tracing 2008, IEEE/EG, pp. 3–10 (2008)
21. Lagae, A., Dutré, P.: Compact, fast and robust grids for ray tracing. In: Computer Graphics Forum (Proceedings of the 19th Eurographics Symposium on Rendering), vol. 27, pp. 1235–1244 (2008)
22. Westermann, R.: A multiresolution framework for volume rendering. In: Symposium on Volume Visualization, pp. 51–58. ACM Press, New York (1994)
23. Bajaj, C., Ihm, I., Park, S.: 3d rgb image compression for interactive applications. ACM Trans. Graph. 20, 10–38 (2001)
24. Lacewell, D., Burley, B., Boulos, S., Shirley, P.: Raytracing prefiltered occlusion for aggregate geometry, pp. 19–26 (2008)

Cartoon Animation Style Rendering of Water

Mi You[1], Jinho Park[2], Byungkuk Choi[1], and Junyong Noh[1]

[1] Graduate School of Culture Technology, KAIST, Republic Of Korea
[2] Namseoul University, Republic Of Korea

Abstract. We present a cartoon animation style rendering method for water animation. In an effort to capture and represent crucial features of water observed in traditional cartoon animation, we propose a Cartoon Water Shader. The proposed rendering method is a modified Phong illumination model augmented by the optical properties that ray tracing provides. We also devise a metric that automatically changes between refraction and reflection based on the angle between the normal vector of the water surface and the camera direction. An essential characteristic in cartoon water animation is the use of flow lines. We produce water flow regions with a Water Flow Shader. Assuming that an input to our system is a result of an existing fluid simulation, the input mesh contains proper geometric properties. The water flow lines can be recovered by computing the curvature from the input geometry, through which ridges and valleys are easily identified.

1 Introduction

With the recent development of computer graphics technology, manually created traditional animation is increasingly being replaced by computer-based cartoon style rendering. However, one problem with non-photorealistic rendering (NPR), in contrast with photorealistic rendering (PR), is excessive sacrifice of the details of individual materials and objects. For instance, a considerable amount of recent research on cartoon shading has focused on rather simple opaque objects [9][10][23]. Transparency, a common characteristic of water, is meanwhile often ignored. This treatment of water as an opaque object fails to meet the standard set by the results created by traditional cartoon artists.

In this work we present methods to draw cartoon style images that properly represent the particular characteristics of water. Cartoon water has abstract optical features such as transparency, reflection, and refraction. Our approach incorporates those features via a *Cartoon Water Shader*. Fig. 1 shows the optical features of water. Unlike previous methods that deal with only ambient and diffuse components [9] (see Fig. 6(a)), the proposed shader also accounts for specular components, which can represent the reflection or refraction effect using ray tracing. Moreover, traditional cartoon animation often shows a timely change between reflection and transparent refraction depending on the position of the viewpoint with respect to the surface of the water. We construct a similar mechanism for automatic selection of appropriate effects based on the angle between the normal vector of the water surface and the camera direction.

G. Bebis et al. (Eds.): ISVC 2009, Part I, LNCS 5875, pp. 67–78, 2009.

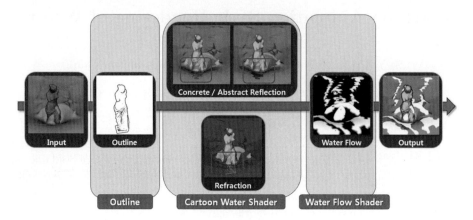

Fig. 1. An overview of our system. From the 3D input geometry, the bold outline is extracted. 'Cartoon Water Shader' is applied to the water input. The user can select from concrete reflection, abstract reflection or refraction. The difference of those effects is shown in the red box. 'Water Flow Shader' extracts the flow lines. These 3 components are combined to produce an output.

When depicting an object that moves across water, animators typically draw explicit flow lines to signify the motion waves generated by the movement of the object. In our approach, we produce water flow lines with a *Water Flow Shader*. The flow regions are geometric features and can be efficiently extracted by the computation of the curvatures on the water surface. Ridge and valley regions determined by the curvature computations represent the water flow regions, as shown in Fig. 1. This water flow region is incorporated into the result of the cartoon water shader.

2   Related Work

Diverse research in cartoon rendering has been reported to date. Todo et al. [23] provided user flexibility by adding localized light and shade using a paintbrush metaphor. Anjyo et al. [1] also proposed an intuitive and direct manipulation method for animating light and shade. In the latter, they mainly focus on the treatment of highlights. Mitchell et al. [17] applied their real-time shading method to commercial games. Barla et al. [2] suggested X-Toon, a shader that expands conventional cartoon shading. Various NPR techniques, including cartoon shading, pencil sketch shading, and stylistic inking, have also been developed [14]. As in cartoon rendering emphasizing abstraction, Gooch et al. [11] proposed a different style of abstraction that relies on interactive technical illustration. The abstraction is applied in [5], in which they develop a soft and abstract style of shadows similar to that seen in fine art.

Cartoon-style rendering of fluids from physical simulation has also attracted recent attention. McGuire and Fein [16] introduced a cartoon-style rendering method for animated smoke particles. Selle et al. [21] proposed a cartoon style rendering technique for smoke simulation. Advected marker particles are rendered as texture-mapped 2D stencils. Instead of relying on a physical simulation, Yu et al. [24] present a template-based approach. They classify cartoon water into different types, and templates of water shapes are designed from the specified types of water. Running water is drawn in a Chinese painting style [25]. From the input videos, they generate the painting structure and the brush stroke.

One of the main characteristics of cartoon animation compared to recent 3D animation rendered by photorealistic rendering methods is the use of bold lines in the object boundary. There are several object-based approaches. Finding ridges and valleys from the geometry can help draw lines that describe the shape. Lee et al. [15] drew lines along tone boundaries around the thin dark areas in the shaded image with a GPU-based algorithm. Decarlo [6] developed a new type of line, a suggestive contour, which is determined by the zero crossings of the radial curvature. Curvature estimation plays an important role in extracting salient lines from the geometry. Judd et al. [12] estimated the view-dependent ridges and valleys from the curvature. Chandra and Sivaswamy [3] analyzed curvature based ridges and valleys represented in digital images. View-independent ridge and valley detection has also been proposed [19]. Meanwhile, demarcating curves, another new class of view-independent curves, are defined on the slopes between ridges and valleys [13].

Our approach is most similar to the work in [9], which involves a cartoon style rendering of liquid animation. They utilize a bold outline, constant color, and oriented texture on the liquid surface. Although their method recovers many of the important features that water possess, they fail to address the artistic side of cartoon rendering such as abstract optical features and water flow lines observed in traditional cartoon animation. The proposed approach therefore tackles this.

3   Methods

The input to our system is a surface from a three-dimensional physically based fluid simulation. The surface mesh inherently contains the geometry information, which is useful in extracting water specific features for cartoon-style rendering. Commercial software, RealFlow is used for the generation of water simulation. However, any particle-based [4] or 3D grid-based solver [22] should work for our purpose.

3.1   Line Drawing

In traditional animation, lines are drawn mostly around the boundaries of objects to distinguish them from the background. Having the same aim as in traditional animation, we find the silhouette from the input meshes to draw desirable lines. We use the method suggested by [20]. Here, two identical polygons are combined into a set with slightly different scales, and each of the two polygons has a

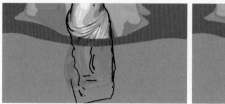 

(a) An image-based line drawing   (b) An object-based line drawing

Fig. 2. Visual comparison of (a) and (b). (b) generates suitably distorted refractions when combined with ray tracing. (a) contains visual artifacts.

different culled face; the first is a front-facing polygon utilized for drawing the object itself, and the second is a back-facing polygon utilized for drawing the outlines. The scale difference between the two polygons creates the border lines around the object, as shown in [18][20]. As our line drawing is based on objects, our line objects also easily generate distorted refractions when combined with ray tracing (Fig. 2).

3.2   Cartoon Water Shader

Modified Phong for Cartoon Style. Adjusting the Lambertian illumination model [14] effectively generates cartoon style shading for scenes with opaque objects. Their cartoon shading equation consists of two main terms, ambient and diffuse terms, which return the largest value between $\overline{L} \cdot \overline{n}$ and 0. This value is used for dividing two colors as texture coordinates. Their cartoon shading equation is:

$$I_{cs} = I_a k_a O_d + I_d[k_d Max(\overline{L} \cdot \overline{n}, 0)] \tag{1}$$

Here, $\overline{L}$ is the normalized direction of light source and $\overline{n}$ is the normalized direction of surface normal.

For opaque objects, their equation is sufficient to create a cartoon style image. However, water exhibits three peculiar characteristics: transparency, refraction, and reflection. Successful incorporation of these features helps convey realism, even in cartoon-style water animation.

To create transparently refractive effects, we adapt the Phong illumination model, modify $\overline{L} \cdot \overline{n}$ to O_d divided by 3 colors and add ray-tracing terms. The final illumination of the Cartoon Water Shader, I_{cws}, is obtained as follows:

$$I_{cws} = I_a k_a O_d + I_d[k_d O_d + k_s O_s (\overline{R} \cdot \overline{V})^n] + k_s I_r + k_t I_t \tag{2}$$

Here, I_a and I_d are the intensity of the ambient and diffuse light. k_a, k_d, and k_s are the ambient, diffuse, and specular coefficient, respectively. O_s is an object's specular color. $\overline{R}$ is the normalized direction of reflection and $\overline{V}$ is the normalized viewpoint direction. $k_s I_r$ and $k_t I_t$ are the intensity of the reflected and transmitted ray for ray tracing. Fig. 3 (a) shows the refraction of Venus under water.

To create a cartoon style shader, the object color is simplified into three levels: bright, medium, and dark. We evaluated different numbers of levels and determined that three levels produced the most convincing results, as too many levels approached photorealistic rendering and too few levels resulted in an indistinguishable appearance. The level is determined by the angle between the surface normal direction and the light direction, as follows:

$$O_d = \begin{cases} \text{Bright Color} & \text{if } \overline{L} \cdot \overline{n} > T_{bc} \\ \text{Medium Color} & \text{if } \overline{L} \cdot \overline{n} > T_{mc} \text{ and } \overline{L} \cdot \overline{n} \leq T_{bc} \\ \text{Dark Color} & \text{Otherwise} \end{cases} \quad (3)$$

Here, T_{bc} and T_{mc} are the threshold for a bright color and a medium color, respectively.

We distinguish two types of reflection. The first is concrete reflection, which shows an ordinary reflection effect commonly observed in the real world. The second is abstract reflection. The latter is also considered important in the context of cartoon animation. We allow both types of reflection in order to serve the artist's intentions depending on the situation. Rendering the two types of reflection is similar to the line drawing method described in Section 3.1. Concrete reflection is generated by setting the front-facing polygons as the reflection target of an object. In contrast, abstract reflection is produced by setting the back-facing polygons as the reflection target. Fig. 3 (b) and (c) show both effects reflected from Venus on a water surface.

Automatic Control of Refraction and Reflection Effect. In traditional animation, for simplicity artists tend to employ either a reflection or refraction effect for a given scene, unlike the real world where both effects coexist. Artists utilize the refraction effect when the distance between the camera and the main character and the angle created by the two are small. Otherwise, the reflection effect is employed.

Dynamic camera movements in a 3D scene may result in frequent transitions between the reflection effect and refraction effect causing unwanted flickering.

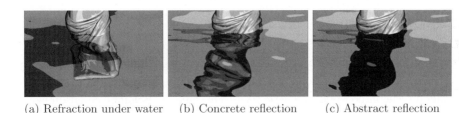

(a) Refraction under water   (b) Concrete reflection   (c) Abstract reflection

Fig. 3. Refraction effect of Venus (a) and reflection effect (b)(c) in the Cartoon Water Shader

Moreover, it would be very time consuming to determine a desirable effect according to the dynamic camera movements. Therefore, a simple but effective interpolation function that automatically determines when to apply reflection or refraction is proposed here. As transparency directly influences the visibility of refraction, we utilize transparency in the following equations.

$$
k_R = \begin{cases} k_{Rmax} & \textbf{if} \quad x < \cos s_{max} \\ k_{Rmin} & \textbf{if} \quad x > \cos s_{min} \\ f(\overline{n} \cdot \overline{V}, k_{Rmax}, k_{Rmin}) & \textbf{Otherwise} \end{cases} \tag{4}
$$

$$
k_T = \begin{cases} k_{Tmin} & \textbf{if} \quad x < \cos s_{max} \\ k_{Tmax} & \textbf{if} \quad x > \cos s_{min} \\ f(\overline{n} \cdot \overline{V}, k_{Tmax}, k_{Tmin}) & \textbf{Otherwise} \end{cases} \tag{5}
$$

Here, $f(\overline{n} \cdot \overline{V}, k_{Rmax}, k_{Rmin})$ and $f(\overline{n} \cdot \overline{V}, k_{Tmax}, k_{Tmin})$ are cubic polynomial interpolation functions. $\overline{n}$ is the normal direction and $\overline{V}$ is the viewpoint direction, respectively. k_{Rmax}, k_{Rmin}, k_{Tmax}, and k_{Tmin} are the maximum reflection, the minimum reflection, the maximum transparency, and the minimum transparency coefficient, respectively. In Fig. 4(a), θ_B is the angle where the critical change occurs and θ_S is the angle interval where interpolation happens. A user can specify k_{Rmax}, k_{Rmin}, k_{Tmax}, k_{Tmin}, θ_B, and θ_S at the key camera positions to produce desirable reflection and refraction effects. k_R and k_T are then computed automatically along the scene compositions. Fig. 4(b) shows each parameter of the control function and the shift of the reflection and transparency coefficients according to $\overline{n} \cdot \overline{V}$.

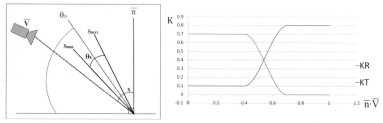

(a) The parameters used in control function

(b) The shift of the reflection and transparency coefficients according to $\overline{n} \cdot \overline{V}$

Fig. 4. The parameters and the behavior of the of control function

Although utilizing well-known physics such as the Fresnel reflection and Brewster's angle can provide a viable solution, the primary goal of our shader is to render a scene in non-photorealistic style. This cubic polynomial equation works well for our purpose.

3.3   Water Flow Shader

In traditional animation, special lines are drawn in the neighborhood of objects to effectively show the interaction between water and objects. In our simulation results, those shapes represent the flowing motions of water and traces of the objects. To emphasize these motions and traces, we introduce a Water Flow Shader. With the Water Flow Shader, the critical regions from flowing motions and traces are determined using the geometrical properties from input meshes. As the shader takes into account important geometric features, our results faithfully reproduce physically persuasive motions and appearances.

Estimation of Geometrical Properties. We compute both the Gaussian curvature and the mean curvature at every vertex of the triangular mesh to identify the critical regions on the mesh surface, as the curvature on the region formed by the movement of objects has a higher value than other flat regions. Using a discretization of the Gauss-Bonet theorem, we approximate the Gaussian curvature as described in [8]. The equation is as follows:

$$K_i = \frac{1}{A}\left(2\pi - \sum_j \theta_j\right) \qquad (6)$$

Here, A is the Voronoi area around X_i, and θ_j is an angle connected with X_i and X_j.

The mean curvature is approximated using a discretization of the Laplace-Beltrami operator also known as the mean curvature normal operator. A detailed explanation can be found in the literature [8][7]. The equation is given as follows:

$$H_i = \frac{1}{4A}\sum_j (\cot \alpha_{ij} + \cot \beta_{ij})(X_i - X_j) \cdot \hat{n}_i \qquad (7)$$

Here, A is the Voronoi area. α_{ij} and β_{ij} are angles adjacent to the specified vertices, and X_j is a neighbor vertex connected to X_i. $\hat{n}_i$ is a normal vector at X_i. Both approximations are applied to the one-ring neighborhood of the vertex.

From the recovered Gaussian curvature and mean curvature, we specify a ridge area connected by the vertices in which the absolute value of the Gaussian curvature is larger than the threshold and the mean curvature is greater than zero. Although our approach to calculate the curvatures and a rendering style are different, our concept of specifying the regions is similar to the principle of demarcating curves [13].

Enhancement. After determining the ridge and valley region, we apply an enhancement algorithm. As the computation of ridges and valleys heavily depends on the input meshes, bad geometry in the meshes can adversely affect the result. For example, if the mesh was created by a particle simulation that represents the details of the fluid well but introduces jiggling surfaces, our method will

(a) A fountain in a forest (b) A boat sailing on the sea (c) An image created by an artist (d) An image created by an artist

Fig. 5. The images rendered by our system and the images used in the experiments. Total rendering time of (a) is 8.0 sec(C.W.S: 8.0 / W.F.S: n/a / number of vertices: 75406) and that of (b) is 1330.0 sec(C.W.S: 5.0 / W.F.S: 1325.0 / number of vertices: 55091).

draw lines erroneously. Through application of the *median filter* to the one-ring neighbor of each vertex, we can effectively suppress these errors.

In traditional animation, special regions to be lined on the surface of water are opaque and monochromic. To generate these images, we divide curvature values into three parts. The first is for ridges, the second for valleys, and the third is the part between ridges and valleys. We paint ridge or valley parts as shaded and the other parts as transparent. As a result, the water flow region is separated from the other parts clearly. As the water flow region can be efficiently represented by simple shading, a simple Lambertian illumination model is sufficient for rendering.

4   Results and Analysis

To demonstrate the effectiveness of our system, the Cartoon Water Shader and the Water Flow Shader were created as a Maya plug-in (version 2009). The plug-in system was implemented in C++ with the Maya API. Moreover, to maximize the efficiency of our plug-in, we provide a user interface.

The images in Fig. 5 show the results rendered by our system. Images (a) and (b) clearly demonstrate that our system can produce high quality cartoon animation water rendering. Additional results are available in demo clips(see http://143.248.249.138/isvc09/paper66/). Rendering times of (a) and (b) are recorded on an Intel Core 2 Quad Q6600 machine with 4GB memory, using a single thread implementation. As the computation time of the Water Flow Shader depends on the number of vertices with the curvature approximations and the enhancement procedure, much more time is required to render the scene via the Water Flow Shader than with the Cartoon Water Shader(see Fig. 5).

To verify the visual enhancement of our method over previous approaches, we compare similar scenes with results rendered by earlier methods. In addition,

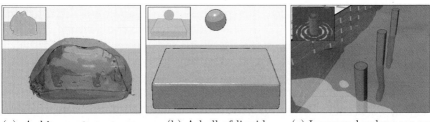

(a) A hippopotamus cov-   (b) A ball of liquid   (c) Immersed poles near an
ered by water   embankment

Fig. 6. Visual comparison with two earlier approaches. Insets are the images from the
previous methods.

an evaluation was conducted to verify that our results are visually equivalent to
those created by an artist.

4.1   Visual Comparison with Previous Approaches

Among the studies on cartoon rendering of fluids, two prominent studies [9][24]
were selected, as their main subject is similar to ours.

Eden A. M. et al. successfully recovered many crucial visual cues generated by
the movement of a fluid [9]. However, their method is absent of optical features
(Fig. 6 (a)). In contrast, our method shows a refracted view of a hippopotamus
inside water. As we render the shading of an object after calculating the surface
normal and the light information, our method can flexibly cope with the light
position. With our method, the middle image in Fig. 6 is shaded following the
light direction. In Eden A. M. et al. [9], lighting information is not considered.

As an approach of Yu j. et al.[24] is template-based, their images are similar
to the drawing of an artist. However, the water types that they can represent
are limited. With this method it is difficult to express the relationship of the
shading of water, which is affected by the environment, with reflections of nearby
objects on the water or transparent refractive shapes under the water. On the
other hand, our image showing immersed poles near an embankment (Fig. 6);
accurately reflects all objects near the water. These reflective images are changed
following the variation of the mesh and the light.

4.2   Evaluation

An experiment was conducted to confirm that our results are visually equivalent
to those created by an artist.

Participants: 30 participants took part in the evaluation. Among the participants, 15 were experts who work in the field of art and graphics and are
considered to have excellent observation capability of images. The remaining
participants were non-specialists who were nevertheless familiar with cartoon
animation.

Material: To collect the images for comparison, a 3D artist was asked to create scenes similar to those in a feature animation. We chose two concepts: a fountain in a forest and a boat sailing on the sea. These selected concepts were visualized as a 3D scene by a 3D artist. Based on the images rendered by the Lambertian illumination model, a cartoon animation artist composed water in the style of cartoon animation. Our instructions were to represent the artist's sense inside the outline of water and to follow the motion of the water mesh. However, we did not provide instructions concerning artistic flavors such as the properties of shading, the color to use, or which points the water flow lines should represent. These restrictions allowed the artist to draw freely, tapping into her own ingenuity. All of the images for the test were organized into 12 drawings per second, as in the style of limited animation usually employed in traditional Japanese animation. Images in the experiments were scaled to 640×480, which is the same scale used for the demo clip.

Analysis: P-values for the two-sided hypothesis testing problem were computed using an independent-samples t-test.

Procedure: Each participant was informed that the only variation between each set of video clips was the shading of water. Participants observed four video clips two times each in a random order. Two clips depicted 'a fountain in a forest'; the first was created by our method, and the second was drawn by the artist. The other two clips depicted clips of 'a boat sailing on the sea'; the first was created by our method, and the second was drawn by the artist. After observation of four video clips, the participants were asked to rate the images from 1(Bad) to 5(Good). The first question was to evaluate the aesthetic impression of the video clips according to the participant's own standards. The second question was to rank the clips according to whether they offered a rich depiction of the water.

Results and Discussion: For the first question, participants answered that the aesthetic impression of Fig. 5 (a) and Fig. 5 (c) were similar to each other as $Mean_{artist} = 3.17$, $Mean_{ours} = 3.07$. However, in the set of 'a boat sailing on the sea' clips, the video clip generated by our method in Fig. 5 (b) was rated to lend a higher aesthetic impression than the results by the artist ($p < 0.01$)($Mean_{artist} = 3.03$ vs $Mean_{ours} = 4.03$).

  For the second question, most participants answered that Fig. 5 (a) and Fig. 5 (c) have similar values, implying that both clips exhibit a rich depiction of the water, as $Mean_{artist} = 3.20$, $Mean_{ours} = 3.21$. In comparison of Fig. 5 (b) and Fig. 5 (d), $M_{ours} = 4.20$ is better than $Mean_{artist} = 3.80$. In conclusion, the evaluation demonstrates that our results match those created by an artist in terms of visual equivalence.

5   Conclusion and Discussion

The contribution of our system includes the proposed Cartoon Water Shader which can convey the essential components of water -transparency, refraction, and reflection- from an artistic point of view using a modified Phong illumination model. Automatic change between refraction and reflection, frequently observed effects in traditional cartoon animation, is also offered in the shader. Furthermore, utilizing the information of geometrical properties, Water Flow Shader is specifically designed for rendering water flow observed in traditional cartoon animation. To demonstrate the advantages of our method, we evaluated rendering results in comparison with those yielded by previous approaches and by an artist.

Although the proposed method can represent many characteristics of cartoon animation style rendering of water, there remain issues that must be addressed in the future. As we assume that the input meshes do not contain spray or bubbles, our current system cannot reproduce them. Another limitation involves the regions that cannot be described by the geometry properties alone in the Water Flow Shader. Although our system provides some thresholds for adjusting the effects to meet the artist's demands, exaggerated expressions of water flow lines are difficult to produce. We believe that these problems should be investigated in the future for a greater audience reception.

References

1. Anjyo, K.I., Wemler, S., Baxter, W.: Tweakable light and shade for cartoon animation. In: NPAR 2006: Proceedings of the 4th international symposium on Non-photorealistic animation and rendering, pp. 133–139. ACM, New York (2006)
2. Barla, P., Thollot, J., Markosian, L.: X-toon: an extended toon shader. In: NPAR 2006: Proceedings of the 4th international symposium on Non-photorealistic animation and rendering, pp. 127–132. ACM, New York (2006)
3. Chandra, S., Sivaswamy, J.: An analysis of curvature based ridge and valley detection. In: Proceedings of IEEE International Conference on Acoustics, Speech and Signal Processing, 2006. ICASSP 2006, p. 2. IEEE Computer Society Press, Los Alamitos (2006)
4. Desbrun, M., Cani, M.-P.: Smoothed particles: A new paradigm for animating highly deformable bodies. In: Boulic, R., Hegron, G. (eds.) Eurographics Workshop on Computer Animation and Simulation (EGCAS), August 1996, pp. 61–76. Springer, Heidelberg (1996) (Published under the name Marie-Paule Gascuel)
5. DeCoro, C., Cole, F., Finkelstein, A., Rusinkiewicz, S.: Stylized shadows. In: NPAR 2007: Proceedings of the 5th international symposium on Non-photorealistic animation and rendering, pp. 77–83. ACM, New York (2007)
6. DeCarlo, D., Finkelstein, A., Rusinkiewicz, S., Santella, A.: Suggestive contours for conveying shape. In: SIGGRAPH 2003: ACM SIGGRAPH 2003 Papers, pp. 848–855. ACM, New York (2003)
7. Desbrun, M., Meyer, M., Schröder, P., Barr, A.H.: Implicit fairing of irregular meshes using diffusion and curvature flow. In: SIGGRAPH 1999: Proceedings of the 26th annual conference on Computer graphics and interactive techniques, pp. 317–324. ACM Press/Addison-Wesley Publishing Co. (1999)
8. Desbrun, M., Meyer, M., Schroder, P., Barr, A.H.: Discrete differential-geometry operators for triangulated 2-manifolds, pp. 35–57. Springer, Heidelberg
9. Eden, A.M., Bargteil, A.W., Goktekin, T.G., Eisinger, S.B., O'Brien, J.F.: A method for cartoon-style rendering of liquid animations. In: GI 2007: Proceedings of Graphics Interface 2007, pp. 51–55. ACM, New York (2007)

10. Gooch, A., Gooch, B., Shirley, P., Cohen, E.: A non-photorealistic lighting model for automatic technical illustration. In: SIGGRAPH 1998: Proceedings of the 25th annual conference on Computer graphics and interactive techniques, pp. 447–452. ACM, New York (1998)

11. Gooch, B., Sloan, P.-P.J., Gooch, A., Shirley, P., Riesenfeld, R.: Interactive technical illustration. In: I3D 1999: Proceedings of the 1999 symposium on Interactive 3D graphics, pp. 31–38. ACM, New York (1999)

12. Judd, T., Durand, F., Adelson, E.: Apparent ridges for line drawing. In: SIGGRAPH 2007: ACM SIGGRAPH 2007 papers, p. 19. ACM, New York (2007)

13. Kolomenkin, M., Shimshoni, I., Tal, A.: Demarcating curves for shape illustration. In: SIGGRAPH Asia 2008: ACM SIGGRAPH Asia 2008 papers, pp. 1–9. ACM, New York (2008)

14. Lake, A., Marshall, C., Harris, M., Blackstein, M.: Stylized rendering techniques for scalable real-time 3d animation. In: NPAR 2000: Proceedings of the 1st international symposium on Non-photorealistic animation and rendering, pp. 13–20. ACM, New York (2000)

15. Lee, Y., Markosian, L., Lee, S., Hughes, J.F.: Line drawings via abstracted shading. In: SIGGRAPH 2007: ACM SIGGRAPH 2007 papers, p. 18. ACM, New York (2007)

16. McGuire, M., Fein, A.: Real-time rendering of cartoon smoke and clouds. In: NPAR 2006: Proceedings of the 4th international symposium on Non-photorealistic animation and rendering, pp. 21–26. ACM, New York (2006)

17. Mitchell, J.L., Francke, M., Eng, D.: Illustrative rendering in team fortress 2. In: SIGGRAPH 2007: ACM SIGGRAPH 2007 courses, pp. 19–32. ACM, New York (2007)

18. Markosian, L., Kowalski, M.A., Goldstein, D., Trychin, S.J., Hughes, J.F., Bourdev, L.D.: Real-time nonphotorealistic rendering. In: SIGGRAPH 1997: Proceedings of the 24th annual conference on Computer graphics and interactive techniques, pp. 415–420. ACM Press/Addison-Wesley Publishing Co., New York (1997)

19. Ohtake, Y., Belyaev, A., Seidel, H.-P.: Ridge-valley lines on meshes via implicit surface fitting. In: SIGGRAPH 2004: ACM SIGGRAPH 2004 Papers, pp. 609–612. ACM, New York (2004)

20. Raskar, R., Cohen, M.: Image precision silhouette edges. In: I3D 1999: Proceedings of the 1999 symposium on Interactive 3D graphics, pp. 135–140. ACM, New York (1999)

21. Selle, A., Mohr, A., Chenney, S.: Cartoon rendering of smoke animations. In: NPAR 2004: Proceedings of the 3rd international symposium on Non-photorealistic animation and rendering, pp. 57–60. ACM, New York (2004)

22. Stam, J.: Stable fluids. In: SIGGRAPH 1999: Proceedings of the 26th annual conference on Computer graphics and interactive techniques, pp. 121–128. ACM Press/Addison-Wesley Publishing Co. (1999)

23. Todo, H., Anjyo, K.-I., Baxter, W., Igarashi, T.: Locally controllable stylized shading. In: SIGGRAPH 2007: ACM SIGGRAPH 2007 papers, p. 17. ACM, New York (2007)

24. Yu, J., Jiang, X., Chen, H., Yao, C.: Real-time cartoon water animation. Comput. Animat. Virtual Worlds 18(4-5), 405–414 (2007)

25. Zhang, S., Chen, T., Zhang, Y., Hu, S., Martin, R.: Video-based running water animation in chinese painting style. Science in China Series F: Information Sciences 52(2), 162–171 (2009)

Deformable Proximity Queries and Their Application in Mobile Manipulation Planning

M. Gissler, C. Dornhege, B. Nebel, and M. Teschner

Computer Science Department, University of Freiburg, Germany

Abstract. We describe a proximity query algorithm for the exact minimum distance computation between arbitrarily shaped objects. Special characteristics of the Gilbert-Johnson-Keerthi (GJK) algorithm are employed in various stages of the algorithm. In the first stage, they are used to search for sub-mesh pairs whose convex hulls do not intersect. In the case of an intersection, they guide a recursive decomposition. Finally, they are used to derive lower and upper distance bounds in non-intersecting cases. These bounds are utilized in a spatial subdivision scheme to achieve a twofold culling of the domain. The algorithm does not depend on spatial or temporal coherence and is, thus, specifically suited to be applied to deformable objects. Furthermore, we describe its embedding into the geometrical part of a mobile manipulation planning system. Experiments show its usability in dynamic scenarios with deformable objects as well as in complex manipulation planning scenarios.

1 Introduction

Proximity queries play an important role in robot motion planning, dynamic simulation, haptic rendering, computer gaming, molecular modeling and other applications [1]. A plethora of papers has been published on different aspects of these queries in computational geometry and other research areas. Furthermore, many systems and libraries have been developed for performing different proximity queries. However, the attention to deformable proximity queries has been of moderate extent when compared to the many techniques capitalizing the special properties of rigid bodies. Regarding motion planning in robotics, distance queries have been used to accelerate the verification of execution paths [2]. If a deformable environment or flexible robots have to be considered in the planning process, an efficient distance computation algorithm is needed in the geometrical part of the motion planner that is capable of handling such environments.

Our Contribution: We describe an approach for the computation of the minimum distance between arbitrarily shaped deformable objects. Objects are represented by triangulated surface meshes. We show how to use GJK for an adaptive decomposition of the meshes. GJK is employed on the sub-mesh pairs to find lower and upper bounds for the minimum distance. The bounds are used for a spatial subdivision scheme that only takes a small part of the domain into account to determine the exact minimum distance between the sub-meshes. The

G. Bebis et al. (Eds.): ISVC 2009, Part I, LNCS 5875, pp. 79–88, 2009.

algorithm does not depend on spatial or temporal coherence. Thus, it is suitable to be applied to deformable objects. We show the usability of the algorithm in a planning system for mobile manipulation. The system is able to find execution plans for complex tasks that require the replacement of objects to reach a specific goal or to take the deformability of the manipulable objects into account.

Organization: The rest of the paper is organized as follows. Section 2 surveys related work on proximity queries and manipulation planning. The proximity query algorithm is described in section 3. The embedding of the proximity queries into the manipulation planning framework is discussed in section 4. The paper concludes by presenting results for the proximity query technique and its application to manipulation problems in section 5.

2   Related Work

Proximity Queries: Proximity query algorithms find their application in many research areas such as computer graphics, physically-based simulation, animation, interactive virtual environments and robotics. They include the query for collision detection, distance computation or penetration depth. Extensive research has produced a variety of specialized algorithms. They may differ in the model representations they are able to process, the type of query they can answer or the specific properties of the environment. Excellent surveys can be found in [3,4,5]. Considering collision detection, many approaches exploit the properties of convex sets to be able to formulate a linear programming problem. Gilbert et al. propose an iterative method to compute the minimum distance between two convex polytopes using Minkowski differences and a support mapping [6]. In contrast, Lin and Canny [7] execute a local search over the Voronoi regions of convex objects to descend to the closest point pair. In dynamic environments, geometric and time coherence can be exploited to employ feature-tracking to improve the efficiency of the algorithms even more [7]. The described techniques can be applied to non-convex objects, if the non-convex objects are either composed of [6,7] or decomposed into [8] convex subparts. The algorithms then work on the convex subparts as usual. However, surface decomposition is a nontrivial and time consuming task and can only be considered as a preprocessing step when applied to undeformable objects.

Apart from the family of feature-tracking algorithms, there is the class of bounding volume hierarchies. For each object, a hierarchy of bounding volumes is computed that encloses the primitives of the object at successive levels of detail. Different types of bounding volumes have been investigated, such as spheres [9,10], axis-aligned bounding boxes [11], k-DOPs [12] or oriented bounding boxes [13]. Further, various hierarchy-updating methods have been proposed [14].

Spatial subdivision is the third family of acceleration techniques for proximity queries. The simplest but also most efficient subdivision would be the use of a regular grid. Only primitives within the same grid cell are then queried for collision. This approach is best suited for n-body collision queries, since it only takes linear time to query the collisions between the n^2 object pairs. Furthermore, it is well-suited for the detection of collisions and self-collisions between deformable objects [15]. An approach that combines the benefits of feature-tracking algorithms and spatial subdivision algorithms is described in [16] and [17].

In the recent years, graphics hardware has been used to accelerate various geometric computations such as collision detection [18,19] or distance field computation [20]. Possible drawbacks of GPU-based approaches are that their accuracy is limited by the frame buffer resolution and the time for reading back the frame buffers. However, in [19] the amount of read-back is reduced with the introduction of occlusion queries for collision detection.

Manipulation Planning: Solving the robotic planning problems in high-dimensional configuration spaces is often addressed using probabilistic roadmap (PRM) planners [21,22]. Such planners randomly generate samples in the configuration space and attempt to connect each newly generated sample to one of the existing samples by means of shortest paths in configuration space. This procedure results in a connectivity graph that spans the configuration space. The sampled nodes and the path segments stored in the graph have to be tested for collision. The validation of a collision-free graph takes up most of the computation time in the construction of the PRM. Schwarzer et al. presented an approach to integrate distance computation algorithms in the PRM framework for a more efficient dynamic collision checking [2]. The problem of computing a measure of distance between two configurations of a rigid articulated model has also been addressed by Zhang et al. [23]. Furthermore, manipulation planning is addressed by building a so-called "manipulation graph". It consists of nodes representing viable grasps and placements. Nodes are connected by transit or transfer paths moving either the manipulator alone or together with a grasped object. Those paths are solved using the above-mentioned PRM planners [24,25]. Integrating symbolic and manipulation planning has been studied in the past. Cambon et al. [26,27] use the FF planner which they modified to integrate roadmap planning into the planner. However, they do not provide a general interface to the domain-independent planner. Therefore, we base our implementation on the work of Dornhege et al. [28] that presents a general domain-independent planner interface to geometric planning. A comprehensive survey on robot planning algorithms can be found in [29].

3   Proximity Queries

In this section, the proximity query algorithm is described. It returns the minimum distance between pairs of arbitrarily shaped objects in three-dimensional space. The objects may be given as closed non-convex triangulated surface meshes. The algorithm can be divided into three stages. The first stage employs a variation of the Gilbert-Johnson-Keerthi algorithm (GJK) [6]. It determines the separation distance between the convex hulls of a pair of non-convex objects. Obviously, the points that define the separation distance lie on the convex hulls of the objects, but not necessarily on their surfaces. Thus, we obtain a

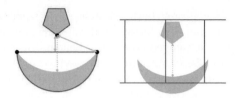

Fig. 1. Step one and two of the distance computation algorithm. Left: Lower and upper distance bounds (gray lines) between the two shapes are derived from GJK. The lower bound is the minimum distance between the two convex hulls (red), the upper bound is the minimum distance between pairs of support points (black dots). The actual separation distance (dotted gray line) lies within these bounds. Right: Twofold culling using spatial hashing: 1. Only the object parts inside the margins (horizontal red lines) are hashed. 2. Only primitives inside the same cell (red rectangle) are considered in the pair-wise primitive test.

lower distance bound for the exact separation distance. Furthermore, in GJK the closest points on the convex hull are expressed by a combination of points on the actual surfaces, the support points. Thus, the closest pair of support points gives an upper distance bound (see the left side of figure 1). If the lower distance bound is greater than zero, i.e. the convex hulls of the two objects do not overlap, the algorithm proceeds with stage two. It employs spatial hashing [15]. All surface triangles are hashed to the cells in the hash table. The cell size $\mathbf{c} = [x, y, z]^T$ of the hash grid is determined using the distance bounds found in the first stage, with dist_{upper} and dist_{lower} being the upper and lower bound, respectively. The grid is aligned to a local coordinate system, which has the z-axis parallel to the vector that connects the closest points on the convex hulls. We define the cell size along the different axes to be: $c_x = c_y := \sqrt{t_x^2 + t_y^2}$ and $c_z = 2 \cdot \mathrm{dist}_{upper} - \mathrm{dist}_{lower}$. Here, $\mathbf{t} = (t_x, t_y, t_z)^T$ is the vector that connects the support points for which $\|\mathbf{t}\| = \mathrm{dist}_{upper}$. Using this scheme, only triangles within the same cell and its neighbors can still contribute to the exact minimum distance. The distance for all other triangle pairs is guaranteed to be greater than the upper distance bound. They are efficiently culled away by the intrinsic properties of the subdivision scheme (see the right side of figure 1). If the convex hulls of the mesh pair overlap, the algorithm proceeds with stage three. In this stage, information computed by GJK is utilized to adaptively decompose the meshes into sub-meshes and pair-wise repeat the process in stage one recursively. In particular, GJK returns extremal points of the objects along a support vector. Planes orthogonal to this support vector and going through the extremal points divide the objects into sub-meshes (see figure 2). The overall minimum distance between the object pair is the minimum of the set of distances computed for all the sub-mesh pairs. In contrast to other approaches, the input data does not need to be pre-processed, i. e. no full surface decomposition is executed and no bounding volume hierarchies are constructed. Instead, decomposition of the surface meshes is only performed if it is required in the separation distance

Fig. 2. Step three of the algorithm is invoked, if the convex hulls of the two shapes overlap. Left: The objects are recursively divided into sub-meshes according to support planes (dashed red lines). Right: The minimum of the set of separation distances (red lines) of the sub-mesh pairs gives the separation distance.

computation. This makes the approach suitable for the simulation of deformable objects.

4   System Overview

In this section, we describe our framework for manipulation planning and the embedding of the proximity query algorithm into this framework. First of all, the manipulation problem is decomposed into a symbolic and a geometrical part. The symbolic planner allows for task specifications and domain descriptions to be given in high-level, human-like language, e. g. task specifications look like *on(box, table)* and domain descriptions like *pick-up(box)* or *put-down(box, table)*, respectively. On the symbolic level, the applicability of actions can be decided by evaluating the conditions of state variables. On the other hand, the geometric planner is used for constraint checking and effect calculation, i. e. the detection of collision free states and execution paths. Therefore, the geometric planner has access to a full domain description that represents the kinematics of the manipulator and a three-dimensional scene description. The decomposition serves to partition the complex manipulation problem into simpler planning problems. The interaction between the two parts is realized by external modules called semantic attachments [28]. They compute the valuations of state variables in the symbolic part by answering question like "Is there a collision-free way to move from point a to point b?" using the geometric part at run-time. Using the semantic attachments, the low-level geometric planner can provide information to the high-level symbolic planner *during* the planning process. However, it is only evoked when it seems relevant to the high-level planner. This is of particular importance, since the low-level planner performs the most time-consuming tasks, the proximity queries. The semantic attachments are implemented as probabilistic roadmaps (PRM) [22]. The roadmaps are connectivity graphs that provide collision-free states and path segments in configuration space. The configuration space is given by the kinematics of the manipulator and the configurations of the objects.The verification of collision-free states and paths in workspace is performed using the proximity query algorithm described in section 3. Depending on the query, the exact distance is returned or a distance threshold is verified. Using distance queries instead of collision queries may seem to be slower in comparison, but only distance queries allow for a fast and safe path verification [2].

Incorporating the possibility to manipulate deformable objects extends the collision-free configuration space in the case of transfer paths - the paths, where the manipulator has grasped an object and moves it along. We require the manipulator to be collision-free. Only the grasped object may collide. If this collision can be resolved by deformation, the current state is verified to be collision-free. The deformation energy of an object is computed using a linear relation between the forces and the displacement of the tetrahedrons in the volume representation. If an object-specific threshold is passed, the state is not-collision-free. Besides the ability for deformable manipulable objects, navigation amongst deformable objects [30,31] could also be realized with the tetrahedral data structure.

5   Experiments

We have staged a series of experiments to evaluate the distance computation approach as well as its application in the manipulation planning framework. In all experiments, the object representation is twofold. Surfaces are represented by triangular meshes to provide the input for the proximity query algorithm, whereas volumes are represented by tetrahedral meshes to provide the input for the force-displacement relation. All run-times were computed as average run-times on an Intel Core2Duo 6400 with 2 GB RAM using a single core.[1]

Proximity Queries: The set of test scenarios for the evaluation of the distance computation approach includes (1) a pair of cows, (2) a pair of horses, (3) a stick and a dragon and (4) a pair of deforming teddies. The objects vary in shape and complexity. Scene complexity and timings are given in table 1. We compare the timings with the ones gathered with the software package SWIFT++ [8]. SWIFT decomposes the surface of a non-convex object into convex pieces, which are stored in a bounding volume hierarchy (BVH). The query is then executed on the hierarchy of convex pieces. Using this data structures, distance queries can be answered very quickly. However, if the scene is considered to be unknown in each time step, surface decomposition and BVH generation has to be included in the total computation time. In comparison, our approach decomposes the objects into a tree of sub-meshes whose convex hulls do not overlap. This is more general when compared to a decomposition into convex pieces. However, it is also more adaptive with respect to the current scene configuration. Therefore, our algorithm achieves lower average computation times. Please note that SWIFT++ is optimized for the application in rigid body simulations. Therefore, the surface decomposition and the construction of the BVH can be executed as preprocessing steps. Thus, they are probably not optimized. Nevertheless, the timings indicate that the decomposition is less suitable for online computations in the context of deformable objects or for single-shot algorithms like the approach proposed here.

---

[1] A video with five exemplary scenarios and plans can be found at:
http://tinyurl.com/p5z82t

Table 1. Results for the test scenarios. The timings resemble the average distance computation time in milliseconds over 1000 consecutive frames.

Scenario	# of triangles	avg. [ms] our algorithm	avg. [ms] SWIFT
(4)	4400	67	1518
(3)	6000	90	1250
(1)	12000	680	1681
(2)	19800	762	2904

Manipulation Planning: We demonstrate our manipulation planning framework on two synthetic test scenarios (see figures 3 and 4). The manipulator used in both scenarios consists of 2400 triangles and 2500 tetrahedrons. The first scenario consists of an additional three tables, with manipulable items placed on top. Triangles and tetrahedrons sum up to 2500 and 2600, respectively. In the second scenario, cubes are arranged to form a small narrow passage. The manipulator platform and a teddy are placed left and right of the passage. Triangles and tetrahedrons sum up to 3000 and 3500, respectively. The two scenarios demonstrate two different problems. In the tables scene, problems are formulated that place objects at the locations of other objects, forcing the planner to detect such situations and plan for them accordingly. The results shown in table 2 indicate that even multiple replacing of objects still results in reasonable runtimes.

In the second scene, a problem is formulated that forces the geometric planner to take the deformability of the manipulable object into account. The execution path depicted in figure 4 required the deformation computation to be executed for 600 configurations. This adds an additional planning time of 100 ms per configuration on average.

Fig. 3. An advanced pick-and-place task. The manipulator is required to pick up the red box and place it to where the green box is located. Therefore, it first has to pick up the green box and place it somewhere else (upper row) and then pick up and move the red box to the final position (lower row).

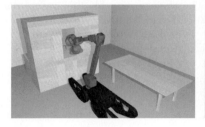 

Fig. 4. A pick-and-place task applied to a deformable movable object. The teddy is picked up from behind the wall and moved trough the small narrow passage (left) to its final position above the table (right). An execution path can only be found, if deformability of the teddy is considered by the geometric planner.

Table 2. Results for the tables scene. The problem instances are separated in three classes: Simple pick-and-place tasks (Class I), problems that require the replacing of object to reach the goal configuration (Class II), and problems that require the replacing of multiple objects (Class III). Various tasks have been posed per class. Runtimes are given in seconds.

Class I	Runtime (s)
01	3.48 ± 1.23
02	6.08 ± 3.49
03	1.47 ± 0.12
04	3.77 ± 0.97
05	4.75 ± 2.36
06	5.27 ± 2.71
07	63.83 ± 7.67
08	5.66 ± 7.50
09	12.48 ± 14.74

Class II	Runtime (s)
01	24.32 ± 8.63
02	24.95 ± 9.25
03	91.87 ± 14.01
04	30.26 ± 9.74

Class III	Runtime (s)
01	37.33 ± 6.85
02	15.50 ± 2.52
03	78.55 ± 45.61

6   Conclusion

We have presented an algorithm for deformable proximity queries. It employs GJK to recursively find sub-mesh pairs whose convex hulls do not overlap. For such pairs, the minimum distance can be efficiently computed using spatial hashing. The overall minimum distance is governed by the minimum distance between the sub-mesh pairs. We have illustrated the efficiency and suitability of the algorithm with regard to deformable objects. Furthermore, we have described and demonstrated the application of the algorithm in a manipulation planning framework. Currently, we are investigating how to improve the runtimes of the framework. An optimized recursive decomposition would speed up the proximity query algorithm. Integration of geometric heuristics in the symbolic planning process could significantly reduce the amount of calls to the geometric planner.

Acknowledgments

This work has been supported by the German Research Foundation (DFG) under contract number SFB/TR-8. We also thank the reviewers for their helpful comments.

References

1. Lin, M., Gottschalk, S.: Collision detection between geometric models: a survey. In: Proc. of IMA Conference on Mathematics of Surfaces, pp. 37–56 (1998)
2. Schwarzer, F., Saha, M., Latombe, J.: Adaptive dynamic collision checking for single and multiple articulated robots in complex environments. IEEE Transactions on Robotics and Automation 21(3), 338–353 (2005)
3. Lin, M.C., Manocha, D.: 35. In: Handbook of Discrete and Computational Geometry, pp. 787–806. CRC Press, Boca Raton (2004)
4. Ericson, C.: Real-Time Collision Detection. Morgan Kaufmann (The Morgan Kaufmann Series in Interactive 3-D Technology) (2004)
5. Teschner, M., Kimmerle, S., Heidelberger, B., Zachmann, G., Raghupathi, L., Fuhrmann, A., Cani, M.-P., Faure, F., Magnenat-Thalmann, N., Strasser, W., Volino, P.: Collision detection for deformable objects. Computer Graphics Forum 24, 61–81 (2005)
6. Gilbert, E., Johnson, D., Keerthi, S.: A fast procedure for computing the distance between complex objects in three-dimensional space. IEEE Transactions on Robotics and Automation 4, 193–203 (1988)
7. Lin, M., Canny, J.: A fast algorithm for incremental distance calculation. In: IEEE Int. Conf. on Robotics and Automation, pp. 1008–1014 (1991)
8. Ehmann, S., Lin, M.: Accurate and fast proximity queries between polyhedra using surface decomposition. In: Computer Graphics Forum (Proc. of Eurographics 2001), vol. 20, pp. 500–510 (2001)
9. Quinlan, S.: Efficient distance computation between non-convex objects. In: IEEE Int. Conf. on Robotics and Automation, vol. 4, pp. 3324–3329 (1994)
10. Hubbard, P.: Approximating polyhedra with spheres for time-critical collision detection. ACM Transactions on Graphics 15, 179–210 (1996)
11. van den Bergen, G.: Efficient collision detection of complex deformable models using AABB trees. J. Graphics Tools 2, 1–13 (1997)
12. Klosowski, J., Held, M., Mitchell, J., Sowizral, H., Zikan, K.: Efficient collision detection using bounding volume hierarchies of k-DOPs. IEEE Transactions on Visualization and Computer Graphics 4, 21–36 (1998)
13. Gottschalk, S., Lin, M., Manocha, D.: OBB-Tree: a hierarchical structure for rapid interference detection. In: SIGGRAPH 1996: Proc. of the 23rd annual conference on Computer graphics and interactive techniques, pp. 171–180. ACM Press, New York (1996)
14. Larsson, T., Akenine-Moeller, T.: Collision detection for continuously deforming bodies. In: Eurographics, pp. 325–333 (2001)
15. Teschner., M., Heidelberger, B., Mueller, M., Pomeranets, D., Gross, M.: Optimized spatial hashing for collision detection of deformable objects. In: Proc. Vision, Modeling, Visualization VMV 2003, Munich, Germany, pp. 47–54 (2003)
16. Gissler, M., Frese, U., Teschner, M.: Exact distance computation for deformable objects. In: Proc. Computer Animation and Social Agents CASA 2008, pp. 47–54 (2008)

17. Gissler, M., Teschner, M.: Adaptive surface decomposition for the distance computation of arbitrarily shaped objects. In: Proc. Vision, Modeling, Visualization VMV 2008, pp. 139–148 (2008)
18. Knott, D., Pai, D.: CInDeR: Collision and interference detection in real-time using graphics hardware. In: Proc. of Graphics Interface, pp. 73–80 (2003)
19. Govindaraju, N., Redon, S., Lin, M., Manocha, D.: CULLIDE: Interactive collision detection between complex models in large environments using graphics hardware. In: HWWS 2003: Proc. of the ACM SIGGRAPH/EUROGRAPHICS conference on Graphics hardware, pp. 25–32 (2003)
20. Sud, A., Govindaraju, N., Gayle, R., Kabul, I., Manocha, D.: Fast proximity computation among deformable models using discrete Voronoi diagrams. ACM Trans. Graph. 25(3), 1144–1153 (2006)
21. Latombe, J.: Robot Motion Planning. Kluwer Academic Publishers, Dordrecht (1991)
22. Kavraki, L., Svestka, P., Latombe, J.-C., Overmars, M.: Probabilistic roadmaps for path planning in high-dimensional configuration spaces. IEEE Transactions on Robotics and Automation 12(4), 566–580 (1996)
23. Zhang, L., Kim, Y.J., Manocha, D.: C-dist: efficient distance computation for rigid and articulated models in configuration space. In: SPM 2007: Proceedings of the 2007 ACM symposium on Solid and physical modeling, pp. 159–169. ACM Press, New York (2007)
24. Alami, R., Laumond, J.P., Siméon, T.: Two manipulation planning algorithms. In: WAFR: Proceedings of the workshop on Algorithmic foundations of robotics, pp. 109–125. A.K. Peters, Ltd., Wellesley (1995)
25. Simeon, T., Cortes, J., Laumond, J., Sahbani, A.: Manipulation planning with probabilistic roadmaps. The International Journal of Robotics Research 23, 729–746 (2004)
26. Cambon, S., Gravot, F., Alami, R.: A robot task planner that merges symbolic and geometric reasoning. In: Proc. of ECAI, pp. 895–899 (2004)
27. Fabien Gravot, S.C., Alami, R.: aSyMov: A planner that deals with intricate symbolic and geometric problems. Springer Tracts in Advanced Robotics 15 (2005)
28. Dornhege, C., Eyerich, P., Keller, T., Trüg, S., Brenner, M., Nebel, B.: Semantic attachments for domain-independent planning systems. In: Proc. of ICAPS (2009) (to appear)
29. LaValle, S.M.: Planning algorithms. Cambridge University Press, Cambridge (2006)
30. Rodriguez, S., Lien, J.M., Amato, N.: Planning motion in completely deformable environments. In: Proc. IEEE Int. Conf. on Robotics and Automation (ICRA), pp. 2466–2471 (2006)
31. Frank, B., Stachniss, C., Schmedding, R., Burgard, W., Teschner, M.: Real-world robot navigation amongst deformable obstacles. In: Proc. IEEE Int. Conf. on Robotics and Automation (ICRA), pp. 1649–1654 (2009)

Speech-Driven Facial Animation Using a Shared Gaussian Process Latent Variable Model

Salil Deena and Aphrodite Galata

School of Computer Science, University of Manchester
Manchester, M13 9PL, United Kingdom
{deenas,agalata}@cs.man.ac.uk

Abstract. In this work, synthesis of facial animation is done by modelling the mapping between facial motion and speech using the shared Gaussian process latent variable model. Both data are processed separately and subsequently coupled together to yield a shared latent space. This method allows coarticulation to be modelled by having a dynamical model on the latent space. Synthesis of novel animation is done by first obtaining intermediate latent points from the audio data and then using a Gaussian Process mapping to predict the corresponding visual data. Statistical evaluation of generated visual features against ground truth data compares favourably with known methods of speech animation. The generated videos are found to show proper synchronisation with audio and exhibit correct facial dynamics.

1 Introduction

Synthesis of a talking face driven driven by speech audio has many applications from cinema, games, virtual enviroments, online tutoring and in devising better Human Computer Interaction (HCI) systems. Humans perceive speech by interpreting both the sounds produced by speech movements and the visual cues that accompany it. Suppression of one channel at the expense of the other results in ambiguities in speech perception as shown by McGurk and McDonald [1]. Moreover, given the high fine tunement in the way humans perceive speech, slight glitches in an animated character are very conspicuous. Thus, an animated character needs to exhibit plausible speech movements without jerks and with proper synchronisation with the audio.

The pioneering work on facial animation was done by Parke [2] where a 3D model of the face was built using a polygon mesh which was texture-mapped and animation was achieved by interpolating between prototypes or keyframes. Facial animation can also be done using anatomical models of the face constrained by the laws of physics [3], [4]. Whilst 3D models of the face offer a high level of flexibility to the animator, they are very labour intensive and fail to achieve very high levels of realism. Data-driven approaches to facial animation seek to use text or audio data to directly synthesise animation with minimal manual intervention. They can be grouped into text-to-visual synthesis [5] and audio-to-visual synthesis [6], [7], [8], [9], [10], [11]. Rendering for data-driven facial

G. Bebis et al. (Eds.): ISVC 2009, Part I, LNCS 5875, pp. 89–100, 2009.

animation can be using 3D graphics-based models of the face [12], [13]; 2D image-based models [6], [7] or through hybrid appearance-based models [8], [9], [10], [11].

The basic unit of spoken language is the phoneme and the corresponding visual unit pertaining to different lip configurations is the viseme. The english language has a total of 41 phonemes [14] and according to the MPEG-4 standard, these are grouped into 14 visemes [15]. Thus, the mapping from phonemes to visemes is many-to-one. Moreover, the visual counterpart of speech is dependent on the context of the speech signal, which means that the same phoneme may produce a different visual output, depending on the phonemes preceeding and following it. This phenomenon is known as coarticualtion.

Our focus is on a data-driven approach to speech animation using machine learning techniques. Because the audio-visual mapping is many-to-one and modelling coarticulation involves taking context into account, regression techniques like artificial neural networks or support vector machines fail to produce appropriate results. Successful techniques that effectively model coarticulation include hidden Markov models [7], Gaussian phonetic models [8] and switching linear dynamical systems [11]. In this work, we make use of the Gaussian Process Latent Variable Model [16] (GPLVM) framework to learn a shared latent space between audio and visual data. The GPLVM is a non-linear dimensionality reduction technique and has recently been applied to multimodal data by learning a shared latent space between human silhouette features and 3D poses [17], [18]. This allows the inferrence of pose from silhouettes. We apply this framework to learn an audio-visual mapping and compare the results with Brand's Voice Puppetry [7].

2   Background and Related Work

We begin by providing some background on the Shared GPLVM (SGPLVM) and refer readers to [16] and [18] for more information.

2.1   The GPLVM

The GPLVM is a probabilistic dimensionality reduction technique that uses Gaussian Processes (GPs) to find a non-linear manifold of some data that seeks to preserve the variance of the data in latent space. The latent space $\mathbf{X} = [x_1, \ldots, x_N]$ is assumed to be related to the mean centered data set, $\mathbf{Y} = [y_1, \ldots, y_N]^T$ through a mapping f that is corrupted by noise:

$$y_n = f(x_n) + \epsilon . \tag{1}$$

By placing a GP prior of the mapping f and marginalising it, the likelihood function (2) is obtained, which is a product of D GPs and Φ are the hyperparameters of the covariance function, which is also referred to as the kernel.

$$p(\mathbf{Y}|\mathbf{X}, \Phi) = \prod_{i=1}^{D} \frac{1}{(2\pi)^{\frac{N}{2}} |\mathbf{K}|^{\frac{1}{2}}} \exp\left(-\frac{1}{2} y_{:,i}^{T} \mathbf{K}^{-1} y_{:,i} \right) . \tag{2}$$

For non-linear mappings, a closed-form solution is not available and the likelihood function is optimised with respect to the latent values $\mathbf{X}$ using conjugate-gradient optimisation. Maximising the marginal likelihood (2) with respect to the latent points and the hyperparameters Φ results in the latent space representation of the GPLVM.

$$\{\hat{\mathbf{X}}, \hat{\Phi}\} = \underset{\mathbf{X}, \Phi}{\arg\max} P(\mathbf{Y}|\mathbf{X}, \Phi) . \tag{3}$$

Back Constraints. The GPLVM, being a mapping from the latent space to the data space, ensures that points that are close on the latent space are found close on the data space. However, it does not ensure the opposite, i.e. points that are close in the data space to be mapped close on the latent space. The aim of the back constraints [19] is to enforce this distance preservation. It is done by using an inverse parametric mapping that maps points from the data space to the latent space. The mapping takes the following form:

$$x_i = g(y_i, \mathbf{W}) . \tag{4}$$

Where $\mathbf{W}$ are the parameters of the back-constraint kernel function, which can be any non-linear kernel such as the Radial Basis Function (RBF) or the Multilayer Perceptron (MLP). The optimisation in (3) is then done with respect to the back constraint parameters $\mathbf{W}$:

$$\{\hat{\mathbf{W}}, \hat{\Phi}\} = \underset{\mathbf{W}, \Phi}{\arg\max} P(\mathbf{Y}|\mathbf{W}, \Phi) . \tag{5}$$

Dynamics. Wang et al. [20] proposed an extension of the GPLVM which produces a latent space that preserves sequential relationships between points on the data space, on the latent space. This is done by specifying a dynamical function over the sequence in latent space:

$$x_t = h(x_{t-1}) + \epsilon_{dyn} . \tag{6}$$

Where $\epsilon_{dyn} \sim N(\mathbf{0}, \beta_{dyn}^{-1}\mathbf{I})$. This is a first-order dynamics kernel that assumes that each latent point x_t is only conditioned on the preceeding frame, x_{t-1}. By placing a Gaussian Process prior over the function $h(x)$ and marginalising this mapping, a new objective function is obtained. Optimising this objective function results in latent points that preserve temporal relationships in the data. The new objective function is given by (7) with $\hat{\Phi}_{dyn}$ being the hyperparameters of the dynamics kernel.

$$\{\hat{\mathbf{X}}, \hat{\Phi}_Y, \hat{\Phi}_{dyn}\} \underset{X, \Phi_Y, \Phi_{dyn}}{\arg\max} P(\mathbf{Y}|\mathbf{X}, \Phi_Y)P(\mathbf{X}|\Phi_{dyn}) . \tag{7}$$

2.2   The SGPLVM

To construct a shared latent space between two sets of variables, $\mathbf{Y}$ and $\mathbf{Z}$ and with a shared latent space $\mathbf{X}$, the likelihood function is taken to the the product of each individual likelihood function, conditioned on a common latent space. This leads to the optimisation of two different sets of hyperparameters for the two kernel functions. The joint likelihood of the two observation spaces is given by:

$$P(\mathbf{Y}, \mathbf{Z}|\mathbf{X}, \Phi_s) = P(\mathbf{Y}|\mathbf{X}, \Phi_Y)P(\mathbf{Z}|\mathbf{X}, \Phi_Z) . \tag{8}$$

Where $\Phi_S = \{\Phi_Y, \Phi_Z\}$ is a concatenation of the two different sets of hyperparameters.

Back-constraints can similarly be integrated, but with respect to only one data space, because in practice, two separate mappings from two different data spaces, that produce a common latent space cannot be defined. Moreover, a dynamics prior can also be placed on the latent space, just like for the GPLVM.

The Shared GPLVM (SGPLVM) used by [17] and [18] has been used to learn a mapping between pose and silhouette data. However, the mapping from silhouette to pose is one-to-many because silhouettes are ambiguous, especially when the figure is turning around. Ek et al. have addressed ambiguity in [18] by putting a back constraint with respect to poses, which forces a one-to-one relationship between the data and latent space. In [21], ambiguity has been catered for by using a Non-Consolidating Components Analysis (NCCA) whereby a private latent space for each of the observation spaces is learnt in addition to the shared latent space. This allows for the disambiguation of human pose estimation given silhouettes because the variance in both data spaces is retained. Thus, the variance from the space pertaining to the test data can used in the inferrence as a discriminant to resolve ambiguities. The same NCCA model has been used in [22] for mapping human facial expression data, represented by facial landmarks to a robotic face. The ambiguity in this case is with respect to robot poses, with multiple robot poses corresponding to a given facial expression vector.

The SGPLVM can be viewed as a non-linear extension of Canonical Correlation Analysis (CCA). CCA learns a correspondence between two datasets by maximising their joint correlation. Theobald and Wilkinson [10] use CCA to learn an audio-visual mapping. Modelling coarticulation is achieved by appending speech features to the right and to the left of each frame. This, however, leads to a combinatorial explosion and requires large amounts of data to provide adequate generalisation ability. Our approach, based on the SGPLVM framework allows for coarticulation to be modelled in two ways. Placing a back constraint with respect to audio features ensures distance preservation of speech features in the latent space, thus ensuring a smooth transition of latent points for test audio data. Moreover, placing a dynamical model on the latent space constraints the optimisation of latent points to respect the data's dynamics both in the training and synthesis phases.

3   Building an Audio-Visual Corpus

The *Democracy Now!* dataset [11] has been used for our experiments. It features an anchor giving news presentations under roughly the same camera and lighting conditions. We use the dissected video sequences mentioned in [11], featuring the anchor speaking sentences delimited by pauses for breath. However, we perform our own parameterisation of the visual and speech data. The video sequences are converted into frames sampled at the rate of 25 frames per second and cropped around the face region. High quality uncompressed audio has also been made available separately by the authors of [11], that match the dissected video sequences. We now detail how a compact parameterisation is obtained for both visual and audio data. A total of 236 video sequences, corresponding to about 20 minutes of video have been used, together with the corresponding uncompressed audio.

3.1 Visual Data Pre-processing

Active Appearance Models (AAMs) [23] have been chosen for facial parameterisation because they capture the statistical variation in shape and texture and provide a generative model to extrapolate novel faces as a linear combination of basis shape and texture vectors. They require a training set of annotated prototype face images where the annotations provide for the shape data and the texture data is sampled from the convex hull spanned by these shape vectors (Figure 1a). AAM training first normalises the shape vectors by removing rotations and translations and aligns the the shape with respect to the mean shape by a piecewise affine warp. This requires a triangulation of the landmarks to be performed (Figure 1b). In our case, 31 landmarks have been used. PCA is then applied to the shape and texture data separately and then further on the concatenation of the PCA parameters for shape and texture. Ater training, AAM parameters can be extracted from novel images by projecting the shape and texture data to the corresponding retained eigenvectors of the PCA and then again on the combined eigenvectors. In addition, given a set of AAM parameters, novel frames can be generated by first reconstructing the shape and texture separately and then warping the texture to the shape (Figure 1d). AAMs can also be used for tracking landmarks on novel facial images (Figure 1c). By retaining 95% of the variance in the shape, texture and combined PCA, a 28-dimensional AAM feature vector is obtained.

3.2 Speech Parameterisation

Speech needs to be parameterised so as to represent the acoustic variability within and between the different phonemes. This is done by extracting features from the speech signal that help distinguish between the phonemes. The most common speech feature extraction techniques are: Linear Predictive Coding (LPC), Mel-Frequency Cepstral Coefficients (MFCC), Line Spectral Frequencies (LSF) and Formants [14]. MFCCs have been chosen for speech feature extraction of our data because of its robustness to noise and also because we do not need accurate reconstructions provided by linear prediction methods such as LPC and LSF. The MFCC features have been computed at 25 Hz in order to match the sampling rate of the image frames.

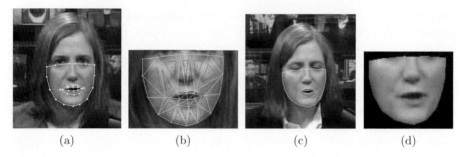

(a)   (b)   (c)   (d)

Fig. 1. (a) Annotations marked on a sample face image. (b) Triangulation of the landmarks for warping. (c) Results of AAM search on a new image. (d) Reconstruction of face from a set of AAM parameters.

4 Audio-Visual Mapping

Taking $\mathbf{Y}$ to be the MFCC feature vector and $\mathbf{Z}$ to be the AAM feature vector, an SGPLVM is learnt between $\mathbf{Y}$ and $\mathbf{Z}$. Whilst in [21] and [22], the data to be synthesised is ambiguous, in our case, we have a many-to-one mapping between audio and visual data. This leads to more flexibility in building the model and the NCCA model of [21] and [22] brings no benefit to our system. However, placing a back-constraint with respect to the audio favours the modelling of coarticulation by constraining similar audio features to be mapped close on the latent space. In addition, it allows the initialisation of latent points from novel audio using the back-constrained mapping. We place an MLP back-constraint with respect to the audio data. An autoregressive dynamics GP is also placed on the latent space. The graphical model of our system is shown in Figure 2. All the parameters of the model are optimised during training. The resulting latent space can be viewed as a non-linear embedding of both audio and visual data that can generate both spaces.

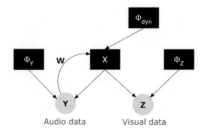

Fig. 2. Graphical model of the shared GPLVM with a back-constraint with respect to the audio and an autoregressive dynamics model on the latent space

4.1   Synthesis

Once the SGPLVM model is trained, audio-visual synthesis proceeds by first extracting MFCC features from test audio. AAM parameters $\hat{\mathbf{Z}}$ can then be synthesised from the test MFCC features $\hat{\mathbf{Y}}$ by first obtaining the corresponding latent points, $\hat{\mathbf{X}}$. The optimisation of latent points is done both with respect to the GP mapping from $\mathbf{X}$ to $\mathbf{Y}$ as well as with respect to the dynamical model, by formulating a joint likelihood given in (9). The likelihood is then optimised using conjugate gradient optimisation to find the most likely latent coordinates for a sequence of audio features.

$$\hat{\mathbf{X}} = \arg\max_{\mathbf{X}_*} P(\hat{\mathbf{Y}}, \mathbf{X}_* | \mathbf{Y}, \mathbf{X}, \Phi_Y, \Phi_{dyn}) \,. \tag{9}$$

Where $\mathbf{X}_*$ is an initialisation of the latent points. Once $\hat{\mathbf{X}}$ is obtained, $\hat{\mathbf{Z}}$ is obtained from the mean prediction of the GP from $\mathbf{X}$ to $\mathbf{Z}$.

$$\hat{\mathbf{Z}} = k(\hat{\mathbf{X}}, \mathbf{X})^T \mathbf{K}^{-1} \mathbf{Z} \,. \tag{10}$$

4.2   Initialisation of Latent Points

Optimisation of (9) is likely to be highly multimodal with multiple local optima. Thus, a proper initialisation of the latent space, $\mathbf{X}_*$ is required to get good results. We use two initialisation techniques for $\mathbf{X}_*$. In the first method, the latent points obtained from the SGPLVM training are taken to be the states of a hidden Markov Model (HMM) and the training audio features are taken to be the observations. The transition log likelihood is computed as the GP point likelihood between each latent point and every other latent point and the observation log likelihood is obtained by computing the GP point likelihood between the test audio vector and each of the training latent points (states of the HMM). The optimal sequence of latent points $\mathbf{X}_*$ is obtained from the Viterbi algorithm in log space. This is analogous to choosing a set of latent points from the training set that best match the test audio. To speed computation when the number of training data points for the SGPLVM is very high, a subset of the points can be randomly chosen instead, for initialisation.

The second method of initialisation is from the back-constrained mapping from the audio space $\mathbf{Y}$ to the latent space $\mathbf{X}$, which can be obtained as follows:

$$\mathbf{X}_* = g(\hat{\mathbf{Y}}, \mathbf{W}) \,. \tag{11}$$

We shall call the method based on the back-constraint initialisation SGPLVM A and the method based on the HMM initialisation SGPLVM B.

4.3   Experiments

GPLVM training is quite expensive and has a complexity $O(N^3)$, where N is the number of data points. Various sparsification methods have been proposed

[24] by making use of a subset of data at a time, called the active set. However, even with sparsification, optimisation of a GPLVM likelihood becomes intractable when the number of data points exceeds a few thousands. This is in contrast to other methods to audio-visual mapping such as HMMs, which can cope with tens of thousands of data points. In our experiments, we have used a repeated random subsampling method for choosing 50 sequences from the 236 audio-visual sequence pairs for training SGPLVM A and SGPLVM B, giving an average of 6000 frames. We fix the dimensionality of the latent space to be 8 as further increasing the dimensionality does not improve the reconstructions of AAM parameters. We then randomly choose 20 sequences for testing, such that the training and testing sets do not overlap. Only the audio features from this test set are used for inferring novel AAM parameters using SGPLVM A and SGPLVM B.

We have used Brand's Voice Puppetry [7] as a benchmark. We train the cross-modal HMMs using the same subsets of audio and visual features as used for the SGPLVM and use the same data for testing. The repeated random subsampling experiment is done ten times for both the SGPLVM and Brand's method.

4.4   Results

We present both quantitative and qualitative results from our experiments. Quantitative results are obtained by finding the Root Mean Square (RMS) error between test AAM feature vectors and ground truth. Figure 3 shows the results

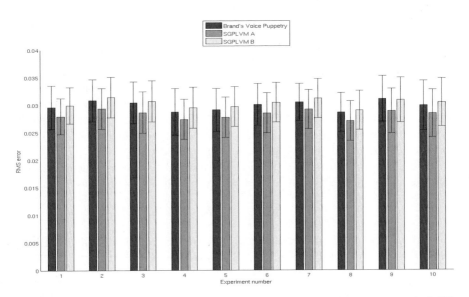

Fig. 3. RMS errors obtained between ground truth AAM feature vectors and 1) SG-PLVM A 2) SGPLVM B and 3) Brand's Voice Puppetry. The plots also include the standard deviation of the errors.

obtained accross the ten runs of the experiment. The results show no statistically significant difference between the errors obtained from Brand's method and the SGPLVM. In general, the errors for SGPLVM B are slightly higher than those for SGPLVM A, mostly due to a smoother latent space obtained from the back-constraint initialisation.

We also compare the trajectories of the first mouth landmark parameter reconstructed from the AAM parameters, of the three approaches against ground truth. Figures 5, 6 and 7 shows the results for Brand's Voice Puppetry, SGPLVM A and SGPLVM B respectively. The results for Brand's method show that the trajectories are smoothed out as compared to the SGPLVM approaches. This is

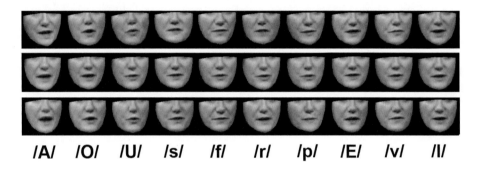

/A/   /O/   /U/   /s/   /f/   /r/   /p/   /E/   /v/   /l/

Fig. 4. Reconstructions from AAM features obtained from: ground truth (top row), Voice Puppetry (middle row) and SGPLVM A (bottom row). The audio used for synthesis contains the sentence: "*House of representatives has approved legislation*". The frames correspond to ten different visemes from the test audio sentence.

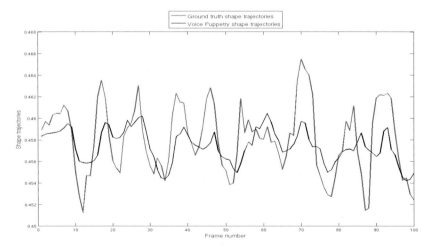

Fig. 5. Shape trajectories obtained from Brand's Voice Puppetry and the corresponding ground truth trajectories

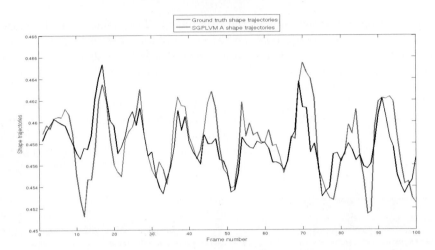

Fig. 6. Shape trajectories obtained from SGPLVM A and the corresponding ground truth trajectories

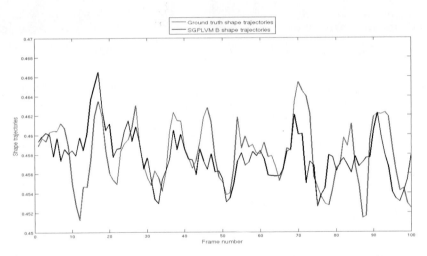

Fig. 7. Shape trajectories obtained from SGPLVM B and the corresponding ground truth trajectories

because Brand's approach involves synthesising AAM parameters from a state sequence, which represents Gaussian clusters, and is thus very approximative. The SGPLVM approaches, on the other hand, bypass this approximation and make use of the full variance of the visual data in synthesis.

Qualitative results are obtained by rendering frames from the AAM parameters in order to visualise the output. The videos show proper lip synchronisation with the audio with smooth lip movements. The results from SGPLVM A appears to be the best whilst SGPLVM B gives proper lip synchronisation but with a few jerks in the animation. The results from Brand's Voice Puppetry are overly smoothed with under articulation. Figure 4 shows ground truth frames as well as frames generated from AAM features obtained from Voice Puppetry and SGPLVM A. The audio contains a sentence which has 12 of the 14 visemes from the MPEG-4 standard [15].

5   Conclusions and Future Work

We have shown how the shared GPLVM can be applied to multimodal data comprising of audio and visual features, in order to synthesise speech animation. The results show that our methods are comparable to Brand's Voice Puppetry in terms of RMS errors of AAM features generated, but with more articulated lip movements.

In future work, a perceptual evaluation of the animation will be carried out where viewers would be asked to asses the realism of the generated videos as well as the intelligibility of the lip movements. Experiments will also be performed with different parameterisations of speech, which favour speaker independence. Moreover we would also investigate delta features in speech to more effectively capture context in speech animation.

References

1. McGurk, H.: MacDonald: Hearing lips and seeing voices. Nature 264, 746–748 (1976)
2. Parke, F.I.: A parametric model of human faces. PhD thesis, University of Utah (1974)
3. Terzopoulos, D., Waters, K.: Analysis and synthesis of facial image sequences using physical and anatomical models. IEEE Trans. on Patt. Anal. and Mach. Intel. 15(6), 569–579 (1993)
4. Kähler, K., Haber, J., Yamauchi, H., Seidel, H.P.: Head shop: generating animated head models with anatomical structure. In: SCA 2002: Proc. of the 2002 ACM SIGGRAPH/Eurographics symposium on Computer animation, pp. 55–63 (2002)
5. Ezzat, T., Poggio, T.: Miketalk: A talking facial display based on morphing visemes. In: Proc. of the Computer Animation Conference (1998)
6. Bregler, C., Covell, M., Slaney, M.: Video rewrite: driving visual speech with audio. In: SIGGRAPH 1997: Proc. of the 24th ACM annual conference on Computer graphics and interactive techniques, pp. 353–360 (1997)
7. Brand, M.: Voice puppetry. In: SIGGRAPH 1999: Proc. of the ACM 26th annual conference on Computer graphics and interactive techniques, pp. 21–28 (1999)
8. Ezzat, T., Geiger, G., Poggio, T.: Trainable videorealistic speech animation. In: SIGGRAPH 2002: Proceedings of the ACM 29th annual conference on Computer graphics and interactive techniques, pp. 388–398 (2002)
9. Cosker, D., Marshall, D., Rosin, P.L., Hicks, Y.: Speech driven facial animation using a hidden Markov coarticulation model. In: ICPR 2004: Proc. of the IEEE 17th International Conference on Pattern Recognition, vol. 1, pp. 128–131 (2004)
10. Theobald, B.J., Wilkinson, N.: A real-time speech-driven talking head using active appearance models. In: AVSP 2007: Proc. of the International Conference on Auditory-Visual Speech Processing (2007)

11. Englebienne, G., Cootes, T.F., Rattray, M.: A probabilistic model for generating realistic lip movements from speech. In: NIPS 2008: Avances in Neural Information Processing Systems 21, pp. 401–408 (2008)
12. Chai, J.X., Xiao, J., Hodgins, J.: Vision-based control of 3D facial animation. In: SCA 2003: Proc. of the ACM SIGGRAPH/Eurographics symposium on Computer animation, pp. 193–206 (2003)
13. Cao, Y., Faloutsos, P., Kohler, E., Pighin, F.: Real-time speech motion synthesis from recorded motions. In: SCA 2004: Proc. of the ACM SIGGRAPH/Eurographics symposium on Computer animation (2004)
14. Huang, X., Acero, A., Hon, H.-W.: Spoken Language Processing: A Guide to Theory, Algorithm and System Development. Prentice Hall PTR, Englewood Cliffs (2001)
15. Tekalp, M., Ostermann, J.: Face and 2-D mesh animation in MPEG-4. Image Communication Journal (1999)
16. Lawrence, N.D.: Probabilistic non-linear principal component analysis with Gaussian process latent variable models. Journal of Machine Learning Research 6, 1783–1816 (2005)
17. Shon, A., Grochow, K., Hertzmann, A., Rao, R.: Learning shared latent structure for image synthesis and robotic imitation. In: NIPS 2005: Advances in Neural Information Processing Systems 18, pp. 1233–1240 (2005)
18. Ek, C.H., Torr, P.H.S., Lawrence, N.D.: Gaussian process latent variable models for human pose estimation. In: Popescu-Belis, A., Renals, S., Bourlard, H. (eds.) MLMI 2007. LNCS, vol. 4892, pp. 132–143. Springer, Heidelberg (2008)
19. Lawrence, N.D., Quinonero-Candela, J.: Local distance preservation in the GP-LVM through back constraints. In: ICML 2006: Proc. of the ACM 23rd International Conference on Machine learning, pp. 513–520 (2006)
20. Wang, J.M., Fleet, D.J., Hertzmann, A.: Gaussian process dynamical models. In: NIPS 2005: Advances in Neural Information Processing Systems 18 (2005)
21. Ek, C.H., Rihan, J., Torr, P.H., Rogez, G., Lawrence, N.D.: Ambiguity modeling in latent spaces. In: Popescu-Belis, A., Stiefelhagen, R. (eds.) MLMI 2008. LNCS, vol. 5237, pp. 62–73. Springer, Heidelberg (2008)
22. Ek, C.H., Jaeckel, P., Campbell, N., Lawrence, N.D., Melhuish, C.: Shared Gaussian process latent variable models for handling ambiguous facial expressions. American Institute of Physics Conference Series (2009)
23. Cootes, T.F., Edwards, G.J., Taylor, C.J.: Active appearance models. In: ECCV 1998: Proc. of the 5th European Conference on Computer Vision-Vol. II, pp. 484–498 (1998)
24. Lawrence, N.D.: Learning for larger datasets with the Gaussian process latent variable model. In: AISTATS 2007: Proc. of of the Eleventh International Workshop on Artificial Intelligence and Statistics (2007)

Extracting Principal Curvature Ridges from B-Spline Surfaces with Deficient Smoothness

Suraj Musuvathy and Elaine Cohen

School of Computing, University of Utah

Abstract. Principal curvature ridges identify characteristic feature curves on a surface that can be used for surface registration, quality control, visualization and various other shape interrogation applications across disciplines such as medical imaging, computer vision, computer-aided design and engineering and geology. Current techniques for accurate extraction of ridges from B-Spline surfaces require C^n, $n \geq 3$ smoothness. In practice, many fitting techniques and modeling systems yield surface representations that may be only C^2, C^1 or C^0 on the knot lines. In this paper, we generalize a robust tracing algorithm to address the problem of extracting ridges from surfaces with lower orders of smoothness to broaden its applicability to a much larger set of surfaces.

1 Introduction

Principal curvature ridges as defined by Porteous [1] are loci of points on a surface where one of the principal curvatures attain an extremum along its respective principal direction. Identifying feature curves of the intrinsic geometry of surfaces, ridges have been found to be very useful in diverse application domains such as computer vision [2,3], medical image analysis [4,5,6], computer aided design and engineering [7], visualization [8], and geology [9].

At any point on a surface $S(u, v) \in R^3$, we denote principal curvatures and principal directions by κ_i and $t_i, i = 1, 2$ respectively.[1] Ridges are defined [2] as locations where $\phi_i(u, v) = < \nabla \kappa_i, t_i > = 0$ and the solutions form a set of 1-manifolds in the parametric domain and on the surface. The ridge condition imposes a third order derivative (C^3) smoothness requirement on $S(u, v)$ since derivatives of the principal curvatures are required to be continuous. Previous works on extracting accurate continuous ridge curves from smooth surfaces assume that the input surfaces have the necessary C^3 smoothness. To the best of our knowledge, there is no existing technique in the literature that addresses the case when the input surfaces have deficient smoothness $(C^2, C^1$ or $C^0)$ along embedded 1-manifolds. Techniques that extract ridges directly from discrete data (grid data, polygonal meshes) approximate principal curvatures and their derivatives by fitting smooth surfaces. Since derivatives are evaluated at discrete locations, it is assumed that the desired smoothness is available at all required points.

---

[1] See [10] for background material on differential geometry of surfaces.

G. Bebis et al. (Eds.): ISVC 2009, Part I, LNCS 5875, pp. 101–110, 2009.

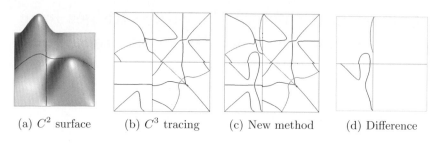

(a) C^2 surface   (b) C^3 tracing   (c) New method   (d) Difference

Fig. 1. (a) A bicubic B-Spline surface that is C^2 across the isoparametric knot curves, shown in black, that divide the parametric domain into four charts. (b) Top view of ridges extracted with the C^3 numerical tracing method. (c) Top view of ridges extracted using the proposed method. κ_1-ridges are black and κ_2-ridges are grey. Ridges are discontinuous at chart boundaries, the horizontal and vertical grey lines. (d) Ridge segments extracted using the new method but missed by the C^3 method.

It is difficult to enforce C^3 continuity while creating smooth surface representations during *ab initio* design as well as from discrete data. Surface derivatives are typically not available with discrete data, and merging surface patches fit to regions of discrete data or designed *ab initio* with C^3 smoothness imposes difficult constraints. The main contribution of this paper is a novel technique for extracting accurate ridges from rational tensor product parametric B-Spline (NURBS) surfaces that do not have C^3 smoothness.

This paper extends a robust tracing algorithm [11] that addresses C^3 surfaces and is computationally less demanding than other methods. The algorithm presented in [11] computes the necessary start points for tracing generic ridges on C^3 surfaces. However, when surfaces are C^2 or lower on knot lines, non-generic ridges may occur that appear as disjoint segments across images of the knot lines on the surface. A new type of start point is required to trace such ridge segments. We address this problem by subdividing the parametric domain into *charts* at knots where pieces of a NURBS surface meet with C^2 or lower smoothness in either of the parametric directions. Within each chart, the surface has smoothness of C^n, $n \geq 3$. In this paper, an approach for computing appropriate start points for tracing all ridge segments in all charts is presented, thereby guaranteeing extraction of all ridge segments. Fig. 1 shows an example of ridges extracted from a C^2 B-Spline surface.

Since B-Splines and NURBS are widely used surface representations for smoothly approximating discrete data as well as for ab initio design, this technique enables the use of smooth ridge extraction for a variety of applications on extended classes of surfaces. Many of the discrete methods do not address ridges around umbilics, which represent important features, and thus do not provide a complete solution. In addition, the ridges obtained using discrete methods are coarse approximations that are not as accurate as those obtained using the method presented in this paper and therefore can also miss ridges and incorrectly report the presence of ridges.

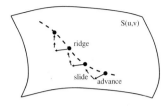

Fig. 2. Tracing ridges by advancing and sliding along principal directions

2   Previous Work on Ridge Extraction

Computing accurate ridges on smooth surfaces is computationally challenging since $\phi_i(u, v)$ is a complex function that is not piecewise rational. Ridges exhibit complex behavior at umbilics and ridge turning points, and these points along with the ridges around these points represent important surface features. Methods for extracting accurate ridges must address issues of computational and topological complexity. This section summarizes previous work on ridge extraction from smooth parametric surfaces. Previous work on ridge extraction from discrete representations may be found in [12,13] and references therein.

Lattice methods (conceptually similar to those used for surface intersection problems [14]) involve computing solutions of the ridge equations ($\{(u, v) : \phi_i(u, v) = 0\}$) on a dense grid of isoparametric curves on the surface and then computing the topology of the obtained ridge points. This approach has been adopted by [15,16] to compute ridges on surfaces represented by single polynomial patches (Bézier patches). This technique handles umbilics as well as ridge turning points but has the drawback of being computationally demanding. Since only single polynomial surfaces are addressed, surfaces with deficient smoothness are not considered.

Sampling methods are discrete techniques that detect isolated ridge points by evaluating $\phi_i(u, v)$ on various types of tessellation of the parametric domain [3,4,17,18] or along principal curvature lines [7]. Although these methods are typically fast, they do not address the topology of ridges at umbilics. Also, their accuracy is limited since the ridges are computed on a tessellation of the domain, and can miss ridges as well as compute false positives. Although C^2 B-Spline surfaces are used in [4], issues related to deficient smoothness are not addressed since derivatives are required at discrete locations only.

The approach presented in [11] builds upon numerical tracing ideas to robustly march along ridge segments using generic properties of ridges. This approach handles piecewise rational surfaces that join with at least C^3 smoothness. On C^3 surfaces, ridges are continuous curves and every ridge passes through at least one umbilic or curvature extremum, so umbilics and principal curvature extrema are computed as start points (seeds) for tracing. Therefore, all continuous ridges are guaranteed to be detected. The algorithm marches along ridge segments by advancing and sliding along local coordinate systems formed by principal directions (Fig. 2). For a κ_1-ridge, at each marching step, the algorithm first advances along the t_2 direction and then slides along the t_1 direction until a ridge point is detected. For a κ_2-ridge, the algorithm advances along the t_1 direction and then slides along the t_2 direction. Principal directions and curvature gradients are recomputed at every advance and slide step. Step sizes are adaptively varied to ensure that the trace does not jump to an adjacent ridge segment or get stuck in local non-zero extrema of $\phi_i(u, v)$. Umbilics and ridge turning points are also handled by the algorithm. Details of the method are found in [11].

In this paper, we have chosen to adopt the numerical tracing approach since it is robust and is computationally suitable for accurate ridge extraction on complex smooth surfaces. We add ideas from lattice methods, adapted to B-Splines, to solve for ridge points at the few specific isoparametric curves of deficient smoothness. In our experiments, a tracing method using ordinary differential equations (ODE) required much smaller step sizes for achieving the same accuracy as the method of [11]. In addition they required computation of singular points of $\phi_i(u, v)$ for robust tracing. Therefore, it seemed that the method presented in [11] is computationally more suitable for tracing than ODE based methods.

3   Tensor Product B-Spline Representation

A tensor product NURBS surface [19] is given by,

$$S(u, v) = \frac{\displaystyle\sum_{i=0}^{m} \sum_{j=0}^{n} w_{ij} R_{ij} B_{i,d^{(u)}}(u) N_{j,d^{(v)}}(v)}{\displaystyle\sum_{i=0}^{m} \sum_{j=0}^{n} w_{ij} B_{i,d^{(u)}}(u) N_{j,d^{(v)}}(v)} \tag{1}$$

R_{ij} are the control vertices and w_{ij} are the weights of $S(u, v)$. $B_{i,d^{(u)}}(u)$ and $N_{j,d^{(v)}}(v)$ are B-Spline basis functions of degrees $d^{(u)}$ and $d^{(v)}$ respectively. Let $T^{(u)} = \{T_p^{(u)}\}_{p=0}^{(m+d^{(u)}+1)}$ and $T^{(v)} = \{T_q^{(v)}\}_{q=0}^{(n+d^{(v)}+1)}$ be the knot vectors in the u and v parametric directions respectively. The support of $B_{i,d^{(u)}}(u) = [T_i^{(u)}, T_{i+d^{(u)}+1}^{(u)})$ and the support of $N_{j,d^{(v)}}(v) = [T_j^{(v)}, T_{j+d^{(v)}+1}^{(v)})$. It should be noted that the upper end of each interval is open. $S(u, v)$ is defined over the interval $[T_{d^{(u)}}^{(u)}, T_{m+1}^{(u)}) \times [T_{d^{(v)}}^{(v)}, T_{n+1}^{(v)})$.

B-Spline basis functions are piecewise polynomial i.e., they are polynomial functions within each knot interval. The smoothness at each knot is determined by its multiplicity. Let $\pi^{(u)}$ and $\pi^{(v)}$ be the set of unique knots (breakpoints) and $\mu^{(u)}$ and $\mu^{(v)}$ be the multiplicities of each breakpoint in the u and v parametric direction respectively. Then, at each breakpoint $\pi_p^{(u)}$, $S(u, v) \in C^{(d^{(u)} - \mu_p^{(u)})}$ in the u direction and at each knot $\pi_q^{(v)}$, $S(u, v) \in C^{(d^{(v)} - \mu_q^{(v)})}$ in the v direction. The derivatives of $S(u, v)$ are right continuous at a knot but $S(u, v)$ fails to be C^l when the lth derivatives are not left continuous at the knot.

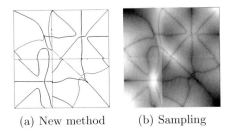

(a) New method   (b) Sampling

Fig. 3. Validation of disjoint ridges. (a) is identical to Fig. 1(c) and is repeated here for comparison with the sampled ridges shown in (b). κ_1-ridges and κ_2-ridges are shown in same color in (b).

4   Characteristics of Ridges on Surfaces with Deficient Smoothness

Generic ridges of a principal curvature on C^3 smooth surfaces either form non-intersecting closed curves on the surface or end at a boundary. They do not begin or end within the surface except at umbilics where a κ_1-ridge changes into a κ_2 ridge. However, these properties do not hold on surfaces with deficient smoothness lower than C^3 as illustrated in Fig. 1 and Fig. 3. Fig. 1(a) shows a bicubic B-Spline surface with single knots at $u = 0.25$ and $v = 0.5$. The surface is therefore C^2 across both knots and the ridge equations $\phi_i(u, v)$, $i = 1, 2$ are discontinuous at those knot lines $((0.25, v)$ and $(u, 0.5))$ since the third derivatives of the surface representation are discontinuous. The ridges on the surface thus are disjoint at the knots as shown in Fig. 1(c) and Fig. 3(a) which is verified by the results obtained by a sampling approach in Fig. 3(b). The C^3 tracing method can therefore end at a knot line and miss ridge segments when curvature extrema and umbilics are not present in the adjacent chart across the knot line. Fig. 1(d) shows the ridge segments missed by the C^3 tracing method (Fig. 1(b)), since umbilics and curvature extrema are not present on the missed segments in the relevant charts, but successfully detected by the method presented in this paper (Fig. 1(c)). Similar problems are encountered when the surface is C^1 or C^0 (See Fig. 6). It is assumed that the surface is at least C^0 i.e., we are concerned with only continuous surfaces. It is possible for the surface to be smooth even in the presence of knots of higher multiplicity. Ridges are continuous curves across such *pseudo* knots (Fig. 7).

Our goal is to trace all segments of all ridges on surfaces with deficient smoothness. We do not attempt to resolve the topology of the segments across the discontinuities. The next section presents extensions to the algorithm presented in [11] to resolve issues for tracing ridges on surfaces with deficient smoothness; viz., 1) ridge segments may be missed if no umbilic or curvature extremum is found in the corresponding chart, 2) ridge segments may or may not be continuous across knot lines and 3) ridge segments that lie exactly along a knot line may be missed.

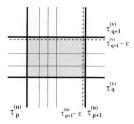

Fig. 4. A chart of a rational B-Spline surface. Chart boundaries are shown as thick black lines. Knots within the chart at which the surface has at least C^3 smoothness are shown as thin black lines.

5   Dealing with Deficient Smoothness

Let $S(u,v) \in R^3$ be a regular ($\|S_u \times S_v\| \neq 0$) tensor product NURBS surface. Within each knot interval a B-Spline basis function is a polynomial and $S(u,v)$ is rational, so all derivatives exist at any point on a surface that does not lie on the image of a knot line in either parametric direction. Since ridge computation is a local problem, the parametric domain is divided into charts between the knots at which $S(u,v)$ has smoothness lower than C^3 determined by $d^{(u)} - \mu_p^{(u)} < 3$ or $d^{(v)} - \mu_q^{(v)} < 3$. Let $\tau^{(u)} \subset T^{(u)}$ and $\tau^{(v)} \subset T^{(v)}$ be the set of knots at which the surface pieces meet with smoothness less than C^3 in the u and v direction respectively. The parametric domain is split into charts $[\tau_p^{(u)}, \tau_{p+1}^{(u)}) \times [\tau_q^{(v)}, \tau_{q+1}^{(v)})$, $p = 0 \ldots (|\tau^{(u)}| - 2), q = 0 \ldots (|\tau^{(v)}| - 2)$ (See Fig. 4) thereby creating $(|\tau^{(u)}| - 1)(|\tau^{(v)}| - 1)$ charts.

Since ridges start or stop at a chart boundary on surfaces with deficient smoothness, ridge seeds are computed along the boundary of each chart in addition to computing curvature extrema and umbilic seeds within every chart. Boundary seeds are computed as zeros of the univariate versions of the ridge equation; $\phi_i(\tau_p^{(u)}, v)$, $\phi_i(\tau_{p+1}^{(u)} - \epsilon, v)$, $\phi_i(u, \tau_q^{(v)})$ and $\phi_i(u, \tau_{q+1}^{(v)} - \epsilon)$. Since the upper end of each knot interval is open (Section 3), the isoparametric knot lines at $\tau_{p+1}^{(u)}$ and $\tau_{q+1}^{(v)}$ are not considered part of the chart and ridge seeds are computed at $\tau_{p+1}^{(u)} - \epsilon$ and $\tau_{q+1}^{(v)} - \epsilon$ instead. Since $\phi_i(u,v)$ is not piecewise rational, the equation $\phi_i(u,v) = 0$ is converted into $\tilde{\phi}_i(u,v) = 0$ by rearranging terms and squaring so that $\tilde{\phi}_i(u,v)$ is piecewise rational. The coefficients of each of the univariate functions are determined using symbolic computation [20] and the zeros are computed using efficient subdivision based rational B-Spline constraint solving techniques [21,22]. Ridges are then traced from all seeds using the technique presented in [11] within each chart independently.

When a trace reaches a chart boundary, surface derivatives are tested for continuity by comparing the vector values at that boundary point in the current chart and the adjacent chart. This is done by comparing the derivatives evaluated at the knot and the limit values when approaching the knot from the lower end of the chart interval in the corresponding parametric direction. If the derivative values do happen to be equal within a numerical tolerance, they are deemed to be continuous across the knot and the tracing continues into the relevant adjacent chart. Otherwise the derivatives are deemed discontinuous and the tracing for that particular segment ends. Ridge segments may coincide with isoparametric knot lines. These segments are traced from seed points on knot lines by stepping along the relevant knot lines in parametric space until $|\phi_i(u, v)|$ is greater than a user specified threshold value. In addition, curves on the surface along C^0 knot lines where the surface normals are discontinuous at each point across the knot lines appear as sharp feature lines on the surface. These ridges are extracted by marching along the length of all C^0 knot lines in parameter space and checking for discontinuities in the surface normals (See Fig. 6).

6   Results and Discussion

In this section, examples of ridge extraction from several tensor product NURBS surfaces with deficient smoothness are presented. Results are validated against

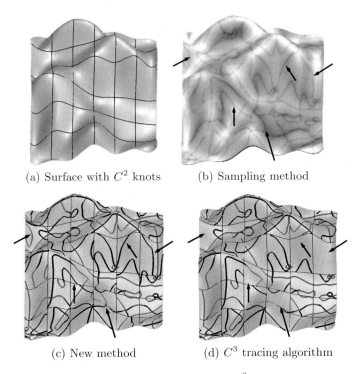

(a) Surface with C^2 knots   (b) Sampling method

(c) New method   (d) C^3 tracing algorithm

Fig. 5. (a) The NURBS surface fit to terrain data is C^2 at knot lines rendered as thin black curves. (b) Implied ridges from the sampling method, with κ_1-ridges and κ_2-ridges indicated by the same color. (c) Ridges extracted using our method. (d) Ridges extracted using the C^3 tracing algorithm. Black curves represent κ_1-ridges and grey curves represent κ_2-ridges. Arrows in (b), (c) and (d) point to ridges extracted by our new method, missed by the C^3 tracing algorithm and validated with the sampling method.

implied ridges computed using a sampling approach on a fine tessellation of the parametric domain. Lattice methods [15,16] address only single polynomial surface representations and are therefore not used for validating results in this paper. The algorithm presented in this paper has been implemented in the IRIT B-Spline modeling and programming environment [23]. Computation time varied from a few seconds to a few minutes for all surfaces used in this paper depending on the surface degree, number of internal knots and the complexity of ridge structures.

Fig. 5 shows ridges extracted from a bicubic C^2 surface that is fit to discrete data on a regular grid. This data is from the GLOBE [24] terrain elevation data set. Fig. 6 illustrates ridges extracted from a NURBS surface (degree 2 in u direction and degree 3 in v direction) with C^1 and C^0 knots. The results show several ridge segments missed by the C^3 tracing algorithm of [11](Figs. 5(d) and 6(d)) but successfully extracted using the method presented in this paper (Figs. 5(c) and 6(c)). Results of the proposed method are validated against a sampling of $|\phi_i(u,v)|$ in the parametric domain in Figs. 5(b) and 6(b). The sampling method can miss ridges as well as compute false positives since the accuracy is limited to the resolution of the tessellation as mentioned in [11].

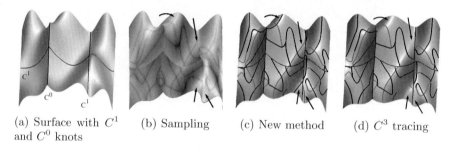

(a) Surface with C^1 (b) Sampling   (c) New method   (d) C^3 tracing
and C^0 knots

Fig. 6. (a) Surface with C^1 and C^0 knots rendered as thin black curves. (b) Implied ridges from the sampling method, with κ_1-ridges and κ_2-ridges indicated by the same color. (c) Ridges extracted using our method. (d) Ridges extracted using the C^3 tracing algorithm. Black curves represent κ_1-ridges and grey curves represent κ_2-ridges. Arrows in (b), (c) and (d) point to ridges extracted by our new method, missed by the C^3 tracing algorithm and validated with the sampling method.

In Fig. 7 we compare ridges extracted from surfaces created with successive B-Spline knot insertion [19]. The surfaces have identical geometry but the repeated knots suggest lower order smoothness (pseudo knots). Only curvature extrema and umbilics were used as seeds for the tracing algorithm. The results are identical for all four surfaces as expected. Even though seeds were not found in all charts, the method traces over a knot boundary since the derivatives are determined to be continuous. We can conclude that the algorithm presented in this paper does not fail when the surface geometry is smooth at repeated knots and computes connected ridge segments.

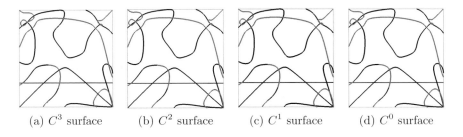

(a) C^3 surface   (b) C^2 surface   (c) C^1 surface   (d) C^0 surface

Fig. 7. Top view of ridges on biquartic surfaces created with successive knot insertions at $v = 0.25$ (horizontal black line)

7   Conclusions, Limitations and Future Work

Generic ridges on C^3 surfaces are continuous curves. Since ridges are defined using third order surface derivatives, they are discontinuous segments on surfaces with lower order smoothness. Previous methods for ridge extraction from parametric surfaces impose a C^3 smoothness requirement. This paper presents a novel technique for extracting all ridge segments on C^2, C^1 and C^0 tensor product NURBS surfaces by extending a previous robust numerical tracing technique. The parametric domain is divided into charts between non-C^3 knot lines and seeds for tracing are computed on the boundaries of each chart in addition to curvature extrema and umbilics within each chart, thereby guaranteeing extraction of every ridge segment. Since univariate versions of the ridge equations are solved only at non-C^3 knots, the technique retains computational suitability and robustness properties of the tracing approach. Since C^2, C^1 or C^0 NURBS are widely used surface representations across several applications, this technique enables access to ridges on smooth surfaces for various shape analysis tasks. This paper does not address resolution of ridge topology at knot intersections and across knot lines. Surface smoothness may also degenerate due to surface irregularities such as the case when multiple control vertices of a B-Spline surface are coincident. These are areas for future work.

Acknowledgement. This work was supported in part by NSF (CCF0541402). All opinions, findings, conclusions or recommendations expressed in this document are those of the author and do not necessarily reflect the views of the sponsoring agencies.

References

1. Porteous, I.: Geometric differentiation: for the intelligence of curves and surfaces. Cambridge University Press, Cambridge (2001)
2. Hallinan, P., Gordon, G., Yuille, A., Giblin, P., Mumford, D.: Two-and three-dimensional patterns of the face. AK Peters, Ltd., Natick (1999)
3. Kent, J., Mardia, K., West, J.: Ridge curves and shape analysis. In: The British Machine Vision Conference 1996, pp. 43–52 (1996)

4. Guéziec, A.: Large deformable splines, crest lines and matching. In: Proceedings of Fourth International Conference on Computer Vision, 1993, pp. 650–657 (1993)

5. Pennec, X., Ayache, N., Thirion, J.-P.: Landmark-based registration using features identified through differential geometry. In: Bankman, I. (ed.) Handbook of Medical Image Processing and Analysis - New edn., pp. 565–578. Academic Press, London (2008)

6. Subsol, G.: Crest lines for curve-based warping. Brain Warping, 241–262 (1999)

7. Hosaka, M.: Modeling of curves and surfaces in CAD/CAM with 90 figures Symbolic computation. Springer, Heidelberg (1992)

8. Interrante, V., Fuchs, H., Pizer, S.: Enhancing transparent skin surfaces with ridge and valley lines. In: Proceedings of the 6th conference on Visualization 1995. IEEE Computer Society, Washington (1995)

9. Little, J., Shi, P.: Structural lines, TINs, and DEMs. Algorithmica 30, 243–263 (2001)

10. O'Neill, B.: Elementary differential geometry, 2nd edn. Academic Press, London (2006)

11. Musuvathy, S., Cohen, E., Seong, J., Damon, J.: Tracing Ridges on B-Spline Surfaces. In: Proceedings of the SIAM/ACM Joint Conference on Geometric and Physical Modeling (to appear, 2009)

12. Ohtake, Y., Belyaev, A., Seidel, H.P.: Ridge-valley lines on meshes via implicit surface fitting. ACM Transactions on Graphics 23, 609–612 (2004)

13. Cazals, F., Pouget, M.: Topology driven algorithms for ridge extraction on meshes (2005); INRIA Technical Report

14. Patrikalakis, N., Maekawa, T.: Shape interrogation for computer aided design and manufacturing. Springer, Heidelberg (2002)

15. Cazals, F., Faugère, J., Pouget, M., Rouillier, F.: Ridges and umbilics of polynomial parametric surfaces. In: Jüttler, B., Piene, R. (eds.) Geometric Modeling and Algebraic Geometry, pp. 141–159 (2007)

16. Cazals, F., Faugère, J., Pouget, M., Rouillier, F.: Topologically certified approximation of umbilics and ridges on polynomial parametric surface (2005); INRIA Technical Report

17. Jefferies, M.: Extracting Crest Lines from B-spline Surfaces. Arizona State University (2002)

18. Morris, R.: Symmetry of Curves and the Geometry of Surfaces. PhD thesis, University of Liverpool (1990)

19. Cohen, E., Riesenfeld, R., Elber, G.: Geometric modeling with splines: an introduction. AK Peters, Ltd., Wellesley (2001)

20. Elber, G.: Free form surface analysis using a hybrid of symbolic and numeric computation. PhD thesis, The University of Utah (1992)

21. Elber, G., Kim, M.: Geometric constraint solver using multivariate rational spline functions. In: Proceedings of the sixth ACM symposium on Solid modeling and applications, pp. 1–10. ACM, New York (2001)

22. Elber, G., Grandine, T.: Efficient solution to systems of multivariate polynomials using expression trees. In: IEEE International Conference on Shape Modeling and Applications, 2008. SMI 2008, pp. 163–169 (2008)

23. Elber, G.: The IRIT modeling environment, version 10.0 (2008)

24. Hastings, D., Dunbar, P., Elphingstone, G., Bootz, M., Murakami, H., Maruyama, H., Masaharu, H., Holland, P., Payne, J., Bryant, N., et al.: The global land one-kilometer base elevation (GLOBE) digital elevation model. Version 1.0. National Oceanic and Atmospheric Administration, National Geophysical Data Center, Boulder, Colorado. National Geophysical Data Center, Digital data base on the World Wide Web (1999), http://www.ngdc.noaa.gov/mgg/topo/globe.html (CD-ROMs)

Adaptive Partitioning of Vertex Shader for Low Power High Performance Geometry Engine*

B.V.N. Silpa[1], Kumar S.S. Vemuri[2], and Preeti Ranjan Panda[1]

[1] Dept of CSE, IIT Delhi
{silpa,panda}@cse.iitd.ac.in
[2] Intel Technology India
kumar.vemuri@intel.com

Abstract. Advances in Computer Graphics have led to the creation of sophisticated scenes with realistic characters and fascinating effects. As a consequence
the amount of geometry per frame is escalating, making the performance of geometry engine one of the major factors affecting the overall performance of a
graphics application. In this paper we present a mechanism to speed-up geometry processing and at the same time reduce the power consumption by reducing
the amount of computation on processing the geometry of a scene. Based on the
observation that large number of triangles are trivially rejected in each frame, we
propose to partition the vertex shader into position-variant and position-invariant
parts and execute the position-invariant part of the shader only on those triangles that pass the trivial reject test. Our main contributions in this work are: (i)
a partitioning algorithm that minimizes the duplication of code between the two
partitions of the shader and (ii) an adaptive mechanism to enable the vertex shader
partitioning so as to minimize the overhead incurred due to thread-setup of the
second stage of the shader. By employing the proposed shader partitioning approach, we have achieved a saving of up to 50% of vertex shader instructions
on games like Unreal Tournament 2004 and Chronicles of Riddick. Depending
on the architecture implementing the pipeline, we expect that this huge saving
on instructions would translate to significant saving of cycles and power of the
geometry engine. Our experiments on ATTILA, a cycle level simulator for modern graphics pipelines, show a promising speed-up of up to 15% on geometry
processing for various games.

1 Introduction

Geometry– the measure of number of objects present in the scene and the detail at which
they are modeled, is one of the most important aspects that determines the complexity
and visual reality of a scene. Increasing emphasis on incorporating even the intricate
details in a scene is leading to an increase in the number of primitives/frame, since
modeling at finer levels of granularity requires the objects to be represented with large
number of smaller primitives. In the older generation graphics systems, the geometry
was processed on CPU and hence, the amount of geometry that could be accommodated
in a scene was constrained by the computational capacity of the CPU. Acceleration of
vertex processing in newer generation graphics cards by programmable vertex shading

---

* This work was partially supported by a research grant from Intel India.

G. Bebis et al. (Eds.): ISVC 2009, Part I, LNCS 5875, pp. 111–124, 2009.

in hardware, facilitates advanced geometry processing, thus paving the way towards the generation of realistic images. In modern workloads used for benchmarking the performance of graphics cards, it has been observed that:

i) There is a surge in the polygon count per frame - The polygon count in 3DMark05 is about a few Million polygons/frame in contrast to 10-30K polygons/frame in yesteryear games like Quake3 or Doom3.

ii) Complexity of vertex shaders is increasing - It is now possible to apply advanced transformations to vertex position, use complex per-vertex lighting models and also render the surfaces with realistic material models.

With increasing vertex counts and vertex shader complexities, vertex shading has become one of the factors that significantly impact the overall performance of a graphics application and the power consumed by it. In this paper, we propose to reduce the amount of computations on geometry and hence reap the benefits on performance gain and power saving.

It has been observed from the simulation of games and benchmarks as shown in Figure 1, that on an average about 50% of primitives are trivially rejected in each frame. Trivial rejects account for the primitives that fall totally outside the viewing frustum and also front/back face culled primitives. Since testing for Trivial Rejection requires only the position information of the vertex, the time spent on processing the non-transformation part of the vertex shader on these vertices is wasteful. Instead, if we partition the vertex shader into position variant (transformation) and position invariant (lighting and texture mapping) parts and defer the position invariant part of the vertex shader, post trivial reject stage of the pipeline, we can achieve significant savings in cycles and energy expended on processing these rejected vertices. An example illustrating vertex shader partitioning is shown in Figure 2.

The changes to be incorporated in the conventional graphics pipeline to introduce partitioned vertex shading are shown in Figure 4. In the modified pipeline, the VS1 stage computes only the position variant part of the vertex shader and rest of the vertex processing is deferred to VS2 stage. The Clipper stage is divided into Trivial Reject and Must Clip stages. Triangles passing through the trivial reject test are sent to the VS2

Benchmark/ Game	% Trivial Rejects
3dMark05	50
Chronicles of Riddick	61
UT2004	54
Prey	47
Quake4	56
Doom3	36

Fig. 1. % TR / frame

```
SHADER PROGRAM
dp4 o0.x, c0, i0
dp4 o0.y, c1, i0
dp4 o0.z, c2, i0
dp4 o0.z, c3, i0
dp4 o6.z, c4, i0
dp4 o6.z, c5, i0
dp4 o7.z, c6, i0
dp4 o7.z, c7, i0
mov o1, i3
```

```
PARTITION 1 (VS1)
dp4 o0.x, c0, i0
dp4 o0.y, c1, i0
dp4 o0.z, c2, i0
dp4 o0.z, c3, i0
```

```
PARTITION 2 (VS2)
dp4 o6.z, c4, i0
dp4 o6.z, c5, i0
dp4 o7.z, c6, i0
dp4 o7.z, c7, i0
mov o1, i3
```

Fig. 2. Ex : VS Partitioning

Shader	% Inst Saved
Hatching [15]	46.3
Cartoon [15]	41
Directional Lighting [16]	32.4
Crystal [17]	39.4
Water [17]	18.2
Bubble [14]	16.12
Vertex Blending Lighting[18]	6.86

Fig. 3. % Instructions Saved

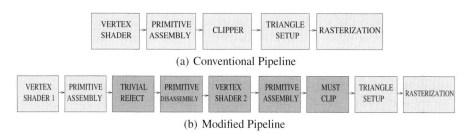

(a) Conventional Pipeline

(b) Modified Pipeline

Fig. 4. Pipeline Modified to support Vertex Shader Partitioning

stage after disassembling them into vertices (since the vertex shader can only work on vertex granularity). These vertices after being processed in VS2 are assembled back to triangles and sent to Must Clip stage. The geometry engine of the pipeline is thus modified and the fragment generation and rendering takes place as it was in the original pipeline.

To illustrate the efficacy of our proposal, we have used our shader partitioning algorithm on a set of commonly used vertex shaders and the saving in instructions achieved are reported in Figure 3. The savings are calculated assuming about 50% trivial reject rate and compiling the shaders to instruction set implemented in ATTILA framework [1]. From the results we see that vertex shader partitioning leads to significant reduction in instruction count, thus motivating the adoption of vertex shader partitioning into a graphics pipeline.

From the discussion so far it might appear that it is most appropriate for the API to support vertex shader partitioning. This would require the application developer to provide two pieces of vertex shader programs, one for transforming the vertices and one for lighting and texturing. But we have observed that such hard partitioning of shaders is not always viable and hence propose the framework for adaptive partitioning of vertex shaders which includes (i) incorporation of vertex shader partitioning pass in the shader compiler (ii) enhancements to the driver so as to dynamically decide if vertex shader is to be partitioned and handle the state for VS2 stage in case the shader is partitioned (iii) incorporation of a fixed-function unit into the architecture for setting up VS2 threads.

2   Related Work

The initial work on VLSI implementation of a programmable geometry engine was presented in [2]. The system was designed for high performance by connecting multiple geometry engines in a pipeline where each of the engine could be configured to perform one of the fixed-functions - transformation, clipping or projection. Similarly in [3], six floating point processors are pipelined to implement a programmable geometry subsystem. In [4], the authors propose a vertex shader implementation for mobile platforms and hence target the power consumption of a vertex shader. They suggest a Fixed point SIMD implementation of the vertex shader in contrast to the floating point shader proposed in [5] or the integer data-path suggested in [6]. They point out that the integer implementation does not provide the required performance and the floating point implementation fails to meet the power budget. In [7], the authors study the performance benefits of a unified shader architecture in comparison to using separate processing units for vertex and pixel shading. This results in performance benefits due to better utilization of processing units by balancing the workload of vertex and pixel shader threads. The authors of [8] aim to speed-up geometry processing by proposing a compressed representation of vertex data that not only reduces the bandwidth requirement for transferring data from CPU to graphics subsystem but also results in high vertex coherence resulting in good hit-rate into the vertex cache. An algorithm to re-order the sequence in which triangles are rendered so as to increase the hit rate into post-TnL vertex cache is presented in [9]. In [10], the authors report performance gain by using a cache to hold vertices pre-TnL and another to hold vertices post-TnL. They indicate that pre-TnL cache enhances the performance by both pre-fetching the vertices and also results in reuse of fetched vertices and the post-TnL cache caters to the reuse of the shaded vertices.

In contrast to the architectural techniques suggested in the literature presented so far, we propose to achieve power and performance improvements by reducing the number of computations in the geometry engine. This is achieved by performing lighting and texturing on only those vertices that fall in the view frustum. In [11], the authors indicate that power savings could be achieved if the API supported lighting and texturing of the vertices after the trivial reject stage. But we observed that such a hard partitioning of the vertex shader is not always beneficial. This could be due to one of the two reasons discussed below.

- Thread setup overhead for the second vertex shader could overshadow the advantage of deferring the position invariant part of the shader code. Hence we propose an adaptive algorithm for vertex shader partitioning, which would take the decision based on a trade-off between the setup overhead and cycles saved due to partitioning.
- Moreover, we observe that there could be a significant number of instructions common to position-variant and position-invariant part of the vertex shader. Our algorithm identifies the set of intermediate values to be transferred from VS1 stage to VS2 so as to minimize the amount of code duplication resulting from partitioning the shader.

Shader partitioning has been studied in [12,13], in the context of code generation for multi-pass shaders for execution on GPUs constrained by the availability of resources. Virtualization of GPU resources is achieved by dividing the shader into multiple smaller programs; the number of such passes is to be minimized to maximize the performance. In our case, the number of code partitions is fixed at two and the aim is to minimize the amount of code duplicated across the two partitions. To the best of our knowledge, ours is the first work that attempts dynamic partitioning of the vertex shader and employs compiler automation for partitioning the shader.

3   Shader Compiler

We propose to include the vertex shader partitioning pass in the code generation phase of shader compilation. The binaries are generated for both partitioned and non-partitioned versions of the shader and passed on to the driver. Since shader compilation is only a one time process, the incorporation of this new pass would result in minimal software overhead. The algorithm used to partition the vertex shader code is explained below.

3.1 Algorithm for Partitioning with Minimum Duplication (PMD)

A DAG (Directed Acyclic Graph) representing the data-flow in the vertex shader program is the input to the partitioning phase. Each node in the graph represents an operation and a directed edge between two nodes exists if there exists a data dependency between them. A principal input (PI) has no incoming edge and a principal output has no outgoing edge. We introduce a source node S which is connected to all the PIs (I0, I1, I2 and I3 in Figure 5) and a destination node T connecting all the POs (O0, O1 and O2 in Figure 5) to this DAG and call this new graph G.

Trivial code partitioning can be done by simple BFS (Breadth First Search) traversal of the graph starting from the destination node and separating all nodes reachable from the PO representing the position attribute (O0 in Figure 5) of the vertex program into VS1 and all the nodes that are reachable from rest of the POs into VS2 (O1 and O2 in Figure 5). However, this partitioning results in duplication of the instructions that are common to both VS1 and VS2 (the shaded nodes in Figure 5). The duplication of code could be minimized by sending some of the auxiliary intermediate values (AUX) to VS2 along with the position output attribute, and representing the VS2 program as a function of the PIs and outputs from VS1 (AUX)

$$VS2 = f(PI, AUX)$$

Let i be the number of PIs to the VS2 stage and k be the number of auxiliaries sent from VS1 to VS2. If a vertex shader program can accept a total of C vertex attributes as inputs,

$$i + k = C$$

Now the problem is to identify the set of k intermediate values to be transferred from VS1 to VS2 from the set of m of them so that, the number of instructions duplicated in VS2 is minimized. A heuristic that uses an iterative approach to solve this problem is presented here.

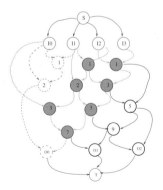

Fig. 5. DAG representing the data-flow in the code

We start with the DAG of the vertex shader G and separate the common code shared by VS1 and VS2 by BFS traversal of graph starting from the leaves (POs) and separating all the nodes that are reachable from the PO of VS1 (position output attribute of the shader program) and the POs of VS2 (rest of the output attributes generated by the shader program) into a graph M. The leaf nodes of this graph represent all the temporaries that are common to both VS1 and VS2 and hence the instructions generating them need duplication in VS2, if not transferred from VS1. We assign to each node a weight equal to one more than the sum of weights of all its parents, starting from the PIs which are given a weight of one. Thus the weight of a node gives the measure of number of instructions required to generate the corresponding temporary value. Since at this stage in compilation, high level instructions are already divided into micro-operations, it is acceptable to assume that each operation takes equal amount of time for execution. In a system with variable instruction latencies, the weight of a node is calculated as sum of weights of its parents and the delay of the instruction represented by the node. We sort the leaf nodes in order of their weight and divide them into two sets A and B such that set A contains k nodes of largest weight. Taking set A as an initial solution, we iteratively replace nodes in set A with nodes from the rest of the graph M, such that each replacement reduces the residual computations required to generate all the leaf nodes from the PIs and the nodes in the solution set. Starting from the leaf nodes, we traverse up the graph until we reach a node(say node R) which is reachable from at least two of the leaf nodes. Consider the following three scenarios.

1. *The node R is reachable from at least two nodes from set A.*
   Consider the scenario shown in Figure 6.
   Instead of sending P and Q, we can send node R and node D. By sending node R we can compute P and Q in VS2 with a cost of 2 instructions, but we have saved a cost of 5 by sending D. If the cost saved by sending D is greater than cost of computing nodes P and Q from R, we replace P and Q from the solution set with nodes R and D. This is captured in the equation below.
   Let P and Q be the nodes of smallest weight from set A that are reachable from R, and let D be the node of largest weight from set B.

$$\text{Condition 1: } W_r + W_d > W_p + W_q - 2W_r$$
$$\text{Action: Replace P and Q from set A with R and D}$$

In case the above condition fails, as shown in the scenario Figure 7, we check if fanout of node R has nodes in set B and compute the decrease in cost of generating these nodes when R is sent to VS2. If this difference is greater than the cost incurred in computing node P from R, we replace node P with node R.

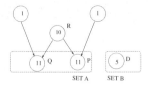

Fig. 6. Case 1

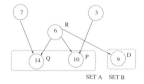

Fig. 7. Case 2

$$Condition\ 2:\ W_r > W_p - W_r$$
$$Action:\ Replace\ node\ P\ with\ node\ R$$

2. *The node R is reachable from only one node P from set A.* We use the Condition 2 for evaluating a possible replacement of node P with node R.
3. *The node R is not reachable from any of the nodes from set A.* - No replacement is possible in this case.

The BFS traversal is done until the root node is reached, and whenever a node is encountered that is common to two or more leaf nodes, we use the conditions enumerated above to find a possible replacement that would improve the solution. The pseudo code for the algorithm discussed above is presented herein.

The algorithm requires three BFS traversals of the graph. In each BFS traversal, the nodes and edges of the graph are visited once and a constant amount of work is done at each node and edge. If e is the number of edges and m is number of nodes in the graph, time complexity of the algorithm is $O(e + m)$.

3.2 Comparison with the Naive Algorithm

Figure 8 shows the reduction in duplicated instructions and hence increase in instructions saved by using our partitioning algorithm over the trivial one used in [11]. The savings are reported assuming a Trivial Reject rate of 50%.

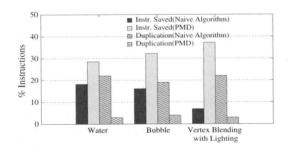

Fig. 8. Comparison of proposed PMD algorithm with the existing one

---

Algorithm 1. PASS 1 : Separate the Duplicated Code

Input: DAG of the vertex shader
Output: DAG of the duplicated code
1: Start BFS traversal of the DAG starting from the destination node. and Color the node corresponding to O0 red and all other outputs blue.
2: If a node is reachable only from a red colored node,color it red.
3: If a node is reachable only from a blue colored node,color it blue.
4: If a node is reachable from both red colored nodes and blue colored nodes color it green.
5: The DAG with green colored nodes represents the duplicated code.

Algorithm 2. PASS 2 : Label the DAG

Input: DAG from PASS1
Output: DAG with nodes labeled with weights
1: BFS traversal of the DAG starting from the source node.
2: Assign all the PIs a weight of 0.
3: On reaching a node,
4: **if** node is green **then**
5:     $Weight\ of\ node \leftarrow 1 + sum\ of\ weights\ of\ its\ parents$
6: If a leaf node is reached, insert it into a max heap (sorted by weight)

Algorithm 3. PASS 3 : Partitioning the DAG

Input: DAG from PASS 2 and the heap of sorted leaf nodes. Each node has fields w1,w2,w3 set
   to zero and n1,n2 pointing to NULL. w1 and w2 hold the weights and n1,n2 point to the two
   highest weight nodes of set A that the node is reachable from. w3 holds the weight of the
   heighest weight node of set B, that the node is rechable from.
Output: Set A containing k intermediate nodes to be transferred from VS1 to VS2
1: The first k nodes de-queued from the heap form set A (solution set) and the remaining nodes
   set B.Insert all the leaf nodes in a an empty queue Q
2: **while** Q not empty **do**
3:     $N \leftarrow dequeue(Q)$, $w \leftarrow weight(N)$
4:     Reach all the parents of N.
5:     **if** N' is reachable from N, such that $N \in A$ **then**
6:         **if** $w(N) < w1(N')$ **then**
7:             $w1(N') \leftarrow w(N)$, $n1(N') \leftarrow N$
8:         **else if** $w(N) < w2(N')$ **then**
9:             $w2(N') \leftarrow w(N)$, $n2(N') \leftarrow N$
10:    **if** N' is reachable from N, such that $N \in B$ **then**
11:        $w3(N') \leftarrow w(N)$
12:    Insert N' into Q, if not already enqueued.
13:    **if** N is not a leaf node **then**
14:        $w' \leftarrow w(N_h)$, where N_h is the head of the heap H
15:        **if** n2(N) is not NULL **then**
16:            **if** $w(N) + w' > w1(N) + w2(N) - 2w(N)$ **then**
17:                replace nodes N1 and N2 from set A, with nodes N and N_h. Delete N_h from the
                   heap.
18:        **else if** n1(N) is not NULL and w3(N) is not zero **then**
19:            **if** $w(N) > w1(N) - w(N)$ **then**
20:                replace node N1 from set A with node N.

3.3   Selective Partitioning of Vertex Shader

Spawning a thread on the Programmable Shader Unit incurs some thread setup overhead, the extent of which is dependent on the micro-architecture of the thread setup unit.
This could include the idle time waiting for the availability of resources, time spent on
loading the inputs, time spent on transmission of outputs, etc. Partitioning of the vertex shader results in VS2 thread setup overhead. Hence it is very important to weigh
the benefit of cycles saved on rejected vertices against the overhead incurred on thread
setup for the vertices that are not rejected.

The cost incurred to process a batch of vertices (B) without vertex shader partitioning is given as

$$Cost_{no-part} = B \times (\text{VS Thread Setup overhead} + \text{Execution time of VS})$$

If we assume C be the rate at which vertices are trivially rejected, then cost incurred to process the batch with partitioning is given as

$$Cost_{part} = B \times (\text{VS1 Thread Setup overhead} + \text{Execution time of VS1}) +$$
$$B \times (1 - C) \times (\text{VS2 Thread Setup overhead} + \text{Execution time of VS2})$$

Vertex shader partitioning is profitable only if $Cost_{part}$ is less than $Cost_{no-part}$ and hence we propose to enhance the driver so that it can take the decision of partitioning dynamically at run time. The driver is provided with the thread setup overhead for VS1 and VS2 stages and the execution time is approximated to the number of instructions in the program. Since the partitioning decision is to be taken prior to the trivial reject stage of the pipeline, we use history based prediction for clip rate. Due to spatial coherence of the frames, we can expect the trivial reject rate of adjacent frames to be comparable. The same has been observed from simulation of various games and shown in Figure 9. Thus, the clip rate of the previous frame can be taken as an approximation for the clip rate of present frame. We have observed that this history based adaptive partitioning algorithm results in attractive performance benefits in comparison to hard partitioning of the vertex shader.

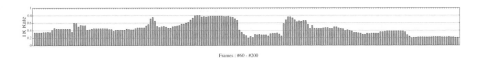

Frames : #60 - #200

Fig. 9. Variation of Trivial Rejects across Frames - From UT2004

4   Simulation Framework

We have used the ATTILA [1] simulator framework for implementing our vertex shader partitioning algorithm. ATTILA is an open source simulation framework that models a modern GPU micro architecture. Though the implementation details of our proposal are with reference to the ATTILA framework, the ideas are generic enough to be incorporated into any micro-architecture with minor variations. The shaded portions of Figure 10 show the additional units added to the existing model. In this section the architectural details pertaining to vertex processing on the Graphics Processor modeled in ATTILA framework are explained.

Command Processor acts as an interface between the driver and the pipeline. It maintains the state of the pipeline and on receiving the commands from the driver, updates the state of the pipeline and issues appropriate control signals to the other units in the pipeline. At every context switch, the command processor does the following tasks:

(i) Loads the control registers : it updates the registers representing the render state.
(ii) Initiates the transfer of data to GPU memory : it sets up the transactions to fill vertex buffer, index buffer, load shader program, load textures etc. from system memory to GPU memory through the memory controller.

Streamer is responsible for reading the vertices from the GPU memory and setting up the vertex shader threads to process them. Vertices and Indices are buffered in FIFOs and the Shader loader spawns a vertex shader thread whenever a Programmable unit is free. When Indexed mode is used to address the vertices, a post shading vertex cache is used for reusing the shaded vertices. The shaded vertices are sent to the primitive assembly unit.

Unified Shader architecture implemented in ATTILA is based on the ISA described in ARB vertex and fragment program OpenGL extensions.The ALU is a 4 way SIMD working on 4 component floating point vectors. The instruction memory is implemented as a scratchpad memory divided into logical partitions to hold the vertex, triangle setup and fragment shader programs. The driver is responsible for loading the shader code into the scratchpad (if not already present) whenever there is a state change. The register file in the shader has four banks, one each to store input attributes, output attributes, constants and intermediate results of the program. The constant register bank is shared by all threads whereas 16 each of input/output and temporary registers are allocated per thread.

The following modules are added to the existing architecture to support vertex shader partitioning.

Vertex Buffer: We introduce a small fully associative cache to buffer the vertex input attributes read at the VS1 stage for reuse at VS2 stage. Since the vertices reach the

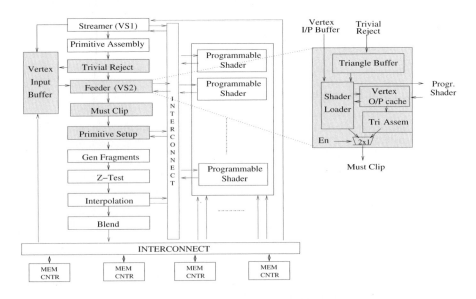

Fig. 10. Modified ATTILA Architecture

VS2 stage in the same order in which they are processed at VS1 stage, we choose to use FIFO replacement policy for the buffer. Since the trivially rejected vertices are not processed at the VS2 stage, a de-allocation signal to the buffer from the trivial reject stage helps free the lines that carry the trivially rejected vertices, thus increasing the hit rate into the cache.

Feeder Unit: The micro-architecture of the feeder unit is as shown in Figure 10. The Triangle buffer holds the triangles accepted after passing through the Trivial Reject stage of the pipeline. A triangle is a set of three vertices, each associated with an index to address it and the position attribute computed by VS1. If vertex shading is disabled for the current shader, the triangles are sent to the 3D clipping unit from the Triangle Buffer. If partitioning is enabled, the triangle is disassembled into vertices and sent to the Shader Loader. The Vertex Output Cache aids in the reuse of the shaded vertices. When the shader loader receives a vertex, it looks up the vertex output cache for the vertex. If the result is a cache miss, the shader loader reads the input attributes of the vertex from the Vertex Input Buffer and spawns a VS2 thread for processing the vertex. The triangle assembly unit receives the position output attribute of the vertices from the Triangle buffer and rest of the attributes from the programmable shader. After the triangle assembly, the triangles are sent to the 3D clipping unit.

5   Experiments and Results

To illustrate the advantages of vertex shader partitioning, we have taken a few frames from the games Unreal Tournament 2004(UT), Chronicles of Riddick(CR) and Quake4 and rendered them on ATTILA modified to support vertex shader partitioning. Some of the sample frames rendered on the modified architecture are shown in Figure 12. The frames rendered by the modified architecture are compared with the frames rendered on the basic architecture for validating the correctness of our implementation.

Figure 11 shows up to 50% saving in vertex shader instructions and about 15 % improvement achieved on cycles spent on geometry processing by adopting our algorithm. The results are compared against those achieved by the naive partitioning algorithm proposed in [11]. We observe that due to hard partitioning of vertex shader, as proposed

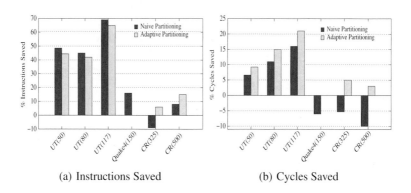

(a) Instructions Saved   (b) Cycles Saved

Fig. 11. Savings Achieved

(a) UT:Frame 50   (b) CR:Frame 500   (c) Quake:Frame 100

Fig. 12. Frames rendered on ATTILA simulator enhanced with Vertex Shader Partitioning

by them, there is a degradation of performance on those frames with smaller clip rate or on those that have so few instructions in VS2 stage that the thread setup overhead overshadows the advantage of partitioning. This is observed in the case of the frame rendered from game Quake4 where most of the effects are achieved by texturing and vertex shaders are used only for simple transformation. In this case, hard partitioning, though resulting in a small saving in instructions, leads to a negative impact on performance. In contrast, our adaptive algorithm has almost negligible effect on performance in such frames. Similarly in frames containing shaders that have a lot of common code between VS1 and VS2 stages, hard partitioning leads to negative impact on performance due to duplication in instructions. This is observed in the frames rendered from Chronicles of Riddick. These frames use shaders for vertex blending and lighting. This shader when partitioned naively would lead to large number of instructions being duplicated as reported in Figure 3. Our approach of reusing the intermediate values generated at VS1 in VS2 leads to better results. In frames having higher trivial rejects and using large number of instructions in VS2 stage, our algorithm gives better performance improvement and the saving in instructions is comparable to that achieved by hard partitioning method. This is observed from the scenes rendered from Unreal Tournament. From the graph we see that number of instructions saved by our algorithm is slightly less than that achieved by the naive method. But this is intentional since we avoid partitioning the shaders that would result in greater thread setup overhead and hence achieve better performance than the naive method.

The technique proposed in this paper works unconditionally on DirectX9 applications and also with OpenGl application without Geometry Shader Extension. For DirectX10 and DirectX11 applications, this technique could be used when the Geometry Shader and Tesselator stages are disabled respectively. When these stages are enabled, we switch back to non-partitioned vertex shader.

6   Conclusion

In contrast to older generation games which were predominantly pixel processing intensive, the newer ones tend to aim towards achieving greater realism by incorporating more and more geometry, thus increasing the load on vertex processing. This is as a result of increasing number of primitives/frame and also increasing size of vertex shader programs. Based on the observation that about 50% of the primitives are trivially rejected in each frame, we proposed a mechanism of partitioning the vertex shader and deferring the positioning invariant part of the shader to the post trivial reject stage of the pipeline. We also identified that such partitioning can have a negative impact on performance in some cases, and proposed an adaptive partitioning scheme that applies partitioning only to scenarios that benefit from it.

From the experiments on ATTILA framework, we observe upto 50% saving in shader instructions due to vertex shader partitioning leading to 15% speed-up and can expect up to 50% power saving on computations in the geometry engine. Since programmable cores are the major consumers of power, we can expect a significant saving on overall power consumption by using vertex shader partitioning.

References

1. del Barrio, V., et al.: Attila: a cycle-level execution-driven simulator for modern gpu architectures. In: IEEE International Symposium on Performance Analysis of Systems and Software, pp. 231–241 (2006)
2. Clark, J.H.: The geometry engine: A vlsi geometry system for graphics. SIGGRAPH Comput. Graph., 127–133 (1982)
3. Akeley, K., Jermoluk, T.: High-performance polygon rendering. In: SIGGRAPH 1988: Proceedings of the 15th annual conference on Computer graphics and interactive techniques, pp. 239–246. ACM, New York (1988)
4. Sohn, J.H., et al.: A programmable vertex shader with fixed-point simd datapath for low power wireless applications. In: HWWS 2004: Proceedings of the ACM SIGGRAPH/EUROGRAPHICS conference on Graphics hardware, pp. 107–114. ACM, New York (2004)
5. Kameyama, M., et al.: 3d graphics lsi core for mobile phone "z3d". In: HWWS 2003: Proceedings of the ACM SIGGRAPH/EUROGRAPHICS conference on Graphics hardware, Eurographics Association, pp. 60–67 (2003)
6. Sohn, J.H., et al.: Low-power 3d graphics processors for mobile terminals. IEEE Communications Magazine, 90–99 (2005)
7. Barrio, V.M.D., et al.: A single (unified) shader gpu microarchitecture for embedded systems. In: Conte, T., Navarro, N., Hwu, W.-m.W., Valero, M., Ungerer, T. (eds.) HiPEAC 2005. LNCS, vol. 3793, pp. 286–301. Springer, Heidelberg (2005)
8. Chhugani, J., Kumar, S.: Geometry engine optimization: cache friendly compressed representation of geometry. In: I3D 2007: Proceedings of the 2007 symposium on Interactive 3D graphics and games, pp. 9–16. ACM, New York (2007)
9. Sander, P.V., et al.: Fast triangle reordering for vertex locality and reduced overdraw. ACM Trans. Graph., 89 (2007)
10. Chung, K., et al.: Vertex cache of programmable geometry processor for mobile multimedia application (2006)
11. Tsao, Y.M., et al.: Low power programmable shader with efficient graphics and video acceleration capabilities for mobile multimedia applications, pp. 395–396 (2006)
12. Chan, E.: Ng, et al.: Efficient partitioning of fragment shaders for multipass rendering on programmable graphics hardware. In: HWWS 2002: Proceedings of the ACM SIGGRAPH/EUROGRAPHICS conference on Graphics hardware, Eurographics Association, pp. 69–78 (2002)

13. Heirich, A.: Optimal automatic multi-pass shader partitioning by dynamic programming. In: HWWS 2005: Proceedings of the ACM SIGGRAPH/EUROGRAPHICS conference on Graphics hardware, pp. 91–98. ACM, New York (2005)
14. http://developer.download.nvidia.com/SDK/9.5/Samples/samples.html
15. http://ati.amd.com/developer/shaderx/ShaderX_NPR.pdf
16. http://attila.ac.upc.edu/wiki/index.php/Traces
17. http://amd-ad.bandrdev.com/samples/Pages/default.aspx
18. http://www.flipcode.com/archives/articles.shtml

Visualized Index-Based Search for Digital Libraries

Jon Scott, Beomjin Kim, and Sanyogita Chhabada

Department of Computer Science
Indiana University-Purdue University
Fort Wayne, IN, U.S.A.
scotjc01@ipfw.edu, kimb@ipfw.edu

Abstract. As the amount of available information continues to expand, traditional text-based searches for digital libraries and similar systems become increasingly cumbersome to the user. Selection of the best result calls upon the user to compare and contrast top results; this can involve investigative reading of each, to determine what quality and amount of the desired topic is present in each. This paper presents an alternative search strategy, utilizing visualization to relate detailed content information obtained through indexes. By providing such information in a visual manner, the aim is to reduce the burden of investigation placed upon in present systems.

1 Introduction

The proliferation of computing technology and the advance of the Internet has greatly changed the way in which people obtain information. In these days, a person can access to the information which is more than we can handle without the limitation of time and location. The amount of available data grows ever rapidly, but the ability for a user to find their desired information has progressed with less vigor. This is particularly true of highly concentrated sources of a broad array of information, such as documents on the Internet and Digital Libraries.

The current library system provides several attributes associated with books as a response to the users' inquiry. The search results include book title, author, publication year, ISBN number, thickness, etc. However, there is often a large expectation left on the user's ability to read through the results. Furthermore, the text-based approach is non-intuitive and inefficient for finding suitable information through comparison of many possible search results [1]. While ranked search results may assist the user in this endeavor, there is still a reliance on the user investigating the top results individually [2, 3]. This will only become more problematic as the information domain they are applied to continues to grow larger and more complex.

Information visualization is an effective tool that can present a large amount of data compactly, but intuitively for easy comparison. By exploiting users' perceptual cognition, studies have shown that the graphical illustration of data has contributed in improving the users' understanding and reviewing speeds [2, 4, 5]. Borner and Chen explained that there are three common usage requirements for visual interfaces to Digital Libraries: First, to support the identification of the composition of retrieval result; second, to understand the interrelation of retrieved documents to one another,

G. Bebis et al. (Eds.): ISVC 2009, Part I, LNCS 5875, pp. 125–134, 2009.

and last, to refine a search, to gain an overview of the coverage of a Digital Library (DL) and to facilitate browsing and to visualize user interaction data in relation to available documents in order to evaluate and improve DL usage [6].

When using physical books, people tend to view multiple at once; to better compare and review information across multiple sources, and to have a better overall understanding of the domain. In their study, Good et al, identify this to be a major weakness in current DL displays [1]. To address related issues, researchers have conducted studies applying visualization techniques for book searches and presenting various forms of search results [7, 8].

The Graphical Interface for Digital Libraries (GRIDL) is a system that displays a hierarchical cluster of the relevant data to a query on two-dimensional display [9]. This system uses a two-dimensional coordinate, the axes of which are selectable from a variety of different attributes. Results were displayed within each cell as a collection of different size icons, color coded by document type. Marks and his colleagues present a similar approach, based on scatter plots, known as ActiveGraph [10]. Because this approach results in much more node clustering and overlap, a logarithmic transformation is provided, along with the ability to filter out user specified documents. ActiveGraph also provides the ability to specify shape, color, and size of nodes representing documents. By allowing users to manipulate the manner in which data is displayed, these visualizations provided a strong ability to reveal patterns within the data that may not typically be apparent.

These studies mainly focused on aiding the user in comparing the search results effectively by presenting book properties through various visual attributes; but they don't express in detail the amount of content related to user interest. Lin proposed a graphical table of contents (GTOC) that showed the dimension of items in the table of contents based on Kohonen's self organizing feature map algorithm [11]. The paper introduces how documents can be organized and then visualized to allow the user easy access of underlying contents. The GTOC prototype describes various interactive tools to assist the user exploring document contents and analyzing relationships among terms in the table of contents.

The main goal of the research presented in this paper is the development of visualization techniques that will make the user's book search effective by exploiting the book index.

2   Methods

The Visualized Index-Based Search (VIBS) system utilizes an Overview + Detail approach for presenting book search results. This is a visualization technique that uses multiple images to display the entire data space, as well as show an up-close, detailed view of the data [12]. Similar to traditional library searches, the overview will present outline of the book search results through graphical illustration. The user interactively selects a subset of visualized icons that will allow them to execute content level exploration. When a user provides search terms of interests, the Detail view presents a rich visualization of the assets of the given query in a book index with other related information. The resulting visualization supports the user for intuitive comparing and contrasting the selections in greater detail.

2.1 Overview Visualization

The Overview allows the user to perform a general search on the data space, similar to traditional library tools. The current prototype utilizes a title based search, although a more robust implementation would make use of additional categorization provided by the environment. To address the inefficiency of space utilization of text-based book search application, VIBS presents the results using a grid based layout. The X and Y axis represent individual book attributes, including author, publication year, number of pages, and review details. These are freely changed via drop-down selection boxes; providing greater control over the result display and assisting with user understanding [9]. Book nodes are located accordingly.

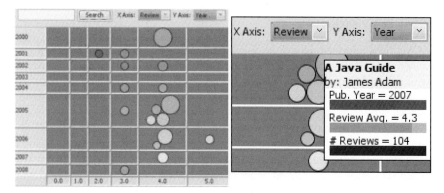

Fig. 1. Overview display (left). Close-up of Overview with tool-tip (right).

Total page count defines the radius of the corresponding circular icon. The books are classified into N categories depending on the number of pages, where each group is mapped to a predefined radius. Within the same category, the radius is linearly interpolated according to the page count of the item. By properly utilizing non-liner and linear transformation, the VIBS can display variable sizes of book volume on the limited screen space while still delivering its magnitude. Through the visual illustration of book dimension, the users can intuitively estimate the amount of content available.

VIBS uses the RGB color model to deliver other attributes to the user. As part of this, the Overview employs collaborative filtering to relate the perceived user value of each result. Collaborative filtering is a type of content-based filtering which utilizes the opinions of users who have already evaluated the quality of information [13]. This is done by collecting user reviews and their fidelity, typically found on most book merchant web-sites. The user's satisfaction with a book corresponds to the green color component of the circular node. The average review score a book receives is mapped to the green intensity, as shown by equation 1, where C_i^G is the green intensity associated with a book i, E_i^G is the average user rating of book i, I_{max} is the brightest hardware intensity of green, and E_{max}^G is the highest available user rating.

$$C_i^G = (E_i^G \cdot I_{max}) / E_{max}^G \qquad (1)$$

The accuracy of collaborative filtering is highly dependent on the number of evaluators. Because of this, the user rating of a book can be unreliable if the number of individual reviews is relatively low [14]. To deliver the fidelity of user ratings of the book, the Overview utilizes the blue color component to represent the number of unique evaluators who rated the book. The red color components relates the publication year, as more recent works are likely to hold more up to date information. These two colors are treated similarly to equation 1. The three color components are blended together to form a color Ω_i which is the final color of the circular node. The end result is that books with high ratings, a large reviewer base, and more recent publication will have a prominent color closer to white. To assist the user, mouse-over tooltips are employed to relate the color data individually along with the title and author (figure 1).

2.2 Detailed Visualization

The Overview interface assists the user to compare attributes associated with books. Meanwhile the Detail visualization focuses on showing the amount of searched contents of selected books through a graphical illustration. The VIBS system utilizes the book index to present the amounts of related contents, search terms distribution, and associated sub-terms to search terms in the index pages.

When a user enters a search term of which information they are looking for, the Detail view displays the corresponding data using a radial tree structure. In the display, terms are represented as circles with the center representing the given search term and sub-terms are displayed as nodes branched off of the center circle (Figure 2). With the similar way used in Overview visualization, the VIBS system makes use of multiple attributes in relaying a range of information that will assists the user's finding a book of their interest.

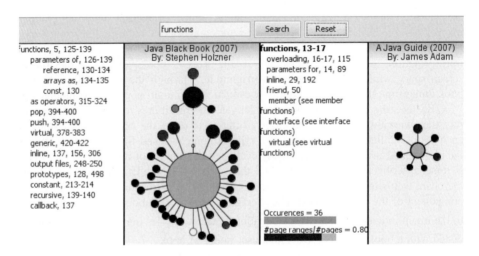

Fig. 2. Detail view showing (left) diagram with expanded sub-term and an overflow node, and (right) context highlighting of text list and bar display

The total number of pages associated with a term in the index could be considered to have strong correlation to the amount of information provided by the work on that subject. In the Detail view, this important measure is naturally mapped to the radius of the term icons. This is a cumulative page total of the node's associated term, and all of the term's subordinates within the index. This graphical illustration allows the user to gain a general understanding of how different book results compare, without having to compare each level of sub-terms in detail.

The VIBS system applies color-coding to the terms as a visual abstraction occurrences and concentration associated with that item. This is intended to provide the user with an understanding of the comparative value of the different entries being displayed. The green intensity represents the total number of occurrence of the term throughout the index. A larger number of occurrences could be indicative of the term having broader, or perhaps more complex coverage within the book. The system applies normalization to the occurrence of all terms in the index that results in the node with the highest occurrence having the brightest green intensity, and the node with the least having the darkest.

Blue represents the concentration of sub-terms appearance within the book. A term which appears in two books with the same number of pages allocated by both indexes may not necessarily indicate equal coverage on the topic. If one index were to have the term listed as a single contiguous section of pages while the other had each page listed independently, the former would be considered to likely have a more meaningful coverage of that term. The ratio of page continuity to all pages containing a sub-term determines the blue intensity of color-code by equation 2, where C_i^b is the blue intensity for sub-term i, $\sum_{j=0}^{n} P_j^C$ is the number of individual page ranges for i, T_i is the total number of pages for the term i, and I_{max} is the brightest intensity of blue possible on the machine. This gives terms with more concentrated information a stronger representation in the visualization.

$$C_i^b = (1 - (\sum_{j=0}^{n} P_j^C / T_i)) * I_{max} \tag{2}$$

Displaying the relationship between search term and its sub-terms is valuable information for content search. A sub-term which has page allocation close to the search term will likely have a stronger correlation than one which is on the opposite end of the book. This relationship is presented by the distance between the parent and sub nodes within the visualization. The magnitude of the distance is average distance between the search term and child page occurrences, normalized across all terms displayed. This is computed with equation 3 where D_j is a normalized distance between the root node and sub-node j, P_j^k is the page number contains sub-term j, $\overline{Root_i}$ is the mean page number having the root word i, and NORM and ABS are a normalize and absolute function respectively.

$$D_j = \text{ABS}(\text{NORM}(\sum_{k=0}^{n} (P_j^k - \overline{Root_i}) / N)) \tag{3}$$

Sub-terms with a larger number of associated pages are deemed more important, and are subsequently given display priority. These children nodes are arranged around the parent top to bottom in decreasing value, ensuring the user can readily identify which items are potentially more valuable for the particular work being displayed. In the event that the space around the parent node can't accommodate any additional terms, the remaining un-drawn entities are collected in a single node, indicated via a red outline. As higher priority terms are drawn first, only lower value items will be present within this 'overflow' node. A user can still investigate these items by expanding the node.

Space is also preserved by displaying only the first level of sub-terms around the root as these are considered higher value than subsequent child nodes. Terms with hidden children are indicated by a jagged, dashed outline. When the users want to explore deeper into the index hierarchy, they can freely expand these elements by clicking on them; clicking the root will expand all such nodes. When a node is expanded, it is drawn further away from the parent in order to avoid overlapping. To maintain the information originally relayed by the branch length, a marker is drawn to indicate the original length.

The Detail view interface supports additional tools to assist the user. This is in the form of an information panel that updates based upon the user mouse activity with the display. By hovering over a node, an information panel will appear, highlight the term in a text list. A histogram reflecting the value of the number of occurrences, and page range ratio is also shown. This will support the user in keeping track of what different topics are available in the different books, while also relating more specific information for the different attributes.

3   Experiment

To investigate the effectiveness of the VIBS system as a search supporting tool, we conducted survey-based usability tests that collected users' opinion to the VIBS over traditional text-based search systems. For the experiments, two testing applications were implemented using Java. A text-based application was constructed simulating the conventional library book search system. The second was implemented as explained in Section 2.

Seventeen students majoring in computer science participated in the experimental sessions. Participants had no problems in color perception, were comfortable using a mouse-driven GUI application, and had no previous exposure to VIBS. Sessions were held in groups of three to five, and started with a brief orientation. This orientation presented the functionality provided by both systems along with an explanation of the visualization. Each participant was then given time to investigate the systems until they were comfortable using it.

Next, the testers were asked to search books by using the text and visualization based approaches independently. Both applications show the overall view first either text-based or graphically forms respectively. From the overall view each student selected desirable books by comparing their different properties while utilizing the interactive functionality of the system. The selection of books from the overall view leads the user to the detail view, where participants are able to perform searches on the indexes. From this view, participants can analyze the index-level information related to their queries which will assist them to understand better about the underlying contents; eventually the better chance to find books has more useful or related information.

Participants were then asked to fill out a post experimental survey. The survey was designed to determine user satisfaction about the proposed visualization over a traditional text-based approach. Comments and suggestions on how the visualization can be practically deployed to fill such a role in the future were also collected.

3.1 Survey Results

The questionnaire has 14 questions which utilize a 5 point Likert scale; 5 indicating the highest level of satisfaction, and 1 the lowest. Table 1 below shows the survey results that summarizes the user's feedback to VIBS over a text-based library search system. 16 out of 17 participants stated that they had previous experiences using text-based book search system. Overall users were satisfied with both the overall and detailed view of the proposed system. Although it was expressed in their comments indicating unfamiliarity of visualization system over the current text-based interface, 53% of participants stated that they are more satisfied using VIBS than a text-based approach, 35% of the participants had a neutral stance, 12% of participants still prefer to use text-based book search interface. 82% of participants responded that the overview visualization improved their ability to identify desirable book. For questions specifically associated with the overall view, determining the overall concentration of information for individual books, as well as comparing multiple books to decide which is most appropriate, was considered a strong benefit provided by the prototype. Participants strongly agreed that presenting a visual abstract of book attributes was meaningful for their search activities. They also expressed that the grid-based layout assisted them in comparing multiple books and the selectable axes were very helpful and assisted them in identifying desirable results.

Table 1. Post experiment survey results as percentages

Questions	Pos	Neut	Neg
VIBS overall was preferable to a text-based search	53	35	12
The Overview improved identification of desirable books	82	6	12
The choice of visualized attributes provided a meaningful search environment.	76	18	6
Use of the selectable axis facilitated a better understanding of a set of books.	71	24	5
The Overview helped in selecting a subset of books.	65	29	6
It was easy to discern book attributes based on node color	53	18	29
The Detailed view was preferable to a text approach	65	24	11
The radial layout of the Detail view was intuitive	65	18	17
The Detail view made concentrations of information easy to identify	76	24	0
It was easy to discern term attributes based on node color	47	24	29
The detail view made relevant book selection easier	88	12	0
The relation between a term and it's child was clear	65	35	0
The Detail view helped identify terms related to the search	94	6	0

The participants also replied positively to utilizing visualization for presenting index information of books. 65% of participants satisfied using the visual interface for presenting book contents, meanwhile 11% of users prefer to explore book indexes in text. Users responded optimistically to questions asking the effectiveness of radial tree visualization that displays hierarchies of index terms and its magnitude. Especially participants strongly agreed on two facts that the detailed visualization helped them to choose books with more relevant information and to understand other term related with the search term to 88% and 94% respectively. Meanwhile in both views, the primary recommendation provided by users was to make the color blending of the visualization more intuitive. The post-experiment interview showed that this stems primarily from the fact that some users were unfamiliar with the RBG color blending scheme, having had little contact with it in their daily lives.

4   Conclusion and Future Work

Overall, feedback was positive toward the VIBS system. The testers found it to be an interesting and robust alternative to more traditional search methodologies. While there were users that still held preference toward a traditional text-based approach, several brought up the topic of familiarity. Although these types of searches have the weaknesses outlined earlier, they benefit in wide-spread usage and familiarity among users. Even though the testers had limited exposure to the VIBS system prior to the experiment; that more than half would prefer it over a traditional text-based search is very promising.

Taking a closer look, the Overview was well liked by most users. Similar comments were made regarding familiarity about this display; the chart-like layout was also readily understood, and was an aspect of the view that users felt improved their understanding of the data. The interaction provided through the selectable axes assisted in this, as users appreciated the ability to tailor the display closer to the characteristics more relevant to his or her interest.

Although the Detail view lacked the familiarity present in the Overview, participants still responded favorably. The ability to compare multiple sources at once in detail was an aspect of the system users were very appreciative of. One important point in particular is the high percentage of users that found this view greatly enhanced the identification of other important terms related to the original query. This exploratory aspect of the system is important, in that it can lead users to more appropriate terminology to refine the search. Additionally, it could expose users to other concepts contained within each work that, while not to original goal, could be something which would sway the user to placing more value in one result over another.

The primary issue highlighted by the experiment is user difficulty with interpreting the color codes of both views. Although around half in both instances were comfortable with this aspect of the system, around 30% had trouble with it. The post-experiment interviews provided two main causes for the diverging opinions. First, the RBG color model was not familiar to some. These users cited heavy exposure to the RBY model as being a source of confusion when interpreting the displays. The increased unfamiliarity left those individuals feeling more comfortable using a text-based search.

The other difficulty reported by users was in determining the relative value of one result with another. This could be, in part, a result of the human eye being more sensitive to some colors rather than others. For example, green-yellow colors have the strongest reception, which could mislead a user into considering a result with this color to have more overall value than another when that may not be the case [15]. Researching and examining alternate color models will be one of the challenging task for future work on the VIBS system.

The other area for future work is with regard to content analysis. As books and similar works move toward electronic rather than traditional hard-copy formats, the use of indexes in the traditional sense may become obsolete. For VIBS to be viable in such a scenario, full content analysis would be an alternative for classifying information. Instead of relying upon pre-existing documentation of content, the system could examine the work in full, determining such items as key words and topics, along with their frequency.

In a similar vein, the inclusion of a thesaurus-like aspect would also be a potentially valuable addition. The current Detail view arranges sub-terms solely by content size, a redundant expression to the size of the nodes. Instead, sub-terms could be grouped according to similarity, which would serve to strengthen the exploratory aspect of the system, and improve the user's experience.

By working toward the solutions and ideas presented here, continued work on the VIBS system will be better able to address some of the difficulties presented through the experiments. This will also serve to improve upon its strengths, and allow for a more comprehensive alternative to more traditional text-based search methodologies.

References

1. Good, L., Popat, A., Janssen, W., Bier, E.: Fluid Interface for Personal Digital Libraries. In: Proceedings of the 9th European Conference on Research and Advanced Technology for Digital Libraries, pp. 162–173 (2005)
2. Veerasamy, A., Heikes, R.: Effectiveness of a graphical display of retrieval results. In: Proceedings of the 20th Annual International ACM SIGIR Conference on Research and Development in Information Retrieval, pp. 236–245 (1997)
3. Dushay, N.: Visualizing Bibliographic Metadata – A Virtual (Book) Spine Viewer. D-Lib Magazine 10(10) (2004)
4. Hawkins, D.T.: Information Visualization: don't tell me, show me! Online 23(1), 88–90 (1999)
5. Kim, B., Johnson, P., Huarng, A.: Colored-sketch of Text Information. Journal of Informing Science 5(4), 163–173 (2002)
6. Borner, K., Chen, C.: Visual interfaces to digital libraries: motivation, utilization, and socio-technical challenges. In: Börner, K., Chen, C. (eds.) Visual Interfaces to Digital Libraries. LNCS, vol. 2539, pp. 1–9. Springer, Heidelberg (2002)
7. Shen, R., Vemuri, N., Fan, W., Torres, R., Fox, E.: Exploring digital libraries: integrating browsing, searching, and visualization. In: Proceedings of the 6th ACM/IEEE-CS joint conference on Digital libraries, pp. 1–10 (2006)
8. Silva, N., Sánchez, A., Proal, C., Rebollar, C.: Visual exploration of large collections in digital libraries. In: Proceedings of the Latin American conference on Human-computer interaction, pp. 147–157 (2003)

9. Shneiderman, B., Feldman, D., Rose, A., Grau, X.: Visualizing Digital Library Search Results with Categorical and Hierarchical Axes. In: Proceedings of the 5th ACM conference on Digital Libraries, pp. 57–66 (2000)

10. Marks, L., Hussell, J., McMahon, T., Luce, R.: ActiveGraph: A Digital Library Visualization Tool. International Journal on Digital Libraries 5(1), 57–69 (2005)

11. Lin, X.: Graphical Table of Contents. In: Proceedings of the first ACM international conference on Digital libraries, pp. 45–53 (1996)

12. Baudich, P., Good, N., Bellotti, V., Schraedley, P.: Keeping Things in context: A comparative Evaluation of Focus Plus Context Screens, Overviews, and Zooming. In: Proceedings of the SIGCHI conference on Human Factors in Computing Systems, pp. 259–266 (2002)

13. Resnick, P., Iacovou, N., Suchak, M., Bergstrom, P., Riedl, J.: GroupLens: An Open Architecture for Collaborative Filtering of Netnews. In: Proceedings of the ACM Conference on Computer Supported Cooperative Work, pp. 175–186 (1994)

14. Allen, R.B.: User Models: Theory, Method, and Practice. International Journal of Man-Machine Studies 32, 511–543 (1990)

15. Foley, J., Van Dam, A., Feiner, S., et al.: Computer Graphics: Principles and Practice, 2nd edn. Addison-Wesley Publishing Company, Reading (1996)

Generation of an Importance Map for Visualized Images

Akira Egawa[1] and Susumu Shirayama[2]

[1] Department of Systems Innovation, Graduate School of Engineering,
The University of Tokyo
[2] RACE, The University of Tokyo
{egawa,sirayama}@race.u-tokyo.ac.jp

Abstract. Visualized images have always been a preferred method of communication of information contained in complex data sets. However, information contained in the image is not always efficiently communicated to others due to personal differences in the way subjects interpret image content. One of the approaches to solving this issue is to determine high-saliency or eye-catching regions/objects of the image and to share information about the regions of interest (ROI) in the image among researchers. In the present paper, we propose a new method by which an importance map for a visualized image can be constructed. The image is first divided into segments based on a saliency map model, and eye movement data is then acquired and mapped into the segments. The importance score can be calculated by the PageRank algorithm for the network generated by regarding the segments as nodes, and thus an importance map of the image can be constructed. The usefulness of the proposed method is investigated through several experiments.

1 Introduction

As complexity of raw data in computational simulations increases, visualization of the data has played an important role in its analysis, and such visualized images have always been a preferred method of communication. On the other hand, interpretation of these visualized images is becoming more difficult, due to the enormous amount of information contained in them. Therefore, clear interpretation of the images is required in order to share useful information. However, information contained in the image is not always efficiently communicated to others due to personal differences in the way subjects interpret image content. Moreover, the need for collaborative evaluations, second opinions, and third party evaluations makes the problem of conveying information to others correctly even more difficult to solve.

Both formative and summative evaluations of the image are often required in order to solve those issues. In these cases, determining how to estimate high saliency or eye-catching regions/objects of the image is one of the most important problems.

A number of methods for conveying information to others and for determining high-saliency regions have been proposed. One such method is to manifest regions of interest (ROI) [1] and to show the ROI according to level of detail (LOD) [2]. In most cases, particular objects or regions are extracted from the image based on the meaning of the content, after which the LOD is designed. Although this is a powerful method, there are two major problems with this approach.

G. Bebis et al. (Eds.): ISVC 2009, Part I, LNCS 5875, pp. 135–146, 2009.

One is the difficulty involved in separating particular objects or regions from the image. In order to separate these objects/regions from the image, the problem of figure-ground separation has been solved [3], and methods to detect the critical points and particular lines/regions have been developed [4][5]. However, it is not always easy to detect the boundaries in an image, or to analyze the figure-ground relationship, and figure-ground separation remains an unsolved problem. Despite this, methods of detecting the critical points and particular lines have been established for images produced by visualizing fields that have a mathematical structure.

The second problem is that if there is no definitive structure, then the ROI and LOD depend strongly on personal interpretation, and the definition of an ROI or LOD becomes subjective. Therefore, it is difficult to assign a certain quantity of interest to a particular region. This is related to the first. In general, it has been pointed out that images have inherently recognizable areas. Considering the visual characteristics of the human eye, most viewers naturally focus their eyes on regions that contain recognizable areas. Therefore, such regions may be regarded as ROI candidates.

In order to detect such regions, we focused on the following two approaches. The first is a model-based approach that consists of computational models that imitate visual attention based on the human cognitive system. One of the most popular computational models was introduced by Itti et al. [6]. They modeled high-saliency regions that attract the human eye in a task-independent manner. Their model is an extension of the visual attention model proposed by Koch et al. [7]. The model of Itti et al. is referred to herein as the saliency map model. Using the saliency map model, eye-catching areas can be extracted from a source image [8][9]. Furthermore, the validity of this model has been confirmed by eye-tracking experiments [10][11] and in applications [12][13][14].

The second is an eye movement analysis. Although most eye movement data is meaningless, as pointed out by Duchowski et al. [15], the overt response of the subject in visual attention processes, tacit interest, and personal skill or knowledge can be estimated based on the fixation and the saccade [16][17][18].

These two approaches for detecting recognizable areas or ROI have been applied to image processing. Maeder [19] reviewed methods for image quality assessment and argued the usefulness of a perceptual modeling framework known as an image importance map. Although he considered the human cognitive system, his goal was to assess the image quality effectively. Lee et al. [20] proposed methods for detecting manipulated images. Their methods involve segmentation, classification, and common-sense reasoning. An importance map constructed by mean-shift image segmentation [21] and the visual attention algorithm of Itti et al. [6] is used in the processes of segmentation and classification. In these studies, an importance map of an image played an important role in processing the image.

In the present paper, a method for conveying quantitative information about both subjective/ personal ROI favored by ordinary or skilled viewers and also an objective view of ROI in a visualized image is considered. For this method, we use the saliency map model and eye movement analysis to extract an ROI from an image and then assign a quantitative index to the ROI. We refer to a distribution of such ROI as an Image Importance Map (IIM). A method for constructing an IIM from the visualized image is described below.

2   Method

First, a source image and eye movement data for a visualized image are provided as the input data. Second, the source image is segmented by the following three steps:

Step 1. A saliency map for the image is created according to the model proposed by Itti et al. [6].

Step 2. Regions of high salience in the map are clustered.

Step 3. The image is segmented based on the clusters produced in Step 2.

Third, two types of Image Importance Map (IIM) are constructed. The IIM is constructed by assigning a quantitative index to the segments. The quantitative index for the first IIM is calculated using the attention shift model of Itti et al. [6]. The index for the second IIM is calculated by eye movement data using the PageRank algorithm [22]. Finally, two IIMs are integrated into one map using the Biased PageRank algorithm [23]. The details of this process are presented in the following sections.

2.1   Segmentation of the Source Image

In a task-dependent manner in particular, it may be considered that conveying information related to subjective/objective ROI in the image to others is based on important objects. In order to identify the objects in an image, an image segmentation method for objects that have discernable or presumed contours has been used [24]. Therefore, object-based segmentation is used in the importance map of Lee et al. [20].

However, considering the visual characteristics of the human eye, it is assumed that both ordinary and skilled viewers naturally train their eyes on regions that contain recognizable areas, even while working to complete a task. Such regions or areas do not always have clearly identified boundaries.

Accordingly, we propose a new image segmentation method based on the saliency map model.

Creating saliency map. First, the image is decomposed into three feature channels: intensity, color (red/green and blue/yellow), and orientation ($0°$, $45°$, $90°$, $135°$). The orientation channel is then calculated using the Gabor filter and nine spatial scales are created from these three channels using Gaussian pyramids.

Second, differences between the fine and coarse scales in each channel are calculated based on a center-surround operation that imitates the human visual system.

Third, for each channel, three conspicuity maps that highlight the parts that differ noticeably from their surroundings are created using across-scale combinations. Finally, a saliency map is created as a linear combination of the three conspicuity maps. The saliency map is composed of $lmax \times mmax$ cells. Let (l, m) be the coordinates of a cell in the map. The saliency at (l, m) is denoted by $s(l, m)$. In our implementation, the scale of the saliency map is 1/16 the size of the source image (one cell on the saliency map corresponds to a 16×16 pixel region in the source image). Figure 1 shows the source image (left figure) and the saliency map (right figure). White and black areas indicate high- and low-saliency regions, respectively.

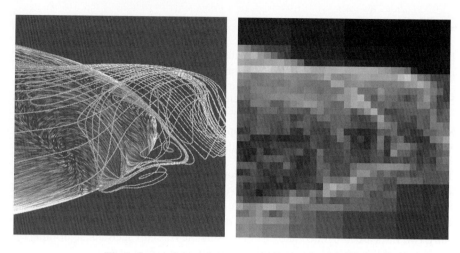

Fig. 1. Source image (left) and its saliency map (right)

Fig. 2. Saliency Cluster Map. Left and right figures show the core-clusters and all of the other clusters, respectively.

Clustering of the saliency map. Regions of high salience in the map are clustered. The proposed procedure consists of the two steps. The pseudocode for the first step is shown in Table 1, where *imax* and *jmax* denote the width and height, respectively, of the source image. A cluster is identified by index K. The index of the cluster at (l, m) is stored in $c(l, m)$. Let c_K be a set of cells in cluster K. The centroid of cluster K is (x_K, y_K), which is computed as $(x_K, y_K) = \frac{1}{N_K}(\sum_{(l,m) \in c_K} l, \sum_{(l,m) \in c_K} m)$, where N_K is the number of cells in cluster K. The distance between a cell $c(l, m)$ and cluster K is calculated as $\sqrt{(l - x_K)^2 + (m - y_K)^2}$. NearestCluster$(l, m)$ is the function that returns the index of the cluster that has the shortest distance between (l, m) and (x_k, y_k) in eight

Table 1. Pseudocode for the first step

```
set 0 to c(l,m)
K ← 0
for l = 0 to [(imax − 1)/16] {
    for m = 0 to [(jmax − 1)/16] {
        if s(l,m) ≥ δ and c(l,m) = 0 {
            if at least one of the eight neighbors of (l,m)
            belongs to any cluster {
                K ← K + 1
                c(l,m) = K
                (xK, yK) = (l,m)
            }
            else {
                k = NearestCluster(l,m)
                c(l,m) = k
                update (xk, yk)
            }
        }
    }
    if s(l,m) < δ { c(l,m) = 0 }
```

neighbors of (l,m). The threshold value δ is given by the user. We refer to the cluster obtained in the first step as the core-cluster.

In the second step, the cells for which $s(l,m) < \delta$ are assigned to one of the clusters which are closed to the cells. All of the cells on the saliency map belong to the clusters after this clustering.

In this way, all of the cells are eventually assigned to the appropriate cluster. These clusters are referred to collectively as the Saliency Cluster Map (SCM). The SCM for Figure 1 is shown in Figure 2. The core-clusters are shown in the left-hand image, and the right-hand image shows all of the other clusters.

Segmentation using the saliency cluster map. The SCM is applied to the image segmentation in a straightforward manner. A segment consists of the pixel regions that correspond to cells belonging to the same cluster. Note that the segment is composed of two types of pixels. One type exists in the core-cluster. We refer to the region of a segment composed of these pixels as the *core-segment*. The other type exists outside the core-cluster. Let the index of a segment be k. The core-segment is then denoted as R_k, and the region outside the core-segment is denoted as R'_k. An example of image segmentation of a source image is shown in Figure 3.

2.2 First Image Importance Map

Our approach is based on the Focus Of Attention (FOA) and FOA shift proposed by Itti et al. algorithm [6].

We define a set R as a collection of all segments, where the total number of elements of R is M. Let the saliency of segment be $s_k(k = 1, ..., M)$. The saliency s_k is set as the

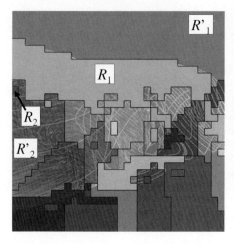

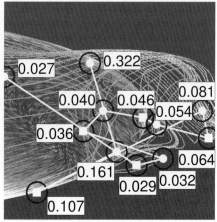

Fig. 3. Example of image segmentation   **Fig. 4.** First Image Importance Map

highest saliency in the cells corresponding to the core-segment R_k. Let (x_k, y_k) be the centroid of segment k, which is defined as the centroid of the cluster corresponding to segment k. In addition, the preferential saliency is denoted as s'_k.

A set F that consists of a sorted FOA is obtained as follows:

Step 1. Let F be an empty set $\emptyset$. Also, let $s'_k = s_k (k = 1, ..., M)$

Step 2. Find the segment that has the highest preferential saliency in $R \setminus F$, and set the index of the segment to i

Step 3. Add i to set F

Step 4. Compute s'_k as follows:

if $\sqrt{(x_i - x_k)^2 + (y_i - y_k)^2} \leq \beta$ {
$s'_k = s_k + \alpha$
}
else {
$s'_k = s_k$
}

Step 5. Repeat Step 2 through 4

where α and β are the parameters of the proximate preference. α and β indicates the strength and range of the proximate preference respectively. We use $\alpha = 50$ and $\beta = 100$ (pixels) in this paper.

We use the set F for the quantitative index assigned to the segment. First, R_k and R'_k are redefined. The value of index k indicates the k-th element of F, i.e., the order of the FOA. Next, the importance of the segment is defined. Let i^1_k be the importance of segment k.

$$i^1_k = \frac{1}{k \cdot \sum_{k=1}^M \frac{1}{k}}. \tag{1}$$

The distribution of the segments that have this importance is referred to as the first Image Importance Map (first IIM). An example of a first IIM is shown in Figure 4. In

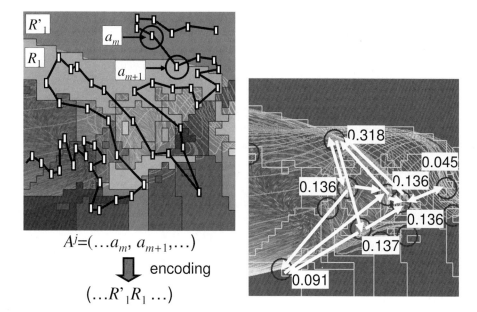

$A^j = (\ldots a_m, a_{m+1}, \ldots)$

⇩ encoding

$(\ldots R'_1 R_1 \ldots)$

Fig. 5. Scheme of encoding the eye movement dataset

Fig. 6. Second Image Importance Map

this figure, the blue circles are the centroids of the segments, and the numerical values indicate the importance of the segments. In addition, the green filled circle denote R_1, and the red lines denote the FOA shift.

2.3 Second Image Importance Map

In the first IIM, top-down (task-dependent manner) factors are not considered. Previous researches have improved this model by adding top-down factors such as eye movement data [25] and a stochastic factor [26]. As in those studies, in the present study, the top-down factors are considered in the construction of the IIM. In our method, we use the same image segmentation of the first IIM and an IIM with a top-down factor is constructed as follows.

First, we obtain eye movement data from the subjects using a head mounted eye-tracking system. Let $a_n = (x_n, y_n, t_n)$ be the n-th eye movement data. The data a_n is composed of (x_n, y_n) (the coordinates in the image) and t_n (the recorded time). An eye movement dataset for person j is denoted by $A^j = (a_1, a_2, \ldots, a_N)$, and N is the total number of eye movement data.

Second, the eye movement dataset A^j is mapped into the segmented source image and encoded using the identifier of the segment that contains the eye position. The scheme for encoding the eye movement dataset is shown in Figure 5.

Third, a network for the eye movements is generated by regarding the segments as nodes. We denote a node as v_i. The number of nodes is the same as the number of segments (M). The node corresponds to the centroid of the segment that contains the eye position. For simplicity, in the present paper, we do not discriminate between the core-segment and the region outside the core-segment. A link (an arc) is generated by the eye movement. The encoded eye movement dataset A^j is transformed into a network.

In this way, one network is generated from one eye movement dataset. This network is represented by a weighted adjacency matrix $\mathbf{A}$ in which the weight is given by the number of links.

The score of importance is obtained by the PageRank algorithm [22]. First, the matrix $\mathbf{A}$ is converted to a transition probability matrix $\mathbf{T}$. For example, in the case of the eye movement dataset $R_2R_4R_1R_1R_1R_2R_4R_3$, the matrix $\mathbf{T}$ is obtained by

$$
\mathbf{T} = \begin{array}{c} \\ v_1 \\ v_2 \\ v_3 \\ v_4 \end{array} \begin{array}{cccc} v_1 & v_2 & v_3 & v_4 \\ \left(\begin{array}{cccc} 2/3 & 1/3 & 0 & 0 \\ 0 & 0 & 0 & 1 \\ 0 & 0 & 0 & 0 \\ 1/2 & 0 & 1/2 & 0 \end{array} \right) \end{array}. \tag{2}
$$

Second, the PageRank of each node is calculated by the following iterative process:

$$
\mathbf{p}^{v+1} = \mathbf{T}'\mathbf{p}^v, \tag{3}
$$

where $\mathbf{p} = (i_1^2, ..., i_i^2, ..., i_M^2)$, i_i^2 represents the PageRank of the v_i, and v is the iterative number.

The importance of the segment k is i_k^2. The distribution of the segments that have this importance is designated as the second Image Importance Map (second IIM). An example of a second IIM is shown in Figure 6. In this figure, the blue circles are the centroids of the segments, and the numerical values indicate the importance of the segments. In addition, the orange arrows indicate the links of the network.

2.4   Integration of Image Importance Maps

A second IIM is obtained for each eye movement dataset. In order to integrate the second IIMs, the weighted adjacency matrix $\mathbf{A}$ is composed of the eye movement datasets that we attempt to integrate.

For the case in which the first IIM and the second IIM are integrated, we use the biased PageRank algorithm [23]. First, a vector $\mathbf{i}$ that denotes the importance of the segment in the first IIM is calculated. Second, the weighted adjacency matrix $\mathbf{A}$ is obtained and converted to the transition probability matrix $\mathbf{T}$.

The first IIM and the second IIM are integrated by the following iterative process:

$$
\mathbf{p}^{v+1} = \omega\mathbf{T}'\mathbf{p}^v + (1-\omega)\mathbf{i}, \tag{4}
$$

where $\mathbf{p} = (i_1^s, ..., i_i^s, ..., i_M^s)$, i_i^s represents the integrated importance of segment i, and ω $(0 \leq \omega \leq 1)$ is the user parameter. In the present paper, we use $\omega = 0.5$.

3   Construction of the Image Importance Maps

First, an image produced by the visualization of instantaneous streamlines in a flow past a spheroid was used for the experiment. The source image and the first IIM are shown in the left part of Figure 1 and in Figure 4, respectively. Three subjects participated in the experiment. One (subject A) was an expert in fluid dynamics, whereas the others (subjects B and C) were novices. The second Image Importance Maps (IIMs) and the integrated IIMs are shown in Figure 7 and Figure 8, respectively.

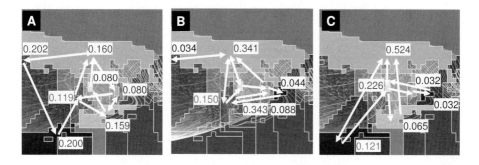

Fig. 7. Three subjests' second IIMs of the first source image

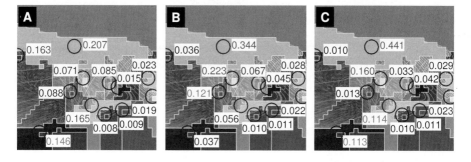

Fig. 8. Three subjests' integrated IIMs of the first source image

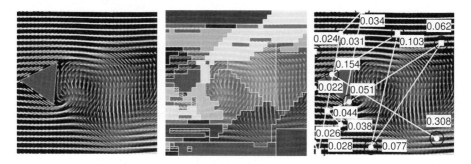

Fig. 9. A source image for the second experiment (left), the corresponding segments of the image (center) and the first IIM (right)

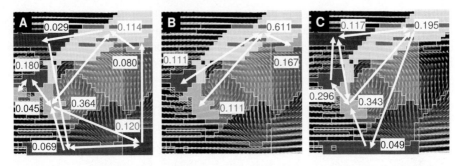

Fig. 10. Three subjests' second IIMs of the second source image

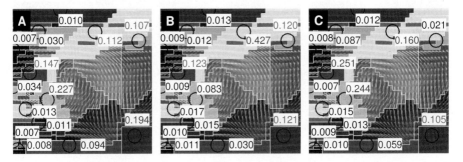

Fig. 11. Three subjests' integrated IIMs of the second source image

Second, a visualized image produced by a tuft visualization for a flow around a trianguler cylinder is used. A source image, the corresponding segments of the image and the first IIM are demonstrated in Figure 9. Figure 10 and Figure 11 shows the second IIMs and integrated IIMs, respectively.

As shown in Figures 7 and 10, all the regions where the expert (subject A) gazed are not always eye-catching for subjects B and C. This means that some important information is lost despite the fact that raw data is visualized and transformed into a recognizable form.

The region which has the highest saliency in the right lower part of the first IIM shown in Figure 9 is remarkable. After the experiment, it was confirmed through interviews with the expert that this region is important to prediction of the far wake region. Actually, the expert (subject A) had focused on the region, and the subject A's score is striking in the integrated IIMs presented in Figure 11. It can be said that the region is inherently recognizable and includes the expert's preference. On the other hand, subjects B and C did not gaze at the region as shown in Figure 10. This also means that some important information will be lost.

From the standpoint of visual communication, it may be considered that the experts' ways of viewing are deliberate and inappropriate since their viewing behaviors are not always correct and are sometimes unnatural. In order to avoid premature judgement, it is important to know the average and personal preferences of the viewers. The integrated IIMs shown in Figures 8 and 11 are considered to give an immediate indication of the average and personal preferences of the viewers since they are created by combining the importance values of the first and second IIMs. The region that has a higher importance value on the integrated IIM is inherently recognizable; it is the focus of the viewer's gaze, whereas regions that have lower importance are not eye-catching, and so are not gazed at by the viewer. The differences in the evaluations of the subjects can be expressed by the quantitative values. Thus, the integrated IIM provides a more accurate indication of the important regions of the image.

4   Conclusion

We have presented a new method by which an importance map of a visualized image can be constructed. We used a saliency map model for image segmentation and quantitative importance values for the importance maps were obtained based on the attention shift model of the saliency map model and eye movement data. The usefulness of the proposed method was investigated through eye-tracking experiments.

Our method enables researchers to share information about ROIs with quantitative values of importance and indications of differences among individuals. Therefore, this method will help to enhance the interpretability and recognizability of visualized images subsequent to their generation, and can be used in such applications as visual communication for education purposes.

Several extensions can be developed to improve the proposed method. For a more detailed treatment of the eye movement dataset, the regions outside the core-segments can be subdivided into a number of segments in an appropriate manner. In addition, although the saliency map model of Itti et al. is effective for use with the proposed method, other saliency map models that simulate human visual attention may be used as alternatives. Finally, object-based image segmentation may be combined with the proposed method in order to consider eye movement in detail and in a task-dependent manner.

Acknowledgements

One of the authors was supported through the Global COE Program, gGlobal Center of Excellence for Mechanical Systems Innovation,h by the Ministry of Education, Culture, Sports, Science and Technology.

References

1. Kim, H., Min, B., Lee, T., Lee, S., Lee, C., Park, J.: Three-Dimensional Digital Subtraction Angiography. IEEE Transactions on Medical Imaging MI-1, 152–158 (1982)
2. Rauschenbach, U.: Progressive Image Transmission using Levels of Detail and Regions of Interest. In: Proc. of IASTED CGIM 1998, pp. 38–41 (1998)
3. Bhandarkar, S., Zeng, X.: Figure-Ground Separation: A Case Study in Energy Minimization via Evolutionary Computing. In: Pelillo, M., Hancock, E.R. (eds.) EMMCVPR 1997. LNCS, vol. 1223, pp. 375–390. Springer, Heidelberg (1997)

4. Schultz, T., Theisel, H., Seidel, H.: Topological Visualization of Brain Diffusion MRI Data. IEEE Transactions on Visualization and Computer Graphics 13, 1496–1503 (2007)
5. Bürger, R., Hauser, H.: Visualization of Multi-variate Scientific Data. In: Proc. of EuroGraphics 2007 State of the Art Reports, pp. 117–134 (2007)
6. Itti, L., Koch, C., Niebur, E.: A Model of Saliency-Based Visual Attention for Rapid Scene Analysis. IEEE Transactions on Pattern Analysis and Machine Intelligence 20, 1254–1259 (1998)
7. Koch, C., Ullman, S.: Shifts in selective visual attention: towards the underlying neural circuitry. Human Neurobiology 4, 219–227 (1985)
8. Itti, L., Kock, C.: Target Detection using Saliency-Based Attention. In: Proc. of RTO/SCI-12 Workshop on Search and Target Acquisition (Unclassified), pp. 3.1–3.10 (1999)
9. Walther, D., Edgington, D., Koch, C.: Detection and Tracking of Objects in Underwater Video. In: Proc. of IEEE Conference on Computer Vision and Pattern Recognition, pp. 544–549 (2004)
10. Itti, L., Koch, C.: Feature combination strategies for saliency-based visual attention systems. Journal of Electronic Imaging 10, 161–169 (2001)
11. Parkhurst, D., Law, L., Niebur, E.: Modeling the role of salience in the allocation of overt visual attention. Vision Research 42, 107–123 (2002)
12. Kim, Y., Varshney, A.: Saliency-guided Enhancement for Volume Visualization. IEEE Transaction on Visualization and Computer Graphics 12, 925–932 (2006)
13. Davies, C., Tompkinson, W., Donnelly, N., Gordon, L., Cave, K.: Visual saliency as an aid to updating digital maps. Computers in Human Behavior 22, 672–684 (2006)
14. Ma, Y., Lu, L., Zhang, H.J., Li, M.: A User Attention Model for Video Summarization. In: Proc. of the 10th ACM international conference on Multimedia, pp. 533–542 (2002)
15. Duchowski, A.: Eye Tracking Methodology, 2nd edn. Springer, Heidelberg (2007)
16. Yarbus, A.: Eye Movements and Vision. Plenum Press, New York (1967)
17. Russo, J.E., Rosen, L.D.: An eye fixation analysis of multialternative choice. Memory and Cognition 3, 167–276 (1975)
18. Martinez-Conde, S., Macknik, S.L.: Windows on the mind. Scientific American Magazine 297, 56–63 (2007)
19. Maeder, A.: The image importance approach to human vision based image quality characterization. Pattern Recognition Letters 26, 347–354 (2005)
20. Lee, S., Shamma, D., Gooch, B.: Detecting false captioning using common-sense reasoning. Digital Investigation 3, 65–70 (2006)
21. Meer, P., Georgescu, B.: Edge Detection with Embedded Confidence. IEEE Transactions on Pattern Analysis and Machine Intelligence 23, 1351–1365 (2001)
22. Brin, S., Page, L.: The anatomy of a large-scale hypertextual web search engine. Computer Networks and ISDN Systems 30, 107–117 (1998)
23. Kamvar, S., Haveliwala, T., Manning, C., Golub, G.: Exploiting the Block Structure of the Web for Computing Pagerank. Stanford University Technical Report 2003-17 (2003)
24. Ko, B., Byun, H.: FRIP: A Region-Based Image Retrieval Tool Using Automatic Image Segmentation and Stepwise Boolean AND Matching. IEEE Transactions on Multimedia 7, 105–113 (2005)
25. Igarashi, H., Suzuki, S., Sugita, T., Kurisu, M., Kakikura, M.: Extraction of Visual Attention with Gaze Duration and Saliency Map. In: Proc. of IEEE International Conference on Control Applications, 562–567 (2006)
26. Pang, D., Kimura, A., Takeuchi, T., Yamato, J., Kashino, K.: A Stochastic Model of Selective Visual Attention with a Dynamic Baysian Network. In: Proc. of ICME 2008, pp. 1073–1076 (2008)

Drawing Motion without Understanding It

Vincenzo Caglioti, Alessandro Giusti, Andrea Riva, and Marco Uberti

Politecnico di Milano, Dipartimento di Elettronica e Informazione
alessandro.giusti@polimi.it

Abstract. We introduce a novel technique for summarizing a short video to a single image, by augmenting the last frame of the video with comic-like lines behind moving objects (*speedlines*), with the goal of conveying their motion. Compared to existing literature, our approach is novel as we do not attempt to track moving objects nor to attain any high-level understanding of the scene: our technique is therefore very general and able to handle long, complex, or articulated motions. We only require that a reasonably correct foreground mask can be computed in each of the input frames, by means of background subtraction. Speedlines are then progressively built through low-level manipulation of such masks. We explore application scenarios in diverse fields and provide examples and experimental results.

1 Introduction

Representing motions and actions in a single image is a well known problem for visual artists. Universally accepted solutions are found in comics, where motion is represented through various abstract graphical devices [1] (Figure 1), and in photography, where object images are often deliberately motion blurred in order to convey their speed (Figure 2). In this paper, we propose a simple, low-level algorithm for representing in a single image the motion occurring in a video, by augmenting the last frame of the video sequence with lines conveying the motion occurred in the previous frames (*speedlines*). In particular, we recover speedlines as the envelope of the objects' apparent contours in time.

Drawing speedlines is mainly a matter of visual style in case the motion of objects in the scene is known in advance; this happens in synthetically modeled scenes, or when motion is previously recovered by, e.g., object/people tracking

Fig. 1. Speedlines are commonplace in comic books

G. Bebis et al. (Eds.): ISVC 2009, Part I, LNCS 5875, pp. 147–156, 2009.
© Springer-Verlag Berlin Heidelberg 2009

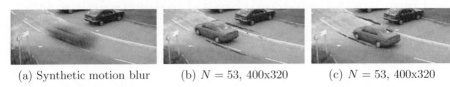

(a) Synthetic motion blur   (b) $N = 53$, 400x320   (c) $N = 53$, 400x320

Fig. 2. Motion blur (a) fails to convey whether the car was moving forwards (b) or backwards (c). Short speedlines in (b,c) can instead represent the motion orientation, and do not cause loss of detail in the foreground.

algorithms; then, a straightforward, albeit visually rude technique for representing such motion could be to simply draw the known object trajectories in the previous frames; several other techniques, more refined and appealing, are presented in literature (see Section 2). Our approach is fundamentally different as it operates at a much lower level, and it does not explicitly try to separate or reconstruct moving objects from the input video; we just require an acceptable foreground/background segmentation for each frame, and assume that the camera is still. Then, speedlines are computed by means of few simple operations on the foreground masks of successive frames; in other words, we do not try to use or infer any information pertaining the scene semantics, and we represent motion without understanding it.

On one hand, this approach has several important advantages: low computational complexity; robustness to segmentation errors; and a remarkable generality, which allows applications in diverse fields without any modification of the algorithm. On the other hand, since they lack any explicit high level, symbolic information about the scene evolution, our results are worthless for any subsequent automated processing step: the only actual user of the output of our algorithm is an human observer.

Such a system can be useful for a number of reasons. A first obvious application is *summarizing a short video segment* in order to convey instantaneous motion in media where video is not available (printed, low-bandwidth devices), or where an image is preferable, as its meaning can be grasped immediately, without explicit attention. In these contexts, a similar goal is commonly achieved by shooting a single image with a longer exposure time, which results in motion blur; however, motion blur corrupts the foreground and does not convey motion orientation (see Figure 2).

Another application, which is commonly overlooked in related works, is *summarizing a longer video segment* in order to represent an extended motion, possibly complex, articulated or inhomogeneous in time. Motion blur is hardly helpful in this case, as it would make the subject too confused; an applicable technique is instead multiple exposure photography (or *strobing*), i.e. a long exposure photograph with stroboscopic lighting. Physical devices frequently used to attain similar goals are smoke trails or ribbons (e.g. in rhythmic gymnastics). In Section 4 we describe a prototype video surveillance application which uses speedlines to summarize long, temporally sparse events, and detail a number of practical advantages of such approach.

(a) 38 frames, 180x144   (b) 473 frames, 320x256

Fig. 3. Speedlines describing long motions

Finally, speedlines can be used to *augment every frame of a video stream*; when combined with refined, good-looking visualizations (which is not a primary focus in this paper), this can be applied for video special effects. In other contexts, speedlines can allow a casual viewer (such as a surveillance operator) to immediately grasp what's happening without having to look at the video stream for a longer time or rewind it, as each single frame visually conveys the scene's past temporal history. Another example is using speedlines for immediately understanding if and how slow objects, such as ships at a distance, are moving; interestingly, this is intuitively achieved by looking at the ship's trail, if visible, which is a real-world counterpart to our speedlines.

The paper is structured as follows: after reviewing existing literature in Section 2, we describe our technique in Section 3: we initially introduce our model and briefly discuss background subtraction, which is given as granted in the rest of the paper; then we detail the core our approach and its possible extensions. In Section 4 we present application examples and experimental results. We conclude with Section 5, which also presents future works, extensions, and additional foreseen application scenarios.

2   Related Works

Few main graphical devices for representing motion in a single image are described and used in literature [2,3]: drawing trailing speedlines (or motion lines) behind the moving object; replicating the moving object's image (such as in strobing) or its contours; introducing deformations on the object in order to convey its acceleration. In this work, we focus on the first (and partially the second, see Section 3.3) of these techniques, whose perceptual foundations are investigated by brain researchers in [4].

Several related works aim at rendering motion cues from synthetic animations; in particular [5] uses the scene's 3D model and animation keyframes as input, whereas [6] presents a more specialized system in the context of human gait representation; in [7] speedlines are synthesized in a framework for nonphotorealistic rendering, by selecting vertexes of the 3D moving objects which will generate speedline trails whenever the object's speed exceeds a given threshold.

Our system radically differs from the cited efforts as we use a video as input, instead of a 3D animated synthetic model.

A semi-automated system with similar goals is introduced in [8], which tracks moving objects in video and applies several graphically-pleasing effects to represent their motion, including short rectilinear speedlines. Straight speedlines can also be generated for translational motion of a tracked object in [9]. A different approach is adopted in [10], where *individual features* are tracked in a manually selected foreground object, and speedlines drawn as their trajectories; high level scene understanding is also a key characteristic of several other works [11], which fully exploit their higher abstraction level by introducing interesting additional effects, such as distortion for representing acceleration. The technique we are proposing differs as it lets speedlines "build themselves" through low-level manipulations of the evolving foreground masks, without requiring any user interaction nor explicitly tracking the objects; also, we do not assume anything on the object's motion, which is not required to be explicitly modeled or recovered: in fact, our system effortlessly handles complex motion of articulated or deformable objects, and also scenes with multiple moving objects which can not be easily separated or tracked. Moreover, our technique directly works on very low resolution video, where moving objects (possibly affected by motion blur) often bear no recognizable features. This also allows us to handle new scenarios and applications, such as surveillance tasks, which, to the best of our knowledge, have never been targeted by previous literature. An interesting relation also holds with [12], which deals with the opposite problem of inferring object movement from a single image affected by motion blur.

3   Technique

3.1   Overview

We work on a sequence of N video frames $I_1..I_N$, uniformly sampled in time. We assume that the camera is fixed, and that one or several objects are moving in the foreground.

We require that for each frame I_i, the foreground can be extracted, so to obtain a binary foreground mask M_i, whose pixels are 1 where frame I_i depicts the foreground, and 0 otherwise; this problem (background subtraction) is extensively dealt with in literature, and many effective algorithms such as [13] are available for application in most operating conditions, including unsteady cameras and difficult backgrounds.

From now on, we will assume that a reasonable segmentation is obtained using any background subtraction technique; Section 4 shows that spatially or temporally local segmentation errors only marginally affect the quality of results.

By processing foreground masks, we finally output an image F as the last frame I_N with speedlines superimposed, conveying the motion of objects in the scene during the considered time frame. From the theoretical point of view, these lines are an approximation of the envelope of the moving objects' apparent contours; we now describe their construction.

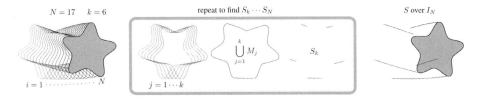

$N = 17$ $k = 6$

repeat to find $S_k \cdots S_N$

S over I_N

$i = 1 \cdots \cdots \cdots N$

$j = 1 \cdots k$

$\bigcup_{j=1}^{k} M_j$

S_k

Fig. 4. Our algorithm applied to $N = 17$ frames of a rototranslating object, with $k = 6$

3.2   Building Speedlines

First, the foreground mask M_i of each input frame I_i $i = 1 \ldots N$ is computed.
The video is then processed by analyzing adjacent foreground masks in small
groups (see Figure 4), whose size is defined by a parameter k, $3 \leq k \leq N$. k is a
parameter controlling the amount of detail in the speedlines in case of complex
motion, and its effects are better described in Section 4.1.

In particular, we compute $N - k + 1$ binary images S_i $k \leq i \leq N$; each S_i represents the atomic pieces of speedlines originating from the k frames $I_{i-k} \ldots I_i$.
S_i is computed as the set of edge pixels of the union[1] of masks $M_{i-k} \ldots M_i$,
which are not also edge pixels of M_{i-k} or M_i:

$$S_i = \text{edge}\left(\bigcup_{j=i-k}^{i} M_j\right) \setminus (\text{edge}(M_{i-k}) \cup \text{edge}(M_i)) \qquad (1)$$

When the trajectories of the objects in frames $i - k \ldots i$ are simple and non
intersecting, S_i approximates the envelopes of the foreground contours during
such time interval.

The $N - k + 1$ atomic pieces of speedlines S_i are then merged through a visualization function $f(\ldots)$, in order to determine the final speedlines S and their
appearance. A simplistic $f(\ldots)$ function just computes the union of all S_i. Alternatively, such function may improve visualization in several ways, some of which
we briefly explored in our implementation, and are introduced in Section 3.3

S is finally composited over I_N, which gives the system's output image F.
The complete algorithm is summarized as follows:

for $i = 1 \ldots N$ **do**
    $M_i \leftarrow \text{foreground}(I_i)$
end for
for $i = k \ldots N$ **do**
    $S_i \leftarrow \text{edge}\left(\bigcup_{j=i-k}^{i} M_j\right) \setminus (\text{edge}(M_{i-k}) \cup \text{edge}(M_i))$
end for
$S \leftarrow f(S_k, S_{k+1} \ldots S_N)$
$F \leftarrow \text{composite } S \text{ over } I_N$

---

[1] We consider the union of binary masks as the boolean OR operation on such masks.

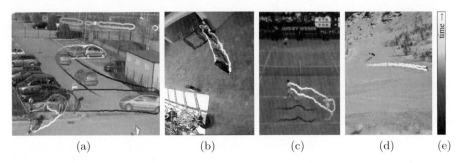

(a)   (b)   (c)   (d)   (e)

Fig. 5. Speedlines for complex motions, with foreground replication; speedline color encodes time (e). In (a) two cars park, and a person comes out from the car below. (a,b) from PETS dataset [16].

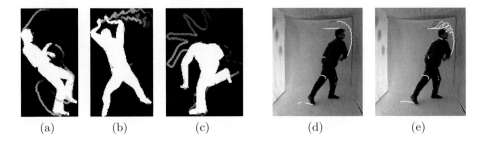

(a)   (b)   (c)   (d)   (e)

Fig. 6. (a,b,c): Speedlines computed directly from 640x480 silhouette data [17]. Note replication of contours in (c) due to fast motion (see Section 4.1). (d,e): the quickly moving hand draws copies of its contour instead of speedlines in (e), due to a coarse frame rate.

3.3   Extensions and Implementation Notes

In order to improve the informational content and polish of speedlines, the function $f(\ldots)$, which combines speedline pieces originating from different frames, can implement several visualization improvements, such as drawing speedline pieces with varying stroke width, color, or transparency.

Also, when applying the algorithm to longer timeframes, the foreground of important frames (*keyframes*) can be semitransparently superimposed over the output image (see Figure 5); in the simplest implementations, keyframes can be regularly sampled in time; as a more powerful alternative, a number of sophisticated techniques for automatically finding meaningful keyframes in video are proposed in the literature on video summarization [14,15].

4   Experimental Results

We implemented the proposed algorithm in a Matlab prototype; we tested several different application scenarios, both with video sequences from external sources such as [17,16] and produced *ad hoc*. We experimented with both short motions (Figures 2, 6, 7 and 6) and long, complex events (Figures 3 and 5); our videos have a wide range of different image resolutions, video quality and frame rates; we also applied our technique to time-lapse image sequences of very slow events (see [18]). We provide source code, original videos and full details in the supplementary material [18].

In our prototype, background subtraction is implemented by simply thresholding the absolute differences of each video frame with respect to a static background model, as our test videos were not very demanding from this point of view. Also, we smooth the recovered foreground by means of a median filter with square support, which helps in reducing artifacts due to background subtraction, and also allows the algorithm to create smoother speedlines. In fact, the exact shape of foreground masks is not fundamental for drawing good-looking, descriptive speedlines, which is also a reason why we tolerate imprecise background subtraction in the first place. Moreover, even macroscopic errors in background subtraction, if limited to few frames, only marginally affect our final results; in fact, such errors would only affect few of the speedline pieces we use to construct the final speedlines.

Computational costs in our unoptimized prototype are currently rather low, which allows us to process low-resolution webcam video in real time. Still, we expect huge improvements from optimized implementation in lower-level languages, as the algorithm has no computationally-intensive steps, at least unless more sophisticated background subtraction comes into play.

4.1 Discussion

As we previously mentioned, parameter k allows us to define how detailed the resulting speedlines appear: in particular, in presence of self-intersecting trajectories, a low k value causes more detailed speedlines to be built; in presence of unreliable masks or very complex motions which do not have to be represented in fine detail, setting an higher k parameter helps in improving the system's robustness and simplicity of the output.

Our system also requires that the framerate of the input video is sufficiently fast w.r.t. the object motion: in fact, in order for speedlines to be smooth, foreground masks in adjacent frames must be mostly overlapping. On one hand, this rules out the use of our system for surveillance video sampled very low frame rates, which is not an uncommon scenario. On the other hand, when the framerate is too coarse, the system does not abruptly fail, but basically replicates the subject's contours instead of drawing its envelopes, which is also a traditionally-employed technique for motion representation [1] (Figures 6 and 7b). In practice, an user can often make sense of the resulting image, and in some scenarios may also consider this a feature rather than a shortcoming, as quickly-moving objects gain a distinctive appearance.

We investigated a specific application of our technique, which is detailed in the Supplementary Material [18]: video from a surveillance camera is analyzed in order to segment simple temporally sparse events; each event, with a typical

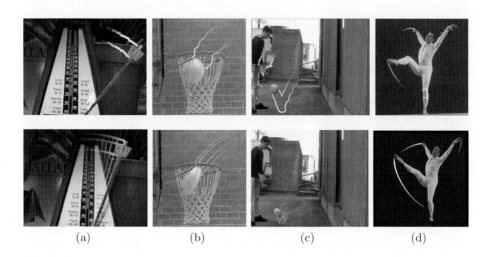

(a)   (b)   (c)   (d)

Fig. 7. Although less graphically pleasant, our results in drawing short speedlines (top) are comparable in informational content to those recovered by the high-level approach in [11] (a,b,c). Note speedlines for hands moving down in third image, as we describe *every* motion in the scene. In the bottom-right example (d) speedlines are determined manually [6], and are remarkably similar to our result.

duration between several seconds and one minute, is summarized in a single image with speedlines and foreground replication. This is useful for a number of reasons:

- the summary image can be easily and cheaply transmitted for immediate review to e.g. a mobile device.[2]
- an operator can quickly review many events by grasping what happened in each event at a glance; if the summary image for an event is judged unclear, unusual or suspect, the operator can click on it in order to see the corresponding video segment. This can potentially dramatically speedup the review process of surveillance video.
- An additional advantage of our approach in a security-related scenario is the transparency of the algorithm due to its low level of complexity. Moreover, speedlines are built in such a way that they encompass any area affected by motion, which is a nice property in this context.

When applied to a large, simple moving object for creating short speedlines, a fundamental disadvantage of our approach with respect to others becomes apparent: our speedlines are in fact only drawn at the extremes of the object's motion, which creates a poor visual effect (e.g. compare Figure 1a to Figure 2b); we are currently investigating other options for dealing with this issue.

---

[2] In several "extreme" scenarios, such as wireless sensor networks with very strict computational and power requirements, storing and/or transmitting a single summary image for an event may often be the only viable option.

5   Conclusions and Future Works

In this paper we presented a low-level algorithm for representing in a single image the motion occurring in a video: the last frame of the video sequence is augmented with speedlines, which convey the motion occurred in the previous frames by tracing the envelopes of the contours of moving objects. We propose an efficient low-level technique for constructing such speedlines, which is robust to temporally or spatially local segmentation errors, and does not require object tracking. Experimental results confirm that the approach is valid and can be applied to many different scenarios.

The main contribution of this paper over the state of the art is twofold.

- Our algorithm is very general: due to its low level of abstraction, it effortlessly deals with long and complex trajectories, multiple moving objects, articulated motion, and low resolution and quality of the input video, which existing algorithms can not handle.
- We explore the use of speedlines for representing long, complex motions, and introduce speedlines as an effective tool for video surveillance applications.

We are currently implementing the algorithm on a DSP-equipped smart camera, and studying a number of possible optimizations for its efficient implementation. Moreover, we are also experimenting other uses, for surveillance and entertainment applications. Finally, we are experimenting with improved visualization techniques, in order to better convey motion without the need of scene understanding.

References

1. McCloud, S.: Understanding Comics: The Invisible Art. Kitchen Sink Press (1993)
2. Strothotte, T., Preim, B., Raab, A., Schumann, J., Forsey, D.R.: How to render frames and influence people. In: Proc. of Eurographics (1994)
3. Kawagishi, Y., Hatsuyama, K., Kondo, K.: Cartoon blur: Non-photorealistic motion blur. In: Computer Graphics International, p. 276 (2003)
4. Kawabe, T., Miura, K.: Representation of dynamic events triggered by motion lines and static human postures. Experimental Brain Research 175 (2006)
5. Masuch, M., Schlechtweg, S., Schulz, R.: Depicting motion in motionless pictures. In: Proc. SIGGRAPH (1999)
6. Chen, A.: Non-photorealistic rendering of dynamic motion (1999), http://gvu.cc.gatech.edu/animation/Areas/nonphotorealistic/results.html
7. Lake, A., Marshall, C., Harris, M., Blackstein, M.: Rendering techniques for scalable real-time 3d animation. In: Proc. of symposium on NPR animation and rendering (2000)
8. Kim, B., Essa, I.: Video-based nonphotorealistic and expressive illustration of motion. In: Proc. of Computer Graphics International (2005)
9. Hwang, W.-I., Lee, P.-J., Chun, B.-K., Ryu, D.-S., Cho, H.-G.: Cinema comics: Cartoon generation from video. In: Proc. of GRAPP (2006)
10. Markovic, D., Gelautz, M.: Comics-like motion depiction from stereo. In: Proceedings of WSCG (2006)

11. Collomosse, J.P., Hall, P.M.: Video motion analysis for the synthesis of dynamic cues and futurist art. Graphical Models (2006)
12. Caglioti, V., Giusti, A.: On the apparent transparency of a motion blurred object. International Journal of Computer Vision (2008)
13. Mahadevan, V., Vasconcelos, N.: Background subtraction in highly dynamic scenes. In: Proc. of CVPR (2008)
14. Gong, Y., Liu, X.: Generating optimal video summaries. In: Proc. of ICME, ICME (2000)
15. Zhao, Z.P., Elgammal, A.M.: Information theoretic key frame selection for action recognition. In: Proc. of BMVC (2008)
16. PETS: Performance evaluation of tracking and surveillance (2008)
17. Ragheb, H., Velastin, S., Remagnino, P., Ellis, T.: Vihasi: Virtual human action silhouette data for the performance evaluation of silhouette-based action recognition methods. In: Proc. of Workshop on Activity Monitoring by Multi-Camera Surveillance Systems (2008)
18. Caglioti, V., Giusti, A., Riva, A., Uberti, M.: Supplementary material (2009), http://www.leet.it/home/giusti/speedlines

Image Compression Based on
Visual Saliency at Individual Scales

Stella X. Yu[1] and Dimitri A. Lisin[2]

[1] Computer Science Department
Boston College, Chestnut Hill, MA 02467, USA
[2] VideoIQ, Inc. Bedford, MA 01730, USA

Abstract. The goal of lossy image compression ought to be reducing entropy while preserving the perceptual quality of the image. Using gaze-tracked change detection experiments, we discover that human vision attends to one scale at a time. This evidence suggests that saliency should be treated on a per-scale basis, rather than aggregated into a single 2D map over all the scales. We develop a compression algorithm which adaptively reduces the entropy of the image according to its saliency map within each scale, using the Laplacian pyramid as both the multiscale decomposition and the saliency measure of the image. We finally return to psychophysics to evaluate our results. Surprisingly, images compressed using our method are sometimes judged to be better than the originals.

1   Introduction

Typical lossy compression methods treat an image as a 2D signal, and attempt to approximate it minimizing the difference (e.g. L_2 norm) from the original. By linearly transforming an image using an orthogonal basis (e.g. Haar wavelets), solutions of minimal difference can be computed by zero-ing out small coefficients [1,2]. As there are many different zero-ing schemes corresponding to the same total difference, various thresholding techniques (e.g. wavelet shrinkage) that aim to reduce visual artifacts have been developed [3,4,5,6].

However, an image is not just any 2D signal. It is viewed by human observers. Lossy image compression should reduce entropy while preserving the perceptual quality of the image. Signal-based methods fall short of both requirements: zero-ing out small coefficients aims at reducing pure signal differences instead of entropy, and reducing signal difference does not guarantee visual quality.

Our work concerns the use of visual saliency to guide compression. This topic has been explored on multiple fronts, such as modifying the JPEG format [7], compressing salient and non-salient regions with separate algorithms [8], and applying saliency-based non-uniform compression to video [9]. Most saliency models yield a location map based on low-level cues [10], or scene context and visual task [11], treating scale like any other primary feature such as orientation, color, and motion. Computer vision algorithms often concatenate measurements at multiple scales into one feature vector without questioning its validity.

G. Bebis et al. (Eds.): ISVC 2009, Part I, LNCS 5875, pp. 157–166, 2009.

We first conduct an eye tracking experiment, and discover that human vision often attends to one scale at a time, while neglecting others (Sec. 2). We then develop a saliency-based compression scheme in which the entropy is reduced at each scale separately, using that scale's saliency map (Sec. 3). We finally validate our approach in another psychophysical experiment where human subjects render their judgement of visual quality between pairs of briefly presented images (Sec. 4). Our compression results not only look better than the signal-based results, but, surprisingly, in some cases even better than the originals! One explanation is that our saliency measure captures features most noticeable in a single glance, while our entropy reduction aggressively suppresses the often distracting background, enhancing the subjective experience of important visual details.

2   Scale and Human Visual Attention

Our inspiration comes from studying *change blindness* [12]: When two images are presented with an interruption of a blank, the blank wipes out the retinal stimulation usually available during natural viewing, making the originally trivial change detection extremely difficult, even with repeated presentations. Using an eye tracker, we discover 3 scenarios between looking and seeing (Fig. 1):

1) Most detections are made after the gaze has scrutinized the change area.
2) If the gaze has never landed upon the area, seeing is unlikely.
3) Sometimes the gaze repeatedly visits the change area, however, the subject still does not see the change.

Our gaze data reveals two further scenarios for the last case of no seeing with active looking. **1)** For 80% of visits to the area of change, the gaze did not stay long enough to witness the full change cycle. As the retina is not receiving sufficient information regarding the change, blindness naturally results. **2)** For the rest 20% of visits which involve 9 of 12 stimuli and 10 out of 11 subjects, the gaze stayed in the area more than a full cycle, yet the change still escaped detection. Those are true instances of looking without seeing [13].

1. looking and seeing   2. no looking, no seeing   3. active looking, no seeing

Fig. 1. Relationship between looking and seeing. A four-image sequence, I, B, J, B, is repeatedly presented for 250ms each. I and J denote images a major difference (the presence of a person in the white circle in this case), and B a blank. Shown here are 3 subjects' gaze density plots as they search for the difference. Red hotspots indicate the locations that are most looked at. Only Subject 1 detected the change.

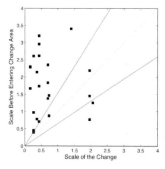 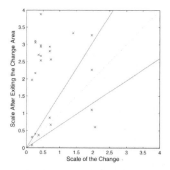

Fig. 2. The scale difference between fixations is more than 1.5 folds (solid lines) in 88% cases. The horizontal axis is for the size of change. The vertical axis is the size of the area examined in the fixation before entering (■) or after exiting (×) the change area. While the size here is determined based on manually outlined focal regions, similar results are obtained with synthetic stimuli varying only in the size dimension.

We examine the retinal inputs fixation-by-fixation. In most true instances of looking without seeing, the areas visited by the eye right before or after the change area tend to have features of a different scale from the change (Fig. 2). If at time $t-1$ the subject is looking at a coarse-scale structure, he is likely to be oblivious to the change in a fine-scale structure at time t, and he tends to continue looking at a coarse-scale structure at time $t+1$. In other words, when the visual system attends to one scale, other scales seem to be neglected.

3   Saliency and Compression

Our experiment suggests that human vision attends to one scale at a time, rather than processing all scales at once. This implies that saliency should be defined on a per-scale basis, rather than aggregated over all scales into a single 2D saliency map, as it is typically done [10]. We use the Laplacian pyramid [1] to define a multi-scale saliency map, and we use range filters [14] to reduce the entropy of each scale, applying more range compression to less salient features (Fig. 3).

We adopt the Laplacian pyramid as both the multiscale signal decomposition and the saliency measure of the image, since the Laplacian image is the difference of images at adjacent scales and corresponds to center-surround filtering responses which indicate low-level feature saliency [10].

Step 1: Given image I and number of scales n, build Gaussian pyramid G and Laplacian pyramid L, where $\downarrow$ = downsampling, $\uparrow$ = upsampling

$$G_{s+1} = \downarrow (G_s * \text{Gaussian}), \qquad G_1 = I, \qquad s = 1 \to n \qquad (1)$$
$$L_s = G_s - \uparrow L_{s+1}, \qquad L_{n+1} = G_{n+1}, \qquad s = n \to 1 \qquad (2)$$

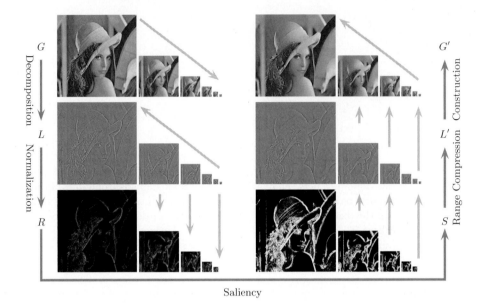

Fig. 3. Our image compression uses the Laplacian pyramid as both a signal representation and a saliency measure at multiple scales

To turn the Laplacian responses into meaningful saliency measures for compression, we first normalize it $(L \to R)$ and then rectify it $(R \to S)$ using sigmoid transform with soft threshold m and scale factor α (Fig. 4). α controls saliency sharpness and is user-specified. We then use binary search to find the optimal m that satisfies the total saliency percentile p: If $S = 1$, $p = 1$, every pixel has to be maximally salient, whereas if $p = 0.25$, about 25% of the pixels are salient.

Step 2: Given percentile p and scaling factor α, compute saliency S from L using a sigmoid with threshold m:

$$S_s = \left(1 + e^{-\frac{R_s - m_s}{\alpha}}\right)^{-1}, \, s = 1 : n \tag{3}$$

$$m_s = \arg\left\{\sum_i S_s(i; m_s) = p \cdot \sum_i 1\right\}, R_s = \frac{|L_s|}{\max(|L_s|)} \tag{4}$$

We modify Laplacian L by range filtering with saliency S. Range compression replaces pixel i's value with the weighted average of neighbours j's, larger weights for pixels of similar values [14]. Formulating the weights W as a Gaussian of value differences, we factor saliency S into covariance Θ: High saliency leads to low Θ, hence high sensitivity to value differences, and value distinction better preserved. The maximal amount of compression is controlled by the range of Laplacian values at that particular scale (Eqn. 7).

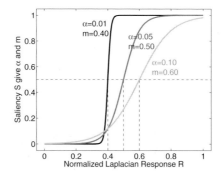

Fig. 4. Sigmoid function rectifies the Laplacian to become a saliency measure. m is a soft threshold where saliency becomes 0.5. α controls the range of intermediate saliency values. Low α forces a binary (0 or 1) saliency map.

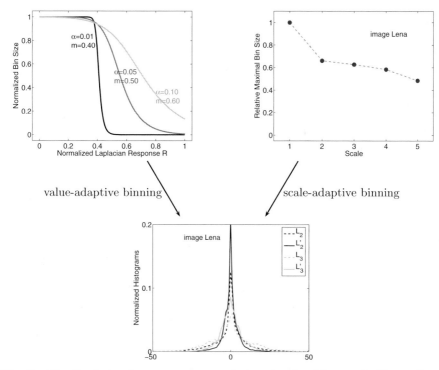

Fig. 5. The Laplacian is range compressed as a signal subject to itself as saliency measure. The reduction in entropy is achieved by implicit adaptive binning in the histograms. We can have some idea of the bin size by examining the standard deviation of W: $\Theta^{\frac{1}{2}} = \sqrt{1-S} \cdot (\max(L) - \min(L))/\beta$. The first factor $\sqrt{1-S}$ makes the bin dependent on the value, whereas the second $(\max(L) - \min(L))/\beta$ makes it dependent on the scale, for the range of L naturally decreases over scale.

Step 3: Given S, neighbourhood radius r and range sensitivity factor β, generate a new Laplacian pyramid L' by spatially-variant range filtering of L:

$$L'_s(i) = \frac{\sum_{j \in N(i,r)} L_s(j) \cdot W_s(i,j)}{\sum_{j \in N(i,r)} W_s(i,j)}, L'_{n+1} = L_{n+1}, s = 1 : n \tag{5}$$

$$W_s(i,j) = e^{-\frac{(L_s(i) - L_s(j))^2}{2\Theta_s(i)}}, \tag{6}$$

$$\Theta_s(i) = (1 - S_s(i)) \cdot \left(\frac{\max(L_s) - \min(L_s)}{\beta}\right)^2 \tag{7}$$

The nonlinear filtering coerces the Laplacian values towards fewer centers (Fig. 5). It can be understood as scale- and value-adaptive binning: As the scale goes up, the bin gets smaller; as the value increases, the saliency increases, and the bin also gets smaller. As the value distribution becomes peakier, the entropy is reduced and compression results. The common practice of zero-ing out small values in Laplacians or wavelets only expands the bin at 0 while preserving signal fidelity, whereas our saliency regulated local range compression adaptively expands the bin throughout the levels while preserving perceptual fidelity.

Finally, we synthesize a new image by collapsing the compressed Laplacian pyramid (Fig. 3 G', L'). Note that L and L' look indistinguishable, whereas nonsalient details in G are suppressed in G'.

Step 4: Construct the compressed image J by collapsing the new Laplacian L':

$$G'_s = L'_s + \uparrow G'_{s+1}, G'_{n+1} = L'_{n+1}, s = n \to 1; J = G'_1 \tag{8}$$

4   Evaluation

Lossy image compression sacrifices quality for saving bits. Given infinite time to scrutinize, one is bound to perceive the loss of details in a compressed image. However, in natural viewing, instead of scanning the entire image evenly, we only dash our eyes to a few salient regions. Having developed an image compression method based on human vision, we now return to it to evaluate our results.

We carry out two-way forced choice visual quality comparison experiments using 12 standard test images (Fig. 8). Using our method, we generate 16 results per image with $\alpha \in \{0.01, 0.1\}$, $p \in \{0.25, 0.5\}$, $r \in \{3, 6\}$ $\beta \in \{5, 10\}$. For each image, we choose 3 compression levels that correspond to minimal, mean and maximal JPEG file sizes. For each level, we find a signal-compressed version of the same JPEG file size but reconstructed from zero-ing out sub-threshold values in the Laplacian pyramid. The threshold is found by binary search.

We first compare our results with signal-based results (Fig. 6). Each comparison trial starts with the subject fixating the center of a blank screen. Image 1 is presented for 1.2s, followed by a gray random texture for 0.5s, image 2 for

signal-based compression:

our perception-based compression:

visual quality comparison:

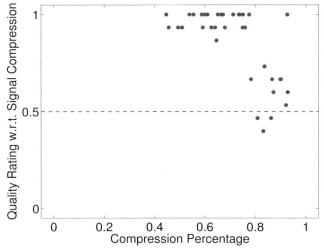

Fig. 6. Comparison of signal-based compression (row 1) and our perception-based compression (row 2). Row 3 shows a plot of quality ratings of our results for different compression ratios. The quality rating is the fraction of subjects who judged our results to be better than those produced by the signal-based algorithm. Each dot in the plot represents a quality rating of the perception-based compression of a particular image resulting in the compression ratio given by the horizontal axis. Our results are better in general, especially with more compression.

wavelet compression:

our compression:

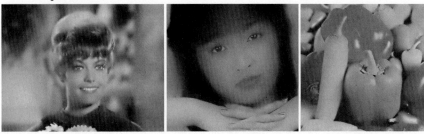

Fig. 7. Our results (row 2) are better than compression by Daubechies 9-tap/7-tap wavelet with level-dependent thresholding (row 1) for the same JPEG file size

1.2s, and random texture again till a keypress indicating which one looks better. The occurrence order within each pair is randomized and balanced over 15 naive subjects, resulting in 30 trials per pair of images. Our quality rating is determined by the percentage of favorable votes for our method: 0.5 indicates that the images from two methods have the same visual quality statistically, whereas a value greater(less) than 0.5 means our results are better(worse). The visual quality of our results is better overall, especially with heavier compression.

We have also computed wavelet compression results with various settings: Haar vs Daubechies 9-tap / 7-tap wavelet, global- vs. level-dependent thresholding via Birge-Massart strategy. They have their own characteristic patterns in quality loss over heavy compression. Our compressed images degrade more gracefully than those as well (Fig. 7).

Finally, we compare our results at the best quality level to the original images (Fig. 8). 1) At a short exposure, our results are entirely indistinguishable from the original; 2) At a medium exposure, ours are better than the original! The enhancement is particularly strong for face images. 3) At a long exposure, our results become slightly worse. Such exposure-dependence in fact supports the validity of our saliency model: Our method captures visual features of the first-order significance at the cost of losing details of the second-order significance.

At low levels of compression, our method produces an air-brush effect which emphasizes strong straight edges and evens out weak and curly edges, lending more clarity to a face image while destroying the natural texture in a pepper image (Fig. 8 bottom). At higher levels of compression (Fig. 7), our method

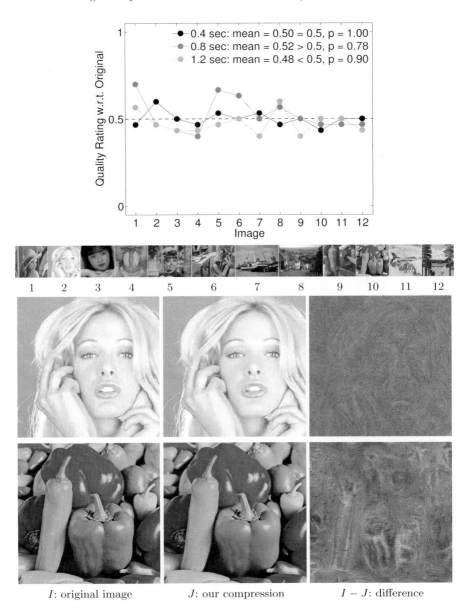

Fig. 8. Comparison between the originals and our compressed results (**row 1**) on 12 test images (**row 2**) at the exposure times of 0.4, 0.8, and 1.2 seconds, with p-values from two-, right-, and left-tailed one-sample t-tests between the means and the equal quality level 0.5. Our results are equally good at the short exposure (too short for anyone to notice any differences), better at the medium exposure, and worse at the long exposure (long enough to notice the distinction in richness). The enhancement at the medium exposure is most positive in image 2 (**row 3**), and most negative in image 10 (**row 4**), where the air-brush effect makes the facial characteristics clearer and the pepper textures disturbingly fake.

creates the soft focus style used by photographer David Hamilton, which blurs the image while retaining sharp edges.

Summary. Our human vision study suggests that a saliency model must treat each scale separately, and compression must preserve salient features *within each scale*. We use the Laplacian pyramid as both signal representation and saliency measures at individual scales. Range compression modulated by saliency not only results in entropy reduction, but also preserves perceptual fidelity. This can be viewed as value- and scale-adaptive binning of the distributions, an elegant alternative to various thresholding strategies used in wavelet compression. Our validation with human viewers indicates that our algorithm not only preserves visual quality better than standard methods, but can even enhance it.

Acknowledgements. This research is funded by NSF CAREER IIS-0644204 and a Clare Boothe Luce Professorship to Stella X. Yu.

References

1. Burt, P., Adelson, E.H.: The Laplacian pyramid as a compact image code. IEEE Trans. Communication 31 (1983)
2. Mallat, S.: A theory for multiresolution signal decomposition: The wavelet representation. PAMI (1989)
3. Simoncelli, E., Adelson, E.: Noise removal via Bayesian wavelet coring. In: IEEE International Conference on Image Processing, vol. 1 (1996)
4. Donoho, D.L.: De-noising by soft-thresholding. IEEE Trans. Information Theory 4, 613–627 (1995)
5. Chambolle, A., DeVore, R.A., Lee, N., Lucier, B.J.: Nonlinear wavelet image processing: variational problems, compression, and noise removal through wavelet shrinkage. IEEE Trans. Image Processing 7, 319–335 (1998)
6. DeVore, R., Jawerth, B., Lucier, B.: Image compression through wavelet transform coding. IEEE Transactions on Information Theory 32 (1992)
7. Golner, M.A., Mikhael, W.B., Krishnang, V.: Modified jpeg image compression with region-dependent quantization. Circuits, Systems, and Signal Processing 21, 163–180 (2002)
8. Lee, S.-H., Shin, J.-K., Lee, M.: Non-uniform image compression using biologically motivated saliency map model. In: Intelligent Sensors, Sensor Networks and Information Processing Conference (2004)
9. Itti, L.: Automatic foveation for video compression using a neurobiological model of visual attention. IEEE Trans. Image Processing 13, 669–673 (2003)
10. Itti, L., Koch, C.: Computational modelling of visual attention. Nature Neuroscience, 194–203 (2001)
11. Torralba, A.: Contextual influences on saliency. In: Itti, L., Rees, G., Tsotsos, J. (eds.) Neurobiology of Attention, pp. 586–593. Academic Press, London (2004)
12. Rensink, R.A.: Change detection. Annual Review of Psychology 53, 4245–4277 (2002)
13. O'Regan, J.K., Deubel, H., Clark, J.J., Rensink, R.A.: Picture changes during blinks: looking without seeing and seeing without looking. Visual Cognition 7, 191–211 (2000)
14. Tomasi, C., Manduchi, R.: Bilateral filtering for gray and color images. In: International Conference on Computer Vision (1998)

Fast Occlusion Sweeping

Mayank Singh[1], Cem Yuksel[1], and Donald House[2]

[1] Texas A&M University
[2] Clemson University
{mayank,cem}@cs.tamu.edu,
dhouse@cs.clemson.edu

Abstract. While realistic illumination significantly improves the visual quality and perception of rendered images, it is often very expensive to compute. In this paper, we propose a new algorithm for embedding a global ambient occlusion computation within the fast sweeping algorithm while determining isosurfaces. With this method we can approximate ambient occlusion for rendering volumetric data with minimal additional cost over fast sweeping. We compare visualizations rendered with our algorithm to visualizations computed with only local shading, and with a ambient occlusion calculation using Monte Carlo sampling method. We also show how this method can be used for approximating low frequency shadows and subsurface scattering.

Realistic illumination techniques used in digitally synthesized images are known to greatly enhance the perception of shape. This is as true for renderings of volume data as it is for geometric models. For example, Qiu et al. [1] used full global illumination techniques to improve visualizations of volumetric data, and Stewart [2] shows how computation of local ambient occlusion enhances the perception of grooves in a brain CT scanned dataset. Tarini et al. [3] observed that perception of depth for large molecules was significantly improved with the use of ambient occlusion as compared to standard direct shading methods even when coupled with other techniques such as depth cueing and shadowing. Recently, a carefully designed experimental study by Weigle and Banks [4] definitively demonstrated that physically-based global illumination is a powerful adjunct to perspective projection in aiding human subjects to understand spatial relationships in a complex volume rendered scene. Despite the strong evidence for its efficacy in conveying spatial information in visualization, the use of global illumination is rare in practical visualization systems. This is most likely due to the high overhead of existing global illumination rendering algorithms.

In this paper, we provide a new solution for ambient occlusion computation that is significantly faster than existing techniques. The method integrates well with a volumetric ray marching algorithm implemented on the GPU. While not a full global illumination solution, ambient occlusion provides a more realistic illumination model than does local illumination, and permits the use of realistic light sources, like skylights. For accelerating our ray marching algorithm, we build a volumetric signed distance field using the *fast sweeping method*, and we embed our ambient occlusion approximation directly into the sweeping algorithm. Thus, our algorithm can produce an ambient occlusion estimate with only a minor computational overhead. We are also able to use

G. Bebis et al. (Eds.): ISVC 2009, Part I, LNCS 5875, pp. 167–178, 2009.

our approach to approximate low-frequency shadows due to direct illumination from certain angles, and to approximate subsurface scattering effects.

1   Background

Since our method combines an ambient occlusion computation with the fast sweeping method, in this section, we briefly overview the fast sweeping method, and ambient occlusion, and review previous work on computing ambient occlusion.

1.1   Fast Sweeping

The aim of fast sweeping is to build a volumetric signed distance field from volume data, for a specified isolevel. This defines a surface in 3D as the zero distance level set of the field, with all cells away from this surface containing the minimum distance from that cell to the surface. This distance field is commonly used as an aid in speeding the process of isosurface rendering using methods such as ray-marching. The fast sweeping method was introduced by Zhao [5] as a linear time alternative to the *fast marching* method [6] for computing a signed distance field.

The fast sweeping method can be divided into two distinct steps, as indicated in Fig. 1. First, the distance values of all grid vertices are initialized to positive infinity. Then, all vertices that participate in grid edges with exact zero-crossings of the isolevel (grey curve in the figure) are updated to their interpolated distance from the isolevel (open vertices in the figure).

The second step conceptually consists of multiple diagonal sweeps that update the distance values from the distance values of neighboring vertices. In Fig. 1, the filled vertices show the order of vertices visited by one diagonal sweep, starting from bottom-left towards the top-right (the scanlines are numbered according to sweeping order).

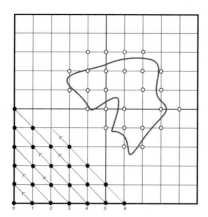

Fig. 1. The fast sweeping method for building a distance field to the zero level set (grey curve) in 2D. The open vertices indicate voxels initialized by direct distance computation. The closed vertices show the sweeping order of voxels for one diagonal sweep.

Table 1. Sweeping Directions and Neighboring Positions

s	Sweep Direction	$\mathbf{P}_s^x$	$\mathbf{P}_s^y$	$\mathbf{P}_s^z$
1	(1, 1, 1)	$\mathbf{P} - (h,0,0)$	$\mathbf{P} - (0,h,0)$	$\mathbf{P} - (0,0,h)$
2	(-1, 1, 1)	$\mathbf{P} + (h,0,0)$	$\mathbf{P} - (0,h,0)$	$\mathbf{P} - (0,0,h)$
3	(1,-1, 1)	$\mathbf{P} - (h,0,0)$	$\mathbf{P} + (0,h,0)$	$\mathbf{P} - (0,0,h)$
4	(-1,-1, 1)	$\mathbf{P} + (h,0,0)$	$\mathbf{P} + (0,h,0)$	$\mathbf{P} - (0,0,h)$
5	(1, 1,-1)	$\mathbf{P} - (h,0,0)$	$\mathbf{P} - (0,h,0)$	$\mathbf{P} + (0,0,h)$
6	(-1, 1,-1)	$\mathbf{P} + (h,0,0)$	$\mathbf{P} - (0,h,0)$	$\mathbf{P} + (0,0,h)$
7	(1,-1,-1)	$\mathbf{P} - (h,0,0)$	$\mathbf{P} + (0,h,0)$	$\mathbf{P} + (0,0,h)$
8	(-1,-1,-1)	$\mathbf{P} + (h,0,0)$	$\mathbf{P} + (0,h,0)$	$\mathbf{P} + (0,0,h)$

This would be followed by a sweep back from the upper-right to the lower-left, and then sweeps in both directions along the other diagonal. 3D fast sweeping uses eight similar diagonal sweeps. In actual implementation, it is usual to arrange the algorithm so that iterations are across the columnar directions in the volume, instead of along diagonals.

During a sweeping operation in 3D, we consider the current distance field values $D(\mathbf{P})$ at three neighboring vertices of each vertex $\mathbf{P}$ in the grid. These three neighboring vertices are the ones that are visited before $\mathbf{P}$ depending on the current sweeping direction. Table 1 shows the sweeping directions and formulations of neighboring vertices $\mathbf{P}_s^x$, $\mathbf{P}_s^y$, and $\mathbf{P}_s^z$ for each sweep s, through a cubic grid whose spacing is h.

For each vertex, the distance field $D(\mathbf{P})$ is updated as the minimum of its current value and a new distance estimate $\hat{d}$, calculated from the existing values $D(\mathbf{P}_s^x)$, $D(\mathbf{P}_s^y)$, and $D(\mathbf{P}_s^z)$ of the indicated three neighbors. To calculate the new estimate, these three distances are sorted into ascending order, and renamed d_1, d_2, and d_3. Then

$$\hat{d} = \begin{cases} s_1 = d_1 + h & \text{if } |s_1| < d_2, \\ s_2 = \frac{d_1 + d_2 + \sqrt{2h^2 - (d_1 - d_2)^2}}{2} & \text{if } |s_2| < d_3, \\ s_3 = \frac{d_1 + d_2 + d_3 + \sqrt{3h^2 - (d_1 - d_2)^2 - (d_1 - d_3)^2 - (d_2 - d_3)^2}}{3} & \text{otherwise.} \end{cases} \tag{1}$$

1.2   Ambient Occlusion

Ambient occlusion was first introduced by Zhukov et al. [7] as a better approximation to ambient light than the constant ambient term used in many local illumination models. Ambient occlusion on a point $\mathbf{P}$ with surface normal $\mathbf{N}$ is defined as

$$A(\mathbf{P},\mathbf{N}) = \int_\Omega V(\mathbf{P},\omega)\, g(\mathbf{N},\omega)\, d\omega\,, \tag{2}$$

where Ω is the unit sphere, $V(\mathbf{P},\omega)$ is the *visibility function* accounting for occlusion and scattering along direction ω, and $g(\mathbf{N},\omega) = \max(\mathbf{N}\cdot\omega, 0)$ is the *geometry term*. In this formulation, ambient occlusion corresponds to illumination due to an isotropic skylight including global shadows cast by surrounding objects (i.e. occlusion). It is the computation of the visibility function for all directions that makes the ambient occlusion computationally intensive.

The typical way of solving Equation 2 is through Monte Carlo integration using raycasting for a binary decision of visibility in a chosen direction (if the ray hits any object visibility is zero in that direction, otherwise it is one). This approach requires many samples to reduce noise and it is very slow, especially when the integration is performed for all pixels of an image. Therefore, for real-time visualizations of static objects, ambient occlusion is often precomputed.

A popular alternative to Monte Carlo integration for ambient occlusion is precomputing shadows from multiple directional light sources (such as in [8,9]). While this method can be faster, depending on the number of light directions and the complexity of the scene, it tends to produce undesired aliasing artifacts instead of noise.

For ambient occlusion in dynamic scenes, researchers have proposed different ways of keeping volumetric ambient occlusion fields around moving objects [10,11,12,13]. However, the initial computation of the ambient occlusion values is handled through a long precomputation step, similar to previous approaches.

For the purpose of ambient occlusion, Bunnell's [14] method represents each vertex in the scene as a planer disk called a *surface element*. The ambient occlusion computation is performed for every vertex by considering the contributions of all surface elements. Since the complexity of this algorithm is $O(n^2)$, where n is the number of surface elements, it is not suitable for scenes with high resolution surfaces.

Shanmugam and Arikan [15] used nearby pixels for computing local occlusion information. They precompute an expensive spherical representation of the model for distant occlusion information. Ritschel et al. [16] also used *only* local occlusion information derived from the nearby pixels in the image space. Their method does not account for global occlusion. Salama [17] proposed a multipass algorithm that precomputes a set of random directions and runs a GPU-based Monte Carlo raycasting.

2   Occlusion Sweeping

We compute an approximate ambient occlusion solution using eight sweeps of the fast sweeping method. During each diagonal sweep s, we compute the ambient contribution for each voxel from the corresponding octant of the sphere Ω_s. The final ambient occlusion is the sum of all of these eight octants, given by

$$A_s(\mathbf{P}, \mathbf{N}) = \int_{\Omega_s} V(\mathbf{P}, \omega) g(\mathbf{N}, \omega) d\omega . \qquad (3)$$

Notice that the only difference between Equation 2 and Equation 3 is the integration domain.

To simplify the computation of this equation, we introduce three theoretical simplifications. First, we approximate the visibility function within an octant by a constant value $V_s(\mathbf{P})$. This implies that all of the light rays arriving at point $\mathbf{P}$ from the same octant are occluded the same amount. Using this simplification, we can take the visibility function outside of the integral and approximate ambient occlusion for the octant as

$$A_s(\mathbf{P}, \mathbf{N}) \approx V_s(\mathbf{P}) G_s(\mathbf{N}) , \qquad (4)$$

where $G_s(\mathbf{N})$ is the integrated geometry term

$$G_s(\mathbf{N}) = \int_{\Omega_s} g(\mathbf{N}, \omega) d\omega . \qquad (5)$$

In this formulation $G_s(\mathbf{N})$ depends solely on the surface normal and can be easily precomputed. Therefore, all we need to compute to find the ambient occlusion is the visibility approximation of the octant for each voxel.

Our second simplification provides a fast estimation to $V_s(\mathbf{P})$ of the voxel at P by approximating it as a combination of the visibility at three neighboring voxels. These voxels, P_s^x, P_s^y, and P_s^z, are the three of the six neighbors of the voxel at P given in Table 1. We multiply the visibility values at these neighboring voxels by the transmittance $\tau(\mathbf{P})$ through these voxels to account for the light that is occluded within these voxels. As a result, our visibility estimation becomes

$$V_s(\mathbf{P}) \approx \frac{1}{3}\left[V_s(\mathbf{P}_s^x)\,\tau(\mathbf{P}_s^x) + V_s(\mathbf{P}_s^y)\,\tau(\mathbf{P}_s^y) + V_s(\mathbf{P}_s^z)\,\tau(\mathbf{P}_s^z)\right] . \tag{6}$$

Finally, we approximate the value of the transmittance function using the distance to the level set surface $D(\mathbf{P})$. We define the transmittance function to be

$$\tau(\mathbf{P}) = \begin{cases} 0 & \text{if } D(\mathbf{P}) \leq -\frac{h}{2} \\ \frac{D(\mathbf{P})}{h} + \frac{1}{2} & \text{if } -\frac{h}{2} < D(\mathbf{P}) < \frac{h}{2} \\ 1 & \text{if } D(\mathbf{P}) \geq \frac{h}{2} \end{cases} . \tag{7}$$

To summarize, in our occlusion sweeping method, we visit each voxel eight times in eight diagonal sweeps. Each sweep computes the visibility function for the octant that is on the opposite side of the sweeping direction, using Equation 6. Table 1 gives the ordering of sweeping directions and the neighboring voxels used in the calculations. The order in which the voxels are visited in a particular sweep is chosen to assure that the visibility functions at the corresponding neighboring voxels of $\mathbf{P}$ are computed before $\mathbf{P}$. This whole computation can be easily implemented as a part of the fast sweeping method for building a signed distance field. Notice that even though we use the distance field values in Equation 7, the distance values within the range $[-\frac{h}{2}, \frac{h}{2}]$ are set during the initialization step. Therefore, we can use the distance field to evaluate $\tau(\mathbf{P})$ while building the distance field itself.

3   Implementation and Results

For demonstration purposes, we implemented our method within a single pass, GPU-based, ray marcher. The volumetric distance field and the ambient occlusion terms are precomputed offline, and stored as 3D textures on the GPU. A single ray is then marched for every pixel on the screen. If the ray hits a surface, hit position $\mathbf{P}$ is returned along with the gradient of the distance field, which is used as the surface normal $\mathbf{N}$. Since $\mathbf{N}$ is computed on the GPU per pixel (as opposed to per voxel of the dataset), Equation 4 needs to be evaluated separately for each octant at runtime. Therefore, we store the eight $V_s(\mathbf{P})$ fields as color channels in two 3D RGBA textures.

Since the eight integrated geometry terms $G_s(\mathbf{N})$ depend only on the surface normal, they can be precomputed once and stored in the color channels of two 2D textures (Fig. 2). The x and y coordinates of $\mathbf{N}$ then serve to determine the texture coordinate for lookup. These are shown in Fig. 2.

Fig. 2. Precomputed geometry term $G(\mathbf{N})$, stored as eight channels in two 2D RGBA textures

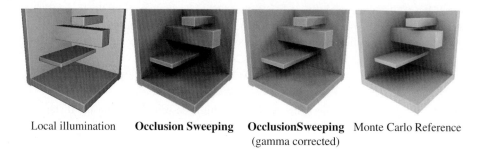

Local illumination   **Occlusion Sweeping**   **OcclusionSweeping**   Monte Carlo Reference
                                                 (gamma corrected)

Fig. 3. A simple $100 \times 100 \times 100$ dataset for demonstrating the validity of our algorithm, where we compare our occlusion sweeping results to local illumination and Monte Carlo reference

Since ambient occlusion does not correctly account for inter-reflected light, as in a full global illumination calculation, images can be overly dark. To overcome the extra darkening, we applied gamma correction directly to our ambient occlusion estimation as a post-processing operation, computed as $A \leftarrow A(\mathbf{P},\mathbf{N})^{1/\gamma}$. We use $\gamma = 2$ for all gamma corrected images in this paper.

To test the validity of our ambient occlusion estimation, we prepared the simple $100 \times 100 \times 100$ voxel dataset shown in Fig. 3, and rendered it using local illumination only, our occlusion sweeping method, and with a Monte Carlo ray tracer. The precomputation of ambient sweeping took only a fraction of a second, while the Monte Carlo reference image was rendered in several minutes. Both the occlusion sweeping and Monte Carlo images are arguably much more effective than the local illumination image in depicting the depth relationships of the various objects in the scene. It can also be seen that occlusion sweeping provides a good estimate for ambient occlusion, in that the dark areas and shadows correspond well to the those in the Monte Carlo reference image.

While occlusion sweeping greatly improves the illumination as compared to local illumination, there are two inaccuracies introduced by our simplifications. Ambient sweeping produces soft diagonal shadows, which are especially visible on the flat walls of the images in Fig. 3. Secondly, ambient sweeping produces a glow effect along sharp edges. While this can be interpreted as a useful visualization feature (as argued by Tarini et al. [18]), it actually originates from our transmittance function $\tau(\mathbf{P})$ in Equation 7. Since $\tau(\mathbf{P})$ is directly computed from the distance field $D(\mathbf{P})$ and it does not depend on the octant, neighboring voxels on a flat surface end up partially occluding each other. While this effect darkens flat surfaces, sharp features are not affected as much and appear brighter.

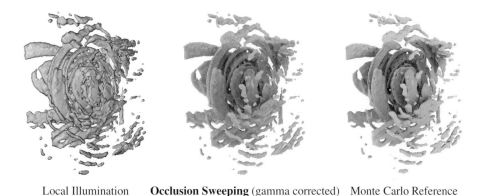

Local Illumination   **Occlusion Sweeping** (gamma corrected)   Monte Carlo Reference

Fig. 4. Comparison of occlusion sweeping to a Monte Carlo reference

Fig. 4, shows a comparison of local illumination to our ambient occlusion solution for a vorticity isosurface within a lightning cloud simulation dataset. As can be seen from this figure, ambient occlusion greatly improves depth perception and the global shape of this complicated surface becomes easier to understand. In Fig. 4, we provide a comparison of the gamma corrected occlusion sweeping to a Monte Carlo reference image for the same dataset. Notice that on a complicated surface like this one, the di-agonal artifacts of ambient sweeping are not visible, while the darkening of flat regions emphasizes sharp local features.

Fig. 5 shows another dataset that contains capillaries from the mouse brain (cerebel-lum). Notice that ambient occlusion with our method significantly improves the percep-tion of the thread-like 3D vascular structure, while using only local illumination makes the image look flat and visually confusing.

In our experiments, we noticed that local features can be further emphasized by dropping the geometry term $G_s(\mathbf{P})$ from Equation 4. In this case the visibility fields of

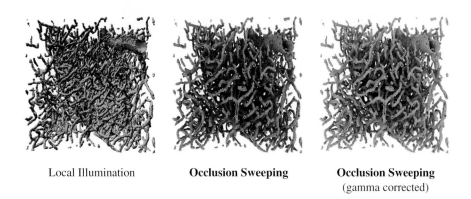

Local Illumination   **Occlusion Sweeping**   **Occlusion Sweeping**
                                                        (gamma corrected)

Fig. 5. Vascular data of capillaries from the mouse brain (cerebellum) with different illumination methods. The dataset is about 0.5mm across and the capillaries occupy less than six percent of the overall tissue volume. The data is sampled on a $256 \times 256 \times 256$ grid.

all eight octants can be combined to form a scalar ambient occlusion field, significantly reducing the memory requirements. Fig. 6 shows such an example where it can be seen that sharp features appear brighter when the geometry term is dropped.

With our occlusion sweeping technique, we can also produce more natural illumination conditions than those produced by an isotropic skylight. For example, in Fig. 7 the surface is illuminated through only top four of the eight octants, emulating the lighting conditions of a cloudy day.

Table 2 provides timing information for our ambient occlusion precomputation, for five different data sets of varying sizes. The timings are computed on an Intel Core2 (2.66 GHz) machine with 4 GB RAM and NVidia GeForce 8800 GTX graphics card.

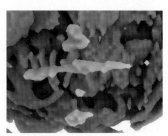

Occlusion Sweeping   **Occlusion Sweeping** without G_s

Fig. 6. Ambient sweeping with and without the geometry term. Notice that the images are very similar. Eliminating the geometry term $G_s(\mathbf{N})$ makes edges and sharp features slightly brighter.

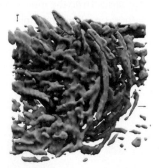

Fig. 7. Occlusion sweeping with lighting from only the top four octants, emulating the lighting conditions of a cloudy day. The image on the right is gamma corrected.

Table 2. Computational time (in seconds) for sweeping the dataset

NX	NY	NZ	Without Occlusion	Combined Occlusion	8 Octant Occlusion
101	101	61	0.72	0.91	1.01
100	100	100	1.06	1.386	1.52
150	150	150	3.86	6.52	7.78
200	200	200	9.56	17.12	23.43
250	250	250	23.26	39.66	46.64

Our method adds an overhead of from 30% to 100% to a standard fast sweeping calculation, which we believe to be bounded by memory access time, and not by computation. This is actually only a minor overhead to the full rendering calculation, as this computation needs to be done only once per viewing session.

4   Other Illumination Effects

4.1   Low-Frequency Shadows

In our method, the result of ambient occlusion from each octant corresponds to low-frequency shadows from an area light source. The assumed directions of the light source is in the opposite direction to the sweeping direction. In this sense, our ambient occlusion solution can be thought of as the combination of low-frequency shadows from eight area light sources.

Fig. 8 shows a comparison of ray traced hard shadows to low-frequency shadows using a single octant of occlusion sweeping. While for simple datasets hard shadows can provide good depth cues, for more complicated datasets they are not as suitable. A complicated example comparing hard shadows to our low-frequency shadows are shown in Fig. 9. Notice that hard shadows of this complicated dataset produces additional high-frequency features, while our low-frequency shadows do not interfere with the actual data.

While the intensities of each of these eight light sources can be adjusted at run time, their positions are attached to the grid. For computing low frequency shadows with our technique from an arbitrary direction, one needs to rotate the grid on which the ambient occlusion is computed. While the ambient occlusion grid can be easily rotated independent of the distance field grid, changing the light direction requires recomputation of the occlusion values with a new sweep. However, the size of an area light source cannot be adjusted as it is a function of the grid resolution.

4.2   Subsurface Scattering Effects

Subsurface scattering accounts for the light that enters a translucent surface and scatters within the object. For visualization purposes, subsurface scattering can be used to provide visual cues to the thickness of an object.

With a slight modification to our fast occlusion sweeping technique, we can achieve a subsurface scattering effect for materials with strong forward scattering. We perform this by replacing the transmittance functions τ in Equation 6 with translucent transmittance functions τ_{trans}, such that

$$\tau_{trans}(\mathbf{P}) = 1 - (1 - \tau(\mathbf{P}))\,\alpha\,, \tag{8}$$

where α is a user defined opacity parameter between 0 and 1. Decreasing α values permit more light to penetrate through the surface, effectively softening the shadows of ambient occlusion and allowing surfaces to be lit from behind via forward scattering.

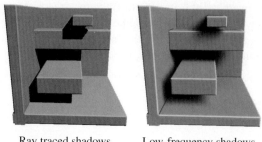

Ray traced shadows   Low-frequency shadows

Fig. 8. Comparison of hard shadows computed using ray marching on the GPU to low-frequency shadows generated by our occlusion sweeping method

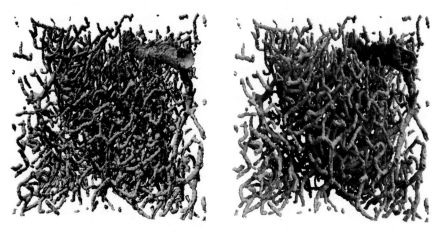

Ray traced shadows   Low-frequency shadows

Fig. 9. Comparison of hard shadows computed using ray marching on the GPU to low-frequency shadows generated by our occlusion sweeping method for the vascular data in Fig. 5

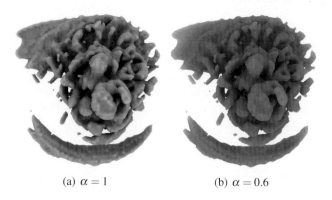

(a) $\alpha = 1$ (b) $\alpha = 0.6$

Fig. 10. With decreasing α values subsurface scattering effects become more prominent

However, the geometry term $g(\mathbf{N}, \omega)$ in Equation 2 and the integral of the geometry term over an octant $G_s(\mathbf{N})$ eliminate the illumination contribution from the opposite direction of the surface normal. Therefore, for backward illumination from forward scattering, we need to modify the geometry term. We can easily achieve this by redefining it as $g(\mathbf{N}, \omega) = |\mathbf{N} \cdot \omega|$.

Fig. 10 shows examples of the subsurface scattering effect computed with our method. Notice that subsurface scattering makes thinner parts of the object brighter giving a visual cue indicating how deep the object is beyond the visible surface.

5   Conclusion

We have presented a new method, which we call *occlusion sweeping*, for the fast computation of ambient occlusion in the rendering of volumetric data. Unlike a full global illumination solution, our method produces approximate ambient occlusion in time that scales linearly with the size of the data set. Since it uses only volumetric calculations, its time complexity is not affected by the geometric complexity of the data. Further, it integrates very easily into the fast sweeping method for determining a signed distance field in a data volume. Therefore, our method should be particularly useful for visualization applications that use ray-marching through a distance field. In addition, our method permits other illumination effects like soft shadows and sub-surface scattering, both of which provide essential visual cues that enhance spatial perception.

As future work, we would like to explore other formulations for the visibility computation and experiment with more accurate transmittance functions to reduce or eliminate the illumination artifacts produced with our current algorithm.

Acknowledgements

This work was supported in part by the National Science Foundation ITR 0326194. We would like to thank David Mayerich for the mouse brain vascular data, and Edward Mansell for the lightning cloud simulation data.

References

1. Qiu, F., Xu, F., Fan, Z., Neophytos, N.: Lattice-based volumetric global illumination. IEEE Transactions on Visualization and Computer Graphics 13, 1576–1583 (2007)
2. Stewart, A.J.: Vicinity shading for enhanced perception of volumetric data. IEEE Visualization (2003)
3. Tarini, M., Cignoni, P., Montani, C.: Ambient occlusion and edge cueing for enhancing real time molecular visualization. IEEE Transactions on Visualization and Computer Graphics 12, 1237–1244 (2006)
4. Weigle, C., Banks, D.: A comparison of the perceptual benefits of linear perspective and physically-based illumination for display of dense 3d streamtubes. IEEE Transactions on Visualization and Computer Graphics 14, 1723–1730 (2008)
5. Zhao, H.: A fast sweeping method for eikonal equations. Mathematics of Computation 74, 603–627 (2005)

6. Sethian, J.A.: A fast marching level set method for monotonically advancing fronts. Proc. Nat. Acad. Sci., 1591–1595 (1996)
7. Zhukov, S., Inoes, A., Kronin, G.: An ambient light illumination model. In: Drettakis, G., Max, N. (eds.) Rendering Techniques 1998. Eurographics, pp. 45–56. Springer, Wien (1998)
8. Pharr, M., Green, S.: Ambient Occlusion. In: GPU Gems, pp. 667–692. Addison-Wesley, Reading (2004)
9. Sattler, M., Sarlette, R., Zachmann, G., Klein, R.: Hardware-accelerated ambient occlusion computation. In: 9th Int'l Fall Workshop Vision, Modeling, And Visualization (VMV), Stanford, California, pp. 119–135 (2004)
10. Sloan, P.P., Kautz, J., Snyder, J.: Precomputed radiance transfer for real-time rendering in dynamic, low-frequency lighting environments. ACM Trans. Graph. 21, 527–536 (2002)
11. Mei, C., Shi, J., Wu, F.: Rendering with spherical radiance transport maps. Computer Graphics Forum 23, 281–290 (2004)
12. Kontkanen, J., Laine, S.: Ambient occlusion fields. In: Proceedings of ACM SIGGRAPH 2005 Symposium on Interactive 3D Graphics and Games, pp. 41–48. ACM Press, New York (2005)
13. Malmer, M., Malmer, F., Assarsson, U., Holzschuch, N.: Fast precomputed ambient occlusion for proximity shadows. Journal of Graphics Tools 12, 59–71 (2007)
14. Bunnell, M.: Dynamic Ambient Occlusion and Indirect Lighting. In: GPU Gems 2, Pearson Education, Inc, London (2005)
15. Shanmugam, P., Arikan, O.: Hardware accelerated ambient occlusion techniques on gpus. In: I3D 2007: Proceedings of the 2007 symposium on Interactive 3D graphics and games, pp. 73–80. ACM, New York (2007)
16. Ritschel, T., Grosch, T., Seidel, H.-P.: Approximating dynamic global illumination in image space. In: I3D 2009: Proceedings of the 2009 symposium on Interactive 3D graphics and games, pp. 75–82. ACM, New York (2009)
17. Salama, C.R.: Gpu-based monte-carlo volume raycasting. In: PG 2007: Proceedings of the 15th Pacific Conference on Computer Graphics and Applications, Washington, DC, USA, pp. 411–414. IEEE Computer Society, Los Alamitos (2007)
18. Tarini, M., Cignoni, P., Montani, C.: Ambient occlusion and edge cueing to enhance real time molecular visualization (2006)

An Empirical Study of Categorical Dataset Visualization Using a Simulated Bee Colony Clustering Algorithm

James D. McCaffrey

Microsoft MSDN / Volt VTE
One Microsoft Way
Redmond, WA 98052 USA
v-jammc@microsoft.com

Abstract. This study investigates the use of a biologically inspired meta-heuristic algorithm to cluster categorical datasets so that the data can be presented in a useful visual form. A computer program which implemented the algorithm was executed against a benchmark dataset of voting records and produced better results, in terms of cluster accuracy, than all known published studies. Compared to alternative clustering and visualization approaches, the categorical dataset clustering with a simulated bee colony (CDC-SBC) algorithm has the advantage of allowing arbitrarily large datasets to be analyzed. The primary disadvantages of the CDC-SBC algorithm for dataset clustering and visualization are that the approach requires a relatively large number of input parameters, and that the approach does not guarantee convergence to an optimal solution. The results of this study suggest that using the CDC-SBC approach for categorical data visualization may be both practical and useful in certain scenarios.

Keywords: Categorical data, category utility, cluster analysis, data visualization, simulated bee colony algorithm.

1 Introduction

This paper presents a study of the use of a biologically inspired meta-heuristic algorithm for processing large datasets composed of categorical data in order to present the data in a useful visual form. The analysis and visualization of datasets which contain categorical data has great practical importance. Examples include examining sales data to forecast consumer purchasing behavior, examining telecommunications data for possible terror-related activity, and examining medical information for various clinical diagnoses. For the sake of concreteness, consider the artificial dataset presented below. The nine tuples in the dataset are based on three attributes: color, size, and temperature. Each of the three attributes can take on a single categorical value: red, blue, green or yellow; small, medium, or large; and hot or cold, respectively. With even this unrealistically small dataset, it is quite difficult for human observers to categorize or group the raw dataset in a meaningful way so that the categorized data can then be presented in some visually descriptive form.

G. Bebis et al. (Eds.): ISVC 2009, Part I, LNCS 5875, pp. 179–188, 2009.

```
001:   Green    Medium   Hot
002:   Blue     Small    Hot
003:   Red      Large    Cold
004:   Red      Medium   Cold
005:   Yellow   Medium   Hot
006:   Green    Medium   Hot
007:   Red      Small    Hot
008:   Red      Large    Cold
009:   Blue     Medium   Hot
```

The primary source of the difficulty of clustering the data presented above is the fact that the attribute values are categorical rather than numerical. It is not so obvious how to compute a meaningful difference or a representative value for categorical tuples such as (red, small, hot) and (blue, large, cold). Data clustering is a widely studied problem domain. A search of the IEEE digital library Web site for the keyword "clustering" returned a list of references to over 36,000 documents

In order to measure the quality of a particular clustering algorithm compared to alternative approaches, some measure of clustering effectiveness must be employed. One approach for evaluating the quality of a clustering algorithm which works on categorical data is to generate a synthetic dataset which is based on some hidden, underlying rule set, run the proposed clustering algorithm against the synthetic dataset, and then gauge the quality of the resulting clusters using some form of similarity or likelihood measure. Examples of similarity measures which can be used to evaluate clustering effectiveness include the Simple Matching coefficient, Jaccard's coefficient, Dice's coefficient, the Cosine coefficient, and the Overlap coefficient [1]. Examples of likelihood measures which can be used to evaluate clustering effectiveness include various forms of entropy functions and the category utility function [2]. The category utility (CU) function is generally attributed to a 1985 paper by Gluck and Corter [3]. The CU function is defined in terms of the bivariate distributions produced by a clustering. Suppose a dataset is composed of t tuples where each tuple is based on A_i attributes (i = 1... m) and where each attribute value, V_{ij}, is a categorical value. If a dataset under analysis is partitioned into a cluster set $C = \{C_k\}$ (k = 1... n), then the category utility function for the clustering scheme is given by the equation:

$$CU(C) = \frac{1}{n}\sum_{k=1}^{n} P(C_k)\left[\sum_i \sum_j P(A_i = V_{ij} \mid C_k)^2 - \sum_i \sum_j P(A_i = V_{ij})^2\right] \qquad (1)$$

The left-hand term in the brackets of equation (1) represents conditional probabilities that each attribute takes on a particular categorical value, given the distribution of that value within a cluster. The right-hand term is similar except that it represents unconditional probabilities of attribute values for the entire dataset. Therefore, the entire term in the square brackets in equation (1) measures the difference of the probabilities of finding attribute values in a cluster purely by chance and the probabilities of finding those values given the clustering scheme.

The fact that quality of categorical data clustering algorithms can be evaluated using the CU function raises the possibility of using the CU function as the basis of a clustering generation mechanism. This is the foundation of the approach used by the algorithm introduced in this study. Except in situations with very small datasets, the

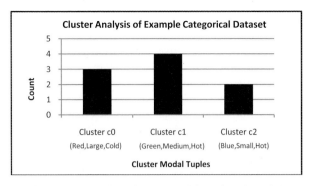

Fig. 1. Visualization of categorical data after clustering

CU function cannot be used to directly generate an optimal clustering of categorical data. Complete analysis of the effectiveness of any clustering algorithm is an NP-complete problem and requires a full enumeration of all possible partitions of the dataset under analysis [2]. Therefore an indirect approach must be employed. The approach introduced by this study is to use a simulated bee colony algorithm (SBC) in conjunction with the category utility function. As will be explained in the following sections of this paper, in essence, the simulated bee colony algorithm intelligently searches the entire solution space of all possible dataset partitions, seeking the partition which has a global maximum category utility value. This algorithm is called CDC-SBC (Categorical Dataset Clustering with a Simulated Bee Colony) to distinguish it from other algorithms in the literature. The resulting clustered categorical data can be visually represented in useful ways, such as the one shown in Fig. 1 which uses modal values from each cluster. The integral relationship between the visualization of categorical datasets and clustering is formalized in a widely cited framework for visualization techniques called the Information Visualization Data State Reference Model (IVDSRM) [4]. The IVDSRM framework is summarized in Fig. 2.

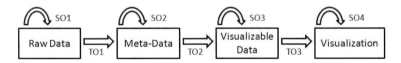

Fig. 2. The Information Visualization Data State Reference Model

The IVDSRM visualization framework models the creation of a visualization as four distinct data stages: raw data (the value stage), meta-data (the analytical abstraction stage), visualizable data (the visualization abstraction stage), and visualization (the view stage). Two types of operators can be applied to each data stage: transformation operators (TO1 through TO3), which create a new data stage, and stage operators (SO1 through SO4) which do not change the underlying structure of data. In the context of the IVDSRM framework, the clustering technique presented in this paper is a Transformation Operator 1 and produces clustering meta-data from the raw categorical dataset data. The clustering meta-data can then be used to produce several different visualizations such as the modal histogram shown in Fig. 1.

2 Algorithms Inspired by Bee Behavior

Algorithms inspired by the behavior of natural systems have been studied for decades. Examples include algorithms inspired by ants, biological immune systems, metallurgic annealing, and genetic recombination. A review of the literature on algorithms inspired by bee behavior suggests that the topic is evolving and that there is no consensus on a single descriptive title for meta-heuristics based on bee behavior. Algorithm names in the literature include Bee System, BeeHive, Virtual Bee Algorithm, Bee Swarm Optimization, Bee Colony Optimization, Artificial Bee Colony, Bees Algorithm, and Simulated Bee Colony.

Common honey bees such as Apis mellifera take on different roles within their colony over time [5]. A typical hive may have 5,000 to 20,000 individuals. Young bees (2 to 20 days old) nurse larvae, construct and repair the hive, guard the entrance to the hive, and so on. Mature bees (20 to 40 days old) typically become foragers. Foraging bees typically occupy one of three roles: active forgers, scout foragers, and inactive foragers. Active foraging bees travel to a food source, gather food, and return to the hive. Roughly 10% of foraging bees in a hive are employed as scouts.

A 1997 study by Sato and Hagiwara used a model of honey bee behavior named Bee System to create a variation of the genetic algorithm meta-heuristic [6]. The algorithm essentially added a model of the behavior of scout bees to introduce new potential solutions and avoid premature convergence to local minima solutions. A 2002 study by Lucic and Teodorvic used a variation of the Bee System model to investigate solving complex traffic and transportation problems [7]. The study successfully used Bee System to solve eight benchmark versions of the traveling salesman problem. A 2004 paper by Nakrani and Tovey presented a honey bee inspired algorithm for dynamic allocation of Internet services [8]. The study concluded that bee inspired algorithms outperformed deterministic greedy algorithms in some situations. A 2005 study by Drias et al. used a meta-heuristic named Bee Swarm Optimization to study instances of the Maximum Satisfiability problem [9]. The study concluded that Bee Swarm Optimization outperformed other evolutionary algorithms, in particular an ant colony algorithm. A 2006 paper by Basturk and Karaboga investigated a bee-inspired algorithm named Artificial Bee Colony to solve five multi-dimensional numerical problems [10]. The paper concluded that the performance of the bee algorithm was roughly comparable to solutions by differential evolution, particle swarm optimization, and evolutionary algorithms. A 2009 study by McCaffrey demonstrated that an algorithm named Simulated Bee Colony outperformed existing deterministic algorithms for generating pairwise test sets, for six out of seven benchmark problems [11].

3 Simulated Bee Colony Algorithm Implementation

There are many ways to map honey bee foraging behavior to a specific algorithm which clusters categorical data in order to create a useful visual presentation. The three primary design features which must be addressed are 1.) design of a problem-specific data structure that simulates a foraging bee's memory and which represents the location of a food source, which in turn represents a dataset clustering scheme, 2.) formulation of a problem-specific function which measures the goodness, or quality, of a candidate partitioning, and 3.) specification of generic algorithm parameters such as the numbers of foraging, scout, and inactive bees in the colony, and the maximum number of times a bee will visit a particular food source. Suppose the dataset to be analyzed contains the data described in the Introduction section of this paper, with attributes of color (red, blue, green, yellow), size (small, medium, large), and temperature (hot, cold), and a cluster size of n = 3 is specified. The screenshots shown in Fig. 3 and Fig. 4 show the result of a sample program run and illustrate many of the implementation details.

```
C:\CDC-SBC\Run\bin\Debug> Run.exe

Begin cluster analysis of categorical data using SBC

Number clusters = 3

Initializing Hive

Number Active bees = 60
Number Inactive bees = 20
Number Scout bees = 20
Maximum number of cycles = 10,000
Maximum cycles without improvement = 10,000
Maximum visits to a food source = 10
Probability waggle dance will convince = 0.9000
Probability a bee accepts a worse source = 0.0100

Hive initialized
```

Fig. 3. Screenshot of initialization phase of the CDC-SBC implementation

The CDC-SBC algorithm implementation used in this study models a bee as an object with four data members. The primary data member is a two-dimensional integer array named MemoryMatrix which corresponds to a bee's memory of the location of a food source, which in turn represents a dataset clustering. A Status field identifies the bee's role (1 = an active forager). A CategoryUtility field is a value which is a measure of the quality of the memory matrix, as described in the Introduction section of this paper. A NumberVisits field is a counter that tracks the number of times the bee object has visited a particular food source. The honey bee colony as a whole is modeled as an array of bee objects. The CDC-SBC algorithm iterates through each bee in the colony and examines the current bee's Status field. If the current bee is an active forager, the algorithm simulates the action of the bee leaving the hive to go to the current food source in memory. Once there, the bee examines a single neighbor food source. A neighbor food source is one which, relative to the current food source, has a single tuple assigned to a different cluster. If the quality of the neighbor food source is superior to the current food source, the foraging bee's memory is updated with the neighbor location and the NumberVisits counter is reset to 0.

After examining a neighbor food source, an active bee returns to the hive. If the returning bee has reached a threshold for the maximum number of visits to its food source in memory, that bee becomes inactive and a randomly selected inactive bee is

```
All cycles completed

Best clustering matrix found is
0  0  1  1  0  0  0  1  0
1  0  0  0  1  1  0  0  1
0  1  0  0  0  0  1  0  0

Corresponding category utility is 0.3971

Cluster c0 =
003 ( Red       Large    Cold     )
004 ( Red       Medium   Cold     )
008 ( Red       Large    Cold     )
----------------------------
mode: Red       Large    Cold

Cluster c1 =
001 ( Green     Medium   Hot      )
005 ( Yellow    Medium   Hot      )
006 ( Green     Medium   Hot      )
009 ( Blue      Medium   Hot      )
----------------------------
mode: Green     Medium   Hot

Cluster c2 =
002 ( Blue      Small    Hot      )
007 ( Red       Small    Hot      )
----------------------------
mode: Blue      Small    Hot

End SBC visualization run
```

Fig. 4. Screenshot of execution and results of the CDC-SBC implementation

converted to an active forager. Otherwise the returning bee performs a simulated waggle dance to all inactive bees in the hive. This dance conveys the goodness of the current food source / clustering in the dancing bee's memory. Inactive bees with food sources in memory which have lower quality than the returning bee's food source will update their memories to the returning bee's memory with probability = 0.90. Scout bees are not affected by the waggle dances of returning foragers. Instead, scouts leave the hive, examine a randomly selected food source, return to the hive, and perform a waggle dance to the audience of currently inactive bees.

4 Results

Two common metrics for measuring the effectiveness of clustering algorithms are precision and recall [12]. Suppose some dataset contains t tuples and some clustering algorithm assigns each tuple to one of n clusters. Let a_i represent the number of tuples correctly assigned to cluster i. Let b_i represent the number of tuples which have been incorrectly assigned to cluster i. And let c_i represent the number of tuples which have been incorrectly rejected from cluster i (and incorrectly assigned to some cluster j where $j \neq i$). Then the precision for cluster i is given by $p_i = a_i / (a_i + b_i)$. The recall for cluster i is given by $r_i = a_i / (a_i + c_i)$. The precision for a given cluster can be thought of as a measure of accuracy, and the recall can be thought of as a measure of completeness. The micro-precision of a clustering result is computed as a whole, across all clusters, using overall numbers of correctly assigned tuples, incorrectly assigned tuples, and incorrectly rejected tuples.

4.1 Experiment #1 – Congressional Voting Data

In order to evaluate the effectiveness of the CDC-SBC algorithm and resulting data visualizations compared to alternative clustering algorithms, the CDC-SBC algorithm was executed against the UCI voting dataset. The voting dataset consists of actual congressional votes from the U.S. House of Representatives on 16 issues in 1984. Results of running the CDC-SBC algorithm against the voting dataset (with party affiliation omitted) and the corresponding values for seven other categorical data clustering algorithms are shown in Table 1.

Table 1. Effectiveness of different clustering algorithms on the benchmark UCI voting dataset

Algorithm	Correct	Precision	CU	CU'
CDC-SBC	383	0.8805	1.4711	2.9422
COBWEB	378	0.8690	1.4506	2.9011
Ahmad-K	377	0.8667	1.4465	2.8929
LIMBO	376	0.8644	1.4424	2.8847
K-Means	376	0.8644	1.4424	2.8847
Huang-K	364	0.8368	1.3931	2.7861
COOLCAT	363	0.8345	1.3890	2.7779
ROCK	345	0.7931	1.3150	2.6300

The values in the column labeled Correct in Table 1 are the number of tuples in the voting dataset which were correctly clustered as Democrat or Republican by each algorithm. The Precision column is the micro-precision value as described above, which in this situation is just the number of tuples which are correctly clustered, divided by the total number of tuples (t = 435) in the voting dataset. The CU column is the category utility of the clustering produced by each algorithm, as defined by equation (1). The CU' column is a slightly different definition of category utility used by some studies, which is not normalized for number of clusters. Because the number of clusters in this situation is 2, the values in the CU' column are simply twice the values in the CU column, and have been included solely to provide a consistent comparison with the reported results of other studies.

The COBWEB clustering algorithm incrementally builds a probabilistic hierarchy tree of clusters from a dataset using category utility to measure clustering effectiveness [13]. The Ahmad-K clustering algorithm is a variation of the Huang-K algorithm, which in turn is based on the simple k-means algorithm [14]. The LIMBO algorithm is based on a concept called the information bottleneck, which is essentially a measure of entropy [12]. The COOLCAT clustering algorithm is an iterative technique that uses a greedy algorithm based combined with an entropy measure [2]. The

ROCK algorithm uses a hierarchical approach in conjunction with a distance measure modeled on graph theory [15].

The data in Table 1 were derived from several sources and should be interpreted somewhat cautiously. Most of the studies represented in Table 1 reported result values for the UCI voting dataset in terms of category utility. In the situations where the number of correct tuples was not reported (COOLCAT, COBWEB, LIMBO), an auxiliary program, developed as part of this study, which computes the number of correct values for a given category utility was employed to produce the values shown in the Correct column of Table 1. The results for the K-Means algorithm were determined by executing the WEKA data analysis tool [16]. Further, published results differ slightly for the ROCK, COOLCAT, and COBWEB algorithms, presumably because of differences in input parameters to the algorithms. In situations where the differences in reported values for these algorithms could not be resolved, the data in Table 1 represent arithmetic means of reported results.

The results of the CDC-SBC algorithm and the seven other algorithms listed in Table 1 represent the best clustering results of the benchmark UCI voting dataset discovered by a comprehensive review of the literature. The data indicates that the CDC-SBC algorithm produced more accurate results than all previously published algorithms for clustering the UCI voting dataset.

4.2 Experiment #2 – Synthetic Datasets

In order to evaluate the efficiency of the CDC-SBC algorithm and its resulting data visualizations, the algorithm was executed against six synthetic datasets. The results are shown in Table 2.

Table 2. Accuracy of the CDC-SBC algorithm on synthetic datasets

Dataset	Attributes	Attribute Values	Tuples	Clusters	Partitions	Precision
DS0	3	(4,3,2)	9	2	$3.02 * 10^3$	1.00
DS1	4	(5,5,5,5)	20	3	$5.81 * 10^8$	1.00
DS2	5	(2,3,4,3,2)	36	4	$1.97 * 10^{20}$	1.00
DS3	6	(3,3,..,3)	50	5	$7.40 * 10^{32}$	1.00
DS4	10	(2,2,...,2)	200	2	$8.03 * 10^{59}$	0.98
DS5	16	(2,2,...,2)	435	2	$4.44 * 10^{130}$	0.95

After the synthetic datasets had been generated, a program which implemented the CDC-SBC algorithm was executed using each synthetic dataset (without cluster values) as input. The micro-precision was computed for each resulting clustering, and is listed in Table 2. For all synthetic dataset inputs, the maximum number of iterations of the main SBC algorithm loop was limited to a count of 10^8 or until a partitioning result with precision of 1.00 was discovered. The column in Table 2 which is labelled Partitions holds the total number of possible partitions for the associated synthetic dataset, computed using Stirling numbers of the second kind, and is a measure of dataset complexity.

The results in Table 2 suggest that the CDC-SBC algorithm is highly effective at clustering datasets which contain self-consistent data. The results also suggest that the CDC-SBC algorithm is at least reasonably effective at clustering datasets which have huge search spaces. The results for dataset D05 are particularly noteworthy; the CDC-SBC algorithm correctly placed 413 out of 435 tuples from a problem domain with over 10^{130} possible partitions.

5   Conclusions

The results of this study demonstrate the feasibility of using a simulated bee colony meta-heuristic algorithm in conjunction with the category utility function to cluster categorical datasets so that the data can be usefully visualized. Because the scope of this study is limited and is for the most part empirical, it is not possible to draw definitive conclusions from the results. However, when taken as a whole the results suggest that categorical data visualization using the CDC-SBC technique is a promising technique which has the potential to outperform existing algorithms in terms of clustering accuracy and accuracy of any resulting visualization format, and that the technique merits further investigation. One disadvantage of the CDC-SBC algorithm compared to alternative approaches is that CDC-SBC requires a relatively large number of generic algorithm parameters such as the numbers and percentages of each type of bee object, and simulation probabilities such as the probability that an active foraging bee will accept a neighbor solution with a lower category utility value than the current CU value. Because algorithms based on bee behavior are relatively unexplored, there are very few guidelines available for selecting input parameters and trial and error is often required to tune the algorithm for better performance. Additionally, because the CDC-SBC algorithm is probabilistic, there is no guarantee that the algorithm will produce an optimal solution to any clustering problem.

In addition to clustering accuracy, an advantage of the CDC-SBC algorithm compared to existing approaches is that CDC-SBC can in principle be applied to arbitrarily large datasets. A promising potential extension of CDC-SBC is to investigate datasets with mixed categorical and numerical data. According to a mathematical analysis of the category utility function by Mirkin, in spite of a significantly different outward appearance compared to traditional numerical clustering measures, the CU function is in fact closely related to the square-error criterion used in numerical clustering [17]. This raises the possibility of adapting the CDC-SBC algorithm to deal with mixed data using a unified form of CU function.

References

1. Liu, Y., Ouyang, Y., Sheng, H., Xiong, Z.: An Incremental Algorithm for Clustering Search Results. In: Proceedings of the 2008 IEEE International Conference on Signal Image Technology and Internet Based Systems, pp. 112–117 (2008)
2. Barbara, D., Li, Y., Couto, J.: COOLCAT: An Entropy-Based Algorithm for Categorical Clustering. In: Proceedings of the 11th International Conference on Information and Knowledge Management, pp. 582–589 (2002)

3. Gluck, M., Corter, J.: Information, Uncertainty, and the Utility of Categories. In: Program of the 7th Annual Conference of the Cognitive Science Society, pp. 283–287 (1985)

4. Chi, E.: A Taxonomy of Visualization Techniques using the Data State Reference Model. In: Proceedings of the IEEE Symposium on Information Visualization, pp. 69–75 (2000)

5. Seeley, T.D.: The Wisdom of the Hive: The Social Physiology of Honey Bee Colonies. Harvard University Press, Boston (1995)

6. Sato, T., Hagiwara, M.: Bee System: Finding Solution by a Concentrated Search. In: Proceedings of the IEEE International Conference on Systems, Man, and Cybernetics, vol. 4, pp. 3954–3959 (1997)

7. Lucic, P., Teodorovic, D.: Transportation Modeling: An Artificial Life Approach. In: Proceedings of the 14th IEEE International Conference on Tools with Artificial Intelligence, pp. 216–223 (2002)

8. Nakrani, S., Tovey, C.: On Honey Bees and Dynamic Server Allocation in Internet Hosting Centers. Adaptive Behavior - Animals, Animats, Software Agents, Robots, Adaptive Systems 12(3-4), 223–240 (2004)

9. Drias, H., Sadeg, S., Yahi, S.: Cooperative Bees Swarm for Solving the Maximum Weighted Satisfiability Problem. In: Cabestany, J., Prieto, A.G., Sandoval, F. (eds.) IWANN 2005. LNCS, vol. 3512, pp. 318–325. Springer, Heidelberg (2005)

10. Basturk, B., Karaboga, D.: An Artificial Bee Colony (ABC) Algorithm for Numeric Function Optimization. In: Proceedings of the IEEE Swarm Intelligence Symposium, pp. 687–697 (2006)

11. McCaffrey, J.: Generation of Pairwise Test Sets using a Simulated Bee Colony Algorithm. In: Proceedings of the 10th IEEE International Conference on Information Reuse and Integration (2009)

12. Andritsos, P., Tsaparas, P., Miller, R., Sevcik, K.: LIMBO: Scalable Clustering of Categorical Data. In: Bertino, E., Christodoulakis, S., Plexousakis, D., Christophides, V., Koubarakis, M., Böhm, K., Ferrari, E. (eds.) EDBT 2004. LNCS, vol. 2992, pp. 123–146. Springer, Heidelberg (2004)

13. Fisher, D.: Knowledge Acquisition via Incremental Conceptual Clustering. Machine Learning 2(2), 139–172 (1987)

14. Ahmad, A., Dey, L.: A k-Mean Clustering Algorithm for Mixed Numeric and Categorical Data. Data Knowledge and Engineering 63(2), 503–527 (2007)

15. Hsu, C., Chen, C., Su, Y.: Hierarchical Clustering of Mixed Data Based on Distance Hierarchy. Information Sciences 177(20), 4474–4492 (2007)

16. Holmes, G., Donkin, A., Witten, I.: WEKA: A Machine Learning Workbench. In: Proceedings of the 2nd Austraila and New Zealand Conference on Intelligent Information Systems, pp. 357–361 (1994)

17. Mirkin, B.: Reinterpreting the Category Utility Function. Machine Learning 45(2), 219–228 (2001)

Real-Time Feature Acquisition and Integration for Vision-Based Mobile Robots

Thomas Hübner and Renato Pajarola

Visualization and MultiMedia Lab, University of Zurich
Binzmühlestr. 14, 8050 Zurich, Switzerland
http://vmml.ifi.uzh.ch

Abstract. In this paper we propose a new system for real-time feature acquisition and integration based on high-resolution stereo images that is suitable for mobile robot platforms with limited resources. We combine a fast feature detection stage with a stable scale-invariant feature description method utilizing optimized spatial matching. Putative image feature matches are used to determine 3D coordinates of feature points and to estimate corresponding view transformations. Experimental results show the advantages of our system in terms of performance and accuracy compared to the standard methods used in the area.

Keywords: Feature acquisition, feature integration, vision, mobile robot.

1 Introduction

Reliable feature localization and mapping from multiple images is an important capability for mobile robots. A major fraction of robotic systems utilizes laser scanners as the primary input sensor to obtain information about the environment, in particular distance values. However, vision-based approaches using single- or multiple camera configurations are able to provide much more information relatively cheap, but also lead to a vast amount of image data that needs to be processed within the speed and latency constraints of the robot platform.

In vision-based approaches natural visual features are exploited as landmarks. These landmarks need to be detected and described by a feature vector to enable tracking and unique view-independent matching. The main computational effort is spent on feature tracking and description.

We address the problem of real-time acquisition and integration of visual features in natural indoor environments based on a rotation sequence of high-resolution stereo images. Our optimized feature detection and matching methods are suitable for online processing on vision-based robot platforms with limited computational resources. Despite focusing mainly on system performance, we achieve excellent accuracy in feature integration even in the presence of outliers or sparse landmarks. Our contributions include:

- Adaptive feature detection based on the *FAST* corner detector
- Improved *SIFT* based feature description for real-time processing

G. Bebis et al. (Eds.): ISVC 2009, Part I, LNCS 5875, pp. 189–200, 2009.

- Fast feature matching exploiting spatial and temporal coherence
- Stable and accurate feature integration from a non-stationary stereo camera
- Experimental comparisons to standard methods used in the area

2   Related Work

The acquisition of landmarks is generally carried out in two steps: (1) Detection of suitable visual features that can be used as landmarks. (2) Description of the features with a feature vector that uses local image neighborhood information. A number of methods for both steps have been proposed in the past. We focus here on the ones primarily used in natural feature acquisition.

KLT developed by Kanade, Lucas and Tomasi [1] extracts image features that are adequate for tracking. *Normalized cross correlation* (NCC) tracks these features in subsequent images. Tracking has the advantage that features can still be followed even after the feature detector ceases to detect them. Thus, feature localization is very accurate, but if the search area is not sufficiently limited NCC exposes a poor performance with increasing image size (see also Section 4). The most popular feature detection and description approach is *SIFT* (Scale-Invariant-Feature-Transform) [2]. Visual features are detected as local extrema of the *Difference of Gaussian* (DoG) over scale space. The SIFT descriptor is rotation and scale invariant. Computationally, SIFT is one of the most expensive descriptors though achieving an excellent invariant feature description. More recently *SURF* (Speeded up Robust Features) [3] has been proposed. In SURF, feature detection is based on the *Hessian matrix* while the descriptor uses sums of 2D Haar wavelet responses. SURF is exposing a similar description quality as SIFT at a better performance [4].

Acquired landmarks are matched by calculating the Euclidian of their descriptors. To avoid a brute-force comparison of the descriptors, k-d trees, spatial hashing, and epipolar matching can be employed. Once putative matches are found, 3D coordinates of the landmarks can be calculated by triangulation. Features are spatially integrated estimating the view transformation from corresponding 2D image points [5,6] or corresponding 3D world points [7,8]. Estimating the view transformation based on 3D point correspondences is inferior as triangulations are much more uncertain in the depth direction. Therefore, the estimation based on 2D image points could give a more precise view transformation.

Our system for the online acquisition and integration of image features avoids expensive computational feature detection and tracking by using the *FAST* corner detector proposed by Rosten et al. [9] combined with a modified reduced SIFT descriptor [2]. Registration and matching of features in real-time is achieved by exploiting optimized spatial hashing [10]. Features are spatially integrated by estimating the view transformation on corresponding 2D image points directly, using a rather simple but stable algorithm. Our system provides very accurate results at comparatively low computational cost that no other previously proposed method is capable of.

3   System Overview

Our mobile robot platform as shown in Figure 1(a) uses a Bumblebee 2 stereo camera from Point Grey Research as the optical imaging sensor that is capable of capturing stereo image frames with a resolution of 1024 × 768 pixels at 20fps. The camera is attached to a pan-and-tilt device permitting an absolute rotation of 130°. Mechanically, the rotation angle can be controlled from −65° to +65°, however, no exact feedback or control is possible for accurate setting of intermediate angles.

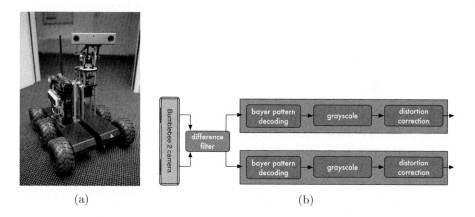

(a)   (b)

Fig. 1. (a) Mobile robotic platform with Bumblebee 2 stereo camera. (b) Pre-processing pipeline.

Feature points are acquired and integrated in real-time from a continuous stream of images. Feature acquisition includes detection, description and matching of features while during integration the transformations between the images are estimated. The consecutive processing steps of our system are outlined in Figure 2 and will be explained in detail below.

Fig. 2. Consecutive steps of real-time feature acquisition and integration

3.1   Pre-processing

Raw stereo image data acquired with the camera needs to be pre-processed before actual feature acquisition can take place. Figure 1(b) shows the pre-processing pipeline.

Captured stereo frames are run initially through a difference filter that limits the number of frames to be processed. For this purpose the filter subtracts two consecutive stereo frames and calculates the mean value of the resulting difference frame. If the mean value is under a certain threshold, the stereo frame is discarded. Image noise defines the threshold for this filter. Subsequently, the frame rate varies between 0-20fps depending on the image differences and thus the robot and camera motion.

Frames that reach the pre-processing stage are decoded from their raw Bayer pattern format to 8-bit grayscale. This is the image format which is used by practically all feature detectors. Camera lens distortions are corrected by a lookup-table and bi-linear interpolation.

The whole pre-processing pipeline is multi-threaded so that both stereo frames are processed in parallel. Use of SSE (Streaming SIMD Extensions) ensures that pre-processing consumes a minimal amount of available computational time before the actual feature detection is executed.

3.2   Feature Detection

Our feature detection is based on the *FAST* corner detector with non-maximum suppression for feature detection [9]. FAST is a high quality corner detector that significantly outperforms other existing algorithms. The principle of FAST is to examine a small patch around a candidate image point to see if it looks like a corner. This approach is efficiently implemented by a segment test algorithm improved through machine learning. We use the 9-point variant of this corner detector for our feature detection stage, as it provides optimal performance.

Regardless of being optimized for performance, the FAST corner detector is invariant to substantial view transformations and independent of the feature type. Its major disadvantage lies in the dependance on a user defined threshold. Nevertheless, this feature detector exposes such a great performance so that it can be used for adaptive feature detection. Instead of defining an image-based threshold we can define a desired feature count. The optimal threshold can then be found in a few iterations using the algorithm shown in Figure 3.

First we pre-define threshold step sizes that proved to be appropriate for fast threshold determination (1). We then iterate until a user defined limit (2), run the feature detector (3), and return the features if they lie within a 10% threshold of the desired feature count (4, 5). If the feature number differs to a greater extent the threshold is adjusted by a step value in the direction of the desired feature count (7, 8). Should we pass the target features we start reducing the step size and repeat steps (2-9). If no appropriate threshold can be found after exceeding the iteration limit, the feature search with the closest count is returned. This way we aim for a constant feature count by adaptive thresholding.

Section 4.1 shows that adaptive feature detection has only a marginal impact on the overall performance, whereas keeping a constant feature count is an important component for feature description and integration. Too sparse features result in an uncertainty in the estimated view transformation, while an excessive number of features increases significantly the time spent on feature description and thus breaking the real-time constraint of the robot platform.

```
1    threshold step = {10, 5, 2, 1};
2    while iteration limit not reached
3        run feature detector;
4        if feature count within 10% of target features
5            return features;
6        (* adjust feature detection threshold *);
7            determine threshold step sign;
8            threshold ± = threshold step;
9            if passed target features
10               begin reducing threshold step;
```

Fig. 3. Adaptive feature detection procedure

3.3   Feature Tracking

Detected features could be tracked between image frames by cross correlation or sum-of-squared differences (SSD) matching. We avoid actual feature tracking because the involved search in high resolution images would be much more time-consuming than our fast feature detection, description and matching. Tracking is eventually achieved by matching features in multiple images as outlined below in Section 3.5.

3.4   Feature Description

Feature detection is directly followed by feature description. Feature description is a fundamental part as it is used to associate landmarks from different views. Wrong associations between landmarks will result in inaccurate feature registration and integration, thus each detected feature needs to be assigned a unique invariant descriptor.

As noted in Section 2, SIFT achieves an excellent invariant feature description at the expense of decreased performance. Though quite fast implementations exist that employ SSE and OpenMP (Open Multi-Processing) claiming speed improvements of a factor of 6 over Lowe's standard approach [11], they are still not sufficient for real-time usage on high-resolution images.

Referring to SIFT, it is always considered in its most expensive variant with a 128-element feature vector. A smaller variant of the descriptor exists that performs only 8% worse in terms of feature association than the full version [2]. This variant uses a 3×3 descriptor array with only 4 gradient orientations. The resulting 36-element feature vector is much faster to compute and suits the real-time constraints of our system.

In detail, we choose a subregion around the feature that is 15×15 pixels wide. This subregion is divided into 5×5 pixel wide regions. Keeping close to the original implementation we calculate the gradient orientation and magnitude for each pixel in the subregion. Weighted by the distance from the region center and the magnitude, gradient orientations are accumulated into 4 gradient orientation histograms (Figure 4). The desriptor is normalized to unit length to reduce effects

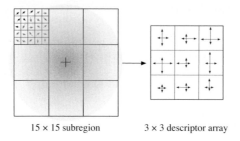

15 × 15 subregion   3 × 3 descriptor array

Fig. 4. 36-element reduced SIFT descriptor creation

```
1    (* Initialization *)
2       pre-calculate square-root lookup table [256, 256];
3       pre-calculate arc-tangent lookup table [512, 512];
4
5    (* Descriptor calculation *)
6       calculate image x - gradients;
7       calculate image y - gradients;
8       for each detected feature do
9          lookup gradient magnitudes in square-root table;
10         lookup gradient orientations in arc-tangent table;
11         accumulate weighted orientations to descriptor array;
12      return descriptors;
```

Fig. 5. Feature description process

of illumination changes. To reduce the influence of large gradient magnitudes, descriptor values are clamped to 0.2 and re-normalized as proposed in [2].

We implemented the descriptor calculation without any specific multi-processing or SSE extensions. Nevertheless, we achieve real-time performance on high resolution images (see Section 4.1). Figure 5 shows the outline of our *real-time SIFT* (RTSIFT) method.

During the initialization of RTSIFT we pre-calculate square root and arc tangent lookup tables (2, 3). Before considering individual descriptors we calculate the x- and y-image gradients once (6, 7). This can be performed more efficiently on the entire image than on separate subimages as feature regions tend to overlap. To describe an image feature we lookup the gradient magnitudes and orientations for each pixel in the feature's subregion (9, 10) and accumulate them after Gaussian weighting into the descriptor array (11). A list of descriptors is eventually returned (12).

3.5   Feature Matching and Outlier Removal

Feature matching associates landmarks from different views that correspond to the same feature. The similarity of two features is defined by the Euclidian distance of their feature descriptors. Comparing feature descriptors using

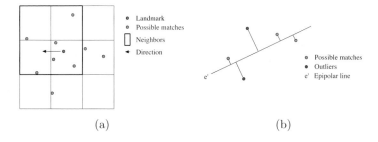

(a)   (b)

Fig. 6. (a) Spatial hashing to limit feature search. (b) Epipolar matching.

a brute-force approach leads to a matching time that is linearly dependent on the number of features. Using spatial and temporal coherence we can avoid the linear dependency and considerably decrease the time spent on feature matching and reduce outliers at the same time.

To limit the number of descriptor comparisons we use *spatial hashing* [10]. Spatial hashing divides the image space into grid cells (Figure 6(a)). Each of these grid cells is assigned with a unique hash value. The grid cell size influences the number of landmarks that will reside in a cell. For matching, only landmarks within the same and neighboring grid cells are considered. The optimal grid cell size depends on the matching type. In stereo matching the grid size is set according to the expected depth range and thus the corresponding disparities. To estimate view transformations we use knowledge about the pixel displacement given by a certain robot or camera movement. We set the grid cell size to the expected maximum disparity or pixel displacement.

Our matching method is outlined in Figure 7. After setting the grid size according to the matching type the hash table is initialized with landmarks. We use a hash function (4) with the prime numbers $p1 = 73856093$, $p2 = 19349663$ and the hash table size n set to the landmark count. For each landmark (3) its grid position and the corresponding hash value are generated (4). The landmark's Euclidian distance is calculated to all landmarks corresponding to the same hash value (5). If a match is not found (6), neighboring grid cells are searched (7). These neighbor grid cells depend on the landmark position and the direction of the pixel displacement. Features are only associated if they are closer than 50% of the second closest match (9).

Spatial hashing contributes greatly to the reduction of outlier matches as the spatial coherence constraint generates fewer mismatches.

Feature matching can be improved further by considering temporal coherence. Robot and camera motion are nearly constant over a short time, thus it is possible to predict a landmark position based on previous matches. This is done by *epipolar matching*. The fundamental matrix F can be determined based on 2D image point correspondances [6]. Having F, we can estimate the epipolar lines along which landmarks should move. Matching is hence reduced to features that are near an epipolar line e' as illustrated in Figure 6(b). But uncertainty in the estimation of the fundamental matrix leads to wrong predictions.

```
1    set grid size according matching type;
2    initialize hash table;
3    for each described feature do
4        generate hash = (x · p1 ⊕ y · p2)  mod n;
5        compare feature descriptors inside grid cell;
6        if feature not found
7            search neighboring grid cells;
8        if match 50% closer than second match
9            associate features;
```

Fig. 7. Feature matching algorithm

As the camera movement of our robotic platform at a single position is limited to rotations, landmarks move either horizontal or vertical in image space, and we can reduce the problem to a simple 1D lookup table. Mismatches between similar features that remain after feature matching are handled during the following feature integration stage.

3.6   Feature Integration

Assuming a calibrated camera with known intrinsic parameter matrix (Equation 1), f being the focal length and X_c, Y_c the camera image center, we can easily triangulate the 3D position of associated landmarks (Equation 2).

$$\begin{pmatrix} f & 0 & X_c \\ 0 & f & Y_c \\ 0 & 0 & 1 \end{pmatrix} \tag{1}$$

The parameter b represents the camera baseline and d the stereo disparity.

$$z = \frac{b \cdot f}{d}, \quad x = \frac{(x_i - X_c) \cdot z}{f}, \quad y = \frac{(y_i - Y_c) \cdot z}{f} \tag{2}$$

As noted in Section 2, estimating the view transformation from 3D points is unreliable due to depth uncertainties. Hence we estimate the camera movement based on 2D points from monocular images at continuous time steps.

Standard algorithms for estimating the camera movement based on the fundamental matrix F [6] showed to be inappropriate for feature integration on our robotic system. The angle and axis of rotation recovered from the fundamental matrix varied strongly with the standard algorithms and in most cases misleadingly a translation was found instead of a proper rotation. Furthermore, the fundamental matrix is very sensitive to slight changes in the landmarks' 2D positions and to outliers even when using RANSAC.

The camera on our robotic platform is rotating in angular steps $\leq 1°$, we therefore need a more robust feature integration approach that is able to reliably estimate even small angles. For the sake of simplicity we consider in the following a camera rotation around the y-axis, as a x-axis rotation can easily be derived from the given equations.

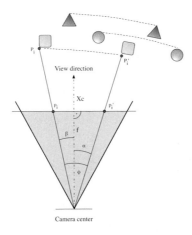

Fig. 8. Relation of camera rotation angles to disparities

For each pair of associated landmarks we calculate the rotation angle ϕ using the cameras intrinsic parameters and trigonometric relations as indicated in Figure 8.

The rotation angle ϕ is found based on the projected image positions (p_i, p_i') of the landmark (P_i, P_i'). According to the projection of p_i and p_i' relative to the camera's image center X_c we determine the intermediate angles α and β as:

$$\alpha = atan(\frac{p_i' - X_c}{f}), \quad \beta = atan(\frac{X_c - p_i}{f}). \tag{3}$$

The angle ϕ is then easily found from

$$\phi = \alpha + \beta. \tag{4}$$

Given a set of individual rotation angles ϕ we need to find a common angle that agrees with most landmarks and takes outliers into account which were not eliminated during the previous feature matching stage.

Our solution to this is the following: First, we find the angle ϕ that corresponds to the majority of angles inside the set within a 10% threshold. This is done by simply testing each angle ϕ against all others. Second, landmarks that are not consistent with ϕ are considered to be outliers and excluded from estimating the rotation angle. The angle ϕ is finally used as starting point for iterative, non-linear least squares Levenberg-Marquardt optimization under the condition that the error $E(\phi)$ becomes minimal.

$$E(\phi) = \sum_{n=0}^{N} [p_i' - f(p_i, \phi)]^2. \tag{5}$$

The angle obtained through this optimization is used for 3D feature integration according to triangulation and the estimation of the cumulative rotation angle (see Section 4.2).

4   Experimental Results

We tested our system on a mobile robot platform that uses a Apple Mac Mini Core 2 Duo (2GHz CPU) for vision processing. The results presented are using two parallel threads unless stated otherwise.

4.1   Performance

To compare the performance of our RTSIFT implementation with adaptive feature detection to different standard methods, a real-world indoor sequence of 100 frames was recorded. On this pre-recorded sequence we compared RTSIFT to KLT [12], Fast SIFT [11], and SURF [13]. Timings are given in Figure 9(a) for the feature detection and description and for three different image resolutions. RTSIFT clearly outperforms any standard method for the feature detection and description on high-resolution images. We achieve 33fps at a resolution of 1024×768, while SURF (4.8fps), Fast SIFT (1.2fps) and KLT (0.8fps) are significantly slower.

(a)

	1024×768	512×384	256×192
RTSIFT	2.98	1.57	1.27
KLT	125.11	23.43	5.5
SIFT	84.31	25.65	7.54
SURF	20.77	7.34	1.99

(b)

	3648×2736	1920×1440	1024×768
FAST	0.21	0.06	0.02
# Features	21817	6587	2582
Description	1.40	0.32	0.18
FAST AFD	0.63	0.21	0.07
# Features	521	460	501
Description	0.49	0.15	0.06

(c)

	$130°$	Error	$65°$	Error	$10°$	Error
RTSIFT	128.88	1.12	63.93	1.07	8.56	1.44
KLT	135.36	5.36	66.35	1.35	9.01	0.99
SIFT	128.92	1.08	63.70	1.3	8.59	1.41
SURF	128.59	1.41	62.69	2.31	8.13	1.87

Fig. 9. (a) Feature detection and description applied to 100 frames. (b) Adaptive feature detection (single-threaded). (c) Accuracy of estimating the cumulative rotation angle. All timings are given in seconds.

We additionally evaluated the influence of the adaptive feature detection (AFD) on the performance of the feature detection, as well as on the number of detected features and the resulting description time. For the adaptive detection the number of target features was set to 500 and the iteration limit to 6. The initial threshold for FAST feature detection was set to 15. Figure 9(b) shows the comparison between the original FAST corner detector and our implementation with adaptive feature detection. With increasing image size the number of detected features increases proportionally when using the original FAST corner detector. While the excessive number of features is not necessarily beneficial, this reduces the performance of the subsequent feature description stage. Our adaptive feature detection keeps the number of features near a given constant and thus guarantees fast computation for feature description.

4.2   Accuracy

The accuracy of the system is an important factor for the feature integration. We tested our method with different pre-defined rotation angles. While rotating the camera, landmarks are continuously acquired and intermediate rotation angles are estimated. The resulting accumulated angle should ideally correspond to the real camera rotation. Figure 9(c) shows the error in estimating the cumulative rotation angle for the different methods. While all methods achieve a rather low error for estimating the rotation angle, using our feature integration, RTSIFT is significantly faster.

In Figure 10 we show an example of the real-world environment with matched features in the left-right stereo images as well as matched features in subsequent frames over time.

(a)   (b)   (c)

Fig. 10. (a) Left-view image with detected landmarks in green. (b) Right-view with matched stereo correspondences. (c) Right image at $t+1$ with matched correspondences over time.

5   Conclusion

In this paper we presented an efficient system for real-time feature acquisition and integration for vision-based mobile robots. A fast adaptive feature detection stage is combined with a SIFT-based stable, scale-invariant and real-time feature description while utilizing spatial hashing and epipolar matching for feature association and outlier removal. The proposed feature integration method showed to be robust in real-world indoor environments with low texture information as well as fast, and thus can be successfully applied for real-time robot navigation and 3D reconstruction.

References

1. Shi, J., Tomasi, C.: Good features to track. In: IEEE Conference on Computer Vision and Pattern Recognition (CVPR 1994), Seattle (1994)
2. Lowe, D.G.: Distinctive image features from scale-invariant keypoints. Int. J. Comput. Vision 60, 91–110 (2004)

3. Bay, H., Tuytelaars, T., Van Gool, L.: Surf: Speeded-up robust features. In: Leonardis, A., Bischof, H., Pinz, A. (eds.) ECCV 2006. LNCS, vol. 3951, pp. 404–417. Springer, Heidelberg (2006)
4. Bauer, J., Sünderhauf, N., Protzel, P.: Comparing several implementations of two recently published feature detectors. In: Proc. of the International Conference on Intelligent and Autonomous Systems (2007)
5. Hartley, R.I., Zisserman, A.: Multiple View Geometry in Computer Vision, 2nd edn. Cambridge University Press, Cambridge (2004)
6. Zhang, Z., Kanade, T.: Determining the epipolar geometry and its uncertainty: A review. International Journal of Computer Vision 27, 161–195 (1998)
7. Besl, P.J., McKay, N.D.: A method for registration of 3-d shapes. IEEE Trans. Pattern Anal. Mach. Intell. 14, 239–256 (1992)
8. Umeyama, S.: Least-squares estimation of transformation parameters between two point patterns. IEEE Trans. Pattern Anal. Mach. Intell. 13, 376–380 (1991)
9. Rosten, E., Drummond, T.: Machine learning for high-speed corner detection. In: Leonardis, A., Bischof, H., Pinz, A. (eds.) ECCV 2006. LNCS, vol. 3951, pp. 430–443. Springer, Heidelberg (2006)
10. Teschner, M., Heidelberger, B., Mueller, M., Pomeranets, D., Gross, M.: Optimized spatial hashing for collision detection of deformable objects. In: Proceedings of Vision, Modeling, Visualization (VMV 2003), pp. 47–54 (2003)
11. OpenSource: Fast sift image features library. Website (2008), http://libsift.sourceforge.net/ (visited on July 14, 2009)
12. Birchfield, S.: Klt: An implementation of the kanade-lucas-tomasi feature tracker. Website (2007), http://www.ces.clemson.edu/stb/klt/ (visited on July 14, 2009)
13. Bay, H.: Surf: Speeded up robust features. Website (2006), http://www.vision.ee.ethz.ch/surf/ (visited on July 14, 2009)

Matching Planar Features for Robot Localization

Baptiste Charmette, Eric Royer, and Frédéric Chausse

LASMEA, université Blaise Pascal, Clermont Universites

Abstract. Localizing a vehicle with a vision based system often requires to match and track landmarks whose position is known. This paper tries to define a new method to track some features in modeling them as local planar patches with a monocular camera. First a learning sequence is recorded to compute the planar features and their orientation around landmarks tracked on several views. Then in the localization part, camera pose is predicted and features are transformed to fit with the scene as seen by the camera. Landmarks can then easily be matched and position is computed more accurately. With this method many features can be tracked on longer sequences than with standard methods, even if the camera is moving away from the learning trajectory. This improves the localization.

1   Introduction

An important problem in mobile robotics consists in localizing the robot at any time with a good precision. To have the robot moving from one place to another, it is very important to keep robot on the expected trajectory and to correct its movement as soon as necessary. Vision-based systems need only a standard camera and use directly the different visible elements in the area to be localized without artificial landmarks.

Such systems consist generally in creating a map of key points which can be found on the current view of the camera. Analyzing how the position of the landmarks on the image evolves lead to the movement of the robot. The map can be generated either in the same process as the localization, with a SLAM (Simultaneous Localization and Mapping) algorithm or in a learning part while the robot is being manually driven. In our case, we consider a learning stage. The vehicle is first driven in the test area to acquire a video sequence. Then the map is computed off-line. After that, the robot can move automatically and use the map to find its position in real-time.

In both situations - SLAM or use of pre-learned map - images features have to be associated with data from the map. Usually to realize this data association some points are detected and matching is done with points in the map. The main problem consists in matching points seen from different viewpoints.

To do that, many researches try to encode landmarks in such a way they do not change with the viewpoint. Classical descriptors such as the SIFT [1] or the SURF [2] use directly information of one image to generate a set of values which are robust to viewpoint change. Other works [3], [4] use several images of

G. Bebis et al. (Eds.): ISVC 2009, Part I, LNCS 5875, pp. 201–210, 2009.

the same point and consider matching in different viewpoint as a classification problem.

Other researches, instead of searching for a viewpoint independent descriptor, try to define a 3D model of the landmarks and adapt it to the viewpoint. Main contributions [5], [6] close to our work consider landmarks as points lying on locally planar surfaces. The features can then be projected in the current view to have the same appearance as the searched landmark.

The main difference between these methods is when two different points look similar but are located in different positions, for example a building with several identical windows. Viewpoint invariant descriptors will probably be approximately the same for all points because all feature are altmost identical and so window from different places could match. On the contrary the planar features adapted to the current view will have a different projection induced by the position change, and even when the features had the same appearance, they do not match. But a constraint of the feature projection is the need of an approximate pose to know the current view. As in our case the robot movement is easily modeled, pose can be predicted, and this constraint is not very annoying. Our work will so use planar feature to model the landmarks. Contrary to [6] our approach uses only one camera to evaluate the planar feature, and instead of using direct computing as in a SLAM context [5], the learning part enables to use more points and more accurate features on every images.

Section 2 presents the initialization part where the planar features are recorded. Localization with the generated map is described in section 3, and results are shown in section 4.

2   Computing the Planar Features

In our system, the robot is first driven manually through the area to record an image sequence with the on-board camera. With this sequence, the 3D map is built with a structure from motion algorithm published in [7]. This algorithm consists in detecting interest points with the Harris detector [8], and tracking them along the image sequence. The essential matrices are then computed and 3D coordinates of the points are found with triangulation. Bundle adjustment described in [9] is finally used to refine the solution.

After this part each pose i is known and Rotation matrix R_i and translation vector T_i are computed as shown on figure 1.

2.1   Computation of the Features Orientation

As shown on figure 1 we consider that each point lies on a locally planar surface. In order to model it, we need to compute the normal N to the surface.

As explained in [10], when two cameras observe a plane from different viewpoint, we know that an homography $H_{1\to2}$ can be defined to link every point with homogeneous coordinates P_1 on the first image by the corresponding point with homogeneous coordinate P_2 on the second image with the equation

$$P_2 = H_{1\to2}P_1 \tag{1}$$

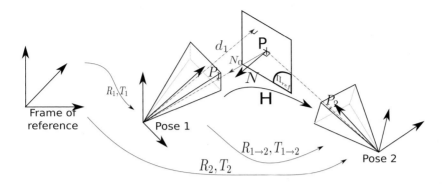

Fig. 1. Patch viewed by two cameras

$H_{1\to2}$ is the homography induced by the plane and can be expressed by

$$H_{1\to2} = d_1 R_{1\to2}(I - \boldsymbol{T_{1\to2}}\boldsymbol{N_1}^t) \tag{2}$$

$R_{1\to2}$ and $\boldsymbol{T_{1\to2}}$ are the rotation matrix and translation vector to transform the pose 1 in pose 2, d_1 is the algebraical distance from the center of camera 1 to the plane, $\boldsymbol{N_1}$ is the normal to the plane expressed in the reference associated to camera 1 and I is the 3×3 identity matrix

The homography $H_{1\to2}$ is computed by minimizing the cost function C:

$$C = \sum_{(i,j)}[I_1(i,j) - I_2(H_{1\to2}(i,j))]^2 + [I_1(H_{1\to2}^{-1}(i,j)) - I_2(i,j)]^2 \tag{3}$$

with (i,j) coordinates in the neighborhood of the analyzed point $\boldsymbol{P}$ - in our case a 20 pixels-side square centered around $\boldsymbol{P}$. This minimization is done with a Levenberg-Marquard method.

Equation 2 shows that the only unknown is the orientation of $\boldsymbol{N}$. The unit vector $\boldsymbol{N_0}$ which comes from $\boldsymbol{P}$ to the center of the camera 1 is set as an initial value and 2 other vectors $\boldsymbol{u_1}$ and $\boldsymbol{u_2}$ are defined to have an orthonormal basis $(\boldsymbol{N_0}, \boldsymbol{u_1}, \boldsymbol{u_2})$. Then the scalar parameters α and β define vector $\boldsymbol{N}$ with expression

$$\boldsymbol{N} = \boldsymbol{N_0} + \alpha\boldsymbol{u_1} + \beta\boldsymbol{u_2} \tag{4}$$

With these 2 parameters, $\boldsymbol{N}$ can have all the possible values in a half sphere in direction to the camera. As the plane has been seen on the image, the direction can indeed not be on the opposite.

Better results can be obtained if more images are used and if the viewpoint changes a lot between the different images. To keep reasonable calculation time, only 3 images are used, for example the first image, the last one and another image in the middle of the sequence to have a wide basis. The cost function, in case of more than 2 images, becomes

$$C = \sum_{n_1 \neq n_2} \sum_{(i,j)}[I_{n_1}(i,j) - I_{n_2}(H_{n_1\to n_2}(i,j))]^2 \tag{5}$$

Here the $H_{n_1 \to n_2}^{-1}$ is not used because it is the same as $H_{n_2 \to n_1}$ and the corresponding cost will be added in the sum.

When we look at the result on the image, we can see that most of the normals calculated with this method seem accurate except in case of bad matching or when the detected points are not lying on a plane. To detect these problems, correctness of the patch has to be evaluated, and it will be done in the next section.

After all these operations, every landmark is associated with its 3D coordinates and the orientation of the local planar surface around it.

2.2   Generation of the Planar Textures

To use the planar features in the localization, the texture of the plane around the point is necessary, in order to compare it with the current view. The goal of this part is to compile all the images where the point was matched from different viewpoints, and use it to find the best texture to apply on the plane. Then it is necessary to control the correctness of the obtained image, and to determine from which viewpoint the image can be seen.

Texture generation. We generate a virtual view Π_{ref} as the projection of the patch in a virtual camera. After that every image is transformed to generate Π_{ref}. The virtual camera defines the patch appearance. The first idea could be to define a fronto-parallel pose but this pose could be very different from the original image, as seen in the figure 2. To avoid important resampling and define Π_{ref} close to the initial view, the center C of the virtual camera is set as the average position of the centers C_i of the original camera.

The virtual pose is then oriented so that P is on the optical axis, to put it in the center of the image. The homography $H_{vir \to i}$ which gives for each pixel (x, y) in the virtual view, its coordinates in the image i can be computed. A simple mean of the pixel value in every image gives the value in Π_{ref} with equation 6

$$\Pi_{ref}(x, y) = \frac{1}{N_{view}} \sum_{i=0}^{N_{view}} I_i(H_{vir \to i}(x, y)) \tag{6}$$

With this method, Π_{ref} is computed using every image where P was found. As the value is computed separately for each pixel, dimensions of Π_{ref} can be chosen arbitrary, for example a 20 pixels-large square.

Evaluation of the correctness. Sometimes, particularly in case of wrong matching, Π_{ref} is blurred and can not be reliable. To eliminate this kind of defaults, it is essential to evaluate the correctness of the patches. A way to perform this evaluation is to try to match Π_{ref} with the original views.

First, we use $H_{vir \to i}$ to generate an image Π_i, projection of the image I_i in the virtual view i. Then a Zero-mean Normal Cross Correlation (ZNCC) is used to compare Π_{ref} and Π_i. A threshold (0.8 in our test) is used to determine if

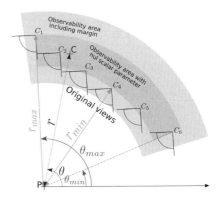

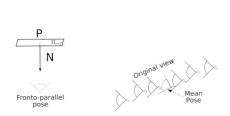

Fig. 2. Position of the pose relatively to the plane in top view. For comparison fronto parallel view is drawn to show how far from the original view it can be.

Fig. 3. Top view of a point P and its observability area defined with the original views

they match together. Then two criterias are used to evaluate the correctness of the patch:

- the number of original views $N_{Matched}$ which match with the point have to be greater than 3
- The ratio of $N_{Matched}$ to the number of view N_{view} to generate Π_{ref} have to be greater than 90%

After that, it is possible to eliminate every unreliable point, in order to avoid bad matches and to reduce calculating time.

Observability area. This area is based on the positions where the patch has been observed initially and we can consider it is not visible far from this place. The observability area is illustrated on figure 3. We consider a camera with center C, this camera can see the point P if C is in the observability area of P. C_i denote the center of each camera i initially observing P. The observability area is more easily defined with spherical coordinates with origin in P and angle θ considered in the horizontal plane. As the vehicle movement is such as the distance from the camera to the ground is not changing, then the ϕ coordinate is not used for the computation. Every C_i has coordinates (r_i, θ_i, ϕ_i). The observability area $Obs(P)$ is defined by expression 7

$$(r, \theta, \phi) \in Obs(P) \Leftrightarrow \begin{cases} r_{min}(1 - \rho_r) - \lambda_r \leq r \leq r_{max}(1 + \rho_r) + \lambda_r \\ \theta_{min}(1 - \rho_\theta) - \lambda_\theta \leq \theta \leq \theta_{max}(1 + \rho_\theta) + \lambda_\theta \end{cases} \quad (7)$$

with r_{min} and r_{max} minimal and maximal value of r_i, θ_{min} and θ_{max} minimal and maximal value of θ_i and ρ_r, λ_r, ρ_θ, λ_θ scalar parameters used to define the margin area to detect the point close to the original view.

Finally a test can quickly be computed to know if a point P can be seen from the position C, considering where the point was seen and a small interval around.

After all these computations the planar features can be stored. To sum up we need for the next step only:

- 3D coordinates of the feature
- Normal orientation
- Planar texture and virtual pose associated
- Observability limits

3   Localization Using Planar Features

Figure 4 shows the global view of the localization algorithm for an iteration i.

The current pose i is predicted using the pose $i - 1$ and a kinematic model of the robot. This prediction enables to generate patch reprojection in the current view and their uncertainty area. As the center of the camera is approximatively known the observability limits can be checked and planar patches which are not visible can be eliminated.

For every visible patch, the homography $H_{img \rightarrow vir}$ to link coordinates in the image to the coordinates in the virtual view is computed and a new patch Π_{img} is generated. After repeating this part on every visible patch, we obtain the patches as seen by the predicted camera and can use it to try to match it in the image.

At the same time, a point detector is applied on the current image to find every interest point IP. The detector is the same as the one used in the structure from motion algorithm [7] described in section 2 to assure that the same points are found.

Then, every interest point IP of the image lying in the uncertainty area is considered as a potential match with the patch. Π_{img} is translated to put the IP in its center. Then a ZNCC is computed to have a matching score S. A threshold is then used to filter every bad match. In our tests, 0.5 was used as threshold, but other values can either save computation time (higher value) or increase accuracy (lower value). Viewpoint invariance is not necessary because the patches are theoretically the same as the image. As only illumination changes can occur in case of weather or lighting conditions evolutions, ZNCC is used because of its robustness to affine change in luminosity.

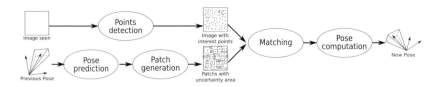

Fig. 4. Chart showing the localization algorithm

After computing the matching score for every patch with all the interest points in its uncertainty area, every matching pairs are compared and if a patch or an IP is used in two different pairs, the pair with the worse score is suppressed. After that, every interest point can be matched with only one patch, and every patch matches with only one IP. With these pairs, the 2D coordinates of the IP can be linked with the 3D coordinates of the planar feature.

Algorithm proposed in [11] is finally used with these coordinates to compute the new camera pose by minimizing the reprojection error.

4   Results

This algorithm has been tested in an urban area with buildings around.

The learning part is made with an images sequence shot while the vehicle moves along the right side of the road. The trajectory is 68 meters long. The whole sequence is made of 509 images. 6145 landmarks were initially tracked, to generate 4033 accurate patches which are tracked on an average of 58 images. The figure 5 shows some images from this sequence and a top view (e) of the trajectory generated by the structure from motion algorithm. In the part (b) and (d), the generated features are all transformed to show the image in the same view as in (a) and (c).

The localization part is computed on an other sequence where the vehicle was moving on a different path in the same area. Figure 6 shows an image from a sequence where the robot was zigzagging from the left to the right side of the road. Image (e) shows the top-view generated by the localization algorithm. The robot never drives further than 3.5 m from the reference trajectory with a

Fig. 5. Images extracted from the learning part: (a) and (c) images viewed by the camera, (b) and (d) generated features in the same view as the image. (e) top view of the trajectory with every pose (black square) and landmarks seen around.

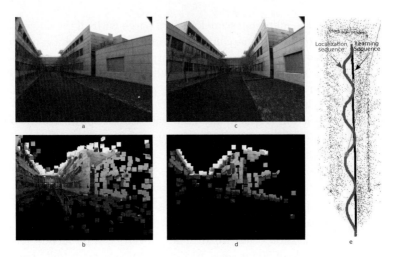

Fig. 6. Images extracted from the localization part: (a) image viewed by the camera. (b) generated features in the same view as the image with all the landmarks. (c) closest image taken in the learning part. (d) features matching the view (e) top view of the trajectory with landmarks.

maximal angular deviation of 30°. The top images show the difference between the current image (a) and the reference (c). A viewpoint change can clearly be evaluated by comparing these images. On the bottom, images (b) and (d) show the planar features transformed to have the same view as the image. Despite the viewpoint change, the transformed features look like the image. This is an improvement of the matching process.

To test the benefit of the system, the number of correct matches on the sequence has been compared with another algorithm. Instead of using generated planar features, the matching is done directly with the reference images. Two methods have been tested, using the ZNCC for matching and with the SURF descriptor ([2]) which is known to be almost viewpoint invariant. The surf descriptor used was computed directly with the library provided by its author. As the SURF descriptor needs the scale factor to be accurate, the detector provided with the SURF library was used instead of the Harris detector. In order to compare the methods with the same data, this detector has been used with all the methods, to have the same detected point and reference point, and the only change is the descriptor used for the matching. Figure 7 shows the number of correct matches on every image of the sequence using the different methods when the vehicle was zigzagging.

The use of planar features increases clearly the number of correct matches compared with the other methods. Particularly when the vehicle is moving far from its first trajectory (around image 100, 300, 550, 650 and 800 on the figure 7), the method based only on the original image and ZNCC descriptor can hardly localize the vehicle because of the lack of correct matching. In this sequence, the direct matching has even lost the position of the robot around the 800th image.

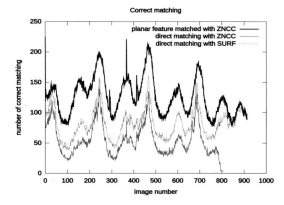

Fig. 7. Number of correct matches along the 915 images sequence where the vehicle is zigzagging on the road. The method with planar features (in thick black line) is compared with method using simple descriptors computed directly with the closest reference image.

On the contrary the use of planar feature enables to localize easily the vehicle with more than 100 correct matches.

At this time the localization algorithm needed around 2 seconds per image with a core 2 duo 2.5 GHz processor. But this C++ code has not been optimized yet. We think that after some improvement, this algorithm can be used in real time with several images per seconds.

However in order to use the planar features we need to predict the camera position first. As the planar patches are projected as seen in the predicted view, they are highly dependent on this prediction. This remains the main drawback and can be a real problem for the initialization with the first image of the sequence, where position has to be estimated from nothing. A way to initialize the position could be to use a low cost GPS, or using global localization algorithm such as Monte Carlo algorithm [12].

The planar features give an other advantage in memory occupation. Instead of saving lots of previous observations, to compare them from different positions, only one patch is saved with coordinates of the normal.

5   Conclusion

This matching method explicitly enables to deal with the change of viewpoint. We have shown that this representation gives many reliable reference lanmarks, which can be matched from different viewpoints. Results are better than the descriptors which are known to be robust to viewpoint changes. Instead of using a lot of different small features this kind of representation allows to keep only one reference as accurate as possible, and uses less memory. Moreover this representation allows to go further from the initial learning trajectory which can be useful for example to avoid some obstacles on the way.

A future issue could be to become more robust to the lighting change. Most of the descriptors are robust to affine illumination change but due to shadows and varying weather conditions, the change is not always affine. An improvement could be made by changing ZNCC in an other descriptor more robust to illumination changes.

References

1. Lowe, D.G.: Object recognition from local scale-invariant features. In: ICCV, pp. 1150–1157 (1999)
2. Bay, H., Tuytelaars, T., Van Gool, L.: Surf: Speeded up robust features. In: Leonardis, A., Bischof, H., Pinz, A. (eds.) ECCV 2006. LNCS, vol. 3951, p. 404. Springer, Heidelberg (2006)
3. Calonder, M., Lepetit, V., Fua, P.: Keypoint signatures for fast learning and recognition. In: Forsyth, D.A., Torr, P.H.S., Zisserman, A. (eds.) ECCV 2008, Part I. LNCS, vol. 5302, pp. 58–71. Springer, Heidelberg (2008)
4. Lepetit, V., Lagger, P., Fua, P.: Randomized trees for real-time keypoint recognition. In: CVPR, pp. 775–781. IEEE Computer Society, Los Alamitos (2005)
5. Molton, N., Davison, A., Reid, I.: Locally planar patch features for real-time structure from motion. In: BMVC (2004)
6. Berger, C., Lacroix, S.: Using planar facets for stereovision SLAM. In: IROS, pp. 1606–1611. IEEE, Los Alamitos (2008)
7. Royer, E., Lhuillier, M., Dhome, M., Lavest, J.M.: Monocular vision for mobile robot localization and autonomous navigation. International Journal of Computer Vision 74, 237–260 (2007)
8. Harris, C., Stephens, M.: A combined corner and edge detector. In: Alvey vision conference, vol. 15, p. 50 (1988)
9. Triggs, B., McLauchlan, P., Hartley, R., Fitzgibbon, A.: Bundle adjustment-a modern synthesis. LNCS, pp. 298–372 (1999)
10. Hartley, R., Zisserman, A.: Multiple View Geometry in Computer Vision, 2nd edn. Cambridge University Press, Cambridge (2004)
11. Araujo, H., Carceroni, R., Brown, C.: A fully projective formulation to improve the accuracy of Lowe's pose-estimation algorithm. Computer vision and image understanding(Print) 70, 227–238 (1998)
12. Thrun, S., Fox, D., Burgard, W., Dellaert, F.: Robust Monte Carlo localization for mobile robots. Artificial Intelligence 128, 99–141 (2001)

Fast and Accurate Structure and Motion Estimation

Johan Hedborg, Per-Erik Forssén, and Michael Felsberg

Department of Electrical Engineering
Linköping University, Sweden
hedborg@isy.liu.se

Abstract. This paper describes a system for structure-and-motion estimation for real-time navigation and obstacle avoidance. We demonstrate a technique to increase the efficiency of the 5-point solution to the relative pose problem. This is achieved by a novel sampling scheme, where we add a distance constraint on the sampled points inside the RANSAC loop, before calculating the 5-point solution. Our setup uses the KLT tracker to establish point correspondences across time in live video. We also demonstrate how an early outlier rejection in the tracker improves performance in scenes with plenty of occlusions. This outlier rejection scheme is well suited to implementation on graphics hardware. We evaluate the proposed algorithms using real camera sequences with fine-tuned bundle adjusted data as ground truth. To strenghten our results we also evaluate using sequences generated by a state-of-the-art rendering software. On average we are able to reduce the number of RANSAC iterations by half and thereby double the speed.

Structure and motion (SaM) estimation from video sequences is a well explored subject [1,2,3]. The underlying mathematics is well understood, see e.g. [1], and commercial systems, such as Boujou by 2d3 [4], are used in the movie industry on a regular basis. Current research challenges involve making such systems faster, more accurate, and more robust, see e.g. [2,3]. These issues are far from solved, as is illustrated by the 2007 DARPA urban challenge [5]. In the end, none of the finalists chose to use the vision parts of their systems, instead they relied soley on LIDAR to obtain 3D structure. Clearly there is still work to be done in the field.

This paper aims to increase the speed and accuracy in structure-and-motion estimation for an autonomous system with a forward looking camera, see figure 1. Although on a smaller scale, this platform has the same basic geometry and motion patterns as the DARPA contenders, and as the vision based collision warning systems developed for automotive applications. In such systems, estimated 3D structure can be used to detect obstacles and navigable surfaces.

When dealing with forward motion there are a number of problems that must be adressed. The effective baseline is on average much smaller than for the sideways motion case, resulting in a more noise sensitive structure estimation. A tracked point feature near the camera often has a short lifespan because it quickly

G. Bebis et al. (Eds.): ISVC 2009, Part I, LNCS 5875, pp. 211–222, 2009.

Fig. 1. Left: Robotic car platform. Upper right: Frame from a forward motion sequence. Lower right: Estimated structure using the proposed algorithm.

moves out of the the visual field. Unfortunately, such points also contain most of the structural information [6]. Forward motion also produces large scale changes in some parts of the image, and this can be a problem for some trackers.

This paper studies the calibrated SaM formulation, which has several advantages over the uncalibrated formulation. In calibrated SaM, estimated cameras and structure will be in Euclidean space instead of a projective space, and we can use more constrained problem formulations [1]. Planar-dominant scenes are not an issue when doing calibrated five point pose estimation [7], which turns out to be a very desirable property when doing autonomous navigation, as these kinds of scenes are quite common. Note also that in autonomous navigation the camera is often fixed, which makes calibration of the camera straightforward.

We should also note that monocular SaM has an inherent scale ambiguity [1]. Despite this, the estimated structure can still be effectively used for obstacle avoidance if *time-to-collision* is used as the metric [8].

The main contributions of this paper are:

1. Introduction of a distance constraint that significantly reduces the number of RANSAC iterations in the five point algorithm, while retaining the pose accuracy. Even when all correspondences in a sample are inliers, their distribution in space makes a big difference. This idea is very easy incorporate. Somewhat surprisingly it does not appear to be described elsewhere.
2. Experimental evaluation of a recently introduced outlier rejection technique for the KLT tracker [9], in the SaM setting. This technique adds an outlier rejection step already in the tracking algorithm. With respect to performance this is used to move calculations from the CPU to a GPU.

We also demonstrate how sophisticated rendering techniques can be used for controlled evaluation of the system.

1   Previous Work

The use of calibrated epipolar geometry in computer vision was pioneered by Longuet-Higgins in his seminal work [10]. The minimal number of point correspondences in the calibrated case is five, and the current state-of-the-art five-point method is the one introduced by Nistér [7]. The exact solution involves cubic constraints, which result in a polynomial with 13 roots. Nistér reduced the number of roots to 10 and provided very efficient solutions to each step of the algorithm. His paper also describes how to add a third view, by solving the perspective-three-point problem [11]. We use this complete three view method in our paper, and refer to it as the Nistér three view method.

Ever since the original RANSAC algorithm was introduced [11], many modifications to the algorithm have appeared. Some methods assume prior information of which points are likely to be inliers, and use this to bias the sampling, e.g. PROSAC. Others estimate the point inlier likelihoods as they go [12], for instance using the point residual distributions [13]. Others discard samples (i.e. groups of points), before scoring them against a model, by comparing the sample points against the model. R-RANSAC [14] and preemptive RANSAC [15], are examples of this. In our setup model estimation is relatively expensive, so it would be better if we could discard a sample even before computing the model. This is exactly what our constraint does, and in this respect, it is similar to NAPSAC [16], which selects points that lie close together when estimating hyperplanes. But, as we will show, for our problem it is on the contrary better to select points that are far apart.

Wu et al.[17] have shown how the KLT tracker can be improved by simultaneously tracking both forwards and backwards in time. Another approach is to simply run the tracker again, backwards in time, and reject trajectories that do not end up at the starting point [9]. We will use the latter approach, and demonstrate its effect in the experiment section.

Rendered 3D scenes as synthetic ground truth has a long history in the field of motion estimation, e.g., the famous Yosemite sequence [18]. This was at the time a very complex scene as it had real 3D structure. Baker et al. argued in [9] that the Yosemite sequence is outdated and they introduce a new set of ground truth data. These new datasets use modern rendering software that can accurately model effects such as shadows, indirect lighting and motion blur. We will use similar data of our own design, in the experiment section.

2   Method

2.1   Overview

The real-time SaM method that we are using consists of two steps:

1. Point correspondences are maintained over time using the Kanade-Lucas-Tomasi (KLT) tracking framework [19,20].

2. The relative pose is estimated for 3 cameras, according to the method described in [7]. In this approach, the relative pose is first found between 2 cameras. Then, triangulation and the perspective-3-point algorithm [11] is used to incorporate the third camera. All of this is done for the minimal 5-point case, inside a RANSAC [11] loop.

We will describe these two steps in detail below, as well as the modifications we have added to each step.

2.2 Tracking

We use the KLT-tracker [20] to maintain point correspondences across time. The KLT-tracker is basically a least-squares matching of rectangular patches that obtains sub-pixel accuracy through gradient search. We use a translation-only model between neighbouring frames. Tracking between neighbouring frames instead of across larger temporal windows improves the stability, especially since it helps us to deal with the scale variations that are present in forward motion. When tracking frame-by-frame, the changes in viewing angle and scale are sufficiently low for tracking to work well.

A simple way to increase the quality of the point correspondences is to add an early outlier rejection step. We do this by running the tracker backwards from the current frame and position and checking if it ends up at its initial position in the previous frame. We will call this procedure *track-retrack* from now on. Adding the track-retrack step doubles the amount of computations. However, since our tracker is running on the GPU, which has cycles to spare, this does not affect the overall performance of the rest of our system.

The KLT tracker has successfully been implemented on a GPU by several authors [21,22,23]. Reference [22] shows a speed increase of more than 20x, and can track thousands of patches in real time. Such a large amount of tracked points is not necessary to estimate the camera motion, and we can thus easily afford to run the tracker a second time. In order to fully utilize the GPU, the number of threads of an implementation must be high. While a CPU can efficiently run 2 threads on a 2 core system, the GPU's core is a simpler version of a processor with very high memory latencies, little or no cache and with many SIMD characteristics. To achieve maximum performance from such a design we need many more threads than processors, and as the current high-end hardware has 240 processors, one often needs more than 5000 threads [24].

2.3 Five Point Pose Estimation

The minimal case for relative pose estimation in the calibrated case is five corresponding points seen in two cameras. Currently, the fastest algorithm for this problem is given by Nistér in [7]. It runs in real-time, and thus we have chosen it as our starting point. In this method, the relative pose estimation is extended to three cameras by doing the following within the RANSAC loop: The essential matrix is estimated from five point correspondences between two cameras.

The relative camera position and rotation are then extracted from the essential matrix. From these, the camera projection matrices are created and used to triangulate the five 3D points. The perspective-3-point algorithm [11] is then used to calculate the third camera from three 3D points and their respective projections onto this camera.

One advantage with using the minimal case is that it imposes all available constraints on the estimation. This is especially important when handling more complicated cases like the forward motion case. If one plane is dominant, the uncalibrated case has several solutions. In the calibrated case, the number of solutions is reduced to two, where one can be discarded (as it has the cameras below ground). It is thus not necessary to switch between the homographic and the full epipolar geometry model. We use the Nistér method here, but note that in principle any five point solver would benefit from the improvements we suggest in this paper.

2.4   Distance Constraint

Forward motion in structure from motion is a notoriously difficult case, because of the much smaller *equivalent baseline*[1] created between two cameras than with other types of motions. The forward motion also gives rise to large scale differences in the point correspondence estimation. The point tracking becomes less accurate under these scale transformations. Most of the time we will also have a singular point (the epipole) lying in the image (in the motion direction). At this point the equivalent baseline is zero, and it increases towards the edges of the frame.

Computation of the relation between two cameras by estimating the essential matrix is quite sensitive to the actual 3D positions of the used correspondences. This is demonstrated by a recent discovery by Martinec and Pajdla [3]. In their paper, they show that bundle adjustment using only four carefully chosen correspondences between each pair of views can be almost as accurate (and much faster) as using all correspondences. These four points are chosen to be maximally distant in the 3D space with metric determined by the data covariance. However, for a direct solution of SaM this is too expensive as it requires triangulated 3D points. Instead, our proposal is to look at the projections in the image plane. The rationale is that points that are distant in the image plane are *likely* to be distant also in 3D space.

The standard approach when solving structure and motion with [7] is to place the minimal case five point solver inside a RANSAC loop. Our proposal is to add a simple distance test inside the loop, before the minimal case solver. With this test we put a minimum distance constraint on the randomly chosen image points, $\mathbf{x}$. Only sets of five points $\mathbf{x}_1,\ldots,\mathbf{x}_5$, that satisfy:

$$||\mathbf{x}_i - \mathbf{x}_j||_2 > T, \ \forall\{(i,j) : i,j \in [1..5], j > i\}, \tag{1}$$

---

[1] By equivalent baseline, we informally mean the distance between the camera centers when projected onto the image plane.

reconstruction of *concrete* sequence   reconstruction of *grass* sequence

Fig. 2. The top four subplots show the used frames from the real world sequences. The indoor sequence (top left) is chosen as it has plenty of occlusions. The three outdoor scenes are captured while driving the robotic car forward on different terrains: concrete, gravel and grass. The two lower images are the three-view reconstructions of two of the sequences (a navigable planar surface is also estimated and textured for illustration).

will be used for pose estimation. For other sets the sampling is run again. We use the threshold $T = 0.1$ on distances in normalized image coordinates. This corresponds to approximately 150 pixels in our sequences (or about 10% of the image diagonal). This value gave a reasonable compromise between the number of resamplings, and the precision obtained. We have not done any extensive tests on the exact value to use.

The computational load of the point pre-selection procedure is small compared to the 5 point solver. It consists of 5 `rand()` function calls, 10 conditional instructions and some simple arithmetical operations. If necessary this can be further optimized by different gridding methods. The average time to compute one sample on our platform (in one thread on an Intel 2.83GHz Q9550 CPU) in standard C++ is 0.1 microseconds. On our datasets this is done on average 2-4 times, for $T = 0.1$. The three-view five-point method is reported to take 140 microseconds in 2004 [7]. Accounting for CPU speedups gives us about 35 microseconds, or 87-175x more than our sampling step.

3   Evaluation

Evaluation was carried out on four real world sequences captured with a Point-Grey Flea2 camera (1280 × 960 at 15 Hz) with pure forward motion, see top

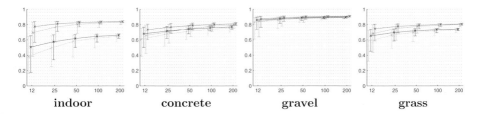

indoor   concrete   gravel   grass

Fig. 3. Statistics of inlier/outlier ratios as a function of number of RANSAC iterations (12, 25, ...). Each graph shows results for one sequence. Each experiment is repeated 500 times, and the median, and 5% and 95% quantiles are plotted. The four curves (in left-right order within each group) show results without distance constraint and without track-retrack (CYAN), with distance constraint and without track-retrack (BLUE) without distance constraint and with track-retrack (GREEN), with distance constraint and with track-retrack (RED).

of figure 2. OpenCV's software library is used to calibrate the camera from a checkerboard pattern [25]. We have also chosen to use OpenCV's KLT implementation in the experiments to make it easier for others to reproduce our results.

Additionally we have generated two synthetic sequences, shown in figure 4. We have used a rendering software called Mental ray that has a wide variety of modelling capabilities such as complex geometry, soft shadows, specular highlights and motion blur. These are effects that impact the performance of the SaM, and we would like to further investigate this in the future. For now we have just used them with settings that give footage similar to the real camera. Besides being used in many movies, Mental ray was also used in [9] to generate image sequences and ground truth for optical flow.

We will use the real-world sequences together with the synthetic sequences to evaluate the efficiency of the distance constraint. We use two measures in the experiments:

- **Inlier Frequency.** We count the number of inliers in the best model found by RANSAC. This is the criterion that RANSAC itself tries to maximise, and thus it demonstrates how much our modifications have assisted RANSAC.
- **Model Precision.** We evaluate the scale normalised position of the third camera. On real sequences, this is done by comparing our estimate against the output of the bundle-adjustment algorithm described in [26]. On the synthetic sequences, we know the exact locations of each camera, and use that as ground truth.

3.1   Inlier Frequency Evaluation

In each sequence, we have used 3 images to compute SaM, and this has been done for 12, 25, 50, 100 and 200 RANSAC iterations. This procedure is run 500 times, and from this we calculate the median and the 5 and 95 percentiles to show where 90% of the estimates lie. This kind of evaluation is used for all graphs in the paper. To evaluate the performance of the distance constraint the same setting is run both with, and without the constraint.

In figure 3 we give graphs of the expected inlier frequency in the best model found by RANSAC, as a function of the number of RANSAC iterations. These graphs clearly show an inlier increase when the distance constraint is used, and this holds both with, and without the track-retrack step. Almost everywhere, the curves without the distance constraint need more than double the amount of iterations to reach the same inlier frequency. In most of the real sequences we could reduce the number of RANSAC iterations by half and still have the same inlier count as when the distance constraint was not used.

The indoors sequence was chosen to demonstrate an important aspect of the track-retrack scheme: As can clearly be seen in the graphs, the improvement caused by track-retrack is much bigger in the indoor sequence than in the others. The reason for this is that track-retrack is very effective at detecting outliers caused by occlusions.

The same test is run for the synthetic test data, see figure 4. The behaviour here is nearly identical to the evaluation with real images, and we can also observe that we can reduce the number of RANSAC steps by approximately half with maintained inlier/outlier ratio when adding the distance constraint.

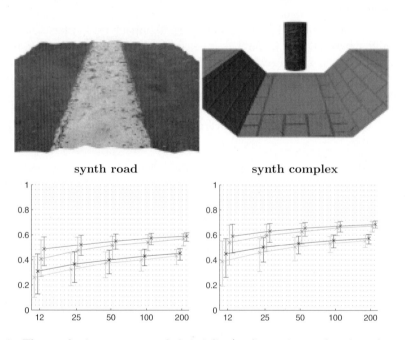

Fig. 4. The synthetic sequences and their inlier/outlier ratios, as function of number of RANSAC iterations. Same legend as figure 3.

The first scene consists of a forward motion on a planar road and the second is also forward motion but on a more complex scene. For each synthetic sequence, we have also generated images of calibration patterns. This allows us to process the synthetic sequences in exactly the same manner as the real ones.

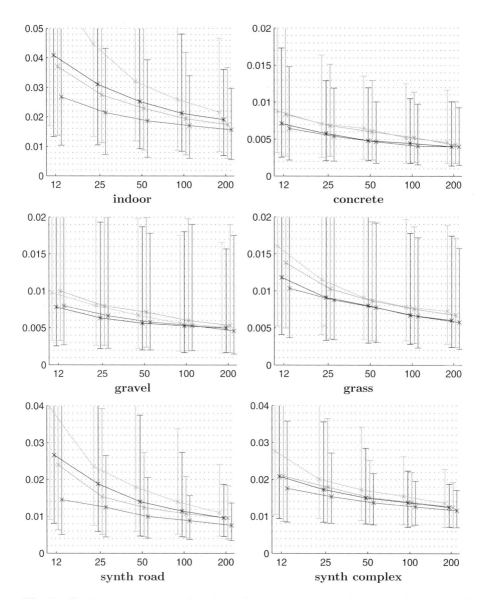

Fig. 5. Position errors on real and synthetic sequences as function of number of RANSAC iterations. The sequences are evaluated against bundle adjustment output on the same data set. The synthetic sequences are evaluated against their ground truth. The four methods are coloured as in figure 3.

3.2 Precision Evaluation

As calibrated monocular reconstructions are only defined up to scale, accuracy evaluation is rather problematic. We have chosen to evaluate the accuracy of the obtained SaM solutions using the position error of the third camera. For this to be possible, we need first to adjust the distance of the second camera to be the same as in the ground truth (here we set this distance to be 1). Only after this normalisation are we able to compare positions of the third camera.

In the absence of real ground truth on the real sequences, we have used the output of the bundle-adjustment (BA) algorithm described in [26]. Bundle-adjustment is the maximum likelihood estimate of SaM, and has been shown to greatly improve the results [27]. On the synthetic sequences we simply save the camera locations used for generating the frames. Note that we get similar results on both real and synthetic sequences, which supports the use of BA for evaluation purposes.

Figure 5 shows the absolute position error of the third camera in each triplet. Here we can see that the position error follows the same trend as the inlier/outlier ratio. If anything the improvement is even more pronounced here.

We have summarised the results of the precision experiments in table 1 and 2. Table 1 shows the speed increase (i.e. reduction in number of RANSAC samplings needed to obtain the same precision) from the distance constraint, when the track-retrack step is disabled, and table 2 shows the speed increase with track-retrack enabled.

The percentages are obtained as follows: For each number of iterations, we look up the precision with the distance constraint enabled. We then estimate how many iterations it would take to archieve the same precision when the constraint is not used. The estimate is found through linear interpolation (note although

Table 1. Speed increase from the distance constraint, with track-retrack disabled (i.e. CYAN curve vs. BLUE)

#iterations	synt. r.	synt. c.	indoor	concrete	gravel	grass
12	180%	190%	230%	198%	216%	193%
25	184%	198%	215%	256%	234%	199%
50	199%	225%	216%	265%	199%	187%
100	185%	194%	>200%	>200%	162%	>200%

Table 2. Speed increase from the distance constraint, with track-retrack enabled (i.e. GREEN curve vs. RED)

#iterations	synt. r.	synt. c.	indoor	concrete	gravel	grass
12	229%	211%	215%	243%	201%	197%
25	198%	191%	242%	277%	244%	195%
50	264%	209%	235%	258%	232%	183%
100	>200%	191%	>200%	>200%	>200%	>200%

the graphs are in log scale, the interpolation is done on a linear scale). The speed increase is now the ratio of the two iteration counts. There are some cases where the intersection point lies after the 200 iteration value, these values could have been extrapolated, but we have instead chosen to just show them as >200.

Note that a speed increase computed in this way does not take the extra overhead of the sampling into account. But, as shown in section 2.4 this overhead is 0.6-1.2% of the total time, and as mentioned, there are ways to reduce this even further.

4   Conclusions

We have introduced a method that significantly speeds up the current state of the art SaM algorithm for calibrated cameras. On average we are able to reduce the number of RANSAC iterations by half and thereby double the speed. The improvement is achieved by adding a distance constraint to the point selection inside the RANSAC loop. We have also added an outlier rejection step (which we call track-retrack) to the KLT tracker. For scenes with high level of occlusion the track-retrack scheme also gives a similar improvement in performance. For scenes with little or no occlusion, however, the difference is negligible. Note also that the two improvements are independent, for scenes where the track-retrack scheme gives and improvement, we will get further improvement by adding the distance constraint.

Acknowledgments

The research leading to these results has received funding from the European Community's Seventh Framework Programme (FP7/2007-2013) under grant agreement no 215078, DIPLECS.

References

1. Hartley, R., Zisserman, A.: Multiple View Geometry in Computer Vision, 2nd edn. Cambridge University Press, Cambridge (2003)
2. Pollefeys, M., et al.: Detailed real-time urban 3D reconstruction from video. International Journal of Computer Vision 78, 143–167 (2008)
3. Martinec, D., Pajdla, T.: Robust rotation and translation estimation in multiview reconstruction. In: IEEE CVPR 2007 (2007)
4. 2d3 Ltd., B. (2009), http://www.2d3.com
5. Various: Special issues on the 2007 DARPA urban challenge, parts I- III. In: Buehler, M., Iagnemma, K., Singh, S. (eds.) Journal of Field Robotics, vol. 25. Wiley, Blackwell (2008)
6. Vedaldi, A., Guidi, G., Soatto, S.: Moving forward in structure from motion. In: IEEE International Conference on Computer Vision (2007)
7. Nistér, D.: An efficient solution to the five-point relative pose problem. IEEE TPAMI 6, 756–770 (2004)

8. Källhammer, J.-E., et al.: Near zone pedestrian detection using a low-resolution FIR sensor. In: Intelligent Vehicles Symposium, IV 2007 (2007)
9. Baker, S., Scharstein, D., Lewis, J.P., Roth, S., Black, M.J., Szeliski, R.: A database and evaluation methodology for optical flow. In: IEEE ICCV (2007)
10. Longuet-Higgins, H.: A computer algorithm for reconstructing a scene from two projections. Nature 293, 133–135 (1981)
11. Fischler, M., Bolles, R.: Random sample consensus: a paradigm for model fitting, with applications to image analysis and automated cartography. Communications of the ACM 24, 381–395 (1981)
12. Tordoff, B., Murray, D.: Guided sampling and consensus for motion estimation. In: Heyden, A., Sparr, G., Nielsen, M., Johansen, P. (eds.) ECCV 2002. LNCS, vol. 2350, pp. 82–96. Springer, Heidelberg (2002)
13. Zhang, W., Kosecka, J.: A new inlier identification scheme for robust estimation problems. In: Robotics Science and Systems RSS02 (2006)
14. Chum, O., Matas, J.: Randomized RANSAC with t(d, d) test. In: BMVC 2002, pp. 448–457 (2002)
15. Nistér, D.: Preemptive RANSAC for live structure and motion estimation. Machine Vision and Applications 16, 321–329 (2005)
16. Myatt, D., et al.: NAPSAC: High noise, high dimensional model parametrisation - it's in the bag. In: BMVC 2002 (2002)
17. Wu, H., Chellappa, R., Sankaranarayanan, A.C., Zhou, S.K.: Robust visual tracking using the time-reversibility constraint. In: IEEE ICCV (2007)
18. Barron, J.L., Fleet, D.J., Beauchemin, S.S.: Performance of optical flow techniques. International Journal of Computer Vision 12, 43–77 (1994)
19. Lucas, B.D., Kanade, T.: An iterative image registration technique with an application to stereo vision. In: IJCAI 1981, pp. 674–679 (1981)
20. Shi, J., Tomasi, C.: Good features to track. In: IEEE Conference on Computer Vision and Pattern Recognition, CVPR 1994, Seattle (1994)
21. Ringaby, E.: Optical flow computation on compute unified device architecture. In: Proceedings of SSBA 2009 (2009)
22. Sinha, S.N., Frahm, J.M., Pollefeys, M., Genc, Y.: GPU-based video feature tracking and matching. Technical report, UNC Chapel Hill (2006)
23. Hedborg, J., Skoglund, J., Felsberg, M.: KLT tracking implementation on the GPU. In: Proceedings SSBA 2007 (2007)
24. Kirk, D., Hwu, W.-m.W.: The CUDA programming model (2007)
25. Open source computer vision software library, O.A. (2009), http://opencv.willowgarage.com/
26. Lourakis, M., Argyros, A.: The design and implementation of a generic sparse bundle adjustment software package based on the Levenberg-Marquardt algorithm. Technical Report 340, Institute of Computer Science - FORTH (2004)
27. Engels, C., Stewénius, H., Nistér, D.: Bundle adjustment rules. In: Photogrammetric Computer Vision, PCV (2006)

Optical Flow Based Detection in Mixed Human Robot Environments*

Dario Figueira, Plinio Moreno, Alexandre Bernardino, José Gaspar,
and José Santos-Victor

Instituto Superior Técnico & Instituto de Sistemas e Robótica
1049-001 Lisboa, Portugal
{dfigueira,plinio,alex,jag,jasv}@isr.ist.utl.pt

Abstract. In this paper we compare several optical flow based features in order
to distinguish between humans and robots in a mixed human-robot environment.
In addition, we propose two modifications to the optical flow computation: (i) a
way to standardize the optical flow vectors, which relates the real world motions
to the image motions, and (ii) a way to improve flow robustness to noise by selecting the sampling times as a function of the spatial displacement of the target
in the world.

We add temporal consistency to the flow-based features by using a temporal-
Boost algorithm. We compare combinations of: (i) several temporal supports, (ii)
flow-based features, (iii) flow standardization, and (iv) flow sub-sampling. We
implement the approach with better performance and validate it in a real outdoor
setup, attaining real-time performance.

1 Introduction

Current trends in robotics research envisage the application of robots within public environments helping humans in their daily tasks. Furthermore, for security and surveillance
purposes, many buildings and urban areas are being equipped with extended networks
of surveillance cameras. The joint use of fixed camera networks together with robots in
social environments is likely to be widespread in future applications.

The long term goal of this work, integrated in the URUS project[1], adopts this vision. The URUS project aims to achieve the interaction of robots with people in urban
public areas, to improve mobility in downtown areas. A key element of the project is a
monitoring and surveillance system composed by a network of fixed cameras that provide information about the human and robot activities. These multi-camera applications
must also consider constraints such as real-time performance and low-resolution images
due to limitations on communication bandwidth. Thus, It is fundamental to be able to
detect and categorize humans and robots using low resolution and fast to compute features. We propose the use of optical flow derived features to address this problem.

Detection of humans in images is a very active research area in computer vision
with important applications such as pedestrian detection , people tracking and human

---

* Research partly funded by the FCT Programa Operacional Sociedade de Informação(POSI) in
the frame of QCA III, and EU Project URUS (IST-045062).

G. Bebis et al. (Eds.): ISVC 2009, Part I, LNCS 5875, pp. 223–232, 2009.

activity recognition. These approaches aim to model the human limbs by using features such as the silhouette [2,3], image gradient [4], color distribution of each limb [5], optic flow [6,7] and combinations of the features just mentioned. Detection of robots in images have become a very popular field of research in the RoboCup[1] framework, which focus on cooperative robot interaction [8,9,10].We address the unexplored issue of discrimination between these two classes, people and robots, which is essential to the development of algorithms that deal with *e.g.,* surveillance, in mixed human-robot environments. Our approach relies on the motion patterns extracted from optical flow, which have been used previously by Viola *et al.* [6] and Dalal *et al.* [4,7] in order to detect pedestrian in images and videos. Viola *et al.* combine the optical flow with wavelet-based features to model the people appearance, while Dalal *et al.* compute histograms of the flow directions. Our work explores on this latter approach, comparing two types of features:

- Histogram of gradients (HOG), which computes the histogram of the optic flow orientation weighted by its magnitude.
- Motion boundary histogram (MBH), computed from the gradient of the optical flow. Similarly to HOG, this feature is obtained by the weighted histogram of the optical flow's gradient.

Using optical flow to separate robots' movement from people's movement is appealing for its independence on people and robot visual appearances (*i.e.,* color, size or shape), allowing it to model individuals with different outlooks, requiring only "different enough" patterns of movement. Since most current robots are rigid, while people tend to not be rigid at all while moving about, this a reasonable assumption to start with. Also, optical flow is not limited to high resolution images, being able to capture enough information from only a limited amount of pixels.

In order to improve the classifier's accuracy we also test the features with standardized flow described in Section 2. Using localized detections on a world reference frame, we scale the flow to its corresponding real-world metric value, creating invariance to the target's distance to the camera. In addition, we don't use consecutive images to compute flow which we refer to as spatial sub-sampled flow. We store a frame, wait for the target to move in the real world, and only after its displacement is larger than a threshold, we compute the optical flow from the stored frame to the present image.

The histogram-based features provide the data samples for the learning algorithm, GentleBoost [11]. This algorithm is a very efficient and robust classifier that adds the response of several base (weak) classifiers. In addition, we consider the temporalGentleBoost [12], a recent modification of GentleBoost that exploits the temporally local similarities of the features.

In the next section we will describe the features employed to represent the targets. Section 3 describes the learning algorithm used to learn how to distinguish people from robots. We then present some results on a real live setting and finish with the conclusions.

---

[1] http://www.robocup.org/

2   Target Representation

In this work we assume that detection, tracking and localization (in the ground plane) of targets in the camera's field-of-view (FOV) is already done by any of the existing algorithms available in the literature. For instance we use background subtraction [13] for detection, nearest-neighbor for tracking and homographies for localization. Our goal in this paper is to discriminate among different classes of targets using motion cues.

For (dense) optical flow computation, we use the implementation of [14][2], an algorithm that introduces a new metric for intensity matching, based on the unequal matching (i.e. unequal number of pixels in the two images can be correspondent to each other). We chose this algorithm for its good balance between computational load and robustness to noise [14].

2.1   Flow Standardization

The optical flow vectors encode the pixel displacement between two images, independent of the corresponding real displacement. This means that an object closer to the camera will have a large pixel displacement, while the same object will have small flow vectors when moving far away. Thus, it is very difficult to match similar motions by using the features computed directly from the optical flow. In order to overcome this limitation we standardize the flow using the world coordinate locations of the moving objects in the scene.

Given the displacement of each object in the world in metric coordinates, and in the image in pixels, we derive a linear scale factor to relate the optical flow, in the image, to the motion in the world. We illustrate this in Figure 1, where the gray arrows display the optical flow in two different detections, the blue arrows represents the movement in the world and the red arrows encode the mean displacement of the detected bounding boxes in the image. The flow magnitude (f pixels), is larger when the object is close, and smaller when farther away. The average displacement (P pixels) follows the same behavior, while both world displacements (M meters) keep the same value. Therefore the optical flows can be scaled to a similar value ($f.M/P$ meters).

The flow standardization just described assumes that the motions of the target's limbs are aligned to the mean displacement of the target. If this assumption is violated, the motions with other directions are projected to the direction aligned with the mean displacement's vector. Since in general, while a person is moving, their limb motions will be parallel to the motion of her body, the assumption will hold for most of the sequences.

The flow standardization described above causes the flow magnitude to be independent to the target's distance to the camera, but still dependent on the target's velocity. If an individual moves faster in some frames and slower in other frames, its displacement will be different for the respective pairs of frames, so the features extracted will be dissimilar.

We implement spatial sub-sampling of the optical flow in order to provide velocity independence to the features. This comprises, for a given target, the selection of the frames to compute the optical flow based on its displacement. The method includes

---

[2] http://www.cs.umd.edu/~ogale/download/code.html

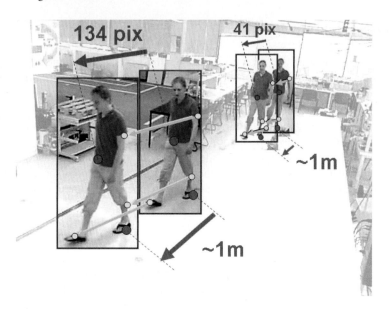

Fig. 1. Smaller grey arrows: Optical flow; Big red arrow: mean pixel displacement in the image; Big blue arrow: meter displacement in the world

these steps: (i) store a frame; (ii) wait for the target to move more than a threshold distance; (iii) compute the optical flow using the stored frame and the present frame. In addition, the spatial sub-sampling provides invariance to changes on the sampling frequency of the cameras.

2.2 Flow-Based Features

We compare two kinds of features: motion boundary histogram (MBH) [7] and histogram of gradients (HOG) [4], considering four kinds of flow data: (i) raw flow, (ii) spatially sub-sampled, (iii) standardized flow and (iv) spatially sub-sampled and standardized flow.

MBH captures the local orientations of motion edges. We do it by considering the two flow components (x and y) as independent images, and taking their gradients. To extract the spatial information of the gradient image, two types of sampling are considered: dividing the image in cartesian or polar regions. HOG is computed in a similar way but directly on the flow vectors, and we also consider the same two sampling types: cartesian and polar. In total, we compare among sixteen different combinations of features, samplings and flow data. In difference to the original MBH and HOG features, that overlap sampling regions, we don't consider overlapping.

3 Learning Algorithm

The Boosting algorithm provides a framework to sequentially fit additive models in order to build a final strong classifier, $H(x_i)$. This is done minimizing, at each round, the weighted squared error, $J = \sum_{i=1}^{N} w_i(y_i - h_m(x_i))^2$, where $w_i = e^{-y_i h_m(x_i)}$ are the weights, N the number of training samples, x_i is a feature and y_i is the correspondent class label. At each round, the weak classifier with lowest error is added to the strong classifier and the data weights adapted, increasing the weight of the misclassified samples and decreasing correctly classified ones [11]. Then, in the subsequent rounds the weak classifier focus on the misclassified samples of the previous round.

In the case of GentleBoost it is common to use simple functions such as regression stumps. They have the form $h_m(x_i) = a\delta\left[x_i^f > \theta\right] + b\delta\left[x_i^f \le \theta\right]$, where f is the number of the feature and δ is an indicator function (i.e. $\delta[condition]$ is one if $condition$ is $true$ and zero otherwise). Regression stumps can be viewed as decision trees with only one node, where the indicator function sharply chooses branch a or b depending on threshold θ and feature x_i^f. To optimize the stump one must find the set of parameters $\{a, b, f, \theta\}$ that minimizes J w.r.t. h_m. The optimal a and b are obtained by closed form and the value of pair $\{f, \theta\}$ is found using an exhaustive search [15].

A recent approach considers the temporal evolution of the features in the boosting algorithm, improving the noise robustness and performance. Ribeiro et al. [12] model temporal consistency of the features, by parameterizing time in the weak classifiers. The Temporal Stumps compute the mean classification output of the regression stump, in a temporal window of size T,

$$h_m(x_i) = a\left(\frac{1}{T}\sum_{t=0}^{T-1}\delta\left[x_{i-t}^f > \theta\right]\right) + b\left(\frac{1}{T}\sum_{t=0}^{T-1}\delta\left[x_{i-t}^f \le \theta\right]\right). \qquad (1)$$

The temporal weak classifier of Eq. 1 can be viewed as the classic regression stump with a different "indicator function". If $T = 1$ it becomes the original regression stump, and for $T > 1$ the indicator function changes. The new indicator functions

$$\Delta_+^T(f, \theta, T) = \frac{1}{T}\sum_{t}^{T-1}\delta\left[x_{i-t}^f > \theta\right], \quad \Delta_-^T(f, \theta, T) = \frac{1}{T}\sum_{t}^{T-1}\delta\left[x_{i-t}^f \le \theta\right], \qquad (2)$$

compute the percentage of points above and below the threshold θ, in the temporal window T and for the feature number f. The indicator functions with temporal consistency in Eq. 2, can take any value in the interval $[0\ 1]$, depending on the length of the temporal window used. For example, if $T = 2$ the functions can take 3 different values, $\Delta_+^T \in \{0,\ 1/2,\ 1\}$, if $T = 3$ can take four values, $\Delta_+^T \in \{0,\ 1/3,\ 2/3,\ 1\}$ and so on. The fuzzy output of the new "indicator function", Δ, represents the confidence of threshold choice to use the data with temporal support T. Thus, at each boosting round, we use a weighted confidence of both branches, instead of choosing only one branch.

Using the weak classifier with temporal consistency of Eq. 1 in the cost function, Ribeiro et al. [12] obtain closed expressions for the parameters a and b that minimize the error J. The optimal f, θ and T are obtained by exhaustive search. The learning algorithm shown in figure 2 is similar to GentleBoost, but optimizing the temporal stump of Eq. (1).

1. Given: $(x_1, y_1), \ldots, (x_N, y_N)$ where $x_i \in X$, $y_i \in Y = \{-1, +1\}$, set $H(x_i) := 0$, initialize the observation weights $w_i = 1/N$, $i = 1, 2, \ldots, N$
2. Repeat for $m = 1, \ldots, M$
   (a) Find the optimal weak classifier h_m^* over (x_i, y_i, w_i).
   (b) Update strong classifier $H(x_i) := H(x_i) + h_m^*(x_i)$
   (c) Update weights for examples $i = 1, 2, \ldots, N$, $w_i := w_i e^{-y_i h_m^*(x_i)}$

Fig. 2. Temporal Gentleboost algorithm

4   Results

We compute the 16 different combinations of flow-based features (Section 2.2) in three scenarios: people walking, people loitering and robot moving. The motion patterns of people walking and robot moving will be properly extracted by optical flow-based features, so they are the nominal classification scenario. People loitering on the other hand, is a difficult situation as it provides small optical flow values. Both people walking and loitering are very common activities, therefore we decide to focus on them in this work. Figure 3 shows the setup of each scenario, which includes video sequences from 10 cameras.

We grabbed five groups of sequences, where each one includes images from 10 cameras. One group with a person walking, another group with a different person walking, two groups with the same pioneer robot moving in two different conditions, and the

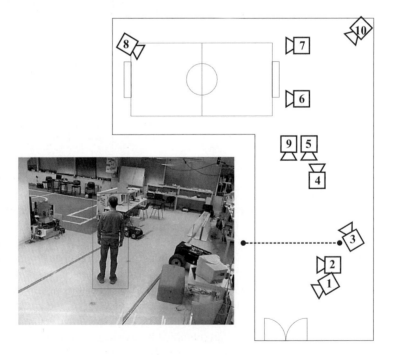

Fig. 3. Experimental setup for training scenario

last group with a third person loitering. The people class videos have a total of 9500 samples of the optical flow and the robot class videos have a total of 4100 samples. The segmentation and tracking of the moving objects in the scene are provided by: - LOTS background subtraction for detection [13] and nearest neighbor for tracking. The LOTS algorithm provides the bounding boxes of the regions of interest and its respective segmented pixels. Nearest neighbor is computed between the center points of the two bounding boxes.

We follow a cross validation approach to compare the classification result of the temporal GentleBoost algorithm. We build two different groups of training and testing sets. The people loitering data is always in the testing set, each person belongs to the training set for one of the experiments, and each pioneer robot sequence belongs to the training set once. The Tables 1, 2 and 3 show the average of the recognition rate for each frame using the two experiments. Each table summarizes the results for a fixed value of temporal support, T, and we notice the large performance improvement brought by the temporal support of the flow-based features when compared to the common GentleBoost ($T = 1$).

Table 1. Recognition rate of several features, using a maximum temporal support $T = 5$ frames of the temporal boost algorithm

Feature	sub-sampled+standardized	standardized	sub-sampled	raw flow
polar flow histogram	76.15	**92.26**	75.96	90.62
cartesian flow histogram	71.90	**87.90**	71.63	87.20
MBH cartesian	90.60	83.13	**91.39**	82.73
MBH polar	93.14	90.60	**93.67**	89.40

Table 2. Recognition rate of several features, without temporal support ($T = 1$ frames) of the temporal boost algorithm

Feature	sub-sampled+standardized	standardized	sub-sampled	raw flow
polar flow histogram	76.71	**87.23**	76.82	85.68
cartesian flow histogram	78.71	**85.18**	78.74	84.13
MBH cartesian	79.79	75.48	**79.98**	74.65
MBH polar	88.80	83.87	**88.94**	81.30

Table 3. Recognition rate of several features, using a maximum temporal support $T = 10$ frames of the temporal boost algorithm

Feature	sub-sampled+standardized	standardized	sub-sampled	raw flow
polar flow histogram	77.23	**95.43**	77.77	93.02
cartesian flow histogram	73.64	**89.32**	74.22	88.52
MBH cartesian	91.68	87.59	**92.25**	85.25
MBH polar	**94.58**	91.41	**94.58**	91.01

Fig. 4. Examples of person and robot training samples (on top) and real-time classification in an outdoors setting (bottom)

We observe three general patterns from the recognition rate:

- The polar sampling of the images performs better than the cartesian counterpart. It seems that the polar sampling is better suited for modeling the motion of the peoples' limbs, so it is easier to discriminate between people and robots.
- The Motion Boundary Histogram (MBH) feature has better performance than the optical flow histogram. The MBH has a richer representation based on two images that extract the first order spatial derivatives of the optical flow, while the flow histogram is a more efficient representation based on only one image, the optical flow.
- The spatial sub-sampling of the optic flow computation has a positive effect on the MBH features, while has a negative impact on the flow-based histogram features. On one hand, it seems that the MBH feature needs optical flow measurements with low levels of noise, which is provided by the spatial sub-sampling for computing the optical flow. On the other hand, the evolution in time of the optical flow histogram is better captured by the computation of the optical flow between consecutive images.

- The standardization of the optical flow has a very small improvement of the recognition rate, because all the features compute normalized histograms that provide a sort of standardization of the features.

From Table 3 we see that the best compromise between accuracy and robustness is provided by the MBH polar using the spatial sub-sampling. Thus, we implemented this feature in a C++ program that distinguishes between people and robots in real-time.

5   Conclusions

In this work we compared several optical flow based features to distinguish people from robots. We propose a way to standardize the optical flow vectors, scaling them to their corresponding metric value in the real-world, and also a more efficient and robust way of computing the optical flow that subsamples the images on time using the spatial displacement of the targets in the world. We used Temporal GentleBoost algorithm for learning, which is able to improve the classification rate by considering previous features' values, thus including a temporal support of the features. We test for several combinations of temporal supports, type of feature and flow standardization in order to verify the combination with better performance and robustness. The application of spatial sub-sampling to the optical flow reduces the computational load of the algorithm while keeping similar results to its counterparts. These computational savings guarantees real-time classification. The Motion Boundary Histogram feature with world spatial sub-sampling of the optical flow and temporal support of 10 frames have a very good trade-off between accuracy and robustness. We implement the combination just mentioned, validating it in a outdoors setting that shows the generalization capabilities of the proper combination of features, classifier and sampling approaches, providing a very good performance.

References

1. Sanfeliu, A., Andrade-Cetto, J.: Ubiquitous networking robotics in urban settings. In: Workshop on Network Robot Systems. Toward Intelligent Robotic Systems Integrated with Environments. Proceedings of 2006 IEEE/RSJ International Conference on Intelligence Robots and Systems (IROS 2006) (October 2006)
2. De Leon, R.D., Sucar, L.: Human silhouette recognition with fourier descriptors. In: Proceedings of 15th International Conference on Pattern Recognition, 2000, vol. 3, pp. 709–712 (2000)
3. Wang, L., Tan, T., Ning, H., Hu, W.: Silhouette analysis-based gait recognition for human identification. IEEE Transactions on Pattern Analysis and Machine Intelligence 25(12), 1505–1518 (2003)
4. Dalal, N., Triggs, B.: Histograms of oriented gradients for human detection. In: CVPR 2005: Proceedings of the 2005 IEEE Computer Society Conference on Computer Vision and Pattern Recognition (CVPR 2005) - Vol.1, Washington, DC, USA, pp. 886–893. IEEE Computer Society, Los Alamitos (2005)
5. Ramanan, D., Forsyth, D.A., Zisserman, A.: Tracking people by learning their appearance. IEEE Transactions on Pattern Analysis and Machine Intelligence 29(1), 65–81 (2007)

6. Viola, P., Jones, M.J., Snow, D.: Detecting pedestrians using patterns of motion and appearance. International Journal of Computer Vision 63(2), 153–161 (2005)
7. Dalal, N., Triggs, B., Schmid, C.: Human detection using oriented histograms of flow and appearance. In: Leonardis, A., Bischof, H., Pinz, A. (eds.) ECCV 2006. LNCS, vol. 3952, pp. 428–441. Springer, Heidelberg (2006)
8. Kaufmann, U., Mayer, G., Kraetzschmar, G., Palm, G.: Visual robot detection in robocup using neural networks. In: Nardi, D., Riedmiller, M., Sammut, C., Santos-Victor, J. (eds.) RoboCup 2004. LNCS (LNAI), vol. 3276, pp. 262–273. Springer, Heidelberg (2005)
9. Mayer, G., Kaufmann, U., Kraetzschmar, G., Palm, G.: Biomimetic Neural Learning for Intelligent Robots. Springer, Heidelberg (2005)
10. Lange, S., Riedmiller, M.: Appearance-based robot discrimination using eigenimages. In: Lakemeyer, G., Sklar, E., Sorrenti, D.G., Takahashi, T. (eds.) RoboCup 2006: Robot Soccer World Cup X. LNCS (LNAI), vol. 4434, pp. 499–506. Springer, Heidelberg (2006)
11. Friedman, J., Hastie, T., Tibshirani, R.: Additive logistic regression: a statistical view of boosting (with discussion and a rejoinder by the authors). Annals of Statistics 28(2), 337–407 (2000)
12. Ribeiro, P.C., Moreno, P., Santos-Victor, J.: Boosting with temporal consistent learners: An application to human activity recognition. In: Proc. of 3rd International Symposium on Visual Computing, pp. 464–475 (2007)
13. Boult, T.E., Micheals, R.J., Gao, X., Eckmann, M.: Into the woods: Visual surveillance of noncooperative and camouflaged targets in complex outdoor settings. Proceedings of the IEEE 89(10), 1382–1402 (2001)
14. Ogale, A.S., Aloimonos, Y.: A roadmap to the integration of early visual modules. International Journal of Computer Vision 72(1), 9–25 (2007)
15. Torralba, A., Murphy, K., Freeman, W.: Sharing visual features for multiclass and multiview object detection. IEEE Transactions on Pattern Analysis and Machine Intelligence 29(5), 854–869 (2007)

Using a Virtual World to Design a Simulation Platform for Vision and Robotic Systems

Om K. Gupta and Ray A. Jarvis

Department of Electrical and Computer Systems Engineering
Monash University, Clayton, VIC 3800, Australia
{Om.Gupta,Ray.Jarvis}@eng.monash.edu.au

Abstract. Virtual worlds are increasingly being used in research and development to provide simulation platforms for testing, debugging and validating proof-of-concepts, as they offer a significant savings in cost, time and other resources. This paper describes the design of a real-time open-source 3D simulation platform based on a commercially available Half Life 2 game engine. It is primarily aimed towards research in mobile robotics, in-game characters manipulation, surveillance related vision application, and high quality synthetic video generation. Along with the illustrations of the platform and the analysis of algorithms, this paper also provides a comprehensive tutorial in developing similar tools for researchers, enthusiasts and commercialists to pursue their interests in creating and applying such tools to their own objectives.

1 Introduction

Although, virtual worlds are primarily being used in the gaming industry for entertainment, they are fast becoming one of the major tools in training and education, and in many areas of research and development. Due to the advantage of having rich visual effects and user-friendly interactive features, it is most appealing to people, irrespective of their age, race, or their technical expertise. The rise in virtual world based applications with multi-million active subscribers provides the proof of virtual worlds' success and popularity. For example, Second Life[1] is a rich three-dimensional interactive environment that goes beyond entertainment and socialization to the arena of learning and creativity, providing a vast range of services for artists, researchers and enthusiasts. Second Life supports features such as virtual classrooms, hosting lectures, distance learning, virtual conferences and webinars. Also several health, medical education and economics research projects have been carried out in Second Life [1] [2].

Similarly, Whyville[2] is another popular example of an educational virtual world for children. Study of historic civilizations such as the twelfth century

---

[1] http://www.secondlife.com/
[2] http://www.whyville.net/

G. Bebis et al. (Eds.): ISVC 2009, Part I, LNCS 5875, pp. 233–242, 2009.

Khmer empire and the city of Angkor in Cambodia [3] is being conducted by generating a model of the city using a virtual world.

Thus, virtual worlds not only provide risk free environments to test and validate proof-of-concepts, but also reduce the cost involved and other resources significantly. There are many simulation environments and synthetic videos generation tools [4] [5] based on virtual worlds providing the platforms for testing, debugging and validating proof-of-concepts before moving on to the real world. In computer vision, such platforms provide a researcher with exact control over staging the desired scene. Changing various lighting conditions, such as dark, bright day-light, foggy and rainy conditions could easily be generated using simulation video. Repeatability of the same scene for various conditions is one of the major advantages offered by such synthetic videos. This is particularly useful in accident research[6], where repeating the same accident conditions and control over the scene is almost impossible and impractical, and may require a strong consideration of occupational health/safety issues, risk analysis, legal issues, permission from council/government, costs involved, time and other resources.

Another reason why researchers have found such virtual worlds alluring is in ground truth generation and annotations. They have been widely used in comparing the quality of processed video frames in applications, such as, foreground segmentation, background subtraction, tracking, surveillance and geo-location. It normally requires days of manual labor and intensive processing to generate the ground truth, going through each frame at a time just for a minute long video. However, ObjectVideo Virtual Video (OVVV) [4] is a tool offered by ObjectVideo Inc. that allows features for generating accurate ground truth and bounding boxes in a 3D virtual world environment at a click of a button.

This paper describes the design and development of a real-time and open-source simulation platform primarily aimed towards research in path planning of a mobile robot, surveillance related computer vision applications, and synthetic video generation and manipulation. It illustrates the development of the tool based on a virtual world and experiments being conducted in the authors' research. Along with the illustration, this paper particularly provides a comprehensive tutorial on developing such tools. Although the paper mainly discusses computer vision and robotics applications, plenty of examples and suggestions are provided for researchers, enthusiasts and commercialists to pursue their interests in creating and applying such tools to their own objectives.

The rest of this paper is organized as follows. Section 2 briefly reviews various existing simulation tools for robotic and vision applications. Section 3 describes the methods involved in the development of the simulation platform which includes designing world maps, scheduling different events in the virtual world, the modeling of different entities, and programming features and characteristics of the entities, physics and artificial intelligence (AI). Section 4 demonstrates the path planning and navigation experiment for a mobile robot in the virtual world using the simulation platform. Finally discussion of future work, extensions and conclusions are given in Section 5.

2   Related Work

There is no doubt that the visual simulation platform has become an invaluable tool for analyzing theoretical concepts, complex algorithms and challenging experiments otherwise difficult and impractical to set up in the real world. Especially in the early stages of development, algorithms inevitably contains bugs and unknown defects that could cause immense loss to researchers both in terms of time and other resources if applied directly to hardware. There are countless platforms in all sectors of academia and industries. Most of them are specific to applications or machines and are very limited in their capability. Webots developed by Cyberbotics Ltd. is one of the well-known and powerful mobile robotics simulation software package that provides a prototyping environment for modeling, programming and simulating mobile robots [7]. It has many impressive features, allowing researchers to simulate and program a wide range of robotic experiments in its virtual environment. However, the graphics of generated videos are not accurate enough compared to those of the real world, making it inadequate for vision based experiments and control. Also a researcher or an enthusiast may find it a bit expensive to get a full version license.

As the design of every robot is different, a single software package cannot accommodate the demand of a very wide diversity of robots. So open source is always an attractive solution. Gamebot [5] is a simulation platform based on a virtual world using the Unreal Tournament game engine modification and Object Video Virtual Video (OVVV) [4] is a synthetic video generator based on a virtual world using the Half-Life 2 game engine. OVVV produces high quality videos compare to Gamebot and supports live streaming directly from the virtual world. It has impressive features for simulating noisy images, controlling camera view, ground truth generation and annotations; however, it is limited to computer vision based research. The proposed platform is inspired by both OVVV for using virtual world Half-Life 2 game engine and Gamebot extending its application beyond vision to a more generic platform for artificial intelligence and robotics.

3   Design

Our design aim is to develop a real-time, general purpose open source simulation platform independent of robot make/model with realistic graphics for vision and robotic systems research. The details of the design are described in the following sub-sections.

3.1   Half Life 2 World

The virtual world used in the design was the Half Life 2 (HL2) game engine developed by Valve Software. The HL2 world is particularly suitable for our objective because it provides a very flexible, powerful and open-source software development kit (SDK) on purchase of the game. It has visually stunning graphics, real world physics (kinematics, inverse kinematics, ropes, etc.) and advanced artificial intelligence (collision detection, vision, hearing, relationships, path finding, etc.). The HL2 engine is also renowned for high quality dynamic lighting and shadows, a scalable high-performance rendering system, environmental effects like fog and rain and realistic-looking reflective water surfaces with refraction.

3.2 Modding

Half Life 2 SDK supports alteration or creation of files for its game engine, which allows it to change the game-play style, graphics, environments, models, etc. The process is known as 'modding', and an altered version of the game is called a 'mod'. It can be single player, multi-player or total conversion. Our mod falls into the single player category, which means it allows one client to connect to the game. Half Life 2 is based on server-client architecture. The server side communicates directly to the game engine. It holds the map information, including the number of entities in the map, their positions, states, velocities and other properties. It also controls artificial intelligence and the physics of the world. On the other hand, the client side holds the player information connected to the game. It manages the health, properties and characteristics of the player, and renders the camera view from the player's eye angle. Our mod involves modification of both the server and client sides. Modding consists of mapping, programming and modeling, and is described in the following subsections.

3.3 Mapping

Mapping, also known as map building, is setting up a scene in the virtual world. This involves creating scenarios and sequencing events very similar to designing a conventional game level. Although there is a bit of learning curve, it is not actually as grueling as it may seem on a first glance. Valve Software provides a free map building and editing tool called 'Hammer' editor as part of their SDK and there are plenty of supports and tutorials available on different discussion forums to get started on this. Figure 1 shows a typical view of a Hammer editor. It can be seen that the window is divided into four views: Camera view, top, front and side views. It is easy to navigate through these views and Valve Software provides plenty of models, entities and features to help with map building.

The 'Hammer' editor was used for map building for our simulation platform. Creating a scenario involved defining the static world geometry (walls, ceilings, terrains), specifying artificial and environmental light sources, adding props (furniture, trees, parked cars), populating the world with different moving entities (random people, animals, police) and defining their states and properties [4]. Sequencing an event could be either scripted or AI controlled. A scripted sequence is a pre-defined path or movements for a dynamic entity to follow. This is usually user defined while mapping with a set of node links. Alternatively, AI control allows intelligent algorithms to handle the dynamics of an entity. This prevents replication of the same movement sequences, combining randomness and reactivity of a map, making it lively, which could be useful in evaluating performance of various algorithms such as tracking, object recognition, etc.

Figure 2 shows the snapshots of various indoor and outdoor maps from the author and other developers that could be used for a range of experimental analysis.

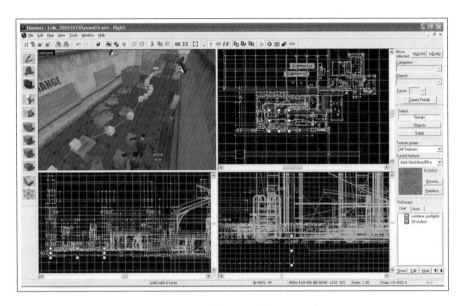

Fig. 1. Map building in Hammer editor

(a) Outdoor scene   (b) Dark street   (c) City square   (d) Water surface

(e) Rainy condition   (f) Indoor scene   (g) Train station   (h) Stadium

Fig. 2. Scenarios from Half Life 2 mods

3.4   Modeling

The process of creating any visible entity is called modeling. All of the models in Half Life 2 were created with Softimage—XSI. A free version Softimage—XSI, called XSI Mod Tool, is available online to download. This is especially tailored for prop and character creation. However, modeling is not obligatory for the virtual world because Half Life 2 comes with thousands of inbuilt models and props, enough for general purpose map building. It is required if one wants to build a specific model, such as a custom robot or a structure. Also, there are many tutorials and models available on different community forums for the HL2 virtual world. The model could be created using other tools like 3DS Max, Blender, etc.

The entities presented in Figure 2 are models for Half Life 2. Modeling was not pursued in depth by the authors except for a few minor features tweaking and using a few models from third-party vendors.

3.5   Programming

Programming in HL2 provides flexible and robust control over entities. Many advanced features and algorithms can be implemented for the virtual world through programming, including artificial intelligence, real world physics, player view control, behavior control, etc. Although there is a very little support available on the aspect of modding, an experienced programmer may follow through the source codes, which is reasonably modular and illustrative. Source codes for Half Life 2 are written in the C/C++ programming language and can be modified in Microsoft Visual Studio Development Environment on an Windows operating system. It has over a thousand C/C++ files and requires a bit of patience investigating, understanding and modifying them.

As we mentioned earlier, the Half Life 2 world is based on server-client architecture. On the server side, we created new definition modules for our robot models and set up their characteristics, such as, size, hull type, move type (walking, flying, rolling), health, classification, etc. Also we established modules for our experimental algorithm for path planning of the robot and an invisible logical entity in the game to link and interface between the algorithm and the robot. The details of the algorithm are explained in Section 4.

On the client side, we rendered the camera view by capturing the default client player's eye angle and placing it at a desired position. The views were saved on a hard disk for performance evaluation or further analysis. Alternatively, the live streaming view could be captured by creating a socket over TCP/IP connection for surveillance and monitoring of live frame from the world. Ground truth frames of the foreground can be created simply by rendering the desired entities (such as humans or moving entities) instead of everything in the player's view.

4   Experimental Results

Various robotics and vision experiments runs were conducted on the virtual world based platform by the authors. Two of these are described in the following

subsections. The design of the platform and the simulations were all carried on a
2.99 GHz Pentium IV machine with 2 Gigabyte of RAM and Microsoft Windows
XP operating system.

4.1 Navigation and Path Planning

The navigation of a mobile robot requires online control of the robot motion in
the virtual world. In the real world, one could face a lot of challenges in setting
up a scene for the robot navigation. Quite often, the motion of a moving entity
in such an environment is unpredictable and can not be perfectly modelled.
The dynamics of a human-centric environment adds an extra complexity to it.
Most often, the real world experiments fail to deliver real-time performance in
such scenarios. A cost-evaluation function based algorithm for path planning was
evaluated on the simulation platform [8]. It minimises a path cost made up of
accumulated weighted mixtures of distance and risk of collision on time-space
Distance Transforms (DT), thus providing efficient and low risk trajectories. The
basic approach of the method is to propagate distance in an incremental manner
out from specified goals, flowing around obstacles, until all of the free space has

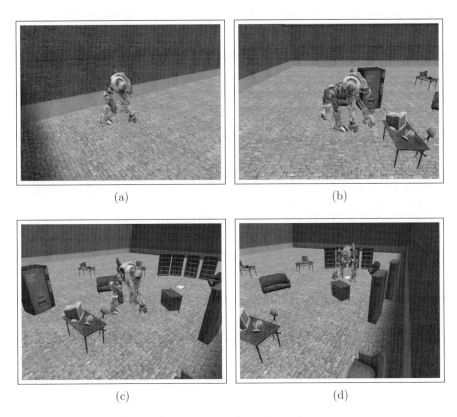

(a) (b)

(c) (d)

Fig. 3. Path planning in static environment

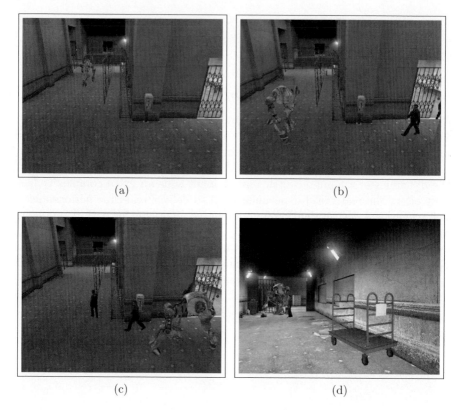

Fig. 4. Path planning in dynamic environment

a distance from the nearest goal parameter associated with it. Moving down the steepest descent path from any free-space cell in this distance transformed space generates the least-cost (in distance/risk terms) path to the nearest goal [9].

A 3D Grid-based map is used to represent the configuration space of a robot projected into an x-y plane with evenly spaced grid cells, known as an Occupancy Grid. Occupancy Grids were originally proposed by Elfes [10] and have a value attached to each grid cell that measures the probabilistic belief that a cell is occupied based on sensor readings. The occupancy map of the system includes permanent obstacles and places which are inaccessible to the robot, such as walls, furniture, holes in a ground, etc.

Tests were conducted in a virtual 3D simulated world with settings for obstacles of different shapes, sizes, locations and velocities. The robot position was assumed to be a point, of about the size of a cell in the map and obstacles were dilated by an appropriate amount in order to avoid collision with the robot.

Figures 3 and 4 represent a collection of key snapshots from the simulator at different time slices for a complex scenario with static and variable speed moving obstacle fields respectively. When the robot encounters the predicted moving obstacles, it re-adjusts its path to avoid obstacles and continue towards

(a) Synthetic image from HL2   (b) Real image from panoramic camera

Fig. 5. Analysis of SIFT algorithm on real and synthetic images

its destination. Also, it was demonstrated that in some circumstances choosing a longer path or waiting for a obstacle to pass can be the optimal approach.

4.2 Scale-Invariant Feature Transform

The scale-invariant feature transform (SIFT) is an algorithm which describes local scale-invariant features in images for correspondence matching and is widely used in computer vision and robotics application for object recognition, mapping, 3D modeling, etc. The algorithm was published by David Lowe [11] and University of British Columbia holds its patent.

We analyzed the algorithm on both a synthetic image from the virtual world and from a panoramic camera mounted on the ceiling of a room. K-nearest neighbourhood kdtree and Random Sample Concensus (RANSAC) were employed for matching and fitting a model of a robot in the images. We tested this algorithm for recognition of specific objects in our environment. Figure 5 shows the result of the algorithm for both synthetic and real images. An average of 15 positive matches were found for both type of images; however, the results were not in real-time, with a processing delay of 5 seconds between the frames. Also, the distortion in the panoramic image affected the performance of the algorithm significantly.

5 Conclusion

A 3D simulation platform based on a Half Life 2 game engine was presented. Because of the versatility of the open-source virtual world engine, the platform can easily be extended for many different applications, such as, following or tracking people, covert path planning for moving sentries, indoor and outdoor navigation, which are the open areas of future research work. Applications of the tool are not limited to our research in robotics and computer vision and are beyond our imagination. For example, it could be used in creating short animation films and games, security and defence related research, monitoring and surveillances, accident researches, modeling and design, architectural engineering, study of crowd dynamics, etc. There is no argument that the meticulous investigations of real-world videos and real-world testings are always required for the validation of any concept or an algorithm. But a simulation platform like this, makes a valuable contribution to the research and development community by lowering the risks, cost factor and other resources.

References

1. Boulos, M., Hetheringtont, L., Wheeler, S.: Second life: an overview of the potential of 3-d virtual worlds in medical and health education. Health Information and Libraries Journal 24, 233–245 (2007)
2. Harris, B., Novobilski, A.: Real currency economies: using real money in virtual worlds. In: Proceedings of 2008 International Conference on Frontiers in Education: Computer Science and Computer Engineering (FECS 2008), pp. 241–246 (2008)
3. Geographic, O.: Angkor - interactive - national geographic magazine (2009), http://ngm.nationalgeographic.com/2009/07/angkor/angkor-animation (accessed, July 10, 2009)
4. Taylor, G.R., Chosak, A.J., Brewer, P.C.: Ovvv: using virtual worlds to design and evaluate surveillance systems. In: Proceedings of the IEEE Computer Society Conference on Computer Vision and Pattern Recognition (2007)
5. Adobbati, R., Marshall, A.N., Scholer, A., Tejada, S., Kaminka, G., Schaffer, S., Sollitto, C.: Gamebots: A 3d virtual world test-bed for multi-agent research. In: Proceedings of the 2nd International Workshop on Infrastructure, MAS, and MAS Scalability (2001)
6. Félez, J., Vera, C., Martínez, M.L.: Virtual reality applied to traffic accident analysis. Computer Networks 30, 1907–1914 (1998)
7. Michel, O.: Cyberbotics ltd.: Webots: Professional mobile robot simulation. Int. Journal of Advanced Robotics Systems 1, 39–42 (2004)
8. Gupta, O.K., Jarvis, R.A.: Optimal global path planning in time varying environments based on a cost evaluation function. In: Wobcke, W., Zhang, M. (eds.) AI 2008. LNCS (LNAI), vol. 5360, pp. 150–156. Springer, Heidelberg (2008)
9. Jarvis, R.A.: Distance transform based path planning for robot navigation. In: Zheng, Y.F. (ed.) Recent Trends in Mobile Robots, pp. 3–31. World 129 Scientific Publishers, River Edge (1993)
10. Elfes, A.: Sonar-based real-world mapping and navigation. IEEE Journal of Robotics and Automation RA(3), 249–265 (1987)
11. Lowe, D.G.: Distinctive image features from scale-invariant keypoints. International Journal of Computer Vision 2, 91–110 (2004)

Accurate and Efficient Computation of Gabor Features in Real-Time Applications

Gholamreza Amayeh[1], Alireza Tavakkoli[2], and George Bebis[1]

[1] Computer Vision Laboratory, University of Nevada, Reno
[2] Computer Science Department, University of Houston–Victoria

Abstract. Gabor features are widely used in many computer vision applications such as image segmentation and pattern recognition. To extract Gabor features, a set of Gabor filters tuned to several different frequencies and orientations is utilized. The computational complexity of these features, due to their non-orthogonality, prevents their use in many real-time or near real-time tasks. Many research efforts have been made to address the computational complexity of Gabor filters. Most of these techniques utilize the separability of Gabor filters by decomposing them into 1-D Gaussian filter. The main issue in these techniques is the efficient pixel interpolation along the desired direction. Sophisticated interpolation mechanisms minimize the interpolation error with the increased computational complicity. This paper presents a novel framework in computation of Gabor features by utilizing a sophisticated interpolation scheme – quadratic spline – without increasing the overall computational complexity of the process. The main contribution of this work is the process of performing the interpolation and the convolution in a single operation. The proposed approach has been used successfully in real-time extraction of Gabor features from video sequence. The experimental results show that the proposed framework improves the accuracy of the Gabor features while reduces the computational complexity.

1   Introduction

Recently, computer scientists have become interested in modeling the human vision systems [1]. It is explained by neuroscientists [2] that receptive fields of the human vision system can be represented as basis functions similar to Gabor filters. In NeoCortical Simulators(NCS), the response profile of neurons in visual cortex area of the human brain is modeled by Gabor features. In the latest version of the NCS (version 5.0), the robot's eye (a tracking pan-tilt-zoom camera) captures the video images from the real world [3]. Then, Gabor features of these images are extracted and uploaded to the brain simulator running on a cluster of computers, executing a pre-specified spiking brain architecture. Real-time extraction of these features from video images (30 frames per second) is important in order to avoid small delays which slow down the entire system. Moreover, inaccurate features can trigger inappropriate neuron regions. As a result, accurate and efficient extraction of Gabor features from video sequences is a crucial task.

G. Bebis et al. (Eds.): ISVC 2009, Part I, LNCS 5875, pp. 243–252, 2009.
© Springer-Verlag Berlin Heidelberg 2009

Gabor features are based on Gabor filter responses to a given input image. The responses over the image are calculated for a set of filters – a *bank* – tuned to various orientations and frequencies. The most straightforward technique to conduct the filtering operation is by performing the convolution in the spatial domain. The complexity of convolution depends directly on the size of the convolution mask. The mask in this case is the Gabor filter. The complexity of calculating the filter response for one point is $O(M^2)$, where M is the width and height of the mask. If the filtering is done on the entire image of size $N \times N$, the complexity becomes $O(M^2N^2)$.

One trivial solution to reduce the computational complexity is to perform the filtering process in the frequency domain [4]. In this approach, the image is first converted to the frequency domain using the Fast Fourier Transform (FFT). Afterwards the FFT transformed image is multiplied by a FFT transformed Gabor filter. Finally, the responses are converted back to the spatial domain using the inverse FFT. For an image of size $N \times N$, the computational complexity of this approach becomes $O(N^2 \log N)$ with a constant multiplier [4]. One of the issues with this method is the fact that the generic FFT formulation is limited to signals of length 2^n. Moreover, the memory requirement of this approach is very high.

Many research efforts have been made to significantly improve the computational complexity of Gabor filtering [5,6,7,8]. Nestares *et al.* in [6] improved the standard convolution with Gabor filters by utilizing the separability of Gabor filters. Ranganathan *et al.* in [5] used symmetry and anti-symmetry characteristics of Gabor filters to reduce their computational complexity. These convolution improvements can reduce the computational complexity of the Gabor filter from $O(M^2N^2)$ to $O(2MN^2)$. Compared to FFT filtering complexity of $O(N^2 \log N)$ it is evident that these techniques are beneficial when $M < \log N$. The main issue is that these methods can be applied only to certain configurations (e.g. $\theta = k\frac{\pi}{4}, k \in Z$), making them merely special cases.

Recently Areekul *et al.* in [8] generalized separable Gabor filters to any orientation. Their method uses three steps. The first step is to define and interpolate consecutive sequences of pixels to form a new image along selected convolutional orientations and their perpendicular directions. They employed an interpolation technique with the least expensive complexity – the linear interpolation of the two nearest pixels. Secondly, two continuous 1-D Gabor filters with suitable parameters are generated and re-sampled with uniform space between pixels. This task resembles the image re-sampling from the first step. Finally, separable convolutions can be performed along any selected orientation using these tessellated and interleaved patterns. In the best cases when $\theta = k\frac{\pi}{4}, k \in Z$ the computational complexity is $O(2MN^2)$ [8]. However, for an arbitrary orientation the re-sampling process plays a critical role if the required pixel is not on the sampling grid. As a result, in the worst case scenarios the computational complexity reaches $O(6MN^2)$. In this method, the main issue is interpolation error resulting in less accurate Gabor features. The accuracy can be improved by employing more sophisticated interpolation schemes (e.g. quadratic spline). Unfortunately, such sophisticated techniques drastically increase the computation cost of the first step – up to 4 times.

In this paper, a new framework is presented to generalize separable Gabor filters for any orientation by integrating the interpolation and the convolution processes in a single step. The proposed approach employs a sophisticated interpolation method – quadratic spline. This, results in a low interpolation error without any increase in the computational complexity. As it was mentioned, to extract Gabor features, a filter bank containing $Q \times S$ Gabor filters in Q directions and S scales is utilized. In this study, the computation complexity of Gabor features is further reduced by applying 1D filters in specific direction for all scales. This paper is organized as follows: In section 2, a review of the Gabor filter is presented. In section 3, the details of separable Gabor filters are described. Section 4 discusses the integration of the interpolation and the convolution processes in a single step. Section 5 shows experimental results of the proposed approach and compares our method with the state-of-the-art. Finally, Section 6 concludes this work and discusses future directions of this study.

2   Gabor Filters

The Gabor filter is a product of an elliptical Gaussian in any rotation and a complex exponential function representing a sinusoidal plane wave [9]. The sharpness of the filter is controlled on the major and minor axes by σ_x and σ_y, respectively. The filter response in spatial domain can be expressed by the following equation [9]:

$$g(x, y, f, \theta) = e^{-\frac{1}{2}\left(\frac{x_\theta^2}{\sigma_x^2} + \frac{y_\theta^2}{\sigma_y^2}\right)} \times e^{j2\pi f x_\theta} \tag{1}$$

where f is the frequency of the sinusoidal plane wave, θ is the orientation of the Gabor filter, σ_x is the sharpness along the major axis, σ_y is the sharpness along the minor axis, $x_\theta = x\cos\theta + y\sin\theta$ and $y_\theta = -x\sin\theta + y\cos\theta$. In most applications, the real part of the filter's impulse response (namely even-symmetric Gabor filter) is considered. As a result, the equation 1 can be rewritten as [9]:

$$g(x, y, f, \theta) = e^{-\frac{1}{2}\left(\frac{x_\theta^2}{\sigma_x^2} + \frac{y_\theta^2}{\sigma_y^2}\right)} \times \cos(2\pi f x_\theta) \tag{2}$$

The normalized Gabor filter in the frequency domain can be represented by [9]:

$$G(u, v, f, \theta) = \frac{1}{2\pi\sigma_u\sigma_v}\left[e^{-\frac{1}{2}\left(\frac{(u_\theta - u_0)^2}{\sigma_u^2} + \frac{(v_\theta - v_0)^2}{\sigma_v^2}\right)} + e^{-\frac{1}{2}\left(\frac{(u_\theta + u_0)^2}{\sigma_u^2} + \frac{(v_\theta + v_0)^2}{\sigma_v^2}\right)}\right] \tag{3}$$

where u_0, v_0 and $\sigma_{u,v}$ are equal to $\frac{2\pi\cos\theta}{f}$, $\frac{2\pi\sin\theta}{f}$ and $\frac{1}{2\pi\sigma_{x,y}}$, respectively. Also $u_\theta = u\cos\theta + v\sin\theta$ and $v_\theta = -u\sin\theta + v\cos\theta$. Figure 1 shows an even-symmetric Gabor filter in the spatial and frequency domains.

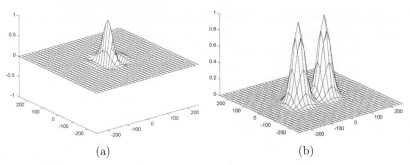

(a)   (b)

Fig. 1. An even-symmetric Gabor filter with $\theta = \frac{\pi}{2}$, $f = 0.01$ and $\sigma_x = \sigma_y = 3$ in (a) spatial domain and (b) frequency domain

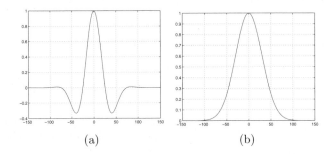

(a)   (b)

Fig. 2. Decomposing the Gabor filter in Fig. 1 into (a) a band-pass Gaussian filter, and (b) a low-pass Gaussian filter

3   Separability of Gabor Filters

A filter g is called separable if it can be expressed as the multiplication of two vectors – $g_{row} \times g_{col}$. For separable filters the convolution can be performed separately with one dimensional filters g_{row} and g_{col}. Employing one dimensional filters decreases the two dimensional filter's computational complexity from $O(M^2N^2)$ to $O(2MN^2)$, where M and N are the width (and height) of the filter mask and the image, respectively.

According to the definition of separable filters, the Gabor filters are parallel to the image axes – $\theta = k\frac{\pi}{2}$, $k = 0, 1, \ldots$ – are separable. In separable Gabor filters, one of the 1-D filters is a sinusoidal function with a Gaussian envelope and the other one is a Gaussian envelope. For example, if $\theta = \frac{\pi}{2}$, equation (1) gives $x_\theta = x$ and $y_\theta = y$. Therefore, equation (2) can be rewritten as:

$$g(x, y, f, \theta) = g_{bp}(x, f) \times g_{lp}(y) = e^{-\frac{x^2}{2\sigma_x^2}} \cos(2\pi f x) \times e^{-\frac{y^2}{2\sigma_y^2}} \tag{4}$$

where g_{bp} is a 1D band-pass Gaussian filter, and g_{lp} is a 1D low-pass Gaussian filter as shown in figure 2.

Separable Gabor filters can be extended to work with $\theta = k\frac{\pi}{4}$ by going through the image along diagonal directions instead of the image axes [6]. In order to implement separable 2D Gabor filters in any direction it should be separated into 1-D low-pass and band-pass filters along the desired orientation and its perpendicular direction, respectively. However, the line formed by pixel sequences along an arbitrary direction $\theta \neq k\frac{\pi}{4}$ is not well defined due to square sampling grid pattern in an image. For example, if we draw a straight line in any chosen direction, it is difficult to pick consecutive pixels in order to form a straight line in that particular direction. Therefore, making Gabor filters separable in arbitrary directions needs a re-sampling process. The re-sampling should be performed to get an exact sequence of missing pixels on the desired orientations.

4   Integrating Interpolation and Convolution

In general, there are several ways to find approximate values for the re-sampled pixels. Linear interpolation, spline interpolation, or sinc interpolation are among the most widely used techniques in image re-sampling. In all of these schemes, there is a trade off between computational complexity and interpolation error.

Areekul *et al.* in [8] proposed a generalized separable Gabor filter for any orientation. Their method has two main steps. The first step is to interpolate consecutive sequences of pixels along an arbitrary direction. The second step performs a separable convolution along the direction. In order to reduce the computational complexity of the interpolation process in [8], missing pixels are linearly interpolated between their two nearest pixels. Although this interpolation scheme has a low complexity $O(2MN^2)$, it suffers from increased interpolation error. Employing more sophisticated interpolation schemes in this approach will increase the computational complexity significantly – e.g. $O(9MN^2)$ for quadratic spline interpolation.

We proposed a novel approach in performing the interpolation and convolution processes required to achieve a separable Gabor filter along an arbitrary direction. The main idea behind our framework is the integration of the interpolation and the convolution processes. To this end, we propose a technique to re-sample an image $f(x, y)$ by an interpolation kernel $k(x, y)$ and then convolve it by a convolution kernel $p(x, y)$. By performing this integration scheme the overall process saves one step by convolving the image by a kernel $q(x, y)$.

Let's define $f_i(x, y)$ to be the image after interpolation but before convolution:

$$f_i(x, y) = \sum_{x_1, y_1} f(x_1, y_1)k(x - x_1, y - y_1) \tag{5}$$

By convolving it with $p(x, y)$ we get the final result $f_p(x, y)$:

$$f_p(x, y) = \sum_{x_2, y_2} f_i(x - x_2, y - y_2)p(x_2, y_2) \tag{6}$$

Substituting equation (5) in (6) results in:

$$f_p(x, y) = \sum_{x_2, y_2} [\sum_{x_1, y_1} f(x_1, y_1)k(x - x_2 - x_1, y - y_2 - y_1)]p(x_2, y_2) \tag{7}$$

After regrouping the sums, we get:

$$f_p(x,y) = \sum_{x_1,y_1} f(x_1,y_1)[\sum_{x_2,y_2} k(x-x_2-x_1, y-y_2-y_1)p(x_2,y_2)] \quad (8)$$

$$= \sum_{x_1,y_1} f(x_1,y_1)q(x-x_1, y-y_1)$$

Therefore, the final result – $f_p(x,y)$ – is the convolution of the image $f(x,y)$ by $q(x,y)$, where $q(x,y) = k(x,y) * p(x,y)$.

The 1D low-pass Gaussian filter – g_{lp} – in the direction θ can be generated with uniform displacement d. In this case d is related to θ by the following equation:

$$d = \begin{cases} \frac{1}{\cos\theta} & \text{if } |\cos\theta| \geq \frac{\sqrt{2}}{2} \\ \frac{1}{\sin\theta} & \text{if } |\sin\theta| > \frac{\sqrt{2}}{2} \end{cases} \quad (9)$$

The same relation is true for the 1-D band-pass filter along the perpendicular direction as well. Therefore, the 1D low-pass and band-pass Gaussian filters become:

$$g_{bp}[n] = e^{-\frac{(nd)^2}{2\sigma_x^2}} \cos(2\pi f n d) \quad (10)$$

$$g_{lp}[n] = e^{-\frac{(nd)^2}{2\sigma_y^2}}$$

When the orientation is $\theta = k\frac{\pi}{4}$ all of $g_{lp}[n]$ and/or $g_{bp}[n]$ are located on the sampling Cartesian grid. However, for an arbitrary direction (figure 3(a)) some of the low- and/or band- pass filtered pixels (i.e. the red circles) may be on the sampling Cartesian grid while others may not. For the pixels which do not lie on the sampling gird their pixel values need to be regenerated by a re-sampling process. We utilize a quadratic B-spline interpolation scheme to estimate these missing pixel values. Moreover, the proposed framework can be extended to other interpolation schemes as well. The B-spline re-sampling process is one of the most commonly used family of spline functions [10]. It can be derived by several self-convolutions of a basis function. Quadratic B-spline interpolation kernel, $k(r)$, can be presented by the following formula [11]:

$$k(r) = \begin{cases} \frac{1}{2}r^2 & \text{if } 0 < |r| \leq \frac{1}{2} \\ -r^2 + r + \frac{1}{2} & \text{if } \frac{1}{2} < |r| \leq 1 \\ \frac{1}{2}(1-r^2) & \text{if } 1 < |r| \leq \frac{3}{2} \\ 0 & \text{otherwise} \end{cases} \quad (11)$$

where r is the distance of an estimated value from a pixel on the sampling Cartesian grid – figure 3(b).

Instead of performing interpolation and convolution separately, we can accomplish them in a single step by defining a mask q, shown in figure 3(c). This mask is the convolution of the interpolation kernel $k(r)$ and the 1D low- and/or

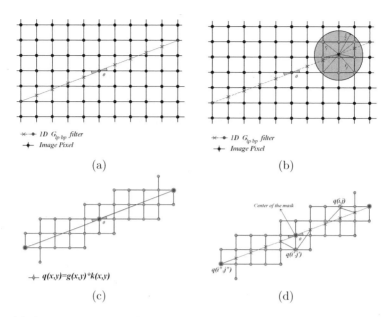

Fig. 3. (a) Some pixels in 1D filtering may not on sampling Cartesian grid; (b) Regenerating missing pixels by Quadratic B-spline interpolation; (c) Building the mask q which integrate 1D convolution and Quadratic B-spline interpolation; (d) Calculating the $q(i,j)$ coefficients

band-pass Gaussian filters – g_{lp} and/or g_{bp}. The size of q is almost $3 \times M$ in the worst case scenario – figure 3(c). As a result, the asymptotic computational complexity of our approach is $O(6M^2N^2)$.

For the implementation purpose, we find all Cartesian grids whose minimum distance to $g_{mis}[n]$ (the red crosses in figure 3(d)) is less than $\frac{3}{2}$. Then we calculate their coefficients by the following formula:

$$q(i,j) = \sum_{l=-\frac{M}{2}, |r_l| \le \frac{3}{2}}^{\frac{M}{2}} k(r_l) g_{mis}[n_l] \tag{12}$$

Some $g_{lp}[n'']$ and/or $g_{bp}[n'']$ positions may be located on the sampling Cartesian grid of the mask q – the $q(i'', j'')$ in figure 3(d). In these cases we need to update the $q(i'', j'')$ by adding $g_{lp}[n'']/g_{bp}[n'']$ to it.

5 Experimental Results

In this section some experiments have been conducted in order to evaluate the performance and efficiency of the proposed framework. The computational complexity of our approach is compared to the traditional techniques in the literature. To extract Gabor features from an acquired image, a Gabor filter bank containing $P \times Q$ Gabor filters in different directions $\theta \in \{\theta_1, \theta_2, ..., \theta_Q\}$ and different frequencies $f \in \{f_1, f_2, ..., f_P\}$, is utilized. From equation (4), in separable Gabor filters, the frequency only affects the 1-D band-pass Gaussian filter. Therefore, in a practical implementation, the 1-D low-pass Gaussian filter is computed once for all P Gabor filters in a specific direction θ_i with different frequencies.

Table 1 shows the computational complexity and the memory requirement of our proposed approach compared to the traditional method of filtering in frequency domain, and the separable technique proposed in[8]. In this experiment we use a specific direction and frequency and a Gabor filter bank containing $P \times Q$ filters. Although the traditional method in spatial domain is the least expensive technique in terms of memory requirement, it is the most computationally complex. As compared to filtering in the frequency domain, $O(12N^2 \log N)$, the spatial domain convolution with separable filters is beneficial when $M < 2 \log N$. Moreover, the generic FFT formulation is limited to working with signals of length 2^n. From table 1, however, the proposed method employs a sophisticated interpolation scheme based on Quadratic B-spline. Compared to the method in [8], it is evident that our approach does not increase the computational complexity and the memory requirements.

In this experiment, we have employed different methods to extract Gabor features from a video sequence with dimensionality of 640×480 and the rate of 30 frames per second. Gabor features were extracted using 12 Gabor filters in 4 different directions $\theta \in \{\frac{\pi}{8}, \frac{3\pi}{8}, \frac{5\pi}{8}, \frac{7\pi}{8}\}$, and 3 different scales $f \in \{1, 3, 5\}$. All methods have been implemented in C/C++ on a 64-bit machine with 2G byte

Table 1. Comparison of different methods in terms of memory requirement and computational complexity

Techniques	Memory space	Computation cost for a Gabor filter	Computation cost for a Gabor bank
Traditional method	$O(N^2)$	$O(M^2N^2)$	$O(PQM^2N^2)$
Filtering in frequency domain	$O(3N^2)$	$O(12N^2 \log N)$	$O(12PQN^2 \log N)$
Separable method[8]	$O(2N^2)$	$O(6MN^2)$	—
Proposed method	$O(2N^2)$	$O(6MN^2)$	$O(3(P+1)QMN^2)$

Table 2. Average number of filtered video frames by a Gabor filter bank (12 filters) in one second using different Gabor filtering methods

Techniques	$\overline{Speed}$ (fps)
Traditional method	1.1
Filtering in frequency domain	12.8
Separable method[8]	30
Proposed method	30

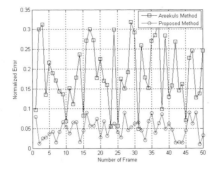

Fig. 4. Comparison of Areekul's method [8] and proposed method in terms of accuracy for the first 50 frames od video sequence

RAM. In the case of filtering in frequency domain, since the dimensionality of video frames were not exactly the length of 2^n, we used FFTW3 library [12]. This technique provides fast solutions for the signal lengths that the original FFT is not suitable for. Table 2 shows that the separable Gabor filter has the best performance among others and, as expected, the traditional method has the worst one. To evaluate the accuracy of both separable approaches, we calculated the normalized error E:

$$E = \frac{\sum_x \sum_y |\dot{f}(x,y) - \ddot{f}(x,y)|^2}{\sum_x \sum_y \dot{f}(x,y)^2} \tag{13}$$

where $\dot{f}$ is the filtered image by traditional method in spatial domain, and $\ddot{f}$ is the filtered image by one of the separable Gabor techniques. Figure 4 shows the average normalized error for all direction and frequencies for the first 50 frames. As you can see, our proposed method has significantly reduced the interpolation error by employing Quadratic B-spline interpolation technique. As it was expected, the method in [8] has considerable error due to the use of linear interpolation of the two nearest pixels.

6   Conclusion

Fast and efficient computation of Gabor features from video frames have become the focus of recent studies of the functionalities of the human brain and visual cognitive systems. The issue of efficient calculation of Gabor filters is of particular interest since the directional properties of these filters makes them inseparable along arbitrary direction.

In this paper we proposed a novel technique to integrate the interpolation and the convolution processes of the Gabor filter. This integration of the two processes makes the 2-D Gabor filter separable along any direction. Moreover, the integration performs the two processes as a single step in real-time. Therefore, our interpolation process employs a sophisticated interpolation technique in order to increase its accuracy while performing the entire process in real-time.

References

1. Serre, T., Wolf, L., Poggio, T.: Object recognition with features inspired by visual cortex. IEEE Conf. on Computer Vision and Pattern Recognition 2, 994–1000 (2005)
2. Olshausen, A.B., Field, D.J.: Emergence of simple-cell receptive field properties by learning a sparse code for natural images. Intl. Jurnal of Nature 381, 607–609 (1996)
3. Goodman, P., Zou, Q., Dascalu, S.: Framework and implications of virtual neurorobotics. Frontiers in Neuroscience 2, 123–128 (2008)
4. Bracewell, R.N. (ed.): The Fourier Transform and Its Applications, 3rd edn. McGraw-Hill, New York (2000)
5. Ranganathan, N., Mehrotra, R., Namuduri, K.: An architecture to implement multiresolution. In: Intl. Conf. on Acoustics, Speech, and Signal Processing, vol. 2, pp. 1157–1160 (1991)
6. Nestares, O., Navarro, R., Portilla, J., Tabernero, A.: Efficient spatial-domain implementation of a multiscale image representation based on gabor functions. Journal of Electronic Imaging 7, 166–173 (1998)
7. Areekul, V., Watchareeruetai, U., Tantaratana, S.: Fast separable gabor filter for fingerprint enhancement. In: Zhang, D., Jain, A.K. (eds.) ICBA 2004. LNCS, vol. 3072, pp. 403–409. Springer, Heidelberg (2004)
8. Areekul, V., Watchareeruetai, U., Suppasriwasuseth, K., Tantaratana, S.: Separable gabor filter realization for fast fingerprint enhancement. In: IEEE Intl. Conf. on Image Processing, vol. 3, pp. 253–256 (2005)
9. Daugman, J.: Uncertainty relation for resolution in space, spatial-frequency, and orientation optimized by two-dimensional visual cortical filters. Journal of Optical Socity America 2, 160–169 (1985)
10. Hou, H.S., Andrews, H.C.: Cubic splines for image interpolation and digital filtering. IEEE Tran. on Acoustic, Speech and Signal Processing 26, 508–517 (1978)
11. Dodgson, N.A.: Quadratic interpolation in image resampling. IEEE Tran. on Image Processing 6, 1322–1326 (1997)
12. Frigo, M., Johnson, S.: The design and implementation of fftw3. Proceedings of IEEE 93, 216–231 (2005)

Region Graph Spectra as Geometric Global Image Features

Qirong Ho, Weimiao Yu, and Hwee Kuan Lee

Bioinformatics Institute, A*STAR, Singapore

Abstract. In quantitative biology studies such as drug and siRNA screens, robotic systems automatically acquire thousands of images from cell assays. Because these images are large in quantity and high in content, detecting specific patterns (phenotypes) in them requires accurate and fast computational methods. To this end, we have developed a geometric global image feature for pattern retrieval on large bio-image data sets. This feature is derived by applying spectral graph theory to local feature detectors such as the Scale Invariant Feature Transform, and is effective on patterns with as few as 20 keypoints. We demonstrate successful pattern detection on synthetic shape data and fluorescence microscopy images of GFP-Keratin-14-expressing human skin cells.

1 Introduction

Sophisticated microscopy and cell culture systems have enabled high-throughput and high-content screens that yield thousands of images. Timely biological discovery using these images requires computational methods that rapidly exploit the voluminous amount of information they contain. In particular, many biological studies involve identification of cells with specific phenotypes (visual appearances), either to identify compounds with a specific biological effect [1] or to perform further data mining on the cell images [2]. The need for computerized phenotype detection becomes evident when one considers the time required for visual inspection of each cell.

Because traditional approaches to medical image analysis have proven less than ideal for this task, the biomedical imaging community has begun to adopt data mining and machine learning methods [3]. Pioneering work by Boland *et al.* [4] demonstrated successful classification of cells into a fixed number of categories, using a neural network classifier trained on global features extracted from manually-labeled cells. If automatic cell detection is provided, then their method becomes applicable to large-scale phenotype detection. Recently, Jones *et. al* [5] have combined cell segmentation, gentle boosting [6] on cell features, and iterative learning into a framework for supervised detection of arbitrary cell phenotypes. Because the features are cell-specific, their framework is reliant on accurate cell segmentation, which may be challenging under certain imaging conditions.

On the other hand, the trend in the computer vision community has been to represent visual objects as unordered distributions of local descriptors produced

G. Bebis et al. (Eds.): ISVC 2009, Part I, LNCS 5875, pp. 253–264, 2009.

by some detector [7]; examples of detectors include Harris detector specializa-
tions [8], the Scale-Invariant Feature Transform (SIFT) [9] and the derivative
Rotation-Invariant Feature Transform (RIFT) [10], as well as Speeded-Up Robust Features (SURF) [11]. Such unordered distributions, also known as *bag-
of-words* representations, have been popular due to their successful application
to natural (real-world) scenes [12]. Because natural scene objects are complex,
containing hundreds or thousands of local descriptors, bag-of-words methods
usually employ clustering to obtain representative descriptors [12]. Conversely,
some bio-image phenotypes are low in complexity and local feature count — for
instance, the biological image set in this paper contains ~25 SIFT descriptors per
exemplar. This reduces the number of meaningful clusters that can be extracted,
which may limit the discriminative capabilities of bag-of-words approaches.

To overcome this problem, we propose a phenotype detection framework
whose key contribution is a *global feature* for visual objects derived from *relationships* between local descriptors, as opposed to the *distribution* of descriptors
that characterizes bag-of-words approaches. This global feature is a vector in
$\mathbb{R}^k$, hence we describe it as *geometric* because it naturally admits the Euclidean
norm as one notion of inter-feature distance. Furthermore, it can be made both
scale- and rotation-invariant for local feature detectors with those properties,
such as SIFT. Additionally, our framework does not rely on segmentation; we
employ a general region-sampling technique that performs well for objects that
fit within cell-shaped regions. We demonstrate our framework's capabilities on
synthetic and biological images, where in both cases the patterns of interest
express few (i.e. tens of) SIFT descriptors.

2   Global Features from Local Feature Detectors

In this section, we develop a geometric global feature for arbitrary closed regions
in an image. This global feature, which we call a *region spectrum*, is derived
from local feature descriptors such as SIFT [9], SURF [11] and RIFT [10]; these
descriptors characterize visually significant points of the image. The general idea
is to find all descriptors within a region of interest, then construct multiple graphs
whose vertices represent descriptors, and whose edge weights are proportional
to similarities between descriptor feature vectors. Collectively, the eigenvalues of
every graph's Laplacian matrix make up the region spectrum, a vector describing
the region's geometric properties.

2.1   Regions and Local Feature Descriptors

Formally, let $\mathcal{D}$ be the set of descriptors generated by a local feature detector
(e.g. SIFT) on an image domain Ω. Each descriptor $\mathbf{d} \in \mathcal{D}$ is associated with
a position $\mathbf{x} = (x, y)$ on the image. In addition, $\mathbf{d}$ also contains features such
as local patch information. Generally, $\mathbf{d}$ is a real valued vector. For any closed
region $R \subseteq \Omega$ in the image, let $\mathcal{D}_R \subseteq \mathcal{D}$ be the set of descriptors whose associated
positions fall within R. This is illustrated in Fig. 1(a).

In our experiments we used SIFT [9] to obtain local feature descriptors. SIFT detects gradient maxima and minima in an image's scale-space representation, generating a descriptor $\mathbf{d} = (\mathbf{x}, \sigma, \theta, \mathbf{h})$ for each detected extremum. Here, $\mathbf{x}$ is the keypoint location associating the descriptor to a position in the original image and σ is the extremum's scale coordinate[1]. θ is the dominant gradient orientation of the region surrounding $(\mathbf{x}, \sigma)$ in scale-space, and $\mathbf{h}$ is a 128-bin histogram of gradients in a θ-oriented window centered at $(\mathbf{x}, \sigma)$. We chose SIFT for its robustness and utility in image registration tasks [9,13], though our method works with any feature detector for which meaningful inter-descriptor distances can be taken [11,10].

2.2 Region Graphs and Region Spectra

A region's descriptor set $\mathcal{D}_R$ can be used to construct one or more graphs $G_{R,C}$, where $C : \mathcal{D}_R \times \mathcal{D}_R \mapsto [0, 1]$ is a "connectivity" function between two descriptors. The vertex set of $G_{R,C}$ is $\mathcal{D}_R$, while its edge weights are $w_{\mathbf{d}_i, \mathbf{d}_j} = C(\mathbf{d}_i, \mathbf{d}_j)$, $\mathbf{d}_i, \mathbf{d}_j \in \mathcal{D}_R$. Refer to Fig. 1(b) for an illustration. For SIFT descriptors, we define three connectivity functions $C_{\mathbf{x}}$, C_σ and C_θ:

$$C_{\mathbf{x}}(\mathbf{d}_i, \mathbf{d}_j) = \exp(-\alpha_{\mathbf{x}} \|\mathbf{x}_i - \mathbf{x}_j\|^2) \tag{1}$$

$$C_\sigma(\mathbf{d}_i, \mathbf{d}_j) = \exp(-\alpha_\sigma |\sigma_i - \sigma_j|^2) \tag{2}$$

$$C_\theta(\mathbf{d}_i, \mathbf{d}_j) = \exp(-\alpha_\theta \min\left[|\theta_i - \theta_j|, 2\pi - |\theta_i - \theta_j|\right]^2) \tag{3}$$

where $\mathbf{x}_i$, σ_i and θ_i denote the $\mathbf{x}$, σ and θ components of $\mathbf{d}_i$ respectively. $\alpha_{\mathbf{x}}$, α_σ and α_θ are scaling coefficients. These connectivity functions indicate descriptor similarity in terms of image Euclidean distance $\mathbf{x}$, scale σ and angle θ. We did not use histogram $\mathbf{h}$ similarities in our experiments, as they do not improve performance on our data sets, yet are relatively expensive to compute. Regarding the coefficients α, our primary concern was avoiding numerical underflow; we set $\alpha_\sigma = \alpha_\theta = 1$ and $\alpha_{\mathbf{x}} = 10/(\text{mean training exemplar area})$.

For each graph $G_{R,C}$, we can generate its Laplacian matrix

$$\mathcal{L} = I - DAD \tag{4}$$

$$A_{i,j} = w_{\mathbf{d}_i, \mathbf{d}_j}$$

$$D_{i,j} = \begin{cases} (\sum_k w_{\mathbf{d}_i, \mathbf{d}_k})^{-1/2} & i = j \text{ and } \sum_k w_{\mathbf{d}_i, \mathbf{d}_k} \neq 0 \\ 0 & \text{otherwise} \end{cases}$$

where I is the identity matrix [14]. Applying an eigendecomposition algorithm to $\mathcal{L}$ yields $\boldsymbol{\lambda}$, its vector of eigenvalues in ascending order. The number of eigenvalues is equal to the number of descriptors in $\mathcal{D}_R$, which we denote by $|\mathcal{D}_R|$. Since each connectivity function $C_{\mathbf{x}}, C_\sigma$ and C_θ gives rise to one graph, we may concatenate the $\boldsymbol{\lambda}$'s resulting from each graph, $\boldsymbol{\lambda}_{\mathbf{x}}$, $\boldsymbol{\lambda}_\sigma$ and $\boldsymbol{\lambda}_\theta$, to obtain the *region spectrum*

---

[1] The SIFT computes σ as $2^{o_i + o_f}$, where o_i and o_f are the integer "octave" and fractional "interval" described in [9].

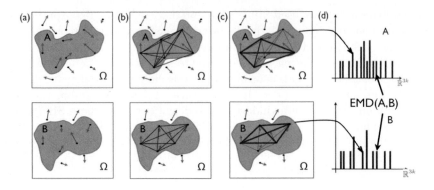

Fig. 1. Example illustration of the steps involved in generating geometric global features. (a) Descriptors of two regions A and B are generated (e.g. using SIFT), (b) Graphs $G_{A,C}$ and $G_{B,C}$ are formed by connecting descriptors within the regions A and B. In this example, $G_{A,C}$ has six vertices while $G_{B,C}$ has five vertices. (c) Sample $k=4$-vertex subsets from each graph to form subgraphs. (d) Each sample of 4 vertices contributes one bin to its region's EMD signature. The distance between regions A and B is the EMD between their signatures.

$\mathbf{s}_R = (\boldsymbol{\lambda}_{\mathbf{x}}, \boldsymbol{\lambda}_\sigma, \boldsymbol{\lambda}_\theta)$, which serves as a global feature vector for regions. Some properties of $\boldsymbol{\lambda}$ bear mentioning. First, $\boldsymbol{\lambda}$ is bounded in $[0, 2]^{|\mathcal{D}_R|}$, and the smallest eigenvalue $\lambda_1 = 0$. Moreover, $\boldsymbol{\lambda}$ bounds invariants of $G_{R,C}$ such as the graph diameter, distances between subgraphs, and random walk stationary distribution convergence times [14]. These properties capture the collective behavior of the set of local features in an image region R, hence turning local features into a geometric global feature $\mathbf{s}_R \in [0, 2]^{3|\mathcal{D}_R|}$. This feature's notion of region similarity is near-cospectrality, or similarity in Laplacian eigenspectra. Region spectra $\mathbf{s}_R$ have several notable properties:

1. $\mathbf{s}_R$ *is rotation invariant if the local feature detector is rotation invariant, and the connectivity function* $C(\mathbf{d}_i, \mathbf{d}_j)$ *remains invariant when the same rotation is applied to* $\mathbf{d}_i$ *and* $\mathbf{d}_j$.
   This is because C's satisfying the latter condition generate isomorphic graphs for any rotation of R, while isomorphic graphs have identical eigenspectra. The three SIFT connectivity functions C_σ, $C_{\mathbf{x}}$ and C_θ are rotation invariant.
2. $\mathbf{s}_R$ *is scale invariant for* C's *and local feature detectors exhibiting scale-invariance.*
   The function C_θ is scale invariant, while C_σ becomes scale-invariant if σ_i and σ_j are substituted with $o_i = \log_2 \sigma_i$ and $o_j = \log_2 \sigma_j$. $C_{\mathbf{x}}$ can also be made scale invariant by substituting $\|\mathbf{x}_i - \mathbf{x}_j\|^2$ with $\frac{\|\mathbf{x}_i - \mathbf{x}_j\|^2}{\sigma_i \sigma_j}$. In practice however, the non-scale-invariant versions of $C_{\mathbf{x}}$ and C_σ (Eq. (1),(2)) yielded better results on our data sets.
3. *The equivalence class of graphs with the same (or similar) eigenspectra is not limited to isomorphisms.*

Region spectra thus enable generalization in ways that are meaningful, yet not immediately obvious from a visual standpoint.

3   Comparing Region Spectra

Despite their advantages, region spectra require further modifications to make them a suitable feature for machine learning. The chief difficulty arises when trying to compare $\mathbf{s}_R$'s with differing $|\mathcal{D}_R|$ and consequently with different vector lengths. This is an issue because many local feature detectors (SIFT included) do not produce a fixed number of descriptors per region.

3.1   Region Signatures

The eigenvalues are related to distances between subgraph partitions of $G_{R,C}$ [14], which are incomparable for graphs with small vertex sets of different sizes due to discretization effects. Hence the eigenvalues of two regions should not be compared directly, except when they contain the same number of descriptors.

To compare regions with different descriptor counts, we make use of k-descriptor combinations from the $|\mathcal{D}_R|$ descriptors in R. There are $\binom{|\mathcal{D}_R|}{k}$ combinations, and for each combination we generate the induced subgraph $S_i \subseteq G_{R,C}, i = 1, \cdots, \binom{|\mathcal{D}_R|}{k}$, followed by the three k-dimensional eigenvalue vectors $\mathbf{s}_R^{k,i} = (\boldsymbol{\lambda}_{\mathbf{x}}^i, \boldsymbol{\lambda}_{\sigma}^i, \boldsymbol{\lambda}_{\theta}^i)$, $i = 1, \cdots, \binom{|\mathcal{D}_R|}{k}$ corresponding to connectivity functions $C_{\mathbf{x}}, C_{\sigma}$ and C_{θ}. Next, we construct an Earth Mover's Distance (EMD) [15] signature $\mathcal{S}_R^k : [0, 2]^{3k} \mapsto \mathbb{R}$ from the $\binom{|\mathcal{D}_R|}{k}$ spectra $\mathbf{s}_R^{k,i}$. Each spectrum $\mathbf{s}_R^{k,i}$ gives rise to one bin in $[0, 2]^{3k}$, and the squared Euclidean distance is used as the ground distance between bins. We refer to the EMD signatures $\mathcal{S}_R^k$ as *region signatures*.

In practice, calculating all $\binom{|\mathcal{D}_R|}{k}$ region spectra will often be computationally prohibitive. We alleviate this problem by estimating the EMD signatures $\mathcal{S}_R^k$, i.e. we randomly sample some number of subsets as a function[2] of $|\mathcal{D}_R|$ and k; random sampling has been shown to be effective in estimating true distributions [16]. Subset sampling and region signature creation are illustrated in Fig. 1(c,d).

3.2   Restoring Information with Neumann Eigenvalues

The eigenvalues of a k-descriptor subset's induced subgraph $S \subseteq G_{R,C}$ do not capture all information encoded in $G_{R,C}$. However, we can restore some information from $G_{R,C}$ by considering the *Neumann eigenvalues* of S. These eigenvalues arise from boundary conditions analogous to a "Neumann random walk", in which agents moving to a vertex $v \notin S$ immediately move to some neighbor of v in S — in other words, they "reflect" off the subgraph boundary [14]. Neumann eigenvalues therefore incorporate information from descriptors near to the k-descriptor subset. Since our graphs $G_{R,C}$ are fully connected, using the

---

[2] $10|\mathcal{D}_R|/k$ subsets for a given R in our experiments.

Neumann eigenvalues of S restores the information encoded in $G_{R,C}$ to a significant extent. Intuitively, few subset samples will be needed to approximate the full signature on all $\binom{|\tilde{D}_R|}{k}$ subsets. The Neumann eigenvalues are obtained via eigendecomposition of the modified Laplacian [14]:

$$\mathcal{L}_N = D_{k \times k} I_{k \times n} (T - A) \mathcal{N} D_{k \times k} \tag{5}$$

$$T_{i,j} = \begin{cases} \sum_{h=1}^{n} w_{\mathbf{d}_i, \mathbf{d}_h} & i = j \\ 0 & \text{otherwise} \end{cases}$$

$$\mathcal{N}_{i,j} = \begin{cases} 1 & i = j \\ 0 & i \leq k \text{ and } i \neq j \\ w_{\mathbf{d}_i, \mathbf{d}_j} / \sum_{h=1}^{k} w_{\mathbf{d}_i, \mathbf{d}_h} & \text{otherwise, i.e. } i > k \end{cases}$$

where vertices have been relabeled so that the k chosen descriptors correspond to matrix indices $\{1, \ldots, k\}$. T is an $n \times n$ diagonal matrix of vertex degrees, $\mathcal{N}$ is an $n \times k$ matrix that redistributes edge weights according to the Neumann random walk, and A is as defined in (4) (taking into account the relabeling). $D_{k \times k}$ is D from (4) (relabeled) but truncated to the upper left $k \times k$ block, while $I_{k \times n}$ denotes a $k \times n$ matrix with 1's on the main diagonal and 0's everywhere else.

4   Pattern Detection Using Region Signatures

We now describe our machine learning framework for pattern detection using region signatures. Given a set of images $\mathcal{I}$, our framework ranks them according to its confidence that the pattern of interest is present, and also provides the approximate center of each detection.

A brief summary of our method follows. We begin by training a ν-SVM classifier [17] on user-provided training *exemplars* — image regions with the pattern of interest — as well as non-exemplars, which can be provided or obtained from a bootstrapping procedure that will be discussed shortly. For each image $\Omega \in \mathcal{I}$, we generate a series of random region masks and subject the mask regions to the ν-SVM classifier, which outputs 1 for the exemplar class and 0 for the non-exemplar class. The classification outcomes are averaged for each pixel to generate a *score landscape* (Figs. 3,4), whose local maxima are detected and sorted in descending score order. By ranking the local maxima lists for all images in lexical descending order, we obtain the required detection confidence ranking for $\mathcal{I}$. Moreover, the local maxima coordinates locate detected patterns of interest.

4.1   Pattern Detection without Segmentation

In order to detect patterns, we could segment each image $\Omega \in \mathcal{I}$ into regions and classify them using the SVM. However, this requires a segmentation algorithm that *a priori* separates the pattern of interest from other image regions. Finding such an algorithm can be a difficult task onto itself.

Instead of segmentation, we employ a general technique we term *masking* that is similar to sliding-window techniques. Given an image $\Omega \in \mathcal{I}$, we start by generating a set of *masks* $\mathcal{M}$, where each mask $M \in \mathcal{M}$ is a set of randomly-generated regions $R \subseteq \Omega$. For each mask M, we compute region signatures $\mathbf{s}_R$ for every $R \in M$. The $\mathbf{s}_R$ are then classified in $\{0, 1\}$ via the trained SVM, where 1 indicates the class of exemplars and 0 otherwise. In our experiments, we generated enough masks to cover each pixel ≥ 50 times. Finally, we generate Ω's *score landscape* $\Phi_\Omega(\mathbf{x})$ (Figs. 3,4), the average classification of all random mask regions R covering pixel $\mathbf{x}$. Local maxima $\mathbf{x}_m$ of Φ_Ω then correspond to detected patterns of interest, where $\Phi_\Omega(\mathbf{x}_m)$ is the detection confidence, a value in $[0, 1]$ with 1 representing perfect confidence and 0 representing no confidence.

Since the score landscape may have numerous local maxima, we employ a DBSCAN-like [18] algorithm to group local maxima into significant clusters. Each cluster is assigned a score equal to the greatest maxima inside it, and clusters are sorted in descending score order to get a *maxima list* for Ω. By sorting all image maxima lists in descending lexical order, we obtain a ranking for all images in $\mathcal{I}$. Comparing this ranking with the ground truth then gives a Receiver Operating Characteristic (ROC) curve.

Ideally, the randomly-generated regions should have shapes that fully contain the pattern of interest, yet contain little in the way of other regions. In our experiments, the patterns of interest are either cell-shaped or fit reasonably well into cell-shaped regions. Thus, we employed the following procedure to generate masks with cell-shaped regions:

1. Initialize a blank image with larger dimensions than Ω.
2. Place random seeds on the image, and convolve with a Gaussian filter.

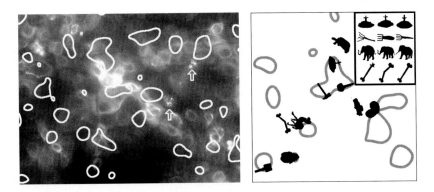

Fig. 2. Example images and masks from the MPEG-7 and skin cell image sets. LEFT: Skin cell image (resolution 696x520) with example mask region overlaid (parameters $(m_\sigma, m_p, m_t) = (16.0, 10^{-3}, 0.9)$). Two exemplars of the phenotype of interest (Keratin-14 aggregates) are indicated by arrows. RIGHT: MPEG-7 montage image (resolution 1000x1000), with example mask region overlaid (parameters $(m_\sigma, m_p, m_t) = (40.0, 2.5^{-4}, 0.9)$). The upper right insert shows the 4 classes to be detected, with 3 exemplars each.

3. Threshold at some (high) percentile, which generates connected components.
4. Place an Ω-sized window at the image center, and extract all connected components in the window. By overlaying these connected components onto Ω, we obtain a mask M.

Three parameters are involved: Gaussian filter standard deviation in pixels m_σ, seed distribution defined as the independent probability m_p that a given pixel will become a seed, and threshold percentile m_t. In our experiments, we adjust them so that the mean region size approximately matches training exemplar region sizes; refer to Fig. 2 for example masks used in each experiment.

Some regions may have too few descriptors to take k-descriptor samples. These regions do not contribute to the score landscape Φ_Ω, except for regions with exactly zero descriptors (such as blank or low-detail regions of the image) — these are assumed to have classification 0. If the proportion of non-contributing regions for some location $\mathbf{x}$ exceeds 0.8, we set $\Phi_\Omega(\mathbf{x})$ to 0. Such locations are not being sampled adequately, hence they are ambiguous and should not be considered during maxima detection.

4.2   Classifying Regions Using Support Vector Machines

We employ a ν-SVM classifier [17] on EMD signatures with $K(\mathcal{S}_{R_1}, \mathcal{S}_{R_2}) = \exp(-\gamma E(\mathcal{S}_{R_1}, \mathcal{S}_{R_2}))$ as the kernel function, where $\mathcal{S}_{R_1}$ and $\mathcal{S}_{R_2}$ are region signatures, $\gamma > 0$ is a scale parameter, and E is the EMD function with the squared Euclidean distance $\|\mathbf{b}_1 - \mathbf{b}_2\|^2$ as the ground distance[3] between signature bins $\mathbf{b}_1$ and $\mathbf{b}_2$. Although we cannot prove that $K(\mathcal{S}_{R_1}, \mathcal{S}_{R_2})$ is positive semidefinite, we did compute the spectra for a large sample of kernel matrices from our experiments, and did not find any negative eigenvalues. We also note that a similar kernel was used in [7], but with the unmodified Euclidean distance (rather than its square) as the ground distance.

The SVM is trained with exemplar regions containing the pattern of interest labeled as class 1, and non-exemplar regions labeled as class 0. If the user does not provide non-exemplar regions, they may be generated by *bootstrapping*, in which images known to lack the pattern of interest are *masked* (Section 4.1).

5   Experiments

We demonstrate our framework's performance on two image sets: the 216-image MPEG-7 CE Shape-1 Part-B database subset used by Sebastian *et al.* in [19], and a 304-image subset of the GFP-Keratin-14-expressing human skin cell fluorescence microscopy images used by Law *et al.* in [20]. Both image sets are single-channel. These choices reflect two distinct applications, namely shape retrieval and bio-image phenotype detection.

---

[3] To be specific, we use the squared *normalized* Euclidean distance, in which every bin dimension is rescaled to have standard deviation 1.0 (over all training data). This allows us to use $\gamma = 1.0$ as a reasonable starting point for parameter tuning. Another possibility would be to use the squared Mahalanobis distance.

Note that we did not use the MPEG-7 image subset as-is; instead we generated 29 montage images with 5-10 randomly placed shape images in each (Fig. 2). None of the 216 images were used more than once. Furthermore, some of the shapes overlap in the montage images. Using montages rather than invididual shapes demonstrates the functionality of our masking technique.

The common experimental setup was as follows:

1. Annotate exemplar regions with the pattern of interest, in the form of closed polygons. Skin cell image exemplars required the consensus of 5 individuals.
2. Divide the images for k-fold cross-validation — 2 folds for MPEG-7 montages, 5 for skin cell images.
3. Train the ν-SVM classifier using the exemplar regions as class 1, while bootstrapping (Section 4.2) to obtain non-exemplar regions as class 0.
4. Apply masking (Section 4.1) to score and rank images. We adjusted masking parameters independently for both experiments (Fig. 2).
5. Use the rankings to compute ROC curves and Area Under Curves (AUCs).

5.1   MPEG-7 Results

The MPEG-7 subset contains 18 classes of 12 shapes each, which we randomly placed into 29 1000x1000 montage images. We chose 4 shape classes for retrieval: fountains, forks, elephants and bones (Fig. 2). For each class 11-12 training

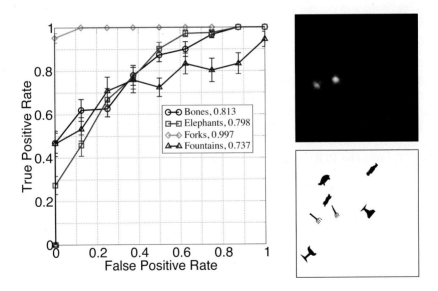

Fig. 3. LEFT: MPEG-7 database subset — average ROC curves and AUCs for detecting fountains, forks, elephants and bones. The curves may be non-monotonic due to binning effects on the small image set. RIGHT: Fork detection example score landscape (top) and original image (bottom). In the score landscape, black represents score 0, white represents score 1.

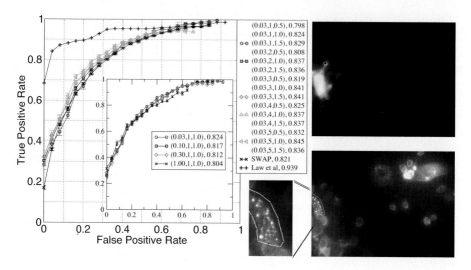

Fig. 4. LEFT: Skin Cell image set — average ROC curves and AUCs for detecting Keratin-14 aggregates. Legends for each ROC curve are given as "(γ, r, α), AUC". **GRAPH INSET:** Tuning over $\gamma = \{1.00, 0.30, 0.10, 0.03\}$, $r = 1$ and $\nu = 0.5$. **MAIN GRAPH:** Tuning over $\gamma = 0.03$, $r = \{1, 2, 3, 4, 5\}$ and $\nu = \alpha/(1 + r)$, $\alpha \in \{0.5, 1.0, 1.5\}$. Also shown are "SWAP", a training/validation-set-exchanged 5-fold cross-validation averaged over 5 trials with $\gamma = 0.03$, $r = 5$, and $\nu = 1/6$, and "Law et al.", the results of Law *et al.*'s deterministic aggregate-specific detector [20]. RIGHT: Example score landscape (top) and original image (bottom) with annotated exemplar. In the score landscape, black represents score 0, white represents score 1.

exemplars were annotated[4], and the montages divided into 2 folds with 5-6 exemplars each. As all exemplars contained 20 to ~50 descriptors each, we set the number of descriptors per subset $k = 20$. The SVM was trained with a $1:5$ ratio of exemplars to non-exemplars (i.e. we bootstrapped 5 non-exemplars for every training exemplar), $\nu = 1/6$ and $\gamma = 0.03$.[5] Masking was carried out with $(m_\sigma, m_p, m_t) = (40.0, 2.5 \times 10^{-4}, 0.9)$ (Fig. 2).

Fig. 3 shows the ROC curves and AUC for each class, averaged over 10 trials of 2-fold cross validation. We took timings using one core of a 3.0GHz Intel Core 2 system running x86-64 Linux; training and bootstrapping took ~10 seconds per fold, while each image's score landscape took 4-5 minutes to generate. The time spent on maxima finding was negligible in comparison to score landscape generation.

5.2   Skin Cell Results

The phenotype (pattern) of interest was Keratin-14 aggregates, which manifest as fields of bright dots (Fig. 2). The image set contained 304 single-channel

---

[4] We omitted occluded shapes from the exemplar set.

[5] These parameters were selected based on Section 5.2's tuning.

images of 696x520 resolution, of which 152 had at least one exemplar. There were 193 exemplar regions in total, containing 25.3 ± 17.4 descriptors on average with significant right skew in the distribution. Based on this, we set the number of descriptors per subset $k = 10$, which required us to discard 19 exemplar regions containing ≤ 10 descriptors. The masking parameters were $(m_\sigma, m_p, m_t) = (16.0, 10^{-3}, 0.9)$ (Fig. 2). ROC curves and AUCs are averaged over 10 trials of 5-fold cross-validation, unless otherwise indicated.

We tuned the SVM kernel width γ for values $\{1.00, 0.30, 0.10, 0.03\}$, using $\nu = 0.5$ and a $1:1$ ratio of exemplars to bootstrapped non-exemplars. Having identified $\gamma = 0.03$ as yielding the optimal AUC, we proceeded to tune the bootstrap ratio and ν. First, we selected ratios of the form $1:r$, $r \in \{1, 2, 3, 4, 5\}$. Then, since ν upper-bounds the fraction of training set outliers [17], for each r we selected $\nu = \alpha/(1 + r)$ for $\alpha \in \{0.5, 1.0, 1.5\}$. Choosing ν in this manner protects the smaller exemplar set from over-penalization by the ν-SVM's regularization.

All AUCs and some ROC curves from our tuning are shown in Fig. 4. To demonstrate the effect of a small training set, we include a 5-fold cross-validation with training and validation sets exchanged. We also include results for Law *et al.*'s spot detector [20]. We emphasize that their detector is application-specific, whereas our method can be trained to recognize arbitrary patterns. Our worst runtimes are from the $1:5$ exemplar-to-non-exemplar ratio: training and bootstrapping took ~ 4 min/fold, while score landscapes took ~ 4.5 minutes each. Again, maxima finding takes negligible time compared to score landscape generation. In comparison, the method of Jones *et al.* requires ~ 2.5 minutes to preprocess each 3-channel 512x512 image, on a 2.4GHz Intel CPU [5].

6   Conclusion

We have described a global geometric image feature for pattern retrieval called a *region signature*. This feature derives the collective behavior of local image descriptors from graphs of their differences. By utilizing the EMD as a distance measure between region signatures, the latter can be used with ν-SVM classifiers and image *masking* to perform pattern detection without segmentation. Our framework demonstrates good performance on synthetic shapes and real biological images — in particular, it retrieves patterns with only tens of local descriptors, a quantity far smaller than typically used [12,7]. We also note that the EMD distance allows region signatures to be employed in image clustering.

Acknowledgements

We would like to thank Ivy Yan Nei Law, Yudistira Mulyadi, Boyang Zheng and Peili Yu for assistance related to this work.

References

1. Yarrow, J.C., Feng, Y., Perlman, Z.E., Kirchhausen, T., Mitchison, T.J.: Phenotypic Screening of Small Molecule Libraries by High Throughput Cell Imaging. Combinatorial Chemistry & High Throughput Screening 6, 279–286 (2003)
2. Bakal, C., Aach, J., Church, G., Perrimon, N.: Quantitative Morphological Signatures Define Local Signaling Networks Regulating Cell Morphology. Science 316, 1753–1756 (2007)
3. Peng, H.: Bioimage informatics: a new area of engineering biology. Bioinformatics 24, 1827–1836 (2008)
4. Boland, M.V., Markey, M.K., Murphy, R.F.: Automated recognition of patterns characteristic of subcellular structures in fluorescence microscopy images. Cytometry 33, 366–375 (1998)
5. Jones, T.R., Carpenter, A.E., Lamprecht, M.R., et al.: Scoring diverse cellular morphologies in image-based screens with iterative feedback and machine learning. PNAS 106, 1826–1831 (2009)
6. Friedman, J., Hastie, T., Tibshirani, R.: Additive logistic regression: a statistical view of boosting. Annals of Statistics 28, 337–374 (2000)
7. Zhang, J., Marszalek, M., Lazebnik, S., Schmid, C.: Local features and kernels for classification of texture and object categories: A comprehensive study. IJCV 73, 213–238 (2007)
8. Schmid, C., Mohr, R.: Local grayvalue invariants for image retrieval. PAMI 19, 530–535 (1997)
9. Lowe, D.: Distinctive image features from scale-invariant keypoints. IJCV 60, 91–110 (2004)
10. Lazebnik, S., Schmid, C., Ponce, J.: A sparse texture representation using local affine regions. PAMI 27(8), 1265–1278 (2005)
11. Bay, H., Tuytelaars, T., Van Gool, L.: Surf: Speeded up robust features. In: Leonardis, A., Bischof, H., Pinz, A. (eds.) ECCV 2006. LNCS, vol. 3951, pp. 404–417. Springer, Heidelberg (2006)
12. Everingham, M., Zisserman, A., Williams, C.K.I., Gool, L.V.: The 2006 pascal visual object classes challenge (voc2006) results. Technical report, University of Oxford (2007)
13. Mikolajczyk, K., Schmid, C.: A performance evaluation of local descriptors. PAMI 27, 1615–1630 (2005)
14. Chung, F.R.K.: Spectral Graph Theory. American Mathematical Society, Providence (1997)
15. Rubner, Y., Tomasi, C., Guibas, L.J.: The earth mover's distance as a metric for image retrieval. IJCV 40, 99–121 (2000)
16. Liu, J.S.: Monte Carlo Strategies in Scientific Computing. Springer, Heidelberg (2002)
17. Schlkopf, B., Smola, A.J., Williamson, R.C., Bartlett, P.L.: New support vector algorithms. Neural Computation 12, 1207–1245 (2000)
18. Ester, M., Kriegel, H.-p., Sander, J., Xu, X.: A density-based algorithm for discovering clusters in large spatial databases with noise. In: International Conference on Knowledge Discovery and Data Mining, pp. 226–231. AAAI Press, Menlo Park (1996)
19. Sebastian, T., Klein, P., Kimia, B.: Recognition of shapes by editing their shock graphs. PAMI 26, 550–571 (2004)
20. Law, Y.N., Ogg, S., Common, J., Tan, D., Lane, E.B., Yip, A.M., Lee, H.K.: Automated protein distribution detection in high-throughput image-based sirna library screens. Journal of Signal Processing Systems 55, 1–13 (2009)

Robust Harris-Laplace Detector by Scale Multiplication

Fanhuai Shi[1,2], Xixia Huang[3], and Ye Duan[1]

[1] Computer Science Department, University of Missouri-Columbia, USA
[2] Welding Engineering Institute, Shanghai Jiao Tong University, China
[3] Marine Technology & Control Engineering Key Lab, Shanghai Maritime University, China

Abstract. This paper proposes a robust Harris-Laplace detector by scale multi-plication. The specific Harris corner measure functions at adjacent scales are multiplied as a product function to magnify the corner like structures, while suppress the image noise and weak features simultaneously. Unlike the contour-based multi-scale curvature product for image corner detection, we detect the corner like features directly in intensity image. Experiments on natural images demonstrate that the proposed method has good consistency of corner detection under different noise levels.

1 Introduction

Corner feature detection is an essential process in image analysis, matching and rec-ognition. Note that, the term *corner* here has a specific meaning: points in the 2D image with high curvature. So far, many techniques have been proposed on corner detection, see [1] for a survey on the state of the art. Currently, there are mainly two kinds of corner detection methods: contour curvature based methods and intensity based methods.

Contour curvature based method was mainly focused on the accuracy of point lo-calization [1]. Early work of detecting dominant points on a digital closed curve is proposed in [2]. They indicate that the detection of dominant points relies primarily on the precise determination of the region of support rather than on the estimation of discrete curvature. Ji and Haralick [3] estimates the parameters of two lines fitted to the two segments neighboring to the corner point. A corner is declared if the parame-ters are statistically significantly different. To explore the corner detection in multi-scale, the curvature scale space analysis was performed to find the local scale of curves [4][5]. More recently, Zhang et al. [6] proposed multi-scale curvature product for image corner detection in curvature scale space. Awrangjeb and Lu [7] proposed a complete corner detection technique based on the chord-to-point distance accumula-tion (CPDA) for the discrete curvature estimation. The CPDA discrete curvature es-timation technique is less sensitive to the local variation and noise on the curve. Moreover, it does not have the undesirable effect of the Gaussian smoothing.

Although theoretically well founded for continuous curves, contour curvature based methods, which are less robust in case of discrete curves [2], are mainly applied to line drawings, piecewise constant regions, and CAD/CAM images rather than more challenging natural scenes. Nowadays, in most practical applications of interest points or corners, the focus is on robust, stable, and distinctive points. There has been less

G. Bebis et al. (Eds.): ISVC 2009, Part I, LNCS 5875, pp. 265–274, 2009.

activity in this area recently (over the past ten years), due to complexity and robustness problems, while methods based directly on image intensity attracted more attention [1]. Methods based on image intensity have only weak assumptions and are typically applicable to a wide range of images. Harris detector [8], proposed by Harris and Stephens, is based on a second moment matrix. A function based on the determinant and trace of that matrix was introduced which took into account both eigenvalues of the matrix. This detector is widely known today as the Harris detector or Plessey detector. Steerable filters can also be used for corner detection [9], however, one need to steer the filters to detect multiple orientations first. It is indirect and iterative. To fully explore the feature detection in multi-scale intensity image, Lindeberg [10] proposed a successful scale selection mechanism for Gaussian filters with a theoretical formulation. More recently, Mikolajczyk and Schmid [11] introduced a scale invariant corner detector, referred to as *Harris-Laplace*, and a scale-invariant blob detector, referred to as *Hessian-Affine*. Benchmark tests show that Harris-Laplace has the best repeatability over large-scale changes among the existing detectors, a very important attribute in the applications of accurate localization.

However, the location of each Harris-Laplace corner is derived from only a single scale, and multi-scale information is used only for scale selection. Therefore, it fails to detect true corners when the image noise is relative big. If a complex image is detected, the conflict between missing true corners versus detecting false corners become more severe.

Meanwhile, in edge detection, edge structures can present observable magnitudes along the scales while the noise decreases rapidly. With this observation, several researchers proposed the idea of scale multiplication for edge detection [6], and it was shown that the scale product could enhance the edge signal and improve detection and localization of the edge. In addition, Zhang et al. [6] successfully adopted scale multiplication for corner detection in edge image.

Inspired by the work of scale multiplication in edge detection and multi-scale curvature product for corner detection, in this paper, we proposed multi-scale product of Harris corner measure and a simple and robust Harris-Laplace detector based on scale multiplication.

2   Harris-Laplace and Harris Corner Measure

In this section, we briefly review the original Harris-Laplace detector and the related Harris corner measure function.

Define a scale-adapted second moment matrix as following:

$$M = \sigma_D^2 g(\sigma_I) * \begin{bmatrix} I_x^2(X,\sigma_D) & I_x(X,\sigma_D)I_y(X,\sigma_D) \\ I_x(X,\sigma_D)I_y(X,\sigma_D) & I_y^2(X,\sigma_D) \end{bmatrix} \tag{1}$$

where σ_I is the integration scale, σ_D is the differentiation scale and I_a is the derivative computed in the a direction. M is the so-called auto-correlation matrix.

Harris-Laplace starts with a multi-scale Harris corner detector as initialization to determine the location of the local features. The commonly used Harris corner measure combines the trace and the determinant of a second moment matrix:

$$F_{cornerness} = \det(M) - \lambda \cdot trace^2(M) \tag{2}$$

where M is the second moment matrix defined in (1) and λ is a constant coefficient. A typical value for λ is 0.04. Local maxima of *cornerness* determine the location of interest points.

In this paper, we will adopt another cornerness measure function, that is Alison Noble measure [12]:

$$F_{cornerness} = \frac{\det(M)}{trace(M)} \tag{3}$$

where the denominator should add a tiny positive number in order to keep it a non-zero value.

After a multi-scale Harris corner detection, the characteristic scale is then determined based on scale selection as proposed by Lindeberg [10]. The idea is to select the characteristic scale of a local structure, for which a given function attains an extremum over scales. The selected scale is characteristic in the quantitative sense, since it measures the scale at which there is maximum similarity between the feature detection operator and the local image structures. The size of the region is therefore selected independently of the image resolution for each point. As the name Harris-Laplace suggests, the Laplacian operator is used for scale selection.

Although Harris-Laplace has a better detection performance, the location of each detected corner is derived from only a single scale, and multi-scale is used only for scale selection. In this paper, we extend the Harris-Laplace by scale multiplication, which can not only magnify the corner like structures, but also suppress the noise and weak features simultaneously.

3   Analysis of Multi-scale Product of Cornerness Measure

In this section, we will define the concept of multi-scale product of cornerness measure.

Let $F(\sigma^i)$ denotes the Harris cornerness measure at scale σ^i ($i=1,2,\ldots$). According to (2) or (3), we can have the multi-scale product of cornerness measure as:

$$P_N = \prod_{i=1}^{N} F(\sigma^i) \tag{4}$$

More generally the multi-scale product of cornerness measure on an arbitrary set of different scales denoted by

$$P = \prod_{\sigma \in \Omega} F(\sigma) \tag{5}$$

where Ω is the set of different scales.

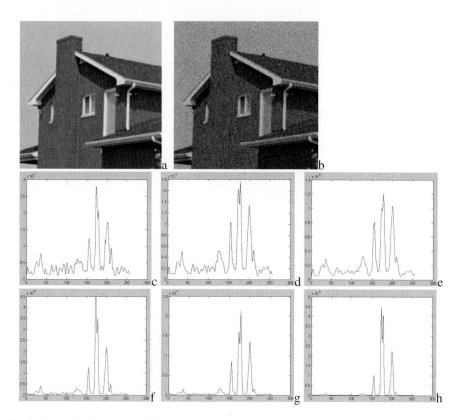

Fig. 1. Example of scale multiplication of Alison Noble corner measure. The house image is a 256 gray-level image and added by Gaussian noise with mean zero and variance 20. (a) original image. (b) noise image. (c)-(e) are first three scales of cornerness measure at image row 128. (f)-(h) shows the cornerness measure product of image row 128 at scales 1,2; at scales 2,3; and at scales 1,2,3 respectively.

An example of the scales product of Alison Noble corner measure is given in Fig 1. We select the 256 gray-level house image that is corrupted by Goussian noise with mean zero and variance 20. For illustration purposes, we only show the cornerness measure of one single image row in the figure. Without loss of generality, the corner-ness measure of image row 128 at scales 1, 2, 3 are shown in Fig. 1(c)-(e) respec-tively. Apparently, the cornerness measure present observable magnitudes along the scales, while the magnitudes of the noise and weak features decrease. As shown is Fig. 1(f)-(h), the weak corner features and the noise are suppressed after the scale multiplication, and the responses of the corners become more salient. Thus we can say that the multi-scale product of cornerness measure enhances the corner peaks while suppressing the noise and weak features.

Further more, as shown in Fig. 1, the product of only two adjacent scales seems to be enough to suppress the noise and weak features. Thus in the proposed scale multi-plication Harris-Laplace, when carrying out the multi-scale Harris corner detection, we adopt the product of Alison Noble corner measure of current scale and previous scale as the current cornerness measure.

3.1 Performance Evaluation of the Proposed Approach

In order to evaluate the performance of the proposed approach under different noise levels, similar to the criteria in [6], we also use the consistency of corner numbers(CCN) and accuracy (ACU) for measuring the robustness of the corner detectors. For convenience, we refer Harris, Harris-Laplace, the proposed method as Harris, HarLap and SMHL.

Let N_0 be the number of corners in original image, N_t be the number of corners in each of the noise images. Consistency of corner numbers is given as follows:

$$CCN = 100 \times 1.01^{-|N_t - N_0|} \tag{6}$$

where CCN stands for consistency of corner numbers and the base of the exponential function is a little different with that in [6].

Remark 1. Since the motivation of scale multiplication of cornerness measure is to suppress the noise, CCN should be insensitive to the image noise. A stable corner detector does not change the corner numbers from original image to noise images, in terms of consistency, the value of CCN for stable corner detector should be close to 100%. This criterion for corner detectors with more false corners is closer to zero.

An example of the consistency of corner numbers with respect to different image noise is given in Fig 2. As can be seen that, the CCN of Harris and Harris-Laplace detector decrease rapidly with the increase of image noise. Instead, the proposed approach presents a good performance.

More importantly, corner locations should keep constant when different noise is added to the image. Therefore, we also need to evaluate the accuracy of the corner detector. Let N_o be the number of the corners in the original image (note that $N_o \neq 0$), N_a the number of the matched corners in the noise image when compared to the original corners. The criterion of accuracy is:

$$ACU = 100 \times \frac{N_a}{N_o} \tag{6}$$

where ACU stands for "accuracy".

Remark 2. We use ACU to describe the accuracy of corner detectors. The value of ACU for accurate corner detectors should be close to 100% as the same as CCN. The matched corners are defined as those corners that both have the same scale and within a specific spatial distance.

An illustration of the accuracy of corner detection with respect to different image noise is given in Fig 3. As can be seen that, the ACU of Harris presents lower performance than that of Harris-Laplace and the proposed approach. Whereas, the ACU of Harris-Laplace and the proposed approach have nearly equal performance, which means that scale multiplication of Harris corner measure has little effect on the performance of ACU.

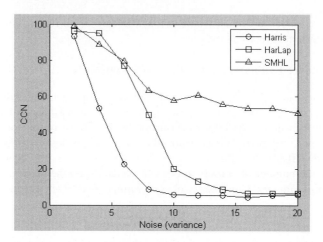

Fig. 2. The consistency of corner numbers for house image with respect to different noise

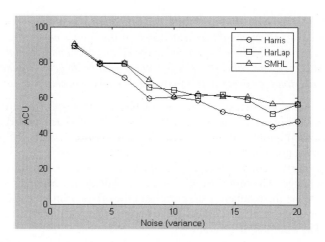

Fig. 3. The accuracy of corner detection for house image with respect to different noise

Table 1. Efficiency evaluation results of Harris-Laplace and scale multiplication Harris-Laplace

Detectors	Harris-Laplace	Scale multiplication Harris-Laplace
Average time per image (seconds)	5.8165	12.6577

In order to compare the running time of Harris-Laplace and scale multiplication Harris-Laplace, we recorded the time spent in detecting the house image. To get more reasonable results, this experiment was carried out with different image noise and then we calculate their average time per image. The experiments run on a notebook

PC with Intel® Core™2 Duo CPU T7250 2.00 GHz, and all the function implementation are in MATLAB R14. The image size is 256×256 and the scale number for scale selection is 7. The comparison of running efficiency is listed in Table 1. From the results we can see that, the time cost for scale multiplication approach is about the double of the original one.

4   Experiment Results and Discussion

In this section, we give the corner detector results based on the method described in Section 3. Also some comparative experiments with Harris-Laplace or Harris are performed. All the tests are carried out under different noise levels on different images. In order to compare the detection result under image noise to the results without noise, we mark all the matched corners (defined in section 3.1) by red color. We will briefly discuss the results in the following.

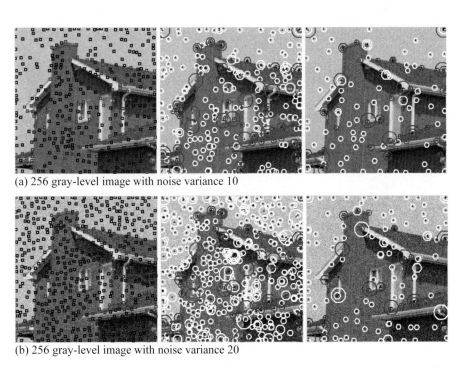

(a) 256 gray-level image with noise variance 10

(b) 256 gray-level image with noise variance 20

Fig. 4. The corner detection for house image with respect to different noise levels. Left: Harris. Center: Harris-Laplace. Right: scale multiplication Harris-Laplace.

First, the output of scale multiplication Harris-Laplace given the popular house image are shown in the right column in Fig. 4. Compare them with the left column and center column, output from Harris and Harris-Laplace respectively. Each detector gives results under two kinds of different image noise. The consistency of corner detection of scale multiplication Harris-Laplace is apparent. While the other two detectors produce more weak corner features or noise corners with the increase of image noise.

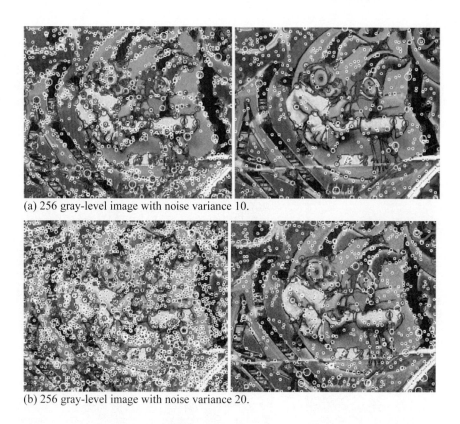

(a) 256 gray-level image with noise variance 10.

(b) 256 gray-level image with noise variance 20.

Fig. 5. The corner detection for graph image with respect to different noise levels. Left: Harris-Laplace. Right: scale multiplication Harris-Laplace.

Next, we give the detection results using the graph and boat image sequences, which are provided by Mikolajczyk[1]. The image size of graph sequence is 800×640 and boat sequence 850×680. Due to the page limit, we only give the results of Harris-Laplace and the proposed approach under two kinds of image noise. Fig. 5. and Fig. 6 demonstrate the corner detection results on graph and boat image respectively. It is obvious in the results that, similar to the corner detection on house image, much more weak features or noise corners were produced by Harris-Laplace.

---

[1] http://www.robots.ox.ac.uk/~vgg/research/affine/

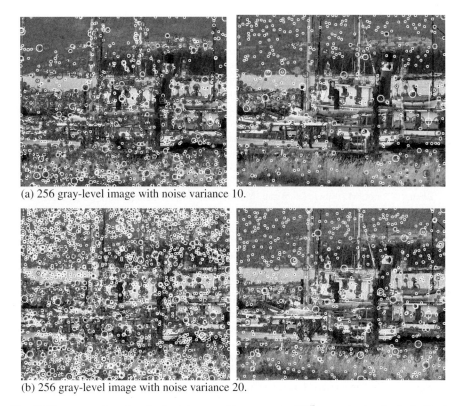

(a) 256 gray-level image with noise variance 10.

(b) 256 gray-level image with noise variance 20.

Fig. 6. The corner detection for boat image with respect to different noise levels. Left: Harris-Laplace. Right: scale multiplication Harris-Laplace.

5   Conclusion

In this paper, the technique of scale multiplication is analyzed in the framework of Harris corner measure. We define the scale multiplication function as the product of the Harris corner measure of adjacent scales, and propose a robust Harris-Laplace detector. The proposed product function of cornerness measure will magnify the corner like structures, while suppresses the image noise and weak features simultaneously. Unlike the contour-based multi-scale curvature product for image corner detection, we detect the corner like features directly in intensity image. Experiments on natural images also demonstrate that the proposed method has the results of good consistency of corner detection to variable noise.

Acknowledgments

This work is supported in part by National Natural Science Foundation of China (No. 60805018).

References

1. Tuytelaars, T., Mikolajczyk, K.: Local Invariant Feature Detectors: A Survey. Foundations and Trends in Computer Graphics and Vision 3(3), 177–280 (2008)
2. Teh, C., Chin, R.: On the detection of dominant points on digital curves. IEEE Trans. Pattern Anal. Mach. Intell. 11, 859–872 (1989)
3. Ji, Q., Haralick, R.: Corner detection with covariance propagation. In: Proc. of CVPR, pp. 362–367 (1997)
4. Mokhtarian, F., Suomela, R.: Robust image corner detection through curvature scale-space. IEEE Trans. Pattern Anal. Mach. Intell. 20(12), 1376–1381 (1998)
5. He, X., Yung, N.: Curvature scale space corner detector with adaptive threshold and dynamic region of support. In: Proc. of ICPR, pp. 791–794 (2004)
6. Zhang, X., Lei, M., Yang, D., Wang, Y., Ma, L.: Multi-scale curvature product for robust image corner detection in curvature scale space. Pattern Recognition Letters 28, 545–554 (2007)
7. Awrangjeb, M., Lu, G.: Robust Image Corner Detection based on the Chord-to-Point Distance Accumulation Technique. IEEE Trans. on Multimedia 10(6), 1059–1072 (2008)
8. Harris, C., Stephens, M.: A combined corner and edge detector. In: Alvey Vision Conference, pp. 147–151 (1988)
9. Jacob, M., Unser, M.: Design of steerable filters for feature detection using Canny-like criteria. IEEE Trans. Pattern Anal. Mach. Intell. 26(8), 1007–1019 (2004)
10. Lindeberg, T.: Feature detection with automatic scale selection. Inter. J. Comput. Vision 30(2), 79–116 (1998)
11. Mikolajczyk, K., Schmid, C.: Scale and affine invariant interest point detectors. Inter. J. Comput. Vision 1(60), 63–86 (2004)
12. Noble, J.: Finding corners. Image and Vision Computing 6(2), 121–128 (1988)

Spatial-Temporal Junction Extraction and Semantic Interpretation

Kasper Broegaard Simonsen[1,2], Mads Thorsted Nielsen[2], Florian Pilz[1],
Norbert Krüger[2], and Nicolas Pugeault[3]

[1] Department of Media Technology, Aalborg University Copenhagen, Denmark
[2] Cognitive Vision Lab,The Maersk Mc-Kinney Moller Institute,
University of Southern Denmark
[3] Centre for Vision, Speech and Signal Processing,
University of Surrey, United Kingdom

Abstract. This article describes a novel junction descriptor that encodes junctions' semantic information in terms incoming lines' orientations, both in 2D and 3D. A Kalman filter process is used to reduce the effect of local noise on the descriptor's error and to track the features. The improvement gained by our algorithm is demonstrated quantitatively on synthetic scenes and qualitatively on real scenes.

1 Introduction

Different kinds of image structures co–exist in natural images. It is common to classify them into edges, texture, junction or homogeneous image structures (see, e.g., [1]). Junctions are rare yet important image features, useful for various higher level vision tasks such as matching of stereo images, object recognition and scene analysis.

A significant amount of work studied the process of junction detection and the extraction of junction descriptors encoding local semantic information as oriented lines intersecting at the junction's center (see, e.g., [2,3,4,5,6]). In [2], both edges and junctions become detected and classified in the first step, and the semantic interpretation needs to be performed a posteriori. In [4], junctions are detected based on Hough lines and the semantic interpretation of junctions is computed in 2D and 3D. In [3], the number of lines is assumed to be known and a junction model is fitted to the data by minimizing an energy function. In [5], a simple method for extracting the semantic representation of junctions is proposed by analyzing orientation histograms. In [6], a biologically motivated approach to junction detection is presented, where junctions are characterized by a high activity for multiple orientations within a single cortical hyper–column – equivalently defined as points in the image where two or more edges join or intersect.

It is known that junctions are stable features for matching [7], and that their semantic interpretation is important for the reconstruction of visual scenes [8] (for example by interfering depth discontinuities from T-junctions). However, the

G. Bebis et al. (Eds.): ISVC 2009, Part I, LNCS 5875, pp. 275–286, 2009.

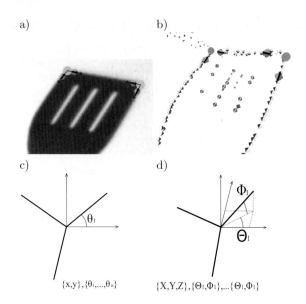

Fig. 1. A figure with 2D and 3D semantic interpretation. a) 2D junction interpretation in the left stereo image. b) 3D junction interpretation corresponding to a). c) Semantic 2D junction interpretation. d) Semantic 3D interpretation.

detection and semantic interpretation of junctions (in particular in 3D) faces the problem that high–level information is inferred from a (noisy) local image patch (or two corresponding images patches in the 3D case).

A suitable semantic representation for junctions in 2D can be formulated in terms of: (i) a center point $\mathbf{x}$; (ii) an integer n indicating the number of rays intersecting in $\mathbf{x}$; and (iii) a vector of n angles $(\theta_1, \ldots, \theta_n)$ indicating the incoming lines' directions (see figure 1c). An equivalent 3D interpretation (see, e.g., [4]) can be formulated in terms of: (i) a 3D center point $\mathbf{X}$; (ii) an integer n; and (iii) a vector of n pairs of angles $((\Theta_1, \Phi_1), \ldots, (\Theta_1, \Phi_1)))$ indicating the 3D lines' direction in polar coordinates (see figure 1d). However, for T-junctions (two of the edges are parallel), an actual intersection in the 3D point $\mathbf{X}$ is not likely due to a high probability of a depth discontinuity being existent at that position and hence this case needs to be detected on the 2D level beforehand.

The central novelty of this article is that it describes a system in which junction detection and the interpretation in 2D and 3D in terms of point as well as orientation information becomes improved by a Kalman filter based spatial–temporal filtering. Hence junctions become understood as spatial-temporal entities with considerable semantic information associated. As a result, junctions are detected more reliably than in individual frames, and their localization and semantic interpretation is more stable and correct. This is quantified on artificial data and results are shown on rather controlled indoor scenes in a robot vision set–up, as well as outdoor scenes with considerable noise level. Moreover, we have introduced some improvements in the actual junction extraction process in 2D and 3D from individual frames.

2   Junction Interpretation as Part of an Early Cognitive Vision System

The junction detection and interpretation described here is part of the Early Cognitive Vision framework described in [9,10]. In this framework, the interpretation of local information both in 2D and 3D, is performed by different experts for all four image structures described above (i.e., textures, junctions, edges and homogeneous image patches, see figure 2).

These interpretations are then disambiguated; both spatial and temporal contexts are used to eliminate outliers and come to a more precise estimation of the semantic parameters. Previous stages focused on the computation, interpretation and disambiguation of edge structures (see, e.g., [11,12]); this paper extends this framework to junctions.

Before extracting a junction semantic descriptor from a local image area, it is required to assess this patch's likelihood to contain a junction. The intrinsic dimension (iD) – see, e.g., [13] – is a suitable classifier in this context. The iD characterizes image patches according to the dimension of the subspace occupied by the local spectral energy. The spectral energy of an intrinsically zero–dimensional signal is concentrated in the origin, whilst the energy of an intrinsically one–dimensional signal spans a line, and the energy of an intrinsically two–dimensional signal varies in more than one dimension. Ideal homogeneous image patches have an intrinsic dimension of zero (i0D), ideal edges are intrinsically one–dimensional (i1D), and junctions and most textures have an intrinsic dimension of two (i2D). Going beyond such a discrete classification [13,14], we use a *continuous* formulation [15] that, based in a compact triangular representation, provides three confidences that express the likelihood of an image patch being either i0D, i1D or i2D.

In this paper, we understand junctions as image patches in which lines intersect. The intrinsic dimensionality is well suited to distinguish junctions from edges and homogeneous image patches; however, textures and junctions are both intrinsically 2D structures, and can not be reliably distinguished by local iD confidences. As shown in [5], junctions can be distinguished from textured areas by looking at the intrinsic dimensionality within a certain neighborhood (which size depends on the scale): junctions have high i1D and i0D values in their immediate neighborhood, whereas most textures have only i2D values (for details, see [16]).

3   2D and 3D Junction Extraction and Interpretation

In this section, we briefly describe the process of extracting a semantic 2D and 3D interpretation of junctions from stereo images. The algorithm used in this work is an improved version of [5]. The process is exemplified by showing 2D junctions and their 3D reconstructions on artificial image sequences. Results on real images are shown in section 5.2.

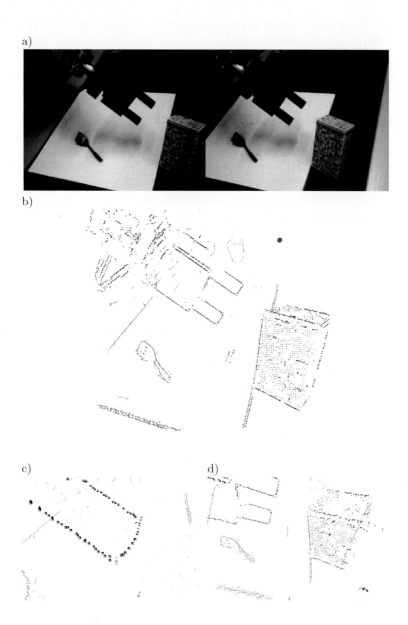

Fig. 2. 3D reconstructed icons according to the classified image patches. Red icons represent texture, green points junctions and the remaining icons edge–like structures; a) left and right input image; b) 3D reconstruction of the entire scene from the camera position; c) zoom-in on the top gripper with reconstructed junction and edges; d) display of the 3D reconstruction showing the correct reconstructed textured plane from an alternative perspective.

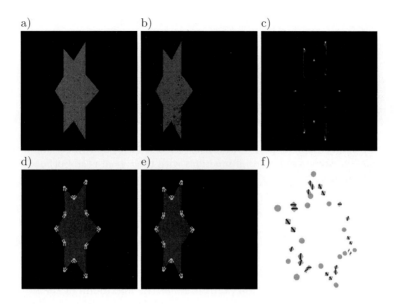

Fig. 3. 2D junction interpretation: a,b) shows the left and right view, c) shows the intersection consistency. d) and e) shows the junctions in the left and the right view and f) shows the 3D reconstruction from one stereo pair with the semantic interpretation.

3.1   2D Interpretation

This section briefly describes the 2D semantic interpretation of junction as proposed by [5], where the interested reader will find a detailed discussion.

A junction is represented by a set of rays r_i, each defining one of the intersecting half–lines, expanding from the junction's center in a direction θ_i [8,5].

Junctions are localized using a continuous formulation of the intrinsic dimensionality (iD), where intrinsically two dimensional image structures ($i2d$) indicate the presence of junctions. This positioning of potential junctions is improved by a local search for the pixel with the highest intersection consistency (iC) [5] (see figure 4b). Intersection consistency defines a more accurate junction position, where each line around the center votes for an intersection position. A pixel's intersection consistency is defined as:

$$\mathrm{ic}(p_c) = (1 - \mathrm{i1D}(p_c)) \sum_{p_x \in N_p} \left(k(p_x)(c_{mag}(p_x))^2 \left(1 - \frac{\mathrm{dist}(l_{p_x}, p_c)}{\mathrm{dist}(p_x, p_c)} \right) \right) \qquad (1)$$

N_p are all pixels in the local neighborhood with a certain distance to p_c (see figure 4c). $i1D(p_x)$ denotes the $i1D$ confidence for a given pixel [15], $c_{mag}(p_x)$ is the magnitude for a given pixel, l_{p_x} is the line going through p_x oriented as the local pixel orientation, $dist(l_{p_x}, p_c)$ is the distance between l_{p_x} and the center pixel p_c, and $dist(p_x, p_c)$ is the distance between the two pixels p_x and p_c. The

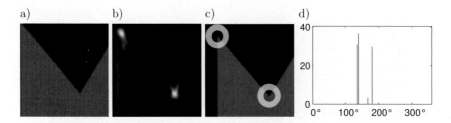

Fig. 4. 2D interpretation on part of star image (as discussed) a) artificial input image b) intersection consistency c) potential junctions with their local neighborhood for which the orientation histogram is computed. d) the orientation histogram of the top-left junction.

expression $k(p_x)$ is 1 if $dist(l_{p_x}, p_c) < maxDist$, zero otherwise. The basic idea behind this formula is that edge pixels in the surrounding vote for an optimal position of the junction. The intersection consistency is displayed in figure 3c and figure 4b. The junction's position (x, y) is set as the maximum of the intersection consistency computed with sub–pixel accuracy. The algorithm by [5] produced large iC in some edge structures; therefore, we introduced the term $(1 - i1D(p_c)$ that enforces low iC value at edge structures. For a detailed discussion of these differences we refer to [17].

The orientation θ_i is extracted for each potential ray r_i of a junction by finding the dominant orientations in its neighborhood (figure 4c); this is coded as an orientation histogram $H(\theta_c)$ (see figure 4d). The number of rays and their orientations is then determined by the clusters in that histogram $H(\theta_c)$. Having the information on the number of rays the semantic interpretation of a junction in 2D is defined as:

$$\pi^j = \{(x, y), n, (\theta_1, ..., \theta_n)\} \tag{2}$$

where x and y is the 2D position, n the number of edges and $(\theta_1, ..., \theta_n)$ the angle associated to each edge.

3.2   3D Interpretation

Junctions are matched in the left and right view according to the epipolar geometry using cross correlation. Additionally, we match the junction rays allowing for a 3D reconstruction in terms of the two angles (Θ, Φ) for each matched pair of rays, indicating the 3D lines' direction in polar co-ordinates (see figure 1d). The algorithm is discussed in more detail in [4]. The matching algorithm is based on the angular distance between 2D rays' orientations in the left and right image. First, the distance between all rays in the left and right view is computed (see figure 5a). The distance is measured by the number of bins, where the total number of bins is determined by the resolution of the orientation histogram (in our case 64 bins, being equal to 360°). If the distance between two edges in the left and right view is small, they are likely to be part of the same 3D edge (see

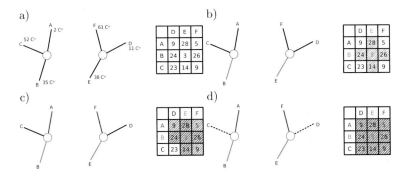

Fig. 5. a) left and right junction with table of orientation distances between their edges. b) The smallest distance is found to be between edge B and E and all cells related to edge B and E are cleared. c) Remaining cells with the smallest gaps are found (edge A and F) and cells for found pairs are cleared d). Remaining angle gaps above a threshold value are are discarded.

figure 5b,c); if the distance is above a threshold the edge is discarded (see figure 5d). For details, we refer again to [4,17].

Besides the 3D reconstruction also the 3D uncertainty is computed needed for the spatial–temporal filtering process using Kalman filters. The 3D reconstruction uncertainty is propagated from the 2D uncertainty as follows [18]:

$$\Lambda_{3D} = \boldsymbol{D}\boldsymbol{f}(\boldsymbol{x})\Lambda_{2D}\boldsymbol{D}\boldsymbol{f}(\boldsymbol{x})^T \tag{3}$$

where Λ_{3D} is the propagated 3D covariance, $\Lambda_{2D} = diag(\Lambda_1, ..., \Lambda_n)$ the uncertainty in 2D and $\boldsymbol{f}(\boldsymbol{x})$ the reconstruction function. For more details of the noise modeling we refer to [17].

Because edges are part of the semantic interpretation, they are also reconstructed as 3D edges. Their position uncertainty is propagated to 3D in the same way as it is done for the 3D position of junctions. An example of reconstructed junctions from a stereo image pair is shown in figure 3f.

4   Spatial–Temporal Filtering

The spatial-temporal filtering process is the combination of three processes. First, a temporal matching of junctions in images; Second a re–evaluation of junctions' confidence depending on how consistently they have been observed in a succession of frames; Third, multiple Kalman filters that correct the junction's position and incoming lines' orientations.

4.1   Temporal Matching

To find the best match for each prediction, it is compared with all the newly detected junctions in the neighborhood. The comparison is done by back projecting the 3D junction to the image planes and matching them with the newly detected 2D junctions.

4.2   Confidence

The junctions' confidence is re–evaluated using a Bayesian scheme wherein 3D junction's likelihood is estimated based on predictions and verifications of their change of 2D position over frames due to the underlying object or ego motion. This process is governed by the equation:

$$C = \frac{\beta^m (1 - \beta)^{n-m} \alpha}{\beta^m (1 - \beta)^{n-m} \alpha + \gamma^m (1 - \gamma)^{n-m} (1 - \alpha)},\qquad(4)$$

where n is the total number of frames since the junction was first detected, m is the number of times it was successfully matched since this time, α is the prior probability that an observed junction is true, β is the probability that a true junction is matched successfully, and γ is the probability that a false junction is matched. For further details see [12].

4.3   Filtering

The junction descriptor's parameters are corrected over time using a collection of independent Kalman filters [19], one for the junction's center and one for each of the intersecting lines. The prediction step is provided by the object motion, that is either known (object manipulation scenario) or computed from feature correspondences (road scenario). An estimation of this prediction uncertainty is gathered either from the process that provides the motion, and will not be discussed here. We use the temporal matching process to pair predicted junctions with the observed ones, and the observation uncertainty is provided by the junction 3D reconstruction process (and therefore is different for each observation). Because both observations and predictions lie in the same space, the Kalman equations simplify to the following:

$$K = \Sigma \cdot (\Sigma + \Sigma^*)^{-1} \qquad (5)$$
$$x' = x + K \cdot (x^* - x) \qquad (6)$$
$$\Sigma' = (I - K) \cdot \Sigma \qquad (7)$$

where K is the Kalman gain, (x, Σ) are the predicted state vector's mean and covariance, (x^*, Σ^*) the observed vector's, and (x', Σ') the corrected vector's.

Note that the filtering process is an extension of a process defined for edge elements in [12].

5   Results

We have tested our algorithm on artificial (section 5) as well as indoor and outdoor stereo sequences (section 5.2).

5.1   Quantification on an Artificial Image Sequences

First, we tested our algorithm in a stereo sequence showing a star rotating around the y axis. The images are created in OpenGL and hence the ground truth of the motion is known, and used for the Kalman filter's prediction step. Also

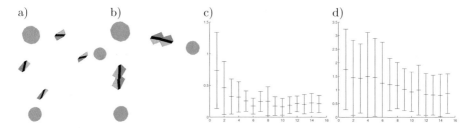

Fig. 6. Accumulation of junctions: a) shows the non accumulated 3D representation of a tip of the star, b) shows the accumulated tip, c) and d) shows the 3D position error (in mm) of junctions and their adjacent edges, respectively, over accumulation iterations

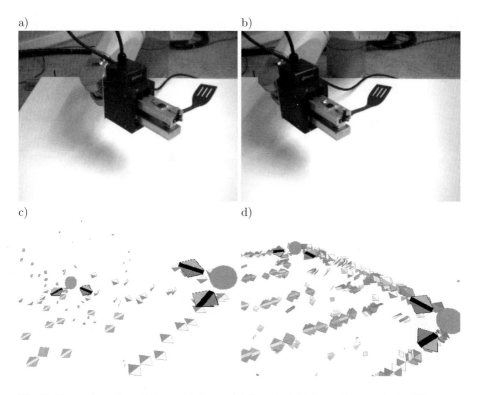

Fig. 7. Reconstruction of the spatula. a,b) left and right input image c) the 3D reconstruction of the spatula c) accumulated junctions after 72 frames. Edges are displayed for better scene understanding.

the junctions' exact 3D positions and angles between incoming rays. Figure 6 exemplifies the convergence process. In a), the reconstruction based on one stereo pair is drawn; in b), the reconstruction after 15 iterations is drawn, showing an improvement in the 3D orientation estimate. This is also quantified in figure 6c, d. In figure 6c the position error mean and variance over all junctions is displayed. In figure 6d, the deviation from the true angles between two rays is shown. The accumulation scheme lead to improvements for both quantities.

5.2   Indoor and Outdoor Scenes

Figure 7 shows results on a scene where the motion is generated by a robot (and hence is known). The junctions are displayed by circles and the orientations of the associated 3D rays are shown as the big rectangular icons. In addition, accumulated edges are shown according to the algorithm described in [12]. Figure 7 a, b) shows the left and right images; figure 7c shows the stereo representation at the first frame; figure 7d shows the accumulated representation. It is clearly visible that the representation improves both in terms of edges and junctions.

Figure 8 shows results on a scene recorded in a driver assistance scenario where the motion is estimated using the algorithm described in [20]. An improvement in positioning for this scene is not as eminent as for the indoor sequence, due to an increased scene depth. However by utilizing spatial temporal filtering, wrong re-constructed junctions have been discarded providing a more robust scene representation.

a)   b)

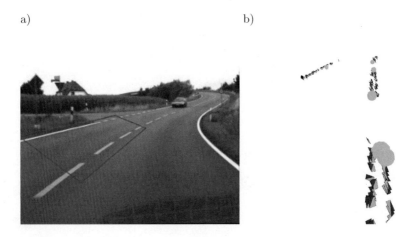

Fig. 8. Results for outdoor scene recorded in a driver assistance scenario. a) left input image b) accumulated junctions after 20 frames. (For a better display, edges are displayed and the display size for both junctions and edges has been increased.

6   Conclusion

We introduced a junction extraction process that encodes not only junctions' positions, but also semantic information in terms of intersecting lines' orientations; this is also extended to in the 3D domain using stereopsis. It is known that the extraction of such high level information can be severely affected by noise. To counteract the effect of noise, we applied a Kalman filter scheme on both the junction's position, and intersecting lines' orientations. The algorithm was evaluated on artificial and real data, and lead to significant improvements in position and orientation estimates. Results were shown quantitatively on artificial data, and qualitatively on real scenes.

The spatial–temporal junction extraction algorithm is part of an Early Cognitive Vision system in which different local symbolic descriptors become computed for different image structures. This system, so far dealing with line segments only, has been used is a large number of applications in computer vision and vision based robotic tasks (e.g., [11,21]). The integration of junctions into the Early Cognitive Vision system is an important step to enhance the system's generality and performance for tasks such as motion estimation, object recognition and pose estimation.

Acknowledgment

This work has been supported by the European Commission - FP6 Project DRIVSCO (IST-016276-2).

References

1. Kalkan, S., Wörgötter, F., Krüger, N.: Statistical analysis of local 3d structure in 2d images, pp. 1114–1121 (2006)
2. Harris, C.G., Stephens, M.: A combined corner and edge detector. In: 4th Alvey Vision Conference, pp. 147–151 (1988)
3. Rohr, K.: Recognizing corners by fitting parametric models, vol. 9, pp. 213–230 (1992)
4. Hahn, M., Krüger, N.: Junction detection and semantic interpretation using hough lines. In: DAGM 2000 (2000)
5. Kalkan, S., Shi, Y., Pilz, F., Krüger, N.: Improving junction detection by semantic interpretation. In: VISAPP (1), pp. 264–271 (2007)
6. Hansen, T., Neumann, H.: Neural mechanisms for the robust representation of junctions. Neural Comput. 16, 1013–1037 (2004)
7. Shen, X., Palmer, P.: Uncertainty propagation and the matching of junctions as feature groupings. IEEE Trans. Pattern Anal. Mach. Intell. 22, 1381–1395 (2000)
8. Parida, L., Geiger, D., Hummel, R.: Junctions: detection, classification and reconstruction, vol. 20, pp. 687–698 (1998)
9. Krüger, N., Lappe, M., Wörgötter, F.: Biologically Motivated Multi-modal Processing of Visual Primitives. Interdisciplinary Journal of Artificial Intelligence & the Simulation of Behaviour, AISB Journal 1, 417–427 (2004)

10. Pugeault, N.: Early Cognitive Vision: Feedback Mechanisms for the Disambiguation of Early Visual Representation. PhD thesis, Informatics Institute, University of Göttingen (2008)
11. Kraft, D., Pugeault, N., Başeski, E., Popović, M., Kragic, D., Kalkan, S., Wörgötter, F., Krüger, N.: Birth of the Object: Detection of Objectness and Extraction of Object Shape through Object Action Complexes. Special Issue on Cognitive Humanoid Robots of the International Journal of Humanoid Robotics 5, 247–265 (2009)
12. Pugeault, N., Wörgötter, F., Krüger, N.: Accumulated Visual Representation for Cognitive Vision. In: Proceedings of the British Machine Vision Conference, BMVC (2008)
13. Zetzsche, C., Barth, E.: Fundamental limits of linear filters in the visual processing of two dimensional signals. Vision Research 30, 1111–1117 (1990)
14. Jähne, B.: Digital Image Processing – Concepts, Algorithms, and Scientific Applications. Springer, Heidelberg (1997)
15. Felsberg, M., Kalkan, S., Krüger, N.: Continuous dimensionality characterization of image structures. Image and Vision Computing 27, 628–636 (2009)
16. Goswami, D., Kalkan, S., Kruger, N.: Bayesian classification of image structures. In: Scandinavian Conference on Image Analysis, SCIA (2009)
17. Simonsen, B., Nielsen, M.: Junctions in an early cognitive vision system. Master's thesis, Cognitive Vision Group, Maersk Institute, University of Southern Denmark (2009)
18. Faugeras, O.: Three–Dimensional Computer Vision. MIT Press, Cambridge (1993)
19. Kalman, R.: A new approach to linear filtering and prediction problems. Journal of Basic Engineering 82 (1960)
20. Pilz, F., Pugeault, N., Krüger, N.: Comparison of point and line features and their combination for rigid body motion estimation. In: Statistical and Geometrical Approaches to Visual Motion Analysis. LNCS, vol. 5604. Springer, Heidelberg (2009)
21. Detry, R., Pugeault, N., Piater, J.: A probabilistic framework for 3d visual object representation. IEEE transactions on Pattern Analysis and Machine Intelligence (in press)

Cross-Correlation and Rotation Estimation of Local 3D Vector Field Patches

Janis Fehr[1,3], Marco Reisert[2,3], and Hans Burkhardt[3]

[1] HCI, University Heidelberg, Germany
[2] Medical Physics, University Hospital Freiburg, Germany
[3] LMB, University Freiburg, Germany
janis.fehr@iwr.uni-heidelberg.de

Abstract. In this paper, we present a method for the fast and accurate computation of the local cross-correlation of 3D vectorial data. Given spherical patches of 3D vector fields, our method computes the full cross-correlation over all possible rotations and allows an accurate estimation of the rotation offset between two patches. Our approach is based on a novel harmonic representation for 3D vector fields located on spherical surfaces which allows us to apply the *convolution theorem* and perform a fast correlation in the frequency domain. The theoretical advances presented in this paper can be applied to various computer vision and pattern recognition problems, such as finding corresponding points on vector fields for registration problems, key point detection on vector fields or designing local vector field features.

1   Introduction

3D vector fields play an important role in a wide range of computer vision and pattern recognition problems. From methods like 3D motion estimation, which have a natural representation in vector fields, to the extraction of 3D gradient fields from volume images, 3D vector fields are a suitable standard data representation for many complex vision problems. Given such 3D vector fields, a common task is to compare and measure similarities between two or more data sets. In this paper, we approach the problem of locally comparing 3D vectorial data. Our method operates on local spherical vector field patches (1) and uses a full correlation over all possible rotations between two patches to compute a rotation invariant similarity measure as well as an accurate rotation offset estimation. Our proposed method could be used to directly transfer some well established methods from the 2D and 3D scalar domain to 3D vector fields: like finding corresponding points on vector fields for registration problems, key point and object detection in vectorial data or for the design of local vector field features.

Parameterization of Local Patches. Given a 3D vector field $\mathbf{X} : \mathbb{R}^3 \rightarrow \mathbb{R}^3$, we parameterize a local spherical patch at position $\mathbf{x} \in \mathbb{R}^3$ as a set of n concentric

G. Bebis et al. (Eds.): ISVC 2009, Part I, LNCS 5875, pp. 287–296, 2009.

spherical surfaces $\mathcal{S}_i[r_i]\,(\mathbf{x})$ with radii r_i and $i \in \{1,\ldots,n\}$ which are centered in $\mathbf{x}$:

$$\mathcal{S}_i[r_i]\,(\mathbf{x}) := \{\mathbf{x}' \in \mathbb{R}^3 \,|\, \|\mathbf{x} - \mathbf{x}'\|_2 = r_i\}. \tag{1}$$

Related Work. The fast correlation of scalar data over all possible cyclic translations in an nD Euclidean space by use of the *Convolution Theorem* is a standard procedure in computer vision and pattern recognition applications. An extension to nD Euclidean vector spaces is also quite common, e.g. a general method for the correlation of vector fields can be found in [10]. Fast correlations over all possible rotations tend to be more difficult. Our approach is motivated by [3], which solves this problem for scalar signals by extending and improving the accuracy of the fast correlation methods originally introduced by [7] and [8]. All of these methods use a harmonic expansion of the scalar signals in Spherical Harmonics to compute the correlation in the frequency domain. We introduce a novel harmonic representation of vectorial signals on a sphere that is largely derived form a more general harmonic tensor representation introduced by [9]. The fast correlation and the rotation estimation are both a direct extension of the scalar methods we presented in [3]. The remainder of this paper is structured as follows: first, we briefly review some mathematical foundations and introduce a common notation in section 1.1 before we derive the harmonic representation of 3D vector fields on spheres in section 2. Then, in sections 3 and 3.1, we present the cross-correlation and rotation estimation algorithms. Finally, we experimentally evaluate the performance of our methods in section 4.

1.1   Mathematical Foundations

Our parameterization of the local vector field patches in form of vectorial signals on concentric spherical surfaces (see section 1) makes it necessary to find a suitable mathematical formulation of such signals. For this purpose, we derive an orthonormal base for vectorial signals on the 2-sphere (see section 2) that provides a sound mathematical representation of the parameterized patches in the harmonic domain. The entire approach is based on widely known methods from *angular momentum theory* [2] such as Spherical Harmonic base functions for scalar valued functions on spheres, *Wigner-D* rotation matrices and the *Clebsch-Gordan Coefficients*. Hence, we start with a brief review of these methods.

Spherical Harmonics. Spherical Harmonics ($\mathcal{SH}$) [5] form an orthonormal base on the 2-sphere. Analogical to the Fourier Transform, any given real valued signal f on a sphere with its parameterization over the angles Θ, Φ (latitude and longitude of the sphere) can be represented by an expansion in its harmonic coefficients:

$$f(\Phi, \Theta) = \sum_{l=0}^{\infty} \sum_{m=-l}^{m=l} \widehat{f}_m^l Y_m^l(\Phi, \Theta), \tag{2}$$

where l denotes the band of expansion, m the order for the l-th band and $\widehat{f}_m^l$ the harmonic coefficients. The harmonic base functions $Y_m^l(\Phi, \Theta)$ are computed as follows:

$$Y_m^l(\Phi, \Theta) = \sqrt{\frac{2l+1}{4\pi}\frac{(l-m)!}{(l+m)!}} \cdot P_m^l(\cos\Theta)e^{im\Phi}, \tag{3}$$

where P_m^l is the associated Legendre polynomial [2].

Rotations in $\mathcal{SH}$. Throughout the rest of the paper we will use the Euler notation in zyz'-convention denoted by the angles φ, θ, ψ with $\varphi, \psi \in [0, 2\pi[$ and $\theta \in [0, \pi[$ to parameterize the rotations $\mathcal{R} \in \mathcal{SO}(3)$ (abbreviation for $\mathcal{R}(\varphi, \theta, \psi) \in \mathcal{SO}(3)$). Rotations $\mathcal{R}$ in the Euclidean space find their equivalent representation in the Spherical Harmonic domain in terms of the so called Wigner D-Matrices, which form an irreducible representation of the rotation group $\mathcal{SO}(3)$ [2]. For each band l, $D^l(\varphi, \theta, \psi)$ (abbreviated as $D^l(\mathcal{R})$) defines a band-wise rotation in the $\mathcal{SH}$ coefficients. Hence, a rotation in the Euclidean space can be computed in the harmonic domain by:

$$\mathcal{R}f = \sum_{l=0}^{\infty}\sum_{m=-l}^{l}\sum_{n=-l}^{l} D_{mn}^l(\mathcal{R})\widehat{f}_n^l Y_m^l. \tag{4}$$

Clebsch-Gordan Coefficients. Clebsch-Gordan Coefficients (CG) of the form $\langle lm|l_1m_1, l_2m_2\rangle$ are commonly used for the representation of direct sum decompositions of $\mathcal{SO}(3)$ tensor couplings [2]. The CG define the selection criteria for such couplings, are only unequal to zero if the constraints $m = m_1 + m_2$ and $|l_1 - l_2| \le l \le l_1 + l_2$ hold and fulfill several useful orthogonality constraints (see [2] for more details).

2   Vectorial Harmonics

Based on the rich and well established methods for scalar signals on spheres, we extend the Spherical Harmonics to vectorial data. The so-called Vectorial Harmonics ($\mathcal{VH}$) inherit most of the favorable properties from their scalar counterparts, while solving the non trivial problem that rotations of vectorial data not only affect the position of a vector, but also its direction. It should be noted that there have been several different previous approaches towards Vectorial Harmonics, like [6] or [1]. All of these methods basically provide the same functionality, but use different parameterizations and notations which are not directly suitable for our purposes. We derive our methods from a very general theory of Tensorial Harmonics [9], which provides expansions for arbitrary real valued tensor functions $\mathbf{f}$ of order d on the 2-sphere:

$$\mathbf{f}[r, d](\Theta, \Phi) := \sum_{l=0}^{\infty}\sum_{k=-d}^{k=d}\sum_{m=-(l+k)}^{m=(l+k)} \widehat{\mathbf{f}}_{km}^l(r)\mathbf{Z}_{km}^l(\Theta, \Phi), \tag{5}$$

where $\widehat{\mathbf{f}}_{km}^l(r)$ is the expansion coefficient of the l-th band of tensor order d and harmonic order m at radius r. The orthonormal Tensorial Harmonic base functions $\mathbf{Z}_{km}^l$ are given as:

$$\mathbf{Z}_{km}^l(\Theta, \Phi) := \mathbf{e}_m^{(l+k)} \circ_l Y_m^l(r)(\Theta, \Phi), \tag{6}$$

where $\mathbf{e}_m^l$ are elements of the standard Euclidean base of $\mathbb{C}^{2d+1}$. $\circ_l$ denotes a bilinear form connecting tensors of different ranks:

$$\circ_d : V_{l_1} \times V_{l_2} \to \mathbb{C}^{2d+1}. \tag{7}$$

$l_1, l_2 \in \mathbb{N}$ have to hold $|l_1 - l_2| \le l \le l_1 + l_2$ and $\circ_l$ is computed as follows:

$$(\mathbf{e}_m^l)^T(\mathbf{v} \circ_l \mathbf{u}) := \sum_{m=m_1+m_2} \langle lm|l_1m_1, l_2m_2\rangle v_{m_1} u_{m_2}. \tag{8}$$

See [9] for details and proofs. If we limit the general form to tensors of order one ($d := 1$), then we obtain our Vectorial Harmonic expansions

$$\mathbf{f}[r, d=1](\Theta, \Phi) := \sum_{l=0}^{\infty} \sum_{k=-1}^{k=1} \sum_{m=-(l+k)}^{m=(l+k)} \widehat{\mathbf{f}}_{km}^l(r) \mathbf{Z}_{km}^l(\Theta, \Phi) \tag{9}$$

with the orthonormal base functions:

$$\mathbf{Z}_{km}^l = \begin{pmatrix} \langle 1 \ 1 \ | l+k \ \ m, l \ \ 1-m\rangle & Y_{1-m}^l \\ \langle 1 \ 0 \ | l+k \ \ m, l \ \ -m\rangle & Y_{-m}^l \\ \langle 1 \ -1 \ | l+k \ \ m, l \ \ -1-m\rangle & Y_{-1-m}^l \end{pmatrix}^T. \tag{10}$$

Refer to figure 1 for an overview of the first two bands of the $\mathcal{VH}$ basis.

$$\mathbf{u} \in \mathbb{C}^3 : \mathbf{u} := \begin{pmatrix} \frac{x-iy}{\sqrt{2}} \\ z \\ \frac{-x-iy}{\sqrt{2}} \end{pmatrix}. \tag{11}$$

Rotations in $\mathcal{VH}$. The analogy of our Vectorial Harmonics to Spherical Harmonics continues also in the case of rotations in the harmonic domain. Complex 3D vector valued signals $\mathbf{f}$ with Vectorial Harmonic coefficients $\widehat{\mathbf{f}}$ are rotated [9] by:

$$\mathcal{R}\mathbf{f} = \sum_{l=0}^{\infty} \sum_{k=-1}^{k=1} \sum_{m=-(l+k)}^{l+k} \sum_{n=-(l+k)}^{l+k} D_{mn}^{l+k}(\mathcal{R})\widehat{\mathbf{f}}_{km}^l \mathbf{Z}_{kn}^l, \tag{12}$$

which is a straight forward extension of (4). One notable aspect is, that we need to combine Wigner-D matrices of the upper $l+1$ and lower $l-1$ bands in order to compute the still band-wise rotation of $\widehat{\mathbf{f}}_{km}^l$. Hence, we rotate $\widehat{\mathbf{f}}_{km}^l$ by $\mathcal{R}(\phi, \theta, \psi)$ via band-wise multiplications:

$$\mathbf{f}' = \mathcal{R}(\phi, \theta, \psi)\mathbf{f} \Rightarrow \widehat{\mathbf{f}'}_{km}^l = \sum_{n=-(l+k)}^{l+k} D_{mn}^{l+k}(\phi, \theta, \psi)\widehat{\mathbf{f}}_{km}^l. \tag{13}$$

Due to the use of the zyz'-convention, we have to handle inverse rotations with some care:

$$\mathbf{f}' = \mathcal{R}^{-1}(\phi, \theta, \psi)\mathbf{f} \Rightarrow \widehat{\mathbf{f}'}_{km}^l$$

$$= \sum_{n=-(l+k)}^{l+k} D_{mn}^{l+k}(-\psi, -\theta, -\phi)\widehat{\mathbf{f}}_{km}^l. \tag{14}$$

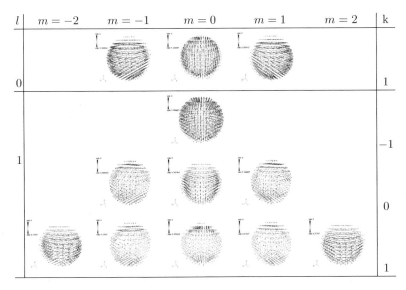

l	$m = -2$	$m = -1$	$m = 0$	$m = 1$	$m = 2$	k
0						1
1						-1
						0
						1

Fig. 1. Real part of the first 2 bands of the Vectorial Harmonic base functions, note that the base functions for $l = 0$ and $k = -1, 0$ do not exist

Band Limitation. A practically very relevant property of the harmonic representation (2) is that due to the band-wise nature of the representation, all properties, including the rotation (3.1), are independent of the actual extent of the expansion. Hence, we can limit the expansion to an arbitrary maximum band b_{max} without changing any of the characteristics. Naturally, a band limitation acts like a low-pass filter, cutting off the information in higher frequencies.

3   Fast and Accurate Correlation in $\mathcal{VH}$

Given our harmonic representation of vectorial signals on spheres (9) and the methods to rotate theses signals (13), we are able to present an algorithm for the fast correlation over all possible rotations. Analogous to the duality given by the *Convolution Theorem*, we derive a correlation method which operates in the Vectorial Harmonic (frequency) domain. This approach is based on a similar method which we have derived for scalar signals expanded in the Spherical Harmonic domain [3].

Correlation of Vectorial Signals on Spheres. We use local dot-products of vectors to define the correlation under a given rotation $\mathcal{R}$ in Euler Angles ϕ, θ, ψ as:

$$\mathcal{C}^{\#}(\mathcal{R}) := \int_{\Phi,\Theta} \langle \mathbf{f}(\Phi, \Theta), \mathcal{R}\mathbf{g}(\Phi, \Theta) \rangle \quad \sin \Theta d\Phi d\Theta, \tag{15}$$

where $\mathcal{C}^{\#}$ is the 3D correlation matrix which holds the correlation results for all possible rotations $\mathcal{R}$ in a (φ, θ, ψ)-space.

Applying the Convolution Theorem. Using the *Convolution Theorem* and substituting $\mathbf{f}$ and $\mathbf{g}$ with their $\mathcal{VH}$ expansions (3.1, 9) leads to

$$
\mathcal{C}^{\#}(\mathcal{R}) = \sum_{l=0}^{l=\infty} \sum_{k=-1}^{k=1} \sum_{mn=-(l+k)}^{mn=(l+k)} \overline{D_{mn}^{l+k}(\mathcal{R})} \widehat{\mathbf{f}}_{km}^{l} \overline{\widehat{\mathbf{g}}_{kn}^{l}}. \tag{16}
$$

We factorize the original rotation $\mathcal{R}(\varphi, \theta, \psi)$ into $\mathcal{R} = \mathcal{R}_1 \mathcal{R}_2$, choosing $\mathcal{R}_1(\xi, \pi/2, 0)$ and $\mathcal{R}_2(\eta, \pi/2, \omega)$ with $\xi = \varphi - \pi/2, \eta = \pi - \theta, \omega = \psi - \pi/2$.
Using the fact that

$$
D_{mn}^{l}(\varphi, \theta, \psi) = e^{-im\varphi} d_{mn}^{l}(\theta) e^{-in\psi}, \tag{17}
$$

where d^l is a real valued and so called "Wigner (small) d-matrix" [2], and

$$
D_{mn}^{l}(\mathcal{R}_1 \mathcal{R}_2) = \sum_{h=-l}^{l} D_{nh}^{l}(\mathcal{R}_1) D_{hm}^{l}(\mathcal{R}_2), \tag{18}
$$

we can rewrite $D_{mn}^{l}(\mathcal{R})$ as

$$
D_{mn}^{l}(\mathcal{R}) = \sum_{h=-l}^{l} d_{nh}^{l}(\pi/2) d_{hm}^{l}(\pi/2) e^{-i(n\xi + h\eta + m\omega)}. \tag{19}
$$

Substituting (19) into (16) provides the final formulation for the correlation function regarding the new angles ξ, η and ω:

$$
\mathcal{C}^{\#}(\xi, \eta, \omega) = \sum_{l=0}^{l=\infty} \sum_{k=-1}^{k=1} \sum_{m,h,m'=-(l+k)}^{m,h,m'=(l+k)} d_{mh}^{l+k}(\pi/2)
$$
$$
d_{hm'}^{l+k}(\pi/2) \widehat{\mathbf{f}}_{km}^{l} \overline{\widehat{\mathbf{g}}_{km'}^{l}} e^{-i(m\xi + h\eta + m'\omega)}. \tag{20}
$$

Obtaining the Fourier Transform of the Correlation Matrix. The direct evaluation of this correlation function is not possible - but it is rather straight forward to obtain the Fourier transform $\widehat{\mathcal{C}^{\#}} := \mathcal{F}(\mathcal{C}^{\#})$ of (20), hence eliminating the missing angle parameters:

$$
\widehat{\mathcal{C}^{\#}}(m, h, m') = \sum_{l=0}^{l=\infty} \sum_{k=-1}^{k=1} d_{mh}^{l+k}(\pi/2) d_{hm'}^{l+k}(\pi/2) \widehat{\mathbf{f}}_{km}^{l} \overline{\widehat{\mathbf{g}}_{km'}^{l}}. \tag{21}
$$

Finally, the correlation $\mathcal{C}^{\#}(\xi, \eta, \omega)$ can be retrieved via inverse FFT of $\widehat{\mathcal{C}^{\#}}$

$$
\mathcal{C}^{\#}(\xi, \eta, \omega) = \mathcal{F}^{-1}(\widehat{\mathcal{C}^{\#}}(m, h, m')), \tag{22}
$$

revealing the correlation values on a sparse grid in a three dimensional (ξ, η, ω)-space. Figure (2) shows a resulting correlation grid for a sample rotation of random input data. Hence, if we are interested in the maximum correlation value, we simply have to search the (ξ, η, ω)-space.

3.1   Rotation Estimation

The second objective of
our method is to not
only compute the max-
imum correlation value,
but also to obtain the rotation offset between two
vectorial signals. The resulting correlation matrix
$\mathcal{C}^{\#}$ from (22) directly allows us to compute the ro-
tation parameters (φ, θ, ψ)
from the (ξ, η, ω)-angles
associated with the posi-

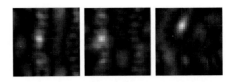

Fig. 2. Ortho-view of a resulting 3D correlation grid in
the (ξ, η, ω)-space with a maximum spherical harmonic
expansion to the 16th band, $\phi = \pi/4, \theta = \pi/8, \psi =
\pi/2$. From left to right: xy-plane, zy-plane, xz-plane.
Clearly, our fast correlation function obtains an isolated
and stable maximum in a single point on the grid.

tion of the maximum correlation peak. However, this direct rotation estimation
approach has a major drawback: the angular resolution of such an estimate directly depends on the maximum expansion band b_{max}, because the parameters
m, m', h in (21) are running from $l = 0, \ldots, l = b_{max}$. Hence, an example expansion to the 16th band would result in a $\mathcal{C}^{\#}$ size of $33 \times 33 \times 33$. Given rotations
up to $360°$, this leaves us in the worst case with an overall estimation accuracy
of less than $15°$.

In general, even if our fast correlation function (22) perfectly estimated the
maximum position in all cases, we would have to expect a worst case accuracy
of

$$Err_{corr} = 2 \cdot \frac{180°}{2l} + \frac{90°}{2l}. \tag{23}$$

Hence, if intended to achieve an accuracy of $1°$, we would have to take the harmonic expansion roughly beyond the 180th band. This would be computationally
too expensive for most applications. Even worse, since we are usually considering
discrete data, the signals on the sphere are band-limited. So for smaller radii,
higher bands of the expansion are actually not carrying any valuable information. To solve this resolution problem we follow an estimation method introduced
by [3], which uses a *Sinc Interpolation* in the frequency domain to obtain high
angular accuracies even if limited to low maximum expansion bands.

Sinc Interpolation. It is easy to see that the (m, h, m')-space in (21) actually
represents a discrete 3D Fourier spectrum. Hence, one can directly perform a
Sinc interpolation by adding a zero padding into the (m, h, m')-space [3]. This
way, the resolution of our correlation function can be increased drastically at very
low additional cost. It has to be noted, that even though the Sinc interpolation
implies some smoothing characteristics to the correlation matrix, the maxima
remain fixed to singular positions in the grid.

Using a pad size of p, we are able to reduce the worst case accuracy even for
low expansion bands:

$$Err_{corr}^{pad} = 2 \cdot \frac{180°}{2l + p} + \frac{90°}{2l + p}. \tag{24}$$

Of course, the padding approach has practical limitations - inverse FFTs are becoming computationally expensive at some point. But as our experiments show, resolutions below one degree are possible even for very low expansions. Finally, we are able to retrieve the original rotation parameters. For a given correlation peak at the grid position $C^\#(x, y, z)$, with maximum harmonic expansion b and padding p, the rotation angles are given by:

$$\phi = \begin{cases} \pi + (2\pi - x\Delta) & \text{for } x\Delta > \pi \\ \pi - x\Delta & \text{otherwise} \end{cases} \tag{25}$$

$$\theta = \begin{cases} (2\pi - y\Delta) & \text{for } y\Delta > \pi \\ y\Delta & \text{otherwise} \end{cases} \tag{26}$$

$$\psi = \begin{cases} \pi + (2\pi - z\Delta) & \text{for } z\Delta > \pi \\ \pi - z\Delta & \text{otherwise} \end{cases} \tag{27}$$

with $\Delta = 2\pi/(b + p)$.

4   Experiments

We conduct a series of experiments on artificial 3D vector field data to evaluate the accuracy and complexity performance of our proposed methods.

Rotation Estimation Accuracy. First, we investigate the performance of the rotation estimation algorithm. We use a sample 3D vector field, which is rotated around the center of one spherical patch parameterized by a single radius of $r = 10$. For each experiment, we evaluate the error statistics of 100 random rotations of this vector field. We generate the rotations over all possible angles $\varphi, \psi \in [0, 2\pi[$ and $\theta \in [0, \pi[$ with a resolution of $0.001 \approx 0.1°$. Note that an

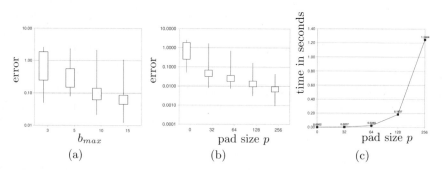

(a)   (b)   (c)

Fig. 3. (a) Accumulated rotation estimation error for increasing b_{max} and without using the Sinc interpolation method ($p = 0$). (b) Accumulated rotation estimation error for increasing pad size p of the Sinc interpolation with $b_{max} = 5$. (c) Computational complexity for increasing pad size p of the Sinc interpolation with $b_{max} = 5$. The experiments were performed on a standard 2GHz PC, using the FFTW [4] implementation of the inverse FFT.

error of $1° \approx 0.017$. All given error rates are the accumulated errors of all three angles. Figure 3 shows the direct effect of the maximum expansion band b_{max} on the rotation estimate. But even for expensive "higher band" expansions, we encounter strong outliers and a rather poor average accuracy. This can be compensated by our Sinc interpolation approach: Figure 3 shows how we can reduce the rotation estimation error well below $1°$, just by increasing the pad size p. The additional computational costs caused by the padding are also given in figure 3. Summarizing these first experiments, are we able to show that our proposed method is able to provide a fast and accurate rotation estimation even for rather low band expansions, i.e. if we choose $p = 64$ and $b_{max} = 5$, we can expect an average estimation error below $1°$ at a computation time of less than 25ms.

Key Point Detection. In a second series of experiments, we evaluate the performance of our methods in a key point (or object) detection problem on artificial data. Figure 4 shows the 3D vector fields of two of our target structures. Our goal is to detect the center of such X- and Y-like shaped bifurcations under arbitrary rotations in larger vector fields. For each target

Fig. 4. Sample target structures for the detection problem: 3D vector fields of X- and Y-like shaped bifurcations

structure, we extract a single patch, parameterized in four different radii with $b_{max} = 3$, at the center of the bifurcations. We extract patches with the same parameterization at each point of the test samples and apply our cross-correlation to detect the target structures in the test vector fields. Figure 5 shows some example test data together with the correlation results.

It should be noted that the test bifurcations are only similar in terms of an X or Y shape, but not identical to the given target structures. We also rotate the test data in a randomized procedure over all angles. Applying a threshold of 0.9

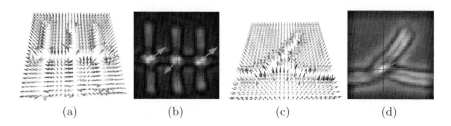

(a)   (b)   (c)   (d)

Fig. 5. (a) Test data for the X-sample. (b) xy-slice of a sample correlation result for the X-bifurcation target. The red cross indicates the position of the maximum correlation value and the green arrows the estimated rotation offset. (c) Test data for the Y-sample. (d) xy-slice of the correlation result for the Y-bifurcation target. The red cross indicates the position of the maximum correlation value and the green arrows the estimated rotation offset.

to the correlation results, we were able to detect the correct target structures in all of our test samples without false positives.

Conclusions and Outlook. We presented a novel, fast and accurate method for the computation of the cross-correlation and the estimation of the rotation offset of local spherical patches extracted from 3D vector fields. We were able to show that on artificial "toy problems", our method achieves very good results for both tasks. For the future, our next step will be to investigate the performance of our method on "real world" data. Given the low-pass property of our Vectorial Harmonics, we expect our method to be even able to cope with noisy data.

References

1. Barrera, R., Estevez, G., Giraldo, J.: Vector spherical harmonics and their application to magnetostatic. Eur. J. Phys. 6, 287–294 (1985)
2. Brink, D., Satchler, G.: Angular Momentum, 2nd edn. Clarendon Press, Oxford (1968)
3. Fehr, J., Reisert, M., Burkhardt, H.: Fast and accurate rotation estimation on the 2-sphere without correspondences. In: Forsyth, D., Torr, P., Zisserman, A. (eds.) ECCV 2008, Part II. LNCS, vol. 5303, pp. 239–251. Springer, Heidelberg (2008)
4. Frigo, M., Johnson, S.G.: The design and implementation of FFTW3. Proceedings of the IEEE 93(2), 216–231 (2005); special issue on "Program Generation, Optimization, and Platform Adaptation"
5. Groemer, H.: Geometric Applications of Fourier Series and Spherical Harmonics. Cambridge University Press, Cambridge (1996)
6. Hill, E.: The theory of vector spherical harmonics. Am. J. Phys. 22, 211–214 (1954)
7. Kovacs, J.A., Wriggers, W.: Fast rotational matching. Acta Crystallogr (58), 1282–1286 (2002)
8. Makadia, A., Daniilidis, K.: Rotation recovery from spherical images without correspondences. IEEE Transactions on Pattern Analysis and Machine Intelligence 28(7) (2006)
9. Reisert, M., Burkhardt, H.: Efficient tensor voting with 3d tensorial harmonics. In: CVPR Workshop on Tensors, 2008, Anchorage, Alaska (2008)
10. Stephens, M.: Vector correlation. Biometrica 66, 41–48 (1979)

Scene Categorization by Introducing Contextual Information to the Visual Words

Jianzhao Qin and Nelson H.C. Yung

Laboratory for Intelligent Transportation Systems Research,
Department of Electrical & Electronic Engineering,
The University of Hong Kong, Pokfulam Road, Hong Kong SAR, China
{jzhqin,nyung}@eee.hku.hk

Abstract. In this paper, we propose a novel scene categorization method based on contextual visual words. In this method, we extend the traditional 'bags of visual words' model by introducing contextual information from the coarser scale level and neighbor regions to the local region of interest. The proposed method is evaluated over two scene classification datasets of 6,447 images altogether, with 8 and 13 scene categories respectively using 10-fold cross-validation. The experimental results show that the proposed method achieves 90.30% and 87.63% recognition success for Dataset 1 and 2 respectively, which significantly outperforms previous methods based on the visual words that represent the local information in a statistical manner. Furthermore, the proposed method also outperforms the spatial pyramid matching based scene categorization method, one of the scene categorization methods which achieved the best performance on these two datasets reported in previous literatures.

1 Introduction

Automatic labeling or classification of an image to a specific scene category (e.g. indoor, outdoor, forest, coast) has been widely applied to multi-disciplines such as image retrieval [1,2] and intelligent vehicle/robot navigation [3,4]. Comparing with object recognition, due to the ambiguity and variability in the content of scene images, the problem of scene categorization becomes more challenging.

In previously published literatures, a popular approach for scene classification is to employ global features to represent the scene. The basic idea of this approach is to take the whole image as an entity then relies on low-level features (e.g. color [2], edges intensity [2], texture, gradient, etc.) to represent the characteristics of the scene. Using global features to represent the scene may be sufficient for separating scenes with significant differences in the global properties. However, if scenes with similar global characteristics (e.g. bedroom vs. sitting room) are to be differentiated, then global features may not be discriminative enough. Thus, features extracted from local regions in a scene have been proposed for classification [5,6]. In the methods proposed by Luo et al [5] and Vogel et al [6], the types of the local regions are man labeled or automatically labeled by a semantic concept classifier based on low-level image features. However, accurate object

G. Bebis et al. (Eds.): ISVC 2009, Part I, LNCS 5875, pp. 297–306, 2009.

recognition remains an unattainable goal at the moment. Recently, representing an image by a collection of local image patches of certain size using unsupervised learning methods [7,8,9] has become very popular and achieved a certain success in visual recognition, image retrieval, scene modeling/categorization, etc., because of its robustness to occlusions, geometric deformations, and illumination variations. In scene categorization, the scene image is represented by the co-occurrence of a large number of visual components or the co-occurrence of a certain number of visual topics (intermediate representation) [10,11]. A comparative study conducted by Bosch et a [12] has pointed out that using visual-word representations jointly with different techniques, such as the probabilistic latent semantic analysis (pLSA) [10,11,12] or latent Dirichlet allocation [10], is the one which obtains the best classification results for scene classification.

The visual words proposed previously, however, only represent local image feature within the region of interest (local visual words). We believe the image features surrounding this region (ROI) also can provide useful information or cue about the ROI. Many recent researches in computer vision have employed contextual information for object recognition [13,14] that achieved better object recognition performance than the methods based solely on local features. Thus, we believe that integrating the contextual information with local features will give a better representative of the ROI. This can help differentiate regions which are only similar locally but have significant difference in its surrounding regions. In previous works, Lazebnik et.al. [15] proposed a spatial pyramid matching based scene categorization method which also can incorporate some contextual information. (This method is one of the methods reporting the best result on the two datasets used in this paper [16]). The method calculates the distribution of the visual words at multi-spatial resolutions then measures the similarity of the images using pyramid matching. The distribution of the visual words in some spatial resolutions is able to introduce some information about the distribution of the visual words surrounding the ROI. However, this surrounding information is just introduced after the stage of visual words creation which forms the visual words based on the local feature. In this manner, if the local regions in a specific spatial resolution are wrongly represented by the local visual words, this may also result in an error surrounding visual words distribution, i.e., we may get an error context representation. Therefore, we suggest introducing the context information in the stage of visual words creation. We believe this can achieve better representation of ROI. The experimental result also confirms the superiority of the proposed context visual words.

Fig. 1 illustrates a sample of the sources of the contextual information in our proposal, i.e. the patches surrounding the ROI at the same scale and the patch at a coarser scale level with respect to the ROI. After incorporating contextual information in the representation of the ROI, we employed our previously proposed category-specific visual words generation strategy [17] to generate the contextual visual words to represent the scene. The proposed method is rigorously evaluated over two scene classification with 8 and 13 scene categories respectively using 10-fold cross-validation. The experimental results show that the proposed

Fig. 1. A sample of the sources of the contextual information in our proposal, i.e. the patches surrounding the ROI at the same scale and the patch at a coarser scale level with respect to the ROI

method achieves 90.30% and 87.63% recognition success for Dataset 1 and 2 respectively, which significantly outperforms previous methods based on the visual words that represent the local information in a statistical manner, which reveals that the contextual information is rather useful for enhancing the representative ability of the local visual words. Additionally, we explored how the weighting of the contextual information would affect the final performance of classification. The results reveal that, generally, an optimal accuracy is achieved when the weighting value is around 0.7 (The range of the weighting value is between 0 to 1), which indicates that the contextual information can be fully utilized if we put appropriate weight on it.

The paper is organized as follows. In Section 2, we first formulate the scene classification problem. This is followed by an overview of the proposed method. It then describes the method and various steps involved in generating the contextual words using the category-specific creation strategy as well as the feature extraction process and the classifier training. Section 3 presents the experimental results. This paper is concluded in Section 4.

2   Proposed Method

2.1   Problem Formulation

The scene classification problem based on visual words representation can be formulated in the following manner: given an image $\mathbf{I} \in^{m \times n}$ and a set of scene categories $\mathbf{c} = \{c_1, c_2, \cdots, c_m\}$, we first represent the image $\mathbf{I}$ by a codebook $\mathbf{V}$ consisting of a set of contextual visual words $\mathbf{V} = \{v_1, v_2, \cdots, v_k\}$. We denote this representation by $R(\mathbf{I})$, which is a vector $\mathbf{r} = R(\mathbf{I})$, $\mathbf{r} \in \mathbf{R}^k$ that indicates the distribution or the presence of the visual words. The problem then becomes the issue of finding a projection:

$$f : R(\mathbf{I}) \rightarrow \mathbf{c}, \tag{1}$$

which projects the visual words representation of the image to the scene category c_i, $i = 1, \cdots, m$ where it belongs.

2.2 Overview

Fig. 2 depicts the overall framework of the proposed method. In the training stage, each training image is divided into regular patches at different scales. Then, their Scale-Invariant Feature Transform (SIFT) features [18] are extracted. Given the SIFT features of a ROI, the SIFT features from the coarser scale level and neighbor regions are combined to describe the ROI. After integrating the contextual information, clustering is performed according to different scales and scene categories to create representative visual words, which are denoted by the centroids of the clusters. The visual words are then entered into a codebook. Then, based on this visual word codebook, for each training image, a feature vector representing the existence of the visual words is extracted, which is used for training the classifier. In the classification of testing images, the unknown image is partitioned into patches at different scales and its SIFT features extracted. As in training, based on the SIFT feature of the ROI, the SIFT feature from coarser scale level and the SIFT features from neighbor regions, a feature in which the contextual information is integrated is formed to describe the ROI, then a list of contextual visual words that best represent the features of the image is selected then a feature vector is compiled according to this list. Finally, the feature vector is classified by the SVM to obtain the scene type.

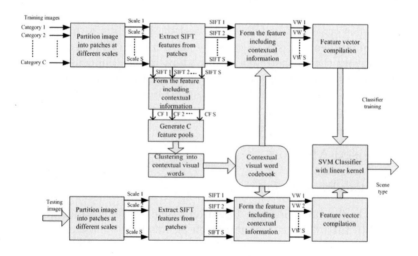

Fig. 2. Framework of the proposed method

2.3   Contextual Visual Words

In this subsection, we introduce the concept of using contextual visual words to represent a scene image. Visual words represent a set of image regions having similar characteristic. This characteristic is usually described using the SIFT feature [10] or the color SIFT feature [16] which can be extracted from the local image region. The previous visual word representation only considers the feature coming from the ROI without taking into consideration the features of the neighboring image regions at the same scale level or the features of the ROI at the coarser scale level. We believe these additional features can provide useful information to identify the ROI and reduce ambiguity. We call these features coming from outside of ROI 'context'. We call the visual word represented in this manner 'contextual visual word'.

Let $\mathbf{P}_L \in \Re^{m_L \times n_L}$ denote the ROI, $\mathbf{P}_C \in \Re^{m_C \times n_C}$ denote the region having the same center as the ROI but at a coarser scale level and $\mathbf{P}_N \in \Re^{m_N \times n_N}$ denotes the neighbor regions of the ROI at the same scale level. For the local visual word, the ROI is represented as $\mathbf{f} = f(\mathbf{P}_L)$ where f denotes the feature extraction function. For the contextual visual word, we represent the ROI as $\mathbf{f} = f(\mathbf{P}_L, \mathbf{P}_C, \mathbf{P}_N)$, that is, we do not only extract the feature from the ROI but also extract the features from the 'context' of the ROI, then combine the features. There are many possible ways to combine the features. In this paper, we linearly combine these features for simplicity. The feature of the ROI is then represented as

$$\mathbf{f} = [f(\mathbf{P}_L), w_C \cdot f(\mathbf{P}_C), w_N \cdot f(\mathbf{P}_N)], \tag{2}$$

where w_C and w_N are the weighting parameters which control the significance of the features from the coarser scale level and the neighbor regions. The range of w_N (w_C) is between 0 and 1. If w_N (w_C) takes 0, the contextual information from neighbor regions (coarser scale level) does not provide any contribution and if w_N (w_C) takes 1, it means that the contribution provided by the neighbors (coarser scale level) is the same as that provided by the ROI.

2.4   Category-Specific Contextual Visual Word

In this subsection, we introduce the category-specific visual word creation strategy to generate the contextual visual word. In this strategy, we firstly generate feature pools from C-category scene images separately. Then, quantize the features to create visual words independently from each feature pool. Finally, the visual words are collated to form the final visual word codebook. The steps for the contextual visual word generation using the category-specific visual word creation strategy are as follows:

Step 1: Divide a subset of the scene images from the training set into patches at different scales.

Step 2: Form the feature including contextual information using equation (2), where the feature extraction function f is the SIFT feature extractor. Then, use principle component analysis (PCA) to reduce the dimension of the feature.

Step 3: Generate C (the number of categories) feature pools at scale s, i.e., $\mathbf{P}_1^s = \{\mathbf{y}_1^1, \mathbf{y}_2^1, \mathbf{y}_3^1, \cdots\}, \mathbf{P}_2^s = \{\mathbf{y}_1^2, \mathbf{y}_2^2, \mathbf{y}_3^2, \cdots\}, \cdots, \mathbf{P}_C^s = \{\mathbf{y}_1^C, \mathbf{y}_2^C, \mathbf{y}_3^C, \cdots\}$. The features in each pool are patch features belonging to the same category at scale s.

step 4: Quantize the features in each pool separately using k-means clustering to create the visual words belonging to category c at scale s, $\mathbf{v}_{1(s)}^c, \mathbf{v}_{2(s)}^c, \mathbf{v}_{3(s)}^c, \cdots, \mathbf{v}_{n(s)}^c, \quad c = 1, \cdots, C$.

step 5: Group the visual words together to form the final codebook.

2.5   Feature Extraction and Classifier Training

This section presents the steps involved in extracting features from the scene image based on the contextual visual words, and in training the classifier.

Given the contextual visual words, a codebook is used to represent the scene image by calculating the presence of the visual words in the image. Assuming that the codebook has n visual words, and a scene image is represented by a n dimensional vector $\mathbf{x}$. The ith element in the vector corresponds to the ith visual word. If the ith visual word exists in the current image, the corresponding i^{th} element of the feature vector is set to 1, otherwise 0. The feature extraction steps are as follows:

Step 1: Given an image $\mathbf{I}$, divide it into n_s patches at scale s for every scale level.

Step 2: Form n_s features for every scale level including the contextual information using equation (2) then reduce the dimension of the features.

Step 3: Set $k = 1$.

Step 4: For the kth feature $\mathbf{y}_k$ at scale s, we calculate its distance, $d_{kj} = \|\mathbf{y}_k - \mathbf{v}_j\|_2$, $j = s_1, \cdots, s_n$, where $s_1, \cdots, s_n$ is the index of visual words in the codebook at Scale s, to each visual word in the codebook from the same scale s. The kth patch can be represented by the lth visual word with the minimum distance to the feature of patch, i.e. $\ell = \min_j \|\mathbf{y}_i - \mathbf{v}_j\|_2$.

Step 5: Set the lth element of $\mathbf{x}$ to 1.

Step 6: If k equals N ($N = \sum_{s=1}^{S} m_s$, the number of patches), terminate the process, otherwise $k = k + 1$ go back to Step 4.

In the training process, images in the training set can be represented as a set of n dimensional features, $\{\mathbf{x}_1, \mathbf{x}_2, \mathbf{x}_3, \cdots, \mathbf{x}_L\}$, where L denotes the number of training images. Based on the features of the labeled training images, a SVM classifier with linear kernel is trained for classification.

3   Experimental Results

This section reports the experimental results of the proposed method including how to set the optimal weighting values of the context information and the comparison of the proposed method with local visual words and spatial pyramid matching based method (The implementation of the spatial pyramid matching based method is based on the LIBPMK toolkit [19]. The parameters setting is the same as the settings in [15].)

Performance of the proposed scene classification method is tested based on two datasets which has been widely used in the previous research [10,20,16,21]. For simplicity sake, we focus our analysis and discussion on Dataset 1, which consists of 2688 color images from 8 categories, whereas we only report the overall results for Dataset 2, which contains 3759 images from 13 categories (Gray version of the images is used for our experiments for these 2 datasets). In our experiments, we perform a 10-fold cross-validation in order to achieve a more accurate performance estimation (The comparision results reported in [20,16] is only based on a training and test set split). Moreover, in order to have a reliable comparison between different methods, we also performed the paired Student t-test on the accuracy rates from 10-fold cross-validation.

We select the best parameters, i.e. w_c and w_n in equation (2) using 10-fold cross-validation and curve fitting. Note that, in the parameter setting stage, the 10-fold cross validation only conduct on the training set. No samples from the test set has been included. Fig. 3(a) and 3(b) depict the accuracy rates versus different weighting values at interval 0.1 for the context from coarser scale level at scale 2 and the context from neighborhood at scale 5 respectively. A quadratic function is fitted to these values to represent the general trend of the data. From Fig. 3 (a) and (b), we can observe that in these two cases, for weights from 0 to 1, the accuracy rate generally increases at the beginning, and it reaches the maximum at around 0.7, then falls back when the value approaches 1, of which the performance is still better than without introducing the contextual information (i.e. weight value equals 0). This indicates that the introduction of contextual information improves the performance of classification. The optimal performance is achieved when we appropriately weights the contextual information. The same situation is also observed at scale 3, 4 and 5 when the context is from coarser

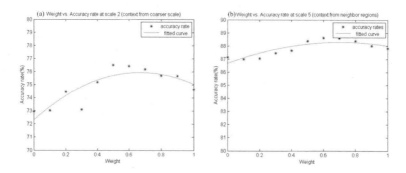

Fig. 3. (a) Accuracy rates versus the weight of the contextual information from coarser scale level at scale 2; (b) Accuracy rates versus the weight of the contextual information from neighbor regions at scale 5

Table 1. Average (standard deviation) classification accuracy rates (%) after combining the 5-scale visual words (Multi-scale local visual words vs. spatial pyramid matching and Contextual visual words)

	Local visual words [17]	spatial pyramid matching	Contextual visual words
Dataset 1	88.81±3.74	88.19±3.46	**90.30**±2.54
Dataset 2	85.05±2.16	84.40±1.90	**87.63**±2.30

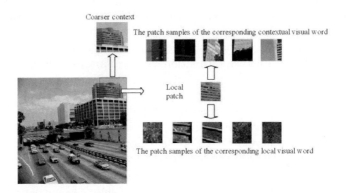

Fig. 4. A sample of a local patch represented by the local visual word and contextual visual word respectively

scale level and at scale 2, 3, 4 when the context is from neighbor regions except for scale 5, where the maximum is when the weight reaches 1. Table 1 shows that in terms of average classification accuracy rate, compared with local visual words, the performance obtained by using contextual visual words is improved by 1.49% (The context is from coarser scale level plus neighbor regions) and 2.58% (The context is from coarser scale level plus neighbor regions) for dataset 1 and 2 respectively. The paired Student t-test shows that this improvement is statistically significant with the p value equals 0.0062. Table 1 also reveals that the proposed method is superior than the spatial pyramid matching based method. Fig. 4 depicts a sample of a local patch represented by the local visual word and contextual visual word respectively. This figure demonstrates a patch which depicts a part of the a building is incorrectly represented by the local visual word which denotes a part of glass land or trees due to the similarity in the texture property. This patch, however, is correctly represented by the contextual visual word which denotes a part of buildings due to the introduction of the context from coarser scale level. In the coarser context, the sky behind the building and the building structure under the ROI provide further information to differentiate regions from part of glass land or trees. The image in this sample was incorrectly classified by using the local visual words but is correctly classified by employing the contextual visual words.

4   Conclusion

In this paper, we have presented a scene categorization approach based on contextual visual words. The contextual visual words represent the local property of the region of interest and the contextual property (from the coarser scale level, neighbor regions or both) simultaneously. By considering the contextual information of the ROI, the contextual visual word gives us a richer representation of the scene image which reduces the ambiguity and error. The reduction in the representation ambiguity improves the visual words representation of the scene image. This improvement further enhances the classification performance. Unlike the spatial pyramid matching, the proposed method utilizes the contextual information in the visual word creation stage. This manner gives a better representation of the contextual information. The experimental result shows the superiority of the proposed method. Additionally, we also explored how the weighting of the contextual information may influence the classification performance. Our experimental results reveal that, generally, an optimal accuracy is around 0.7, which indicates that we cannot fully utilize the contextual information if we put too light or too heavy weight on it. As it is, in our proposed method, the contextual information is linearly combined with the local information. It would be appropriate to consider nonlinear combination or combination that is determined by the feature of the patch in our future research.

Acknowledgments

The authors would like to thank Antonio Torralba, Fei-Fei Li for providing their data sets, Dr. Kwan Wing Keung from Computer Center of the University of Hong Kong for providing high performance computing support and the anonymous reviewers for providing useful feedback.

References

1. Wang, J.Z., Jia, L., Wiederhold, G.: Simplicity: semantics-sensitive integrated matching for picture libraries. IEEE Transactions on Pattern Analysis and Machine Intelligence 23(9), 947–963 (2001)
2. Vailaya, A., Figueiredo, M., Jain, A., Zhang, H.J.: Content-based hierarchical classification of vacation images. In: Figueiredo, M. (ed.) IEEE International Conference on Multimedia Computing and Systems, vol. 1, pp. 518–523 (1999)
3. Siagian, C., Itti, L.: Gist: A mobile robotics application of context-based vision in outdoor environment. In: Itti, L. (ed.) 2005 IEEE Computer Society Conference on Computer Vision and Pattern Recognition, vol. 3, pp. 1063–1069 (2005)
4. Manduchi, R., Castano, A., Talukder, A., Matthies, L.: Obstacle detection and terrain classification for autonomous off-road navigation. Autonomous Robots 18(1), 81–102 (2005)
5. Luo, J., Savakis, A.: Indoor vs outdoor classification of consumer photographs using low-level and semantic features. In: Savakis, A. (ed.) 2001 International Conference on Image Processing, vol. 2, pp. 745–748 (2001)

6. Vogel, J., Schiele, B.: A semantic typicality measure for natural scene categorization. In: Rasmussen, C.E., Bülthoff, H.H., Schölkopf, B., Giese, M.A. (eds.) DAGM 2004. LNCS, vol. 3175, pp. 195–203. Springer, Heidelberg (2004)

7. Fergus, R., Fei-Fei, L., Perona, P., Zisserman, A.: Learning object categories from google's image search. In: Fei-Fei, L. (ed.) Tenth IEEE International Conference on Computer Vision, vol. 2, pp. 1816–1823 (2005)

8. Agarwal, S., Awan, A., Roth, D.: Learning to detect objects in images via a sparse, part-based representation. IEEE Transactions on Pattern Analysis and Machine Intelligence 26(11), 1475–1490 (2004)

9. Sivic, J., Zisserman, A.: Video google: a text retrieval approach to object matching in videos. In: Ninth IEEE International Conference on Computer Vision, pp. 1470–1477 (2003)

10. Fei-Fei, L., Perona, P.: A bayesian hierarchical model for learning natural scene categories. In: IEEE Computer Society Conference on Computer Vision and Pattern Recognition, vol. 2, pp. 524–531 (2005)

11. Quelhas, P., Monay, F., Odobez, J.M., Gatica-Perez, D., Tuytelaars, T., Van Gool, L.: Modeling scenes with local descriptors and latent aspects. In: Tenth IEEE International Conference on Computer Vision, vol. 1, pp. 883–890 (2005)

12. Bosch, A., Munoz, X., Marti, R.: Which is the best way to organize/classify images by content? Image and Vision Computing 25(6), 778–791 (2007)

13. Torralba, A.: Contextual priming for object detection. International Journal of Computer Vision 53(2), 169–191 (2003)

14. Torralba, A., Murphy, K.P., Freeman, W.T.: Contextual models for object detection using boosted random fields. In: Saul, L.K., Weiss, Y., Bottou, L. (eds.) Adv. in Neural Information Processing Systems 17 (NIPS), pp. 1401–1408. MIT Press, Cambridge (2005)

15. Lazebnik, S., Schmid, C., Ponce, J.: Beyond bags of features: Spatial pyramid matching for recognizing natural scene categories. In: IEEE Computer Society Conference on Computer Vision and Pattern Recognition, 2006, vol. 2, pp. 2169–2178 (2006)

16. Bosch, A., Zisserman, A., Muoz, X.: Scene classification using a hybrid generative/discriminative approach. IEEE Transactions on Pattern Analysis and Machine Intelligence 30(4), 712–727 (2008)

17. Qin, J., Yung, N.H.C.: Scene categorization with multi-scale category-specific visual words. Optical Engineering (to appear, 2009)

18. Lowe, D.G.: Object recognition from local scale-invariant features. In: The Proceedings of the Seventh IEEE International Conference on Computer Vision, vol. 2, pp. 1150–1157 (1999)

19. Lee, J.J.: Libpmk: A pyramid match toolkit. Technical Report MIT-CSAIL-TR-2008-17, MIT Computer Science and Artificial Intelligence Laboratory (April 2008)

20. Bosch, A., Zisserman, A., Munoz, X.: Scene classification via plsa. In: Leonardis, A., Bischof, H., Pinz, A. (eds.) ECCV 2006. LNCS, vol. 3954, pp. 517–530. Springer, Heidelberg (2006)

21. Oliva, A., Torralba, A.: Modeling the shape of the scene: A holistic representation of the spatial envelope. International Journal of Computer Vision 42(3), 145–175 (2001)

Edge-Preserving Laplacian Pyramid

Stella X. Yu

Computer Science Department
Boston College, Chestnut Hill, MA 02467, USA

Abstract. The Laplacian pyramid recursively splits an image into local averages and local differences using a fixed Gaussian interpolation function. We propose a spatially variant interpolation function that is adaptive to curvilinear edges in the image. Unlike the signal-based multiscale analysis where a step edge is multiply represented at all scales, our perception-based multiscale analysis preserves the edge at a single scale as much as possible. We demonstrate that our average pyramid retains boundaries and shading at lower spatial and tonal resolutions, whereas our difference pyramid refines edge locations and intensity details with a remarkably sparse code, delivering an image synopsis that is uncompromising between faithfulness and effectiveness.

1   Introduction

An image of a natural scene is not a collection of random numbers. Pixels nearby often have similar values, yet it is their differences that give away shapes and textures. We propose an edge-preserving Laplacian pyramid that provides an image synopsis which removes spatial redundancy, retains perceptually important structures such as boundaries and textures, and refines the representation over scale (Fig. 1). As the synopsis adopts a larger size, boundaries become more precisely localized, textures more elaborated. These synopses can be related using the smallest synopsis and a series of sparse differences to refine it (Fig. 2).

image   synopsis towards a smaller size ⟶

Fig. 1. Image synopsis should be effective, faithful and progressive. The original image (285×288) is represented at $\frac{1}{2}$, $\frac{1}{4}$ and $\frac{1}{8}$ of its size respectively, all shown at the full size (with obvious pixelization in the rightmost 36×36 image). Perceptually important features, such as shape boundaries, material texture, highlights and cast shadows, remain visible throughout the synopsis reduction process.

G. Bebis et al. (Eds.): ISVC 2009, Part I, LNCS 5875, pp. 307–316, 2009.
© Springer-Verlag Berlin Heidelberg 2009

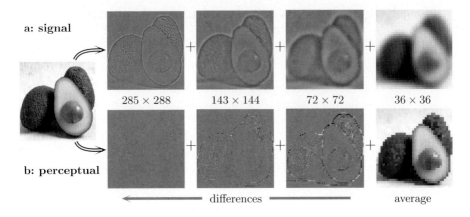

Fig. 2. Multiscale analysis of an image. **a:** The signal-based Laplacian pyramid parses an image into an average image of lower frequency and a series of difference images of higher frequencies. In the average, boundaries are fuzzy and textures are smoothed out. In the differences, a step edge is represented as multiple smooth transitions with artificial halos (*Gibbs phenomenon*). **b:** Our edge-preserving Laplacian pyramid has an average image that retains boundaries and overall textural variations and a set of differences that successively refine edge locations and intensity details.

Multiscale analysis of an image is a well traversed area in signal processing, e.g. the Gaussian-Laplacian pyramid [1] and wavelets [2]. The basic idea is that every pixel value can be decomposed into a neighbourhood average component and a local difference component. This process can be recursively applied to the average, producing a frequency subband decomposition of the image (Fig. 2a). Signal-based multiscale analysis methods vary in their choices of filters to compute the neighbourhood average, yet they share one commonality: the filter is the same everywhere in the image, whether the pixel is on a boundary or inside a region. Signal frequencies matter; perceptual structures do not matter. Consequently, signal-based multiscale analysis is great for blending images across frequencies [3], but as image synopsis it is neither effective nor faithful.

We develop a Laplacian pyramid that adapts the neighbourhood average computation according to edges. Since the average maximally preserves edges within a single scale, edges are no longer repeatedly represented at multiple levels of the pyramid. In fact, there is little perceptual structure left in the difference images, other than sparse correction near edges due to inevitable loss of spatial resolution at a coarser scale (Fig. 2b). We demonstrate that our new Laplacian pyramid is effective at both coding and representing the image.

2   Edge-Preserving Multiscale Analysis

The Laplacian pyramid is developed from the idea that the intensity of pixel p in a real image I can be largely predicted by its local context $\bar{I}$:

$$I(p) \approx \bar{I}(p) = \sum_{q=p'\text{s local neighbour}} W(p,q)I(q) \qquad (1)$$

where weight $W(p,q)$ describes how neighbour q contributes to predicting p's intensity. In the original formulation [1], W is pre-chosen and fixed over the entire image, which has nothing to do with the image I itself. There is no guarantee that the prediction $\bar{I}$ is a good synopsis of I, or the residue $I(p) - \bar{I}(p)$ is small. In our formulation, W adapts to I and varies across the image, with $\bar{I}$ maximally preserving the edges in I while making $I - \bar{I}$ as small as possible.

Our multiscale analysis follows the same procedure as in [1]:

Step 1: An image is decomposed into an average and a difference.
Step 2: The average is smoother and thus reduced in size.
Step 3: Repeat Steps 1 and 2 to the average.

This process recursively splits an image into an average and a difference, resulting in a difference pyramid that can be used to synthesize the image.

Multiscale Analysis:

Given image I and number of scales n, construct average pyramid A and difference pyramid D, where $\downarrow(\cdot, W_\triangledown)$ is downsampling with *analysis weights* $W_\triangledown$, $\uparrow(\cdot, W_\triangle)$ is upsampling with *synthesis weights* $W_\triangle$. The sampling factor is 2.

$$A_1 = I, \qquad\qquad A_{s+1} = \downarrow(A_s, W_\triangledown), \qquad s = 1 \to n \qquad (2)$$
$$D_{n+1} = A_{n+1}, \qquad D_s = A_s - \uparrow(A_{s+1}, W_\triangle), \qquad s = n \to 1 \qquad (3)$$

Multiscale Synthesis:

Given difference pyramid D, reconstruct average pyramid A and image I.

$$A_{n+1} = D_{n+1}, \qquad A_s = D_s + \uparrow(A_{s+1}, W_\triangle), \qquad s = n \to 1; \qquad I = A_1 \qquad (4)$$

Two functions, $\downarrow(\cdot, W_\triangledown)$ and $\uparrow(\cdot, W_\triangle)$ need to be defined. In the Laplacian pyramid, the analysis weights $W_\triangledown$ and the synthesis weights $W_\triangle$ are not only identical but also spatially invariant. They are entirely determined by the distance between pixels, regardless of what and where these pixels are in the image:

$$W_\triangledown(p,q) = W_\triangle(p,q) = G(\|\overrightarrow{p} - \overrightarrow{q}\|; \sigma) \qquad (5)$$
$$G(d; \sigma) = \exp\left(-\frac{d^2}{2\sigma^2}\right) \qquad (6)$$

where $\overrightarrow{p}$ is p's 2D image coordinates, $\|\cdot\|$ a vector's L_2 norm, and $G(d;\sigma)$ the un-normalized 1D Gaussian function with mean 0 and standard deviation σ.

However, a quick examination of Eqns. 2–4 reveals that $W_\triangledown$ and $W_\triangle$ can be **independently** defined and in fact **arbitrary** without jeopardizing a perfect reconstruction. In our multiscale analysis, not only $W_\triangledown \neq W_\triangle$, but both of them also vary according to perceptual structures at each pixel.

We characterize the perceptual structure in terms of pixel proximity and edge geometry. Our new weight W for Eqn. 1 is a product of these two factors. Pixel p itself always contributes with the maximal weight 1, while pixel q contributes with the minimal weight 0 if it is separated (by boundaries) or far from p.

Edge-Preserving Averaging:
Given image I and neighbourhood radius r, compute the local average $\bar{I}$ using spatial proximity kernel K_s and edge geometry kernel K_g.

$$\bar{I}(p) = \frac{\sum_{k=1}^{r}\sum_{q \in N(p,k)} W(p,q;I)I(q)}{\sum_{k=1}^{r}\sum_{q \in N(p,k)} W(p,q;I)}, N(p,k) = p\text{'s neighbours at radius } k \quad (7)$$

$$W(p,q;I) = K_s(p,q;r) \cdot K_g(p,q;I,r) \quad (8)$$

$$K_s(p,q;r) = G(\|\overrightarrow{p} - \overrightarrow{q}\|; \frac{r}{3}) \quad (9)$$

The geometry kernel K_g describes curvilinear edges with pairwise pixel grouping relationships, with edges first localized at zero-crossings of 2nd-order derivatives.

The edge magnitude E and phase P of image I are the size of the 1st-order derivative and the sign of the 2nd-order derivative along the gradient direction respectively. The magnitude measures the maximal contrast, whereas the binary phase indicates on which side the pixel lies with respect to zero-crossings [4].

Zero-crossings alone are not sufficient to characterize boundaries [5,6]. We formulate K_g based on the idea of intervening contour (IC) affinity in segmentation [7,8] and the idea of bilateral extension in contour completion [6].

The IC affinity $C(p,q)$ between pixels p and q is inversely proportional to the maximal edge magnitude encountered along the line connecting them. For adjacent pixels, it is 0 if they are on the opposite sides of an edge, and 1 otherwise (Fig. 3a, Eqn. 10, Eqn. 11 line 1). K_g is the gap-completed version of C from bilateral extension: Either two pixels are separated by an edge, or their neighbours at *both* ends are separated by an edge (Fig. 3b). This curvilinearity operation of K_g can be modeled as minimax filtering of C along angular directions (Eqn. 12).

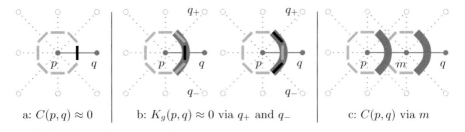

a: $C(p,q) \approx 0$ b: $K_g(p,q) \approx 0$ via q_+ and q_- c: $C(p,q)$ via m

Fig. 3. Boundaries are characterized by curvilinear edges. **a:** IC affinity C at radius 1 checks if there is an edge (black line) between adjacent pixels. **b:** K_g checks if there is a curved boundary (thick gray curve) between two pixels: Either they (left) or their neighbours at both ends (right) are separated by edges. **c:** IC affinity C at radius 2 checks if there is a curved boundary between successive pairs of adjacent pixels.

To extend the boundary characterization from radius 1 to radius 2, we first establish affinity C for pixels at distance 2 from that between successive pairs of adjacent pixels (Fig. 3c, Eqn. 11 line 2). K_g is subsequently obtained as bilateral extension of C to complete boundary gaps.

Formally, we first define K_g for $N(p, 1)$ and then propagate it to $N(p, 2)$. This process (Eqn. 11-Eqn. 12) is recursively applied at an increasing radius to fill in K_g values within a log-polar neighbourhood: 0 if two pixels are separated by boundaries, and 1 otherwise. K_g is sparse. The space and time complexity is linear to the number of pixels and to the number of neighbours per pixel.

Edge Geometry Kernel:

Given edge magnitude E and phase P, edge parameter σ_g, $N(p, r)$ denoting the set of pixels at radius r from p and along 8 directions, compute geometry K_g which describes boundaries enclosing a pixel at an increasing radius (Fig. 3).

$$L(p, q) = \begin{cases} \min(E(p), E(q)), & P(p) \neq P(q) \\ 0, & P(p) = P(q) \end{cases} \tag{10}$$

$$C(p, q) = \begin{cases} G(L(p, q); \sigma_g), & q \in N(p, 1) \\ \min(K_g(p, m), K_g(m, q)), \ \overrightarrow{m} = \frac{\overrightarrow{p} + \overrightarrow{q}}{2}, & q \in N(p, 2) \end{cases} \tag{11}$$

$$K_g(p, q) = \min(C(p, q), \max_{o \in \{q_+, q_-\}} C(p, o)), |\angle q_{\pm} pq| = 45^\circ, q, q_{\pm} \in N(p, r) \tag{12}$$

Downsampling is trivial since we only need to perform decimation on the average $\bar{I}$. Upsampling requires boundary estimation at subpixel locations. To relate subpixels to original pixels, we first interpolate edge magnitudes and phases at subpixel locations using the Gaussian function with standard deviation $\frac{1}{3}$. We then compute the affinity C and hence K_g between subpixels and their 8 immediate original pixels (Fig. 4a). Using K_g between original pixels, we propagate weights from subpixel locations to original pixels at a farther radius (Fig. 4b).

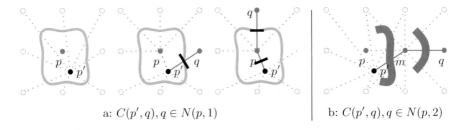

a: $C(p', q), q \in N(p, 1)$ b: $C(p', q), q \in N(p, 2)$

Fig. 4. Geometry kernel relating subpixel locations to original pixels. **a:** It starts with establishing IC affinity C between p' and its 8 immediate original pixels (left). There are two scenarios. If p' is closer to q than p, then C checks the edge between p' and q directly (middle). Otherwise, both edges intersecting p' and p, p and q are checked (right). Minimax filtering on C with neighbouring directions gives rise to geometry kernel $K(p', q)$. For example, the left thick gray curve in b) illustrates $K(p', m)$. **b:** IC affinity at radius 2 checks if there are boundaries (thick gray curves) intersecting the line connecting p' and q via mid-point m.

Edge Geometry Kernel at Subpixel Locations:

Given edge L, parameter σ_g, subpixel displacement $\vec{d}$ where $\|d\| < 1$, compute geometry kernel K_g between location $\vec{p'} = \vec{p} + \vec{d}$ and original pixel q.

$$C(p',q) = \begin{cases} G(L(p',q);\sigma_g), & \|\vec{p'} - \vec{q}\| \leq \|\vec{p} - \vec{q}\|, q \in N(p,1) \\ G(\max(L(p',p),L(p,q));\sigma_g), & \|\vec{p'} - \vec{q}\| > \|\vec{p} - \vec{q}\|, q \in N(p,1) \\ \min(K_g(p',m),K_g(m,q)), & \vec{m} = \frac{\vec{p}+\vec{q}}{2}, & q \in N(p,2) \end{cases} \tag{13}$$

$$K_g(p',q) = \min(C(p',q), \max_{o \in \{q_+,q_-\}} C(p',o)), |\angle q_{\pm} pq| = 45°, q, q_{\pm} \in N(p,r) \tag{14}$$

Our analysis and synthesis weights realize weight W in Eqn. 8 on a downsampled grid and an upsampled grid respectively.

Edge-Preserving Analysis and Synthesis Weights:

We apply Eqn. 8 to downsample and upsample image I with:

$$W_{\triangledown}(p,q) = K_s(p,q;2) \cdot K_g(p,q) \tag{15}$$

$$W_{\triangle}(p',q) = K_s(p',q;1) \cdot K_g(p',q), \quad \vec{p'} = \vec{p} + \vec{d}, \vec{d} \in \{0,0.5\} \times \{0,0.5\} \tag{16}$$

Our weight formula appears similar to bilateral filtering [9] based on spatial proximity and intensity similarity. We replace the intensity similarity with our geometry kernel which characterizes boundaries.

Our approach also shares the same anisotropic diffusion principle as many partial differentiation equation formulations [10,5]. We adapt weights according to local image structures, yet they are neither low-level signal quantifiers such as gradients [10], nor high-level hidden causes such as perceptual boundaries with smoothness priors imposed [5], but mid-level characterization of boundaries in terms of curvilinear zero-crossing edges.

Consequently, our method does not create flat zones and artificial boundaries inside smooth regions (so-called *staircasing effect*) as local [9,10] or non-local [11] neighbourhood filtering approaches. The local average operator does not need to be upgraded to linear regression in order to dissolve the staircasing artifacts[12].

We decompose an image into scales just like the Laplacian pyramid [1] and its elaborated wavelet version [2]. However, instead of expanding the wavelet basis to accommodate edges, e.g., ridgelets, wedgelets, curvelets [13,14,15], we create a local structure adaptive basis at each pixel to acknowledge the discontinuity, avoiding artificial oscillations that are inevitable in harmonic analysis methods.

Our synthesis weight formula expands boundaries to a higher resolution with local pairwise pixel grouping relationships. It can be used for single image super-resolution without relying on massive image patch comparisons [16,17,18].

3   Experimental Results

We evaluate our perceptual multiscale analysis over the following signal-based multiscale analysis methods: the traditional Laplacian pyramid, i.e., Gaussian interpolation, nearest neighbour, bilinear, and bicubic interpolation methods. For the Gaussian, we use $\sigma = 2$. For the nearest, bilinear, and bicubic, we use MATLAB built-in function *imresize.m* with a image size factor of 2 and the default anti-aliasing setting. For our method, we set $\sigma_g = 0.05, r = 2, n = 5$.

We first compare the average image as an image representation. Fig. 5 shows that our interpolation preserves corners and contrast as well as the nearest neighbour interpolation, and refines curves and gradation as well as the bicubic interpolation. Fig. 6 further demonstrates that our results have neither excessive blur and halos around edges as the Gaussian, bilinear, bicubic methods, nor magnified pixelation artifacts in textures as the nearest neighbour method. Our method thus provides a faithful image synopsis at a much reduced size.

We then compare the difference images as an image code on a set of standard test images (Fig. 7a). In the signal-based multiscale analysis, a single sharp

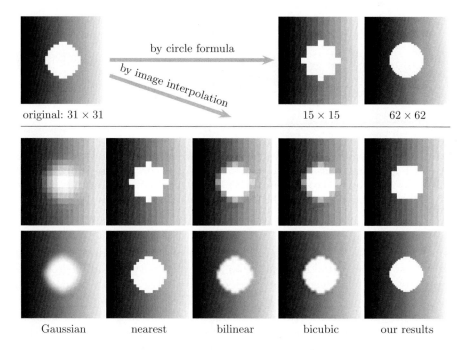

Fig. 5. Comparison of interpolation methods. The scene consists of a circle on a shaded background. **Row 1** shows three images generated by the same formula at scales 1, 0.5, and 2 respectively. **Row 2** shows downsampling results, **Row 3** upsampling results, all interpolated from the original image using 5 interpolation methods. Gaussian: boundaries dissolving over blur. Nearest: spikes at downsampling, jagged boundaries and streaked shading at upsampling. Bilinear and bicubic: finer shading at upsampling, smoother boundaries at the cost of blur and halos. Our results: boundaries smoothed and shading refined without damaging sharp corners and contrast, approximating the images generated by formula but without spikes at scale 0.5 and rounding at scale 2, both unforeseeable from the original image.

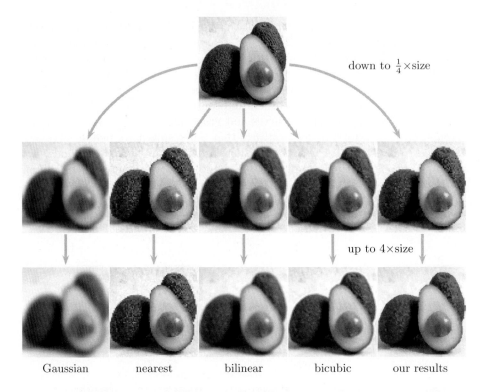

down to $\frac{1}{4}\times$size

up to 4×size

| Gaussian | nearest | bilinear | bicubic | our results |

Fig. 6. Comparison of the average image as a synopsis representation. For the 285×288 image in Fig. 2, we downsample it twice to 72 × 72 (shown at the full size), then unsample twice to bring the size back to 285 × 288. Our results have neither excessive blur as the Gaussian, bilinear, and bicubic, nor pixelation artifacts as the nearest.

discontinuity in the intensity is decomposed into smooth transitions at multiple frequencies. As the spatial frequency goes up, the intensity oscillation grows relatively large near the edge. Our perception-based multiscale analysis encodes an edge within a single scale as much as possible. There is no intensity overshooting, and the difference needed to refine the edge location is at most 2 pixels wide, creating a sparser representation. Since most information is concentrated in the average image of the smallest size (Fig. 7b), the reduction of entropy in the difference images of larger sizes leads to significant savings in the number of bits (Fig. 7c,d). While parsing an image into frequency bands can save 0.25 bits per pixel over the original, parsing it into perceptual multiscale can save 1.45 bits per pixel. That is a 5-time increase in the lossless compression performance.

Among signal-based approaches, as an image code, multiscale with the simplest nearest neighbour interpolation is far more efficient than the widely known Laplacian pyramid; as an image representation, multiscale with bicubic interpolation has a better trade-off between clarity and smoothness. Our edge-preserving Laplacian pyramid can yet outperform these signal-based multiscale approaches

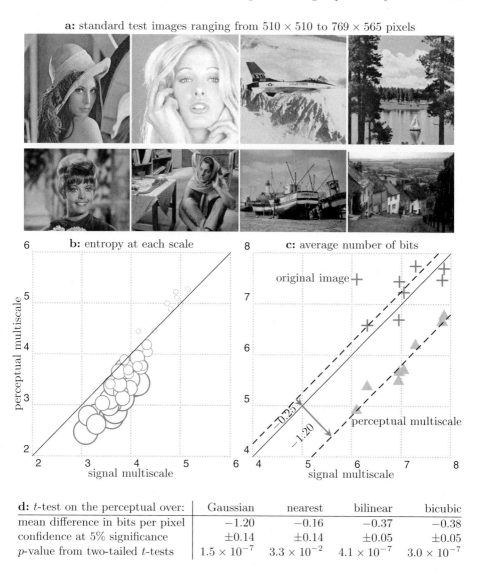

a: standard test images ranging from 510×510 to 769×565 pixels

b: entropy at each scale

c: average number of bits

d: t-test on the perceptual over:	Gaussian	nearest	bilinear	bicubic
mean difference in bits per pixel	-1.20	-0.16	-0.37	-0.38
confidence at 5% significance	± 0.14	± 0.14	± 0.05	± 0.05
p-value from two-tailed t-tests	1.5×10^{-7}	3.3×10^{-2}	4.1×10^{-7}	3.0×10^{-7}

Fig. 7. Lossless compression performance comparison. **a)** Test images. **b)** Entropy at each scale for the Gaussian (x-axis) and our method (y-axis). The circle size reflects the image size. While our average images have higher entropy than the Gaussian averages (shown as the smallest circles above the diagonal line), most difference images, have lower entropy than the Laplacian differences (shown as the rest circles). **c)** Average number of bits per pixel for the original Laplacian pyramid on the x-axis, and the original $+$ and our edge-preserving Laplacian pyramid ▲ on the y-axis. Linear relations that fit through $+$'s and ▲'s respectively are shown as dashed lines. On average, 0.25 bits are saved with the Laplacian pyramid, while 1.45 bits are saved with our pyramid. **d)** t-test results on the average numbers of bits per pixel between our method and the other four methods. Our method saves more bits than any other method.

on either account: The average images retain boundaries and shading at lower spatial and tonal resolutions, whereas the difference images refine edge locations and intensity details with a remarkably sparse code. It delivers an image synopsis that is uncompromising between faithfulness and effectiveness.

Acknowledgements. This research is funded by NSF CAREER IIS-0644204 and a Clare Boothe Luce Professorship.

References

1. Burt, P., Adelson, E.H.: The Laplacian pyramid as a compact image code. IEEE Transactions on Communication COM-31, 532–540 (1983)
2. Mallat, S.: A theory for multiresolution signal decomposition: The wavelet representation. IEEE Trans. on PAMI 11(7), 674–693 (1989)
3. Burt, P., Adelson, E.H.: A multiresolution spline with application to image mosaics. ACM Transactions on Graphics 2(4), 217–236 (1983)
4. Marr, D.: Vision. Freeman, CA (1982)
5. Mumford, D., Shah, J.: Boundary detection by minimizing functionals. In: IEEE Conference on Computer Vision and Pattern Recognition, pp. 22–26 (1985)
6. Williams, L.R., Jacobs, D.W.: Local parallel computation of stochastic completion fields. Neural computation 9, 859–881 (1997)
7. Malik, J., Belongie, S., Leung, T., Shi, J.: Contour and texture analysis for image segmentation. International Journal of Computer Vision (2001)
8. Yu, S.X.: Segmentation induced by scale invariance. In: IEEE Conference on Computer Vision and Pattern Recognition (2005)
9. Tomasi, C., Manduchi, R.: Bilateral filtering for gray and color images. In: International Conference on Computer Vision (1998)
10. Perona, P., Malik, J.: Scale-space and edge detection using anisotropic diffusion. IEEE Trans. on PAMI 33(7), 629–639 (1990)
11. Buades, A., Coll, B., Morel, J.-M.: A review of image denoising algorithms, with a new one. SIAM Journal on Multiscale Modeling and Simulation 4, 490–530 (2005)
12. Buades, A., Coll, B., Morel, J.-M.: The staircasing effect in neighbourhood filters and its solution. IEEE Transactions on Image Processing 15(6), 1499–1505 (2006)
13. Candes, E.: Ridgelets: Theory and Applications. PhD thesis, Stanford University (1998)
14. Candes, E., Donoho, D.L.: Curvelets: A surprisingly effective nonadaptive representation of objects with edges. In: Curves and Surfaces. Vanderbilt University Press (1999)
15. Candes, E.J., Guo, F.: New multiscale transforms, minimum total variation synthesis: Applications to edge-preserving image reconstruction. Signal Processing 82, 1519–1543 (2002)
16. Ebrahimi, M., Vrscay, E.R.: Solving the inverse problem of image zooming using 'self-examples'. In: International Conference on Image Analysis and Recognition, pp. 117–130 (2007)
17. Protter, M., Elad, M., Takeda, H., Milanfar, P.: Generalizing the nonlocal-means to super-resolution reconstruction. IEEE Trans. on Image Processing 18 (2009)
18. Glasner, D., Bagon, S., Irani, M.: Super-resolution from a single image. In: International Conference on Computer Vision (2009)

Automated Segmentation of Brain Tumors in MRI Using Force Data Clustering Algorithm

Masoumeh Kalantari Khandani[1], Ruzena Bajcsy[2], and Yaser P. Fallah[2,3]

[1] School of Engineering Science, Simon Fraser University, Canada
mka47@sfu.ca
[2] Electrical Engineering and Computer Sciences, University of California, Berkeley, USA
[3] Institute of Transportation Studies, University of California, Berkeley, USA

Abstract. In this paper, we present a novel automated method for detecting tumor location in brain magnetic resonance (MR) images, and identifying the tumor boundary. This method employs an unsupervised learning algorithm called Force for coarse detection of the tumor region. Once tumor area is identified, further processing is done in the local neighborhood of the tumor to determine its boundary. The Force method, which is based on the rules of electrostatics, is used for finding spatial clusters of high intensity in the 2D space of MR image. Further analysis of the identified clusters is performed to select the cluster that contains the tumor. This method outperforms many existing methods due to its accuracy and speed. The performance of the proposed method has been verified by examining MR images of different patients.

Keywords: Tumor detection, segmentation, data clustering, brain MRI.

1   Introduction

Automated detection of the abnormalities in medical images is an important and sometimes necessary procedure in medical diagnostics, planning, and treatment. While detection of abnormalities such as tumors is possible by experts, manual segmentation is usually tedious and time consuming [5][9], and subject to error [6]. There are many methods that find a tumor in MR images (MRI) semi-automatically. In such methods, human intervention is required, which again makes the process time consuming and expensive. The critical problem is finding the tumor location automatically and later finding its boundary precisely. The objective of this work is to present an automated unsupervised method for finding tumor (high-grade gliomas) in slices of T2 FLAIR MRI of Brain (no enhancements by contrast agent). In such images the tumor needs to be identified amongst brain soft tissues, white matter (WM), gray matter (GM) and cerebrospinal fluid.

An important factor in detecting tumor from healthy tissues is the difference in intensity level. However, relying only on the intensity level is usually not enough. The spatial information available in clusters of pixels that form a tumor should also be used in the detection process. In this work we propose a new method that is comprised of three tasks of preprocessing, coarse detection of tumor area, and fine detection of tumor boundary. The coarse detection is done using an enhanced version of the

G. Bebis et al. (Eds.): ISVC 2009, Part I, LNCS 5875, pp. 317–326, 2009.

Force clustering algorithm [11]. 'Force' is applied to a data set that is created from a preprocessed slice of brain MRI. In 'Force', the rules of electrostatics are used to determine clusters of pixels with higher intensity. Once these clusters are found, the algorithm identifies the cluster/region of the brain that contains the tumor. In the last step, the tumor cluster is further analyzed and the tumor boundaries are determined. The proposed algorithm is designed to be robust to variations in MR images and is able to efficiently and accurately identify tumor boundaries in different brain MRIs; this is shown through comparing this method by ground truth manually produced by experts.

1.1 Related Work

There have been significant efforts to develop automated computer algorithms for locating tumors in brain MRI. A review of pattern recognition methods for MRI segmentation is presented in [1], and methods and applications of MRI segmentation can be found in [3]. We describe few notable algorithms in this section.

Among supervised methods, the work in [6] combines information from a registered atlas template and user input to train a supervised classifier. The method in [7] detects tumors based on outlier detection and uses affine transformation for the registration. However, this method fails in case of large tumors. The method described in [8] is based on training on healthy brain images instead of training on pathology. To recognize deviations from normalcy, a multi-layer Markov random field is used which is computationally expensive. In the work reported in [4], the authors employ an atlas based pathological segmentation using affine transformation. They assume tumor growth has a radial expansion from its starting point. All of the above methods are time consuming, and also need expert input for large set of data. Supervised pattern recognition methods have exhibited problems with reproducibility [2], due to significant intra and inter-observer variance introduced over multiple training trials.

The unsupervised method reported in [9] divides the T2 weighted images into few blocks, and calculates the number of edges, the intensity and the contrast parameter in each block. It assumes the abnormalities occupy less that 10% of all pixels, and that the blocks containing tumor pixels exhibit fewer edge pixels. However, tumor may fall in different blocks, making parts of the tumor undetectable. In another method, presented in [10], color-based clustering is used. The MR image is translated to RGB, and RGB to L*a*b* planes. K-means clustering [12] is used on a* and b* planes to find thresholds and mark the tumor. The issue with such methods is that they rely on intensity level classification, which is susceptible to misclassification.

In our paper, a new approach toward tumor detection and segmentation is introduced. Our method is based on the use of a new unsupervised data clustering algorithm called 'Force' [11] for initial tumor detection. The unsupervised nature of this method avoids the problems of observer variability found in supervised methods; therefore, the results of our method are reproducible. The proposed method can also be combined with prior methods to enhance them in initial tumor location detection. Section 2 provides a brief background on "Force". Section 3 describes the proposed method. The evaluation of the proposed method is presented in section 4.

2 Background: Force Algorithm

In [11] we presented a numerical unsupervised clustering algorithm, Force, which is inspired by the laws of electrostatics. Clustering itself means grouping data points of a set into different classes according to some similarities between points. The Force algorithm allows efficient and robust clustering of multi dimensional (MD) data sets and always converges to the same solution under different conditions (initial guesses and noise). In the Force algorithm, data points are assumed to be negative electrical charges (with charge -1) scattered and fixed in a MD space. To cluster these charges, a number of positive charges (with variable magnitude) are randomly released in the space; these charges are allowed to move. The positive charges are the centroids. Their movement direction is determined by the electrostatic field that affects them in the space. We adjust the positive charge of centroids in each iteration to ensure they end up in the center of the clusters of negative charges. In each iteration, the charge of each centroid is set equal to the total charge of its current cluster, with opposite sign. When the balance is achieved, positive charges will be at true centers of the found clusters. Under the balance condition the centroids do not move anymore. The total force affecting centroid j is calculated for each step as:

$$\vec{F}_j = \sum_{i \neq j, i \in D \cup C} \frac{Q_j Q_i}{R_{ij}^2} \frac{(c_j - p_i)}{\left\| c_j - p_i \right\|}. \tag{1}$$

where c_j and p_i are vectors describing the positions of centers and data points, R_{ij} is the distance between charges i and j. Q_i is the charge of each data point or centroid. D and C denote the set of data points and centroids, respectively. Here, the direction of the force is the only parameter that is used for determining the centroid's new position. Therefore the new position of each centroid, in each step, is calculated as:

$$c_j^{(\tau+1)} = c_j^{(\tau)} + \eta . \vec{F}_j / \left\| \vec{F}_j \right\| \tag{2}$$

where $c_j^{\tau+1}$ is the new position of the centroid, c_j^{τ} is the previous position, and $F_j / \| F_j \|$ is a unit vector of force providing the movement direction. Knowing the direction, a fixed step size η is utilized for moving the centroid. Other variants of Force, including one with adaptive step size, are presented in [11]. One of the benefits of Force is that the found centers, after different runs of the algorithm with different initial parameters, are at most different by 2η. Also, the algorithm performs a globalized search, contrary to local searches done by methods such as k-means. A 2-D example of how Force moves the centroids to cluster centers is shown in Fig. 1.

2.1 Using Force for Tumor Detection

In existing methods, data clustering is used for finding different clusters of intensity level for tumor and brain tissues. This is done by clustering image histograms to find the right thresholds. However, this method is prone to misclassification of intensity levels. Besides, valuable spatial information is usually ignored in the clustering stage.

In our approach, instead of classification based on the image histogram, both intensity level and spatial information are used for clustering. Also, our use of clustering/classification is not to directly determine intensity thresholds for segmentation;

rather, we use Force to find the region of interest (ROI), or coarsely find the tumor location. Here, the idea is to use the fact that tumors are spatial clusters of pixels with higher intensity levels than soft tissues. In our approach, an extended version of Force is used. In particular, each pixel of the image is assumed a negative charge with magnitude set to the intensity of the pixel raised to the power of k (instead of charge -1 as in original Force). Exponentiation is used to exaggerate the intensity difference between tumor and soft tissue, helping faster and more accurate convergence of Force.

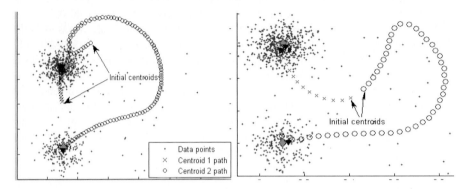

Fig. 1. Force centroid path for different initial centriod positions; diamonds are final positions

3 Brain Tumor Detection

Since tumors are more condensed, they produce brighter reflections. Therefore, tumor detection mainly relies on finding clusters of pixels with different color intensity than their surroundings. In our approach, we first try to find the region containing the tumor, and then more precisely find the tumor borders. Thus, our approach consists of two main parts: coarse detection and fine detection. Coarse detection is done using Force. Utilizing 'Force' helps avoiding the common intensity analysis and thresholding problems. Fine detection uses histogram analysis, thresholding, and region growing. The following explains the 3 steps of the algorithm:

 Step 1: Preprocessing for Clustering:
  o Skull removal and coarse removal of un-necessary information (GM, WM)
 Step 2: Coarse Detection: Finding the region of Interest (ROI)
  o Use 'Force' data clustering to find N clusters
  o Determine which cluster is ROI
 Step 3: Fine Detection: Finding the boundary of the tumor in the ROI
  o histogram analysis
  o region growing to identify tumor pixels

3.1 Preprocessing

Before coarse detection step, preprocessing is done which consists of removing the skull and removing some part of non tumor pixels as shown in Fig. 2 and explained in this section. Since, skull consists of very high intensity values in T2 images, such that it suppresses other intensity levels, it has to be removed to enhance the next steps of tumor detection. Fortunately, skull is geometrically identifiable and can be efficiently removed. In our method, first the edges of the skull are found using Canny method [13]. Later, the area belonging to the skull which is contained within two strong edges are removed as in Fig. 3.

To be able to detect the ROI more precisely by 'Force', our experiments have shown that it is better to remove some parts of the image which are easily identified as non tumor pixels. These points are found with histogram analysis and are removed by thresholding. In this step, pixels below the histogram peak will be removed, since tumor is always brighter than GM and WM in T2 images. Similar gross separation of tumor and non-tumor pixels is also reported in other works [5]. After removing the obvious non-tumor pixels, an exponential transformation is applied to the remainder of the image (right most plot in Fig. 3). Here we raise the intensity value of each pixel to the power k, where k can be in the range of 2-5 (we use 4 here). The processed image of this step is passed to the Force algorithm for spatial-intensity clustering. To ensure robustness to imperfections of preprocessing, the remaining steps of the algorithm, after clustering, use the original image and not to the preprocessed image.

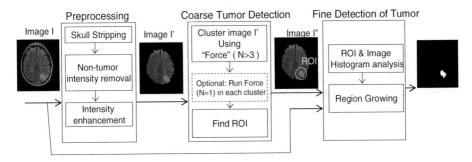

Fig. 2. Proposed tumor detection procedure

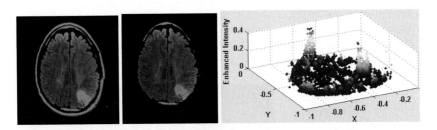

Fig. 3. Left to right: original image, after skull stripping, enhanced intensity levels after removing some non-tumor pixels

3.2 Coarse Detection of Tumor

Clustering the Image using "Force". The resulting image from the preprocessing step is used to find clusters of high intensity pixels using Force. For this purpose, we release N (> 3) centroids randomly in the image, some will move to clusters of brighter pixels. The extra centroids that can not be associated with clusters are repelled and may stop in locations where there are no visible clusters. If enough centroids are used, the result of running Force is that each cluster attracts one centroid. If there are large scattered masses of data (which do not look like clusters), some centroids will associate with them; the remaining centroids will be placed almost outside the areas where data points are (Fig. 4). After reaching equilibrium, each centroid will have a charge equal to the total charge of its cluster. This information is used to remove centroids that are outside the data area and are not associated with many data points. Such centroids will have small final charges.

An example of running Force on the preprocessed image is shown in Fig. 4. Here 4 centroids were released. At the beginning, the centroids were in random positions; they were gradually attracted toward data masses of either high number of data points or high charge or both. In Fig. 4, two initial centroids were attracted toward the tumor which was high in number and charge. Later the centroid that moved faster toward the tumor position repelled the other one. Since there was not a big data mass nearby the repelled centroid stopped outside the brain area. Two other centroids were attracted toward other data masses. As shown in Fig. 4 one of the centroids stops inside the tumor; this centroid gives an idea about the area where the tumor is located. Nevertheless, the algorithm does not require the centroid to be exactly inside the tumor and a rough indication of the area is enough.

Finding Region of Interest. After clustering, the next step is to identify which of the N centroids or clusters includes the tumor. We have tested few methods for this purpose. A simple method is to pick the center that is located at a position with the highest intensity. The assumption is that tumor area has the highest intensity in the preprocessed image. The problem of this method is that sometimes the tumor is not convex shaped, or some other mass nearby causes the center to be slightly outside the tumor; also it is possible that some small but very high intensity pixels remain after preprocessing and attract one of the centroids.

The second method is to pick the centroid with the highest charge. The centroid's charge value is proportional to the total intensity of its cluster. It is expected that the region containing tumor have the highest charge due to its brightness and size. However, sometimes scattered big clusters of low intensity pixels (soft tissues) form a cluster with total charge higher than the tumor. The final and best method that we use is to pick the centroid with highest total intensity in a limited spatial neighborhood, 5-10% image dimensions. This ensures that only a real limited size cluster is considered.

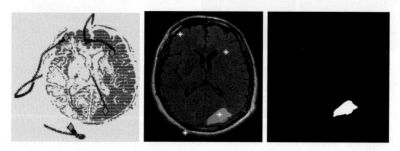

Fig. 4. Left to right: clustering result on preprocessed image, centroids on original image; tumor boundary after region growing

An optional operation to enhance the ROI detection task is to run the Force algorithm for each region again, but with one centroid, to ensure that each center will be located in the middle of the data mass (the tumor for the ROI) of its region, free from the effect of other pixels or centers.

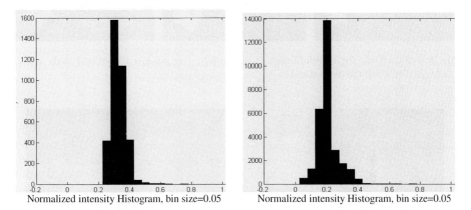

Fig. 5. Left to right: rough histogram of the tumor, histogram of the brain tissues

3.3 Fine Tumor Detection

Tumor pixels are usually geometrically connected, but tumor edges are of lower intensity than its center. To identify the boundary of the tumor in the ROI, we use region growing to detect tumor pixels. For this purpose, we need to find the threshold between the tumor and its surrounding. In this step, histogram analysis is used. Since we already roughly know the tumor area, the histogram analysis is significantly less prone to misclassification of intensities than the histogram analysis of the original image. First, we find a rough histogram of the tumor by observing histogram of a neighborhood (5% image dimension) of the centroid in the ROI. We find a rough histogram of the soft tissues by looking at image histogram (most pixels are GM or WM pixels). Fig. 5 shows the histograms found in this step. We then find the difference between the peaks of the two histograms and choose the mid point as the threshold. Following identifying the threshold, region growing is performed from the centroid location. The resulting area is declared as the tumor, and tumor boundary is consequently identified; this is shown in the right most plot of Fig. 4.

4 Evaluation

The performance of our method has been evaluated on different MRI slices of 13 different patients (~ 100 slices). We depict some of the results in Fig. 6. Despite the difficult cases, our algorithm is able to successfully identify the tumor pixels. Comparing our result with ground truth, we can see the ability of the proposed algorithm in finding the tumor and then identifying its area, even for small parts of the tumor in areas where the difference of tumor and the surroundings intensity is low. The results show the robustness of our algorithm in handling different image intensity levels and dynamic ranges and different tumor shapes and sizes (Fig. 6).

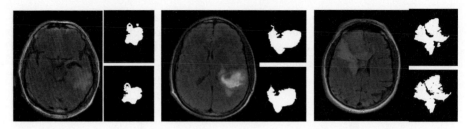

Fig. 6. For each set, the original image is on the left; Top mask is the ground truth, bottom mask is the result of the our method

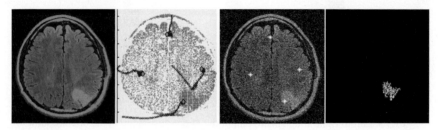

Fig. 7. Left to right: original image, Force result, centroids on noisy image, tumor pixels

4.1 Robustness with Regard to Noise

To further evaluate the performance of our method, we have considered noisy images. Gaussian white noise with mean zero and variance 0.02 has been added to the images. One example is shown in Fig. 7. The tumor is in a cluttered area and it is hard to see its boundary. Force is able to detect the tumor area, as one of the centroids moves inside the tumor. Next steps of the algorithm determine the tumor area more precisely.

4.2 Comparison of Force with Ground Truth

To see the accuracy of our tumor detection method, we have compared our result with ground truth (manually marked tumors by experts) for different slices of different patients. We selected slices with different shapes and sizes of tumors and from different patients. The results presented in Table 1 are selected from 100 slices to be representation of different conditions, and show that in all cases our method correctly finds the tumor (the coarse detection); the identified tumor boundaries are also relatively accurate. In this table, three quantities are shown: true positive, false positive and false negative. The fact that a high percentage of true positive is reported verifies the success of the algorithm in correctly finding the tumor. The presence of some false negative or positive pixels (e.g., Slice24p9 or Slice24p13) shows that the fine detection step can be enhanced with more complex methods. Nevertheless, the robustness of the proposed method in identifying tumor is a significant advantage of the algorithm.

Table 1. Results of our proposed method compared to ground truth (ratios, out of 1)

Image	T. pos	F. pos.	F. neg.	Image	T. pos.	F. pos.	F. neg.
Slice28p1	0.920	0.056	0.079	Slice33p7	0.873	0.117	0.127
Slice35p1	0.900	0.065	0.099	Slice24p8	0.947	0.135	0.052
Slice25p2	0.876	0.007	0.124	Slice24p9	0.778	0.055	0.221
Slice35p3	0.901	0.144	0.098	Slice31p10	0.938	0.051	0.061
Slice24p4	0.939	0.110	0.098	Slice27p11	0.920	0.006	0.079
Slice18p5	0.959	0.041	0.040	Slice31p12	0.815	0.028	0.184
Slice26p6	0.967	0.072	0.032	Slice35p13	0.827	0.027	0.172
Slice38p6	0.933	0.024	0.066	Slice24p13	0.964	1.539	0.035

5 Concluding Remarks

In this paper, a fully automated tumor detection method based on Force algorithm is proposed. "Force" is an unsupervised data clustering method which has been modified here to be used as a method to find the region that contains tumor in brain MRI. The proposed method has three steps of preprocessing, coarse tumor detection using Force, and fine tumor detection. The performance of this method has been evaluated; and it is shown that it is robust and able to find the tumor boundaries in MR images with different intensity, dynamic range, noise, tumor size and shape. The proposed clustering algorithm can be employed in 3D space with no change. Therefore, the entire tumor detection approach is extendable to 3D. The clustering method can also be applied to other medical images, e.g., breast or liver MRI.

References

1. Bezdek, J.C., Hall, L.O., Clarke, L.P.: Review of MR image segmentation techniques using pattern recognition. Medical physics 20(4), 1033–1048 (1993)
2. Held, K., Rota, K.E., Krause, B.J., Wells III, W.M., Kikinis, R., Muller-Gartner, H.W.: Markov random field segmentation of brain MR images. IEEE Trans. Med. Imaging, 878–886 (1997)
3. Clarke, L., et al.: MRI segmentation: Methods and applications. Magnetic Resonance Imaging 13(3), 343–368 (1995)
4. Cuadra, M.B., Gomez, J., Hagmann, P., Pollo, C., Villemure, J.G., Dawant, B.M., Thiran, J.: Atlas-based segmentation of pathological brains using a model of tumor growth. In: Medical Image Comp.& Computer-Assisted Intervention MICCAI (2002), pp. 380–387. Springer, Heidelberg (2002)
5. Clark, M.C., Hall, L.O., Goldgof, D.B., Velthuizen, R., Murtagh, F.R., Silbiger, M.S.: Automatic tumor-segmentation using knowledge-based techniques. IEEE Trans. on Medical Imaging 17, 187–201 (1998)
6. Kaus, M.R., Warfield, S.K., Nabavi, A., Chatzidakis, E., Black, P.M., Jolesz, F.A., Kikinis, R.: Segmentation of meningiomas and low grade gliomas in MRI. In: Taylor, C., Colchester, A. (eds.) MICCAI 1999. LNCS, vol. 1679, pp. 1–10. Springer, Heidelberg (1999)
7. Prastawa, M., Bullitt, E., Ho, S., Gerig, G.: A Brain Tumor Segmentation Framework Based on Outlier Detection. Medical Image Analysis 8, 275–283 (2004)

8. Gering, D., Eric, W., Grimson, L., Kikinis, R.: Recognizing Deviations from Normalcy for Brain Tumor Segmentation. Medical Image Computing and Computer-Assisted Intervention, 388–395 (September 2002)

9. Phooi Yee Lau Voon, F.C.T., Ozawa, S.: The detection and visualization of brain tumors on T2-weighted MRI images using multi parameter feature blocks. In: Engineering in Medicine and Biology Society. IEEE-EMBS, pp. 5104–5107 (2005)

10. Wu, M.-N., Lin, C.-C., Chang, C.-C.: Brain Tumor Detection Using Color-Based K-Means Clustering Seg. In: IEEE Computer Society, IIH-MSP 2007, pp. 245–250 (2007)

11. Kalantari Khandani, M., Saeedi, P., Fallah, Y.P., Khandani, M.K.: A Novel Data Clustering Algorithm Based on Electrostatic Field Concepts. In: IEEE Symposium on Computational Intelligence and Data Mining (IEEE CIDM 2009), pp. 232–237 (2009)

12. Mc Queen, J.: Some methods for classification and analysis of multivariate observations. In: Proc. of the 5th Berkeley Symp. on Mathematical Statistics and Probability, pp. 281–297 (1967)

13. Canny, J.F.: A computational approach to edge detection. IEEE Trans. Pattern Analysis and Machine Intelligence, 679–698 (1986)

Top-Down Segmentation of Histological Images Using a Digital Deformable Model

F. De Vieilleville[1], J.-O. Lachaud[1], P. Herlin[2],
O. Lezoray[3], and B. Plancoulaine[2]

[1] Laboratoire de Mathématiques, UMR CNRS 5127
Université de Savoie, 73776 Le-Bourget-du-Lac, France
francois.de-vieilleville@univ-savoie.fr,
jacques-olivier.lachaud@univ-savoie.fr
[2] GREYCAN, Centre François Baclesse
Avenue du Général Harris 14076 Caen cedex 5, France
p.herlin@baclesse.fr, b.plancoulaine@baclesse.fr
[3] GREYC 6 Boulevard du Maréchal Juin
14050 Caen Cedex, France
olivier.lezoray@unicaen.fr

Abstract. This paper presents a straightforward top-down segmentation method based on a contour approach on histological images. Our approach relies on a digital deformable model whose internal energy is based on the minimum length polygon and that uses a greedy algorithm to minimise its energy. Experiments on real histological images of breast cancer yields results as good as that of classical active contours.

1 Introduction

Breast cancer may be one of the oldest known forms of cancerous tumors in humans. Worldwide, breast cancer is the second most common type of cancer after lung cancer and the fifth most common cause of cancer death. Prognostic and diagnosis largely depend on the examination of stained tissue images by expert pathologists, which is time consuming and may lead to large variations. Therefore, it is essential to develop Image Decision Guided Systems to assist prognostic, diagnostic and early detection of cancer by automatically analyzing images of pathological tissue samples. One important prognostic factor for pathologist is the assessment of cellular proliferation by calculation of a mitotic grade [1]. To that aim, histological slides are stained by Immunohistochemistry that stains cells in proliferation in brown and other cells in blue. To establish an accurate mitotic grade, one has, in addition to mitosis detection, to properly detect cancer cells clusters (clusters of tumoral cells) to evaluate the mitotic grade only in tumor areas of the tissue. Fig 1 presents cancer cell clusters with mitotic and non mitotic cells.

In this paper, we focus on the segmentation of cancer cell clusters in breast histological images. In literature, such a task is performed with machine learning

G. Bebis et al. (Eds.): ISVC 2009, Part I, LNCS 5875, pp. 327–336, 2009.

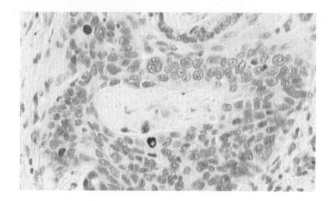

Fig. 1. Cancer cell clusters with cells in proliferation stained in brown

methods [2,3]. Alternatively, we propose a top-down contour based approach using a digital deformable model [4]. On digital curves, using a direct analog of the classic energies of [5] is difficult for the expression of the geometrical quantities, mainly length and curvature estimation, suffer many drawbacks, see discussion in [6] about the curvature estimation. As a result, the internal energy term which usually monitors the smoothness of the curve does not behave as expected. Even when considering digital estimators that are asymptotically convergent on digital curves (see [7]), they often lead the evolution process to non-significant local minimum. As a result we here consider a digital deformable model that benefits of an internal energy based on the minimum length polygon (see [8,9]) yielding a convex functional. As a result, a descent on the internal energy ensures a global minimum. On an open 4-connected simple path this global minimum is very close to the simplest digital straight segment linking the endpoints of the path.

The paper is written as follow, we first recall the definition of the digital deformable model (Section 2). Later-on we elaborate on our Top-Down approach on histological images (Section 3). Experiments illustrate the use of our digital deformable model on real histology images, in particular we compare our model with another internal energy and a with one of the classical active contour formulation using a greedy algorithm for the energy minimisation(Section 4). Eventually, we conclude on the benefits of the proposed approach and possible future works.

2   Digital Deformable Model

We here recall briefly the definition and main properties of a digital analog of active contours, (see [4]). The geometry of our model is that of an simple oriented digital 4-connected open path Γ with its endpoints being fixed, says A and B. The endpoints of the path are not allowed to be moved, while the remaining points of the curve can be deformed using elementary local transformations. Those separates into three types : *flips*, *bumps* and *flats*. All of these can only be applied on points with the proper corresponding geometry such as inside or outside corners, flat parts or inside or outside bumps. Those features can be easily listed when the contour is read as a succession of straight moves, left moves and right moves. See Fig. 2 for an illustration of possible deformations on some specific contours. The admissible deformations of Γ are chosen such that they preserve its topology when applied and such that they are always reversible.

As an analog of active contour, the evolution of the model is monitored and highly dependant of the energy associated to Γ. This energy divides into two terms, one for the geometrical constrain over the curve (internal energy term E_{int}) and one for the fit to the underlying datas (image energy term E_{image}).

$$E^D_{DM}(\Gamma) = E^D_{int}(\Gamma) + E^D_{image}(\Gamma).$$

In the case of parametrized curves on the Euclidean plane constrain energies are usually based of the length and the integration of the normal and curvature along the curve, that is $\int_0^1 E_{int}(v(s), v'(s), v''(s))ds$. In digital space, it is difficult to find good curvature estimators on open contour, although there exists approaches which accurately estimate the curvature on closed digital contour such as the GMC in [10] (which is based on an optimisation scheme), or the Brunet-Malgouyres estimator in [11] (which uses binomial convolutions), these approaches are not suited to make a reliable estimation on the border of open curves. Moreover, as noticed by many authors (see for instance the geometric active contours of [12]), the internal energy is in fact the length of the curve. Our digital analog to E_{int} is therefore defined as the estimation of the length of Γ. This estimation is based on the euclidean length of the constrained minimum length polygon (CMLP for short) of Γ, that is the minimum length polygon in

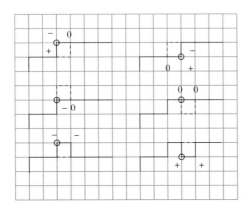

Fig. 2. Example of some local deformations used by the digital deformable model, solid line for Γ and dashed line for Γ after the local deformation, the points where the deformation is applied are circled. The + symbol stands for a left turn, the − symbol stands for a right turn and the 0 symbol stands for a straight ahead move. From left to right and top to bottom : flip of an outside corner, flip of an inside corner, inside bump of a flat part, outside bump of a flat part, flat of an inside bump and flat of an outside bump.

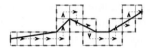

Fig. 3. A digital open path Γ whose Freeman code word is EEEENNESESEENENE. In dashed lines we represent the one pixel band along Γ. The thick line is the constrained minimum length polygon of Γ.

a one pixel wide band along Γ passing through its constrain points. The first and last vertices of the $CMLP$ are respectively the first and last point of Γ. See Fig. 3 for an illustration of the CMLP.

Thus our internal energy becomes:

$$E_{int}^{D}(\Gamma) = \alpha \mathcal{L}_2(CMLP(\Gamma))$$

One important property of this energy is that it is convex (see proof in [4]), that is, a descent on this energy always leads to a global minimum. This minimum is such that the CMLP of Γ is exactly the euclidean straight segment between the endpoints of Γ. The image energy term favors strong gradients and is defined as:

$$E_{image}^{D}(\Gamma) = \sum_{p \in \Gamma} \left(\max_{c \in Image} (\|\nabla I(c)\|) - \|\nabla I(p)\| \right),$$

This energy term being positive everywhere, the tuning of the α parameter is eased with respect to length term.

3   Top-Down Segmentation for Histology Images

The minimisation of the digital deformable model relies on a greedy approach (see Alg. 1) being costly because of it requires to compute many times the CMLP of Γ. As a consequence, a top-down approach seems particularly suited. We here consider three levels of resolution, the size of the image being multiplied by four at the next level. At each level there are two phases: an initialisation and a minimisation based on energy criteria.

At the first level, the initialisation is done via a binary mask image, from which we extract the initialisation paths. In fact we have several digital deformable modelsn to minimize at each level. Although we do not strictly preserve the topology of the initialisation, we prevent the paths to collide with one another. The initialisation at the other levels is obtained by scaling the paths resulting from the minimisation phase at the previous level, see Fig. 4 for an illustration.

The minimisation phase is straightforward: for a given path, we try all the deformations and apply the one that brings the smallest energy, we repeat this scheme as long as smaller energies can be found.

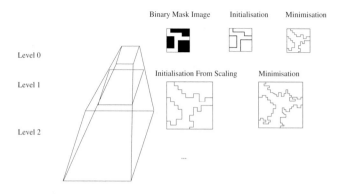

Fig. 4. The three levels used in our Top-Down approach

Algorithm 1: Greedy1 algorithm: extracts the deformation that brings the lowest energy among all possibles ones

---

1 **Function** Greedy1(**In** Γ, **Out**) : boolean ;
  Input: Ω: an Interpixel Partition Deformable Model
  Output: Ω': when returning *true*, elementary deformation of Ω with
        $E_{DP}^{D}(\Omega') < E_{DP}^{D}(\Omega)$, otherwise $\Omega' = \Omega$ and it is a local minimum.
  Data: Q : Queue of (Deformation, double) ;
2 $E_0 \leftarrow E_{DM}^{D}(\Gamma)$;
3 **foreach** *valid Deformation d on Γ* **do**
4 $\quad$ Γ.applyDeformation(d);
5 $\quad$ Q.push_back($d, E_{DM}^{D}(\Gamma)$);
6 $\quad$ Γ.revertLastDeformation();
7 **end**
8 (d, E_1) = SelectDeformationWithLowestEnergy (Q);
9 $\Gamma' \leftarrow \Gamma$;
10 **if** $E_1 < E_0$ **then** Γ'.applyDeformation(d);
11 **return** $E_1 < E_0$;

---

4 Experiments

The binary masks at low resolution are obtained with a coarse segmentation algorithm. The latter performs an automatic binary thresholding by entropy on a simplified version of the image with a morphological opening by reconstruction, see [13] for a similar method.

Our first examples use the constrained minimum length polygon as internal energy. As the model explicitly uses a weighting coefficient, we have run our experiments with various values, as shown on Fig. 5,6 and 7. This coefficient is such that the smaller, the less the internal energy monitors the deformation, consequently for α equal to 400 the length penalisation overcome the data term.

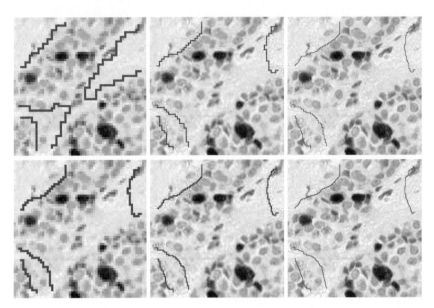

Fig. 5. Results of the top-down segmentation process using the value 400 as balance term, the internal energy term is based on the CMLP. From left to right and top to bottom: image at level 0,1 and 2 with initialisation, image at level 0,1 and 2 after minimisation. The high value of α penalise the length of the contours, smoothing the contours too much, only top-left contour seems to be correctly delineated.

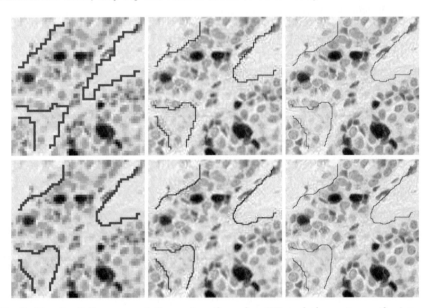

Fig. 6. Result of the segmentation using the value 150 as balance term, the internal energy term is based on the CMLP. From left to right and top to bottom: image at level 0,1 and 2 with initialisation, image at level 0,1 and 2 after minimisation. Top-left and bottom-left contours are correctly delineated.

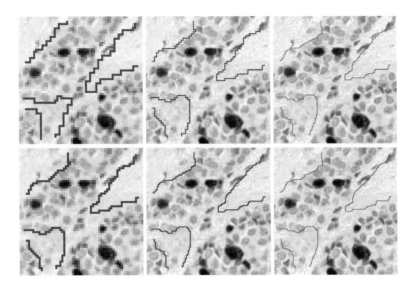

Fig. 7. Results of the top-down segmentation process using the value 100 as balance term, the internal energy term is based on the CMLP. From left to right and top to bottom: image at level 0,1 and 2 with initialisation, image at level 0,1 and 2 after minimisation. Global delineation of contours is correct.

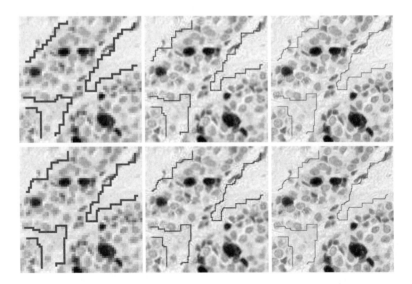

Fig. 8. Results of the top-down segmentation process using the value 100 as balance term, the internal energy term is based on the length of the freeman code of the path, other positive values for the balance term bring very similar contours. From left to right and top to bottom: image at level 0,1 and 2 with initialisation, image at level 0,1 and 2 after minimisation. Although the global delineation seems correct, the obtained contours are not smoothed at all.

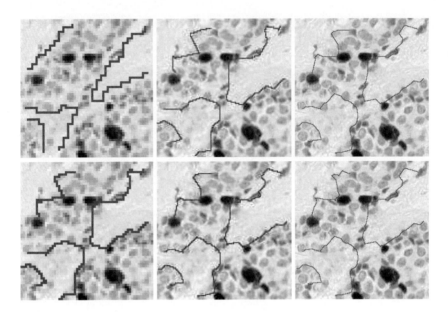

Fig. 9. Results of the top-down segmentation process using the deformable model described in [6]. Parameters are such that (α, β) equal $(0.1, 0.45)$, the neighborhood is chosen as a 3×3 square, and minimisation is done until no smaller energy can be found. From left to right and top to bottom: image at level 0,1 and 2 with initialisation, image at level 0,1 and 2 after minimisation. Top-right contour and middle contours are well delineated, top left and bottom left are partially correctly delineated.

In order to compare our results we have also run the same experiments using another internal energy. We have used a simpler internal energy which also leads to a global minimum. This internal energy is simply defined as the sum of the length of the freeman chain code constituting the path, that is the number of points of the path minus one. Using various positive coefficients as α, results are very similar and the paths are much less smoothed than with the CMLP, as illustrated on Fig. 8.

Finally we have run the experiments using one of the classic active contour method in the literature, [6]. In this the curve is approached by a polygon whose vertices have integer coordinates. The minimisation process uses a greedy algorithm, which is closer to our optimisation scheme than a variational method. The quantity being minimized by this approach is:

$$E = \int (\alpha E_{cont} + \beta E_{curv} + \gamma E_{image}) ds.$$

At each iteration, and for each vertex a fixed-sized neighborhood around the point is considered and the point giving the smallest value becomes the new vertex. The energy terms are such that small values of E_{cont} causes the polygon points to be more equidistant and E_{curv} is the estimation of the curvature using finite differences. The γ coefficient is chosen equal to one and $E_{image}(p)$ is chosen as $\max(||\nabla I||) - ||\nabla I(p)||$ to ease the comparison with our approach. Between each level we also double the number of points of the model, obtained results are good considering that this method allows the end points of the polygon to move according to the minimisation process, as illustrated on Fig. 9.

5 Conclusion

In this paper we have used a top-down straightforward approach using a digital deformable model to segment cancer cluster cells in histological images. The behaviour of our digital deformable model was shown to behave as the classic continuous active contours and shown to give similar results. Future work will focus on integrating our digital deformable model with topological map (following [14]) so as to ensures that no topological changes may occur during the deformation of the contours. Once this step is achieved, top-down approaches on irregular pyramids such as [15] would be able to mix region and contour based segmentation process.

References

1. Cheryl, E.: Assessment of cellular proliferation by calculation of mitotic index and by immunohistochemistry. In: Metastasis Research Protocols. Methods in Molecular Medicine, vol. 57, pp. 123–131. Humana Press (2001)
2. Doyle, S., Agner, S., Madabhushi, A., Feldman, M.D., Tomaszewski, J.: Automated grading of breast cancer histopathology using spectral clusteringwith textural and architectural image features. In: ISBI, pp. 496–499 (2008)
3. Signolle, N., Plancoulaine, B., Herlin, P., Revenu, M.: Texture-based multiscale segmentation: Application to stromal compartment characterization on ovarian carcinoma virtual slides. In: Elmoataz, A., Lezoray, O., Nouboud, F., Mammass, D. (eds.) ICISP 2008 2008. LNCS, vol. 5099, pp. 173–182. Springer, Heidelberg (2008)
4. de Vieilleville, F., Lachaud, J.O.: Digital deformable model simulating active contour. In: Proceedings of 15th International Conference on Discrete Geometry for Computer Imagery (accepted, 2009)
5. Kass, M., Witkin, A., Terzopoulos, D.: Snakes: Active contour models. International Journal of Computer Vision 1, 321–331 (1988)
6. Williams, D.J., Shah, M.: A fast algorithm for active contours and curvature estimation. CVGIP: Image Underst 55, 14–26 (1992)
7. Lachaud, J.O., Vialard, A., de Vieilleville, F.: Fast, accurate and convergent tangent estimation on digital contours. Image and Vision Computing 25, 1572–1587 (2007)
8. Sloboda, F., Stoer, J.: On piecewise linear approximation of planar jordan curves. J. Comput. Appl. Math. 55, 369–383 (1994)
9. Klette, R., Yip, B.: The length of digital curves. Machine Graphics Vision 9, 673–703 (2000); Also research report CITR-TR-54, University of Auckland, NZ (1999)
10. Kerautret, B., Lachaud, J.-O.: Curvature estimation along noisy digital contours by approximate global optimization. Pattern Recognition (in Press, 2009) (Corrected Proof)

11. Malgouyres, R., Brunet, F., Fourey, S.: Binomial convolutions and derivatives estimation from noisy discretizations. In: Coeurjolly, D., Sivignon, I., Tougne, L., Dupont, F. (eds.) DGCI 2008. LNCS, vol. 4992, pp. 370–379. Springer, Heidelberg (2008)
12. Caselles, V., Catte, F., Coll, T., Dibos, F.: A geometric model for active contours. Numerische Mathematik 66, 1–31 (1993)
13. Elie, N., Plancoulaine, B., Signolle, J., Herlin, P.: A simple way of quantifying immunostained cell nuclei on the whole histological section. Cytometry 56 A, 37–45 (2003)
14. Dupas, A., Damiand, G.: First results for 3d image segmentation with topological map. In: Coeurjolly, D., Sivignon, I., Tougne, L., Dupont, F. (eds.) DGCI 2008. LNCS, vol. 4992, pp. 507–518. Springer, Heidelberg (2008)
15. Goffe, R., Brun, L., Damiand, G.: A top-down construction scheme for irregular pyramids. In: Proceedings of the Fourth International Conference On Computer Vision Theory And Applications (VISAPP 2009), pp. 163–170 (2009)

Closing Curves with Riemannian Dilation: Application to Segmentation in Automated Cervical Cancer Screening

Patrik Malm and Anders Brun

Centre for Image Analysis, Box 337, SE-751 05, Uppsala, Sweden
{patrik,anders}@cb.uu.se

Abstract. In this paper, we describe a nuclei segmentation algorithm for Pap smears that uses anisotropic dilation for curve closing. Edge detection methods often return broken edges that need to be closed to achieve a proper segmentation. Our method performs dilation using Riemannian distance maps that are derived from the local structure tensor field in the image. We show that our curve closing improve the segmentation along weak edges and significantly increases the overall performance of segmentation. This is validated in a thorough study on realistic synthetic cell images from our Pap smear simulator. The algorithm is also demonstrated on bright-field microscope images of real Pap smears from cervical cancer screening.

1 Introduction

Cervical cancer is the second most common type of cancer among women. During 2005 it caused a quarter of a million deaths worldwide, of which 80% occurred in developing countries [1]. Current screening programmes mainly use the staining and visual inspection method developed by Papanicolaou during the 1940s. Images from the Papanicolaou (Pap) test, commonly know as the Pap smear, are mostly analyzed manually by experts, which is a costly procedure [2]. In this paper we describe an automatic segmentation algorithm for cell nuclei in such images. In particular we introduce and study the use of anisotropic dilation for curve closing to achieve better segmentations.

When using edge detection algorithms, e.g., the Canny edge detector [3], it is common that the resulting edge map has gaps in the object borders. To achieve a segmentation based on the edge detection result, such flaws need to be corrected. The method described in this paper generates a non-Euclidean distance transform of the edge map, derived from local gradients in the image. The distance map is then used to perform anisotropic dilation, which we call Riemannian dilation since the distances are derived from geodesic distances in a Riemannian manifold [4]. Locally, the metric we use is related, but not identical, to the local structure tensor field [5,6,7] in the image. In this metric, the geodesic disc has an elongated appearance that follows edges and lines in the image. When it is used as structuring element in dilation, it repairs the gaps in the edge map. In

G. Bebis et al. (Eds.): ISVC 2009, Part I, LNCS 5875, pp. 337–346, 2009.

this paper, we show that by modeling the image as a Riemannian manifold in this manner, we achieve better performance for weak edges compared to using ordinary isotropic dilation.

1.1 Related Work

A thorough discussion regarding the concepts of spatially variant morphology can be found in [8]. In [9] a similar approach is described, where the structure tensor field is also used to perform spatially varying morphology in images.

2 Segmentation

The segmentation process can be thought of as consisting of two related processes, recognition and delineation. Recognition is the task of roughly deciding where in an image an object is, whereas delineation is the process of determining the precise spatial extent and point-by-point composition of the object [10].

The segmentation method described in this paper is mainly aimed at recognizing the location of nuclei in images. This is achieved by using the Canny edge detector followed by a series of morphological processes aimed at refining and closing detected edges and thereby producing recognizable objects (see Fig. 1).

The basic way to perform a closing operation ($\bullet$) is to do a binary morphological dilation ($\oplus$) followed by an erosion ($\ominus$),

$$A \bullet B = (A \oplus B) \ominus B, \qquad (1)$$

where A represents the set and B the structuring element used [11]. If the structuring element B is disk shaped the same dilation can also be achieved by thresholding a distance transform generated using the binary object as a seed. However,

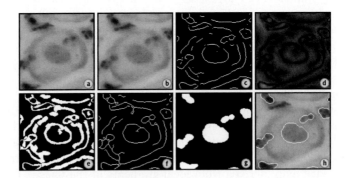

Fig. 1. The individual steps of the segmentation method when used to segment a nucleus (big roundish object in the center of the image): a) initial image, b) median filtering, c) Canny edge result, d) resulting distance map using the weighted measure, e) thresholded distance map, f) skeleton of threshold, g) final segmentation result achieved through filling of closed areas and opening, h) final segmentation overlayed on the original image

both methods are limited in that they dilate a binary object equally in all directions and everywhere in the image, without regard to the local image structure.

An alternative approach to curve closing is to perform an edge linking procedure in which matching pairs of endpoints are found and connected [12]. However, image content is still not taken into account. The two matching endpoints are connected using a straight line, making it possible to create boundaries that exclude part of the object or include background in the object segmentation.

Our method uses an image adaptive dilation, meaning that, from a given seed point, it propagates edges with regards to certain features in the image. This makes it possible to close edges in a way that better corresponds to the underlying image data.

2.1   Method Overview

Figure 1 shows the individual steps of the described method. The initial image (a) is first median filtered to remove noise (b). Canny edge detection is then applied (c) and the result is used as a seed image for a tensor weighted distance transform, implemented[1] from [5], that will be further described in the following section (d). The distance map is thresholded (e) and the subsequent binary image cleaned up via skeletonization (f), filling of closed areas and finally an isotropic opening (binary erosion followed by dilation) in order to get the final segmentation result (g, h).

2.2   Tensor Weighted Distances

We model the image as a chart of a 2D Riemannian manifold. The distance transforms we compute are based on estimates of geodesic distance in this curved Riemannian manifold. Since we only approximate geodesic distances, there is a possibility that the triangle inequality is not fulfilled and that the corresponding space we create is not a metric. For the task at hand, however, this possibility is of less practical importance.

The concept of looking at an image as a landscape, where the gray-level values correspond to height, is well known. How distances can be calculated in such a landscape is also a previously studied concept [13,14]. A 2D image $f(x, y)$ is in this case represented by the embedding of a surface in $\mathbb{R}^3$:

$$\chi = (x, y, \alpha f(x, y)), \tag{2}$$

where α is positive. From Eq. 2 we derive the pullback metric from χ in the image plane $(x, y) \in \mathbb{R}^2$:

$$\mathbf{B} = I + \alpha^2 \nabla f \nabla f^T = \begin{bmatrix} 1 & 0 \\ 0 & 1 \end{bmatrix} + \alpha^2 \begin{bmatrix} f_x f_x & f_x f_y \\ f_y f_x & f_y f_y \end{bmatrix}. \tag{3}$$

---

[1] Toolbox Fast Marching by Gabriel Peyré
http://www.mathworks.com/matlabcentral/fileexchange/6110

The scalar product of two vectors $\mathbf{u}, \mathbf{v} \in \mathbb{R}^2$, in a particular point in the image, is then $\langle \mathbf{u}, \mathbf{v} \rangle_B = \mathbf{u}^T \mathbf{B} \mathbf{v}$. With this metric, the distance between two neighboring pixels, separated by a vector $\mathbf{u}$ (Fig. 2(a)), is approximately

$$\|\mathbf{u}\| = \sqrt{\langle \mathbf{u}, \mathbf{u} \rangle_B} = \sqrt{\mathbf{u}\mathbf{u} + \alpha^2 (\nabla f \cdot \mathbf{u})^2}. \tag{4}$$

The effect of the α parameter on the resulting distance transform is illustrated in Fig. 2(b). When $\alpha = 0$ the distance is reduced to the Euclidean distance in the image plane, while for larger values of α, steps taken in the direction of the gradient are penalized. Equation 3 is known as the Beltrami framework [15] and for cell images it means that distances increase faster perpendicular to the nucleus border. As pointed out in [16], Eq. 3 may be generalized by replacing the outer product of the gradient with the structure tensor [17,18,19]

$$\mathbf{S} = I + \alpha^2 \nabla f \nabla f^T * G_\sigma = \begin{bmatrix} 1 & 0 \\ 0 & 1 \end{bmatrix} + \alpha^2 \begin{bmatrix} f_x f_x & f_x f_y \\ f_y f_x & f_y f_y \end{bmatrix} * G_\sigma, \tag{5}$$

where $*G_\sigma$ is the convolution by a Gaussian with σ standard deviation. The use of the structure tensor extends the Beltrami framework to penalize both image gradients (edges) and ridges (lines). It is a generalization because when $\sigma \to 0$ then $\mathbf{S} \to \mathbf{B}$. It is also different from the pure structure tensor because of the inclusion of the unit metric and the α parameter.

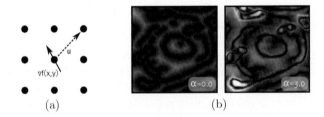

(a)   (b)

Fig. 2. a) Illustration showing distance vector from a pixel with gradient ∇f. b) Effect of the α parameter on the distance transform.

3   Data

Our algorithms were applied on two types of data: realistic synthetic cell images and minimum intensity projections of focus stacks taken of cervical smears with a bright-field microscope.

3.1   Realistic Synthetic Cell Images

For validation and parameter tuning purposes we have created a program that generates synthetic cell images. This program is a part of ongoing research. Here we briefly describe the image formation in the Pap smear simulator.

The images used for this paper have dimensions 1000×1000 pixels. The dynamic range is set to [0,255] in order to match the real data which is being mimicked. An overview over the different steps of the image generation can be seen in Fig. 3.

The process starts with an empty image to which two levels of Perlin noise [20] is added to simulate general lighting inconsistencies and noise due to glass imperfections (Fig. 3a). The image area is then populated with cells.

Each generated cell is unique. The cytoplasm shape is created by selecting a field from a Voronoi diagram and then distorting the shape using Perlin noise. Texture generated from normally distributed noise and shadow created by adding a blurred layer are used to increase realism when drawing the finished shape in the image (Fig. 3b, c).

The nucleus shape, a randomly distorted circle, is created using a truncated Fourier series. Texturing of the nucleus includes two types of noise, a normally distributed noise to simulate chromatin texture and a Perlin noise to simulate staining imperfections (Fig. 3d, e).

Debris, e.g., blood cells, is generated as clusters containing three to ten small roundish objects. These are then added to the image as two layers with different levels of blurring (Fig. 3f, g).

The final step of creating the synthetic image is the addition of a small Gaussian blur ($\sigma = 0.8$) throughout the entire image. This ensures that all components are blended together in a way that simulates the point spread function of the microscope (Fig. 3h). The resulting images are hard to visually distinguish from real microscope images (Fig. 4).

3.2 Bright-Field Microscopy Images

The biological samples used in this study were prepared using the standard Papanicolaou protocol. We obtained focus stacks of the Pap smear samples using

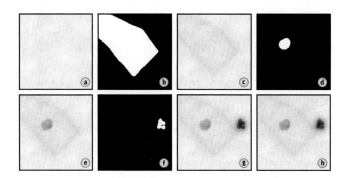

Fig. 3. The key steps in the synthetic image creation: a) initial plane with noise, b) finished cytoplasm mask, c) cytoplasm subtracted from image, d) nucleus mask, e) nucleus subtracted from image, f) debris mask, g) debris subtracted from image, h) final image after blurring

Fig. 4. Synthetic image (left) with corresponding ground truth segmentation of nuclei (middle) compared to the minimum intensity projection of a focal stack taken of a Pap smear sample (right)

a MIRAX MIDI system (3DHISTECH Ltd., Budapest, Hungary) equipped with a 20× / 0.8 Plan-Apochromat objective. The images used have 1376×1032 pixels and a dynamic range in $[0, 255]$. For the application described in this paper, the focus stacks were reduced to a single image using minimum intensity value projection.

4   Evaluation

The focus of the evaluations performed in this paper is to obtain a high rate of nuclei recognition, rather than perfect delineation of every individual cell nuclei. For this reason we evaluate the segmentation result in terms of true positives (found nuclei) and false negatives (missed nuclei). An efficacy metric which compares the segmentation results with the ground truth for the synthetic images was implemented. The metric is based on theory described in [10], which compares two segmentation results of an object i, τ_{i_1} and τ_{i_2}, using the equation

$$r_i = \frac{|\tau_{i_1} \cap \tau_{i_2}|}{|\tau_{i_1} \cup \tau_{i_2}|}, r_I = \sum_{i \in I} r_i. \tag{6}$$

Here r_i represents the common part of the two segmentation results as a fraction of their combined area. If a ground truth is used as one of the segmentations in Eq. 6 then r_i represent the precision of the segmentation for each nucleus i. The global score r_I can be used to rate the performance over the entire image I.

We consider an object correctly segmented when r_i exceeds a threshold ρ. With the segmentation result given as a series of binary outcomes we are able to perform a statistical analysis on the results to compare two segmentation methods. We use a single sided McNemar test [21] that uses the values p_{TF}, the number of objects found by method 1 but not by method 2, and p_{FT}, the number of objects found by method 2 but not by method 1, to compare methods.

In the case of the synthetic data, the ground truth was known for all nuclei, making it possible to apply the efficacy metric described in Eq. 6 to evaluate segmentation results. The ratio threshold, ρ, was set to 0.7 for all evaluations. The results could then be statistically analyzed using McNemar's test.

Lacking a ground truth, the segmentation results for the Pap smear images were analyzed based on an expert's visual inspection. The McNemar test was then performed using the exact binomial distribution, instead of the χ^2 approximation, because of the small sample sizes in the real datasets.

5   Experiments

The experiments in this paper were focused at comparing the curve closing ability of the anisotropic dilation scheme to a standard isotropic one. The isotropic dilation method based on the distance transform, described in Section 2, was chosen for the comparison. It was implemented using the same framework as the anisotropic dilation only with the σ and α parameters set to 0. This means that dilation becomes fully dependent on the threshold, λ. The Canny edge parameters were heuristically selected prior to the evaluation process and then kept fixed throughout.

Algorithm development was done in the MATLAB^{TM} programming environment (The MathWorks Inc., Natick, MA) using the DIPimage toolbox version 2.0.1 [22].

5.1   Synthetic Data

A quantitative evaluation was performed using synthetic data sets generated as described in Section 3.1. A total of 160 images were used, each containing approximately 60 nuclei (not counting those touching the image border). The total number of fully segmentable nuclei was 9750.

Prior to evaluation the parameters of the dilation schemes were optimized based on the criterion described in Section 4 using a brute force approach. The parameter tuning was performed by running the segmentation algorithm on three synthetic images with a range of parameters. The dilation schemes were rated based the image score of Eq. 6. The optimal parameters were then chosen to be $\alpha = 8$, $\sigma = 6$ and $\lambda = 210$ for the anisotropic dilation and $\lambda = 6.5$ for isotropic dilation. In Fig. 5 maximum value projections illustrating the relationships between the three parameters σ, α and λ for the anisotropic parameter optimization are shown.

Using the optimized parameters all 160 images were then analyzed. The results are illustrated in Table 1(a) in the form of a contingency matrix. The results were analyzed using a single sided McNemar test, mentioned in Section 4: H_0 : $p_{TF} = p_{FT}$ and the alternative hypothesis H_1 : $p_{TF} > p_{FT}$. The test showed that the anisotropic method was significantly better, $p \ll 10^{-6}$.

5.2   Bright-Field Microscope Images

Because of the limited number of Pap smears available, the experiments using real data were focused towards a qualitative analysis. We used two Pap smear images, one where the cells were less densely packed and with a smaller amount

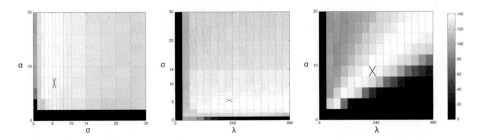

Fig. 5. A illustration of the evaluated parameter space for the anisotropic dilation scheme. The figure shows the maximum value projections along the λ (left), α (middle) and σ axis (right). The 'X' marks the global optimum.

Table 1. Tables showing the resulting contingency matrices for the experiments performed on the synthetic (a) and the real (b, c) datasets. The results corresponding to the two real images have been divided with respect to the type of cellular distribution. The matrices show the distribution of correct (cor.) and incorrect (in.) segmentations.

(a)			(b)			(c)		
Synthetic data			Real data (dense)			Real data (sparse)		
	Aniso.			Aniso.			Aniso.	
	cor.	in.		cor.	in.		cor.	in.
Iso. cor.	6444	419	Iso. cor.	26	0	Iso. cor.	37	0
in.	1487	1400	in.	2	6	in.	0	5

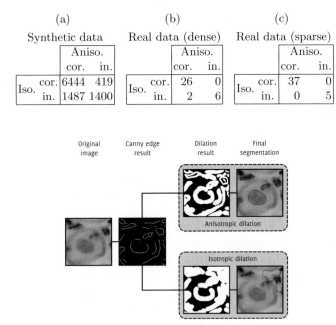

Original image Canny edge result Dilation result Final segmentation

Anisotropic dilation

Isotropic dilation

Fig. 6. The difference in curve closing ability between an anisotropic dilation and a standard isotropic dilation

of debris and one we consider to have a more normal composition. In total the two images contained 76 nuclei. No ground truth existed for the segmentation so parameter tuning was performed manually, starting from optimal values found for synthetic images (described in Section 5.1). We used the previously found optimal values for α and σ, but changed λ slightly.

We experienced that both dilation schemes found the same nuclei in the less cluttered image but that the anisotropic approach was able to segment two more nuclei in the cluttered image. This behavior is demonstrated in Fig. 6. The contingency matrices for the experiments are shown in Table 1(b) and 1(c). The results are encouraging but not significant due to the small sample size: The single side McNemar test did not reject the zero hypothesis, $p \leq 0.25$, for the results in the cluttered image.

6   Conclusions

We can conclude that Riemannian dilation performs better than isotropic dilation in a realistic biomedical setting where it is used for curve closing. Even though there is a need for a more complete validation on real data our extensive study on realistic simulated data shows the benefits of the anisotropic framework.

Through the use of our Pap smear simulator we were able to tune the parameters optimally by an exhaustive search in the relevant part of the parameter space. In the future, we expect to reduce the number of parameters in the segmentation by exploiting the almost linear dependency shown in Fig. 5 (right). When more Pap smear images become available in our project, we will have the option to auto-tune the algorithm by including a cell classification step, combined with a performance criteria, in a closed loop. We also expect the Pap smear simulator to be helpful in the future, to test image processing methods, and we plan to develop it further.

Acknowledgments

Ewert Bengtsson, Gunilla Borgefors, Cris Luengo, Robin Strand and Carolina Wählby are all acknowledged for their valuable input and support. The project "3D analysis of chromatin texture in cell nuclei to improve cervical cancer screening" is funded by the Swedish Research Council (2008-2738).

References

1. WHO: Comprehensive cervical cancer control: A guide to essential practice. WHO Press (2006)
2. Grohs, H., Husain, O. (eds.): Automated Cervical Cancer Screening. IGAKU-SHOIN Medical Publishers, Inc. (1994)
3. Canny, J.: A computational approach to edge detection. IEEE Trans. Pattern Analysis and Machine Intelligence 8, 679–698 (1986)
4. Do Carmo, M.: Riemannian Geometry. Birkhäuser, Boston (1992)
5. Prados, E., Lenglet, C., et al.: Control theory and fast marching techniques for brain connectivity mapping. In: Proceedings of the IEEE Conference on Computer Vision and Pattern Recognition, New York, NY, vol. 1, pp. 1076–1083. IEEE, Los Alamitos (2006)

6. Jeong, W.K., Fletcher, P., et al.: Interactive visualization of volumetric white matter connectivity in dt-mri using a parallel-hardware hamilton-jacobi solver. IEEE Transactions on Visualization and Computer Graphics 13, 1480–1487 (2007)
7. Tsai, R., Zhao, H., Osher, S.: Fast sweeping algorithms for a class of hamilton-jacobi equations. SIAM journal on numerical analysis 41, 673–694 (2004)
8. Bouaynaya, N., Charif-Chefchaouni, M., Schonfeld, D.: Theoretical foundations of spatially-variant mathematical morphology part 1: Binary images. IEEE Transactions on Pattern Analysis and Machine Intelligence 30, 823–836 (2008)
9. Breuss, M., Burgeth, B., Weickert, J.: Anisotropic continuous-scale morphology. In: Pattern Recognition and Image Analysis, Pt 2, Proc., Berlin, Germany, vol. 2, pp. 515–522 (2007)
10. Udupa, J., LeBlanc, V., et al.: A framework for evaluating image segmentation algorithms. Computerized Medical Imaging and Graphics 30, 75–87 (2006)
11. Gonzalez, R., Woods, E.: 9. In: Digital Image Processing, 3rd edn., pp. 633–635. Pearson Education, London (2008)
12. Ghita, O., Whelan, P.: Computational approach for edge linking. Journal of Electronic Imaging 11, 479–485 (2002)
13. Ikonen, L.: Pixel queue algorithm for geodesic distance transforms. In: Andrès, É., Damiand, G., Lienhardt, P. (eds.) DGCI 2005. LNCS, vol. 3429, pp. 228–239. Springer, Heidelberg (2005)
14. Verbeek, P., Verwer, B.: Shading from shape, the eikonal equation solved by grey-weighted distance transform. Pattern Recognition Letters 11, 681–690 (1990)
15. Sochen, N., Kimmel, R., Malladi, R.: A general framework for low level vision. IEEE Transactions on Image Processing 7, 310–318 (1998)
16. San Jose Estepar, R.: Local Structure Tensor for Multidimensional Signal Processing. Applications to Medical Image Analysis. PhD thesis, University of Valladolid, Spain (2005)
17. Knutsson, H.: A tensor representation of 3-D structures. Poster presentation of 5th IEEE-ASSP and EURASIP Workshop on Multidimensional Signal Processing, Noordwijkerhout, The Netherlands (1987)
18. Bigün, J., Granlund, G.H.: Optimal orientation detection of linear symmetry. In: IEEE First International Conference on Computer Vision, London, Great Britain, pp. 433–438 (1987)
19. Förstner, W., Gülch, E.: A fast operator for detection and precise location of distinct points, corners and centres of circular features. In: Proc. ISPRS Intercommission Workshop, pp. 281–304 (1987)
20. Perlin, K.: An image synthesizer. In: SIGGRAPH 1985: Proceedings of the 12th annual conference on Computer graphics and interactive techniques, New York, USA, vol. 19, pp. 287–296 (1985)
21. Dietterich, T.: Approximate statistical tests for comparing supervised classification learning algorithms. Neural Computation 10, 1895–1923 (1998)
22. Luengo Hendriks, C.L., van Vliet, L.J., Rieger, B., van Ginkel, M.: DIPimage: a scientific image processing toolbox for MATLAB. Computer Program (1999)

Lung Nodule Modeling – A Data-Driven Approach

Amal Farag, James Graham, Aly Farag, and Robert Falk*

Department of Electrical and Computer Engineering, University of Louisville
aafara02@louisville.edu
www.cvip.uofl.edu

Abstract. The success of lung nodule detection depends on the quality of nodule models. This paper presents a novel approach for nodule modeling which is data-driven. Low dose CT (LDCT) scans of clinical chest data are used to create the required statistics for the models based on modern computer vision techniques. The new models characterize the tissue characteristics of typical lung nodules as specified by human experts. These models suit various machine learning approaches for nodule detection including simulated annealing, genetic algorithms, SVM, AdaBoost, and Bayesian methods. The quality of the new nodule models are studied with respect to parametric models, and are tested on clinical data showing significant improvements in sensitivity and specificity.

1 Introduction

Lung cancer screening studies date back to the 1950s and 1960s when a variety of screening protocols that combined sputum analysis and chest radiography where undertaken for several studies. Studies were designed as either uncontrolled or controlled but not randomized, and employed various screening time intervals. The Philadelphia Pulmonary Neoplasm Research Project was the most widely publicized study, which was a prospective study of 6,027 older men. The Philadelphia Pulmonary Neoplasm Research Project started in 1951 and continued to follow-up on each patient for ten years. The study performed semi-annual screening with 70-mm chest photofluorograms and questionnaires concerning their symptoms, to study the natural history of bronchogenic carcinoma. In this study 6 of 94 patients with lung cancer detected at screening survived more than 5 years [1][2][3]. As technology techniques developed into more refined developments in the 1960s for chest radiography and sputum analysis, the limitations of methods used in the earlier studies led to the perception that a more thorough lung cancer screening study design could be beneficial. Thus in the 1970s three randomized controlled studies were initiated among male smokers: Mayo Clinic [4], Memorial Sloan-Kettering cancer Center and the John Hopkins Medical Institutions. Due to the failure of these studies and others conducted to demonstrate a mortality advantage for lung cancer screening with either sputum cytology or chest radiography led most organizations to not recommend routine screening. Reassessments and re-examinations of the studies have been proposed by

---

* Medical Imaging Division, Jewish Hospital, Louisville, KY, USA.

G. Bebis et al. (Eds.): ISVC 2009, Part I, LNCS 5875, pp. 347–356, 2009.

some investigators since these studies were not specifically designed to assess the effectiveness of chest radiographic screening, but achieved greater survival rates. Renewed interest in lung cancer would come in the 1990s with the development of sophisticated new imaging and non-imaging techniques in combination with lingering questions from the lung screening studies in the 1970s and 1980s. Some of the technologies developed in recent years include low-dose spiral CT (LDCT), digital radiography, advanced sputum analysis and autofluorescence and virtual bronchoscopy. LDCT has played an important technological role in the assessment of patients with clinically proven and suspected bronchogenic carcinoma.

Lung Cancer in the United States account for 30% of all cancer-related deaths, resulting in over 160,000 deaths per year [5], which is more than the annual deaths for colon, breast, prostate, ovarian and pancreatic cancers combined. In 2008, there were 1,437,180 new cases of cancer reported in the United States: (does not include non-melanoma skin cancers), and 565,650 cancer related deaths. The survival of lung cancer is strongly dependent of diagnosis [6][7]. Early detection of lung cancer is the hope for improved survival rate, thus research studies to reach an optimal detection rate is important. Should the use of LDCT scans become a standard clinical practice an automatic way to analyze the scans will lend great benefit for the entire healthcare system; e.g., [8]-[11] and extensive surveys in [12][13].

The generalized framework of Computer-Assisted Diagnosis (CAD) system we use consists of four main steps: 1. Filtering and normalization of the LDCT scans. 2. Segmentation of the lung regions (parenchyma) from the surrounding tissue. 3. Detection of the lung nodules and 4. Classification of the nodules as benign or malignant. The authors of this paper are involved in developing a comprehensive database of nodules that would allow for rigorous analysis of features versus pathology, thus the classification into certain pathology will be possible. This paper will focus on the novel approach for generating non-parametric templates used for nodule detection. The ELCAP dataset [14] is used for nodule design and testing and the sensitivity and specificity of the template matching approach in terms of detection is studied in comparison to our parametric template approach which consists of using circular and semi-circular template for various radii and orientation (in the semi-circular case) in the detection of candidate nodules.Categorization for the various approaches used for automated pulmonary nodule detection since the early 90's are as follows: density-based and model-based approaches. Template matching is one technique for model-based approaches which exploit a priori information of the size, intensity and shape of the nodules. Density-based approaches use the fact that the lung parenchyma has relatively lower CT attenuation (density) than those of the lung nodules, thus they utilize image processing algorithms that rely on multiple thresholding [9], region growing and clustering. The techniques employed for filtering scan artifacts and segmentation of the lung tissue will be briefly mentioned, since the components of the CAD system in Fig. 1 are serial.

This paper is organized as follows: section 2 briefly discussess filtering of the scanning artifacts and segmentation of the lung region from the surrounding tissues; section 3 discusses the novel approach for template modeling and generation of the intelligent nodule templates; section 4 discusses performance evaluation; and section 5 concludes the paper.

2 LDCT Processing

Filtering: Various filtering techniques can be used but three main requirements must be fulfilled before choosing which filtering approach to use: 1. Removal of noise, artifacts, etc. with minimal information loss by preserving object boundaries and detailed structures 2. Sharpening discontinuities for enhancement of morphological definition, and 3. efficiently remove noise in regions of homogeneous physical properties. In our work we used both the Weiner and anisotropic diffusion filters [15][23].

Image Segmentation: This step is important in image analysis and as the name connotes, this consists of separating or dividing the information content in an image (or volume of images) into recognizable classes. It would be a futile effort to survey the approaches developed in the past 50 years, which runs into tens of thousands of papers in the technical literature, [12][13] are just a couple technical surveys the reader can be referred to. The method used in this paper on the ELCAP database is a simplified image segmentation approach that exploits the intensity characteristics of lung CT scans (e.g., [22][23]).

To decrease the sensitivity of the segmentation result to the structuring element diameter, we apply it to the inner and outer lung region contour. After segmentation was completed small nodules attached to the pleural surface were found to no longer exist since these nodules were segmented as not belonging to the lung parenchyma. This operation resulted in 6.5% of the ground truth nodules to be excluded from further experimentations.

3 Novel Nodule Modeling and Simulation

3.1 Pulmonary Nodule Definitions

CT images are analyzed by radiologist during the screening process to locate pulmonary nodules on patient CT images. Small nodules can be over-looked due to several main reasons: nodule characteristics (density, size and location), scanning technique (radiation dosage and slice thickness) and human error. The enhancement of CT imaging with respect to resolution, dose and scanning approaches are behind the increased interest in large scale screening studies, and produce enormous data that has motivated researchers to design fully automated computer-aided diagnosis (CAD) systems for optimum nodule detection (sample work can be found in [9]-[13]). A pulmonary nodule usually has a spherical shape; however, it can be perplexed by surrounding anatomical structures such as vessels and the pleural surface. Nodules, as observed in a CT slice, may appear at various locations in the lung tissues, and may take various size and shape. There is no standard nodule definition in the literature, and no agreement among radiologists about the main features of nodules. Kostis et al. [16] classify nodules into four classes: 1) well-circumscribed where the nodule is located centrally in the lung without being connected to vasculature; 2) vascularized where the nodule has significant connection(s) to the neighboring vessels while located centrally in the lung; 3) pleural tail where the nodule is near the pleural surface, connected by a thin structure; and 4) juxta-pleural where a significant portion of the

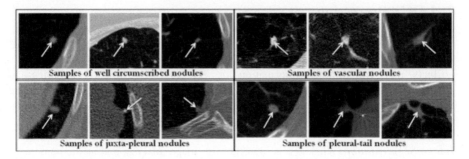

Fig. 1. Nodule types per the classification of Kostis et al [16][17]

nodule is connected to the pleural surface. Fig. 1 illustrates these types of nodules. These definitions will be adopted in this paper and the image analysis methods are developed and tested based on these nodule types.

3.2 Nodule Simulation

In a CT scan the nodules can take various shapes and topologies, but the common characteristic amongst the nodules is the density distribution that tends to be concentrated around a region with an exponential decay. To illustrate this behavior, Fig. 2 shows the image intensity or Hounsfield Units (HU) vs. radial distance for the juxta-pleural nodule type in the ELCAP study. This distance was calculated by summing up the intensity values on concentric circles of various radii centered at the nodules centroid. Fig. 3 shows the average distribution of HU for the juxta-pleural nodule. All nodule types in the ELCAP dataset used in this paper possess the same characteristics of the radial distance which was observed in earlier studies, using different datasets, by Hu et al. (2001) [10] and Farag et al. (2006) [18]. That is, the HU or density decays exponentially with respect to the radial distance from the nodule's centroid.

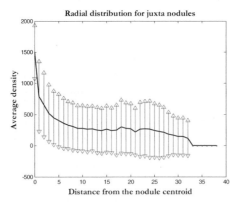

Fig. 2. Plot of the gray level density vs. radial distance from the centroid of the Juxta-Pleural nodules. The bars are one standard deviation off the mean values. This exponential pattern has also been confirmed for the other nodule types.

Furthermore, the decay of the HU is quite significant past a radial distance of 5 pixels. Hence, in designing a nodule template, we may use a bounding box of size 10 pixels (corresponding to physical dimensions of 5mm, which is the range of interest for radiologists). In our experimentations we used templates of size 21x21 pixels.

For parametric templates (Fig. 4), e.g., circular, given the range of nodule density distribution q_{min} and q_{max}, the HU, at a distance r from the centroid, can be estimated by the following equations (e.g., Farag et al. (2005) [19]).

$$q(r) = q_{max}e^{-(r/\rho)^2}, \qquad 0 \le r \le R \qquad (1)$$

$$\rho = R\left(\ln(q_{max}) - \ln(q_{min})\right)^{-1/2} \qquad (2)$$

Where q_{max} and q_{min} are obtained from the density distribution of each nodule type, and R is the radius of the template.

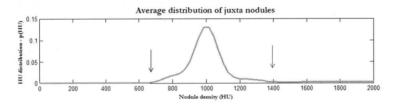

Fig. 3. Probability density of the Radial distance of the Juxta-Pleural nodule; Arrows show q_{min} and q_{max} of the range of densities

Fig. 4. An ensemble of generated circular and semi-circular templates with various orientations

3.3   Statistical Nodule Modeling

The major disadvantage of parametric nodule models is the low sensitivity and unreliable specificity of the detected lung nodules. The detection approach (e.g., template matching) is usually carried out using various template sizes (e.g., diameters of circular templates) and orientations (in case of non-isotropic templates). Optimization approaches, such as genetic algorithms, have been used to carryout the matching process (e.g., Lee et al. (2001) [11]) and Farag et al. (2004) [20]), which reduced the computational time, yet the results are very hard to decipher with respect to the ground truth. Hence, we would like to decouple the nodule model from the detection approach. This will help interpret the performance of the overall nodule detection algorithm in terms of the physics of the problem; i.e., the appearance of typical nodules in LDCT scans.

In the remainder of this section, we describe the nodule design approach from general statistical prospective. First, we created ensembles for each nodule type using the definitions in Kostis et al. [16]. Based on our analysis in the previous section, we chose the bounding box around the location of the nodule to be of size 21x21 pixels

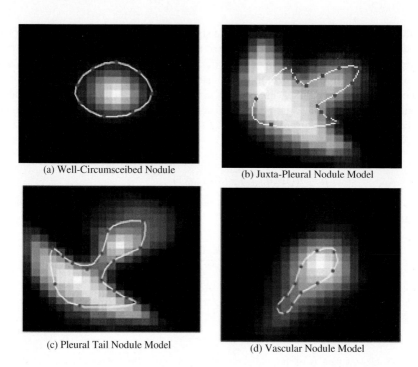

(a) Well-Circumsceibed Nodule

(b) Juxta-Pleural Nodule Model

(c) Pleural Tail Nodule Model

(d) Vascular Nodule Model

Fig. 5. The data-driven nodule models for the well-circumscribed, juxta-pleural, pleural tail and vascular nodule types. These models bear a great similarity to the true nodules.

(this corresponds to spatial resolution of 1.5cm x 1.5 cm; thus covering the basic nodule sizes expected in LDCT scans). The ensemble of nodules contains variations in intensity distribution, shape/structural information and directional variability. The cropped regions inside the bounding boxes will maintain such variations. Second, we co-register these regions and obtain an average nodule, which conveys the statistics of the ensemble. This average nodule is the "intelligent template" which is used in the template matching process. We generate one such template per nodule type. Various statistical methods may be employed in building the template process, including Principle Component Analysis (PCA), mutual information (MI) approach shape-based registration (e.g., Abdelmunim and Farag (2007) [21]) or similar methods.

Fig. 5 shows the resultant nodule model for each nodule type in the ELCAP study. Note that these models possess the major characteristics of the real nodules in terms of shape and texture. This provides the clue for the enhanced performance in the detection process over using parametric nodule models.

In the implementation of the detection process, e.g., using template matching, we could use various orientations of the templates in Fig. 5.

4   Performance Evaluation

To test the effectiveness of the data-driven nodule models (Fig. 5) with respect to the parametric models (Fig. 4), we implemented a basic nodule detection approach using the template matching, in which the nodule model (template) is swept across the scan (2D slices or the 3D volume) in a raster fashion and a similarity measure is calculated between the intensity (or HU) of the template and the region of the CT data underneath. Among the widely used similarity measures is the normalized cross correlation (NCC), which has a maximum value of unity that occurs if and only if the image function under the template exactly matches the template. The normalized cross-correlation of a template, $t(x,y)$ with a sub-image $f(x,y)$ is:

$$NCC = \frac{1}{n-1} \sum_{x,y} \frac{(f(x,y) - \overline{f})(t(x,y) - \overline{t})}{\sigma_f \sigma_t}, \tag{3}$$

where n is the number of pixels in template $t(x,y)$ and sub-image $f(x,y)$ which are normalized by subtracting their means and dividing by their standard deviations.

The probability density functions (pdf) of nodule and non-nodule pixels are computed using the normalized cross correlation coefficients resulted from templates with varying parameters (shape, size and orientation if applicable). Based on the Bayesian classification theory, the intersection between the pdf's of the two classes is chosen as the threshold separating the correlation coefficients resulted from nodule and non-nodule pixels.

Fig. 6 shows the pdf of the NCC for parametric nodule models (left) and the data-driven nodule models (right). Obviously the NCC is more robust with data-driven nodule models, which results in better sensitivity and specificity than the parametric nodule models.

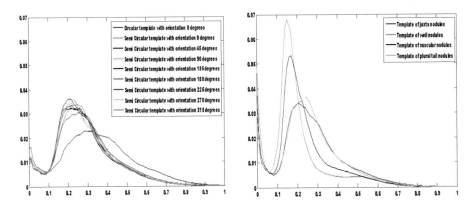

Fig. 6. Distribution of the NCC (i.e., histogram of the NCC values as the templates are swept across the image in a raster fashion) for parametric (left) and data-driven (right) nodule models

Tables 1-7 provide the results of extensive evaluation of nodule detection using the parametric and data-driven nodule models. For parametric models, both size and orientation were examined. For the data-driven models, various orientations were considered. For both nodule model types, nodules were detected using a hard decision rule by setting NCC = 0.5 or above to denote nodules.

Table 1. Performance of data-driven nodule models during template matching

	All no-dule types	Juxta-Pleural	Well-Circumscribed	Plural Tail	Vascular
Sensitivity	85.22%	94.78%	69.66%	95.65%	80.49%
Specificity	86.28%	86.54%	87.10%	83.33%	87.08%

Table 2. Performance of parametric nodule models during template matching for radius size 10 and single orientation angle (0°) for semi-circular models

	All no-dule types	Juxta-Pleural	Well-Circumscribed	Plural Tail	Vascular
Sensitivity	72.16%	83.48%	49.44%	89.13%	70.73%
Specificity	80.95%	79.59%	81.72%	79.33%	84.17%

Table 3. Performance of data-driven nodule models during template matching with changes in nodule models orientation 0°-360° with step size 90°

	All no-dule types	Juxta-Pleural	Well-Circumscribed	Plural Tail	Vascular
Sensitivity	80.07%	93.04%	62.92%	93.48%	65.85%
Specificity	74.89%	73.16%	76.20%	75.53%	75.53%

Table 4. Performance of parametric nodule models during template matching for radius size of 10 and orientation angle for semi-circular models 0°-360° with step size 90° for semi-circular models

	All no-dule types	Juxta-Pleural	Well-Circumscribed	Plural Tail	Vascular
Sensitivity	78.01%	92.17%	51.69%	95.65%	75.61%
Specificity	54.48%	47.61%	61.16%	57.14%	56.19%

Table 5. Performance of parametric templates for templates with radius size from 1-20 circular and radius size 1-20 with orientations 0°-360° w/step size 45° for semi-circular

	All no-dule types	Juxta-Pleural	Well-Circumscribed	Plural Tail	Vascular
Sensitivity	96.22%	95.65%	94.38%	100%	97.56%
Specificity	61.61%	60.53%	63.23%	60.05%	62.33%

Table 6. Performance of parametric templates for template matching with template radius size 10 circular and radius size 10 with orientations 0°-360° w/step size 45° for semi-circular

	All no-dule types	Juxta-Pleural	Well-Circumscribed	Plural Tail	Vascular
Sensitivity	80.41%	93.04%	58.43%	97.83%	73.17%
Specificity	73.18%	71.80%	75.55%	72.81%	72.36%

Table 7. Performance of data-driven nodule models for template matching with orientation $0°$ -$360°$ with step size $45°$

	All no-dule types	Juxta-Pleural	Well-Circumscribed	Plural Tail	Vascular
Sensitivity	87.63%	95.65%	73.03%	97.83%	85.37%
Specificity	63.94%	62.23%	66.34%	61.98%	64.41%

5 Conclusions

In this paper, a data-driven approach was devised to model and simulate typical lung nodules. We studied the effect of template shape on detection of different nodules types. From our extensive experimentation we can concluded that the new data-driven models for template matching yielded an overall higher sensitivity and specificity rate then our previously used parametric templates. In the parametric case where we tested on all radii sizes between 1 and 20 pixels the sensitivity was higher but the specificity in comparison to the data driven nodule templates were still lower.

The overall performance depends on template shape and nodule type. The pleural-tail nodule for both the parametric and non-parametric template matching cases was the most sensitive nodule type, while the well-circumscribed nodule was the least sensitive. The well-circumscribed nodule emphasizes the greatest improvement when our data-driven models were used since the sensitivity rate nearly doubled without increasing the specificity, which in-turn can mean more false positives. Overall, the new data-driven models yielded promising results as shown in tables 1-7 yielding overall high sensitivity and specificity rates for all nodule types using the ELCAP database. Current efforts are directed to constructing and testing the new data-driven modeling approach on a large clinical data and extend this work into the 3D space.

Acknowledgments

This research has been supported by the Kentucky Lung Cancer Program. The first author was sponsored by a NASA graduate fellowship. Assistance of Dr. Asim Ali, Melih Aslan, Shireen Elhabian and Ham Rara at the CVIP-Lab is acknowledged.

References

1. Kloecker, G., et al.: Lung Cancer in the US and in Kentucky. KMA 105, pp. 159–164 (2007)
2. Henschke, C.I., et al.: Early lung cancer Action Project: Overall Design and Findings from baseline Screening. Cancer. 2000 (suppl.) 89(11), 2474–2482 (2000)
3. Boiselle, P., White, C.: Lung Cancer Screening: Past, Present and Future. In: New Techniques in Thoracic Imaging, ch.1
4. Swenson, S.J., et al.: Lung cancer screening with CT: Mayo Clinic experience. Radiology 226, 756–761 (2003)
5. National Institute of Health, http://www.nih.gov
6. Gajra, A., et al.: Impact of tumor size on survival in stage IA non-small cell lung cancer: a case for subdividing stage IA disease. Lung Cancer 42, 51–57 (2003)

7. Zaho, B., Gamsu, G., Ginsberg, M.S., Jiang, L., Schwartz, L.H.: Automatic Detection of small lung nodules on CT utilizing a local density maximum algorithm. Journal of Applied Clinical Medical Physics 4 (2003)
8. Mountain, C.F.: Revisions in the international system for staging lung cancer. Chest 111, 1710–1717 (1997)
9. Armato III, S.G., Giger, M.L., Moran, C.J., Blackburn, J.T., Doi, K., MacMahon, H.: Computerized detection of pulmonary nodules on CT scans. Radio Graphics 19, 1303–1311 (1999)
10. Hu, S., Hoffman, E.A., Reinhardt, J.M.: Automatic lung segmentation for accurate quantitation of volumetric X-ray CT images. IEEE Transactions on Medical Imaging 20, 490–498 (2001)
11. Lee, Y., Hara, T., Fujita, H., Itoh, S., Ishigaki, T.: Automated Detection of Pulmonary Nodules in Helical CT Images Based on an Improved Template-Matching Technique. IEEE Transactions on Medical Imaging 20 (2001)
12. Ginneken, B., Romeny, B., Viergever, M.: Computer-Aided Diagnosis in Chest Radiography: A Survey. IEEE Transactions on Medical Imaging 20 (2001)
13. Sluimer, I., Schilham, A., Prokop, M., van Ginneken, B.: Computer Analysis of Computed Tomography Scans of the Lung: A Survey. IEEE Transactions on Medical Imaging 25(4), 385–405 (2006)
14. ELCAP public lung image database,
    http://www.via.cornell.edu/databases/lungdb.html
15. Grieg, G., Kubler, O., Kikinis, R., Jolesz, F.A.: Nonlinear Anisotropic Filtering of MRI Data. IEEE Transactions on Medical Imaging 11(2), 221–232 (1992)
16. Kostis, W.J., Reeves, A.P., Yankelevitz, D.F., Henschke, C.I.: Three dimensional segmentation and growth-rate estimation of small pulmonary nodules in helical CT images. Medical Imaging IEEE Transactions 22, 1259–1274 (2003)
17. Kostis, W.J., et al.: Small pulmonary nodules: reproducibility of three-dimensional volumetric measurement and estimation of time to follow-up. Radiology 231, 446–452 (2004)
18. Farag, A., El-Baz, A., Gimel'farb, G.L., Falk, R., Abou El-Ghar, M., Eldiasty, T., Elshazly, S.: Appearance Models for Robust Segmentation of Pulmonary Nodules in 3D LDCT Chest Images. In: Larsen, R., Nielsen, M., Sporring, J. (eds.) MICCAI 2006. LNCS, vol. 4190, pp. 662–670. Springer, Heidelberg (2006)
19. Farag, A.A., El-Baz, A., Gimel'farb, G.: Quantitative Nodule Detection in Low Dose Chest CT Scans: New Template Modeling and Evaluation for CAD System Design. In: Proc. of International Conference on Medical Image Computing and Computer-Assisted Intervention (MICCAI 2005), Palm Springs, California, USA, October 26-29, 2005, pp. 720–728 (2005)
20. Farag, A.A., El-Baz, A., Gimelfarb, G., Falk, R., Hushek, S.G.: Automatic detection and recognition of lung abnormalities in helical CT images using deformable templates. In: Barillot, C., Haynor, D.R., Hellier, P. (eds.) MICCAI 2004. LNCS, vol. 3217, pp. 856–864. Springer, Heidelberg (2004)
21. Abdelmunim, H., Farag, A.A.: Shape Representation and Registration using Vector Distance Functions. In: Proc. of IEEE Conference on Computer Vision and Pattern Recognition (CVPR 2007), Minneapolis, MN, June 18-23, pp. Y1–Y8 (2007)
22. Farag, A.A., Elhabian, S.Y., Elshazly, S.A., Farag, A.A.: Quantification of Nodule Detection in Chest CT: A Clinical Investigation Based on the ELCAP Study. In: Proc. of Second International Workshop on Pulmonary Image Processing in conjunction with MICCAI 2009, September 2009, pp. 149–160 (2009)
23. Farag, A.A.: Lung Nodule Modeling and Detection for Computerized Image Analysis of Low Dose CT Imaging of the Chest, Master of Engineering Thesis, CVIP Lab, University of Louisville (April 2009)

Concurrent CT Reconstruction and Visual Analysis Using Hybrid Multi-resolution Raycasting in a Cluster Environment

Steffen Frey, Christoph Müller, Magnus Strengert, and Thomas Ertl

Visualisierungsinstitut der Universität Stuttgart

Abstract. GPU clusters nowadays combine enormous computational resources of GPUs and multi-core CPUs. This paper describes a distributed program architecture that leverages all resources of such a cluster to incrementally reconstruct, segment and render 3D cone beam computer tomography (CT) data with the objective to provide the user with results as quickly as possible at an early stage of the overall computation. As the reconstruction of high-resolution data sets requires a significant amount of time, our system first creates a low-resolution preview volume on the head node of the cluster, which is then incrementally supplemented by high-resolution blocks from the other cluster nodes using our multi-resolution renderer. It is further used for graphically choosing reconstruction priority and render modes of sub-volume blocks. The cluster nodes use their GPUs to reconstruct and render sub-volume blocks, while their multi-core CPUs are used to segment already available blocks.

1 Introduction

CT scanners using modern flat-panel X-ray detectors are popular in industrial applications. They are capable of acquiring a set of high-resolution 2D X-ray images from a huge number of different angles at rapid pace. However, the reconstruction of a volumetric data set on a Cartesian grid from these images is very time consuming as the commonly used reconstruction method by Feldkamp et al. [1] has a runtime complexity of $O(N^4)$. Subsequently, oftentimes a computationally also expensive segmentation algorithm is run to support analysis which in total results in a long delay until the examination can be started.

In this work, we focus on industrial applications, where engineers require a high-resolution reconstruction and segmentation, while it is often critical to have the results of e. g. a non-destructive quality test at hand as early as possible. We therefore propose to distribute the reconstruction and segmentation processes on a cluster equipped with CUDA-enabled GPUs and multi-core CPUs. Employing our hybrid mult-resolution renderer, finished full-resolution parts are successively displayed in the context of a low-resolution preview volume. The low-resolution data set can be created quickly by a single GPU within several seconds on the front-end node. It is also used for prioritising blocks for reconstruction and render-mode selection. The full-resolution volume is progressively created by back-end nodes, which also segment and render their respective blocks.

G. Bebis et al. (Eds.): ISVC 2009, Part I, LNCS 5875, pp. 357–366, 2009.

2   Related Work

In this work, we use the reconstruction algorithm for 3D cone beam computer tomography that was developed by Feldkamp et al. [1]. Turbell [2] gives
extensive and detailed overview over varitions of this method, as well as fundamentally different approaches for CT reconstruction. The use of graphics hardware for computer tomography was first investigated by Cabral et al. [3] on
non-programmable, fixed function SGI workstations. Xu et al. [4] introduced
a framework that implements the Feldkamp algorithm using shaders. Scherl et
al. [5] presented a comparison between their Cell and a CUDA implementation.

Distributed volume rendering has been investigated for a long period of time
and a magnitude of publications can be found on this issue. Most of the existing
systems fit either into the *sort-first* or *sort-last* category according to Molnar
et al.'s classification [6]. Recent systems use GPU-based raycasting [7] with a
single rendering pass [8] since GPUs support dynamic flow control in fragment
programs. Dynamic load balancing issues in such systems have been addressed
by Wang et al. [9] using a hierarchical space-filling curve as well as by Marchesin
et al. [10] and Müller et al. [11], who both use a kd-tree in order to dynamically
reorganise the data distribution in a cluster.

Multiresolution rendering is an LOD approach enabling the interactive visualisation of large data sets on a GPU. Different data representations are used
depending on various parameters (e. g. the view point or estimated screen-space
error) resulting in a quality/performance trade-off. Most frequently, tree data
structures are used in combination with raycasting [12]. Ljung et al. [13] address the interpolation between blocks with different LOD in detail, while Guthe
et al. [14] employ texture-based volume rendering using a compressed hierarchical wavelet representation, which is decompressed on-the-fly during rendering.

The analysis of industrial workpieces using the segmentation of CT data was
discussed by Heinzl [15], and the integration of segmentation information in a
raycaster was discussed by Bullitt and Aylward [16] in a medical context.

3   Architecture Overview

Our reconstruction and visualisation system consists of two classes of nodes:
a single front-end node and one or more back-end nodes. The front-end node
exposes the user interface and displays the volume rendering of the data set
that is being reconstructed. Furthermore, it allows the user to influence the
order of reconstruction and the rendering mode of high-resolution sub-volume
blocks on the back-end nodes. The back-end nodes in turn perform the actual
reconstruction and segmentation of the high-resolution sub-volume blocks and
additionally render images of these blocks, which are used by the front-end node
to generate the final rendering.

Figure 1 illustrates how the distributed program is controlled by the user.
The first step is to provide the input, most importantly the X-ray images. Our
approach is not limited to reading the projection images from disk, but could also

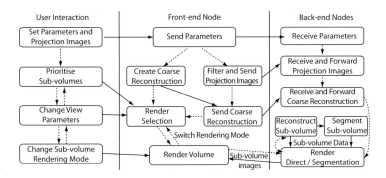

Fig. 1. Control (dashed) and data flow between the threads and processes of the system and the user. The back-end process may run simultaneously on multiple nodes.

handle images being streamed directly from the scanner while performing the incremental reconstruction. We do not elaborate further on that in the following as the time for image acquisition is in the order of minutes while high-resolution reconstruction can take hours hours in our field of application. Besides, industrial CT scanners often perform modifications (e. g. center displacement correction) after the actual acquisition.

While distributing the parameter set is completed quickly using a single synchronous broadcast operation, scattering the projection images can take a long time due to disk I/O and network bandwidth limitations. We address this problem firstly by distributing the images asynchronously while the front-end node completes a low-resolution preview reconstruction, and secondly by daisy-chaining the back-end nodes for the distribution process. This approach avoids unnecessary disk load on the front-end and the back-end nodes, which is the bottleneck in case of a high-speed interconnect like InfiniBand. Additionally, projection images are kept in main memory on the back-end nodes as long as possible to avoid I/O slowing down the reconstruction process. They are only swapped to disk if the system comes under memory pressure.

As the reconstruction of the coarse preview volume usually completes by far earlier than the distribution of the projection images, the user can start investigating the data set at an early state of the overall procedure. The preview rendering can be superimposed by a grid showing the sub-volume blocks that are reconstructed on the back-end nodes, which allows the user to specify the order of reconstruction that is then communicated from the front-end node to the back-end nodes. The visualisation of the reconstructed data set is compiled and presented on the front-end node. At the beginning, the low-resolution preview reconstruction volume data is exclusively used for local rendering. As a detailed sub-volume block has been reconstructed on a back-end node, it is rendered on the same node and included in the final rendering on the front-end node as a kind of pre-integrated sub-volume block using our hybrid raycasting and compositing approach. Additionally, the sub-volume is queued for segmentation on the multicore CPU of the respective node.

4 Implementation Details

Our system consists of three basic modules: the CT reconstruction module (Section 4.1) uses a CUDA implementation of the Feldkamp algorithm. The segmentation module (Section 4.2) is a fully automatic 3D flood-fill variant designed for distributed operation. The volume rendering module (Section 4.3) is used for image generation on the front-end and back-end nodes. Additionally, there is a sub-volume selection and picking rendering mode. It displays the coarse volume and allows the precise selection of sub-volume blocks by mouse click for moving these in the reconstruction priority queue or for choosing the visualisation mode.

4.1 Reconstruction of Sub-volume Blocks

The Feldkamp cone beam reconstruction algorithm works for industrial CT scanners which move the X-ray source on a circular trajectory shooting rays on a detector, which diverge as a cone through the object of interest. The algorithm can be subdivided into two phases: the preparation of the projection images and their subsequent depth-weighted backprojection into the volume. The preparation consists of weighting and filtering each image with a filter kernel derived from a ramp filter. The computationally most expensive part of the reconstruction is the backprojection, on which we will concentrate in the following. It is commonly implemented by determining for each volume element which projection image value it corresponds to by projecting it along the X-ray from the source to the detector. The depth-weighted sum of the respective pixels from all projection images yields the reconstructed voxel value.

This can be accomplished – even for large data sets as we focus on – by only considering one sub-volume block for reconstruction at a time such that just subsets of the projection images are needed. The dimensions of the sub-volumes are determined in a preprocessing step to cover the volume with a minimal amount of blocks considering the graphics memory available.

All projection images are cropped and stored in a single container texture, similar to the storage of renderings for the front-end raycaster (Section 4.3). The coordinates to access this texture for each voxel and every projection image are computed by projecting the eight corners of each sub-volume along the X-rays on the detector plane. These coordinates are subsequently linearly interpolated on the CPU to get the coordinate values $\boldsymbol{p}_{xyw}$ for a considered slice. Afterwards, the window position $\boldsymbol{w}_{xy}$ of the sub-image with respect to the whole image and its coordinates $\boldsymbol{i}_{xy}$ in the container texture need to be applied to the projection coordinates $\boldsymbol{p}_{xy}$. Further, $\boldsymbol{p}_{xy}$ needs to be weighted with the projective component $\boldsymbol{p}_w$ to yield $\boldsymbol{q}_{xyw}$ that is uploaded to a texture: $\boldsymbol{q}_{xy} = \boldsymbol{p}_{xy} + \boldsymbol{p}_w(\boldsymbol{i} - \boldsymbol{w}); \boldsymbol{q}_w = \boldsymbol{p}_w$. Weighting with $\boldsymbol{p}_w$ is required to counter the effect of the projective division $\boldsymbol{c} = (\frac{\boldsymbol{p}_x}{\boldsymbol{p}_w}, \frac{\boldsymbol{p}_y}{\boldsymbol{p}_w})^T$ that takes place after the bilinear interpolation on the GPU. This finally yields the coordinates $\boldsymbol{c}$ for accessing the value of one projection image in the container texture that is backprojected on the considered voxel.

4.2   Volume Segmentation

For segmentation, we use a fully automatic 3D flood fill variant that leverages a multi-core CPU. While the GPUs are reconstructing sub-volume blocks, multiple CPU threads grow regions around a user-defined number of randomly distributed seed points in the already completed blocks. The decision on whether to add a voxel to a region is based on a user-defined threshold (e. g. the maximum range of values allowed in a region) and a gradient-based criterion.

Volume segmentation also requires some communication – between different threads and nodes – due to the fact that the algorithm must merge segments that have been created on different CPU cores or cluster machines. For merging two sub-volumes' regions that belong to different cluster nodes, the region IDs at the face they meet always has to be transmitted. Additional data needs to be transferred depending on the region criteria chosen, e. g., for a gradient-based method using central differences, it suffices to transmit a sub-volume face while value-based methods require extra region information to combine the regions.

4.3   Hybrid Multiresolution Volume Rendering Incorporating Sub-volume Blocks

Each back-end node renders images of the high-resolution sub-volume blocks it has reconstructed and segmented on behalf of the front-end node. The renderer only raycasts the pixels that lie within the image-space footprint of the current sub-volume with respect to the camera parameters transmitted by the head node. As the reconstruction of a sub-volume can be interrupted by rendering requests, the renderer must be able to handle sub-volumes for which high-resolution images are only available up to a certain part. It therefore substitutes the missing high-resolution slices with coarse volume data that has been reconstructed by the front-end node during the initialisation phase. In order to avoid dynamic branching on the GPU and to achieve more efficient texture fetching, the coarse volume data of resolution l is appended to the texture of high resolution h. Thus, the sampling coordinates s on the ray must be scaled to yield the texture coordinates c for the coarse volume past the boundary b in z-direction to low-resolution data: $c = (s_x \cdot \tau_x, s_y \cdot \tau_y, (s_z - b) \cdot \tau_z + b)^T$ with $\tau = (\frac{l_x}{h_x}, \frac{l_y}{h_y}, \frac{l_z}{h_z})^T$ for $s_z > b$ and $\tau = (1, 1, 1)^T$ otherwise (note that $b = h_z$ for a completed block).

The renderer on the front-end node combines high-resolution imagery from the back-end nodes with the coarse volume data that are available on this node into a volume rendering by integrating compositing in the raycasting loop of the front-end node. High-resolution images are raycasted by the front-end node as a kind of pre-integrated voxels. All pre-rendered images are stored in one colour texture on the graphics card, similar to the container texture of the reconstruction algorithm. Images are placed next to each other until the end of the texture is reached and then a new row of images is started at the base level of the tallest image of the previous row (Figure 2).

The information on whether a high-resolution rendering for a sub-volume exists respectively the coordinates to access it are uploaded to the graphics

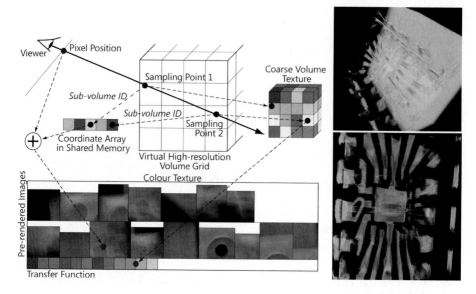

Fig. 2. Left: When rendering the final image, it is determined for each sampling point whether there is pre-rendered high-resolution data available (sampling point 1) or not (sampling point 2). As the case may be, different coordinates are used to access the colour map. Right: Renderings of the IC datset resulting from this technique.

card's shared memory for efficient access. Two 16 bit integers per sub-volume are utilised to determine the texture coordinates of a pre-rendered pixel for the sample point of a ray. These coordinates have already been pre-modified such that the pixel position of a ray only has to be added to fetch a pre-rendered pixel. Due to the use of shared memory and the overloading of the colour map access, no actual branching is required and the amount of texture memory accesses in the sampling loop is the same as of a standard single pass raycaster: one volume texture fetch requesting a scalar density value and one colour transfer texture lookup retrieving a 4D vector. Yet, the colour texture lookup is here also used for accessing a rendered high-resolution image, depending on the sub-volume the respective sample is in.

4.4   Communication and Data Exchange

The communication patterns of our application differ in two different phases: At the beginning, the input parameters and projection images are distributed synchronously, the latter in a daisy-chain from one back-end node to the other. This alleviates disk I/O load on the front-end node, which initially stores the input data, and introduces only a small latency in the distribution pipeline.

After initialisation, both node classes enter a message loop for asynchronous communication. For the front-end node, this is equivalent to the message loop of the window system, which starts as soon as the preview volume has been reconstructed. The back-end nodes have a second message loop for the communication of the segmentation subsystem in order to decouple this process completely from the reconstruction and rendering tasks. The message handling of communication with the front-end can interrupt the reconstruction process, which is running whenever no more important requests must be served, e. g. requests for high-resoluting renderings. It may therefore take a significant amount of time from issuing a rendering request until all sub-images are available. Hence, images are received asynchronously and replace parts of the local preview rendering as they come in.

The assignment of reconstruction tasks to back-end nodes is carried out by the front-end node as it must be possible to re-prioritise blocks on user request. Load-balancing is implicitly achieved by the back-end nodes polling for new tasks once they have completed a sub-volume block.

5   Results

We tested our system on an eight node GPU cluster with an InfiniBand interconnect. Each node was equipped with an Intel Core2 Quad CPU, 4 GB of RAM, an NVIDIA GeForce 8800GTX and a commodity SATA hard-disk. One node acted as front-end creating a 256^3 voxel preview volume, while the remaining ones reconstructed a 1024^3 volume from 720 32-bit X-ray images with a resolution of 1024^2 pixels (Figure 4 shows the volumes). The calculated sub-volume size was 352^3 resulting in a total of 27 sub-volumes.

The time from program start until the preview volume is reconstructed on the frontend-node and rendered is around 29 s, of which the most part is required for I/O caused by projection image downsampling that runs in parallel with the data distribution, while the actual reconstruction on the GPU takes only 1.3 s. The determination of the sub-volume dimension that takes a few seconds runs concurrently. Figure 3 (left) shows the data distribution and reconstruction times measured on the front-end and the back-end nodes. The times for the front-end node also include communication overhead and show the span between program start and the availability of the complete high-resolution volume. In contrast, the numbers for the back-end nodes comprise only the longest computation. So although the average reconstruction time on the back-end nodes quickly decreases with an increasing number of nodes, the observed time on the front-end node declines more slowly, because this timing includes the input distribution and other communication. Input distribution takes slightly longer the more nodes are involved, because the measurement on the front-end node includes the time from reading the file from disk until the last node received the data. Thus the last node in the daisy chain must wait longer for its data.

The rendering times depicted in Figure 3 (left) indicate the time span between the moment the front-end node requests new sub-volume images and the moment the first respectively the last remotely generated image is used in the visualisation. Images from the back-end nodes can either be sent in batches or as separate messages. In our measurements, we let the batch requests – in

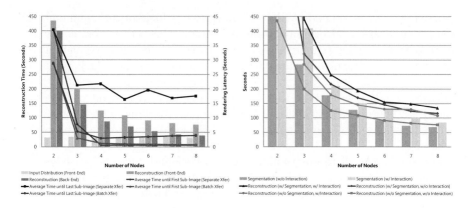

Fig. 3. Left: Timing results for several cluster configurations. The bar chart shows initialisation and reconstruction times on the front-end and back-end nodes (units on the left), while the lines show the average time from a remote rendering request until the reception of the first/last sub-volume image on the front-end node (units on the right). Right: Mutual influence of reconstruction, segmentation and user interaction during the segmentation. The results for only one back-end node are clamped for clarity.

contrast to the separate requests – interrupt the reconstruction not only after a sub-volume block but already after a sub-volume slice has been completed. For the reconstruction, this means that after rendering projection images have to be re-uploaded to the graphics card, resulting in a slightly worse rendering performance than when interrupts are prohibited. For the separate transfer of sub-images there is only little latency between the request and display of an image on the front-end node which gives the impression of a more fluid interaction, while in our test setup the time until the last image is received on the head node is much longer. This is partly due to the decrease in network throughput and the potential interruption of the image transfer by higher priority messages. But more significantly, a node potentially has to wait the reconstruction time of up to 14 s for a sub-volume block to be completed by the reconstruction until rendering can be started. This happens more often the more nodes are involved and hinders scaling with the cluster size of the average latency until the last image has been received. Please note, however, that the system always remains responsive as it can use the coarse volume data that is available from the beginning.

Figure 3 (right) points out the mutual influence of the reconstruction, the segmentation and user interaction during the reconstruction phase. Frequently interrupting the reconstruction by manipulating the scene increases overall reconstruction time and subsequently also segmentation time. The latter is caused by the fact that all sub-volumes have to be read for rendering from disk resulting in reduced I/O performance of the segmentation threads. The same holds true for reconstruction performance, but in this case both computations additionally conflict in the usage of the GPU.

When reconstructing a 2048^3 volume from 1440×2048^2 projection images on eight nodes, data distribution takes ~ 13.5 min due to the considerable need for

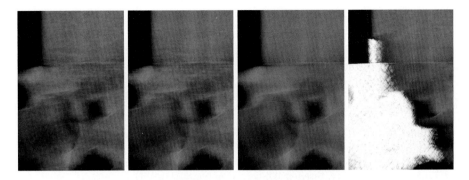

Fig. 4. From left to right: The coarse 256^3 volume, a partially reconstructed volume, the fully reconstructed high-resolution 1024^3 volume and an intermediate state when the left part of the volume has already been segmented

disk swapping already in this first phase. The reconstruction needs $\sim 103\,\mathrm{min}$, which is about 84 times longer than reconstructing the 1024^3 volume. The scaling behaviour is severly hindered by I/O performance in this case due to the excessive need for data swapping, which affects primarily the storage/access of the input data (23 GB versus 4 GB of RAM). Sub-volume sizes are further limited to 256^3 due to GPU memory restrictions, which requires a total of 512 sub-volumes to cover the complete volume – resulting in 19 times more accesses to projection images and sub-volumes. This significantly adds to the permanent memory pressure as the ratio of projection images and sub-volumes that can be stored in memory is already eight times worse in comparison to the 1024^3 case.

6   Conclusion and Future Work

We introduced a distributed software architecture for the 3D reconstruction of computer tomography data sets while continuously visualising and segmenting. Exploiting the computational power of GPUs, our system provides a fast preliminary reconstruction and visualisation, which allows for prioritising interesting sub-volume blocks. While the reconstruction is in progress, the visualisation is continuously kept up-to-date by integrating all available high-resolution blocks immediately into the rendering. For that, we use a hybrid CUDA-based volume raycaster that can replace low resolution blocks with pre-rendered images. We also leverage multi-core CPUs for parallel volume segmentation, which is often utilized for advanced processing steps as well as the support of visual analysis. As this is executed in parallel with the reconstruction on the GPU, the user can switch the render mode to segmentation on a per block base display shortly after the sub-volume has been reconstructed. In our tests, we could see a 256^3 preview of a 1024^3 data set after 30 seconds and the full resolution volume after less than three minutes. A large 2048^3 volume could be reconstructed in 103 minutes, but the process was hampered by the limited I/O performance and GPU and main memory. Testing our system with a bigger cluster and a faster distributed filesystem therefore remains for future work. Our architecture could be further extended by (semi-) automatically assisting the user during the sub-volume selection by pointing out regions that could be of high interest, e. g. by identifying areas with large gradients in the coarse volume.

References

1. Feldkamp, L.A., Davis, L.C., Kress, J.W.: Practical cone-beam algorithm. J. Opt. Soc. Am. 1, 612–619 (1984)
2. Turbell, H.: Cone-Beam Reconstruction Using Filtered Backprojection. PhD thesis, Linköping University, Sweden, Dissertation No. 672 (2001)
3. Cabral, B., Cam, N., Foran, J.: Accelerated Volume Rendering and Tomographic Reconstruction using Texture Mapping Hardware. In: Proceedings of the Symposium on Volume Visualization, pp. 91–98 (1994)
4. Xu, F., Mueller, K.: Accelerating popular tomographic reconstruction algorithms on commodity pc graphics hardware. IEEE Transactions on Nuclear Science, 654–663 (2005)
5. Scherl, H., Keck, B., Kowarschik, M., Hornegger, J.: Fast gpu-based ct reconstruction using the common unified device architecture (cuda). SIAM Journal of Applied Mathematics (2007)
6. Molnar, S., Cox, M., Ellsworth, D., Fuchs, H.: A sorting classification of parallel rendering. IEEE Computer Graphics and Applications 14, 23–32 (1994)
7. Krüger, J., Westermann, R.: Acceleration Techniques for GPU-based Volume Rendering. In: Proceedings of IEEE Visualization 2003, pp. 287–292 (2003)
8. Stegmaier, S., Strengert, M., Klein, T., Ertl, T.: A Simple and Flexible Volume Rendering Framework for Graphics-Hardware–based Raycasting. In: Proceedings of the International Workshop on Volume Graphics 2005, pp. 187–195 (2005)
9. Wang, C., Gao, J., Shen, H.W.: Parallel Multiresolution Volume Rendering of Large Data Sets with Error-Guided Load Balancing. In: Eurographics Symposium on Parallel Graphics and Visualization, pp. 23–30 (2004)
10. Marchesin, S., Mongenet, C., Dischler, J.M.: Dynamic Load Balancing for Parallel Volume Rendering. In: Eurographics Symposium on Parallel Graphics and Visualization, Eurographics Association, pp. 51–58 (2006)
11. Müller, C., Strengert, M., Ertl, T.: Optimized Volume Raycasting for Graphics-Hardware-based Cluster Systems. In: Eurographics Symposium on Parallel Graphics and Visualization, Eurographics Association, pp. 59–66 (2006)
12. Crassin, C., Neyret, F., Lefebvre, S., Eisemann, E.: Gigavoxels: Ray-guided streaming for efficient and detailed voxel rendering. In: ACM SIGGRAPH Symposium on Interactive 3D Graphics and Games (I3D), Boston, MA, Etats-Unis. ACM Press, New York (to appear, 2009)
13. Ljung, P., Lundström, C., Ynnerman, A.: Multiresolution interblock interpolation in direct volume rendering. In: Santos, B.S., Ertl, T., Joy, K.I. (eds.) EuroVis, pp. 259–266. Eurographics Association (2006)
14. Guthe, S., Strasser, W.: Advanced techniques for high quality multiresolution volume rendering. In: Computers and Graphics, pp. 51–58. Elsevier Science, Amsterdam (2004)
15. Heinzl, C.: Analysis and visualization of industrial ct data (2009)
16. Bullitt, E., Aylward, S.R.: Volume rendering of segmented image objects. IEEE Transactions on Medical Imaging 21, 200–202 (2002)

Randomized Tree Ensembles for Object Detection in Computational Pathology

Thomas J. Fuchs[1,4], Johannes Haybaeck[2], Peter J. Wild[3], Mathias Heikenwalder[2], Holger Moch[3,4], Adriano Aguzzi[2], and Joachim M. Buhmann[1,4]

[1] Department of Computer Science, ETH Zürich, Zürich, Switzerland
{thomas.fuchs,jbuhmann}@inf.ethz.ch
[2] Department of Pathology, Institute for Neuropathology, University Hospital, Zürich
[3] Department of Pathology, Institute of Surgical Pathology, University Hospital Zürich
[4] Competence Center for Systems Physiology and Metabolic Diseases, ETH Zürich

Abstract. Modern pathology broadly searches for biomarkers which are predictive for the survival of patients or the progression of cancer. Due to the lack of robust analysis algorithms this work is still performed manually by estimating staining on whole slides or tissue microarrays (TMA). Therefore, the design of decision support systems which can automate cancer diagnosis as well as objectify it pose a highly challenging problem for the medical imaging community.

In this paper we propose Relational Detection Forests (RDF) as a novel object detection algorithm, which can be applied in an off-the-shelf manner to a large variety of tasks. The contributions of this work are twofold: (i) we describe a feature set which is able to capture shape information as well as local context. Furthermore, the feature set is guaranteed to be generally applicable due to its high flexibility. (ii) we present an ensemble learning algorithm based on randomized trees, which can cope with exceptionally high dimensional feature spaces in an efficient manner. Contrary to classical approaches, subspaces are not split based on thresholds but by learning relations between features.

The algorithm is validated on tissue from 133 human clear cell renal cell carcinoma patients (ccRCC) and on murine liver samples of eight mice. On both species RDFs compared favorably to state of the art methods and approaches the detection accuracy of trained pathologists.

1 Introduction

Clear Cell Renal Cell Carcinoma: Renal cell carcinoma (RCC) is one of the ten most frequent malignancies in Western societies and can be diagnosed by histological tissue analysis. Current diagnostic rules rely on exact counts of cancerous cell nuclei which are manually counted by pathologists. The prognosis of renal cancer is poor since many patients suffer already from metastases at first diagnosis. The identification of biomarkers for prediction of prognosis (prognostic marker) or response to therapy (predictive marker) is therefore of utmost importance to improve patient prognosis. Various prognostic markers have been suggested in the past, but conventional estimation of morphological parameters is still most useful for therapeutical decisions. Clear cell RCC (ccRCC) is the most common subtype of renal cancer and it is composed of cells

G. Bebis et al. (Eds.): ISVC 2009, Part I, LNCS 5875, pp. 367–378, 2009.

with clear cytoplasm and typical vessel architecture. ccRCC shows an architecturally diverse histological structure, with solid, alveolar and acinar patterns. The carcinomas typically contain a regular network of small thin-walled blood vessels, a diagnostically helpful characteristic of this tumor.

Murine Tissue Specimen Analysis: Inflammatory disorders of the liver can be classified by histological analysis where exact counts of proliferating organ specific cells is important for various studies. We propose a completely automated image analysis pipeline to count cell nuclei that are indicated by a proliferation marker (MIB-1) based on the analysis of immunohistochemical staining of mouse tissues. Most laboratories that are dealing with evaluation of immunohistochemically stained tissue specimens are confronted with very tedious, time consuming and thereby prone to error analysis. Current image analysis software requires extensive user interaction to properly identify cell populations, to select regions of interest for scoring, to optimize analysis parameters, and to organize the resulting raw data. Due to these facts in current software, typically pathologists manually assign a composite staining score for each spot during many microscopy sessions over a period of several days. Manual scoring also introduces a possible bias when investigations are not performed on the same day or for too many hours in one session.

Motivation: The absence of an objective ground truth requires to generate a gold standard by combining the knowledge of expert pathologists and computerized systems enabling scientists to count cells in a high throughput fashion. These counts are indispensable for the training of classifiers, for their validation and it is highly non-trivial to acquire them from a technical as well as statistical viewpoint. To facilitate the labelling procedure for trained pathologists, we developed dedicated labelling tools for tissue specimens. The software enables the user to view single tissue areas and it provides zooming and scrolling capabilities. It is possible to annotate the image with vectorial data in SVG (support vector graphics) format and to mark cell nuclei of different cell types which are recognized by their shape, vessels and other biological structures. An additional requirement to the software is the usability on a tablet PC so that a scientist/pathologist can perform all operations with a pen alone in a simple an efficient manner. In the domain of cytology, especially blood analysis and smears, automated analysis is already established [1]. The big difference to histological tissues is the homogeneous background on which the cells are clearly distinguishable and the absence of vessels and connection tissue. The isolation of cells simplifies the detection and segmentation process of the cells significantly. A similar simplification can be seen in the field of immunofluorescence imaging [2]. Only the advent of high resolution scanning technologies in recent years made it possible to consider an automated analysis of histological slices. Cutting-edge scanners are now able to scan slices with resolution, comparable to a $40\times$ lens on a light microscope. In addition the automated scanning of staples of slices enables an analysis in a high throughput manner.

2   Methods

2.1   Tissue Preparation and Scanning

Murine Tissue: The tissue blocks were generated in a trial at the Department of Pathology from the University Hospital Zürich. Murine hepatic tissue from independent experiments were formalin fixed and paraffin embedded. Sections were cut at a thickness of $2\mu m$, stained with the MIB-1 (Ki-67) antigen and stored at 4° Celsius till use. Slices from the murine tissue block and the RCC TMA were scanned on a Nanozoomer C9600 virtual slide light microscope scanner from HAMAMATSU. The magnification of $40\times$ resulted in a per pixel resolution of $0.23\mu m$. Finally 11 patches of size 2000×2000 pixel were randomly sampled from whole tissue slides of each of the 8 mice.

Human Renal Cell Carcinoma: The tissue microarray block was generated in a trial from the University Hospital Zürich. The TMA slides were immunohistochemically stained with the MIB-1 (Ki-67) antigen. Scanning was performed as above and the tissue microarry was tiled into 133 single spots of size 3000×3000 pixel, representing one patient each.

2.2   Voronoi Sampling

It is common in object detection frameworks [3,4,5] to randomly choose negative background samples. Therefore first, points are uniformly sampled from an interval for each coordinate and second, points which are closer to a positive sample than a predefined threshold are discarded. On images with dense packing of multiple object, as it is the case in most cell detection tasks, this procedure leads to two main problems: (i) Uniform sampling with rejection results on one hand in a higher percentage of negative samples in areas with sparse object distribution and on the other hand in very few samples in dense packed areas. Hence negative samples are lacking especially there, where differentiation between object and background is difficult. (ii) Without specific presorting or local sampling algorithms, uniform sampling can be cumbersome and slow in domains with dens packed objects.

To overcome these drawbacks we propose a simple sampling algorithm which we term *Voronoi Sampling*. A Voronoi diagram or Dirichlet tessellation is a decomposition of a metric space determined by distances to a specified discrete set of objects in this space. In a multiple object detection scenario these mathematical objects are the centers of the queried objects.

To create a Voronoi diagram first a Delaunay triangulation is constructed, second the circumcircle centers of the triangles is determined, and third these points are connected according to the neighborhood relations between the triangles. In the experiments in Section 3 the Voronoi tessellation is created based on the joint set of nuclei from both pathologists to prevent the use of good and ambiguous nuclei as negative instances for learning. In contrast to uniform sampling, using a tessellation has the additional advantage that the negative samples are concentrated on the area of tissue and few samples are spent on the homogeneous background. Figure 1 shows a comparison of Voronoi and uniform rejection sampling.

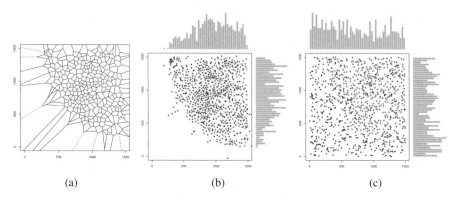

(a)	(b)	(c)

Fig. 1. Voronoi Sampling. (a) Voronoi tessellation of the input space based on labeled nuclei. (b) Sampling negative instances from nodes of the Voronoi tessellation (red dots). Cell nuclei are marked with blue crosses. (c) Uniform rejection sampling in the whole input space. The marginal histograms show the frequency of negative samples. In contrast to uniform sampling, using a tessellation has the additional advantage that negative samples are concentrated on the area of tissue and few samples are spent on the homogeneous background.

2.3   Feature Space

In this framework we consider the following simple feature base. The discrete coordinates of two rectangles R_1 and R_2 are sampled uniformly within a predefined window size w:

$$R_i = \{c_{x1}, c_{y1}, c_{x2}, c_{y2}\}, \quad c_i \sim U(x|0, w)$$

For each rectangle the intensities of the underlying gray scale image are summed up and normalized by the area of the rectangle. The feature $f_{R_1,R_2}(s)$ evaluates to a boolean value by comparing these quantities:

$$f_{R_1,R_2}(s) = I\left(\frac{\sum_{i|x_i \in R_1} x_i}{\sum_{i|x_i \in R_1} 1} < \frac{\sum_{i|x_i \in R_2} x_i}{\sum_{i|x_i \in R_2} 1}\right),$$

where x_i is the gray value intensity of pixel i of sample $s = \{x_1, x_2, \ldots, x_n\}$. From a general point of view this definition is similar to generalized Haar features except the fact that not the difference between rectangles is calculated but the relation between them. For example, in the validation experiments a window size of 65×65 pixels was chosen. Taking into account that rectangles are flipping invariant this results in $\left((64^4)/4\right)^2 \approx 2 \cdot 10^{13}$ possible features.

Putting this into perspective, the restriction of the detector to windows of size 24×24 leads to $\sim 6.9 \cdot 10^9$ features which are significantly more than the $45,396$ Haar features from classical object detection approaches [3].

For such huge feature spaces it is currently not possible to exhaustively evaluate all features while training a classifier. Approaches like AdaBoost [6] which yield very good results for up to hundreds of thousands of features are not applicable any more. In Section 2.4 we propose a randomized approach to overcome these problems.

Considering relations between rectangles instead of cut-offs on differences leads to a number of benefits:

Illumination Invariance: A major problem in multiple object detection in general and on microscopic images of human tissue in special are vast differences of illumination within a single image. In natural images this can be due to shadows or fog and in histopathology it is mainly due to varying thickness of the slide or imperfect staining. Taking into account only the relation between rectangles lead to illumination invariant features which give the same response for high and low contrast patches as long as the shape of the object is preserved. It has to be noted, that due to the directionality of the relation they fail if the image is inverted. In general, illumination invariance speeds up the whole analysis process because neither image normalization nor histogram equalization are required.

Fast Evaluation: The most time intensive step in the induction of tree classifiers is the repeated evaluation of features. In the classical textbook approach as followed by [7,6,3] at each node all possible cut-offs on feature values are evaluated for every feature exhaustively. and testing each intermediate split. In the worst case the number of feature evaluations is as high as the number of samples at a given node. Recently [8] proposed extremely randomized trees to overcome this problem. In their approach only a small number of cut-offs is randomly sampled to reduce the number of tests. Instead of ordering all feature values only the maximum and minimum have to be calculated and the cut-offs are drawn uniformly. The only assumption necessary for this approach is, that the feature values for all samples increase linearly. Given an exponential or logarithmic behavior a low number of sampled cut-offs are not able to capture the power of the feature.

In contrast to that the proposed features require only one single evaluation. In practice this allows for testing hundreds or thousands of features per split with the same number of CPU cycles classical approaches needed to test all thresholds of a single feature. Analyzing single features, the use of relations instead of cut-offs on differences seems to lead to lower variance and higher bias.

Ensemble Diversity: One of the main concerns in ensemble learning is to induce sufficiently diverse base classifiers. Increased diversity leads to increased performance. Randomized algorithms as described in Section 2.4 can benefit from the enormous size of the described feature space. Leaning trees by randomly selecting features from this large set guarantees small similarity between base classifiers. Even if thousands of features are evaluated at each split, the chances that two trees are built of the same features are negligible.

2.4   Tree Induction

The base learners for the ensemble are binary decision trees, designed to take advantage of large feature spaces as described in Section 2.3. Tree learning follows loosely the original approach for random forests described in [7]. A recursive formulation of the learning algorithm is given in procedure `LearnTree`. The sub procedure `SampleFeature` returns a feature consisting of two rectangles uniformly sampled within a predefined window as described in Section 2.3. In accordance with [7] the

---

Procedure `LearnTree`

    Input: set of samples $S = \{s_1, s_2, \ldots, s_n\}$

    Input: depth d; max depth d_{max}; features to sample $mTry$

1 **Init**: $\widehat{label} = null$; $g = -\inf$; $N_{left} = null$; $N_{right} = null$

2 **if** $(d = d_{max})$ OR $(\texttt{isPure}(S))$ **then**

3      $\widehat{label} = \begin{cases} T \textbf{ if } & |\{s_j = T\}| > |\{s_j = F\}|; \quad j = 1, \ldots, |S| \\ F \textbf{ else } \end{cases}$

4 **else**

5   **for** $i = 0, i < mTry, i++$ **do**

6          $f_i = \texttt{SampleFeature}()$

7          $S_L = \{s_j | f_i(s_j) = T\}$; $S_R = \{s_j | f_i(s_j) = F\}$; $j = 1, \ldots, |S|$

8          $g_i = \widehat{\Delta G}(S_L, S_R)$

9   **if** $g_i > g$ **then**

10            $f^* = f_i$; $g = g_i$

11   **end**

12   **end**

13     $N_L = \texttt{LearnTree}(\{s_j | f^*(s_j) = T\})$

14     $N_R = \texttt{LearnTree}(\{s_j | f^*(s_j) = F\})$

15 **end**

Gini Index is used as splitting criterion, i.e. the Gini gain is maximized. At a given node, the set $S = s_1, \ldots, s_n$ holds the samples for feature f_j. For a binary response Y and a feature f_j the Gini Index of S is defined as:

$$\widehat{G}(S) = 2\frac{N_{false}}{|S|}\left(1 - \frac{N_{false}}{|S|}\right), \qquad N_{false} = \sum_{s_i} I\left(f_j(s_i) = false\right),$$

where $|S|$ is the number of all samples at the current node and N_{false} denotes the number of samples for which f_j evaluates to $false$. The Gini indices $\widehat{G}(S_L)$ and $\widehat{G}(S_R)$ for the left and right subset are defined similarly. The Gini gain resulting from splitting S into S_L and S_R with feature f_j is then defined as:

$$\widehat{\Delta G}(S_L, S_R) = \widehat{G}(S) - \left(\frac{|S_L|}{|S|}\widehat{G}(S_L) + \frac{|S_R|}{|S|}\widehat{G}(S_R)\right),$$

where $S = S_L \cup S_R$. From that follows, that the larger the Gini gain, the larger the impurity reduction. Recently [9] showed that the use of Gini gain can lead to selection bias because categorical predictor variables with many categories are preferred over those with few categories. In the proposed framework this is not an obstacle due to the fact that the features are relations between sampled rectangles and therefore evaluate always to binary predictor variables.

2.5 Multiple Object Detection

For multiple object detection in a gray scale image every location on a grid with step size δ is considered as an independent sample s which is classified by the ensemble.

Therefore each tree casts a binary vote for s being and object or background. The whole ensemble predicts probability of being class 1: $RDF(s) = \sum_{i|t_i(s)=1} 1/\sum_i 1$, where t_i is the ith tree. This procedure results in an accumulator or probability map for the whole image.

The final centroids of detected objects are retrieved by applying weighted mean shift clustering with a circular box kernel of radius r. During shifting, the coordinates are weighted by the probabilities of the accumulator map. While this leads to good results in most cases, homogeneous ridges in the accumulator can yield multiple centers with a pairwise distance smaller than r. To this end we run binary mean shift on the detection from the first run until convergence. The radius is predefined by the average object size. If the objects vary largely in size the whole procedure can be employed for different scales. To this end, in accordance with [3], not the image but the features respectively the rectangles are scaled.

2.6 Staining Classification

To differentiate a stained cell nucleus from a non-stained nucleus a simple color model was learned. Based on the labeled nuclei color histograms were generated for both classes based on the pixels within the average cell nuclei radius. To classify a nucleus on a test image the distance to the mean histograms of the both classes is calculated.

2.7 Survival Analysis

The main goal of tissue microarray analysis for the proliferation marker MIB-1 is the search for subgroups of patients which show different survival outcomes. Therefore, the patients are split in two (1/2 : 1/2) groups based on the estimated percentage of cancerous nuclei which express MIB-1. Then the Kaplan-Meier estimator is calculated for each subgroup. This involves first ordering the survival times from the smallest to the largest such that $t_1 \le t_2 \le t_3 \le \ldots \le t_n$, where t_j is the jth largest unique survival time. The Kaplan-Meier estimate of the survival function is then obtained as

$$\hat{S}(t) = \prod_{j:t_{(j)} \le t} \left(1 - \frac{d_j}{r_j} \right)$$

where r_j is the number of individuals at risk just before t_j, and d_j is the number of individuals who die at time t_j .

To measure the goodness of separation between two or more groups, the log-rank test (Mantel-Haenszel test) is employed which assesses the null hypothesis that there is no difference in the survival experience of the individuals in the different groups. The test statistic of the log-rank test (LRT) is χ^2 distributed: $\chi^2 = [\sum_{i=1}^m (d_{1i} - \hat{e}_{1i})]^2 / \sum_{i=1}^m \hat{v}_{1i}$ where d_{1i} is the number of deaths in the first group at t_i and $e_{1i} = n_{1j} \frac{d_i}{n_i}$ where d_i is the total number of deaths at time $t_{(i)}$, n_j is the total number of individuals at risk at this time, and n_{1i} the number of individuals at risk in the first group.

2.8 Performance Measure

One way to evaluate the quality of the nuclei detection is to consider true positive (TP), false positive (FP) and false negative (FN) rates. The calculation of these quantities

is based on a matching matrix where each boolean entry indicates if a machine extracted nucleus matches a hand labeled one or not within the average nucleus radius. To
quantify the number of correctly segmented nuclei, a strategy is required to uniquely
match a machine detected nucleus to one identified by a pathologist. To this end we
model this problem as bipartite matching problem, where the bijection between extracted and gold-standard nuclei is sought inducing the smallest detection error [10].
This prevents overestimating the detection accuracy of the algorithms. To compare the
performance of the algorithms we calculated precision ($P = TP/(TP + FP)$) and
recall ($R = TP/(TP + FN)$).

2.9   Implementation Details

The ensemble learning framework was implemented in C# and the statistical analysis
was conducted in R [11]. Employing a multi threaded architecture tree ensembles are
learned in real time on a standard dual core processor with 2.13 GHz. Inducing a tree
for 1000 samples with a maximum depth of 10 and sampling 500 features at each split
takes on average less than $500ms$. Classifying an image of 3000×3000 pixels on a grid
with $\delta = 4$ takes approximately ten seconds using the non optimized C# code.

3   Experiments and Results

3.1   Generating a Gold Standard

The absence of an objective ground truth requires generating a gold standard by combining the knowledge of expert pathologists. These labels are indispensable for the
training of classifiers, for their validation and it is highly non-trivial to acquire these
labels from a technical as well as statistical point. Although studies were conducted on
a global estimation of staining on TMA spots [12,13], to our knowledge this is the first
in depth study for tissue microarrays which incorporates expert labeling information
down to the detail and precision of single cell nuclei.

   To facilitate the labeling process for trained pathologists, we developed a special
labeling tool, dedicated for TMA spots. The software allows the user to view single
TMA spots and it provides zooming and scrolling capabilities. It is possible to annotate
the image with vectorial data in SVG (support vector graphics) format and to mark cell
nuclei, vessels and other biological structures. An additional demand to the software
was the usability on a tablet PC so that a pathologist can perform all operations with a
pen alone in a simple an efficient manner.

   Two trained pathologists and experts in renal cell carcinoma from the University
Hospital Zürich used the software to annotate TMA spots of 9 different patients. They
marked the location of each cell nucleus and its approximates size. In total each pathologist has detected more than 2000 cell nuclei on these images. This tedious process
demanded several days of work and was performed independently of each other. Therefore, the detection results of the two pathologists differ for approximately 15% of the
nuclie, which also depicted in Figure 2.

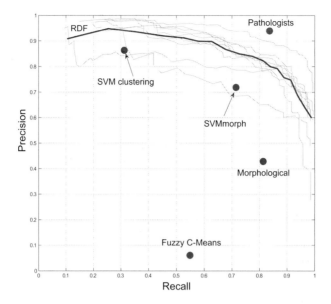

Fig. 2. Precision/Recall plot of cross validation results on the renal clear cell cancer (RCC) dataset. For Relational Detection Forests (RDF) curves for the nine single patients and their average (bold) are depicted. RDF with the proposed feature base outperforms previous approaches based on SVM clustering [14], mathematical morphology and combined methods [15]. The inter pathologist performance is depicted in the top right corner.

3.2   Clear Cell Renal Cell Carcinoma (ccRCC)

Detection Accuracy: Three fold cross validation was employed to analyze the detection accuracy of RDFs. The nine completely labeled patients were randomly split up into three sets. For each fold the ensemble classifier was learned on six patients and tested on the the other three. During tree induction, at each split 500 features were sampled from the feature generator. Trees were learned to a maximum depth of 10 and the minimum leave size was set to 1. The forest converges after 150 to an out of bag (OOB) error of approximately 2%. Finally, on the test images each pixel was classified and mean shift was run on a grid with $\delta = 5$.

Figure 2 shows precision/recall plot for single patients and the average result of the RDF object detector. The algorithm is compared to point estimates of several state of the art methods: SVM clustering was successfully employed to detect nuclei in H&E stained images of brain tissue by [14]. SVMmorph [15] is an unsupervised morphological [16] approach for detection combined with an supervised support vector machine for filtering. The marker for the pathologists shows the mean detection accuracy if alternately one expert is used as gold standard. On average the pathologists disagree on 15% of the nuclei.

Although only grayscale features were used for RDF it outperforms all previous approaches which also utilize texture and color. This observation can be a cue for further research that the shape information captured in this framework is crucial for good detection results.

Survival Estimation: The only objective and undisputed target in the medical domain relates to the survival of the patient. The experiments described in Section 1 show the large disagreement between pathologists for the estimation of staining. Therefore, the adaption of an algorithm to the estimates of one pathologist or to a consensus voting of a cohort of pathologist is not desirable. Hence we validate the proposed algorithm against the right censored clinical survival data of 133 patients. In addition these results were compared to the estimations of an expert pathologist specialized on renal cell carcinoma. He analyzed all spots in an exceptional thorough manner which required him more than two hours. This time consuming annotation exceeds the standard clinical practice significantly by a factor of 10-20 and, therefore the results can be viewed as an excellent human estimate for this dataset.

Figure 3 shows Kaplan-Meier plots of the estimated cumulative survival for the pathologist and RDF. The father the survival estimated of the two groups are separated the better the estimation. Quantifying this difference with log-rank test shows that the proposed algorithm is significantly ($p = 0.0113$) better than the trained pathologist ($p = 0.0423$).

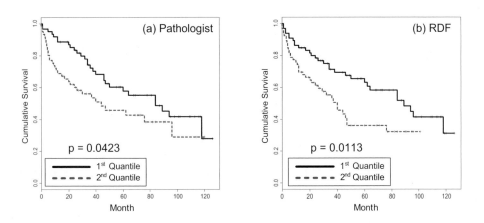

Fig. 3. Kaplan-Meier estimators showing significantly different survival times for renal cell carcinoma patients with high and low proliferating tumors. Compared to the manual estimation from the pathologist (a) ($p = 0.04$), the fully automatic estimation from the algorithm (b) performs better ($p = 0.01$) in terms of survival prediction on the partitioning of patients into two groups of equal size.

3.3   Proliferation in Murine Liver Tissue

To validate the performance of the proposed algorithm on diverse tissue samples and different species we conducted an experiment with liver tissue from eight mice. In addition this experiment was designed to learn the difference between non parenchymal cells and parenchymal, organ specific cell types in order to distinguish between organ prone proliferation and inflammatory cell derived positivity for MIB-1. Figure 4 shows a comparison of MIB-1 staining in 88 images of murine liver tissue between case and control groups. Although the algorithm is as well capable as the pathologist in differentiating between case and control ($p < 0.001$) the absolute estimates of stained nuclei are much higher. This is mostly due to wrong hits in area of high lymphocyte density as shown in the tissue image in Figure 4.

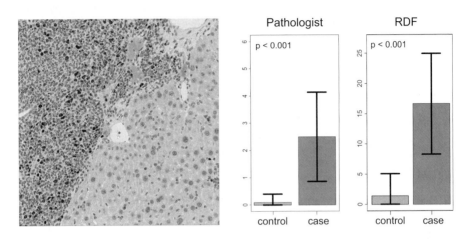

Fig. 4. Left: Murine liver tissue with parenchymal nuclei of interest in the bottom right part of the image. The goal is to estimate the number of stained nuclei of this specific type. Detections are shown as pink dots. Most of the inflammatory cells in the top left part are correctly not detected. **Right:** Comparison of MIB-1 staining in 88 images of mouse liver tissue between case and control groups. Although the algorithm is as well capable as the pathologist in differentiating between case and control ($p < 0.001$) the absolute estimates of stained nuclei are much higher.

4   Conclusion

We presented a framework for learning Relational Detection Forests (RDF) for object detection. A comprehensive study was conducted on two different species to investigate the performance of the algorithm in terms of detection accuracy and survival estimation.

The proposed framework is characterized by the following properties: (i) **Simplicity:** It can be used off-the-shelf to train object detectors in near real time for large variety of tasks. (ii) **Novel Feature Basis:** The introduced relational features are able to capture shape information, they are illumination invariant and extremely fast to evaluate. (iii) **Randomization:** The randomized tree induction algorithm is able to handle the intractable large feature space and to take advantage of it by increasing diversity of the ensemble. (iv) **Real World Applicability:** We successfully applied the proposed RDF algorithm to real world problems in computational pathology. We are convinced that the availability of an off-the-shelf object detection framework is of immense benefit for medical research where fast and accurate adaption to a large number of cancer types is indispensable.

References

1. Yang, L., Meer, P., Foran, D.J.: Unsupervised segmentation based on robust estimation and color active contour models. IEEE Transactions on Information Technology in Biomedicine 9, 475–486 (2005)
2. Mertz, K.D., Demichelis, F., Kim, R.,, Schraml, P., Storz, M., Diener, P.-A., Moch, H., Rubin, M.A.: Automated immunofluorescence analysis defines microvessel area as a prognostic parameter in clear cell renal cell cancer. Human Pathology 38, 1454–1462 (2007)
3. Viola, P., Jones, M.: Rapid object detection using a boosted cascade of simple features (2001)
4. Grabner, H., Leistner, C., Bischof, H.: Semi-supervised on-line boosting for robust tracking. In: Forsyth, D., Torr, P., Zisserman, A. (eds.) ECCV 2008, Part I. LNCS, vol. 5302, pp. 234–247. Springer, Heidelberg (2008)
5. Dalal, N., Triggs, B.: Histograms of oriented gradients for human detection. In: CVPR 2005: Proceedings of the 2005 IEEE Computer Society Conference on Computer Vision and Pattern Recognition (CVPR 2005), Washington, DC, USA, vol. 1, pp. 886–893. IEEE Computer Society, Los Alamitos (2005)
6. Freund, Y., Schapire, R.: Experiments with a new boosting algorithm. In: Machine Learning: Proceedings of the Thirteenth International Conference, pp. 148–156 (1996)
7. Breiman, L.: Random forests. Machine Learning 45, 5–32 (2001)
8. Geurts, P., Ernst, D., Wehenkel, L.: Extremely randomized trees. Machine Learning 63, 3–42 (2006)
9. Strobl, C., Thomas Augustin, A.L.B.: Unbiased split selection for classification trees based on the gini index. Computational Statistics & Data Analysis 52, 483–501 (2007)
10. Kuhn, H.W.: The hungarian method for the assignment problem. Naval Research Logistic Quarterly 2, 83–97 (1955)
11. R Development Core Team: R: A Language and Environment for Statistical Computing. R Foundation for Statistical Computing, Vienna, Austria (2009) ISBN 3-900051-07-0
12. Hall, B., Chen, W., Reiss, M., Foran, D.J.: A clinically motivated 2-fold framework for quantifying and classifying immunohistochemically stained specimens. In: Ayache, N., Ourselin, S., Maeder, A. (eds.) MICCAI 2007, Part II. LNCS, vol. 4792, pp. 287–294. Springer, Heidelberg (2007)
13. Yang, L., Chen, W., Meer, P., Salaru, G., Feldman, M.D., Foran, D.J.: High throughput analysis of breast cancer specimens on the grid. Med. Image Comput. Comput. Assist. Interv. Int. Conf. Med. Image Comput. Comput. Assist Interv. 10, 617–625 (2007)
14. Nikiforidis, G.: An image-analysis system based on support vector machines for automatic grade diagnosis of brain-tumour astrocytomas in clinical routine. Medical Informatics and the Internet in Medicine 30, 179–193 (2005)
15. Fuchs, T.J., Lange, T., Wild, P.J., Moch, H., Buhmann, J.M.: Weakly supervised cell nuclei detection and segmentation on tissue microarrays of renal cell carcinoma. In: Rigoll, G. (ed.) DAGM 2008. LNCS, vol. 5096, pp. 173–182. Springer, Heidelberg (2008)
16. Soille, P.: Morphological Image Analysis: Principles and Applications. Springer, Secaucus (2003)

Human Understandable Features for Segmentation of Solid Texture

Ludovic Paulhac[1], Pascal Makris[1], Jean-Marc Gregoire[2], and Jean-Yves Ramel[1]

[1] Université François Rabelais Tours, Laboratoire Informatique (EA2101)
[2] UMR INSERM U930, CNRS ERL 3106, équipe 5, Université François Rabelais Tours
{ludovic.paulhac,pascal.makris,jean-marc.gregoire,
jean-yves.ramel}@univ-tours.fr

Abstract. The purpose of this paper is to present new texture descriptors dedicated to segmentation of solid textures. The proposed texture attributes are inspired by the human description of texture and allows a general description of texture. Moreover it is more convenient for a user to understand features signification particularly in a man-aided application. In comparison with psychological measurements for human subjects, our characteristics gave good correspondences in rank correlation of 12 different solid textures. Using these texture features, segmentation results obtained with the classical K-means method on solid textures and real three-dimensional ultrasound images of the skin are presented and discussed.

Keywords: Texture features, Segmentation, Solid textures, 3D Ultrasound Images.

1 Introduction

Texture analysis is an important topic of image analysis and computer vision. It is used to identify a given texture or to divide an image in several regions with similar characteristics. A great number of methods have been proposed to analyze textures. They can be classified in four categories [1], that is to say statistical [2,3], geometrical [4], frequential [5] and model based methods [6].

The disadvantage of the major part of theses methods is that they do not have a general applicability and can not identify some classes of texture. For example some of these approaches are not able to describe the granularity properties. In comparison, the human visual system is efficient for almost all types of texture and allows outstanding performance even without a good context. For instance, the difficulty to give a general definition of texture is well-known. Indeed, a texture is an abstract thing, hard to define literally. Nevertheless, a texture can be described using human understandable adjectives. Usually, humans use and are able to quantify textural properties like directionality, coarseness, contrast, granularity, shape etc. In order to have general texture measures, some authors [7,8] made some research in this direction proposing a description of texture using textural properties understandable by humans.

With some classical texture analysis methods, it is not always obvious to describe what a given feature allows to measure. For a man-aided application, it is better to have

G. Bebis et al. (Eds.): ISVC 2009, Part I, LNCS 5875, pp. 379–390, 2009.

a set of features which corresponds to the one used by human for the description of texture. Then it is more convenient for a user to select which features are interesting to use in a given application.

Texture analysis methods have been mainly developed and experimented on 2D texture images. Recently, some of these methods have been investigated to analyze solid texture [9]. In the medical domain, there is an increasing use of three-dimensional acquisition technology that needs three-dimensional segmentation or analysis methods. Medical images like volumetric ultrasound images are complex and texture analysis is very efficient for tissue classification and segmentation [10].

This paper presents 3D texture features, inspired by human description of texture, in order to have general texture measures. Section 2 presents a detailed description of the features and the assigned computation method. Section 3 shows the correspondance between the proposed features and the human vision. To validate our proposition, our texture descriptors are used to segment synthetic solid textures and 3D ultrasound images. Section 4 describes the segmentation process and section 5 presents a comparison of texture segmentation between our method and the grey co-occurrence matrix of Haralick. Section 6 shows segmentation results on 3D ultrasound images. To conclude, we provide a discussion about our work and introduce different prospects.

2   Understandable Features for Segmentation of Solid Textures

In this section, a set of 3D texture features, inspired by the human way to describe a texture, is proposed. The set of characteristics has been chosen using analysis results from previous works [7,8]. The chosen characteristics are the following: Granularity, which can be represented by the number of three-dimensional patterns constituting the texture, shape information about these patterns, regularity of these patterns, contrast and roughness of the image, which are also important information.

2.1   A Multiresolution Schema for Texture Segmentation

The proposed approach allows us to give a description of an image using texture features for several resolutions (Figure 1). To do so, we use a 3D discrete wavelet transform. In [5], Mallat proposes a decomposition scheme using filters: a highpass filter, which allows to obtain detail coefficients and a lowpass filter which gives approximation coefficients. Roughness is computed using detail coefficients whereas the other proposed features are computed using approximation coefficients.

During the segmentation process, the proposed features are computed for each voxel and for several resolutions. In order to obtain a segmentation of the start image, a step of feature upsampling is necessary for each resolution. Then, a voxel of the initial image is described by a vector containing $6n$ different features with n the number of resolutions and 6 the number of proposed features. At last, the K-means algorithm [11] allows to generate a segmentation using the set of computed vectors.

2.2   Geometric Study of 3D Textures

We have chosen to describe the geometric structure of the textures with the help of the three-dimensional connected components which can be viewed as the patch patterns in

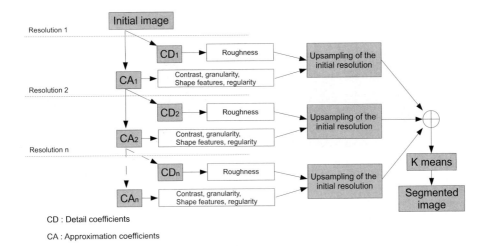

CD : Detail coefficients

CA : Approximation coefficients

Fig. 1. Multiresolution segmentation

the texture. To compute connected components, we propose a similar method to the one presented by Shoshany [12], that is to say a gray-level textured image is decomposed into a progressive sequence of binary textures in order to study patterns and their evolution. In our approach, a clustering algorithm applied on the voxels of the initial 3D image allows us to identify a set of thresholds used to compute the sequence of binary versions of the image. Then, connected components are computed for each binary textures of the produced sequence. Figure 2 presents a binarized image with examples of connected components. If we consider two points A and B included in a sub-set S of an image I, these two points are connected in S if and only if there exists a connecting path in S which links A and B. The connected components of an image are obtained by associating to each connected voxel the same label. To do so there are several algorithms for two-dimensional images and the main ones are presented by Chassery and Montanvertin in [13]. Among these methods, we have chosen to adapt to three dimensions an algorithm which only needs two scans to process an image. With this algorithm, the complexity depends on the size of the image whereas with a sequential algorithm, the number of iterations depends on the complexity of the objects. Connected components represent the basic objects inside binary textures. Their analysis can provide important geometrical and volume information and it allows computation of features like granularity which corresponds to the number of patterns per volume unit, the volume and compacity of each connected components providing information about the shape of texture patches. The regularity can also be estimated using the variance of these patterns. In our case, the number of patterns corresponds to the number of connected components ($nbCC$) per volume unit. For a given texture, if the number of connected component is important for the resolution β then there is an important number of patterns and the granularity (f_{gran_β}) of the texture is significant.

Besides the number of connected components, we compute shape characteristics with the average volume and the average compacity of connected components. Like

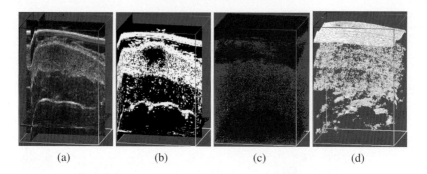

(a)   (b)   (c)   (d)

Fig. 2. a) An ultrasound image, b) The same ultrasound image after a binarization, [c-d] Connected components extracted from different binary images

the number of connected components, the volume is an additional information to identify the fineness of a texture. The average volume of connected components is computed as follows:

$$f_{vol_\beta}(x, y, z) = (\sum_{i=1}^{nbCC_\beta} V_{i,\beta})/nbCC_\beta \tag{1}$$

where $V_{i,\beta}$ is the volume of the connected component i for the resolution β. The considered connected components are located in a cube of size N^3 centered at the coordinates (x, y, z).

Compacity of connected components gives information about pattern shape. A texture with an important compacity is a texture with compact patterns. Otherwise this is a texture with elongate shapes. This characteristic is invariant by translation, rotation but also to scale changes [14]. It has been used to texture characterization by Goyal et al in [15]. The compacity of a connected component can be computed as follows:

$$C_{i,\beta} = \frac{S_{i,\beta}^2}{V_{i,\beta}} \tag{2}$$

where $S_{i,\beta}$ is the surface and $V_{i,\beta}$ is the volume of connected component i for the resolution β. It is then possible to compute the average compacity f_{comp_β} :

$$f_{comp_\beta}(x, y, z) = \frac{1}{nbCC_\beta} \sum_{i=1}^{nbCC_\beta} C_{i,\beta} \tag{3}$$

We can also obtain information about the regularity of a texture from the study of the connected components. Therefore, we decided to use the compacity variance. We have seen that this characteristic is invariant by any transformation [14]. The shape of patterns is the only element that affects the variance feature. A low variance of the compacity indicates an important regularity of the connected components whatever their spatial organization is.

$$f_{reg_\beta}(x, y, z) = E(C_\beta^2) - (E(C_\beta))^2 \tag{4}$$

where E is the expected value.

2.3 Roughness Measure Using the Discrete Wavelet Transform

The surface of a rough texture presents a high number of asperities. In an image, roughness can be described as a set of quick spatial transitions with varying amplitude. From a frequential point of view, the image asperities in the spatial domain correspond to the presence of high frequencies. Detail coefficients give a description of high frequencies in an image and this in several directions. It is then possible to have an estimation of the texture roughness for a specific resolution.

After the decomposition process, we have a set of subband which can be used to compute the features. Generally researchers [16,17] use first order statistics like average, energy, variance etc. Energy is one of the most used statistics:

$$E(x, y, z) = \sum_{i=1}^{N} \sum_{j=1}^{N} \sum_{k=1}^{N} |c(i, j, k)| \tag{5}$$

where $c(i, j, k)$ is the set of wavelet coefficients in a cube of size N^3 corresponding to voxels in a subband at coordinates (x, y, z). To estimate the roughness of three-dimensional textures, we use the detail coefficients of the wavelet decomposition. Indeed these coefficients allow the identification of high frequencies. For a given resolution, we propose to compute this texture attribute:

$$f_{rgh_\beta}(x, y, z) = \sum_{\alpha=1}^{M} (\sum_{i=1}^{N} \sum_{j=1}^{N} \sum_{k=1}^{N} |w_{\alpha,\beta}(i, j, k)|)/M \tag{6}$$

where f_{rgh_β} is the roughness at the resolution β, $w_{\alpha,\beta}(i, j, k)$ corresponding to the set of detail coefficients in a cube of size N^3 centered at a voxel of a subband α at the coordinates (x, y, z) and M being the number of detail coefficient subbands for a resolution.

2.4 Statistical Measure for Contrast

In [2,3], Haralick proposes a measure to estimate contrast using second order statistics. To do so, the moment of inertia is computed from the main diagonal of the co-occurrence matrix. Nevertheless, the construction of a co-occurrence matrix, only to obtain an estimation of the contrast, can be expensive in computing time. In [7], Tamura et al claim that four factors are supposed to influence the contrast difference between two textures. They are the dynamic range of gray-levels, the polarization of the distribution of black and white on the gray-level histogram or ratio of black and white areas, the sharpness of edges, and the period of repeating patterns. They propose to approximate the contrast with a measure incorporating the two first factors. We use this approximation in our work. To obtain a measure of polarization they use the kurtosis α_4. It allows a measurement of the disposition of probability mass around their center.

$$\alpha_{4,\beta} = \frac{\mu_{4,\beta}}{\sigma_\beta^4} \tag{7}$$

where μ_4 is the fourth moment about the mean and σ^2 the variance of gray-levels for the resolution β. In order to take into account the dynamic range of gray-levels, they combine the kurtosis with the standard deviation as follows:

$$f_{cont_\beta}(x, y, z) = \frac{\sigma_\beta}{\alpha_{4,\beta}^n} \qquad (8)$$

where n is a positive value. In their paper, Tamura et al make comparisons between psychological experiments and their operators before concluding that the value $n = 1/4$ gives the best approximation. At last, the values of σ_β and $\alpha_{4,\beta}^n$ are computed in a cube of size N^3 centered at the coordinates (x, y, z).

3 Psychological Experiments

The purpose of these experiments is to show that there is a strong correspondance between the proposed features and the human vision. Thus, we propose to construct psychometric prototypes and to compare them to our texture measures.

The set of textures presented in Figure 3 has been used in our experiments. These textures have been constructed using methods presented in [18,19] except for textures (j) and (l) that are subsets of ultrasound images. Each of them is a volumetric texture

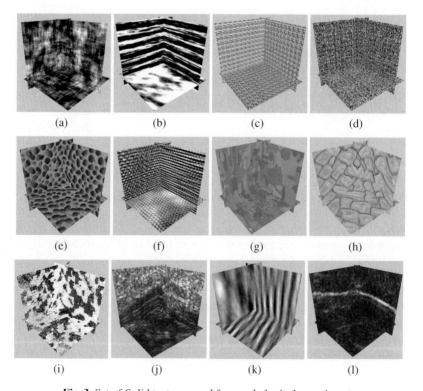

Fig. 3. Set of Solid textures used for psychological experiments

(texture in the 3D domain) of size 128^3 with 256 gray-levels. They have been printed using a printer HP Color LaserJet 3700 and presented to human subjects.

Our group of human subjects comprises 15 men and 11 women and the majority of them has no knowledge in image and texture analysis. We distributed to each one of them a questionnaire containing the set of textures (Figure 3) and an explanation of the texture features used in our model. For each feature, textures are classified in descending order that is to say from the most rough to the most smooth, from the most regular to the most irregular, etc. Using these questionnaires, we construct a ranking of these textures for each texture attributes. For a given characteristic, a score is assigned to a texture according to its ranking. For exemple, The most rough takes the value $+12$ (for the roughness feature), the second one $+11$, the last one $+1$, and this for all the questionnaires. The addition of the questionnaire scores for each texture allows to obtain a final ranking for a given texture feature.

Using the proposed texture attributes, we also generate a feature ranking. A vector of 6 features is computed for each texture in the questionnaire. Here only the first resolution is considered ($\beta = 1$) because this is the one which corresponds the best to the observation of textures by human subjects through questionnaires.

3.1   Comparison between Human and Computational Ranking

To compare human and feature ranking, the degree of correspondances between them has been determined. In this respect, we choose to use the well-known Spearman's coefficient of rank correlation which is given as:

$$r_s = 1 - \frac{6}{n^3 - n} \sum_{i=1}^{n} d_i^2 \tag{9}$$

where n is the number of individuals, and d_i is the difference between the ranks assigned to the ith object in the two measurements. The value of this coefficient is between -1 and 1. Value 1 corresponds to the complete agreement of the two rankings whereas value -1 indicates complete disagreement. Table 1 presents coefficients of rank correlations between human and feature ranking.

These results show a huge correlation between human and feature ranking. For the volume, the Spearman's coefficient indicates that there is a link between the two measurements with a confidence rate included between 95 and 98 percent. The compacity feature gives the best result with a confidence rate which tends toward 100 percent. The volume feature has the smallest correlation results. It can be supposed that it is sometimes difficult for our human subjects to visualise the volume of patterns because of the 3D.

Table 1. Coefficients of rank correlations between human and feature ranking

f_{gran}	f_{comp}	f_{vol}	f_{reg}	f_{rgh}	f_{cont}
0.83	0.9	0.61	0.82	0.75	0.65

4 Process of Segmentation Using 3D Texture Features

Features presented previously are computed for each voxel of the image being processed. A voxel is then described by a vector containing several texture attributes. A vector is composed of a maximum of $6n$ different features with n the number of chosen resolution and 6 the number of proposed feature: Granularity(f_{gran}), shape information of patterns(f_{vol} and f_{comp}), regularity of these patterns(f_{reg}), contrast(f_{con}) and roughness(f_{rgh}). In our approach, texture features have a correspondence with the human description of texture. It is then easier for a user to make a selection of attributes to process an image. To produce a clustering of the voxels, we used the K-means algorithm [11]. It allows to classify voxels in subsets according to their texture features. The main advantages for this approach are its fastness and its low memory cost. Indeed, the size of the processed images is 300^3 which represents a huge number of individuals (voxels). In terms of performance, K-means algorithm does not guarantee the return of a global optimum but it allows an efficient clustering of voxels in a low execution time. Likewise, this method requires the user to choose the number of classes needed for a segmentation. To provide a solution to this problem, our software gives the possibility to merge the different classes generated by the K-means algorithm. After the visualisation, the user chooses a large number of classes to obtain a first segmentation. Then, using a user friendly interface and by merging different classes, he can choose the best segmentation according to his goals. The merging of classes is achieved by using an ascendant hierarchical classification and the two most similar regions are merged at each step (Figure4). The distance between regions is computed using the texture features of the centroid of each class. This information is also stored in a xml file and can be given to the user if needed.

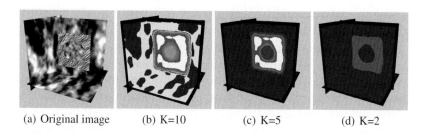

(a) Original image   (b) K=10   (c) K=5   (d) K=2

Fig. 4. Examples of segmentation from $K = 10$ to $K = 2$

5 Evaluation of Our Method of Texture Segmentation

To evaluate our human understandable features (HUF) method, we made a comparison with the classical grey level co-occurrence matrix (GLCM) method of Haralick [2,3].

Segmentation results have been produced using 15 different volumetric texture images (Figure 5): 5 images with 2 classes of textures, 5 images with 3 classes of textures and 5 images with 4 classes of textures. For the GLCM method we computed the following features: the angular second moment, the variance, the contrast, the correlation,

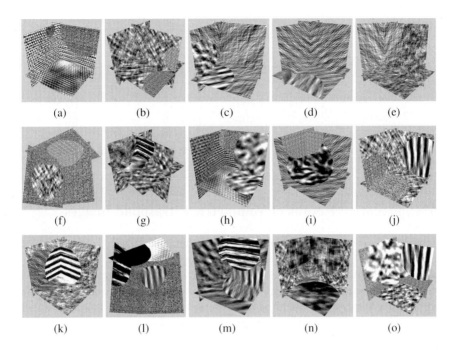

(a)   (b)   (c)   (d)   (e)

(f)   (g)   (h)   (i)   (j)

(k)   (l)   (m)   (n)   (o)

Fig. 5. Solid texture images: [a-e] 2 classes of textures, [f-j] 3 classes of textures, [k-o] 4 classes of textures

the entropy, the homogeneity, and the uniformity. To obtain these characteristics, we computed a co-occurence matrix according to 13 directions in 3D space, and with a distance $d = 1$ between a pair of voxels. With our method, we used all the proposed texture features for one level of decomposition. For these two methods, the considered neighborhood around each voxel is a cube of size 7^3. To give an evaluation of the produced segmentation, a metric based on the comparison of regions (voxel classification) has been used: the normalized partition distance [20]. In [21], Gusfield defines the partition distance as follows :

Definition 1: Given two partitions P and Q of S, the partition distance is the minimum number of elements that must be deleted from S, so that the two induced partitions (P and Q restricted to the remaining elements) are identical.

Table 2, shows the normalized partition distance for each solid texture segmentation. An ideal segmentation will have a partition distance of 0 whereas an inverse segmentation will generate the value 1. Except for image (e) of 2 classes, our method gives better segmentation results than Haralick's gray level co-occurrence matrix. Moreover, our approach rarely exceeds the value $1,5$ for the partition distance. At last, some images like (m) and (o) of 4 classes are segmented with difficulty with Haralick features whereas the proposed features allow to obtain a good segmentation.

Table 2. Comparison of texture segmentation methods using the normalized partition distance

Number of classes	Texture methods	Image (a) to (o)				
2 classes	GLCM	0.0131	0.0276	0.0462	0.0185	0.2698
	HUF	0.0100	0.0065	0.0206	0.0107	0.3000
3 classes	GLCM	0.1317	0.0917	0.2518	0.3536	0.2534
	HUF	0.0447	0.0387	0.1305	0.1373	0.1710
4 classes	GLCM	0.0984	0.1744	0.3546	0.4767	0.3221
	HUF	0.0561	0.1050	0.1200	0.3883	0.1298

6   Segmentation Results of Volumetric Ultrasound Images

Sonography of the skin allows the visualization of tumor (cyst, nevus, melanoma, basal cell carcinoma (BCC) etc.), inflammatory pathologies, scars. The discrimination between the different lesions is not always obvious but cutaneous sonography is an important help for detection and diagnosis. The possibility to segment and characterize a lesion in three dimensions would be very useful in the establishment of therapeutic strategies. The three-dimensional sonography of the skin is rarely used because of the lack of three-dimensional image analysis tools but the recent evolution of three-dimensional probes should allow the emergence of new technics. With a three-dimensional acquisition, it is possible to obtain features that are inaccessible using two dimensions. Moreover three-dimensional sonography is well adapted to the supervision of the evolution of a structure or a lesion notably using volume measures. To evaluate our system on realistic images and applications, we have provided some dermatologists with our software in order to help them to collect 3D information about pathologies. Figure 6 presents segmentation results for different skin ultrasound images. Using a segmentation, it is possible to extract different things like lesion, tendon, layers of the skin etc. Figure 6[a-b] shows two nevi, Figure 6[c-d] shows two Histiocytofibroma, Figures 6[a-d] contains at the left the original three-dimensional ultrasound image, at the center an image of classified voxels and on the right a mesh of the lesion built using the segmented image. With these results, it is then possible to perform measures like volume and depth, in order to help specialists in their diagnostic, to track pathologies evolution or a more precise excision etc. With a clustering, all the voxels of an image are classified and it is possible to make a visualization of the different layers of the skin (figure 6e). For a specialist, this visualization can be interesting. Indeed, layers of the skin evolve according to the age and it can be interesting to supervise the healing of a burned skin. This visualization can also be interesting for the evaluation of the effects of cosmetic products. To evaluate our results, the classified images and their corresponding meshes have been presented to specialists in ultrasound images. The evaluation is only a qualitative one at the moment. Producing a ground truth for a two-dimensional ultrasound image is not very restrictive. It is possible to obtain different ground truths for the same image produced by different specialists. But, for a three-dimensional ultrasound image, it is very difficult and costs a lot of time to create a ground truth. Indeed, it is necessary to produce an expert segmentation for each two-dimensional cut (for example in z-axis direction) of a three-dimensional ultrasound image. With this two-dimensional ground truth it could be possible to construct a three-dimensional one but the gathering of the two-dimensional images could generate holes and then problems of precision. It is then difficult or even impossible to obtain a fine resolution for the evaluation.

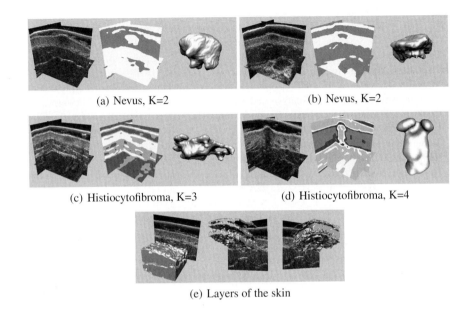

(a) Nevus, K=2   (b) Nevus, K=2

(c) Histiocytofibroma, K=3   (d) Histiocytofibroma, K=4

(e) Layers of the skin

Fig. 6. Segmentation of three-dimensional echographic images of the skin

7   Conclusion

This paper presents a set of understandable features for solid texture analysis. These characteristics are comprehensible by human and allow a better use for a man-aided application. The following texture attributes have been proposed: the granularity, shape information with the volume and the compacity of patterns, the regularity with the compacity variance of patterns, roughness and contrast. To prove the strong correspondance between the proposed features and human vision, psychological experiments have been presented. To validate, in a quantitative manner, our proposition, our texture descriptors have been used to segment synthetic solid textures and 3D ultrasound images. Synthetic solid textures allow us to make a comparison of texture segmentation between our method and Haralick's gray level co-occurrence matrix. The evaluation of segmentation results obtained using the normalized partition distance shows a better performance with our approach. Eventually, we have seen that it is difficult to construct a ground truth to evaluate three-dimensional ultrasound images. It could be interesting to discuss with specialists in order to identify and propose metrics for an unsupervised evaluation to obtain a quantitative and objective evaluation of our proposition. So far, our results have been validated, in a qualitative way, by dermatologists.

References

1. Tuceryan, M., Jain, A.K.: 2.1. In: Texture Analysis. The Handbook of Pattern Recognition and Computer Vision, pp. 207–248 (1998)
2. Haralick, R.M.: Statistical and structural approaches to textures. Proceedings of the IEEE 67(5), 786–804 (1979)
3. Haralick, R.M., Shanmugam, K., Dinstein, I.: Texture features for image classification. IEEE Transactions on Systems, Man and Cybernetics 3(6), 610–621 (1973)
4. Tuceryan, M., Jain, A.K.: Texture segmentation using voronoi polygons. IEEE Transactions On Pattern Analysis And Machine Intelligence 12, 211–216 (1990)
5. Mallat, S.G.: A theory for multiresolution signal decomposition: the wavelet representation. IEEE transaction on Pattern Analysis and Machine Intelligence 11, 674–693 (1989)
6. Chellappa, R., Jain, A.K.: Markov Random Fields Theory and Application. Academic Press, London (1993)
7. Tamura, H., Mori, S., Yamawaki, T.: Texture features corresponding to visual perception. IEEE transaction on Systems, Man, Cybernetics 8(6), 460–473 (1978)
8. Amadasun, M., King, R.: Texture features corresponding to textural properties. IEEE transactions on Systems, Man and Cybernetics 19(5), 1264–1274 (1989)
9. Suzuki, M.T., Yoshitomo, Y., Osawa, N., Sugimoto, Y.: Classification of solid textures using 3d mask patterns. In: ICSMC 2004: International Conference on Systems, Man and Cybernetics (2004)
10. Noble, J.A., Boukerroui, D.: Ultrasound image segmentation: A survey. IEEE Transactions on Medical Imaging 25(8), 987–1010 (2006)
11. Coleman, G., Andrews, H.: Image segmentation by clustering. Proceedings of the IEEE, 773–785 (1979)
12. Shoshany, M.: An evolutionary patch pattern approach for texture discrimination. Pattern Recognition 41, 2327–2336 (2008)
13. Chassery, J.M., Montanvert, A.: Géométrie discrète en analyse d'images (1991)
14. Zhang, J., Tan, T.: Brief review of invariant texture analysis methods. Pattern Recognition 35, 735–747 (2002)
15. Goyal, R., Goh, W., Mital, D., Chan, K.: Scale and rotation invariant texture analysis based on structural property. In: IECON 1995: Proceedings on the International Conference on Industrial Electronics, Control, and Instrumentation (1995)
16. Unser, M.: Texture classification and segmentation using wavelet frames. IEEE Transactions on Image Processing 4, 1549–1560 (1995)
17. Muneeswaran, K., Ganesan, L., Arumugam, S., Soundar, K.R.: Texture classification with combined rotation and scale invariant wavelet features. Pattern Recognition 38, 1495–1506 (2005)
18. Kopf, J., Fu, C.-W., Cohen-Or, D., Deussen, O., Lischinski, D., Wong, T.-T.: Solid texture synthesis from 2d exemplars. In: SIGGRAPH 2007: Proceedings of the 34th International Conference and Exhibition on Computer Graphics and Interactive Techniques (2007)
19. Paulhac, L., Makris, P., Ramel, J.Y.: A solid texture database for segmentation and classification experiments. In: VISSAPP 2009: Proceedings of the Fourth International Conference on Computer Vision Theory and Applications (2009)
20. Cardoso, J.S., Corte-Real, L.: Toward a generic evaluation of image segmentation. IEEE Transaction on Image Processing 14(11), 1773–1782 (2005)
21. Gusfield, D.: Partition-distance: A problem and class of perfect graphs arising in clustering. Information Processing Letters 82(9), 159–164 (2002)

Exploiting Mutual Camera Visibility in Multi-camera Motion Estimation

Christian Kurz[1], Thorsten Thormählen[1], Bodo Rosenhahn[2], and Hans-Peter Seidel[1]

[1] Max Planck Institute for Computer Science (MPII), Saarbrücken, Germany
[2] Leibniz University Hannover, Institut für Informationsverarbeitung

Abstract. This paper addresses the estimation of camera motion and 3D reconstruction from image sequences for multiple independently moving cameras. If multiple moving cameras record the same scene, a camera is often visible in another camera's field of view. This poses a constraint on the position of the observed camera, which can be included into the conjoined optimization process. The paper contains the following contributions: Firstly, a fully automatic detection and tracking algorithm for the position of a moving camera in the image sequence of another moving camera is presented. Secondly, a sparse bundle adjustment algorithm is introduced, which includes this additional constraint on the position of the tracked camera. Since the additional constraints minimize the geometric error at the boundary of the reconstructed volume, the total reconstruction accuracy can be improved significantly. Experiments with synthetic and challenging real world scenes show the improved performance of our fully automatic method.

1 Introduction

Simultaneous estimation of camera motion and 3D reconstruction from image sequences is a well-established technique in computer vision [1–3]. Often this problem is referred to as Structure-from-Motion (SfM), Structure-and-Motion (SaM), or Visual Simultaneous Location and Mapping (Visual SLAM). This paper investigates the special scenario of multiple independently moving cameras that capture the same scene. In such a scenario it is often the case that a camera can be observed by another camera. This puts an additional constraint on the position of the observed camera. The additional constraint can be exploited in the estimation process in order to achieve more accurate results.

Multi-camera systems, e.g., stereo cameras, light field capturing systems [4], and markerless motion capturing setups [5] employing multiple cameras, are very common in computer vision. Until now, almost all of these camera setups have been static.

For static cameras, Sato [6] analyzed the epipolar geometry for cases where multiple cameras are projected into each other's images. In these cases, the epipoles are directly given by the projection of the camera centers. Therefore, the epipolar geometry can be calculated from less feature correspondences between the images.

Sometimes static setups are mounted on a moving platform [7], e.g., Stewénius and Åström investigated the structure-and-motion problem for multiple rigidly moving cameras in an autonomous vehicle [8], and Frahm et al. [9] mounted several rigidly coupled cameras on a moving pole.

G. Bebis et al. (Eds.): ISVC 2009, Part I, LNCS 5875, pp. 391–402, 2009.

Recently, Thormählen et al. [10] presented a solution for multiple independently moving cameras that capture the same scene. This scenario frequently occurs in practice, e.g., multi-camera recordings of TV shows or multi-camera shots in movie productions. Camera motion estimation and 3D reconstruction is performed independently for each sequence with a feature-based single camera structure-and-motion approach. The independent reconstructions are then merged into a common global coordinate system, followed by a conjoined bundle adjustment [2, 11] over the merged sequences.

This paper adapts a similar approach and extends it for the case where a moving camera is located in the field of view of another moving camera. The following two contributions are made:

– A detection and tracking algorithm is used to determine the projection of a camera center in the image of another camera. Thereby, the user has the choice between a fully automatic and a semi-automatic approach. For the fully automatic approach, the cameras have to be retrofitted with a color pattern. For the semi-automatic approach the user manually defines the position of the camera center projection in the first image where the camera is visible. For both approaches the camera center is automatically tracked in the subsequent images, whereby the tracking algorithm is guided by the available initial camera center estimates.
– A sparse bundle adjustment algorithm is presented that allows incorporating the additional constraints given by the tracked camera centers. These constraints minimize the geometric error at the boundary of the reconstructed volume, which is usually the most sensitive part for reconstruction. Consequently, the total reconstruction accuracy can be improved significantly.

2   Scene Model

Consider a total number of N moving cameras, which capture the image sequences S_n, with $n = 1, \ldots, N$, consisting of K images $I_{k, n}$, with $k = 1, \ldots, K$, each. The cameras are synchronized, so that images $I_{k, n}$ for all n are recorded at the same point in time k. Let $\mathsf{A}_{k, n}$ be the 3×4 camera matrix corresponding to image $I_{k, n}$. A set of J 3D object points $\mathbf{P}_j = (P_x, P_y, P_z, 1)^\top$, with $j = 1, \ldots, J$, represents the static scene geometry, where the individual 3D object points are visible in at least a subset of all the images. In addition, the 2D feature points corresponding to $\mathbf{P}_j$, as seen in image $I_{k, n}$, are given by $\mathbf{p}_{j, k, n} = (p_x, p_y, 1)^\top$. This notation is clarified by Fig. 1.

Let $\mathbf{C}_{k, n} = (C_x, C_y, C_z, 1)^\top$ be the center of camera n at time k. The 2D image position of $\mathbf{C}_{k, n}$, as seen from camera $\tilde{n}$, with $n \neq \tilde{n}$, is now defined as $\mathbf{c}_{k, n, \tilde{n}} = (c_x, c_y, 1)^\top$. Likewise, the position of the projection of $\mathbf{C}_{k, n}$ in $I_{k, \tilde{n}}$ is defined as $\hat{\mathbf{c}}_{k, n, \tilde{n}} = \mathsf{A}_{k, \tilde{n}} \mathbf{C}_{k, n}$. Note that, in an ideal noise-free case $\hat{\mathbf{c}}_{k, n, \tilde{n}} = \mathbf{c}_{k, n, \tilde{n}}$; however, in real situations, it can usually be observed that $\hat{\mathbf{c}}_{k, n, \tilde{n}} \neq \mathbf{c}_{k, n, \tilde{n}}$.

3   Unconstrained Reconstruction

In a first step, synchronization of the N individual image sequences S_n is achieved using a method similar to the one presented by Hasler et al. [12]. This method analyzes

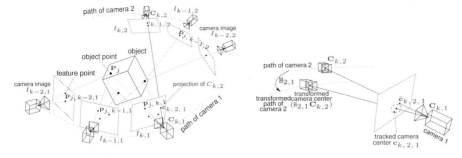

Fig. 1. Multiple cameras observe the same object. The camera center $\mathbf{C}_{k,2}$ of camera 2 is visible in image $I_{k,1}$ of camera 1, and vice versa.

Fig. 2. Compensation of the systematically erroneous path of camera 2 by applying a similarity transformation $\mathtt{H}_{2,1}$

the audio data, which is recorded simultaneously with the video data. A synchronization offset of at most half a frame is usually achieved. This approach allows the application of standard consumer cameras; a hard-wired studio environment is not required and the recordings can take place at arbitrary sets, including outdoor locations.

In a second step, each camera sequence is processed independently with a standard structure-from-motion algorithm. This establishes initial estimates for every single camera matrix $\mathtt{A}_{k,n}$ and every 3D object point $\mathbf{P}_j$ of the rigid scene. The 2D feature points $\mathbf{p}_{j,k,n}$ are detected and tracked through the image sequences. For each tracked 2D feature point $\mathbf{p}_{j,k,n}$ a corresponding 3D object point $\mathbf{P}_j$ is estimated. The applied algorithms are robust against outlier feature tracks introduced by moving objects, repetitive structures, or illumination changes. Intrinsic camera parameters are determined by self-calibration [3]. The estimation is finalized by a bundle adjustment.

In a third step, a similar approach as in [10] is employed to register the independent reconstructions into a common global coordinate system. The required similarity transformation for each individual reconstruction is estimated from corresponding feature tracks found via wide baseline matching between the image sequences. A conjoined bundle adjustment over all N reconstructions is performed to achieve equal distribution of the residual error over the whole scene. This minimization problem requires finding

$$\underset{\mathtt{A},\mathbf{P}}{\arg\min} \ \sum_{n=1}^{N}\sum_{j=1}^{J}\sum_{k=1}^{K} \mathrm{d}(\mathbf{p}_{j,k,n}, \mathtt{A}_{k,n}\mathbf{P}_j)^2, \tag{1}$$

where $\mathrm{d}(\dots)$ denotes the Euclidean distance. It is solved using the sparse Levenberg-Marquardt (LM) algorithm, as described in [2].

After these processing steps, an initial reconstruction of the scene has been established, which will be referred to as *unconstrained reconstruction* henceforth. Though the residual error is usually small, the inhomogeneous distribution of the corresponding feature tracks found by wide baseline matching may lead to estimation results not accurately reflecting the true structure of the scene. These inhomogeneities can arise because reliable merging candidates can usually be found more easily at the center of the reconstructed volume where the individual camera's fields of view overlap.

4   Detection and Tracking of Camera Centers

The unconstrained reconstruction can be improved by exploiting the visibility of the camera center in the field of view of another camera. To incorporate this additional constraint into the bundle adjustment, the determination of the 2D image positions of the visible camera centers $c_{k,n,\tilde{n}}$ is necessary. The user has the choice to either use a fully automatic or a semi-automatic approach. The fully automatic approach comprises the detection and tracking of the camera centers, whereas the semi-automatic approach requires the user to provide the positions of the projection of the camera centers for the first image they appear in.

4.1   Detection

The automatic detection of the camera centers requires the image of the cameras to be descriptive. One possibility would be to use a learning-based approach trained on the appearance of the camera. However, as small consumer cameras are used, reliable detection is challenging. As a consequence, the cameras were retrofitted with descriptive color patterns to facilitate the automatic detection.

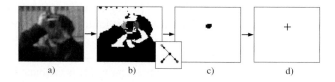

Fig. 3. Steps of the detection algorithm: a) input image detail, b) image after the conversion to HSV color space and color assignment, c) pixels that pass the geometric structure evaluation, d) detected camera center

Fig. 3 summarizes the automatic detection process and also shows the used color pattern, which consists of three patches with different colors. The pattern colors red, green, and blue were chosen, as they can easily be separated in color space. Since the front of the cameras is usually visible, the camera lens serves as additional black patch.

At first, the image is converted from RGB to HSV color space. All the pixels are then either assigned to one of the three pattern colors, black, or the background based on their proximity to the respective colors in HSV color space. Thereby, the value parameter (V) of the HSV color space model is ignored to achieve illumination invariance. For each black pixel the geometric structure of the pixels in a window around the pixel is examined. To be more specific, for each red pixel in the neighborhood of the black pixel, a green pixel is required to lie in the exact same distance in the opposite direction. Furthermore, a blue pixel must be located in the direction perpendicular to the connection line between the red and the green pixel. Again, the distance of the blue pixel from the black pixel must be exactly the same as the distance from the red to the black pixel. In addition, it must lie on the correct side of the connection line (see Fig. 3). Since there are usually multiple detections per camera, the centers of the clusters yield the desired positions of the camera centers.

4.2 Tracking

If fully automatic detection is not used, e.g., because color patterns for the camera are not available, the user is required to input the initial positions of the camera centers for the first image the camera appears. To simplify the notation, it is assumed for a moment that all other cameras are visible in the first image of every camera. Thus, the positions $c_{1,n,\tilde{n}}$ are now determined, either through user input or automatic detection.

For the tracking of the camera positions through the image sequences, a tracking algorithm based on Normalized Cross Correlation (NCC) matching is employed. A special feature of the algorithm is the guided matching process, which relies on the known initial estimates for the camera positions given by the unconstrained reconstruction to improve the robustness of the tracking.

Starting from $c_{k,n,\tilde{n}}$ in image $I_{k,\tilde{n}}$, the algorithm searches for $c_{k+1,n,\tilde{n}}$. This is done by calculating NCC matching scores for a window around $c_{k,n,\tilde{n}}$ in image $I_{k+1,\tilde{n}}$. The results of this operation are stored in a sorted list with positions producing the highest matching scores at the front.

Starting with the best match, it is checked whether the NCC score is above a user-defined threshold t_0 or not. If no matches with sufficiently high score are present, the algorithm aborts. In case of a valid match, the solution is cross-checked by calculating a second NCC score, between the current best match and the initial camera position $c_{1,n,\tilde{n}}$ (assuming the initial position to originate from $I_{1,\tilde{n}}$).

If the score for the second NCC matching is below another user-defined threshold t_1, instead of terminating, the algorithm simply processes the match with the next-lower score in the list. The cross-check reduces the effects of slow deviation of the feature point's description over time, since it assures that the original position can be found by reverse tracking.

Albeit performing very well and producing results of high tracking accuracy, this unguided tracking fails under certain conditions. Mismatches can occur due to similar image regions in the search window. Moreover, camera centers can leave and reenter the camera's field of view, or might get occluded by foreground objects, which causes traditional unguided tracking algorithms to lose the target.

Therefore, an additional constraint is introduced. As stated before, a set of good initial estimates for the camera projection matrices $A_{k,n}$ is available from the unconstrained reconstruction. These estimates contain estimates for the camera centers $C_{k,n}$, since $A_{k,n}C_{k,n} = 0$.

Due to registration errors, the tracked positions of the camera centers $c_{k,n,\tilde{n}}$ and the positions resulting from reprojection of the camera centers $\hat{c}_{k,n,\tilde{n}} = A_{k,\tilde{n}}C_{k,n}$ systematically deviate from each other (see Fig. 2).

These registration errors are compensated by estimating a common similarity transformation for the camera centers $C_{i,n}$, with $i = 1, \ldots, k$, represented by a 4×4 matrix $H_{n,\tilde{n}}$. More formally, it is required to find

$$\underset{H_{n,\tilde{n}}}{\arg\min} \; \sum_{i=1}^{k} d(c_{i,n,\tilde{n}}, \, A_{i,\tilde{n}} H_{n,\tilde{n}} C_{i,n})^2. \tag{2}$$

The similarity transformation $H_{n,\tilde{n}}$ allows for 7 degrees of freedom (3 for translation, 3 for rotation, and 1 for scale), and therefore a minimum of 4 measurements $c_{i,n,\tilde{n}}$

is needed to prevent ambiguities. For that reason it is clear that this approach cannot be applied to the first 3 images after the initial one, but starting from the fourth one it provides a sophisticated means of determining whether the match is a false positive or not, as described in the following.

The projection $(A_{k,\tilde{n}} H_{n,\tilde{n}} C_{k,n})$ gives a quite accurate estimate of the true $\hat{c}_{k,n,\tilde{n}}$ that can be used to determine if $c_{k,n,\tilde{n}}$ lies within a certain distance t_2 from the estimated projection of the camera center (see Fig. 2). The expectation of the residual error of Eq. (2) is given by $\epsilon_{res} = \sigma(1-(d/M))^{1/2}$, where $d = 7$ is the number of parameters of $H_{n,\tilde{n}}$ and $M = 2k$ is the number of measured $c_{i,n,\tilde{n}}$ (see Hartley and Zisserman [2] for details on the expectation of residual errors). Using the relation $t_2 = \epsilon_{res} + b$, with user-defined values for standard deviation σ and bias b, t_2 can be changed adaptively. The bias value accounts for systematic errors, which cannot be compensated by the similarity transformation.

If the currently best match from the sorted list does not fulfill the requirements, the next match in the list is processed. Once a match is accepted, the transformation $H_{n,\tilde{n}}$ is refined and the algorithm moves on to the next image.

If a tracked camera center leaves the camera's field of view or gets occluded by foreground objects, the remainder of the image sequence is checked for possible reappearance of the camera. The reappearance point can be predicted with $(A_{k,\tilde{n}} H_{n,\tilde{n}} C_{k,n})$ using the last transformation $H_{n,\tilde{n}}$ that was estimated before the track was lost. This prediction is then used to reinitialize the NCC matching process.

5 Sparse Bundle Adjustment with Additional Camera Center Constraints

Given tracked positions of camera centers $c_{k,n,\tilde{n}}$, Eq. (1) is expanded to accommodate for the additional constraints:

$$\underset{A,P}{\arg\min} \sum_{n=1}^{N}\sum_{j=1}^{J}\sum_{k=1}^{K} d(\mathbf{p}_{j,k,n}, A_{k,n}\mathbf{P}_j)^2 + w \sum_{n=1}^{N}\sum_{k=1}^{K}\sum_{\tilde{n}=1}^{N} d(\mathbf{c}_{k,n,\tilde{n}}, A_{k,\tilde{n}}\mathbf{C}_{k,n})^2$$

$$(3)$$

for $n \neq \tilde{n}$, with w being a user-defined weight factor.

As in the unconstrained case, this minimization problem can be solved by the sparse LM algorithm, as derived in the following. A similar notation as in the book by Hartley and Zisserman [2] is used.

The measurement vector $\tilde{\mathbf{p}} = (\bar{\mathbf{p}}^\top, \bar{\mathbf{c}}^\top)^\top$ is assembled from the vector $\bar{\mathbf{p}}$ of all 2D feature points $\mathbf{p}_{j,k,n}$ placed one after another in a single column, and the vector $\bar{\mathbf{c}}$ constructed alike from all tracked camera centers $c_{k,n,\tilde{n}}$.

In a similar fashion, the parameter vector $\mathbf{q} = (\mathbf{a}^\top, \mathbf{b}^\top)^\top$ can be obtained by assembling a parameter vector $\mathbf{a}$ denoting the corresponding set of parameters describing the cameras, and parameter vector $\mathbf{b}$ denoting the corresponding set of parameters describing the points.

In each step of the LM algorithm the following linear equation system needs to be solved:

$$J\boldsymbol{\delta} = \boldsymbol{\epsilon} \tag{4}$$

with the Jacobian matrix $J = \partial\tilde{p}/\partial q$, the residual vector ϵ, and the update vector δ of the LM algorithm, which is the solution to the least squares problem. The residual vector $\epsilon = \left(\epsilon_p{}^\top, \epsilon_c{}^\top\right)^\top$ is assembled from the residual vector of the 2D feature points ϵ_p and the residual vector of the camera centers ϵ_c.

The Jacobian matrix J has a block structure

$$J = \begin{bmatrix} \bar{A} & \bar{B} \\ \bar{C} & 0 \end{bmatrix} \quad, \quad \text{where} \quad \bar{A} = \begin{bmatrix} \dfrac{\partial\bar{p}}{\partial a} \end{bmatrix} \quad, \quad \bar{B} = \begin{bmatrix} \dfrac{\partial\bar{p}}{\partial b} \end{bmatrix} \quad \text{and} \quad \bar{C} = \begin{bmatrix} \dfrac{\partial\bar{c}}{\partial a} \end{bmatrix}. \quad (5)$$

The linear equation system of Eq. (4) evaluates to

$$[\tilde{A}|\tilde{B}] \begin{pmatrix} \delta_a \\ \delta_b \end{pmatrix} = \epsilon \quad, \quad \text{with} \quad \tilde{A} = \begin{bmatrix} \bar{A} \\ \bar{C} \end{bmatrix} \quad \text{and} \quad \tilde{B} = \begin{bmatrix} \bar{B} \\ 0 \end{bmatrix}. \quad (6)$$

The normal equations corresponding to Eq. (4) are given as

$$J^\top \Sigma^{-1} J\delta = J^\top \Sigma^{-1} \epsilon \quad, \quad \text{with} \quad \Sigma = \begin{bmatrix} \Sigma_p & 0 \\ 0 & \Sigma_c \end{bmatrix}, \quad (7)$$

where Σ_p is the covariance matrix of the 2D feature points, and Σ_c the covariance matrix of the tracked camera centers. In absence of other knowledge, the matrix Σ_c is chosen to be the identity matrix. The normal equations evaluate to

$$\begin{bmatrix} \tilde{A}^\top \Sigma^{-1} \tilde{A} & \tilde{A}^\top \Sigma^{-1} \tilde{B} \\ \tilde{B}^\top \Sigma^{-1} \tilde{A} & \tilde{B}^\top \Sigma^{-1} \tilde{B} \end{bmatrix} \begin{pmatrix} \delta_a \\ \delta_b \end{pmatrix} = \begin{pmatrix} \tilde{A}^\top \Sigma^{-1} \epsilon \\ \tilde{B}^\top \Sigma^{-1} \epsilon \end{pmatrix}, \quad (8)$$

which can be simplified by back-substitution:

$$\begin{bmatrix} \bar{A}^\top \Sigma_p^{-1} \bar{A} + \bar{C}^\top \Sigma_c^{-1} \bar{C} & \bar{A}^\top \Sigma_p^{-1} \bar{B} \\ \bar{B}^\top \Sigma_p^{-1} \bar{A} & \bar{B}^\top \Sigma_p^{-1} \bar{B} \end{bmatrix} \begin{pmatrix} \delta_a \\ \delta_b \end{pmatrix} = \begin{pmatrix} \bar{A}^\top \Sigma_p^{-1} \epsilon_p + \bar{C}^\top \Sigma_c^{-1} \epsilon_c \\ \bar{B}^\top \Sigma_p^{-1} \epsilon_p \end{pmatrix}. \quad (9)$$

The corresponding block structure is

$$\begin{bmatrix} U^* & W \\ W^\top & V^* \end{bmatrix} \begin{pmatrix} \delta_a \\ \delta_b \end{pmatrix} = \begin{pmatrix} \epsilon_A \\ \epsilon_B \end{pmatrix}, \quad (10)$$

where U^* denotes U augmented by multiplying its diagonal entries by a factor of $1 + \lambda$, and V^* likewise. Left multiplication with $\begin{bmatrix} I & -WV^{*-1} \\ 0 & I \end{bmatrix}$, where I is the identity matrix, yields

$$\begin{bmatrix} U^* - WV^{*-1}W^\top & 0 \\ W^\top & V^* \end{bmatrix} \begin{pmatrix} \delta_a \\ \delta_b \end{pmatrix} = \begin{pmatrix} \epsilon_A - WV^{*-1}\epsilon_B \\ \epsilon_B \end{pmatrix}. \quad (11)$$

The equation

$$\left(U^* - WV^{*-1}W^\top\right) \delta_a = \epsilon_A - WV^{*-1}\epsilon_B \quad (12)$$

can be used to find δ_a, which may be back-substituted to get δ_b from

$$V^* \delta_b = \epsilon_B - W^\top \delta_a. \quad (13)$$

These derivations are closely related to those of standard sparse bundle adjustment. It is thus very easy to incorporate the modifications into existing implementations without introducing significant additional computational overhead.

This *constrained bundle adjustment* can improve the unconstrained reconstruction of Sec. 3. At first, a projective constrained bundle adjustment is performed (12 parameters per 3×4 camera matrix A). Afterwards, a new self-calibration and a metric constrained bundle adjustment with 7 parameters per camera view (3 for translation, 3 for rotation, and 1 for focal length) is executed.

6   Results

In this section, experiments with synthetic and real scenes are shown. The experiments on real scenes are also presented in the video provided with this paper, which can be found at http://www.mpi-inf.mpg.de/users/ckurz/

6.1   Experiments with Synthetic Data

To evaluate if the constrained sparse bundle adjustment of Sec. 5 achieves higher accuracy than the standard bundle adjustment used for the generation of the unconstrained reconstruction of Sec. 3, a comparison with synthetic data is performed.

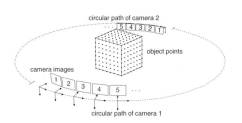

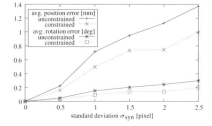

Fig. 4. Setup of the scene to generate synthetic measurement values with known ground truth camera parameters

Fig. 5. Average absolute position error and average absolute rotation error of the estimated camera motion over standard deviation σ_{syn}

Fig. 4 shows the setup of the scene to generate synthetic measurement values for the 2D feature points $\mathbf{p}_{j,k,n}$ and the tracked camera centers $\mathbf{c}_{k,n,\tilde{n}}$. Two virtual cameras with known ground truth camera parameter are observing the same 296 object points $\mathbf{P}_j$, which are placed in a regular grid on the surface of a cube with an edge length of 100 mm. The two virtual cameras have an opening angle of 30 degrees and rotate on a circular path with a radius of 300 mm around the object points. Each virtual camera generates 20 images, and the camera centers are mutually visible in every image. Using the ground truth camera parameters and ground truth positions of object points, ground truth measurements are generated for the 2D feature points $\mathbf{p}_{j,k,n}$ and the tracked camera centers $\mathbf{c}_{k,n,\tilde{n}}$. These measurements are then disturbed with Gaussian noise with a standard deviation σ_{syn}. Furthermore, 20 percent of the 2D feature points are disturbed with a very large offset to simulate outliers. The structure-from-motion algorithm is then applied with constrained and standard bundle adjustment for 50 times, each time with different randomly disturbed measurements.

Each resulting reconstruction is registered to the ground truth reconstruction by aligning both with an estimated similarity transformation.

In Fig. 5 the average absolute position error and average absolute rotation error for the estimated camera motion for different standard deviations are shown. It can be observed that the constrained always outperforms the standard bundle adjustment; e.g., for a standard deviation of $\sigma_{syn} = 1.0$ pixel, the average absolute position error is reduced by 30.0 percent and the average absolute rotation error by 38.7 percent.

6.2 Experiments with Real Scenes

The presented approach is applied to several real image sequences. These image sequences are first processed by a standard bundle adjustment, resulting in an unconstrained reconstruction as described in Sec. 3. Afterwards, the camera positions are obtained with the described detection and tracking algorithm of Sec. 4, and the constrained sparse bundle adjustment including an updated self-calibration is applied to the sequences.

In accordance with Eq. (3), two different error measures are introduced: The root-mean-squared residual error of the tracked camera centers

$$r_1 = \left(\frac{1}{C} \sum_{n=1}^{N} \sum_{k=1}^{K} \sum_{\tilde{n}=1}^{N} \mathrm{d}(\mathbf{c}_{k,n,\tilde{n}} , \mathsf{A}_{k,\tilde{n}} \, \mathbf{C}_{k,n})^2 \right)^{\frac{1}{2}} \tag{14}$$

with C the total number of all tracked camera centers, and the root-mean-squared residual error of the 2D feature points

$$r_2 = \left(\frac{1}{P} \sum_{n=1}^{N} \sum_{j=1}^{J} \sum_{k=1}^{K} \mathrm{d}(\mathbf{p}_{j,k,n} , \mathsf{A}_{k,n} \, \mathbf{P}_j)^2 \right)^{\frac{1}{2}} , \tag{15}$$

where P is the total number of all 2D feature points.

Obviously, the introduction of additional constraints restrains the bundle adjustment, so that a value for r_2, as obtained in the unconstrained case, can usually not be achieved in the constrained case. Therefore, if r_1 is significantly reduced and the value of r_2 increases only slightly, it can be assumed that a plausible solution was found.

Both r_1 and r_2 are first evaluated for the unconstrained reconstruction. Then the constrained bundle adjustment is applied and r_1 and r_2 are measured again.

In the following paragraphs results for three scenes are presented. The image sequences of these scenes have a resolution of 1440×1080 pixel and were recorded by 4 moving HDV consumer cameras. The lengths of the sequences in the first scene are 80 images per camera (320 images total), 400 images per camera for the second scene (1600 images total), and 400 images per camera for the third scene. The parameters $\sigma = 3$ pixel, $b = 2$ pixel, $t_0 = 0.8$, and $t_1 = 0.6$ are used for the tracking algorithm.

For the first scene, depicting a runner jumping over a bar, the weight factor $w = 0.1 \cdot P/C$ is used for the constrained bundle adjustment. The error measures for this scene can be found in Table 1 and the result is shown in Fig. 6.

The second scene depicts a skateboard ramp. The wide baseline matching used for the generation of the unconstrained reconstruction finds mainly corresponding feature tracks at the center of the reconstruction volume. As can be verified in Fig. 7, this leads to acceptable results in the center of the reconstruction volume but results in large deviations at the borders of the reconstruction volume. This becomes evident because the overlay geometry in the center (green rectangle) does fit but the projections of the camera centers show large errors. In contrast, the constrained bundle adjustment with $w = 0.001 \cdot P/C$ is able to guide the estimate parameters to a solution, which generates plausible results for the whole reconstruction volume. In particular, the self-calibration benefits from improved estimates, as can be verified by the overlay geometry in Fig. 8, where the perpendicular structure is slightly off in the unconstrained reconstruction and fits well after the constrained bundle adjustment. Table 1 shows the results.

Table 1. Results for r_1 and r_2 of the three scenes

Scene	"Running"		"Ramp"		"Statue"	
Method	r_1 [pixel]	r_2 [pixel]	r_1 [pixel]	r_2 [pixel]	r_1 [pixel]	r_2 [pixel]
unconstrained	16.85	1.67	99.37	0.77	123.78	0.88
constrained	6.26	1.75	5.92	0.98	4.43	0.93

The third scene shows an art statue in a park. For this scene the automatic camera detection algorithm is employed, whereas for the two previous examples only the automatic tracking with manual initialization is used. The automatic detection works reliably and similar results as for the previous examples are achieved. Results for a weight factor $w = 0.001 \cdot P/C$ are shown in Figs. 8 and 9, and Table 1).

To evaluate the automatic detection algorithms, it is applied to all images of the sequence and the result is checked manually. In spite of severe illumination changes due to the grazing incidence of the sunlight, the detection algorithm reliably determines all camera center positions without producing any false positives.

7   Conclusion

An algorithm for multi-camera motion estimation is presented, which takes advantage of mutual visibility of cameras. An automatic detection and tracking algorithm using color patterns, NCC matching, and homography estimation is proposed that is capable of tracking the camera positions through the image sequences. Furthermore, a constrained bundle adjustment is introduced, which allows to include the additional constraints for the tracked camera centers. It is an extended version of the widely used sparse Levenberg-Marquardt algorithm for bundle adjustment. Despite the introduction of additional constraints, the sparse matrix structure of the equation systems is preserved, so that the computational effort does not increase.

Fig. 6. Image sequence "Running" recorded with 4 moving cameras: Example images of camera 2 (the 4 leftmost images) and camera 4 (the 4 rightmost images) are presented. The two left images of camera 2 depict the path of camera 3 and 4 (in blue), prior to (left) and after constrained optimization (right), and the two left images of camera 4 depict the path of camera 2 in a similar fashion. The two right images show detail magnifications in each case. The detected camera positions are indicated by a red circle. Deviations of the estimated camera positions from the actual positions are depicted by red lines and are clearly visible in the magnifications.

Fig. 7. Image sequence "Ramp" recorded with 4 moving cameras: Example images of camera 1 and camera 3 are presented, showing the path of camera 3 and camera 1, respectively

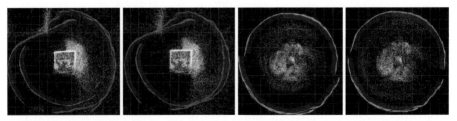

Fig. 8. Top views of the reconstructions for scenes "Ramp" (leftmost images) and "Statue" (rightmost images): Comparison between the unconstrained reconstruction (left) and the result of the constrained bundle adjustment (right). The estimated camera positions and orientations are depicted by small coordinate systems (optical axis, horizontal image direction, and vertical image direction, in blue, red, and green, respectively). The 3D object points are displayed as white dots.

Fig. 9. Image sequence "Statue" recorded with 4 moving cameras: Example images of camera 2 and camera 3 are presented, showing the path of camera 3 and camera 2, respectively

Evaluations have been conducted on both, synthetic and real-world image sequences. On the synthetic sequences the average absolute error could be reduced by approximately 30 percent. For the real world image sequences, a very significant improvement of the estimated camera parameters could be observed. It turns out that the additional constraints minimize the geometric error at the boundary of the reconstructed volume and thereby can also ameliorate the self-calibration process.

An obvious drawback is the necessity of the cameras to be at least visible in a subsequence of frames, to allow our algorithm to generate estimation results with improved accuracy. However, during recording image sequences, it turned out to be quite hard to avoid situations where other cameras are visible. Therefore, the presented algorithm can find numerous applications in multi-camera computer vision. A current limitation is that the tracking algorithm does determine only the position of the camera lens and not the true mathematical camera center point. However, as the projection of the camera is small in the images, this is a good approximation.

Future work will address the automatic determination of the weighting factor w of the camera center constraints as well as the inclusion of other constraints to further improve the accuracy and robustness of camera motion estimation.

References

1. Gibson, S., Cook, J., Howard, T., Hubbold, R., Oram, D.: Accurate camera calibration for off-line, video-based augmented reality. In: ISMAR, Darmstadt, Germany (2002)
2. Hartley, R.I., Zisserman, A.: Multiple View Geometry. Cambridge University Press, Cambridge (2000)
3. Pollefeys, M., Gool, L.V., Vergauwen, M., Verbiest, F., Cornelis, K., Tops, J., Koch, R.: Visual modeling with a hand-held camera. IJCV 59, 207–232 (2004)
4. Wilburn, B., Joshi, N., Vaish, V., Talvala, E.V., Antunez, E., Barth, A., Adams, A., Horowitz, M., Levoy, M.: High performance imaging using large camera arrays. Proceedings of Siggraph 2005, ACM Trans. Graph. 24, 765–776 (2005)
5. Rosenhahn, B., Schmaltz, C., Brox, T., Weickert, J., Cremers, D., Seidel, H.-P.: Markerless motion capture of man-machine interaction. In: CVPR, Anchorage, USA (2008)
6. Sato, J.: Recovering multiple view geometry from mutual projections of multiple cameras. IJCV 66, 123 (2006)
7. Jae-Hak, K., Hongdong, L., Hartley, R.: Motion estimation for multi-camera systems using global optimization. In: CVPR, Anchorage, AK, USA (2008)
8. Stewénius, H., Åström, K.: Structure and motion problems for multiple rigidly moving cameras. In: Pajdla, T., Matas, J(G.) (eds.) ECCV 2004. LNCS, vol. 3023, pp. 252–263. Springer, Heidelberg (2004)
9. Frahm, J.-M., Köser, K., Koch, R.: Pose estimation for multi-camera systems. In: 26th DAGM Symposium, Tübingen, Germany, pp. 27–35 (2004)
10. Thormählen, T., Hasler, N., Wand, M., Seidel, H.-P.: Merging of unconnected feature tracks for robust camera motion estimation from video. In: CVMP, London, UK (2008)
11. Triggs, B., Mclauchlan, P.F., Hartley, R.I., Fitzgibbon, A.W.: Bundle adjustment – a modern synthesis. In: Triggs, B., Zisserman, A., Szeliski, R. (eds.) ICCV-WS 1999. LNCS, vol. 1883, p. 298. Springer, Heidelberg (2000)
12. Hasler, N., Rosenhahn, B., Thormählen, T., Wand, M., Gall, J., Seidel, H.P.: Markerless motion capture with unsynchronized moving cameras. In: CVPR, Miami Beach, FL, USA (2009)

Optical Flow Computation from an Asynchronised Multiresolution Image Sequence

Yusuke Kameda[1], Naoya Ohnishi[2], Atsushi Imiya[3], and Tomoya Sakai[3]

[1] Graduate School of Advanced Integration Science, Chiba University
[2] School of Science and Technology, Chiba University
[3] Institute of Media and Information Technology, Chiba University
Yayoicho 1-33, Inage-ku, Chiba, 263-8522, Japan

Abstract. We develop a method for the optical flow computation from a zooming image sequence. The synchronisation of image resolution for a pair of successive images in an image sequence is a fundamental requirement for optical flow computation. In a real application, we are, however, required to deal with a zooming and dezooming image sequences, that is, we are required to compute optical flow from a multiresolution image sequence whose resolution dynamically increases and decreases. As an extension of the multiresolution optical flow computation which computes the optical flow vectors using coarse-to-fine propagation of the computation results across the layers, we develop an algorithm for the computation of optical flow from a zooming image sequence.

1 Introduction

In this paper, we develop an optical flow computation method for zooming image sequences. In the traditional optical flow computation methods [5,7,9,10,12], the resolution of a pair of successive images in a sequence is synchronised, that is, all images in an image sequence are assumed to be observed in the same resolution. However, if we observe an image sequence using a zooming and dezooming camera, the resolution of images in an image sequence, respectively, increase and decrease, that is, the resolution in a pair of successive images in an image sequence is asynchronised.

Optical flow is an established method for motion analysis in computer vision [5,7,9,10,12] and introduced to many application areas such as cardiac motion analysis [1], robotics [3]. In a real application, we are required to deal with an image sequence whose image resolution is time dependent, that is, we are required to compute optical flow from a multiresolution image sequence whose resolutions increases and decreases temporally. For instance, for the robust tracking of moving object using a camera mounted on the robot, the robot is required to have the geometric depth to the object to observe clear images for the accurate computation of geometrical features. Without geometric depth to the object from the camera, the camera mounted on the robot captures a defocused image sequence. Therefore, while tracking of the object, the robot would observe a zooming image sequence.

G. Bebis et al. (Eds.): ISVC 2009, Part I, LNCS 5875, pp. 403–414, 2009.

Therefore, for the computation of the optical flow vectors from a zooming image sequence, we develop a method to synchronise the resolution of a pair of successive images in this zooming image sequence. To synchronise the resolution of a pair of images, we construct an estimation method of the zooming operation from images in the observed image sequence. Furthermore, we clarify a convergence property of a numerical scheme for optical flow computation from asynchronised image sequence using the unimodality of the variational functionals for optical flow computation.

Zooming of the lower resolution image and dezooming of the higher resolution image are a methods to yield a pair of images of synchronised resolution from zooming images. We use the second method. For the achievement of this dezooming operation, we are required to estimate the zooming operation from images in the image sequence. In this paper, assuming that the zooming and dezooming operations are the reverse operations, we develop a method to identify this zooming operation.

The proposing method to compute optical flow from this zooming image sequence is an extension of the multiresolution optical flow computation, [15,16,17,14,4] since our method locally utilises tow different resolution images. The common assumption on refs. [14,4] is that the motion in coarse resolution image can be accepted as an initial estimator to the motion in finer resolution images. This assumption comes from the heuristics that the global motion in a region is an average of local motion. We also accept the same assumption on the motion in the image sequence.

2   Multiresolution Optical Flow Computation

2.1   Optical Flow Computation

We assume that our images are elements of the Sobolev space $H^2(\mathbf{R}^2)$. For a spatiotemporal image $f(\boldsymbol{x}, t)$, $\boldsymbol{x} = (x, y)^\top$, the optical flow vector $\boldsymbol{u} = (u, v)^\top$ of each point $\boldsymbol{x} = (x, y)^\top$ is the solution of the singular equation

$$f_x u + f_y v + f_t = \nabla f^\top \boldsymbol{u} + \partial_t f = 0. \tag{1}$$

To solve this equation, the regularisation method [5,6,10,12] which minimises the criterion

$$J(\boldsymbol{u}) = \int_{\mathbf{R}^2} \left\{ (\nabla f^\top \boldsymbol{u} + \partial_t f)^2 + \kappa P(\boldsymbol{u}) \right\} d\boldsymbol{x} \tag{2}$$

is employed, where $P(\boldsymbol{u})$ is a convex prior of $\boldsymbol{u}$ such that

$$\lambda P(\boldsymbol{u}) + (1 - \lambda)P(\boldsymbol{v}) \geq P(\lambda \boldsymbol{u} + (1 - \lambda)\boldsymbol{v}), \ \ 0 \leq \lambda \leq 1. \tag{3}$$

The minimisation of eq. (2) is achieved by solving a dynamic system

$$\frac{\partial \boldsymbol{u}}{\partial \tau} = -\nabla_{\boldsymbol{u}} J(\boldsymbol{u}), \tag{4}$$

where $\nabla_{\boldsymbol{u}}$ is the gradient with respect to $\boldsymbol{u}$. If

$$P(\boldsymbol{u}) = tr\nabla\boldsymbol{u}\nabla\boldsymbol{u}^{\top} = |\nabla u|^2 + |\nabla v|^2, \tag{5}$$

$$P(\boldsymbol{u}) = tr\boldsymbol{H}\boldsymbol{H}^{\top} = |u_{xx}|^2 + 2|u_{xy}|^2 + |u_{yy}|^2 + |v_{xx}|^2 + 2|v_{xy}|^2 + |v_{yy}|^2, \tag{6}$$

$$P(\boldsymbol{u}) = |\nabla u| + |\nabla v|, \tag{7}$$

where $\boldsymbol{H}$ is the Hessian of $\boldsymbol{u}$, we have the Horn-Schunck method [5,9,10], deformable model method [7,8], and total variational method [11,12], respectively for optical flow computation.

2.2   Multiresolution Expression

For a function $w(x,y)$ which is nonnegative in a finite closed region Ω in $\mathbf{R}^2$ and

$$\int_{-\infty}^{\infty} \int_{-\infty}^{\infty} w(x,y)dxdy = 1, \tag{8}$$

we deal with a linear transformation

$$g(x,y,t) = Rf(x,y,t) = \int_{-\infty}^{\infty} \int_{-\infty}^{\infty} w(u,v)f(x-u,y-v,t)dudv = w *_2 f, \tag{9}$$

where $*_2$ stands for the two-dimensional convolution. Equation (9) defines a smoothing operation on $\mathbf{R}^2$, since

$$\int_{-\infty}^{\infty} \int_{-\infty}^{\infty} |g(x,y,t)|^2 dxdy \leq \int_{-\infty}^{\infty} \int_{-\infty}^{\infty} |f(x,y,t)|^2 dxdy. \tag{10}$$

Equation (9) generates a pair of hierarchical multiresolution images

$$f^n(x,y,k) = R^n f(x,y,k), \quad f^n(x,y,k+1) = R^n f(x,y,k+1) \tag{11}$$

and the sequence of hierarchical multiresolution image sets

$$F(k) = \{f^n(x,y,k)\}_{n=0}^{N} = \{f_k^n\}_{n=0}^{N}, \quad k = 0, 1, 2, \cdots, K. \tag{12}$$

From a successive pair of hierarchical multiresolution image sets $F(k)$ and $F(k+1)$, Algorithm 1 computes $\boldsymbol{u} := \boldsymbol{u}_k^0$.

Algorithm 1. The Multiresolution Algorithm

Data: $f_k^N \cdots f_k^0$
Data: $f_{k+1}^N \cdots f_{k+1}^0$
Result: optical flow u_k^0
$n := N$;
while $n \neq 0$ **do**
$\quad | \quad \boldsymbol{u}_k^n := I(u(f_k^n, f_{k+1}^n)\boldsymbol{u}_k^{n+1}))$;
$\quad | \quad n := n - 1$
end

The operation $I(u(f_k^i, f_{k+1}^i), u_k^{i+1})$ in Algorithm 1 expresses computation of optical flow from a pair of images $f_k^n = f^n(x, k)$ and $f_k^n(x - u_k^{n+1}, k)$. We call this algorithm the Gaussian Multiresolution Algorithm [14,15,18] This is general expression of the multiresolution optical flow computation, which is an extension of the Lucas-Kanade with pyramid transform [13,14].

Let u^k be the optical flow computed from $R^n f(x, y, t)$ and $R^n(x, y, t + 1)$. If optical flow field is locally stationary, u^n lies in an neighbourhood of u^{n-1}. Therefore, by replacing f_k^n to $f^n(x - u_k^{n+1}, k)$, the solutions u^k converges to bmu.

3   Zooming and Optical Flow

The operator R satisfies the relations $g_x = Rf_x$ and $g_y = Rf_y$ for $g_t = Rf_t$. The solution of eq. (1) is

$$u = -\frac{\partial_t f}{|\nabla f|^2} \nabla f + c \nabla f^\perp \tag{13}$$

where $\nabla f^\perp = (-f_y, f_x)^\top$ for a real constant c, since $\nabla f^\top \nabla f^\perp = 0$. In the same way, for $g = Rf$ the solution of $\nabla g^\top v + \partial_t g = 0$ is

$$v = -\frac{\partial_t g}{|\nabla g|^2} \nabla Rf + c' \nabla g^\perp \tag{14}$$

where $\nabla g^\perp = (-g_y, g_x)^\top = \nabla Rf^\perp$ for a real constant c'. Therefore,

$$Ru = R\left(-\frac{\partial_t f}{|\nabla f|^2} \nabla f\right) + c \nabla f^\perp \neq -\frac{\partial_t g}{|\nabla g|^2} \nabla g + c' \nabla f^\perp = v. \tag{15}$$

Equation (15) implies the next lemma.

Lemma 1. *Since $Ru \neq v$, setting where O to be the operation to compute optical flow from an image sequence, the relation $RO \neq OR$ is satisfied.*

The relations
    Setting $R^\dagger$ to be the Moor-Penrose inverse operation of R, Lemma 1 implies the relation $R^\dagger ROR^\dagger \neq R^\dagger ORR^\dagger$. Here $R^\dagger$ symbolically express zooming operation. Therefore, we have the next lemma.

Lemma 2. *The operation $R^\dagger$ satisfies the relation $R^\dagger O \neq OR^\dagger$.*

From lemmas 1 and 2, we have the next theorem.

Theorem 1. *The operator O which computes optical flow is noncommutative with both R, dezooming, and $R^\dagger$, zooming.*

4   Optical Flow Computation of Zooming Sequence

4.1   Zooming Flow

A zooming multiresolution image sequences is modelled as

$$f(x, y, t) = \begin{cases} R^{N-t} f_0(x, y, t), & 0 \le t \le N \\ f_0(x, y, t), & t \ge N+1 \end{cases} \quad (16)$$

assuming that $f_0(x, y, t)$ is the clear zoomed image sequence. We develop the algorithm to compute $\boldsymbol{u}(x, y, t)$ from this zooming sequence.

Theorem 1 implies that synchronisation of the resolution between $f(x, y, t)$ and $f(x, y, t+1)$ is required for the optical flow computation from a zooming sequence. Therefore, we accept the solution $\boldsymbol{u}(x, y, t)$ of the equation

$$\nabla f(x, y, t)^\top \boldsymbol{u} + (Rf(x, y, t+1) - f(x, y, t)) = 0 \quad (17)$$

as the optical flow field at time t for $t < n$, since the resolution of $Rg(x, y, t+1)$ and $f(x, y, t)$ is equivalent. Equation (17) derives the following procedure for the computation of optical flow $\boldsymbol{u}(x, y, t)$ for $t < n$.

1. *Compute* $\nabla f(x, y, t) = (f_x f_y)^\top$
2. *Set* $f_t = Rf(x, y, t+1) - f(x, y, t)$
3. *Compute the solution of eq. (2).*

This procedure derives Algorithm 2.

Algorithm 2. The Dynamic Multiresolution Algorithm

Data: $f_k^N f_{k+1}^{N-1} \cdots f_{k+N}^0 \quad 0 \le k \le N$
Result: optical flow u_N^0
$i := n$;
$k := 0$;
while $n \ne 0$ **do**
$\quad \boldsymbol{u}_k^{n-1} := I(u(f_k^{n-1}, f_{k+1}^{n-1}), \boldsymbol{u}_{k-1}^{n(\infty)})$;
$\quad n := n - 1$;
$\quad k := k + 1$;
end

Figure 1 shows the charts on the relation between the resolution of images and the frame numbers of image sequences both for Algorithms 1 and 2. The Multiresolution Dynamic Algorithm propagates the flow vectors to the next successive frame pairs. The classical algorithm computes optical flow from all resolution images of fixed successive times as shown in Fig. 1 (b).

4.2   Mathematical Property of Algorithm

For the energy

$$E(\boldsymbol{u}_t^n) = \int_{\mathbf{R}^2} \left\{ ((\nabla f^n)^\top \boldsymbol{u}^n + \partial_t f^n)^2 + \kappa_n P(\boldsymbol{u}^n)) \right\} d\boldsymbol{x}, \quad (18)$$

we set $\boldsymbol{u}_t^n = \text{argument} \left(\min E(\boldsymbol{u}_t^n) \right)$.

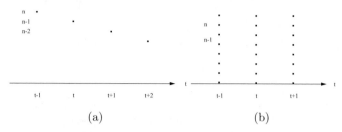

Fig. 1. Layer-Time Charts of Algorithms. (a) The dynamic algorithm propagates the flow vectors to the next successive frame pairs. (b) The traditional algorithm computes optical flow from all resolution images of fixed times.

Since $E(\boldsymbol{u}_t^n)$ is a convex functional, this functional satisfies the inequality $E(\boldsymbol{u}_t^n) < E(\boldsymbol{u}(t))$, for $\boldsymbol{u}_t^n \neq \boldsymbol{u}(t)$. Therefore, we have the relation

$$E(\boldsymbol{u}_t^{n-1}) \leq E(\boldsymbol{u}_t^n). \tag{19}$$

This relation implies that it is possible to generate a sequence which reaches to $E(\boldsymbol{u}_{t+N}^0)$ from $E(\boldsymbol{u}_t^N)$, since

$$E(\boldsymbol{u}_t^N) \geq E(\boldsymbol{u}_{t+1}^{N-1}),$$
$$E(\boldsymbol{u}_{t+1}^{N-1}) \geq E(\boldsymbol{u}_{t+2}^{N-2}),$$
$$\vdots \tag{20}$$
$$E(\boldsymbol{u}_{t+(N-2)}^2) \geq E(\boldsymbol{u}_{t+(N-1)}^1),$$
$$E(\boldsymbol{u}_{t+(N-1)}^1) \geq E(\boldsymbol{u}_{t+N}^0),$$

setting $\boldsymbol{u}_{t+1}^{n-1} = \boldsymbol{u}_t^n + \boldsymbol{d}_k^n$ for $|\boldsymbol{d}^{n-1}| \ll |\boldsymbol{u}^n|$.

5   Numerical Examples

For performance evaluation of the Algorithm 2, we adopt $P(\boldsymbol{u}) = tr\nabla\boldsymbol{u}\nabla\boldsymbol{u}^\top$. Furthermore, we assume our zooming sequence is

$$f^{n-k}(x, y, k) = G(x, y, \sigma_{n-k}) *_2 f(x, y, k), \tag{21}$$

where

$$G(x, y; \sigma) = \frac{1}{2\pi\sigma} \exp\left(-\frac{x^2 + y^2}{2\sigma^2}\right), \ \sigma_k = s \times k \tag{22}$$

for an appropriate positive constant s, and the initial image $G(x, y, \sigma_n) *_2 f_0(x, y, 0)$.

5.1   Optical Flow Computation

Figures 2 and 3 show numerical results of optical flow computation from zooming image sequences generated from Marbled Block 1 sequence and Yosemite sequences. In these examples, the number of layers N is selected as 6, so that he scale parameters are 9, 7, 5, 3, 1, 0. The regularisation parameter κ is 1000 for each layer, and the iteration is 300 for each layer. These results show that our algorithm for the optical flow computation from zooming image sequence accurately computes the optical flow vectors.

Let $l_n = |\hat{\boldsymbol{u}}_n - \boldsymbol{u}_n|$, $a_n = \angle[\hat{\boldsymbol{u}}_n, \boldsymbol{u}_n]$, and $r_n = \frac{|\hat{\boldsymbol{u}}_n - \boldsymbol{u}_n|}{|\boldsymbol{u}_n|}$ be the displacement, the angles, and residual of between two optical flow vectors $\{\boldsymbol{u}_n\}_{i=1}^{N}$ and $\{\hat{\boldsymbol{u}}_n\}_{i=1}^{N}$ which are computed using Algorithm 1 and Algorithm 2, respectively. We evaluated the maximum, the mean, and the variance of l_n, a_n, and r_n In the table we denote them L, A, and R, respectively.

In Tables 1 and 2, the maximums of differences of angles are almost 180deg. This geometrical property of the resolution means that the flow vectors computed by the two methods are numerically in the opposite directions. This geometrical configuration happens on in the background of image sequences at the points where the gradient of images is zero and the structure tensor is zero tensor. However, the length of these anti-directional vectors are short. Therefore, these anti-directional configuration of vectors do not effect to the mean and variance of the lengths. If we add an operation to remove these anti-directional vectors by evaluating the structure tensor of each point, the results would be refined. However, considering this configuration of flow vectors in the computed results, the statistical analysis shows that Algorithm 2 which computes optical flow from a zooming image sequence numerically derives acceptable results.

Table 1. L, A, and R for the Gaussian Multiresolution Dynamic Algorithm and the Gaussian Multiresolution Algorithm

	L			A			R		
sequence	max	mean	variance	max	mean	variance	max	mean	variance
block1	0.162	0.027	0.001	172.0	0.108	0.174	5.943	0.068	0.0333
yosemite	0.2112	0.018	0.001	131.0	4.409	5.960	1.558	0.099	0.045

Table 2. L, A, and R for the Gaussian Multiresolution Dynamic Algorithm and the Horn and Schunck Method

	L			A			R		
sequence	max	mean	variance	max	mean	variance	max	mean	variance
block1	0.761	0.116	0.020	172.0	1.614	2.313	481.520	1.454	83.0218
yosemite	0.812	0.091	0.0204	172.0	8.0353	10.331	5900.0	67.282	157000.0

(a)   (b)   (c)   (d)

(e)   (f)   (g)   (h)

Fig. 2. Image sequences: Form left to right for $\sigma = 9, 5, 3, 1$

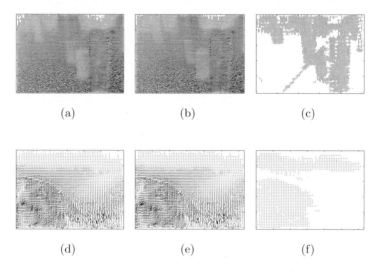

(a)   (b)   (c)

(d)   (e)   (f)

Fig. 3. Results: From left to right: The Gaussian Multiresolution Algorithm, The Gaussian Multiresolution Dynamic Algorithm and Differences of the results computed these two algorithms. Differences of the results of two methods show that our method computes optical flow field which is compatible to the classical method from a zooming image sequence.

5.2   Estimation of Zooming Operation

Let the Fourier transforms of $f^\sigma(x, y, t)$ and $G_\sigma(x, y)$ be $F_\sigma(m, n, t)$ and $\bar{G}_\sigma = e^{-2\pi^2\sigma^2(m^2+n^2)}$, respectively, for

$$F(m, n) = \int_{-\infty}^{\infty} \int_{-\infty}^{\infty} f(x, y)e^{-2\pi i(xm+yn)} dxdy. \tag{23}$$

For $f^0(x, y, t)$ and $f^\sigma(x, y, t)$, we have

$$-2\pi^2\sigma^2(m^2 + n^2) = \log \frac{|F^0(m, n, t)|}{|F^\sigma(m, n, t)|}, \tag{24}$$

since $|F^\sigma(m, n, t)| = \bar{G}_\sigma|F^0(m, n, t)|$. Therefore, we can compute σ as

$$\sigma^2 = \frac{1}{-2\pi^2(m^2 + n^2)} \log \frac{|F^0(m, n, t)|}{|F^\sigma(m, n, t)|}. \tag{25}$$

We can describe this relation as

$$\sigma = E(F^0(m, n, t), F^\sigma(m, n, t)) = \sqrt{\frac{1}{-2\pi^2(m^2 + n^2)} \log \frac{|F^0(m, n, t)|}{|F^\sigma(m, n, t)|}}. \tag{26}$$

If we observe $f^\sigma(x, y, t)$ and $f^{\sigma+1}(x, y, t)$, we can have have the estimation in the form as

$$\sigma = S(F^\sigma(m, n, t), F^{\sigma+1}(m, n, t)) = \frac{1}{2}(E^2(F^\sigma(m, n, t), F^{\sigma+1}(m, n, t)) - 1) \tag{27}$$

We derive a method to estimate σ from a pair of observatins $f^{\sigma+k}(x, y, t - k)$ and $f^\sigma(x, y, t)$. It is possible to express

$$f^0(x, y, t + 1) = f^0(x, y, t) + h(x, y, t + 1). \tag{28}$$

Furthermore, for f, we can assume the property

$$|h(x, y, t + 1)|^2 \ll |f^\sigma(x, y, t + 1)|^2, \quad |f^\sigma(x, y, t)|^2, \tag{29}$$

Therefore, similar to eq.(27) we have the relations

$$\sigma = S(F^\sigma(m, n, t), F^{\sigma+1}(m, n, t)) \simeq \frac{1}{2}(E^2(F^\sigma(m, n, t), F^{\sigma+1}(m, n, t - 1)) - 1). \tag{30}$$

We show the results for the image sequences Marbled Block 1 sequence and Yosemite sequence in Fig. 4. The σ is estimated from image sequences of Fig. 5. Results in Fig. 4 lead to the conclusion that this method can estimate the Gaussian blurring parameter from a zooming image sequence.

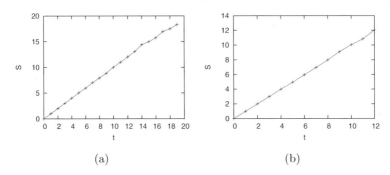

Fig. 4. Estimation of σ using $S(F^k(t), F^{k+1}(t-1))$ for Marbled Block 1 (a) and Yosemite sequences (b), respectively

(a) $f^{15}(x, y, t - 15)$ (b) $f^{10}(x, y, t - 10)$ (c) $f^5(x, y, t - 5)$ (d) $f^0(x, y, t)$

(e) $f^9(x, y, t - 9)$ (f) $f^6(x, y, t - 6)$ (g) $f^3(x, y, t - 3)$ (h) $f^0(x, y, t)$

Fig. 5. Zooming image sequences. Top: Marbled Block 1. Bottom Yosemite.

6   Conclusions

In this paper, we developed an optical flow computation algorithm for zooming image sequence, assuming that the image blurring operation is approximated by the Gaussian kernel convolution. We adopted the Horn and Schunck method for the optical flow computation in each pair of successive images.

In early '90, "Active Vision Paradigm" has been proposed in computer vision. The main methodology of the paradigm is to unify visual information observed actively controlled sensors. An example is a reconstruction of 3D object from images captured by the various directions, various depths, various resolutions, various magnitudes, and various colour spectra. This idea have mainly come from the state of arts of the imaging systems at that time, that is, it had been believed that by integrating various visual information, it is possible to improve the *quality of information*, which is defined qualitatively, for 3D understanding from an image/ image sequence. Today, this is a basic methodology in robot vision, since autonomous robot observes the environment using mount-on vision system. From the viewpoint of active vision, the method introduced in this paper achieves the optical flow computed from an actively measured blurred image sequence, that is, the method derives a mathematical framework for the optical flow computation in active vision methodology.

Since the system of inequalities of eq. 20 is satisfied for any convex functionals, for the optical flow computation in each step of Algorithm 2, we can accept

$$\boldsymbol{u} = arg\min \left[\int_{\mathbf{R}^2} \left\{ (\nabla f^\top \boldsymbol{u} + \partial_t f)^p + \left\{ \frac{1}{2}(\boldsymbol{H}\boldsymbol{u} + \nabla f_t) \right\}^q + \kappa P(\boldsymbol{u}) \right\} d\boldsymbol{x} \right],$$

$$(31)$$

for $f_t(x, y, t) = Rf(x, y, t+1) - f(x, y, t)$, where $p, q \geq 1$ and $\boldsymbol{H}$ is the Hessian of the image f [11]. Furthermore, equation (29) implies that

$$w(x, y) \simeq \int_{-\infty}^{\infty} \int_{-\infty}^{\infty} W(m, n) e^{2\pi i(xu + yv)} dudv.$$

$$(32)$$

setting $W = \frac{F_{k+1}}{F_k}$ where F_k is the Fourier transform of f_k.

This research was supported by Grants-in-Aid for Scientific Research@founded by Japan Society of the Promotion of Sciences, Research Fellowship for Young Scientist founded by Japan Society of the Promotion of Sciences, and Sciences, and Research Associate Program of Chiba University.

References

1. Zhou, Z., Synolakis, C.E., Leahy, R.M., Song, S.M.: Calculation of 3D internal displacement fields from 3D X-ray computer tomographic images. In: Proceedings of Royal Society: Mathematical and Physical Sciences, vol. 449, pp. 537–554 (1995)
2. Kalmoun, E.M., Köstler, H., Rüde, U.: 3D optical flow computation using parallel vaiational multigrid scheme with application to cardiac C-arem CT motion. Image and Vision Computing 25, 1482–1494 (2007)
3. Guilherme, N.D., Avinash, C.K.: Vision for mobile robot navigation: A survey. IEEE Trans. on PAMI 24, 237–267 (2002)
4. Ruhnau, P., Knhlberger, T., Schnoerr, C., Nobach, H.: Variational optical flow estimation for particle image velocimetry. Experiments in Fluids 38, 21–32 (2005)
5. Horn, B.K.P., Schunck, B.G.: Determining optical flow. Artificial Intelligence 17, 185–204 (1981)
6. Nir, T., Bruckstein, A.M., Kimmel, R.: Over-parameterized variational optical flow. IJCV 76, 205–216 (2008)
7. Suter, D.: Motion estimation and vector spline. In: Proceedings of CVPR 1994, pp. 939–942 (1994)

8. Grenander, U., Miller, M.: Computational anatomy: An emerging discipline. Quarterly of applied mathematics 4, 617–694 (1998)

9. Weickert, J., Schnörr, C.: Variational optic flow computation with a spatio-temporal smoothness constraint. Journal of Mathematical Imaging and Vision 14, 245–255 (2001)

10. Weickert, J., Bruhn, A., Papenberg, N., Brox, T.: Variational optic flow computation: From continuous models to algorithms. In: Proceedings of International Workshop on Computer Vision and Image Analysis, IWCVIA 2003 (2003)

11. Papenberg, N., Bruhn, A., Brox, T., Didas, S., Weickert, J.: Highly accurate optic flow computation with theoretically justified warping. International Journal of Computer Vision 67, 141–158 (2006)

12. Werner, T., Pock, T., Cremers, D., Bischof, H.: An unbiased second-order prior for high-accuracy motion estimation. In: Rigoll, G. (ed.) DAGM 2008. LNCS, vol. 5096, pp. 396–405. Springer, Heidelberg (2008)

13. Bouguet, J.-Y.: Pyramidal implementation of the Lucas Kanade feature tracker: Description of the algorithm, Microsoft Research Labs, Tech. Rep. (1999)

14. Hwan, S., Hwang, S.-H., Lee, U.K.: A hierarchical optical flow estimation algorithm based on the interlevel motion smoothness constraint. Pattern Recognition 26, 939–952 (1993)

15. Weber, J., Malik, J.: Robust computation of optical flow in a multi-scale differential framework. Int. J. Comput. Vision 14, 67–81 (1995)

16. Battiti, R., Amaldi, E., Koch, C.: Computing optical flow across multiple scales: An adaptive coarse-to-fine strategy. Int. J. Comput. Vision 2, 133–145 (1991)

17. Condell, J., Scotney, B., Marrow, P.: Adaptive grid refinement procedures for efficient optical flow computation. Int. J. Comput. Vision 61, 31–54 (2005)

18. Amiz, T., Lubetzky, E., Kiryati, N.: Coarse to over-fine optical flow estimation. Pattern recognition 40, 2496–2503 (2007)

Conditions for Segmentation of Motion with Affine Fundamental Matrix

Shafriza Nisha Basah[1], Reza Hoseinnezhad[2], and Alireza Bab-Hadiashar[1]

[1] Faculty of Engineering and Industrial Sciences, Swinburne University of Technology
[2] Melbourne School of Engineering, The University of Melbourne,
Victoria, Australia
sbasah@swin.edu.au, rezah@unimelb.edu.au, abab-hadiashar@swin.edu.au

Abstract. Various computer vision applications involve recovery and estimation of multiple motions from images of dynamic scenes. The exact nature of objects' motions and the camera parameters are often not known a priori and therefore, the most general motion model (the fundamental matrix) is applied. Although the estimation of a fundamental matrix and its use for motion segmentation are well understood, the conditions governing the feasibility of segmentation for different types of motions are yet to be discovered. In this paper, we study the feasibility of separating a motion (of a rigid 3D object) with affine fundamental matrix in a dynamic scene from another similar motion (unwanted motion). We show that successful segmentation of the target motion depends on the difference between rotation angles and translational directions, the location of points belonging to the unwanted motion, the magnitude of the unwanted translation viewed by a particular camera and the level of noise. Extensive set of controlled experiments using synthetic images were conducted to show the validity of the proposed constraints. The similarity between the experimental results and the theoretical analysis verifies the conditions for segmentation of motion with affine fundamental matrix. These results are important for practitioners designing solutions for computer vision problems.

1 Introduction

Recovering structure-and-motion (SaM) from images of dynamic scenes is an indispensable part of many computer vision applications ranging from local navigation of a mobile robot to image rendering in multimedia applications. The main problem in SaM recovery is that the exact nature of objects' motions and the camera parameters are often not known a priori. Thus, any motion of 3D object needs to be modelled in the form of a fundamental matrix [1] (if all moving points are in the same plane, the motion can be modelled as a homography). Motion estimation and segmentation based on the fundamental matrix are well understood and solved in the established work summarised in (Chapters.9–12,[2]). Soon after that work, researchers resumed to the more challenging multibody structure-and-motion (termed MSaM by Schindler and Suter in [3]) where multiple objects in motions need to be concurrently estimated and

G. Bebis et al. (Eds.): ISVC 2009, Part I, LNCS 5875, pp. 415–424, 2009.
© Springer-Verlag Berlin Heidelberg 2009

segmented. However, the conditions governing the feasibility of segmentation involving MSaM for different types of motions are yet to be established. These conditions are important as they provide information on the limits of current MSaM methods and would provide useful guidelines for practitioners designing solutions for computer vision problems.

Well known examples of previous works in motion segmentation using the fundamental matrix are by Torr [4], Vidal et.al [5] and Schindler and Suter [3]. Torr uses the fundamental matrix to estimate an object in motion and cluster it in a probabilistic framework using the Expectation Maximisation algorithm [4]. Vidal.et.al propose to estimate the number of moving objects in motion and cluster those motions using the multibody fundamental matrix; the generalization of the epipolar constraint and the fundamental matrix of multiple motions [5]. Schindler and Suter have implemented the geometric model selection to replace degenerate motion in a dynamic scene using multibody fundamental matrix [3].

The focus of this paper is to study the feasibility of the detection and segmentation of an unknown motion (of a rigid 3D object) with affine fundamental matrix in a dynamic scene (using images taken by an uncalibrated camera). Motion with affine fundamental matrix consists of rotation around Z axis of the image plane and translation parallel to the image plane (no translation in Z direction) [1]. This work is an extension to the study of conditions for motion-background segmentation and segmentation of translational motions in [6,7]. In Section 2, we derive the conditions for the detection and segmentation of motion with affine fundamental matrix and provide quantitative measures for detection using theoretical analysis. Section 3 details the Monte Carlo experiments using synthetic images conducted to verify the theoretical analysis and the proposed conditions for successful segmentation of motion with affine fundamental matrix. Section 4 concludes the paper.

2   Segmentation of Motion with Affine Fundamental Matrix

Consider $n = n_i + n_o$ feature points $[X_i, Y_i, Z_i]^\top$ belonging to two 3D objects undergoing motion-a and motion-b ($i = 0, 1 \ldots n_i$ denote the points belonging to motion-a and $i = n_i + 1, n_i + 2 \ldots n$ denote points belonging to motion-b). Each motion consists of rotations around Z axis followed by a non-zero translation in the $X - Y$ plane denoted by θ_a and T_a for motion-a and θ_b and T_b for motion-b where $T_a = [T_{xa}, T_{ya}, T_{za}]^\top$ and $T_b = [T_{xb}, T_{yb}, T_{zb}]^\top$ with $T_{za} = T_{zb} = 0$. The location $[X_i, Y_i, Z_i]^\top$ before and after the translations are visible in the image plane and are denoted by $m_{1i} = [x_{1i}, y_{1i}]^\top$ and $m_{2i} = [x_{2i}, y_{2i}]^\top$. All points in the image plane are perturbed by measurement noise e assumed to be independent and identically distributed (i.i.d) with Gaussian distribution:

$$x_{1i} = \underline{x}_{1i} + e^1_{ix}, \quad y_{1i} = \underline{y}_{1i} + e^1_{iy}, \quad x_{2i} = \underline{x}_{2i} + e^2_{ix}, \text{ and } y_{2i} = \underline{y}_{2i} + e^2_{iy}, \quad (1)$$

where $e^1_{ix}, e^1_{iy}, e^2_{ix}$ and $e^2_{iy} \sim N(0, \sigma_n^2)$ and σ_n is the unknown scale of noise. The underlined variables denote the true or noise-free locations of points in image plane. The relationship between all noise-free matching points in the image plane and world coordinate points as viewed by a camera (with focal length f and principle points $[P_x, P_y]$) are:

$$x_{1i} = \frac{fX_i}{Z_i} + P_x, \quad x_{2i} = x_{1i}\cos\theta - y_{1i}\sin\theta + \frac{fT_x}{Z_i} + \widetilde{P_x},$$

$$y_{1i} = \frac{fY_i}{Z_i} + P_y, \quad y_{2i} = x_{1i}\sin\theta + y_{1i}\cos\theta + \frac{fT_y}{Z_i} + \widetilde{P_y}, \tag{2}$$

where $\widetilde{P_x} = P_x(1-\cos\theta) + P_y\sin\theta$ and $\widetilde{P_y} = P_y(1-\cos\theta) - P_x\sin\theta$. The symbols θ, T_x and T_y in (2) denote the motion parameters where $\theta = \theta_a$, $T_x = T_{xa}$ and $T_y = T_{ya}$ for motion-a and $\theta = \theta_b$, $T_x = T_{xb}$ and $T_y = T_{yb}$ for motion-b respectively. We aim to segment the matching points belonging to motion-a from the mixture of matching points belonging to motion-a and motion-b in two images, thus the points undergoing motion-a are considered the inliers and the points undergoing motion-b would be the outliers or the unwanted motion.

The fundamental matrix of motion-a is computed using:

$$F = A^{-T}[T]_x R A^{-1}, \tag{3}$$

where R is the rotation matrix of the motion and $[T]_x$ is the skew symmetric matrix of translation T [8,2,9]. For motion-a ($\theta = \theta_a$, $T_x = T_{xa}$ and $T_y = T_{ya}$), equation (3) yields:

$$F_a = \frac{1}{f}\begin{pmatrix} 0 & 0 & T_{ya} \\ 0 & 0 & -T_{xa} \\ T_{xa}\sin\theta_a - T_{ya}\cos\theta_a & T_{ya}\sin\theta_a + T_{xa}\cos\theta_a & Q \end{pmatrix}, \tag{4}$$

where $Q = (T_{ya}\cos\theta_a - T_{xa}\sin\theta_a - T_{ya})P_x + (T_{xa} - T_{ya}\sin\theta_a - T_{xa}\cos\theta_a)P_y$. We assume that, a perfect estimator provides the true fundamental matrix given in (4). If F_a is known, the Sampson distances can be computed using [10,1,11]:

$$d_i = \frac{m_{2i}^T F m_{1i}}{\sqrt{\left[(\partial/\partial x_{1i})^2 + (\partial/\partial y_{1i})^2 + (\partial/\partial x_{2i})^2 + (\partial/\partial y_{2i})^2\right] m_{2i}^T F m_{1i}}}. \tag{5}$$

Substitution of real plus noise forms in (1) and F_a in (4) into equation (5) yields:

$$d_i = (2(T_{ya}^2 + T_{xa}^2))^{-\frac{1}{2}}[(x_{1i} + e_{ix}^1)(T_{xa}\sin\theta_a - T_{ya}\cos\theta_a) + (y_{1i} + e_{iy}^1)\cdots \\ (T_{ya}\sin\theta_a + T_{xa}\cos\theta_a) + (x_{2i} + e_{ix}^2)T_{ya} - (y_{2i} + e_{iy}^2)T_{xa} + Q]. \tag{6}$$

For points undergoing motion-a ($i = 0, 1 \ldots n_i$), the above expression without noise terms equals zero because the true F_a is used to compute d_i's. Thus, equation (6) can be simplified to:

$$d_i = (2(T_{ya}^2 + T_{xa}^2))^{-\frac{1}{2}}[e_{ix}^1(T_{xa}\sin\theta_a - T_{ya}\cos\theta_a) + e_{iy}^1(T_{ya}\sin\theta_a + \cdots \\ T_{xa}\cos\theta_a) + e_{ix}^2 T_{ya} - e_{iy}^2 T_{xa}]. \tag{7}$$

Distances d_i's of the points belonging to motion-a in (7) turn out to be a linear combination of the i.i.d. noises therefore, they are also normally distributed with zero mean. The variance of d_i's $(i = 0, 1 \ldots n_i)$ also equals σ_n^2 as the numerator and denominator cancel each other. Therefore, the distribution of d_i's of motion-a are the same as the noise e which are $N(0, \sigma_n^2)$.

The points belonging to motion-a are to be separated from the points belonging to motion-b. The distances of points belonging to motion-b $(i = n_i + 1, n_i + 2 \ldots n)$ with respect to F_a are calculated using (6) in which $[\underline{x}_{2i}, \underline{y}_{2i}]$ are replaced with the terms in (1) for motion-b, yields:

$$
\begin{aligned}
d_i = & (2(T_{ya}^2 + T_{xa}^2))^{-\frac{1}{2}}[\underline{x}_{1i}(T_{xa}\sin\theta_a - T_{ya}\cos\theta_a) + \underline{y}_{1i}(T_{ya}\sin\theta_a + \cdots \\
& T_{xa}\cos\theta_a) + \underline{x}_{2i}T_{ya} - \underline{y}_{2i}T_{xa} + Q] + e,
\end{aligned}
\tag{8}
$$

where $e \sim N(0, \sigma_n^2)$ based on the distribution of d_i's in (7). By combining equation (8) with the image-world points relationship for motion-b in (2), and expressing it in term of directions of translation (ϕ_a for T_a and ϕ_b for T_b) and magnitude of T_b, we obtain (after manipulations using several trigonometric identities):

$$
d_i = \sqrt{2}\sin\frac{\Delta\theta}{2}(\acute{x}_{1i}\cos\Theta + \acute{y}_{1i}\sin\Theta) + \frac{K}{Z_i}\sin\Delta\phi + e,
\tag{9}
$$

where $\acute{x}_{1i} = \underline{x}_{1i} - P_x$, $\acute{y}_{1i} = \underline{y}_{1i} - P_y$, $\Delta\theta = \theta_a - \theta_b$, $\Delta\phi = \phi_a - \phi_b$, $\Theta = \phi_a - \frac{\theta_a + \theta_b}{2}$ and $K = \frac{f\|T_b\|}{\sqrt{2}}$. By using the harmonic addition theorem [12], equation (9) is simplified to:

$$
d_i = G_i \sin\frac{\Delta\theta}{2}\cos\breve{\Theta}_i + \frac{K}{Z_i}\sin\Delta\phi + e,
\tag{10}
$$

where $G_i = \sqrt{2(\acute{x}_{1i}^2 + \acute{y}_{1i}^2)}$ and $\breve{\Theta}_i = \Theta + \tan^{-1}(-\acute{x}_{1i}/\acute{y}_{1i}) + \beta$ (the value of $\beta = 0$ if $\acute{x}_{1i} \geq 0$ or $\beta = \pi$ if $\acute{x}_{1i} < 0$). Since the distribution of d_i's of motion-a is always $N(0, \sigma_n^2)$ (according to equation (7)), the feasibility of identification and segmentation of points belonging to motion-a using a robust estimator depends on the distribution of d_i's of motion-b. If the population of d_i's from points belonging to motion-b overlap with population of d_i's from points belonging to motion-a, they would not be separable. However if both populations (d_i's from motion-a and b) do not overlap, the population of d_i's of each motion will be separable. Thus, a robust estimator should be able to correctly identify and segment the points belonging to motion-a. In order to ensure that both populations do not overlap, the following conditions must be satisfied[1]:

$$
\underline{d}_i \geq 5\sigma_n \text{ or } \underline{d}_i \leq -5\sigma_n \quad \text{when} \quad i = n_i + 1, n_i + 2 \ldots n,
\tag{11}
$$

where $\underline{d}_i$'s are the noise-free d_i's given by all terms in equation (8) except for the noise term e. If the conditions in (11) is satisfied 99.4% of points belonging

---

[1] Since the d_i's of motion-a is distributed according to $N(0, \sigma_n^2)$ and d_i's of motion-b are also perturbed by measurement noise of $N(0, \sigma_n^2)$, from probability theory if the maximum or minimum value of d_i's of motion-b are at least $5\sigma_n$ away from the mean of d_i's of motion-a, then only about 0.6% of d_i's of each population would overlap.

to motion-a will be correctly segmented. Since the term G_i's in (10) are always positive, the range of the term $G_i \cos \breve{\Theta}_i$ in (10) are:

$$-\widehat{G} \leq G_i \cos \breve{\Theta}_i \leq \widehat{G} \quad \text{for all } i\text{'s}, \tag{12}$$

where $\widehat{G}$ is the maximum values of G_i's depending on the locations of the objects undergoing motion-b. Thus, the range of $\underline{d}_i$'s from equations (10) and (12) are:

$$-\widehat{G} \sin \frac{\Delta\theta}{2} + \frac{K}{Z_i} \sin \Delta\phi \leq \underline{d}_i \leq \widehat{G} \sin \frac{\Delta\theta}{2} + \frac{K}{Z_i} \sin \Delta\phi. \tag{13}$$

Combining the inequalities in (11) and (13), the conditions for successful segmentation of points belonging to motion-a are expressed as:

$$\widehat{G} \sin \frac{\Delta\theta}{2} + \frac{K}{Z_i} \sin \Delta\phi \leq -5\sigma_n \quad \text{or} \quad -\widehat{G} \sin \frac{\Delta\theta}{2} + \frac{K}{Z_i} \sin \Delta\phi \geq 5\sigma_n. \tag{14}$$

Solving for $\Delta\theta$, the inequalities in (14) are expressed as:

$$\begin{aligned}
\frac{\Delta\theta}{2} &\leq \sin^{-1} \frac{-5\sigma_n \pm \frac{K}{Z_i} \sin \Delta\phi}{\widehat{G}} \quad && \text{for } \Delta\theta \geq 0, \\
\frac{\Delta\theta}{2} &\geq -\sin^{-1} \frac{-5\sigma_n \pm \frac{K}{Z_i} \sin \Delta\phi}{\widehat{G}} \quad && \text{for } \Delta\theta < 0.
\end{aligned} \tag{15}$$

In most computer vision problems, the distance between the camera and the object in motion is roughly known. Therefore, the term Z_i in equation (15) can be expressed in term of average distance $\bar{Z}$ between camera and objects in motions or $Z_i \approx \bar{Z}$.

We assume that an accurate estimate for F_a is obtained by minimising the cost function of a robust estimator. Having F_a, the distances (d_i's) of all matching points can be computed. Then d_i^2's for all points are used as residuals for segmentation to identify and segment points belonging to motion-a using a robust estimator. Here, we use the Modified Selective Statistical Estimator (MSSE) [13] as it has been shown to outperform other robust estimation techniques in term of its consistency [14]. In MSSE, the residuals are sorted in an ascending order and the scale estimate given by the smallest kth distances is calculated using [13] for a particular value of k:

$$\sigma_k^2 = \frac{\sum_{i=1}^{k} d_i^2}{k-1}. \tag{16}$$

While incrementing k, the MSSE algorithm is terminated when d_{k+1} is larger than 2.5 times the scale estimate given by the smallest k distances:

$$d_{k+1}^2 > 2.5^2 \sigma_k^2. \tag{17}$$

With the above threshold, at least 99.4% of the inliers will be segmented if there are normally distributed [13].

From our analysis, the separability of motion with affine fundamental matrix depends on the difference between rotation angles and translational directions

($\Delta\theta$ and $\Delta\phi$), the location of points belonging to motion-b ($\widehat{G}$), the magnitude of T_b viewed by a particular camera ($\frac{K}{Z}$) and the level of noise (σ_n) presented in equation (15). We verified these conditions using Monte Carlo experiments and the results are presented in the next section. The correctness of these conditions are verified by studying the variance of the result of the Monte Carlo experiments. The conditions for segmentation for more general motions (including $T_z \neq 0$ and rotation around other axes) are too complex to be derived theoretically. However, the derived condition for segmentation of motion with affine fundamental matrix is served as the basis of approximation for more general motions when T_z and rotation around other axes are very small or close to zero.

3   Monte Carlo Experiments

The Monte Carlo experiments with synthetic images have two parts. The first part was conducted to verify the conditions for segmentation in (15) for separating motion-a from motion-b. The second part of the experiments was designed to examine how the conditions change when the inlier ratio ϵ was varied.

In each iteration in the Monte Carlo experiments, 2000 pairs of points in the world coordinate according to motion-a were mixed with the pairs of matching points according to motion-b (the number of matching points belonging to motion-b depends on the inlier ratio ϵ). All X and Y coordinates of the matching points were randomly generated while Z coordinates were uniformly distributed according to $U(\bar{Z}-\sigma_Z, \bar{Z}+\sigma_Z)$ where $\bar{Z} = 10\text{m}$ and $\frac{\sigma_Z}{Z} = 10\%$. Then, all matching points were projected to two images using a synthetic camera (with $f = 703$ pixels, $[P_x, P_y]=[320,240]$ and image size of 640×480 pixels). Random noise with the distribution of $N(0, \sigma_n^2)$ was added to all points. We assumed that the image points ($\acute{\underline{x}}_{1i}$'s and $\acute{\underline{y}}_{1i}$'s) belonging to each motion (motion-a and b) were clustered together, since in many computer vision applications the objects in motions are rigid. The points belonging to motion-b were assumed to be within $20\%\times20\%$ width and length of the image size and according to $\widehat{G}$, while the points belonging to motion-a could be anywhere in the image plane (since its d_i's will always be $N(0, \sigma_n^2)$ according to (7)). The segmentation was performed using MSSE with d_i^2's (calculated based on the true F_a) as the segmentation residuals. The ratio of the number of segmented inliers over the actual inliers ζ was calculated and recorded. Each experiment consists of 1000 experimental iterations and the mean and standard deviation of 1000 ζ's were recorded (denoted by $\bar{\zeta}$ and $\sigma\zeta$). These experiments were then repeated for various ϵ.

In the first part of the experiments, we consider two scenarios; Scenario-I with parameters $\frac{K}{Z}=50$ (corresponding to $\|T_b\| = 1\text{m}$), $\sigma_n = 0.5$ and $\widehat{G} = 0.75G_{\text{max}}$ while Scenario-II with parameters $\frac{K}{Z}=40$ (corresponding to $\|T_b\| = 0.8\text{m}$), $\sigma_n = 1$ and $\widehat{G} = 0.75G_{\text{max}}{}^2$. The conditions for segmentation for Scenario-

---

2 The term G_{max} is the value of G_i when $[\acute{\underline{x}}_{1i}, \acute{\underline{y}}_{1i}]$ are maximum and in this case [320,240] according to a camera with image size 640×480 with $[P_x, P_y]=[320,240]$. Generally smaller $\widehat{G}$ means that the points moved by motion-b were closer to $[P_x, P_y]$ of the image.

I and II were generated from (15) and shown in Fig.2(c)-(d) and Fig.2(e)-(f) respectively, where the shaded area denote the area where motion-a will be successfully segmented from motion-b (in this area, both population of d_i's will not overlap). This analysis were also performed in five different cases with $\epsilon = 30\%$. The motion parameters were selected from the magnitudes of $\Delta\theta$'s and $\Delta\phi$'s from the shaded region (Case-1 and 2) and unshaded region (Case-3, 4 and 5) in Fig.2(c)-(d). In all cases, the histogram of d_i's of all image points were plotted and ζ's were recorded. For five instances of the data samples generated in Case 1 to 5 for Scenario-I (in Fig.2(c)-(d)), the histogram of d_i's for all points are plotted in Fig.1. These figures show that in Case-1 ($\Delta\theta = 1°$ and $\Delta\phi = 10°$) and Case-2 ($\Delta\theta = 5°$ and $\Delta\phi = 30°$), the points belonging to motion-a were correctly segmented, denoted by $\zeta = 0.98$ for both cases ($\zeta = 1$ indicates perfect segmentation). Successful segmentation was expected as the population of d_i's of points belonging to motion-a and motion-b were not overlapping, thus there were separable as shown in Fig.1(a) and 1(b). As the magnitudes of $\Delta\theta$ and $\Delta\phi$ were selected to be outside the shaded region in Fig.2(c)-(d) in Case-3 ($\Delta\theta = 2°$ and $\Delta\phi = 1°$) and Case-4 ($\Delta\theta = 10°$ and $\Delta\phi = 4°$), the points belonging to motion-a were incorrectly segmented denoted by $\zeta = 1.92$ and 1.77. The failure is due to the overlap of the population of d_i's of both motions and thus there were very little distinction between them as shown in Fig.1(c) and 1(d). However, in Case-5 ($\Delta\theta = 10°$ and $\Delta\phi = 4°$) it was observed that the points belonging to motion-a was correctly segmented ($\zeta = 0.99$ and non overlapping d_i's in Fig.1(d)) even though the magnitudes of $\Delta\theta$ and $\Delta\phi$ were the same as in Case-4. These experiment results were consistent with the conditions for segmentation derived in equation (15), where population of d_i's of both motions are not overlap in the shaded region of Fig.2(c)-(d). Thus the points belonging to motion-a will be correctly segmented when the magnitudes of $\Delta\theta$ and $\Delta\phi$ are in this region. However, when the magnitudes of $\Delta\theta$ and $\Delta\phi$ are outside the shaded region in Fig.2(c)-(d), the are no guarantee that points belonging to motion-a will be correctly segmented since there is a chance that the population of d_i's of both motions would overlap.

In the second part of the experiments, the effect of varying ϵ from 30% to 80% to the conditions for segmentation of Scenario-I and II (the shaded regions) in Fig.2(c) to (e) were examined. The mean and sigma of 1000 ζ's (denoted as $\bar{\zeta}$ and $\sigma\zeta$) were recorded for each pair of $\Delta\theta$'s and $\Delta\phi$'s in the experiments. Both $\Delta\theta$ and $\Delta\phi$ in the experiments were varied from $0°$ to $90°$ with the increment of $2.5°$. Fig.2(a) and 2(b) show $\bar{\zeta}$ and $\sigma\zeta$ versus $\Delta\theta$ and $\Delta\phi$ for Scenario-I. It was observed that for small $\Delta\theta$ and $\Delta\phi$ (both $< 5°$), points from motion-a were mixed with points from motion-b and segmented ($\zeta > 1$). In such cases, an inaccurate inlier-outlier dichotomy would result in an incorrect motion estimation and segmentation. As both $\Delta\phi$ and $\Delta\theta$ increased to $90°$, the magnitudes of $\bar{\zeta}$'s approaching 0.99 and $\sigma\zeta$'s reduced to around 0.02. From Fig.2(a) and 2(b), there are areas when $\bar{\zeta} = 0.99$ and $\sigma\zeta < 0.01$ (when $\Delta\theta$ between $0°$ to $12°$ and $\Delta\phi$ from

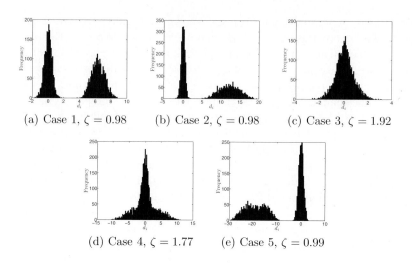

(a) Case 1, $\zeta = 0.98$ (b) Case 2, $\zeta = 0.98$ (c) Case 3, $\zeta = 1.92$

(d) Case 4, $\zeta = 1.77$ (e) Case 5, $\zeta = 0.99$

Fig. 1. Histogram for d_i's for all points for Case-1 to 5 in Scenario-I

$6°$ to $90°$) indicating correct and consistent segmentation of motion-a. Then the magnitudes of $\Delta\theta$'s and $\Delta\phi$'s when $\bar{\zeta} \leq 0.994$ and $\sigma\zeta \leq 0.01$ were extracted from Fig.2(a) and 2(b) and compared to the analytical conditions for segmentation as shown in Fig.2(c). The magnitudes of $\bar{\zeta} \leq 0.994$ and $\sigma\zeta \leq 0.01$ were selected to make sure that the segmentation of motion-a was correct and consistent for all iterations (1000 iterations for each $\Delta\theta$ and $\Delta\phi$) in the experiments[3]. The extracted $\Delta\theta$'s and $\Delta\phi$'s when $\bar{\zeta} \leq 0.994$ and $\sigma\zeta \leq 0.01$ for different inlier ratio ϵ and Scenario-II are shown in Fig.2(c) to 2(f). When the value of ϵ is increased from 30% to 80%, the region where the magnitudes of $\bar{\zeta} \leq 0.994$ and $\sigma\zeta \leq 0.01$ are slightly expanded as shown in Fig.2(d) and 2(f). In addition, we observed that when $\Delta\theta > 20°$ for Scenario-I and $\Delta\theta > 45°$ for Scenario-II, the points belonging to motion-a were also correctly and consistently segmented ($\bar{\zeta} \leq 0.994$ and $\sigma\zeta \leq 0.01$). This is because, when ϵ was higher (more points of motion-a than motion-b) and high values on $\Delta\theta$, the density of d_i's for points belonging to motion-b were not widely spread. Thus, the likelihood of both populations of d_i's to overlap decreased. Hence, expanding the region for correct and consistent segmentation of points belonging to motion-a. The similarity between the experimental and analytical results for the magnitude of $\Delta\theta$'s and $\Delta\phi$'s to achieve correct and consistent segmentation ($\bar{\zeta} \leq 0.994$, $\sigma\zeta \leq 0.01$ and both population of d_i's do not overlap) in Fig.2(c) to 2(f) verifies the segmentation analysis in Section.2.

---

[3] From probability theory, the standard deviation of a uniformly distributed variables B's is according to $\frac{B_{max} - B_{min}}{\sqrt{12}}$, where B_{max} and B_{min} are the maximum and minimum values of B's [15]. Thus, if ζ's was uniformly distributed with $\sigma\zeta = 0.01$, the values of ζ's were between $\bar{\zeta} \pm 0.017$.

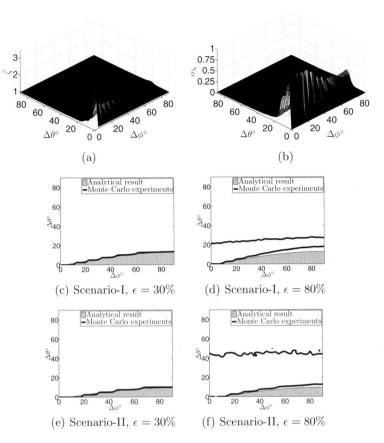

(a)   (b)

(c) Scenario-I, $\epsilon = 30\%$ (d) Scenario-I, $\epsilon = 80\%$

(e) Scenario-II, $\epsilon = 30\%$ (f) Scenario-II, $\epsilon = 80\%$

Fig. 2. $\bar{\zeta}$ and $\sigma\zeta$ versus $\Delta\theta$ and $\Delta\phi$ for Scenario-I when $\epsilon = 30\%$ in (a) and (b). The analytical and the extracted magnitudes of $\Delta\theta$'s and $\Delta\phi$'s when motion-a will be correctly and consistently segmented in (c)-(d) for Scenario-I and (e)-(f) for Scenario-II (the boundary of the region where $\bar{\zeta} \leq 0.994$ and $\sigma\zeta \leq 0.01$ from Monte Carlo experiments were plotted instead of the shaded region for illustration purposes).

4   Conclusions

The conditions for segmentation of motions with affine fundamental matrix were proposed in terms of the difference between rotation angles ($\Delta\theta$) and translational directions ($\Delta\phi$), the location of points belonging to the unwanted motion ($\widehat{G}$), the magnitude of the unwanted translation viewed by a particular camera ($\frac{K}{Z}$) and the level of noise (σ_n). If this conditions are satisfied, it is guaranteed that the population of distances belonging to multiple affine motions do not overlap, and therefore the target motion can be successfully segmented. The proposed conditions were both studied by theoretical analysis and verified by Monte Carlo experiments with synthetic images. The performance of these conditions for various inlier ratio was also examined. The magnitudes of $\Delta\theta$'s and

$\Delta\phi$'s (for particular values of $\frac{K}{Z}$, $\widehat{G}$ and σ_n) for correct and consistent segmentation did not changed significantly when the inlier ratio was varied. However when the inlier ratio was high ($\epsilon = 80\%$), the segmentation was also successful when the magnitude of $\Delta\theta$ was large ($\Delta\theta > 20°$ for Scenario-I and $\Delta\theta > 45°$ for Scenario-II). This is explained by the fact that, less contamination of the unwanted motion and high magnitude of $\Delta\theta$ reduced the likelihood for both populations of distances to overlap resulting in the successful segmentation of the target object.

References

1. Torr, P.H.S., Zisserman, A., Maybank, S.J.: Robust detection of degenerate configurations while estimating the fundamental matrix. Vision Computing and Image Understanding 71, 312–333 (1998)
2. Hartley, R., Zisserman, A.: Multiple View Geometry in Computer Vision, 2nd edn. Cambridge University Press, Cambridge (2003)
3. Schindler, K., Suter, D.: Two-view multibody structure-and-motion with outliers through model selection. IEEE Transactions on Pattern Analysis and Machine Intelligence 28, 983–995 (2006)
4. Torr, P.H.S.: Motion Segmentation and Outlier Detection. Phd thesis, Department of Engineering Science, University of Oxford (1995)
5. Vidal, R., Ma, Y., Soatto, S., Sastry, S.: Two-view multibody structure from motion. International Journal of Computer Vision 68, 7–25 (2006)
6. Basah, S.N., Hoseinnezhad, R., Bab-Hadiashar, A.: Limits of motion-background segmentation using fundamental matrix estimation. In: DICTA 2008, pp. 250–256 (2008)
7. Basah, S.N., Bab-Hadiashar, A., Hoseinnezhad, R.: Conditions for segmentation of 2d translations of 3d objects. In: ICIAP 2009. LNCS, vol. 5716, pp. 82–91. Springer, Heidelberg (2009)
8. Armangué, X., Salvi, J.: Overall view regarding fundamental matrix estimation. Image and Vision Computing 21, 205–220 (2003)
9. Zhang, Z.: Determining the epipolar geometry and its uncertainty: A review. International Journal of Computer Vision 27(2), 161–195 (1998)
10. Torr, P.H.S., Murray, D.W.: The development and comparison of robust methodsfor estimating the fundamental matrix. International Journal of Computer Vision 24, 271–300 (1997)
11. Weng, J., Huang, T., Ahuja, N.: Motion and structure from two perspective views: Algorithms, error analysis, and error estimation. IEEE Transactions on Pattern Analysis and Machine Intelligence 11, 451–476 (1989)
12. Weisstein, E.W.: Harmonic addition theorem (From MathWorld-A Wolfram Web Resource), http://mathworld.wolfram.com/HarmonicAdditionTheorem.html
13. Bab-Hadiashar, A., Suter, D.: Robust segmentation of visual data using ranked unbiased scale estimate. Robotica 17, 649–660 (1999)
14. Hoseinnezhad, R., Bab-Hadiashar, A.: Consistency of robust estimators in multi-structural visual data segmentation. Pattern Recognition 40, 3677–3690 (2007)
15. Evans, M., Hastings, N., Peacock, B.: Statistical Distributions, 3rd edn. Wiley, Chichester (2000)

Motion-Based View-Invariant Articulated Motion Detection and Pose Estimation Using Sparse Point Features

Shrinivas J. Pundlik and Stanley T. Birchfield

Clemson University, Clemson, SC USA
{spundli,stb}@clemson.edu

Abstract. We present an approach for articulated motion detection and pose estimation that uses only motion information. To estimate the pose and viewpoint we introduce a novel motion descriptor that computes the spatial relationships of motion vectors representing various parts of the person using the trajectories of a number of sparse points. A nearest neighbor search for the closest motion descriptor from the labeled training data of human walking poses in multiple views is performed. This observational probability is fed to a Hidden Markov Model defined over multiple poses and viewpoints to obtain temporally consistent pose estimates. Experimental results on various sequences of walking subjects with multiple viewpoints demonstrate the effectiveness of the approach. In particular, our purely motion-based approach is able to track people even when other visible cues are not available, such as in low-light situations.

1 Motivation for Articulated Human Motion Analysis

The detection of articulated human motion finds applications in a large number of areas such as pedestrian detection for surveillance, or traffic safety, gait/pose recognition for human computer interaction, videoconferencing, computer graphics, or for medical purposes. Johansson's pioneering work on moving light displays (MLDs) [1] has enabled researchers to study the mechanism and development of human visual system with a different perspective by decoupling the motion information from all other modalities of vision such as color and texture. One compelling conclusion that can be drawn from these studies is that motion alone captures a wealth of information about the scene. Others have made a similar observation [2,3].

Figure 1 shows some examples of humans walking as seen from multiple angles along with their motion trajectories. Even though the appearance features (shape, color, texture) can be discriminative for detection of humans in the sequence, the motion vectors corresponding to the point features themselves can be used to detect them. The motion of these points becomes even more compelling when viewed in a video, as the human visual system fuses the information temporally to segment human motion from the rest of the scene. It is common knowledge that in spite of having a separate motion, each body part moves in a particular pattern. Our goal is to exploit the motion properties of the sparse points attached to a human body in a top-down approach for human motion analysis. More specifically, our attempt is to answer the question: If provided

G. Bebis et al. (Eds.): ISVC 2009, Part I, LNCS 5875, pp. 425–434, 2009.

Fig. 1. Two examples of human walking motion at different viewing angles, and the motion vectors of the tracked feature points

only with the motion tracks (sparse point trajectories) and no appearance information, how well can an algorithm detect, track, and estimate the pose of a walking human in a video?

Previous work related to human motion detection and analysis can be loosely classified into three categories: pedestrian detection for surveillance, pose estimation, and action recognition. The nature of the algorithms dealing with the different categories varies significantly due to the differences in the input image sequences. Approaches for pedestrian detection are either appearance-based [4,5,6], use both appearance and stereo [7], or are based on modeling the periodic motion [8]. In contrast to pedestrian detection, human pose estimation [9,10,11,12,13,14,15,16,3,17,18,19] requires greater detail of the human motion to be captured, with a model that accounts for the disparate motions of the individual body parts. A related area of research is human action recognition [20,21], in which the objective is to classify the detected human motion into one of several predefined categories using off-line training data for learning these action categories.

Even while considering only a single action category such as walking, human motion analysis remains a challenging problem due to various factors such as pose, scale, viewpoint, and scene illumination variations. Most approaches use appearance cues to perform human motion analysis, but these will not work when appearance information is lacking (e.g., at night in poorly lit areas). The few approaches that are predominantly motion based [3,18] are limited in terms of viewpoint and lighting variations. In this paper, using only the sparse motion trajectories and a *single gait cycle* of 3D motion capture data points of a walking person for training, we demonstrate detection and pose estimation of articulated motion on various sequences that involve viewpoint, scale, and illumination variations, as well as camera motion. Our focus is on a top-down approach, where instead of learning the motion of individual joints and limbs as in [3], we learn the short-term motion pattern of the entire body in multiple pose and viewpoint configurations. Pose estimation can then be performed by a direct comparison of the learned motion patterns to those extracted from the candidate locations. The advantage of using such a top-down approach is that it greatly simplifies the learning step, facilitating one-shot learning. At the same time, the learned motion patterns can be reliably used to estimate the pose and the viewpoint in the presence of noise.

2   Learning Models for Multiple Poses and Viewpoints

An overview of the proposed approach is shown in Figure 2. Given an image sequence our goal is to segment, track, and determine the configuration of the walking human subject (2D pose and viewpoint) using only the sparse motion vectors corresponding to the feature points in the sequence. The primary reason for using sparse optical flow obtained from the tracked point features instead of a dense flow field for motion representation is efficiency of computation. The point features are detected and tracked using the Lucas-Kanade algorithm. Since there is a significant amount of self-occlusion, many point features representing the target are lost. Therefore, we use only short term feature trajectories between two consecutive frames. Let $V_t = \left(\mathbf{v}_1^{(t)}, \ldots, \mathbf{v}_k^{(t)} \right)$ be the tuple that describes the velocities of the k feature points at frame t, $t = 0, \ldots, T$, where $T + 2$ is the total number of frames in the sequence. The configuration of the subject in the current frame is denoted by $c_t = (m_t, n_t)$, where m_t and n_t are the 2D pose and view at time t, respectively. We assume that the viewpoint stays the same throughout the sequence. The configuration in the current frame is dependent not only on the motion vectors in the current frame but also on the configuration in the previous time instants. For determining c_t, the Bayesian formulation of the problem is given by

$$p(c_t|V_t, c_{0:t-1}) \propto p(V_t|c_{0:t})p(c_t|c_{0:t-1}), \tag{1}$$

where $p(V_t|c_{0:t})$ is the likelihood of observing the particular set of motion vectors given the configurations up to time t, and $p(c_t|c_{0:t-1})$ is the prior for time instant t that depends on previosu configurations. Assuming a Markov process, we can write the above equation as

$$p(c_t|V_t, c_{0:t-1}) \propto p(V_t|c_t)p(c_t|c_{t-1}). \tag{2}$$

The estimate of the configuration at time t is $\hat{c}_t$, and our goal is to estimate configurations over the entire sequence, $\mathcal{C} = (\hat{c}_0, \ldots, \hat{c}_T)$. Learning the motion patterns of the multiple poses and viewpoints involves obtaining a set of motion descriptors that

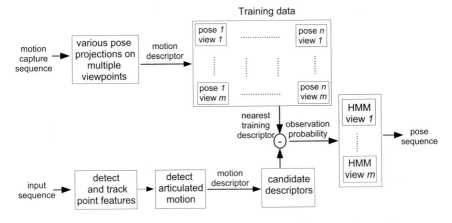

Fig. 2. Overview of the proposed approach to extract human motion models

describe each pose in each viewpoint first in the training data. The test data is then processed in a similar manner to obtain motion descriptors that are compared with the training data to obtain the likelihood of observing a particular pose and viewpoint configuration.

2.1 Training Data

For training, we used a single sequence from the CMU Motion Capture (mocap) data[1] in which the human subject is walking. A single gate cycle was extracted from the sequence. The obtained marker locations associated with the joints and limbs were projected onto simulated image planes oriented at various angles with respect to the subject for each pose (i.e., gait phase), and the corresponding motion vectors were obtained. A similar multi-view training approach was also adopted in [18]. The advantage of using the 3D data is that a single sequence provides a large amount of training data. Note that even though the motion capture data were obtained by calibrated cameras, our technique does not require any calibration since standard cameras have near unity aspect ratio, zero skew, and minimal lens distortion.

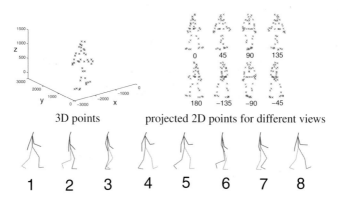

Fig. 3. Top: 3D Motion capture data and its projection onto various planes to provide multiple views in 2D. Bottom: Stick figure models for a sequence of poses (gait phases) for the profile view.

All possible views and poses are quantized to a finite number of configurations. Let M be the number of poses and N the number of views. Let $\mathbf{q}_m^{(i)} = (q_x^{(i)}, q_y^{(i)}, q_z^{(i)})^T$, be the 3D coordinates of the ith point obtained from the mocap data for the mth pose, $i = 1, \ldots, l$. Then the projection of this point onto the plane corresponding to the nth view angle is given by $\mathbf{p}_{mn}^{(i)} = \mathbf{T}_n \mathbf{q}_m^{(i)}$. Here $\mathbf{T}_n$ is the transformation matrix for the nth view angle which is the product of the 2×3 projection matrix and the 3×3 rotation matrix about the vertical axis. Let $\mathcal{P}_{mn} = \left(\mathbf{p}_{mn}^{(1)}, \ldots, \mathbf{p}_{mn}^{(l)} \right)$ be the tuple of 2D points representing the human figure in phase m and view n and $\mathcal{V}_{mn} = \left(\mathbf{v}_{mn}^{(1)}, \ldots, \mathbf{v}_{mn}^{(l)} \right)$ be their corresponding 2D motion vectors. Note that $\mathcal{V}$ denotes motion vectors obtained

[1] http://mocap.cs.cmu.edu

from the training data while V represents the motion vectors obtained from the test sequences. Figure 3 shows the multiple views and poses obtained from the 3D marker data. In this work we use 8 views and 8 poses.

2.2 Motion Descriptor

It is not possible to compare the sets of sparse motion vectors directly using a technique like PCA [18] because there is no ordering of the features. Instead, we aggregate the motion information in spatially local areas. Given the training data of positions $\mathcal{P}_{mn}$ and velocities $\mathcal{V}_{mn}$, we define the motion descriptor ψ_{mn} for pose m and view n as an 18-element vector containing the magnitude and phase of the weighted average motion vector in nine different spatial areas, where the weight is determined by an oriented Gaussian centered in the area. More precisely, the jth bin of the motion descriptor is given by

$$\psi_{mn}(j) = \sum_{i=1}^{l} \mathbf{v}_{mn}^{(i)} G_j(\mathbf{p}_{mn}^{(i)}),$$

(3)

where G_j is a 2D oriented Gaussian given by

$$G_j(\mathbf{x}) = \frac{1}{2\pi|\Sigma_j|^{1/2}} \exp\left(-\frac{1}{2}\left(\mathbf{x} - \mu_j\right)^T \Sigma_j^{-1}\left(\mathbf{x} - \mu_j\right)\right),$$

(4)

with μ_j and Σ_j being the mean and covariance matrix of the jth Gaussian, precomputed with reference to the body center.

Figure 4 shows the nine spatial ellipses used in computing the motion descriptor, along with their Gaussian weight maps. The discriminative ability of the motion descriptor is illustrated in the rest of the figure. The confusion matrix shows the pseudocolored Euclidean distance between the motion descriptors of all pairs of 64 configurations, with zero values along the diagonal. It is clear from this matrix that motion alone carries sufficient information to discriminate between the various poses and views in nearly all situations. The bottom row of the figure shows the descriptor bin values for two cases: three different views of the same pose, and the same view of three different poses. Because they capture the motion of the upper body, the first several bins have similar values, while the last several bins representing the lower body show a larger degree of variation. It is this larger variation in the lower part of the body that gives the descriptor its discriminatory power.

3 Pose and Viewpoint Estimation

Hidden Markov Models (HMMs) are well suited for the estimation of human gait over time. HMMs are statistical models consisting of a finite number of states which are not directly observable (hidden) and which follow a Markov chain, i.e., the likelihood of occurrence of a state at the next instant of time conditionally depends only on the current state. Each discrete pose for each viewpoint can be considered as a hidden state of the model. Assuming that the pose of a human walking is a Markov process, the observation probabilities can be computed from the image data using the motion of the

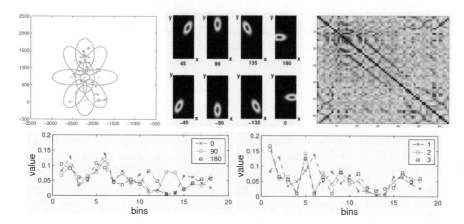

Fig. 4. TOP: The proposed motion descriptor (left), weight maps (middle) of all but the central Gaussian used for computing the motion descriptor, and the 64×64 confusion matrix (right) for 8 poses and 8 views. BOTTOM: The motion descriptor bin values for different views of the same pose (left), and for the same view of different poses (right).

limbs, and the state transition probabilities and priors can be determined beforehand. The goal is then to determine the hidden state sequence (pose estimates and viewpoint) based on a series of observations obtained from the image data.

Let $\lambda = (A, B, \pi)$ be the HMM, where A is the state transition probability matrix, B is the observational probability matrix, and π is the prior. Let the configuration c_t represent the hidden state of the model at time t, and let O_t be the observation at that time. There is a finite set of states $\mathcal{S} = \{(1, 1), \ldots, (M, N)\}$ corresponding to each pose and view angle. The state transition probability is $A(i, j) = P(c_{t+1} = s_j | c_t = s_i)$, $s_i, s_j \in \mathcal{S}$, i.e., the probability of being in state s_j at time $t + 1$ given that the current state is s_i. The observation probability is given by $B(j, t) = P(O_t | c_t = s_j)$, i.e., the probability of observing O_t at time t given that the current state is s_j. Given the HMM $\lambda = (A, B, \pi)$, and series of observations $\mathcal{O} = \{O_0, \ldots, O_T\}$, our goal is to find the sequence of states $\mathcal{C} = \{c_0, \ldots, c_T\}$ such that the joint probability of the observation sequence and the state sequence given the model $P(O, \mathcal{C} | \lambda)$ is maximized.

The state transition probability between two states $s_i = (m_i, n_i)$ and $s_j = (m_j, n_j)$ is predefined to be

$$ p(s_j | s_i) = \begin{cases} \phi_{next} & \text{if } n_i = n_j \text{ and } m_j = m_i + 1 \\ \phi_{remain} & \text{if } n_i = n_j \text{ and } m_j = m_i \\ 0 & \text{otherwise} \end{cases} \tag{5} $$

where $\phi_{next} = 0.51$ is the probability of transitioning to the next pose, and $\phi_{remain} = 0.43$ is the probability of remaining in the same pose. Note that, as mentioned earlier, the transition probability from one view to another view is zero, creating effectively a disconnected HMM. The observation probability is given by a normalized exponential of the Euclidean distance between the test and training motion descriptors. The optimum state sequence $\mathcal{C}$ for the HMM is then computed using the Viterbi algorithm.

4   Experimental Results

Our approach was tested on a variety of sequences of walking humans from different viewpoints, scales, and illumination conditions. The detection of articulated bodies is performed by computing the motion descriptor to each pixel of the image at three different scales and projecting the descriptor onto a line to determine the similarity with respect to a model of human motion. A strength map is generated indicating the probability of a person being at that location and scale, and the maximum of the strength map

Fig. 5. Articulated motion detection for various viewpoints: right profile, left profile, at an angle, and frontal. In the bottom row, the camera is moving.

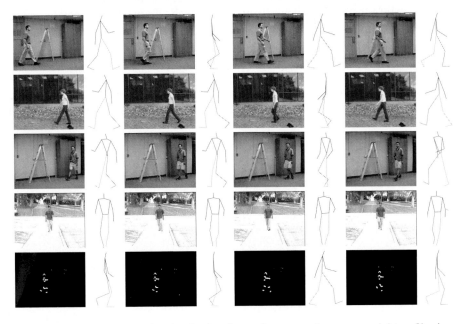

Fig. 6. Top to bottom: Pose estimation for four frames from several sequences: right profile view, left profile view, angular view, frontal view, and profile view at night with reflectors

is used as the location and scale of the target. Figure 5 shows human detection based on this procedure. The unique characteristics of human motion when compared to other motions present in natural scenes is clear from the ability of such a simple procedure to detect the people. Using only motion information, the person is correctly detected in each sequence, even when the camera is moving, because only differences between motion vectors are used. Once the person has been detected, Lucas-Kanade point features are tracked through the image sequence, and the location and scale of the person is updated using the tracked points attached to the detected target. The entire process is fully automatic.

Figure 6 shows the pose estimation results for sequences captured from various viewpoints. Each sequence covers an entire gait cycle. The stick figure models correspond to the nearest configuration found in the training data by the HMM. It is important to keep in mind that point feature tracks are not very accurate in sequences such as these involving non-rigid motion and large amounts of occlusion, and a large number of point features belonging to the background cause noise in the data, especially when the camera is moving. Moreover, when the person walks toward or away from the camera (frontal view), the pose estimation is difficult due to the ambiguity in motion. Nevertheless, the estimated poses are qualitatively correct.

The last row of the figures shows a sequence captured at night by an infrared camera. The person is wearing a special body suit fitted with reflectors that reflect the light emitted by the headlights of an oncoming vehicle. This suit has been used in psychological studies of the effectiveness of reflectors for pedestrian safety by exploiting the biomotion capabilities of the human visual system of automobile drivers [22]. The utility of a purely motion based approach can be especially seen in this sequence, in which no appearance information is available. Even without such information, the motion vectors are highly effective within the current framework for estimating the pose. To provide quantitative evaluation, Figure 7 shows the estimated knee angles at every frame of the right profile view and the frontal view sequences, along with the ground truth.

As can be seen from these results, our approach offers several advantages over previous motion-based approaches [3,18,17,21]. First, it is invariant to scale and viewpoint, and it is able to deal with noisy video sequences captured from a moving camera. In

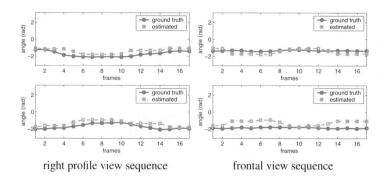

right profile view sequence   frontal view sequence

Fig. 7. Estimated and ground truth knee angles for two sequences. The top row shows the right knee, while the bottom row shows the left knee.

contrast, many of the previous algorithms rely on a static camera, tightly controlled imaging conditions, and/or a particular walking direction (e.g., profile view). Another advantage of our approach is that it is easy to train, requiring only a small amount of training data since there is no need to account for all the variations in appearance that occur in real imagery. Since the estimated poses of our approach are necessarily tied to the training data, it is not possible to recover arbitrary body poses not seen in the training data. Nevertheless, it may possible to train a similar detector to handle various other actions such as running or hand waving with appropriate data.

5   Conclusion

Motion is a powerful cue that can be effectively utilized for biological motion analysis. We have presented a motion-based approach for detection, tracking, and pose estimation of articulated human motion that is invariant of scale, viewpoint, illumination, and camera motion. In this spirit of one-shot learning, the approach utilizes only a small amount of training data. The spatial properties of human motion are modeled using a novel descriptor, while temporal dependency is modeled using an HMM. A clear advantage of using a purely motion based approach is demonstrated in pose estimation in nighttime sequences where no appearance information is available. In demonstrating the effectiveness of motion information alone, our intention is not to discount the importance of appearance information but rather to highlight the effectiveness of this particular cue. Future work involves exploring ways of articulated motion detection in the presence of noise, allowing the subjects to change viewpoints as they are tracked, combining the bottom-up and top-down approach for more accurate pose estimation, and incorporating appearance information for increased robustness.

Acknowledgments

We would like to thank Dr. Rick Tyrrell for graciously providing the nighttime sequence.

References

1. Johansson, G.: Visual perception of biological motion and a model for its analysis. Perception and Psychophysics 14, 201–211 (1973)
2. Brostow, G.J., Cipolla, R.: Unsupervised Bayesian detection of independent motion in crowds. In: Proceedings of the IEEE Conference on Computer Vision and Pattern Recognition (CVPR), pp. 594–601 (2006)
3. Daubney, B., Gibson, D., Campbell, N.: Real-time pose estimation of articulated objects using low-level motion. In: CVPR (2008)
4. Viola, P., Jones, M., Snow, D.: Detecting pedestrians using patterns of motion and appearance. In: ICCV (2003)
5. Papageorgiou, C., Poggio, T.: A trainable system for object detection. IJCV 38, 15–33 (2000)
6. Wu, B., Nevatia, R.: Detection and tracking of multiple partially occluded humans by Bayesian combination of edgelet based part detectors. IJCV 75, 247–266 (2007)

7. Gavrila, D., Munder, S.: Multi-cue pedestrian detection and tracking from a moving vehicle. IJCV 73, 41–59 (2007)
8. Cutler, R., Davis, L.: Robust real-time periodic motion detection, analysis, and applications. PAMI 22, 781–796 (2000)
9. Agarwal, A., Triggs, B.: Tracking articulated motion using a mixture of autoregressive models. In: ECCV, pp. 54–65 (2004)
10. Sigal, L., Bhatia, S., Roth, S., Black, M., Isard, M.: Tracking loose limbed people. In: CVPR, pp. 421–428 (2004)
11. Urtasun, R., Fleet, D., Hertzman, A., Fua, P.: Priors for people from small training sets. In: ICCV, pp. 403–410 (2005)
12. Ramanan, D., Forsyth, D.: Finding and tracking people from bottom-up. In: CVPR, pp. 467–474 (2003)
13. Song, Y., Goncalves, L., Perona, P.: Unsupervised learning of human motion. PAMI 25, 814–827 (2003)
14. Lee, M., Nevatia, R.: Human pose tracking using multiple level structured models. In: Leonardis, A., Bischof, H., Pinz, A. (eds.) ECCV 2006. LNCS, vol. 3953, pp. 368–381. Springer, Heidelberg (2006)
15. Bregler, C.: Learning and recognizing human dynamics in video sequences. In: CVPR, pp. 568–575 (1997)
16. Lan, X., Huttenlocher, D.: A unified spatio-temporal articulated model for tracking. In: CVPR, pp. 722–729 (2004)
17. Fathi, A., Mori, G.: Human pose estimation using motion exemplars. In: ICCV, pp. 1–8 (2007)
18. Fablet, R., Black, M.: Automatic detection and tracking of human motion with a view based representation. In: Heyden, A., Sparr, G., Nielsen, M., Johansen, P. (eds.) ECCV 2002. LNCS, vol. 2350, pp. 476–491. Springer, Heidelberg (2002)
19. Lipton, A.: Local applications of optic flow to analyse rigid versus non-rigid motion. In: ICCV Workshop on Frame-Rate Applications (1999)
20. Niebles, J., Fei-Fei, L.: A hierarchical model of shape and appearance for human action classification. In: CVPR, pp. 1–8 (2007)
21. Bobick, A., Davis, J.: The recognition of human movement using temporal templates. PAMI 23, 257–267 (2001)
22. Wood, J.M., Tyrrell, R.A., Carberry, T.: Unsupervised learning of human motion. Human Factors 47, 644–653 (2005)

Robust Estimation of Camera Motion Using Optical Flow Models

Jurandy Almeida[1], Rodrigo Minetto[1], Tiago A. Almeida[2],
Ricardo da S. Torres[1], and Neucimar J. Leite[1]

[1] Institute of Computing, University of Campinas, Brazil
{jurandy.almeida,rodrigo.minetto,rtorres,neucimar}@ic.unicamp.br
[2] School of Electrical and Computer Engineering, University of Campinas, Brazil
tiago@dt.fee.unicamp.br

Abstract. The estimation of camera motion is one of the most important aspects for video processing, analysis, indexing, and retrieval. Most of existing techniques to estimate camera motion are based on optical flow methods in the uncompressed domain. However, to decode and to analyze a video sequence is extremely time-consuming. Since video data are usually available in MPEG-compressed form, it is desirable to directly process video material without decoding. In this paper, we present a novel approach for estimating camera motion in MPEG video sequences. Our technique relies on linear combinations of optical flow models. The proposed method first creates prototypes of optical flow, and then performs a linear decomposition on the MPEG motion vectors, which is used to estimate the camera parameters. Experiments on synthesized and real-world video clips show that our technique is more effective than the state-of-the-art approaches for estimating camera motion in MPEG video sequences.

1 Introduction

Advances in data compression, data storage, and data transmission have facilitated the way videos are created, stored, and distributed. The increase in the amount of video data has enabled the creation of large digital video libraries. This has spurred great interest for systems that are able to efficiently manage video material [1, 2, 3].

Making efficient use of video information requires that the data be stored in an organized way. For this, it must be associated with appropriate features in order to allow any future retrieval. An important feature in video sequences is the temporal intensity change between successive video frames: apparent motion. The apparent motion is generally attributed to the motion caused by object movement or introduced by camera operation. The estimation of camera motion is one of the most important aspects to characterize the content of video sequences [4].

G. Bebis et al. (Eds.): ISVC 2009, Part I, LNCS 5875, pp. 435–446, 2009.

Most of existing techniques to estimate camera motion are based on analysis of the optical flow between consecutive video frames [5,6,7,8,9,10,11]. However, the estimation of the optical flow, which is usually based on gradient or block matching methods, is computationally expensive [12].

Since video data are usually available in MPEG-compressed form, it is desirable to directly process the compressed video without decoding. A few methods that directly manipulate MPEG compressed video to extract camera motion have been proposed [13,12,4,14]. These approaches use MPEG motion vectors[1] as an alternative to optical flow which allows to save high computational load from two perspectives: full decoding the video stream and optical flow computation [4].

The most popular models in estimating camera motion from MPEG motion vectors are the four parameter [14] and the six parameter [4] affine model. However, the affine parameters are not directly related to the physically meaningful camera operations.

In this paper, we propose a novel approach for the estimation of camera motion in MPEG video sequences based on optical flow models. The proposed method generates the camera model using linear combinations of prototypes of optical flow produced by each camera operation.

In order to validate our approach, we use a synthetic test set and real-world video sequences including all kinds of camera motion and many of their possible combinations. Further, we have conducted several experiments to show that our technique is more effective than the affine model-based approaches for estimating camera motion in MPEG video sequences.

The remainder of the paper is organized as follows. In Section 2, we review three existing approaches used as reference in our experiments. Section 3 presents our approach for the estimation of camera motion. The experimental settings and results are discussed in Section 4. Finally, Section 5 presents conclusions and directions for future work.

2   Related Work

In this section, we review three approaches used as reference in our experiments. These methods were implemented and their effectiveness are compared in Section 4.

Kim et al. [4] have used a two-dimensional affine model to detect six types of motion: panning, tilting, zooming, rolling, object motion, and stationary. Beforehand, motion vector outliers are filtered out by a simple smoothing filter. The camera parameters for the model are estimated by using a least squares fit to the remaining data.

Smolic et al. [14] have used a simplified two-dimensional affine model to distinguish between panning, tilting, zooming, and rolling. They use the M-estimator

---

[1] In video compression, a motion vector is a two-dimensional vector used for inter prediction that provides an offset from the coordinates in the decoded picture to the coordinates in a reference picture.

approach [15] to deal with data corrupted by outliers. It is basically a weighted least square technique, which reduces the effect of outliers by using an influence function.

Gillespie et al. [12] have extended such approach in order to improve its effectiveness by using a robust Least Median-of-Squares (LMedS) [15] to estimate the camera parameters and minimize the influence of outliers.

3   Our Approach

The previous approaches simply find the best-fit affine model to estimate camera motion by using the least squares method. However, the affine parameters are not directly related to the physically meaningful camera operations.

In this sense, we propose a novel approach for the estimation of camera motion based on optical flow models. The proposed method generates the camera model using linear combinations of prototypes of optical flow produced by each camera operation. It consists of three main steps: (1) feature extraction; (2) motion model fitting; and (3) robust estimation of the camera parameters.

3.1   Feature Extraction

MPEG videos are composed by three main types of pictures: intra-coded (I-frames), predicted (P-frames), and bidirectionally predicted (B-frames). These pictures are organized into sequences of groups of pictures (GOPs) in MPEG video streams.

A GOP must start with an I-frame and can be followed by any number of I and P-frames, which are usually known as anchor frames. Between each pair of consecutive anchor frames can appear several B-frames. Figure 1 shows a typical GOP structure.

An I-frame does not refer to any other video frame. On the other hand, the encoding of a P-frame is based on a previous anchor frame, while the encoding of a B-frame can be based on two anchor frames, a previous as well as a subsequent anchor frame.

Each video frame is divided into a sequence of non-overlapping macroblocks. For each macroblock, a motion vector which points to a similar block in an anchor frame is estimated. Motion estimation algorithms try to find the best block match in terms of compression efficiency. This can lead to motion vectors that do not represent the camera motion at all [17].

The motion vectors are extracted directly from the compressed MPEG stream. Only the motion vectors from P-frames are processed in our approach. They were chosen due to the following reasons. First, usually each third until fifth frame in a MPEG video is a P-frame, and thus, the temporal resolution is suficient for most applications. Further, both the prediction direction and the temporal distance of motion vectors are not unique in B-frames, resulting in additional computational complexity.

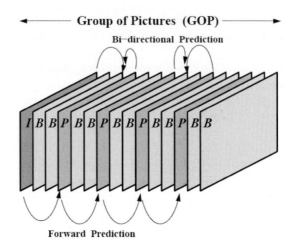

Fig. 1. A typical group of pictures (GOP) in MPEG video sequences [16]

3.2   Motion Model Fitting

A camera projects a 3D world point into a 2D image point. The motion of the camera may be limited to a single motion such as rotation, translation, or zoom, or some combination of these three motions. Such camera motion can be well categorized by few parameters.

If we consider that the visual field of the camera is small, we can establish ideal optical flow models, which are noise-free, by using a numerical expression for the relationship of the MPEG motion vectors, and creating prototypes of optical flow models.

Hence, we can approximate the optical flow by a weighted combination of optical flow models:

$$f = P \cdot p + T \cdot t + Z \cdot z + R \cdot r, \tag{1}$$

where p, t, z, and r are the prototypes generated by panning, tilting, zooming, and rolling, respectively.

The parameter-estimation problem is now to obtain an estimate for the parameters P, T, Z, and R, based on a set of measured motion vectors $\{\hat{f}_i\}$. Since the measurements are not exact, we can not assume that they will all fit perfectly to the model. Hence, the best solution is to compute a least-squares fit to the data. For this, we define the model error as the sum of squared norm of the discrepancy vectors between the measured motion vectors $\hat{f}_i$ and the motion vectors obtained from the model:

$$E = \sum_i \|(P \cdot p_i + T \cdot t_i + Z \cdot z_i + R \cdot r_i) - \hat{f}_i\|^2, \tag{2}$$

where P, T, Z, and R represent the motion induced by the camera operations of panning (or tracking), tilting (or booming), zooming (or dollying), and rolling, respectively.

To minimize the model error E, we can take its derivatives with respect to the motion parameters

$$\frac{\partial E}{\partial P} = \sum_i 2\, p_i^T\, (P \cdot p_i + T \cdot t_i + Z \cdot z_i + R \cdot r_i - \hat{f}_i),$$

$$\frac{\partial E}{\partial T} = \sum_i 2\, t_i^T\, (P \cdot p_i + T \cdot t_i + Z \cdot z_i + R \cdot r_i - \hat{f}_i),$$

$$\frac{\partial E}{\partial Z} = \sum_i 2\, z_i^T\, (P \cdot p_i + T \cdot t_i + Z \cdot z_i + R \cdot r_i - \hat{f}_i),$$

$$\frac{\partial E}{\partial R} = \sum_i 2\, r_i^T\, (P \cdot p_i + T \cdot t_i + Z \cdot z_i + R \cdot r_i - \hat{f}_i),$$

and set them to zero, giving

$$\sum_i (P\, p_i^T p_i + T\, p_i^T t_i + Z\, p_i^T z_i + R\, p_i^T r_i - p_i^T \hat{f}_i) = 0,$$

$$\sum_i (P\, t_i^T p_i + T\, t_i^T t_i + Z\, t_i^T z_i + R\, t_i^T r_i - t_i^T \hat{f}_i) = 0,$$

$$\sum_i (P\, z_i^T p_i + T\, z_i^T t_i + Z\, z_i^T z_i + R\, z_i^T r_i - z_i^T \hat{f}_i) = 0,$$

$$\sum_i (P\, r_i^T p_i + T\, r_i^T t_i + Z\, r_i^T z_i + R\, r_i^T r_i - r_i^T \hat{f}_i) = 0,$$

which can be written in matrix form as

$$
\begin{bmatrix}
\sum_i \langle p_i, p_i \rangle & \sum_i \langle p_i, t_i \rangle & \sum_i \langle p_i, z_i \rangle & \sum_i \langle p_i, r_i \rangle \\
\sum_i \langle t_i, p_i \rangle & \sum_i \langle t_i, t_i \rangle & \sum_i \langle t_i, z_i \rangle & \sum_i \langle t_i, r_i \rangle \\
\sum_i \langle z_i, p_i \rangle & \sum_i \langle z_i, t_i \rangle & \sum_i \langle z_i, z_i \rangle & \sum_i \langle z_i, r_i \rangle \\
\sum_i \langle r_i, p_i \rangle & \sum_i \langle r_i, t_i \rangle & \sum_i \langle r_i, z_i \rangle & \sum_i \langle r_i, r_i \rangle
\end{bmatrix}
\begin{pmatrix} P \\ T \\ Z \\ R \end{pmatrix}
=
\begin{pmatrix}
\sum_i \langle p_i, \hat{f}_i \rangle \\
\sum_i \langle t_i, \hat{f}_i \rangle \\
\sum_i \langle z_i, \hat{f}_i \rangle \\
\sum_i \langle r_i, \hat{f}_i \rangle
\end{pmatrix}, \quad (3)
$$

where $\langle u, v \rangle = u^T v$ is the inner product between the vectors u and v.

Here, we define the optical flow model for panning (p), tilting (t), zooming (z), and rolling (r), respectively, as:

$$p(x, y) = \begin{pmatrix} -1 \\ 0 \end{pmatrix}, \quad t(x, y) = \begin{pmatrix} 0 \\ -1 \end{pmatrix}, \quad z(x, y) = \begin{pmatrix} -x \\ -y \end{pmatrix}, \quad r(x, y) = \begin{pmatrix} y \\ -x \end{pmatrix},$$

where (x, y) is the sample point whose coordinate system has origin at the center of the image.

Figure 2 represents the prototypes which consist of optical flow models generated by panning, tilting, zooming, and rolling, respectively. These optical flow models express the amount and direction of the camera motion parameters, respectively.

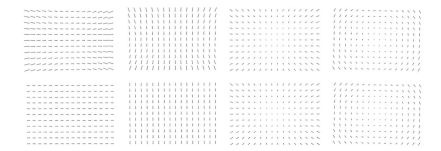

Fig. 2. The optical flow (top) and the prototype (bottom) generated by panning, tilting, zooming, and rolling, respectively (left to right)

3.3 Robust Estimation of the Camera Parameters

The direct least-squares approach for parameter estimation works well for a small number of outliers that do not deviate too much from the correct motion. However, the result is significantly distorted when the number of outliers is larger, or the motion is very different from the correct camera motion. Especially if the video sequence shows independent object motions, a least-squares fit to the complete data would try to include all visible object motions into a single motion model.

To reduce the influence of outliers, we apply a well-known robust estimation technique called RANSAC (RANdom SAmple Consensus) [18]. The idea is to repeatedly guess a set of model parameters using small subsets of data that are drawn randomly from the input. The hope is to draw a subset with samples that are part of the same motion model. After each subset draw, the motion parameters for this subset are determined and the amount of input data that is consistent with these parameters is counted. The set of model parameters with the largest support of input data is considered the most dominant motion model visible in the image.

4   Experiments and Results

In order to evaluate the performance of the proposed method for estimating camera motion in MPEG video sequences, experiments were carried out on both synthetic and real-world video clips.

4.1 Results with Noise-Free Synthetic Data

First, we evaluate our approach on synthetic video sequences with known ground-truth data. For this, we create a synthetic test set with four MPEG-4 video clips[2]

---

[2] All video clips and ground-truth data of our synthetic test set are available at
http://www.liv.ic.unicamp.br/~minetto/videos/

(a)   (b)   (c)

Fig. 3. The POV-Ray scenes of a realistic office model used in our synthetic test set

$(640 \times 480$ pixels of resolution) based on well textured POV-Ray scenes of a realistic office model (Figure 3), including all kinds of camera motion and many of their possible combinations. The main advantage is that the camera motion parameters can be fully controlled which allows us to verify the estimation quality in a reliable way.

The first step for creating the synthetic videos is to define the camera's position and orientation in relation to the scene. The world-to-camera mapping is a rigid transformation which takes scene coordinates $p_w = (x_w, y_w, z_w)$ of a point to its camera coordinates $p_c = (x_c, y_c, z_c)$. This mapping is given by [19]

$$p_c = R p_w + T, \tag{4}$$

where R is a 3×3 rotation matrix that defines the camera's orientation, and T defines the camera's position.

The rotation matrix R is formed by a composition of three special orthogonal matrices (known as *rotation matrices*)

$$R_x = \begin{bmatrix} \cos(\alpha) & 0 & -\sin(\alpha) \\ 0 & 1 & 0 \\ \sin(\alpha) & 0 & \cos(\alpha) \end{bmatrix}, R_y = \begin{bmatrix} 1 & 0 & 0 \\ 0 & \cos(\beta) & \sin(\beta) \\ 0 & -\sin(\beta) & \cos(\beta) \end{bmatrix}, R_z = \begin{bmatrix} \cos(\gamma) & \sin(\gamma) & 0 \\ -\sin(\gamma) & \cos(\gamma) & 0 \\ 0 & 0 & 1 \end{bmatrix},$$

where α, β, γ are the angles of the rotations.

We consider the motion of a continuously moving camera as a trajectory where the matrices R and T change according to the time t, in homogeneous representation,

$$\begin{bmatrix} p_c \\ 1 \end{bmatrix} = \begin{bmatrix} R(t) & T(t) \\ 0 & 1 \end{bmatrix} \begin{bmatrix} p_w \\ 1 \end{bmatrix}. \tag{5}$$

Thus, to perform camera motions such as tilting (gradual changes in R_x), panning (gradual changes in R_y), rolling (gradual changes in R_z), and zooming (gradual changes in focal distance f), we define a function $F(t)$ which returns the parameters (α, β, γ, and f) used to move the camera at the time t. We use a smooth and cyclical function

$$F(t) = \mathcal{M} * \frac{1 - \cos(2\pi t/\mathcal{T})(0.5 - t/\mathcal{T})}{0.263}, \tag{6}$$

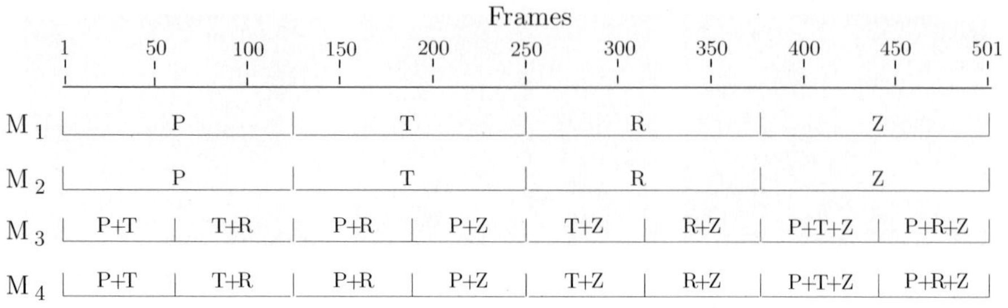

	Frames									
	1	50	100	150	200	250	300	350	400	450	501

M$_1$		P			T			R			Z	
M$_2$		P			T			R			Z	
M$_3$	P+T	T+R	P+R	P+Z	T+Z	R+Z	P+T+Z	P+R+Z
M$_4$	P+T	T+R	P+R	P+Z	T+Z	R+Z	P+T+Z	P+R+Z

Fig. 4. The main characteristics of each video sequence (M$_i$) in our synthetic test set

where $\mathcal{M}$ is the maximum motion factor and $\mathcal{T}$ is the duration of camera motion in units of time. We create all video clips using the maximum motion factor $\mathcal{M}$ equals to 3° for tilting (α), 8° for panning (β), 90° for rolling (γ), and 1.5 for zooming (f).

Figure 4 shows the main characteristics of each resulting video sequence (M$_i$). The terms P, T, R, and Z stand for the motion induced by the camera operations of panning, tilting, zooming, and rolling, respectively. The videos M$_3$ and M$_4$ have combinations of two or three types of camera motions. In order to represent a more realistic scenario, we modify the videos M$_2$ and M$_4$ to have occlusions due to object motion.

Furthermore, we change the artificial illumination of lamps and the reflection of the sunrays in some parts of the scene according to the camera motion. In addition, all the objects present in the scene have complex textures, which are very similar to the real ones.

Moreover, we are very severe in the intensity of the camera movements. Fast camera motions and combinations of several types of motion at the same time are rare to occur. Our goal in these cases is to measure the response of our algorithm in adverse conditions, and not only with simple camera operations.

We assess the effectiveness of the proposed method using the well-known Zero-mean Normalized Cross Correlation (ZNCC) metric [20], defined by

$$\text{ZNCC}(\mathcal{F}, \mathcal{G}) = \frac{\sum_t ((\mathcal{F}(t) - \bar{\mathcal{F}}) \times (\mathcal{G}(t) - \bar{\mathcal{G}}))}{\sqrt{\sum_t (\mathcal{F}(t) - \bar{\mathcal{F}})^2 \times \sum_t (\mathcal{F}(t) - \bar{\mathcal{G}})^2}} \quad (7)$$

where $\mathcal{F}(t)$ and $\mathcal{G}(t)$ are the estimate and the real camera parameters, respectively, at the time t. It returns a real value between -1 and $+1$. A value equals to $+1$ indicates a perfect estimation; and -1, an inverse estimation.

Tables 1, 2, 3, and 4 compare our approach with the techniques presented in Section 2. Clearly, the use of optical flow models for estimating camera motion in MPEG video sequences is more effective than the affine model-based approaches.

Despite MPEG motion vectors improve the runtime performance, they often do not model real motion adequately [17]. Note that the effectiveness achieved by all methods is reasonably reduced for tilting operations in presence of several types of camera motions at the same time.

Table 1. Effectiveness achieved by all approaches in video clip M_1

Method	Tilting	Panning	Rolling	Zooming
Our Approach	**0.981419**	**0.996312**	**0.999905**	**0.964372**
Gillespie et al.	0.970621	0.987444	0.999830	0.958607
Smolic et al.	0.950911	0.994171	0.999199	0.949852
Kim et al.	0.649087	0.912365	0.994067	0.858090

Table 2. Effectiveness achieved by all approaches in video clip M_2

Method	Tilting	Panning	Rolling	Zooming
Our Approach	**0.981029**	**0.995961**	**0.999913**	**0.965994**
Gillespie et al.	0.972189	0.988062	0.999853	0.959516
Smolic et al.	0.936479	0.991438	0.999038	0.949367
Kim et al.	0.633559	0.821266	0.986408	0.865052

Table 3. Effectiveness achieved by all approaches in video clip M_3

Method	Tilting	Panning	Rolling	Zooming
Our Approach	**0.587136**	**0.950760**	**0.999624**	**0.956845**
Gillespie et al.	0.575178	0.931957	0.999521	0.954215
Smolic et al.	0.559669	0.940782	0.999037	0.951701
Kim et al.	0.501764	0.942563	0.997240	0.942588

Table 4. Effectiveness achieved by all approaches in video clip M_4

Method	Tilting	Panning	Rolling	Zooming
Our Approach	**0.592071**	**0.949922**	**0.999659**	**0.956440**
Gillespie et al.	0.577467	0.932568	0.999545	0.954286
Smolic et al.	0.557849	0.940886	0.998920	0.951640
Kim et al.	0.498081	0.941956	0.997334	0.944102

4.2 Results with Real-World Video Sequences

We also evaluate our technique over four real-world video sequences[3]. These video clips were shot with a hand-held consumer-grade DVR (Canon Optura 40) with variable zoom. They were recorded in MPEG format at 320×240 resolution, 14.98 frames per second.

Table 5 summarizes the main characteristics of each resulting real-world video sequence (R_i). All videos clips were affected by natural noise. The videos R_3 and R_4 have occlusions due to object motion.

---

[3] All real-world video sequences are available at http://www.liv.ic.unicamp.br/~minetto/videos/

Table 5. The main characteristics of each real-world video sequence (R_i)

Video	Frames	Camera Operations
R_1	338	P,T,R,Z
R_2	270	P,T,R,Z
R_3	301	P,T,R,Z
R_4	244	P,T,R,Z

In these experiments, we analyze the effectiveness of motion vector-based techniques in relation to the well-known optical flow-based estimator presented in [9]. Each video clip (R_i) takes less than 1 second to process the whole sequence using a motion vector-based approach on a Intel Core 2 Quad Q6600 (four cores running at 2.4 GHz), 2GB memory DDR3. It is important to realize that the optical flow-based method requires a magnitude of almost one second per frame.

Tables 6, 7, 8, and 9 compare our approach with the techniques presented in Section 2. In fact, the use of optical flow models for estimating camera motion in MPEG video sequences outperforms the affine model-based approaches.

Note that the optical flow models identify the camera operations better than the affine parameters. For instance, considering zooming operations in the video

Table 6. Effectiveness achieved by all approaches in video clip R_1

Method	Tilting	Panning	Rolling	Zooming
Our Approach	**0.986287**	**0.986294**	**0.987545**	**0.982227**
Gillespie et al.	0.982345	0.978892	0.980464	0.964398
Smolic et al.	0.984085	0.976381	0.977135	0.966526
Kim et al.	0.982998	0.884470	0.795713	0.944286

Table 7. Effectiveness achieved by all approaches in video clip R_2

Method	Tilting	Panning	Rolling	Zooming
Our Approach	**0.914379**	**0.954113**	**0.929268**	**0.684219**
Gillespie et al.	0.863166	0.931218	0.899512	0.357249
Smolic et al.	0.874244	0.952316	0.919447	0.611227
Kim et al.	0.899520	0.901673	0.846316	0.670006

Table 8. Effectiveness achieved by all approaches in video clip R_3

Method	Tilting	Panning	Rolling	Zooming
Our Approach	**0.964425**	**0.960878**	**0.957735**	**0.454204**
Gillespie et al.	0.949270	0.931442	0.927145	0.379836
Smolic et al.	0.957662	0.953751	0.956303	0.444741
Kim et al.	0.954097	0.912121	0.924798	0.368722

Table 9. Effectiveness achieved by all approaches in video clip R_4

Method	Tilting	Panning	Rolling	Zooming
Our Approach	**0.976519**	**0.958020**	**0.927920**	**0.577974**
Gillespie et al.	0.948314	0.902511	0.851247	0.308588
Smolic et al.	0.969314	0.956417	0.903442	0.523507
Kim et al.	0.969613	0.938639	0.839906	0.474439

R_4, our method is more than 10% ($\approx$ 5 percentual points) better than the best affine model-based one.

5 Conclusions

In this paper, we have presented a novel approach for the estimation of camera motion in MPEG video sequences. Our technique relies on linear combinations of optical flow models. Such models identify the camera operations better than the affine parameters.

We have validated our technique using synthesized and real-world video clips including all kinds of camera motion and many of their possible combinations. Our experiments have showed that the use of optical flow models for estimating camera motion in MPEG video sequences is more effective than the affine model-based approaches.

Future work includes an extension of the proposed method to distinguish between translational (tracking, booming, and dollying) and rotational (panning, tilting, and rolling) camera operations. In addition, we want to investigate the effects of integrating the proposed method into a complete MPEG system for camera motion-based search-and-retrieval of video sequences.

Acknowledgment

The authors thank the financial support of Microsoft EScience Project, CAPES/COFECUB Project (Grant 592/08), and Brazilian agencies FAPESP (Grants 07/52015-0, 07/54201-6, and 08/50837-6), CNPq (Grants 311309/2006-2, 142466/2006-9, and 472402/2007-2), and CAPES (Grant 01P-05866/2007).

References

1. Chang, S.F., Chen, W., Meng, H.J., Sundaram, H., Zhong, D.: A fully automated content-based video search engine supporting spatio-temporal queries. IEEE Trans. Circuits Syst. Video Techn. 8, 602–615 (1998)
2. Hampapur, A., Gupta, A., Horowitz, B., Shu, C.F., Fuller, C., Bach, J.R., Gorkani, M., Jain, R.: Virage video engine. In: Storage and Retrieval for Image and Video Databases (SPIE), pp. 188–198 (1997)

3. Ponceleon, D.B., Srinivasan, S., Amir, A., Petkovic, D., Diklic, D.: Key to effective video retrieval: Effective cataloging and browsing. In: ACM Multimedia, pp. 99–107 (1998)
4. Kim, J.G., Chang, H.S., Kim, J., Kim, H.M.: Efficient camera motion characterization for mpeg video indexing. In: ICME, pp. 1171–1174 (2000)
5. Dufaux, F., Konrad, J.: Efficient, robust, and fast global motion estimation for video coding. IEEE Trans. Image Process. 9, 497–501 (2000)
6. Park, S.C., Lee, H.S., Lee, S.W.: Qualitative estimation of camera motion parameters from the linear composition of optical flow. Pattern Recognition 37, 767–779 (2004)
7. Qi, B., Ghazal, M., Amer, A.: Robust global motion estimation oriented to video object segmentation. IEEE Trans. Image Process. 17, 958–967 (2008)
8. Sand, P., Teller, S.J.: Particle video: Long-range motion estimation using point trajectories. IJCV 80, 72–91 (2008)
9. Srinivasan, M.V., Venkatesh, S., Hosie, R.: Qualitative estimation of camera motion parameters from video sequences. Pattern Recognition 30, 593–606 (1997)
10. Zhang, T., Tomasi, C.: Fast, robust, and consistent camera motion estimation. In: CVPR, pp. 1164–1170 (1999)
11. Minetto, R., Leite, N.J., Stolfi, J.: Reliable detection of camera motion based on weighted optical flow fitting. In: VISAPP, pp. 435–440 (2007)
12. Gillespie, W.J., Nguyen, D.T.: Robust estimation of camera motion in MPEG domain. In: TENCON, pp. 395–398 (2004)
13. Tiburzi, F., Bescos, J.: Camera motion analysis in on-line MPEG sequences. In: WIAMIS, pp. 42–45 (2007)
14. Smolic, A., Hoeynck, M., Ohm, J.R.: Low-complexity global motion estimation from p-frame motion vectors for mpeg-7 applications. In: ICIP, pp. 271–274 (2000)
15. Rousseeuw, P.J., Leroy, A.M.: Robust Regression and Outlier Detection. John Wiley and Sons, Inc., Chichester (1987)
16. Tan, Y.-P., Saur, D.D., Kulkarni, S.R., Ramadge, P.J.: Rapid estimation of camera motion from compressed video with application to video annotation. IEEE Trans. Circuits Syst. Video Techn. 10, 133–146 (2000)
17. Ewerth, R., Schwalb, M., Tessmann, P., Freisleben, B.: Estimation of arbitrary camera motion in MPEG videos. In: ICPR, pp. 512–515 (2004)
18. Fischler, M.A., Bolles, R.C.: Random sample consensus: A paradigm for model fitting with applications to image analysis and automated cartography. Commun. ACM 24, 381–395 (1981)
19. Ma, Y., Soatto, S., Kosecka, J., Sastry, S.S.: An Invitation to 3-D Vision: From Images to Geometric Models. Springer, Heidelberg (2003)
20. Martin, J., Crowley, J.L.: Experimental comparison of correlation techniques. In: Int. Conf. on Intelligent Autonomous Systems (1995)

Maximum Likelihood Estimation Sample Consensus with Validation of Individual Correspondences

Liang Zhang, Houman Rastgar, Demin Wang, and André Vincent

Communications Research Centre Canada
3701 Carling Avenue, Ottawa, Ontario, K2H 8S2 Canada

Abstract. This paper presents an extension of the maximum likelihood estimation sample consensus (MLESAC) by introducing an online validation of individual correspondences, which is based on the Law of Large Numbers (LLN). The outcomes of the samples, each considered a random event, are analyzed for useful information regarding the validities of individual correspondences. The information from the individual samples that have been processed is accumulated and then used to guide subsequent sampling and to score the estimate. To evaluate the performance of the proposed algorithm, the proposed method was applied to the problem of estimating the fundamental matrix. Experimental results with the Oxford image sequence, *Corridor*, showed that for a similar consensus the proposed algorithm reduced, on average, the Sampson error by about 13% and 12% in comparison to the RANSAC and the MLESAC estimator, while the associated number of samples decreased by about 14% and 15%, respectively.

1 Introduction

The fundamental matrix describes a very important image relation between two images. It is required for many applications such as image rectification and augmented reality. One common problem in image or video processing is to estimate the fundamental matrix from a set of correspondences that are acquired from two images. To reduce the negative effect of mismatched correspondences (outliers), robust estimation techniques are required to overcome problems with noise, which can originate from image capture, and also from errors associated with feature matching. Amongst the methods presented so far in the literature, random sampling consensus (RANSAC) algorithms provide the best solution [9]. RANSAC is essentially a hypothesis and verification algorithm, where solutions are repeatedly generated (hypothesis) from the minimal sets of correspondences randomly selected from the data. The algorithm proceeds by testing each solution for support (consensus) from the complete set of correspondences to determine the consensus for the model hypothesis to be estimated.

RANSAC-like algorithms usually employ a hypothesis scoring technique to evaluate the support of a generated model hypothesis from the complete set of correspondences. The standard RANSAC algorithm [2] counts the number of inliers (as a measure of consensus) for each generated model hypothesis by binarizing the errors with a given threshold. The MSAC (M-estimator sample consensus) estimator [10] measures the quality of a hypothesis in such a way that outliers are given a fixed

G. Bebis et al. (Eds.): ISVC 2009, Part I, LNCS 5875, pp. 447–456, 2009.
© Springer-Verlag Berlin Heidelberg 2009

penalty while inliers are scored on how well they fit the data. The MLESAC (maximum likelihood estimation sample consensus) algorithm [11] evaluates the likelihood of the model hypothesis instead of using heuristic measures. It requires the estimation of a parameter, which represents the proportion of valid correspondences and is solved by an EM (expectation maximization) algorithm. All the methods mentioned so far takes into account equal constant validities of correspondences. The G-MLESAC (guided maximum likelihood estimation sample consensus) algorithm [8] extends the MLESAC algorithm by adding prior validity information for individual correspondences. The prior validities of correspondences are however calculated only from the information gathered from the feature matcher and are kept constant while estimating the parameters of the motion model. In the absence of meaningful matching scores this algorithm is reduced to the same performance as the MLESAC. Recently, a model-driven method was proposed to validate individual correspondences [6]. It makes use of the sample that is considered the best one in the previous samples and, thus, the validation could be unstable because it is highly dependent on a single outcome. Several techniques have also been proposed to speed up the verification phase of the standard RANSAC algorithm. For instance, Chum and Matas [1] made use of randomized sequential sampling to enable the early termination of the hypothesis evaluation. Nister presented a preemptive RANSAC method [5] to efficiently select, with a predefined confidence, the best hypothesis from a fixed number of generated hypotheses.

One key attribute of the conventional RANSAC-like algorithms described above is that each sample is processed independently. No information from the previous samples is exploited to provide guidance in the subsequent samples. An individual sample of the RANSAC-like algorithm can be viewed as a random event and, according to probability theory, the outcomes of the random events will exhibit certain statistical patterns that can be predicted. This motivated us to study the possibility of exploiting the statistical patterns provided by previous samples for analyzing subsequent samples.

This paper is organized as follows. After the Introduction, Section 2 reviews the formulation of the maximum likelihood (ML) estimation of the fundamental matrix. Section 3 describes the online validation of correspondences according to the Law of Large Numbers (LLN). Section 4 describes the proposed algorithm. Experimental results with the Oxford test sequence, *Corridor*, are presented in Section 5. The last section concludes the paper.

2 ML Formulation

The fundamental matrix represents the epipolar geometrical constraint between two images and is applicable for general motion and structure attributes with uncalibrated cameras. Let { $\tilde{x}_i^j$, $i=1$, ..., n and $j=1,2$} be the sets of contaminated homogenous image points, which are the image projections from an object, as viewed in the first and second image. F is the fundamental matrix. $\varepsilon_i(F)$ is an error generated by the ith correspondence { $\tilde{x}_i^1$, $\tilde{x}_i^2$ } by a given F. Then, we have

$$\tilde{x}_i^{2^T} F \tilde{x}_i^1 = \varepsilon_i(F).$$

(1)

To determine the fundamental matrix using maximum likelihood estimation is to find an estimate $\hat{F}$ so that

$$\hat{F} = \max_F \left\{ \prod_{i=1}^{n} p(\varepsilon_i \mid F) \right\}, \tag{2}$$

where $p(\varepsilon_i \mid F)$ is a likelihood function that describes how well the ith correspondence $\{ \tilde{x}_i^1, \tilde{x}_i^2 \}$ matches, when a fundamental matrix F is given. Torr and Zisserman [10] model the errors of correct matches with a Gaussian distribution and the errors of mismatched correspondences with a uniform distribution. In practice, it is unknown whether the ith correspondence $\{ \tilde{x}_i^1, \tilde{x}_i^2 \}$ is a mismatch or not. To quantify the match, let us define an indicator variable v_i, which asserts that the ith correspondence $\{ \tilde{x}_i^1, \tilde{x}_i^2 \}$ is a correct match, and $P(v_i)$ be the probability of it being a correct match (inlier). Taking into consideration both matched and mismatched cases [8], the ML estimation is formulated as follows

$$\hat{F} = \max_F \prod_{i=1}^{n} \left(\left[\frac{1}{\sqrt{2\pi\sigma^2}} e^{-\|\varepsilon_i\|^2 / 2\sigma^2} \right] P(v_i) + \frac{1}{w}(1 - P(v_i)) \right). \tag{3}$$

To implement the ML estimate as shown in Eq. (3), the probability $P(v_i)$ has to be determined. MLESAC algorithm assumes that $P(v_i)$ is constant within the set of all correspondences, e.g. $P(v_i) = P(v)$, and $P(v)$ is calculated using the expectation-maximization (EM) algorithm. In contrast, G-MLESAC algorithm considers that each correspondence has a different $P(v_i)$ and determines $P(v_i)$ based on similarity measures of a feature matcher. In addition, $P(v_i)$ is fixed during hypothesis generation (samples).

3   Validation of Point Correspondences

An individual sample is a random and independent event. The probability theory indicates that if an event is repeated many times the sequences of the random events will exhibit certain statistical patterns, which can be studied and predicted. This suggests that $P(v_i)$ could be estimated from the outcomes of the previously completed samples.

Recall that for each sample the error $\varepsilon_i(F)$ of the ith correspondence for each generated hypothesis F is calculated. Against a pre-defined error threshold T_{thr} the complete set of correspondences are classified into two subsets, namely one for inliers and the other for outliers, i.e,

$$the\ i^{th}\ correspondence\ is \begin{cases} an\ inlier,\ if\ \|\varepsilon_i(F)\| < T_{thr} \\ an\ outlier,\ if\ \|\varepsilon_i(F)\| \geq T_{thr} \end{cases} \tag{4}$$

In the current experiment, the value of T_{thr} was chosen to be 1.96σ with $\sigma=1$ as suggested in [9] for the case of the fundamental matrix, where σ stands for the standard deviation of the error. This inlier-outlier classification is one of the outcomes of the event and provides the information related to the validity of the correspondence. From the viewpoint of probability theory, the value of $P(v_i)$ can be determined from this inlier-outlier classification performed in the previously completed samples.

We accumulate the inlier-outlier classification obtained from successful random samples as an estimate of $P(v_i)$ for the ith correspondence. The successful random sample is defined as the one that reaches a score better than any one before. Let C_m^i be the inlier-outlier classification result of the ith correspondence at the mth independent successful random sample, i.e.

$$C_m^i = \begin{cases} 1, & inlier \\ 0, & outlier \end{cases} \tag{5}$$

Suppose that there are a total of m independent successful random samples after the kth independent sample. $C_1^i, C_2^i, ..., C_m^i$ are the outcomes of the ith correspondence from those m independent successful random samples, with an expected finite value of $\mu_i=E(C^i)$ and a finite variance of $\sigma_i^2 =V(C^i)$. Let

$$S_m^i = C_1^i + C_2^i + ... + C_m^i . \tag{6}$$

From the theorem of the Law of Large Numbers (LLN) [3], for any $\delta>0$, we have

$$P\left(\left|\frac{S_m^i}{m} - \mu_i\right| \geq \delta\right) \to 0 \tag{7}$$

as $m\to\infty$. Equivalently,

$$P\left(\left|\frac{S_m^i}{m} - \mu_i\right| < \delta\right) \to 1 \tag{8}$$

as $m\to\infty$.

As an estimate of the validity of a correspondence, we propose that

$$P(v_i) \leftarrow \frac{S_m^i}{m} \tag{9}$$

constitute the validity of the ith correspondence and be exploited for subsequent samples. Based on the law of large numbers, this estimate approaches the ground-truth validity value of the ith correspondence as the number of successful random samples increases. This estimation is also in accordance with a common assumption that the consensus corresponding to the best score so far is more likely an inlier set at the instance. As the number of successful samples increases, the consensus approaches to real one. It is important to note that the estimates of $P(v_i)$ are *not* exploited for the subsequent samples until at least a certain percent of the total number of samples has been completed because successful but unreliable samples are likely to occur at the beginning of the process.

4   Algorithm

The proposed MLESAC algorithm with validation of individual correspondences as described in Section 3 is named the *LLN-MLESAC* algorithm. The entire LLN-MLESAC algorithm is shown in Fig. 1 and explained as follows. At the beginning of the estimation, all values of $P(v_i)$ are assumed to be equal and chosen to be 0.5. For each sample, a minimal subset S_f of l correspondences is randomly selected by the Monte-Carlo method according to $P(v_i)$ [8]. This minimal subset S_f gives an estimate of the fundamental matrix F. After that, the Sampson error $\varepsilon_i(F)$ of each correspondence is calculated and the score of this hypothesis is determined according to (3). Whenever a new best hypothesis F is determined, the whole data set is classified into inlier and outlier subsets according to (4). The classification results of each correspondence are summed up according to (5) and (6). Meantime, the expected fraction r of inliers is determined by

$$r = \frac{1}{n}\sum_{i=1}^{n} C_m^i .\tag{10}$$

and the upper limit on the number of samples I_{max} is adjusted based on

$$I_{max} = \frac{\log(1-p)}{\log(1-(1-r)^l)},\tag{11}$$

LLN-MLESAC algorithm

1: Initiate each value of $P(v_i)$ to 0.5.
2: for j=1 to I_{max} samples do
3:   Select minimal set S_f of l correspondences using $P(v_i)$.
4:   Derive motion hypothesis F.
5:   for all correspondences i do
6:   Find residual Sampson error ε_i.
7:   end for
8:   Find the score using (3).
9:   if the current score is largest so far then
10:   Retain F,
11:   for all correspondences i do
12:   Classify the correspondence using (4) and keep the consensus,
13:   Accumulate the outcomes using (5) and (6).
14   Set I_r be equal to $0.15 \times I_{max}$
15:   if j is larger than the sampled number I_r then
16:   Update the $P(v_i)$ using (9);
17:   end if
18:   end for
19:   Find expected inlier fraction r using (10); adjust I_{max} using (11).
20:   end if
21: end for
22: Re-derive motion hypothesis F from the consensus using normalized
*   eight-point algorithm.*

Fig. 1. The LLN-MLESAC algorithm

where p is the confidence level that at least one of the random samples of l correspondences is free from outliers. As usual, p is chosen to be 0.99. According to (9), the new values of $P(v_i)$ replace the old ones only if the current number of samples is larger than the pre-defined number I_r. In the experiment, I_r is chosen to be $0.15 \times I_{max}$. The updated values of $P(v_i)$ are exploited for the subsequent samples. This process will continue until the upper limit on the number of samples is reached. The consensus with the highest score is kept for deriving the final estimate of the fundamental matrix using the normalized eight-point algorithm [4].

Note that the error defined in eq.(1) was the algebraic error. However, we replaced the algebraic error in eq.(3) with the Sampson error in our implementation because it is better [8].

5   Experiments

The proposed LLN-MLESAC algorithm was applied to the problem of estimating the fundamental matrix F that specifies the relation between corresponding features in epipolar geometry. We tested the algorithm's performance with pure synthetic data sets and a real image sequence *Corridor* (Oxford vision image database, retrieved from http://www.robots.ox.ac.uk/~vgg/data/data-mview.html). Given that the results from both the synthetic and real data show similar improvements over comparative methods, in order to save space, we only show the experiment results from the real test sequence *Corridor*.

The real image sequence *Corridor* consists of 10 pairs of images, each with a spatial resolution of 512×512 pixels, for testing the estimation of two-view geometry. The number of ground-truth correspondences for each image pair is shown in Table 1.

The proposed algorithm was compared to the conventional RANSAC and MLESAC estimators. To evaluate the accuracy of the estimated fundamental matrix $\hat{F}$ after optimal fitting to the inliers, the Sampson error was calculated as follows:

$$\frac{1}{n} \sum_{i=1}^{n} \frac{\left(\vec{x}_i^{2^T} \hat{F} \vec{x}_i^{1}\right)^2}{\left(\hat{F}\vec{x}_i^{1}\right)_1^2 + \left(\hat{F}\vec{x}_i^{1}\right)_2^2 + \left(\hat{F}^T \vec{x}_i^{2}\right)_1^2 + \left(\hat{F}^T \vec{x}_i^{2}\right)_2^2}, \tag{12}$$

Table 1. Number of ground-truth correspondences for 10 image pairs

Image Pair	1,2	2,3	3,4	4,5	5,6	6,7	7,8	8,9	9,10	10,11
Number of correspondences	409	409	350	350	388	388	292	292	260	260

where n is the total noise-free correspondent pairs (see Table 1 for the exact number for each test image pair), $(\vec{x}_i^1, \vec{x}_i^2)$ are the coordinates of the i-th noise-free point correspondent pair in the first and second image, respectively, and $\left(\hat{F}\vec{x}_i\right)_k^2$ is the square

Fig. 2. The first two images of the test sequence *Corridor,* superimposed with black lines indicating 89 correspondences that were detected by SUSAN [7] feature detectors

of the k-th entry of the vector $\hat{F}\vec{x}_i$. Note that, instead of the corrupted data ($\tilde{x}_i^1$, $\tilde{x}_i^2$), we used the noise-free correspondent pairs ($\vec{x}_i^1$, $\vec{x}_i^2$) to calculate the Sampson errors for the evaluation of the accuracy of the estimated fundamental matrix.

In the first test, the noise-free correspondences, provided by Oxford University, were corrupted by noise in order to test the performance of the proposed robust estimator. All correspondences were first corrupted by a zero-mean Gaussian noise with a predetermined noise variance. The Gaussian noise was simply added to the image coordinates of the corresponding points. Then, a randomly selected subset of correspondences, to be considered as outliers in the test, was further corrupted by a uniform noise. The parameters of this distribution were based on the minimum and maximum displacements amongst all noise-free correspondences. Keep in mind that the uniform noise was added to the coordinates of the corresponding points in only one image. In contrast, Gaussian noise was added to the coordinates in both images.

To perform a reliable comparison, in each test we generate noise addition as described previously and randomly select correspondences as the outliers. Such a test was repeated 200 times for every given ratio of outlier ranging from 5% to 65%. The noise variances are chosen and are set at 1 and 2 pixels. We calculated the Sampson error, the associated number of samples actually required for the algorithm to reach that Sampson error with a 99% confidence level, the consensus represented by the estimates of the inlier ratios, and the ratios of the correctly classified inliers over total number of the actual estimated inliers.

The experiment results confirm that the proposed LLN-MLESAC algorithm increases the estimate accuracy meanwhile it reduces the actually required number of samples in comparison to the RANSAC and MLESAC algorithms. Figure 3 shows the results averaged out from 10 image pairs. It can be seen that the LLN-MLESAC algorithm outperforms the conventional RANSAC and MLESAC estimators for various outlier ratios. When the noise variance is 1, the LLN-MLESAC on average reduced not only the Sampson error by 13.77% and 13.06% in comparison to RANSAC and

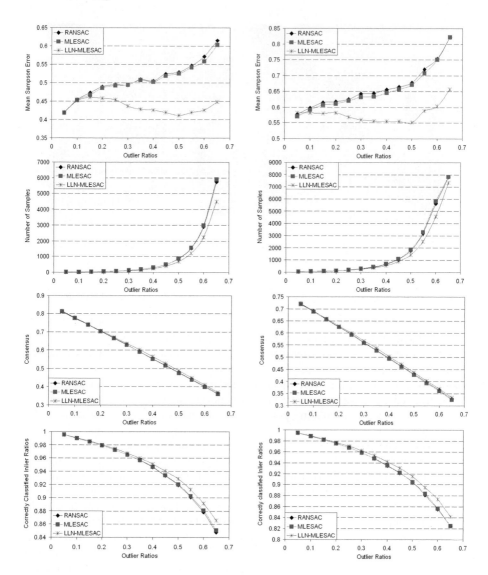

Fig. 3. Experimental results with the sequence *Corridor*. The left column depicts the results obtained with a noise variance of 1 pixel and the right column depicts the results obtained with a variance of 2 pixels. From the top row to the bottom, the graphs show the results with respect to the Sampson error, the actual number of samples, the consensus, and the ratio of correctly classified inliers.

MLESAC, but also the associated number of samples by 14.59% and 15.68%, respectively. This reduction on Sampson errors is the result of the increased consensus and also the result of the improved quality of the classified inliers. The LLN-MLESAC increases the consensus by 1.21% and 1.52% compared to the RANSAC and MLE-SAC, while the percentage of the ground truth inliers that were correctly identified is also slightly increased by 0.63% and 0.55%, respectively. The comparison results are the same, when the noise variance is 2. The LLN-MLESAC reduced Sampson error by 12.13% and 11.14%, cut the number of samples by 13.49% and 15.01%, increased the consensus by 1.01% and 1.52%, and improved the percentage of the ground truth inliers by 0.78% and 0.75%, respectively, when compared to the RANSAC and MLESAC.

Note that for the LLN-MLESAC algorithm the Sampson error does not monotonously go up when the outlier ratio increases. The reason for that can be explained as follows. A small number of samples generate an immature validity measure, which makes the LLN-MLESAC perform similar to other algorithms. After a certain number of samples are reached, the gain from the improved validity measure overcompensates the loss from the increased outlier ratio. This makes the Sampson error of the LLN-MLESAC algorithm decrease. When the validity measure becomes mature, the error increases again as the outlier ratio goes up.

In the second test, we use the SUSAN detector [7] to find the features of *Corridor* sequence and determine the correspondences using the cross-correlation as the criterion. Fig. 2 shows 89 correspondences obtained from the first two images. The estimated fundamental matrix $\hat{F}$ after optimal fitting to the inliers will be evaluated using the noise-free data, provided by Oxford University. Table 2 shows the results averaged over 200 tests, where error variances are calculated to measure the variation of Sampson error over the tests. It can be seen that at comparable consensus the proposed LLN-MLESAC algorithm is superior to the RANSAC and MLESAC algorithms in terms of Sampson error and trail number.

Table 2. Experimental results with the first image pair of the test sequence Corridor

	RANSAC	*MLESAC*	*LLN-MLESAC*
Sampson Error	0.29137	0.2891	0.27798
Error Variance	0.00274	0.00175	0.00169
Trial Number	28.505	29.06	28.2
Consensus	0.77028	0.76983	0.77028

6 Conclusion

This paper proposed a novel method for validating individual correspondences to improve the RANSAC-like robust estimator. The novelty resides in the idea that an analysis of the outcomes of previous random samples can benefit subsequent samples of RANSAC-like robust estimators. This idea is founded on the Law of Large Numbers (LLN) that states if an event is repeated many times the outcome of the random events will exhibit a certain statistical pattern that can be studied and predicted. In addition, this proposed algorithm differs from the G-MLESAC in two aspects. First, G-MLESAC determines $P(v_i)$ based on similarity measures, whereas the proposed algorithm determines $P(v_i)$ from the inlier-outlier classification according to the LLN. Second, for G-MLESAC $P(v_i)$ is known before the sampling procedure starts and is fixed during all the samples. In contrast, in the proposed algorithm $P(v_i)$ is estimated online during each sampling.

Based on the experimental results and analysis, the conclusion of this paper is that information accumulated from the individual samples that have been processed is useful in guiding subsequent sampling. It can improve the performance of robust estimators. The experimental results obtained with the Oxford image sequence, *Corridor*, showed that for a similar consensus the proposed LLN-MLESAC algorithm on average reduced the Sampson error by about 13% and 12% in comparison to the RANSAC and MLESAC algorithms, respectively. In conjunction, the associated number of samples decreased by about 14% and 15%, respectively.

References

1. Chum, O., Matas, J.: Optimal Randomized RANSAC. IEEE Trans. on Pattern Analysis and Machine Intelligence 30(8), 1472–1482 (2008)
2. Fischler, M.A., Bolles, R.C.: Random sample consensus: a paradigm for model fitting with application to image analysis and automated cartography. Comm. of the ACM 24, 381–395 (1981)
3. Grinstead, C.M., Snell, J.L.: Introduction to Probability. The American Mathematical Society, Providence (1997)
4. Longuet-Higgins, H.C.: A computer algorithm for reconstructing a scene from two projections. Nature 293, 133–135 (1981)
5. Nister, D.: Preemptive RANSAC for live structure and motion estimation. In: Proc. Int'l Conf. Computer Vision (ICCV 2003), vol. 1, pp. 199–206 (2003)
6. Rastgar, H., Zhang, L., Wang, D., Dubois, E.: Validation of correspondences in MLESAC robust estimation. In: 2008 Int'l Conf. on Pattern Recognition, TuAT10.11, Tampa, Florida (2008)
7. Smith, S.M., Brady, J.M.: SUSAN - a new approach to low level image processing. International Journal of Computer Vision 23(1), 45–78 (1997)
8. Tordoff, B.J., Murray, D.W.: Guided-MLESAC: fast image transform estimation by using matching priors. IEEE Trans. on Pattern Analysis and Machine Intelligence 27(10), 1523–1535 (2005)
9. Torr, P.H.S., Murray, D.W.: The development and comparison of robust methods for estimating the fundamental matrix. Int. J. Computer Vision 24(3), 271–300 (1997)
10. Torr, P.H.S., Zisserman, A.: Robust computation and parameterization of multiple view relations. In: ICCV, Bombay, India, pp. 727–732 (1998)
11. Torr, P.H.S., Zisserman, A.: MLESAC: a new robust estimator with application to estimating image geometry. Computer Vision and Image Understanding 78(1), 138–156 (2000)

Efficient Random Sampling for Nonrigid Feature Matching

Lixin Fan and Timo Pylvänäinen

Nokia Research Center
Tampere, Finland
fanlixin@ieee.org,
timo.pylvanainen@iki.fi

Abstract. This paper aims to match two sets of nonrigid feature points using random sampling methods. By exploiting the principle eigenvector of correspondence-model-linkage, an adaptive sampling method is devised to efficiently deal with non-rigid matching problems.

1 Introduction

The goal of this work is to simultaneously recover unknown *correspondence* and *motion* models between two sets of feature points. The problem is of interest to many computer vision applications, e.g. structure from motion [1,2,3], image registration [4,5,6,7] and object recognition [8,9].

The problem can be broadly categorized as two sub-problems, i.e. *rigid* and *non-rigid* matching. Under the assumption of rigid world, random sampling methods [10, 2, 11, 12] have been used to seek the optimal motion model that minimizes the reprojection error between two sets of feature points. For nonrigid matching with unknown correspondence, however, the sheer number of required random samples becomes intractable. Instead, graph matching and softassign methods [13, 14, 6, 8, 9, 15, 16] have been used to approximate the optimal solution, by relaxing binary correspondence constraint and adding pairwise compatibility into cost functions. However, the exact definition of such cost terms varies from methods to methods, and is often tailored to specific motion models in an ad-hoc manner. Methods [5, 17, 18] are proposed to learn the optimal forms of complex cost functions.

This paper presents a novel approach that casts non-rigid matching as *optimization via random sampling*. The method is novel in two aspects. First, we reveal an intriguing *linkage* between feature correspondences and random models, i.e. a candidate correspondence is linked with a number of random models and vice versa (Section 3). Second, the principle eigenvector of correspondence-model linkage is used to adapt random model sampling, so that high likelihood models are visited more frequently. In this way, the proposed adaptive sampling method is order of magnitudes more efficient than standard sampling methods, and thus effectively solves non-rigid matching problem within hundreds of iterations (Section 4).

G. Bebis et al. (Eds.): ISVC 2009, Part I, LNCS 5875, pp. 457–467, 2009.

2   Previous Work

Our work shares the *random optimization* nature with random sampling methods [10, 2, 11] and is most similar to the adaptive sampling method [12], in which data points were weighted based on previously evaluated models. As compared with these methods, our key contribution is the fast computation of eigenlinks and using it for adaptive sampling. In terms of application, traditional random sampling methods often assume known feature correspondence and rigid world. We extend the random optimization framework to matching problems with unknown correspondence and thin-plate spline warping (having > 40 control points). To our best knowledge, no similar results have been reported before.

Shape context [19] and feature descriptors such as SIFT and SURF [20, 21] are commonly used to match feature points for different vision tasks. Approximations of the optimal solution are obtained by using these feature descriptors as unitary correspondence confidence score and solving correspondence problem with *linear assignment* i.e. Hungarian algorithm [22]. However, matching results obtained this way can be further improved by imposing geometrical compatibilities between different correspondences.

Graph matching has long been used to solve correspondence problem. Various graph matching approaches approximate the optimal solution by imposing quadratic terms on cost functions [13, 8, 9, 15, 16], and learning methods [5, 17, 18] are used to determine the optimal forms of such terms. Our approach complements graph matching methods in different ways. First, we propose a novel correspondence compatibility measure that is rooted in the linkage between correspondence and random models. Second, we propose an incremental power method to solve the principle eigenvector for large matrices. Third, the proposed adaptive sampling is essentially an online "learning" process taking advantages of previously evaluated models.

Our method constantly updates the confidence score in correspondences and models, which is akin to soft assignment approaches [14, 6]. Compared with these softassign methods, the proposed method is more robust to high noisy data and outliers (see Section 5). An integer optimization method [7] is used to solve both correspondences and outlier rejection in a single step. EM algorithm [1] formulates the problem as iterative correspondence-motion estimation.

In the context of webpage structure analysis, linkage matrices are constructed from hyperlinks between webpages, *hub* and *authoritative* pages are then identified by using power method to recover the principle eigenvector of linkage matrices [23, 24].

3   Problem Formulation

We formally define feature correspondence and random model as follows. A candidate correspondence between two sets of image points is a quadruple $m_{ab} = (x_a^1, y_a^1, x_b^2, y_b^2)$, where $(x_a^1, y_a^1), a \in \{1, ...K_1\}$ and $(x_b^2, y_b^2), b \in \{1, ...K_2\}$ are coordinates of matched feature points in images 1 and 2 respectively. In total, there are $M (= K_1 * K_2)$ candidate correspondences in the *correspondence set*, denoted as $\mathcal{M} = \{m_{ab}, a = 1...K_1, b = 1...K_2\}$. Two arbitrary correspondences $m_{ab}, m_{a'b'}$ satisfy the *uniqueness constraint* iff $a \neq a'$ and $b \neq b'$. In the rest of the paper, we abuse the notation by saying m_i and m_j in $\mathcal{M} = \{m_i, i = 1...M\}$ satisfy the uniqueness constraint.

A general motion relation is an implicit function $\mathcal{G}(m, \theta) = 0$, where $m \in \mathcal{M}$ and θ is a parametric motion model that agrees with m. For a subset of s correspondence, there exists a unique motion model θ for which all s correspondences fit perfectly[1]. We refer to such a set as a minimal set and s is the model dimensionality. For instance, $s = 3$ for affine image transformation and $s = 8$ for eight-point fundamental matrix estimation. A permissible model is the one defined by a minimal set, and all permissible models form a finite discrete model set Θ. Note that the combinatorial search space Θ defined this way is data dependent, and for realistic data, permissible models are often unevenly distributed.

Optimization via Random Sampling: Given a model fitting threshold ε, the optimal model $\hat{\theta} \in \Theta$ is the one that has the maximum number of inlier correspondences that satisfy the uniqueness constraint:

$$[\hat{\theta}] = \arg\max_{\theta} |I_U(\theta)| \tag{1}$$

where $|.|$ is the cardinality of a set, and the set of inlier correspondences $I_U(\theta) = \{m \in \mathcal{M} || \mathcal{G}(m, \hat{\theta})| \leq \varepsilon$ and $\forall i, j, m_i, m_j$ satisfy the uniqueness constraint$\}$.

Random sampling methods search for the optimal model by repeatedly drawing models θ from Θ and evaluating $|I_U(\theta)|$ for each model. In order to reduce the search effort, one should *sample more frequently* those models with large number of inliers $|I_U(\theta)|$. We illustrate such an efficient adaptive sampling method in the rest of the paper.

3.1 Linkage between Feature Correspondences and Models

Given a correspondence $m \in \mathcal{M}$, let $A(m) = \{\theta \in \Theta || \mathcal{G}(m, \theta)| \leq \varepsilon\}$ denote the set of *association models* that fit m within the threshold ε. A likelihood function ℓ_1 is defined over Θ:

$$\ell_1(\theta|m) = \begin{cases} 1 & \text{if } \theta \in A(m), \\ 0 & \text{otherwise.} \end{cases} \tag{2}$$

The compatibility between two known correspondences m_1, m_2 is then measured by the inner product of two likelihood functions:

$$c(m_1, m_2) = \sum_{\theta \in \Theta} \ell_1(\theta|m_1)\ell_1(\theta|m_2) = |A(m_1) \cap A(m_2)|, \tag{3}$$

where $|.|$ is the cardinality of a set. $c(m_1, m_2)$ measures the likelihood that both m_1, m_2 agree with a single model, regardless of the model being selected. This definition of compatibility is independent of any specific motion relation $\mathcal{G}$. Specially, the auto-compatibility is $c(m_1, m_1) = |A(m_1)|$.

Now, let $I(\theta) = \{m \in \mathcal{M} || \mathcal{G}(m, \theta)| \leq \varepsilon\}$ denote the set of inlier correspondences that fit a given θ. In contrast to $I_U(\theta)$, there is no uniqueness constraint imposed on $I(\theta)$. A likelihood function ℓ_2 is defined over all candidate correspondences $\mathcal{M}$:

$$\ell_2(m|\theta) = \begin{cases} 1 & \text{if } m \in I(\theta), \\ 0 & \text{otherwise.} \end{cases} \tag{4}$$

---

[1] Without loss of generality, we assume s correspondences are non-degenerated.

And similarly,

$$c(\theta_1, \theta_2) = \sum_{m \in \mathcal{M}} \ell_2(m|\theta_1)\ell_2(m|\theta_2) = |I(\theta_1) \cap I(\theta_2)|. \tag{5}$$

Specially, $c(\theta_1, \theta_1) = |I(\theta_1)|$ is the Hough Transform [25, 26] of model θ_1.

3.2   Eigenlink

Direct computing of (3) is challenging, due to the sheer number of random models in Θ. Section 4 will show how to approximate (3) by adaptive model sampling. For now, let us assume the availability of a set of models, denoted by $\theta = \{\theta_j, j = 1...K\}$, that are evaluated over all correspondences. The "linkage" between correspondences m_i and random models θ_j can be represented by an M by K linkage matrix L:

$$L = \begin{bmatrix} \ell_2(m_1|\theta_1) & \dots & \ell_2(m_1|\theta_K) \\ \vdots & \ddots & \vdots \\ \ell_2(m_M|\theta_1) & \dots & \ell_2(m_M|\theta_K) \end{bmatrix}, \tag{6}$$

where columns of L represent the likelihood given by (4). Note that $\ell_2(m_i|\theta_j) = \ell_1(\theta_j|m_i)$, so equivalently, rows of L represent the likelihood given by (2).

Now assign a *soft assignment* x_i to each correspondence, and denote them as a column vector $\mathbf{x} = (x_1, x_2, ...x_M)^T$. Similarly, denote *confidence scores* of models as a column vector $\mathbf{t} = (t_1, t_2, ...t_K)^T$. Through L, $\mathbf{x}$ and $\mathbf{t}$ mutually enhance each other by applying operators:

$$\mathbf{x} = \frac{1}{Z_x}L\mathbf{t}, \qquad \mathbf{t} = \frac{1}{Z_t}L^T\mathbf{x}, \tag{7}$$

where Z_x and Z_t are normalization constants so that $\mathbf{1}_M \cdot \mathbf{x} = 1$ and $\mathbf{1}_K \cdot \mathbf{t} = 1$. Note that this normalization is different from the soft assignment used in [14, 6].

By simple substitution, and multiplication with $Z_x Z_t$:

$$Z_x Z_t \mathbf{x} = LL^T\mathbf{x}, \qquad Z_x Z_t \mathbf{t} = L^T L\mathbf{t}. \tag{8}$$

The equilibrium solution $\mathbf{x}^*$ and $\mathbf{t}^*$ to (7) are eigenvectors of symmetric matrices LL^T and $L^T L$ respectively, having the same eigenvalue $Z_x Z_t$. Therefore, $\mathbf{x}^*$ and $\mathbf{t}^*$ obtained this way are called *eigenlinks* of correspondence and random models.

One can interpret LL^T from a graph matching point of view. Following (6) we have

$$LL^T = \begin{bmatrix} \tilde{c}(m_1, m_1) & \dots & \tilde{c}(m_1, m_M) \\ \vdots & \ddots & \vdots \\ \tilde{c}(m_M, m_1) & \dots & \tilde{c}(m_M, m_M) \end{bmatrix}, \tag{9}$$

where off-diagonal elements $\tilde{c}(m_a, m_b) = \sum_{j=1}^{K} \ell_1(\theta_j|m_a)\ell_1(\theta_j|m_b)$ are pairwise correspondence compatibilities defined in (3), and diagonal elements $\tilde{c}(m_a, m_a) = |A(m_a)|$ are the fitness of individual correspondences.

$\mathbf{x}^*$ maximizes the total inter-cluster score $\mathbf{x}^T L L^T \mathbf{x}$ [9]:

$$\mathbf{x}^* = \arg\max_{\mathbf{x}} \mathbf{x}^T L L^T \mathbf{x} = \arg\max_{\mathbf{x}} \left(\sum_{i=1}^{M} x_i^2 \tilde{c}(m_i, m_i) + \sum_{i=1}^{M} \sum_{j=1, j \neq i}^{M} x_i x_j \tilde{c}(m_i, m_j) \right).$$

(10)

Notice that we are not attempting to maximize (10) as an objective function. Rather, eigenlink $\mathbf{x}^*$ obtained this way is used to adapt random sampling (see Section 4).

3.3 Incremental Power Method

Solving all eigenvectors for large matrices is memory and time extensive. In this work, we adopt a modification of iterative power method [27] to compute $\mathbf{x}^*$ in an incremental manner.

Denote each column of (6) as column vectors $\delta_j, j = 1...k$, and the $(k+1)$th step matrix $L_{k+1} = [L_k \delta_{k+1}]$. Following (7), the incremental updating equation is then given by:

$$\mathbf{t}_{k+1} = \frac{1}{Z_t^{k+1}} \left(\begin{matrix} Z_t^k \mathbf{t}_k \\ \delta_{k+1}^T \mathbf{x}_k \end{matrix} \right), \qquad \mathbf{x}_{k+1} = \frac{1}{Z_x^{k+1}} \left(\frac{(Z_t^k Z_x^k)}{Z_t^{k+1}} \mathbf{x}_k + \frac{(\delta_{k+1}^T \mathbf{x}_k)}{Z_t^{k+1}} \delta_{k+1} \right), \quad (11)$$

where Z_t^k and Z_x^k are the kth step normalization constants. Using this incremental power method brings about two advantages. First, there is no need to store and compute L, LL^T and $L^T L$ explicitly. Only δ_{k+1} is needed, thus the memory cost is small. Second, operation (11) iterates just once, whenever the linkage matrix L is appended with one column vector δ_{k+1}. This way, eigenlinks are continuously updated, yet without repeatedly solving the eigenvalue problem for many iterations.

4 Adaptive Random Sampling

In order to improve sampling efficiency, we take an adaptive sampling approach that exploits eigenlink $\mathbf{x}^*$ obtained from previously sampled models. The proposed **AdaLink** algorithm demonstrates superior robustness and efficiency, in dealing with a variety of motion models (see next page for algorithm outline).

Figure 1 illustrates typical non-rigid matching results and how eigenlinks are updated during the course of optimization. AdaLink algorithm as such shows a tendency to explore the entire solution space in early iterations and gradually lock on a stabilized matching in later iterations. The stabilized solution corresponds to highly peaked eigenlink $\mathbf{x}^*$, which in turn leads to more random models being sampled around the stabilized solution.

Model dimensionality s : For nonrigid matching, s is the number of thin plate spline control points, which is set to be half of the number of points in template sets. For different test datasets in our experiments, s ranges from 25 to 52. For fundamental matrix estimation, $s = 8$.

Algorithm 1. AdaLink

- Input: All candidate correspondences $\mathcal{M}$, threshold ε, model dimensionality s and $\gamma \in (0, 1)$.
- Output: the best model $\hat{\theta}$.

1. Initialize $\mathbf{x} = \mathbf{1}/M$ and $\ell_o(\hat{\theta}) = 0$;
2. (a) With probability γ, randomly draw s correspondences using $\mathbf{x}$ (see text).
   (b) With probability $1 - \gamma$, draw s correspondence using priori knowledge (see text).
3. Evaluate new model θ_j according to (1) and denote model likelihood as $\ell_o(\theta_j)$.
4. With probability $\frac{\ell_o(\theta_j)}{\ell_o(\hat{\theta})}$, accept model θ_j and keep fitting outcomes as δ_j.
5. If θ_j accepted, update $\mathbf{t}$ and $\mathbf{x}$ using (11).
6. If $\ell_o(\theta_j) > \ell_o(\hat{\theta})$, set $\ell_o(\hat{\theta}) = \ell_o(\theta_j)$,
7. Repeat steps 2-6 until stopping criterion has been reached (see text).

Threshold ε is related to the scale of additive inlier noise, which is known a priori in our experiments. Alternatively, it can be automatically estimated. We refer readers to [12] for detailed treatment of robust scale estimation.

Step 2 (a): In order to select high confidence correspondences and impose one-one uniqueness constraint on selected correspondences, we first rearrange column vector $\mathbf{x}$ as a $K_1 \times K_2$ matrix, where K_1 and K_2 are numbers of feature points in two images. Maximal elements along both matrix rows and columns are then selected as *pivotal correspondences*, from which s minimal set correspondences are randomly selected to define the motion model[2].

Step 2 (b): If features matching methods such as shape-context [19], SIFT [20] and SURF [21] are used, pairwise feature similarities can be stored in $K_1 \times K_2$ matrix. Maximal elements along both matrix rows and columns are selected as *pivotal correspondences*, from which s minimal set correspondences are randomly selected. If no prior knowledge is available, we select s minimal set correspondences uniformly distributed from all candidate correspondences, subject to the uniqueness constraint.

Mixing probability γ is related to the adaptivity of sampling. In this work, we set $\gamma = 0.5$ to draw balanced samples.

Stopping criterion adopted by AdaLink is as follows: whenever a model with higher likelihood is found and let i_{best} denote the number of current iterations, then the maximal number of iterations is updated as $i_{\text{max}} = \max(\alpha \times i_{\text{best}}, i_{\text{burn\_in}})$. In our experiments, we set $\alpha = 2.0$ and $i_{\text{burn\_in}} = 100$. With this setting, many matching experiments terminate in hundreds of iterations (see Section 5).

Computational complexity of the algorithm depends on (a) the number of required iterations and (b) the computation cost for each iteration. For different matching problems, AdaLink terminates in hundreds to thousands of iterations. Table 1 summarizes matlab profile report of running AdaLink for a non-rigid point matching problem. Time

---

[2] If the number of pivotal correspondence is smaller than s, additional correspondences are randomly drew from all candidate correspondences.

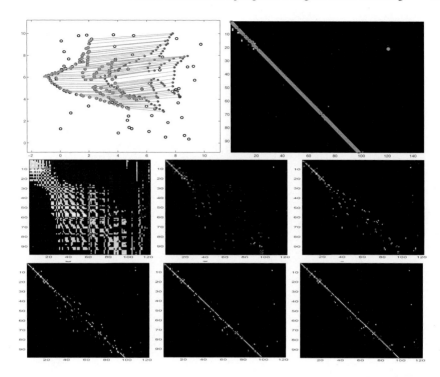

Fig. 1. Up Left: Non-rigid matching of 98 template points (red dots) with 122 noisy points (blue circles), in which 24 points (≈ 0.2) are outliers. $s = 49$ control points are used for thin plate spline transformation. **Up right**: Eigenlink **x** rearranged as 98×122 matrix. White pixels represent high confidence correspondences. Red circles are ground truth correspondences, green dots are correspondences selected by AdaLink. **Middle left:** pairwise shape-context similarities used in Step 2(b). **The rest**: Eigenlinks correspond to iterations $100, 505, 598, 625$ and 807 respectively.

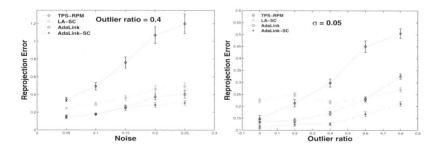

Fig. 2. Nonrigid point matching using different methods (see Section 5.1). **Left**: x-axis represents noise levels increasing from $(0.05 - 0.25)$. y-axis represents the reprojection error of all matched points. Error bars indicate standard deviation. Outlier ratio is fixed as 0.4. **Right**: outlier ratio increases from 0 to 0.8, and noise level $\sigma = 0.05$.

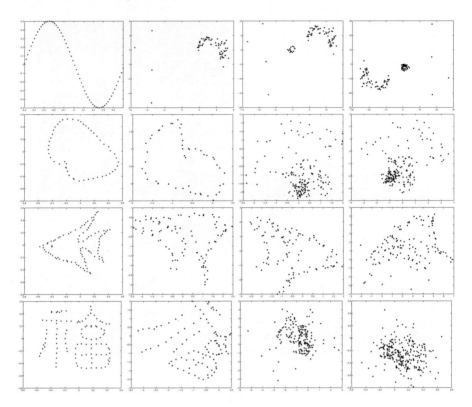

Fig. 3. Random datasets for nonrigid point matching. Left column: four template points - *wave*, *heart*, *fish* and *blessing-Chinese*. The rest are example noisy point sets to be matched.

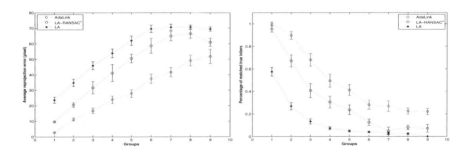

Fig. 4. Image matching using epipolar geometry. Left: average reprojection error w.r.t. different baseline groups. Right: average percentage of matched true inliers.

cost of model evaluation step (3) depends on model dimensionality s and the number of candidate correspondences M. This step spends more than twice as much time as of adaptive sampling step (2). Incremental eigenlinks updating step (5) is fast, which amounts to merely 3% of total computational cost. Overall time cost for 1000 iterations is 6.53s.

Table 1. Computational cost per iteration, for non rigid matching with 30 control points. Matlab profile report on a Pentium 4 Duo core 1.6GHz laptop.

Steps:	2	3	6	Others	**Total**
Time (ms)	1.65	3.89	0.2	0.78	**6.53**
(percent.)	25%	60%	3%	12%	

5   Experiments

5.1   Non-rigid Point Matching

In this experiment, we compare the proposed AdaLink algorithm with two existing nonrigid matching methods. The first method, denoted as *LA-SA*, uses shape-context [19] to measure pairwise point distances and uses *linear assignment* i.e. Hungarian algorithm [22] to solve correspondence. No motion model is defined for this method. The second method, *TPS-RPM* [6], is essentially a graph matching method that utilizes the softassign [14] and deterministic annealing for the correspondence and the thin plate spline for the non-rigid mapping. Two AdaLink variants, respectively, use random sampling (*AdaLink*) and pairwise shape-context similarities (*AdaLink-SC*) in step 2 (b).

Four point sets i.e. *wave*, *heart*, *fish* and *blessing-Chinese*, adopted from [6], are subject to non-rigid warping (see Figure 3). On top of that, we add random transformations, outliers and Gaussian noise to already warped data points. We test different methods with 100 random transformation of each data set, which gives us a total of 400 data sets. Due to their random nature, *AdaLink* and *AdaLink-SC* run 10 times for each data set.

For all three methods, we measure average *reprojection errors* of matched points. Figure 2 summarizes matching results using different methods. When noise ($\sigma = 0.05$) and outlier ratio (≤ 0.2) are low, *TPS-RPM* outperforms *LA-SC*, which, on the other hand, is more robust to high noise and outlier ratio. *AdaLink* is compared in favor of *LA-SC*, with the only exception when $\sigma = 0.05$ and outlier ratio = 0.6. This performance degradation is mitigated by integrating shape-context into adaptive sampling step 2(b). Overall, the average errors of *AdaLink-SC* is about 0.22, which amounts to 61% and 26% of average errors of *LA-SC* and *TPS-RPM*, respectively.

5.2   Two View Image Matching

In this experiment, we test *AdaLink* method with image matching problem assuming rigid motion models. We use *Dinosaur* dataset consisting of 36 views, for which ground truth correspondence of 4838 points are available[3]. A subset of 557 reliable points that are viewed by at least 5 views are selected for our experiment. Total number of 228 image pairs, which have at least 10% overlapped points, are used for performance comparison. According to the baseline between image pairs, we categorize them into 9 groups with increasingly wider baseline. Group 1 has 35 subsequent frame pairs i.e.

---

[3] http://www.robots.ox.ac.uk/ vgg/data/data-mview.html from Visual Geometry Group, Oxford University.

1-2, ... 35-36, which are easy to match. Group 9 has 4 difficult pairs 16-25, 17-26, 18-27 and 19-28, in which object rotates more than 90 degrees.

We first extract SURF feature descriptor [21] at each feature points to compute pairwise feature similarity matrix, which is used by all three methods below. The first method, denoted as *LA*, uses *linear assignment* i.e. Hungarian algorithm [22] to solve correspondence based on SURF feature similarity matrix. No motion model is defined for this method and correspondence are selected based on photometric similarity only. The second method, denoted as *LA-RANSAC+*, refines the outcome of *LA* method by imposing geometrical constraints on correspondences. In the experiment, *LA-RANSAC+* is set to run for a fixed 5000 iteration. Our method, *AdaLink*, uses SURF feature similarity in adaptive sampling step 2(b). Both *LA-RANSAC+* and *AdaLink* run 10 times for each image pair.

Figure 4 summarizes performance comparison of different methods, using epipolar geometry to relate two views. A standard eight-point algorithm [3] is used to estimate fundamental matrix. For group 1, Adalink has reprojection error < 3 pixels and > 0.99 inliers matched, which is comparable to the state-of-the-art. For groups 2-9, it is observed that *AdaLink* consistently outperforms other two methods. Average reprojection errors of *AdaLink* method are 7 (70%) to 20 (30%) pixels lower than that of *LA-RANSAC+* for different groups. Average percentage of matched true inliers are in general 10% higher. Nevertheless, groups 2-9 are included to compare different matching methods only, one should always use group 1 for 3D reconstruction.

6   Conclusion

Simultaneous recovering of correspondence and unknown motion is a long standing problem. In this work, we recast it within an random optimization framework, in which a linkage between feature correspondence and random model is revealed. Consequently, a correspondence compatibility measure that is independent of underlying motion models is proposed. Using the principle eigenvector of correspondence-model-linkage leads to an efficient sampling method, so that one can extend random sampling to deal with unknown correspondence nonrigid matching problem. To our best knowledge, AdaLink is the first random sampling method that solves such a problem. Robustness and efficacy of AdaLink are demonstrated with both nonrigid point matching and two view image matching.

References

1. Dellaert, F., Seitz, S.M., Thorpe, C.E., Thrun, S.: Structure from motion without correspondence. In: CVPR, pp. 2557–2564 (2000)
2. Torr, P.H., Davidson, C.: IMPSAC: Synthesis of importance sampling and random sample consensus. IEEE Transactions on Pattern Analysis and Machine Intelligence 25, 354–364 (2003)
3. Hartley, R., Zisserman, A.: Multiple view geometry in computer vision, 2nd edn. Cambridge University Press, Cambridge (2003)
4. Bujnak, M., Sára, R.: A robust graph-based method for the general correspondence problem demonstrated on image stitching. In: ICCV, pp. 1–8 (2007)

5. Caetano, T., Cheng, L., Le, Q., Smola, A.: Learning graph matching. In: IEEE International Conference on Computer Vision, ICCV (2007)

6. Chui, H.: A new algorithm for non-rigid point matching. In: CVPR, pp. 44–51 (2000)

7. Maciel, J., Costeira, J.P.: A global solution to sparse correspondence problems. IEEE Transactions on pattern Analysis and Machine Intelligence 25, 187–199 (2003)

8. Berg, A.C., Berg, T.L., Malik, J.: Shape matching and object recognition using low distortion correspondence. In: CVPR, pp. 26–33 (2005)

9. Leordeanu, M., Hebert, M.: A spectral technique for correspondence problems using pairwise constraints. In: International Conference of Computer Vision (ICCV), vol. 2, pp. 1482–1489 (2005)

10. Fischler, M.A., Bolles, R.C.: Random sample consensus: A paradigm for model fitting with applications to image analysis and automated cartography. Communications of the ACM 24, 381–395 (1981)

11. Tordoff, B.J., Murray, D.W.: Guided-MLESAC: Faster image transform estimation by using matching priors. IEEE Transactions on Pattern Analysis and Machine Intelligence 27, 1523–1535 (2005)

12. Fan, L., Pylvänäinen, T.: Robust scale estimation from ensemble inlier sets for random sample consensus methods. In: Forsyth, D., Torr, P., Zisserman, A. (eds.) ECCV 2008, Part III. LNCS, vol. 5304, pp. 182–195. Springer, Heidelberg (2008)

13. Gold, S., Rangarajan, A.: A graduated assignment algorithm for graph matching. IEEE Trans. Patt. Anal. Mach. Intell. 18, 377–388 (1996)

14. Gold, S., Rangarajan, A., ping Lu, C., Mjolsness, E.: New algorithms for 2d and 3d point matching: Pose estimation and correspondence. Pattern Recognition 31, 957–964 (1998)

15. Caetano, T., Caelli, T., Schuurmans, D., Barone, D.: Graphical models and point pattern matching. IEEE Transactions on pattern Analysis and Machine Intelligence 10, 1646–1663 (2006)

16. Cour, T., Srinivasan, P., Shi, J.: Balanced graph matching. In: Schölkopf, B., Platt, J., Hoffman, T. (eds.) Advances in Neural Information Processing Systems, vol. 19, pp. 313–320. MIT Press, Cambridge (2007)

17. McAuley, J.J., Caetano, T.S., Smola, A.J.: Robust near-isometric matching via structured learning of graphical models. In: NIPS, pp. 1057–1064 (2008)

18. Torresani, L., Kolmogorov, V., Rother, C.: Feature correspondence via graph matching: Models and global optimization. In: Forsyth, D., Torr, P., Zisserman, A. (eds.) ECCV 2008, Part II. LNCS, vol. 5303, pp. 596–609. Springer, Heidelberg (2008)

19. Belongie, S., Malik, J., Puzicha, J.: Shape matching and object recognition using shape contexts. IEEE Transactions on Pattern Analysis and Machine Intelligence 24, 509–522 (2002)

20. Lowe, D.: Object recognition from local scale-invariant features. In: Proceedings of the International Conference on Computer Vision, pp. 1150–1157 (1999)

21. Bay, H., Tuytelaars, T., Van Gool, L.: SURF: Speeded Up Robust Features. In: Proceedings of the ninth European Conference on Computer Vision (2006)

22. Kuhn, H.W.: Variants of the hungarian method for assignment problems. Naval Research Logistic Quarterly 3, 253–258 (1956)

23. Kleinberg, J.M.: Authoritative sources in a hyperlinked environment. Journal of the ACM 46, 668–677 (1999)

24. Ng, A.Y., Zheng, A.X., Jordan, M.I.: Link analysis, eigenvectors and stability. In: IJCAI, pp. 903–910 (2001)

25. Duda, R.O., Hart, P.E.: Use of the hough transformation to detect lines and curves in pictures. Comm. ACM 15, 11–15 (1972)

26. Kultanen, P., Xu, L., Oja, E.: Randomized hough transform. In: Proceedings of the International Conference on Pattern Recognition, pp. 631–635 (1990)

27. Golub, G.H., van Loan, C.F.: Matrix Computations, 3rd edn. Johns Hopkins University Press, Baltimore (1996)

RiverLand: An Efficient Procedural Modeling System for Creating Realistic-Looking Terrains

Soon Tee Teoh

Department of Computer Science, San Jose State University

Abstract. Generating realistic-looking but interesting terrains quickly is a great challenge. We present RiverLand, an efficient system for terrain synthesis. RiverLand creates a realistic-looking terrain by first generating river networks over the land. Then, the terrain is created to be consistent with the river networks. In this way, the terrains created have a proper drainage basin, an important feature lacking in many existing procedural terrain methods. The terrains generated by RiverLand are also widely varied, with rolling hills, river valleys, alpine mountains, and rocky cliffs, all seamlessly connected in the same terrain. Since RiverLand does not use complex fluid simulations, it is fast, and yet is able to produce many of the erosion features generated by the simulation methods.

1 Introduction

Computer-generated terrains are in high demand due to the popularity of 3D games, online 3D fantasy worlds, and CGI movies. These virtual 3D terrains are desirable because they can be aesthetically pleasing and visually dramatic. However, efficiently generating realistic-looking terrains remains a huge challenge. The attractiveness of virtual terrains is that they can be creatively designed, so that they are not identical to real-world terrains. Yet they have to bear similarities to real-world terrains because users desire realism in virtual landscapes.

Creating realistic terrains is challenging because terrains on earth are sculpted by various complex forces. The two most influential physical processes are *tectonics* and *water erosion*. Plate tectonic movements create large features such as volcanoes, fold mountains and block mountains. River or glacial erosion then modify the land to create canyons, plains and river valleys. Other forces such as wind, rain and freeze-thaw action also alter the physical landscape. The simplest and fastest procedural terrain modeling algorithms make use of fractals. Although these algorithms can produce terrains with mountains and valleys with interesting shapes, they are not realistic, because there is no river network in the resulting terrain. In fact, creating a river network on a fractal terrain is difficult because rivers must flow from higher ground to lower ground, while a fractal terrain often does not have a natural path for a river to flow. Therefore, complex river simulation methods have been proposed. However, these methods tend to be significantly slower.

We have developed a new system, RiverLand, that generates a procedural terrain by first generating the river network. The method used by RiverLand

G. Bebis et al. (Eds.): ISVC 2009, Part I, LNCS 5875, pp. 468–479, 2009.

generates rivers with tributaries, and then creates mountains and other physical features on the land. In this way, a realistic-looking terrain is created. RiverLand is versatile; it allows users the option of controlling the terrain by specifying ridges and mountains, and it can produce many interesting features including alpine mountains, rolling hills, cliffs, plains and valleys. Furthermore, these different features compose a single terrain in a seamless manner.

In this paper, we examine the existing procedural terrain methods, and describe the advantage of RiverLand compared to these methods. We also describe how RiverLand accomplishes the challenges of terrain modeling mentioned in these papers. RiverLand is also a fast system, generating a 256×256 height-map in 0.2 seconds.

2   Related Works

Early works [1,2,3,4] in procedural terrain synthesis were based on fractals and fractional Brownian motion. These foundational algorithms can efficiently create landscapes with mountains and valleys. However, as pointed out by later works, a major weakness of these fractal terrains is that they have no global erosion features, and are therefore limited in their realism.

Noticing this defect in fractal approaches, Kelley et al. [5] presented a method for simulating stream networks. Musgrave et. al [6] then added local control to fractal terrains, and then simulated various erosion processes on these terrains. Other simulation-based methods [7] [8] [9] are also able to produce realistic-looking terrain. Improving on speed, Neidhold et al. [10] are able to run fluid dynamics simulations on an initial terrain, and generate eroded terrain in near real-time (a 256x256 terrain in 0.25 seconds on a 2.4 GHz machine). Recent works [11] [12] implemented fluid dynamics computations on the GPU, also improving on speed. They are able to simulate a $1K \times 1K$ terrain with 50 iterations in 11.2 seconds (compared to 67 seconds using the CPU). These simulation-based methods are limited because they require an initial starting terrain, and they also tend to generate uniform-looking terrains because the same erosion processes and parameters are used throughout the entire terrain.

Prusinkiewicz and Hammel [13] proposed a fractal algorithm to generate a river, and corresponding mountains. A weakness of this method is that it produces rivers with assymetric valleys and no tributaries.

Belhadj and Audiber [14] proposed first generating ridge lines using *Ridge Particles* and then growing river networks by dropping *River Particles* on top of the ridges. A midpoint displacement method is then used to fill in the rest of the terrain. This method is relatively fast, producing a 512×512 terrain in 0.45 seconds on a 3.0 GHz machine. In comparison, our system starts with only the river network, and is able to generate more varied terrains. We also allow users to optionally influence the terrain by specifying desired ridges and mountains.

More recently, Schneider et al. [15] presented a system for real-time editing, synthesis, and rendering of infinite landscapes using the GPU. However, like earlier methods, this system produces no river network and is limited in realism.

Recently, several works [16] [17] [18] have taken a new approach. Observing that it is difficult for most existing systems to generate terrain with a desired style, these methods use actual real-world Digital Elevation Maps (DEM) as examples, to synthesize new terrain. Brosz et al. [17] extracts high-resolution details from existing terrain to add details to a low-resolution terrain. Zhou et al. [18] allows the user to sketch a path on an empty canvas, and provide a sample terrain style DEM dataset, for example the DEM of the Grand Canyon. They then produce a terrain with a canyon traced around the sketch provided by the user. The images produced look realistic, compared to previous methods. However, these example-based methods have several important limitations. First, they have difficulty in producing realistic heterogeneous terrain, merging different terrain styles seamlessly. Second, they are unable to generate novel terrain styles that are not identical to any existing style in their database. Third, the resulting terrain may not be physically realistic, since the river networks they produce are arbitrary, and not based on physical simulation.

There are also several excellent commercial and freely available procedural terrain systems. Mojoworld (www.pandromeda.com), for example, allows the user to create entire planets, including terrain, vegetation and atmosphere. Terragen (www.planetside.co.uk) also provides photorealistic scenery rendering, with tools available to create terrains. It allows the user to "paint" the shape of a landscape to position mountains and valleys; and automatically shapes the rest of the terrain with random fractal methods and physical-based methods such as simulation of glaciation. Bryce (www.Daz3D.com) allows the user to create virtual worlds, including both natural and urban landscapes.

Gain et al. [19] presented a novel system that allows the user to sketch the silhouette of a mountain on a 3D view of the terrain, and optionally modify shadow and boundary lines. The program then automatically generates a landscape that fits the user sketch.

3   RiverLand Overview

RiverLand begins with an empty 2D canvas which represents the top view of the land. The user is allowed to paint regions on the canvas. Within each region, the user is allowed to specify ridge lines, and the heights and widths of various points on the ridge lines.

The procedural terrain synthesis method begins by creating random river seed points on the region boundary. From these seed points, meandering rivers are grown into the region. Tributaries are also spawned from these rivers. Rivers are not allowed to cross user-specified ridges. Once inside the influence of a user-defined ridge, a river grows up in the direction towards the ridge line. Once the river network is established, mountains are grown on the rest of the terrain. Fractal methods are used to introduce interesting variations into the terrain. Figure 1 shows an overview of this process.

Compared to previous methods, RiverLand has the following advantages: (1) It produces more heterogenous landscapes, with steep ravines, alpine mountains,

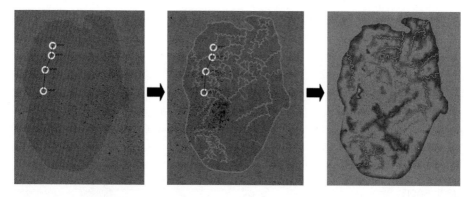

Fig. 1. RiverLand Terrain Synthesis Process. Left: User optionally paints region and ridges. Middle: Algorithm generates fractal ridges and rivers. Right: Terrain topography is fitted onto map.

rolling hills, flat river valleys and cliffs within the same terrain; (2) it produces a globally physically consistent river network, a feature only present in a few of the existing algorithms; (3) it is near real-time, faster than most of the existing algorithms; and (4) it allows users to influence the terrain, a feature found only in some existing methods.

4   Creating the River Network

For each region painted by the user, RiverLand assumes that its boundary is flat land. Terrain synthesis begins by setting n random boundary cells to be river mouths, each mouth being at least d cells apart. By adjusting the parameters n and d, the user can influence the eventual appearance of the terrain.

From each river mouth, a river network is grown. The initial direction vector of the river is the vector perpendicular to the boundary. Going forwards on this vector for distance *SegmentLength* $+ r$ (where r is a random number), set a *SegmentPoint*. From this new SegmentPoint, repeatedly rotate the vector by a slight random angle, go forwards by a distance, and set a new SegmentPoint, until one of the following conditions is reached: (1) It goes with a distance d from the river or region boundary, or (2) The target t number of SegmentPoints has been set, where d and t are set by the user. By adjusting the limits of angle rotation of the river vector, and parameters such as SegmentLength, each river can have different characteristics. For example, large straight rivers can be made by increasing SegmentLength and decreasing the rotation angle, and also increasing the target number of SegmentPoints. By generating random rivers with different parameters, rivers of different characteristics can be created in the same region.

Next, a meandering river is fit through the SegmentPoints. In nature, most rivers do not travel in a straight path; instead they alternately curve left and right, forming the familiar meanders. This is done by fitting a curve between

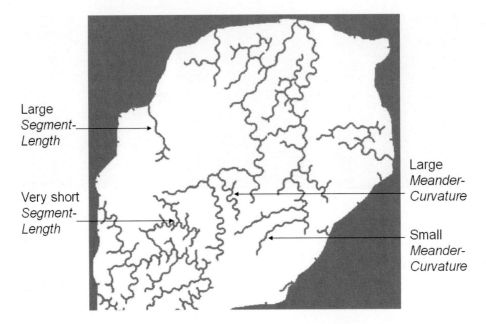

Large
Segment-
Length

Very short
Segment-
Length

Large
Meander-
Curvature

Small
Meander-
Curvature

Fig. 2. Different values of parameters SegmentLength and MeanderCurvature create rivers with different characteristics, producing a more varied terrain

each SegmentPoint. A random curvature is set for each segment. Each river has a *MeanderCurvature* parameter that controls the mean meander curvature of the river. Figure 2 shows how different parameters for *SegmentLength* and *MeanderCurvature* influence the characteristics of a river. RiverLand generates rivers with different parameters to create a more varied terrain.

Once the meandering river has been fit, new seed points are placed randomly on the concave banks of the river. Tributaries can be grown outwards from these new seed points. These new seed points are inserted into the list of all seed points. At each iteration, a seed point is randomly selected, and a river or tributary is grown from it, until a target number of rivers or tributaries have been created, or until there are no seed points left. This completes the creation of the river networks in the region.

After the river network is created, a height is assigned to each river cell. At all river mouths, the height is assumed to be zero. For each river, a parameter *AverageSlope* is set. Going from one river cell to the next, the height is increased by AverageSlope, plus some random offset. When the algorithm reaches a river cell that is a distance d (set by the user) from the source, the river is allowed to be steeper. This is because real-world rivers tend to be steeper near the source. Due to the deposit of eroded sediments, rivers tend to form large, flat river plains downstream. RiverLand can also simulate water-flow so that rivers get wider as they flow downstream, and as more tributaries join them.

If the user has defined some ridges, the river network algorithm has to take them into consideration. When a river SegmentPoint is set within the influence of a user-defined ridge, the river vector is constrained to go straight up the user-defined ridge, within a narrow angle. Also, the river is not allowed to grow past the user-defined ridge line. This ensures that the generated river network is consistent with the user-defined ridges.

5   Generating Terrain Height

Once the height of each river cell is set, the next step is to generate the heights of all the cells in the terrain. First, the medial axis (skeleton) of the river and boundary cells is found. This medial axis consists of all the cells that are of maximal distance from a river or boundary cell. These medial axis cells are called "Maximal Cells". To find the Maximal Cells, first add all the river and boundary cells into a list, and set their distance d to 0. For each cell C_1 in the list, consider all its adjacent cells to the list. Let C_2 be an adjacent cell to C_1. Then, the candidate distance of C_2 from C_1 is $d_{C2} = d_{C1} + slope \times dist$, where $dist$ is the horizontal distance between C_1 and C_2, and $slope$ is a parameter set by the user. For each river or region boundary cell, a $slope$ parameter is set to control the slope of the terrain from this cell. To create gentle slopes, set the value to less than 1, say 0.75 or 0.5. Allowing the $slope$ variable allows the system to generate assymetric ridges, where one side is steeper than the other. If this candidate distance d_{C2} is less than the existing distance $d_{C2existing}$ from C_2 to the river, set $d_{C2existing}$ to d_{C2}, and set the $PathToRiver$ pointer of C_2 to point to C_1. Then, remove C_1 from the list.

When the list becomes empty, all the cells that have no incoming $PathToRiver$ pointers are the Maximal Cells. Each Maximal Cell has an associated River Cell or Boundary Cell from which it came. Next, a M̂inimumHeight and $IdealHeight$ of each Maximal Cell is assigned. The MinimumHeight of a Maximal Cell is the height of its associated River Cell, since the path from the Maximal Cell to the River Cell is considered the drainage basin of the River Cell, and so cannot be lower than the River Cell. The $IdealHeight$ of the Maximal Cell is set to $MinimumHeight + slope \times distancetoriver$. Next, the $IdealHeights$ of all Maximal Cells are smoothed by taking the average $IdealHeights$ of all neighboring Maximal Cells. The actual height of each Maximal Cell is set to its $IdealHeight$ unless it is less than the $MinimumHeight$, which happens only rarely. Finally, the heights of all the cells along the path from the Maximal Cell to the River Cell are linearly interpolated. Other adjustments are later made to vary the heights, as well as to add the influence of user-defined ridges.

The algorithm described can produce many cliffs along maximal cells. This is because maximal cells mark the boundary between the drainage basins of different rivers. Since the rivers have different heights, the maximal cells associated with each river would also have different heights. The solution is to use a cliff-removal step to produce a more natural-looking terrain. First, a cliff threshold c is defined. Any cell that has a height difference greater than c with any adjacent cell is considered a cliff. The cliff-removal algorithm adds all the lower cliff cells into a list L, and raises each lower cliff cell to within c of its tallest adjacent cell, to remove the cliff. Because raising this cell may create new cliffs, the iterative step examines every neighbor N of every cell C in list L, and if that neighbor N's height $h_N < h_C - c$ (where h_C is C's height), N is added to list L, and h_N is raised to $h_C - c$. This is repeated until list L is empty. After this step, all the cliffs created on the river drainage basin boundaries are removed.

So far, all the heights increase linearly from the river/coast to the top of the hills. To generate a more interesting terrain, the slope profile can be changed. Using the default linear function, the height h of a cell is set to $h = sd$, where s is the slope, and d is the distance of the cell to the river or coast. To make the terrain more interesting, different functions can be used, so that $h = f(d)$. Currently, a few hard-coded functions are available in RiverLand. Figure 3 shows an example of creating cliffs by using a non-linear function for the slope profile.

6   User-Defined Ridges

The user is allowed to define and edit ridge lines to influence the terrain, before the algorithm computes the river network. The user clicks with the left mouse button to place control points for ridges, and has the option of selecting and moving control points with the right mouse button. For each control point, the user can enter the desired height and width. RiverLand scan-converts the ridge lines to find all the cells along the lines. For each cell between the control points, RiverLand uses the random midpoint offset method to assign its width and height. According to the random midpoint offset method, the height h of a point midway between two other points P_{left} and P_{right} is given by $h = avg(h_{Left}, h_{Right}) + srd$, where s is a user-defined roughness factor, r is a Gaussian random variable, and d is the horizontal distance between P_{left} and P_{right}. The same formula to used to calculate the width. Starting with a pair of control points, the midpoint height and width are calculated, and then the rest of the points are recursively calculated.

Once the height of each cell on the ridge line is calculated, the height of all the cells within the influence of the ridge is set by linearly interpolating between the ridge cells and the boundary cells, whose height is set to 0. This is done by finding the path from each ridge line cell to the ridge boundary, using a method similar to the method described in Section 5 used to set terrain heights from river cells.

After that, a random fractal method is used to roughen the terrain. Then, the user-defined terrain is merged with the rest of the terrain height-map, which is created by setting the heights of cells from the algorithm-generated river network (see Section 5). For each cell, take the maximum of the height from the river network and the user-defined ridge. The result is shown in Figure 4.

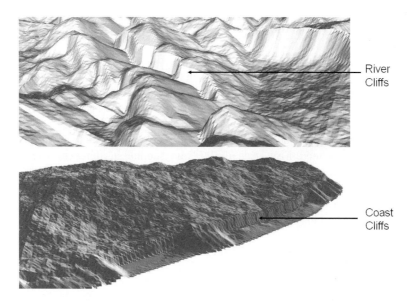

Fig. 3. Changing the slope profile to form river cliffs and coast cliffs

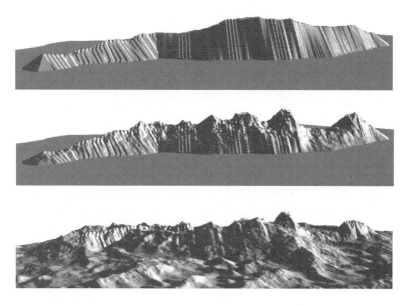

Fig. 4. Top: 3D view of user-defined ridge. Middle: Heights adjusted by random offsets to roughen terrain. Bottom: User-defined ridge shown with the rest of the terrain.

7   Randomized Fractal Overlay

The random midpoint offset algorithm was introduced by Miller [4] to produce fractal terrains. Although this is an efficient method to quickly produce large terrains which look realistic in small patches, the resulting terrains are uniform and do not have realistic erosion features found in real-world terrains. However, this method is useful for adding roughness to the terrain.

In RiverLand, we use the random midpoint displacement algorithm in several ways already mentioned. First, it is used to vary the height along the user-defined ridge lines. It is also used to vary the width along the user-defined ridges.

Next, it is also used to vary the height on all the cells covered by the user-defined ridges. It is also used to vary the height of all the cells created from the river network. This is done by first creating a fractal terrain "overlay". For example, if the terrain is $n \times n$ cells, we create another fractal terrain that is at least $n \times n$ cells. The fractal terrain is created this way. First a coarse $k \times k$ grid of random height values is created. Each cell of the coarse grid is then subdivided using the random midpoint displacement mehtod into $m \times m$ fine grid points, where m is a power of 2. Finally, there would be $(k - 1)m + 1 \times (k-1)m + 1$ cells. The roughness of this fractal terrain can be controlled by the user by controlling the range of height values of the initial coarse grid, and also by controlling the roughness parameter in the random midpoint displacement algorithm.

To prevent the fractal terrain from altering the correctness of the river network, river cells' heights are not adjusted by the fractal terrain overlay. This is to preserve the rule that rivers' heights are non-increasing as they flow towards the mouth. The height change of a cell is therefore $dh = f \times d_r$ where f is a constant set in the program, and d_r is the distance of a cell from a river or region boundary cell.

In addition, on the 2D map, RiverLand also allows the user to paint over an area called a "Modifier Patch", to modify the terrain to add rocky outcrops. Many examples of famous scenery in the real world, such as Guilin (China), the Three Gorges (China), Yosemite Valley (USA) and Rio De Janeiro (Brazil), are rocky outcrops, because they make the landscape interesting. Rocky outcrops can take different shapes, and this is allowed in RiverLand by setting different parameters for a Modifier Patch: r, the average radius of a rock, h the average height, d the average distance between outcrops, and s the steepness. Rocky outcrops are then generated within the Modifier Patch and added to the fractal overlay. Since the rocky outcrop modifies the fractal terrain overlay, it does not modify the final terrain directly. In that way, it will not change the course and height of a river, so that the river networks remain correctly downward flowing.

8   Performance

RiverLand is a fast system; it is able to generate a 256×256 terrain height-map in 0.2 seconds, a 512×512 terrain in 0.9 seconds, and a 1024×1024 terrain in 4 seconds on a 2 GHz machine with 2 GB RAM. This compares well with other recent methods. This does not include the time for generating the fractal overlay, which is done only once during set-up, and can be re-generated by user request.

9   Conclusions

We have presented RiverLand, a system for the fast procedural modeling of terrains. Existing terrain synthesis methods can be generalized into two main types: (1) fractal-based, and (2) simulation-based. In general, fractal-based terrains are faster to generate, but are limited in realism, whereas simulation-based methods are much slower, and their results vary: some terrains look more realistic than others. RiverLand is fundamentally a fractal-based procedural modeling method, but the process is designed to produce terrains consistent with river networks, an important feature lacking in most existing faster fractal-based methods.

One attractive feature of RiverLand is that it allows the user to influence the terrain if the user has preferences. First, by choosing different parameters, the user can influence the style of the terrain. For example, increasing the number and length of rivers makes the river network cover more of the terrain. Decreasing the minimum distance between rivers results in higher river density. Users can also control the roughness and slope of the terrain. The user can also draw ridge lines, and for each control point on the ridge line, the user specifies the desired height and width. RiverLand produces a river network and terrain height-map consistent with the ridge lines defined with the user. The user can paint Modifier Patches to create rocky outcrops, to make the terrain more interesting. Most of the existing terrain synthesis methods do not allow the user such extensive control of the terrain, and yet generate a terrain that is consistent over different areas. RiverLand can create a terrain without any user input at all, but can also create a terrain with a significant amount of user input.

Another advantage of RiverLand is that the terrains produced contain a wide variety of interesting physical features found in real-world terrains, such as cliffs, gentle river valleys, deeper ravines, alpine mountains, rocky outcrops, and rolling hills. All these features can be found in the various example images shown in this paper. Figures 5 shows more examples. This figure also shows an example of automatic adjustment of river width according to water flow simulation by RiverLand.

One important desirable physical feature mentioned in other papers is ridges, which are not generated by the basic fractal methods, but occur in real-world terrains. The terrains produced by RiverLand produce ridges, even if no ridge lines are specified by the user. The terrains generated by RiverLand contain a diversity of features fitting naturally in the same terrain. Also, since RiverLand is not an example-based terrain synthesis program, it can create terrains which are not replicas of real-world examples, and so is less limited in range. By adjusting different parameters, users can also create a terrains in a variety of different styles, and so RiverLand is less rigid, compared to example-based methods.

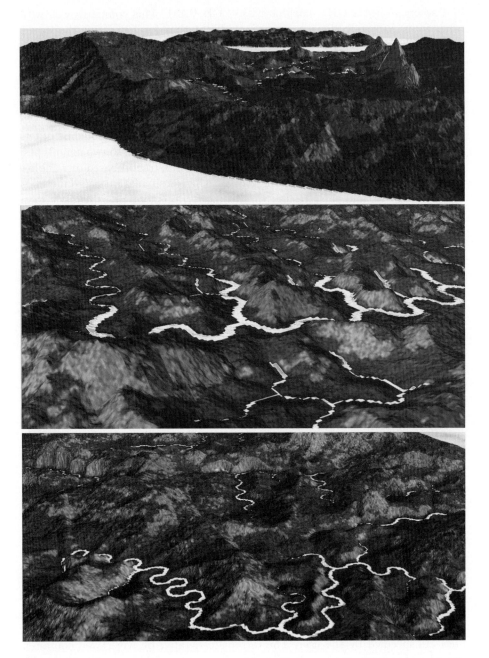

Fig. 5. 3D view of terrains enhanced by color, showing terrain in green, river cells in blue, steep cliffs in grey, and high elevation in white. RiverLand allows automatic adjustment of river width according to water flow simulation, shown in the bottom terrain.

References

1. Carpenter, L.: Computer rendering of fractal curves and surfaces. In: Proceedings of the 7th Annual Conference on Computer Graphics and Interactive Techniques, p. 109 (1980)
2. Mandelbrot, B.B.: The Fractal Geometry of Nature. W.H. Freeman and Co., New York (1982)
3. Fournier, A., Fussell, D., Carpenter, L.: Computer rendering of stochastic models. Communications of the ACM 25, 371–384 (1982)
4. Miller, G.S.P.: The definition and rendering of terrain maps. In: Proceedings of the 13th Annual Conference of Computer Graphics and Interactive Techniques (SIGGRAPH 1986), pp. 39–48 (1986)
5. Kelley, A., Malin, M., Nielson, G.: Terrain simulation using a model of stream erosion. Computer Graphics 22, 263–268 (1988)
6. Musgrave, F., Kolb, C., Mace, R.: The synthesis and rendering of eroded fractal terrains. Computer Graphics 23, 41–50 (1989)
7. Roudier, P., Perrin, B.: Landscapes synthesis achieved through erosion and deposition process simulation. Computer Graphics Forum 12, 375–383 (1993)
8. Nagashima, K.: Computer generation of eroded valley and mountain terrains. The Visual Computer 13, 456–464 (1997)
9. Chiba, N., Muraoka, K., Fujita, K.: An erosion model based on velocity fields for the visual simulation of mountain scenery. The Journal of Visualization and Computer ANimation 9, 185–194 (1998)
10. Neidhold, B., Wacker, M., Deussen, O.: Interactive physically based fluid and erosion simulation. In: Eurographics Workshop on Natural Phenomena (2005)
11. Nguyen, H., Sourin, A., Aswani, P.: Physically based hydraulic erosion simulation on graphics processing unit. In: Proceedings of the 5th international conference on Computer graphics and interactive techniques in Australia and Southeast Asia, pp. 257–264 (2007)
12. Krištof, P., Beneš, B., Křivánek, J., Šťava, O.: Hydraulic erosion using smoothed particle hydrodynamics. Computer Graphics Forum (Proceedings of Eurographics 2009) 28 (2009)
13. Prusinkiewicz, P., Hammel, M.: A fractal model of mountains with rivers. In: Proceedings of Graphics Interface, pp. 174–180 (1993)
14. Belhadj, F., Audiber, P.: Modeling landscapes with ridges and rivers. In: Proceedings of the ACM Symposium on Virtual Reality Software and Technology (VRST 2005), pp. 151–154 (2005)
15. Schneider, J., Boldte, T., Westermann, R.: Real-time editing, synthesis, and rendering of infinite landscapes on GPUs. In: Vision, Modeling and Visualization 2006 (2006)
16. Chiang, M.-Y., Huang, J.-Y., Tai, W.-K., Liu, C.-D., Chiang, C.-C.: Terrain synthesis: An interactive approach. In: Proceedings of the International Workshop on Advancaed Image Tehnology, pp. 103–106 (2005)
17. Brosz, J., Samavati, F., Sousa, M.: Terrain synthesis by-example. In: Advances in Computer Graphics and Computer Vision International Conferences VISAPP and GRAPP 2006, Revised Selected Papers, vol. 4, pp. 58–77 (2006)
18. Zhou, H., Sun, J., Turk, G., Rehg, J.: Terrain synthesis from digital elevation models. IEEE Transactions on Visualization and Computer Graphics 13, 834–848 (2007)
19. Gain, J., Marais, P., Strasser, W.: Terrain sketching. In: Proceedings of the ACM Symposium on Interactive 3D Graphics and Games (2009)

Real-Time 3D Reconstruction for Occlusion-Aware Interactions in Mixed Reality

Alexander Ladikos and Nassir Navab

Chair for Computer Aided Medical Procedures (CAMP)
Technische Universität München
Boltzmannstr. 3, 85748 Garching, Germany

Abstract. In this paper, we present a system for performing real-time occlusion-aware interactions in a mixed reality environment. Our system consists of 16 ceiling-mounted cameras observing an interaction space of size 3.70 m x 3.20 m x 2.20 m. We reconstruct the shape of all objects inside the interaction space using a visual hull method at a frame rate of 30 Hz. Due to the interactive speed of the system, the users can act naturally in the interaction space. In addition, since we reconstruct the shape of every object, the users can use their entire body to interact with the virtual objects. This is a significant advantage over marker-based tracking systems, which require a prior setup and tedious calibration steps for every user who wants to use the system. With our system anybody can just enter the interaction space and start interacting naturally. We illustrate the usefulness of our system through two sample applications. The first application is a real-life version of the well known game Pong. With our system, the player can use his whole body as the pad. The second application is concerned with video compositing. It allows a user to integrate himself as well as virtual objects into a prerecorded sequence while correctly handling occlusions.

1 Introduction

The integration of virtual objects into a real scene and their interaction with real objects is one of the key aspects in mixed reality. However, this is a non-trivial problem. To create the illusion of actually belonging to the scene, a virtual object has to behave properly in the face of occlusion. Many existing systems are not capable of handling this case, leading to unconvincing augmentations, where the virtual object appears in front of the occluder. Another important aspect for a convincing presentation is the ability to interact with virtual objects. For instance, this would allow the user to pick up a virtual object and place it at another position. In this paper we present a system which is capable of addressing both the occlusion and the interaction issue to create a convincing mixed reality environment. The user can interact with virtual objects without requiring any additional tools or external tracking while at the same time occlusions are seamlessly handled. Our system is based on the reconstruction of the 3D shape of objects inside an interaction space. To this end we equipped our laboratory

G. Bebis et al. (Eds.): ISVC 2009, Part I, LNCS 5875, pp. 480–489, 2009.

with 16 cameras which are mounted on the ceiling. Each camera creates a fore-ground/background segmentation which is used to construct the visual hull [1] of all the objects in the scene. This gives us a 3D representation for every object, which in turn allows us to convincingly add virtual objects into the scene and to handle occlusions automatically. One of the key advantages of such a system over more traditional tracking-based systems is that we do not require any a priori information about the objects in the scene. We also do not need any prior setup or calibration for someone to use our system. There can even be multiple people in the interaction space at the same time. This makes our system a good candidate for use in real environments where people can just enter the inter-action space, start to interact naturally with the virtual scene and then leave, without having to put on any special equipment. This significantly lowers the barrier to try the system and makes it attractive for presenting it to a wider audience, for instance in museums. We implemented two exemplary applications which highlight the aspects of interaction and occlusion handling respectively. The first application is loosely based on the game Pong. The goal of the game is to prevent a virtual ball which is bouncing between the user and a wall from leaving the interaction space, by placing oneself in its path (see figure 1). This application shows the interaction between real and virtual objects. The second application is more focused on providing correct occlusion handling. This is done in the context of video compositing. We record several sequences in the interac-tion space at different points in time and use the depth map computed from the 3D reconstruction to join the sequences while correctly handling occlusions.

2   Related Work

In recent years, several real-time 3D reconstruction systems which explicitly recover the visual hull have been proposed [2,3,4,5,6,7]. However, only [6,5,7] actually run at frame-rates which allow interactivity. Other researchers have fo-cused on implicitly computing the visual hull [8,9,10,11,12]. The main difference to the explicit systems is that they only generate an image of the visual hull from a novel viewpoint without recovering an explicit 3D reconstruction. This is ac-ceptable in some cases, but does not allow any form of interaction which requires the full 3D shape (e.g. taking the volume of the object into account). However, it is still possible to use it for collision detection [12,11]. As is to be expected these systems run faster than comparable systems performing an explicit 3D reconstruction. However, today the explicit reconstruction systems reach real-time performance, so that there is no drawback to making use of the additional information.

Some early work on real-time 3D content capture for mixed reality, was pre-sented in [10]. In this paper a novel view generation system was used to insert 3D avatars of real objects into a virtual environment. The system runs at ap-proximately 25 fps using 15 cameras. However, the aspect of interaction between real and virtual objects was not considered. In [11] the authors present a col-lision detection scheme which extends the work in [12] allowing the interaction

Fig. 1. Our system allows the user to interact with virtual objects using natural movements due to a real-time 3D reconstruction. The images placed around the center show some of the input views while the center and the left side show a orthographic and a perspective view of the scene respectively.

between real and virtual objects. However, they are also not using an explicit 3D reconstruction. In addition their system is running at only 10 fps using 7 cameras which is rather low for real interactivity.

Our system recovers the explicit 3D reconstruction of all objects in the scene at a real-time frame rate of 30 Hz using 16 cameras. This allows us to also perform interactions with objects which are occluded by other objects and would therefore not be visible in a system based on an implicit reconstruction.

3   Real-Time 3D Reconstruction System

3.1   System Architecture

Hardware. Our system consists of 4 PCs used for the reconstruction, 1 PC used for visualization and 16 cameras mounted on movable aluminum profiles on the ceiling (see figure 2). The cameras have an IEEE 1394b interface and provide color images at a resolution of 1024x768 and a frame rate of 30 Hz. To cover a big working volume we use wide angle lenses with a focal length of 5 mm. The cameras are externally triggered to achieve synchronous image acquisition. Groups of four cameras are connected to one PC using two IEEE 1394b adapter cards. There are four PCs (slaves) dedicated to capturing the images

Fig. 2. Lab setup for our real-time 3D reconstruction system. The cameras are mounted on movable profiles to allow an easy reconfiguration of the camera setup.

and computing the visual hull and one PC (master) dedicated to visualizing the result and controlling the acquisition parameters. The four PCs used for image acquisition and reconstruction are equipped with an Intel 2.6 GHz Quad-Core CPU (Q6700), 2 GB of main memory and a NVIDIA 8800 GTX graphics board with 768 MB of memory. The master PC uses an Intel 3.0 GHz Dual-Core CPU (E6850), 2 GB of main memory and a NVIDIA 8800 GTS graphics board with 640 MB of memory. The PCs are connected through a Gigabit Ethernet network.

Software. To achieve real-time performance the reconstruction process (running on the slave PCs) is implemented as a four stage pipeline consisting of image acquisition, silhouette extraction, visual hull computation and volume encoding and transmission. Each pipeline step is realized as a thread and will be described in detail in the following sections. On the master PC the processing is also distributed into several steps. There is a separate thread for handling network communication, compositing the partial reconstructions, visualizing the result and performing the application-specific logic, such as the interactions. Figure 3 gives an overview of the processing steps in the system.

3.2 Calibration

In order to perform the reconstruction the cameras have to be calibrated. The calibration is performed using the multi-camera calibration method proposed in [13]. The method relies on point correspondences between the cameras created

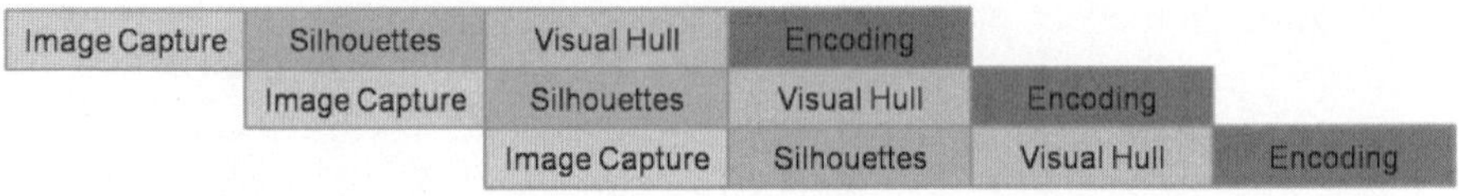

Image Capture	Silhouettes	Visual Hull	Encoding		
	Image Capture	Silhouettes	Visual Hull	Encoding	
		Image Capture	Silhouettes	Visual Hull	Encoding

Fig. 3. Reconstruction pipeline on the slave PCs. Each row represents the processing steps taken for each group of simultaneously acquired input images. The processing times are 8 ms, 15 ms, 15 ms and 10 ms respectively. This leads to a very low latency on the slave PCs, which is pushed to about 100 ms when considering image exposure time and computations on the master PC. Due to the pipelining the whole system can run at a frame rate of 30 Hz or higher.

by means of a point light source such as an LED. First, the lighting in the room is dimmed, so that it becomes easier to extract the point created by the LED in the camera images. We run the cameras with a short exposure time and a low frame rate (1 fps) to obtain well-defined light points. By moving the light source through the reconstruction volume a large number of correspondences is created which is then used in a factorization-based algorithm to determine the camera intrinsic and extrinsic parameters. This requires synchronized cameras to make certain that the point seen in each image is created by the same physical point. The method is robust to occlusions of the points in some cameras. The computed camera coordinate system is registered to the room coordinate system by using a calibration target at a known position in the room.

3.3   Reconstruction

Silhouette Extraction. The silhouettes are computed using a robust background subtraction algorithm [14] working on color images. Before the system is used background images are acquired. During runtime the images are first corrected for illumination changes using a color mapping table which is built using the color distributions of corresponding non-foreground regions in the current image and the background image. After applying this mapping to the background image, a thresholding is applied to extract the foreground pixels in the current image. Small holes in the segmentation are filled using morphological operations.

Handling Static Occluders. While our system seamlessly handles multiple persons in the scene which can occlude each other, one problem which has to be addressed during silhouette extraction is the presence of static occluders. Static occluders are objects inside the working volume which cannot be removed, such as tables mounted to the floor. Hence static occluders are also present in the background images. The assumption during background subtraction, however, is that all foreground objects are located in front of the background. This is not the case in the presence of an occluder because a foreground object could move behind the occluder and effectively disappear from the silhouette image. This will result in the partial or complete removal of the object from the reconstruction.

To overcome this problem we use a method which is similar to the one proposed in [15]. The areas in the silhouette images corresponding to the static occluder have to be disregarded during the visual hull computation. We achieve this goal by building a 3D representation of the object and projecting it into the cameras or by manually segmenting the object in the reference images. This gives us a mask for every camera in which the static occluder is marked as foreground. This mask is then added (logical OR) to the silhouette images computed during runtime.

Visual Hull Computation. Using the silhouette images the object shape is reconstructed using the GPU-based visual hull algorithm described in [7]. In order to increase the working volume we also reconstruct regions which are only seen by at least four cameras instead of only using the overlapping region of all 16 cameras. This allows us to avoid the use of extreme wide angle lenses for covering a big area, which also results in a higher spatial resolution of the camera images. To reconstruct the non-overlapping regions, one has to consider the handling of voxels which project outside of the image in other cameras. The traditional approach is to just mark these voxels as empty. Instead, we do not consider the contribution of the images in which the voxels are not visible, thereby also reconstructing regions only seen by a few cameras. To avoid the introduction of artifacts due to a small number of cameras, we only use regions which are seen by at least four cameras. This is implicitly accomplished in our system by performing an unconstrained reconstruction on the slave PCs which also reconstructs the regions seen by only one camera. On the master PC the local reconstructions are combined using a logical AND operator, which will remove any regions which have not been observed by at least one camera at each of the four slave PCs.

3.4 Visualization

For visualization the voxel representation is converted to a mesh representation using a CUDA-based marching cubes implementation. A CPU-based implementation was not able to generate meshes at the desired frame rate. The meshes are visualized on the master PC on a grid showing the origin and the extent of the reconstruction volume. At this time we do not texture the resulting meshes online. However, it is possible to use an offline texturing step to perform this task when required. This is also not a major concern for our system, since we only need the geometrical and depth information to achieve convincing interaction and occlusion handling results.

4 Applications

4.1 Pong

The idea of using mixed reality to create a game in which users interact with virtual objects has already been introduced with the ARHockey system [16]. The

ARHockey system used a lot of tracking devices and HMDs to enable the illusion of having a virtual puck which is controlled by the hands of the users. Picking up on this idea we used our system to implement a game which is loosely based on the game Pong. In the original game two players each control a pad and pass a ball between each other. If a player fails to catch the ball his opponent gains a point. We modified the game so that one player is playing against a wall. The goal is to keep the ball from exiting the scene. The player has a certain amount of life points and has to try to keep the ball in the game for as long as possible.

We use a video projector to display the reconstruction of the interaction space on a wall. The user can see himself moving in 3D and he has to position himself, such that the ball is reflected off of him (see figure 4). The collision test is performed between the virtual object and the visual hull. There are two modes. In the first mode we only use the bounding box of the visual hull to perform the collision test. This has the advantage that it is easier for the user to hit the ball, because there is a bigger interaction area. Using the bounding box also allows children to easily capture the ball, because they can extend their arms to compensate for their lesser body size. The second mode performs the collision test directly between the visual hull and the virtual object. This leads to a more

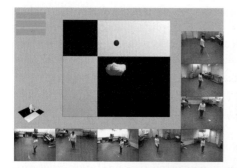
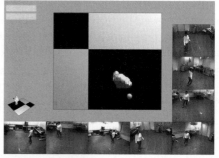

Fig. 4. Our system allows the user to play a game by interacting with a virtual ball. The images placed around the center show some of the input views while the center and the left side show a orthographic and a perspective view of the scene respectively. In the upper left corner the player's remaining life points are shown.

natural interaction, because it is very intuitive. However, the problem here is that it is hard for the user to estimate the height of the ball, so that it might happen that he extends his arm, but the ball passes below it. This problem can be reduced by showing several views of the reconstruction. Optimally a stereoscopic HMD would allow the most natural interaction.

It is also possible for multiple people to play the game at the same time. Due to the use of our reconstruction system the player can use his whole body to catch the ball. The interaction is very natural and people intuitively know how to move to catch the ball. Their movement is not hindered by any additional equipment as would be the case in a tracking-based solution. Even when using a tracking-based solution it would be quite complex to correctly compute the extents of the body. Due to its easy usability and the fact that no setup or training phase is necessary for the user, our system is well suited for use in a real environment, for instance in a museum.

4.2   Video Compositing

As a second application we implemented a video composition system which properly handles occlusions. This is an important topic in mixed/augmented reality

Fig. 5. Video compositing. We created a new sequence by composing the same video six times in 1 second intervals. Note the correct occlusion handling with respect to the virtual ball and the different time steps of the original sequence.

[17,18]. Using our system we took a sequence of a person walking inside the interaction volume. We subsequently used the reconstruction to compute the depth map for one of the cameras. By using both the information from the depth map and the segmentation we created a sequence which shows the same scene at six time steps with an interval of one second in between at the same time (see figure 5). The effect is that instead of one person you can see a queue of 6 copies of the same person walk inside the room. Due to the use of the depth map we correctly handle the occlusion effects. In addition, we added a virtual bouncing ball to the scene which also correctly obeys the occlusion constraints. For creating the composited scene we currently do not apply any image-based refinement on the silhouette borders, but this could be easily added into the system.

These compositing results would be very hard to achieve using purely image-based techniques which do not consider any information about the 3D structure of the scene. It would require a (manual) segmentation of the objects of interest in the entire sequence which is extremely time consuming especially for long sequences. With our solution the segmentation and the depth information is automatically recovered without any additional intervention from the user.

5   Conclusion

We presented a real-time 3D reconstruction system for occlusion-aware interactions in mixed reality environments. Our system consists of 16 cameras and reconstructs the contents of the interaction volume at a frame rate of 30 Hz. The reconstruction is used to allow users to interact naturally with virtual objects inside the scene while correctly handling the problem of occlusions in the augmentation. This is an important aspect in mixed and augmented reality. We demonstrated the results of our system in two application scenarios. The first application is an interactive game which focuses on the interaction aspect, while the second application is a video compositing task which focuses on occlusion handling. In the future we plan to also integrate a tracked HMD into the system, to create an even more immersive experience for the user.

References

1. Laurentini, A.: The visual hull concept for silhouette-based image understanding. IEEE Transactions on Pattern Analysis and Machine Intelligence 16, 150–162 (1994)
2. Cheung, G., Kanade, T., Bouguet, J.Y., Holler, M.: A real-time system for robust 3d voxel reconstruction of human motions. In: IEEE Conference on Computer Vision and Pattern Recognition (2000)
3. Borovikov, E., Sussman, A., Davis, L.: A high performance multi-perspective vision studio. In: ACM International Conference on Supercomputing (2003)
4. Wu, X., Takizawa, O., Matsuyama, T.: Parallel pipeline volume intersection for real-time 3d shape reconstruction on a pc cluster. In: IEEE International Conference on Computer Vision Systems (2006)

5. Allard, J., Menier, C., Raffin, B., Boyer, E., Faure, F.: Grimage: Markerless 3d interactions. In: SIGGRAPH - Emerging Technologies (2007)

6. Hasenfratz, J.M., Lapierre, M., Sillion, F.: A real-time system for full body interaction with virtual worlds. In: Eurographics Symposium on Virtual Environments, pp. 147–156 (2004)

7. Ladikos, A., Benhimane, S., Navab, N.: Efficient visual hull computation for real-time 3d reconstruction using cuda. In: Proceedings of the 2008 Conference on Computer Vision and Pattern Recognition Workshops (2008)

8. Matsuik, W., Buehler, C., Raskar, R., Gortler, S., McMillan, L.: Image-based visual hulls. In: SIGGRAPH (2000)

9. Li, M., Magnor, M., Seidel, H.: Improved hardware-accelerated visual hull rendering. In: Vision, Modeling, and Visualization (2003)

10. Prince, S., Cheok, A., Farbiz, F., Williamson, T., Johnson, N., Billinghurst, M., Kato, H.: 3D live: Real time captured content for mixed reality. In: ISMAR 2002: Proceedings of the 1st IEEE/ACM International Symposium on Mixed and Augmented Reality (2002)

11. Decker, B.D., Mertens, T., Bekaert, P.: Interactive collision detection for free-viewpoint video. In: GRAPP 2007: International Conference on Computer Graphics Theory and Applications (2007)

12. Lok, B., Naik, S., Whitton, M., Brooks, F.: Incorporating dynamic real objects into immersive virtual environments. In: I3D 2003: Proceedings of the 2003 symposium on Interactive 3D graphics (2003)

13. Svoboda, T., Martinec, D., Pajdla, T.: A convenient multi-camera self-calibration for virtual environments. Presence: Teleoperators and Virtual Environments 14, 407–422 (2005)

14. Fukui, S., Iwahori, Y., Itoh, H., Kawanaka, H., Woodham, R.: Robust background subtraction for quick illumination changes. In: Chang, L.-W., Lie, W.-N. (eds.) PSIVT 2006. LNCS, vol. 4319, pp. 1244–1253. Springer, Heidelberg (2006)

15. Guan, L., Sinha, S., Franco, J.-S., Pollefeys, M.: Visual hull construction in the presence of partial occlusion. In: 3DPVT 2006: Proceedings of the Third International Symposium on 3D Data Processing, Visualization, and Transmission, 3DPVT 2006(2006)

16. Ohshima, T., Satoh, K., Yamamoto, H., Tamura, H.: Ar2 hockey: A case study of collaborative augmented reality. In: Proceedings of the IEEE VRAIS 1998 (1998)

17. Berger, M.: Resolving occlusion in augmented reality: a contour based approach without 3d reconstruction. In: IEEE Conference on Computer Vision and Pattern Recognition (1997)

18. Kim, H., Yang, S., Sohn, K.: 3d reconstruction of stereo images for interaction between real and virtual worlds. In: ISMAR 2003: Proceedings of the 2nd IEEE/ACM International Symposium on Mixed and Augmented Reality (2003)

Augmenting Exercise Systems with Virtual Exercise Environment

Wei Xu, Jaeheon Jeong, and Jane Mulligan

Department of Computer Science, University of Colorado at Boulder, Boulder,
Colorado 80309-0430 USA
{Wei.Xu,Jaeheon.Jeong,Jane.Mulligan}@Colorado.edu

Abstract. Adhering to an exercise program is a challenge for everybody who wants to build a healthier body and lifestyle through physical exercise. We have developed an Virtual Exercise Environment (VEE) that augments stationary exercise equipment with virtual reality techniques to make exercising more enjoyable. Our VEE system consists of a recording system to capture video, distance and incline data about real trails, and a playback system which "displays" both video and terrain data in the form of video speed and resistance. Trails are played back according to the speed the user generates on the stationary exercise equipment. The system uses commodity capture and display devices and supports standard interfaces for existing exercise equipment. User studies have shown that users enjoy the ability to guage their progress and performance via their progress through trail playback in the VEE.

1 Introduction

Sticking to exercise programs can be challenging for most of us, especially using stationary exercise machines at home or at the gym . Studies show that an external focus of attention helps improve enjoyment and intensity of workouts [1], but typically the only distraction available is a TV or mp3 player. To address this deficiency we have developed an add-on virtual reality system to augment standard exercise equipment, such as treadmills, stationary bikes or arm ergometers. Our augmentation takes the form of a Virtual Exercise Environment (VEE) that simulates real, vivid, outdoor exercise trails for exercising in an indoor environment.

The VEE system consists of a target exercise machine, immersive video displays and a workstation to drive the displayed percepts. What distinguishes our system from others that integrate displayed games and graphics for exercise equipment (e.g., [2]), is that we have developed a capture phase which records both appearance and terrain of outdoor real trails using a panoramic camera head and special purpose boards and electronic sensors. We have also focused on low cost solutions to make the system affordable to most of people.

G. Bebis et al. (Eds.): ISVC 2009, Part I, LNCS 5875, pp. 490–499, 2009.

2   Related Work

To the best of our knowledge, there is no previous work that addresses all the features of our VEE system. However, there is separate work in various related research communities, including computer vision, image processing, computer graphics and virtual reality community, that gives us hints on how to build our system.

The Virtual Environments group at the University of Utah has produced a number of interesting results on *locomotion interfaces* [3,4]. These are interfaces which cause the user to expend energy as they simulate unconstrained activities such as walking or running in virtual reality (VR) in limited space. The particular system they use combines a special purpose treadmill with immersive visual displays to study perception action couplings. Their locomotion display includes a large Sarcos treadmill with an active mechanical tether. The tether applies inertial forces and emulates slope changes. The standard exercise equipment which forms the locomotion display in our VEE system can not match the devices used in these studies, but concepts such as the importance of matching the visual percept to walking speed [3], are highly relevant.

Some navigation-based VR systems such as the early Aspen Movie Map [5] and the more recent Google Street View [6] have also some similarity to our VEE system in the aspect that they also visually record and play back panoramic frames of real outdoor trails. However, these systems do not have a locomotiion display part comparing to our VEE system — there is no monitoring and association of the user's walking speed to the playback of recorded panoramic frames. And the terrain changes of the trails are not recorded and thus there is no way to play them back via a locomotion display in these systems.

A number of multi-camera systems have been proposed for capturing surround video. Point Grey Research packages a six camera proprietary spherical camera head [7]. Foote and Kimber [8] and Nanda and Cutler [9] describe 8 and 5 camera systems respectively, applied to recording office meetings. These multi-camera systems are carefully designed, manufactured and calibrated, but are usually inaccessible or unaffordable to our target community. We instead built our system upon low-cost commodity board cameras that are affordable to most of people. In order to achieve the overlap required to produce cylindrical video frames while using as few cameras as possible, we also used extremely wide angle lenses (1.7mm focal length) that have significant radial distortions (see Figure 4(a)). We will describe how we solve these practical problems in detail in the following sections.

3   The VEE System

Our VEE system is composed of a trail recording phase, a data processing and integration phase, and a trail playback phase. The trail recording sub-system can record both appearance and terrain of a real outdoor trail. The records are then processed and integrated by the data processing and integration sub-system, and

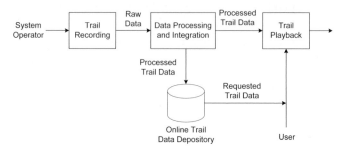

Fig. 1. VEE system overview

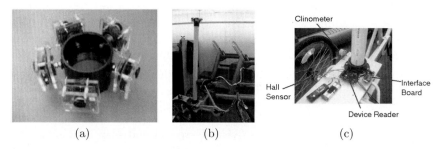

(a)   (b)   (c)

Fig. 2. (a) Low cost cylindrical capture head. (b) Camera head mounted on trail capture vehicle. (c) Terrain recording hardware mounted on capture trike.

the processed data are uploaded into a trail data distribution website from which the public can download the data for free. With downloaded trail data, a user can play the trail back using an immersive display device and standard exercise equipment. Figure 1 gives an overview of the VEE system.

3.1   Trail Recording

The trail recording sub-system includes a panoramic camera head consisting of five Unibrain firewire board cameras mounted evenly on a cylinder (Figure 2(a)), an altered adult trike used to carry the camera head to move around (Figure 2(b)), plus special purpose boards and electronic sensors (Figure 2(c)) integrated with the trike which measure tilt and odometry (distance travelled). All measurements are recorded simultaneously by a control program running on a laptop computer as the trike is ridden along scenic bike trails.

Distance measurements are provided by a Hall effect transistor which detects the presence or absence of three strong magnets spaced evenly around one of the rear wheels of the trike as the wheel rotates. As the magnets come into proximity to the Hall sensor it turns on, and as the magnets move away it turns off, which generates a tick. The distance traveled D_{curr} is computed as $D_{curr} = (N_{tick} - 1) * U_{dist}$, where N_{tick} is the count of ticks and U_{dist} is the

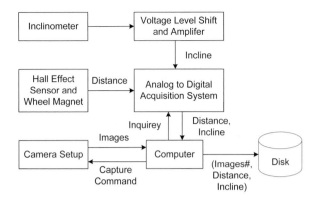

Fig. 3. Trail recording and data processing and integration process

unit distance traveled between two successive ticks. With three sensors on the 24" wheel of the trike, U_{dist} is approximately two feet.

The incline measurement utilizes a Schaevitz Accustar clinometer [10] attached to the base of of the trike. This is an electronic device that puts out analog voltages based on tilt — when tilted in up-hill direction the voltage is positive and when tilted in down-hill direction the voltage is negative. Since the analog data acquisition system we use to collect the clinometer signals only measures positive voltages, an inverting amplifier is used to convert negative voltages into positive voltage. Another thresholding amplifier is used to collect the positive signals only (by converting all of the negative voltages into zero). By comparing these two serials of positive signals, the original serial of signals that contains both positive and negative values can be restored:

$$T^{(t)} = \begin{cases} (T_1^{(t)} + T_2^{(t)})/2 & \text{if } T_2^{(t)} > 0 \\ -T_1^{(t)} & \text{if } T_2^{(t)} = 0 \end{cases}$$

where $T^{(t)}$ is the original signal of tilt at a time tick t, $T_1^{(t)}$ and $T_2^{(t)}$ are corresponding signals collected vis the inverting amplifier and via the thresholding amplifier respectively.

Procedure of the trail recording process is shown in Figure 3.

3.2 Panoramic Video Stitching

The individual videos captured by the five cameras need to be stitched into a 360° panoramic video for immersive playback in future. This includes several operations considering the special capturing device we use. First, the cameras used in our system are equipped with low-grade wide angle lenses (with $\approx 75°$ horizontal FoV) which suffer from substantial radial distortions. We use the Brown-Conrady model [11,12] to model the radial distortions by a low-order polynomial: $x_d = x_u(1 + \kappa_1 r^2 + \kappa_2 r^4)$ and $y_d = y_u(1 + \kappa_1 r^2 + \kappa_2 r^4)$, where

$r^2 = x^2 + y^2$ and κ_1 and κ_2 are called the radial distortion parameters [13], (x_d, y_d) is distorted (image) coordinate and (x_u, y_u) is corresponding undistorted coordinate. Here we omit the tangential (decentering) distortions [11], since they are usually negligible for consumer-level stitching [14,13]. Our calibration using a checkerboard(Figure 4(a)) estimates that $\kappa_1 \approx 1.6 \times 10^{-5}$ and $\kappa_2 \approx 2.0 \times 10^{-11}$. This estimate is applicable for all of the five cameras.

Next, we register the calibrated images and stitch them into a panorama. We assume the five cameras consisting our panoramic camera head approximately follow a rotational geometrical model, i.e. they are co-centered and their poses are only different to each other by a 3D rotation. Under this model, the stitching is simply warping the images into cylindrical coordinates and then using a pure translational model to align them [15]. The formula for converting a rectangular coordinate (x, y) into the corresponding cylindrical coordinate (x_c, y_c) is as follows: $x_c = s\theta = s \arctan \frac{x}{f}$ and $y_c = s\rho = s\frac{y}{\sqrt{x^2+f^2}}$, where s is an arbitrary scaling factor that determines the size of the resulting image. Then, the parameters of the translational model between the i'th pair of adjacent views, $(\Delta x, \Delta y)$, are estimated from robust SIFT feature matches (where RANSAC is used to remove the outlier matches). Also, because what we are stitching is a $360°$ panorama, we need to make sure the left and right ends of the panorama matches with each other. This is achieved by minimizing the global registration error E_G:

$$E_G = \sum_{i=1}^{5} err_i(\Delta x_i, \Delta y_i) = \sum_{i=1}^{5} \sum_{j=1}^{n_i} (\hat{x}_{ij} - x_{ij} - \Delta x_i)^2 + (\hat{y}_{ij} - y_{ij} - \Delta y_i)^2$$

where $err_i(\cdot)$ is the re-projection error and n_i is the number of feature matches between the i'th pair of adjacent views. $(\hat{x}_{ij}, \hat{y}_{ij})$ and (x_{ij}, y_{ij}) are a pair of matched points. After the views are registered in cylindrical surface, a linear image blending technique [16] is used to remove appearance difference and seams in the overlap areas of adjacent views.

Figure 4 illustrates the image calibration and projection operations through an example. And figure 5 shows an example panorama as the final output of the whole processing. In practice, we use direct mapping based on look-up tables to speed up the processing. We learn the geometric transformation and image blending parameters from the stitching of several carefully selected frames. We then compute the direct mapping from pixels on the original images to pixels on the final composite panorama and save the mapping into a static look-up table which is used to stitch all the remaining frames. Direct mapping avoids both step-wise image warping and backward interpolation [17] and saves the time and space for storing/loading intermediate results. Details about the design and implementation of the look-up table can be found in [18].

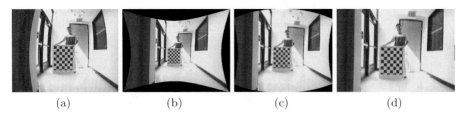

| (a) | (b) | (c) | (d) |

Fig. 4. Image calibration and projection process. (a) Original image frame captured by one of the five cameras. (b) Frame after removal of radial distortions. (c) Calibrated frame projected onto the cylindrical surface. (d) Cropped cylindrical frame.

Fig. 5. An example panorama

3.3 Data Integration

In order to provide a virtual sensation of strolling down a path, the video image needs to play back at a speed that is consistent with the pace of the person walking down the path [3]. At the same time the amount of effort that is expended needs to be consistent with the grade (incline) of the path. This means ideally image and terrain data acquisition would occur simultaneously, where one image set is collected for each distance and incline data measurement. However, in reality images are acquired more often than tics from the hall sensor, so after data collection distances are interpolated over the number of images between tics to produce a file with one distinct distance and incline mark associated with each frame. The timestamps of the data records were used to guide the association.

4   Trail Playback

The result of trail recording and data processing and integration is a dense sequence of panoramic video frames labeled with recorded incline and distance traveled along the trail. The last phase of the VEE system is to play back this sequence using an immersive video display (either a tracked HMD or a surrounding monitor cluster) and a locomotion display consisting of a standard piece of stationary exercise equipment such as a treadmill or bike. We have developed a multi-threaded control program to coordinate the playback on these two kinds of displays. The program is based on three threads running concurrently. One thread pre-fetches images and stores them in a queue, another thread continually

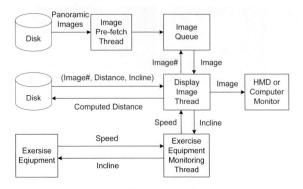

Fig. 6. Trail playback process

requests speed and data from the exercise equipment and updates the distance
traveled measure, and a third thread monitors this distance metric and based
on this data retrieves images from the queue to be displayed on the immersive
video display. This multi-threaded processing could smoothly play back a trail
at a higher speed than single-process progressive processing and it never got any
lag problems in its practice with the several testing trails that we have under
proper parameter settings (e.g., the image queue size). Figure 6 is a sketch map
of the playback process.

4.1 Immersive Video Display

The immersive video display device used in our experiments is a single integrated
HMD, the eMagin z800(Figure 7(a)). Cylindrical panoramic images computed
from the captured multi-image sequences are displayed in the HMD according to head position computed by the head tracker. In this way the user can
"look around" the trail surroundings as he moves through the virtual trail world
(Figure 7(b)). To avoid lags due to loading image files into memory, panoramic
images are pre-fetched and stored in a queue in local memory for display as
needed. Figure 7(c) shows an early non-immersive playback system using a single forward-facing monitor, but including the locomotion display playback on
the treadmill.

4.2 Locomotion Display

The locomotion display devices used in our system include various kinds of exercise equipment, such as treadmills, stationary exercise bikes and arm ergometers.
The exercise equipment itself acts as a physical display, by "playing back" an
approximation of the recorded terrain using the speed and slope or resistance
settings of the equipment. The thread that plays back the trail on the exercise
equipment calculates the distance traveled by the user that is exercising and uses
this information to determine which frame of the panorama sequence should be

(a)   (b)   (c)

Fig. 7. (a) Head Mounted Display (HMD) (b) Recorded surround video and terrain features are played back on synchronized HMD and exercise bike. (c) Playback using synchronized computer monitor and treadmill.

rendered on the immersive video display. The playback effectively runs "on a rail" since currently no navigation interface is provided and only fixed recorded traversal of the trail is available. Some of the exercise equipment does not have a fine enough distance resolution for retrieving continuous image frames for rendering. For example, the distance resolution of the treadmill used by our system (Table 1) is 0.01 miles (52.8 feet) which causes jumping artifacts in the rendered video. So, instead our system uses the instantaneous speed reported by the exercise equipment to calculate the accumulated distance and uses this distance to retrieve the appropriate panorama frame for rendering:

$$curState.dist = curState.dist + rd * \text{mph2fps} * curState.speed$$

where $curState$ is a global structural variable that records the run-time status of the machine. $rd = dispTime.runningDuration()$ records the running duration from last computation of $curState.dist$ to now, and $dispTime$ is a global timer whose count resets after each update of $curState.dist$. mph2fps = 1.4667 is a scaling factor that converts the unit from mph (mile per hour) to fps (frame per second). In our current implementation only the speed information reported by the exercise equipment is used to calculate the distance traveled, but in future versions reported distance might also be used to dynamically calibrate the calculation.

The incline information is physically played back on the locomotion display. Different exercise equipment has different usage of this information. In our system, we use the incline information to control the slope of the deck of treadmills, and the resistance level of arm ergometers or stationary exercise bikes. For machines without controllable incline or resistance, we emulate a low gear on a bicycle, where the user must pedal faster (proportional to the measured trail incline) to travel the same distance. Table 1 shows a list of exercise equipment onto which we can successfully display the incline information. We use either the Communications Specification for Fitness Equipment (CSAFE) protocol [19] or machine associated SDK to control these equipments. The CSAFE protocol is a

Table 1. The exercise equipment experimented in our system, their properties used to play back the incline information, and their communication interfaces

Name	Category	Manufacturer	Property	Interface
FlexDeck Shock Abso. Sys.	treadmill	Life Fitness	slope	CSAFE
LifeCycle 9500 HR	exercise bike	Life Fitness	resistance	CSAFE
SciFit Pro I	arm ergometer	Sci-Fit	resistance	CSAFE
PCGamerBike	arm ergometer	3D Innovations	resistance	FitXF SDK

public standard on exercise equipment manufacturing that is supported by most of today's exercise equipment manufactures. Our playback control program has fully implemented it and thus can communicates with any exercise machines that supports this protocol.

5   Pilot Study and Results

The VEE system we developed was evaluated by researchers at Assistive Technology Partners in Denver, CO, USA. A male participant was recruited for a pilot study of the technology using a treadmill. This individual completed a week long baseline phase in which he logged his daily step count as usual, followed by an intervention phase during which he exercised on our VEE-outfitted treadmill a total of 11 times over a period of four weeks. At the end of the study, the participant was asked to provide feedback on his VEE experience. He described the technology as compelling, stating that he had found it interesting and enjoyable to use. The participant's average daily number of steps during the baseline phase of the study 2065. His average daily step count during the intervention phase was 6151. This included a low of 4433 per day (third week of intervention) to a high of 7926 (fourth week of intervention). Throughout the intervention phase, the participant completed an average of 4563 steps per intervention session. A final pleasant outcome was the patient's body weight, which decreased 5 lbs from the baseline phase to the end of the intervention phase.

6   Concluding Remarks

In this paper we have described a complete system for recording of real outdoor trail environments and playing these recordings back in an immersive Virtual Exercise Environment (VEE). We have developed a low-cost capture device which records both appearance and terrain of outdoor real trails using a panoramic camera head and special purpose boards and electronic sensors. We have also developed a collection of computer programs for controlling the trail recording and playback phases, for rapidly stitching panoramic videos, and for driving and monitoring various kinds of immersive video display and locomotion display devices. Efficacy of the VEE system has been verified by human pilot study.

References

1. Porcari, J.P., Zedaker, J.M., Maldari, M.M.: Real increased exercise performance results when members have more to engage their attention than their own effort and fatigue. Fitness Management Magazine 14, 50–51 (1998)
2. Cybex: Cybex Trazer (2007), http://www.cybextrazer.com
3. Thompson, W.B., Creem-Regehr, S.H., Mohler, B.J., Willemsen, P.: Investigations on the interactions between vision and locomotion using a treadmill virtual environment. In: Proc. SPIE/IS&T Human Vision and Electronic Imaging Conference, vol. 5666, pp. 481–492 (2005)
4. Hollerbach, J.M.: Locomotion interfaces. In: Handbook of Virtual Environments Technology
5. Naimark, M., Clay, P., Mohl, B.: Aspen Movie Map (1978), http://www.naimark.net/projects/aspen.html
6. Google: Google Street View (2007), http://maps.google.com/help/maps/streetview
7. PointGreyResearch: Ladybug (2006), http://www.ptgrey.com/products/spherical.asp
8. Foote, J., Kimber, D.: Flycam: practical panoramic video and automatic camera control. In: Proc. IEEE International Conference on Multimedia and Expo., vol. 3, pp. 1419–1422 (2000)
9. Nanda, H., Cutler, R.: Practical calibrations for a real-time digital omnidirectional camera. In: Technical Sketches IEEE International Conference on Computer Vision and Pattern Recognition (2001)
10. Measurement Specialties, I.: Schaevitz accustar clinometer (2007), http://www.meas-spec.com/myMeas/sensors/schaevitz.asp
11. Brown, D.C.: Decentering distortion of lenses. Photogrammetric Engineering 32, 444–462 (1966)
12. Conrady, A.: Decentering lens systems. Monthly Notices of the Royal Astronomical Society 79, 384–390 (1919)
13. Brown, D.C.: Close-range camera calibration. Photogrammetric Engineering 37, 855–866 (1971)
14. Devernay, F., Faugeras, O.: Straight lines have to be straight: automatic calibration and removal of distortion from scenes of structured environments. Machine Vision and Applications 1, 14–24 (2001)
15. Chen, S.E.: Quicktime vr c an image-based approach to virtual environment navigation. In: Computer Graphics (SIGGRAPH 1995), pp. 29–38 (1995)
16. Goshtasby, A.A.: 2-D and 3-D Image Registration. John Wiley & Sons, Hoboken (2005)
17. Wolberg, G.: Digital Image Warping. IEEE Computer Society Press, Los Alamitos (1990)
18. Xu, W., Penners, J., Mulligan, J.: Recording real worlds for playback in a virtual exercise environment. Computer Science CU-CS 1013-06, University of Colorado at Boulder, Boulder, CO (2006)
19. FitLinxx: Communications SpecificAtion for Fitness Equipment, CSAFE (2004), http://www.fitlinxx.com/CSAFE/

Codebook-Based Background Subtraction to Generate Photorealistic Avatars in a Walkthrough Simulator

Anjin Park[1], Keechul Jung[2], and Takeshi Kurata[1]

[1] Center for Service Research, National Institute of Advanced Industrial
Science and Technology, Japan
[2] Department of Digital Media, Soongsil University, Korea
{anjin.park,t.kurata}@aist.go.jp, kcjung@ssu.ac.kr

Abstract. Foregrounds extracted from the background, which are intended to
be used as photorealistic avatars for simulators in a variety of virtual worlds,
should satisfy the following four requirements: 1) real-time implementation, 2)
memory minimization, 3) reduced noise, and 4) clean boundaries. Accordingly,
the present paper proposes a codebook-based Markov Random Field (MRF)
model for background subtraction that satisfies these requirements. In the proposed method, a codebook-based approach is used for real-time implementation
and memory minimization, and an edge-preserving MRF model is used to eliminate noise and clarify boundaries. The MRF model requires probabilistic measurements to estimate the likelihood term, but the codebook-based approach
does not use any probabilities to subtract the backgrounds. Therefore, the proposed method estimates the probabilities of each codeword in the codebook
using an online mixture of Gaussians (MoG), and then MAP-MRF (MAP: Maximum A-Posteriori) approaches using a graph-cuts method are used to subtract
the background. In experiments, the proposed method showed better performance than MoG-based and codebook-based methods on the Microsoft DataSet
and was found to be suitable for generating photorealistic avatars.

1 Introduction

Before constructing buildings, details such as the suitability of the floor layout and
navigation signs and whether users will feel comfortable in the buildings should be
considered. Virtual reality techniques are used to investigate virtual structures in
detail. However, since information is generally displayed through monitors and a
keyboard or mouse is used to navigate the structure, it is difficult to evaluate the relationship between the details of the structure and the sense of absolute direction of the
user. Therefore, we are developing simulation environment, referred as to *Walk-
through Simulator* (WTS), in order to enable subjects to navigate virtual constructions
from the perspective of the customer. The hexahedral-shaped device shown in the
center of Fig. 1 is the WTS. The virtual building is displayed inside the device using
multi-projectors, as shown on the right-hand side of Fig. 1.

In some buildings, such as public institutions, guides provide instructions to customers or visitors to help them reach their destination. In the virtual building, guides
are displayed as avatars. In the present study, we use a *Photorealistic Avatar*, in
which the appearance of an actual person is used as CG texture, as a guide. The

G. Bebis et al. (Eds.): ISVC 2009, Part I, LNCS 5875, pp. 500–510, 2009.

present paper focuses on displaying the photorealistic avatar in virtual buildings. The image of the person that is used to create the photorealistic avatar is extracted by the camera in front of a remote computer connected via a network with the WTS, as shown in Fig. 1(A), and the photorealistic avatar is displayed in a fixed location in the virtual world inside the simulator, as shown in Fig. 1(B).

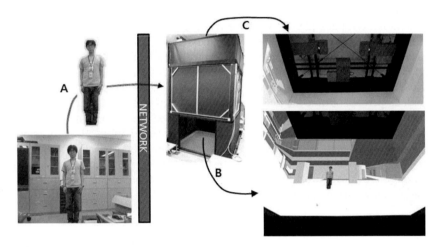

Fig. 1. Schematic diagram of the WTS: (A) photorealistic avatar extracted from the modeled background, (B) photorealistic avatar integrated into the virtual building, and (C) multi-projectors

In the present paper, it is assumed that the moving foreground in front of the fixed camera of the remote computer is an individual whose image will be used to generate the photorealistic avatar, and the actual person who will take a role of the guide can stand on any places, for example, a room with complex backgrounds. The present paper uses background subtraction to extract the appearance of the guide from images captured by a camera. There are four requirements for the background subtraction: 1) extraction must be performed in real time, 2) memory consumption must be limited, 3) the image must be extracted with little noise, and 4) the boundaries of the avatar must be clear.

The proposed method integrates a codebook-based approach, which helps to perform extraction in real time and reduces the required memory, and an edge-preserving MRF model, which can eliminate noise and generate clear boundaries. Although the codebook-based algorithm [10] can model an adaptive and compact background over a long period of time with limited memory, it cannot be used as the likelihood term in the edge-preserving MRF, because the similarity (rather than the probability) is used to compare input pixels with the modeled background. Therefore, online mixture of Gaussians (MoG) is used to estimate the probabilities for all codewords in the codebook. In addition, the proposed method models the prior term using the codebook-based method in order to substantially reduce extraction errors caused by high-contrast edges in cluttered backgrounds, thereby reducing errors on the boundaries of extracted foregrounds.

2   MRF Modeling for Background Subtraction

2.1   Related Research

The simplest background model assumes that pixel values can be modeled by a single Gaussian distribution [1]. However, this basic model cannot handle multiple backgrounds, such as trees moving in the wind. The MoG has been used to model non-static backgrounds [2]. However, it is difficult to detect sudden changes in the background when the learning rate is low, and slowly moving foreground pixels will be absorbed into the background model when the learning rate is high [7]. Sheikh and Shah [4] proposed a MAP-MRF framework, which results in clear boundaries without noise by enforcing spatial context in the process, but this technique [4] cannot be used when long periods of time are needed to sufficiently model the background, primarily due to memory constraints, because they used a kernel density estimation technique [7]. In order to address the memory constraint problem, Kim et al. [7] proposed a codebook background subtraction algorithm intended to model pixel values over long periods of time, without making parametric assumptions. However, since this algorithm did not evaluate probabilities, but only calculated the distance from the cluster means, it is hard to extend this algorithm to the MAP-MRF framework.

2.2   Energy Function

In the present paper, background subtraction is considered as an MRF framework. The MRF is specified in terms of a set of sites S and a set of labels $\mathcal{L}$. Consider a random field consisting of a set of discrete random variables $\mathbf{F} = \{F_1, \dots, F_n\}$ defined on the set S, such that each variable F_s takes a value f_s in $\mathcal{L}$, where s is index of the set of sites. For a discrete label set $\mathcal{L}$, the probability that random variable F_s takes the value f_s is denoted as $P(F_s = f_s)$, and the joint probability is denoted as $P(\mathbf{F} = \mathbf{f}) = (F_1 = f_1, \dots, F_n = f_n)$, abbreviated as $P(\mathbf{f})$, where $\mathbf{f} = \{f_1, \dots, f_n\}$. Here, $\mathbf{f}$ is a configuration of $\mathbf{F}$.

If each configuration, $\mathbf{f}$ is assigned a probability $P(\mathbf{f})$, then the random field is said to be an MRF [11] with respect to a neighborhood $N = \{N_s | s \in S\}$, where N_s is the set of sites neighboring s, if and only if it satisfies the following two conditions: the positivity property $P(\mathbf{f}) > 0, \forall \mathbf{f} \in \mathbf{F}$ and the Markovian property $P(f_s | f_{S-\{s\}}) = P(f_s | f_{N_s})$, where $f_{N_s} = \{f_{s'} | s' \in N_s\}$ denotes the set of labels at the sites neighboring s.

Since $\mathbf{F}$ is generally not accessible, its configuration $\mathbf{f}$ can only be estimated through an observation obs. The conditional probability $P(\mathbf{f} | obs)$ is the link between the configuration and the observation. A classical method of estimating the configuration $\mathbf{f}$ is to use MAP estimation. This method aims at maximizing the posterior probability $P(\mathbf{f} | obs)$, which is related to the Bayes rule as follows: $P(\mathbf{f} | obs) = \frac{P(obs | \mathbf{f}) P(\mathbf{f})}{P(obs)}$.

Since the problem lies in maximizing the previous equation with respect to $\mathbf{f}$, which $P(obs)$ does not act on, the MAP problem is equivalent to

$$P(\mathbf{f} | obs) = \text{argmax}_{\mathbf{f} \in \mathbf{F}} \left(\sum_{s \in S} D_s(f_s) + \sum_{\{s,s'\} \in N} V_{s,s'}(f_s, f_{s'}) \right), \tag{1}$$

in an energy function, where $P(\mathbf{f})$ is the Gibbs distribution, and pairwise cliques are considered. For more information on the MAP-MRF, please refer to the paper by Geman and Geman [5].

In the present paper, $D_s(f_s)$ in Eq. 1 is referred to as the likelihood term derived from the modeled background, which reflects how each pixel fits into the modeled data given for each label, and $V_{s,s'}(f_s, f_{s'})$ is referred to as the a prior term that encourages spatial coherence by penalizing discontinuities between neighboring pixels s and s'. In addition, $V_{s,s'}(f_s, f_{s'})$ is replaced by $V_{s,s'} \cdot \delta(f_s, f_{s'})$, where $\delta(f_s, f_{s'})$ denotes the delta function defined by 1 if $f_s \neq f_{s'}$, and otherwise denotes the delta function defined by 0. Thus, this is a penalty term when two pixels are assigned different labels.

2.3 Graph Cuts

To minimize the energy function (Eq. 1), we use a graph-cuts method [8], because this method showed the best performance among the conventional energy minimization algorithms [9]. The procedure for energy minimization using the graph-cuts method includes building a graph, wherein each cut defines a single configuration, and the cost of a cut is equal to the energy of its corresponding configuration [9].

For the graph-cuts method, a graph $G = \langle v, \varepsilon \rangle$ is first constructed with vertices corresponding to the sites. Two vertices, namely, *source (Src)* and *sink (Sin)*, also referred to as terminals, are needed in order to represent two labels, and each vertex has two additional edges, $\{s, Src\}$ and $\{s, Sin\}$. Therefore, the sets of vertices v and edges ε are $v = \mathcal{S} \cup \{Src, Sin\}$ and $\varepsilon = N \cup_{s \in \mathcal{S}} \{\{s, Src\}, \{s, Sin\}\}$, where N are referred to as *n-links* (neighboring links) and $\{s, Src\}$ and $\{s, Sin\}$ are referred to as *t-links* (terminal links). The weights of the graph are set for both *n-links* and *t-links*, where the *t-links* connecting each terminal and each vertex correspond to the likelihood term and the *n-links* connecting neighboring vertices correspond to the prior term.

Note that the background subtraction problem can be solved by finding the least energy consuming configuration of the MRF among the possible assignments of the random variables **F**. Minimizing the energy function defined in Eq. 1 is equivalent to finding the cut with the lowest cost, because the costs of two terms are assigned to the weights of the graph. Specific labels are then assigned to two disjointed sets connected by *Src* and *Sin* by finding the cut with the lowest cost in the graph. The minimum-cost cut of the graph can be computed through a faster version of max-flow algorithm, proposed in [9]. The obtained configuration corresponds to the optimal estimate of $P(\mathbf{f}|obs)$.

3 Proposed Energy Function

3.1 Likelihood Term

The likelihood term is derived from the modeled background data to measure the cost of assigning the label f_p to the pixel p, and $D_p(f_p)$ is defined as follows:

$$
\begin{cases}
D_p(f_p = \text{foreground}) = 1, & D_p(f_p = \text{background}) = 0, & \text{if } P(p) < T_f, \\
D_p(f_p = \text{foreground}) = 0, & D_p(f_p = \text{background}) = 1, & \text{if } P(p) > T_b, \\
D_p(f_p = \text{foreground}) = T_b^p, & D_p(f_p = \text{background}) = P(p), & \text{otherwise,}
\end{cases}
$$

where T_f and T_b are thresholds for hard constraints [10] in constructing graphs, T_b^p is a threshold to extract moving objects from the background, and $P(p)$ is the probability that a pixel p is included in the background. In the present paper, the codebook-based algorithm and MoGs are used to estimate the probabilities for the background.

The codebook algorithm is used to construct a background model from long input sequences and adopts a quantization technique to minimize the required memory. For each pixel, the codebook algorithm builds a codebook consisting of one or more codewords. Samples at each pixel are quantized into a set of codewords based on color and brightness information. The background is then encoded on a pixel-by-pixel basis.

Fig. 2 shows algorithm to construct codebook. Let X be a training sequence for a single pixel consisting of n_x RGB-vectors: $X = \{x_1, ..., x_{n_x}\}$, and let C be the codebook for a pixel consisting of n_c codewords. Each pixel has a different codebook size based on its sample variation. Each codebook $c_i, i = 1, ..., n_c$ consists of an RGB vector $v_i = (\bar{R}_i, \bar{G}_i, \bar{B}_i)$ and a 7-tuple $\mathbf{aux}_i = \langle \check{I}_i, \hat{I}_i, \check{\mathbf{T}}_i, \hat{\mathbf{T}}_i, \tau_i, q_i, f_i \rangle$, where $\check{I}_i$ and $\hat{I}_i$ denote the minimum brightness and maximum brightness, respectively, of the ith codeword, $\check{\mathbf{T}}_i$ and $\hat{\mathbf{T}}_i$ denote the thresholds for the RGB vector v_i, τ_i denotes the maximum negative run-length (MNRL), which is defined as the longest interval during the training period in which the codeword did not recur, q_i denotes the last access time at which the codeword occurred, and f_i is the frequency with which the codeword occurs.

1	$n_c \leftarrow 0, C \leftarrow \emptyset$ (empty set)
2	**For** $t = 1$ to n_x do
2.1	$x_t = (R, G, B), I \leftarrow \sqrt{R^2 + G^2 + B^2}$
2.2	Find the codeword c_m in $C = \{c_i | 1 \leq i \leq n_c\}$ that matches x_t based on two conditions $\check{\mathbf{T}}_i \leq x_t \leq \hat{\mathbf{T}}_i$ and $\check{I}_i \leq I \leq \hat{I}_i$
2.3	If $C = \emptyset$, or, if there is no match, then $n_c = n_c + 1$. Create a new codeword c_{n_c} by setting $v_{n_c} \leftarrow (R, G, B)$ and $\mathbf{aux}_{n_c} = \langle I - t_I, I + t_I, v_{n_c} - t_v, v_{n_c} + t_v, n_c - 1, n_c, 1 \rangle$
2.4	Otherwise, update the matched codeword c_m, consisting of $v_m = (\bar{R}_m, \bar{G}_m, \bar{B}_m)$ and $\mathbf{aux}_m = \langle \check{I}_m, \hat{I}_m, \check{\mathbf{T}}_m, \hat{\mathbf{T}}_m, \tau_m, q_m \rangle$, by setting $$v_m \leftarrow \left(\frac{f_m \bar{R}_m + R}{f_m + 1}, \frac{f_m \bar{G}_m + G}{f_m + 1}, \frac{f_m \bar{B}_m + B}{f_m + 1} \right) \text{ and }$$ $\mathbf{aux}_m = \langle \min\{I, \check{I}_m\}, \max\{I, \hat{I}_m\}, \min\{v_m, \check{\mathbf{T}}_m\}, \max\{v_m, \hat{\mathbf{T}}_m\}, \max\{\tau_m, t - q_m\}, t, f_m + 1 \rangle$
	end for

Fig. 2. Algorithm for codebook construction

After construction, the codebook may be sizeable because it contains all of the codewords that may include moving foreground objects and noise. Therefore, the codebook is refined by eliminating the codewords that contain moving foreground objects. The MNRL in the codebook is used to eliminate the codewords that include moving objects, based on the assumption that pixels of moving foreground objects appear less frequently than moving backgrounds. Thus, codewords having a large τ are eliminated by the following equation: $\mathbb{C} = \{c_m | c_m \in C \wedge \tau_m \leq T_\mathbb{C}\}$, where $\mathbb{C}$ denotes the background model, which is a refined codebook, and $T_\mathbb{C}$ denotes the threshold value. In the experiments, $T_\mathbb{C}$ was set to be equal to half the number of training frames.

In the case of using codebook-based algorithms, it is difficult to use an MRF because the MRF does not evaluate probabilities, but rather calculates the distance from the RGB vectors and the brightness of the codewords.

To evaluate the probabilities from the codebooks, a mixture of K Gaussian distributions proposed by Stauffer and Grimson [2] is chosen to model the recent history of each pixel, which is included in the same codewords. The probability of observing the current pixel value $\mathbf{x_t}$ is $P(\mathbf{x_t}) = \sum_{i=1}^{K} w_{i,t} * \eta(\mathbf{x_t}, \mathbf{\mu}_{i,t}, \mathbf{\Sigma}_{i,t})$, where K is the number of distributions, $\mathbf{\mu}_{i,t}$ is an estimate of the weight of the ith Gaussian in the mixture at time t, $\mathbf{\mu}_{i,t}$ and $\mathbf{\Sigma}_{i,t}$ are the mean value and covariance matrix, respectively, of the ith Gaussian in the mixture at time t, and η is a Gaussian probability density function. In the experiments, K is determined by the number of frames used for background modeling, and the covariance matrix is assumed to be of the following form: $\mathbf{\Sigma}_{k,t} = \sigma_k^2 \mathbf{I}$.

3.2 Prior Term

Since a common constraint is that the labels should vary smoothly almost everywhere while preserving sharp discontinuities that may exist, e.g., at boundaries [8], the costs of the smoothness are assigned for discontinuity-preservation between two neighboring pixels, and we use a generalized Potts model [8]. As such, $V_{p,p'}$ is defined as follows:

$$V_{p,p'} = dis(p,p')^{-1} e^{(-\beta \cdot \|p-p'\|^2)}, \tag{2}$$

where the contrast term $\|p - p'\|^2$ denotes the dissimilarity between two pixels p and p', and $dis(\cdot)$ is the Euclidean distance between neighboring pixels in the image domain. When $\beta = 0$, the smoothness term is simply the Ising model, which promotes smoothness everywhere. However, it has been shown that it is more effective to set $\beta > 0$, because this relaxes the tendency to smooth regions of high contrast. The constant β is chosen to be $\beta = (\langle \|p - p'\|^2 \rangle)^{-1}$, where $\langle \cdot \rangle$ denotes the expectation over an image. This choice of β ensures that the exponential term in Eq. 2 switches appropriately between high and low constants.

However, when the scene contains a cluttered background, notable segmentation errors often occur around the boundary, which generates flickering artifacts in the final results displayed in the virtual world [6]. These errors occur because the MRF model contains two terms for color and two terms for contrast. A straightforward idea is to subtract the contrast of the background image from the current image [6]. However, since only one background image is used for this approach, the nonstationary background motion that is ubiquitous in the real world cannot be modeled.

To overcome this problem, the contrast of the background is modeled using the codebook-based algorithm described in Section 3.1. The difference is that the codebook for the smoothness terms uses $V_{p,p'}$ instead of I as input and does not use $\mathbf{x_t}$. This means modeling contrasts between adjacent pixels. After modeling the contrasts, if the contrasts of the input frame are within the ranges $\check{V}_m$ and $\hat{V}_m$ of any codeword m, then the contrast is considered to be background contrast, and $V_{s,s'}$ is set 0. Otherwise, $V_{s,s'}$ is set as the value of an input frame. This approach helps not only to eliminate the flickering artifacts but also facilitates the use of the generalized Potts model.

4   Experimental Results

Background subtraction was used to generate a photorealistic avatar in the virtual world for the WTS. Section 4.1 presents the resultant images displayed in the WTS, and Section 4.2 presents a quantitative evaluation to verify the effectiveness of the proposed method. All of the experiments were carried out on a 2.40-GHz Pentium 4 CPU.

4.1   Simulated Environment

The proposed method was based on the codebook-based method [7]. Images resulting from use of the codebook-based method (Fig. 3(a)) and the proposed method are shown in Fig. 3(b) and (b), respectively. As shown in Fig. 3, the results of the code-book-based method include some noise and holes in extracted regions. However, by applying an edge-preserving MRF framework, the proposed method includes no noise or holes and has clean boundaries. The photorealistic avatar, based on the resultant images presented in Fig. 3(b), was integrated into the WTS as shown in Fig. 4. In addition, MoG-based [2] and codebook-based [7] methods were compared to the proposed method, as shown in Fig. 5.

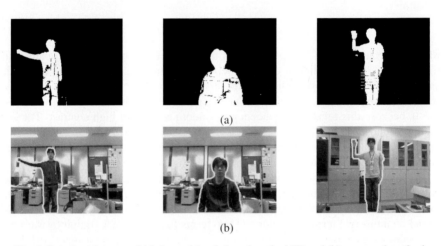

(a)

(b)

Fig. 3. Resultant images of (a) the codebook-based method [7] and (b) proposed method

Fig. 4. Photorealistic avatar integrated into a virtual building

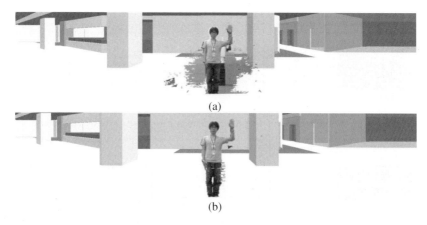

(a)

(b)

Fig. 5. Guide representation in a virtual building using (a) MoG- [2] and (c) codebook-based [7] methods

4.2 Qualitative Analysis

We tested four data sets described in [3]: Waving Trees, Camouflage, Time of Day, and Moved Object. We chose these four sets because the background images to be modeled might include nonstationary background motion, as in the Waving Trees and Camouflage sets, and because the sequential background images might change gradually as a result of changing light conditions throughout the day, as in the case of the Time of Day set. In the present study, in contrast to [3], moving objects are considered to be in the foreground, because the photorealistic avatar can use objects to express information to the subject. In the experiments, codebook-based [7] and MoG-based [2] results were compared with the results of the proposed method, and the results of these tests are shown in Figs. 6–9.

(a)   (b)   (c)   (d)

Fig. 6. Sample results for Waving Trees obtained using (a) the MoG-based method, (b) the graph-cuts method using MoG, (c) the codebook-based method, and (d) the proposed method

(a)   (b)   (c)   (d)

Fig. 7. Sample results for Camouflage obtained using (a) the MoG-based method, (b) the graph-cuts method using MoG, (c) the codebook-based method, and (d) the proposed method

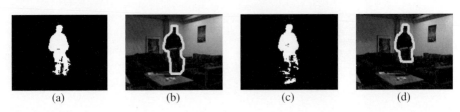

Fig. 8. Sample results for Time of Day obtained using (a) the MoG-based method, (b) the graph-cuts method using MoG, (c) the codebook-based method, and (d) the proposed method

Fig. 9. Sample results for Moved Object obtained using (a) the MoG-based method, (b) the graph-cuts method using MoG, (c) the codebook-based method, and (d) the proposed method

The accuracy rates were evaluated by two criteria: the number of false positives and the number of false negatives. The number of false positives is the number of foreground pixels that were misidentified, and the number of false negatives is the number of background pixels that were identified as foreground pixels. As shown in Table 1, the proposed method had the best performance, except in the case of the Time of Day data set, as shown in Fig. 8. Since brightness values were used to deal with shadows, the proposed approach worked poorly in dark areas of images. Therefore, the leg regions of the human were not extracted by the proposed method. On the other hand, since shadow regions are included in the results shown in Fig. 7, the proposed method had better performance than the MoG-based methods. The processing times for each step of the proposed method are presented in Table 2. Approximately nine frames per second could be extracted using the proposed method.

Table 1. False positives and false negatives (%)

		MoG [3]	Codebook [7]	MoG + graph cuts	Proposed method
Fig. 6	F. Positive	6.89	3.17	3.08	0.02
	F. Negative	2.43	2.55	0.18	0.04
Fig. 7	F. Positive	16.13	9.11	5.88	0.0
	F. Negative	38.09	1.15	19.27	0.93
Fig. 8	F. Positive	5.21	9.80	2.18	11.2
	F. Negative	1.38	0.54	0.91	0.09
Fig. 9	F. Positive	11.10	12.34	5.11	1.57
	F. Negative	9.75	8.27	9.33	8.12

Table 2. Processing times for each step of the proposed method (msec)

Resolution	Codebook construction	MoG	Graph construction	Graph cuts
160×120	8	5	7	25
320×240	20	10	20	60

5 Conclusions

In the present paper, we proposed a codebook-based MRF model for background subtraction to generate a photorealistic avatar displayed in the virtual world. Although an edge-preserving MRF can eliminate the noise and generate suitable object boundaries, the MRF depends on how the likelihood terms in the energy function are estimated. The proposed method uses a codebook-based method to estimate the likelihood term, which not only reduces the required memory and enables real-time implementation. Moreover, the proposed method used online MoG to estimate the probability for each codeword, which resulted in minimization of the required memory, reduced noise, and clean boundaries. In addition, the proposed method enabled the photorealistic avatar to be displayed clearly in the virtual world, as compared with previously proposed methods, such as codebook and MoG.

However, the proposed method was not able to extract the foreground in dark regions because brightness values were used to handle shadows. Therefore, in future studies, we intend to investigate how to extract the foreground in the dark regions more effectively. Moreover, we intend to extend the proposed method to extract the foreground inside a WTS that contains non-static backgrounds due to virtual world displayed inside the WTS.

Acknowledgement. This research was partially supported by the Ministry of Economy, Trade and Industry (METI), and partially supported by the Japan Society for the Promotion of Science (JSPS).

References

1. Wren, C., Azarbayejani, A., Darrell, T., Pentland, A.: Pfinder: Real-Time Tracking of the Human Body. IEEE Transactions on Pattern Analysis and Machine Intelligence 23(7), 780–785 (1997)
2. Stauffer, C., Grimson, W.: Learning Patterns of Activity using Real-Time Tracking. IEEE Transactions on Pattern Analysis and Machine Intelligence 22(8), 747–757 (2000)
3. Toyama, K., Krumm, J., Brumitt, B., Meyers, B.: Wallflower: Principles and Practice of Background Maintenance. In: Proceedings of International Conference on Computer Vision, pp. 255–261 (1999)
4. Sheikh, Y., Shah, M.: Bayesian Modeling of Dynamic Scenes for Object Detection, IEEE Transactions on Object Detection. IEEE Transactions on Pattern Analysis and Machine Intelligence 27(11), 1778–1792 (2005)
5. Geman, S., Geman, D.: Stochastic Relaxation, Gibbs Distributions, and the Bayesian Restoration of Image. IEEE Transactions on Pattern Analysis and Machine Intelligence 6(6), 721–741 (1984)

6. Sun, J., Zhang, W., Tang, X., Shum, H.-Y.: Background Cut. In: Proceedings of European Conference on Computer Vision, Part II, pp. 628–641 (2006)
7. Kim, K., Chalidabhongse, T.H., Harwood, D., Davis, L.: Real-Time Foreground-Background Segmentation using Codebook Model. Real-Time Imaging 11(3), 172–185 (2005)
8. Boykov, Y., Veksler, O., Zabih, R.: Fast Approximation Energy Minimization via Graph Cuts. IEEE Transactions on Pattern Analysis and Machine Intelligence 23(11), 1222–1239 (2001)
9. Szeliski, R., Zabih, R., Scharstein, D., Veksler, O., Kolmogorov, V., Agarwala, A., Tappen, M., Rother, C.: A Comparative Study of Energy Minimization Methods for Markov Random Fields. IEEE Transactions on Pattern Analysis and Machine Intelligence 30(6), 1068–1080 (2008)
10. Boykov, Y., Kolmogorov, V.: An Experimental Comparison of Min-Cut/Max-Flow Algorithms for Energy Minimization in Vision. IEEE Transactions on Pattern Analysis and Machine Intelligence 26(9), 1124–1137 (2004)
11. Li, S.Z.: Markov Random Field Modeling in Computer Vision. Springer, Heidelberg (2001)

JanusVF: Adaptive Fiducial Selection

Malcolm Hutson and Dirk Reiners

University of Louisiana at Lafayette
malcolm@louisiana.edu,
dirk@louisiana.edu

Abstract. Critical limitations exist in the currently available commercial tracking technologies for fully-enclosed virtual reality (VR) systems. While several 6DOF solutions can be adapted for fully-enclosed spaces, they still include elements of hardware that can interfere with the users visual experience. JanusVF introduced a tracking solution for fully-enclosed VR displays that achieves comparable performance to available commercial solutions but without artifacts that obscure the users view. In this paper we extend JanusVF by exploring two new methods for selecting the fiducials to be displayed. The first method creates an additional offset grid of fiducials for each original resolution. The algorithm selects fiducials from the grid that yields more visible markers. The second method selects multiple resolutions of markers per wall by intersecting the cameras view frustum with the display surface. The resulting area is populated with fiducials by a quadtree subdivision. Comparison results show that the positional error has been reduced.

1 Introduction

A tracking system for a fully-enclosed space, such as a six-sided CAVE [1,2], should have as minor of an impact on the user's experience as possible: Sensor hardware inside the VR space can obscure the user's view of the display surface, which could lead to a break in presence. In a head-mounted display, this requirement is totally non-existent as no tracking device will ever obscure the user's view of his display because it is fixed to his head. In a fish tank or wall display, there are multiple options for placement of the tracking hardware so that they are totally out of the user's view such as behind or above the user. Often, the fixed component of the tracking device can simply be placed above or to the side of the display surface. These assumptions regarding where a user cannot or will not look are not possible in a spatially immersive display. By definition, the device allows the user to look in any direction.

1.1 Previous Work

Unfortunately, most common commercial tracking solutions do not conform to these stricter requirements, requiring a portion of the hardware to remain visible. Acoustic solutions such as the InterSense IS-900 require that the ultrasonic

G. Bebis et al. (Eds.): ISVC 2009, Part I, LNCS 5875, pp. 511–520, 2009.

emitters be placed inside the display surfaces because they cannot permeate the screen surfaces. While options exist for low-profile surface mounted emitters instead of the traditional sensor bars, these devices can still be identified by their shadow on a display surface.

Magnetic tracking solutions such as the Polyhemus FastTrack or Ascension MotionStar are often chosen for fully enclosed VR spaces because the world-fixed magnetic transmitter can be mounted outside the display walls. If the entire virtual environment is made of non-metallic materials, magnetic trackers can be reasonably accurate up to a moderate distance. With extensive calibration, the tracker can compensate for static noise and distortions in the magnetic fields, but transient changes from devices like CRT monitors or the movement of dense objects near the tracker can have erratic effects.

In order to implement standard optical feature tracking [3] or IR tracking [4,5], the camera hardware must be placed within the VR space so as to fully observe the user's motions. These hardware artifacts make these types of systems far less attractive options for use in fully-enclosed VR spaces.

Another recent optical tracking system is the HiBall [6]. It is an inside-looking-out design with multiple cameras that track LED arrays mounted on the ceiling. This design is obviously not conducive to use inside a fully-enclosed space where the LED arrays would obscure the view of the ceiling.

The Hedgehog [7] project presents a novel tracking system expressly designed for fully enclosed virtual reality spaces. A head mounted device projects calibrated laser beams in a specific pattern onto the interior of the projection screens. Cameras mounted outside the projection space view the resulting dots on the display walls. The algorithm can filter and triangulate the user's position and orientation. However, it is possible for the user to see the laser dots on the display surface. This could lead to a break in presence by highlighting the true display surface.

2   JanusVF

JanusVF [8] has successfully been demonstrated as a tracking system for fully enclosed, active stereo spaces that satisfies the original goals of Hedgehog but while showing no visible artifacts within the view of the user. The system determines the head position and orientation by optically tracking fiducial markers. Images captured by a head-worn camera are processed by ARToolkitPlus [9] and filtered by a SCAAT [10] filter. Our system exploits the fact that the virtual reality space, if considered by the camera to be the world space, can be directly manipulated. Thus, the virtual fiducial markers can be made to only appear directly behind the user's field of view as part of the rendered scene (see Figure 1). Unlike most tracking systems, JanusVF is a low-cost tracking solution involving the cooperation of the display system with the tracking hardware.

Fig. 1. Example of how fiducials are displayed in the VR space by the original JanusVF. Worn on the user's head, the camera faces exactly opposite the the user's point of view to track fiducials that are displayed on the wall behind the user.

3   Fiducial Selection

3.1   Previous Work

In order that there be no synchronization required between the fiducial drawing and fiducial recognition components of the JanusVF solution, it was necessary that the fiducials be statically positioned in space. If markers were not statically assigned their translations in world space, it would be necessary for the drawing application to constantly inform the recognition application about each assignment as they changed, placing unnecessary requirements on our commodity system.

In the original implementation of JanusVF, each display surface was automatically pre-allocated several grids of markers. Each static grid had a resolution of between 3 to 10 fiducials per edge. During the run-time of the application before drawing each frame, the optimum resolution and fiducial size was chosen based on the camera's normal distance to the display surface. The algorithm attempted to maximize the size of the marker within the camera's view without clipping any corners. The selection thresholds defining the optimum grid size for distance were determined empirically by testing the stability of the tracking algorithm for each grid size at several distances. For these tests, the camera was oriented perpendicularly to the display surface.

Several limitations of this marker allocation and selection method were discovered during use of the JanusVF system. To overcome these limitations, we have developed two alternative marker selection routines. Both new implementations maintain our requirement that the fiducial layout remain static and pre-allocated. Thus, there is no need for the display software to communicate with the tracker about the changing locations of fiducials, nor does there need to be any synchronization as there would be if fiducials were not unique.

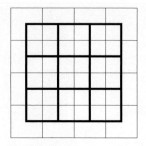

Fig. 2. Example diagram of a 4 x 4 grid (thin stroke), overlaid with its corresponding offset grid (bold stroke). The offset grid has one less marker per edge, but each marker remains the same size.

3.2 Offset Grids

The range of sizes of markers created during the pre-allocation phase of the original implementation was chosen such that the largest marker fills the camera's view when the camera is positioned at a maximum practical distance from the display surface, and similarly that the smallest marker fills the camera's view when it is at the shortest practical distance from the display surface. However, each grid is laid out from the origin, such that it fully populates a square display surface. Thus, for any distance and associated optimum marker size, it may be a common occurrence that the camera is positioned such that even though the chosen size of marker is reasonable, the camera is posed such that it views only an intersection between four markers and therefore registers none of them. In the original implementation, we compensated for this limitation by displaying fiducials that were smaller than optimal to improve our probability of registering full markers in every observation.

Our improved offset grid algorithm generates two set of grids. The first set is inherited from the original generation algorithm: A grid of markers that fully populates the square display surface from edge to edge is created for each resolution from 3 x 3 to 10 x 10. In a second pass, the offset algorithm adds a pair to each member of the first set where the size of each marker is held constant but the count of markers per edge is reduced by one. The grid is positioned in the center of the display surface, such that the new offset grid has a marker placed upon each four-way intersection of the larger grid that it is paired with. Figure 2 shows a conceptual example of how a 4 x 4 grid and its offset would look if they were overlaid.

To take advantage of the offset grid, the JanusVF Graphics Hooks that run alongside the user's VR application were extended to make two fiducial selection passes. After determining the optimum marker size based on the camera's distance to the display surface, the algorithm first checks each fiducial within the standard grid of the chosen grid size. Then, it checks each fiducial in the standard grid's complementing offset grid. Because the two grids potentially conflict with each other by overlapping, results can only be accepted from one grid per frame. The algorithm determines which grid has the most area of visible fiducials by comparing the count of visible markers in each grid. It accepts the grid with the highest number of visible markers.

In situations where the user pauses upon a border between where an offset or the original grid may become optimal, some flickering between states may occur. Filtering of the choice of grids by averaging previous choices reduces this behavior, but it does not appear to be necessary.

3.3   Quadtree Subdivision

While the original algorithm chooses logical sizes of markers when the camera is oriented perpendicularly to the display surface, the selections become increasingly more inappropriate as the angle of inclination decreases. When perpendicular to the surface, the camera's distance to each fiducial is relatively similar, so it is reasonable to select markers that are all of the same size. However, at lower angles of inclination, the camera's depth of field upon the screen becomes larger. For these cases, recognition could be improved if the system could select multiple resolutions of markers and display them simultaneously. In the near field, markers would be smaller than in the far field.

Grundhofer et al. [11] explored the use of fiducials for surface tracking. Their primary interests focused on the imperceptibility of markers that are within view and required interleaving frames containing fiducials between the normally visible frames. Synchronized cameras were required to separate the visible and fiducial frames. Unfortunately, the active stereo in a CAVE display already demands the full frame rate of standard projectors, and the existing keying of the stereo frames is tightly coupled with the active stereo glasses. We have however adapted their use of quadtrees for populating markers within the camera's view.

The fiducial pre-allocation phase is modified to generate marker resolutions of 2×2, 4×4, 8×8, 16×16, and 32×32. A quadtree map is built, linking each marker to the four smaller markers that occupy its space in the next resolution.

During the selection phase, a view frustum of the camera is calculated based on current prediction of the camera's pose and parameters of the camera's lens. Effectively, an intersection is computed between this view frustum and the display surface, and the area is subdivided by a quadtree.

In our implementation, the algorithm first tests if all four corners in world space of each of the four large markers in the 2×2 grid are within volume that defines the camera's view frustum. If a marker has all four corners visible, it is selected. If only a subset of corners are visible, the algorithm recursively tests the children of the current marker. If a marker has some corners visible but has no children, or if no corners are visible, the recursion stops.

This selection routine yields a set of fiducials that are adaptively fit to the camera's perspective. Figure 3 shows two examples of how the quadtree selected markers appear when rendered. This method allows JanusVF to select multiple resolutions of markers per wall from a statically allocated set of markers while guaranteeing that the selected markers will not overlap each other. When the camera views the surface at low angles of inclination, the viewable surface area

(a) High Angle   (b) Low Angle

Fig. 3. Two example photographs from a third person perspective of how a set of markers appear on the display surface after the camera's view frustum has been populated with fiducials by quadtree subdivision. In the first, the camera is nearly orthogonal to the surface but rotated along the Z axis. In the second, the camera is posed such that it has a low angle of inclination to the display surface.

is populated such that each marker appears at an optimum size on the camera's image plane: Nearer markers are drawn smaller and farther markers drawn larger on the display surface.

4 Testing

Testing was performed in a 6-sided CAVE that measures 3 meters on each edge and is driven by a six node cluster running VRJuggler on Linux. Each node is equipped a dual-core 2.0 GHz AMD Opteron, 2 gigabytes of RAM, and an NVIDIA Quadro 4500. It is outfitted with an InterSense IS-900 hybrid acoustic-inertial tracker. Four transmitters per edge are embedded in the edges of the space where the ceiling meets the side walls.

It was hypothesized that original selection algorithm would perform best when the camera views the display surface from an orthogonal angle because it was at this angle that the original selection algorithm was optimized. As the angle of inclination to the surface is reduced, we expected that the stability of the original algorithm would decline and that the improved selection algorithms would remain more stable.

4.1 Comparison Tests

During these comparison tests, our JanusVF tracker was rigidly attached to an Intersense IS-900 head tracker. A single transform was added to the JanusVF output so that its eye pose aligned with that of the IS-900. The IS-900 was used in tethered mode to remove possible latencies incurred in its native wireless communication. A logging application was written on an independent machine to poll and record the outputs of the VRPN servers of both trackers simultaneously.

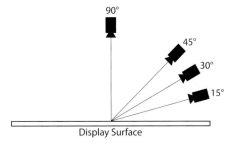

Fig. 4. Tracker placement for the four rotational accuracy tests varying the angle of inclination while maintaining distance to the focus point

4.2 Viewing Angle Tests

The accuracy of the JanusVF tracker as compared to the Intersense IS-900 tracker was measured by statically orienting the tracker pair by using the IS-900 as a reference. At each position chosen, we measured the difference between the reported orientation of JanusVF with the precisely placed IS-900. The first orientation test was taken from the true center of the space, facing the front wall with no pitch, roll, or yaw. This pose resembles the poses used during the optimization of the original JanusVF fiducial selection routine.

At three more positions, we varied the camera's angle of incidence to the display surface. As noted from experience, the fiducial analysis is often weak at discerning rotations around a point on the surface from a strafe along the surface. As the angle upon a fiducial grows smaller, camera pixel error increases as each pixel accounts for greater distances along the fiducial's edges.

Each position maintained the distance from the camera to its center of focus on the screen at 1.5 meters, varying only in the angle of inclination. The test was conducted at the angles of 90, 45, 30, and 15 degrees. The state acquisition was reset between each position. The trackers were placed into position before being enabled so that knowledge of a previous state would not affect their ability to resolve the new state. Figure 4 shows a plot of the locations chosen.

The entire test was conducted once each for the original fiducial selection algorithm, the offset grid algorithm, and the quadtree subdivision algorithm. Figure 5 shows how the camera's view appeared at a low angle of inclination for each of the selection methods.

Table 1 shows the results of the viewing angle experiments. As anticipated, the accuracy of the original algorithm degrades quickly as the angle of inclination becomes lower. The offset selection algorithm provided no significant improvement over the original selection algorithm. While in some poses the offset grid was preferred over the basic grid, it provided no means to compensate for the stretched depth of field upon the markers. In contrast, the quadtree selection showed a pronounced strength through the low angles. Where in these poses the previous two algorithms struggled to fuse noisy data from suboptimal markers, the camera's view in the quadtree selection test was populated such that more

Table 1. The RMS delta between the IS-900 tracker the JanusVF tracker in their reported position for each of the marker selection methods at each of the angles described in Figure 4

Angle (deg)	Delta (mm RMS)		
	Original	Offset	Quadtree
90	0.488	0.465	0.477
45	4.932	5.180	0.882
30	8.148	8.051	2.452
15	20.327	19.258	5.233

markers were optimally sized. Performance did decrease as the angle became lower, but the magnitude of the error was much smaller.

4.3   Free Motion Tests

To gauge the average accuracy of the JanusVF tracker with each of the proposed selection algorithms, a subject navigated through a model of an interior space while wearing both trackers. Over the 5 minute trial, we recorded the outputs of both tracking systems. We then calculated the error in the positions reported by the InterSense and all three versions of JanusVF. To accomplish this, the test was run three separate times by the same user, intentionally trying to reproduce the same motions each time. This is obviously no longer an exact comparison, but over the length of the trial, we expect that individual differences will be smoothed by average the results over the course of the entire run. Note that the InterSense in this case is not necessarily ground truth, as its setup with only four transmitters in the ceiling edges is not optimal. While the IS-900's output is consistently precise, its output is often inaccurate when measured manually. These slight variations, possibly due to echos or suboptimal transmitter placement, are not overtly noticeable during normal use, but they do account for some of the statistical error during our comparison tests.

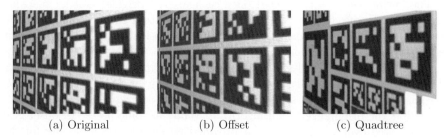

(a) Original   (b) Offset   (c) Quadtree

Fig. 5. The JanusVF camera's view of the display surface when placed at a low angle of inclination for each of the selection and display methods. In each view, the near field markers are within an optimum range for the recognition algorithm. In the original 5(a) and offset 5(b) selection methods, the distant markers are too small.

Table 2. The RMS delta between the IS-900 tracker the JanusVF tracker in their reported position for each of the marker selection methods during a free motion test

	Delta (cm RMS)
Original	2.446
Offset	2.163
Quadtree	1.108

Results for this test are meant to be representative of a typical use, but they cannot be considered as an absolute performance metric. Table 2 shows the results of the free motions tests as the RMS delta between the IS-900 tracker's positions and the JanusVF tracker's position for each of the selection methods.

The results of testing the original JanusVF algorithms were consistent with those reported in the first JanusVF publication. During the free motion tests, we observed an RMS delta between the original JanusVF tracker and the IS-900 of 2.446 cm.

When the offset selection algorithm was employed, we noticed a negligible improvement in the positional resolution. The result suggests that the condition where the original algorithm would select a probable resolution under which the camera could not view a single complete fiducial occurs infrequently or is only a small contributor to the error in the system.

The quadtree selection algorithm showed a significant improvement over the original selection algorithm, reducing the dispersion by over half. While the offset algorithm only offers improvements in special cases, the quadtree selection algorithm allows JanusVF to select multiple resolutions of fiducials per surface, potentially improving every camera view.

5   Conclusion

The advantages provided by the offset grid fiducial selection algorithm were limited in scope. This very simple two-pass addition does resolve a known problem where the original algorithm can select properly sized fiducials that may happen to be arranged such that the camera is oriented towards an intersection of four corners and thus cannot resolve any complete fiducial. As evidenced by the static tests, it does have some positive effect, but its benefits are seen only in a small percentage of cases.

The quadtree algorithm differs significantly from the original and offset grid selection methods because it is capable of selecting multiple sizes of markers per wall, quickly and efficiently. By populating the camera's view with markers that have been selected by a quadtree subdivision of the viewable surface area, we place markers that are optimally large, especially in the case of low angles. In these cases where the depth of field is very long, the algorithm, by nature of the quadtree subdivision, places larger markers in the distant field and smaller markers in the near field. Whereas the previous two selection algorithms only consider the camera's normal distance to the surface, the quadtree selection algorithm considers and adapts to the camera's perspective upon the surface. This adds a great deal of robustness to the fiducial recognition as shown in the static rotation tests. This ability also translated to a reduction in the magnitude of the error during the free motion tests, suggesting that this algorithm provides a benefit in a significant percentage of situations.

Acknowledgements

The authors wish to thank Dr. Carolina Cruz-Neira and her team at LITE for the use of their facilities to carry out this research. The Cathedral model in Figure 1 was used courtesy of Fraunhofer IGD.

References

1. Cruz-Neira, C., Sandin, D.J., DeFanti, T.A., Kenyon, R.V., Hart, J.C.: The CAVE: audio visual experience automatic virtual environment. Commun. ACM 35, 64–72 (1992)
2. Robinson, M., Laurence, J., Zacher, J., Hogue, A., Allison, R., Harris, L., Jenkin, M., Stuerzlinger, W.: IVY: The Immersive Visual environment at York (2002)
3. Jiang, B., You, S., Neumann, U.: Camera tracking for augmented reality media. In: IEEE International Conference on Multimedia and Expo, 2000. ICME 2000, vol. 3, pp. 1637–1640 (2000)
4. Rasmussen, N.T., Strring, M., Moeslund, T.B., Granum, E.: Real-time tracking for virtual environments using SCAAT Kalman filtering and unsynchronized cameras. In: VISAPP 2006: International Conference on Computer Vision Theory and Applications (2006)
5. Foursa, M.: Real-time infrared tracking system for virtual environments. In: VR-CAI 2004: Proceedings of the 2004 ACM SIGGRAPH international conference on Virtual Reality continuum and its applications in industry, pp. 427–430. ACM, New York (2004)
6. Welch, G., Bishop, G., Vicci, L., Brumback, S., Keller, K., Colucci, D.: High-Performance Wide-Area Optical Tracking: The HiBall Tracking System. Presence: Teleoper. Virtual Environ. 10, 1–21 (2001)
7. Vorozcovs, A., Hogue, A., Stuerzlinger, W.: The Hedgehog: a novel optical tracking method for spatially immersive displays. In: Proceedings of Virtual Reality, 2005. VR 2005, pp. 83–89. IEEE, Los Alamitos (2005)
8. Hutson, M., White, S., Reiners, D.: JanusVF: Accurate Navigation Using SCAAT and Virtual Fiducials. In: Virtual Reality Conference, 2009. VR 2009, pp. 43–50. IEEE, Los Alamitos (2009)
9. Wagner, D., Schmalstieg, D.: ARToolKitPlus for Pose Tracking on Mobile Devices. In: CVWW 2007: Proceedings of 12th Computer Vision Winter Workshop (2007)
10. Welch, G., Bishop, G.: SCAAT: incremental tracking with incomplete information. In: SIGGRAPH 1997: Proceedings of the 24th annual conference on Computer graphics and interactive techniques, pp. 333–344. ACM Press/Addison-Wesley Publishing Co., New York (1997)
11. Grundhofer, A., Seeger, M., Hantsch, F., Bimber, O.: Dynamic adaptation of projected imperceptible codes. In: 6th IEEE and ACM International Symposium on Mixed and Augmented Reality, 2007. ISMAR 2007, pp. 181–190 (2007)

Natural Pose Generation from a Reduced Dimension Motion Capture Data Space

Reza Ferrydiansyah and Charles B. Owen

Media and Entertainment Technologies Laboratory
Computer Science Department
Michigan State University
{ferrydia,cbowen}@msu.edu

Abstract. Human animation from motion capture data is typically limited to whatever movement was performed by the actor. A method to create a wider range of motion in the animation utilizes the motion capture database to synthesize new poses. This paper proposes a method to generate original natural poses based on the characteristics of natural poses based on motion capture data. Principal Component Analysis is used to transform the data into a reduced dimensional space. An unconstrained pose data set is created by calculating the position of the human skeleton based on the reduced dimensional space. Constrained pose data can be created using interpolation and iteration on the unconstrained pose data. We show some example results of the generated poses and compare these poses to poses created with iterative inverse kinematics methods. Results show that our method is more accurate and more natural than iterative inverse kinematics methods.

1 Introduction

Graphical objects representing humans should appear life-like in both physical and dynamic appearance. In particular, they should move in ways that appear natural. Animation using kinematic methods finds a possible solution to a motion by calculating the correct angles for each bone that achieves desired constraints. Motion capture data is based on motion captured from real human movement [1].

Using motion capture data, a new animation is created by merging and blending existing frames that create smooth motion ([2], [3], and [4]). Motion capture can determine joint angles from real humans both statically and dynamically, but is only useful if data exists for exactly the pose desired or the application can wait for appropriate data to be acquired. Capturing all movement with all the exact poses needed for every possible animation is impossible. New natural poses that are vastly different from original poses can be used as a basis for animation through keyframing or optimization such as utilized by Abe et al. [5].

In many animation applications, especially interactive ones such as games, the problem is to determine joint angles that place a particular body part at a specified location. Inverse kinematics (IK) is the numerical solution that determines joint angles that achieve the proper positioning of an end effector in space [6]. IK with a very small number of degrees of freedom can be solved algebraically. But, in most cases, a

G. Bebis et al. (Eds.): ISVC 2009, Part I, LNCS 5875, pp. 521–530, 2009.
© Springer-Verlag Berlin Heidelberg 2009

numerical method is needed to solve an inverse kinematics problem. These numerical methods tend to use iterative approaches that converge to the target but are not sensitive to a naturalistic appearance.

We propose a method that uses statistical characteristics of the captured data to generate new poses which appear natural. Both unconstrained and constrained poses are generated. The first step is to create a set of *unconstrained poses* which are natural and novel for use in an animation. Principal Component Analysis (PCA) is used to create a reduced dimensionality space capturing the characteristics of natural data. This new *pose space* is used to generate new pose data. Our method to create *constrained poses* uses interpolation of existing novel unconstrained poses. Iterative inverse kinematics is then used to refine the pose to meet the constraint.

2   Related Work

Often used for creating key frame poses, Inverse Kinematics (See [6] for details) is a technique for computing joint angles in order to place a particular body part at a particular position. IK requires the use of iterative methods such as the popular iterative Jacobian method or Cyclic Coordinate Descent. A variety of methods are discussed in Fedor [7].

Meredith and Maddock [8] show how to weigh the effect of inverse kinematics on different joints at every step of the animation. Their method is mainly used for retargeting motions to other characters. Komura [9] computes weights from motion capture data based on the position and direction of movements in the database.

Capturing human data and using that data directly is a good solution to finding natural poses for animation, particularly when animation requirements are known well in advance. However, it is impossible to capture the entire range of human motion. Thus animation of general human motion must be synthesized based on data. A simple method to do this is by joining frames together from different captured sequence ([2], [3], and [4]).

The creation of new poses from motion data has been proposed by various authors. Abe et al. creates a limited number of poses which have the same physical requirements using transformation (by rotation or translation) of motion capture data [5]. Grochow et al [10] compute poses via inverse kinematic methods based on data learned from captured motion. They calculate the likelihood of a particular pose using a probabilistic model. New poses are synthesized using an optimization algorithm in which the objective function depends on the learned poses. Similarly, in Yamane [11], inverse kinematics calculation of joint angles is performed using a constrained optimization algorithm. Captured data is stored and used as soft constraints. Motion is created by the smoothing of various results of IK computations over multiple positions.

PCA has been used in conjunction with motion capture databases in order to reduce dimensionality and capture relations between features. Arikan uses PCA to create an algorithm that compresses motions in the databases [12]. Safonova's [13] work creates simulated movement based on an optimization on an objective function that uses the torques, joint angle trajectories, and the reduced space. Our work focuses on creating a generalized space consisting of natural poses and utilizing this space to speed up the process for finding new natural poses.

3 A Reduced Space Based on Natural Pose Characteristics

In a system with j degrees of freedom, a j-dimensional space can represent all possible values for each joint angle. Due to joint limitations, not all points in that space are correct poses. Of the possible poses, some are physically implausible and some are simply unnatural. These poses should be removed from the set of available poses. Unfortunately, it is very difficult to automatically differentiate which poses are natural and which are not.

Our method attempts to characterize and describe natural poses by reducing the j-dimensional space to a lower dimensionality space that reasonably describes natural poses. We attempt to find an m-dimensional space, where m < j, and any point in a certain range in this m-space can be transformed back to a natural pose.

Motion data were selected from the CMU MOCAP database [14]. We used 10000 frames for training. Data were taken from different subjects and were chosen based solely on the description to reduce any bias. We assume that poses are interchangeable between different subjects, The motion data that were selected had primarily walking motions with some additional motions (jump, reach, etc.).

Each frame is represented as a vector x with each dimension corresponding to a degree of freedom in 18 dimensions. The Karhunen-Loeve transform method (PCA) is used to calculate the reduced dimension data. Human natural motion is highly correlated; the movement of the forearms for example, will imply movement of the upper arm in natural ways. PCA allows us to find these correlations among the data and reduce the dimensionality and, thus, the system complexity.

Each pose from a motion frame correspond to a point in the reduced dimensional pose space (referred to as pose space in this paper) created by our analysis. Based on this new space, a set of poses can be created. Any point in the space represents a single pose which will be shown to be close to natural poses. These other points are synthesized poses that are significantly different from existing poses.

4 Unconstrained Poses

Unconstrained poses are simply poses created by selecting points in the pose space. This point is then transformed to an actual pose by using the inverse of the original transformation matrix. The original transformation matrix T is orthonormal, and therefore the inverse of this matrix is simply the transpose. The dimension of the pose space is less than the original dimension. Therefore, only the first n columns of the inverse of the transformation matrix are used.

Multiple points in the original space may map to the same point in the reduced space. Inverting the point in the reduced space returns a point with the least mean squared-error to all possible points that map to that point. It is not uncommon that some degrees of freedom may exhibit values that slightly exceed the specified DOF limits. If any of the angles of X is outside the bounds of the joint ($AMin_i...AMax_i$) the angle is adjusted so as to be bounded by the limit on the joint angles.

The motion capture data utilizes skeletons consisting of 29 joints and 59 degrees of freedom. In this work we limit the method to finding poses for the right hand. Additional work not described herein is examining the problem of multiple simultaneous constraints such as multiple targeted hand locations. As we are only concerned with arm pose location, we only utilize the angles from bones that connect the root bone to the right hand bone. The right hand bone acts as the end effector. There are 9 bones and 18 degrees of freedom between the right hand bone and the root bone.

For this data, 98% of the arm data variability can be described by only 7 dimensions. A transform matrix T is created from the eigenvectors that transforms the 18 angle vector to a 7 dimensional space. There is a many-to-one correspondence between points in the 18 dimension space and those in the 7 dimensional reduced space. A sample point x consists of 7 values $(x_1...x_7)$, each corresponding to one vector element. The range for x is between the minimum and maximum value for the ith dimension based on the training data.

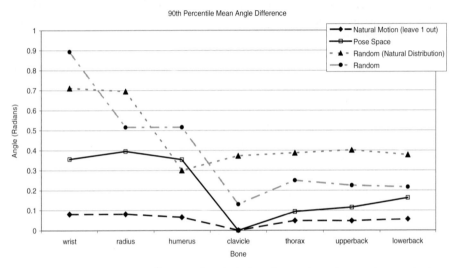

Fig. 1. Comparison of the 90th Percentile calculation of naturalness from various methods

To evaluate the naturalness of poses in this space, sample points were randomly selected and compared to a significant portion of the MOCAP database. Comparison was done on a bone by bone basis. Given two poses, one a randomly selected sample point, and a pose from the MOCAP database, the quaternion dot product between each bone angles was computed. The dot product can be used to measure the angle needed to rotate from one angle to another. As the angle becomes more similar, the rotation needed decreases, and the dot product approaches a value of 1. Equation 1 shows the similarity measure between two poses, each having k bones,

$$\delta = \sum_{k}(1-(\alpha_{1k} \cdot \alpha_{2k}))$$ (1)

We compare this method with two random methods of generating poses. The first random method simply selects a set of random angles (within the DOF limits) to generate a pose. The second random method creates random numbers using the distribution of the angles in the training data. The method is also compared to actual natural pose taken from the MOCAP database. A random pose is taken from the MOCAP database, the similarity distance is calculated over all poses not in the same motion.

The reason for this is that poses in the same motion tend to be close together (especially a pose which comes before or after the reference pose in a motion).

500 poses were generated for each method. The 90th percentile of the distance is calculated. Resampling is used to find a 95% confidence interval of this percentile. The data was resampled 1000 times. The mean angle in radians of the 90th percentile for each degree of freedom is shown in figure 1. The method proposed creates poses that are quite natural compared to the random algorithm.

Aside from naturalness, the generated poses must also be significantly different than existing poses. By having poses that are different, a wider range of motion can be created. Figure 2 shows the Cartesian coordinates of the end effector from the original poses and 8000 generated poses. The graphs are cutaway views of a combination of two axes. The grey points are the end effector coordinates of the created sample points, while the black points represent the end effector coordinates of the original data points. This graph shows that this method creates new poses that are not in the original motion captured data.

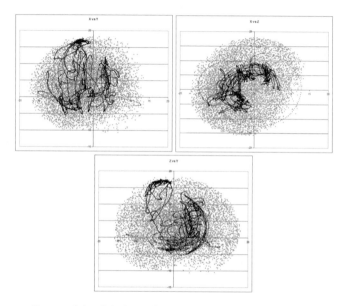

Fig. 2. The coordinates of the right hand from both the original data (black) and the natural pose samples (grey). The top-left picture is the cutaway x-y coordinate view, the top-right, x-z and the bottom z-y.

5   Constrained Poses

Constrained poses require the pose to meet some predefined conditions. In our experiments, the goal is to place the end effector (hand) at a particular, specified, position. The starting pose of the virtual human is all the same. Bones from the hand to the root bone (lower back) are considered; other bones are ignored. The algorithm proposed is called the Constrained Pose Generator (CPG) algorithm.

Table 1. Accuracy comparison of various pose generation algorithms

Method	Accuracy	Accuracy for Reachable Constraints
CPG (CCD)	0.7280 ± 0.0276	0.8125 ± 0.0256
CPG (Jacobian)	0.8090 ± 0.0244	0.9029 ± 0.0194
CCD	0.6380 ± 0.0298	0.7121 ± 0.0296
Jacobian	0.6030 ± 0.0303	0.6730 ± 0.0307

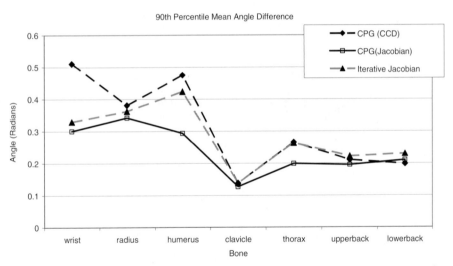

Fig. 3. The mean angle difference for each degree of freedom based on different generation algorithms

Sample points were predetermined by choosing points on a grid superimposed on the space. For each of the seven dimensions, points are sampled on the grid, starting from the minimal value to the maximum value. The lower dimension of PCA captures more variability than the higher ones. Therefore the lower dimensions were sampled at a higher rate. The total number of points used as a seed for this method is 53000. These points are stored in database, indexed by the end effector position to facilitate fast searching.

Given a starting pose P_0, our method seeks to determine a set of DOFs that place the end effector (the right hand) at a desired target position T(x, y, and z). We find a sample point R_i having the end effector position S in the pose space which is the best match for that pose. The criteria to find R_i is based on the distance of the end effector as well as a function of the naturalness score δ (equation 2).

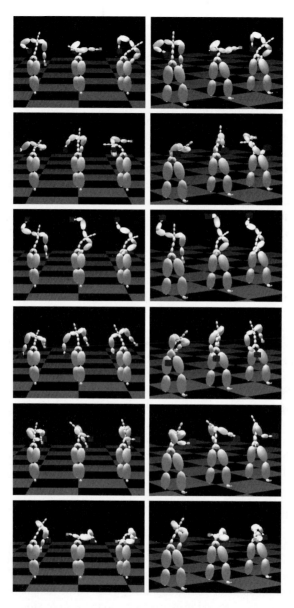

Fig. 4. Each frame shows the result of running our algorithm for each of 6 poses. The pictures on the left side are the front view of the pose; the pictures of the right side are is from an angled view from the right side of the animated humans. The leftmost pose in each frame was created using the NSPA algorithm followed by the CCD algorithm, the middle pose was calculated using CCD, and the rightmost pose was calculated using the iterative Jacobian method.

$$\arg\min_i \|T_i - S_i\| + f(\delta) \tag{2}$$

Once the algorithm determines a candidate pose that is nearest to achieving the desired constraint, either Coordinate Cyclical Descent (CCD) [7] or the iterative Jacobian is used to refine the pose so as to accurately achieve the constraint.

We performed a search on 1000 random single constraint problems (the constraint was on the position of the right hand) to determine accuracy of algorithm as well as naturalness of results. The algorithms tested are the CPG using CCD, CPG using the Jacobian, CCD algorithm, and Iterative Jacobian algorithm.

Table 1 shows how accurate the various algorithms were at finding a solution. Out of the 1000 constraints given, 104 constraints were never found by any of the algorithm. This could mean that the constraints were out of reach range of the virtual human. The two scores in table 1 show the accuracy for all constraints, and accuracy for only the reachable constraints (with a confidence interval of 95%). Based on this table it is clear that in terms of accuracy in finding the correct pose for a given constraint, using the pose space is an improvement to using CCD or Iterative Jacobian.

In some of the poses, both the Jacobian and CCD algorithm results in a pose where the body is twisted and awkward. There are of course exceptions. One of the main problems with the Jacobian and the CCD methods is that all joint angles are changed, even though in natural human motion not all joint angles change to achieve a pose.

The naturalness measurement described in equation 1 is used again to determine the naturalness of poses created by the various algorithms. Figure 3 shows the naturalness comparison of the CPG algorithms (using CCD and Jacobian) versus pure Jacobian. The figure shows that the CPG algorithm with Jacobian creates significantly more natural poses than Jacobian alone.

Figure 4 illustrates the ending poses according to three algorithms. The algorithm was tested in a side by side testing, simultaneously calculating poses by running the CPG algorithm (using CCD) and comparing it to a CCD and an iterative Jacobian solution. Objects are placed at either a comfortable reaching distance, extended reaching distance (where the human must extend hands fully), or twist position where the location is usually behind the virtual human.

6   Conclusion and Future Work

We describe a method that statistically summarizes motion capture and creates a natural space for motions involving the right hand. The new reduced dimension space is used to create sample poses. We show that the sample poses are in fact quite natural compared to other methods of generating new poses. These sample poses can be used in an animation sequence created using motion capture data or simple keyframes. It is perhaps possible to create an animation by traversing points in the pose space.

We propose the CPG method to find a pose with a constraint, is where the end effector position must be at certain coordinates. Our method uses unconstrained generated poses to finds a starting pose having the closest effector position to the target. The CPG method is more accurate than iterative methods, it is also more natural.

In the future the use of multiple constraints as well as multiple pose spaces corresponding to different body parts will be addressed. One possible method would be to use PCA on the whole body to create a reduced space. Another method is to divide the whole body into hierarchically organized parts, each with its own natural space, and then join these natural spaces together to create a composite natural space.

Currently the bounds of the reduced dimension space used to create new poses are based on the minimum and maximum value for each dimension. Other methods such as SVM and mixture of Gaussians are being explored in order to create a better boundary on the reduced space.

The reduced space can also be used in other ways. We plan to animate characters based on multiple poses in the reduced space. These poses can be smoothed out and interpolated to create an animation.

Acknowledgement

The data used in this project was obtained from mocap.cs.cmu.edu. The database was created with funding from NSF EIA-0196217.

References

1. Bodenheimer, B., et al.: The Process of Motion Capture: Dealing with the Data. In: Computer Animation and Simulation 1997, Eurographics Animation Workshop (September 1997)
2. Gleicher, M., et al.: Snap-together motion: assembling run-time animations. In: Proceedings of the 2003 symposium on Interactive 3D graphics, pp. 181–188. ACM Press, Monterey (2003)
3. Arikan, O., Forsyth, D.A.: Interactive motion generation from examples. In: Proceedings of the 29th annual conference on Computer graphics and interactive techniques, pp. 483–490. ACM Press, San Antonio (2002)
4. Lee, J., Lee, K.H.: Precomputing avatar behavior from human motion data. In: Proceedings of the 2004 ACM SIGGRAPH/Eurographics symposium on Computer animation, pp. 79–87. ACM Press, Grenoble (2004)
5. Abe, Y., Liu, C.K., Popovic, Z.: Momentum-based parameterization of dynamic character motion. In: Proceedings of the 2004 ACM SIGGRAPH/Eurographics symposium on Computer animation 2004, Eurographics Association, Grenoble, France, pp. 173–182 (2004)
6. Watt, A., Watt, M.: Advanced Animation and Rendering Techniques Theory and Practice. Addison-Wesley, New York (1992)
7. Fedor, M.: Application of inverse kinematics for skeleton manipulation in real-time. In: Proceedings of the 19th spring conference on Computer graphics, pp. 203–212. ACM Press, Budmerice (2003)
8. Meredith, M., Maddock, S.: Adapting motion capture data using weighted real-time inverse kinematics. Comput. Entertain. 3(1), 5 (2005)
9. Komura, T., et al.: An inverse kinematics method for 3D figures with motion data. In: Computer Graphics International (2003)
10. Grochow, K., et al.: Style-based inverse kinematics. In: ACM SIGGRAPH 2004 Papers, pp. 522–531. ACM, Los Angeles (2004)

11. Yamane, K., Kuffner, J.J., Hodgins, J.K.: Synthesizing animations of human manipulation tasks. ACM Trans. Graph. 23(3), 532–539 (2004)
12. Arikan, O.: Compression of motion capture databases. In: ACM SIGGRAPH 2006 Papers, pp. 890–897. ACM, Boston (2006)
13. Safonova, A., Hodgins, J.K., Pollard, N.S.: Synthesizing physically realistic human motion in low-dimensional, behavior-specific spaces. ACM Trans. Graph. 23(3), 514–521 (2004)
14. CMU, CMU Graphics Lab Motion Capture Database (2008)

Segmentation of Neural Stem/Progenitor Cells Nuclei within 3-D Neurospheres

Weimiao Yu[1], Hwee Kuan Lee[1], Srivats Hariharan[2], Shvetha Sankaran[2], Pascal Vallotton[3], and Sohail Ahmed[2]

[1] Bioinformatics Institute, #07-01, Matrix,
30 Biopolis Street, Singapore 138671
[2] Institute of Medical Biology, #06-06, Immunos,
8A Biomedical Grove, Singapore 138648
[3] CSIRO Mathematical and Information Sciences,
Locked Bag 17, North Ryde, NSW, 1670, Australia

Abstract. Neural stem cells derived from both embryonic and adult brain can be cultured as neurospheres; a free floating 3-D aggregate of cells. Neurospheres represent a heterogenous mix of cells including neural stem and progenitor cells. In order to investigate the self-renewal, growth and differentiation of cells within neurospheres, it is crucial that individual nuclei are accurately identified using image segmentation. Hence effective segmentation algorithm is indispensible for microscopy based neural stem cell studies. In this paper, we present a seed finding approach in scale space to identify the center of nuclei in 3-D. Then we present a novel segmentation approach, called "Evolving Generalized Voronoi Diagram", which uses the identified centers to segment nuclei in neurospheres. Comparison of our computational results to mannually annotated ground truth demonstrates that the proposed approach is an efficient and accurate segmentation approach for 3-D neurospheres.

1 Introduction and Background

The dogmatic concept, "adult brains are unable to make new neurons", has dominated neuroscience thinking for centuries[1] until the first challenge from Altman J. and Das G.[2], who saw cells that appeared to be newly born neurons in 1960's. Later in 1980s, Goldman S. and Nottebohm F.[3] found solid evidence that canaries give birth to waves of new brain neurons seasonally in a particular area of their brains. The discovery of *adult* neural stem cells (NSCs) opens the door to potential treatments for neurological diseases, such as Parkinsons and Alzheimers, by endogenous repair. To be able to utilize NSCs for therapy and investigations of neurodevelopment we need to have a better understanding of their cell biology. There are no definitive markers for NSCs and they are normally followed by functional critiera[5] including; (a) self-renewal ability to passage for many generations, (b) multipotency ability to generate neurons, astrocytes and oligodendrocytes and (c) to regenerate brain tissue. Our understanding of NSCs and their therapeutic potential relies on propagating these cells *in vitro*. One

G. Bebis et al. (Eds.): ISVC 2009, Part I, LNCS 5875, pp. 531–543, 2009.

of the most popular functional assays for NSCs is the *Neurosphere Formation Assays* (NFAs). It provides a relatively simple and robust means to investigate the propagation of NSCs. Although the exact relationship between NSCs and NFAs is unclear yet[4], NFAs remain a good evidence of the presence of NSCs[5].

Fluorescence microscopy is a common, probably preeminent, tool to understand the cell biology of NSCs. Automated analysis of the acquired 3-D images is critical to identify the stem cells within neurospheres based on their activities and behaviors. In order to do so, we need to accurately identify and segment each nucleus from the neurosphere, however, it is challenging since the nuclei are morphologically diverse and usually clumpy with each other. Using simple approaches, such as thresholding or watershed, will cause severe under-segmentation or over-segmentation. One way to solve this segmentation problem is to first find the centers of the nuclei, which we call *seed finding* in this paper, and then perform some kind of flooding or region-growing from the seeds to segment the nuclei. Traditionally, the centers of the nuclei are identified as local maxima in the distance transform of binary or gray level images. This approach works well only when the shapes of nuclei are ellipsoidal or spherical and their sizes are homogeneous. However, these two conditions are satisfied in few biological experiments. More sophisticated approaches are needed, for example, fast radial symmetry transform[6] and phase symmetry approach[7,8]. More recently, an iterative voting approach of radial symmetries based on the gradient direction is presented in [9] for inferring the center of objects. This approach is applied to identify the center of closely packed cells in 2-D images in [10] and then it is extended to identify the center of nuclei in 3-D mammosphere images in [11]. Another seed detection approach based on Laplacian of Gaussian filter is applied on 2-D gray level images in [12].

Seed finding is only the first step of nuclei segmentation. Many segmentation approaches have been reported. The distance transform and a modified watershed approach are applied in [14] to segment the nuclei in 3-D microscopy images of *C.elegans*. Most of the nuclei can be correctly segmented by the proposed approach, however post-processing may be needed to further split or merge objects[14]. A gradient flow and local adaptive thresholding approach for 3-D nuclei segmentation is presented[15] and tested on both synthetic images and real images. The combination of level set and watershed approach is also popular, for example [16]~[17]. In order to overcome over-segmentation and under-segmentation, different topological constraints are exploited, such as topological dependence[18,19], simple point concept[20] and topology-preserving model[21]. An important concept is proposed in [18]: image intensity and geometrical shape of the objects are both important cues for an accurate segmentation. It is difficult to include all important work in this short paper. Other related work can be found, such as the flexible contour model[22] and ellipse detector[23].

For the rest of the paper, we describe the preparation of biological samples and the microscope configurations in Section 2. In Section 3, we first present the pre-processing of 3-D images. Then a seed finding approach is presented to identify the center of nuclei by searching the local maxima in scale space representations of the distance transform of binary images. Section 4 proposes a novel algorithm,called "Evolving Generalized Voronoi Diagram", to segment the clumpy nuclei with irregular shapes. The experimental results and validation are presented in Section 5 followed by discussions and conclusions in Section 6.

2   Image Acquisition

The cells in our study are Neural Stem/Progenitor Cells derived from mouse embryo stage (E 14.5). Cells were cultured in growth media containing Epidermal Growth Factor and N2 (a growth supplement). Cells were nucleofected with plasmid Plasma Membrane Targeted-YFP (PMT-YFP) and allowed to grow in an incubator at $37°C$ and 5% CO_2 for 4-5 days to form neurospheres. At the end of 5 days, Hoechst was added to stain the nuclei and incubated for 10-15 mins before imaging. We used the Olympus confocal FV1000 for imaging. 488nm laser set at 5% power (0.86mw) was used to image PMT-YFP and 405nm laser set at 2% power (0.12mw) was used to image Hoechst. Images were acquired with a z step of $0.2\mu m$. The resolution of x and y axis is $0.25\mu m$. The photomultiplier tube voltage in the confocal was set based on the signal from the sample. A $60\times$ water immersion lens with a numerical aperture of 1.2 was used. The 3-D images contain two channels: green for PMT-YFP and blue for Hoechst. A representative neurosphere image from two different viewing angles is shown in Fig. 1.

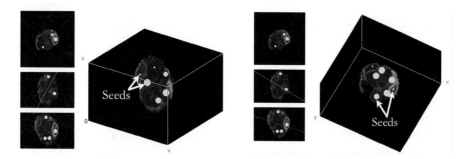

Fig. 1. Original 3-D Neurosphere Image and Seed Finding in Scale Space. Two different views of the original 3-D image are illustrated. The identified seeds are annotated by the arrows. The seeds are dilated by a ball structuring element of 7 voxel radius for the purpose of visualization. The apparent size of dot indicates its relative depth to the observer.

3   Image Pre-processing and Seed Finding

3.1   Image Pre-processing

The images in this paper are defined on a finite subset of three-dimensional Euclidean space, $\Theta \subset \mathbb{R}^3$. A point $\vec{r}$ in Θ is represented by Cartesian coordinate, e.g. $\vec{r} = (x, y, z) \in \Theta$. $f^n(\vec{r}) : \Theta \mapsto \mathbb{R}$ and $f^c(\vec{r}) : \Theta \mapsto \mathbb{R}$ represent the intensities of Hoechst (blue) and PMT-YFP (green) at $\vec{r}$, respectively. The superscripts "n" and "c" indicate "nucleus" and "cell". The image intensities are normalized such that $f^n(\vec{r}) \in [0, 1]$ and $f^c(\vec{r}) \in [0, 1]$.

Both $f^n(\vec{r})$ and $f^c(\vec{r})$ contain important information for the 3-D nuclei segmentation. In order to avoid photobleaching and phototoxicity for the live cells, fast scanning speed is applied, $i.e.$ 2ms/voxel, which limits the achievable signal-to-noise ratio. Image enhancements such as histogram equalization and contrast improvement are applied to $f^n(\vec{r})$ and $f^c(\vec{r})$ and produce $\tilde{f}^n(\vec{r})$ and $\tilde{f}^c(\vec{r})$, respectively. As shown in Fig. 1, the value of $f^c(\vec{r})$ is relatively high (bright) at the cell boundary. Thus, we combine these two images into one image:

$$f(\vec{r}) = \tilde{f}^n(\vec{r}) - \tilde{f}^c(\vec{r}) \tag{1}$$

The combination will make $f(\vec{r})$ darker near the cell boundaries. This facilitates the subsequent processing. Then $f(\vec{r})$ is converted to a binary image according to Otus threshold[24]. The binary image contains two regions, the background Ω_b and the foreground Ω_f with $\Omega_b \cup \Omega_f = \Theta$ and $\Omega_b \cap \Omega_f = \emptyset$. Then the distance transform is applied:

$$D(\vec{r}) = \min_{\vec{r'} \in \Omega_b} |\vec{r} - \vec{r'}| \tag{2}$$

where the distance between the two point $\vec{r}$ and $\vec{r'}$ is the Euclidean distance: $|\vec{r} - \vec{r'}| = \sqrt{(x - x')^2 + (y - y')^2 + (z - z')^2}$. It is obvious that if $\vec{r} \in \Omega_b$, $D(\vec{r}) = 0$, otherwise, $D(\vec{r}) > 0$.

3.2   Seed Finding in Scale Space

Segmentation of the nuclei in neurospheres is very challenging, because they are closely packed and touch each other. Hence, finding the centers of the nuclei, known as *seed finding*, is a critical step to assist our subsequent processing. The distance transform given in Eq.(2) can identify the seeds nicely, provided the sizes of the nuclei are similar and their shapes are spherical or ellipsoidal. Unfortunately, these two conditions can not be satisfied in our study. The results of using the distance function directly are very sensitive to the thresholding value, which is used to identify the local maxima. If smaller nuclei are detected, then the seeds of bigger nuclei may merge due to irregular shapes; on the other hand, if bigger nuclei are successfully separated, it is very likely that some smaller nuclei are undetected.

In order to overcome these challenges, we present a robust method based on scale space theory[25], which can be used to identify centers of objects with different sizes. It is also applied to identify the stable key points in [13]. It has been shown by [25] and [26] that under a variety of reasonable assumptions, Gaussian kernel is the only scale-space kernel. The 3-D Gaussian kernel of width σ is given by:

$$G(\vec{r}, \sigma^2) = \frac{1}{\sqrt{(2\pi\sigma)^3}} \exp\left(-\frac{|\vec{r}|^2}{2\sigma^2}\right) \tag{3}$$

where $|\vec{r}| = \sqrt{x^2 + y^2 + z^2}$.

For a given scale σ, the scale-space representation of distance function $D(\vec{r})$ is given as:

$$L(\vec{r}, \sigma^2) = D(\vec{r}) * G(\vec{r}, \sigma^2) \tag{4}$$

where "$*$" means convolution. A straightforward way to obtain a multi-scale detector with automatic scale selection is to consider the normalized scale Laplacian operator. Lindeberg T. showed that the normalization of the Laplacian with the factor σ^2 is required for true scale invariance[25]. In detailed experimental comparisons[27] , it is found that the maxima and minima of scale-normalized Laplacian of Gaussian produce the most stable image features. In practice, the difference-of-Gaussian (DoG) provides a good approximation of the scale-normalized Laplacian of Gaussian. The DoG can be efficiently calculated according to:

$$
\begin{aligned}
DoG(\vec{r}, \sigma^2) &= \frac{1}{2\Delta\sigma^2}(G(\vec{r}, \sigma^2 + \Delta\sigma^2) - G(\vec{r}, \sigma^2 - \Delta\sigma^2)) * D(\vec{r}) \\
&= \frac{1}{2\Delta\sigma^2}(L(\vec{r}, \sigma^2 + \Delta\sigma^2) - L(\vec{r}, \sigma^2 - \Delta\sigma^2))
\end{aligned}
\tag{5}
$$

where $2\Delta\sigma^2$ is a positive normalization factor, which is essential to achieve the scale invariant representation of the DoG. In order to identify the seeds, we find the local maxima of $DoG(\vec{r}, \sigma^2)$:

$$(\tilde{\vec{r}}, \tilde{\sigma}^2) = arg\ \text{local-max}_{(\vec{r}, \sigma^2)}\left[DoG(\vec{r}, \sigma^2)\right] \tag{6}$$

The identified local maximum voxel $\tilde{\vec{r}}$ is the center of the nuclei, $i.e.$ the $seeds$. We denote them by s_i ($i = 1, 2, ..N$), where N is the number of seeds. The convolution of $D(\vec{r})$ at different scales will also partially solve the problem associated with irregular nuclei shape. Two different views of an original image are illustrated in Fig. 1. As shown in Fig. 1, the dots annotated by the arrows illustrate the detected seeds using our approach. Note that the size of dot indicates its relative depth to the observer. The detected seeds can successfully represent the center of nuclei of different sizes and irregular shapes.

4   Nuclei Segmentation Based on Evolving Generalized Voronoi Diagram

The nuclei segmentation is not only to simply separate the nuclei from the background, but also separate them from each other. Inspired by the concept of topological dependence[18,19], we present our algorithm of "Evolving Generalized Voronoi Diagram" (EGVD) to segment the 3-D nuclei. In this algorithm, we evolve the level set function using Chan-Vese method introduced in [18]. At each iteration of level set evolution, we prevent splitting and merging of the objects using Generalized Voronoi Diagram. We followed the idea introduced in [18,19] and modified the distance transform $D(\vec{r})$ based on the identified seeds:

$$\bar{D}(\vec{r}) = \begin{cases} 1 & \text{if } \vec{r} \in \bigcup_{i=1}^{N} s_i \\ \frac{D(\vec{r})}{\max(D(\vec{r}))} & \text{otherwise} \end{cases} \tag{7}$$

Then initialization of the level set is given by [18,19]:

$$\phi^{t=0}(\vec{r}) = \bar{D}(\vec{r}) - 1 \tag{8}$$

such that active contours are initialized at the found seeds in Section 3.2. We evolve the level set function based on the formulation given in [18,19]:

$$\phi^{t+\Delta t} = \phi^t + \Delta t \cdot \delta_\epsilon(\phi^t)[-\lambda_1(D(\vec{r}) - c_1)^2 + \lambda_2(D(\vec{r}) - c_2)^2] \tag{9}$$

where the constants c_1 and c_2 are the mean values of the background and foreground.

For numerical stability reasons, the level set function is usually reinitialized to be the distance function after a few iterations. This is particularly important when the level set curvature term $div(\frac{\nabla\phi}{|\nabla\phi|})$ is present in the level set updating. In this paper, we did not do so, but still achieved numerical stabilities for the following two reasons. Firstly, the length parameter ν is zero in Eq. (9) by [18,19] so that the level set curvature term is absent. Secondly, using a regularized delta function $\delta_\epsilon(\phi^t)$ with a large ϵ, ($\epsilon = 1.0$) contributed to maintaining the numerical stability.

Evolving the level set using Eq.(9) is insufficient to segment the nuclei correctly. We develop the EGVD algorithm to prevent objects from splitting and merging. Before we present our algorithm, two important definitions are needed: *Generalized Voronoi Diagram*(GVD) and *Choice Function*. Let $\omega_i^{n,t}$ ($i = 1, 2, ...N$) denote nuclei segments at artificial time step t. They are defined as follows:

Generalized Voronoi Diagram: *Given a set of disjoint connected regions $\omega_i^{n,t}$ for $i = 1, 2, ...N$ with $\omega_i^{n,t} \cap \omega_j^{n,t} = \emptyset \ \forall i \neq j$, define the Generalized Voronoi Diagram (GVD) as $V_i(\omega_1^{n,t}, \omega_2^{n,t}, ...\omega_N^{n,t})$ corresponding to each $\omega_i^{n,t}$:*

$$V_i(\omega_1^{n,t}, \omega_2^{n,t}, ...\omega_N^{n,t}) = \left\{ \vec{r} \in \Theta | \min_{\vec{s} \in \omega_i^{n,t}} |\vec{r} - \vec{s}| < \min_{\vec{s}' \in \bigcup_{j \neq i}^{N} \omega_j^{n,t}} |\vec{r} - \vec{s}'| \right\} \tag{10}$$

Choice Function: *Given a connected region s_i as seeds, and a set of points Γ. Γ may consist of several connected regions. Define the choice function, also known as selector, $C(\Gamma|s_i)$ that chooses the connected region from Γ which contains s_i . $C(\Gamma|s_i)$ returns empty set $\emptyset$, if $s_i \not\subset \Gamma$*

Based on the above definitions, our Evolving Generalized Voronoi Diagram algorithm is given as follows:

1. Find the seeds to obtain s_i, $(i = 1, 2, ...N)$ according to Eq. 6.
2. Initialize the level set function for the nuclei segmentation according Eq. (8).
3. Update the level set function $\phi^t \rightarrow \phi^{t+\Delta t}$ using Eq. (9). Then update the GVD regions iteratively at each time step $t + \Delta t$ to obtain $w_i^{n,t+\Delta t}$ as follows:
   (a) Let $\Omega^{n,t+\Delta t} = \{\vec{r} \in \Theta | \phi^{t+\Delta t} \geq 0\}$.
   (b) Define $\bar{w}_{i,k=0}^{n,t+\Delta t} = w_i^{n,t}$, for $i = 1, 2, ...N$. k is used to index successive estimates of nuclei segments at $t + \Delta t$.
   (c) For each $i = 1, 2, ...N$, calculate GVD: $V_i(\bar{w}_{1,k}^{n,t+\Delta t}, \bar{w}_{2,k}^{n,t+\Delta t}, ...\bar{w}_{N,k}^{n,t+\Delta t})$ and update the nuclei segment using the choice function:

$$\bar{w}_{i,k+1}^{n,t+\Delta t} = \mathcal{C}(\Omega^{n,t+\Delta t} \cap V_i(\bar{w}_{1,k}^{n,t+\Delta t}, \bar{w}_{2,k}^{n,t+\Delta t}, ...\bar{w}_{N,k}^{n,t+\Delta t})|s_i) \qquad (11)$$

   and then set $k \rightarrow k + 1$.
   (d) Iterate step 3(c) until convergence, i.e. $\bar{w}_{i,k+1}^{n,t+\Delta t} = \bar{w}_{i,k}^{n,t+\Delta t}$
   (e) Set $w_i^{n,t+\Delta t} \leftarrow \bar{w}_{i,k+1}^{n,t+\Delta t}$. This completes the update of GVD and nuclei segments at $t + \Delta t$.
4. Repeat step 3 until convergence of the level set function.

In Fig. 2, we present an illustration in 2-D for better understanding. Our algorithm consists of an outer loop for level set evolution and an inner loop for EGVD. GVDs define the boundary of nucleus segments where two level set segments might merge. Essentially, EGVD algorithm involves a series of fine adjustment of the intermediate GVDs. Suppose at time t, $w_i^{n,t}$ are obtained and V_i is calculated based on Eq.(10). At the next time step $t + \Delta t$, the level set function is updated and a new V_i is needed to give correct $w_i^{n,t+\Delta t}$. We first use the V_i calculated at the previous time step, illustrated by shaded area in Fig. 2(a), as initial GVD for the inner loop iteration. Given GVD and s_i, we use the choice function to calculate a connected region for each nucleus, as shown by the dotted regions in Fig. 2(b). Then a new GVD is calculated based on these dotted regions, as shown by the shaded area in Fig. 2(c). The process of evolving GVD is performed iteratively until GVD does not change anymore, which will be the final GVD in this time step. As we can see from Fig. 2(d), the nucleus in V_j has converged in this iteration. It can be proven that the EGVD algorithm converges, while we shall leave its rigorous mathematical proof for further publication due to the limitation of space.

Comparing with existing approaches, EGVD algorithm has a few advantages. First of all, EGVD is conceptually very simple. Comparing with the maximum common boundary criterion in [18,19], it does not require the considerations of many different cases in which the level set function split and merge. EGVD is also fast and efficient. It is not necessary to determine whether a given point is a simple point[20]. EGVD is more flexible than the formulation introduced in [21], in which any topological change is forbidden. EGVD only disallows splitting and merging, while it tolerates other topological changes, such as adding a hole. Lastly, it is trivial to extend EGVD to other dimensions.

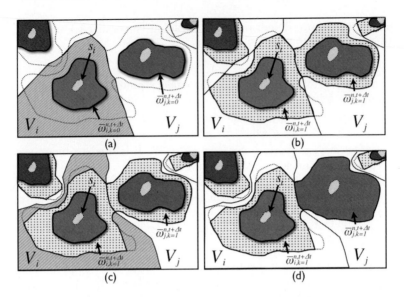

Fig. 2. Evolving the Generalized Voronoi Diagram. Central regions annotated by s_i are the identified seeds. The shaded regions in (a) and (c) are corresponding V_i for $\bar{\omega}_{i,k=0}^{n,t+\Delta t}$ and $\bar{\omega}_{i,k=1}^{n,t+\Delta t}$. The dotted regions represent the different $\bar{\omega}_{i,k}^{n,t+\Delta t}$ at different iterations before convergence.

5   Experimental Results

Two of our segmentation results are shown in Fig. 3. We use random colors to represent different nuclei. Although the nuclei are clumpy and touch each other in the neurosphere, our EGVD algorithm is able to segment them satisfactorily.

Quantitative validation is an important and necessary step to evaluate the accuracy of algorithms. We select eight 3-D Neurosphere images and manually create the ground truth using a touch screen laptop and "Segmentation Editor" in Fiji [1]. The boundary of each nucleus is labelled manually from the top to bottom. Segmentation Editor has the function of 3-D interpolation and we don't need to draw the boundary at each image slice. The interval of the boundary drawing is 3~7 slices depending on the shape of the nuclei. A few slices of a ground-truth image are shown in Fig. 4. The masks are the created nucleus segments. The boundaries of a nucleus, annotated by the arrows in Fig. 4 (b)~(h), are drawn manually according to best human perception.

Based on the ground truth, there are *246* nuclei in eight 3-D images. For a given image, let $\omega_i^g, i = 1, 2, ...N$ denote the objects of the ground truth and $\omega_i^s, i = 1, 2, ...M$ denote the computational objects given by EGVD approach. We define a score α_i to describe accuracy of the computational results, which is

---

[1] Fiji package with Segmentation Editor is available at: http://pacific.mpi-cbg.de

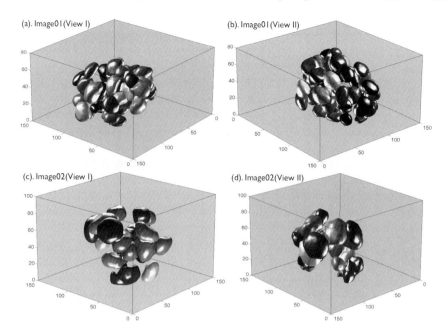

Fig. 3. Segmented Nuclei in 3-D. (a) and (b) display two different views of the segmented nuclei in Image01. (c) and (d) illustrate the segmentation results of Image02. Random color is selected to represent each nucleus.

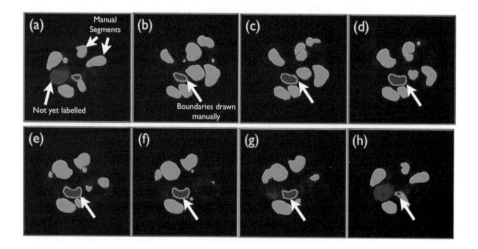

Fig. 4. The Procedure of Ground Truth Labeling. The nuclei segments are the manually created. The contours annotated by the arrows represent boundaries of a nucleus. They are drawn manually according to the best human perception. The boundaries in the slices between them will be interpolated by the Segmentation Editor.

$$\alpha_i = \max_j \left(\frac{|\omega_i^s \cap \omega_j^g|}{|\omega_i^s \cup \omega_j^g|} \right) \tag{12}$$

where $|\cdot|$ means the volume of the given connected component. It is obvious that $\alpha_i \in (0,1)$. If a computational object perfectly matches a ground truth object, α_i is 1.0; while $\alpha_i = 0$ when ω_i^s does not overlap any ω_j^g. On the other hand, it is also possible that some ω_j^g does not overlap any ω_i^s, which we called Missing Segments.

In order to test the performance of EGVD when seeds are correctly provided, we use the geometrical center of the ground truth objects as seeds and then apply EGVD approach to segment nuclei. We also use the proposed approach in scale space with different σ to detect the seeds and then compare our computational results with ground truth. The probability distribution of score $p(\alpha)$ and $P(\alpha) = \int_\alpha^1 p(\alpha')d\alpha'$ are shown in Fig. 5. From this figure, we can see that the Missing Segments caused a small peaks near $\alpha = 0$ in $p(\alpha)$, while the falsely detected seeds caused under-segmentation or over-segmentation and thus produced the small peaks near $\alpha = 0.5$ in $p(\alpha)$. From the curves of $p(\alpha)$ given different σ, we know there is a optimal value of σ for nuclei segmentation. The numbers of seeds identified by different σ are indicated in legend. It is clear that the number of detected seeds does not necessarily imply better performance, because some of them might be positioned inaccurately. The strong peaks near $\alpha = 0.9$ in the curves of $p(\alpha)$ indicate that majority of the nuclei are satisfactorily segmented. The mean segmentation accuracy of EGVD given real seeds is about 75% and it is about 70% given $\sigma = 2.1$ for seed finding.

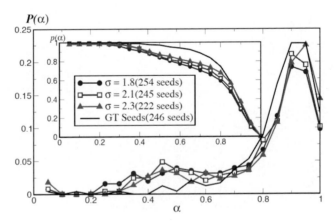

Fig. 5. Performances of EGVD Algorithm. The probability distribution function $p(\alpha)$ and corresponding $P(\alpha)$ are illustrated. If seeds are correctly provided, EGVD can segment majority of nuclei nicely indicated by the peak near $\alpha = 0.9$. The average segmentation accuracy is 75%. Using our seed finding approach, the falsely detected seeds caused the small peaks near $\alpha = 0.5$ in $p(\alpha)$.

6 Discussions and Conclusions

A seed finding approach in scale space and a novel segmentation algorithm, Evolving Generalized Voronoi Diagram (EGVD), are proposed to segment the nuclei in neurosphere. Our study combines careful image acquisition with dual dyes for nuclear and membrane labelling, and robust image analysis to use the data optimally. In order to quantitatively evaluate our proposed approaches, we create the ground truth of eight 3-D images containing *246* nuclei.

The computational results demonstrate that the EGVD algorithm can satisfactoryly segment the nuclei in neurospheres if the seeds are provided accurately. Our seed finding approach is validated by comparing the computational results with the ground truth. The results demonstrate our seed finding approach can identify most of the seeds correctly, though there are still some errors, which might cause under-segmentation and over-segmentation, as indicated by the weak peaks near $\alpha = 0.5$ in $p(\alpha)$ in Fig. 5. A more complete study on the parameter tuning of σ in the seed finding algorithm is needed to optimize the segmentation performance. Another possible improvement is that instead of simply using the PMT-YFP signal as in Eq. (1), we should utilize this important information more sufficiently. However, more sophisticated techniques are needed to extract the cell boundaries, since the PMT-YFP signal is not uniformly distributed near the cell boundaries and sometimes is really weak.

We expect our methods to be generally applicable to other stem cell assemblies, such as mamospheres, and more generally in the area of embryology. In the future, we plan to use the tools described to investigate in detail neurosphere development, including long range cell motion, apoptosis, and cavity formation, using 4D microscopy.

References

1. Barinaga, M.: Newborn Neurons Search for Meaning. Science 299, 32–34 (2003)
2. Altman, J., Das, G.D.: Post-natal origin of microneurones in the rat brain. Nature 207, 953–956 (1965)
3. Goldman, S.A., Nottebohm, F.: Neuronal Production, Migration, and Differentiation in a Vocal Control Nucleus of the Adult Female Canary Brain. PNAS 80, 2390–2394 (1983)
4. Reynolds, B.A., Rietze, R.L.: Neural Stem Cells and Neurospheres – Re-evaluating the Relationship. Natural Methods 2, 333–336 (2005)
5. Louis, S.A., Rietze, R.L., Deleyrolle, L., Wagey, R.E., Thomas, R.E., Eaves, R.E., Reynolds, B.A.: Enumeration of Neural Stem and Progenitor Cells in the Neural Colony-forming Cell Assay. Stem Cell 26, 988–996 (2008)
6. Loy, G., Zelinsky, A.: Fast Radial Symmetry for Detecting Points of Interest. IEEE Trans. on PAMI 25, 959–973 (2003)
7. Kovesi, P.: Image Features From Phase Congruency. Videre: A Journal of Computer Vision Research 1, 2–26 (1999)
8. Kovesi, P.: Phase Congruency: A Low-level Image Invariant. Psychological Research Psychologische Forschung 64, 136–148 (2000)

9. Yang, Q., Parvin, B.: Perceptual Organization of Radial Symmetries. In: Proceedings of CVPR, vol. 1, pp. 320–325 (2004)

10. Raman, S., Parvin, B., Maxwell, C., Barcellos-Hoff, M.H.: Geometric Approach to Segmentation and Protein Localization in Cell Cultured Assays. In: Bebis, G., Boyle, R., Koracin, D., Parvin, B. (eds.) ISVC 2005. LNCS, vol. 3804, pp. 427–436. Springer, Heidelberg (2005)

11. Han, J., Chang, H., Yang, Q., Barcellos-Hoff, M.H., Parvin, B.: 3D Segmentation of Mammospheres for Localization Studies. In: Bebis, G., Boyle, R., Parvin, B., Koracin, D., Remagnino, P., Nefian, A., Meenakshisundaram, G., Pascucci, V., Zara, J., Molineros, J., Theisel, H., Malzbender, T. (eds.) ISVC 2006. LNCS, vol. 4291, pp. 518–527. Springer, Heidelberg (2006)

12. Althoff, K., Degerman, J., Gustavsson, T.: Combined Segmentation and Tracking of Neural Stem-Cells. In: Kalviainen, H., Parkkinen, J., Kaarna, A. (eds.) SCIA 2005. LNCS, vol. 3540, pp. 282–291. Springer, Heidelberg (2005)

13. Lowe, D.G.: Distinctive Image Features from Scale-Invariant Keypoints. International Journal of Computer Vision 60, 91–110 (2004)

14. Long, F., Peng, H., Myers, E.: Automatic Segmentation of Nuclei in 3D Microscopy Images of C. elegans. In: Proceedings of ISBI 2007, pp. 536–539 (2007)

15. Li, G., Liu, T., Tarokh, A., Nie, J., Guo, L., Mara, A., Holley, S., Wong, S.T.C.: 3D Cell Nuclei Segmentation Based on Gradient Flow Tracking. BMC Cell Biology 8 (2007), http://www.biomedcentral.com/1471-2121/8/40

16. Tai, X., Hodneland, E., Weickert, J., Bukoreshtliev, N.V., Lundervold, A., Gerdes, H.: Level Set Methods for Watershed Image Segmentation. In: Sgallari, F., Murli, A., Paragios, N. (eds.) SSVM 2007. LNCS, vol. 4485, pp. 178–190. Springer, Heidelberg (2007)

17. Yan, P., Zhou, X., Shah, M., Wang, S.T.C.: Automatic Segmentation of High-throughput RNAi Fluorescent Cellular Images. IEEE Transaction on Information Technology in Biomedicinel 12, 109–117 (2008)

18. Yu, W.M., Lee, H.K., Hariharan, S., Bu, W.Y., Ahmed, S.: Level Set Segmentation of Cellular Images Based on Topological Dependence. In: Bebis, G., Boyle, R., Parvin, B., Koracin, D., Remagnino, P., Porikli, F., Peters, J., Klosowski, J., Arns, L., Chun, Y.K., Rhyne, T.-M., Monroe, L. (eds.) ISVC 2008, Part I. LNCS, vol. 5358, pp. 540–551. Springer, Heidelberg (2008)

19. Yu, W.M., Lee, H.K., Hariharan, S., Bu, W.Y., Ahmed, S.: Quantitative Neurite Outgrowth Measurement Based on Image Segmentation with Topological Dependence. Cytometry Part A 75A, 289–297 (2009)

20. Xiao, H., Chenyang, X., Jerry, L.P.: A Topology Preserving Deformable Model Using Level Sets. In: Proceeding of CVPR, vol. 2, pp. 765–770 (2001)

21. Le Guyader, C., Vese, L.A.: Self-Repelling Snakes for Topology-Preserving Segmentation Models. IEEE Transactions on Image Processing 17(5), 767–779 (2008)

22. Clocksin, W.F.: Automatic Segmentation of Overlapping Nuclei with High Background Variation Using Robust Estimation and Flexible Contour Model. In: Proceedings of ICIAP, vol. 17, pp. 682–687 (2003)

23. Yap, C.K., Lee, H.K.: Identification of Cell Nucleus Using a Mumford-Shah Ellipse Detector. In: Bebis, G., Boyle, R., Parvin, B., Koracin, D., Remagnino, P., Porikli, F., Peters, J., Klosowski, J., Arns, L., Chun, Y.K., Rhyne, T.-M., Monroe, L. (eds.) ISVC 2008, Part I. LNCS, vol. 5358, pp. 582–593. Springer, Heidelberg (2008)

24. Otsu, N.: A Threshold Selection Method from Gray-level Histograms. IEEE Transactions on Systems, Man & Cybernetics 9, 62–66 (1979)

25. Lindeberg, T.: Scale-space Theory: A Basic Tool for Analysing Structures at Different Scales. Journal of Applied Statistics 21(2), 224–270 (1994)
26. Koenderink, J.J.: The Structure of Images. Biological Cybernetics 50, 363–396 (1984)
27. Mikolajczyk, K., Schmid, C.: An Affine Invariant Interest Point Detector. In: Heyden, A., Sparr, G., Nielsen, M., Johansen, P. (eds.) ECCV 2002. LNCS, vol. 2350, pp. 128–142. Springer, Heidelberg (2002)

Deconvolving Active Contours for Fluorescence Microscopy Images

Jo A. Helmuth and Ivo F. Sbalzarini*

Institute of Theoretical Computer Science and Swiss Institute of Bioinformatics,
ETH Zurich, Switzerland

Abstract. We extend active contours to constrained iterative deconvolution by replacing the external energy function with a model-based likelihood. This enables sub-pixel estimation of the outlines of diffraction-limited objects, such as intracellular structures, from fluorescence micrographs. We present an efficient algorithm for solving the resulting optimization problem and robustly estimate object outlines. We benchmark the algorithm on artificial images and assess its practical utility on fluorescence micrographs of the Golgi and endosomes in live cells.

1 Introduction

Active contours are among the most important frameworks for image segmentation. In the original formulation by Kass et al. [1], a contour is defined as a (closed or open) parametric curve in the image domain that minimizes an energy functional. Closed active contours can also be represented implicitly as level sets [2]. This is particularly beneficial when the number of objects to be segmented is not known a priori since it allows for topology changes during energy minimization. In both representations, the energy functional consists of two terms: (1) an *external energy* that depends on image data, and (2) an *internal energy* that solely depends on the geometry of the contour. While the former defines an unconstrained image segmentation problem, the latter provides regularization, helps overcoming local minima, and allows bridging regions with little information in the image. Many extensions of active contours have been proposed over the last two decades, including active masks [3,4], active surfaces [5], and stochastic active contours (STACS) [6,7]. For implicit active contours, very efficient algorithms have been developed based on narrow-band level sets [8] or graph cuts to minimize the energy functional [9].

Active contours are widely used in biological light-microscopy imaging. Their application has, however, so far been restricted to images of objects well above the resolution limit of the imaging equipment. If the size of the object becomes comparable to the width of the point spread function (PSF) of the microscope, the objects are under-resolved and active contour segmentations can no longer

---

* This work was supported by ETH Research Commission grant TH-1007-1 and the Swiss SystemsX.ch initiative, evaluated by the Swiss National Science Foundation.

G. Bebis et al. (Eds.): ISVC 2009, Part I, LNCS 5875, pp. 544–553, 2009.

be considered unbiased estimates of the object's geometry [10]. The only exception are objects of dimension 0 or 1, imaged with a symmetric PSF. This includes particles modeled as points [11] and filaments modeled as lines [12]. For extended objects of co-dimension zero, deconvolution can be used to estimate their geometry from an image blurred by the PSF. Deconvolution is, however, an inverse problem that is known to be ill posed [13,14]. Moreover, direct linear deconvolution is not feasible for sub-cellular structures [15]. Therefore, constrained iterative methods have to be used.

In this paper, we extend the active contour framework to constrained iterative deconvolution by including models of the objects and the imaging process. This involves replacing the external energy functional with a negative log-likelihood function and optimizing it under the constraint of the internal energy. This optimization involves simulating the forward imaging process. We propose an efficient minimization algorithm that uses a domain decomposition approach and exploits the linearity of the convolution operator. The resulting method is an iterative deconvolving active contour that is constrained by the object model and the imaging model. We demonstrate the accuracy and precision of deconvolving active contours on synthetic benchmark images of sub-resolution objects. We show that the present framework allows unbiasing the estimation of object geometries from fluorescence micrographs. We further demonstrate the applicability of the proposed algorithm to images of the Golgi complex and endosomes in live cells. The resulting estimated outlines allow biological observations that were not possible before.

2   Energy Functional

We seek a parametric description of a set S of outlines of objects, supported by an error-corrupted digital image $I_m = I + \varepsilon$. Given an imaging model $I(S)$, parameters $\boldsymbol{\Theta}$ of the set of N objects $S = \{\boldsymbol{\Theta}^k\}_{k=1}^{N}$ have to be found that best explain the measured image I_m. The imaging model $I(S)$ predicts the image I of the set S of objects in the absence of noise. This parameter estimation problem can be rephrased in the explicit active contour framework. Hereby, the outline of an object k is represented by a piece-wise linear spline $\boldsymbol{\Theta}^k = [x_1^k, y_1^k, \ldots, x_{n_k}^k, y_{n_k}^k]^T$. As described at the end of Sec. 3, the computational cost of the algorithm is not significantly influenced by the number of control points n_k used. We minimize the sum of internal and external energy

$$E(I_m, S) = \sum_{\boldsymbol{\Theta} \in S} \left(E_b(\boldsymbol{\Theta}) + E_s(\boldsymbol{\Theta}) \right) + E_{\text{ext}}(I_m, I(S)), \tag{1}$$

where the external energy is given by the similarity between the model image $I(S)$ and the real image I_m, quantifying the likelihood that the objects S have indeed created the observed image I_m. The internal energy comprises regularizations for bending and stretching of the contour as:

$$E_b(\boldsymbol{\Theta}) = \beta \sum_{i=1}^{n} \| x_{i+1} - 2x_i + x_{i-1} \|^2 \text{ and}$$
$$E_s(\boldsymbol{\Theta}) = \alpha \sum_{i=1}^{n} \| x_i - x_{i-1} \|^2 \tag{2}$$

with $\boldsymbol{x}_i = (x_i, y_i)$ of the spline $\boldsymbol{\Theta}$. While the "bending stiffness" β limits undulations of the outlines, the "stretching stiffness" α constrains the shrinking of outlines to significant image energy gradients. We define $I(S)$ based on a function $O(i, j)$ over the image pixels $\{x_i\} \times \{y_j\}$. This *object intensity function O* represents the pre-imaging objects up to a multiplicative constant. In fluorescence microscopy, O is proportional to the concentration of fluorophores at each point in the focal plane. Formation of the model image (Fig. 2B) is done by convolving the object intensity function O with the (measured) PSF P of the microscope:

$$I = O * P, \tag{3}$$

where $*$ denotes the discrete convolution operator. In practice, we sample $O(i, j)$ at higher spatial resolution (two, three, or four-fold) than the measured image in order to include sub-pixel information. This requires down-sampling of I before comparison to I_m. The object intensity function $O(i, j)$ is defined from the set S of outlines by setting to non-zero values only the pixels (i, j) close to or enclosed by one of the outlines $\boldsymbol{\Theta}^k$:

$$O(i, j) = \begin{cases} c^k & \text{if } (x_i, y_j) \text{ enclosed by } \boldsymbol{\Theta}^k \\ (1 - d) c^k & \text{if } d = D((x_i, y_j), S) < 1 \\ 0 & \text{else} \end{cases} \tag{4}$$

where D is the distance to the closest spline $\boldsymbol{\Theta}^k$ and c^k the constant intensity of object k. Similar to the simplified Mumford-Shah functional [2], this object intensity function is piecewise constant, but with linearly decaying intensities at the boundaries. We favor this piecewise linear functional over more complex models as it increases the robustness of the estimator on noisy data. The external energy E_{ext} is given by:

$$E_{\text{ext}} = \sum_i \sum_j R(i, j) \left(I_m(i, j) - I(i, j) \right)^2 . \tag{5}$$

The weight matrix R allows including a model for the distribution of the imaging noise ε. In the absence of knowledge about ε, or for Gaussian white noise, R is the identity matrix.

Minimizing E over the $\boldsymbol{\Theta}^k$ yields an estimate of $O(i, j)$. Direct estimation of O based on I_m amounts to direct linear deconvolution, a problem known to be ill posed. The present framework can thus be interpreted as a constrained, iterative deconvolution [13,14] with the constraints defined in Eqs. 2 and 4.

3   Minimization of the Energy Functional

We assume that initial estimates of the outlines are provided by a suitable pixel-based segmentation procedure and that they represent the correct topology of the objects. Further, the 2D PSF of the imaging device is assumed to be known. Minimizing Eq. 1 could then be done using any general purpose optimizer. A specialized procedure exploiting the structure of this high-dimensional problem

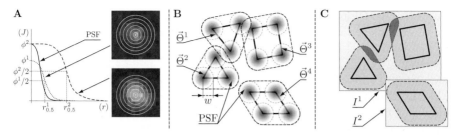

Fig. 1. (A) Radial intensity profile of a small (top image, dotted line) and a large (bottom image, dashed line) object. The intensity profile of the small object and the PSF (solid line) have a similar shape. (B) The width w of the PSF defines the influence regions (dashed lines) of the objects Θ^k (solid lines). (C) Objects with overlapping influence regions (dark gray areas) are grouped together.

is, however, favorable. Our algorithm consists of three steps: (1) estimating the object intensities c^k, (2) decomposing the image into smaller parts, and (3) estimating the precise outlines Θ^k of all objects.

The **object intensities** c^k are separately **estimated** in order to use them in Eq. 4 and as a biologically relevant readout. For objects that are far larger than the width of the PSF, the object intensity is approximately equal to the intensity ϕ in the center of the image of the object (bottom image in Fig. 1A). Smaller objects have a central intensity that is reduced by a factor κ. In order to estimate κ^k for a given outline Θ^k, we analyze the radial intensity profile $J(r)$ of the object (Fig. 1A), found by averaging interpolated intensities along concentric circles around the intensity centroid. The same procedure is also used to measure the radially symmetric 2D PSF of the microscope from images of point-like sources such as fluorescent beads. The half width at half maximum (HWHM) $r_{0.5}$ of $J(r)$ serves as a size parameter (Fig. 1A). Since the dependence $\kappa = f(r_{0.5})$ is not explicitly known, we empirically calibrate it on synthetic images, generated by convolving circular objects of different sizes and known intensities c^i with the PSF. For each synthetic image I^i we measure $r_{0.5}^i$ and the central intensity ϕ^i, and then compute the calibration function as $\bar{\kappa}^i(r_{0.5}^i) = c^i/\phi^i$. Based on this function, the object intensity c^k of an experimentally observed object k can later be estimated as $c^k = \phi^k \kappa(r_{0.5}^k)$ using (linear) interpolation. For point-like objects, $r_{0.5}$ converges to the HWHM of the PSF and does no longer vary with object size. The analysis of sizes and shapes is, therefore, restricted to objects with $r_{0.5}$ above an empirically determined threshold of 1.1–1.5 HWHM. Smaller objects are treated as circles with centers at the observed intensity centroid and object intensities c found by least squares regression on Eqs. 3 and 5.

In order to accelerate the computations and estimate the outlines of different objects on different computer processor cores in parallel, we **decompose the image** into smaller, independent parts. This is possible since the PSF P is of limited spatial extent and the influence of objects on distant pixels in the image can be neglected. We define the radius of the region of influence around each outline Θ^k as the width w where the PSF has decayed below 1% of its peak (Fig. 1B). Objects with overlapping influence regions are grouped together and their influence regions are merged. This decomposes the image into (not necessarily disjoint) rectangular sub-images I^l containing the merged influence regions (Fig. 1C). Eq. 1 is then evaluated independently for each sub-image and we no longer consider pixels that are in no sub-image.

Estimating the outlines of the objects to sub-pixel resolution is done by minimizing the energy functional in Eq. 1 with respect to the Θ^k. This amounts to implicit deconvolution. The energies of different sub-images are minimized independently in parallel. We use the optimizer proposed by Kass et al. [1] with an explicit Euler method of adaptive step size. This requires approximating the partial derivatives of E with respect to the x and y positions of all spline points (x_i^k, y_i^k), $i = 1, \ldots, n_k$. We use a finite-difference approximation to the derivatives with respect to a point (x_i^k, y_i^k) on the outline Θ^k by displacing the point by a dynamically adapted Δx. This yields a deformed hypothetical outline Θ_*^k in a new set of outlines S_*. The derivative of the external energy E_{ext}^l in sub-image l is then approximated as:

$$\frac{\partial E_{\text{ext}}^l}{\partial x_i^k} \approx \frac{E_{\text{ext}}^l \left(I_m^l, I^l(S_*) \right) - E_{\text{ext}}^l \left(I_m^l, I^l(S) \right)}{\Delta x} . \tag{6}$$

The derivatives with respect to other points and in the y-direction are found similarly. The hypothetical sub-image $I^l(S_*)$ needs to be computed once per iteration of the minimization procedure. Since convolution is a linear operation, the image $I^l(S_*)$ can be expressed as the sum of $I^l(S)$ and a change ΔI^l caused by the deformation of Θ^k. ΔI^l is found by computing $O(i,j)_*^k$ from Θ_*^k, subtracting it from $O(i,j)^k$, and convolving this difference $\Delta O(i,j)^k$ with P. This drastically reduces the number of compute operations and the run-time. The computational cost of the algorithm is dominated by the cost of the convolutions; the cost of the active contour updates is insignificant. The total cost of all convolutions is proportional to the total number of non-zero entries in *all* $\Delta O(i,j)^k$, which scales linearly with the length of the outline. Reducing the number of control points n_k would thus not lead to significant computational savings since the total length of the outline (and hence the computational cost of the convolutions) remains the same. After computing all partial derivatives of the image error, one Euler step is performed. Minimization starts from the initial outline estimates and ends when the L_∞-norm of the change of outlines between two iterations, $\|\Theta_{i+1} - \Theta_i\|_\infty$, is below a user-defined threshold. This implies that the gradient-descent minimizer has converged to a (local) minimum. Local minima can possibly be escaped by increasing α.

4   Benchmarks

We quantify the accuracy and precision of the presented algorithm on synthetic benchmark images of diffraction-limited objects. Images are generated as illustrated in Fig. 2A to C using a real, measured PSF with full width at half maximum FWHM = 322 nm. We consider images of a circular object with diameter

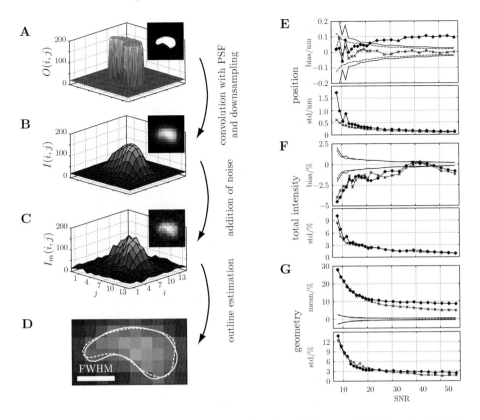

Fig. 2. (A–C) Generation of the synthetic benchmark images. (D) True (solid line) and estimated (dashed line) outline for SNR = 12.5 (geometry error = 14%). (E–G) Benchmark results for a pear-shaped (solid lines with diamonds) and circular (dashed lines with crosses) object. Lines without markers delimit the ± 1 standard error interval.

1.5 FWHM and of a 0.6 to 1.2 FWHM wide and 2.8 FWHM long pear-shaped object. The ground-truth $O(i,j)$ is generated from the true outlines according to Eq. 4. The object intensity c is set to 200 and a background level of $b = 20$ is added (Fig. 2A). Different amounts of noise yield signal-to-noise ratios SNR $= (c-b)/\sigma_c$ in the range of 7 to 56, where σ_c is the noise level in the center of the objects. For each object and SNR, 250 test images are generated, object outlines are estimated, and compared to ground truth.

We quantify accuracy and precision of the estimated position, total intensity, and geometry (Fig. 2E to G). The position error is defined as the difference between the true and estimated x-position of the intensity centroid[1]. The total intensity error is given by the difference in total intensity (sum over all $O(i,j)$ enclosed by the outline) between the estimated object and ground truth, divided by ground truth. The geometry error is defined as the sum of non-overlapping

---

[1] We prefer this over the Euclidean distance in (x,y) since it enables correlating shape asymmetries with the position errors in the different directions.

areas of the true and estimated outlines, normalized by the area enclosed by the true outline. For both shapes, precision and accuracy of the estimated position are in the range of a few nanometers (Fig. 2E). The errors in the y-direction are comparable for the pear-shaped object and identical for circular shape (data not shown). The position bias for the circular shape is always within an interval of $\pm$ the standard error, and hence not significant. As expected, a small systematic position bias can be observed for the pear-shaped object. Since this shape is not symmetric, the underestimation of high curvatures due to E_b causes a small shift in the position estimate toward the less curved side. For both shapes the standard deviation of the relative total intensity error drops below 5% for SNRs larger than 10 (Fig. 2F). The bias is larger than the standard error, almost always negative, and converges to about -1%. This is due to E_s favoring shorter outlines and thus decreasing the enclosed object intensity. The means and standard deviations of the geometry errors of both shapes converge to values of less than 10% and 3%, respectively (Fig. 2G). The bending and stretching energies E_b and E_s prevent the mean geometry error from converging to zero, and the underestimation of high curvatures causes an additional bias for the more complex pear-shaped object. Nevertheless, we observe that the estimated outlines visually reproduce well the essential characteristics of the true outlines, even at SNRs below 15 and, therefore, error levels >10% (Fig. 2D).

Adjusting the bending stiffness β (Eq. 2) allows trading the accuracy of the outline estimation against its robustness by limiting undulations of the contour. Low values lead to a higher noise sensitivity (less regularization), but allow better estimation of high curvatures. In order to qualitatively assess this trade-off, we apply the algorithm to synthetic images of a triangle (Fig. 3), generated as described above. As expected, we observe that low SNRs favor high values of β, and vice versa. The algorithm is robust over two orders of magnitude of β. Only the most extreme case ($\beta = 0.02$, SNR $= 5$) exhibits significant shape instabilities. The stretching stiffness α has much less influence on the final contour. Higher values lead to faster convergence of the algorithm and better escape from local minima. At the same time, however, they bias the outlines to shorter, more contracted contours. We find a value of $\alpha = 0.005$ to be sufficient to overcome local minima and speed up convergence.

5   Application to Real Data

We demonstrate the utility of the present algorithm on fluorescence microscopy images of different intracellular structures. First, we apply it to an image of the Golgi complex in HeLa cells labelled by fluorescent giantin antibodies (Fig. 4A and B). The Golgi is a complex-shaped intracellular organelle composed of membrane stacks of about $5\,\mu$m size. The same image was also used to demonstrate active mask segmentation [3]. We show how such a coarse, pixel-level segmentation can be refined by the present implicit deconvolution method. We start from a rough manual segmentation obtained from Fig. 11d of Ref. [3]. Since no information about the PSF was available, we model it by a Gaussian with

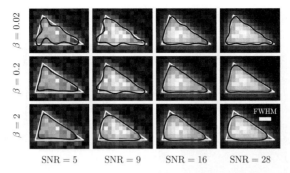

Fig. 3. Estimation of a highly curved outline at different SNRs and values of β ($\alpha = 0.005$ in all cases). White lines: true outlines; black lines: estimated outlines.

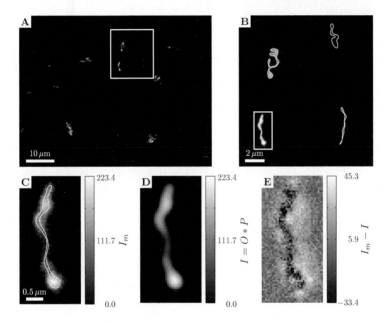

Fig. 4. Outline estimation of the Golgi complex in HeLa cells [16]. (A) Original image containing multiple cells. Magnification of a single cell (B) and a single Golgi segment (C) with estimated outlines (white line); model image (D) and residual error (E).

$\sigma = 150\,\text{nm}$. This is a conservative choice for the imaging set-up used (spinning disk confocal, NA $= 1.4$, oil immersion). The final deconvolving active contours (Fig. 4B and C) capture well the morphological characteristics of the Golgi and the model image I (Fig. 4D) is remarkably close to the real image (Fig. 4C). The estimated outline shows no obvious signs of over-fitting. The residual error $I_m - I$ (Fig. 4E) shows that the model image trends to be too bright in the center of the object and under-represents the blur around it. As shown below, this is likely due to a too narrow model PSF.

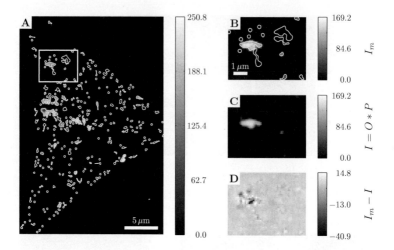

Fig. 5. Outline estimation (white lines) of endosomes in a live HER 911 cell (A). (B) Magnification, (C) model image, and (D) residual error in the magnified region.

The second application considers live HER 911 cells expressing EGFP-tagged Rab5, a protein marker for endosomes. With diameters of about 500 nm, endosomes are much smaller than the Golgi and they appear more compact. Initialized with watershed segmentation, the present algorithm estimates the outlines as shown in Fig. 5A and B. The PSF (FWHM = 322 nm) of the microscope (spinning disk confocal, NA = 1.35, oil immersion) was measured from images of sub-diffraction objects as described in Sec. 3. Fig. 5B shows complex-shaped endosome outlines in close vicinity. The outlines follow well the subjective contours in the images, even for very dim objects. The correspondence between the real and the model image (Fig. 5C) is remarkable. Also, unlike in the Golgi case, there is no clear trend in the residual error (Fig. 5D), highlighting the advantage of using the true, measured PSF. Except for slight over-estimation of the central intensity of the large object on the left, the residual error is dominated by detector noise.

6   Conclusions and Discussion

We have introduced deconvolving active contours, extending explicit active contours to iterative constrained deconvolution by replacing the image-based external energy with a model-based likelihood function that includes prior knowledge about the imaging process. The algorithm iteratively refines an initial image segmentation using regularized optimization. Optimizing the likelihood function is computationally more involved than optimizing a classical pixel-based energy. We have thus introduced a special-purpose algorithm that uses domain decomposition parallelism and exploits the linearity of the convolution operator. Runtimes on a desktop computer are on the order of seconds for individual objects (e.g. 1.2 s for the object in Fig. 2D). The presented algorithm enables estimating the outlines of diffraction-limited, asymmetric objects to sub-pixel accuracy. The benchmarks demonstrated localization precision in the nanometer range (better than 0.01 FWHM) and fluorescence intensity estimation to within a few %. We have further demonstrated the practical utility of deconvolving active contours on images of the Golgi complex and endosomes in live cells.

References

1. Kass, M., Witkin, A., Terzopoulos, D.: Snakes: Active contour models. Int. J. Computer Vision, 321–331 (1988)
2. Chan, T.F., Vese, L.A.: Active contours without edges. IEEE Trans. Image Process. 10, 266–277 (2001)
3. Srinivasa, G., Fickus, M.C., Guo, Y., Linstedt, A.D., Kovačević, J.: Active mask segmentation of flourescence microscope images. IEEE Trans. Image Process. 18, 1817–1829 (2009)
4. Srinivasa, G., Fickus, M.C., Gonzalez-Rivero, M.N., Hsieh, S.Y., Guo, Y., Linstedt, A.D., Kovačević, J.: Active mask segmentation for the cell-volume computation and Golgi-body segmentation of HeLa cell images. In: Proc. IEEE Int. Symp. Biomed. Imaging, pp. 348–351 (2008)
5. Dufour, A., Shinin, V., Tajbakhsh, S., Guillen-Aghion, N., Olivo-Marin, J.C., Zimmer, C.: Segmenting and tracking fluorescent cells in dynamic 3-D microscopy with coupled active surfaces. IEEE Trans. Image Process. 14, 1396–1410 (2005)
6. Pluempitiwiriyawej, C., Moura, J.M.F., Wu, Y.J.L., Ho, C.: STACS: New active contour scheme for cardiac MR image segmentation. IEEE Trans. Med. Imaging 24, 593–603 (2005)
7. Coulot, L., Kirschner, H., Chebira, A., Moura, J.M.F., Kovačević, J., Osuna, E.G., Murphy, R.F.: Topology preserving STACS segmentation of protein subcellular location images. In: Proc. IEEE Int. Symp. Biomed. Imaging, pp. 566–569 (2006)
8. Shi, Y., Karl, W.: Real-time tracking using level sets. In: Proc. IEEE Conf. CVPR, vol. 2, pp. 34–41 (2005)
9. El-Zehiry, N., Elmaghraby, A.: An active surface model for volumetric image segmentation. In: Proc. IEEE Int. Symp. Biomed. Imaging, pp. 1358–1361 (2009)
10. Streekstra, G.J., van Pelt, J.: Analysis of tubular structures in three-dimensional confocal images. Network: Comput. Neural Syst. 13, 381–395 (2002)
11. Sbalzarini, I.F., Koumoutsakos, P.: Feature point tracking and trajectory analysis for video imaging in cell biology. J. Struct. Biol. 151, 182–195 (2005)
12. Li, H., Shen, T., Smith, M.B., Fujiwara, I., Vavylonis, D., Huang, X.: Automated actin filament segmentation, tracking and tip elongation measurements based on open active contours. In: Proc. IEEE Int. Symp. Biomed. Imaging, pp. 1302–1305 (2009)
13. Sibarita, J.B.: Deconvolution microscopy. Adv. Biochem. Eng. Biot. 95, 201–243 (2005)
14. Sarder, P., Nehorai, A.: Deconvolution methods for 3-D fluorescence microscopy images. IEEE Signal Process. Mag. 23, 32–45 (2006)
15. Meijering, E., Smal, I., Dzyubachyk, O., Olivo-Marin, J.C.: Time Lapse Imaging. In: Microscope Image Processing, ch. 15. Academic Press, London (2008)
16. Guo, Y., Linstedt, A.D.: COPII-Golgi protein interactions regulate COPII coat assembly and Golgi size. J. Cell Biol. 174, 53–63 (2006)

Image Registration Guided by Particle Filter

Edgar R. Arce-Santana*, Daniel U. Campos-Delgado, and Alfonso Alba**

Facultad de Ciencias, Diagonal Sur S/N, Zona Universitaria, C.P. 78290,
San Luis Potosi, S.L.P., Mexico

Abstract. Image Registration is a central task to different applications, such as medical image analysis, stereo computer vision, and optical flow estimation. One way to solve this problem consists in using
Bayesian Estimation theory. Under this approach, this work introduces
a new alternative, based on Particle Filters, which have been previously
used to estimate the states of dynamic systems. For this work, we have
adapted the Particle Filter to carry out the registration of unimodal and
multimodal images, and performed a series of preliminary tests, where
the proposed method has proved to be efficient, robust, and easy to
implement.

1 Introduction

The goal of Image Registration is to find an optimal geometric transformation
between corresponding image data [1],[2], where the criterion for optimality depends on a specific application. This task is very important to many applications
involving image processing or analysis such as medical-evaluation, biomedical
systems, image guidance, depth estimation, and optical flow. In the past 10
years, many methods have been published; an extensive and comprehensive survey can be found in [3],[4].

Image Registration Methods can be classified as global or local. In the global
approaches one searches a model, often a parametric one such as rigid, affine,
projective, or curved transformation, explaining the similarities between two
images. Local methods or dense registration seek individual correspondences for
each pixel in both images. A special kind of registration is called Multimodal
Image Registration, in which two or more images coming from different sources
are aligned; this process is very useful, for example, in computer aided visualization in the medical field, since it allows one to find correspondences between
functional and anatomical images.

In the literature, the more common global methods are those based on intensity changes [5],[6],[7]. Within these methods, there is a very popular one
based on Mutual Information (MI), proposed independently by Viola and Wells

* The author was supported by Grant PROMEP/103.5/04/1387 and Fac. de Ciencias,
UASLP, Mexico.
** The author was supported by Grant PROMEP/103.5/09/2416 and Fac. de Ciencias,
UASLP, Mexico.

G. Bebis et al. (Eds.): ISVC 2009, Part I, LNCS 5875, pp. 554–563, 2009.

[8],[9] and Collignon and Maes [10],[11]. Even though it is theoretically robust, it requires to find the maximum mutual information, for example, using optimization techniques such as hill-climbing, making the methods very sensitive to its initial parameters and the derivative calculations. Other related work is found in [12], where an affine transformation between images is modeled locally using a linear approximation (first order Taylor's expansion), but due to this approximation, the algorithm must use a differential multiscale framework in order to approximate large geometric displacements with a series of shorter ones. Other global approaches that have demonstrated to be very efficient to estimate affine transformation parameters are those based on Bayesian Estimation. Generally, in these methods, Image Registration is solved by estimating a statistic, for example the mean or mode, from a cost function, given a posterior probability [13],[14],[15]. A commonly used estimator is the maximum a posteriori (MAP), which can be found by minimizing an energy function; unfortunately, these functions are highly nonlinear with respect to the geometrical parameters, requiring complex optimization methods and demanding considerable computing time.

Under the same Bayesian approach, in the present work, we propose a new algorithm based on a method that has been used to estimate states of nonlinear dynamics systems, known as Particle Filter [16],[17]. This method has been adapted to solve the problem of image registration, showing that the methodology behind Particle Filter can be used in a very efficient way to solve these kinds of image processing tasks. The paper is organized as follows: Section 2 describes the basis of the Particle Filter; in Section 3, we describe how the filter is adapted to the image registration problem; Section 4 shows some experiments and results; and finally, in Section 5, some conclusions are presented.

2 Particle Filter Basis

Particle Filter is a method based on Bayesian Estimation that uses a Monte Carlo algorithm to estimate probability density statistics [16],[17]. The key idea is to obtain a posterior density from a set of random samples with associated weights, which allows one to estimate state variables defining a dynamic system.

In order to establish the problem, consider the evolution of a state sequence, given by

$$s_k = f_k(s_{k-1}, v_{k-1}), \tag{1}$$

where $f_k(.)$ is a function, possibly nonlinear, of the system state vector s_{k-1}, at time k-1, and v_{k-1} is independent and identically distributed (iid) noise. Thus, the goal is recursively to estimate the state vectors s_k from the measurements

$$z_k = h_k(s_k, n_k), \tag{2}$$

where $h_k(.)$ is a possibly nonlinear function, and n_k are iid noise samples. In particular, we want to estimate the distribution $p(s_k|z_{1:k})$ from a set of measurements $z_{1:k} = \{z_i, i = 1, ..., k\}$. Suppose that we have access to the pdf

$p(s_{k-1}|z_{1:k-1})$ at time *k-1*. The method consists in a prediction stage to approximate the pdf of the state system at time k, via the Chapman-Kolmogorov equation

$$p(s_k|z_{1:k-1}) = \int p(s_k|s_{k-1})p(s_{k-1}|z_{1:k-1})ds_{k-1}. \qquad (3)$$

Notice that, we have used the fact that $p(s_k|s_{k-1}, z_{1:k}) = p(s_k|s_{k-1})$, which describes a Markovian process. At time k, a measurement z_k is available, and using the Bayes' theorem, we can predict (update stage) the posterior density

$$p(s_k|z_{1:k}) = \frac{p(z_k|s_k)p(s_k|z_{1:k-1})}{p(z_k|z_{1:k-1})}, \qquad (4)$$

where the normalizing constant is

$$p(z_k|z_{1:k-1}) = \int p(z_k|s_k)p(s_k|z_{1:k-1})ds_k. \qquad (5)$$

Equations (3) and (4) provide a form to estimate the posterior probability $p(s_k|z_{1:k})$ in a recursive way. This structure, in many cases, cannot be analytically established, particularly when $f_k(.)$ and/or $h_k(.)$ are nonlinear. One way to solve the problem is applying Particle Filter. In this method, the distribution $p(s_k|z_{1:k})$ is represented by samples (particles) $\{(s_k^i, w_k^i) : i = 1, ..., N_s\}$, where s_k^i are the particles' values, at time k, and w_k^i are the associated weights such that $\sum_i w_k^i = 1$.

The Particle Filter algorithm is by nature iterative, and is composed of two stages: A) Prediction stage, in which each of the state variables (particles) is modified according to equation (1); and B) Update stage, where the particles' weights are recalculated using the measurements information described by equation (2). It is convenient to mention that it is common to take $p(s_k|s_{1:k-1}) = p(s_k|s_{k-1})$, which means that the actual state variables depend on their previous values (1st order Markov process). Also the likelihood function $p(z_k|s_{1:k}) = p(z_k|s_k)$ depends only on the latest state.

In the prediction stage, particles at time *k-1* are propagated to generate new particles at time *k*. A common drawback in this process is a phenomenon known as the Degeneracy Problem [18], where after a few iterations, all but one particle will have negligible weights. One way to resolve this problem is by *resampling*, in order to eliminate particles having small weights and concentrate in particles having large contribution: the particles that represent $p(s_{k-1}|z_{k-1})$ at time *k-1* are used to obtain new particles (*resampling*); next, the state of each of these particles is modified according to (1)(prediction stage); finally, their weights w_k^i are updated using the likelihood function $p(z_k|s_k)$ in order to obtain representative samples of $p(s_k|z_k)$.

3   Image Registration Guided by Particle Filter

In the present work, we adapted the Particle Filter algorithm to solve the image registration problem between two images I_1 and I_2, which are related by the observation model $I_1(x, y) = F(I_2(T(x, y)))$, where $F(.)$ is an intensity transfer function, depending on the gray values of I_2, and $T(.)$ is the affine transformation:

$$T([x, y]) = [x, y, 1] \begin{bmatrix} \lambda_x \cos\theta & -\sin\theta & 0 \\ \sin\theta & \lambda_y \cos\theta & 0 \\ d_x & d_y & 1 \end{bmatrix}. \qquad (6)$$

In this equation, θ represents the rotation angle of one image with respect to the other one, λ_x is the scale factor on the x-axis, λ_y on the y-axis, d_x is the translation on x, and d_y on y-axis.

To carry out the registration, we assume as state variables the vector $(s_k^i) = \left[\theta_k^i, \lambda_{x k}^i, \lambda_{y k}^i, d_{x k}^i, d_{y k}^i\right]^T$, the geometric transformation parameters in (6); and define the state equation, which in this case is a simple random walk:

$$\begin{bmatrix} \theta_k^i \\ \lambda_{x k}^i \\ \lambda_{y k}^i \\ d_{x k}^i \\ d_{y k}^i \end{bmatrix} = \begin{bmatrix} \theta_{k-1}^i \\ \lambda_{x k-1}^i \\ \lambda_{y k-1}^i \\ d_{x k-1}^i \\ d_{y k-1}^i \end{bmatrix} + \begin{bmatrix} v_\theta \\ v_{\lambda_x} \\ v_{\lambda_y} \\ v_{d_x} \\ v_{d_y} \end{bmatrix}. \qquad (7)$$

Notice that the prediction equation is very simple: it is the preceding value disturbed by the noise vector $\left[v_\theta, v_{\lambda_x}, v_{\lambda_y}, v_{d_x}, v_{d_y}\right]^T$. Here we assumed Gaussian and independent noise for each parameter with zero mean and standard deviations $\sigma_\theta, \sigma_\lambda, \sigma_d$, respectively.

In the update stage, we want to know how well each new particle value s_k^i fits the observation model

$$F(I_2(T(x, y))) = I_1(x, y) + \gamma(x, y), \qquad (8)$$

where $\gamma(x, y)$ is iid noise, with zero mean and standard deviation σ_γ. In order to accelerate the process, we only consider a set of m uniformly distributed pixels $P = \{(x_i, y_i); i = 1...m\}$, in the images I_1, and I_2, so that the likelihood function may be expressed as

$$p(z_k|s_k^i) = \prod_{(x,y) \in P} p(I_1(x, y), F(I_2(T(x, y)))|s_k^i). \qquad (9)$$

Notice that the affine transformation is applied to the coordinates of one of the images, meaning that the observation model is highly nonlinear with respect to the transformation parameters. Next, we detail the image registration algorithm guided by the Particle Filter.

3.1   Algorithm: Image Registration Guided by Particle Filter

Given a set of particles at time k-1, $\{(s_{k-1}^i, w_{k-1}^i) : i = 1, ..., N_s\}$, where the state values are given by the transformation parameters

$$s_{k-1}^i = \left[\theta_{k-1}^i, \lambda_{x k-1}^i, \lambda_{y k-1}^i, d_{x k-1}^i, d_{y k-1}^i\right]^T \qquad (10)$$

1. For each particle, compute the cumulative probability as

$$c_{k-1}^0 = 0 \tag{11}$$

$$c_{k-1}^i = c_{k-1}^{i-1} + w_{k-1}^i, \text{for } i = 1, ..., N_s. \tag{12}$$

2. For each particle s_{k-1}^i, do the resampling as follows:
   - Generate a uniform random value $u \in [0, 1]$.
   - Find the smallest j such that $c_{k-1}^j \geq u$.
   - Select the state $\widehat{s}_{k-1}^i = s_{k-1}^j$.
3. Obtain the new set of samples at time k (Prediction stage), using the equation: $s_k^i = \widehat{s}_{k-1}^i + v_k$.
4. For each new state s_k^i, compute the corresponding weights $w_k^i = p(z_k|s_k^i)$; that is the likelihood function (Update stage):

$$p(z_k|s_k^i) = \prod_{(x,y)\in P} p(I_1(x, y), I_2(T(x, y))|s_k^i). \tag{13}$$

5. Normalize the weight, such that $\sum_i w_k^i = 1$.
6. Once the particles' weights have been computed, we may evaluate the mean for each affine transformation parameter to compute the estimations:

$$\begin{aligned}
E[\theta_k|z_k] &= \sum_{i=1}^{N_s} w_k^i \theta_k^i, \quad E[\lambda_{xk}|z_k] = \sum_{i=1}^{N_s} w_k^i \lambda_{xk}^i, \\
E[\lambda_{yk}|z_k] &= \sum_{i=1}^{N_s} w_k^i \lambda_{yk}^i, \quad E[d_{xk}|z_k] = \sum_{i=1}^{N_s} w_k^i d_{xk}^i, \\
E[d_{yk}|z_k] &= \sum_{i=1}^{N_s} w_k^i d_{yk}^i.
\end{aligned} \tag{14}$$

4   Experiments and Results

In this section, we present two different kinds of results. The first one corresponds to an observation model which tries to match similar gray values in order to carry out the image registration. The model used in the experiments was

$$I_2(T(x, y)) = I_1(x, y) + \gamma(x, y). \tag{15}$$

In this case, the function $F(.)$ was the identity. In order to avoid numerical instability, due to the likelihood $p(z_k|s_k^i)$ being computed as the product of the individual measurement of pixels in the set P, it is appropriated to choose a robust γ function [19], having the maximum at the mean, and tails quickly reaching nonzero values. A function that satisfies both requirements is the following one:

$$\gamma(x, y) = \frac{1}{1 + \frac{|I_1(x,y) - I_2(T(x,y))|}{\sigma_\gamma^2}} \tag{16}$$

where σ_γ^2 is a parameter that depends on the standard deviation of the noise.

Using (15) and (16), the following experiment consisted on registering 256×256 pixel images. For the experiment, we used as initial parameter values the

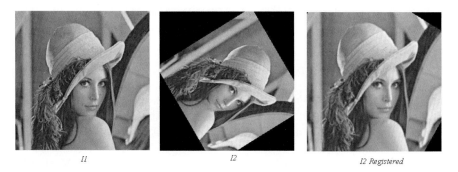

II   *I2*   *I2 Registered*

Fig. 1. Image registration. Left: reference image; middle: image to register; right: registered image I_2.

identity $(\theta = 0, \lambda_x = 1, \lambda_y = 1, d_x = 0, d_y = 0)$, 100 particles, and 256 pixels for computing the likelihood. The image to register was obtained by applying an artificial affine transformation with $\theta = -45$, $\lambda_x = 0.6$, $\lambda_y = 0.6$, $d_x = 10$, and $d_y = 15$. The estimated parameters, after 5 seconds, were $\theta = -44.9276$, $\lambda_x = 0.59821$, $\lambda_y = 0.59822$, $d_x = 9.937$, and $d_y = 15.015$; Figure 1 shows the obtained registered image.

Next, we tested the robustness of the algorithm with respect to noise. We added normal random values, with standard deviations from 0 to 20, to the image I_2 in Fig. 1, and used the same initial values that the previous experiment. The Box-Plot of the True Relative Mean Error (TRME) of the affine parameters are shown in panel $A)$ of Fig. 2; the mean and standard deviation are shown in

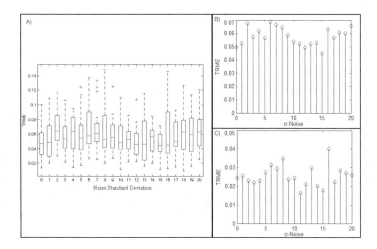

Fig. 2. TRME results for noise standard deviation from (0,....,20): A) TRME-Box-Plot; B)TRME-Mean; C) TRME-SD

panels B) and C). This error function has the advantage of independently taking into account the unit scales of the quantities to evaluate, and it is computed as follows:

$$TRME = \frac{\sum_{k=1}^{5} \left| \frac{s^*(k)-s(k)}{s^*(k)} \right|}{5}, \tag{17}$$

where $s^*(k)$ corresponds to the k-th true parameter value, and $s(k)$ to the estimated one by the algorithm. Notice that, no matter the noise level, the median and mean are between 4% and 7%, and the standard deviation is between 1% and 4%.

In the second set of experiments, we used as measure model the Mutual Information (MI) defined as

$$\mathcal{I}(I_1, I_2) = \mathcal{H}(I_1) + \mathcal{H}(I_2) - \mathcal{H}(I_1, I_2), \tag{18}$$

where $\mathcal{H}(I_1)$ is the entropy of the gray value distribution of the image I_1, and $\mathcal{H}(I_1, I_2)$ is the joint entropy of the images I_1 and I_2. This is a more general measure, since it allows one to register Multi-modal images, coming from different sensor types. An example of such images is in panels A) and B) of Fig. 3, where the first one is a CT-Image and the other one is a RM-Image. This measure is more general in the sense that if I_2 is the result of a 1-1 intensity function mapping $F(I_1)$, then $\mathcal{I}(I_1, I_2) = \mathcal{I}(I_1, F(I_1)) = \mathcal{I}(I_1, I_1) = \mathcal{H}(I_1)$. Also, we know that this measure is bounded: $0 \leq \mathcal{I}(I_1, I_2) \leq \min(\mathcal{I}(I_1, I_1), \mathcal{I}(I_2, I_2))$. Hence, we may define as likelihood function

$$\gamma(\mathcal{H}(I_1), \mathcal{I}(I_1, T(I_2)); \sigma) = \frac{1}{\sqrt{2\pi}\sigma} \exp\left\{ -\frac{(\mathcal{H}(I_1) - \mathcal{I}(I_1, T(I_2)))^2}{2\sigma^2} \right\}, \tag{19}$$

for a given σ, and an affine transformation $T(.)$, since we have defined as reference image I_1. Note that (19) has a gaussian shape and reaches its maximum when $I_1(x, y) = F(I_2(T(x, y)))$. In Figure 3.$C$), we used a checker-board in

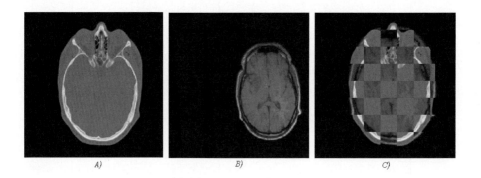

A)   B)   C)

Fig. 3. Multimodal image registration. Left: CT-image; center: MR-image; right: CT and MR registration.

order to appreciate the registration details, in which each square corresponds, alternatively, to the image in A) and to the registered image in B).

Multi-modal image registration is very important in the medical field, since, for example, it allows one to find correspondences between functional and anatomical images, or align a set of MR-slices to reconstruct a 3D brain volume. In this work, we present an example of 3D volume reconstruction in which we took 27 axial MR-slices, each one of 128×128 pixels. For each one of these slices, we applied an artificial relative geometrical transformation with respect to the previous one. For the translation parameters, random values were generated between -10 to 10 pixels, and for the rotation between -10 to 10 degrees; panel A) in Fig. 4 shows the volume obtained from these transformed slices. In order to realign the images, the affine transformation between two consecutive images I_i and I_j was found, and it was propagated from I_j until the last image; the process was repeated from slice I_2 to I_{27}. Panel B) in Fig. 4 shows the 3D reconstructed volume. To verify numerically how well the displacement and rotation transformation were estimated, we compute the squared error for each parameter at each pair of adjacent slices. Figure 5 shows the histograms of these estimations. Notice that the maximum translation error on the x-axis is about 1.6 pixels, that corresponds only to one

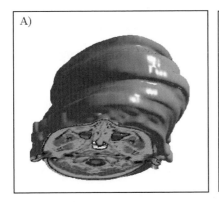
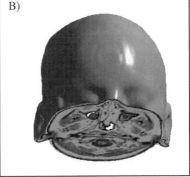

Fig. 4. 3D-Reconstruction. A) Non-aligned volume; B) Aligned volume.

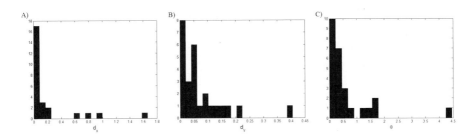

Fig. 5. Error histograms: A) displacement on x-axis; B) displacement on y-axis; C) rotation

Table 1. Mean and SD of the MSE

Parameter	Mean	SD
d_x	0.1979	0.3923
d_y	0.0700	0.0901
θ	0.0105	0.0162

slice; on the y-axis the maximum error is less than 0.5 pixels; and only one slice had a rotation error of 4.5 degrees, the rest had less than 2 degrees. We found that these maximum errors correspond to the last slices, from bottom to top in Fig. 4, this is because the whole volume is composed only by 27 slices from which the last ones are the less similar (top of head), and thus are more difficult to align. Finally, Table 1 shows the mean and standard deviation of the MSE of the estimated geometrical transformations.

5   Conclusions and Future Work

In the present work, we described a new algorithm for Image Registration based on Particle Filter method used to estimate states variables of dynamic systems. This method was adapted to carry out affine image registration. Although Particle Filter is by nature stochastic, we presented experiments where the algorithm showed to be very efficient and accurate to estimate geometric transformation parameters. Finally, we showed that the algorithm is easy to implement and robust with respect to noise, and it is possible to include complex likelihood expressions, in contrast to others optimization methods which complexity depends on this term.

Some immediate perspectives for future research include: a generalization of the proposed methodology for the registration of anatomic and functional 3D brain images, and implementing the algorithm in a parallel computer architecture.

References

1. Modersitzki, J.: Numerical Methods for Image Registration. Oxford University Press, Oxford (2004)
2. Hajnal, J.V., Hill, D.L.G., Hawkes, D.J. (eds.): Medical Image Registration. CRC Press, Boca Raton (2001)
3. Zitova, B., Flusser, J.: Image Registration Methods: a survey. Image and Vision Computing 21, 977–1000 (2003)
4. Brown, L.G.: A survey of image registration techniques. ACM Computing Survey 24, 326–376 (1992)
5. Hill, D.L.G., Hawkes, D.J.: Across-modality registration using intensity-based cost functions. In: Bankman, I. (ed.) Handbook of Medical Imaging: Processing and Analysis, pp. 537–553. Academic, New York (2000)
6. Ding, E., Kularatna, T., Satter, M.: Volumetric image registration by template matching. In: Medical Imaging 2000, pp. 1235–1246. SPIE, Bellinham (2000)

7. Shekhar, R., Zagrodsky, V.: Mutual Information-based rigid and non-rigid registration of ultrasound volumes. IEEE Transactions on Medical Imaging 21, 9–22 (2002)
8. Viola, P.A., Wells III, W.M.: Alignment by maximization of mutual information. In: Proc. 5th Int. Conf. Computer Visiom, Cambridge, MA, pp. 16–23 (1995)
9. Wells III, W.M., Viola, P.A., Atsumi, H., Nakajima, S., Kikinis, R.: Multi-modal Volumen Registration by Maximization of Mutual Information. In: Medical Image Analysis, vol. 1, pp. 35–51 (1996)
10. Collignon, A., Maes, F., Delaere, D., Vadermeulen, D., Suetens, P., Marchal, G.: Automated multimodality medical image registration using information theory. In: Bizais, Y., Barillot, C., Di Paola, R. (eds.) Proc. 14th Int. Conf. Process Med. Imag., Ile de Berder, France, pp. 263–274 (1995)
11. Maes, F., Collignon, A., Vadermeulen, D., Marchal, G., Suetens, P.: Multimodality image registration by maximization of mutual information. IEEE-Trans. Med. Image. 16(2), 187–198 (1997)
12. Periaswamya, S., Farid, H.: Medical image registration with partial data. Medical Image Analisys 10(3), 452–464 (2006)
13. Arce-Santana, E.R., Alba, A.: Image registration using Markov random coefficient and geometric transformation fields. Pattern Recognition 42, 1660–1671 (2009)
14. Arce-Santana, E.R., Alba, A.: Image Registration Using Markov Random Coefficient Fields. In: Brimkov, V.E., Barneva, R.P., Hauptman, H.A. (eds.) IWCIA 2008. LNCS, vol. 4958, pp. 306–317. Springer, Heidelberg (2008)
15. Marroquin, J.L., Arce, E., Botello, S.: Hidden Markov measure field models for image segmentation. IEEE Transactions on Pattern Analysis and Machine Intelligence 25, 1380–1387 (2003)
16. Sanjeev Arulampalam, M., Maskell, S., Gordon, N., Clapp, T.: A Tutorial on Particle Filters for Online Nonlinear/Non-Gaussian Bayesian Tracking. IEEE Transaction on Signal Processing 50, 174–188 (2002)
17. Ristic, B., Arulampalam, S., Gordon, N.: Beyond the Kalman Filter. In: Particle Filter for Tracking Applications. Artech House (2004)
18. Doucet, A.: On sequential Monte Carlo methods for Bayesian filtering, Dept. Eng., Univ. Cambridge, UK, Tech. Rep. (1998)
19. Winkler, G.: Image Analysis, Random Fields and Markov Chain Monte Carlo Methods: A Mathematical Introduction, 2nd edn. Springer, New York (2003)

Curve Enhancement Using Orientation Fields

Kristian Sandberg

Computational Solutions, LLC
1800 30th St. Suite 210B, Boulder, CO 80301

Abstract. We present a new method for enhancing the contrast of curve-like structures in images, with emphasis on Transmission Electron Microscopy tomograms of biological cells. The method is based on the Orientation Field Transform, and we introduce new techniques for generating directions and weights of the orientation field. The new method for generating the orientation field focuses on analyzing local asymmetries in the image. We demonstrate that analyzing geometric attributes such as orientations and symmetries results in a robust method that is relatively insensitive to poor and non-uniform contrast.

1 Introduction

Transmission Electron Microscopy (TEM) is a powerful tool to better understand structure and functionality of biological cells [1]. By imaging a specimen from multiple angles, tomographic reconstructions (tomograms) provide 3D images of cellular structures. In order to build easily viewable models of cellular structures, structures of interest are often segmented and rendered as surfaces.

Automating the segmentation process has proved difficult, and in most cases the user has to rely on manual segmentation tools such as IMOD [2]. Building 3D models using manual segmentation tools is often a slow and tedious process, sometimes requiring months of manual identification of cellular structures.

Due to the anisotropic resolution of TEM tomograms of cells [3], segmenting a 3D tomogram is often done slice-wise in the plane of highest resolution. In such slices, the cross section of structures often appear as curve-like structures.

In this paper we consider the problem of enhancing the contrast of curve-like structures in slices of 3D tomograms. Once the contrast of such structures has been enhanced, one can use thresholding and thinning operations to extract contours. This paper will focus on the contrast enhancement step while referring to the extensive literature for the thresholding/thinning problem (see [4] and references therein).

More specifically, we consider the following problem: *Given a 2D image and a scale parameter r, generate an image where the contrast of curve-like structures of thickness $\sim r$ are enhanced, while the contrast of non-curve-like structures are decreased.*

A major obstacle for adaptation of automated segmentation algorithms for cell biology is "the curse of parametrization". Many segmentation algorithms rely on several parameters to be tuned in order to obtain satisfactory results.

G. Bebis et al. (Eds.): ISVC 2009, Part I, LNCS 5875, pp. 564–575, 2009.

Tuning such parameters can be both time consuming and frustrating, especially in cases where the parameters lack intuitive meaning. We therefore consider minimizing the number of parameters of the segmentation process a priority.

Popular automated segmentation methods include Active Contours [5], Level Set methods [6], and the Watershed Transform [7]. Although these methods quite successfully detect compartment-like structures, they are less suitable for detecting curve-like objects [8]. In recent years there has also been an increasing interest in so-called eigenvector techniques [9],[10], which are flexible and can be tuned to detect curve-like structures.

Despite significant research in automatic segmentation techniques, few methods have proved useful for TEM tomograms of cells since these often suffer from low and non-uniform contrast, low signal to noise ratio, and also the presence of interfering structures [11],[8]. In this paper, interfering structures refer to high contrast structures of different shape attributes than the ones the segmentation algorithm targets (see Figure 1).

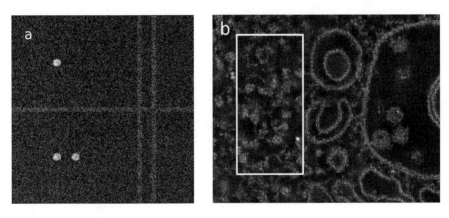

Fig. 1. a) A synthetically generated example where the goal is to detect the horizontal line and the two vertical lines. The dots are considered interfering structures. b) Slice of a tomogram of a Trypanosome. The structures inside the rectangle are examples of what we will refer to as interfering structures.

To address the problem of low and non-uniform contrast while enhancing curve-like structures, *orientation fields* have proved useful, particularly for finger print segmentation and matching [12], and for segmenting slices of tomograms of cells [11]. The orientation field is an assignment of a weighted line segment to each pixel in the image (see Figure 2). Each orientation has two attributes: an angle (ranging between 0 and 180 degrees), and a weight indicating the importance of the orientation. The orientation field can show remarkable uniformity even in cases where the contrast is highly non-uniform (see Figure 3 in [11]).

When using orientation fields for enhancing curve-like objects there are two main questions that have to be addressed: 1) How to generate the orientation field, and 2) how to detect curve-like objects in the orientation field. In this paper, we will focus on the first question and use the Orientation Field Transform (OFT) [11] to detect curve-like objects once the orientation field has been generated.

A common method for generating the orientation field is to compute eigenvectors of the structure tensor [13], or some type of gradient [14], often combined with denoising. In 3D data sets, convolution-based edge detection methods have been described in, e.g., [15] and [16]. In [11], the authors used an approach based on line integration rather than differentiation, which turns out to be relatively stable for the challenging properties of TEM tomograms. However, the approach in [11] has several shortcomings, particularly in the presence of interfering structures.

In this paper, we develop new techniques for generating more robust weights for the orientation field that will properly reflect the reliability of the orientation. We also develop a new robust method for generating orientation directions. These improvements will lead to a robust method for enhancing curve-like objects, while leaving only one parameter, the scale r, for the user to adjust.

Although we will focus on segmentation of curve-like biological structures of TEM images, we note that our approach is general and can be used to enhance curve-like objects in any image. However, we have found TEM tomograms of cells to provide particularly challenging test data sets and therefore an excellent testing environment for image enhancement algorithms.

We review the definition of orientation fields and the OFT in Section 2 and also illustrate where the method for generating the orientation field in [11] fails. In Section 3 we present new techniques for generating a more robust orientation field, and enhance curve-like structures of a synthetic example and a slice from a real TEM tomogram in Section 4. We conclude the paper with a discussion of our approach and mention some of its limitations and future research.

2   Review of Orientation Fields and the OFT

Throughout this paper, we let $I(\mathbf{x})$ denote the image to be processed such that $I(\mathbf{x})$ gives the (gray scale) intensity at location $\mathbf{x} = (x, y)$. We will assume that targeted structures have a locally higher intensity than the background.

We define an orientation $\mathcal{F}$ as the tuple $\{w, \theta\}$ where θ is a direction that ranges between 0 and 180° and w is a positive weight indicating the importance of the direction. An orientation field $\mathcal{F}(\mathbf{x}) = \{w(\mathbf{x}), \theta(\mathbf{x})\}$ is the assignment of an orientation to each location in the image. We can illustrate the orientation field of an image by drawing a line segment at each pixel location, and vary its intensity to indicate the weights (see Figure 2).

To generate the orientation field, we consider the line integral operator

$$R[I](x, y, \theta) \equiv \int_{-\frac{r}{2}}^{\frac{r}{2}} I(x + s \cos\theta, y + s \sin\theta) \, ds, \qquad (1)$$

where r is a scale parameter. This line integral computes the total intensity along a straight line of length r and direction θ centered at pixel (x, y).

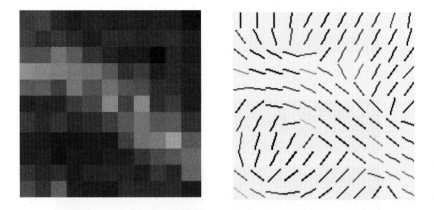

Fig. 2. The orientation field (right) of the image (left)

In [11], the orientation field was generated by

$$\mathcal{F}(x,y) = \left\{ \max_{\theta_k} R[I](x,y,\theta_k), \ \arg\max_{\theta_k} R[I](x,y,\theta_k) \right\} \tag{2}$$

for a sequence of equally spaced angles $\{\theta_k\}_{k=1}^{N_\theta}$ in the range $[0, \pi)$, where r was chosen as approximately two times the width of a typical curve-like structure in the image.

This definition has the advantage of being relatively insensitive to noise compared to gradient based methods for orientation field generation [17],[8]. However, it has some disadvantages that we will now discuss.

First, the orientation field near a structure tends to align with the structure, rather than perpendicularly to the structure, see Figure 3a and b. This happens since the integral along a line that cuts a structure diagonally will give a larger response than an integral along a line that cuts a structure perpendicularly. However, we shall see below that when using the OFT to detect curve-like structures in the orientation field, it is crucial that orientations align *parallel* to the curve for locations *inside* the curve, and align *perpendicular* to the curve for locations near but *outside* curves.

Secondly, the weights of the orientation field may be large even when located at or near a point-like structure, as illustrated in Figure 3d. Ideally, the orientation weights should be close to zero in such case, since these orientations are not associated with a curve-like structure.

In order to to detect curve-like structures in the orientation field, we will use the *Orientation Field Transform* (OFT). To this end, we first define the alignment integral operator Ω of the orientation field $\mathcal{F}(x,y)$ as

$$\Omega[\mathcal{F}](x,y,\alpha) = \tag{3}$$

$$\int_{-\frac{r}{2}}^{\frac{r}{2}} w(x + s\cos\alpha, y + s\sin\alpha)\cos\left(2(\theta(x + s\cos\alpha, y + s\sin\alpha) - \alpha)\right) \ ds$$

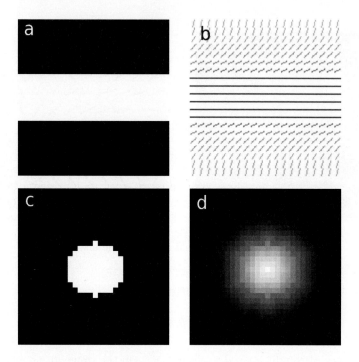

Fig. 3. a) A synthetically generated line-like structure. b) Orientation field for the structure in a) generated by the method in [11]. c) A synthetically generated point-like structure. d) Density plot of the orientation field weights for the structure in c) generated by the method in [11].

where α is an angle between 0 and 180°. This operator integrates the weights of the orientation field along straight lines through each pixel, multiplied by the alignment factor $\cos(2(\theta - \alpha))$. This alignment factor attains its maximum value 1 when θ and α are equal (parallel alignment), and attains it minimum value -1 when θ and α differ by 90° (perpendicular alignment).

To compute the OFT, we first search for the direction in which we have the strongest alignment (parallel or perpendicular), followed by evaluating the alignment integral Ω along the direction of strongest alignment. Formally, we define the OFT, $\mathcal{O}$, of the orientation field $\mathcal{F}(\mathbf{x}) = \{w(\mathbf{x}), \rho(\mathbf{x})\}$ as

$$\mathcal{O}[\mathcal{F}](\mathbf{x}) = \Omega[\mathcal{F}](\mathbf{x}, \tilde{\theta}), \quad \tilde{\theta} = \arg\max_{\alpha_k} |\Omega[\mathcal{F}](\mathbf{x}, \alpha_k)| \ .$$

We note that this operator generates a large negative response near a curve where the orientations (ideally) are aligned perpendicular to the curve, and large positive response inside a curve where the orientations (ideally) are aligned parallel along the curve. Since negative response indicates the exterior of a curve, we can therefore set $\mathcal{O}[\mathcal{F}](\mathbf{x})$ to zero at locations where the OFT response is negative. For examples and details, see [11].

3   A New Method for Generating Orientation Fields

In this section we address the shortcomings listed in Section 2 above. To this end, it will be convenient to introduce operators W and Θ, which extract the weight and direction from an orientation, that is, $\Theta(\mathcal{F}) = \theta$ and $W(\mathcal{F}) = w$.

3.1   Stable Generation of Orientation Field Direction

To generate directions that point perpendicular to a nearby curve, we introduce the notion of an "average" orientation. In order to compute the average of a collection of orientations, we need an addition algorithm for orientations that is associative (order independent). One way to do this, is to map each orientation to a vector in 2D represented in polar coordinates by doubling the direction angle such that $\{w, \theta\}$ is mapped to $\{w, 2\theta\}$. This mapping provides an invertible mapping between orientations and vectors in the plane. Since vector addition in the plane is associative, we can use the following algorithm for adding orientations $\{w_1, \theta_1\}$ and $\{w_2, \theta_2\}$:

1. Map orientations to 2D vectors:
   $\{w_1, \theta_1\} \mapsto \{w_1, 2\theta_1\} \equiv \mathbf{v_1}$ and $\{w_2, \theta_2\} \mapsto \{w_2, 2\theta_2\} \equiv \mathbf{v_2}$
2. Compute $\mathbf{v_{sum}} = \mathbf{v_1} + \mathbf{v_2}$ using the usual rules for vector addition and write the vector $\mathbf{v_{sum}}$ in its polar representation $\{w_{sum}, \theta_{sum}\}$.
3. Map the vector $\mathbf{v_{sum}}$ to an orientation: $\{w_{sum}, \theta_{sum}\} \mapsto \{w_{sum}, \theta_{sum}/2\}$

In particular, we see that two orientations with directions 90° apart and identical weights, add to a zero orientation (orientation with $w = 0$). For an alternative averaging algorithm (which can be generalized to higher dimensions), see [17].

Using the rules for orientation addition, we generate the direction of the orientation field of the image I as

$$\theta(x, y) = \Theta \left(\sum_{k=1}^{N_\theta} \{R[I](x, y, \theta_k),\ \theta_k\} \right) \tag{4}$$

where R is the line integral operator defined in (1) above. This definition differs to the one used in [11], by averaging over the response in different directions, rather than looking for the direction of maximum response. This means that outside a curve, the response will average (because of symmetry), to generate a net orientation perpendicular to the edge (see Figure 4).

3.2   Generation of Orientation Weights

In order to address the problems with orientation weights discussed in Section 2, we will consider two attributes that we refer to as *reliability* and *asymmetry alignment*. We will then use a simple fuzzy system to combine these two attributes to a single weight. One could add even more attributes, but we limit ourselves to only two since these two attributes are fairly general and should therefore work for a broad class of images.

Fig. 4. The orientation field generated by (4) for the structure in Figure 3a

The Reliability Measure. The reliability measure, which we denote as $w_r(\mathbf{x})$, is given by extracting the weight from the sum computed for generating the direction in Section 3.1, that is,

$$w_r(\mathbf{x}) = W\left(\sum_{k=1}^{N_\theta} \{R[I](\mathbf{x}, \theta_k),\ \theta_k\}\right). \tag{5}$$

This expression is best understood by considering the response at the center of a radially symmetric point, in which case the response of the line integral $R[I](\mathbf{x}, \theta)$ is the same in all directions. By symmetry, the sum over orientations will therefore sum to the zero orientation, which we illustrate in Figure 5a.

The Asymmetry Alignment Measure. To prevent large orientation weights nearby point-like objects we observe that the absolute value of the 2D Fourier transform of an image of a radially symmetric point is symmetric, whereas the 2D Fourier transform of a line is highly asymmetric. By computing a local Fourier transform of an area centered at each pixel and analyzing the asymmetry of the resulting Fourier transformed data, we can measure "how curve-like" the neighborhood is.

To formalize these ideas, we first define the following data set

$$\tilde{I}(x, y, \xi, \eta) = \left| \int_{x-\frac{r}{2}}^{x+\frac{r}{2}} \int_{y-\frac{r}{2}}^{y+\frac{r}{2}} I(x', y')e^{-2\pi i(x'\xi + y'\eta)}\ dx'\ dy' \right|.$$

This operator computes the absolute value of the 2D Fourier transform of an r-by-r neighborhood of each pixel (x, y).

The asymmetry alignment measure w_a is defined by measuring the asymmetry of $\tilde{I}$ with respect to the (ξ, η) variables, and measuring the alignment of this asymmetry with the orientation direction:

$$w_a(x, y) = \sum_{k=1}^{N_\theta} \left(\int_{-\frac{r}{2}}^{\frac{r}{2}} \tilde{I}(x, y, s\cos\phi_k, s\sin\phi_k)\ ds \right) \cos(2(\phi_k - \theta(x, y) - \frac{\pi}{2})) \tag{6}$$

where $\{\phi_k\}_{k=1}^{N_\theta}$ is a set of equally spaced angles in the interval $[0, \pi)$.[1]

---

[1] The shift $\frac{\pi}{2}$ is needed since the 2D Fourier transform of a curve orientated at θ degrees in the space domain, will be oriented at $\theta - \frac{\pi}{2}$ degrees in the Fourier domain.

The purpose of this asymmetry measure is twofold. First, it measures the "strength" of the asymmetry.[2] Secondly, it measures the alignment of the asymmetry with the orientation direction computed in (4).

As an example, consider the value of $w_a(\mathbf{x})$ at a line structure with orientation θ. The line integral $\int_{-\frac{r}{2}}^{\frac{r}{2}} \tilde{I}(\mathbf{x}, s\cos\phi_k, s\sin\phi_k)\, ds$ attains it maximum value for $\phi_k = \theta - \pi/2$, for which the alignment factor $\cos(2(\phi_k - \theta - \frac{\pi}{2}))$ attains it maximum value. Hence, w_a will be large and positive at a line structure.

As a second example consider the response at (or near) a radially symmetric point. Since the 2D Fourier transform of a radially symmetric dot is radially symmetric, all terms in (6) cancel out. Hence, w_a is zero at or near a point structure.

As a final example, consider the response outside a straight line. Since we are measuring the absolute value of the asymmetry in the Fourier domain, the measure is independent of spatial shifts of the structure. Hence, the response of $\int_{-\frac{r}{2}}^{\frac{r}{2}} \tilde{I}(\mathbf{x}, s\cos\phi_k, s\sin\phi_k)\, ds$ attains it maximum value for the same ϕ_k as if located at the line structure. However, the orientation θ near but outside a line structure will be perpendicular to the line structure (Figure 4). Hence, $\phi_k = \theta_k$, for which the alignment factor $\cos(2(\phi_k - \theta - \frac{\pi}{2}))$ attains it minimum value (-1). Hence, w_a will be negative outside a line structure.[3]

In Figure 5b we display the weights w_a for the image in Figure 1a.

Combining the Reliability and the Asymmetry Alignment Measures. In order for a pixel to be associated with a curve-like object, we require large reliability response (w_r) and large asymmetry alignment response (w_a). Although there are many ways of combining these measures into a single weight, we have chosen the following criteria for simplicity: We first rescale $w_r(\mathbf{x})$ and $w_a(\mathbf{x})$ to the interval $[0, 1]$ by $w_r(\mathbf{x}) = \frac{w_r(\mathbf{x}) - \min_{\mathbf{x}} w_r(\mathbf{x})}{\max_{\mathbf{x}} w_r(\mathbf{x}) - \min_{\mathbf{x}} w_r(\mathbf{x})}$ and similarly for w_a. We then define $w(\mathbf{x}) = w_a(\mathbf{x}) w_r(\mathbf{x})$ (element-wise multiplication when the weights are represented as matrices), which can be thought of as a simple fuzzy system [18]. We choose this criteria because of its simplicity, and since it does not introduce any additional parameters. In Figure 5c we display the weights w for the image in Figure 1a.

3.3   Summary of the Algorithm

We summarize the curve enhancement algorithm as follows:[4]

---

[2] One can obtain a more intensity independent measure by first setting $\tilde{I}(x, y, 0, 0) = 0$, that is setting the DC component (or zero frequency) to zero, before computing the sum in (6).

[3] Since negative response indicates the exterior of a curve, we can therefore set $w_a(\mathbf{x})$ to zero at locations where the response of (6) is negative.

[4] Note that all integrals are assumed to be approximated by sums.

1. Generate the orientation field:
   (a) For each location $\mathbf{x}$:
       i. Generate the directions $\theta(\mathbf{x})$ by using (4).
       ii. Generate the weights $w_r(\mathbf{x})$ by using (5).
       iii. Generate the weights $w_a(\mathbf{x})$ by using (6). If $w_a(\mathbf{x}) < 0$, set $w_a(\mathbf{x}) = 0$.
   (b) Rescale $w_r(\mathbf{x})$ and $w_a(\mathbf{x})$ to $[0, 1]$.
   (c) For each location $\mathbf{x}$: Compute $w(\mathbf{x}) = w_r(\mathbf{x})w_a(\mathbf{x})$.
2. Compute the OFT. For each location $\mathbf{x}$:
   (a) Compute the OFT by using (3).
   (b) If $\mathcal{O}[I](\mathbf{x}) < 0$, set $\mathcal{O}[I](\mathbf{x}) = 0$.

4   Results

We first verify the consistency of our algorithm by enhancing the line structures in Figure 1a. Although this is obviously a synthetic example, it provides valuable verification of our algorithm's ability to handle noise, and also of its ability to enhance the weak line-like structures while decreasing the contrast of the strong point-like structures as shown in Figure 5d. We used a scale parameter r corresponding to approximately 1.5 times the thickness of the line-like structures.

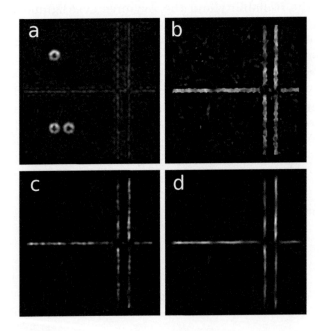

Fig. 5. Illustration of the orientation field weights for the image in Figure 1a displayed as density plots. a) The weights w_r. b) The weights w_a. c) The final weights. d) Result of applying the algorithm in Section 3.3 to the image in Figure 1a.

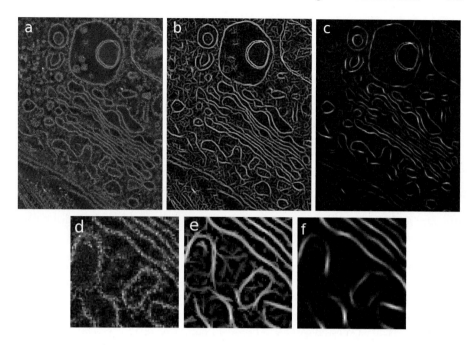

Fig. 6. Enhancement of a tomogram slice of a Trypanosome. a) Original image b) Result using the orientation field generated by the algorithm [11]. c) Result using the algorithm in Section 3.3. d) Closeup of the original image. e) Closeup of the result in b). f) Closeup of the result in c).

We next apply our algorithm to a slice from a real TEM tomogram[5] (Figure 6). In the left column we show the original image, in the center column the result when using the orientation field generated by the algorithm in [11], and in the right column the result of using our algorithm. We used a scale parameter r that corresponds to approximately 1.5 times the thickness of a typical membrane.

5   Discussion

Whereas many traditional segmentation methods focus on detecting edges and local correlation in texture, attributes which are known to be sensitive to nonuniform contrast and noise, the OFT detects correlations in geometrical attributes. However, in order for the OFT to be robust, it is essential for the orientation field to be based on attributes that are insensitive to contrast variations.

The method for generating the orientation field in [11] is relatively insensitive to noise, but still highly dependent on intensity and therefore sensitive to the presence of strong point-like structures. This problem can be partially remedied by locally smoothing the orientation field, thresholding, and allowing different

[5] Tomogram of Trypanosome, courtesy of Mary Morphew, the Boulder Laboratory for 3D Electron Microscopy of Cells, University of Colorado at Boulder.

scale parameters r_1 and r_2 to be used in (2) and (3), respectively. However, this requires more parameters to be tuned.

The methodology in this paper generates a significantly less contrast dependent orientation field by focusing more on local asymmetries than local intensities, and only requires one parameter to be set.

In order for a segmentation algorithms to be used routinely in a laboratory, experience shows that it is essential to minimize the number of parameters for the user to tune, and ensure that existing parameters have intuitive interpretations. The scale parameter used for the algorithm in this paper is easy to estimate as it is directly related to the thickness of a target structure.

We also point out that the suggested algorithm can be extended with more parameters. For example, one can introduce more attributes for generating the weights, and combine these using fuzzy logic, possibly within a neural network framework to train the fuzzy system.

In order to extend the current algorithm to detect objects with varying thickness and curvature, one should introduce a multiscale methodology by simultaneously process the data for a range of scale parameters r combined with some criteria on how to locally select r. We also note that the OFT currently uses a family of straight lines of fixed length to search for correlation in the orientation field. A more sophisticated version can use a family of curves of varying curvature as well, which should improve the accuracy for finding structures with large curvatures.

Finally, we plan on extending the current work to detect curves and planes in 3D data sets.

6   Conclusion

We have refined an earlier suggested method for enhancing the contrast of curve-like structures in TEM tomograms. The method is based on the Orientation Field Transform, but uses a more robust technique to generate the orientation field of an image compared to earlier suggested methods. The resulting method is stable both with respect to noise and presence of high contrast point-like objects. Furthermore, the algorithm only requires one parameter to be set by the user, and is therefore easy to use.

Acknowledgment

The author would like to thank Alejandro Cantarero for his valuable comments and suggestions.

References

1. Richard McIntosh, J. (ed.): Methods in Cell Biology: Cellular Electron Microscopy, vol. 79. Elsevier Inc., Amsterdam (2007)
2. Kremer, J.R., Mastronarde, D.N., McIntosh, J.R.: Computer visualization of three-dimensional image data using IMOD. J. Struct. Biol. 116, 71–76 (1996)

3. Penczek, P., Marko, M., Buttle, K., Frank, J.: Double-tilt Electron Tomography. Ultramicroscopy 60, 393–410 (1995)
4. Gonzales, R.C., Woods, R.E.: Digital Image Processing, 3rd edn. Prentice Hall, Englewood Cliffs (2007)
5. Chan, T.F., Vese, L.A.: Active Contours Without Edges. IEEE Trans. Image Processing 10(2), 266–277 (2001)
6. Osher, S., Fedkiw, R.: Level Set Methods and Dynamic Implicit Surfaces. Springer, Heidelberg (2003)
7. Volkmann, N.: A Novel Three-dimensional Variant of the Watershed Transform for Segmentation of Electron Density Maps. J. Struct. Biol. 138, 123–129 (2002)
8. Sandberg, K.: Methods for Image Segmentation in Cellular Tomography. In: McIntosh, J.R. (ed.) Methods in Cell Biology: Cellular Electron Microscopy, vol. 79, pp. 769–798. Elsevier Inc., Amsterdam (2007)
9. Frangakis, A.S., Hegerl, R.: Segmentation of Two- and Three-dimensional Data from Electron Microscopy Using Eigenvector Analysis. Journal of Structural Biology 138, 105–113 (2002)
10. Coifman, R.R., Lafon, S., Lee, A.B., Maggioni, M., Nadler, B., Warner, F., Zucker, S.W.: Geometric Diffusions as a Tool for Harmonic Analysis and Structure Definition of Data, Part I: Diffusion Maps. Proc. Natl. Acad. Sci. USA 102(21), 7426–7431 (2005)
11. Sandberg, K., Brega, M.: Segmentation of Thin Structures in Electron Micrographs Using Orientation Fields. J. Struct. Biol. 157, 403–415 (2007)
12. Gu, J., Zhou, J.: A Novel Model for Orientation Field of Fingerprints. In: Proceedings of the 2003 IEEE Computer Society Conference on Computer Vision and Pattern Recognition, vol. 2, pp. 493–498 (2003)
13. Weickert, J.: Anisotropic Diffusion in Image Processing. Teubner-Verlag (1998)
14. Kass, M., Witkin, A.: Analyzing Oriented Pattern. In: Computer Vision, Graphics and Image Processing, vol. 37, pp. 362–397 (1987)
15. Zucker, S.W., Hummel, R.A.: A Three-dimensional Edge Operator. IEEE Trans. Pattern Analysis 3(3), 324–331 (1981)
16. Pudney, C., Kovesi, P., Robbins, B.: Feature Detection Using Oriented Local Energy for 3D Confocal Microscope Images. In: Chin, R., Naiman, A., Pong, T.-C., Ip, H.H.-S. (eds.) ICSC 1995. LNCS, vol. 1024, pp. 274–282. Springer, Heidelberg (1995)
17. Brega, M.: Orientation Fields and Their Application to Image Processing. Master's thesis, University of Colorado at Boulder (2005)
18. Ross, T.J.: Fuzzy Logic with Engineering Applications, 2nd edn. Wiley, Chichester (2004)

Lighting-Aware Segmentation of Microscopy Images for In Vitro Fertilization

Alessandro Giusti[1], Giorgio Corani[1], Luca Maria Gambardella[1],
Cristina Magli[2], and Luca Gianaroli[3]

[1] Dalle Molle Institute for Artificial Intelligence (IDSIA), SUPSI and University of Lugano, Switzerland
alessandrog@idsia.ch
[2] International Institute for Reproductive Medicine (IIRM), Lugano, Switzerland
[3] INFERGEN, Lugano, Switzerland

Abstract. We present a practical graph-based algorithm for segmenting circular-shaped structures from Hoffman Modulation Contrast images of human zygotes. Hoffman Modulation Contrast is routinely used during In Vitro Fertilization procedures, and produces images with a sidelit, 3D-like appearance; our algorithm takes advantage of such peculiar appearance in order to improve the robustness of segmentation. The task is not straightforward due to the complex appearance of the objects of interest, whose image is frequently affected by defocus, clutter, debris and other artifacts. We show applications of our technique to the unsupervised segmentation of the zygote oolemma and to the subsequent supervised segmentation of its pronuclei. Experiments are provided on a number of images with different characteristics, which confirm the algorithm's robustness with respect to clutter, noise and overexposure.

1 Introduction

During In Vitro Fertilization (IVF) procedures, biologists observe fertilized ova at different times in order to assess their quality and select the ones maximizing the implantation success rate; this decision-making process is guided by a number of criteria, usually requiring subjective classifications, which are widely discussed in the related literature [1,2,3]; Beuchat, et al. show in [4] that computer-based morphological measurements on zygotes, providing quantitative rather than qualitative data, has the potential to improve the accuracy of implantation predictions.

In order to provide such quantitative measurements, we consider the problem of segmenting circular-shaped structures in zygote images; in particular, our technique is useful for segmenting the zygote cell's oolemma (excluding the surrounding zona pellucida), and pronuclei in the zygote cell (see Figure 1).

This allows us to readily compute a number of quantitative measures (apparent size, simple shape descriptors, relative positions) for each of these objects, which are not easily judged otherwise. The obtained segmentation may also be

G. Bebis et al. (Eds.): ISVC 2009, Part I, LNCS 5875, pp. 576–585, 2009.

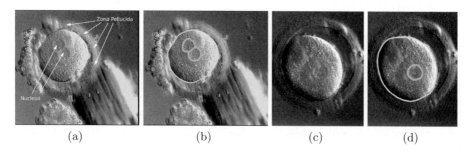

(a)   (b)   (c)   (d)

Fig. 1. Zygote cell (a,c) and its segmentation (b,d). Cell contour in yellow, pronucleus contour in cyan. Note debris in (a) and non-elliptic shape in (b).

applied for other tasks, such as driving an automated microscope for unattended imaging of zygotes, or providing a robust, precise initialization for subsequent (automatic or user-assisted) analysis algorithms, such as those introduced in [5,6] for the segmentation of the zona pellucida.

Our algorithm is designed to operate on Hoffman Modulation Contrast (HMC) images[1]: HMC is an imaging technique converting optical slopes in variations of the light intensity: it is routinely used in IVF labs for observing zygotes, as it provides a large amount of contrast for transparent specimens and eases human observation as the objects appear three-dimensional and side-lit, as if a light source was illuminating them from a side (apparent lighting direction). After locating the center of the structure, we compute a transformed representation of the image in polar coordinates; we take advantage of said lighting peculiarities by only considering edges whose orientation matches the expected sign of the intensity gradient, while penalizing at the same time most unrelated edges. The segmentation problem is finally efficiently solved as a minimum-cost problem on a directed acyclic graph built on the transformed image, which implicitly enforces shape priors.

The main contribution of our technique over the state of the art lies in our simple method for taking advantage of HMC lighting, whose effectiveness is quantitatively evaluated in Section 5. The complete system has shown to be robust and efficient: an image is processed in less than one second, with little effect of debris, noise, overexposure or defocus, and with no need of parameter tuning. This allowed us to easily integrate the technique in the routine of an IVF laboratory, in order to provide precise and effortless zygote measurements.

We briefly review related works in the following Section, then describe our technique in Section 3 in the context of zygote segmentation, and in Section 4 for pronuclei segmentation. Experimental results are shown and discussed in Section 5. Section 6 concludes the paper and presents ongoing work.

---

[1] A technique delivering visually similar results is Differential Interference Contrast (DIC), which is also a likely application scenario for our technique.

2   Related Work

Classical region-based segmentation algorithms, including watersheds [7], are not applicable in this context because of the complex appearance of the cell, including the surrounding zona pellucida, clutter, and artifacts; this also hinders the application of straightforward edge-based segmentation algorithms, as many spurious contours are detected.

Iterative energy minimization methods such as active contours [8] and level sets are frequently employed in biomedical imaging: in this context, their application is not straightforward because debris are likely to generate several local minima in the energy function, which makes quick and robust convergence problematic; for example, in [5] active contours are used for measuring the thickness of the zona pellucida in embryo images, but only after a preprocessing step aimed at removing debris and other artifacts.

In [4] a semisupervised technique for measuring various zygote features is used, where the cell shape is approximated by an ellipse: in our case, instead, we recover the actual shape of the cell, which is often not well approximated by an ellipse.

The technique we are presenting includes a global energy minimization step, and may be classified as a specialized graph-cut [9] approach, where: *a)* priors on the cell shape are accounted for by operating on a spatially-transformed image and searching for a minimum-cost path on a directed acyclic graph, and *b)* priors on the contour appearance due to HMC lighting are directly integrated in the energy terms.

Interestingly, several previous works handled the particular lighting in HMC and DIC images as an obstacle to segmentation [10], and adopted preprocessing techniques for removing it, whereas we actually exploit such appearance for improving robustness. Preliminary results on the zygote contouring have been presented in [11].

3   Segmentation of a Zygote Cell

We now describe the algorithm in the context of zygote cell segmentation. The application of the algorithm to the segmentation of pronuclei is given in Section 4.

We divide the segmentation process in two sequential steps: first, we find the approximate location (c_x, c_y) of the cell center; in doing this, we assume that a single zygote is visible in the image, which is always the case as zygotes are kept in separate wells. Then, the image of the cell is transformed to polar coordinates, the lighting parameters are estimated (if unknown), and a shortest-path formulation is used in order to recover the actual zygote contour. We briefly introduce the former part, which we consider of lesser importance and interest, in Section 3.1. The main focus is instead on the latter part, described in Section 3.2.

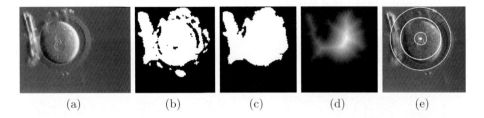

(a)   (b)   (c)   (d)   (e)

Fig. 2. Approximate localization of the cell center. (a): original image. (b): binary mask obtained after thresholding the modulo of the gradient. (c): largest connected component with holes filled. (d): distance transform. (e) the maximum of the distance transform is considered as the approximate center of the cell. Note that the large artifact on the left does not significantly displace the maximum of the distance transform.

3.1 Approximate Localization of Zygote Center

In order to find a rough location for the cell centroid, the modulo of the image gradient is first computed at every pixel, then subsampled to a smaller image, which is automatically thesholded and regularized by means of median filtering (see Figure 2 a,b). The largest connected component is isolated and its holes filled (Figure 2 c); for each point inside the resulting region, the minimum distance to the region boundary is computed by means of the distance transform; the point with the maximum distance is finally chosen as the approximate centroid of the cell.

This preliminary analysis phase is not critical for the quality of results, as the subsequent processing tolerates quite large displacements of the detected centroid; nonetheless, this simple algorithm proves to be quite robust also in presence of large artifacts attached to the cell; this is mainly due to the distance transform, which implicitly cancels or reduces the effect of any non-convex artifact protruding off the border of the cell.

3.2 Detailed Recovery of the Cell Contours

Once the zygote centroid (c_x, c_y) is detected, a circular corona around such point is transformed using bilinear interpolation to an image J in polar coordinates:

$$J(\theta, \rho) = I\left(c_x + \rho \cos(\theta), c_y + \rho \sin(\theta)\right) \qquad 0 \leq \theta < 2\pi \quad \rho' \leq \rho \leq \rho''. \quad (1)$$

In order to account for variations in the cell shape and errors in the centroid location, the range $[\rho' \div \rho'']$ of ρ values is very conservatively set to $[0.3r \div 1.5r]$, where r represents the expected cell radius; this is a quite large range (see Figure 2e), which allows for large variations in the actual radius of the zygote, and for displacements of the estimated centroid (c_x, c_y). ρ and θ values are uniformly sampled in ρ_n and θ_n intervals, respectively, which correspond to rows and columns of image J. We use $\rho_n = 80$, $\theta_n = 180$ in the following.

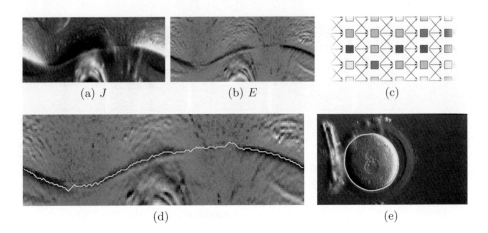

Fig. 3. Contour localization. (a) image in Figure 2 transformed to polar coordinates. (b): energy E computed according to (2), and its graph structure (c). (d): minimum-cost path. (e): the resulting segmentation.

Image J (Figure 3a) is then processed in order to associate an energy to each pixel, which will drive the following graph-based formulation. Let α be the direction of apparent lighting due to HMC, which only depends on the optical setup and can be assumed known in most scenarios (if it's not, it can be easily estimated); we define the energy for each pixel as:

$$E(\theta, \rho) = P\left(\overbrace{\cos(\theta - \alpha) \cdot G_\rho(J)} + \overbrace{\sin^2(\theta - \alpha) \cdot |G_\rho(J)|}\right) \tag{2}$$

$$P(x) = \left(1 + e^{\frac{x}{k}}\right)^{-1} \tag{3}$$

where G_ρ denotes the gradient operator along the ρ axis, and $P(\cdots)$ is a simple decreasing sigmoid function which conditions the energy values to lie in the $[0 \div 1]$ interval; the scaling parameter k is not critical, and can be safely set to $1/5$ of the image's dynamic range.

The first term in (2) dominates where the contour is orthogonal to the apparent light direction, i.e. where the cell is expected to appear significantly lighter $(\theta - \alpha \simeq 0)$ or darker $(\theta - \alpha \simeq \pm\pi)$ than the surroundings; large gradient values with a sign consistent with this assumption lead to lower energies. The second term takes account for the unpredictability of the contour appearance where the contour is parallel to the apparent light direction, and just associates lower energies to large absolute values for $G_\rho(J)$.

A directed acyclic graph is built over J, by considering a node for each pixel, and arcs connecting each pixel to its three 8-neighbors at the right (see Figure 3c). The cost of each arc is set to the energy of the source node. The minimum-cost path is computed from every pixel in the first column to the corresponding pixel in the last column; the cheapest of these paths is chosen as the actual contour

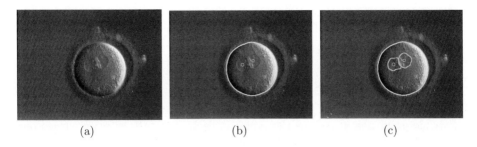

(a)   (b)   (c)

Fig. 4. (a): Original image. (b): Cell segmentation (yellow) and manual initialization of pronuclei contouring (light blue). (c): Segmentation of pronuclei allows to derive relative size measures.

(see Figure 3d), then mildly smoothed for cosmetic reasons and brought back to cartesian coordinates by using the inverse transform to (1). The interior of the resulting polygon defines the computed binary mask M.

Larger values for ratio θ_n/ρ_n allow more freedom to the path built over J, which translates to better accomodation of an irregular cell shape or a displaced centroid (c_x, c_y); at the same time, this reduces the robustness of the approach, as shape priors are less strongly enforced. We found any ratio between 1.5 and 3.0 to be acceptable, although we keep with $\theta_n/\rho_n = 180/80 = 2.25$ in the following.

4   Segmentation of Pronuclei in a Zygote Image

The same technique described in the previous section can be easily adapted for segmenting other circular-shaped structures of particular interest during IVF procedures, which exhibit the same lighting peculiarities we described in Section 3. In particular, we describe in this section the supervised segmentation procedure of pronuclei in a zygote image.

Pronuclei are sometimes not well visible in the zygote images (see Figure 4), which makes the problem somewhat harder than we described previously for the cell's oolemma: still, they are often bounded by (weak) edges which exhibit the same appearance we described in Section 3.2, which also characterizes the whole zygote cell.

In case of pronuclei, however, the apparent lighting direction is the opposite, due to their different optical properties with respect to the surrounding medium[2].

Therefore, the definition of energy E provided in (2) can be successfully used in order to segment pronuclei, exactly as described in Section 3.2, as soon as the α angle is inverted and the expected radius r is adjusted. As the shape of pronuclei is less irregular than the shape of the whole cell, we improve results by

---

[2] HMC images result from rather complex optical phenomena, which are necessary in order to make the transparent cell visible; the resulting lighting is therefore only apparent and not due to the reflectivity of the structures.

using a reduced θ_n/ρ_n ratio of 1.50, which improves robustness by more strongly enforcing shape priors, while being less tolerant about displacements of the given center.

4.1 User-Assisted Segmentation

We developed a number of heuristics in order to automatically detect the number of visible pronuclei, as well as the center of each. The task however turned out to be extremely challanging, due to a number of reasons:

- although most of the cases show two pronuclei approximately at the center of the zygote, the actual number and positions of pronuclei in the image can not be assumed known;
- the edges of the pronuclei are weak, and their internal region is very similar to external texture; moreover, pronuclei images may partially or totally overlap, and/or appear out-of-focus;
- some structures, such as vacuoli, are extremely similar to pronuclei, but have a completely different clinical meaning.

Therefore, we implemented a supervised segmentation technique, in which the user is required to click at the center of each pronucleus, which is then segmented. Still, overlap and weak edges cause localized segmentation errors in a small number of cases. We handle these by allowing the user to adjust the segmentation by clicking on a point on the expected contour; the contour is immediately rerouted after setting the energy of the corresponding point to 0, then recomputing minimum-cost paths on the modified graph; this can be iterated if necessary.

5   Experimental Results

In order to quantitatively evaluate the effectiveness of the technique, we applied the presented algorithms to the tasks described in the paper. The images were acquired with a 0.35 megapixel JVC camera attached to an Olympus IX-51 inverted microscope equipped with a 20x or 40x objective and HMC.

Ground truth is obtained by manually segmenting the structures in each image, thus obtaining a binary mask T representing the interior of each structure.

For a given computed segmentation, represented by a binary mask M, we consider the Jaccard quality metric $q = \frac{|T \cap M|}{|T \cup M|}$ $0 \leq q \leq 1$, which approaches 1 for better segmentations. We also measure the average distance d_a and maximum distance d_m between the true boundary and the computed one, as well as the relative error in the measured area e_a, and absolute error in the measured eccentricity e_e.

We evaluate the results obtained by minimizing the energy term E presented in (2). In addition, in order to test the effectiveness of the lighting priors, we also show results obtained by using two alternative energy terms, which do not include lighting cues:

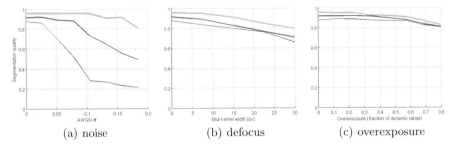

(a) noise   (b) defocus   (c) overexposure

Fig. 5. Average segmentation quality q for zygote segmentation, with varying amounts of degradation. Using enegy E (yellow line), E_ρ (thick red line), E_G (thin red).

$E_\rho(\theta, \rho) = P\left(|G_\rho\left(J\left(\theta, \rho\right)\right)|\right)$, which only considers gradients in the radial direction, without discriminating the sign (i.e. disregarding the expected appearance of the edge); and
$E_G(\theta, \rho) = P\left(|G\left(J\left(\theta, \rho\right)\right)|\right)$, which considers the gradient magnitude only, disregarding both its sign *and* its direction.

5.1 Zygote Segmentation

The dataset is composed by 78 zygote images. The results are shown in the table below:

	q	d_a		d_m		e_a	e_e
		pixels	fraction of R	pixels	fraction of R		
E	0.9613	2.7065	0.0208	4.0438	0.0311	0.0131	0.0491
E_ρ	0.9201	6.8936	0.0530	8.5208	0.0655	0.0757	0.0645
E_G	0.8665	11.4400	0.0880	13.4744	0.1036	0.1589	0.0594

Moreover, we stress-tested our technique by simulating variable amounts of additive white gaussian noise, overexposure, and defocus[3] blur (Figure 5 a,b,c).

The results demonstrate that even in challanging conditions, the zygote segmentation can be performed completely automatically, robustly and with a good accuracy. This also illustrated in Figure 6.

5.2 Segmentation of Pronuclei

For pronuclei segmentation, we used the same dataset described in the previous section, excluding 9 images where no pronuclei were present.

Our software provides a zoomed view of the zygote after the previously-described segmentation phase, and the user is required to click on the center

---

[3] Defocus blur is simulated by convolution with a disk-shaped kernel, which differs from what would occur in a real microscope, where different parts of the cell would come into focus.

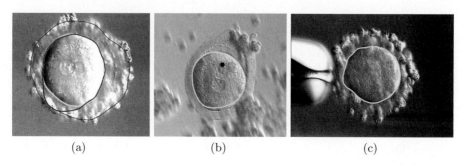

(a)   (b)   (c)

Fig. 6. (a) shows the segmentation obtained using the lighting-aware energy measure E (yellow), versus measures E_ρ (thick red) and E_G (thin red), which are easily misled by artifacts surrounding the cell. (b) shows a failure of our technique (about 15 pixels of error at the bottom of the cell); the failure is due to the recovered approximate cell center (blue dot) being quite displaced. A larger θ_n/ρ_n ratio would fix this problem. (c) shows our technique working on an ovocyte (not a zygote), which however has a regular enough appearance to be detected and correctly segmented.

of each of the pronuclei; we measured an average error in this phase of about 6 image pixels (which correspond to about 30 screen pixels in the zoomed view), when the user is not required to be particularly careful. In our validation, the user initialization for the pronucleus center is simulated by randomly choosing a pixel within a 12 image-pixel distance from the actual centroid; the operation is repeated 5 times for each of the 69 images available.

We consider that a pronucleus is segmented correctly when the Jaccard coefficient is larger than 0.9. In 86% of the images, both pronuclei are segmented correctly; in 8% of the images, a correct segmentation of both pronuclei is obtained after one or two additional clicks, in order to correct segmentation errors (see Section 4). The remaining 6% of the images required more than two clicks on the observed contour of the pronuclei, in order to obtain a correct segmentation of both.

6   Conclusions

We presented a practical edge-based technique for precisely segmenting zygotes, by taking advantage of the peculiar appearance of HMC lighting, which significantly increases the robustness of the system. The technique is easily implemented and not computationally expensive.

The technique is not only useful for directly measuring relevant features: in fact, we are currently using this for automatically initializing iterative techniques for solving more complex problems, such as detection of nucleoli inside pronuclei, or measurements on the zona pellucida around the zygote contour; also, our technique is useful for automatically detecting the presence of the zygote in the microscope field of view, which is an important prerequisite for unattended zygote imaging within an automated microscopy system.

References

1. Kurzawa, R., Głabowski, W., Baczkowski, T.: Methods of embryo scoring in vitro fertilization. Reproductive Biology 4(1), 5–22 (2004)
2. Gianaroli, L., Magli, M.C., Ferraretti, A.P., Lappi, M., Borghi, E., Ermini, B.: Oocyte euploidy, pronuclear zygote morphology and embryo chromosomal complement. Human Reproduction 22, 241–249 (2007)
3. Magli, M.C., Gianaroli, L., Ferraretti, A.P., Lappi, M., Ruberti, A., Farfalli, V.: Embryo morphology and development are dependent on the chromosomal complement. Fertility Sterility 87, 534–541 (2007)
4. Beuchat, A., Thévenaz, P., Unser, M., Ebner, T., Senn, A., Urner, F., Germond, M., Sorzano, C.: Quantitative morphometrical characterization of human pronuclear zygotes. Human Reproduction 23, 1983–1992 (2008)
5. Morales, D., Bengoetxea, E., Larrañaga, P.: Automatic segmentation of zona pellucida in human embryo images applying an active contour model. In: Proc. of MIUA (2008)
6. Karlsson, A., Overgaard, N.C., Heyden, A.: A Two-Step Area Based Method for Automatic Tight Segmentation of Zona Pellucida in HMC Images of Human Embryos. In: Kimmel, R., Sochen, N.A., Weickert, J. (eds.) Scale-Space 2005. LNCS, vol. 3459, pp. 503–514. Springer, Heidelberg (2005)
7. Soille, L., Vincent, P.: Watersheds in digital spaces: an efficient algorithm based onimmersion simulations. IEEE Transactions on Pattern Analysis and Machine Intelligence (1991)
8. Xu, C.: Snakes, shapes, and gradient vector flow. IEEE Transactions on Image Processing (1998)
9. Zabih, R., Kolmogorov, V.: Spatially coherent clustering using graph cuts. In: Proc. of CVPR, vol. 2, pp. II–437–II–444 (2004)
10. Kuijper, A., Heise, B.: An automatic cell segmentation method for differential interference contrast microscopy. In: Proc. of ICPR (2008)
11. Giusti, A., Corani, G., Gambardella, L.M., Magli, C., Gianaroli, L.: Segmentation of human zygotes in hoffman modulation contrast images. In: Proc. of MIUA (2009)

Fast 3D Reconstruction of the Spine Using User-Defined Splines and a Statistical Articulated Model

Daniel C. Moura[1,2], Jonathan Boisvert[5,6], Jorge G. Barbosa[1,2],
and João Manuel R.S. Tavares[3,4]

[1] Lab. de Inteligência Artificial e de Ciência de Computadores, FEUP, Portugal
[2] U. do Porto, Faculdade de Engenharia, Dep. Eng. Informática, Portugal
{daniel.moura,jbarbosa}@fe.up.pt
[3] Instituto de Engenharia Mecânica e Gestão Industrial, Campus da FEUP, Portugal
[4] U. do Porto, Faculdade de Engenharia, Dep. de Eng. Mecânica
Rua Dr. Roberto Frias, 4200-465 Porto, Portugal
tavares@fe.up.pt
[5] National Research Council Canada, 1200, Montreal Rd, Ottawa, K1A 0R6, Canada
[6] École Polytechnique de Montréal, Montréal, QC H3T 1J4, Canada
jonathan.boisvert@nrc-cnrc.gc.ca

Abstract. This paper proposes a method for rapidly reconstructing 3D models of the spine from two planar radiographs. For performing 3D reconstructions, users only have to identify on each radiograph a spline that represents the spine midline. Then, a statistical articulated model of the spine is deformed until it best fits these splines. The articulated model used on this method not only models vertebrae geometry, but their relative location and orientation as well.

The method was tested on 14 radiographic exams of patients for which reconstructions of the spine using a manual identification method where available. Using simulated splines, errors of 2.2±1.3mm were obtained on endplates location, and 4.1±2.1mm on pedicles. Reconstructions by non-expert users show average reconstruction times of 1.5min, and mean errors of 3.4mm for endplates and 4.8mm for pedicles.

These results suggest that the proposed method can be used to reconstruct the human spine in 3D when user interactions have to be minimised.

1   Introduction

Three-dimensional reconstructions of the spine are required for evaluating spinal deformities, such as scoliosis. These deformities have a 3D nature that cannot be conveniently assessed by planar radiography. Clinical indexes that may only be measured with 3D models include, for example, the maximum plane of curvature [1], but there are several others that cannot be accurately quantified using

G. Bebis et al. (Eds.): ISVC 2009, Part I, LNCS 5875, pp. 586–595, 2009.

planar radiography, such as the axial rotation of vertebrae. On the other hand, 3D imaging techniques (i.e. Computer Tomography (CT) and Magnetic Resonance Imaging (MRI)) are not suitable because they require patients to be lying down, which alters the spine configuration. Additionally, they are more expensive and, in the case of CT, the doses of radiation required for a full scan are too high to be justified for multiple follow-up examinations. For all those reasons, 3D reconstructions of the spine are usually done using two (or more) planar radiographs.

Three-dimensional reconstruction of the spine using radiographs is usually done by identifying a predefined set of anatomical landmarks in two or more radiographs. In [2] a set of 6 stereo-corresponding points per vertebra is required to be identified, and in [3] this set is increased even more to enable identifying landmarks that are visible in only one of the radiographs. These methods require expert users and, additionally, they are very time-consuming, error-prone and user-dependent. In [4] the time required by a user to reconstruct a 3D model was decreased to less than 20min by requiring a set of 4 landmarks per vertebra in each radiograph. However, this amount of user interaction is still high for clinical routine use.

Very recently, new methods are arising that try to reduce user interaction by requesting the identification of the spine midline on two radiographs, and make use of statistical data for inferring the shape of the spine. This is the case of [5] where, besides the splines, two additional line segments are needed for achieving an initial reconstruction. Then, several features are manually adjusted that are used along with the initial input for producing the final model. This processes requires an average time of 2.5min, although users may refine reconstructions, increasing interaction time to 10min. Kadoury et al. also proposed using a statistical approach to obtain an initial model of the spine from two splines, which is then refined using image analysis [6]. However, this study only uses the spine midline as a descriptor to get the most probable spine shape for that midline. While this is acceptable, there may be a range of spine configurations for the same spline. Additionally, in both studies, the authors do not make a complete use of the user input, since the control points that the user marks for identifying the spine midline are ignored, while they may carry information about the location of some vertebrae. Finally, the statistical models of this studies do not conveniently explore the inter-dependency of position and orientation between vertebrae.

In this paper, we propose a method for reconstructing the spine from its midline that uses an articulated model [7] for describing anatomical variability. This model effectively represents vertebrae inter-dependency and it has already demonstrated capabilities for inferring missing information [8]. The model is then deformed using an optimisation process for fitting the spine midline while making use of all the information that the user inputs, that is, the location of the control points that define the midline are used for controlling the deformation of the statistical model.

2   Methods

2.1   Articulated Model of the Spine

Statistical models of anatomical structures are often composed by a set of land-marks describing their geometry. In the case of the spine, this could be done by using a set of landmarks for each vertebra. However, the spine is a flexible structure. The position and orientation of the vertebrae are, therefore, not inde-pendent. Capturing the spine as a cloud of points does not differentiate vertebrae and, consequently, information is lost that may be important to capture spinal shape variability and vertebrae inter-dependencies.

For tackling this problem, Boisvert *et al.* proposed the use of articulated mod-els [7]. These models capture inter-vertebral variability of the spine geometry by representing vertebrae position and orientation as rigid geometric transforma-tions from one vertebra to the other along the spine. Only the first vertebra (e.g. L5) has an absolute position and orientation, and the following vertebrae are dependent from their predecessors. This may be formalised as:

$$T_i^{abs} = T_1 \circ T_2 \circ \cdots T_i, \quad \text{for } i = 1..N, \tag{1}$$

where T_i^{abs} is the absolute geometric transformation for vertebra i, T_i is the geometric transformation for vertebra i relative to vertebra $i - 1$ (with the ex-ception of the first vertebra), $\circ$ is the composition operator, and N is the number of vertebra represented by the model.

In order to include data concerning vertebrae morphology, a set of landmarks is mapped to each vertebra in the vertebrae coordinate system, which has its origin at the vertebral body centre. The absolute coordinates for each landmark may be calculated using the following equation:

$$p_{i,j}^{abs} = T_i^{abs} \star p_{i,j}, \quad \text{for } i = 1..N, j = 1..M, \tag{2}$$

where $p_{i,j}^{abs}$ are the absolute coordinates for landmark j of vertebrae i, $p_{i,j}$ are the relative coordinates, $\star$ is the operator that applies a transformation to a point, and M is the number of landmarks per vertebra.

The method proposed on this paper uses an articulated model of the spine composed by $N = 17$ vertebrae (from L5 to T1) with $M = 6$ landmarks per vertebra (centre of superior and inferior endplates ($j = 1..2$), and the superior and inferior extremities of the pedicles ($j = 3..6$)).

2.2   User Input

User input is limited to identifying the spine midline in two different views using parametric splines like illustrated by figure 1. Both splines should begin at the centre of the superior endplate of vertebra T1 and should end at the centre of the inferior endplate of L5. These are the only landmarks that must be present in both radiographs, all the other control points may be identified in only one radiograph without having a corresponding point in the other.

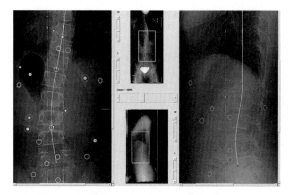

Fig. 1. Graphical user interface designed for defining the splines

In order to impose restrictions concerning vertebrae location, users are asked to mark all the control points in the centre of the vertebral bodies (with the exception of T1 and L5). In fact, there is a natural tendency for users to place control points on the centre of vertebral bodies, even when not asked to, and this way splines also carry information about the location of some vertebrae.

2.3   Fitting the Articulated Model to the Splines

For fitting the model to the splines, an optimisation process is used that iteratively deforms the articulated model and projects the anatomical landmarks on both radiographs simultaneously, towards reducing the distance between the projected landmarks of the model and the splines. Principal Components Analysis (PCA) is used for reducing the number of dimensions of the articulated model, while capturing the main deformation modes. Using PCA in a linearised space, a spine configuration $x = [T_1, \ldots, T_N, p_{1,1}, \ldots, p_{N,M}]$, may be generated using the following equation:

$$x = \bar{x} + \gamma d, \tag{3}$$

where $\bar{x}$ is the Frechét mean of a population of spines represented by articulated models [7], γ are the principal components coefficients (calculated using the covariance matrix of the same population), and d is the parameter vector that controls deformations. The absolute position of every landmark for any configuration x may be obtained by applying equation 1 (for calculating the absolute transformations) and then equation 2.

The goal of the optimisation process is finding the values of d that generate the spine configuration that best fits the splines of both radiographs. For calculating the distance between the articulated model and the splines, we propose projecting to both radiographs the landmarks of the articulated model that define the spine midline. This midline may be represented as a subset of p^{abs} where only the landmarks of the centres of endplates are used:

$$q = \{p_{i,j}^{abs} : \forall i, j \in \{1, 2\}\} . \tag{4}$$

From q, projections of the midline landmarks, q_1^{2D} and q_2^{2D}, may be calculated for the two calibrated radiographs respectively. This is illustrated on figure 2 where it is possible to see the user-identified splines and the projections of the spine midline of the articulated model for the two radiographs.

At this stage, the user input and the articulated model are in the same dimensional space (2D), and both may be represented by splines. However, it is not straightforward to quantify the distance between the two since the user splines may have a different length than the spine midlines of the articulated model (q_k^{2D}). For tackling this, we first project q_k^{2D} to the one dimensional space defined by the spline that has q_k^{2D} as control points and normalise this 1D projections to the spline length. This may be formalised in the following way:

$$\alpha_{k,l} = \frac{length(q_k^{2D}, 1, l)}{length(q_k^{2D}, 1, 2N)} \quad \text{for } k = 1..2, l = 1..2N, \tag{5}$$

where $length(q_k^{2D}, a, b)$ is the function that calculates the length of the segment of spline q_k^{2D} delimited by control points a and b. Vector α is independent of the spline length and, thus, it may be used for mapping the projected spine midlines of the articulated model with the splines identified by the user (figure 2a). This enables to define the following cost function:

$$C = \sum_{k=1}^{2} \sum_{l=1}^{2N} \left\| q_{k,l}^{2D} - (s_k \diamond \alpha_{k,l}) \right\|^2, \tag{6}$$

where s_k represents the user spline for radiograph k, and $\diamond$ is the operator that maps 1D normalised coordinates of a given spline to 2D coordinates. This function is minimised using Matlab's implementation of the trust-region-reflective method [9] for nonlinear least-squares problems.

2.4   Optimising Vertebrae Location

The fitting process just presented only enables to capture the spine shape by placing vertebrae in their probable location along the spine midline, which may not be the correct one. For improving this issue without requesting additional information to the user, we make use of the location of the control points of the splines, which are placed at the centre of vertebral bodies (with the exception of the first and the last control points). Using this specific anatomical position enables us to know that there is a vertebra at the control point location. However, the vertebra identification is unknown, and, therefore, it is not possible to know for sure which vertebra in the articulated model should be attracted by a given control point.

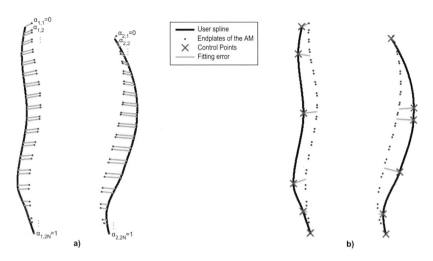

Fig. 2. Fitting the articulated model to the splines: **a)** The spine midline of the 3D articulated model (AM) is projected to both radiographs (frontal and lateral). The 1D normalised coordinates (α) of the landmarks that compose the 2D spine midline are used for mapping the articulated model with the user splines. **b)** Control points attract the nearest vertebra of the articulated model (each vertebra is represented by the centre of its endplates, with the exception of T1 and L5).

For tackling this issue, after a first optimisation with the previously presented process, for each control point, the two nearest vertebrae of the articulated model are selected as candidates. Then, the nearest candidate is elected if the level of ambiguity is low enough. This may be formalised on the following way:

$$\frac{2d_{m,1}}{d_{m,1} + d_{m,2}} \leq \omega, \tag{7}$$

where $d_{m,1}$ is the distance of control point m to the nearest candidate of the articulated model, $d_{m,2}$ is the distance to the second nearest candidate, and ω is a threshold that defines the maximum level of ambiguity allowed. Ambiguity takes the value of 1 when the candidates are equidistant to the control point (total ambiguity), and takes the value of 0 when the nearest candidate is in the exact location of the control point (no ambiguity).

After determining the set of elected candidates, E, the optimisation process is repeated, but now including a second component that is added to equation 6 that attracts the elected vertebrae of the articulated model towards their correspondent control points (figure 2b):

$$C = \sum_{k=1}^{2} \sum_{l=1}^{2N} \left\| q_{k,l}^{2D} - (s_k \diamond \alpha_{k,l}) \right\|^2 + \sum_{m \in E} \|d_{m,1}\|^2. \tag{8}$$

When the second optimisation finishes, the vertebrae location of the articulated model should be closer to their real position, and some of the ambiguities may be solved. Therefore, several optimisation processes are executed iteratively while the number of elected candidates, E, increases.

Concerning the value of ω, using a low threshold of ambiguity may result in a considerable waste of control points due to an over-restrictive strategy. On the other hand, a high threshold of ambiguity may produce worst results, especially when there are control points placed on erroneous locations. For overcoming this issue, a dynamic thresholding technique is used that begins with a restrictive threshold ($\omega = 0.60$), and when no candidates are elected the threshold is increased (by 0.10) up to a maximum threshold of ambiguity ($\omega = 0.80$). If there are any control points still ambiguous at this stage, they are considered to be unreliable.

3   Experiments and Results

For all experiments a data set of 14 *in vivo* exams of scoliotic patients was used. All exams were composed of at least one posterior-anterior (PA) and one lateral (LAT) radiograph. The 6 anatomical landmarks were previously identified on both PA and LAT radiographs by an expert and 3D reconstructions of the landmarks were available. These reconstructions were obtained using a previously validated method [10] and were used as reference on this study.

The articulated model that was used in all experiments was built using 291 exams. The principal components that explained 95% of the data variability were extracted, resulting in a model with 47 dimensions.

3.1   Evaluation Using Simulated Splines

The first experiment consisted of evaluating the proposed method using perfectly marked splines for determining the expected error in ideal conditions. For this, splines with 6 control points were built using the available 2D hand-marked points. In particular, the centre of the superior endplate of T1 (first control point of splines) and the centre of the inferior endplate of L5 (last control point) were copied from the original 2D points. Then the centre of the vertebral body of the remaining vertebrae was calculated using the average location of the centres of both endplates. Finally, for each exam, and for each radiograph, 4 vertebral body centres were selected as control points of the spline in order to best fit all the endplates' centres that were manually identified by an expert. Reconstructions with these splines resulted on an error of 2.2±1.3mm for the endplates and 4.1±2.1mm for pedicles (mean ± S.D.).

For determining the effect of identification errors of the control points on the quality of reconstructions, a second experiment was made where Gaussian noise with standard deviation up to 8 pixels was introduced to the 2D coordinates of the control points. Results of this experience are presented in figure 3.

3.2   Evaluation Using Manually Identified Splines

The last experience concerned evaluating the method performance when using splines manually identified by non-experts. For conducting it, 2 volunteers with

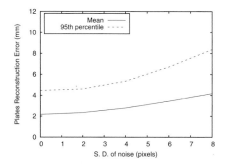 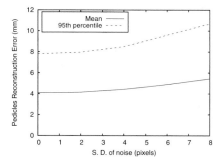

Fig. 3. Effect of simulated Gaussian noise added to the 2D location of the control points of the splines (reconstruction errors for: left – endplates; right: – pedicles)

Table 1. Results for the experiment with splines identified by non-experts with 20min of training. Mean values ($\pm$ S.D. for input and reconstruction errors). Number of C.P. is the average number of control points used for identifying the splines.

| User | Number of C.P. | Input error (pixels) | | Reconstruction Time (min:s) | Reconstruction Error (mm) | |
		Spline	C.P.		Endplates	Pedicles
A	6.1	6.3±5.7	5.9±5.4	1:22	3.4±1.9	4.8±2.5
B	5.5	5.2±5.0	8.7±8.6	1:32	3.4±2.0	4.8±2.6

very limited knowledge of the spine radiological landmarks marked the same 14 exams that were used in the previous experiments. Both of them only had 20min of training with the software tool before performing the experiment. Results of this experiment are presented in table 1, including the average time needed for a reconstruction. The presented times are dominated by user interaction since the average processing time is approximately 10s (on a Pentium Dual Core of 1.86 GHz). Additionally, the input error of each spline was calculated in two ways: a) Spline error: the distance between every endplate centre of the 2D reference data and the nearest point of the spline, and b) Control Points error: the distance between the control points coordinates and their ideal location. Figure 4 shows reconstructions of the proposed and reference methods for an average case.

4   Discussion and Conclusion

The method proposed here achieves 3D reconstructions of the spine by only requiring the identification of two splines by an user. Results with simulated splines composed by 6 control points were compared with the reference method, which needs 102 points per radiograph to be manually identified. The proposed method showed errors near of the *in vitro* accuracy of the reference method [10] for the endplates (2.2±1.3mm vs 1.5±0.7mm), but higher errors for pedicles

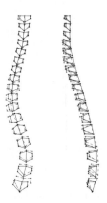

Fig. 4. Comparison of a reconstruction using the proposed method (solid line) with the reference method (dashed line) for an average case of manual identification by non-experts (endplates error– 3.2mm; pedicles error – 4.8mm)

(4.1±2.1mm vs 1.2±0.7mm). We believe that this happens because there is no direct input from the user concerning the localisation of pedicles, and therefore pedicles position and vertebrae axial rotation are completely inferred from the spine curvature.

Simulations considering artificial noise (figure 3) show a robust behaviour with a standard deviation of the noise up to 4 pixels. After this, the curve becomes steeper, especially on the 95th percentile of the endplates error. It is also visible that identification errors have an higher impact on the accuracy of the endplates position than on pedicles, which is natural since the endplate position is much more dependent of the user input. Nevertheless, reconstruction errors of the endplates with noise of 8 pixels are comparable to pedicle errors with no noise.

Experiments with non-expert users revealed an average user time inferior to 1.5min, which is an advance in comparison with the previous generation of reconstruction algorithms. Reconstruction errors are mainly influenced by the quality of the user input either on the accuracy of the spline midline (higher on user A), or in the identification of the precise location of the control points (mainly on user B). In spite of this variability on the dominant source of error, reconstruction errors were comparable for both users. These errors are not comparable with reconstructions performed by experts, but show that a rough initial estimation is at the reach of ordinary users. This is especially visible on figure 4 where it is noticeable that despite the visible input error, vertebrae location is quite accurate given the small amount of user input. Additionally, using more than one user on this experiment gives more credibility to the obtained results, and the fact that reconstruction errors were very similar for both users makes us believe that these results may be generalised. Future work will include thorough tests with even more users to reinforce these conclusions.

We believe that in future assessments of the method with an expert user, the results will be closer to the results achieved with simulated splines. Additionally, using more control points may improve the location of endplates, which creates a tradeoff between accuracy and user interaction. However, additional information will be needed for accurately locating pedicles. Future work includes determining the impact of knowing the location of small sets of pedicles landmarks, which may be either identified by users or by image segmentation algorithms.

Acknowledgements

The first author thanks Fundação para a Ciência e a Tecnologia for his PhD scholarship (SFRH/BD/31449/2006), and Fundação Calouste Gulbenkian for granting his visit to École Polytechnique de Montréal. The authors would like to express their gratitude to Lama Séoud, Hassina Belkhous, Hubert Labelle and Farida Cheriet from Hospital St. Justine for providing access to patient data and for their contribution on the validation process.

References

1. Stokes, I.A.: Three-dimensional terminology of spinal deformity a report presented to the scoliosis research society by the scoliosis research society working group on 3-d terminology of spinal deformity. Spine 19, 236–248 (1994)
2. Aubin, C.E., Descrimes, J.L., Dansereau, J., Skalli, W., Lavaste, F., Labelle, H.: Geometrical modelling of the spine and thorax for biomechanical analysis of scoliotic deformities using finite element method. Ann. Chir. 49, 749–761 (1995)
3. Mitton, D., Landry, C., Vron, S., Skalli, W., Lavaste, F., De Guise, J.: 3d reconstruction method from biplanar radiography using non-stereocorresponding points and elastic deformable meshes. Med. Biol. Eng. Comput. 38, 133–139 (2000)
4. Pomero, V., Mitton, D., Laporte, S., de Guise, J.A., Skalli, W.: Fast accurate stereoradiographic 3d-reconstruction of the spine using a combined geometric and statistic model. Clin. Biomech. 19, 240–247 (2004)
5. Humbert, L., Guise, J.D., Aubert, B., Godbout, B., Skalli, W.: 3d reconstruction of the spine from biplanar x-rays using parametric models based on transversal and longitudinal inferences. Med. Eng. Phys. (in press, 2009)
6. Kadoury, S., Cheriet, F., Labelle, H.: Personalized X-Ray 3D Reconstruction of the Scoliotic Spine From Hybrid Statistical and Image-Based Models. IEEE Trans. Med. Imaging (in press, 2009)
7. Boisvert, J., Cheriet, F., Pennec, X., Labelle, H., Ayache, N.: Geometric variability of the scoliotic spine using statistics on articulated shape models. IEEE Trans. Med. Imaging 27, 557–568 (2008)
8. Boisvert, J., Cheriet, F., Pennec, X., Labelle, H., Ayache, N.: Articulated spine models for 3-d reconstruction from partial radiographic data. IEEE Trans. Biomed. Eng. 55, 2565–2574 (2008)
9. Coleman, T.F., Li, Y.: An interior trust region approach for nonlinear minimization subject to bounds. SIAM J. Optim. 6, 418–445 (1996)
10. Aubin, C., Dansereau, J., Parent, F., Labelle, H., de Guise, J.A.: Morphometric evaluations of personalised 3d reconstructions and geometric models of the human spine. Med. Biol. Eng. Comput. 35, 611–618 (1997)

High-Quality Rendering of Varying Isosurfaces with Cubic Trivariate C^1-Continuous Splines

Thomas Kalbe, Thomas Koch, and Michael Goesele

GRIS, TU Darmstadt

Abstract. Smooth trivariate splines on uniform tetrahedral partitions are well suited for high-quality visualization of isosurfaces from scalar volumetric data. We propose a novel rendering approach based on spline patches with low total degree, for which ray-isosurface intersections are computed using efficient root finding algorithms. Smoothly varying surface normals are directly extracted from the underlying spline representation. Our approach is using a combined CUDA and graphics pipeline and yields two key advantages over previous work. First, we can interactively vary the isovalues since all required processing steps are performed on the GPU. Second, we employ instancing in order to reduce shader complexity and to minimize overall memory usage. In particular, this allows to compute the spline coefficients on-the-fly in real-time on the GPU.

1 Introduction

The visualization of discrete data on volumetric grids is a common task in various applications, e.g., medical imaging, scientific visualization, or reverse engineering. The construction of adequate non-discrete models which fit our needs in terms of visual quality as well as computational costs for the display and preprocessing is an interesting challenge. The most common approach is tri-linear interpolation [1,2], where the tensor-product extension of univariate linear splines interpolating at the grid points results in piecewise cubic polynomials. A sufficiently smooth function is approximated with order two, but in general, reconstructions are not smooth and visual artifacts, like stair-casing or imperfect silhouettes, arise. However, the simplicity of this model has motivated its widespread use. Tri-quadratic or tri-cubic tensor-product splines can be used to construct smooth models. These splines lead to piecewise polynomials of higher total degree, namely six and nine, which are thus more expensive to evaluate.

In order to alleviate these problems, we use cubic trivariate C^1-splines [3] for interactive visualizations of isosurfaces from volumetric data with ray casting. The low total degree of the spline pieces allows for efficient and stable ray-patch intersection tests. The resulting spline pieces are directly available in Bernstein-Bézier–form (BB-form) from the volumetric data by simple, efficient and local averaging formulae using a symmetric and isotropic data stencil from the 27-neighborhood of the nearest data value. The BB-form of the spline pieces has several advantages: well-known techniques from CAGD, like de Casteljau's algorithm and *blossoming,* can be employed for efficient and stable evaluation of

G. Bebis et al. (Eds.): ISVC 2009, Part I, LNCS 5875, pp. 596–607, 2009.

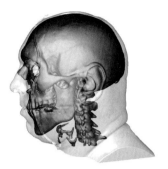

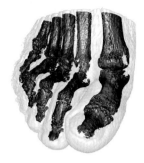

Fig. 1. Blending of different isosurfaces from real-world data sets. *From left to right*: *VisMale* (256^3 voxels), *Tooth* ($256^2 \times 161$ voxels), and *Foot* (256^3 voxels). VisMale and Tooth are smoothed with a Gaussian filter on the GPU. Choosing the desired isosurfaces and smoothing to an appropriate degree is an interactive process.

the splines. The derivatives needed for direct illumination are immediately available as a by-product. The convex hull property of the BB-form allows to quickly decide if a given spline patch contributes to the final surface.

A first GPU implementation for visualization with cubic trivariate splines has been given by [4]. Data streams of a fixed isolevel were prepared on the CPU and then sent to the GPU for visualization. This pre-process, which takes a couple of seconds for medium-sized data sets ($\geq 256^3$ data values), has to be repeated for each change of isolevel. However, in many applications it is essential to vary the isosurface interactively in order to gain deeper understanding of the data (see Fig. 1). In this work, we significantly accelerate the pre-process using NVIDIA's *CUDA* framework and achieve reconstruction times which are even below the rendering times of a single frame. In addition, current innovations of the graphics pipeline in Shader Model 4.0, like *instancing*, allow us to compute all necessary spline coefficients on-the-fly directly in the vertex shader. Therefore, we do not need to inflate the data prior to visualization, but merely store the volume data as a texture on the GPU. In addition, geometry encoding is simplified and memory overhead is reduced. Combining these contributions, we significantly improved the usability of high-quality trivariate splines in real-world applications.

2   Related Work

Techniques for visualizations of gridded scalar data can be categorized into two general classes. *Full volume rendering*, where the equations of physical light transport are integrated throughout the volume, commonly along viewing rays, and the somewhat less complex *isosurfacing*. In the latter case, we are interested in the zero contour of a continuous implicit function which approximates or interpolates the discrete values given at the grid points. We can classify isosurfacing further into methods that obtain discrete representations of the surfaces, e.g., triangle meshes. A standard approach in this area is *marching cubes* [5]. Alternatively, we are only interested in visualizations of the isosurfaces, which is often done by ray casting, where the first intersection of each viewing ray with the surface is determined for later illumination. The recent development of graphics processors has been a massive impulse for interactive volume graphics on consumer hardware (see [6] for a survey). Interactive techniques exist for full volume rendering, isosurface visualization and reconstruction, e.g, [7,8,9,10]. Still, most of these approaches are based on tri-linear interpolation and therefore trade visual quality in favor of rendering speed. Gradients can be pre-computed at the grid points, at the cost of increased memory and bandwidth consumption. Alternatively, the gradients are computed on-the-fly using central differences, which is an expensive operation. Either way, the obtained gradients are not smooth, and visual artifacts arise. To circumvent these problems, higher-order filter kernels, e.g., smooth tri-cubic or tri-quartic box splines, have been proposed [2,11]. One of the few successful implementations of interactive isosurface visualization with higher order filtering has been given by [12]. These splines lead to polynomials of total degree nine, for which no exact root finding algorithms exist. Furthermore, data stencils are large (usually the 64-neighborhood), and important features might be lost resulting from the large support of the filter kernels.

We use smooth trivariate splines defined w.r.t. uniform tetrahedral partitions. Here, the filter kernels are small and isotropic. Since the total degree of the polynomial pieces does not exceed three, we can choose suitable starting values for an iterative root finding algorithm, such that precise intersections with the isosurface are obtained in a stable and efficient way. No further refinements, e.g., near the surface's silhouette, are needed. For an example see Fig. 7, right. An approach for interactive visualization using trivariate splines, has been given by [4]. While this work was the first to allow for real-time rendering of up to millions of smoothly connected spline patches simultaneously, it is based on the common principles described by, e.g, [13,14,15]. These methods project the bounding geometry of the polynomials in screen space and perform intersection tests and illumination during fragment processing. [4] rely on a CPU preprocessing of the data for each change of isovalue, which can be done only off-line. Furthermore, memory requirements for the storage of spline coefficients are substantial. In this paper, we shift the preprocessing to the GPU in order to allow for an interactive change of isosurface. To do that, we use *parallel prefix scans* as described in [16,17]. In addition, we show how to reduce memory overhead to a minimum using current innovations in the graphics pipeline.

3   Trivariate Splines and the BB-Form

In this section, we give a brief outline of the basic terminology and mathematical background of trivariate splines in BB-form on tetrahedral partitions, along with a description of the calculation of the spline coefficients. These coefficients can be directly obtained from the volume data by simple averaging formulae.

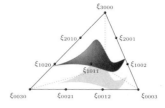

Fig. 2. *Left*: the cube partition $\diamondsuit$ of the domain Ω. *Middle*: the type-6 partition is obtained by first subdividing each cube into six pyramids, which are then further split into four congruent tetrahedra each. *Right*: the zero contour of a single cubic trivariate spline patch $s|_T$ within it's bounding tetrahedron T. The domain points $\xi_{ijk\ell}$ associated with the BB-coefficients $b_{ijk\ell}$ of the front-most triangle are shown.

3.1   Preliminaries and Basic Notation

For $n \in \mathbb{N}$ let $\mathcal{V} := \{\mathbf{v}_{ijk} = (ih, jh, kh) : i, j, k = 0, \ldots, n\}$ be the cubic grid of $(n + 1)^3$ points with grid size $h = 1/n \in \mathcal{R}$. We define a cube partition $\diamondsuit = \{Q : Q = Q_{ijk}\}$ of the domain Ω, where each $Q_{ijk} \in \diamondsuit$ is centered at $\mathbf{v}_{ijk}$ and the vertices of Q_{ijk} are $(2i \pm 1, 2j \pm 1, 2k \pm 1)^t \cdot h/2$, see Fig. 2, left.

We consider trivariate splines on the *type-6 tetrahedral partition* Δ^6, where each Q is subdivided into 24 congruent tetrahedra. This is done by connecting the vertices of Q_{ijk} with the center $\mathbf{v}_{ijk}$. Each of the resulting six pyramids is then further split into four tetrahedra, see Fig. 2, right. The space of cubic trivariate C^1 splines on Δ^6 is defined by $\mathcal{S}_3^1(\Delta^6) = \{s \in C^1(\Omega) : s|_T \in \mathcal{P}_3, \text{ for all } T \in \Delta^6\}$, where $C^1(\Omega)$ is the set of continuously differentiable functions on Ω, $\mathcal{P}_3 := \text{span}\{x^\nu y^\mu z^\kappa : 0 \leq \nu + \mu + \kappa \leq 3\}$ is the 20-dimensional space of trivariate polynomials of total degree three, and T is a tetrahedron in Δ^6. We use the BB-form of the polynomial pieces, i.e.

$$s|_T = \sum_{i+j+k+\ell} b_{ijk\ell} B_{ijk\ell}, \ i + j + k + \ell = 3,$$

where the $B_{ijk\ell} = \frac{3!}{i!j!k!\ell!} \phi_0^i \phi_1^j \phi_2^k \phi_3^\ell \in \mathcal{P}_3$ are the cubic *Bernstein polynomials* w.r.t a tetrahedron $T = [\mathbf{v}_0, \mathbf{v}_1, \mathbf{v}_2, \mathbf{v}_3] \in \Delta^6$. For each T, we set $\mathbf{v}_0$ to the center of it's cube Q, $\mathbf{v}_1$ to the centroid of one of the faces of Q and $\mathbf{v}_2, \mathbf{v}_3$ to the vertices of Q sharing a common edge. The *barycentric coordinates* $\boldsymbol{\phi}(\mathbf{x}) = (\phi_0(\mathbf{x}), \phi_1(\mathbf{x}), \phi_2(\mathbf{x}), \phi_3(\mathbf{x}))^t$ of a point $\mathbf{x} = (x, y, z, 1)^t$ w.r.t a non-degenerate T are the linear trivariate polynomials determined by $\phi_\nu(\mathbf{v}_\mu) = \delta_{\nu,\mu}$, $\nu, \mu = 0, \ldots, 3$, where $\delta_{\nu,\mu}$ is Kronecker's symbol. They are given by the linear system of equations

$$\boldsymbol{\phi}(\mathbf{x}) = \begin{pmatrix} \mathbf{v}_0 & \mathbf{v}_1 & \mathbf{v}_2 & \mathbf{v}_3 \\ 1 & 1 & 1 & 1 \end{pmatrix}^{-1} \cdot \mathbf{x}. \tag{1}$$

The *BB-coefficients* $b_{ijk\ell} \in \mathcal{R}$ are associated with the 20 domain points $\xi_{ijk\ell} = (i\mathbf{v}_0 + j\mathbf{v}_1 + k\mathbf{v}_2 + \ell\mathbf{v}_3)/3$, see Fig. 2, right, and we let $\mathcal{D}(\Delta^6)$ be the union of the sets of domain points associated with the tetrahedra of Δ^6. As pointed out

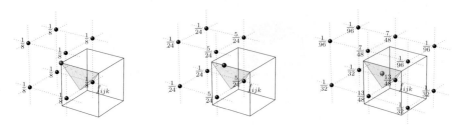

Fig. 3. The masks for the coefficients associated with the domain points ξ_{0003} (left), ξ_{0021} (middle) and ξ_{0111} (right) for the shaded tetrahedron in Q_{ijk}. The remaining coefficients of $\Gamma_{Q_{ijk}}$ follow from symmetry and rotations. Black dots denote data values.

by e.g. [18], the BB-form is especially useful for defining smoothness conditions between neighboring polynomial pieces. Let $T, \tilde{T}$ be two neighboring tetrahedra sharing a common face $F = T \cap \tilde{T} = [\mathbf{v}_0, \mathbf{v}_1, \mathbf{v}_2]$, then a cubic spline s on $T \cup \tilde{T}$ is *continuous* ($s \in \mathcal{S}_3^0$) if $s|_T(\mathbf{x}) = s|_{\tilde{T}}(\mathbf{x})$, $\mathbf{x} \in F$. Using the BB-form, we have $s \in \mathcal{S}_3^0$ if $b_{ijk0} = \tilde{b}_{ijk0}$ and a continuous spline s is uniquely defined by the coefficients $\{b_\xi : \xi \in \mathcal{D}(\Delta^6)\}$. Furthermore, s is C^1-continuous across F iff

$$\tilde{b}_{ijk1} = b_{i+1,j,k,0}\, \phi_0(\tilde{\mathbf{v}}_3) + b_{i,j+1,k,0}\, \phi_1(\tilde{\mathbf{v}}_3) + b_{i,j,k+1,0}\, \phi_2(\tilde{\mathbf{v}}_3) + b_{i,j,k,1}\, \phi_3(\tilde{\mathbf{v}}_3), \qquad (2)$$

where $i + j + k = 2$ and $\tilde{\mathbf{v}}_3$ is the vertex of $\tilde{T}$ opposite to F. Smoothness of the splines on Δ^6 is thus easily described when considering only two neighboring polynomial pieces. The complexity of the spline spaces arises from the fact that smoothness conditions have to be fulfilled not only between two neighboring patches, but across all the interior faces of Δ^6 simultaneously.

We can evaluate a spline patch $s|_T(\mathbf{x})$ with de Casteljau's algorithm. Set $b_{ijk\ell}^{[0]} = b_{ijk\ell}$, a de Casteljau step computes $b_{ijk\ell}^{[\eta]}$, $i+j+k+\ell = 3-\eta$, as the inner product of $(\phi_0(\mathbf{x}), \phi_1(\mathbf{x}), \phi_2(\mathbf{x}), \phi_3(\mathbf{x}))^t$ and $(b_{i+1,j,k,\ell}^{[\eta-1]}, b_{i,j+1,k,\ell}^{[\eta-1]}, b_{i,j,k+1,\ell}^{[\eta-1]}, b_{i,j,k,\ell+1}^{[\eta-1]})^t$ with $s|_T(\mathbf{x}) = b_{0000}^{[3]}$. In addition, the $(3-1)$th step provides the four independent directional derivatives

$$\frac{\partial s|_T(\mathbf{x})}{\partial \phi_\nu(\mathbf{x})} = b_\nu^{[2]}, \quad \nu = 0, \ldots, 3, \qquad (3)$$

where $\boldsymbol{\nu} \in \mathbb{N}_0^4$ is the vector with a 1 at position ν and 0 everywhere else. With $s|_T = p$ and using the chain rule, we have

$$\frac{\partial p(\mathbf{x})}{\partial x_\iota} = \sum_\nu \frac{\partial s|_T(\mathbf{x})}{\partial \phi_\nu(\mathbf{x})} \cdot \frac{\partial \phi_\nu(\mathbf{x})}{\partial x_\iota}, \quad \iota \in 1, 2, 3, \qquad (4)$$

with the gradient $\nabla s|_T(\mathbf{x}) = (\partial p/\partial x_1, \partial p/\partial x_2, \partial p/\partial x_3)^t$. Since each ϕ_ν is a linear polynomial, the $\partial \phi_\nu / \partial x_\iota$ are scalar constants characterized by the barycentric coordinates of the ιth Cartesian unit vector $\mathbf{e}_\iota$ w.r.t. T.

Trivariate *blossoming* [19] is a generalization of de Casteljau's algorithm, where the arguments may vary on the different levels. For any $\mathbf{x}_\eta$, $\eta \in 1, 2, 3$

with $\phi_\eta = \phi(\mathbf{x}_\eta)$, we denote the blossom of $s|_T$ as $\mathrm{bl}[\phi_1, \phi_2, \phi_3]$, meaning that the first step of de Casteljau's algorithm is carried out with ϕ_1, the second step with ϕ_2 and the third with ϕ_3. It is easy to see that $\mathrm{bl}[\phi_\eta, \phi_\eta, \phi_\eta] = s|_T(\mathbf{x}_\eta)$. In addition, the blossom is *multi-affine*

$$\mathrm{bl}[\ldots, \alpha \cdot \phi_\eta + (1-\alpha) \cdot \bar{\phi}_\eta, \ldots] = \alpha \cdot \mathrm{bl}[\ldots, \phi_\eta, \ldots] + (1-\alpha) \cdot \mathrm{bl}[\ldots, \bar{\phi}_\eta, \ldots], \quad (5)$$

for $\alpha \in \mathbb{R}$, and symmetric, i.e., for a permutation $\sigma = (\sigma(1), \sigma(2), \sigma(3))$ we have

$$\mathrm{bl}[\phi_1, \phi_2, \phi_3] = \mathrm{bl}[\phi_{\sigma(1)}, \phi_{\sigma(2)}, \phi_{\sigma(3)}].$$

These properties of the blossom enable us to find intersections of rays with a spline patch in an efficient way, see Sect. 4.4.

3.2   The Approximating Scheme

For smooth approximations of the given data values $f(\mathbf{v}_{ijk})$, associated with the grid points $\mathbf{v}_{ijk}$, we use quasi-interpolating cubic splines as described in [3], which approximate sufficiently smooth functions with order two. The BB-coefficients b_ξ for each tetrahedron are directly available from appropriate weightings of the data values in a symmetric 27-neighborhood of the centering data value $f(\mathbf{v}_{ijk})$,

$$b_\xi = \sum_{i_0, j_0, k_0} \omega_{i_0 j_0 k_0} f(\mathbf{v}_{i+i_0, j+j_0, k+k_0}), \quad i_0, j_0, k_0 \in \{-1, 0, 1\},$$

where the $\omega_{i_0 j_0 k_0} \in \mathcal{R}$ are constant and positive weights. A *determining set* $\Gamma \subseteq \mathcal{D}(\Delta^6)$ is a subset of the domain points with associated BB-coefficients, from which the remaining coefficients for each $s|_T$ can be uniquely identified from the smoothness conditions. We use a symmetric determining set Γ_Q for each $Q \in \Diamond$, formed by the coefficients associated with the domain points $\xi_{00k\ell} \bigcup \xi_{0111}$, where $k + \ell = 3$. For a tetrahedron, ξ_{0030} and ξ_{0003} are vertices of Q, ξ_{0021} and ξ_{0012} are on the outer edges of Q, and ξ_{0111} corresponds to the centroid of the face $\mathbf{v}_1, \mathbf{v}_2, \mathbf{v}_3$. We show the weights for the coefficients of the determining set in Fig. 3. The weights for the remaining coefficients follow from the smoothness conditions (see Eq. 2) and can also be found in [3,4].

4   Trivariate Splines – GPU Visualization

In this section, we give an overview of our GPU-algorithm for efficient visualization of varying isosurfaces, see also Fig. 4, left. In the first part, we use a set of CUDA kernels, which are invoked for each change of isolevel or data set. The kernels determine all the tetrahedra which can contribute to the final surface and prepare the appropriate data structures. The second part uses vertex and fragment programs for the visualization of the surface in a combined rasterization / ray casting approach. For a 2-dimensional example of the visualization principle see Fig. 4, right top. For each *active* tetrahedron, i.e., tetrahedra contributing to the surface, the bounding geometry is processed in the OpenGL pipeline. The vertex programs initialize various parameters, such as the viewing rays, the BB-coefficients and appropriate barycentric coordinates. The fragment programs then perform the actual ray-patch intersection tests.

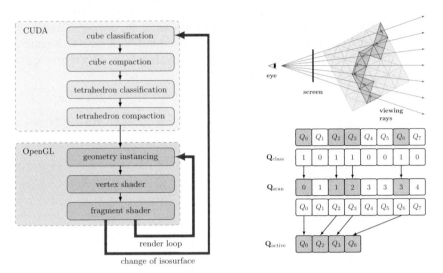

Fig. 4. *Left*: Overview of our our GPU algorithm. The CUDA part specifies the relevant cubes and tetrahedra, respectively, when the isovalue or data set is modified. The visualization is using the rendering pipeline. *Right top*: 2-dimensional illustration of the ray casting principle. On each triangle satisfying the convex hull property (dark shaded), the corresponding BB-curve is intersected with the viewing rays. *Right bottom*: parallel stream compaction with prefix scans. For each entry with a 1 in $\mathbf{Q}_{\text{class}}$, $\mathbf{Q}_{\text{scan}}$ contains an unique address into the compressed array $\mathbf{Q}_{\text{active}}$. The sum of the last entries of $\mathbf{Q}_{\text{class}}$ and $\mathbf{Q}_{\text{scan}}$ gives the size of $\mathbf{Q}_{\text{active}}$. Illustration based on [17].

4.1   The CUDA Kernels

For each $Q \in \Diamond$, we first start a kernel thread that computes the coefficients for the determining set Γ_Q, from which the remaining coefficients can be found quickly using simple averaging. The *cube classification* tests if the coefficients of Γ_Q are either below or above the isolevel. In this case, it follows from the convex hull property of the BB-form that the patches in Q cannot contribute to the surface and we can exclude Q from further examination. Otherwise, Q contains at least one tetrahedron with a visible patch. The result of the classification is written in the corresponding entry of a linear integer array $\mathbf{Q}_{\text{class}}$ of size $(n+1)^3$: we write a 1 in $\mathbf{Q}_{\text{class}}$ if Q passes the classification test and a 0 otherwise. From this unsorted array $\mathbf{Q}_{\text{class}}$, we construct a second array $\mathbf{Q}_{\text{scan}}$ of the same size using the parallel prefix scan from the *CUDA data parallel primitives* (CUDPP) library. For each *active* cube, i.e., cubes with a 1 in $\mathbf{Q}_{\text{class}}$, $\mathbf{Q}_{\text{scan}}$ then contains an unique index corresponding to the memory position in a compacted array $\mathbf{Q}_{\text{active}}$. Since we use an exclusive prefix scan, the sum of the last entries of $\mathbf{Q}_{\text{class}}$ and $\mathbf{Q}_{\text{scan}}$ gives the number of active cubes a for the surface. In the *compaction* step, we reserve memory for the array $\mathbf{Q}_{\text{active}}$ of size a. For each active Q_{ijk}, we write the index $i + j \cdot (n+1) + k \cdot (n+1)^2$ in $\mathbf{Q}_{\text{active}}$ at the position given by the corresponding entry of $\mathbf{Q}_{\text{scan}}$, see Fig. 4, right bottom.

Similarly, we perform a compaction for the active tetrahedra. Since we use instancing for later rendering of the tetrahedra, we reserve 24 arrays $\mathbf{T}_{\text{class},i}$, where each $\mathbf{T}_{\text{class},i}$ has size a and corresponds to one of the 24 different orientations of tetrahedra in Δ^6. For each active cube Q, a kernel thread performs the classification for $T_0, T_1, \ldots, T_{23} \in Q$, now using the convex hull property of the BB-form on the tetrahedra. Note that the BB-coefficients for Q have to be calculated only once and are then assigned to the corresponding domain points on the tetrahedra using a constant lookup table. The following stream compaction works exactly as described above, except that here we use 24 distinct arrays $\mathbf{T}_{\text{scan},i}$, and the compaction is performed by a set of 24 kernels (one for each type of tetrahedron), which write their results into the arrays $\mathbf{T}_{\text{active},i}$. These arrays are interpreted as pixel buffer objects, which can be directly used to render the bounding geometry of the spline patches.

4.2   Geometry Instancing Setup

The 24 arrays $\mathbf{T}_{\text{active},i}$ give us all the information needed to visualize the surface. In order to encode the necessary bounding geometries, i.e., the active tetrahedra, in the most efficient way, we construct a triangle strip for each *generic* tetrahedron of Δ^6 in the unit cube $[-0.5, 0.5]^3$. These 24 triangle strips are then stored as separate vertex buffer objects (VBOs). Each VBO is used to draw all the active tetrahedra of it's type with a call to `glDrawArraysInstanced`, where the number of tetrahedra is given by the size of $\mathbf{T}_{\text{active},i}$.

4.3   Vertex Shader Computations

We use 24 different shader sets, one for each of the different types of tetrahedra. Since each tetrahedron has it's dedicated vertex and fragment programs, we can avoid conditional branches. For every vertex $\mathbf{v}_\mu, \mu = 0, 1, 2, 3$, of a tetrahedron T, the vertex program first determines T's displacement $\mathbf{v}_Q$ from a texture reference into $\mathbf{T}_{\text{active},i}$. For later computation of the ray-patch intersection (see Sect. 4.4), we find the barycentric coordinates $\phi_{\nu,\mu}, \nu = 0, 1, 2, 3$ as $\phi_{\nu,\mu}(\mathbf{v}_\mu) = \delta_{\nu,\mu}$. In addition, a second set of barycentric coordinates $\bar{\phi}_{\nu,\mu}$, corresponding to the unit length extension of the vector defined by $\mathbf{v}_\mu + \mathbf{v}_Q$ and the eye point $\mathbf{e}$ is computed as $\bar{\phi}_{\nu,\mu}(\bar{\mathbf{v}}) = \bar{\phi}_{\nu,\mu}(\mathbf{v}_\mu + (\mathbf{v}_\mu - (\mathbf{e} - \mathbf{v}_Q))/\|\mathbf{v}_\mu - (\mathbf{e} - \mathbf{v}_Q)\|)$. To do this, we use Eq. 1, where the matrices are pre-computed once for each generic tetrahedron. The barycentric coordinates $\phi_{\nu,\mu}, \bar{\phi}_{\nu,\mu}$ are interpolated across T's triangular faces during rasterization. Finally, we have to determine the 20 BB-coefficients of $s|_T$. This can be done in the following way: first, we read the data values $f(\mathbf{v}_Q)$ and it's neighbors from the volume texture. For one patch, only 23 of the 27 values have to be fetched, corresponding to 23 texture accesses. Then, we can directly obtain the $b_{ijk\ell}$ according to Sect. 3.2. Alternatively, we can pre-compute the determining set for each Q in a CUDA kernel and store them as textures. Then, only the remaining coefficients for $s|_T$ need to be computed. This method is less memory efficient, but leads to a slightly improved rendering performance. For a comparative analysis, see Sect. 5.

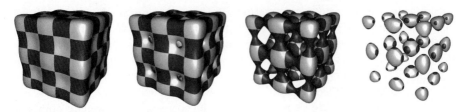

Fig. 5. Varying isosurfaces for a synthetic function of *Chmutov* type, $f(x, y, z) = x^{16} + y^{16} + z^{16} - \cos(7x) - \cos(7y) - \cos(7z)$, sampled from a sparse grid (64^3 data points) with real-time reconstruction and rendering times

4.4   Fragment Shader Computations

For every fragment of a front-facing triangle, the fragment program performs the actual ray-patch intersection. To do this, we need an univariate representation of $s|_T$ restricted along the viewing ray. Using trivariate blossoming, we obtain an univariate cubic BB-curve which can be easily intersected. In addition, we can re-use intermediate results from the blossoms for quick gradient calculation and do not need to determine the exit point of the ray w.r.t. T. Using the interpolated barycentric coordinates $\boldsymbol{\phi} = (\phi_0, \phi_1, \phi_2, \phi_3)$ and $\bar{\boldsymbol{\phi}} = (\bar{\phi}_0, \bar{\phi}_1, \bar{\phi}_2, \bar{\phi}_3)$ obtained from rasterization, we can read off the univariate BB-coefficients directly from the blossoms, setting $b_{30} = \mathrm{bl}[\boldsymbol{\phi}, \boldsymbol{\phi}, \boldsymbol{\phi}]$, $b_{21} = \mathrm{bl}[\boldsymbol{\phi}, \boldsymbol{\phi}, \bar{\boldsymbol{\phi}}]$, $b_{12} = \mathrm{bl}[\bar{\boldsymbol{\phi}}, \bar{\boldsymbol{\phi}}, \boldsymbol{\phi}]$, and $b_{03} = \mathrm{bl}[\bar{\boldsymbol{\phi}}, \bar{\boldsymbol{\phi}}, \bar{\boldsymbol{\phi}}]$, see Sect. 3. Since intermediate results can be re-used, the first step of de Casteljau's algorithm, which accounts for ten inner products each, has to be performed for $\boldsymbol{\phi}$ and $\bar{\boldsymbol{\phi}}$ only once. We proceed in the same way for the second de Casteljau step with $b_{ijk\ell}^{[1]}(\boldsymbol{\phi})$ (resulting from the first step with $\boldsymbol{\phi}$) and $\boldsymbol{\phi}$, as well as $b_{ijk\ell}^{[1]}(\bar{\boldsymbol{\phi}})$ and $\bar{\boldsymbol{\phi}}$, where $i + j + k + \ell = 2$, using in total eight inner products. Finally, the blossoms are completed with four additional inner products using $b_{ijk\ell}^{[2]}(\boldsymbol{\phi})$ and $b_{ijk\ell}^{[2]}(\bar{\boldsymbol{\phi}})$, $i + j + k + \ell = 1$, which correspond to the remaining de Casteljau steps on the last level.

Next, the monomial form of the BB-curve, $\sum_{i=0}^{3} x_i \cdot t^i$, is solved for the ray parameter t. There exist several ways to find the zeros of a cubic equation. [14] choose an analytic approach, whereas [15] apply a recursive BB-hull subdivision algorithm. Since the first method involves trigonometric functions and the second does not converge very quickly, we opt for an iterative Newton approach. As starting values we choose $t_1^{(0)} = -x_0/x_1$, $t_1^{(1)} = (\frac{1}{4}(x_3 + x_2) - x_0)/(\frac{3}{4} \cdot x_3 + x_2 + x_1)$ and $t_1^{(2)} = (2 \cdot x_3 + x_2 - x_0)/(3 \cdot x_3 + 2 \cdot x_2 + x_1)$. Note that this corresponds to the first Newton iteration starting with $0, 1/2$, and 1, respectively. Four additional iterations with $t_{j+1}^{(\mu)} = ((t_j^{(\mu)})^2(x_2 + 2 \cdot t_j^{(\mu)} x_3) - x_0)/(t_j^{(\mu)}(2 \cdot x_2 + 3 \cdot t_j^{(\mu)} x_3) + x_1)$, $\mu = 0, 1, 2$, suffice to find precise intersections without notable artifacts. For each solution $t \in t_5^{(\mu)}$, the associated barycentrics $\boldsymbol{\phi}(\mathbf{x}(t))$ are found by a simple linear interpolation with $\boldsymbol{\phi}$ and $\bar{\boldsymbol{\phi}}$. We take the first valid zero t, where all the components of $\boldsymbol{\phi}(\mathbf{x}(t))$ are positive, if it exists, and discard the fragment otherwise. From the multi-affine property of the blossom (see Eq. 5), it follows that the directional derivatives $b_{\boldsymbol{\nu}}^{[2]}(\boldsymbol{\phi}(\mathbf{x}(t)))$ (see Eq. 3) are obtained by a linear interpolation of $b_{\boldsymbol{\nu}}^{[1]}(\boldsymbol{\phi})$

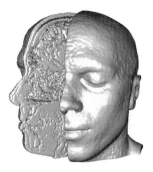
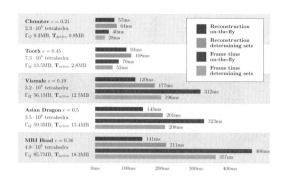

Fig. 6. *Left*: MRI scan ($192^2 \times 126$ voxels) volume clipped with the plane $x = 0$. *Right*: timings of the reconstruction process and the visualization (using pre-computed determining sets as well as on-the-fly spline coefficient computations) for selected data sets and isovalues c. For each data set, the number of active tetrahedra, the size of the determining set Γ_Q and for the geometry encoding T_{active}, respectively, are given.

and $b_\nu^{[1]}(\bar{\phi})$, with the ray parameter t, followed by a de Casteljau step on the second level using $\phi(\mathbf{x}(t))$ Finally, we calculate the gradient for later illumination according to Eq. 4 with three additional scalar products. Here, the $\partial\phi_\nu/\partial x_\iota$ are pre-computed for each of the 24 different generic tetrahedra.

5   Results and Discussion

We demonstrate our results with a series of data sets: *Tooth, VisMale* and *Foot* (see Fig. 1) are publicly available from the Universities of Tübingen and Erlangen, and the US Library of Medicine. Fig. 5 is an example of synthetic data obtained from a sparsely sampled smooth function. Fig. 6, left, shows a MRI scan of the head of one of the authors. Finally, the *Asian Dragon* (Fig. 7) is generated from a signed distance function on the original triangle mesh. All results demonstrate the high visual quality and smooth shading of our method.

Note that the low total degree of the spline patches allows us to obtain precise intersections, even for the objects' silhouettes, without resorting to interval refinements or similar approaches (see Fig. 7, right). Precise intersections are also needed for procedural texturing (see Fig. 5 and 7), and for volume clipping with arbitrary planes and surfaces (see Fig. 6, left). Furthermore, the obtained intersections are exact w.r.t. z-buffer resolution, which allows us to combine raycasted isosurfaces with standard object representations, i.e., triangle meshes.

The table in Fig. 6, right, summarizes the performance for the chosen data sets and lists typical isovalues c, the number of active tetrahedra, as well as the size of the determining set Γ_Q and the geometry encoding $\mathbf{T}_{\text{active}}$. Timings were recorded on a GeForce GTX 285 and CUDA 2.2. For each data set, the first two bars show the timings of our GPU kernels (see Sect. 4.1), which are invoked when the surface needs to be reconstructed. The on-the-fly computation of coefficients is slightly faster than preparing the determining sets, since less data is

Fig. 7. Ray casted isosurface of the *Asian Dragon* head (256^3 voxels) with noise-based procedural texturing. *Right*: close-up into the Dragons' mouth where the C^1-continuous boundary curves on the outside of each cube of $\Diamond$ are shown in black.

written. The most expensive part in the reconstruction is the classification, i.e., the determination of the array $\mathbf{Q}_{\text{class}}$. This could be improved by using appropriate spatial data structures, e.g., min/max octrees, but the recursive nature of these data structures makes an efficient implementation challenging. In addition, the data structures have to be rebuilt if the data itself changes over time, which is not necessary in our simpler approach. Still, for our largest data sets the reconstruction times are in a range of a few hundred ms and in most cases even below the rendering times of a single frame. Compared to former optimized CPU reconstruction based on octrees [4], we achieve significant speed-ups of up to two orders of magnitude.

The per-frame rendering times in Fig. 6, right, are given for a 1280×1024 view port with the surface filling the entire screen. This corresponds to a worst-case scenario, where all active tetrahedra need to be processed and the number of tetrahedra is the limiting factor in both approaches (on-the-fly coefficients and using determining sets). The bottleneck is then determined by the vertex shader complexity. The fact that our on-the-fly vertex programs have about 1/3 more instructions than the version using determining sets is thus directly reflected in the rendering times. An analysis of our fragment programs with NVIDIA's tool *ShaderPerf* yields a peak performance of more than 430 Mio. fragments per second. Thus, fragment processing is already very efficient and further improvements should concentrate on the vertex programs, load balancing, and the reduction of processed geometry during rendering. E.g., for close inspections of the surface, significant speed ups can be achieved from hierarchical view frustum culling, where whole areas of the domain can be omitted and less tetrahedra are processed in the pipeline. This requires splitting up the arrays $\mathbf{T}_{\text{active}}$ based on a coarse spatial partition of $\Diamond$. As [4] have shown, in this case we can expect that frame-rates increase by an order of magnitude.

6   Conclusion

We have shown that interactive and high-quality visualization of volume data with varying isosurfaces can be efficiently performed on modern GPUs. Both, isosurface reconstruction and rendering, are hereby performed using a combined CUDA and graphics pipeline. Our approach benefits strongly from the mathematical properties of the splines. Memory requirements for geometry encoding are significantly reduced using instancing. The method scales well with the fast developing performance of modern graphic processors, and will directly benefit from increased numbers of multiprocessors and texture units. The proposed algorithm can be used for an interactive variation of isolevels, as well as for applications where the data itself varies over time, e.g., simulations and animations.

References

1. Bajaj, C.L.: Data Visualization Techniques. John Wiley & Sons, Chichester (1999)
2. Marschner, S., Lobb, R.: An evaluation of reconstruction filters for volume rendering. In: IEEE Vis., pp. 100–107 (1994)
3. Sorokina, T., Zeilfelder, F.: Local Quasi-Interpolation by cubic C^1 splines on type-6 tetrahedral partitions. IMJ Numerical Analysis 27, 74–101 (2007)
4. Kalbe, T., Zeilfelder, F.: Hardware-Accelerated, High-Quality Rendering Based on Trivariate Splines Approximating Volume Data. In: Eurographics, pp. 331–340 (2008)
5. Lorensen, W.E., Cline, H.E.: Marching cubes: A high resolution 3D surface construction algorithm. In: SIGGRAPH 1987, vol. 21(5), pp. 79–86 (1987)
6. Hadwiger, M., Ljung, P., Rezk-Salama, C., Ropinski, T.: GPU-based Ray Casting. In: Annex Eurographics (2009)
7. Krüger, J., Westermann, R.: Acceleration techniques for GPU-based volume rendering. In: IEEE Vis., pp. 287–292 (2003)
8. Mensmann, J., Ropinski, T., Hinrichs, K.: Accelerating Volume Raycasting using Occlusion Frustum. In: IEEE/EG Symp. on Vol. & Point-Based Gr., pp. 147–154 (2008)
9. NVIDIA CUDA Compute Unified Device Architecture. NVIDIA Corp. (2008)
10. Tatarchuk, N., Shopf, J., DeCoro, C.: Real-Time Isosurface Extraction Using the GPU Programmable Geometry Pipeline. In: SIGGRAPH 2007 courses, pp. 122–137 (2007)
11. Barthe, L., Mora, B., Dodgson, N., Sabin, M.A.: Triquadratic reconstruction for interactive modelling of potential fields. In: Shape Modeling Int., pp. 145–153 (2002)
12. Hadwiger, M., Sigg, C., Scharsach, H., Bühler, K., Gross, M.: Real-Time Ray-Casting and Advanced Shading of Discrete Isosurfaces. CGF 24(3), 303–312 (2005)
13. Reis, G.: Hardware based Bézier patch renderer. In: Proc. of IASTED Visualization, Imaging, and Image Processing (VIIP), pp. 622–627 (2005)
14. Loop, C., Blinn, J.: Real-time GPU rendering of piecewise algebraic surfaces. ACM Trans. on Graphics 25(3), 664–670 (2006)
15. Seland, J., Dokken, T.: Real-Time Algebraic Surface Visualization. In: Geometric Modelling, Numerical Simulation, and Optimization, pp. 163–183. Springer, Heidelberg (2007)
16. Blelloch, G.: Vector Models for Data-Parallel Computing. The MIT Press, Cambridge (1990)
17. Harris, M., Sengupta, S., Owens, J.D.: Parallel Prefix Sum (Scan) with CUDA. In: GPU Gems 3, pp. 677–696. Addison-Wesley, Reading (2007)
18. Lai, M.-J., Schumaker, L.L.: Spline functions on Triangulations. Cambridge University Press, Cambridge (2007)
19. Ramshaw, L.: Blossoming: A Connect-the-Dots Approach to Splines (1987)

Visualizing Arcs of Implicit Algebraic Curves, Exactly and Fast

Pavel Emeliyanenko[1], Eric Berberich[2], and Michael Sagraloff[1]

[1] Max-Planck-Institut für Informatik, Saarbrücken, Germany
{asm,msagralo}@mpi-sb.mpg.de
[2] School of Computer Science, Tel-Aviv University, Tel-Aviv, Israel
ericb@post.tau.ac.il

Abstract. Given a Cylindrical Algebraic Decomposition [2] of an implicitly defined algebraic curve, visualizing distinct curve arcs is not as easy as it stands because, despite the absence of singularities in the interior, the arcs can pass arbitrary close to each other. We present an algorithm to visualize distinct arcs of algebraic curves efficiently and precise (at a given resolution), irrespective of how close to each other they actually pass. Our hybrid method inherits the ideas of subdivision and curve-tracking methods. With an adaptive mixed-precision model we can render the majority of curves using machine arithmetic without sacrificing the exactness of the final result. The correctness and applicability of our algorithm is borne out by the success of our web-demo[1] presented in [11].

Keywords: Algebraic curves, geometric computing, curve rendering, visualization, exact computation.

1 Introduction

In spite of the fact that the problem of rasterizing implicit algebraic curves has been in research for years, the interest in it never comes to an end. This is no surprise because algebraic curves have found many applications in Geometric Modeling and Computer Graphics. Interestingly enough, the task of rasterizing separate curve arcs,[2] which, for instance, are useful to represent "curved" polygons, has not been addressed explicitly upon yet. We first give an overview of existing methods to rasterize complete curves. For an algebraic curve $\mathcal{C} = \{(x, y) \in \mathbb{R}^2 : f(x, y) = 0\}$, where $f \in \mathbb{Q}[x, y]$, algorithms to compute a curve approximation in a rectangular domain $\mathcal{D} \in \mathbb{R}^2$ can be split in three classes.

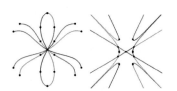

Fig. 1. "Spider": a degenerate algebraic curve of degree 28. The central singularity is enlarged on the left. Arcs are rendered with different colors.

---

[1] http://exacus.mpi-inf.mpg.de
[2] A curve arc can informally be defined as a connected component of an algebraic curve which has no singular points in the interior; see Section 2.

G. Bebis et al. (Eds.): ISVC 2009, Part I, LNCS 5875, pp. 608–619, 2009.
© Springer-Verlag Berlin Heidelberg 2009

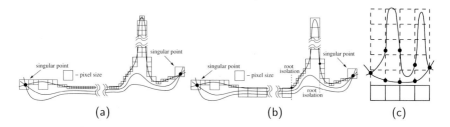

(a)   (b)   (c)

Fig. 2. (a) an attempt to cover the curve arc by a set of xy-regular domains results in a vast amount of small boxes; (b) our method stops subdivision as soon as the direction of motion along the curve is uniquely determined; (c) lifting curve points using root isolation

Space covering. These are numerical methods which rely on *interval analysis* to effectively discard the parts of the domain not cut by the curve and recursively subdivide those that might be cut. Algorithms [9, 18] guarantee the geometric correctness of the output, however they typically fail for singular curves.[3] More recent works [1,5] subdivide the initial domain in a set of xy-regular subdomains where the topology is known and a set of "insulating" boxes of size $\leq \varepsilon$ enclosing possible singularities. Yet, both algorithms have to reach a worst-case root separation bound to guarantee the correctness of the output. Altogether, these methods alone cannot be used to plot distinct curve arcs because no continuity information is involved.

Continuation methods are efficient because only points surrounding a curve arc are to be considered. They typically find one or more seed points on a curve, and then follow the curve through adjacent pixels/plotting cells. Some algorithms consider a small pixel neighbourhood and obtain the next pixel based on sign evaluations [6, 21]. Other approaches [16, 17, 19] use Newton-like iterations to compute the point along the curve. Continuation methods commonly break down at singularities or can identify only particular ones.

Symbolic methods use projection techniques to capture topological events – tangents and singularities – along a sweep line. This is done by computing Sturm-Habicht sequences [8,20] or Gröbner bases [7]. There is a common opinion that knowing exact topology obviates the problem of curve rasterization. We disagree because symbolic methods disrespect the size of the domain $\mathcal{D}$ due to their "symbolic" nature. The curve arcs can be "tightly packed" in $\mathcal{D}$ making the whole rasterization inefficient; see Figure 2 (a). Using *root isolation* to lift the curve points in a number of fixed positions also does not necessarily give a correct approximation because, unless x-steps are adaptive, high-curvature points might be overlooked, thereby violating the Hausdorff distance constraint; see Figure 2 (c).

Given a Cylindrical Algebraic Decomposition (CAD) [2] of $\mathcal{C}$, for each curve arc we output a sequence of pixels which can be converted to a polyline *approximating this arc within a fixed Hausdorff distance*.

---

[3] By geometrically-correct approximation we mean a piecewise linear approximation of a curve within a given Hausdorff distance $\varepsilon > 0$.

The novelty of our approach is that it is hybrid but, unlike [19], the roles of subdivision and curve-tracking are interchanged – curve arcs are traced in the original domain while subdivision is employed in tough cases. Also, note that, the requirement of a complete CAD in most cases can be *relaxed*; see Section 3.5. We start with a "seed point" on a curve arc and trace it in 2 opposite directions. In each step we examine 8 neighbours of a current pixel and choose the one crossed by the arc. In case of a tie, the pixel is subdivided recursively into 4 parts. Local subdivision *stops* as soon as a certain threshold is reached and all curve arcs appear to leave the pixel in one unique direction. From this point on, the arcs are traced collectively until one of them goes apart. When this happens, we pick out the right arc using a real root isolation (e.g., [14]); see Figure 2 (b).

According to our experience, the algorithm can trace the majority of curves without resorting to exact computations even if root separation bounds are very tight. To handle exceptional cases, we switch to more accurate interval methods or increase the arithmetic precision.

2 Preliminaries

Arcs of algebraic curves. For an algebraic curve $\mathcal{C} = \{(x, y) \in \mathbb{R}^2 : f(x, y) = 0\}$ with $f \in \mathbb{Q}[x, y]$, we define its gradient vector as $\nabla f = (f_x, f_y) \in (\mathbb{Q}[x, y])^2$ where $f_x = \frac{\partial f}{\partial x}$ and $f_y = \frac{\partial f}{\partial y}$. A point $\mathbf{p} \in \mathbb{R}^2$ is called *x-critical* if $f(\mathbf{p}) = f_y(\mathbf{p}) = 0$, similarly $\mathbf{p}$ is *y-critical* if $f(\mathbf{p}) = f_x(\mathbf{p}) = 0$ and *singular* if $f(\mathbf{p}) = f_x(\mathbf{p}) = f_y(\mathbf{p}) = 0$. Accordingly, *regular* points are those that are not singular.

We define a *curve arc* as a single connected component of an algebraic curve which has no singular points in the interior bounded by two not necessarily regular end-points. Additionally, an *x-monotone* curve arc is a curve arc that has no x-critical points in the interior.

Interval analysis. We consider only advanced interval analysis (IA) techniques here; please refer to [13] for a concise overview. The First Affine Form (AF1) [15] is defined as: $\hat{x}_{AF1} = x_0 + \sum_{i=1}^{n} x_i \varepsilon_i + x_{n+1}\tilde{\varepsilon}$, where x_i are real coefficients fixed and $\varepsilon_i \in [-1, 1]$ represent unknowns. The term $x_{n+1}\tilde{\varepsilon}$ stands for a cumulative error due to approximations after performing non-affine operations, for instance, multiplication. Owing to this feature, the number of terms in AF1, unlike for a classical affine form, does not grow after non-affine operations. Conversion between an interval $[\underline{x}, \overline{x}]$ and an affine form $\hat{x}$ proceeds as follows:

$$\text{Interval} \rightarrow \text{AF1:} \quad \hat{x} = (\underline{x} + \overline{x})/2 + [(\overline{x} - \underline{x})/2]\varepsilon_k, \quad \tilde{\varepsilon} \equiv 0,$$

$$\text{AF1} \rightarrow \text{Interval:} \quad [\underline{x}, \overline{x}] = x_0 + \left(\sum_{i=1}^{n} x_i \varepsilon_i + x_{n+1}\tilde{\varepsilon}\right) \times [-1, 1],$$

here k is an index of a new symbolic variable (after each conversion k gets incremented). Multiplications on AF1 are realized as follows:

$$\hat{x} \cdot \hat{y} = x_0 y_0 + \sum_{i=1}^{n} (x_0 y_i + y_0 x_i)\varepsilon_i + \left(|x_0|y_{n+1} + |y_0|x_{n+1} + \sum_{i=1}^{n+1} |x_i| \sum_{i=1}^{n+1} |y_i|\right)\tilde{\varepsilon}.$$

The Quadratic Form (QF) is an extension of AF1 that adds two new symbolic variables $\varepsilon^+ \in [0,1]$ and $\varepsilon^- \in [-1,0]$ to attenuate the error when an affine form is raised to even power, and a set of square symbolic variables ε_i^2 to capture quadratic errors:

$$\hat{x}_{QF} = x_0 + \sum_{i=1}^{n} x_i \varepsilon_i + x_{n+1}\tilde{\varepsilon} + x_{n+2}\varepsilon^+ + x_{n+3}\varepsilon^- + \sum_{i=1}^{n} x_{i+n+3}\varepsilon_i^2,$$

where $\varepsilon_i^2 \in [0,1]$. For reasons of space we refer to [15] for a complete set of arithmetic operations on AF1 and QF.

Modified Affine Arithmetic (MAA) [12] is more precise than AF1 and QF. We consider the 1D case here as the only relevant to our algorithm. To evaluate a polynomial $f(x)$ of degree d on $[\underline{x}, \overline{x}]$, we denote $x_0 = (\underline{x} + \overline{x})/2$, $x_1 = (\overline{x} - \underline{x})/2$ and $D_i = f^{(i)}(x_0)x_1^i/i!$. The interval $[\underline{F}; \overline{F}]$ is obtained as follows:

$$\underline{F} = D_0 + \sum_{i=1}^{\lceil d/2 \rceil} \left(\min(0, D_{2i}) - |D_{2i-1}| \right), \quad \overline{F} = D_0 + \sum_{i=1}^{\lceil d/2 \rceil} \left(\max(0, D_{2i}) + |D_{2i-1}| \right).$$

In our implementation we use all three interval methods with AF1 as a default one and QF/MAA used in tough cases; see Section 3.5.

In order to further shrink the interval bounds we exploit the *derivative* information [13]. To evaluate a polynomial $f(x)$ on $X = [\underline{x}, \overline{x}]$, we first evaluate f' on X using the *same* interval method. If the derivative is non-zero, f is strictly monotone on X and the exact bounds are obtained as follows:

$$[\underline{F}; \overline{F}] = [f(\underline{x}), f(\overline{x})] \text{ for } f' > 0, \qquad [\underline{F}; \overline{F}] = [f(\overline{x}), f(\underline{x})] \text{ for } f' < 0.$$

The same approach can be applied recursively to compute the bounds for f', f'', etc. Typically, the number of recursive derivatives in use is limited by a threshold chosen empirically.

3   Algorithm

3.1   Overview

We begin with a high-level overview of the algorithm which is a further development of [10]. After a long-term practical experience we have applied a number of optimizations aimed to improve the performance and numerical stability of the algorithm. In its core the algorithm has an *8-way* stepping scheme introduced in [6]; see Figure 3 (a). As the evidence of the correctness of our approach we use the notion of a *witness* (sub-)pixel.

A "*witness*" (sub-)pixel is a box whose boundaries intersect only twice with an arc to be plotted and do not intersect with any other arc. We implicitly assign a witness (sub-)pixel to each pixel in the curve trace. If we connect the witness (sub-)pixels by imaginary lines, we obtain a piecewise linear approximation of a curve arc within a *fixed Hausdorff distance*; see Figure 3 (b).

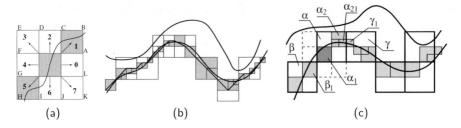

Fig. 3. (a) the 8-pixel neighbourhood with numbered directions, plotted pixels are shaded; (b) adaptive approximation of a curve arc and a polyline connecting witness (sub-)pixels (shaded); (c) more detailed view

Given a set of x-monotone curve arcs, we process each arc independently. The algorithm picks up a "seed point" on an arc and covers it by a witness (sub-)pixel such that the curve arc leaves it in 2 different directions. We trace the arc in both directions until the end-points. In each step we examine an 8-pixel neighbourhood of a current pixel; see Figure 3 (a). If its boundaries are crossed only twice by the arc, we say that the *neighbourhood test* succeeds (see Section 3.2). In this case, we step to the next pixel using the direction returned by the test. Otherwise, there are two possibilities: **1.** the current pixel is *itself* a witness (sub-)pixel: we subdivide it recursively into 4 even parts until the test succeeds for one of its sub-pixels or we reach the maximal subdivision depth.[4] **2.** the current pixel has an assigned witness (sub-)pixel: we proceed with tracing from this witness (sub-)pixel. In both situations tracing at a sub-pixel level is continued until the pixel boundary is met and we step to the *next* pixel. The last sub-pixel we encounter becomes a *witness* of a newly found pixel. Details on the subdivision are given in Section 3.4 in terms of the algorithm's pseudocode.

Suppose we start with a witness (sub-)pixel marked by α_1 in Figure 3 (c), its 8-pixel surrounding box is depicted with dashed lines. The pixel it belongs to, namely α, is added to the curve trace. Assume we choose the direction 1 from α_1 and proceed to the next sub-pixel α_2. The test fails for α_2. Thus, we subdivide it into 4 pieces, one of them (α_{21}) intersecting the arc is taken.[5] We resume tracing from α_{21}, its neighbourhood test succeeds and we find the next "witness" (sub-)pixel (γ_1), its corresponding pixel (γ) is added to the curve trace. The process terminates by reaching one of the arc's end-points. Then, we trace towards another end-point from a saved sub-pixel β_1.

In Section 3.3 we present a technique to stop the local subdivision earlier even if the number of arcs in the pixel neighourhood is not one. Finally, in Section 3.5 we discuss the numerical accuracy issues and the termination criteria of the algorithm. As a possible alternative in the design process we considered using

---

[4] In this situation the algorithm restarts with increased precision; see Section 3.5. We define a subdivision depth k as the number of pixel subdivisions, that is, a pixel consists of 4^k depth-k sub-pixels.

[5] To choose such a sub-pixel we evaluate a polynomial at the corners of α_2 since we know that there is only one curve arc going through it.

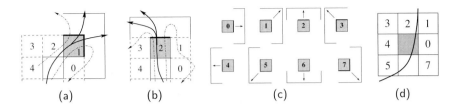

Fig. 4. (a) and (b): Depending on the incoming direction, the curve can leave the shaded pixel's neighbourhood along the boundary depicted with solid lines. Dash-dotted curves show prohibited configurations; (c) boundaries to be checked for all incoming directions; (d) an arc passes exactly between two pixels.

an *4-way* stepping scheme (with the steps in diagonal directions forbidden). However, this scheme has a number of disadvantages (see [10]), among them the main ones is the necessity to use *2D range analysis* (which is more costly and less accurate than its 1D counterpart) and the fact that the tracing is as twice as slow due to the forbidden diagonal directions.

3.2 Counting the Number of Curve Arcs

In this section we discuss the neighbourhood test. Due to the x-monotony constraint, there can be no closed curve components inside a box enclosing an 8-pixel neighbourhood. Hence, the boundary intersection test suffices to ensure that only one curve arc passes through this box. We rely on the following consequence of Rolle's theorem:

Corollary 1. *If for a differentiable function $f(x)$ its derivative $f'(x)$ does not straddle 0 in the interval $[a;b]$, then $f(x)$ has at most one root in $[a;b]$.*

First, we sketch the basic version of the neighbourhood test, and then refine it according to some heuristic observations. The test **succeeds** if the procedure CHECK_SEGMENT given below returns "one root" for exactly 2 out of 9 sub-segments $AB, BC, \ldots, LA$, and "no roots" for the remaining ones; see Figure 3 (a). The test **fails** for all other combinations, resulting in pixel

```
1: procedure CHECK_SEGMENT([a, b] : Interval, f : Polynomial)
2:     if 0 ∉ {[F̲, F̅] = f([a, b])} then        ▷ evaluate f at [a, b] and test for 0 inclusion
3:         return "no roots"                      ▷ interval does not include 0 ⇒ no roots
4:     if sign(f(a)) = sign(f(b)) then                  ▷ test for a sign change at end-points
5:         return "uncertain"                     ▷ no sign change ⇒ even number of roots
6:     loop
7:         if 0 ∉ {[F̲, F̅] = f'([a, b])} then                ▷ check interval for 0 inclusion
8:             return "one root"                    ▷ f' does not straddle 0 ⇒ one root
9:         Split [a, b] in 2 halves, let [x, y]⁺ be the one at which f(x) has no sign change
10:        if 0 ∈ {[F̲, F̅] = f([x, y]⁺)} then              ▷ check interval for 0 inclusion
11:            return "uncertain"       ▷ f straddles 0 ⇒ nothing can be said for sure
12:        Let [x, y]⁻ be the half at which f(x) has a sign change
```

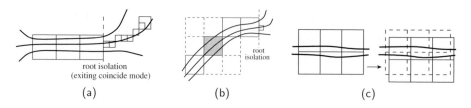

Fig. 5. (a) Tracing arcs collectively in "coincide mode"; (b) diagonal "coincide mode": the neighbourhood test succeeds for the shaded pixel even though 3 arcs are crossing its boundary; (c) pixel grid perturbation

13:        $[a, b] \leftarrow [x, y]^-$ ▷ restart with the refined interval
14:    **end loop**
15: **end procedure**

subdivision. The search space can be reduced because we know the direction of an incoming branch. In Figure 4 (a), the algorithm steps to the shaded pixel in a direction "1" relative to the previous pixel. Hence, a curve must cross a part of its boundary marked with thick lines. The curve can leave its neighbourhood along a part of the boundary indicated by solid lines. Configurations shown by dash-dotted curves are impossible due to x-monotonicity. In Figure 4 (b) we enter the shaded pixel in direction "2", dash-dotted curves are again prohibited since, otherwise, an 8-pixel neighbourhood of a previous pixel (the one we came from) would be crossed more than twice by the curve resulting in a subdivision. Figure 4 (c) lists parts of the boundary being checked for all possible incoming directions. Thus, the neighbourhood test succeeds if CHECK_SEGMENT returns "one root" for exactly one of the "enabled" sub-segments respecting the incoming direction (and "no roots" for the remaining enabled ones).

If an arc passes exactly between two pixels as in Figure 4 (d), we can choose either direction (1 or 2 in the figure). This case is detected by an additional exact zero testing.

3.3   Identifying and Tracing Closely Located Arcs, Grid Perturbation

To deal with tightly packed curve arcs, we modified the neighbourhood test in a way that we allow a pixel to pass this test once a new direction can uniquely be determined *even* if the number of arcs over a sub-segment on the boundary is more than one. We will refer to this as tracing in *coincide mode*. In other words, the test reports the coincide mode if CHECK_SEGMENT returns "uncertain" for one sub-segment and "no roots" for all the rest being checked (as before, the test fails for all other combinations leading to subdivision). From this point on, the arcs are traced *collectively* until one of them goes apart. At this position we *exit* the coincide mode by picking up the right arc using *root isolation* (see Figure 5)(a, b), and *resume* tracing with subdivision. The same applies to seed points – it is desireable to start already in coincide mode if the arcs at this position are too close.

Typically, we enable the coincide mode by reaching a certain subdivision depth which is an evidence for tightly packed curve arcs (depth 5 works well).

Observe that, we need to know the *direction of motion* (left or right) to be able to isolate roots on exiting the coincide mode. Yet, it is absolutely unclear if we step in a vertical direction only. Then, we obtain the direction using the tangent vector to $\mathcal{C}$ evaluated at the "seed" point.

In Figure 5 (c) the arcs are separated by the pixel grid thereby prohibiting the coincide mode: for example, the curve $f(x, y) = y^2 - 10^{-12}$ (two horizontal lines). A simple remedy against this is to shift the grid origin by an arbitrary fraction of a pixel (grid perturbation) from its initial position before the algorithm starts.

Remark that, the coincide mode *violates* our definition of a witness (sub-)pixel. Nevertheless, the approximation is correct because an arc is guaranteed to lie within an 8-pixel neighbourhood even though this neighbourhood does not necessarily contain a single arc.

3.4   Algorithm's Pseudocode

We present the pseudocode in terms of a procedure STEP computing the next pixel in a curve trace. The procedure is applied until tracing reaches one of the end-points (this test is not shown for the sake of simplicity). The neighbourhood test (test_pix) returns a new direction $(0 - 7)$ in case of success (it can also set the flag coincide_mode switching to "coincide mode") or -1 otherwise. The step_pix advances the pixel coordinates with respect to a given direction.

```
 1: procedure STEP(pix: Pixel, witness: Pixel, d: Direction)
 2:     new_d : Direction ← test_pix(pix, d)            ▷ check the pixel's neighbourhood
 3:     if new_d ≠ −1 then
 4:         p:Pixel ← step_pix(pix, new_d)              ▷ step to the next pixel and return it
 5:         return {p, p, new_d}        ▷ a pixel, its witness sub-pixel and a new direction
 6:     if coincide_mode then        ▷ curve branches go apart ⇒ need a new seed point
 7:         coincide_mode = false                     ▷ we exit the "coincide mode"
 8:         {p, new_d} ← get_seed_point(pix, d)    ▷ get a new seed point and a direction
 9:         return {get_pixel(p), p, new_d}      ▷ a pixel, its witness and a new direction
10:     if witness = pix then    ▷ witness sub-pixel is a pixel itself ⇒ perform subdivision
11:         {p: Pixel, new_d} ← SUBDIVIDE(pix, d)
12:     else                       ▷ otherwise continue tracing from the witness sub-pixel
13:         {p: Pixel, new_d} ← {witness, d}
14:     while get_pixel(p) = pix do      ▷ iterate until we step over the pixel's boundary
15:         new_d ← test_pix(p, new_d)                ▷ check the pixel's neighbourhood
16:         if new_d = −1 then
17:             {p,new_d} ← SUBDIVIDE(p, new_d)
18:         p ← advance_pixel(p, new_d)
19:     end while
20:     return {get_pixel(p), p, new_d}▷ a pixel, its witness sub-pixel and a new direction
21: end procedure
22:
23: procedure SUBDIVIDE(pix: Pixel, d: Direction)
24:     sub_p: Pixel ← get_subpixel(pix, d)        ▷ get one of the four sub-pixels of 'pix'
25:     new_d: Direction ← test_pix(sub_p, d)
```

```
26:     if new_d = −1 then
27:         return SUBDIVIDE(sub_p, d)          ▷ neighbourhood test failed ⇒ subdivide
28:     return {sub_p, new_d}
29: end procedure
```

3.5   Addressing Accuracy Problems and Relaxing CAD Assumption

Numerical instability is a formidable problem in geometric algorithms because, on the one hand, using exact arithmetic is unacceptable in all cases while, on the other hand, an algorithm must handle all degeneracies adequately if one strives for an exact approach. To deal with it, we use a *mixed-precision* method, that is, high precision computations are applied only when necessary. The following situations indicate the lack of numerical accuracy:

- reached the maximal subdivision depth (typically 12). It is reasonable to restart with increased precision because we are "stuck" at a single pixel;
- subpixel size is too small, that is, beyond the accuracy of a number type;
- no subpixel intersecting a curve arc was found during subdivision. This indicates that the polynomial evaluation reported a wrong sign.

When any of this occurs, we restart the algorithm using a more elaborate IA method (QF and then MAA; see Section 2). If this does not work either, the arithmetic precision is increased according to a three-level model given below.

Level 0: all operations are carried out in double-precision arithmetic. Polynomial coefficients are converted to intervals of doubles. If a polynomial evaluation results in an interval including zero, the quantity is *reevaluated* using rational arithmetic.

Level 1: all operations are performed in bigfloat arithmetic. Accordingly, polynomial coefficients are converted to bigfloats. As before, the quantity is reevaluated with rationals if the evaluation with bigfloats is not trustful.

Level 2: rational arithmetic is used in all computations. At this level arbitrary subdivision depths are allowed and no constraints are imposed on a sub-pixel size.

In the course of practical application we have not been faced with any particular instance that cannot be handled by this model. The correctness of the algorithm rests on the fact that at a very tiny scale any curve arc can be replaced by a straight line, hence, given the absence of singularities, the tracing always terminates.

Recalling the introduction, unless we deal with isolated singularities, our algorithm can proceed having only x-coordinates of x-critical points of $\mathcal{C}$ (resultant of f and f_y'). Hence, an expensive **lifting phase** of a symbolic CAD algorithm can be avoided. However, since y-coordinates of end-points are not explicitly given, we exploit the x-monotony to decide where to stop tracing. Namely, the tracing terminates as soon as it reaches a (sub-)pixel containing an x-isolating interval of an end-point, and there exists such a box inside this (sub-)pixel that the curve crosses its vertical boundaries only. The last condition

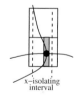

x–isolating
interval

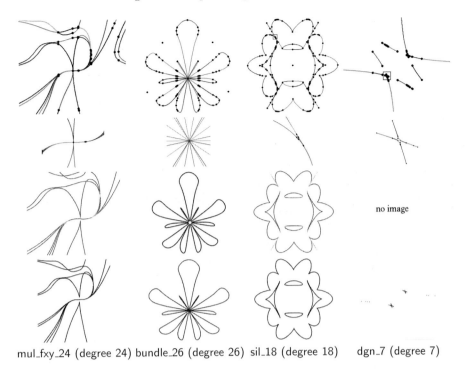

mul_fxy_24 (degree 24) bundle_26 (degree 26) sil_18 (degree 18)   dgn_7 (degree 7)

Fig. 6. First two rows: curves rendered using our method with zooming at singularities; **3rd row:** plots produced in Axel; **4th row:** plots produced in Maple. Curve degrees are given w.r.t. y variable.

is necessary to prevent a premature stopping alarm for arcs with a decent slope; see figure to the right.

4   Results and Conclusion

Our algorithm is implemented in the context of CGAL (Computational Geometry Algorithms Library, www.cgal.org) as part of *Curved_kernel_via_analysis_2* package [3]. The CGAL's development follows a *generic programming paradigm.* This enabled us to parameterize our algorithm by a number type to be able to increase the arithmetic precision without altering the implementation.

We tested our algorithm on 2.2 GHz Intel Core2 Duo processor with 4 MB L2 cache and 2 GB RAM under 32-bit Linux platform. Multi-precision number types were provided by CORE with GMP 4.3.1 support.[6] The CAD of an algebraic curve was computed using [8]. We compared our visualization with the ones provided by Axel and Maple 13 software.[7] Axel implements the algorithm given in [1]. Due to the lack of implementation of the exact approach, we compared

---

[6] CORE: http://cs.nyu.edu/exact; GMP: http://gmplib.org
[7] Axel: http://axel.inria.fr; Maple: www.maplesoft.com

Table 1. Running times in seconds. **Analyze:** the time to compute the CAD; **Render:** visualization using our method; and visualization using Axel and Maple 13 respectively.

Curve	Analyze	Render	Axel	Maple	Curve	Analyze	Render	Axel	Maple
mul_fxy_24	22.2	4.1	2.93	3.35	bundle_26	60.5	2.8	2.11	1.88
sil_18	35.2	6.4	265	2.22	dgn_7	0.61	0.82	timeout	3.1

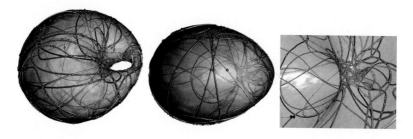

Fig. 7. Rendering curves on a surface of Dupin Cyclide (about 64000 rendered points)

with a subdivision method. We varied the accuracy parameter ε from $5 \cdot 10^{-5}$ to 10^{-8} depending on the curve. The "feature size" *asr* was set to 10^{-2}. In Maple we used implicitplot method with *numpoints* $= 10^5$. Figure 6 depicts the curves[8] plotted with our method, Axel and Maple respectively. Notice the visual artifacts nearby singularities. Moreover, in contrast to our approach, the algorithms from Axel and Maple cannot visualize the arcs selectively. Table 1 summarizes the running times. Rendering the curve dgn_7 in Axel took more than 15 mins and was aborted, this clearly demonstrates the advantages of using coincide mode.

Figure 7 depicts an intersection of a Dupin Cyclide with 10 algebraic surfaces of degree 3 computed using [4]. Resulting arrangement of degree 6 algebraic curves was rendered in tiles with varying resolutions and mapped onto the Dupin Cyclide using rational parameterization. Visualization took 41 second on our machine.

To conclude, we have identified that the interplay of a symbolic precomputation and a numerical algorithm delivers the best performance in practice, because, once the exact solution of $f = 0 \wedge f_y = 0$ is computed, rendering proceeds fast for *any* resolution. In contrast, the subdivision methods have to recompute the topological events for every new domain $\mathcal{D}$ due to the lack of "global" information. Moreover, they can often report a wrong topology if ε is not chosen small enough. Finally, the amount of symbolic computations required by our algorithm can be reduced substantially; see Section 3.5. Yet, the current implementation is still based on a complete CAD, thus we have not been able to evaluate this in practice which is an object of future research.

---

[8] Visit our curve gallery at: http://exacus.mpi-inf.mpg.de/gallery.html

References

1. Alberti, L., Mourrain, B.: Visualisation of Implicit Algebraic Curves. In: Pacific Conference on Computer Graphics and Applications, pp. 303–312 (2007)
2. Arnon, D.S., Collins, G.E., McCallum, S.: Cylindrical algebraic decomposition I: the basic algorithm. SIAM J. Comput. 13, 865–877 (1984)
3. Berberich, E., Emeliyanenko, P.: Cgal's Curved Kernel via Analysis. Technical Report ACS-TR-123203-04, Algorithms for Complex Shapes (2008)
4. Berberich, E., Kerber, M.: Exact Arrangements on Tori and Dupin Cyclides. In: Haines, E., McGuire, M. (eds.) SPM 2008, pp. 59–66. ACM, Stony Brook (2008)
5. Burr, M., Choi, S.W., Galehouse, B., Yap, C.K.: Complete subdivision algorithms, II: isotopic meshing of singular algebraic curves. In: ISSAC 2008, pp. 87–94. ACM, New York (2008)
6. Chandler, R.: A tracking algorithm for implicitly defined curves. IEEE Computer Graphics and Applications 8 (1988)
7. Cheng, J., Lazard, S., Peñaranda, L., Pouget, M., Rouillier, F., Tsigaridas, E.: On the topology of planar algebraic curves. In: SCG 2009, pp. 361–370. ACM, New York (2009)
8. Eigenwillig, A., Kerber, M., Wolpert, N.: Fast and exact geometric analysis of real algebraic plane curves. In: ISSAC 2007, pp. 151–158. ACM, New York (2007)
9. Elber, G., Kim, M.-S.: Geometric constraint solver using multivariate rational spline functions. In: SMA 2001, pp. 1–10. ACM, New York (2001)
10. Emeliyanenko, P.: Visualization of Points and Segments of Real Algebraic Plane Curves. Master's thesis, Universität des Saarlandes (2007)
11. Emeliyanenko, P., Kerber, M.: Visualizing and exploring planar algebraic arrangements: a web application. In: SCG 2008, pp. 224–225. ACM, New York (2008)
12. Huahao Shou, I.V., Martin, R., et al.: Affine arithmetic in matrix form for polynomial evaluation and algebraic curve drawing. Progress in Natural Science 12(1), 77–81 (2002)
13. Martin, R., Shou, H., Voiculescu, I., Bowyer, A., Wang, G.: Comparison of interval methods for plotting algebraic curves. Comput. Aided Geom. Des. 19, 553–587 (2002)
14. Mehlhorn, K., Sagraloff, M.: Isolating Real Roots of Real Polynomials. In: ISSAC 2009, pp. 247–254. ACM, New York (2009)
15. Messine, F.: Extensions of Affine Arithmetic: Application to Unconstrained Global Optimization. Journal of Universal Computer Science 8, 992–1015 (2002)
16. Möller, T., Yagel, R.: Efficient Rasterization of Implicit Functions. Tech. rep., Department of Computer and Information Science, Ohio State University (1995)
17. Morgado, J., Gomes, A.: A Derivative-Free Tracking Algorithm for Implicit Curves with Singularities. In: ICCSA, pp. 221–228 (2004)
18. Plantinga, S., Vegter, G.: Isotopic approximation of implicit curves and surfaces. In: SGP 2004, pp. 245–254. ACM, New York (2004)
19. Ratschek, H., Rokne, J.G.: SCCI-hybrid Methods for 2d Curve Tracing. Int. J. Image Graphics 5, 447–480 (2005)
20. Seidel, R., Wolpert, N.: On the exact computation of the topology of real algebraic curves. In: SCG 2005, pp. 107–115. ACM, New York (2005)
21. Yu, Z.S., Cai, Y.Z., Oh, M.J., et al.: An Efficient Method for Tracing Planar Implicit Curves. Journal of Zhejiang University - Science A 7, 1115–1123 (2006)

Fast Cube Cutting for Interactive Volume Visualization

Travis McPhail, Powei Feng, and Joe Warren

Rice University

Abstract. Visualizing 3D volume datasets has received a great deal of attention in many areas: medical imaging, geoscience, and astrophysics are a few such areas. With improvements of commodity graphics cards, texture-based visualization methods have become popular. Within these methods, intersecting a series of planes with a cube is a common problem. While there are standard methods for approaching this problem, visualizing large data sets in real-time require faster approaches. We present a sweeping algorithm that determines the sequence of topology changes (STC) of the intersection of a cube with a plane as it moves through the cube. We use this algorithm to construct a table of these topological changes that maps view vectors to an STC. With this sequence of topology changes, we generate the polygonal intersections via vector addition. Finally, we use our approach in an octree-based, empty-space culling framework to further reduce the rendering time for large volumes.

1 Introduction

Comprehending the geometric and physiological content of volume data has applications in a variety of fields: medical imaging, geoscience, astrophysics, molecular visualization, and fluid dynamics. There are two primary means for visualizing volume data: polygon rendering (or indirect volume rendering) and direct volume rendering. Polygon rendering involves extracting a surface mesh from grid data and using conventional rendering pipeline to display the surface. These meshes correspond to level-sets of the implicit function given by the volume data. Most well-known works in this area are Marching Cubes and, more recently, Dual Contouring [1, 2]. Direct volume rendering (DVR) is the class of techniques that draws volumetric data directly without constructing polygons that approximate the data [3]. DVR methods can provide internal structure information by using techniques such as transfer function. This extends beyond the capabilities of typical indirect volume rendering methods.

In the history of direct volume rendering techniques, ray-casting, shear-warp, splatting, and texture mapping stand out as being most used and researched [4]. Raycasting and splatting techniques are high-quality but non-interactive methods, whereas shear-warp and texture mapping are fast, hardware-supported methods but lack in fidelity without extra care [5–8]. We will focus on texture-mapping techniques in our work.

G. Bebis et al. (Eds.): ISVC 2009, Part I, LNCS 5875, pp. 620–631, 2009.

Because of the wide availability of hardware support, texture-hardware-based volume rendering, has been a popular topic for research. The technique of texture-mapping requires the generation of a set of view- or object-aligned "proxy" geometries. These geometries are assigned texture coordinates at their vertices, and texture-mapping from these coordinates is either done on the hardware or the software. Cullip et al., Wilson et al., and Cabral et al. were among the first to suggest the use of texture-mapping hardware with volume rendering [8–10]. Much of the earlier work focused on using object-aligned 2D-textures. Lacroute and Levoy used object-aligned slices to render volumes, but they did not rely on hardware acceleration for rendering [7]. Westermann and Ertl described methods for drawing shaded iso-surface using OpenGL extensions to exploit 3D texturing capabilities on the hardware [11]. Rezk-Salama et al. described a method for shaded iso-surfaces using 2D textures [12]. Engel et al. improved the quality of texture-based rendering with a pre-integration technique [13]. Other works on the subject span a wide range of techniques that leveraged performance against quality of rendering [14–16].

Our effort addresses the problem of generating proxy geometries in texture-hardware-based volume rendering. In general, this involves trimming a set of view-aligned planes to a cube, as illustrated in Figure 1. We found few works that are related in this respect. In particular, Dietrich et al. proposed a plane-sweep algorithm for plane-cube intersection that is similar to our work [17]. However, their description is brief, their application is different, and it is unclear how geometries are generated in their algorithm. Rezk-Salama and Kolb have described a method for computing proxy geometries on the vertex shader, where each plane/cube intersection is computed independently [18]. Additionally, six vertices are passed to the GPU for every proxy plane, which is redundant in many cases. In contrast, our method pre-computes all unique topologies from plane/cube intersections and stores the information in a table. This table enables us to extract the exact polygonal intersection on the fly. We will compare our method against Rezk-Salama and Kolb's work to show that a plane-sweeping

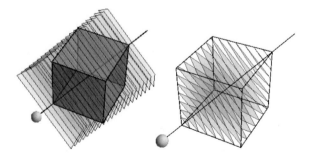

Fig. 1. Comparison between series of quadrilaterals (left) that span the volume of interest versus polygons trimmed to the exact boundary (right). In the right figure, note the amount of superfluous texels that do not contribute to the image.

method is more efficient in the case of multiple plane-cube intersections. Furthermore, we also compare with the use of hardware clipping planes and show that our method performs well in comparison.

Our contributions are as follow:

- Develop a data structure to tabularize all topological changes.
- Generate exact intersections for a cube and series of parallel planes using the data structure.
- Explore the use of this algorithm in an octree-based empty-space culling framework.

2   Method

2.1   Overview

As noted in the introduction, the key to texture-based hardware rendering of 3D volumetric data is generating a set of proxy geometries for mapping volume densities into 3D rendering space. Typically, this process involves generating a set of intersections between a single cube and a sequence of parallel planes. These planes are usually perpendicular to a given viewing vector and equally-spaced. These proxy geometries tile the volume so that the composite of their texture-mapped images produce a 3D rendering of the data. Figure 2 and Figure 1 shows 2D and 3D examples of this tiling processing, respectively.

This paper is built on two key observations. The first observation is that, given a fixed view vector, the cube can be partitioned into a finite set of *wedges*. These wedges are formed by projecting the vertices of the cube onto the view vector and building a set of planes perpendicular to the view vector that pass through these projected points. Figure 2 illustrates this process in 2D.

Inside a given wedge, any plane perpendicular to the view vector intersects the cube in the same set of cube edges. If one wishes to sweep this plane through

Fig. 2. Illustration of wedges with respect a view vector. We start with a view vector and cube (left). We project the vertices on the view vector(middle). Then we define the regions between the projected vertices as *wedges*.

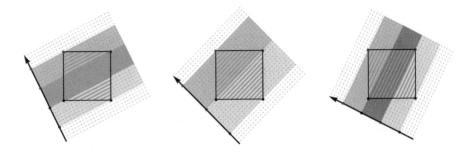

Fig. 3. Series of images that illustrate the occurrence of a change in the order of the vertices. The middle image shows that two vertices have the same projection on the view vector when the ray between them is perpendicular to the view vector.

the wedge, the only updates that need to be done are incremental updates of the positions of the edge intersection points. (Note that no topology calculations need be done.) Thus, the intersection of the cube and a swept sequence of planes perpendicular to the view vector be computed extremely efficiently inside a single wedge.

More generally, if one could pre-compute the set of edge intersections of each wedge, a simple plane sweep algorithm could be used to tile the entire cube. As a plane normal to the view vector is swept through the cube, an outer loop tests whether the plane had moved from one wedge to the next wedge. If not, a second inner loop continues the sweep through a given wedge.

The second key observation is that varying the view vector slightly normally does not affect the algorithm described above. The set of intersected edges associated with each wedge remains unchanged unless *the order of the projected vertices changes*. Figure 3 shows an example of three different vertex orderings and their associated sequence of wedges for the 2D case.

Our approach will be to pre-compute a *sequence of topology changes* (STC) for each possible distinct ordering of projected vertices. An STC will be indexed by its vertex ordering and consist of a list of edge intersection sets associated with each wedge. In the 2D case, there are only eight distinct STCs associated with all possible view vectors. Figure 4 depicts these eight case and the STCs associated with two of these cases. This approach is viable in 3D since the numbers of distinct vertex orderings is also relatively small (96) (see Figure 5 for the 3D partition of the view vector space).

We will briefly describe how to apply our method. The table containing STCs is invariant of the rendering parameters. As long as our bounding polyhedron is a cube, we only need to compute this table one time and use it to supplement the texture-based volume rendering algorithm. During rendering phase, for every change in view direction, the user queries the table of STCs using the view vector.

The query returns an STC for the current view direction. Using this STC, user then applies the plane sweep algorithm, which outputs a set of cube-trimmed polygons. These polygons are sent to the graphics card for 3D volume texturing using standard texture-based DVR. Hence, our method relies on CPU trimming to relieve the GPU of excessive processing.

In the next two subsections, we discuss:

- A method for constructing the table of STCs associated with each distinct vertex ordering as well as a fast method for determining vertex ordering,
- The resulting plane sweep algorithm for generating trimmed polygons.

2.2   Table Structure

Our motivation for building a look-up table for STCs is based on the observation that the number of all vertex orderings is small. Akin to the sign-change table in Marching Cubes, we use table look up to avoid the need to do topological calculations while generating the intersection of the swept planes and the cube.

As noted earlier, the look-up table has one entry for each distinct ordering of the projected cube vertices. A naive approach would be to consider all 8! possible vertex orderings of the eight vertices of the cube. However, this approach is unwieldy. In practice, most of the vertex orderings are impossible. For example, in Figure 4, the ordering $\langle 4, 2, 1, 3 \rangle$ is impossible.

A more precise method for constructing the set of possible vertex orderings is to note that the two vertices of the cube project onto the same point on the view vector if their perpendicular bisecting plane is normal to the view vector. This observation leads to the following algorithm for constructing the look-up table.

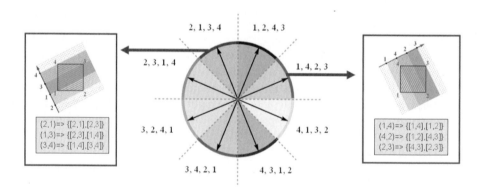

Fig. 4. 2D depiction of the partition of the view vectors for a 2D cube and two example sequence of topological changes (STC). The STC is represented as a list of vertex orderings followed by an edge list.

- Given a unit cube centered at the origin, for each pair of distinct vertices c_i, c_j of the cube, construct its perpendicular bisecting plane P_{ij},
- Use these planes P_{ij} to partition the unit sphere centered at the origin into a set of spherical patches S_k. (Note that there are multiple copies of the same bisecting planes.)
- Points on each patch S_k correspond to the set of view vectors for which the ordering of the projected cube vertices is the same. For each patch S_k, compute this vertex ordering and its associated STC.

Figure 4 shows the partition of the unit circle into eight circular arcs by the four distinct perpendicular bisecting planes associated with the square. In 3D, there are 13 perpendicular bisecting planes; 3 distinct planes for edge adjacent vertices, 6 distinct planes for face adjacent vertices and 4 distinct planes for diagonally opposite pairs of vertices. These planes then partition the unit sphere into 96 spherical patches shown in Figure 5.

To complete our table lookup method, we need to construct a method that, given the view vector, computes the index k for the patch S_k. Our method is a simple one based on BSP trees [19]. We note that the 3 distinct bisecting planes associated with edge adjacent cube vertices partition the unit sphere into 8 octants. (One octant is highlighted in Figure 5). Given a view vector, the top three levels of our BSP tree use these three planes to split the unit sphere into eight octants. Next, note that the six distinct bisecting planes associated with the face adjacent pairs of vertices partition this octant into six triangular patches. (In fact, only three planes even intersect the patch in the dashed edges). The next three levels of the BSP tree use these three planes and partition each octant into six triangular patches. Finally, we note that only one of the four distinct bisecting planes associated with diagonally opposite pairs of cube vertices intersects a particular triangular patch. The last level of our BSP tree uses this plane to partition this triangular patch into two patches, one of which is our desired spherical patch S_k.

Thus, this BSP tree of depth seven allows one to compute the vertex ordering and associated STC using only seven dot products. Given this information, we now present our plane sweep method in detail.

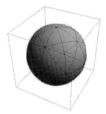 

Fig. 5. Partition of the view vectors for a 3D cube. The highlighted portion of the right figure represents a octant.

2.3   Plane Sweeping

Given a view vector v, we want to generate a set of proxy geometries, which are the intersection of view parallel planes and a cube. Our algorithm utilizes the BSP-tabularized result given in the previous section for fast look-up of the topological changes with respect to v. Also stored within the BSP tree is the order of the cube vertices projected onto the view vector.

We compute the plane-cube intersection by starting from one end of the cube and sweep through the cube in the direction of the view vector. We use P to denote the current plane of the sweep, and P is always normal to the view vector. Let d be the step size of the plane sweep; this is a user given parameter that determines that number of trimmed geometries. We let the wedges be defined as $W_1, W_2, \ldots, W_N$, and note that the number of wedges varies with respect to the view vector (wedges are the regions between dotted blue lines in Figure 6). Let $T_1, T_2, \ldots, T_N$ be the topology (a set of edges) associated with each wedge. (In Figure 6 the topology of the first wedge is $\{\{1,2\}\{1,4\}\}$). We write $u_{ij} = c_i - c_j$ where (c_i, c_j) is an edge in T_k. (These vectors are oriented in the same direction as v; in Figure 6, they are the red arrows).

The algorithm proceeds thus: we initialize P to contain the vertex closest to the eye point. Then we step through each wedge W_k, where for each edge, we compute the intersection of P with the edges of T_k. These intersections are denoted as m_{ij} for each edge $(c_i, c_j) \in T_k$. These intersections will be the starting vertices for the current wedge. After finding the starting vertices, we generate the coordinates of the following trimmed geometry by vector addition. We add to the geometries of the previous iteration by the scaled vector

$$y_{ij} = \frac{d}{u_{ij} \cdot v} u_{ij}$$

for each of the edges in T_k. We also advance P by a fixed distance d along the view vector in every iteration when generating the new geometries (tick marks in Figure 6 denote the plane's advancement). While P is still in W_k, we know that all trimmed geometries will have the same topology. When P crosses over

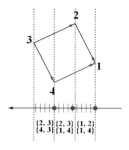

```
k   ←   1
P   ←   plane containing cube vertex closest to eye
while k ≤ N
      ∀(i, j) ∈ T_k, m_ij  ←   intersect P with edge (c_i, c_j)
      while P is in wedge W_k
            ∀(i, j) ∈ T_k, m_ij  ←   m_ij + y_ij
            P  ←  advance P in the direction of v by d
      increment k
```

Fig. 6. The left figure is an example of the plane-sweep algorithm in 2D. The right figure shows the pseudo-code of the algorithm. The notations are detailed in the following section.

into W_{k+1}, then we need to again find the starting vertices for T_{k+1}. (Instances of where this computation is necessary are marked as red circles in Figure 6). This process iterates until P crosses the last wedge. The pseudo-code is given in Figure 6.

The running time of this algorithm is linear in the number planes specified by the user. We cut down on the constant factor of the asymptotic running time by using a table look-up and by replacing edge-plane intersections with vector additions. The initializing step for each wedge presents the heaviest load of computation. However, this step only occurs a small number of times in a sweep.

3   Results

We ran our method against two other methods for comparison. The first is the software implementation of Rezk-Salama and Kolb's method (denoted as RSK) [18]. The second method for comparison uses hardware clipping planes to trim the cubes; this is a straightforward GPU implementation with minimal computation on the CPU. We include results from volume-rendering with no plane trimming (No Op) to establish a baseline of comparison. This is a naive method that sends untrimmed, view-perpendicular, and screen-filling quads to the graphics card (See left image of Figure 1). Although the "No Op" method incurs the very little CPU computation, it introduces wasteful pixel processing for empty regions.

Our primary test machine is a Pentium Xeon machine with two 2.66GHz CPUs and 4GB of memory. The graphics card is NVIDIA Geforce 8800 GTX with 768MB of texture memory.

We compare our method against RSK by just counting the number of planes that can be trimmed in a fixed amount of time. RSK showed a speed of 650,000 planes trimmed per second, and our method posted a speed of 2,500,000 planes per second. Although our method is 3 to 4 times faster than that of RSK, the empirical rendering results we gathered only indicate marginal speed-up. This is due to the fact that the rendering time is mostly dominated by filling pixels as oppose to plane-cube intersections.

3.1   Single Cube

The results from Figure 7 suggest that all three methods perform two to three times better than volume rendering without any optimization. First, for the 500×500 tests, our method has higher frames-per-second than the other two methods. We observe that as the rendering resolution increases from 500×500 to 1000×1000, pixel fills becomes the bottleneck.

We ran tests on a second machine to establish the trade-off between CPU vs. GPU in terms of hardware differences. Our second test machine is a Pentium 4 machine running at 2.8GHZ with 1GB of memory. Its graphics card is NVIDIA Geforce 6200 with 256MB of texture memory. In Figure 8, the hardware

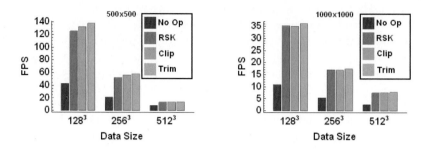

Fig. 7. The left graph records the frames-per-second for each method where the rendering window is 500 × 500 pixels. The right graph has results for 1000 × 1000 pixels. "No Op" stands for volume rendering with no plane-trimming; "RSK" stands for the method by Rezk-Salama and Kolb as described in [18]; "Clip" stands for the hardware-clipping-planes method; and "Trim" is our cube-trimming algorithm.

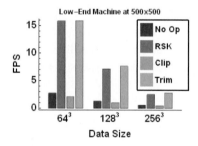

Fig. 8. Results of all the methods running on a machine with limited graphics hardware

clipping-plane method has the lowest number of frames-per-second. However, its performance in the high-end machine tests is highly competitive. Given this trend, we believe that as more powerful graphics hardware become available, the clipping-planes method can out-perform either of the two software implementations on a desktop machine. In the next section, we will discuss a case where the performance issue cannot be addressed with better hardware.

3.2 Empty-Space Culling Using Octree

We use a simple heuristic for empty-space culling by decomposing the volume into an octree where each cell stores the min and max values of the contained subvolume. This is a simple scheme based on ideas from Lamar et al [15]. With the optimization, we can avoid rendering regions that are not within our range of interest. We apply the three methods on each octcell and render the resulting planes or polygons.

We used a largely empty volume as our test example. The results of Figure 9 indicate that empty-space skipping can provide significant speed-up. However,

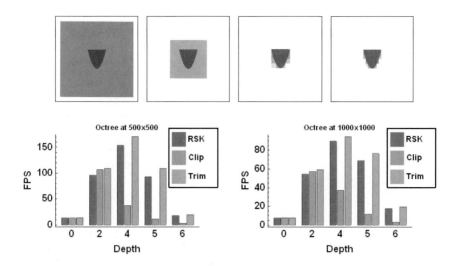

Fig. 9. Results of the three methods running with octree decomposition enabled. "Depth" represents the depth of the octree decomposition. The top four figures show the top-down views of the largely empty volume used in this test. The gray box is the bounding octcell in the octree. The four different levels of decompositions are 0, 2, 4, and 5 (from left to right).

we see the gain from performing empty-space skipping by octree-decomposition diminishes as the octree traversal depth increases beyond level 4. This is an expected result as the decomposition at level 4 is refined enough such that the waste due to rendering empty-spaces is minimized.

In Figure 9, we see a drop in frame rate for the clipping-plane method for levels beyond 2. This is due to the high number planes that have to be passed to and trimmed by the GPU. Since the GPU has to handle both the volume clipping and rendering, its performance staggers as the number of planes increases. We conclude that the hardware approach is not suitable for cases with high number of planes, such is the case with empty-space culling using octrees. Again, note that our cube-trimming method outperforms RSK.

4   Conclusion

We have provided a method for generating the intersection of multiple planes defined by a view vector with cubes. For a given vector, we use a sweeping algorithm uses the vertex connectivity of a cube to create a sequence of topology changes for the plane as it moves through the cube. We tabularize all possible sequences of topology changes for a given cube into a BSP tree. We use this table to quickly map a view vector to a sequence of topology changes and perform a simple sweeping algorithm that generates the polygonal intersections via simple vector addition which is computationally more efficient than regular plane-edge intersection tests. We use this quick polygon generating approach in an octree-based empty-space culling framework. We have shown results that indicate our method is more efficient than previous methods, and in the case of empty-space culling, it outperforms the pure hardware implementation.

For future works, we would like to examine moving some of the computation on to the graphics hardware. As implied in the results section, load-balancing between CPU and GPU requires careful analysis; furthermore, this process is highly dependent on the hardware specification. We think that a careful implementation can make use of the increasingly faster GPU and transfer some of the work-load to the GPU. This direction is especially relevant given the recent development of geometry shaders.

Acknowledgements

We thank Lavu Sridhar for his assistance and the reviewers for helpful suggestions. This work was partially supported by NIH grant R01GM079429.

References

1. Lorensen, W.E., Cline, H.E.: Marching cubes: A high resolution 3d surface construction algorithm. In: SIGGRAPH 1987: Proceedings of the 14th annual conference on Computer graphics and interactive techniques, pp. 163–169. ACM Press, New York (1987)
2. Ju, T., Losasso, F., Schaefer, S., Warren, J.: Dual contouring of hermite data. In: SIGGRAPH 2002: Proceedings of the 29th annual conference on Computer graphics and interactive techniques, pp. 339–346. ACM, New York (2002)
3. Drebin, R.A., Carpenter, L., Hanrahan, P.: Volume rendering. In: SIGGRAPH 1988: Proceedings of the 15th annual conference on Computer graphics and interactive techniques, pp. 65–74. ACM, New York (1988)
4. Huang, J., Mueller, K., Crawfis, R., Bartz, D., Meissner, M.: A practical evaluation of popular volume rendering algorithms. In: Huang, J., Mueller, K., Crawfis, R., Bartz, D., Meissner, M. (eds.) IEEE Symposium on Volume Visualization, 2000. VV 2000, pp. 81–90 (2000)
5. Levoy, M.: Efficient ray tracing of volume data. ACM Trans. Graph. 9, 245–261 (1990)
6. Laur, D., Hanrahan, P.: Hierarchical splatting: a progressive refinement algorithm for volume rendering. SIGGRAPH Comput. Graph. 25, 285–288 (1991)
7. Lacroute, P., Levoy, M.: Fast volume rendering using a shear-warp factorization of the viewing transformation. In: SIGGRAPH 1994: Proceedings of the 21st annual conference on Computer graphics and interactive techniques, pp. 451–458. ACM, New York (1994)
8. Cullip, T.J., Neumann, U.: Accelerating volume reconstruction with 3d texture hardware. Technical report, Chapel Hill, NC, USA (1994)
9. Wilson, O., VanGelder, A., Wilhelms, J.: Direct volume rendering via 3d textures. Technical report, Santa Cruz, CA, USA (1994)
10. Cabral, B., Cam, N., Foran, J.: Accelerated volume rendering and tomographic reconstruction using texture mapping hardware. In: VVS 1994: Proceedings of the 1994 symposium on Volume visualization, pp. 91–98. ACM, New York (1994)

11. Westermann, R., Ertl, T.: Efficiently using graphics hardware in volume rendering applications. In: SIGGRAPH 1998: Proceedings of the 25th annual conference on Computer graphics and interactive techniques, pp. 169–177. ACM, New York (1998)

12. Rezk-Salama, C., Engel, K., Bauer, M., Greiner, G., Ertl, T.: Interactive volume on standard pc graphics hardware using multi-textures and multi-stage rasterization. In: HWWS 2000: Proceedings of the ACM SIGGRAPH/EUROGRAPHICS workshop on Graphics hardware, pp. 109–118. ACM, New York (2000)

13. Engel, K., Kraus, M., Ertl, T.: High-quality pre-integrated volume rendering using hardware-accelerated pixel shading. In: HWWS 2001: Proceedings of the ACM SIGGRAPH/EUROGRAPHICS workshop on Graphics hardware, pp. 9–16. ACM, New York (2001)

14. Boada, I., Navazo, I., Scopigno, R.: Multiresolution volume visualization with a teuxture-based octree. The Visual Computer 17, 185–197 (2001)

15. LaMar, E., Hamann, B., Joy, K.I.: Multiresolution techniques for interactive texture-based volume visualization. In: VIS 1999: Proceedings of the conference on Visualization 1999, pp. 355–361. IEEE Computer Society Press, Los Alamitos (1999)

16. Wilson, B., Ma, K.-L., McCormick, P.S.: A hardware-assisted hybrid rendering technique for interactive volume visualization. In: IEEE Symposium on Volume Visualization and Graphics, pp. 123–130 (2002)

17. Dietrich, C.A., Nedel, L.P., Olabarriaga, S.D., Comba, J.L.D., Zanchet, D.J., da Silva, A.M.M., de Souza Montero, E.F.: Real-time interactive visualization and manipulation of the volumetric data using gpu-based methods. In: SPIE, vol. 5367, pp. 181–192 (2004)

18. Rezk-Salama, C., Kolb, A.: A Vertex Program for Efficient Box-Plane Intersection. In: Proc. Vision, Modeling and Visualization (VMV), pp. 115–122 (2005)

19. Naylor, B., Amanatides, J., Thibault, W.: Merging bsp trees yields polyhedral set operations. In: SIGGRAPH 1990: Proceedings of the 17th annual conference on Computer graphics and interactive techniques, pp. 115–124. ACM, New York (1990)

A Statistical Model for Daylight Spectra

Martyn Williams and William A.P. Smith

Department of Computer Science, The University of York, UK
martyn.williams@gmail.com, wsmith@cs.york.ac.uk

Abstract. In this paper we present a statistical model, learnt from empirical data, which captures variations in the spectra of daylight. We demonstrate two novel techniques which use the model to constrain vision and graphics problems. The first uses the model generatively to render scenes using spectra which are constrained to be plausible. The second uses the model to solve the ill-posed problem of estimating high dimensional illumination spectra from one or more tristimulus images, such as might be observed over the course of a day.

1 Introduction

The analysis of spectra, both reectance and illuminant, is of considerable importance to both computer vision and computer graphics. In vision applications this additional information, lost in the transformation to tristimulus spaces, can help in the identification of objects. In computer graphics, knowledge of how materials respond to specific wavelengths can improve the quality of rendering.

While this additional information is useful, traditional hyperspectral and multispectral cameras require high precision, expensive, optics. There are numerous on going studies working towards alternative, cheaper systems to capture the additional spectral information using COTS tristimulus cameras. Common techniques involve the use of filters to change the response of the camera [1] or the use of modulated light sources [2]. By choosing the optimum filters or illuminants it is hoped that higher dimeinterpreted from a small number of measurements.

In order to convert the observed radiance values of pixels into an object's reflectance, knowledge of the illuminant is required. Typically, an in scene calibration material of known reectance is used to determine the illuminant [1,3,4]. However, in some situations it may not be possible or convenient to place a calibration material in the scene.

Nieves et al. [5] have proposed an unsupervised system capable of recovering the illuminant from a scene without the need of a calibration material, using just a COTS Red, Green, Blue (RGB) camera. This technique creates a direct transform from pixel intensities to the illuminant spectra, by use of a linear pseudo-inverse method. This system must be trained independently for the camera to be used for the recovery. This removes some generality from the system, since it must be retrained if a new camera is used.

In this paper we propose an error minimisation approach to recovering the illuminant spectra in a scene. To do this we first construct a linear model for

G. Bebis et al. (Eds.): ISVC 2009, Part I, LNCS 5875, pp. 632–643, 2009.

daylight illumination spectra. This model is then used to constrain the error minimisation problem of illuminant recovery. We use a number of tests to determine the validity of this solution, including both ground truth and realworld data. The model created will also be used to demonstrate plausible spectral daylight relighting of multispectral images.

1.1   Related Work

The response of a single pixel P in a RGB sensor is a function of the reflective characteristics of the surface imaged at the pixel R, the illuminant incident on the surface I and the response of the sensor to light S. Each of these values are themselves dependent on the wavelength of the incident light. The pixel responses for red P_R, green P_G and blue P_B are given by three integrals Equation 1.

$$P_R = \int_{\lambda_{min}}^{\lambda_{max}} I(\lambda)R(\lambda)S_R(\lambda)\,d\lambda$$

$$P_G = \int_{\lambda_{min}}^{\lambda_{max}} I(\lambda)R(\lambda)S_G(\lambda)\,d\lambda \qquad (1)$$

$$P_B = \int_{\lambda_{min}}^{\lambda_{max}} I(\lambda)R(\lambda)S_B(\lambda)\,d\lambda$$

Previous studies have shown that use of Principal Components Analysis (PCA) can significantly lower the dimensionality of both reflectance [6] and daylight spectra [7,8,9]. These allow for the I and R terms to be represented with far fewer parameters than would be required if the spectra were considered in a raw form.

In most vision tasks it is the reflectance function for a surface which is of interest, not the irradiance values for pixels. It is desirable to know the incident illuminant spectrum to be able to recover the reflectance functions. In most scenarios it is infeasible to have a spectrometer in the scene to record the illuminant. In scene calibration materials can be used to convert the radiance values to reflectance values. This technique has been shown to be effective in both satellite based observations [4] and ground based material identification [3,10].

These in scene calibration methods still require a multi-spectral camera to be used to capture the images. With growing interest in using sensors capturing less spectral data and inferring higher dimensionality, having a robust method to estimate the scene illuminant has become critical.

Chiao et al. [11] attempted to recover the spectral illuminant in forest scenes using only an RGB camera and a white calibration panel. A training set of 238 images of the panel and corresponding ground truth spectra was used to create a direct transformation matrix from RGB values to spectra.

An on going study is being performed at the University of Grenada into recovery of spectra from RGB images. Initial work was performed proving the feasibility of using linear models to describe daylight spectra [8,9,12]. Studies were performed determining the optimum sensors for daylight recovery [13]. It was demonstrated that a standard RGB camera could determine the spectral daylight illuminant in both simulated and real data [14]. Two methods were demonstrated, a pseudo-inverse method based on the eigenvectors in an earlier study [9] and a direct transformation, trained on RGB spectra pairs. The direct transform out performed the eigenvector based approach, however some ambiguities in the use of training and test data may cast a shadow over these results. Finally, a fully unsupervised approach, mentioned earlier, was shown which was capable of determining the daylight illuminant in RGB images without the requirement of any calibration materials [5].

In all previous studies a direct transform was created to take an RGB value to a plausible daylight spectra. This transform implicitly contained the S terms in Equation 1. However, such transforms are camera specific and cannot be generalised to cameras with arbitrary spectral response.

2   Statistical Modelling

Our data was collected during November and December of 2008 using a BWTEK `BSR111E-VIS` spectroradiometer, between 350 and 760 nm at approximately 1/4 nm intervals. This device allows calibrated measurements to be taken, providing data in units of $\mu W/cm^2/nm$. A total of 72 daylight spectra were collected. To provide the most comparability with past reports [6,7,8,9], the captured spectra were down sampled to run from 400 to 700nm at 5nm intervals. All spectra were also normalised to unit length. This prevents the model from fitting to the changes in intensity, instead allowing it to react to the changes in the shape of the spectra.

For measuring spectra similarity a metric is required. A particularly thorough review of available metrics and their applicability was undertaken by Imai et al. [15]. We only regard spectra purely for their physical characteristics and not their effects on human vision, as such the most suitable metrics are the Goodness of Fit Coefficient (GFC) of Romero et al. [8] and the Root Mean Square Error (RMSE). The GFC comes with guidance on its interpretation [8]: a GFC > 0.99 is deemed acceptable, > 0.999 represents a very good representation and > 0.9999 is an exact spectra.

We used PCA to construct a linear statistical model of the data. Any spectra which lay more than 4 standard deviations from the mean in this space were pruned from the data set. This process was repeated until there were no outliers in the data set. In total 4 outlying spectra were removed.

To test the models ability to generalise to out of sample data, a leave-one-out testing strategy was used. A model was trained using all but one spectra in the training set. This model was then used to reconstruct the one spectra not in the model, this process was repeated for every spectra in the training set. Figure 1 shows the effect of varying the number of dimensions in the model on the average GFC values for all the spectra. As expected, using more dimensions improves the reconstructions. Using more than 3 dimensions is sufficient for all spectra in our sample to be represented to at least a very good level, and at least half the spectra to an exact level.

Figure 2 shows the distribution of GFC values for reconstructions with 3 dimensions. It can be seen that 82% of samples could be modelled exactly and all samples could be modelled to a very good or better level.

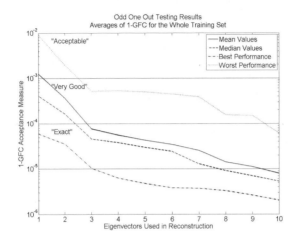

Fig. 1. 1-GFC errors for odd one out analysis against different number of model dimensions. The calculated GFC value is the average for all the spectra captured.

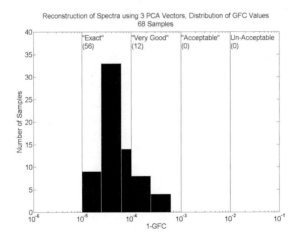

Fig. 2. 1-GFC errors for odd one out analysis. These results were calculated using 3 dimensions.

3   Rendering

We begin by visualing the modes of variation of the statistical model described above. Figure 3 shows 3 images from the Finlayson et al. [16] multi-spectral database, relit using illuminants created by varying the value of the first model parameter. The images were created by applying the rendering equation (Equation 1) to each pixel independently. The top images have not had any white balancing applied, the images shown are how the raw CCD image would look. This gives a slight green tinge to all the colours, due to the integral of the camera response over 400 - 700nm being greater for the green channel than the other channels. The images in the lower half of the figure have been white balanced by dividing the value for each channel by the integral of its corresponding response curve.

A data sample $\mathbf{x}$ can be expressed in terms of the matrix of model eigenvectors, $\mathbf{P}$, the mean, $\overline{\mathbf{x}}$ and a vector of n parameters, $\mathbf{b}$, as $\mathbf{x} = \mathbf{M}(\mathbf{b}) = \mathbf{Pb} + \overline{\mathbf{x}}$. The model provides a constraint on the length of parameter vectors, which is that the square of the Mahalanobis distance from the mean will follow a chi-squared distribution with mean n. Hence, the expected length is:

$$\sum_{i=1}^{n} \left(\frac{b_i}{\sqrt{\lambda_i}} \right)^2 = n, \tag{2}$$

where λ_i is the variance of b_i. This is equivalent to constraining parameter vectors to lie on a hyperellipsoid in parameter space, the diameters of which are defined by the eigenvalues of the PCA. This fact has been used to relight the images with plausible illuminants (Figure 4) created by randomly selecting the parameters of the model and then scaling them such that Equation 2 is satisfied. This demonstrates that our statistical model could be used generatively to provide plausible daylight spectra for rendering.

4   Illuminant Spectra Estimation

The system of Nieves et al. [5] uses the brightest pixels in the scene to interpret the illuminant spectra, these pixels have the highest likelihood of being the colour of the illuminant. This is similar to the max-RGB colour constancy algorithms. Another approach to the colour constancy problem is to use the gray world theorem [17]. This states that the the mean reflectance of objects in an image will be gray, thus the mean colour of an image will be indicative of the light source in an image. We use this fact to generate a colour to be used to reconstruct the illuminant from. The mean RGB value for an image is given by:

$$P_{\{R,G,B\}}^{Ave} = \frac{1}{wh} \sum_{i=1}^{w} \sum_{j=1}^{h} P_{\{R,G,B\}}(i,j), \tag{3}$$

where w is the image width and h is the image height. Substituting this value into Equation 1 and given that the average reflectance will be gray, the following integral is obtained:

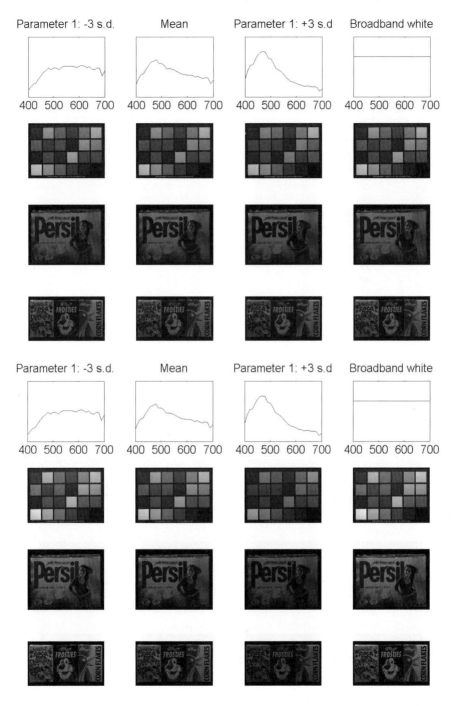

Fig. 3. Testing relighting of multispectral images. The top images are equivalent to a the raw CCD image, the bottom have been white balanced to better reflect the eyes response.

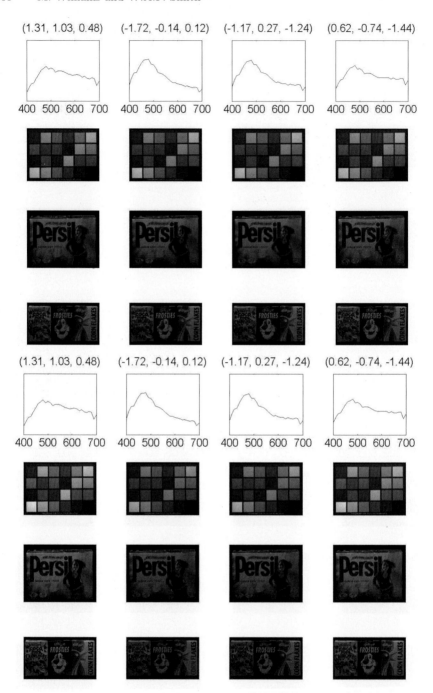

Fig. 4. Relighting to random illuminants. The top images are equivalent to the raw CCD image, the bottom have been white balanced to better reflect the eyes response. The 3 model parameters (in standard deviations from mean) are shown at the top of each column.

$$P^{Ave}_{\{R,G,B\}} = \int_{\lambda_{min}}^{\lambda_{max}} I(\lambda)\mu S_{\{R,G,B\}}(\lambda)\,d\lambda. \tag{4}$$

Here $\mu \in (0,1]$ is a constant for all wavelengths, it is the average gray reflectance. In this integral there is now only one unknown, the incident light spectral function. As a spectrometer takes n discrete samples across λ, these integrals can be replaced with sums over i where $\lambda(i)$ is the wavelength of light for sample $i < n$. For convenience, define $\tilde{\mathbf{I}}$, $\tilde{\mathbf{S}}_{\mathbf{R}}$, $\tilde{\mathbf{S}}_{\mathbf{G}}$ and $\tilde{\mathbf{S}}_{\mathbf{B}}$ to be n-dimensional vectors such that the element i of each corresponds to the value of $I(\lambda(i))$, $S_R(\lambda(i))$, $S_G(\lambda(i))$ and $S_B(\lambda(i))$ respectively. This gives

$$P^{Ave}_{\{R,G,B\}} = \sum_{i=0}^{n} I(\lambda(i))\mu S_{\{R,G,B\}}(\lambda(i)) = \mu\tilde{\mathbf{I}}.\tilde{\mathbf{S}}_{\{\mathbf{R},\mathbf{G},\mathbf{B}\}}. \tag{5}$$

To estimate $\tilde{\mathbf{I}}$ we define the error in an approximation of a spectrum to be

$$\epsilon(\tilde{\mathbf{I}}) = (c\tilde{\mathbf{I}}.\tilde{\mathbf{S}}_{\mathbf{R}} - P^{Ave}_R)^2 + (c\tilde{\mathbf{I}}.\tilde{\mathbf{S}}_{\mathbf{G}} - P^{Ave}_G)^2 + (c\tilde{\mathbf{I}}.\tilde{\mathbf{S}}_{\mathbf{B}} - P^{Ave}_B)^2, \tag{6}$$

with c some unknown scaling, which minimises $c\tilde{\mathbf{I}}.\tilde{\mathbf{S}} - P^{Ave}$ for a given $\tilde{\mathbf{I}}$. c represents all the linear unknowns in the system which cannot be modelled, including, among others: our μ value, unknowns in the camera such as lens transmittance and internal amplification, all scalings to the spectra and scalings in the RGB values as the only true quantities are the ratios between channels.

4.1   A Statistical Constraint

An approximation to the incident spectrum is given when ϵ is minimised. Without further knowledge, this problem is heavily under constrained. To make the problem well-posed, we reduce the problem from estimating a 61 dimensional spectra to estimating the 2 or 3 model parameters of our statistical model described above. Substituting $\mathbf{M}(\mathbf{b})$ for $\tilde{\mathbf{I}}$ in Equation 6 gives

$$\epsilon(\mathbf{b}) = (c\mathbf{M}(\mathbf{b}).\tilde{\mathbf{S}}_{\mathbf{R}} - P^{Ave}_R)^2 + (c\mathbf{M}(\mathbf{b}).\tilde{\mathbf{S}}_{\mathbf{G}} - P^{Ave}_G)^2 + (c\mathbf{M}(\mathbf{b}).\tilde{\mathbf{S}}_{\mathbf{B}} - P^{Ave}_B)^2. \tag{7}$$

This can now be minimised to solve for $\mathbf{b}$. We constrain the solution of this minimisation such that only plausible spectra can be generated. To do so, hard constraints are placed on each of the model parameters to ensure they lie within 3 standard deviations of the mean. This constraint means that the minimisation is non-linear and we solve it using the Levenberg-Marquadt technique.

5   Experimental Evaluation

In this section we present the results of applying our statistical image-based spectra estimation technique to real world and synthetic data.

5.1   Data Collection

For real world data, two time lapse sequences were captured along with images
of di?ering scenes. In each case, images were taken with a Nikon D70 and corresponding ground truth spectra captured. All images were captured in RAW
mode which yields 12-bit measurements with linear response.

The first time lapse sequence comprises images of an outdoor scene and spectra
recorded every 15 minutes between 1105 and 1650. Sunset on this day was at
1642 [18], so the later images will be a?ected by this transition. The second
time lapse sequence comprises images taken indoors (but where the principal
light source is the sun) and spectra recorded every 15 minutes between 1145 and
1745. On this day sunset was at 1716 [18]. In this sequence, the images included a
Macbeth Colour Checker chart which comprises 24 patches of known reectance.

5.2   Experimental Results

An odd one out analysis was performed on all the collected spectra. In this
analysis the odd spectra out was used to generate the associated RGB colour for
a perfect reector and the camera response curves. The illumination recovery was
then performed on this RGB value and the estimated spectra compared against
the ground truth spectra. The left hand plot of Figure 5 shows the resultant
GFC values when varying the dimensions used. It can be seen that the best
results occur when 2 dimensions are used. It seems likely that using additional
dimensions results in overfitting to spurious data.

The right hand plot of Figure 5 shows the distributions of GFCs for reconstructions using 2 dimensions. Approximately 13% of spectra can be reconstructed exactly when convolved through the camera response curves. 83% can

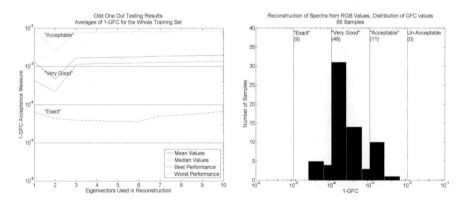

Fig. 5. 1-GFC errors for odd one out analysis against different number of dimensions
used in the reconstruction. The calculated GFC value is the average for all the spectra
captured. Right: 1-GFC errors for odd one out analysis. These results were calculated
using 2 dimensions for reconstruction.

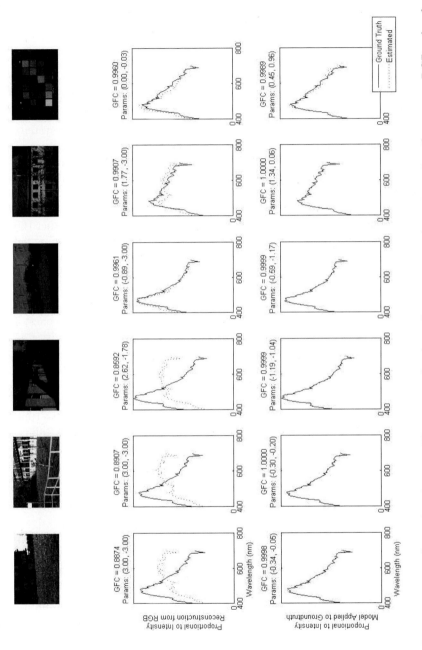

Fig. 6. Reconstructions for various images. The top graphs show the reconstructed spectra using the average RGB value from each image. The bottom graphs show the reconstruction after applying the model to the ground truth spectra, these represent the best case estimation for these images. All reconstructions were performed with 2 dimensions.

Table 1. Distributions of GFC values for illuminant recovery

	Best case (odd one out)	Time Lapse 1	Time Lapse 2	All images
Median	0.9998	0.9978	0.9929	0.9960
Standard Deviation	0.0005	0.0157	0.0028	0.0302

be reconstructed to at least a very good level and all can be reconstructed acceptably.

The results of the odd one out analysis represent the best case for this system. The RGB values tested are those that would be seen if the illuminant were to reflect off a perfect reector. In a true system deducing the RGB value indicative of the light is a non-trivial task, as has been shown by ongoing research into colour constancy.

Figure 6 shows the reconstructed spectra for different images, both using the mean RGB value for the reconstruction and using the model to reconstruct the ground truth spectra. The lower plots demonstrate just how effective the model is at fitting to illumination spectra. The upper plots show the results of image-based spectra estimation, using the average RGB value to fit to. It is clear that for images which deviate largely from the grayworld assumption, such as the first three images (where two contain predominantly grass and the other is largely shadowed), our method performs poorly. However, for the other three scenes, a snowy car park, a building and a macbeth chart the reconstructions register as acceptable by the Romero et al. GFC value.

Table 1 shows the median and standard deviations of Goodness of Fit Co-efficient (GFC) values for the various tests performed. The best case, given by reconstructing the spectra directly from its simulated RGB value, gives a median GFC value just below the level required for exact reconstructions. The two time lapse sequences gave acceptable median reconstructions. Finally, when taking the median of all the images from the time lapses, and of the various situations in Figure 6 the median reconstruction was again acceptable.

6   Conclusions

We have presented a statistical model for daylight spectra and shown how it can be used to relight multispectral images with plausible illuminants. We have also shown how the model can be used to constrain the process of estimating illuminant spectra from RGB images under the gray world assumption. The method shows promise, however its performance is determined by the validity of the gray world assumption.

In future work, we will investigate simultaneously estimating the spectral reflectance of ob jects in the scene alongside the illuminant spectra. We also intend to compare our method with the only other published approach to estimating illuminant spectra from RGB images [5].

References

1. Valero, E., Nieves, J., Nascimento, S., Amano, K., Foster, D.: Recovering spectral data from natural scenes with an RGB digital camera and colored filters. Color Research and Application 32, 352–359 (2007)
2. Park, J.I., Lee, M.H., Grossberg, M., Nayar, S.: Multispectral imaging using multiplexed illumination. In: IEEE 11th International Conference on Computer Vision, 2007. ICCV 2007, pp. 1–8 (2007)
3. Blagg, A.S., Bishop, G.J., Killey, A., Porter, M.D.: Temporal performance of spectral matched filtering techniques. In: Proceedings of the SPIE, vol. 6741 (2007)
4. Karpouzli, E., Malthus, T.: The empirical line method for the atmospheric correction of ikonos imagery. International Journal of Remote Sensing 24, 1143–1150 (2003)
5. Nieves, J., Plata, C., Valero, E., Romero, J.: Unsupervised illuminant estimation from natural scenes: an RGB digital camera suffices. Applied Optics 47, 3574–3584 (2008)
6. Parkkinen, J.P.S., Hallikainen, J., Jaaskelainen, T.: Characteristic spectra of munsell colors. Journal of the Optical Society of America 6, 318–322 (1989)
7. Judd, D.B., MacAdam, D.L., Wyzecki, G.: Spectral distribution of typical daylight as a function of correlated color temperature. Journal of the Optical Society of America 54, 1031–1040 (1964)
8. Romero, J., García-Beltrán, A., Hernández-Andrés, J.: Linear bases for representation of natural and artificial illuminants. Journal of the Optical Society of America 14, 1007–1014 (1997)
9. Hernández-Andrés, J., Romero, J., Nieves, J.L., Raymond, L., Lee, J.: Color and spectral analysis of daylight in southern europe. Journal of the Optical Society of America 18, 1325–1335 (2001)
10. Bishop, G.J., Killey, A., Porter, M.D., Blagg, A.S.: Spectral tracking of objects in real time. In: Proceedings of the SPIE, vol. 7119 (2008)
11. Chiao, C.C., Osorio, D., Vorobyev, M., Cronin, T.W.: Characterization of natural illuminants in forests and the use of digital video data to reconstruct illuminant spectra. Journal of the Optical Society of America 17, 1713–1721 (2000)
12. Hernández-Andrés, J., Romero, J., Raymond, L., Lee, J.: Colorimetric and spectroradiometric characteristics of narrow-field-of-view clear skylight in granada, spain. Journal of the Optical Society of America 18, 412–420 (2001)
13. Hernández-Andrés, J., Nieves, J.L., Valero, E.M., Romero, J.: Spectral-daylight recovery by use of only a few sensors. Journal of the Optical Society of America 21, 13–23 (2004)
14. Nieves, J.L., Valero, E.M., Nascimento, S.M.C., Hernández-Andrés, J., Romero, J.: Multispectral synthesis of daylight using a commercial digital ccd camera. Applied Optics 44, 5696–5703 (2005)
15. Imai, F., Rosen, M., Berns, R.: Comparative study of metrics for spectral match quality. In: Proceedings of the First European Conference on Colour in Graphics, Imaging and Vision, pp. 492–496 (2002)
16. Finlayson, G.D., Hordley, S.D., Morovic, P.: Using the spectracube to build a multispectral image database. In: Proceedings of IS&T Conference on Color, Graphics, Imaging and Vision (2004)
17. Buchsbaum, G.: A spatial processor model for object colour perception. Journal of the Optical Society of America 310, 1–26 (1980)
18. Morrisey, D.: Sunrise, sunset, moonrise and moonset times (2009), http://www.sunrisesunsetmap.com/

Reducing Artifacts between Adjacent Bricks in Multi-resolution Volume Rendering

Rhadamés Carmona[1], Gabriel Rodríguez[1], and Bernd Fröhlich[2]

[1] Universidad Central de Venezuela, Centro de Computación Gráfica,
1041-A, Caracas- Venezuela
rhadames.carmona@ciens.ucv.ve, gaborodriguez@gmail.com
[2] Bauhaus-Universität Weimar, Fakultät Medien, 99423 Weimar, Germany
bernd.froehlich@uni-weimar.de

Abstract. Multi-resolution techniques are commonly used to render volumetric datasets exceeding the memory size of the graphics board, or even the main memory. For these techniques the appropriate level of detail for each volume area is chosen according to various criteria including the graphics memory size. While the multi-resolution scheme deals with the memory limitation, distracting rendering artifacts become noticeable between adjacent bricks of different levels of detail. A number of approaches have been presented to reduce these artifacts at brick boundaries, including replicating or interpolating data between adjacent bricks, and inter-block interpolation. However, a visible difference in rendering quality around the boundary remained, which draws the attention of the users to these regions. Our ray casting approach completely removes these artifacts by GPU-based blending of contiguous levels of detail, which considers all the neighbors of a brick and their level of detail.

1 Introduction

During the past years multi-resolution hardware-accelerated volume rendering has been an important research topic in the scientific visualization domain. Due to the progressive improvements of imaging devices such as tomographs and magnetic resonators, the size of volume datasets continuously increases. Such datasets often exceed the available memory of graphics processing units (GPU), and thus multi-resolution techniques need to be employed to guarantee interactive frame rates for GPU-based rendering approaches. Out-of-core techniques are required for even larger datasets exceeding the main memory capabilities of regular desktop computers, which are generated e.g. by scientific projects such as the visible human ® [1] and the time-dependent turbulence simulation of Richtmyer-Meshkov [2].

While multi-resolution approaches in combination with out-of-core techniques deal with the memory limitations, distracting rendering artifacts between adjacent blocks of different level of detail occur (see Fig. 1). The source of these visual artifacts is the interpolation process since coarser data generates different samples during interpolation, which may be mapped to different colors by classification and during integration along the ray. The presence of these kinds of artifacts in the resulting image subconsciously draws the attention of the user to these regions of the volume instead of allowing the user to focus on the actual data.

G. Bebis et al. (Eds.): ISVC 2009, Part I, LNCS 5875, pp. 644–655, 2009.

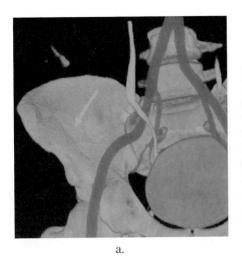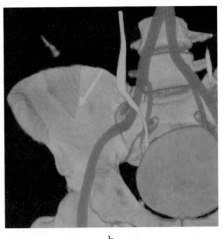

a.   b.

Fig. 1. Angiography dataset with aneurism. Images have been generated by using (a) 30MB of texture memory and (b) 4 MB of texture memory. In both cases, disturbing artifacts are noticeable at levels of detail transitions.

We developed an approach for effectively removing the rendering artifacts related to the quality difference between adjacent bricks of different level of detail. The bricks of the current cut through the octree are interpolated with their representation at the next coarser LOD, such that the resolutions are identical on boundaries between adjacent bricks. This GPU-based interpolation results in an imperceptible transition between adjacent bricks. Our approach requires a restricted octree, where adjacent bricks differ only by one LOD. We consider the resolution of all the neighbors of a brick of the cut during the generation of a volume sample. The interpolation coefficients for a sampling point in the volume can be efficiently generated on the fly from a small pre-computed 3D texture.

Previous research on reducing these multi-resolution rendering artifacts [3], [10], [11] focused on replicating or interpolating boundary voxels between adjacent bricks. While these schemes produce a smooth transition only for the boundary voxels between adjacent bricks, the abrupt change in visual quality around the boundary is still quite noticeable similar to Fig. 1. Our work is inspired by LaMar et al. [12], who achieved imperceptible transitions between LODs for an oblique clipping plane through a multi-resolution volume.

The main contribution of this paper is an effective and efficient approach for removing multi-resolution volume rendering artifacts between adjacent bricks of different level of detail. We integrated our technique into a GPU-based volume ray casting system, which also supports pre-integrated rendering [24], [25]. Our experiments show that the typical memory overhead introduced by our approach is about 10 percent while the increase in computation time is about 20 percent. Our approach extends to CPU-based rendering as well, is easily integrated in any multi-resolution volume rendering system and has the potential to become a standard technique for multi-resolution volume rendering.

2 Related Work

Hardware-accelerated rendering is the current standard for real-time rendering of volume datasets. It was introduced by Akeley [13], who suggested considering the volume dataset as a hardware-supported 3D texture. Various implementations of hardware-accelerated volume rendering have been popular during the past decade, including view-port aligned polygons [14], spherical shells [3], and GPU-based ray casting [15]. These schemes have also been adapted to render large volume datasets [3], [4], [16]. The term "large" is used for volumes, which do not fit into texture memory, and can potentially also exceed the available main memory. To deal with these limitations, the volume can be divided into sub-volumes or bricks, generally of equal size, such that each brick can be rendered independently, and easily swapped with another one [17]. However, the limited bandwidth between the main memory and the GPU still represents a bottleneck for large datasets on desktop computers, limiting the interactivity. Distributed architectures using multiple GPUs have been recently evaluated to alleviate the texture memory and bandwidth limitations [27].

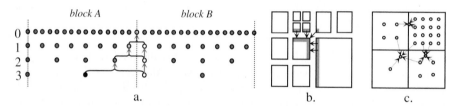

Fig. 2. Consistent interpolation between bricks or blocks of different LODs. (a) The voxels left and right of the boundary between the blocks A and B are used to replace voxels of that boundary for the next LOD. The right boundary voxel is copied into the right boundary voxel of the next coarser LOD, and the average voxel is copied into the left voxel of the next coarser LOD. Notice that when blocks A and B with different LODs are selected for rendering, a consistent interpolation between samples is obtained at boundaries. (b) Samples are replaced only in half of the boundary faces. The voxels of the right ($xmax$), up ($ymax$) and back ($zmax$) faces are replaced, taking the boundary data from adjacent bricks. (c) Interblock interpolation. In general, each sample to be reconstructed at brick boundaries is obtained by weighting up to 4 voxels (8 voxels for the 3D case).

Artifacts at brick boundaries have been reduced with various techniques. LaMar et al. [3] share the boundary voxels between adjacent bricks in each LOD. In this case, the artifacts are only removed between adjacent bricks of the same LOD. Weiler et al. [10] obtain a consistent interpolation between contiguous LODs by letting the boundary voxels between blocks interpolate the scalar field of the coarser LOD. Although this idea is initially outlined for transitions of contiguous LODs, it can be applied iteratively to achieve higher order transitions. Fig. 2a illustrates this process for one-dimensional textures. For the 3D case, this process has to consider the six faces of the blocks. Subsequently, Guthe et al. [7] use a similar concept for octrees, but they replicate voxels of only 3 faces of each brick (see Fig. 2b). Finally, Ljung et al. [11] partition the volume into blocks with a local multi-resolution hierarchy, but boundary voxels are not shared between adjacent blocks. During rendering, they perform the interpolation between blocks of arbitrary resolutions in a direct way (see Fig. 2c), i.e. without replicating voxels or pre-calculating intermediate samples by interpolation.

LaMar et al. [12] presented a multi-resolution technique for interactive texture-based rendering of arbitrarily oriented cutting planes. They achieved smooth transitions on a clipping plane by blending contiguous LODs. We extend this idea to 3D volume rendering considering the volumetric blending of bricks and show how it can be efficiently implemented. Their approach requires also that adjacent bricks along the clipping plane differ in at most one LOD. A similar constraint has been previously introduced for terrain visualization (restricted quad-trees) [18], to guarantee consistent triangulation and progressive meshing during roaming.

3   Multi-resolution Approach

We use a multi-resolution scheme to deal with large volume datasets. During pre-processing, the volume is downsampled into LODs, and partitioned into bricks, to build an octree-based multi-resolution hierarchy [3], [5], [6]. For each frame, a sub-tree of the whole octree is chosen according to a priority function $P(x)$, which may include data-based metrics [9] and image-based metrics [4]. The LOD of each volume area is chosen by a greedy-style subdivision process, considering the priority $P(x)$ [5], [8]. It starts by inserting the root node into a priority queue. Iteratively, the node with highest priority (head node) is removed from the queue, and its children are re-inserted into the queue. This process continues until the size limit is reached, or the head node represents a leaf in the octree hierarchy. We adapt this greedy-style selection algorithm such that the difference between adjacent bricks does not exceed one LOD. Also, the memory cost of the parent bricks has to be considered. Our rendering algorithm implements GPU-based ray casting, with pre-integrated classification.

The multi-resolution approach is presented in the following subsections. We first introduce the core of the approach, which is based on blending of contiguous LODs. Then, the selection algorithm is described, which considers the priority function and the constraint of the levels of detail between adjacent bricks. Finally, we include implementation details of the pre-integrated ray casting system with out-of-core support.

3.1   Blending

Fig. 3a shows a basic one-dimensional example of the blending approach. Consider brick B located at LOD i, and its adjacent brick CD located at the next coarser LOD (level $i-1$). Brick B is gradually blended with its representation at the next coarser LOD (brick AB) such that the coarser representation is reached at the boundary to brick CD. A weight t varying from 0 to 1 along B is required to perform the blending.

For the 3D case, any selected brick x located at the i-th LOD is adjacent to other bricks (26 bricks for the general case), which can be located at level $i-1$, level i or level $i+1$. Weights are assigned to each vertex of the brick; a weight of 0 is assigned if the vertex is only adjacent to bricks located at finer or the same LOD (level $i+1$ or level i), indicating that the resolution in this vertex corresponds to level i. Otherwise, its weight is 1 (the resolution of that vertex corresponds to level $i-1$). Fully transparent adjacent bricks are not considered during the weighting process, since they are

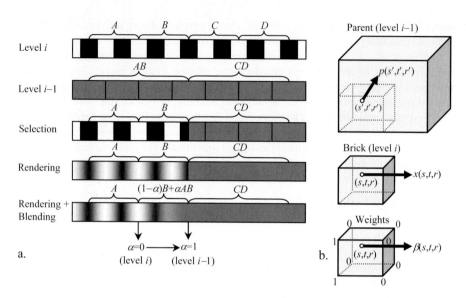

Fig. 3. Interpolation between LODs of adjacent bricks. (a) Levels i and $i-1$ are two subsequent LODs partitioned into bricks. Adjacent bricks of the same LOD share half a voxel at boundaries. The selection criterion selects bricks A, B and CD for rendering. During the rendering, the voxel data is interpolated to generate a volume sample. The difference between LODs of adjacent bricks generates visible artifacts. These artifacts are removed by means of gradual interpolation from the finer level to its next coarser representation, such that the respective LOD matches at the brick boundary. (b) Blending a brick located at level i with its parent at level $i-1$. The blending factor β of any position (s,t,r) inside the brick is computed by a tri-linear interpolation of the weights assigned to each brick vertex.

excluded from rendering. If the weight of every vertex is 0, then its parent brick is not required for blending. In any other case, the blending is performed for each volume sample with texture coordinates (s,t,r) in brick x. Thus, the volume sample $x(s,t,r)$ and the corresponding volume sample $p(s',t',r')$ of its parent are linearly interpolated using equation (1):

$$blend(\beta) = (1-\beta) \cdot x + \beta \cdot p , \qquad (1)$$

where β is computed by tri-linear interpolation of vertex weights (see Fig. 3b). Notice that the blended volume sample $blend(\beta)$ is obtained by quad-linear interpolation since x and p are reconstructed via tri-linear interpolation. Particularly, $blend(\beta)=x$ (level i) if the interpolated weight is $\beta=0$, and $blend(\beta)=p$ (level $i-1$) if $\beta=1$. For any other value of β in $(0,1)$, the level of detail of the resulting volume sample is $i-\beta$, i.e. an intermediate level of detail between the levels $i-1$ and i. This also shows that the rendering achieves voxel-based LODs, guaranteeing a smooth transition between adjacent bricks.

For fast tri-linear interpolation of the vertex weights, a single 3D texture of 2x2x2 voxels (corresponding to the weights of brick vertices) can be used for each brick. However, downloading one extra 3D texture into texture memory for each brick to blend is not appropriate. Similarly to the marching cubes approach [26], this table could be reduced to 15 cases by transforming the texture coordinates (s,t,r) appropriately. However, this would increase the overhead for the fragment program and we opted for using a larger 3D texture instead. This 3D texture of 2x2x512 weights is downloaded only once into texture memory (see Fig. 4). Therefore, a simple index indicating which combination corresponds to the brick being rendered is required for accessing the correct weights for each volume sample.

0   1   2   3   16   17   254   255

Fig. 4. Look-up table with all possible combinations of vertex weights

3.2 Priority Function

In our implementation, the priority of brick x, $P(x)$, combines two metrics similar to [20]: the distortion $D(x)$ in the transfer function domain, and the importance level $I(x)$, based on the distance from x to the viewpoint and region of interest. We present a short summary of the computation of the priority function. A more detailed explanation of the priority function and the real-time updating is found in [20].

$D(x)$ is the distortion of approximating the source voxels (s_i) by the voxels x_i of x. It can be written as (2):

$$D(x) = \sum_{i=1}^{n_s} D(s_i, x_i), \qquad (2)$$

where $D(s_i, x_i)$ is computed in CIELUV color space after applying the transfer function to the voxels s_i and x_i, and n_s is the number of voxels of the source data approximated by x [4]. The importance level $I(x)$ of the brick x is computed considering the distance from brick x to the region of interest (ROI) and to the viewer:

$$I(x) = (1-t)\frac{length(x)}{length(x) + d(x, ROI)} + t\frac{length(x)}{length(x) + d(x, eye)}, \qquad (3)$$

where $diag(x)$ is the diagonal length of the brick x in object space, $d(x,eye)$ is the minimum Euclidean distance from the brick x to the eye, and $d(x,ROI)$ is the average between the minimum distance between x and the ROI and the minimum distance between x and the ROI center. The value of t is used to weigh the distance to the ROI versus the distance to the eye. In our tests, we set $t=0.25$, giving more priority to the distance to the ROI. The distortion level and the importance level are multiplied to define the priority function $P(x)$ as (4):

$$P(x) = D(x) \cdot I(x). \qquad (4)$$

3.3 Selecting the Bricks

Before rendering, the set of bricks for representing the volume under the texture memory constraint have to be selected. We use a greedy-style algorithm, which selects the nodes with highest priority $P(x)$ for splitting. It uses a priority queue PQ to perform this process. Starting by inserting the root node into the queue, the selection process consists of splitting the node of PQ with highest priority [4], [5], [7], i.e. removing the head of PQ and re-inserting its non-fully transparent children into PQ. We use a min-max octree [22] to discard fully transparent bricks. During the refinement process, the following constraints are considered in this approach:

- Splitting a node must not violate the adjacency constraint. This constraint indicates that the difference with respect to the LOD between adjacent bricks selected for rendering must not exceed one level. It suggests building a restricted octree, which can be constructed using an adaptation of the restricted quad-tree algorithm for terrain rendering [18].
- Every node $x \in PQ$ requires blending with its own parent node during rendering, if at least one of its adjacent bricks is coarser than x.
- The number of bricks used for rendering (including parent nodes) is limited by a hardware constraint or user-defined texture size (N bricks).

The adjacency constraint is respected by evaluating the LOD of *adjacent* bricks before splitting a node $x \in PQ$. Each coarser node y adjacent to x needs to be split before splitting the selected node x. Notice that the adjacency constraint has to be evaluated again before splitting any adjacent node, and so on. This suggests using a recursive procedure or an auxiliary queue to perform this task. In addition, for each node x in PQ, we keep the list of its adjacent nodes $A(x)$. When an individual split is performed, the adjacency list of each child is created with the union of its brother nodes and a subset of $A(x)$; also, the adjacency list of each adjacent node is updated, replacing the entry x by the corresponding subset of *children(x)*.

The algorithm keeps track of the number of nodes selected for rendering, *including* the parent bricks. Each node x in PQ has a flag, indicating if such a node requires its parent for blending. One node requires its parent if any of its adjacent bricks is coarser than itself. If a node x is split, the flag of x and the flag of its adjacent bricks need to be re-evaluated.

The refinement process stops if a split operation exceeds the texture memory constraint, or no more refinement is possible.

3.4 Rendering

The selected bricks are rendered in front-to-back order, using GPU-based ray casting with pre-integrated classification. They are composited with the under operator [19]. The pre-integrated table is incrementally generated using the $O(n^2)$ incremental algorithm of Lum et al. [24], which requires about 0.06 seconds for $n=256$. Each brick is stored in an individual 3D texture object; thus, loading and rendering can be performed in an interleaved fashion, and potentially in parallel [21]. For each brick, the front faces of its bounding box are rasterized [15], interpolating the texture coordinates of the vertices, and the viewing vector. Therefore, each fragment obtained from rasterization contains the viewing ray and the entry point into the brick in texture space.

The ray is sampled with constant step length. For each pair of consecutive samples x_f and x_b, the 2D pre-integration table is fetched at $(s,t)=(x_f,x_b)$ to retrieve the corresponding color integrals $RGB(x_f,x_b)$ and the opacity $\alpha(x_f,x_b)$ in the interval $[x_f,x_b]$. If the brick requires blending with its parent, the parent is sampled at constant steps as well. The front sample x_f of the brick x is blended with the corresponding front sample p_f of its parent (1), obtaining the blended front sample f. Also, the back samples are blended, obtaining the blended back sample b. Thus, the pre-integration table is only fetched at location (f,b), i.e. there is only one fetch of the pre-integration table per ray-segment. The last ray-segment inside the brick is clipped at the brick boundaries. In this case, the segment length is used to scale the color integrals, and also to correct the opacity [10].

3.5 Caching and Out-of-Core Support

A texture memory buffer containing 3D texture objects (bricks) is created to store the octree cut in texture memory. To exploit frame-to-frame coherence, we keep a list of used bricks in texture memory, and another list of unused bricks. Every time the list of bricks requested for rendering changes between frames, unused bricks are replaced by new bricks and moved to the used list, according to an LRU (replace the least recently used page) scheme [6]. Since the number of new bricks can be eventually large, it may influence the frame rate significantly. A better approach is suggested in [20], which incrementally updates the previous cut through the octree towards the new cut on a frame-by-frame basis, limiting the number of bricks that are exchanged between frames.

For large datasets fitting only partially into main memory, out-of-core techniques have to be considered. In our system, a simple paging on demand is implemented. A main memory cache is used to hold the bricks required for rendering, and also to keep some other bricks that can be used in future frames. The paging process is running in a separate thread to avoid stalling the rendering process [6], [23]. A simply LRU scheme is also used to replace unused bricks by new bricks [7], [23].

The multi-resolution dataset is stored in a single huge file, and the bricks are loaded in groups of at most 8 bricks (which share the same parent), since the disk bandwidth increases by loading contiguous data [20], [23]. In our tests, for brick sizes varying between 16^3 and 64^3 samples, loading blocks containing 8 bricks doubles the disk bandwidth, in comparison to loading independent bricks. While the requested data is not available in main memory, the rendering thread continues rendering the last available data, which keeps the system interactive.

4 Implementation and Results

The system prototype was developed and evaluated under 32-bit Windows XP using Visual C++ 2005 with OpenGL® support. The hardware platform used for the tests is a desktop PC with a 2.4GHz quad core Intel® processor, 2 GB of main memory, an NVidia® Geforce[tm] 8800 graphics card with 640MB on-board memory, and a SATA II hard disk of 7200 rpm. Medical datasets (See Table 1, and Fig. 5) have been selected for testing: computer tomography of the visible female from the Visible Human project ® [1] (VFCT), angiography with aneurism (Angio), and grayscale-converted photos of the visible female from the Visible Human project ® [1] (VF8b).

Table 1. Test Datasets

Attribute	VFCT	VF8b	ANGIO	Measurement	VFCT	VF8b	ANGIO
Width (v=voxels)	512	2048	512	FPS naive approach	20.01	15.00	22.09
Height (v)	512	1216	512	FPS blended approach	16.7	11.72	17.80
Slices (v)	992	5186	1559	% Blending overhead	16.54%	21.87%	19.42%
Source Size (GB)	0.85	12.03	0.76	(a) Nr. Selected bricks	3865	2761	3255
Brick Size (v)	16^3	32^3	16^3	(b) Nr. Parents used	426	308	426
Bits per voxel	16	8	16	Total bricks = (a)+(b)	4291	3069	3681
Tex. Cache (MB)	35	100	30	% Parent bricks	9.93%	10.04%	11.57%
Ram Cache (MB)	140	400	120	% Blended bricks	65.74%	54.04%	70.26%
Block Size (KB)	64	256	64	(c) LOD ave. naive	5.39	5.67	5.48
Bandwidth (MB)	10	33	10	(d) LOD ave. blended	5.21	5.53	5.30
LODs	0..7	0..8	0..8	LOD difference (c) – (d)	0.17	0.14	0.18

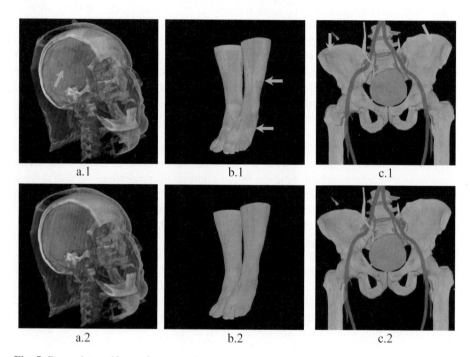

a.1   b.1   c.1

a.2   b.2   c.2

Fig. 5. Removing artifacts of our test datasets. (a) VFCT: Head of the visible female, obtained from CT. (b) VF8b: Feet of the visible female, from full color images converted to grayscale. (c) Angio: Angiography with aneurism. Upper images (*.1) are rendered using the naive approach; bottom images (*.2) are rendered with blending. Each brown arrow points to a visual artifact between adjacent bricks.

For each dataset, a texture memory size is set for caching. A further budget is reserved in main memory to page bricks from disk. We use four times the texture memory size for caching bricks in main memory. During pre-processing, datasets are split into bricks and downsampled to build the multi-resolution hierarchy. Each set of nodes sharing the same parent is grouped into a single block. All blocks are stored in a single binary file. Datasets of 12 bits per sample are scaled to 16 bits to increase the interpolation accuracy [23].

Results are shown in Table 1 and Fig. 5. The transfer function used for VFCT and VF8b is illustrated in Fig. 6. The blended version reduces the frames rate (FPS) rate by about 20%, although more than 54% of the bricks require blending with their parent during rendering. Notice that only about 10% additional bricks (parent bricks) are required for our blending approach. In theory, up to $1/8^{th}$ of the bricks in the cut may be required for a full octree. However, not every selected brick requires its parent for rendering.

We estimated the average LOD for the generated images, both for the naive approach as well as for the blended approach. The average LOD for the naive approach is calculated as the sum of the LODs of each selected brick, weighted by the volume ratio represented by each brick. Let T be the list of nodes selected for rendering (without parent bricks). The average level of detail is defined by equation (5).

$$LOD(T) = \frac{\sum\limits_{x \in T} Volume(x) \cdot LOD(x)}{\sum\limits_{x \in T} Volume(x)} . \qquad (5)$$

The average LOD of the blended approach is also calculated by equation (5). However, the level of detail of brick x $(LOD(x))$ is estimated by averaging the level of detail assigned to each brick vertex. Due to the blending of some bricks with the next coarser LOD, the average LOD of the blended approach is lower. However, it is only about 0.18 levels coarser than the naive approach for our examples.

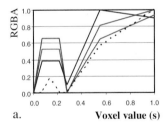
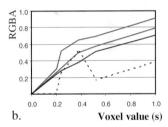

Fig. 6. Normalized transfer functions for the visible female. (a) VFCT, (b) VF8b. R, G and B channels are shown with the corresponding color. Absorption function is stored in the alpha channel A, denoted by a black dotted line.

5   Conclusions and Future Work

We introduced an efficient and effective technique to remove disturbing artifacts between adjacent bricks for direct multi-resolution volume rendering. Our approach blends a brick of the cut with its next coarser representation in such a way that the resolutions match at boundaries with its adjacent bricks. Thus, the LOD is gradually reduced inside a brick instead of simply displaying bricks of different level of detail next to each other or blending a few boundary pixels. Our results show the effectiveness of this technique, which only increases the rendering time by about 20% while requiring only about 10% of texture memory overhead.

Our approach can be extended to work in a multi-resolution framework, which supports roaming through a volume [6], [20]. In this case the cut is often updated from frame to frame, which may incur popping artifacts if the LOD changes in a certain area of the volume. Our technique can be used to generate an animated transition between the previous cut and the current cut. However, the cuts should be nowhere more different than one LOD and the animated transition becomes a 5D interpolation, since each cut requires already a 4D interpolation to perform the blending. Fortunately, the 3D interpolation part is directly hardware-supported.

References

1. Visible Human Project® (2009),
   http://www.nlm.nih.gov/research/visible/visible_human.html
2. Mirin, A., Cohen, R., Curtis, B., Dannevik, W., Dimits, A., Duchaineau, M., Eliason, D., Schikore, D., Anderson, S., Porter, D., Woodward, P., Shieh, L., White, S.: Very High Resolution Simulation of Compressible Turbulence on the IBM-SP System. In: Proc. ACM/IEEE Supercomputing Conference 1999, vol. 13(18), p. 70 (1999)
3. LaMar, E., Hamman, B., Joy, K.I.: Multiresolution Techniques for Interactive Texture-Based Volume Visualization. In: Proc. IEEE Visualization 1999, pp. 355–362 (1999)
4. Wang, C., García, A., Shen, H.-W.: Interactive Level-of-Detail Selection Using Image-Based Quality Metric for Large Volume Visualization. IEEE Transactions on Visualization and Computer Graphics 13(1), 122–134 (2007)
5. Boada, I., Navazo, I., Scopigno, R.: Multiresolution Volume Visualization with Texture-based Octree. The Visual Computer 17, 185–197 (2001)
6. Plate, J., Tirtasana, M., Carmona, R., Froehlich, B.: Octreemizer: A Hierarchical Approach for interactive Roaming through Very Large Volumes. In: Proc. EUROGRAPHICS/IEEE TCVG Symposium on Visualization 2002, pp. 53–60 (2002)
7. Guthe, S., Wand, M., Gonser, J., Straßer, W.: Interactive Rendering of Large Volume Data Sets. In: Proc. IEEE Visualization 2002, pp. 53–60 (2002)
8. Ljung, P., Lundström, C., Ynnerman, A., Museth, K.: Transfer Function Based Adaptive Decompression for Volume Rendering of Large Medical Data Sets. In: Proc. IEEE Symposium on Volume Visualization and Graphics 2004, pp. 25–32 (2004)
9. Guthe, S., Straßer, W.: Advanced Techniques for High-Quality Multi-Resolution Volume Rendering. Computers & Graphics 28(1), 51–58 (2004)
10. Weiler, M., Westermann, R., Hansen, C., Zimmerman, K., Ertl, T.: Level-Of-Detail Volume Rendering via 3D Textures. In: Proc. IEEE Symposium on Volume Visualization and Graphics 2000, pp. 7–13 (2000)
11. Ljung, P., Lundström, C., Ynnerman, A.: Multiresolution Interblock Interpolation in Direct Volume Rendering. In: Proc. EUROGRAPHICS/ IEEE-VGTC Symposium on Visualization 2006, pp. 256–266 (2006)
12. LaMar, E., Duchaineau, M., Hamann, B., Joy, K.: Multiresolution Techniques for Interactive Texture-based Rendering of Arbitrarily Oriented Cutting Planes. In: Proc. EUROGRAPHICS/IEEE TVCG Symposium on Visualization 2000, pp. 105–114 (2000)
13. Akeley, K.: Reality Engine Graphics. In: Proc. annual conference on Computer graphics and interactive techniques SIGGRAPH 1993, pp. 109–116 (1993)

14. Dachille, F., Kreeger, K., Chen, B., Bitter, I., Kaufman, A.: High-Quality Volume Rendering Using Texture Mapping Hardware. In: Proc. ACM SIGGRAPH/EUROGRAPHICS Workshop on Graphics Hardware 1998, pp. 69–76 (1998)
15. Krüger, J., Westermann, R.: Acceleration techniques for GPU-based Volume Rendering. In: Proc. IEEE Visualization 2003, pp. 287–292 (2003)
16. Ljung, P.: Adaptive Sampling in Single Pass, GPU-based Ray Casting of Multiresolution Volumes. In: Proc. EURO-GRAPHICS/IEEE International Workshop on Volume Graphics 2006, pp. 39–46 (2006)
17. Grzeszczuk, R., Henn, C., Yagel, R.: Advanced Geometric Techniques for Ray Casting Volumes. Course Notes No. 4. In: Annual Conference on Computer Graphics - SIGGRAPH 1998 (1998)
18. Pajarola, R., Zürich, E.: Large Scale Terrain Visualization Using the Restricted Quadtree Triangulation. In: Proc. Visualization 1998, pp. 19–26 (1998)
19. Max, N.: Optical Models for Direct Volume Rendering. In: Visualization in Scientific Computing, pp. 35–40. Springer, Heidelberg (1995)
20. Carmona, R., Fröhlich, B.: A Split-and-Collapse Algorithm for Interactive Multi-Resolution Volume Rendering. Elsevier Computer & Graphics (Submitted for publication, 2009)
21. Rezk-Salama, C.: Volume Rendering Techniques for General Purpose Graphics Hardware. Thesis dissertation, Erlangen-Nürnberg University, Germany (2001)
22. Lacroute, P.G.: Fast Volume Rendering Using Shear-Warp Factorization of the Viewing Transformation. Technical Report CSL-TR-95-678, Stanford University (1995)
23. Ljung, P., Winskog, C., Persson, A., Lundström, K., Ynnerman, A.: Full Body Virtual Autopsies using a State-of-the-art Volume Rendering Pipeline. IEEE Transactions on Visualization and Computer Graphics 12(5), 869–876 (2006)
24. Lum, E., Wilson, B., Ma, K.L.: High-Quality Lighting and Efficient Pre-Integration for Volume Rendering. In: Proc. EUROGRAPHICS/IEEE TCVG Symposium on Visualization 2004, pp. 25–34 (2004)
25. Engel, K., Kraus, M., Ertl, T.: High Quality Pre-Integrated Volume Rendering Using Hardware-Accelerated Pixel Shading. In: Proc. ACM SIGGRAPH/EUROGRAPHICS Workshop on Graphics Hardware 2001, pp. 9–16 (2001)
26. Lorensen, W., Cline, H.: Marching Cubes: A high resolution 3D surface construction algorithm. Computer Graphics 21(4), 320–327 (1987)
27. Monoley, B., Weiskopf, D., Möller, T., Strengert, M.: Scalable Sort-First Parallel Direct Volume Rendering with Dynamic Load Balancing. In: Eurographics Symposium on Parallel Graphics and Visualization 2007, pp. 45–52 (2007)

Height and Tilt Geometric Texture

Vedrana Andersen[1], Mathieu Desbrun[2], J. Andreas Bærentzen[1],
and Henrik Aanæs[1]

[1] Technical University of Denmark
[2] California Institute of Technology

Abstract. We propose a new intrinsic representation of geometric texture over triangle meshes. Our approach extends the conventional height field texture representation by incorporating displacements in the tangential plane in the form of a normal tilt. This texture representation offers a good practical compromise between functionality and simplicity: it can efficiently handle and process geometric texture too complex to be represented as a height field, without having recourse to full blown mesh editing algorithms. The height-and-tilt representation proposed here is fully intrinsic to the mesh, making texture editing and animation (such as bending or waving) intuitively controllable over arbitrary base mesh. We also provide simple methods for texture extraction and transfer using our height-and-field representation.

1 Introduction

The advent of laser scanners, structured light scanners, and other modalities for capturing digital 3D models from real objects has resulted in the availability of mesh with complex geometric details at a wide range of scales. Handling this geometric complexity has brought numerous challenges. In this paper, we address the problem of representation and editing of the finest level details known as *geometric texture*. It is important to distinguish this use of the word *texture* from *texture mapping* where an image is mapped onto a shape via parametrization. In recent years the use of texture mapping has expanded greatly, and one application of texture mapping is to map geometric texture onto a smooth base shape by means of height map images. This approach often performs adequately, but geometric texture such as thorns, scales, bark, and overhangs simply cannot be described by height fields: a single valued height field is insufficient for these common types of geometric texture, see Fig. 1.

Tangential displacements could be included alongside normal (height) displacements. However, there is no simple canonical basis in which to encode tangent vectors. To produce a basis one might use the partial derivative of a map from parameter domain to the surface, or choose one outgoing edge from each vertex. Unfortunately, these obvious methods are not intrinsic to the shape, requiring either an added parametrization, or an ordering of the edges, and further editing of the geometric texture may suffer from artifacts accordingly.

To deal with full 3D texture, researchers have proposed cut-and-paste [1] and example-based [2] methods, as well as approaches that stretch and fit patches

G. Bebis et al. (Eds.): ISVC 2009, Part I, LNCS 5875, pp. 656–667, 2009.

Fig. 1. Limitations of the height field representation of the geometric texture. Of the two textures only the *left* one can be described as the texture superimposed on the a shape.

Fig. 2. Examples from our height-and-tilt geometric texture representation. *Left:* A lychee fruit scan is modified to wrap the spikes. *Middle:* Geometric texture applied after a deformation of the base shape. *Right:* A synthetic texture over plane is transferred onto an arbitrary object.

of 3D texture to create complex geometric textures [3]. These methods are also capable of handling weaved textures, or textures of high topological genus. They do not, however, offer intrinsic representations of the texture on the surface, but increase the geometric complexity of the object instead, making use of full-blown mesh editing methods [4].

Contributions. We propose an intermediate type of geometric texture representation, compact and practical, offering a compromise richer than displacement field textures but much simpler than full 3D textures. We will assume that small-scale surface details are easily separable from the base surface, but are not necessarily representable as height fields over the base surface. Our representation adds a *tilt* field to the conventional height field texture representation, with this tilt field being stored using one scalar per edge in a coordinate-free (intrinsic) manner. A resulting height-and-tilt texture model can be used for extraction, synthesis and transfer of a large family of geometric textures. Additionally, we demonstrate that dividing a texture into a height field and a tilt field offers new and intuitive mesh editing and animation possibilities without the computational complexity associated with global mesh editing methods, see Fig. 2.

Related Work. Texture is often an important feature of 3D objects, explaining the abundance and variety of methods proposed to synthesize texture on surfaces [5,6,7]. The main goal of most texture synthesis algorithms is to synthesize a texture (color, transparency, and/or displacement) onto an arbitrary surface resembling a sample texture patch [8,9]. Common to these methods is the limitation to textures represented by an image or a scalar displacement field.

While height fields defined over surfaces have been used for many years, newer and richer representations have only started to appear recently. In [10] for instance, fur was modeled through the addition of a tangential displacement to rotate a discrete set of hair strands away from the normal direction. A similar idea based on vector-based terrain displacement maps to allow for overhangs was also proposed for gaming [11].

Tangent fields have also recently been used to control texture growth directions [12,13]. A convenient, intrinsic representation of tangent vector fields was even proposed in [14], along with vector field processing directly through edge value manipulations.

To overcome the limitations of conventional heightfield-based texture representations, we model geometric texture as a locally tilted height field over the base shape. By storing the height field as scalars over mesh vertices (i.e. discrete 0-forms [15]), and storing the tilt field as scalars over mesh edges (i.e. discrete 1-forms), we obtain an intrinsic, coordinate-free representation of fairly complex geometric textures.

2   Background on Tangent Vector Fields as One-Forms

As we make heavy use of representing tangent vector fields as discrete 1-forms, we briefly review the mathematical foundations proposed in [15,14].

From vector fields to 1-forms. From a vector field defined in the embedding space, one can encode its tangential part $\mathbf{t}$ to a surface mesh by assigning a coefficient c_{ij} to each edge $\mathbf{e}_{ij}$. This coefficient represents the *line integral* of the tangent vector field $\mathbf{t}$ along the edge. The set of all these values on edges offers an intrinsic representation (i.e. needing no coordinate frames) of the tangent vector field.

From 1-forms to vector fields. From the edge values, a tangent vector field can be reconstructed using, for instance, a vertex-based piecewise-linear vector field. The value of the vector field at a vertex is computed from the coefficients of the incident edges: the contribution of one face f_{ijk} (see Fig. 3, *left*) to the field at the vertex v_i is

$$\mathbf{t}_{ijk}(v_i) = \frac{1}{2A_{ijk}}(c_{ij}\mathbf{e}_{ki}^{\perp} - c_{ki}\mathbf{e}_{ij}^{\perp}) , \tag{1}$$

where c_{ij} and c_{ki} are coefficients on edges $\mathbf{e}_{ij}$ and $\mathbf{e}_{ki}$ respectively, and $\mathbf{e}_{ij}^{\perp}$ and $\mathbf{e}_{ki}^{\perp}$ are edges $\mathbf{e}_{ij}$ and $\mathbf{e}_{ki}$ (as 3D vectors) rotated for $\pi/2$ in the plane of f_{ijk} (see the discussion about Whitney edge basis functions in [15]). Averaging these contributions from all incident triangles thus provides a 3D vector per vertex of the mesh.

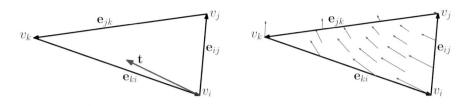

Fig. 3. *Left:* The contribution of the face f_{ijk} to the tangent field at vertex v_i. *Right:* Piecewise linear interpolation of the tangent field.

Least-squares 1-form assignment. The averaging process used in the reconstruction makes the encoding of vector fields by 1-forms lossy: a piecewise-vector field converted into a 1-form may not be exactly recovered once converted back. To provide the best reconstruction of the field from edge coefficients, we do not compute the edge coefficients locally, but proceed instead through a global least squares fit. If we store in **M** the reconstruction coefficients defined in (1) (that depend only the connectivity and vertex coordinates of the mesh, not on the tangent vector field), we find the set of edge coefficients **C** by solving the linear system

$$\mathbf{M\,C} = \mathbf{V} \ , \tag{2}$$

where the vector **V** contains the coordinates of the input vector field at vertices. Each vertex contributing three equations while there is only one unknown per edge, this system is slightly overdetermined (depending on the genus), and solving it in a least squares fashion yields a very good representation of a tangent vector field over the triangular mesh with little or no loss.

Tangent vector field reconstruction. To transfer a tangential field from one mesh to another we need to evaluate the field on arbitrary point of the mesh surface. For a point P on the face f_{ijk} with barycentric coordinates $(\alpha_i, \alpha_j, \alpha_k)$ associated with vertices i, j and k we get

$$\mathbf{t}(P) = \frac{1}{2A_{ijk}} \left((c_{ki}\alpha_k - c_{ij}\alpha_j)\mathbf{e}_{jk}^\perp + (c_{ij}\alpha_i - c_{jk}\alpha_k)\mathbf{e}_{ki}^\perp + (c_{jk}\alpha_j - c_{ki}\alpha_i)\mathbf{e}_{ij}^\perp \right) \ ,$$

which amounts to evaluating the face contribution to each of the vertices, as in (1), and linearly interpolating those using barycentric coordinates, as illustrated in Fig. 3, *right*.

3   Texture Representation

In the texture representation proposed here, we make the usual assumption that the finely-tesselated textured object comes from a smoother base shape, onto which a small-scale geometric texture is superimposed without affecting the topology of the base shape. We will first describe how to establish our discrete representation before introducing applications.

3.1 Texture Extraction

Given a finely-tesselated textured object, we must first decide what constitutes geometry (base shape) and what constitutes small-scale texture (displacement from base shape, see Fig. 4, *left*). While this is a notoriously ill-posed problem, many good practical methods have been proposed. In fact, any approach that proceeds through a smoothing of the textured surface while minimizing the tangential drift throughout the process is appropriate in our context. For example, a few steps of mean curvature flow [16] provides a good vertex-to-vertex correspondence between the original textured surface and a smoother version, used as base shape. For more intricate geometries, a multiresolution smoothing strategy such as [17] or a spectral approach such as [18] are preferable (see Fig. 4, *middle*). Alternatively, defining or altering the base shape by hand might be appropriate if specific texture effects are sought after or if the condition mentioned in Fig. 4, *left*, is significantly violated.

3.2 Pseudo-height and Tilt

From displacements to heights and tilts. With a base shape available, the displacement of vertex v_s is simply defined as

$$\mathbf{d} = \mathbf{v}_0 - \mathbf{v}_s \ ,$$

where $\mathbf{v}_s$ is a position of the vertex v_s on the base (smoother) shape, and $\mathbf{v}_0$ is the position of the corresponding vertex on the textured surface (see Fig. 4, *right*). Storing this displacement as a vector would require either using three coordinates, or defining and maintaining an explicit two-dimensional local coordinate frame over the surface. Instead we split the displacement into two fields: a pseudo-height and a tilt, both of which can be represented in a coordinate-free way based on discrete differential forms [15], [14].

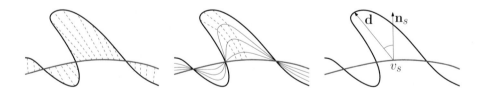

Fig. 4. *Left:* Geometry texture superimposed on objects's base shape in form of vector displacements. The points at the intersections of the textured surface and the base shape have zero displacements. *Middle:* One way of obtaining the base shape in case of non-heightfield texture would be to use the multiresolution hierarchies as in [17] and trace the points through a sufficient number of levels. *Right:* The displacement $\mathbf{d}$ of the vertex v_s can be described in terms of displacement length and the rotation from the normal vector.

The *Pseudo-height* field h represents the signed length of the displacement

$$h = \text{sign}(\mathbf{d} \cdot \mathbf{n}_s)\|\mathbf{d}\| \ ,$$

where $\mathbf{n}_s$ is the normal on the surface at v_s. Our pseudo-height is thus analogous to a typical height field, with values sampled at *vertices* then linearly interpolated across triangles. However we also define a *tilt* field: this is a vector field that defines the tilt (rotation) of the displacement direction with respect to the base normal direction. More precisely, the tilt $\mathbf{t}$ is induced from the displacement $\mathbf{d}$ and the base normal $\mathbf{n}_s$ as

$$\mathbf{t} = \frac{\mathbf{d}}{\|\mathbf{d}\|} \times \mathbf{n}_s \ .$$

Notice that the tilt is a vector in the *tangent space* of the base shape: its direction is the rotation axis for a rotation that transforms the displacement direction into the normal direction and the magnitude of the tilt is the sine of the rotation angle. Therefore, we encode the tilt using the *edge-based* discretization reviewed in Sect. 2. Using the tilt instead of the tangential displacement offers an intuitive description of the texture: the height truly represents the magnitude of the displacement, while the tilt indicates the local rotation of the normal field. We will see that this particular decomposition allows for very simple editing of geometric textures.

In summary, we converted a displacement field into an intrinsic, coordinate-free geometric texture representation

$$\text{texture} = (\text{h}, \mathbf{t}) \ ,$$

consisting of two terms, the pseudo-height h stored as a single scalar per vertex, and the tilt $\mathbf{t}$ stored as a single scalar per edge.

Continuity of height and tilt. Notice that if the condition explained in Fig. 4 is satisfied, our height-and-tilt representation is continuous: the height field vanishes when the textured surface crosses the base shape, while the tilt field approaches the same value on both sides of the surface. However, in practice, one cannot exclude the possibility of having some points that have displacement only in tangential direction, which creates a discontinuity in the height field. To avoid loosing texture information (the "height" of the tangential drift), we use a non-zero sign function in our implementation.

3.3 Texture Reconstruction

Given a base shape and the height-and-tilt texture representation as described above, we can easily reconstruct the textured object. The tilt field $\mathbf{t}$ is calculated first from the edge coefficients, as explained in in Sect. 2. To obtain the direction of the surface displacement, we then simply need to rotate the base shape normal $\mathbf{n}_s$ around the axis $\mathbf{n}_s \times \mathbf{t}$ by the angle α satisfying

$$\sin \alpha = \|\mathbf{t}\|, \qquad \cos \alpha = \text{sign}(h) \sqrt{1 - \|t\|^2} \ .$$

Our height-and-tilt texture can also be transferred from a source shape to a target shape. We need to define a mapping between the two shapes and sample both the height field and the target shape. Typically, such a mapping between two shapes uses a small number of patches as flat as possible [19], and a mapping between each pair of patches is achieved through, for instance, conformal parametrization of small circular patches. Once such a mapping has been established, our pseudo-height field can be copied from source to target through simple resampling (using, e.g., barycentric coordinates). The tilt can also be transferred efficiently: for each of the target edges, we sample the edge at a number of locations (5 in our implementation), evaluate the tilt vector field (as covered in Sect. 2) at these samples from the map we have between the source and the target, and integrate the dot product of the linearly interpolated vector field over the edge.

Fig. 5 and 6 show three examples of transferring a non-heightfield texture patch to the target mesh by the means of simple resampling.

Fig. 5. The texture of the scanned lychee fruit (edited to achieve the whirl effect, *left*) is extracted from the base shape (*middle*) and transferred to the base shape of the avocado fruit (*right*)

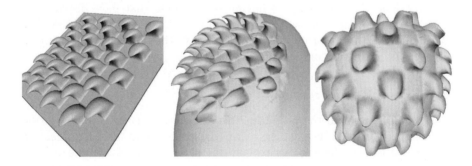

Fig. 6. The synthetical texture (*left*) transferred to the shape of an avocado (*middle*) and to the shape of a lychee fruit (*right*)

4   Applications

We present two types of applications of our height-and-tilt texture representation. For editing and animation, the shape of the object is held constant while the texture on the shape is altered; for deformation and resizing, the shape is deformed and the texture is simply reapplied to it.

4.1   Editing and Animation

Our height-and-tilt texture representation is amenable to a number of simple editing functions. Height and tilt fields can be modified together or separately, which results in new possibilities for geometric texture editing and animation. For instance, we can simulate the effect of spikes swaying on the surface (as if moved by the wind) by changing the texture fields in time. Fig. 7 demonstrates

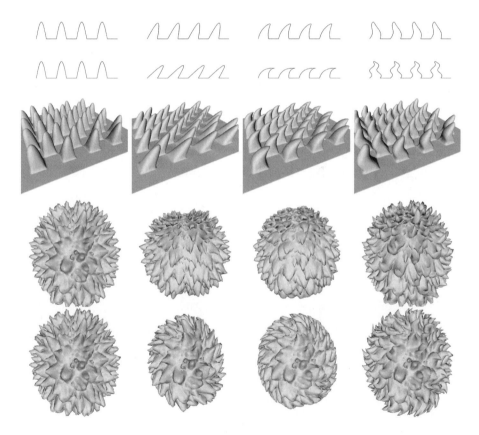

Fig. 7. An example of simple operations on height-and-tilt fields. *Left to right:* A tilt-free texture, set tilt operation, wrap operation and wave operation. *Up to down:* The effect on 2D synthetic texture for two different parameters, on 3D synthetic texture, and on a scan of a lychee fruit for two different directions.

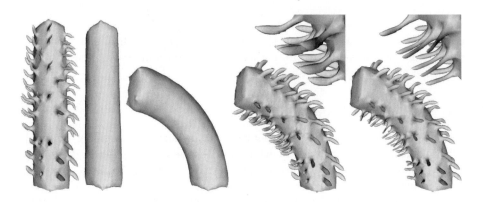

Fig. 8. Our texture representation allows us to extract the texture of the tentacle stick using a (given) base shape. After bending the shape, we can reapply the texture to the shape. On far *right* is the result of applying the space deformation directly to the textured shape. Notice on the enlarged detail that our method does *not* deform texture elements.

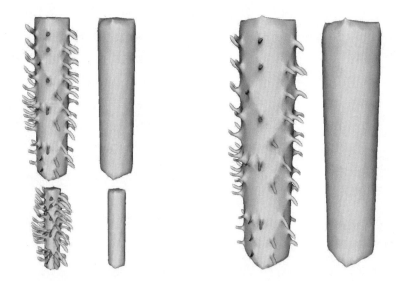

Fig. 9. The original tentacle stick and its (given) shape, *left up*. The shape is then resized (grown by the factor of 1.5 on *right*, shrunk to half size on *left down*) and the texture is put back on it. Due to the texture elements being represented as heights and tilts the size and the shape of the tentacles is not significantly affected by resizing.

a few examples, such as *set tilt*, which fixes the tilt of the texture; *wrap*, which wraps (bends) the texture spikes; and *wiggle*, which creates a wave-like effect on the spikes. While these operations may not be visually relevant on all textures, they are very effective on spiky textures.

4.2   Deformations and Resizing

Combined with base shape deformation, our representation can also handle a wide range of effects. Fig. 8 exhibits some of the benefits of our approach, where a non-purely height field texture is extracted using a given base shape. The base shape is then deformed, and the texture can be added back in a realistic way. However, since our representation is normal-based, it will still exhibit distortion artifacts for severe bending (i.e. large compared to the scale of the texture). The simplicity of our method cannot (and in fact, is not designed to) handle very complex shape deformation that much more costly Laplacian-based editing methods can [20]. Nevertheless, it alleviates the limitations of height field texture methods while keeping their computational efficiency. In an another example shown in Fig. 9 the base shape has been scaled, but the height-and-tilt texture representation preserves the size and shape of the texture elements.

5   Discussion and Conclusion

We presented a height-and-tilt texture representation to efficiently encode and process small-scale geometric textures over fine meshes. As an extension of heightfield-based textures, they share their simplicity (texture editing is achieved only via local computations) and and their intrinsic nature (i.e. they are coordinate-free). Thanks to the added tilt field, a rich spectrum of geometric textures can be stored, edited, animated, as well as transferred between surfaces.

One has to bear in mind some of the present limitations of our method. Firstly, we rely on existing methods to separate texture from geometry. As our notion of texture is richer than the usual height field approach, it is likely that better methods to provide base shapes can be derived. Second, since our representation is normal based, the texture extraction can be sensitive to the smoothness of the base shape. This can be addressed by additional smoothing of the normal field of the base shape prior to texture extraction in our implementation. Additionally, storing the tilt in the tangent field may be, for some applications, inappropriate if the tilt field does not vary smoothly over the surface. To be more robust to non-smoothly varying tilt fields, we utilize the fact that tilt field has maximal magnitude one and constrain the least squares system (2) so that an edge coefficient is not larger than the edge length.

The obvious extension of height-and-tilt texture representation is to synthesize (grow) geometric texture on arbitrary meshes, possibly using the tilt field to control the direction of the growth. Another future endeavor could be to investigate whether we can provide a high fidelity geometric texture with fewer base vertices through field and surface resampling.

Also note the our texture representation is simple enough that a GPU implementation would be fairly easy, allowing for real-time animation of objects displaced with non-heightfield geometric texture or, perhaps more importantly, a system for real time editing of 3D objects with complex geometric texture.

Acknowledgments. This research was partially funded by the NSF grant CCF-0811373.

References

1. Sharf, A., Alexa, M., Cohen-Or, D.: Context-based surface completion. In: SIGGRAPH 2004: ACM SIGGRAPH 2004 Papers, pp. 878–887. ACM, New York (2004)
2. Bhat, P., Ingram, S., Turk, G.: Geometric texture synthesis by example. In: SGP 2004: Proceedings of the 2004 Eurographics/ACM SIGGRAPH symposium on Geometry processing, pp. 41–44. ACM, New York (2004)
3. Zhou, K., Huang, X., Wang, X., Tong, Y., Desbrun, M., Guo, B., Shum, H.Y.: Mesh quilting for geometric texture synthesis. In: SIGGRAPH 2006: ACM SIGGRAPH 2006 Papers, pp. 690–697. ACM, New York (2006)
4. Sorkine, O., Cohen-Or, D., Lipman, Y., Alexa, M., Rössl, C., Seidel, H.-P.: Laplacian surface editing. In: SGP 2004: Proceedings of the Eurographics/ACM SIGGRAPH symposium on Geometry processing, pp. 175–184. ACM, New York (2004)
5. Wang, L., Wang, X., Tong, X., Lin, S., Hu, S., Guo, B., Shum, H.Y.: View-dependent displacement mapping. In: SIGGRAPH 2003: ACM SIGGRAPH 2003 Papers, pp. 334–339. ACM, New York (2003)
6. Wei, L.Y., Levoy, M.: Fast texture synthesis using tree-structured vector quantization. In: SIGGRAPH 2000: Proceedings of the 27th annual conference on Computer graphics and interactive techniques, pp. 479–488. ACM Press/Addison-Wesley Publishing Co., New York (2000)
7. Hertzmann, A., Jacobs, C.E., Oliver, N., Curless, B., Salesin, D.H.: Image analogies. In: SIGGRAPH 2001: Proceedings of the 28th annual conference on Computer graphics and interactive techniques, pp. 327–340. ACM, New York (2001)
8. Ying, L., Hertzmann, A., Biermann, H., Zorin, D.: Texture and shape synthesis on surfaces. In: Proceedings of the 12th Eurographics Workshop on Rendering Techniques, London, UK, pp. 301–312. Springer, Heidelberg (2001)
9. Wei, L.-Y., Levoy, M.: Texture synthesis over arbitrary manifold surfaces. In: SIGGRAPH 2001: Proceedings of the 28th annual conference on Computer graphics and interactive techniques, pp. 355–360. ACM Press, New York (2001)
10. Angelidis, A., McCane, B.: Fur simulation with spring continuum. The Visual Computer: International Journal of Computer Graphics 25, 255–265 (2009)
11. McAnlis, C.: Halo wars: The terrain of next-gen. In: Game Developers Conference (2009)
12. Praun, E., Finkelstein, A., Hoppe, H.: Lapped textures. In: Proceedings of ACM SIGGRAPH 2000, pp. 465–470 (2000)
13. Magda, S., Kriegman, D.: Fast texture synthesis on arbitrary meshes. In: EGRW 2003: Proceedings of the 14th Eurographics workshop on Rendering, Aire-la-Ville, Switzerland, pp. 82–89. Eurographics Association (2003)
14. Fisher, M., Schröder, P., Desbrun, M., Hoppe, H.: Design of tangent vector fields. In: SIGGRAPH 2007: ACM SIGGRAPH 2007 papers, p. 56. ACM, New York (2007)

15. Desbrun, M., Kanso, E., Tong, Y.: Discrete differential forms for computational modeling. In: SIGGRAPH 2006: ACM SIGGRAPH 2006 Courses, pp. 39–54. ACM, New York (2006)
16. Desbrun, M., Meyer, M., Schröder, P., Barr, A.H.: Implicit fairing of irregular meshes using diffusion and curvature flow. In: SIGGRAPH 1999: Proceedings of the 26th annual conference on Computer graphics and interactive techniques, pp. 317–324. ACM Press/Addison-Wesley Publishing Co., New York (1999)
17. Kobbelt, L., Vorsatz, J., Seidel, H.P.: Multiresolution hierarchies on unstructured triangle meshes. Computational Geometry: Theory and Applications 14, 5–24 (1999)
18. Vallet, B., Lévy, B.: Spectral geometry processing with manifold harmonics. In: Computer Graphics Forum (Proceedings Eurographics) (2008)
19. Desbrun, M., Meyer, M., Alliez, P.: Intrinsic parameterizations of surface meshes. In: Eurographics conference proceedings, pp. 209–218 (2002)
20. Botsch, M., Sumner, R.W., Pauly, M., Gross, M.: Deformation transfer for detail-preserving surface editing. In: Vision, Modeling and Visualization 2006, pp. 357–364 (2006)

Using Coplanar Circles to Perform Calibration-Free Planar Scene Analysis under a Perspective View

Yisong Chen

Key Laboratory of Machine Perception (Ministry of Education), Peking University

Abstract. Conics have been used extensively to help perform camera calibration. In this paper, we present a lesser-known result that conics can also be applied to achieve fast and reliable calibration-free scene analysis. We show that the images of the circular points can be identified by solving the intersection of two imaged coplanar circles under projective transformation and thus metric planar rectification can be achieved. The advantage of this approach is that it eliminates the troublesome camera calibration or vanishing line identification step that underlies many previous approaches and makes the computation more direct and efficient. Computation of the vanishing line becomes a by-product of our method which produces a closed form solution by solving the intersection of two ellipses in the perspective view. Different root configurations are inspected to identify the image of the circular points reliably so that 2D Euclidean measurement can be directly made in the perspective view. Compared with other conic based approaches, our algorithm successfully avoids the calibration process and hence is conceptually intuitive and computationally efficient. The experimental results validate the effectiveness and accuracy of the method.

1 Introduction

Conics are widely accepted as one of the most fundamental image features in computer vision due to their elegant mathematical forms and popularities [20]. The study of conics is vital to developing computer vision systems and has been widely employed in computer vision applications to perform camera calibration and pose estimation [14,6,10].

Although recent research based on conics has come up with many fruitful achievements, most work has concentrated on the field of camera calibration while the potential of conics in other vision applications has to a large extent been neglected. In addition, conic based calibration work often involves multiple views, restricted patterns and complicated computations [13]. For example, the work in [17] employs a specially designed pattern with a circles and a pencil of straight lines passing through its center and needs at least three different views to conduct calibration. The work in [15] makes use of planar concentric circles and requires at least two views. The work of [3] uses only a single view but needs complicated computations and iterative optimizations to estimate the direction

G. Bebis et al. (Eds.): ISVC 2009, Part I, LNCS 5875, pp. 668–677, 2009.

and the center of a circle in the perspective view. These constraints make the approaches not particularly attractive to real-world applications.

We point out that it is the process of scene analyses, rather than the estimation of camera parameters, that helps us to understand and to interpret the scene content. Therefore, it is very attractive if we could avoid or by-pass the calibration step and directly go into the work of scene analysis with as few views and computations as possible. Single view metrology [5] and planar rectification [16] are both representative attempts based on this idea. Calibration-free scene analysis represents the most attractive trait of these approaches.

Planar rectification plays a very important role in calibration-free scene analysis. Generally speaking, metric rectification of a perspective view of a world plane is possible once the circular points or the absolute conic is identified on the image plane. This is achieved through either stratified or unstratified planar rectification methods [22,7]. Vanishing line identification is one of the most important steps in planar rectification and there have been many approaches about how to identify the vanishing line on the image plane [2,23,11].

Interestingly, although conics have been extensively used to help perform camera calibration, they have received limited attentions in the field of calibration-free scene analysis. In fact, conics can play as important a role in single view metrology as in camera calibration. An interesting attempt to conduct rectification with the help of conics is made in [4], where the image of the absolute conic (DIAC) is identified by algebraically solving the intersection of the vanishing line and the image of a circle. The advantage of this approach is a rapid and closed form solution for the image of the absolute conic. However, this method still requires the vanishing line to be identified on the image plane.

In this paper, we reap the potential of planar conics in the context of calibration-free scene analysis by exploiting the interaction of multiple coplanar circles. Our approach exhibits the following advantages. First, we use only a single image to achieve good planar scene analysis. Second, we use planar conics to exploit metric properties under 2D homography directly from perspective views and avoid the complicated calibration process. Third, we eliminate the requirement of vanishing line identification that is necessary in most of the related work, and make calibration-free planar scene analysis possible under many difficult cases. Last but not least, our formulation leads to concise closed-form solutions. The simplicity of the underlying mathematics means only minimal knowledge is required to understand our algorithm and put it into practice. Specifically, we propose an alternative planar rectification method that is able to perform quick and reliable planar measures directly from the perspective view with neither camera calibration nor homography parameter estimation. We show that the circular points can actually be identified algebraically by solving the intersection of the images of two circles. The vanishing line and the image of the absolute conic can be calculated later from the two identified imaged circular points.

The major contribution of our work is a simple and elegant algebraic framework to solve the images of the circular points, which eliminates the troublesome but mandatory step of vanishing line identification in many related approaches.

2   Solving the Images of the Circular Point

It is a well known result in projective geometry that every circle on the plane intersects the line at infinity at two fixed complex circular points $I = (1, i, 0)'$ and $J = (1, -i, 0)'$. These two points are fixed under any similarity transformation. The fact that every circle passes through the two circular points implies that the circular points can be identified by solving the intersection of two circles. Although this fact is obvious and can be easily deduced from most projective geometry textbooks, to the best of our knowledge it has been rarely exploited to help perform scene analysis. Note that intersection is an invariant property under projective transformation. Therefore the images of the circular points can be identified on the image plane by identifying the two appropriate intersection points of the images of two world-plane circles, which are in general two ellipses on the image plane.

Suppose that two world-plane circles are mapped to two ellipses on the image plane by 2D non-degenerate homography, which have the forms of equation (1).

$$a_1 x^2 + b_1 xy + c_1 y^2 + d_1 xw + e_1 yw + f_1 w^2 = 0$$
$$a_2 x^2 + b_2 xy + c_2 y^2 + d_2 xw + e_2 yw + f_2 w^2 = 0 \tag{1}$$

The above equations are not as easily solved as the case in [4]. Additionally, unlike in many applications, we are interested only in the complex roots of the equations. Hence the powerful numerical methods searching for real roots are of little help here. Fortunately, such equations have been thoroughly studied in mathematics and there are ready-made algorithms and tools to help solve it [8,1,21]. It is easy to integrate any of the solvers to the above equations into our planar rectification framework. Here we focus on the implications of the solutions and how they can be exploited in calibration-free scene analysis.

After equations (1) are solved we obtain the images of the circular points as in equation (2).

$$I' = (x_0, y_0, 1) = (Re(x_0) + i \cdot Im(x_0), Re(y_0) + i \cdot Im(y_0), 1)$$
$$J' = (\overline{x_0}, \overline{y_0}, 1) = (Re(\overline{x_0}) + i \cdot Im(\overline{x_0}), Re(\overline{y_0}) + i \cdot Im(\overline{y_0}), 1) \tag{2}$$

Since the incidence of the circular points and the line at infinity is invariant under homography, the vanishing line can now be computed by

$$l'_\infty = I' \times J' = (y_0 - \overline{y_0}, \overline{x_0} - x_0, x_0 \overline{y_0} - \overline{x_0} y_0) \tag{3}$$

The image of the absolute conic, ω, can also be computed from I' and J' using the following equation (4) due to the dual relationship between the circular points and the absolute conic.

$$\omega' = I'J' + J'I' \tag{4}$$

We stress that ω is a real symmetric matrix although I' and J' are complex. After ω is identified on the image plane the angle between two world-plane lines l and m can be calculated from their perspective images with equation (5):

$$cos(\theta) = l'\omega'm'/\sqrt{(l'\omega'l')(m'\omega'm')} \qquad (5)$$

Equation (5) is an equivalent expression of the Laguerre formula under 2D homography and can be used to compute the cosine value of any angle between two world-plane lines directly from the image plane without explicitly recovering the world plane [18]. This means that metric rectification is achieved without solving the homography parameters.

Finally, it is worthwhile to make an insightful comparison between our work and that of [3] which is a recent work and closely related to the approach presented here. The two methods look similar in that both approaches work on a single perspective view of two coplanar circles. However, as we have stressed in Section 1, the two approaches are different in their intrinsic motivation. The work in [3] is from the point of view of calibration while ours is from the point of view of calibration-free scene analysis. Although the work in [3] can finally achieve metric rectification, complicated optimization and computations have to be first conducted to estimate the camera parameters and the solution is confronted with the risk of instability in the presence of image noise. By contrast, in our approach we show that it is actually not necessary to perform calibration to achieve metric rectification. Euclidean measurement under projective transformation can be directly reached with much more concise and simple computations. This makes our approach very practical and efficient.

3   Root Configurations

In general there are four pairs of roots to equations (1). So we have to be careful in selecting the appropriate circles and roots because an incautious choice may significantly ruin the final result. We will give some discussions in this section based on several different root configurations of the equations (1).

3.1   Two Intersecting Circles

This is the simplest case. Because the two circles intersect on the world plane, we know immediately that the equations have two pairs of real roots and two pairs of complex roots. In addition, the two pairs of complex roots are conjugate to each other because they are the roots of a real coefficient equations set. Accordingly, the two pairs of complex roots act as the identified images of the circular points.

3.2   Two Non-intersecting, Non-concentric Circles

This case is a little more complicated. The reason is that two such circles on the world plane will have two pairs of complex roots other than the images of the circular points. So a bit of attention should be paid to make sure that no wrong roots are selected. Fortunately this is not a tough task. The incorrect roots can generally be distinguished from the correct ones because they often lead to obviously wrong vanishng line or measure results. Therefore they can be easily removed from the candidates. This will be depicted in more detail in Section 4.

3.3   Two Concentric Circles

As a matter of fact, this is the most popular case in real-world scenes due to the fact that a ring on the world plane naturally defines two concentric circles with its inner edge and outer edge. This looks like an exciting observation because a single ring in the scene seems sufficiently powerful to achieve planar rectification. Unfortunately this is only partially true. In fact, we have to pay special attention here because most degeneracy occurs in this case. We give a brief interpretation in this subsection.

Unlike the aforementioned two cases there are in theory only two pairs of roots to equations (1) for two concentric circles. At first glance this is good news because it seems that no effort is needed to pick up the true images of the circular points from the outliers. Nevertheless this rarely happens in practice. In the presence of noise the error is unavoidably introduced during photo taking and ellipse fitting. As a result we always get four pairs of different roots. Moreover, if the two circles are close in distance then none of the four pairs are actually reliable. This is not surprising because two circles near each other denote a degenerate case and fail to provide adequate information to identify the circular points. Fortunately, as long as the two circles are not very close to each other the degeneracy will NOT occur, even if they are concentric. Therefore, two non-close concentric circles still help to identify the circular points quite well as long as necessary steps are taken to prevent degradation. This is verified by our experiments.

4   Experimental Results

In our experiments, ellipse detection and fitting is conducted by some robust feature extraction and regression algorithms [11,9,19]. One point to stress is that our experiments are done in normalized image coordinates to ensure a steady order of magnitude during the solving process and to achieve a better precision [12].

4.1   Synthetic Scene

In the first experiment, we make a synthetic planar scene with an Olympic logo and several simple geometric shapes on it, as shown in Figure 1(a). The virtue of the Olympic logo is that we can take different circle combinations to test different root configurations addressed in section 3. From figure 1(a) we can see that it is not easy to identify the vanishing line reliably without adequate cues. Nevertheless, the circles in the figure help to avoid this difficulty and make planar rectification possible. The distribution of the roots in the complex plane is intuitively drawn in Figure 2 for five different circle combinations. Figure 2 shows that all but one degenerate case (the inner and the outer edge of the green ring) generate a pair of roots very close in distance (the bold cells in Table 1). This pair of roots is exactly the images of the circular points associated with the corresponding circle combinations.

Fig. 1. (a) The original image. (b) The rectified image.

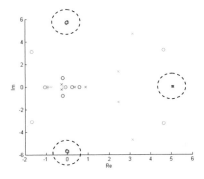

Fig. 2. Distribution of the roots solved by 5 different circle combinations. The "o"s denote the x roots and the "x"s denote the y roots. Different colors denote different circle combinations listed in Table 1. Under all non-degenerate cases (red, blue, cyan, black) the images of the circular points are very close to each other (almost overlapping), which are approximately encircled by 3 dashed circles. In contrast, the roots generated by the degenerate case (green) are all far from the expected positions and none of them is reliable.

Table 1. Circular points computation and selection with different circle combinations

	Red-Green	Yellow-Blue	Red-yellow	Green-innergreen
x1,x2	**-0.0229±5.7100i**	**0.0129±5.7256i**	**-0.0154±5.6756i**	*-1.6811±3.1004i*
y1,y2	**5.0964±0.0183i**	**5.0892±0.0171i**	**5.0587±0.0063i**	*2.4467±1.3476i*
x3,x4	-1.0465,-0.0750	0.5904,0.2470	-0.2024±0.7884i	*4.6277±3.2340i*
y3,y4	-0.9129,-0.8062	0.3617,0.8781	-0.2509±0.2187i	*3.1346±4.6727i*
l'	(-0.003,1.000,-5.096)	(-0.003,1.000,-5.098)	(-0.001,1.000,-5.059)	NA

Table 2. Angle estimating results for the Olympic logo scene (in degrees)

	AD/AB	EG/FG	HK/JK	LM/MN	IJ/LN	BD/AB	JK/MN	AD/FG
Red-green	89.81	57.71	60.34	45.09	0.16	29.96	15.51	75.27
Yellow-blue	89.78	57.63	60.44	44.93	0.11	29.97	15.45	75.26
Red-yellow	89.82	57.70	60.28	45.03	0.18	29.91	15.54	75.23
Red-innergreen	89.54	57.80	60.61	44.90	0.09	29.98	15.51	75.48
Ground truth	**90.00**	**57.62**	**60.25**	**45.00**	**0.00**	**30.25**	**15.25**	**75.00**

After the images of the circular points are recovered the Laguerre formula is employed directly on the image plane to make 2D measurements. As shown in Table 2, in all non-degenerate cases the results are very good. The maximum angle deviation is only about 0.5 degrees. Figure 1-b gives the rectified image for comparison.

4.2   Real World Scene

We use a very challenging plaza scene to test the cases of concentric circles. Although this scene is much noisier compared with the previous one, the distinct magenta color of the two rings in the scene allows robust detection of both edges of the inner ring and the inner edge of the outer ring. The three identified circles are indexed from inner to outer and are highlighted in Figure 3(a), together with several feature lines. All three combinations of the three circles are tested and the results are given in Figure 4 and Table 3. Again except for the degenerate case (circle1-circle2) the images of the circular points are successfully estimated. An interesting observation here is that in the two non-degenerate cases, the two sets of conjugate roots are near each other in the complex number field and both sets can produce reasonable results as the images of the circular points. This result is expected because theoretically these two sets of roots should be identical. The difference is just caused by the impact of the noise. In our experiment, we obtain the final result by averaging the corresponding roots in the two sets and an even better performance is achieved. The rectified image is given in Figure 3(b) and several measure results are listed in Table 4.

The vanishing line can be computed from the images of the circular points by equation (3). The result is also given in Table 3. For comparison we also give the

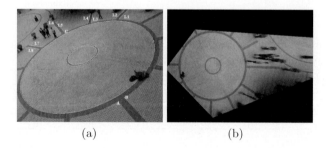

(a)   (b)

Fig. 3. (a) The plaza photo. (b) The rectified plaza image.

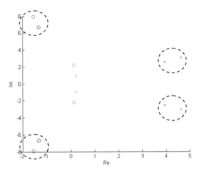

Fig. 4. Distributions of the roots solved by 3 different circle combinations. The "o"s denote the x roots and the "x"s denote the y roots. Different colors denote different circle combinations described in Table 3. Under the two non-degenerate cases (red, blue) the images of the circular points are sufficiently close to each other, which are approximately encircled by 4 dashed circles. In contrast, the roots generated by the degenerate case (green) are far from the expected positions.

Table 3. Circular points computation and selection for the plaza scene

	Circle1-Circle3	Circle2-Circle3	Circle1-Circle2
x1,x2	**-1.5581±7.9239i**	**-1.3254±6.6795i**	*0.1136±2.1698i*
y1,y2	**4.6410±3.0492i**	**3.9511±2.5537i**	*0.2238±0.9403i*
l_∞(estimated)	**(-0.385,1.000,-5.241)**	**(-0.382,1.000,-4.458)**	*(-0.433,1.000,-0.175)*
l_∞(by fitting)	(-0.403,1.000,-5.535)	(-0.403,1.000,-5.535)	(-0.403,1.000,-5.535)

Table 4. Angle estimating results for the plaza scene (in degrees)

	L1/L2	L1/L3	L1/L5	L1/L7	L3/L5	L3/L7	L5/L7
Circle1-circle3	0.124	30.662	61.251	89.178	30.611	59.164	29.549
Circle2-circle3	0.313	30.611	60.901	89.374	30.028	59.645	29.477
Ground truth	**0.00**	**30.00**	**60.00**	**90.00**	**30.00**	**60.00**	**30.00**

vanishing line computed by conventional method with the help of several sets of parallel lines. Table 3 shows that the two approaches give similar results. This validates the feasibility of our approach.

The image of the absolute conic can be computed with the combination of circle3 and either of circle1 and cirlce2. Both results do well in metric rectification. The results in table 4 verify that each neighboring stripe pair forms a 30 degree angle. The largest deviation from the expected value is only about 1.3 degrees. The rectified image in Figure 3-b shows a satisfactory rectification.

There is one point that deserves attention here. To employ equation (5) to make Euclidean measures effectively, we strongly recommend that all straight lines should be fitted through as many points as possible that can be detected from the image. This is due to the fact that statistically more measures will determine the line more accurately and thus generate smaller estimation error [19]. Our experiment reveals that careful line fitting will improve the precision of angle measurement result by as large as 1.0 degrees. Therefore, Line fitting is not an optional but an essential step and should be considered whenever possible.

Finally, it is worth noting that measures can be made directly from the image plane through equation (5) and explicit image rectification is not necessary. The rectified images in Figure 1-b and Figure 3-b are given just for comparison.

All the experiments are conducted through our C++ implementation of the algorithm running on a PC workstation equipped with a 3.0G Pentium4 processor and 1GB of RAM. Feature extraction routine is the dominant time consumer in our experiments. For a photograph of the size 640*480, line and ellipse feature extraction spend around 1.00 seconds. The work of line fitting, ellipse fitting and IAC solving takes less than 0.02 seconds.

5   Conclusion

In this paper, we show that calibration-free planar scene analysis can be performed efficiently by solving the intersections of two world-plane circles on the image plane algebraically. This algorithm achieves metric rectification in the absence of vanishing line and can in turn determine the vanishing line. The idea is implemented by an efficient algorithm for solving the equations set together with a roots selection strategy. Effective length and shape measures can be made directly on the image plane without having to explicitly perform camera calibration, solve the homography parameters, or recover the world plane. Our work exhibits three distinctive advantages: First, It outperforms calibration-mandatory approaches in computational cost in that it is calibration-free. Second, it is more practical than many other calibration-free approaches in that it avoids the vanishing line identification process. Third, it is extremely simple in concept and easy to understand and implement.

Acknowledgement

The work in this paper was supported by Chinese national high-tech R&D program grant (2007AA01Z318, 2007AA01Z159, 2009AA01Z324), national basic research program of China (2004CB719403, 2010CB328002), national natural science foundation of China (60973052, 60703062, 60833067).

References

1. de Almeida Barreto, J.P.: General central projection systems: modeling, calibration and visual servoing. PhD. Thesis, University of Coimara (2003)
2. Caprile, B., Torre, V.: Using vanishing points for camera calibration. International journal of computer vision 4(2), 127–140 (1990)

3. Chen, Q., Hu, H., Wada, T.: Camera calibration with two arbitrary coplanar circles. In: Pajdla, T., Matas, J(G.) (eds.) ECCV 2004. LNCS, vol. 3023, pp. 521–532. Springer, Heidelberg (2004)
4. Chen, Y., Ip, H.: Planar metric rectification by algebraically estimating the image of the absolute conic. In: Proc. ICPR 2004, vol. 4, pp. 88–91 (2004)
5. Criminisi, A., Reid, I., Zisserman, A.: Single View Metrology. International Journal of Computer Vision 40(2), 123–148 (2000)
6. Dhome, M., Lapreste, J., Rives, G., Richetin, M.: Spatial localization of modeled objects of revolution in monocular perspective vision. In: Faugeras, O. (ed.) ECCV 1990. LNCS, vol. 427, pp. 475–485. Springer, Heidelberg (1990)
7. Faugeras, O.: Stratification of 3-D vision: projective, affine, and metric representations. Journal of the Optical Society of America A 12(3), 465–484 (1995)
8. Ferri, M., Managli, F., Viano, G.: Projective pose estimation of linear and quadratic primitive in monocular computer vision. CVGIP: Image understanding 58(1), 66–84 (1993)
9. Fitzgibbon, A., Pilu, M., Fisher, R.: Direct least-square fitting of ellipses. IEEE Trans. PAMI (June 1999)
10. Forsyth, D., et al.: Invariant descriptors for 3-D object recognition and pose. IEEE Trans. PAMI 13(10), 250–262 (1991)
11. Hartley, R., Zisserman, A.: Multiple view geometry in computer vision. Cambridge University, Cambridge (2003)
12. Hartley, R.: In defence of the 8-point algorithm. In: Proc. 5th ICCV, Boston, MA, June 1995, pp. 1064–1070 (1995)
13. Jiang, G., Quan, L.: Detection of concentric circles for camera calibration. In: Proc. ICCV 2005, vol. 1, pp. 333–340 (2005)
14. Kanitani, K., Wu, L.: Interpretation of conics and orthogonality. CVGIP: Image understanding 58, 286–301 (1993)
15. Kim, J., Gurdjos, P., Kweon, I.: Geometric and algebraic constraints of projected concentric circles and their applications to camera calibration. IEEE Trans. PAMI 27(4), 637–642 (2005)
16. Liebowitz, D., Zisserman, A.: Metric Rectification for Perspective Images of Planes. In: Proc. CVPR, June 1998, pp. 482–488 (1998)
17. Meng, X., Hu, Z.: A new easy camera calibration techniques based on circular points. Pattern Recognition 36, 1155–1164 (2003)
18. Mundy, J., Zisserman, A.: Geometric invariance in computer vision. MIT Press, Cambridge (1992)
19. Press, W.H., et al.: Numerical recipes in C: the art of scientific computing, 2nd edn. Cambridge University Press, Cambridge (1997)
20. Semple, J.G., Kneebone, G.T.: Algebraic projective Geometry. Clarendon Press, Oxford (1998)
21. Stahl, S.: Introductory modern algebra: a historical approach. J. Wiley, New York (1997)
22. Sturm, P.F., Maybank, S.J.: Flexible Camera Calibration by Viewing a Plane from Unknown Orientations. In: Proc. CVPR, pp. 432–437 (1999)
23. Zhang, Z.: On plane-based camera calibration: a general algorithm, singularities, applications. In: Proc. ICCV, pp. 666–673 (1999)

Parallel Poisson Surface Reconstruction

Matthew Bolitho[1], Michael Kazhdan[1], Randal Burns[1], and Hugues Hoppe[2]

[1] Department of Computer Science, Johns Hopkins University, USA
[2] Microsoft Research, Microsoft Corporation, USA

Abstract. In this work we describe a parallel implementation of the Poisson Surface Reconstruction algorithm based on multigrid domain decomposition. We compare implementations using different models of data-sharing between processors and show that a parallel implementation with distributed memory provides the best scalability. Using our method, we are able to parallelize the reconstruction of models from one billion data points on twelve processors across three machines, providing a nine-fold speedup in running time without sacrificing reconstruction accuracy.

1   Introduction

New scanning and acquisition technologies are driving a dramatic increase in the size of datasets for surface reconstruction. The Digital Michelangelo project [1] created a repository of ten scanned sculptures with datasets approaching one billion point samples each. New computer vision techniques [2] allow three dimensional point clouds to be extracted from photo collections; with an abundance of photographs of the same scene available through online photo sites, the potential for truly massive datasets is within reach. Processing such large datasets can require thousands of hours of compute time. Recent trends in microprocessor evolution show a movement toward parallel architectures: Multi-core processors are now commonplace among commodity computers, and highly parallel graphics hardware provides even higher performance per watt. Traditional single threaded algorithms will no longer benefit from Moore's law, introducing a new age in computer science in which efficient parallel implementations are required.

This paper presents an efficient, scalable, parallel implementation of the Poisson Surface Reconstruction algorithm [3]. The system is designed to run on multi-processor computer systems with distributed memory, allowing the reconstruction of some of the largest available datasets in significantly less time than previously possible. We begin our discussion with a brief review of both serial and parallel surface reconstruction algorithms in Section 2. We then provide a more in-depth review of the Poisson Surface Reconstruction algorithm on which our work is based, presenting a review of the original implementation in Section 3, and its adaptation to a streaming implementation in Section 4. We describe our parallel reconstruction algorithm in Section 5 and evaluate its effectiveness in terms of both accuracy and efficiency in Section 6. Finally, we conclude by summarizing our work in Section 7.

G. Bebis et al. (Eds.): ISVC 2009, Part I, LNCS 5875, pp. 678–689, 2009.

2   Related Work

Surface reconstruction has been a well studied problem within the field of Computer Graphics. The work can be roughly divided into two categories: Computational geometry based methods; and function fitting methods.

Computational Geometry: Computational geometry based methods use geometric structures such as the Delaunay triangulation, alpha shapes or the Voronoi diagram [4,5,6,7,8,9] to partition space based on the input samples. Regions of space are then classified as either 'inside' or 'outside' the object being reconstructed and the surface is extracted as the boundary between interior and exterior regions. As a consequence of using these types of structures, the reconstructed surface interpolates most or all of the input samples. When noise is present in the data, the resulting surface is often jagged and must be refit to the samples [7] or smoothed [5] in post-processing.

Function Fitting: The function fitting approaches construct an implicit function from which the surface can be extracted as a level set. These methods can be broadly classified as global or local approaches.

Global fitting methods commonly define the implicit function as the sum of radial basis functions centered at each of the input samples [10,11,12]. However, the ideal basis functions, poly-harmonics, are globally supported and non-decaying, so the solution matrix is dense and ill-conditioned. In practice such solutions are hard to compute for large datasets.

Local fitting methods consider subsets of nearby points at a time. A simple scheme is to estimate tangent planes and define the implicit function as the signed distance to the tangent plane of the closest point [13]. Signed distance can also be accumulated into a volumetric grid [14]. For function continuity, the influence of several nearby points can be blended together, for instance using moving least squares [15,16]. A different approach is to form point neighborhoods by adaptively subdividing space, for example with an octree. Blending is possible over an octree structure using a multilevel partition of unity, and the type of local implicit patch within each octree node can be selected heuristically [17]. Since local fitting methods only consider a small subset of the input points at a time, the solutions are more efficient to compute and handle large datasets with greater ease than global methods. The greatest challenge for local fitting methods is how to choose the subset of points to consider at any given point in space. These heuristic partitioning and blending choices make local fitting methods less resilient to noise and non-uniformity in the input samples.

Parallel Surface Reconstruction. Despite the increasing presence of commodity parallel computing systems, there has been comparatively little work on parallel surface reconstruction. The work of [18] implements the Poisson method on the GPU, achieving significant speedups for small datasets. A significant limitation of the implementation is that it requires the entire octree, dataset and supplemental lookup tables to reside in GPU memory, limiting the maximum size of reconstructions possible. To simplify the lookup of neighbor nodes in the octree and reduce the total number of node computations required, the implementation also uses first-order elements. While this allows a more GPU-friendly implementation, the lower-degree functions make the method more susceptible to noise and other anomalies in the input data.

Some other surface reconstruction algorithms lend themselves to efficient parallel implementations. Many local implicit function fitting methods can be at least partially parallelized by virtue of the locality of most data dependencies. Global implicit function fitting methods often have complex data dependencies that inhibit parallelism. Finally, computational geometry approaches can leverage parallel processing by computing structures such as the Delaunay triangulation in parallel (e.g. [19]).

3   Poisson Surface Reconstruction

The Poisson Surface Reconstruction method [3] uses a function fitting approach that combines benefits from both global and local fitting schemes. It is global and therefore does not involve heuristic decisions for forming local neighborhoods, selecting surface patch types, and choosing blend weights. Yet, the basis functions are associated with the ambient space rather than the data points, are locally supported, and have a simple hierarchical structure that allows the resulting linear system to be solved efficiently.

The Poisson Idea: To solve the surface reconstruction problem, the Poisson method reconstructs the indicator function, a function that has value one inside the surface and zero outside. The key idea is to leverage the fact that an oriented point set can be thought of as a sampling of the gradient of the indicator function. This intuition is formalized by using the discrete point set to define a continuous vector field V representing the gradient field of the (smoothed) indicator function. Solving for the indicator function χ then amounts to finding the scalar function whose gradient best matches V, a variational problem that is optimized by solving the Poisson equation: $\Delta\chi = \nabla \cdot V$.

Function Representation: Since the indicator function (and therefore its gradient) only contains high-frequency information around the surface of the solid, an adaptive, multi-resolution basis is used to represent the solution. Specifically, an octree $\mathcal{O}$ is adapted to the point samples and then a function space is defined by associating a tri-variate B-spline F_o to each octree node $o \in \mathcal{O}$. The B-spline F_o is translated to the node center and scaled by the size of the node, and the span $\mathcal{F}$ of the translated and scaled B-splines defines the function-space over which the Poisson equation is solved.

Solving the Poisson Equation: To solve the Poisson equation, a finite-elements approach is used, with the system discretized by using the elements F_o as test functions. That is, the system is solved by finding the function χ in $\mathcal{F}$ such that: $\langle \Delta\chi, F_o \rangle = \langle \nabla \cdot V, F_o \rangle$ for all $o \in \mathcal{O}$.

4   Streaming Implementation

To enable the reconstruction of models larger than memory, the algorithm was adapted to operate out-of-core, using a streaming implementation to maintain only a small subset of the data into memory at any given time [20]. The key observation in performing Poisson Surface Reconstruction in an out-of-core manner is that the computations required for each step of the process are local, due to the compact support of the basis functions F_o. Specifically, each step can be described by evaluating a *node function* N across all nodes in $\mathcal{O}$, where the result of $N(o)$ for each $o \in \mathcal{O}$ is dependent only on the nodes $o' \in \mathcal{O}$ where $\langle F_o, F_{o'} \rangle \neq 0$ (i.e whose base functions overlap). Spatially, this corresponds to a small neighborhood immediately surrounding o and all its ancestors in $\mathcal{O}$.

In one dimension, the most efficient streaming order is a simple left to right traversal of the tree nodes: in this configuration, neighborhoods are always compact and contiguous in the data streams. This ordering is generalized to higher dimensions by grouping nodes into 2D slices (i.e. all nodes at a certain depth with the same x-coordinate), and streaming the data on a slice-by-slice basis. Since a node must remain in main memory from the first to the last time it is referenced by other nodes, and since coarse levels of the tree have longer lifetimes than finer levels of the tree, a multi-level streaming approach is used.

5   Parallel Surface Reconstruction

When designing the streaming implementation, one of the primary concerns was minimizing the effect of the I/O required for out-of-core processing. In particular, this motivated the streaming approach (since streaming I/O is highly efficient) and the minimization of the number of passes required through the data (minimizing the total amount of I/O performed). When considering a parallel implementation, a different set of design concerns prevail: minimizing data sharing and synchronization.

5.1   Shared Memory Implementation

In this section, we consider a simple, shared-memory parallelization of the reconstruction algorithm. Although this is not the model we use for our final implementation, analyzing the shared-memory implementation provides valuable insight regarding the key properties a parallel solver must satisfy to demonstrate good speed-up when parallelized across numerous processors.

The most straightfoward parallelization of the serial streaming implementation is to evaluate each node function N_i in parallel, since the restrictions placed on data dependencies for efficient streaming allow the function to be evaluated in any order within a slice. With slices in large reconstructions typically containing tens or hundreds of thousands of nodes, there appears to be ample exploitable parallelism. A data decomposition approach can be used, with each processing slice in the octree partitioned into a coarse regular grid. Since the data dependencies of a node function are compact, the only shared data between partitions within the same level of the tree resides around the perimeter of each partition. To execute a node function across a slice, each processor is assigned a number of data partitions for processing in a way that best balances the workload. Although this approach provides a straightforward implementation, it has two significant scalability issues.

First, for distributive functions, data are shared not only within a level of the tree, but across all depths of the tree. At the finest levels, the contention for shared data is very low – since a very small portion of each partition is shared, the probability of two processors needing concurrent access to a datum are low. At the coarser levels of the tree, however, the rate of contention becomes very high – and the data associated with the coarsest levels of the tree are updated by each processor for every computation. We found that this problem persisted even when we used an optimistic, lock-free technique that implemented an atomic floating-point accumulation (i.e. A+ = x) using the compare-and-set (CAS) instruction found in most modern processor architectures.

Second, scalability is limited by the frequency of global synchonization barriers required to evaluate multiple functions correctly. Each streaming pass, P, across the octree is a pipeline of functions $P = \{N_1, N_2, ..., N_n\}$ that are executed in sequence. Although the data dependencies are such that the evaluation of $N_i(o)$ cannot depend on the result of $N_i(o')$, it is possible that N_i may depend on N_j if $i > j$. The implication of this in a parallel setting is that function N_i cannot be evaluated for a particular slice until N_j has completed processing in dependent slices *on all processors*, requiring a global synchronization barrier between *each function*, for *each processing slice*. For all but the very largest reconstructions, the resulting overhead is prohibitive. Although this synchronization frequency can be reduced by processing functions over slabs of data formed from multiple octree slices, the associated increase in in-core memory usage results in an undesirable practical limitation on the reconstruction resolution.

5.2   Distributed Memory Implementation

To address the scalability issues that arise from using a shared memory, multi-threaded architecture, we instead implement our solver using a distributed memory model. In this model, each processor shares data explicitly through message passing, rather than implicitly through shared memory. The advantages of this model over the shared memory approach are as follows:

1. Each processor maintains a private copy of all data it needs. Thus, data writes during computation can be performed without the need for synchronization. Data modified on more than one processor can be easily and efficiently reconciled at the end of each computation pass.
2. Without the need for shared memory space, the system can be run on computing clusters, offering the potential for greater scalability, due to the increased memory and I/O bandwidth, as well as number of processors.

Furthermore, we adapt the streaming implementation by implementing each of the functions as a separate streaming pass through the data. While this increases

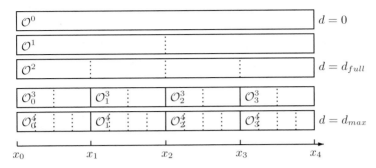

Fig. 1. An illustration of the way data partitions are formed from the tree with $p = 4$ processors. All nodes in $\mathcal{O}^0$, $\mathcal{O}^1$ and $\mathcal{O}^2$ are shared amoungst all processors, and together form the data partition $\mathcal{O}^{full}$. The nodes in remaining depths are split into spatial regions defined by the x-coordinates $\{x_0, x_1, x_2, x_3, x_4\}$, to form the partitions $\mathcal{O}_i^d$.

the amount of I/O performed, it alleviates the need for global, inter-slice, synchronization barriers that are required to allow multiple functions to be evaluated correctly.

Data Partitioning. Instead of fine-grained, slice-level parallelism, the distributed system uses a coarse-grained approach: given p processors, the reconstruction domain is partitioned into p slices along the x-axis given by the x-coordinates $X = \{x_0, x_1, x_2, ..., x_p\}$. The nodes from depth d in the octree are split into partitions $\mathcal{O}^d = \{\mathcal{O}_1^d, \mathcal{O}_2^d, ..., \mathcal{O}_p^d\}$ where $\mathcal{O}_p^d$ are all nodes $o \in \mathcal{O}$ such that $x_p \leq o.x < x_{p+1}$ and $o.d = d$.

Since the coarse nodes in the tree are frequently shared across all processors, we designate the first d_{full} levels in the tree to be part of its own data partition $\mathcal{O}^{full}$, which is not owned by a particular process, and whose processing is carried out *in duplicate* by all processors. Since the total data size of $\mathcal{O}^{full}$ is small, the added expense of duplicating this computation is significantly less than the cost of managing consistent replication of the data.

Figure 1 summarizes the decomposition of the octree into partitions. A processor P_i is assigned to own and process the nodes in $\mathcal{O}_i^*$ in a streaming manner. To allow for data sharing across slabs, processor i has a copy of data in partitions $\mathcal{O}_{i-1}^*$ and $\mathcal{O}_{i+1}^*$ from the result of the previous pass through the data, as well its own copy of $\mathcal{O}^{full}$. Since only a very small portion of data in $\mathcal{O}_{i-1}^*$ and $\mathcal{O}_{i+1}^*$ are ever read or written from P_i (only the data in slices immediated adjacent to $\mathcal{O}_i^*$), the neighboring data partitions are sparsely populated minimizing the amount of redundant storage required.

Since each function is implemented in a separate streaming pass, the execution of a function N_i in one data partition can no longer depend on the execution of a function N_j in another partition, and a global synchronization is only required between the different streaming passes. In practice, we have found that the arithmetic density of most functions is such that the I/O bandwidth required to perform a streaming pass is more than an order of magnitude less that the bandwidth that modern disk drives can deliver, so processing only a single function per pass does not noticeably affect performance.

Load Balancing. Because the octree is an adaptive structure, its nodes are non-uniformly distributed over space. This presents a challenge when choosing the partition bounds X in order to most optimally allocate work across all processors. To minimize workload skew, each partition $\mathcal{O}_i^d$ should be approximately the same size (assuming that the processing time of each node is, on average, constant).

Because we wish to perform the allocation of nodes to partitions before the tree has been created, we use the input point-set to estimate the density of nodes in the tree. Since an octree node may not straddle two data partitions, the partition bounds X must be chosen such that each x_i is a multiple of $2^{-d_{full}}$ (i.e. the width of the coarsest nodes in the high-resolution tree). We use a simple greedy algorithm to allocate X: Given an ideal partition size of $N_{ideal} = \frac{N}{p}$ we grow a partition starting at $x = 0$ until the partition size would exceed N_{ideal}. We then over-allocate or under-allocate the partition depending on which minimizes $|N_i - N_{ideal}|$. The procedure is continued along the x-axis until all partition sizes have been determined.

Replication and Merging of Shared Data. Once data have been modified by a processor, the changes need to be reconciled and replicated between processors. A majority of the shared updates performed by the reconstructor are of the form $o.v = o.v + v$; that is, accumulating some floating point scalar or vector quantity into tree nodes. The merge process for a process P_i is as follows:

1. If P_i has written to $\mathcal{O}_{i-1}$ and $\mathcal{O}_{i+1}$, send data to P_{i-1} and P_{i+1} respectively.
2. If P_{i-1} and P_{i+1} have modified data in $\mathcal{O}_i$, wait for all data to be received.
3. Merge the received data blocks with the data in $\mathcal{O}_i$ (an efficient vector addition operation).

Once data has been reconciled, the updated data can then be redistributed to other processes as follows:

1. If P_i has been updated and is needed by P_{i-1} or P_{i+1} in the next pass, send $\mathcal{O}_i$ to the neighboring processors.
2. If P_{i-1} and P_{i+1} have modified data needed for the next pass, wait for all updated data blocks.

Because each processor streams through the data partitions, changes made to data can be sent asynchronously to other processing nodes as each block in the stream is finalized, rather than after the pass is complete, thereby hiding the latency involved in most message passing operations.

In addition to the accumulation-based data reconciliation, there are two important steps in the reconstruction process that cannot be merged and replicated as efficiently.

Tree Construction. To maximize the parallel processing capability of our system, the construction of the octree itself is performed in parallel. The input point-set P is partitioned during pre-processing into segments $\mathcal{P} = \{\mathcal{P}_1, ..., \mathcal{P}_p\}$ where $\mathcal{P}_i$ contains all points $x_i \leq p.x < x_{i+1}$ (where x_i is the partitioning bounds separating the domain of process P_{i-1} from process P_i).

The first challenge presented in the construction of the tree is the different topological structure created in $\mathcal{O}^{full}$ by each processor. To facilitate efficient merging of data in later steps, it is desirable to have a consistent coarse resolution tree. Although it is possible to merge each of the coarse resolution trees after the first pass, we take a simpler approach: because the coarse resolution tree is small, we pre-construct it as a fully refined octree of depth d_{full}.

The second challenge is that in the initial phases of the reconstruction, a point in partition P_i may affect the creation of nodes outside of $\mathcal{O}_i$ (since the B-splines are supported within the 1-ring of a node). Although this problem could be resolved by allowing processors to generate nodes outside their partition and then merging the nodes at the end of the streaming pass, we have opted for a simpler solution. Recognizing that the points that can create nodes and update data in O_i are in the bounds $x_i - \delta_x \leq p.x < x_{i+1} + \delta_x$, (where $\delta_x = 2^{-d_{full}}$ is the width of the finest-level nodes in the full octree $\mathcal{O}^{full}$) we have processor P_i process this extended subset of points and only perform the associated updates of nodes in $\mathcal{O}_i$. In practice, this adds a small computational cost by processing overlapping point data partitions, but greatly simplifies the creation of the tree.

Solving the Laplacian. To solve the Poisson equation correctly in a parallel setting, we use an approach inspired by domain decomposition methods [21]. In the serial implementation, the linear system is solved in a streaming manner using a block Gauss-Siedel solver, making a single pass through the data. Although we can still leverage this technique within each data partition, the regions of the linear system that fall near the boundaries need special treatment to ensure that the solution across partitions is consistent and correct. To avoid the need for the solver in $\mathcal{O}_i$ to depend on a solution being computed in $\mathcal{O}_{i-1}$, each processor P_i solves a linear system that extends beyond the bounds of $\mathcal{O}_i$ by a small region of padding, and once solutions have been computed by all processors, the solution coefficents in overlapping regions are linearly blended together to give a solution which is consistent across partition boundaries.

6   Results

To evaluate our method, we designed two types of experiments. In the first, we validate the equivalence of our parallel implementation to the serial one, demonstrating that correctness is not sacrificed in the process of parallelization. In the second, we evaluate the scalability of our parallel implementation.

Correctness. We wish to ensure that the surface generated by the parallel implementation is equivalent in quality to the serial implementation. In particular, we want to ensure that the model doesn't significantly change as the number of processors increases, and that any differences that do exist do not accumulate near partition boundaries. To test this, we ran an experiment using the distributed implementation, reconstructing the Stanford Bunny model at depth $d = 9$ using 1, 2, 4, and 8 processors. We then compared the model generated with only one processor M_{serial}, to the models generated with multiple processors M_i by computing an error value δ at each vertex of M_i that is the Euclidean distance to the nearest point on the triangle mesh of M_{serial}. The units of δ are scaled to represent the resolution of the reconstruction so that $1.0\delta = 2^{-d}$ (the width of the finest nodes in the tree).

The table in Figure 2 presents the results of this experiment. Some differences in the output are expected between different numbers of processors because of the lack of commutativity of floating point arithmetic. The results show that in all cases, the average error is low, and the maximum error is bounded within the size of the finest tree nodes. It also shows that error does not change significantly as the number of processors increases. The image in Figure 2 shows the distribution of error across the mesh for $p = 8$, and is typical of all multiple processor results. The image highlights that error is evenly distributed across the mesh, and that the only significant error occurs along the shape crease along the bottom of the bunny's back leg. These errors are the result of a different choice in triangulation along the edge.

Scalability. One of the most desirable properties of a parallel algorithm is scalability, the ability of an algorithm to run efficiently as the number of processors increases. Table 1 shows the running times and Figure 3 shows the speedup of both the shared memory and distributed memory implementations on up to 12 processors when reconstructing the Lucy dataset from 94 million points, and the David dataset from 1 billion points. The shared memory implementations were run on a dual quad core workstation. The distributed memory implementation was run on a three machine cluster with quad core processors and a gigabit

Procs.	Verts.	Tris.	Max δ	Average δ
1	320,944	641,884	-	-
2	321,309	641,892	0.73	0.09
4	321,286	641,903	0.44	0.06
8	321,330	641,894	0.98	0.12

Fig. 2. An analysis of correctness: A comparison of several different reconstructions of the Bunny dataset at depth $d = 9$ created with the distributed implementation. The table summarizes the size of each model, and the maximum and the average distance of each vertex from the ground-truth. The image on the right shows the distribution of error across the $p = 8$ model. The color is used to show δ values over the surface with $\delta = 0.0$ colored blue and $\delta = 1.0$ colored red.

Table 1. The running time (in minutes), aggregate disk use (in MB), and peak memory use (in MB) of the shared memory and distributed memory implementations of the Parallel Poisson Surface Reconstruction algorithms for the Lucy dataset at depth $d = 12, d_{full} = 6$ and the David dataset at depth $d = 14, d_{full} = 8$, running on one through twelve processors. It was not possible to run the shared memory implementation on more than eight processors.

| | Shared Memory | | Distributed Memory | | | | | |
| | Lucy | | Lucy | | | David | | |
Processors	Lock Time	Lock-Free Time	Time	Disk	Memory	Time	Disk	Memory
1	183	164	149	5,310	163	1,970	78,433	894
2	118	102	78	5,329	163	985	79,603	888
4	101	68	38	5,368	164	505	81,947	901
6	102	61	26	5,406	162	340	84,274	889
8	103	58	20	5,441	166	259	86,658	903
10	-	-	18	5,481	163	229	88,997	893
12	-	-	17	5,522	164	221	91,395	897

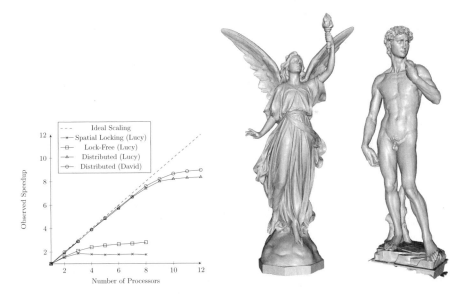

Fig. 3. Analysis of scalability: The speedup of three different parallel Poisson Surface Reconstruction algorithms for the Lucy dataset at depth $d = 12$ and the David dataset at depth $d = 14$ running on one through twelve processors. The Spatial Locking and Lock-Free methods use a shared memory based implementation with two different locking techniques to resolve shared data dependencies. The distributed method uses data replication and message passing to resolve shared data dependencies.

ethernet interconnect. Two variations of the shared memory implementation are examined: one which uses fine-grained spatial locking to manage concurrent updates, and the other using the lock-free update procedure. The lock-free technique is faster and offers greater scalability than the spatial locking scheme, but the scalability is still limited when compared to the distributed implementation. One significant factor affecting the performance is the way in which both the spatial locking and lock free techniques interact with architectural elements of the underlying hardware. When locking shared data between processors, data that was kept primarily in fast on-chip memory caches has to be flushed and shared through main memory each time it is modified to keep separate caches coherent. This forces frequently shared data to be extremely inefficient to access, with no cache to hide high latency memory access. Because the distributed implementation doesn't need to coordinate writes to the same data, the computation is far more efficient, and cleanly scales with increasing numbers of processors. The reduced scalability as the number of processors increases is due to the complete occupancy of all processors on each machine, causing the algorithm to become memory bandwidth bound. Table 1 also lists the peak in-core memory use and aggregate disk use of the distributed algorithm. Since the in-core memory use is related to the size of the largest slices and each data partition is streamed independently, peak memory use is consistent across all degrees of parallelism. Because of the replication of across processors, the disk use grows as the number of processors increases. A majority of the extra data storage is from $\mathcal{O}^{full}$, whose size grows as d_{full} is increased. For the Lucy model, with $d_{full} = 6$, the size of $\mathcal{O}^{full}$ is 18MB, whereas for the David model, with $d_{full} = 8$, it is 1160MB.

7   Conclusion

We have presented an implementation of the Poisson Surface Reconstruction algorithm that is specifically designed for parallel computing architectures using distributed memory. We have demonstrated both its equivalence to the serial implementation and efficient execution on commodity computing clusters with a nine-fold speedup in running time on twelve processors. One avenue we intend to persue in future work is support for parallel processing on a GPU-based cluster.

Acknowledgements

We would like to acknowledge the Stanford 3D Scanning Repository for generously distributing their data. The authors would also like to express particular thanks to Szymon Rusinkiewicz and Benedict Brown for providing non-rigid body aligned Lucy and David scan data [22]. This work was supported in part by an NSF CAREER Award. The authors would also like to thank NVIDIA Corporation for support.

References

1. Levoy, M., Pulli, K., Curless, B., Rusinkiewicz, S., Koller, D., Pereira, L., Ginzton, M., Anderson, S., Davis, J., Ginsberg, J., Shade, J., Fulk, D.: The Digital Michelangelo project: 3D scanning of large statues. In: SIGGRAPH 2000 (2000)
2. Snavely, N., Seitz, S.M., Szeliski, R.: Modeling the world from Internet photo collections. Int. J. Comput. Vision 80, 189–210 (2008)
3. Kazhdan, M., Bolitho, M., Hoppe, H.: Poisson surface reconstruction. In: SGP, pp. 61–70 (2006)
4. Boissonnat, J.D.: Geometric structures for three-dimensional shape representation. ACM TOG 3, 266–286 (1984)
5. Kolluri, R., Shewchuk, J.R., O'Brien, J.F.: Spectral surface reconstruction from noisy point clouds. In: SGP, pp. 11–21 (2004)
6. Edelsbrunner, H., Mücke, E.P.: Three-dimensional alpha shapes. ACM TOG 13, 43–72 (1994)
7. Bajaj, C.L., Bernardini, F., Xu, G.: Automatic reconstruction of surfaces and scalar fields from 3d scans. In: SIGGRAPH, pp. 109–118 (1995)
8. Amenta, N., Bern, M., Kamvysselis, M.: A new voronoi-based surface reconstruction algorithm. In: SIGGRAPH, pp. 415–421 (1998)
9. Amenta, N., Choi, S., Kolluri, R.K.: The power crust, unions of balls, and the medial axis transform. Comp. Geometry 19, 127–153 (2000)
10. Muraki, S.: Volumetric shape description of range data using "blobby model". In: SIGGRAPH, pp. 227–235 (1991)
11. Carr, J.C., Beatson, R.K., Cherrie, J.B., Mitchell, T.J., Fright, W.R., McCallum, B.C., Evans, T.R.: Reconstruction and representation of 3d objects with radial basis functions. In: SIGGRAPH, pp. 67–76 (2001)
12. Turk, G., O'brien, J.F.: Modelling with implicit surfaces that interpolate. ACM TOG 21, 855–873 (2002)
13. Hoppe, H., DeRose, T., Duchamp, T., McDonald, J., Stuetzle, W.: Mesh optimization. In: SIGGRAPH, pp. 19–26 (1993)
14. Curless, B., Levoy, M.: A volumetric method for building complex models from range images. In: SIGGRAPH, pp. 303–312 (1996)
15. Alexa, M., Behr, J., Cohen-Or, D., Fleishman, S., Levin, D., Silva, C.T.: Point set surfaces. In: IEEE VIS, pp. 21–28 (2001)
16. Shen, C., O'Brien, J.F., Shewchuk, J.R.: Interpolating and approximating implicit surfaces from polygon soup. In: SIGGRAPH, pp. 896–904 (2004)
17. Ohtake, Y., Belyaev, A., Alexa, M., Turk, G., Seidel, H.P.: Multi-level partition of unity implicits. ACM TOG 22, 463–470 (2003)
18. Zhou, K., Gong, M., Huang, X., Guo, B.: Highly parallel surface reconstruction. Technical Report 53, Microsoft Research (2008)
19. Hardwick, J.C.: Implementation and evaluation of an efficient parallel delaunay triangulation algorithm. In: Proceedings of the 9th Annual ACM Symposium on Parallel Algorithms and Architectures, pp. 23–25 (1997)
20. Bolitho, M., Kazhdan, M., Burns, R., Hoppe, H.: Multilevel streaming for out-of-core surface reconstruction. In: SGP, pp. 69–78 (2007)
21. Smith, B., Bjorstad, P., Gropp, W.: Domain Decomposition: Parallel Multilevel Methods for Elliptic Partial Differential Equations. Cambridge University Press, Cambridge
22. Brown, B.J., Rusinkiewicz, S.: Global non-rigid alignment of 3-D scans. ACM TOG 26 (2007)

3D Object Mapping by Integrating Stereo SLAM and Object Segmentation Using Edge Points

Masahiro Tomono

Chiba Institute of Technology,
Narashino, Chiba 275-0016, Japan
tomono@furo.org

Abstract. This paper presents a method of 3D object mapping using a binocular stereo camera. The method employs edge points as map element to represent detailed shape and applies a variant of ICP algorithm to 3D mapping. A SIFT descriptor is attached to each edge point for object recognition and segmentation. The 3D map is segmented into individual objects using training images of target objects with different backgrounds based on SIFT-based 2D and 3D matching. In experiments, a detailed 3D map was built by the stereo SLAM and a 3D object map was created through the segmentation of the 3D map into 36 object instances.

1 Introduction

Environment recognition is essential for the robot to perform tasks. Robotic mapping including SLAM (Simultaneous Localization and Mapping) is a key technology for this purpose, and have been studied intensively for decades. Most of the maps created by the conventional mapping schemes are composed of grid cells or geometric elements such as points and polygons for the purpose of representing free space and landmarks for robot navigation. However, more structured maps are necessary in order for the robot to perform high-level tasks such as object manipulation and carrying. Typically, it is desirable that the map describes the shape and pose (location and orientation) of objects relevant to the tasks. If the map are composed of individual objects and have the features to recognize them, the robot can recognize objects in the environment and move to the target object to handle it. We refer to such a map as *object map*.

In this paper, we propose a method of 3D object mapping with a binocular stereo camera. First, we build a 3D map from stereo images based on stereo SLAM. Then, we segment the map into individual objects using training images. The training images have only to contain the objects to be modeled, and need no annotations except object name and object existing region. The 3D map built by the stereo SLAM have the whole shape of the environment, and the object map created from the 3D map contains not only objects but also unstructured regions which can be used in path planning and obstacle detection.

For stereo SLAM, we employ the method proposed by [21]. The method utilizes edge points as map element to represent objects because detailed object

G. Bebis et al. (Eds.): ISVC 2009, Part I, LNCS 5875, pp. 690–699, 2009.

shape is crucial for robotic applications. Edge points can represent more detailed object shape than corner points, which are used for most vision-based SLAM schemes. Moreover, edge points are suitable for recognition of non-textured objects, in which corner points can not be detected sufficiently. The stereo SLAM scheme computes 3D points from the edge points detected by the Canny detector [3] in a stereo image pair, and then estimates the camera motion by matching the consecutive stereo image with the 3D points. The ICP (Iterative Closest Points) algorithm [2] is applied to the registration process.

The object segmentation scheme is based on the co-occurrence of the edge points in the 3D map and training images [20]. Our scheme segments the 3D map into individual objects by extracting the edge points which are contained in both the 3D map and training images. The training images are stored with local descriptors in an image database. The system retrieves a training image that matches with a query stereo image from the image database using local descriptors attached on edge points. We employ a variant of the SIFT (Scale Invariant Feature Transform) descriptor [10] for edge point. Then, we find the 3D points which match with edge points in the training image by 3D-2D matching. The extracted 3D and 2D points from the 3D map are expected to be used as object model for object recognition from a monocular image and task planning using knowledge attached on the target object.

Fig. 1 shows the block diagram of the proposed method. The inputs to the system are stereo images and training images. The stereo images can also be used as training image if there are multiple instances of the target object. Note we need no motion sensors, but some motion sensors could improve the performance of the stereo SLAM. The outputs of the stereo SLAM are a set of 3D points and a camera motion. The outputs of the object segmentation are an object map and object models. The object map consists of 3D points with an object label and 2D points with a SIFT descriptor. An object model is a subset of the 3D and 2D points in the 3D map.

The main contribution of this paper is the integration of stereo SLAM and object segmentation to build a 3D object map and object models. An advantage is that the system provides detailed object shape represented by 3D edge points. Another advantage is the method builds an object map in very weak supervised manner in the sense that no detail annotations for training images are necessary. Our method is based on specific-object recognition technique not category recognition. Category recognition is a powerful scheme, but most of the proposed schemes require supervised training, in which a number of ground truth data are given by human. Our method does not needs such training phase, but just collects training images which contain target objects.

The rest of this paper is organized as follows. Section 2 presents the related work. Section 3 explains 3D mapping by a stereo camera, and Section 4 presents object mapping and object segmentation. Section 5 gives experimental results followed by conclusions.

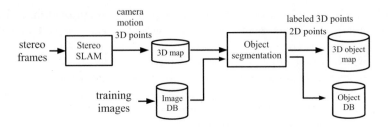

Fig. 1. Block diagram of the proposed method

2   Related Work

In the early stages of object mapping, 3D object maps are built manually using CAD models or simple elements such as lines and polygons. In the last decade, semi-automatic object modeling and object recognition have been utilized for object mapping. 3D object mapping using a laser scanner with functional object recognition rules was proposed by [16]. In general, the laser-based approach provides accurate 3D maps, but is less suitable for object recognition than camera images. 3D object mapping using a monocular camera and a laser scanner was proposed by [19]. This approach performs 3D object modeling and object recognition from monocular images, but it needs a 2D global map created with a laser scanner. 3D scene modeling using camera images were presented in [14,18]. These methods can provide semantic information of the environment but not provide detailed 3D maps.

3D mapping by stereo vision has been studied intensively in the last decade [4,6,17,7]. The most popular approach in recent years is the corner-like feature based one, in which the camera motion is estimated with corner-point matching between consecutive frames, and 3D point clouds are generated based on the estimated camera motion. However, the corner-like features cannot represent detailed object shape, and the corner-based mapping is not suitable for precise object mapping. Recently, edge-point based stereo SLAM has been developed by [21], which provides a detailed 3D map.

Object segmentation methods have been developed due to the highly discriminative features [8,11,15,13]. Object regions can be extracted by clustering the features that co-occur in images. Many of the object segmentation methods aim at category recognition, and their main goal is learning object categories to discover and segment a large variety of objects in 2D images. The precise segmentation of specific 3D objects in real environments was proposed by [20]. The method proposed in this paper is based on the similar idea, and it is applied to object mapping.

3   Stereo SLAM Using Edge Points

Our stereo SLAM is based on the method proposed by [21] since the detailed shape of the objects can be represented using edge points.

Stereo Reconstruction. Edge points are detected using the Canny detector [3]. We refer to a pair of left and right images as *stereo frame* (*frame*, for short). Intra-frame reconstruction (i.e., between the left and right images) is performed based on the epipolar geometry in parallel stereo. We search the matching pair of edge points between left and right images along the scanline since epipolar lines are horizontal for parallel binocular stereo cameras. The matching criterion is the normalized correlation of a small window around the edge point. Also, the orientation of the image gradient at the edge point is optionally used to reduce outliers.

Camera Motion Estimation. A key to stereo SLAM is the estimation of the camera motion between frames. Unlike intra-frame reconstruction, epipolar geometry is unknown in inter-frame reconstruction. Thus, edge point correspondences are much harder to obtain.

The camera motion from time $t-1$ to t can be estimated by matching the edge points in frame I_{t-1} and those in frame I_t. The scheme employs 3D-2D matching, in which the 3D points reconstructed from I_{t-1} are matched with the 2D points detected in I_t. The 3D-2D matching is more stable than 3D-3D matching since errors in depth have less influence on the registration accuracy due to perspective projection [12].

The registration is performed using a variant of ICP algorithm [2] on the image plane. Let r_t be the camera pose at t, P^i_{t-1} be the i-th 3D edge point reconstructed at $t-1$, and p^i_{t-1} be the projected point of P^i_{t-1} onto image I_t. Then, p^i_{t-1} can be written as $p^i_{t-1} = h(P^i_{t-1}, r_t)$, where $h()$ is the perspective transformation. Let q^i_t be the image edge point at t which corresponds to p^i_{t-1}. A cost function $F()$ is defined as follows.

$$F(r_t) = \frac{1}{N} \sum_{i=1}^{N} d(q^i_t, p^i_{t-1}) \tag{1}$$

Here, $d(q^i_t, p^i_{t-1})$ is the perpendicular distance between p^i_{t-1} and the edge segment on which q^i_t lies.

Camera motion r_t and edge point correspondences are searched by minimizing $F(r_t)$ using the ICP. The initial value of r_t is set to r_{t-1}, and the initial correspondence q^i_t of p^i_{t-1} is set to the edge point which is the closest to p^i_{t-1} in terms of Euclidean distance. By repeating the minimization of $F(r_t)$ and edge point matching, the optimal r_t and edge point correspondences are obtained.

The scheme employs a coarse-to-fine approach to improve efficiency. At the first step, the ICP is performed using a small number of edge points, and then the ICP is repeated increasing edge points step by step.

Map Building. Based on the obtained camera pose r_t, a 3D map is built by transforming the intra-frame 3D points from the camera coordinates to the world coordinates. Let P^i_c be the i-th 3D point in the camera coordinates. The location of 3D point P^i can be written as $P^i = f(P^i_c, r_t)$, where $f()$ is the coordinate transformation based on the rotation and translation components of r_t.

4   Object Mapping

The object mapping phase generates a 3D object map from the 3D map created by the stereo SLAM and training images. The training images are stored beforehand in an image database. Each training image contains a target object captured from a typical view. The entry of a training image in the database consists of an object name, a rectangle roughly indicating the object existing region given by human and the edge points with SIFT descriptors [10] extracted from the image.

4.1   Retrieval of Training Images

We choose keyframes from the image sequence used for the stereo SLAM, for example, one keyframe for 10 to 20 stereo frames. For each keyframe, training images are retrieved from the image database based on edge-point matching. The basic scheme is similar to that in [20]. A SIFT descriptor is attached to each edge point for edge-point matching. We re-detect edge points from the keyframe with scale-space analysis [9] in order to find edge points invariant to scale change. Note the scale-space analysis is not performed for stereo images in the stereo SLAM because it is time consuming. Since the detected scale is proportional to object size in the image, the SIFT descriptor is invariant to object size. Using the scale-invariant descriptors, edge-point matching is performed robustly even when the object size in the training image is different from that in the stereo image.

We perform 2D recognition to find the training image which contains the object in the stereo image from the image database by matching edge points between the images. The SIFT descriptor of each edge point is utilized for nearest neighbor indexing using a kd-tree [1]. Then, we choose the training images having a large number of matched edge points. These candidate training images have false correspondences, many of which are caused by coincidentally matching edge points in the background with those in the stereo image. We eliminate them based on the geometric constraint using a pose clustering technique in terms of similarity transform [20]. Then, the training images which have a high clustering score are selected as good candidate.

4.2   Matching between 3D Map and Training Images

We extract the 3D points of the retrieved object from the 3D map. This is performed by matching 3D points in the 3D map with 2D edge points in a training image. The scheme proposed in [20] needs several training images to separate the target object clearly. We improve this scheme to separate the target object using only one training image.

First, we find the correspondences of edge points between the 3D map and the training image. This is done based on the correspondences of edge points between the 3D map and the stereo images obtained by the stereo SLAM, and also based on the correspondences of edge points between the stereo image and the training image obtained by 2D matching. Then, we calculate the camera pose relative to the 3D map, at which the object in the training image is matched with the 3D map. The camera pose is estimated by minimizing the average reprojection errors of the 3D edge points onto the training image. This non-linear minimization problem is solved using a gradient descent method. The initial value given to the method is the camera pose of the selected stereo image.

To cope with outliers in this process, we employ the RANSAC (Random Sample Consensus) algorithm [5]. The ICP used in the stereo SLAM is not suitable for this purpose since the camera pose of the stereo frame can be distant from that of the training image while the camera poses of two consecutive frames in the stereo SLAM is very close. If we simply use the ICP, it will easily fall into local minima. RANSAC is applicable in this case. However, RANSAC is a randomized approach and sometimes provides a solution (best score sample) with large matching errors due to variation in random sampling. This increases false matches between edge points in the training image and 3D edge points. To address this problem, we use not only the best sample in RANSAC but also all the good samples.

We define a score for each edge point for a RANSAC sample. Based on the estimated camera pose of the sample, we find the correspondences of edge points between the 3D map and the training image I. Let $\mathbf{E}$ be the set of 2D edge points detected from I, and $\mathbf{P}$ be the set of 3D edge points in the 3D map. Let p_j be the reprojected point of $P_j \in \mathbf{P}$ to I under camera pose r, and q_j be the edge point in the stereo image that corresponds with p_j. We check correspondence between $e_i \in \mathbf{E}$ and $P_j \in \mathbf{P}$ based on the locations of e_i and p_j and the similarity of the SIFT descriptors of e_i and q_j. We define a score for indicating that 3D point P_j belongs to the target object.

$$g(P_j, \mathbf{E}, r) = \begin{cases} 1, \text{ if } d_1(e_i, p_j) \leq th_1 \ \wedge \ d_2(e_i, q_j) \geq th_2 \\ \quad \text{for } \exists e_i \in \mathbf{E} \\ 0, \text{ otherwise.} \end{cases} \tag{2}$$

Here, $d_1(e_i, p_j)$ is the Euclidean distance between e_i and p_j, and $d_2(e_i, q_j)$ is the normalized correlation between the SIFT descriptors of e_i and q_j. th_1 and th_2 are thresholds, which are determined empirically. In implementation, $th_1 = 2 \ [pixel]$ and $th_2 = 0.8$.

To cope with the abovementioned problem in using RANSAC, we calculate the average score S for the RANSAC samples $\{r_m\}$ ($m = 1 \ to \ M$) which have inliers more than a threshold th_3. In the experiment, th_3 is set to 50% of the number of the inliers in th best sample.

$$S(P_j, \mathbf{E}) = \frac{1}{M} \sum_{m=1}^{M} g(P_j, \mathbf{E}, r_m) . \tag{3}$$

Then, we attach an object label to the 3D edge points such that $S(P_j, \mathbf{E})$ exceeds a given threshold (0.2 in the experiment).

The integration of the scores for multiple training images is useful to refine object segmentation [20]. However, the abovementioned scheme provides relatively good results even if we use only one training image.

5   Experiments

We conducted an experiment of building an object map of a room. The size of the room is approximately 7.8[m] × 6.1[m]. 485 stereo frames were captured manually with Point Grey Research's binocular camera Bumblebee2. The baseline distance is 120 [mm]. The image size was reduced to 320×240 pixels. No motion sensors were used. Fig.2 shows the 3D map generated by the stereo SLAM and some of the captured images. The maximum error compared with the handmade map is approximately 300 [mm]. Fig.3 shows a hand-made map of the room.

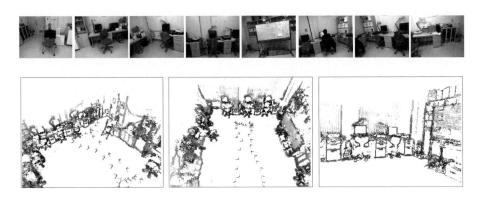

Fig. 2. The map generated by the stereo SLAM. Top: some of the images used for the stereo SLAM. Bottom: the 3D map from several views.

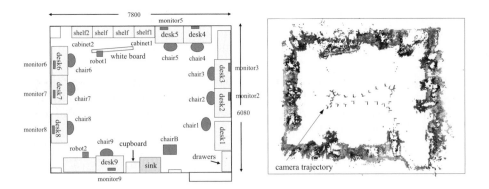

Fig. 3. The floor map of the room. Left: a handmade map of the room. Right: top view of the map generated by the stereo SLAM.

We chose 36 instances of the objects of 11 kinds in the room for object segmentation. 9 desks, 9 chairs (type A), 1 chair (type B), 2 cabinets, 2 book shelves, 7 PC monitors, 1 cupboard, 1 kitchen sink, 1 unit of drawers, 1 white board, 2 small robots. Actually, there are 4 cabinets and 4 book shelves in the room, but two of them are not visible behind the white board.

The image database contains 30 training images for the objects of 11 kinds: 6 images for desk, 12 for chairA, 3 for chairB, 1 for cabinet, 1 for book shelve, 1 for PC monitor, 1 for cupboard, 1 for kitchen sink, 1 for drawers, 1 for white board and 2 for small robot. The chairA needs many training images for the following reason. We prepared training images for its upper part (seat) and its lower part (leg) separately because the upper part turns. When either part is recognized, the region of the other part can be restricted using the relative position between the parts to improve recognition performance. This is useful because the lower part of chairA is easy to recognize but the upper part of chairA is hard.

Fig.4 shows examples of the training images. The rightmost three images at the bottom row were obtained from commercial catalogs, and have no background. All the other images were taken in different environments using the same kinds of objects, or were taken by moving the same object instance if it is movable (e.g., chairs, white board, small robot). The purpose of this is to get training images with different background than the stereo frames. We roughly determined each object region manually as designated by the red rectangles in the images. These object regions still contain some background clutters.

Object recognition was performed using keyframes in the stereo frames. Fig. 5 shows some of the results. In this experiment, 32 of the 36 object instances were recognized with the top recognition score. Three of the 36 instances were recognized with the 2nd score, and one of the 36 instances was recognized with the 3rd score.

Fig. 6 shows a 3D object map created by segmenting the 3D map into the objects of 11 kinds using the result of the object recognition. The colored points represents the separated objects. The 32 instances with the top score were separated directly using the recognition results. For the four instances with 2nd or 3rd score, the correct ones were selected by human from the recognition candidates.

Fig. 4. Image database. From top left to bottom right: desk, book shelf and cabinet, chairA, chairB, PC monitor, small robot, white board, drawers, kitchen sink, cupboard.

Fig. 5. Results of object recognition and segmentation. From top left to bottom right: PC monitor, desk, chairA's leg, book shelf, chairB, and kitchen sink. The right of each image pair shows an input stereo image, and the left image shows the training image retrieved. The red dots in the right image are the 3D points labeled as the object. The red dots in the left image are the 3D map reprojected onto the image based on the estimated camera pose.

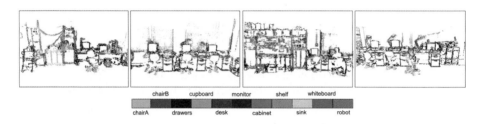

Fig. 6. Object map created by segmenting the 3D map

6   Conclusions

The paper has presented a method of 3D object mapping using a binocular stereo camera. We employ edge points as map element to represent detailed shape and apply a variant of ICP algorithm to build a 3D map. A SIFT descriptor is attached to each edge point for object recognition. The 3D map is segmented using training images of target objects with different backgrounds. In experiments, a detailed 3D map was built by the stereo SLAM and the 3D map was segmented into 36 object instances.

References

1. Beis, J.S., Lowe, D.G.: Shape Indexing Using Approximate Nearest-Neighbor Search in High-Dimensional Spaces. In: Proc. of CVPR, pp. 1000–1006 (1997)
2. Besl, P.J., Mckay, N.D.: A Method of Registration of 3-D Shapes. IEEE Trans. on PAMI 14(2), 239–256 (1992)
3. Canny, J.: A Computational Approach to Edge Detection. IEEE Trans. PAMI 8(6), 679–698 (1986)

4. Elinas, P., Sim, R., Little, J.J.: $\sigma\ SLAM$: Stereo Vision SLAM Using the Rao-Blackwellised Particle Filter and a Novel Mixture Proposal Distribution. In: Proc. of ICRA 2006, pp. 1564–1570 (2006)

5. Fischler, M., Bolles, R.: Random Sample Consensus: a Paradigm for Model Fitting with Application to Image Analysis and Automated Cartography. Communications ACM 24, 381–395 (1981)

6. Garcia, M.A., Solanas, A.: 3D Simultaneous Localization and Modeling from Stereo Vision. In: Proc. of ICRA 2004, pp. 847–853 (2004)

7. Herath, D.C., Kodagoda, K.R.S., Dissanayake, G.: Stereo Vision Based SLAM: Issues and Solutions," Vision Systems. In: Obinata, G., Dutta, A. (eds.) Advanced Robotic Systems, pp. 565–582 (2007)

8. Lazebnik, S., Schmid, C., Ponce, J.: Beyond Bags of Features: Spatial Pyramid Matching for Recognizing Natural Scene Categories. In: Proc. of CVPR 2006, pp. 2169–2178 (2006)

9. Lindberg, T.: Feature Detection with Automatic Scale Selection. Int. J. of Computer Vision 30(2), 79–116 (1998)

10. Lowe, D.G.: Distinctive Image Features from Scale-Invariant Keypoints. Int. J. of Computer Vision 60(2), 91–110 (2004)

11. Marszalek, M., Schmid, C.: Spatial Weighting for Bag-of-Features. In: Proc. of CVPR 2006, pp. 2118–2125 (2006)

12. Nistér, D., Naroditsky, O., Bergen, J.: Visual Odometry. In: Proc. of CVPR 2004, pp. 652–659 (2004)

13. Parikh, D., Chen, T.: Unsupervised Identification of Multiple Objects of Interest from Multiple Images: dISCOVER. In: Proc. of ACCV 2007, pp. 487–496 (2007)

14. Ranganathan, A., Dellaert, F.: Semantic Modeling of Places using Objects. In: Proc. of RSS 2007 (2007)

15. Russell, B.C., Efros, A.A., Sivic, J., Freeman, W.T., Zisserman, A.: Using Multiple Segmentations to Discover Objects and their Extent in Image Collections. In: Proc. of CVPR 2006, pp. 1605–1614 (2006)

16. Rusu, R.B., Marton, Z.C., Blodow, N., Dolha, M.E., Beetz, M.: Functional Object Mapping of Kitchen Environments. In: Proc. of IROS 2008, pp. 3525–3532 (2008)

17. Se, S., Lowe, D., Little, J.: Vision-based Mobile Robot Localization And Mapping using Scale-Invariant Features. In: Proc. of ICRA 2001, pp. 2051–2058 (2001)

18. Sudderth, E.B., Torralba, A., Freeman, W.T., Willsky, A.S.: Depth from Familiar Objects: A Hierarchical Model for 3D Scenes. In: Proc. of CVPR 2006, pp. 2410–2417 (2006)

19. Tomono, M.: 3-D Object Map Building Using Dense Object Models with SIFT-based Recognition Features. In: Proc. of IROS 2006, pp. 1885–1890 (2006)

20. Tomono, M.: 3D Object Modeling and Segmentation Based on Edge-Point Matching with Local Descriptors. In: Bebis, G., Boyle, R., Parvin, B., Koracin, D., Remagnino, P., Porikli, F., Peters, J., Klosowski, J., Arns, L., Chun, Y.K., Rhyne, T.-M., Monroe, L. (eds.) ISVC 2008, Part I. LNCS, vol. 5358, pp. 55–64. Springer, Heidelberg (2008)

21. Tomono, M.: Robust 3D SLAM with a Stereo Camera Based on an Edge-Point ICP Algorithm. In: Proc. of ICRA 2009, pp. 4306–4311 (2009)

Photometric Recovery of Ortho-Images Derived from Apollo 15 Metric Camera Imagery

Taemin Kim[1], Ara V. Nefian[2], and Michael J. Broxton[1,2]

[1] NASA Ames Research Center, Moffett Field, CA, 94035
[2] Carnegie Mellon University

Abstract. The topographical and photometric reconstruction of the moon from Apollo metric data has gained attention to support manned mission planning since the NASA has been working on return to the moon in 2004. This paper focuses on photometric recovery of the moon surface from Apollo orbital imagery. The statistical behavior of photons generates the scene radiance which follows a continuous Poisson distribution with the mean of surface radiance. The pixel value is determined by the camera response of sensor exposure which is proportional to scene radiance and exposure time. The surface radiance, exposure time and camera response are estimated by the maximum likelihood method for sensor exposure. The likelihood function is highly nonlinear and we were unable to find an estimator in closed form. Grouping the three sets of parameters (surface radiance, exposure time, and camera response), an EM-like juggling algorithm is proposed to determine the one family of parameters from the others. The photometric recovery of otho-images derived from Apollo 15 metric camera imagery was presented to show the validity of the proposed method.

1 Introduction

The Lunar Mapping Modeling Project (LMMP) has been actively carried out to develop maps and tools to benefit the Constellation Program (CxP) lunar planning. It will provide common, consistent and useful access to this information for lunar exploration and science communities. One of the requirements for LMMP is to construct geo-registered global and local albedo (visible image) base maps of the Moon from the digital stereo pair scans collected by Apollo era lunar missions (Figure 1). These scans, despite their high quality, are affected by noise inherent to the scanning process: the presence of film grain, dust and lint particles. Attenuating the effect of these scanning artifacts and estimating the surface radiance from Apollo orbital imagery are the central focus of this paper.

More than ever, scanned images are used as texture maps for geometric models. When a picture of a scene is taken and digitized to obtain "brightness" values, these values are rarely true measurements of relative radiance in the scene. There is usually a nonlinear mapping that determines how radiance in the scene becomes pixel values in the image [1]. The image acquisition pipeline shows how the nonlinear mapping composite the each component in a digital image formation (Figure 2).

G. Bebis et al. (Eds.): ISVC 2009, Part I, LNCS 5875, pp. 700–709, 2009.

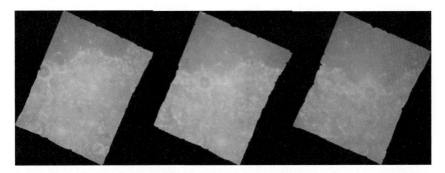

Fig. 1. Examples of Ortho-images from Apollo 15 Metric Camera Imagery

In this paper, the scene radiance is modeled as a continuous Poisson distribution with surface radiance due to the statistical behavior of photons. The pixel value is determined by the camera response of sensor exposure which is proportional to scene radiance and exposure time. The surface radiance, exposure time and camera response are estimated by the maximum likelihood method for sensor exposure. The likelihood function of all parameters is highly nonlinear and an estimator was unable to be found in in closed form. An EM-like juggling algorithm is proposed to determine the one family of parameters from the others. Finally, the reconstructed radiance map from lunar orbital imagery is presented.

2   Image Formation

Scene radiance becomes pixel values through several linear and nonlinear transformations as seen in the image acquisition pipeline (Figure 2). These unknown nonlinear mapping scan occur during exposure, development, scanning, digitization, and re-mapping. The camera response function is the aggregate mapping from sensor exposure X to pixel values Z. We estimate it from a set of sufficiently overlapped images with different exposure times, as described in [1].

After the development, scanning and digitization processes, we obtain an intensity value Z, which is a nonlinear function of the original exposure X at the pixel. Let us call

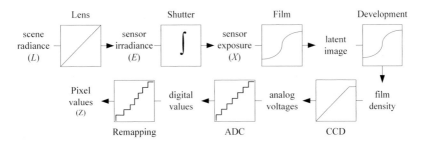

Fig. 2. Image Acquisition Pipeline

this function f, which is the composition of the characteristic curve of the film as well as all the nonlinearities introduced by the later processing steps. We write down the film reciprocity equation as:

$$Z = f(X).\tag{1}$$

Since we assume f is monotonic, it is invertible, and we can rewrite (1) as:

$$X = g(Z).\tag{2}$$

where $g = f^{-1}$.

A continuous Poisson distribution is adopted to model the scene radiance and sensor exposure. Several sources of image noise are listed in [2], but the photon noise dominates the other components in CCD or CMOS cameras. The other noise can be reduced by appropriate design of manufacturer and negligible. Photon production is governed by the laws of quantum physics which restrict us to consider an average number of photons within a give observation window. The probability distribution of p photons during seconds is known to be discrete Poisson [2,3]:

$$P(p \mid \rho,t) = \frac{(\rho t)^p e^{-\rho t}}{p!},\tag{3}$$

where ρ is the rate or intensity parameter measured in photons per second. By the continuous nature of measurement, the sensor exposure X is represented by the continuous Poisson distribution:

$$P(x \mid r,t) = \frac{(rt)^x e^{-rt}}{\Gamma(x+1)},\tag{4}$$

where r is the scene radiance.

3   Radiance Maps from Apollo Imagery

The input to our algorithm is n digitized photographs taken from the same vantage point with different known exposure durations t_j ($j = 1, 2, \cdots, n$). Let the pixel values be uniformly quantized. In this case, 256 gray levels ($z = 0, 1, \cdots, 255$). We will assume that the scene is static and that lighting changes can be safely ignored. For brevity and simplicity, one-dimensional illustrations of images will be presented which can be easily extended to two-dimensional images (Figure 3).

Suppose that we have perfectly aligned images on a regular grid (Figure 4). It can then be assumed that each pixel value on a grid point comes from the same radiance value on that point. We denote i by a spatial index over pixels and j by an image index. Let r_i be the radiance value on ith grid point and t_j be the exposure time of jth image. The inverse function of camera response is represented by a vector. We will denote sensor exposure and pixel values by x_{ij} and z_{ij}, respectively. From (2) we can write

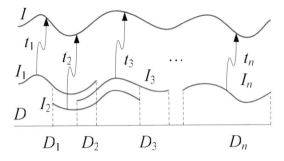

Fig. 3. Radiance Map Recovered from Orbital Imagery

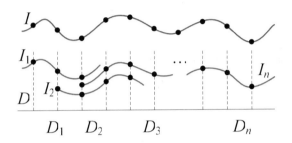

Fig. 4. Aligned Image Set in Regular Grid

$$x_{ij} = g(z_{ij}). \tag{5}$$

All parameters such as surface radiance, exposure time, and camera response are estimated by the maximum likelihood method of the continuous Poisson distribution. Let g, r, and t be the parameterized vectors for camera response, sensor irradiance and exposure time, respectively. We then have their likelihood of the continuous Poisson distribution:

$$
\begin{aligned}
l(g,r,t) &= \prod_{j=1}^{n}\prod_{i\in D_j} p\big(g(z_{ij})\,|\,r_i t_j\big) \\
&= \prod_{j=1}^{n}\prod_{i\in D_j} \frac{(r_i t_j)^{g(z_{ij})}\,e^{-r_i t_j}}{\Gamma\big(g(z_{ij})+1\big)}.
\end{aligned}
\tag{6}
$$

subject to

$$0 \le g(0) \le g(1) \le \cdots \le g(254) \le g(255). \tag{7}$$

Taking the natural logarithm of (6), we have:

$$
\begin{aligned}
L(g,r,t) &= \ln l(g,r,t) \\
&= \sum_{j=1}^{n}\sum_{i\in D_j}\big\{g(z_{ij})\ln r_i t_j - r_i t_j - \ln\Gamma\big(g(z_{ij})+1\big)\big\}.
\end{aligned}
\tag{8}
$$

Taking the derivative with respect to r_i, t_j and $g(k)$, we have:

$$\frac{\partial L}{\partial r_i} = \sum_{i \in D_j} \left\{ \frac{g(z_{ij})}{r_i} - t_j \right\},$$

$$\frac{\partial L}{\partial t_j} = \sum_{i \in D_j} \left\{ \frac{g(z_{ij})}{t_j} - r_i \right\}, \tag{9}$$

$$\frac{\partial L}{\partial g(k)} = \sum_{k=z_{ij}} \left\{ \ln r_i t_j - \Psi(g(k)+1) \right\},$$

where Ψ is the digamma function.

4   Juggling Algorithm

Since we were unable to find a close form solution to make (9) zeros, we determine them iteratively as in the expectation maximization (EM) method. Fortunately, we have the closed-form solution for scene radiance and exposure times from (9):

$$r_i = \frac{\sum_{i \in D_j} g(z_{ij})}{\sum_{i \in D_j} t_j}, \tag{10}$$

$$t_j = \frac{\sum_{i \in D_j} g(z_{ij})}{\sum_{i \in D_j} r_i}, \tag{11}$$

We have the closed-form solution for the camera response function from (9) as long as it satisfies the increasing property (7):

$$g(k) = \Psi^{-1} \left(\frac{\sum_{k=z_{ij}} \ln r_i t_j}{\sum_{k=z_{ij}} 1} \right) - 1. \tag{12}$$

In practice, we have to optimize (8) directly with linear constraints of (7). This optimization is stable in that the objective function and domain are convex. Still, (12) is useful to provide a good initial guess of the optimization. The following approximation is also useful unless we have a built-in function code of the inverse digamma function:

$$g(k) = \exp\left(\frac{\sum_{k=z_{ij}} \ln r_i t_j}{\sum_{k=z_{ij}} 1} \right). \tag{13}$$

This is based on the fact that the digamma function has the same value with the logarithm function asymptotically:

$$\Psi(x+1) \simeq \ln x .\qquad(14)$$

Iteratively we can update all parameters from an initial guess to the convergence. We call it juggling algorithm as the one family of parameters are fully determined by the others. It is reasonable to choose exposure times uniformly because the images were taken continuously. The camera response function is initialized linearly because most cases it follows the gamma correction function.

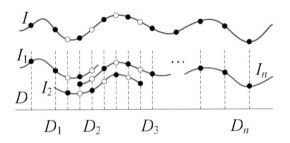

Fig. 5. Unaligned Image Set with Virtual Pixels

The extension to the case of the unaligned images is straightforward for the exposure times and the camera response function in that they have sufficient number of real observations to be determined. However, scene radiance should be determined by a single point of interest because there is no corresponding observation in the other overlapping images. This eliminates the robustness of the estimation. To avoid single observations, virtual observations for all real observations are generated in all other images. Figure 5 illustrates that the virtual observations (hollow points) are interpolated bilinearly to real observations (solid points). But, it is sufficient to calculate exposure times and camera response function from the real observations as in the regular case. The virtual observations are used to calculate the radiance:

$$r_i = \frac{\sum\limits_{i\in D_j} \tilde{g}(z_{ij})}{\sum\limits_{i\in D_j} t_j},\qquad(15)$$

where $\tilde{g}(z_{ij})$ is the estimated sensor exposure for real observation or interpolated values for virtual observations.

5 Experimental Results

The National Aeronautics and Space Administration (NASA) Exploration Systems Mission Directorate (ESMD) has been charged with producing cartographic products via LMMP for use by mission planners and scientists in NASA's Constellation program. As part of the LMMP, we have produced 70 preliminary Digital Terrain Models

(a) Original Image Mosaic

(b) Photometric Recovery Map

Fig. 6. Radiance Maps of Subset Images

(a) Original Image Mosaic

(b) Photometric Recovery Map

Fig. 7. Radiance Maps of Full Set

(DTMs) and ortho-images derived from Apollo 15 Metric Camera (AMC) orbit 33 imagery using the Ames Stereo Pipeline (ASP); a software tool that generates high quality DTMs from orbital imagery using a fully automated process. Given a pair in Apollo Metric Imagery, the reference image is projected onto the reconstructed DTM from the pair by ASP and then the ortho-image is reconstructed by orthographic projection. The whole image set consists of 66 ortho-images and has significant overlap between adjacent frames (80%) so that it is well-suited for photometric recovery.

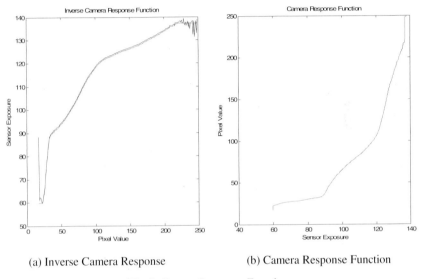

(a) Inverse Camera Response   (b) Camera Response Function

Fig. 8. Camera Response Functions

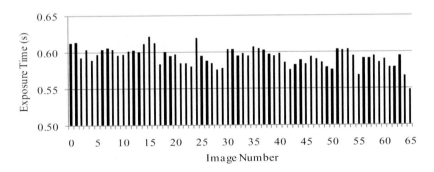

Fig. 9. Exposure Times

The photometric recovery program is implemented based on the NASA Vision Workbench (VW). The NASA VW is a general purpose image processing and computer vision library developed by the Intelligent Robotics Group in the Intelligent Systems Division at the NASA Ames Research Center. Figure 6 shows the original mosaic image constructed from ortho-images and photometrical radiance maps from the consecutive 12 images. The relative radiance is adjusted to be consistent with the remaining images. As you can see in the figure, the original image mosaic shows the vertical seams on the overlapping boundaries. However, the proposed method provides the seamless radiance map. Figure 7 shows the whole image mosaics of original images and the proposed method. The aspect ratio is adjusted because of limited space.

The camera response is shown in Figure 8. The initial guess by (13) is much closer to the optimal solution. The camera response function is obtained by the inverse function shown in Figure 8b. The estimated exposure times of each image are shown in Figure 9.

6   Conclusion

The photometric radiance map of the moon was successfully reconstructed from Apollo 15 metric camera imagery. The pixel value is determined by the camera response of sensor exposure which is proportional to scene radiance and exposure time. The statistical behavior of photons was considered and the maximum likelihood function of the parameters was derived. The surface radiance, exposure time and camera response are estimated by the maximum likelihood method. The likelihood function is highly nonlinear so that the three sets of parameters (surface radiance, exposure time, and camera response) are iteratively optimized. A juggling algorithm is proposed to determine the one family of parameters from the others. The experimental results show the validity of the proposed method.

A residual analysis would be desirable to provide a quantitative measure of the proposed method. A parametric representation of surface radiance would also be valuable to enhance the recovery resolution and robustness of the algorithm.

Acknowledgements

This work was funded by the NASA Lunar Advanced Science and Exploration Research (LASER) program grant #07-LASER07-0148, NASA Advanced Information Systems Research (AISR) program grant #06-AISRP06-0142, and by the NASA ESMD Lunar Mapping and Modeling Program (LMMP). The first author conducted the research under the Visiting Researcher Agreement between the NASA and the Korea Advanced Institute of Science and Technology.

References

1. Debevec, P.E., Malik, J.: Recovering high dynamic range radiance maps from photographs. In: Whitted, T. (ed.) SIGGRAPH 1997 Conference Proceedings. Annual Conference Series, ACM SIGGRAPH, pp. 369–378. Addison Wesley, Reading (1997)
2. Young, I., Gerbrands, J., van Vliet, L.: Fundamentals of image processing, Delft University of Technology (1995)
3. Hwang, Y., Kim, J.-S., Kweon, I.-S.: Sensor noise modeling using the Skellam distribution: Application to the color edge detection. In: IEEE International Conference on Computer Vision and Pattern Recognition, CVPR (2007)

4. Lawrence, S.J., Robinson, M.S., Broxton, M., Stopar, J.D., Close, W., Grunsfeld, J., Ingram, R., Jefferson, L., Locke, S., Mitchell, R., Scarsella, T., White, M., Hager, M.A., Watters, T.R., Bowman-Cisneros, E., Danton, J., Garvin, J.: The Apollo Digital Image Archive: New Research and Data Products. In: Proc of the NLSI Lunar Science Conference, p. 2066 (2008)
5. Broxton, M.J., Moratto, Z.M., Nefian, A., Bunte, M., Robinson, M.S.: Preliminary Stereo Reconstruction from Apollo 15 Metric Camera Imagery. In: 40th Lunar and Planetary Science Conference (2009)
6. Triggs, W., McLauchlan, P., Hartley, R., Fitzgibbon, A.: Bundle Adjustment: A Modern Synthesis. In: Triggs, B., Zisserman, A., Szeliski, R. (eds.) ICCV-WS 1999. LNCS, vol. 1883, pp. 298–373. Springer, Heidelberg (2000)

3D Lunar Terrain Reconstruction from Apollo Images

Michael J. Broxton[1,2], Ara V. Nefian[2], Zachary Moratto[3], Taemin Kim[1],
Michael Lundy[3], and Aleksandr V. Segal[4]

[1] NASA Ames Research Center. Moffet Field, CA, 94035
[2] Carnegie Mellon University
[3] Stinger-Ghaffarian Technologies Inc.
[4] Stanford University

Abstract. Generating accurate three dimensional planetary models is
becoming increasingly important as NASA plans manned missions to return to the Moon in the next decade. This paper describes a 3D surface
reconstruction system called the *Ames Stereo Pipeline* that is designed to
produce such models automatically by processing orbital stereo imagery.
We discuss two important core aspects of this system: (1) refinement of
satellite station positions and pose estimates through least squares bundle adjustment; and (2) a stochastic plane fitting algorithm that generalizes the Lucas-Kanade method for optimal matching between stereo pair
images.. These techniques allow us to automatically produce seamless,
highly accurate digital elevation models from multiple stereo image pairs
while significantly reducing the influence of image noise. Our technique
is demonstrated on a set of 71 high resolution scanned images from the
Apollo 15 mission.

1 Introduction

Accurate, high resolution Lunar 3D maps will play a central role in NASA's
future manned and unmanned missions to the moon. These maps support landing site selection and analysis, lunar landing simulation & training efforts, and
computer assisted landing systems. Furthermore, 3D digital elevation models
(DEMs) provide valuable information to scientists and geologists studying lunar
morphology.

Several recent recent lunar satellite missions, including NASA's Lunar Re-
connaissance Orbiter, have returned stereo pairs with unparalleled resolution
and image quality. However, historical data collected during the Apollo era still
provide some of the best lunar imagery available today [1]. In fact, the Apollo
Metric Camera system collected roughly 8,000 images covering roughly 20% of
the lunar equatorial zone at a resolution of 10-m/pixel (Figure 1). The extensive coverage and relatively high resolution of this camera makes this data set
extremely relevant in modern lunar data processing.

In this paper, we introduce the Ames Stereo Pipeline, a C++ software framework for automated stereogrammetric processing of NASA imagery. We begin

G. Bebis et al. (Eds.): ISVC 2009, Part I, LNCS 5875, pp. 710–719, 2009.

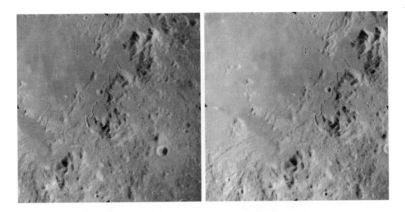

Fig. 1. Adjacent Apollo Metric Camera frames (e.g. AS15-M-1135 and AS15-M-1136 shown here) overlap by 80%. This combined with the relatively wide field of view of the camera (74 degrees) results in ideal stereo angles between successive images.

with an overview of this stereo reconstruction framework in Section 2. Then, specific attention is given to two core components of the system: Section 3 describes the bundle adjustment approach for correcting extrinsic camera parameters and co-registering overlapping images; and Section 4 describes our sub-pixel accurate stereo correlation technique. Finally, in Section 5 we present the results of processing Apollo Metric Camera imagery.

2   The Ames Stereo Pipeline

The entire stereo correlation process, from raw input images to a point cloud or DEM, can be viewed as a multistage pipeline as depicted in Figure 2.

The process begins with least squares Bundle Adjustment, which is described in Section 3, below. This produces corrected extrinsic camera parameters that are utilized by various camera modeling steps.

Then, the left and right images are aligned using interest points or geometric constraints from the camera models. This step is often essential for performance because it ensures that the disparity search space is bounded to a known area. Next, a prepossessing filter such as the Sign of the Laplacian of the Gaussian filter is used, which has the effect of producing images that are somewhat invariant to differences in lighting conditions [2].

Following these pre-processing steps, we compute the disparity space image $DSI(i, j, d_x, d_y)$ that stores the matching cost between a left image block centered around pixel (i, j) and a right image block centered at position $(i - d_x, j - d_y)$. At this stage, the quality of the match is measured as the normalized cross correlation [3] between two 15x15 pixel image patches. We employ several optimizations to accelerate this computation: (1) a box filter-like accumulator that reduces duplicate operations in the calculation of DSI [4]; (2) a coarse-to-fine pyramid based approach where disparities are estimated using low resolution

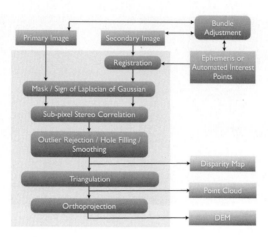

Fig. 2. Flow of data through the Ames Stereo Pipeline

images, and then successively refined at higher resolutions; and (3) partitioning of the disparity search space into rectangular sub-regions with similar values of disparity determined in the previous lower resolution level of the pyramid [4].

The DSI estimate just described efficiently computes $integer$ estimates of disparity between the two images. These estimates are subsequently refined to sub-pixel accuracy using the technique described in Section 4. Finally, in conjunction with the bundle adjusted camera models, the sub-pixel disparity estimates are used to triangulate the location of 3D points as the closest point of intersection of two forward-projected rays emanating from the centers of the two cameras through the matched pixels.

3   Bundle Adjustment

The Apollo-era satellite tracking network was highly inaccurate by today's standards with errors estimated to be 2.04-km for satellite station positions and 0.002 degrees for pose estimates in a typical Apollo 15 image [5]. Such errors propagate through the stereo triangulation process, resulting in systematic position errors and distortions in the resulting DEMs (see Figure 3). These errors can be corrected using least-squares bundle adjustment.

In bundle adjustment the position and orientation of the camera are determined jointly with the 3D position of a set of image tie-points points chosen in the overlapping regions between consecutive images. Tie-points are automatically extracted using the SURF robust feature extraction algorithm [6]. Outliers are rejected using the RANSAC method and trimmed to 1000 matches that are spread evenly across the images.

Our bundle adjustment approach follows the method described in [7] and determines the best camera parameters that minimize the projection error given by $\epsilon = \sum_k \sum_j (I_k - I(C_j, X_k))^2$ where I_k are feature locations on the image

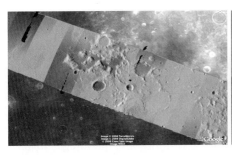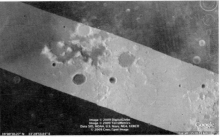

Fig. 3. Bundle adjustment is illustrated here using a color-mapped, hill-shaded DEM mosaic from Apollo 15 Orbit 33 imagery. (a) Prior to bundle adjustment, large discontinuities exist between overlapping DEMs. (b) After bundle adjustment, DEM alignment errors are no longer visible.

plane, C_j are the camera parameters, and X_k are the 3D positions associated with features I_k. $I(C_j, X_k)$ is an image formation model (i.e. forward projection) for a given camera and 3D point. The optimization of the cost function uses the Levenberg-Marquartd algorithm. Speed is improved by using sparse methods described in [8].

To eliminate the gauge freedom inherent in this problem, we add two addition error metrics to this cost function to constrain the position and scale of the overall solution. First, $\epsilon = \sum_j (C_j^{initial} - C_j)^2$ constrains camera parameters to stay relatively close to their initial values. Second, a small handful of 3D *ground control points* are chosen by hand and added to the error metric as $\epsilon = \sum_k (X_k^{gcp} - X_k)^2$ to constrain these points to known locations in the lunar coordinate frame. In the cost functions discussed above, errors are weighted by the inverse covariance of the measurement that gave rise to the constraint.

4   Sub-pixel Stereo Correlation

Apollo images are affected by two types of noise inherent to the scanning process: (1) the presence of film grain and (2) dust & lint particles. The former gives rise to noise in the DEM values that wash out real features, and the latter causes incorrect matches or hard to detect blemishes in the DEM. Attenuating the effect of these scanning artifacts while simultaneously refining the integer disparity map to sub-pixel accuracy has become a critical goal of our system, and is necessary for processing real-world data sets such as the Apollo Metric Camera data.

A common technique in sub-pixel refinement is to fit a parabola to the correlation cost surface in the 8-connected neighborhood around the integer disparity estimate, and then use the parabola's minimum as the sub-pixel disparity value. This method is easy to implement and fast to compute, but exhibits a problem known as pixel-locking: the sub-pixel disparities tend toward their integer estimates and can create noticable "stair steps" on surfaces that should be smooth [9], [10]. One way of attenuating the pixel-locking effect is through the use of a symmetric cost function [11] for matching the "left" and "right" image blocks.

To avoid the high computational complexity of these methods another class of approaches based on the Lucas-Kanade algorithm [12] proposes an asymmetric score where the disparity map is computed using the best matching score between the left image block and an optimally affine transformed block from the right image. For example, the sub-pixel refinement developed by Stein et. al. [9] lets $I_R(m, n)$ and $I_L(i, j)$ be two corresponding pixels in the right and left image respectively, where $i = m + d_x$, $j = n + d_y$ and d_x, d_y are the integer disparities. They develop a linear approximation based on the Taylor Series expansion around pixel (i, j) in the left image

$$I_L(i + \delta_x, j + \delta_y) \approx I_L(i, j) + \delta_x \frac{dI_L}{d_x}(i, j) + \delta_y \frac{dI_L}{d_y}(i, j) \tag{1}$$

where δ_x and δ_y are the local sub-pixel displacements. Let $e(x, y) = I_R(x, y) - I_L(i + \delta_x, j + \delta_y)$ and W be an image window centered around pixel (m, n). The local displacements are not constant accross W and they vary according to:

$$\delta_x(i, j) = a_1 i + b_1 j + c_1$$
$$\delta_y(i, j) = a_2 i + b_2 j + c_2. \tag{2}$$

The goal is to find the parameters $a_1, b_1, c_1, a_2, b_2, c_2$ that minimize the cost function

$$\mathbf{E}(m, n) = \sum_{(x,y) \in W} (e(x, y) w(x, y))^2 \tag{3}$$

where $w(x, y)$ are a set of weights used to reject outliers. Note that the local displacements $\delta_x(i, j)$ and $\delta_y(i, j)$ depend on the pixel positions within the window W. In fact, the values $a_1, b_1, c_1, a_2, b_2, c_2$ that minimize $\mathbf{E}$ can be seen as the parameters of an affine transformation that best transforms the right image window to match the reference (left) image window.

The shortcoming of this method is directly related to the cost function that it is minimizing, which has a low tolerance to noise. Noise present in the image will easily dominate the result of the squared error function, giving rise to erroneous disparity information. Recently, several statistical approaches (e.g. [13]) have emerged to show how stochastic models can be used to attenuate the effects of noise. Our sub-pixel refinement technique [14] adopts some of these ideas, generalizing the earlier work by Stein et. al. [9] to a Bayesian framework that models both the data and image noise.

In our approach the probability of a pixel in the right image is given by the following Bayesian model:

$$P(I_R(m, n)) = \prod_{(x,y) \in W} \mathcal{N}(I_R(m, n) | I_L(i + \delta_x, j + \delta_y), \frac{\sigma_p}{\sqrt{g_{xy}}}) P(z = 0) + \tag{4}$$
$$+ \mathcal{N}(I_R(m, n) | \mu_n, \sigma_n) P(z = 1)$$

The first mixture component ($z = 0$) is a normal density function with mean $I_L(i + \delta_x, j + \delta_y)$ and variance $\frac{\sigma_p}{\sqrt{g_{xy}}}$:

$$P(I_R(m, n)|z = 0) = \mathcal{N}(I_R(m, n)|I_L(i + \delta_x, j + \delta_y), \frac{\sigma_p}{\sqrt{g_{xy}}}) \tag{5}$$

The $\frac{1}{\sqrt{g_{xy}}}$ factor in the variance of this component has the effect of a Gaussian smoothing window over the patch. With this term in place, we are no longer looking for a single variance over the whole patch; instead we are assuming the variance increases with distance away from the center according to the inverted Gaussian, and are attempting to fit a global scale, σ_p. This provides formal justification for the standard Gaussian windowing kernel.

The second mixture component ($z = 1$) in Equation 5 models the image noise using a normal density function with mean μ_n and variance σ_n:

$$P(I_R(m, n)|z = 1) = \mathcal{N}(I_R(m, n)|\mu_n, \sigma_n) \tag{6}$$

Let $\mathbf{I}_R(m, n)$ be a vector of all pixels values in a window W centered in pixel (m, n) in the right image. Then,

$$P(\mathbf{I}_R(m, n)) = \prod_{(x,y) \in W} P(I_R(x, y)) \tag{7}$$

The parameters $\lambda = \{a_1, b_1, c_1, a_2, b_2, c_2, \sigma_p, \mu_n, \sigma_n\}$ that maximize the model likelihood in Equation 7 are determined using the Expectation Maximization (EM) algorithm. Maximizing the model likelihood in Equation 7 is equivalent to maximizing the auxiliary function:

$$\mathbf{Q}(\theta) = \sum_k P(k|\mathbf{I}_R, \lambda_t) \log P(\mathbf{I}_R, k, \underline{\delta}|\lambda)$$

$$= \sum_k \sum_{x,y} P(k|I_R(x, y), \lambda_t) \log P(I_R(x, y)|k, \lambda) P(k|\lambda) \tag{8}$$

Note that the M step calculations are similar to the equation used to determine the parameters $a_1, b_1, c_1, a_2, b_2, c_2$ in the method presented in [9], except here the fixed set of weights is replaced by the a posteriori probabilities computed in the E step. In this way, our approach can be seen as a generalization of the Lucas-Kanade method. The complete algorithm is summarized in the following steps:

- **Step 1:** Compute $\frac{dI_L}{d_x}(i, j)$, $\frac{dI_L}{d_y}(i, j)$ and the $I_R(x, y)$ values using bilinear interpolation. Initialize the model parameters λ.
- **Step 2:** Compute iteratively the model parameters λ using the EM algorithm (see [14] for details).
- **Step 3:** Compute $\delta_x(i, j)$ and $\delta_y(i, j)$ using Equation 2.
- **Step 4:** Compute a new point $(x', y') = (x, y) + (\delta_x, \delta_y)$ and the $I_R(x', y')$ values using bilinear interpolation.
- **Step 5:** If the norm of (δ_x, δ_y) vector falls below a fixed threshold the iterations converged. Otherwise, go to step 1.

Like the computation of the integer DSI, we adopt a multi-scale approach for sub-pixel refinement. At each level of the pyramid, the algorithm is initialized with the

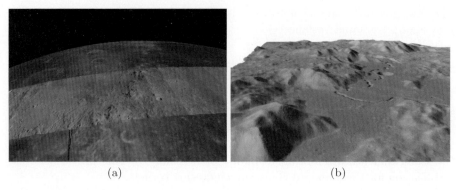

Fig. 4. Hadley Rille and the Apollo 15 landing site derived from Apollo Metric Camera frames AS15-M-1135 and AS15-M-1136. (a) superimposed over the USGS Clementine base map, (b) oblique view.

disparity determined in the previous lower resolution level of the pyramid. This allows the subpixel algorithm to shift the results of the integer DSI by many pixel if a better match can be found using the affine, noise-adapted window.

5   Results

The 3D surface reconstruction system described in this paper was tested by processing 71 Apollo Metric Camera images from Apollo 15. Specifically, we chose frames from orbit 33 of the mission, which includes highly overlapping images that span approximately 90 degrees of longitude in the lunar equatorial region. This exercised our algorithms across a wide range of different terrain and lighting conditions. Figure 4 shows the final results in the vicinity of Hadley Rille: the Apollo 15 landing site.

Tests were carried out on a 2.8-GHz, 8-core workstation with 8-GB of RAM. Stereo reconstruction for all 71 stereo pairs took 2.5 days. In the end, the results were merged into a DEM at 40-m/pixel that contained 73,000 x 20,000 pixels.

5.1   Bundle Adjustment

Bundle adjustment was carried out as described in Section 3. Initial errors and results after one round of adjustment are shown in columns two and three of Table 1, respectively. Subsequently, any tie-point measurements with image-plane residual errors that were greater than 2 standard deviations from the mean residual error were thrown out. Bundle adjustment was run a second time yielding slightly improved results shown in column four of the table.

To constrain the scale and absolute position of the solution, 7 ground control points were selected in a triangle wave pattern across the extent of the orbit to tie specific image pixels to known positions in the lunar coordinate frame. The sigma weights for ground control points were based on the resolution of the underlying base map from which ground control points were derived. These were 300-m on the surface and 500-m normal to the surface. Furthermore, the camera station position and pose estimates were constrained to stay close to their initial values based on radio tracking data. Sigma weights for camera parameters were 2-km for position, and 0.01 radians for pose. These values were drawn from historical estimates of Apollo tracking network accuracy as previously discussed.

Table 1. Residual error at various stages of bundle adjustment. Residual error in the image plane decreases as image tie-point constraints are satisfied. This improvement is made possible as residual "error" for camera position, orientation and ground control points increase to compensate. Triangulation is a measure of the average distance between the closest point of intersection of two forward projected rays for a set of tie-points. Its decrease indicates a substantial improvement in the self-consistency of the DEMs in the data set.

Residual Reconstruction	Initial	After Round 1	After Round 2
Image Plane	0.444-mm	0.012-mm	0.0075-mm
Camera Position	0-km	1.31-km	1.31-km
Camera Orientation	0-mrad	9.0-rad	9.1-mrad
Ground Control Point	0-m	481-m	465-m
Triangulation Error	911-m	24.1-m	15.58-m

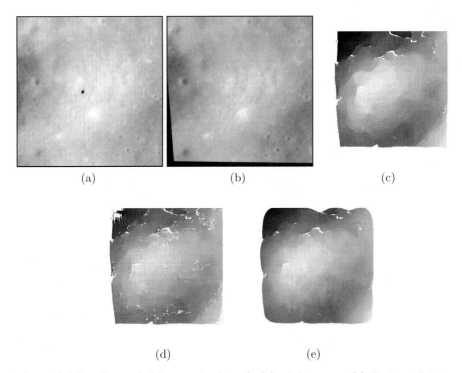

(a)   (b)   (c)

(d)   (e)

Fig. 5. (a) Left Image (with a speck of dust), (b) Right Image, (c) Horizontal integer disparity map, (d) Horizontal disparity map using the parabola method, (e) Horizontal disparity map using the Bayesian approach

5.2   Subpixel Correlation

Film grain and the dust particles are inherent to the scanning process and can significantly limit the accuracy of the stereo processing system. One example where dust particle noise occurs in one of the stereo pair images is shown in detail in Figure 5 (a) and (b). Figure 5 (c) illustrates the integer disparity map obtained by running the fast discrete correlation method described in Section 2. Figure 5 (d), and (e) compares the horizontal sub-pixel disparity maps obtained using the parabola method and the Lucas-Kanade method with the Bayesian approach we introduced in Section 4. The Bayesian approach reduces the "stair-stepping" artifacts apparent in the results from the parabola method. It also demonstrates a degree of immunity to the noise introduced by the speck of dust in (a).

6   Conclusions and Future Work

This paper has introduced a novel statistical formulation for optimally determining stereo correspondence with subpixel accuracy while simultaneously mitigating the effects of image noise. Furthermore, we have successfully demonstrated a significant improvement to the geometric consistency of the results after using least squares bundle adjustment. These techniques were successfully used to process a large, real-world corpus of images and were found to produce useful results. However, this was a preliminary demonstration of these capabilities, and much work remains to quantify residual errors and characterize the degree of noise immunity in our new correlation algorithm. Further research will be directed towards building a comprehensive error model for the 3D surface reconstruction process that can be used to systematically test our system under a variety of different conditions and inputs. Ultimately, we hope to use our technique to process the full collection of over 8,000 Apollo Metric Camera stereo pairs.

Acknowledgments

We would like to thank Mark Robinson and his team at Arizona State University for supplying high resolution scans of the Apollo Metric Camera images. This work was funded by the NASA Lunar Advanced Science and Exploration Research (LASER) program grant #07-LASER07-0148, NASA Advanced Information Systems Research (AISR) program grant #06-AISRP06-0142, and by the NASA ESMD Lunar Mapping and Modeling Program (LMMP).

References

1. Lawrence, S.J., Robinson, M.S., Broxton, M., Stopar, J.D., Close, W., Grunsfeld, J., Ingram, R., Jefferson, L., Locke, S., Mitchell, R., Scarsella, T., White, M., Hager, M.A., Watters, T.R., Bowman-Cisneros, E., Danton, J., Garvin, J.: The Apollo Digital Image Archive: New Research and Data Products. In: Proc of the NLSI Lunar Science Conference, vol. 2066 (2008)

2. Nishihara, H.: PRISM: A Practical real-time imaging stereo matcher. Optical Engineering 23, 536–545 (1984)
3. Menard, C.: Robust Stereo and Adaptive Matching in Correlation Scale-Space. PhD thesis, Institute of Automation, Vienna Institute of Technology, PRIP-TR-45 (1997)
4. Sun, C.: Rectangular Subregioning and 3-D Maximum-Surface Techniques for Fast Stereo Matching. International Journal of Computer Vision 47 (2002)
5. Cameron, W.S., Niksch, M.A.: NSSDC 72-07: Apollo 15 Data User's Guide (1972)
6. Bay, H., Ess, A., Tuytelaars, T., Gool, L.V.: SURF: Speeded Up Robust Features. Computer Vision and Image Understanding (CVIU) 110, 346–359 (2008)
7. Triggs, B., Mclauchlan, P., Hartley, R., Fitzgibbon, A.: Bundle adjustment – a modern synthesis (2000)
8. Hartley, R.I., Zisserman, A.: Multiple View Geometry in Computer Vision, 2nd edn. Cambridge University Press, Cambridge (2004)
9. Stein, A., Huertas, A., Matthies, L.: Attenuating stereo pixel-locking via affine window adaptation. In: IEEE International Conference on Robotics and Automation, pp. 914–921 (2006)
10. Szeliski, R., Scharstein, D.: Sampling the Disparity Space Image. IEEE Transactions on Pattern Analysis and Machine Intelligence (PAMI) 26, 419–425 (2003)
11. Nehab, D., Rusinkiewicz, S., Davis, J.: Improved sub-pixel stereo correspondences through symmetric refinement. In: IEEE International Conference on Computer Vision, vol. 1, pp. 557–563 (2005)
12. Baker, S., Gross, R., Matthews, I.: Lucas-Kanade 20 Years On: A Unifying Framework. International Journal of Computer Vision 56, 221–255 (2004)
13. Cheng, L., Caelli, T.: Bayesian stereo matching. In: CVPRW 2004, Conference on Computer Vision and Pattern Recognition Workshop, 2004, p. 192 (2004)
14. Nefian, A., Husmann, K., Broxton, M., To, V., Lundy, M., Hancher, M.: A Bayesian formulation for sub-pixel refinement in stereo orbital imagery. In: International Conference on Image Processing (2009)

Factorization of Correspondence and Camera Error for Unconstrained Dense Correspondence Applications

Daniel Knoblauch[1], Mauricio Hess-Flores[2], Mark Duchaineau[3], and Falko Kuester[1]

[1] University of California, San Diego
[2] University of California, Davis
[3] Lawrence Livermore National Laboratory

Abstract. A correspondence and camera error analysis for dense correspondence applications such as structure from motion is introduced. This provides error introspection, opening up the possibility of adaptively and progressively applying more expensive correspondence and camera parameter estimation methods to reduce these errors. The presented algorithm evaluates the given correspondences and camera parameters based on an error generated through simple triangulation. This triangulation is based on the given dense, non-epipolar constraint, correspondences and estimated camera parameters. This provides an error map without requiring any information about the perfect solution or making assumptions about the scene. The resulting error is a combination of correspondence and camera parameter errors. An simple, fast low/high pass filter error factorization is introduced, allowing for the separation of correspondence error and camera error. Further analysis of the resulting error maps is applied to allow efficient iterative improvement of correspondences and cameras.

1 Introduction

The main challenges in tracking, structure from motion and other applications that make use of dense correspondences are attributable to faulty correspondences and the estimated camera parameters. These challenges result from different lighting conditions, occlusions, and moving objects within the scene, which introduce uncertainty to the correspondence algorithm. This makes it desirable to be able to iteratively improve these correspondences based on an error metric. To the knowledge of the authors there has been no work evaluating correspondences and camera pose without knowledge of the ground truth. This paper introduces a novel, simple error evaluation based on the triangulation error, without ground truth knowledge.

The usual approach for epipolar-constrained applications is based on the following steps: (1) Find a small number of reliable correspondences between the two images. (2) Estimate camera poses with calculated correspondences. (3) Calculate dense correspondences and scene structure with the help of epipolar constraints.

This project does it by calculating general dense, non epipolar-constrained correspondences and the camera pose estimation from a subset of these correspondences. Based on these two steps, this paper introduces an error metric based on a 3D geometric error. The main contribution of this paper is the factorization of the error into the two main error sources, the camera parameter error and the correspondence error. This error

G. Bebis et al. (Eds.): ISVC 2009, Part I, LNCS 5875, pp. 720–729, 2009.

metric opens up the possibility of automatically performing feedback on both correspondence and camera parameter calculation given a general, non epipolar-constrained, dense correspondence algorithm. While this is not the paper that introduces such a feedback loop, it lays the fundamentals for it.

Further analysis of the extracted errors is performed to allow a quantitative error evaluation. This analysis allows a more systematic decision in which previous steps, correspondence calculation or camera parameter estimation, need further exploration.

The approach presented in this project was chosen because of improvements in hierachical dense correspondence algorithms, which allows efficient general correspondence calculation without knowledge of epipolar geometry. The main reason for this approach is the potential to incrementally improve the correspondences and camera poses by the proposed feedback loop. This is possible because correspondence errors for non-epipolar constraint dense correspondences are independent of the epipolar mapping. The freedom in unconstrained correspondence calculation allows the introduction of geometrical error extraction. By the knowledge of the authors this is the first time that an error factorization for correspondence and camera error in dense correspondence algorithms is possible.

2   Previous Work

Image registration, establishing the correspondence between two images, is a major step towards extracting model geometry. There are different approaches to evaluate pixels in two images representing the same object. Harris and Stephens [1] introduced a motion analysis algorithm based on corners and edges. This approach is only suitable for image motion analysis where objects of interest have to be tracked. Other approaches base the correspondence search on epipolar constraints as shown in [2]. To exploit the epipolar constraints the camera poses have to be known in advance or have to be calculated with a subset of reliable correspondences. The algorithm used in this project is based on dense, non-epipolar constraint correspondences. Due to this constraint a direct method solving correspondences coarse-to-fine on 4-8 mesh image pyramids, with a 5x5 local affine motion model as outlined by Duchaineau et al. [3] has been introduced. The algorithm guarantees that every destination pixel is used only once and, if possible, every pixel gets a correspondence pixel in the destination frame. All the correspondences are calculated without any knowlegde of the camera pose or epipolar constraints. This leads to a very flexible but still reliable correspondence calculation, which fits our newly proposed iterative correspondence calculation.

The camera pose estimation from two corresponding images has been extensively studied in Computer Vision. Hartley [4] introduced the eight-point algorithm that requires at least eight correspondences to evaluate the relative poses of the cameras. Nistér introduced a five-point algorithm in [5]. According to the literature, this algorithm is considered to be more robust than the eight-point algorithm. The five-point algorithm, embedded in RANSAC [6] is used in this project.

The fundamental question is how to quantify the quality of correspondences and calculated camera pose. There has been a vast amount of work in the stereo vision community in error and quality analysis for correspondence and camera pose estimation

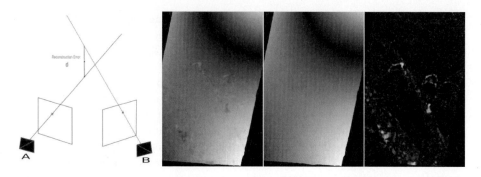

Fig. 1. Triangulation based on camera pose and correspondences. Error metric is defined by d. Resulting Error Maps: Total Error (left), Camera Error (middle), Correspondence Error (right). All error maps are normalized.

algorithms. Rodehorst et al. [7] introduced an approach to evaluate camera pose estimation based on ground truth data. For correspondence evaluation Seitz et al. [8] introduced a comparison and evaluation platform for reconstructions from stereo, termed the Middlebury Stereo Evaluation. This approach is based on reconstructing scenes that are known exactly and comparing the reconstructions against the ground truth data. Mayoral et al. [9] introduced an approach to evaluate the best matching algorithm by introducing a disparity space image based on matching errors. Xiong and Matthies [10] analyse and correct major error sources, based on matching errors, for a certain scene type, in this case a cross country navigation of an autonomous vehicle. All these approaches are based on matching errors in epipolar-constrained correspondence algorithms. To the knowledge of the authors there is no work covering error analysis for correspondences and camera pose at the same time. This paper on the other hand presents a novel technique based on non-epipolar constrained correspondences and a geometric error extraction to evaluate correspondence and camera errors on the fly, without the prerequisite of ground truth data or assumptions about the scene, and lays the fundamentals for an iterative correspondence and camera pose improvement.

3   Factorization of Correspondence and Camera Pose Errors

The error factorization is based on two preceding steps not further covered in this paper, the general dense correspondence calculation and the camera pose estimation. The main achievement of this paper is the introduction of a measure for correspondence and camera quality without knowledge of the perfect solution, any information, or assumption of the scene. This opens the way for automatic iterative correspondence and camera pose calculation.

The error metric is based on triangulation. The basic idea is to intersect rays coming from both cameras which go through corresponding pixels. In order to calculate the direction of these rays we have to take the extrinsic and intrinsic parameters of the cameras into consideration. The camera pose is defined by R, the camera rotation, and

Fig. 2. Aerial source images (left, middle) and the resulting triangulation from a different view point(right)

T, the camera translation, which are given through camera pose estimation. The intrinsic parameters K are given through a one time calibration of the cameras. With these parameters and the correspondences the ray directions D_A and D_B can be calculated; x and y are defined as the pixel coordinates in the base image or the corresponding pixel in the destination image.

$$D_i = R_i * K_i^{-1} * \begin{pmatrix} x \\ y \\ 1 \end{pmatrix} \tag{1}$$

Having the directions of the rays calculated and knowing the start points T_A and T_B, which are based on camera pose estimation, the shortest distance between the two rays can be calculated. The points P_A and P_B on the rays that correspond to the nearest distance points on the rays are defined in the following equations, where t_A and t_B define how far in the given direction the points P_A and P_B are from the camera locations T_A and T_B.

$$P_A = T_A + t_A * D_A \tag{2}$$
$$P_B = T_B + t_B * D_B \tag{3}$$

Knowing P_A and P_B the length of the shortest distance between the two rays is defined by d. In general these rays will not intersect because of noise and errors in the camera pose and the correspondences. Based on this knowledge the length of the nearest distance between the two rays is introduced as the error metric d. Figure 1 illustrates the error calculation. The error is considered to be directional for further calculations. In order to get a direction, the cross product of the two direction rays is evaluated and the resulting vector direction is considered to be the positive direction.

This error calculation is performed for every given correspondence pair across the pair of images. Figure 1 shows the resulting error map on the left. This map consists of a smooth global error superimposed by high frequency errors. Knowing that the main error sources are the camera parameters and the correspondences it can be said that the camera parameters would have to introduce a smooth overall error and the correspondences have a local high frequency error. A closer look at error d results in equations (4) and (5) respectively for P_A and P_B. For this analysis the calculations are done in the coordinate system of camera A. This means that the rotation and translation between camera A and B are relative.

$$P_A = t_A K_{E_A}^{-1} \begin{pmatrix} x + x_{E_d} \\ y + y_{E_d} \\ 1 \end{pmatrix} \tag{4}$$

$$P_B = R_E t_B K_{E_B}^{-1} \begin{pmatrix} x_c + x_{E_d} + x_{E_c} \\ y_c + y_{E_d} + y_{E_c} \\ 1 \end{pmatrix} + T_E \tag{5}$$

Camera parameter errors consist of error in relative rotation R_E, relative translation T_E, the intrinsic errors K_{E_A} and K_{E_B}, and the radial distortion errors x_{E_d} and y_{E_d}. The errors introduced by the inaccuracies in correspondence calculation are represented by x_{E_c} and y_{E_c}. All errors introduced by the camera parameters are global and influence the reconstruction in a smooth manner, resulting in the smooth parts of the total error map. The correspondence errors on the other side are local and therefore result in high frequency errors in the error map. To separate the two error sources from each other the camera error is first estimated. The error map shown in Figure 1 shows the absolute values of the error. The direction of the error is taken into account as it is possible that the crossing rays change their spatial order. This can be seen in the upper right corner of the total error map in Figure 1. The black circle corresponds to an area where the sign of the error changes.

To extract the camera error, a least-square B-spline approximation to the total error height field is introduced. This approximation consists of a 5x5 support point grid and is a special case called Bézier Curve. The goal is to filter the smooth camera error out. The correspondence error is defined as the difference between the total error and the camera error. The resulting smooth camera error can be seen in Figure 1 in the middle. The correspondence error can now be calculated by subtracting the camera error from the total error in each pixel. Figure 1 shows the correspondence error (on the right) based on the image pair in Figure 2. The latter figure shows the triangluation based on the given correspondences and the calculated camera pose. In areas of high error in the correspondence error map the resulting reconstruction shows artifacts. Also the camera error map represents the smooth reconstruction displacements seen in the upper right corner of the reconstruction. This error is produced by internal and external errors of the camera.

4   Error Metric Analysis

4.1   Signal-to-Noise Ratio (SNR) Analysis

As discussed in the previous section, the camera error is modeled as a deterministic function. Correspondence error, on the other hand, appears to be more like 'noise', due to its high-frequency and non-deterministic nature. Thus, a convenient way to examine their relationship is by applying the signal-to-noise ratio (SNR) concept, which is commonly used in image and signal processing. It must be mentioned that the two errors are assumed to be independent, since only an inlier subset of the correspondences chosen by RANSAC are used to compute the camera pose, which are not necessarily representative of the entire set of correspondences. It is also important to take into account that correspondence error is a signal in itself, despite its treatment as noise here for our purposes. Thus, a range of low SNR values, uncommon in normal signal and processing applications, is permissive here, whenever the influence of camera error is less than that of the correspondence error. Our formulation fo the SNR is given by eq. 6, where μ_s and μ_n are respectively the average camera and correspondence errors, while σ_n is the standard deviation of the correspondence error.

$$SNR = \frac{\mu_s - \mu_n}{\sigma_n} \tag{6}$$

A high SNR indicates numerically that the camera error is dominant, and that some algorithm should be applied to overcome this deficiency. On the other hand a SNR smaller than one suggests that the correspondences are the main error source and that the main focus should be on global or local correspondence improvement.

5   Results

In this section the results of the presented approach are discussed and evaluated by using different data sets to show the flexibility of the approach. These tests have been conducted on a machine with Quad Core CPU @2.66 Mhz and 4 GB of RAM. All results were achieved in a few seconds depending on the size of the input images.

5.1   Aerial Imagery

The first data set consists of aerial images taken from different viewing angles of a downtown district. Figure 2 shows the image pair and resulting reconstruction. Figure 1 displays the error maps resulting from the introduced approach. It can be concluded that the largest correspondence errors appear in occlusion areas and in areas where there is not enough texture for the correspondence algorithm to lock down the best

Fig. 3. Source images (left pair), 3D reconstruction with calculated camera pose and correspondences (middle) / perfect camera pose and perfect correspondences (right)

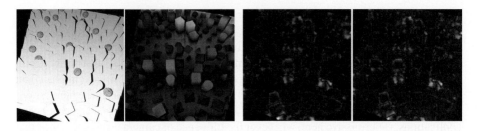

Fig. 4. Left pair: Calculated camera pose/perfect correspondences: camera error (left)/correspondence error (right). Right pair: Perfect camera pose/calculated correspondences: total error (left)/correspondence error (right).

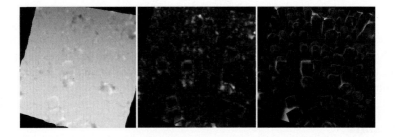

Fig. 5. Estimated camera pose and calculated correspondences. Total error map (left), correspondence error map (middle) and ground truth correspondence error map (right).

correspondences. There are also high errors on the reconstruction of the static scene, such as the streets, where movers appear. The problem is that these objects move from one frame to the other and therefore the correspondences are incorrect. These results demonstrate that problem areas are found by the introduced correspondence error map.

5.2   Artificial Data

Further tests have been completed with artificial data, where the perfect camera positions and correspondences are known. Figure 3 shows the used camera views and the resulting triangulation with calculated and perfect correspondences. The goal of this test is to prove that the assumption of the smooth camera error is correct and that the extraction of correspondence errors results in a reliable error map. The algorithm was run with the perfect camera pose and the calculated correspondences. The resulting correspondence error map can be seen in Figure 4 on the right. This error map illustrates that the main errors in the correspondences are around occlusions and repetitive textures on the cylinders. Considering that we have perfect camera poses the camera error is very small overall, which is supported by the resulting SNR of -0.77. This explains the similarity of the total error map and the correspondence error map. Figure 5 shows the extracted error maps with calculated camera poses and calculated correspondences. This demonstrates that the resulting correspondence error map (middle) is up to normalization just like the one with the perfect camera poses. This shows that the assumed

Fig. 6. Middlebury data set. Base camera view (left), depth map extracted after reconstruction (middle) and Correspondence Error Map (right).

interaction of camera error and correspondence error is correct. In this case the SNR result is 28.31 which implies that the camera error is dominant.

The next test has been conducted to prove that in case of perfect correspondences the total error corresponds to the camera error. Figure 4 shows the results of this test (left). It can be seen that the camera error represents the entire error. The SNR of value 292.96 implies that all the error is in the camera and correspondence errors are negligable. This fulfills the assumed error relation.

A ground truth correspondence error map is introduced to support the extracted correspondence error map. The ground truth correspondence error is given by the distance of the two 3D points based on the perfect camera pose and the perfect correspondences or calculated correspondences respectively. In areas of occlusions no ground truth data can be produced as no perfect correspondences exist. Figure 5 shows that the error areas in the ground truth (right) and the calculated correspondence error (middle) maps are similar up to scale. By taking into account that the error estimation is done without knowledge about the scene it can be said that the results are conclusive.

The introduced approach has additionally been tested with the 'Rocks2' data set from the Middlebury Stereo Evaluation data sets [11]. The resulting correspondence error map can be seen in Figure 6. The comparison of extracted depth map, input image and the correpondence error map shows that problem areas for the correspondences are detected.

5.3 Signal-to-Noise Ratio

The SNR based on the error maps gives information about the relative importance relation between camera and correspondence errors. To underline the benefit of this analysis, tests with the artifical data were conducted. The results in table 1 show the changes in SNR, based on different camera poses and correspondences. The correspondences are reduced in quality from left to right. In the calculated correspondences the number of iterations the algorithm runs are restricted to get different correspondence qualities. For testing purposes the camera poses are optimized based on bundle adjustment (BA) [12]. It can be seen that the better the camera poses get, the lower the SNR is. The worse the correspondences are, the lower the SNR is too. The SNR gives us a measure to estimate which error sources are relatively more dominant. A SNR of approximately one implies that both errors have about the same influence. The SNR value for perfect camera pose

Table 1. SNR values based on different camera poses (vertical) and different correspondences (horizontal)

SNR	Perfect	Calculated (5x)	Calculated (1x)
Perfect	6.69695	-0.789583	-0.770614
Calculated (BA)	278.659	28.5515	27.3323
Calculated	292.969	30.4882	28.3113

and perfect correspondences results because the intrinsic parameters in this data set are not perfect, which shows in the camera error map. These tests were run with all the used data sets and the results comply.

At this point a simple feedback loop can be introduced. An initial camera pose based on dense correspondence calculation is calculated and the resulting SNR is 28.31. This implies that the camera pose is the dominant error source. By using BA the extrinsic parameters can be improved, which shows in the smaller SNR value 27.33. Despite the correction the camera error is still dominant, which implies that most of the remaining camera error is based on internal camera parameters and distortion. To improve this, further parameters could be added to the camera refinement. On the other side a simple global correspondence improvement would be to run the correspondence algorithm with more iterations. This leads to a higher SNR, which implies that the correspondences get globally relatively better. A more advanced way to improve correspondences would be to use the correspondence error map to locally improve bad correspondences with more expensive fitting algorithms. This discussion shows the proof of concept for an iterative correspondence and camera pose estimation algorithm, based on error analysis and separation, though this is out of the scope of this particular paper.

5.4 Zero Crossings

A closer look at the correspondence error map reveals thin lines of 'no-error' inbetween high error regions. The same regions in the total error map reveal that this occurs where total error changes from being smaller to being bigger than the estimated camera error. This is a result of the taken assumptions as we try to calculate the error without comparison to the perfect solution. This artifact is acceptable as it is only a small part of the error and only introduces false positives. Used in an iterative correspondence calculation algorithm these areas will show up as errors in the next iterations.

6   Conclusion

This paper introduces an automated correspondence and camera error metric based on triangulation for general, non-epipolar constrained, dense correspondence applications. The goal of this error metric is to find faulty camera poses and correspondences and lay foundation for feedback to allow updates with more sophisticated and expensive algorithms. To solve for this error metric a triangulation based on correspondences and camera poses is executed. The length of the nearest distance between the resulting triangulation rays is used as the error for this approach. Based on the assumption that the error introduced by the camera is smooth over an image the camera error can be extracted with a least squares B-spline approximation of the total error. The correspondence error, which is considered to be local and represented by high frequency errors is the difference between camera error and total error. Further SNR analysis of the errors reveals if correspondence or camera parameter errors are dominant and helps to iteratively improve the weaker link. An overall test demonstrates the usefulness of the SNR value towards identifying which source of error is relatively dominant, and in case it is the correspondence error the problem areas identified are consistent with ground-truth error maps, such that posterior local corrections can be applied in only these regions.

References

1. Harris, C., Stephens, M.: A combined corner and edge detector. In: Alvey Vision Conference, vol. 15, p. 50 (1988)
2. Pollefeys, M., Van Gool, L., Vergauwen, M., Verbiest, F., Cornelis, K., Tops, J., Koch, R.: Visual Modeling with a Hand-Held Camera. International Journal of Computer Vision 59, 207–232 (2004)
3. Duchaineau, M., Cohen, J., Vaidya, S.: Toward Fast Computation of Dense Image Correspondence on the GPU. In: Proceedings of HPEC 2007, High Performance Embedded Computing, Eleventh Annual Workshop, Lincoln Laboratory, Massachusetts Institute of Technology, pp. 91–92 (2007)
4. Hartley, R.: In defense of the eight-point algorithm. IEEE Transactions on Pattern Analysis and Machine Intelligence 19, 580–593 (1997)
5. Nistér, D.: An Efficient Solution to the Five-Point Relative Pose Problem. IEEE Transactions On Pattern Analysis And Machine Intelligence, 756–777 (2004)
6. Fischler, M., Bolles, R.: Random sample consensus: a paradigm for model fitting with applications to image analysis and automated cartography. Communications of the ACM 24, 381–395 (1981)
7. Rodehorst, V., Heinrichs, M., Hellwich, O.: Evaluation of relative pose estimation methods for multi-camera setups. In: ISPRS 2008, pp. B3b: 135 (2008)
8. Seitz, S., Curless, B., Diebel, J., Scharstein, D., Szeliski, R.: A comparison and evaluation of multi-view stereo reconstruction algorithms. In: Int. Conf. on Computer Vision and Pattern Recognition, pp. 519–528 (2006)
9. Mayoral, R., Lera, G., Perez Ilzarbe, M.: Evaluation of correspondence errors for stereo. IVC 24, 1288–1300 (2006)
10. Xiong, Y., Matthies, L.: Error analysis of a real time stereo system. In: CVPR 1997, pp. 1087–1093 (1997)
11. Hirschmuller, H., Scharstein, D.: Evaluation of cost functions for stereo matching. In: IEEE CVPR, pp. 1–8 (2007)
12. Triggs, B., McLauchlan, P., Hartley, R., Fitzgibbon, A.: Bundle adjustment-a modern synthesis. In: Triggs, B., Zisserman, A., Szeliski, R. (eds.) ICCV-WS 1999. LNCS, vol. 1883, pp. 298–372. Springer, Heidelberg (2000)

Natural Facial Expression Recognition Using Dynamic and Static Schemes

Bogdan Raducanu[1] and Fadi Dornaika[2,3]

[1] Computer Vision Center, 08193 Bellaterra, Barcelona, Spain
bogdan@cvc.uab.es
[2] IKERBASQUE, Basque Foundation for Science
[3] University of the Basque Country, San Sebastian, Spain
fadi_dornaika@ehu.es

Abstract. Affective computing is at the core of a new paradigm in HCI and AI represented by human-centered computing. Within this paradigm, it is expected that machines will be enabled with perceiving capabilities, making them aware about users' affective state. The current paper addresses the problem of facial expression recognition from monocular videos sequences. We propose a dynamic facial expression recognition scheme, which is proven to be very efficient. Furthermore, it is conveniently compared with several static-based systems adopting different magnitude of facial expression. We provide evaluations of performance using Linear Discriminant Analysis (LDA), Non parametric Discriminant Analysis (NDA), and Support Vector Machines (SVM). We also provide performance evaluations using arbitrary test video sequences.

1 Introduction

There is a new paradigm in Human-Computer Interaction (HCI) and Artificial Intelligence focused on human-centered computing [1]. From the HCI perspective, computers will be enabled with perceptual capabilities in order to facilitate the communication protocols between people and machines. In other words, computers must use natural ways of communication people use in their everyday life: speech, hand and body gestures, facial expression. In the past, a lot of effort was dedicated to recognize facial expression in still images. For this purpose, many techniques have been applied: neural networks [2], Gabor wavelets [3] and active appearance models [4]. A very important limitation to this strategy is the fact that still images usually capture the apex of the expression, i.e., the instant at which the indicators of emotion are most marked. In their daily life, people seldom show apex of their facial expression during normal communication with their counterparts, unless for very specific cases and for very brief periods of time. More recently, attention has been shifted particularly towards modelling dynamical facial expressions [5, 6]. This is because that the differences between expressions are more powerfully modelled by dynamic transitions between different stages of an expression rather than their corresponding static key frames. This is a very relevant observation, since for most of the communication act, people rather use 'subtle' facial expressions than showing deliberately exaggerated expressions in order to convey their

G. Bebis et al. (Eds.): ISVC 2009, Part I, LNCS 5875, pp. 730–739, 2009.

message. In [7], the authors found that subtle expressions that were not identifiable in individual images suddenly became apparent when viewed in a video sequence.

Dynamical classifiers try to capture the temporal pattern in the sequence of feature vectors related to each frame such as the Hidden Markov Models (HMMs) and Dynamic Bayesian Networks [8]. In [9], parametric 2D flow models associated with the whole face as well as with the mouth, eyebrows, and eyes are first estimated. Then, mid-level predicates are inferred from these parameters. Finally, universal facial expressions are detected and recognized using the estimated predicates. Most proposed expression recognition schemes rely on the use of image raw brightness changes, which may require fixing the same imaging conditions for training and testing. The recognition of facial expressions in image sequences featuring significant head motions is a challenging problem. However, it is required by many applications such as human computer interaction and computer graphics animation [10] as well as training of social robots [11].

In this paper we propose a novel scheme for dynamic facial expression recognition that is based on the appearance-based 3D face tracker [12]. Compared to existing dynamical facial expression methods our proposed approach has several advantages. First, unlike most expression recognition systems that require a frontal view of the face, our system is view independent since the used tracker simultaneously provides the 3D head pose and the facial actions. Second, it is texture independent since the recognition scheme relies only on the estimated facial actions—invariant geometrical parameters. Third, its learning phase is simple compared to other techniques (e.g., the HMM). As a result, even when the imaging conditions change, the learned expression dynamics need not to be recomputed. It is worth noting that the proposed expression recognition schemes are only depending on the facial shape deformations (facial actions) and not on the image rawbrightness. Certainly, the shape deformations are retrieved using the rawbrightness of the sequence using the 3D face tracker based on the flexible Online Appearance Models [12].

The proposed approach for dynamic facial expression recognition has been compared afterwards against static frame-based recognition methods, showing a clear superiority in terms of recognition rates and robustness. The paper presents comparisons with several static classifiers that take into account the magnitude of facial expressions. We provide evaluations of performance using Linear Discriminant Analysis (LDA), Non parametric Discriminant Analysis (NDA), and Support Vector Machines (SVM).

The rest of the paper is organized as follows. Section 2 briefly presents the proposed 3D face and facial action tracking. Section 3 describes the proposed recognition schemes. In section 4 we report some experimental results and method comparisons. Finally, in section 5 we present our conclusions.

2   3D Facial Dynamics Extraction

2.1   A Deformable 3D Face Model

In our work, we use the 3D face model *Candide* [13]. This 3D deformable wireframe model was first developed for the purpose of model-based image coding and computer animation. The 3D shape of this wireframe model is directly recorded in coordinate form. It is given by the coordinates of the 3D vertices $\mathbf{P}_i, i = 1, \ldots, n$ where n is the number of vertices. Thus, the shape up to a global scale can be fully described by the $3n$-vector $\mathbf{g}$; the concatenation of the 3D coordinates of all vertices $\mathbf{P}_i$. The vector $\mathbf{g}$ is written as:

$$\mathbf{g} = \mathbf{g}_s + \mathbf{A}\,\boldsymbol{\tau}_\mathbf{a} \tag{1}$$

where $\mathbf{g}_s$ is the static shape of the model, $\boldsymbol{\tau}_\mathbf{a}$ the animation control vector, and the columns of $\mathbf{A}$ are the Animation Units. The static shape is constant for a given person. In this study, we use six modes for the facial Animation Units (AUs) matrix $\mathbf{A}$. We have chosen the following AUs: lower lip depressor, lip stretcher, lip corner depressor, upper lip raiser, eyebrow lowerer, and outer eyebrow raiser. These AUs are enough to cover most common facial animations. Moreover, they are essential for conveying emotions. Thus, for every frame in the video, the state of the 3D wireframe model is given by the 3D head pose parameters (three rotations and three translations) and the internal face animation control vector $\boldsymbol{\tau}_\mathbf{a}$. This is given by the 12-dimensional vector $\mathbf{b}$:

$$\mathbf{b} = [\theta_x, \theta_y, \theta_z, t_x, t_y, t_z, \boldsymbol{\tau}_\mathbf{a}^T]^T \tag{2}$$

where:

- θ_x, θ_y, and θ_z represent the three angles associated with the 3D rotation between the 3D face model coordinate system and the camera coordinate system.
- t_x, t_y, and t_z represent the three components of the 3D translation vector between the 3D face model coordinate system and the camera coordinate system.
- Each component of the vector $\boldsymbol{\tau}_\mathbf{a}$ represents the intensity of one facial action. This belongs to the interval $[0, 1]$ where the zero value corresponds to the neutral configuration (no deformation) and the one value corresponds to the maximum deformation. In the sequel, the word "facial action" will refer to the facial action intensity.

2.2 Simultaneous Face and Facial Action Tracking

In order to recover the facial expression one has to compute the facial actions encoded by the vector $\boldsymbol{\tau}_\mathbf{a}$ which encapsulates the facial deformation. Since our recognition scheme is view-independent these facial actions together with the 3D head pose should be simultaneously estimated. In other words, the objective is to compute the state vector $\mathbf{b}$ for every video frame.

For this purpose, we use the tracker based on Online Appearance Models (OAMs)—described in [12]. This appearance-based tracker aims at computing the 3D head pose and the facial actions, i.e. the vector $\mathbf{b}$, by minimizing a distance between the incoming warped frame and the current *shape-free* appearance of the face. This minimization is carried out using a gradient descent method. The statistics of the *shape-free* appearance as well as the gradient matrix are updated every frame. This scheme leads to a fast and robust tracking algorithm. We stress the fact that OAMs are more flexible than Active Appearance Models which heavily depend on the imaging conditions under which these models are built.

3   Facial Expression Recognition

Learning. In order to learn the spatio-temporal structures of the actions associated with facial expressions, we have used a simple supervised learning scheme that consists in

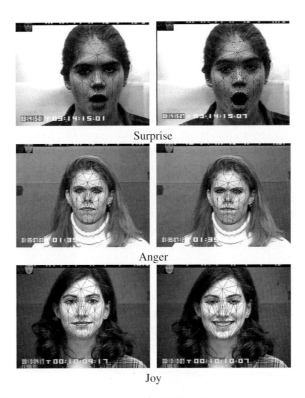

Surprise

Anger

Joy

Fig. 1. Three video examples associated with the CMU database depicting surprise, anger, and joy expressions. The left frames illustrate the half apex of the expression. The right frames illustrate the apex of the expression.

two stages. In the first stage, continuous videos depicting different facial expressions are tracked and the retrieved facial actions τ_a are represented by time series. In the second stage, the time series representation of all training videos are registered in the time domain using the Dynamic Time Warping technique. Thus, a given example (expression) is represented by a feature vector obtained by concatenating the registered τ_a.

Video sequences have been picked up from the CMU database [14]. These sequences depict five frontal view universal expressions (surprise, sadness, joy, disgust and anger). Each expression is performed by 70 different subjects, starting from the neutral one. Altogether we select 350 video sequences composed of around 15 to 20 frames each, that is, the average duration of each sequence is about half a second. The learning phase consists of estimating the facial action parameters τ_a (a 6-element vector) associated with each training sequence, that is, the temporal trajectories of the action parameters. The training video sequences have an interesting property: all performed expressions go from the neutral expression to a high magnitude expression by going through a moderate magnitude around the middle of the sequence. Therefore, using the same training set we get two kinds of trajectories: (i) an entire trajectory which models transitions from the neutral expression to a high magnitude expression, and (ii) a truncated trajectory (the second half part of a given trajectory) which models the transition from small/moderate magnitudes (half apex of the expression) to high magnitudes (apex of the expression). Figure 1 show the half apex and apex facial configurations for three expressions: surprise, anger, and joy. In the final stage of the learning all training trajectories are aligned using the Dynamic Time Warping technique by fixing a nominal duration for a facial expression. In our experiments, this nominal duration is set to 18 frames.

Recognition. In the recognition phase, the 3D head pose and facial actions are recovered from the video sequence using the appearance-based face and facial action tracker. We infer the facial expression associated with the current frame t by considering the estimated trajectory, i.e. the sequence of vectors $\tau_{\mathbf{a}(t)}$ within a temporal window of size 18 centered at the current frame t. This trajectory (feature vector) is then classified using classical classification techniques that rely on the learned examples. We have used three different classification schemes: (i) Linear Discriminant Analysis, (ii) Non-parametric Discriminant Analysis, and (iii) Support Vector Machines with a Radial Basis Function.

4   Experimental Results

In our experiments, we used a subset from the CMU facial expression database, containing 70 persons who are displaying 5 expressions: surprise, sadness, joy, disgust and anger. For training and testing we used the truncated trajectories, that is, the temporal sequence containing 9 frames, with the first frame representing a "subtle" facial expression (corresponding more or less with a "half apex" state, see the left column of Figure 1) and the last one corresponding to the apex state of the facial expression (see the right column of Figure 1). We decided to remove in our analysis the first few frames (from initial, "neutral" state to "half-apex") since we found them irrelevant for the purposes of the current study.

It is worth noting that the static recognition scheme will use the facial actions associated with only one single frame, that is, the dimension of the feature vector is 6. However, the dynamic classifier use the concatenation of facial actions within a temporal window, that is, the feature vector size is $6 \times n$ where n is the number of frames within the temporal window. In the sequel, n is set to 9.

4.1   Classification Results Using the CMU Data

The results reported in this section are based on the "leave-one-out" cross-validation strategy. Several machine learning techniques have been tested: Linear Discriminant Analysis (LDA), Non-parametric Discriminant Analysis (NDA) and Support Vector Machines (SVM). For LDA and NDA, the classification was based on the K Nearest Neighbor rule (KNN). We considered the following cases: K=1, 3 and 5.

In order to assess the benefit of using temporal information, we performed also the "static" facial expression recognition. Three static classifier schemes have been adopted. In the first scheme, training and test data are associated to the apex frames. In the second scheme, training and test data are associated to the half-apex frames. In the third schemes, we considered all the training frames in the 9-frame sequence belonging to the same facial expression, but with different magnitudes. However, during

Table 1. LDA - Overall classification results for the dynamic and static classifiers

Classifier type	K=1	K=3	K=5
Dynamic	94.2857%	88.5714%	82.8571%
Static (apex)	91.4286%	91.4286%	88.5714%
Static (half-apex)	85.7143%	82.8571%	80.0000%
Static (all frames)	84.1270%	91.4286%	89.5238%

Table 2. NDA - Overall classification results for the dynamic and static classifiers

Classifier type	K=1	K=3	K=5
Dynamic	88.5714%	88.5714%	85.7143%
Static (apex)	85.7143%	88.5714%	91.4286%
Static (half-apex)	82.8571%	80.0000%	80.0000%
Static (all frames)	90.7937%	90.1587%	91.1111%

Table 3. SVM - Overall classification results for the dynamic and static classifiers

C	Dynamic	Apex	Half-apex	All frames
5	94.2857%	97.1428%	82.8571%	87.9364%
10	97.1428%	100.0000%	85.7142%	88.8888%
50	100.0000%	94.2857%	94.2857%	86.6666%
100	97.1428%	94.2857%	94.2857%	86.3491%
500	97.1428%	94.2857%	94.2857%	87.3015%
1000	97.1428%	94.2857%	91.4285%	88.5714%

testing every frame is recognized individually and the recognition rate concerns the recognition of individual frames.

The whole results (dynamic and static) for LDA and NDA are reported in tables 1 and 2, respectively. The SVM results for the dynamic classifier are reported in table 3. The kernel was a radial basis function. Thus, the SVM used has two parameters to tune 'C' and 'g' (gamma). The first parameter controls the number of training errors, and the second one controls the RBF aperture. In general, gamma is taken as the inverse of the feature dimension, that is, it is set to $1/dim(vector) = 1/54$ for the dynamic classifier and to $1/dim(vector) = 1/6$ for the static classifier. In this case we wanted to see how the variation of the parameters 'C' (cost) affects the recognition performance. We considered six values for 'C'.

To conclude this part of the experimental results, we could say that, in general, the dynamic recognition scheme has outperformed all static recognition schemes. Moreover, we found out that the SVM clearly outperforms LDA and NDA in classification accuracy. Moreover, by inspecting the recognition results obtained with SVM we can observe that the dynamic classifiers and the static classifiers based on the apex frames are slightly more accurate than the static classifiers (half-apex) and (all frame) (third and fourth columns of Table 3). This can be explained by the fact that these static

Table 4. LDA - Cross-check validation results for the static classifier. Minor: train with half-apex frames and test with apex. Major: train with apex frames and test with half-apex.

Static classifier	K=1	K=3	K=5
Minor	82.8571%	85.7143%	85.7143%
Major	57.1429%	65.7143%	62.8571%

Table 5. NDA - Cross-check validation results for the static classifier. Minor: train with half-apex frames and test with apex. Major: train with apex frames and test with half-apex.

Static classifier	K=1	K=3	K=5
Minor	94.2857%	88.5714%	85.7143%
Major	65.7143%	62.6571%	60.0000%

Table 6. SVM - Cross-check validation results for the static classifier. Minor: train with half-apex frames and test with apex. Major: train with apex frames and test with half-apex.

C	Minor	Major
5	80.0000%	60.0000%
10	85.7142%	51.4285%
50	85.7142%	45.7142%
100	80.0000%	48.5714%
500	82.8571%	48.5714%
1000	82.8571%	48.5714%

classifiers are testing separately individual frames that may not contain high magnitude facial actions.

4.2   Cross-Check Validation Using the CMU Data

Besides the experiments described above, we performed also a cross-check validation. In the first experiment, we trained the static classifier with the frames corresponding to half-apex expression and use the apex frames for test. We refer to this case as 'minor' static classifier. In a second experiment, we trained the classifier with the apex frames and test it using the half-apex frames ('major' static classifier). The results for LDA, NDA and SVM are presented in the tables 4, 5 and 6, respectively. By analyzing the obtained results, we could observe that the 'minor' static classifier has comparable results to the static half apex classifier. This was confirmed by the three classification methods: LDA, NDA, and SVM. This means that a learning based on data featuring half apex expressions will have very good generalization capabilities since the tests with both kinds of data (half-apex and apex expressions) have a high recognition rate. Also, one can notice that the recognition rate of the minor static classifier is higher than that of the major static classifier.

This result may have very practical implications assuming that training data contain non-apex expressions, specially for real-world applications. In human-computer

Fig. 2. Four snapshots from the second video sequence

interaction scenarios, for instance, we are interested in quantifying human reaction based on its natural behavior. For this reason, we have to acquire and process data online without any external intervention. In this context, it is highly unlikely to capture automatically a persons apex of the facial expression. Most of the time we are tempted to show more subtle versions of our expressions and when we indeed show apex, this is in very specific situations and for very brief periods of time.

4.3 Dynamic vs. Static Recognition on Non-aligned Videos

In order to assess the robustness of our method, we also tested the recognition schemes on three arbitrary video sequences. The length of the shortest one is 300 frames and that of the longest is 1600 frames. Figure 2 shows four snapshots associated with the second test video sequence. These sequences depicted unseen subjects displaying a variety of different facial expressions. For training, we employed all the videos from the CMU database used in the previous sections (for which the dynamic expressions are represented by aligned 9-frame sequences). It is worth mentioning that the CMU videos and these three test videos are recorded at different frame rates. Moreover, the displayed expressions are not so similar to those depicted in the CMU data.

We compare the recognized expressions by the static and dynamic classifiers with the ground-truth displayed expressions. Since the test videos are not segmented, we perform the dynamic and static recognition only at some specific frames of the test videos. These keyframes correspond to significant facial deformations and are detected using the heuristic developed in [15]. These keyframes does not correspond to a specific frame in the time domain (onset, apex, offset of the expression). As a result of this, the task of the dynamic classifier will be very hard since the temporal window of 9 frames centered at this detected keyframe will be matched against the learned aligned trajectories. The static recognizer will not be so affected since the recognition is based on comparing the attributes of the individual detected keyframe with those of a set of learned individual frames depicting several amplitudes of the expression.

In the tables 7 and 8, we present the results for the dynamic and static classifiers, respectively. The static scheme has outperformed the dynamic scheme for these three

Table 7. Recognition results for the dynamic classifier on arbitrary non aligned video sequences

Sequence name	LDA	NDA	SVM
Data_1	43.4783%	34.7826%	60.8696%
Data_2	60.0000%	40.0000%	60.0000%
Data_3	61.1111%	55.5556%	66.6667%

Table 8. Recognition results for the static classifier on the three arbitrary non aligned video sequences. We considered only the keyframes.

Sequence name	LDA	NDA	SVM
Data_1	69.5652%	65.2174%	65.2174%
Data_2	80.0000%	80.0000%	60.0000%
Data_3	66.6667%	66.6667%	72.2222%

sequences. This confirms that the dynamic classifiers need better temporal alignment. As can be seen, the recognition rates obtained with both recognition schemes are lower than those obtained with a cross validation test based on the same database. This is due to the fact that the test was performed only on two subjects displaying arbitrary facial expressions.

5   Conclusions and Future Work

In this paper, we addressed the dynamic facial expression recognition in videos. We introduced a view and texture independent scheme that exploits facial action parameters estimated by an appearance-based 3D face tracker. We represented the universal expressions by time series associated with learned facial expressions. Facial expressions are recognized using several machine learning techniques. In order to show even better the benefits of employing a dynamic classifier, we compared it with static classifiers, built on half-apex, apex, and all frames of the corresponding facial expressions.

In the future, we want to further explore the results obtained in this paper by focusing on two directions: trying to discriminate between a fake and a genuine facial expression, and solving simultaneously the alignment and recognition.

Acknowledgements

B. Raducanu is supported by MEC Grant TIN2006-15308-C02 and CONSOLIDER-INGENIO 2010 (CSD2007-00018), Ministerio de Educación y Ciencia, Spain.

References

1. Lisetti, C., Schiano, D.: Automatic facial expression interpretation: Where HCI, AI and cognitive science intersect. Pragmatics and Cognition 8, 185–235 (2000)
2. Tian, Y., Kanade, T., Cohn, J.: Recognizing action units for facial expression analysis. IEEE Trans. on Patt. Anal. and Machine Intell. 23, 97–115 (2001)

3. Bartlett, M., Littlewort, G., Lainscsek, C., Fasel, I., Movellan, J.: Machine learning methods for fully automatic recognition of facial expressions and facial actions. In: Proc. of IEEE. Int'. Conf. on SMC, The Hague, The Netherlands, vol. I, pp. 592–597 (2004)

4. Sung, J., Lee, S., Kim, D.: A real-time facial expression recognition using the staam. In: Proc. of Int'l. Conf. on Pattern Recognition, Hong Kong, PR China, vol. I, pp. 275–278 (2006)

5. Shan, C., Gong, S., McOwan, P.: Dynamic facial expression recognition using a bayesian temporal manifold model. In: Proc. of British Machine Vision Conference, Edinburgh, UK, vol. I, pp. 297–306 (2006)

6. Yeasin, M., Bullot, B., Sharma, R.: Recognition of facial expressions and measurement of levels of interest from video. IEEE Trans. on Multimedia 8, 500–508 (2006)

7. Ambadar, Z., Schooler, J., Cohn, J.: Deciphering the enigmatic face: the importance of facial dynamics to interpreting subtle facial expressions. Psychological Science 16, 403–410 (2005)

8. Zhang, Y., Ji, Q.: Active and dynamic information fusion for facial expression understanding from image sequences 27, 699–714 (2005)

9. Black, M., Yacoob, Y.: Recognizing facial expressions in images sequences using local parameterized models of image motion. Int'l. Journal of Comp. Vision 25, 23–48 (1997)

10. Pantic, M.: Affective computing. In: Pagani, M., Pagani, M.E.A. (eds.) Encyclopedia of Multimedia Technology and Networking, vol. I, pp. 8–14. Idea Group Publishing, USA (2005)

11. Breazeal, C.: Sociable machines: Expressive social exchange between humans and robots. Ph.D. dissertation, Dept. Elect. Eng. & Comput. Sci., MIT, Cambridge, US (2000)

12. Dornaika, F., Davoine, F.: On appearance based face and facial action tracking. IEEE Trans. on Circuits and Systems for Video Technology 16, 1107–1124 (2006)

13. Ahlberg, J.: Model-based coding: extraction, coding and evaluation of face model parameters. Ph.D. Thesis, Dept. of Elec. Eng., Linköping Univ., Sweden (2002)

14. Kanade, T., Cohn, J., Tian, Y.: Comprehensive database for facial expression analysis. In: Proc. IEEE Intl. Conf. on Automatic Face and Gesture Recognition, Grenoble, France, pp. 46–53 (2000)

15. Dornaika, F., Raducanu, B.: Inferring facial expressions from videos: Tool and application. Signal Processing: Image Communication 22, 769–784 (2007)

Facial Shape Recovery from a Single Image with an Arbitrary Directional Light Using Linearly Independent Representation

Minsik Lee and Chong-Ho Choi

School of Electrical Engineering and Computer Science
Seoul National University, Seoul, Korea
{cutybug,chchoi}@csl.snu.ac.kr

Abstract. By the assumption that a face image under an arbitrary point light source is a linear combination of three *linearly independent random vectors*, we propose a novel statistical Shape From Shading (SFS) algorithm which can recover 3-D facial shape irrespective of the illumination direction, unlike most other statistical SFS algorithms. The *scaled surface normal vectors*, which are the products of albedos and surface normal vectors, can be represented by three linearly independent random vectors if we assume that human face is Lambertian. Thanks to this linearly independent representation, 3-D facial shape reconstruction can be accomplished by a few matrix multiplication under an arbitrary point light source. The experimental results show that the proposed algorithm shows good performance under various light conditions at low computational cost.

1 Introduction

Shape From Shading (SFS) [1] has been an active research area in 3-D reconstruction of images, because of the advantage that SFS requires only a single image. SFS is an ill-posed problem having more unknowns than equations, and early works have focused on finding plausible constraints or approximations that make it possible to find a reasonable solution [1]. These approaches, however, do not work well because of some unrealistic assumptions like the uniform albedo assumption, and some ambiguities, such as the generalized bas-relief ambiguity or convex-concave ambiguity [2], and are hardly applicable to practical applications.

A lot of alternative approaches have been proposed to overcome these shortcomings, most of them try to either restrict the object of interest or to incorporate a statistical model about the object [3, 4, 5, 6, 7, 8, 9, 10]. This narrows down the searching space and guarantees a unique solution, which gives relatively good results. Atick et al. [3] proposed a parameter estimation method based on 'eigenheads' derived by Principal Component Analysis (PCA) using the depth information in the cylindrical coordinate. Morphable Model (MM) [4], which generates 3-D face shape by fitting a face image to a pre-built statistical model of face shape and texture. Lei et al. [6] used Canonical Correlation Analysis (CCA)

G. Bebis et al. (Eds.): ISVC 2009, Part I, LNCS 5875, pp. 740–749, 2009.

mapping to find the correspondence between image and depth in conjunction with tensor decomposition [11] to preserve the 2-D structure of images. Smith et al. [8] used the statistical model combined with the azimuthal equidistant projection and Lambert's law to estimate the surface normal directions. Biswas et al. [10] applied nonstationary stochastic image estimation framework [12] to estimate the albedo and illuminance, irrespective of the illumination condition.

These statistical methods, however, still have some drawbacks, such as requiring face images to be in frontal pose under frontal light source, being computationally expensive, or showing poor performance, etc. In this paper, we propose an efficient 3-D facial shape recovery algorithm, which does not have any requirements on the light direction, by using the linearly independent property of the *scaled surface normal vectors* which are the surface normal vectors multiplied by albedos. If we assume that the human face is Lambertian and the scaled surface normal vectors are linearly independent, a face image under an arbitrary point light source is just a linear combination of the scaled surface normal vectors, which can be uniquely determined up to scale. We also show that the 3-D shape of human face satisfies a certain condition, under which these vectors can be determined without scale ambiguity along with the light direction. The algorithm can be implemented with just a few matrix multiplications, so the computational cost is low.

The rest of the paper is organized as follows: We explain the algorithm of facial shape recovery in Section 2. The performance evaluation follows in Section 3, and finally we conclude the paper in Section 4.

2   Facial Shape Recovery

Let us assume that the surface of a human face exhibits Lambertian reflectance. The well-known equation of Lambertian reflectance is

$$z(x,y) = \rho(x,y)\mathbf{s}^T\mathbf{n}(x,y), \tag{1}$$

where $z(x,y)$ is the brightness, $\rho(x,y)$ is the albedo, $\mathbf{n}(x,y)$ ($\in \mathbb{R}^3$) is the surface normal vector of the pixel at (x,y), and $\mathbf{s}$ ($\in \mathbb{R}^3$) is the light source vector. This equation can be expressed as the following form.

$$z(x,y) = \mathbf{s}^T\left(\rho(x,y)\mathbf{n}(x,y)\right) = \mathbf{s}^T\mathbf{x}(x,y). \tag{2}$$

Here we define $\mathbf{x}(x,y) = \rho(x,y)\mathbf{n}(x,y)$ ($\in \mathbb{R}^3$) as the *scaled surface normal vector* of the pixel at (x,y). Let s_i and $x_i(x,y)$ be the ith elements of $\mathbf{s}$ and $\mathbf{x}(x,y)$, respectively. Also let $\mathbf{z}$ be the vector corresponding to the image $\{z(x,y)\}$, and $\mathbf{x}_i$ be the vector corresponding to the image $\{x_i(x,y)\}$. Then (2) can be represented as

$$\mathbf{z} = s_1\mathbf{x}^1 + s_2\mathbf{x}^2 + s_3\mathbf{x}^3 = \sum_i s_i\mathbf{x}^i. \tag{3}$$

Table 1. Minimum principal angles (degree)

		Set 1		
		$S^{(1)}$	$S^{(2)}$	$S^{(3)}$
Set 2	$S^{(i)}$	0.93	2.03	0.58
	$S^{(j_1,j_2)}, j_k \neq i$	6.60	4.92	4.26

Table 2. Minimum angle between $\overline{\mathbf{X}}^i$ and $\widetilde{S}^{(i)}$ (degree)

1	2	3
10.53	16.24	18.56

Let us assume that $\mathbf{z}$ and $\mathbf{x}_i$s are the realizations of random vectors $\mathbf{Z}$ and $\mathbf{X}^i$ for human face, then

$$\mathbf{Z} = \sum_i s_i \mathbf{X}^i. \tag{4}$$

Hence $\mathbf{Z}$ is a linear combination of vectors $\mathbf{X}^i$.

Let us define $S^{(i_1,i_2,\ldots,i_n)}$ be the the linear subspace formed with the linear combination of all possible $\mathbf{X}^{i_1}, \mathbf{X}^{i_2}, \ldots, \mathbf{X}^{i_n}$ of human faces and $\widetilde{S}^{(i_1,i_2,\ldots,i_n)}$ be the the linear subspace formed with the linear combination of all possible $\widetilde{\mathbf{X}}^{i_1}, \widetilde{\mathbf{X}}^{i_2}, \ldots, \widetilde{\mathbf{X}}^{i_n}$ of human faces, where $\overline{\mathbf{X}}^i = E\left[\mathbf{X}^i\right]$ and $\widetilde{\mathbf{X}}^i \triangleq \mathbf{X}^i - \overline{\mathbf{X}}^i$. Also assume two things : 1) $S^{(i)}$ is linearly independent with $S^{(j_1,j_2)}$, $j_k \neq i$. 2) all possible $\widetilde{S}^{(i)}$ is linearly independent with $\overline{\mathbf{X}}^i$. Then the following can be derived.

$$\mathbf{Z} = \sum_i s_i \mathbf{X}^i = \sum_i s_i \left(\overline{\mathbf{X}}^i + \widetilde{P}^i \widetilde{\mathbf{Y}}^i\right)$$

$$= \left[\overline{\mathbf{X}}^1 \ \widetilde{P}^1 \ \overline{\mathbf{X}}^2 \ \widetilde{P}^2 \ \ldots \ \overline{\mathbf{X}}^m \ \widetilde{P}^m\right] \left[s_1 \ s_1(\widetilde{\mathbf{Y}}^1)^T \ s_2 \ s_2(\widetilde{\mathbf{Y}}^2)^T \ \ldots \ s_m \ s_m(\widetilde{\mathbf{Y}}^m)^T\right]^T$$

$$\triangleq \mathcal{P}\mathbf{V}, \tag{5}$$

where $\widetilde{P}^i$ is the matrix whose columns are the bases of $S^{(i)}$, $\widetilde{\mathbf{Y}}^i$ is the random vector that satisfies $\widetilde{\mathbf{X}}^i = \widetilde{P}^i \widetilde{\mathbf{Y}}^i$. Because of the assumption, $\mathcal{P}$ has a full column rank. Therefore, we can reconstruct $\widehat{\mathbf{x}^{i,k}}$ as the following.

$$\left[\begin{array}{c} \widehat{s_i^k} \\ \widehat{s_i^k \widetilde{\mathbf{y}}^{i,k}} \end{array}\right] = E^i \widehat{\mathbf{v}^k} = E^i \mathcal{P}^+ \mathbf{z}^k,$$

$$\widehat{\mathbf{x}^{i,k}} = \overline{\mathbf{X}}^i + \widetilde{P}^i \widehat{s_i^k \widetilde{\mathbf{y}}^{i,k}} / \widehat{s_i^k}, \tag{6}$$

where $E^i = \left[0_{n_i \times \left(\sum_{j=1}^{i-1} n_j\right)} \ I_{n_i} \ 0_{n_i \times \left(\sum_{j=i+1}^{m} n_j\right)}\right]$, $\widehat{(\cdot)}$ is the estimate of $(\cdot)$, $\mathbf{v}^k$, $\mathbf{z}^k$, $\mathbf{x}^{i,k}$, $\mathbf{y}^{i,k}$, and s_i^k are the kth sample of $\mathbf{V}$, $\mathbf{Z}$, $\mathbf{X}^i$, $\mathbf{Y}^i$, and s_i.

It has to be verified first whether the above assumptions holds for $\mathbf{X}_i$s before going through facial shape recovery. We reconstructed 3-D face shapes from PF07 Face Database [13] by photometric stereo [14], and calculated the maximum canonical correlations [15] to check the assumption. 184 subjects were divided into two sets with 92 subjects each, and the minimum principal angles, which are the arccosine of maximum canonical correlations between the two sets, were calculated as in Table 1. Here we can see that the minimum principal angles are smaller for the same index is. Ideally, the angle should be zero for the same index is and nonzero for the others, but the angle for the same index i usually has some small value due to noise or small-sample-size. We assume that $S^{(i)}$s are linearly independent based on the fact that the angles are larger for different index is. Additionally, Table 2 shows that the minimum angle between the mean of $\mathbf{X}^i$ and $\widetilde{S}^{(i)}$ are relatively large, which means that they are independent.

If there is a large shadow area in a face image, then the result of (6) will be erroneous. To mitigate the error, we reformulate the problem as the following.

$$\underset{\mathbf{v}^k}{\text{minimize}} \quad \left\| \mathbf{w}^k \circ (\mathcal{P}\mathbf{v}^k - \mathbf{z}^k) \right\|^2 \tag{7}$$

Here $\mathbf{w}^k$ is a weight vector whose elements are one if the corresponding element of $\mathbf{z}^k$ is not less than a small number ϵ (> 0) and is zero otherwise, $\| \cdot \|$ is the Euclidean norm, and $\circ$ is the Hadamard product, or element-wise product. The objective function is just a quadratic function of $\mathbf{v}^k$ and its solution is

$$\widehat{\mathbf{v}^k} = \left(W^k \mathcal{P} \right)^+ W^k z^k, \tag{8}$$

Where W^k is a diagonal matrix whose elements are the same as the elements of $\mathbf{w}^k$.

Before applying the proposed algorithm, an affine transform $(G^k, \mathbf{t}^k)$ is applied to $z^k(x, y)$ and $\mathbf{x}^{i,k}(x, y)$s for each k so that the centers of eyes and mouth are located at the designated locations respectively, which mitigates small-sample-size problem.

$$z'^k(x', y') = z^k(x, y),$$
$$\mathbf{x}'^{i,k}(x', y') = \mathbf{x}^{i,k}(x, y),$$
$$\begin{bmatrix} x' \\ y' \end{bmatrix} = G^k \begin{bmatrix} x \\ y \end{bmatrix} + \mathbf{t}^k, \tag{9}$$

where x' and y' are the coordinates after the affine transform and z'^k and $\mathbf{x}'^{i,k}$ are the brightness and the scaled surface normal vector of the transformed images. Active Appearance Model (AAM) [16] can also be used as an alternative for this purpose. After retrieving $\widehat{\mathbf{x}'^{i,k}}(x', y')$s from the transformed image $z'^k(x', y')$ for reconstruction, the inverse transform is applied to find $\widehat{\mathbf{x}^{i,k}}(x, y)$s. Note that the depth recovery should not be applied before the inverse transform, because the relationship between the surface normal vector $\mathbf{n}(x, y)$ and the depth $h(x, y)$

changes after the affine transformation. Let us define the transformed depth function $h'(x', y') = h(x, y)$, then the corresponding surface normal vectors are

$$\mathbf{n}(x, y) = \frac{\left[-h_x(x, y) \ -h_y(x, y) \ 1\right]^T}{\sqrt{h_x(x, y)^2 + h_y(x, y)^2 + 1}},$$

$$\mathbf{n}'(x', y') = \frac{\left[-h'_{x'}(x', y') \ -h'_{y'}(x', y') \ 1\right]^T}{\sqrt{h'_{x'}(x', y')^2 + h'_{y'}(x', y')^2 + 1}} = \frac{\left[-h_{x'}(x, y) \ -h_{y'}(x, y) \ 1\right]^T}{\sqrt{h_{x'}(x, y)^2 + h_{y'}(x, y)^2 + 1}}$$

$$= \frac{\left[- \left[h_x(x, y) \ h_y(x, y)\right] G \ 1\right]^T}{\sqrt{h_{x'}(x, y)^2 + h_{y'}(x, y)^2 + 1}},$$

$$\mathbf{n}'(x', y') = \begin{bmatrix} G^T & 0 \\ 0 & 1 \end{bmatrix} n(x, y) \sqrt{\frac{h_x(x, y)^2 + h_y(x, y)^2 + 1}{h_{x'}(x, y)^2 + h_{y'}(x, y)^2 + 1}}.$$

$$(10)$$

The rightmost term of the last equation is nonlinear, which confirms the statement given in the above. Hence the order of reconstruction should not be changed.

There is one more thing to be considered, when the light direction is 'singular'. If one or two elements of $\mathbf{s}^k$s are very small, e.g. $s = [0\,0\,1]^T$, we will not be able to reconstruct the depth function or the reconstruction error will be very large because $\mathbf{x}^{i,k}$ will not be retrievable for the s_i^k with a small value. To avoid this, we should prepare several $\mathcal{P}^l$s by rotating the coordinates with several different rotation matrix R^l ($\in \mathbb{R}^{3\times3}$), $l = 1, 2, \cdots$. Here $\mathcal{P}^l$ is the matrix $\mathcal{P}$ in (5) with $\mathbf{X}^{i,l}$ instead of $\mathbf{X}^i$ for a rotation matrix R^l, where

$$\begin{bmatrix} (\mathbf{X}^{1,l})^T \\ (\mathbf{X}^{2,l})^T \\ (\mathbf{X}^{3,l})^T \end{bmatrix} = R^l \begin{bmatrix} (\mathbf{X}^1)^T \\ (\mathbf{X}^1)^T \\ (\mathbf{X}^3)^T \end{bmatrix}. \qquad (11)$$

If some of s_is are very small for $\mathcal{P}^1$, (6) should be computed with different $l \neq 1$, to obtain full 3-D information.

After finding $\widehat{\mathbf{x}^{i,k}}$s, the surface normal vectors can be easily calculated by normalizing $\widehat{\mathbf{x}^k}(x, y)$. Since the reconstructed surface may not be integrable, we use the Frankot-Chellappa's method [17] to calculate the depth map. The overall procedure for 3-D facial shape recovery is summarized as follows.

[Modeling]

1. Apply an affine transform to training samples $z^k(x, y)$ and $\mathbf{x}^{i,k}(x, y)$, $i = 1, 2, 3$ for all k so that the centers of eyes and mouth are located at their designated coordinates, and transform them into vectors $\mathbf{z}^k$ and $\mathbf{x}^{i,k}$.
2. Let $\widetilde{\mathbf{x}'}^{i,k} = \mathbf{x}'^{i,k} - \overline{\mathbf{x}}^i$, where $\overline{\mathbf{x}}^i = \frac{1}{N} \sum_k \mathbf{x}'^{i,k}$. Apply PCA to $\widetilde{\mathbf{x}'}^{i,k}$ in order to obtain the basis matrices $\widetilde{P}^i$.
3. Construct $\widetilde{P}$ as in (5).
4. Repeat Steps 2 and 3 with $\mathbf{x}'^{i,k,l}$ in (11) for all l.

[Reconstruction]

1. Apply the affine transform to a test sample $z^k(x, y)$, and transform the resultant image into a vector $\mathbf{z}^k$.

2. Set $l = 1$.

3. Find $\widehat{s_i^{k,l}}$ and $\widehat{s_i^{k,l}\widetilde{\mathbf{y}}_i^{k,l}}$ for all i by evaluating $\begin{bmatrix} \widehat{s_i^{k,l}} \\ \widehat{s_i^{k,l}\widetilde{\mathbf{y}}^{i,k,l}} \end{bmatrix} = E^i(\mathcal{P}^l)^+ \mathbf{z}^k$.

4. Compute $\mathbf{s}^{l'} = R^{l'}(R^l)^T \mathbf{s}^l$. If $l^* = \arg\min_{l'}\{\text{std}(\mathbf{s}^{l'})\} \neq l$ and $\text{std}(\mathbf{s}^l) - \text{std}(\mathbf{s}^{l^*})) > \gamma$, set $l = l^*$ and go back to Step 3, otherwise go to Step 5. Here γ is a small nonnegative real number, and $\text{std}(\mathbf{s}^l) = \sqrt{\frac{1}{3}\sum_i \left(s_i^l - \frac{1}{3}\sum_j s_j^l\right)^2}$.

5. Find $\widehat{\mathbf{x}'^{i,k,l}}$ for all i by evaluating $\widehat{\mathbf{x}'^{i,k,l}} = \overline{\mathbf{x}}'^{i,l} + \mathcal{P}^{i,l}\widehat{s_i^{k,l}\widetilde{\mathbf{y}}^{i,k,l}}/\widehat{s_i^{k,l}}$. Then find $\widehat{\mathbf{x}'^{i,k}} = (R^l)^T\widehat{\mathbf{x}'^{i,k,l}}$.

6. Transform $\widehat{\mathbf{x}'^{i,k}}$ into an image and apply the inverse transform of the affine transform in Step 1 to obtain $\widehat{\mathbf{x}^{i,k}}(x, y)$ in the original coordinate. Calculate the albedo and the surface normal vectors by evaluating $\widehat{\rho^k}(x, y) = \left\|\widehat{\mathbf{x}^{i,k}}(x, y)\right\|$ and $\widehat{\mathbf{n}^k}(x, y) = \frac{\widehat{\mathbf{x}^{i,k}}(x,y)}{\widehat{\rho^k}(x,y)}$.

7. Reconstruct the depth map from $\widehat{\mathbf{n}^k}(x, y)$ using the Frankot-Chellappa's method.

Note that the reconstruction algorithm is composed of a few matrix multiplications, and the loop in Steps 3 and 4 usually takes two or three iterations. Therefore the computational cost is low.

3   Experimental Results

For lack of laser-scanned 3-D face databases, we used the photometric stereo results of the PF07 Face Database [13] as a ground truth data. The PF07 Face DB is a large database which contains the true-color face images of 200 people, 100 men and 100 women, representing 320 various images (5 pose variations × 4 expression variations × 16 illumination variations) per person. Among the 16 illumination variations, 15 corresponds to various directional+ambient lights and the last one corresponds to ambient light only. We have used our own variant of photometric stereo to make 3-D models, and the detailed procedure is not included in this paper due to space limitation.

In the experiments, the images with frontal poses of 185 subjects, who did not wear glasses, in PF07 Face DB were used. All the images were transformed by the affine transformation based on the eyes and mouth coordinates, and were cropped to 100×120 pixels. 100 subjects were used for training and 85 subjects were used for test. The MATLAB implementation of the proposed algorithm took 3.09 seconds on average for reconstruction, using Intel Pentium 4 3.0 GHz PC. PCA basis vectors were selected from the eigenvectors with the largest

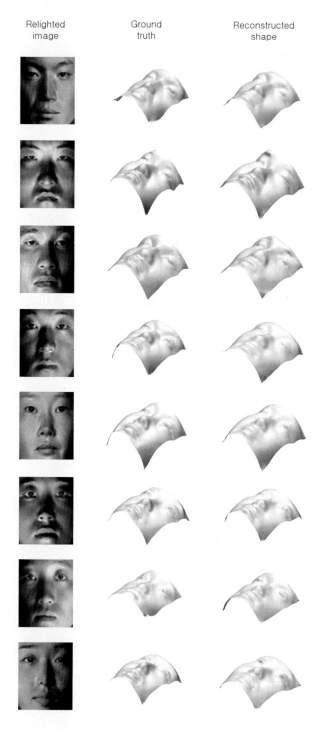

Fig. 1. Reconstructed shapes using the proposed method under various light conditions

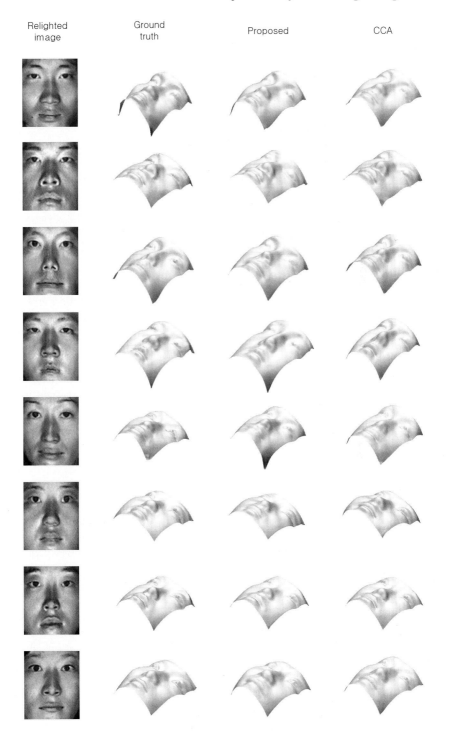

Fig. 2. Reconstructed shapes (frontal light)

eigenvalues to make up 95% of the total sum of eigenvalues. We calculated $\mathcal{P}^l$ for 8 different rotations. We compared the proposed method with Tensor-based CCA mapping [6], which is denoted as 'CCA'. Performance was evaluated in terms of mean absolute angle error E_{angle} and mean absolute depth error E_{depth} of surface normal vectors,

$$E_{angle} = \frac{1}{n} \sum_{x,y} \cos^{-1}(|\mathbf{n}^r(x,y)^T \mathbf{n}^t(x,y)|). \tag{12}$$

$$E_{depth} = \frac{1}{n} \sum_{x,y} |h^r(x,y) - h^t(x,y)|, \tag{13}$$

Here $\mathbf{n}^r$ and h^r are the ground truth surface normal vector and depth, and $\mathbf{n}^t$ and h^t are the reconstructed surface normal vector and depth. These two error measures are not necessarily proportional to each other.

Since the proposed method can handle illumination variation, its performance has been tested under various light directions using the synthetic images based on 3-D models. The detailed results are not included in this paper due to space limitation. But the results shows that the proposed method exhibits good performance evenly under various directional light conditions, giving E_{angle} in the range of 10.1-12.7 degrees and E_{depth} of 3.1-4.4 voxels. CCA can be applied only for the frontal light source, and in this case the proposed method performs better than CCA in E_{angle} by 15 % (10.40 degrees for the proposed, and 11.99 degrees for CCA), but worse in E_{depth} by 10 % (3.08 voxels for the proposed, and 2.78 voxels for CCA). This is because the proposed scheme is trained based on the the scaled surface normal vectors while CCA is based on the depth data, and error is introduced in the process of converting surface normal vectors to depth map or vice versa. Figures 1 and 2 shows some examples of reconstructed shapes.

4   Conclusion

We proposed an effective statistical SFS method that can be applied to a face image with an unknown point light source. By using the disjoint property of scaled surface normal vectors, the problem becomes a simple linear algebra problem. The experimental results showed that the proposed algorithm can reconstruct a facial shape quite accurately with low computational cost under various light conditions. There are some issues left for future work: 1) The current version of the algorithm needs several $\mathcal{P}^l$s to handle full range of light directions, requiring large memory space, which is not preferable. A more simple and convenient representation should be exploited. 2) A way to handle the pose variations should be found.

Acknowledgment

This work has been supported by the Researcher Support Program 2009, Ministry of Education, Science and Technology, Korea.

References

1. Zhang, R., Tsai, P.S.: Shape from Shading: A Survey. IEEE Transactions on Pattern Analysis and Machine Intelligence 21, 690–706 (1999)
2. Belhumeur, P.N., Kriegman, D.J., Yuille, A.L.: The Bas-Relief Ambiguity. International Journal of Computer Vision 35, 33–44 (1999)
3. Atick, J.J., Griffin, P.A., Redlich, A.N.: Statistical Approach to Shape from Shading: Reconstruction of Three-Dimensional Face Surfaces from Single Two-Dimensional Images. Neural Computation 8, 1321–1340 (1996)
4. Blanz, V., Vetter, T.: A Morphable Model For The Synthesis of 3D Faces. In: SIGGRAPH (1999)
5. Samaras, D., Metaxas, D.: Incorporating Illumination Constraints in Deformable Models for Shape from Shading and Light Direction Estimation. IEEE Transactions on Pattern Analysis and Machine Intelligence 25, 247–264 (2003)
6. Lei, Z., Bai, Q., He, R., Li, S.Z.: Face Shape Recovery from a Single Image Using CCA Mapping between Tensor Spaces. In: IEEE International Conference on Computer Vision and Pattern Recognition (2008)
7. Kemelmacher, I., Basri, R.: Molding Face Shapes by Example. In: European Conference on Computer Vision (2006)
8. Smith, W.A., Hancock, E.R.: Recovering Facial Shape Using a Statistical Model of Surface Normal Direction. IEEE Transactions on Pattern Analysis and Machine Intelligence 28, 1914–1930 (2006)
9. Castelán, M., Smith, W.A., Hancock, E.R.: A Coupled Statistical Model for Face Shape Recovery From Brightness Images. IEEE Transactions on Pattern Analysis and Machine Intelligence 16, 1139–1151 (2007)
10. Biswas, S., Aggarwal, G., Chellappa, R.: Robust Estimation of Albedo for Illumination-Invariant Matching and Shape Recovery. IEEE Transactions on Pattern Analysis and Machine Intelligence 31, 884–899 (2009)
11. Vasilescu, M.A.O., Terzopoulos, D.: Multilinear Subspace Analysis of Image Ensembles. In: IEEE International Conference on Computer Vision and Pattern Recognition (2003)
12. Kuan, D.T., Sawchuk, A.A., Strand, T.C., Chavel, P.: Adaptive Noise Smoothing Filter for Images with Signal-Dependent Noise. IEEE Transactions on Pattern Analysis and Machine Intelligence 7, 165–177 (1985)
13. Lee, H.S., Park, S., Kang, B.N., Shin, J., Lee, J.Y., Je, H., Jun, B., Kim, D.: The POSTECH Face Database (PF07) and Performance Evaluation (2008)
14. Woodham, R.J.: Photometric Method for Determining Surface Orientation From Multiple Images. Optical Engineerings 19, 139–144 (1980)
15. Kim, T.K., Kittler, J., Cipolla, R.: Discriminative Learning and Recognition of Image Set Classes Using Canonical Correlations. IEEE Transactions on Pattern Analysis and Machine Intelligence 29, 1005–1018 (2007)
16. Cootes, T.F., Edwards, G.J., Taylor, C.J.: Active Appearance Models. IEEE Transactions on Pattern Analysis and Machine Intelligence 23, 681–685 (2001)
17. Frankot, R.T., Chellappa, R.: A Method for Enforcing Integrability in Shape from Shading Algorithms. IEEE Transactions on Pattern Analysis and Machine Intelligence 10, 439–451 (1988)

Locating Facial Features and Pose Estimation Using a 3D Shape Model

Angela Caunce, David Cristinacce, Chris Taylor, and Tim Cootes

Imaging Science and Biomedical Engineering, The University of Manchester, UK

Abstract. We present an automatic method for locating facial features and estimating head pose in 2D images and video using a 3D shape model and local view-based texture patches. After automatic initialization, the 3D pose and shape are refined iteratively to optimize the match between the appearance predicted by the model, and the image. The local texture patches are generated using the current 3D pose and shape, and the locations of model points are refined by neighbourhood search, using normalized cross-correlation to provide some robustness to illumination. A key aspect is the presentation of a large-scale quantitative evaluation, comparing the method to a well-established 2D approach. We show that the accuracy of feature location for the 3D system is comparable to that of the 2D system for near-frontal faces, but significantly better for sequences which involve large rotations, obtaining estimates of pose to within 10° at headings of up to 70°.

1 Introduction

There are many potential applications which require the location of facial features in unseen images - from in-car safety to crime-prevention. One of the major challenges stems from the fact that the pose of the head, relative to the camera, is often unknown. Although 2D statistical model-based approaches have proved quite successful, they do not deal well with large variations in pose, because the models lose specificity when significant pose variation is included in the training set [1]. Some authors [2, 3] have attempted to augment a 2D approach with a 3D shape model, and, in recent years, other authors have begun to experiment with fully 3D matching algorithms (see [4] for a review).

We present our 3D matching approach and with it attempt to progress two areas. The first is to provide a comprehensive quantitative evaluation of performance in both feature detection and pose estimation. The second is to show that the 3D approach performs as well as a well-developed 2D system on large datasets of near frontal images [5, 6], and surpasses it on large rotations.

1.1 Comparison to Other Methods

The first stage in any 3D modelling approach is building the model. In some work the 3D model is generated from multiple 2D search results [3]. Some authors use an artificial head model [7] or prior knowledge of face deformation [2]. Others find

G. Bebis et al. (Eds.): ISVC 2009, Part I, LNCS 5875, pp. 750–761, 2009.

correspondences between 3D head scans [4, 8] or generate artificial examples [9]. We use manual markups as the basis for our model thus overcoming the correspondence problem.

In early work, authors have used manual methods to show that a good initialisation leads to good pose estimation accuracy [10, 11]. Some authors therefore use automated means to not only locate the face beforehand, but also to make some estimate of the pose before searching begins [3, 9, 12, 13]. Also, integrating the search into a tracking strategy [3, 7, 13] enables systems to deal with the larger rotations without the need for complex initialisations on every image. In our experiments, we use a face detector [14] on images where an independent initialisation is required, and a tracking strategy on sequences.

To compensate for illumination variation, some approaches use illumination models [11, 12]. We use normalised view-based local texture patches similar to Gu and Kanade [9], but continuously updated to reflect the current model pose.

In summary, our approach uses a sparse 3D shape model [15] for pose invariance, and continuously updated view-based local patches for illumination compensation. On images with small rotations the system can locate the features well with a face detector and no specialized pose initialisation. For larger rotations the system works best when integrated into a tracking strategy and we successfully tackle images at headings of up to 70° to within 10° accuracy.

2 Shape Model

We built a 3D statistical shape model [15] from 923 head meshes. Each mesh was created from a manual markup of photographs of an individual. The front and profile shots of each person were marked in detail and the two point sets were combined to produce a 3D representation for that subject (Figure 1 top). A generic mesh, with known correspondence to the 3D points, was warped [16] to fit the markup giving a mesh for each individual (Figure 1 bottom). Since the same mesh was used in each case the vertices are automatically corresponded across the set.

Any subset of vertices from this mesh can be used to build a sparse 3D shape model. We used 238 points (Figure 1 right) which are close to features of interest such as eyes, nose, mouth, etc.

Each example is represented as a single vector in which the 3D co-ordinates have been concatenated:

$$(x_1, \ldots x_n, y_1, \ldots y_n, z_1, \ldots z_n)^T \tag{1}$$

Principle Component Analysis is applied to the point sets to generate a statistical shape model representation of the data. A shape example x_i can be represented by the mean shape $\overline{x}$ plus a linear combination of the principle modes of the data concatenated into a matrix P:

$$\mathbf{x_i} = \overline{\mathbf{x}} + \mathbf{P}\mathbf{b_i} \tag{2}$$

where the coefficients $\mathbf{b_i}$ are the model parameters for shape $\mathbf{x_i}$. We established that the model performance improved when the number of modes was restricted. The results here are quoted for a model with 33 columns in $\mathbf{P}$ which accounts for approximately 93% of the variation in the training data. None of the subjects used in training was present in any of the images or videos used in the experiments.

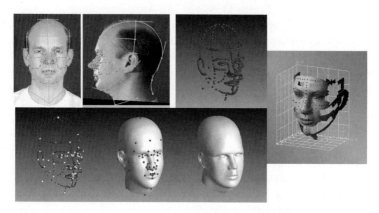

Fig. 1. The front and profile markups are combined to create a 3D point set (*top*). Using known correspondences between the markup and a generic head mesh, an individual mesh can be created for each subject (*bottom*). Only a subset of the mesh vertices are used to build the statistical shape model (*right*).

3 View-Based Local Texture Patches

The local patches are sampled from an average texture generated from 913 subjects. The individual examples are in the form of faces 'unfolded' from the meshes described in Section 2. Because all the vertices of the meshes have the same (UV) coordinates into the texture, all the unwrapped examples correspond directly pixel for pixel and it is easy to obtain the mean (Figure 2). Variation in the texture was not modelled for these experiments.

In order to successfully match the 3D shape model to the face in a 2D image, a texture patch is required at each point for comparison to the image. This patch is always the same size and shape throughout the matching process (5x5 pixels) but changes content at every iteration. It is updated based on the surface normal of the point and the current orientation of the model, and represents the view of the texture at that point (Figure 3). It is assumed for this purpose that the head is a globe and the texture lies tangential to the surface at each point with its major (UV) axes aligned to the lines of latitude and longitude. Black pixels are substituted outside the texture. To reduce speed, only a subset of 155 points are considered in the search. Many of the points excluded are from around the outside of the face where there is less information. Further, of the 155, only points which have surface normals currently facing forwards (less than 90 degrees to the view axis) are actually used to search at each iteration.

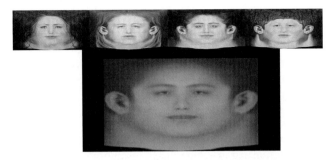

Fig. 2. The texture patches are sampled from a mean texture (*bottom*) averaged over a set of faces 'unwrapped' from the head meshes. Some example faces are shown (*top*).

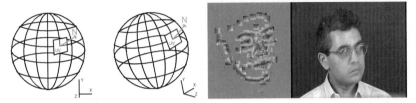

Fig. 3. The local texture patches are generated based on the current pose of the model. The UV axes of the texture are assumed to follow the lines of latitude and longitude of the head at 0° rotation (*left*). During matching the patches are a 'window' onto the texture oriented by the estimated pose of the head. Only forward facing points, determined from the surface normal are used to search.

4 Locating the Features

The model is initialised using the Viola-Jones (V-J) face detector [14]. The detector returns the location of a box, bounding the most likely location of a face in the image. The 3D shape model is placed within the box adopting its default (mean) shape and facing forwards (0° rotation).

The view-based patches are normalized and compared to the image using an exhaustive neighbourhood search. This is done for several iterations at each of a series of resolutions of both the model and the target image. Beginning with the lowest, the search is completed at each resolution before moving on to the next, and the shape and pose parameters are inherited at each resolution from the previous one. As the resolution increases (x2 at each step) the neighbourhood is increased by 2 pixels in each direction, which gradually concentrates the search.

The method begins at the lowest resolution and, at each point, a match value is calculated for all surrounding pixels in a 9x9 neighbourhood using normalised correlation. The best value gives the new target for each point. The targets are weighted in importance by the improvement in match value from the current position. Greater improvements are weighted more strongly. Once each point has a new 2D target location the z-component is estimated as the current z co-ordinate of the point. This assumes an orthogonal projection. Finally, the shape model is fitted in 3D to give a new estimate of the shape and pose parameters. This is a 2 stage process extended from the 2D case [15]. Firstly the points are rigidly aligned (rotation, scale, and translation) to minimise the sum of squared distances between matched points, then the shape model parameters (**b** in (2)) are updated using a least squares approximation.

4.1 Search Summary

- Initialise the model to the mean shape with 0° rotation using the V-J face detector.
- At the lowest resolution of the model find the best matching image resolution.
- At each model resolution and matching image resolution.
  - ○ For a number of iterations.
    - ▪ For each forward point construct a patch based on the current model pose.
    - ▪ Search the neighbourhood around each point for the best match using normalised correlation to get a target point position.
    - ▪ Estimate pose and shape parameters to fit to the target points.

5   3D to 2D Comparison

In order to test the efficacy of the 3D in 2D search, it was compared to an implementation of a well-developed 2D shape matching approach: the Constrained Local Model (CLM) [17]. The two search methods were applied to images from two large publicly available datasets, neither of which contain large variations in pose:

- XM2VTS [6]: We used 2344 images of 295 individuals at 720x576 pixels.
- BioID [5]: We used 1520 images of 23 individuals at 384x286 pixels.

Both of these sets have manual markups but not the same features are located in each. Because of this, and the difference in model points from the 2D and the 3D model, only a small subset of 12 points was used for evaluation. The points chosen are located on the better defined features, common to all sets: the ends of the eyebrows (least well localised); the corners of the eyes (well localised); the corners of the mouth (well localised); and the top and bottom of the mouth (moderately well localised).

  Both the 2D and 3D systems are initialised using the Viola-Jones face detector. We assessed the detector's performance by comparing the box returned by the algorithm to the 12 points of interest in the manual markup. If any points fell outside the box the detection was considered a failure. It was found that the detector failed on 8% of the BioID data set. These examples were excluded from the analysis, since both methods require initialisation in the location of the face.

6   Comparison Results

Figure 5 shows the cumulative distribution of the average point-to-point accuracy for the two methods and Table 1 provides a summary of these results. Due to the wide variation in size of the faces in the images, particularly in the BioID data, the errors are presented as a percentage of the inter-ocular (between pupils) distance and those in the table are the median of the average errors in each example. Also shown are the numbers of poorly located results. This is defined by a median average error of over 15%. The table distinguishes between the average results over all 12 points and the results just for the eyes, which are an easy feature to localise when marking manually. From the Figure and Table it can be seen that the 3D model results are generally better than those of the 2D CLM. Figure 8 shows some sample results for the 3D system.

7   Pose Handling and Estimation

The images in the data sets used in the experiments of section 5 are mainly near front facing. Although still challenging, it would not be expected that either system would dramatically outperform the other. However the advantages of a 3D model search are more apparent when dealing with larger rotations, both in terms of performance and pose estimation. To test this, a series of artificial images were generated with known poses.

Using the full mesh statistical model of the head described in Section 2, and a texture model built from the unwrapped textures described in Section 3, 20 synthetic subjects were generated. These were posed against a real in-car background to generate the artificial images. Figure 4 shows some examples. Feature marking was done automatically by extracting the 2D positions of selected mesh vertices.

The heads were posed as follows (Figure 4):

- Heading +/- (*r/l*) 90° in 10 degree intervals (right and left as viewed)
- Pitch +/- (*d/u*) 60° in 10 degree intervals
- Roll +/- (*r/l*) 90° in 10 degree intervals

For each rotation direction, the images were presented to the 2D and 3D systems as a sequence starting at zero each time. To perform tracking, the 3D system uses its latest result, if successful, to initialise the next search. Otherwise, the search is initialised as in section 4, using the V-J detector. The success of a search, for this purpose, is measured in 2 ways: by the final scaling of the model with respect to the image size; and by the average matching value over all the texture patches. The upper limits for both tests were fixed for all sequences.

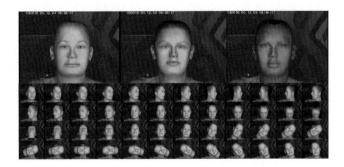

Fig. 4. Some of the synthetic subjects (*top*). And the 48 poses (in addition to zero).

8 Pose Results

Figure 5 shows the cumulative distribution of the point-to-point distances for the two methods on the artificial images. Table 1 summarises the results. The point-to-point errors are presented as pixels (inter-ocular distance is approximately 100 pixels). The automatic markup process described in the previous section failed in 3 cases therefore the results are reported on 977 images.

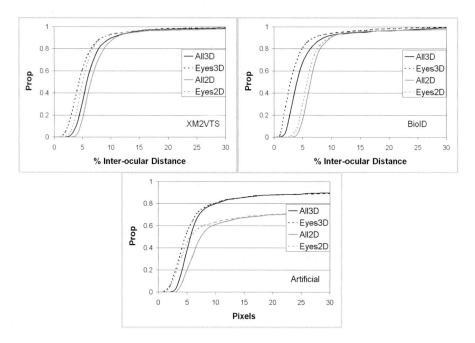

Fig. 5. The cumulative distribution of average point-to-point accuracy as % of inter-ocular distance for the real datasets (*top*), and as pixels for the artificial driver (*bottom*)

It can be seen that the 3D system out performs the 2D system and has a much lower failure rate. The graphs of Figure 6 indicate, as might be expected, that this is related to the larger rotations.

In addition to feature location the 3D model also provides an estimate of pose. Figure 7 shows the accuracy of the pose estimation at each rotation and Table 2 shows the ranges for which the median estimate lies within 10 degrees.

The pose estimation is returned as a quaternion which represents an angle of rotation about an axis and takes the form:

$$Q = (w, x, y, z); \qquad w = \cos\left(\frac{\theta}{2}\right); \qquad (x, y, z) = \sin\left(\frac{\theta}{2}\right)(x', y', z') \quad (3)$$

Where θ is the angle of rotation and (x', y', z') is the axis. The graphs of Figure 7 show two error values for the angle and the axis (angle from actual rotation axis). It can be seen that at smaller rotations the axis error is larger than that at higher rotations. This is due to the ambiguity in head pose close to zero. It is compensated for by the shape model in feature detection.

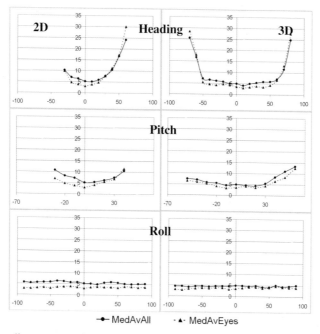

Fig. 6. The median average pixel errors at each rotation for the 2D system (*left*) and the 3D system (*right*). Each graph shows the angle across the horizontal and the pixel error on the vertical.

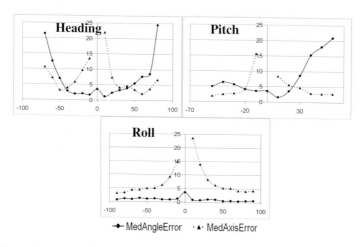

Fig. 7. The median angle errors for the pose estimation at each rotation. Each graph shows the rotation angle across the horizontal and the error, in degrees, on the vertical.

Table 2 indicates the ranges handled by each system where median average errors are within 15 pixels and median angle estimation is within 10°. Both systems handled all roll angles within these limits so these are not included. The 3D system has difficulty estimating pitch at positive rotations above 30°. This is probably due to the disappearance of the nostrils and mouth as the head rotates downwards. In contrast these features can be seen in larger upward rotations.

Figure 8 shows some sample search results from the artificial images for the 3D model.

9   Tracking Video Sequences

The 3D model was used to track features in 3 real-world in-car video sequences of 2000 frames each. The camera is located behind the steering wheel and below the head, therefore the model was initialized with a 40 degree upward pitch. Tracking was performed as described in section 7. Every 10th frame was manually marked but, due to occlusion from the steering wheel, not all 200 frames were suitable for inclusion. Table 1 shows the number of frames used for each assessment and the median average % errors and failures (>15% error). These sequences present difficult challenges because, as well as changing pose and expression, the illumination changes quite dramatically, and there are harsh shadows partially over the face at times. Figure 8 shows some search results. The errors are comparable to those on the datasets of section 5 although the failure rates are somewhat higher.

Table 1. Median average point to point errors for all data sets, presented as % inter-ocular distance or pixels, as indicated. The values are shown for all 12 pts and for just the eyes.

			All	Eyes	All	Eyes
					Fails (>15% Error)	
Data Set	Images	Model	Med. Av. Error %		As % of set [No.]	
XM2VTS	2344	2D	6.44	5.03	4.56 [107]	3.58 [84]
		3D	5.80	4.44	3.75 [88]	3.41 [80]
BioID	1398	2D	6.33	5.75	4.51 [63]	3.86 [54]
		3D	3.98	2.83	5.01 [70]	3.22 [45]
Video 1	150		8.11	6.37	16.00 [24]	14.67 [22]
Video 2	156	3D	8.85	5.68	24.36 [38]	24.36 [38]
Video 3	136		5.57	4.45	14.71 [20]	13.97 [19]
			Med. Av. Error (Pixels)		Fails (>15 pixels Error) As % of set [No.]	
Artificial Driver	977	2D	7.47	5.54	32.86 [321]	32.24 [315]
		3D	5.51	4.61	14.53 [142]	14.33 [140]

Table 2. The ranges handled by each system to within the tolerance shown

	Point to Point Median Average Error <15 pixels		Pose Estimation Median Error <10 degrees	
	Heading	Pitch	Heading	Pitch
2D	-30 to 40	-30 to 40		
3D	-50 to 70	-50 to 60	-50 to 70	-50 to 30

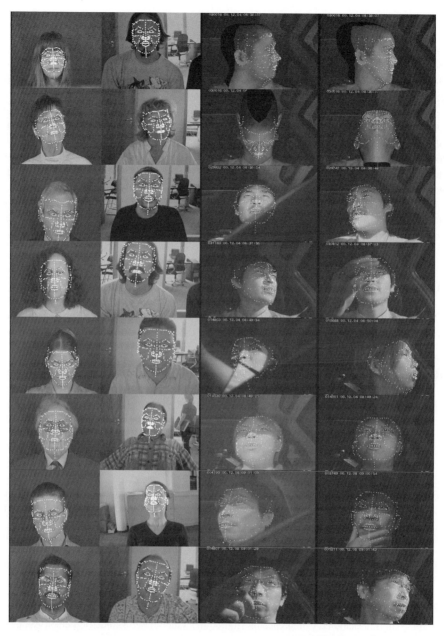

Fig. 8. Some sample 3D search results from: XM2VTS (*1ˢᵗ column*); BioID (*2ⁿᵈ column*); Artificial images (*4 top right*); Video sequences (*rest*). The two images at top left illustrate the failure modes of the search which generally result from the mouth confused with a moustache or the nose. In some cases the ears are not well shaped. This is because only forward facing points are used in the search and therefore the backs of the ears are generally not used. The model deals well with occlusion, glasses, variable illumination, low contrast, and, in many cases, facial hair.

10　Discussion and Future Work

On large datasets of near frontal images the 3D model has been shown to be comparable to a well developed 2D shape matching method. In addition, it has proved superior when handling large rotations and can provide an estimate of pose, critical for gaze dependant applications such as in-car safety.

On the XM2VTS dataset we achieved a median feature detection error of less than 6% inter-ocular distance with only 3.75% of examples falling outside a distance limit of 15%. On artificial images, with known poses, the 3D search exhibited similarly low errors at up to 50° headings and handled rotations of up to 70° with <15 pixels median average error. The system was able to estimate the pose in these images to within a median of 10° for rotations up to 70° right (as viewed) and 50° up.

Currently, the 3D system is initialised using a detector tuned to frontal faces and is instantiated in a frontal pose. One of the key ways that this system may be improved is by developing a more versatile initialisation for unseen sequences, which may not conform to these assumptions.

Acknowledgements

This project is funded by Toyota Motor Europe who provided the driver videos. We would like to thank Genemation Ltd. for the 3D data markups and head textures.

References

1. Matthews, I., Xiao, J., Baker, S.: 2D vs. 3D Deformable Models: Representational Power, Construction, and Real-Time Fitting. International Journal of Computer Vision 75, 93–113 (2007)
2. Vogler, C., Li, Z., Kanaujia, A., Goldenstein, S., Metaxas, D.: The Best of Both Worlds: Combining 3D Deformable Models with Active Shape Models. In: International Conference on Compute Vision, pp. 1–7 (2007)
3. Xiao, J., Baker, S., Matthews, I., Kanade, T.: Real-Time Combined 2D+3D Active Appearance Models. In: Conference on Computer Vision and Pattern Recognition, vol. 2, pp. 535–542 (2004)
4. Romdhani, S., Ho, J., Vetter, T., Kriegman, D.J.: Face Recognition Using 3-D Models: Pose and Illumination. Proceedings of the IEEE 94, 1977–1999 (2006)
5. Jesorsky, O., Kirchberg, K.J., Frischholz, R.W.: Robust Face Detection Using the Hausdorff Distance. In: International Conference on Audio- and Video-based Biometric Authentication, Halmstaad, Sweden, pp. 90–95 (2001)
6. Messer, K., Matas, J., Kittler, J., Jonsson, K.: XM2VTSDB: The Extended M2VTS Database. In: International Conference on Audio- and Video-based Biometric Person Authentication, Washington DC, USA (1999)
7. Dornaika, F., Ahlberg, J.: Fitting 3D face models for tracking and active appearance model training. Image and Vision Computing 24, 1010–1024 (2006)
8. Blanz, V., Vetter, T.: A Morphable Model for the Synthesis of 3D Faces. In: SIGGRAPHH, pp. 187–194 (1999)
9. Gu, L., Kanade, T.: 3D Alignment of Face in a Single Image. In: International Conference on Computer Vision and Pattern Recognition, New York, vol. 1, pp. 1305–1312 (2006)

10. Blanz, V., Vetter, T.: Face Recognition Based on Fitting a 3D Morphable Model. IEEE Transactions on Pattern Analysis and Machine Intelligence 25, 1063–1074 (2003)
11. Ishiyama, R., Sakamoto, S.: Fast and Accurate Facial Pose Estimation by Aligning a 3D Appearance Model. In: International Conference on Pattern Recognition, vol. 4, pp. 388–391 (2004)
12. Romdhani, S., Vetter, T.: 3D Probabilistic Feature Point Model for Object Detection and Recognition. In: Conference on Computer Vision and Pattern Recognition, Minneapolis, USA, pp. 1–8 (2007)
13. Zhang, W., Wang, Q., Tang, X.: Real Time Feature Based 3-D Deformable Face Tracking. In: European Conference on Computer Vision, vol. 2, pp. 720–732 (2008)
14. Viola, P., Jones, M.J.: Robust Real-Time Face Detection. International Journal of Computer Vision 57, 137–154 (2004)
15. Cootes, T.F., Taylor, C.J.: Active Shape Models - 'Smart Snakes'. In: British Machine Vision Conference, pp. 266–275 (1992)
16. Bookstein, F.L.: Principal Warps: Thin-Plate Splnes and the Decomposition of Deformations. IEE Transactions on Pattern Analysis and Machine Intelligence 11, 567–585 (1989)
17. Cristinacce, D., Cootes, T.: Automatic feature localisation with constrained local models. Pattern Recognition 41, 3054–3067 (2007)

A Stochastic Method for Face Image Super-Resolution

Jun Zheng and Olac Fuentes

Computer Science Department,
University of Texas at El Paso,
El Paso, Texas, 79968, U.S.A.

Abstract. In surveillance applications, cameras are usually set up with wide fields of view to capture as much of the scene as possible. This normally results in low-resolution images of the objects of interest. Since most image analysis applications require high or medium resolution inputs, the development of approaches aiming at improving the quality of these image regions has been an active research area in the last few years. A new family of approaches, based on statistical machine learning, aims at analyzing large data sets of images of a particular class of objects and learning the mapping from low-quality to high-quality images of that class. This enables them to infer, for example, the most likely high-resolution face image depicting the same person as a low-resolution image given as input. These super-resolution algorithms are time-consuming, due to the need for exhaustive search in a database of models. This work improves the efficiency of face image super-resolution using stochastic search for local modeling. Experimental results show that the proposed algorithm generates high-quality face images from low-resolution inputs while reducing the computation time dramatically.

1 Introduction

In digital image analysis applications, high-resolution (HR) images are often required. However, in surveillance systems, the regions of interest are often impoverished or blurred due to the large distance between the camera and the objects, or the low spatial resolution of the sensing devices. Figure 1 illustrates an image collected from a surveillance video. In this image, people at a large distance appear very small and their faces cover a small number of pixels, without enough detail to enable analysis by humans or automated face recognition programs. In these type of applications, a way to enhance these low-resolution (LR) images is needed.

Image super-resolution should provide an improvement in the perceived detail content compared to that of the original images. This typically involves restoration of the high-frequency content, which in turn requires an increase in pixel density. Furthermore, in the process of capturing digital images, several problems can affect the quality of the sensed images, such as CCD variations due to different responses of different cells to identical light intensities, scattering due to the medium through which the light beams pass, motion blur due to limited shutter speed, and quantization effects. Hence, the images obtained from digital cameras are distorted. Thus, image super-resolution is closely related to image restoration, which aims to enhance a degraded image without changing its size.

Existing methods for image super-resolution can be divided into two categories: multiple-frame super-resolution and single-frame super-resolution. In multiple-frame

G. Bebis et al. (Eds.): ISVC 2009, Part I, LNCS 5875, pp. 762–773, 2009.

Fig. 1. Faces in surveillance images

super-resolution, the LR frames typically depict the same scene. This means that LR frames are distorted as well as shifted with subpixel precision. If the LR frames contain different subpixel shifts from each other, then the new information contained in each LR frame can be used to construct an HR frame. Through motion analysis from frame to frame, a super-resolution image can be inferred by combining these LR frames into a single image [1].

Single-frame super-resolution aims to estimate missing high-resolution details from a single input low-resolution image. The problems under this category can be generic or object-specific. Generic image super-resolution techniques, such as interpolation, band-pass filtering, and unsharp masking, can be applied on any images. However, they commonly result in blurring of sharp edges, introduction of blocking artifacts, and an inability to generate high frequency components or fine details of semantically important structures [2,3,4,5].

Object-specific super-resolution assumes that only images of a certain type are input. Most approaches to object-specific super-resolution are based on statistical machine learning and work by analyzing large data sets of images of a particular class, for example faces, and learning the mapping from low-resolution to high-resolution images of that class. This enables them to infer, for example, the most likely high-resolution image depicting the same object as a low-resolution image given as input. When applied to face images, this process is known as face hallucination, first proposed by Baker and Kanade [6,7], and has been an active research area at the intersection of computer vision and computer graphics for the last decade [5,8,9,10,11,12,13,14].

We introduce a fast object-specific super-resolution approach. The approach combines separate global and local modeling stages. Global modeling, which provides the general structure of the face image, is done by eigentransformation [13,14], while local modeling, which provides high frequency details, is performed using a novel stochastic search algorithm that efficiently finds near-optimal local patches in a training set of face images. Our experimental results show that our approach provides high-quality results while significantly reducing running times relative to other state-of-the-art methods.

2 Related Work

Baker and Kanade [6,7] first developed a face hallucination method based on a prior on the spatial distribution of the image gradient for frontal face images. It infers the high-frequency component from a parent structure by recognizing the local features of the training set, and aims to recover extremely high-quality HR images of human faces from LR images. For example, given a LR image of 12×16 pixels only, which could barely be recognized as a face, face hallucination can synthesize an HR image of 96×128 pixels.

A successful face super-resolution algorithm must meet the global constraints, which means that the results must have common human characteristics, and the local constraints, which means that the results must have specific characteristics of a particular face image [12]. To fulfill these two constraints, Liu et al. [10,12,11] introduce a two-step statistical hybrid modeling approach that integrates both a global parametric model and a local non-parametric model. The first step is to derive a global linear model to learn the relationship between HR face images and the corresponding smoothed down-sampled LR ones. The second step is to model the residue between an original HR image and the reconstructed HR image by a non-parametric Markov network. Then by integrating both global and local models, they generate the photo-realistic face images.

However the above methods use probabilistic models and are based on an explicit resolution reduction function, which is sometimes difficult to acquire in practice [14]. Instead of using a probabilistic model, Wang et al. [14,13] propose a face hallucination model using PCA to represent the structural similarity of face images. They render the new hallucinated face image by mapping between the LR and HR training pairs of face images. In the PCA representation, different frequency components are independent. By selecting the number of eigenfaces, they extract the maximum amount of facial information from the low-resolution face image and remove the noise.

Motivated by the fact that belief propagation converges quickly to a solution of the Markov network, Freeman et al. [9,5] explore a simpler and faster one-pass algorithm, which uses the same local relationship information as the Markov networks but requires only a nearest-neighbor search in the training set for a vector derived from each patch of local image data. Their algorithms are an instance of a general-training-based approach that can be useful for image processing or graphics applications. It can be applied to enlarge images, remove noise, and estimate 3D surface shapes.

3 Framework

As shown in Figure 2, the proposed method work consists of two steps. The first step uses eigentransformation to infer global faces I_H^g, which we call global modeling. Principal Component Analysis (PCA) is used to fit the input face images as a linear combination of the LR face images in the training set. The HR images are then inferred

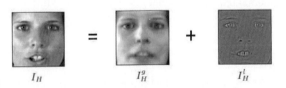

$$I_H \qquad\qquad I_H^g \qquad\qquad I_H^l$$

Fig. 2. Face hallucination framework

by replacing the LR training images with HR ones, while retaining the same combination coefficients [14]. At the second step, high-frequency contents of the HR images, I_H^l, are captured by a patch-based one-pass algorithm, [9], which we call local modeling. To improve the efficiency of searching the most compatible patch, we introduce stochastic search into the one-pass algorithm, which is a probabilistic method that iteratively propagates the targets' position using Bayes' rule. Finally, the super-resolution image, I_H, is the sum of global face and local face, $I_H = I_H^g + I_H^l$. The details of these algorithms are given in the next three sections.

4   Global Modeling

In global modeling we use an algorithm originally introduced by Wang [14], which is called eigentransformation. The eigentransformation is a simple and powerful technique for image enhancement based on principal component analysis (PCA). It assumes that we have a training set of pairs of images $\langle (L_1, H_1), \ldots, (L_n, H_n) \rangle$, where each pair (L_i, H_i) contains a low resolution face image L_i and its corresponding high-resolution counterpart H_i. The eigentransformation allows to represent any image as a linear combination of images in the training set. When given a low resolution image L, it finds the vector of coefficients $[c_1, \ldots, c_n]$ so that

$$L = \Sigma_{i=1}^n c_i L_i + \mu_L$$

where μ_L is the mean low-resolution face.

Given the vector $[c_1, \ldots, c_n]$, the approximate high resolution image H can be computed by

$$H = \Sigma_{i=1}^n c_i H_i + \mu_H$$

where μ_H is the mean high-resolution face image.

Because the coefficients are not computed from the HR training data, some non-face-like distortion may be introduced. To reduce the distortion, we apply constraints by bounding the projection onto each eigenvector by its corresponding eigenvalue, then the synthesized face image is reconstructed from these constrained coefficients.

5   Local Modeling

Given a global face, to construct the corresponding local face, we first filter the global face with a Gaussian high-pass filter, and then subdivide the filtered global face into patches, which we call the low-frequency patches of the HR faces, by scanning a window across the image in raster-scan order. Similarly, we also filter and subdivide the HR faces in the training set into patches which we call high-frequency patches of the training HR faces.

To construct a local face, for each low-frequency patch, a high-frequency patch of the training HR face is selected by a nearest neighbor search from the training set based on local low-frequency details and adjacent, previously determined HR patches. The selected high-frequency patch should not only come from a location in the training images that has a similar corresponding low-frequency appearance, but it should also match at the edges of the patch with the overlapping pixels, which we call high-frequency overlap, of its previously determined high-frequency neighbors to ensure that the high-frequency patches are compatible with those of the neighboring high-frequency patches.

In this work we compute the local faces with an algorithm that is an extension of the one-pass algorithm, proposed by Freeman et al. [9,5]. In the one pass algorithm, we first concatenate the pixels in the low-frequency patch and the high-frequency overlap to form a search vector. The training set also contains a set of such vectors. Then we search for a match by finding the nearest neighbor in the training set. When we find a match we extract the corresponding high-frequency patch from training data set and add it to the initial global face to obtain the output image.

Mathematically, this process can be described as follows. Suppose we have a training data set

$$\{\langle (x^{(i,j,k)}, y^{(i,j,k)}), z^{(i,j,k)} \rangle,$$

$$i = 1, 2, \ldots, l; j = 1, 2, \ldots, m; k = 1, 2, \ldots, n\}$$

where $x^{(i,j,k)}$ is the low-frequency patch at the ith row and j^{th} column of the kth training HR face image, $y^{(i,j,k)}$ is the corresponding high-frequency overlap and $z^{(i,j,k)}$ is the corresponding high-frequency patch of the training HR face image, l is the number of rows of patches in a training image, m is the number of columns of patches in a training image and n is the number of training images,

Given an input LR patch $\bar{x}$, we need to find an HR patch $z^{(i',j',k')}$ such that

$$(i', j', k') = \arg\min_{(i,j,k)} (d(\bar{x}, x^{(i,j,k)}) + \alpha * (d(y^{(i,j,k)}, y_N^{(i,j,k)}))$$

where $d(x, y)$ is the Euclidean distance between x and y, $y_N^{(i,j,k)}$ is the overlap of $z^{(i,j,k)}$ with the adjacent, previously determined high-frequency patches $z^{(i-1,j,k)}$ and $z^{(i,j-1,k)}$, α is a user-controlled weighting factor, and $z^{(i',j',k')}$ is the selected high-frequency patch.

6   Stochastic Search in Local Modeling

This work improves the efficiency of face image super-resolution using stochastic search for local modeling by exploiting the fact that face patches maintain relatively tight distributions for shape at successive iterations. For example, suppose we found the best match at iteration t, which is a patch from a left eye taken from training image T; intuitively, there is a high probability that the best match at iteration $t + 1$ should also belong to a left eye and come also from T, or from a training image that is similar to T. Therefore the position of the most compatible high-frequency patch at iteration t, z_t, and its history $Z_t = z_1, z_2, \ldots, z_{t-1}$ form a temporal Markov chain, so that the new position is conditioned directly only on the immediately preceding state, independent of its earlier history.

$$P(z_t|z_t) = P(z_t|z_{t-1})$$

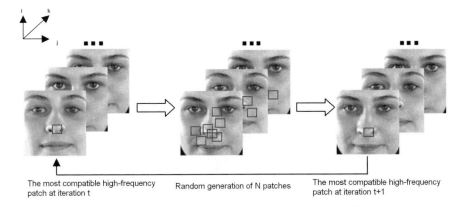

The most compatible high-frequency patch at iteration t   Random generation of N patches   The most compatible high-frequency patch at iteration t+1

Fig. 3. Local modeling with stochastic search

Therefore, based on z_{t-1}, we could estimate the most likely positions of z_t and we just need to search these positions instead of the exhaustive search performed by the conventional one-pass algorithm.

The key idea of the proposed algorithm is shown graphically in Figure 3, where we show how a stochastic search is performed using candidate patches taken from locations and images that are similar to the most recently found patch.

Let $z_t^{(i,j,k)}$ be the most compatible patch found at iteration t (for $t = 0$, an exhaustive search needs to be performed). To find the best patch at iteration $t + 1$, we generate a set of N candidate patch locations $\{(i', j', k')^1, ..., (i', j', k')^N\}$ around location (i, j) in images that are similar to image k.

Each location $(i', j', k')^q$ will be randomly generated according to the following distributions:

$$\begin{cases} i' = i + \alpha \\ j' = j + \beta \\ k' = \gamma(A(k)) \end{cases}$$

where α and β are normally distributed random variables, $A(k)$ is a list of face images that are similar to image k (which includes k itself) and $\gamma(.)$ is a sampling function that randomly selects face images from $A(k)$ with a probability that is directly proportional to their similarity with face image k. In a preprocessing stage, we build a directed graph where every vertex contains a face image and there is an edge $e(k, v)$ if vertices k and v contain similar face images, according to a mean-squared distance metric, as illustrated in Figure 4. Thus $A(k)$ is the set of vertices that are adjacent to k in this graph.

After generating the N candidate patches, we select as the next patch the best match within the generated point set and repeat the process again.

The proposed algorithm generates a set of N candidates around the point (i, j, k). Suppose we have n images in our training data set and each image has k patches, for exhaustive search, the running time for constructing one local face is $O(nk^2)$; for stochastic search, the running time is $O(kn + kN)$.

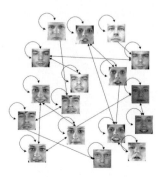

Fig. 4. Similar faces are connected in a graph structure

7 Experimental Results

For experiments we use the BioID data set which consists of 1521 gray-level face images of 384×288 pixels with a frontal view of 23 different people under a high variety of lighting conditions, backgrounds and face sizes. As the variations of human faces, such as glasses and beard, may greatly affect the performance of face hallucination in global modeling, we construct class-based subsets that contain images of small variations as training dataset. In preprocessing we crop and normalize, and then register the face images by 23 manually selected facial feature points (Figure 5), so that we can assume that the same parts of faces appear in roughly the same parts of the images. The face image size is fixed to 128×128 pixels and we use them as the training HR face images. The training LR face images are down-sampled from the training HR face images by averaging the neighborhood pixels. In our experiments, the down-sample factor is 8.

To construct our training dataset, we generate global faces from the training LR face images and filter them with a Gaussian high-pass filter. Then we subdivide the filtered global face images into low-frequency patches by scanning a 4×4 pixel window across the image in raster-scan order. Then we again filter and subdivide the training HR face images into 4×4 pixel high-frequency patches. At each step we also get a 9-bit overlap of each high-frequency patch with the high-frequency patches above and to the left. Then we create our training vectors by concatenating the low-frequency patches and corresponding high-frequency overlaps. In practice, the size of low-frequency patches and high-frequency patches is not necessarily the same. The parameter α, which con-

Fig. 5. 23 manually selected facial feature points

Table 1. Quantitative evaluation

N	MSE	SSIM
300	37.2095	0.9319
500	30.1520	0.9387
1200	28.3502	0.9386
2000	29.6137	0.9396
exhaustive	31.2918	0.9403

Table 2. Computational time

Algorithm	Image size	Mean time
Stochastic	128×128	$16s$
Exhaustive	128×128	$99s$

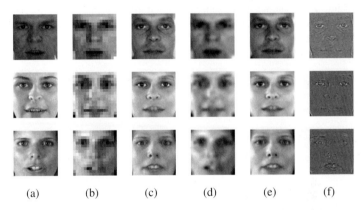

(a)   (b)   (c)   (d)   (e)   (f)

Fig. 6. (a) Original high-resolution face images. (b) Input low-resolution face images. (c) Generated super-resolution face images. (d) Results of bicubic interpolation. (e) Global faces. (f) Local faces. This figure shows the super-resolution results of exhaustive search. Compared with the input low-resolution images and the bicubic interpolation results, the super-resolution face images have much clearer detailed features.

trols the trade-off between matching the low-frequency patches and finding the most compatible high-frequency patches, is set to 0.2.

Figure 6 shows the face image super-resolution results of exhaustive search. The LR inputs are 16×16 pixel face images. The super-resolution face images are 128×128 pixel face images. Compared with the input images and the bicubic interpolation results, the super-resolution face images have much clearer features.

Figure 7 to 10 show the examples of face image super-resolution results of stochastic search. In Figure 7, the size of the randomly generated search set is 300, which is too small, so that the quality of local face is low. In Figure 8, the size of the randomly generated search set is 1200. In Figure 9, the size of the randomly generated search set is 500. In Figure 10, the size of the randomly generated search set is 2000. The results show that a larger generated search set provides a better-quality local face.

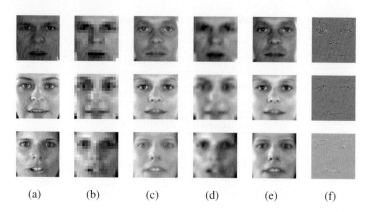

   (a)   (b)   (c)   (d)   (e)   (f)

Fig. 7. (a) Original high-resolution face images. (b) Input low-resolution face images. (c) Generated super-resolution face images. (d) Results of bicubic interpolation. (e) Global faces. (f) Local faces. This figure shows the results of super-resolution using stochastic search. The size of randomly generated search set is set to 300. However, this is too small, so that the quality of local face decreases.

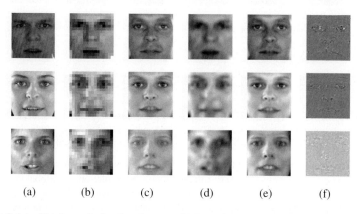

   (a)   (b)   (c)   (d)   (e)   (f)

Fig. 8. (a) Original high-resolution face images. (b) Input low-resolution face images. (c) Generated super-resolution face images. (d) Results of bicubic interpolation. (e) Global faces. (f) Local faces. In this figure, the size of randomly generated search set is set to 1200.

In Table 1, we report the Mean-Squared-Error (MSE) between the super-resolution face images and the ground truth. Although the MSE is a physically meaningful metric for signal reconstruction, it does not necessarily reflect perceived visual quality by humans [15]. Thus we also use the mean Structural Similarity (SSIM) Index, which aims to primarily measure the structural changes between a reference image and its distorted version [16]. These results confirmed the results observed from Figure 7 through 10 that generally a better-quality local face can be generated with a larger search set.

In Table 2 we report our computation time testing results. The time to synthesize a local face of 128×128 pixels using the stochastic search, with a randomly generated set of size 2000, is about $16s$, while using the exhaustive search takes about $99s$, on a 2.4 GHz PC in Matlab.

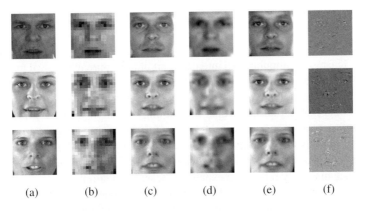

Fig. 9. (a) Original high-resolution face images. (b) Input low-resolution face images. (c) Generated super-resolution face images. (d) Results of bicubic interpolation. (e) Global faces. (f) Local faces. In this figure the size of randomly generated search set is set to 500.

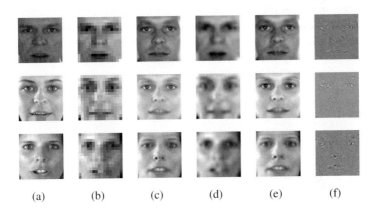

Fig. 10. (a) Original high-resolution face images. (b) Input low-resolution face images. (c) Generated super-resolution face images. (d) Results of bicubic interpolation. (e) Global faces. (f) Local faces. In this figure, the size of randomly generated search set is set to 2000. From Figure 7 to 10, we can see that stochastic local modeling with a larger generated search set would have better quality of local faces but take a longer time. Therefore, a tradeoff between quality and efficiency need to be found.

From Table 1 and 2, and Figure 7 to 10, we can see that stochastic local modeling with a larger generated search set generates better-quality local faces but takes a longer time. Therefore, a tradeoff between quality and efficiency has to be considered.

8 Conclusions and Future Work

In this work, we presented a framework for face image super-resolution integrating global modeling and local modeling. In global modeling, we infer the global face of the input LR face image with a linear combination of PCAs from the HR faces in the training set. In local modeling, we use a stochastic patch-based one-pass algorithm to infer the local face. The final super-resolution image is the sum of the global face and the local face. The most computational intensive part of this approach is local modeling, the computation time of which is proportional to the size of training dataset. We introduce a stochastic local search into the one-pass algorithm that constraints the search space to a fixed size and makes real-time super-resolution possible. If we have n images in our training data set, each image has k patches, and the generated set contains N patches, our method reduces the running time for constructing one local face from $O(nk^2)$ in exhaustive search to $O(kn + kN)$ in stochastic search. Experimental results show that the difference in quality relative to exhaustive search is negligible.

For future work, we will extend this algorithm to generic object super-resolution, for which we need a very large training dataset that contains a subset for each specific object. We are also interested in approaches to exploit super-resolution for object recognition, both by computers and by people.

To further reduce the running time of the super-resolution algorithm, we will use stream processing to parallelize the execution. Stream processing permits the execution of data-parallel algorithms with stream processors such as graphic processing units (GPUs), while using the central processing unit (CPU) for other purposes simultaneously. This would enable a conventional PC to run the image super-resolution algorithm in real time.

References

1. Park, S.C., Park, M.K., Kang, M.G.: Super-resolution image reconstruction: a technical overview. IEEE Signal Processing Magazine 20, 21–36 (2003)
2. Dodgson, N.: Quadratic interpolation for image resampling. IEEE Transactions on Image Processing 6, 1322–1326 (1997)
3. Atkins, C.B., Bouman, C.A., Allebach, J.P.: Optimal image scaling using pixel classification. In: 2001 International Conference on Image Processing, pp. 864–867 (2001)
4. Greenspan, H., Anderson, C., Akber, S.: Image enhancement by nonlinear extrapolation in frequency space. IEEE Transactions on Image Processing 9, 1035–1048 (2000)
5. Freeman, W.T., Pasztor, E.C., Carmichael, O.T.: Learning low-level vision. International Journal on Compter Vision 40, 25–47 (2000)
6. Baker, S., Kanade, T.: Hallucinating faces. In: Fourth IEEE International Conference on Automatic Face and Gesture Recognition, Grenoble, France (2000)
7. Baker, S., Kanade, T.: Limits on super-resolution and how to break them. IEEE Transactions on Pattern Analysis and Machine Intelligence 24, 1167–1183 (2002)
8. Dedeoğlu, G.: Exploiting Space-Time Statistics of Videos for Face "Hallucination". PhD thesis, The Robotics Institute, Carnegie Mellon University, Pittsburgh, Pennsylvania (2007)
9. Freeman, W.T., Jones, T.R., Pasztor, E.C.: Example-based super-resolution. IEEE Computer Graphics and Applications 22, 56–65 (2002)
10. Liang, L., Liu, C., Xu, Y., Guo, B., Shum, H.Y.: Real-time texture synthesis by patch-based sampling. ACM Transactions on Graphics 20, 127–150 (2001)
11. Liu, C., Shum, H.Y., Zhang, C.S.: A two-step approach to hallucinating faces: global parametric model and local nonparametric model. In: Proceedings of the 2001 IEEE Conference on Computer Vision and Pattern Recognition (CVPR 2001), pp. 192–198 (2001)

12. Liu, C., Shum, H.Y., Freeman, W.T.: Face hallucination: Theory and practice. International Journal of Computer Vision (IJCV) 75, 115–134 (2007)
13. Wang, X., Tang, X.: Face hallucination and recognition. In: Proceedings of the Fourth International Conference on Audio- and Video-Based Personal Authentication (IAPR), University of Surrey, Guildford, U.K., pp. 486–494 (2003)
14. Wang, X., Tang, X.: Hallucinating face by eigentransformation. IEEE Transactions on Systems, Man and Cybernetics, Part C: Applications and Reviews 35, 425–434 (2005)
15. Girod, B.: What's wrong with mean-squared error? In: Digital images and human vision, pp. 207–220. MIT Press, Cambridge (1993)
16. Wang, Z., Bovik, A.C., Sheikh, H.R., Simoncelli, E.P.: Image quality assessment: From error visibility to structural similarity. IEEE Transactions on Image Processing 13, 600–612 (2004)

A Framework for Long Distance Face Recognition Using Dense- and Sparse-Stereo Reconstruction

Ham Rara, Shireen Elhabian, Asem Ali, Travis Gault, Mike Miller,
Thomas Starr, and Aly Farag

CVIP Laboratory, University of Louisville, KY, USA
{hmrara01,syelha01,amali003,travis.gault,wmmill06,tlstar01,
aafara01}@louisville.edu

Abstract. This paper introduces a framework for long-distance face recognition using both dense- and sparse-stereo reconstruction. Two methods to determine correspondences of the stereo pair are used in this paper: (a) dense global stereo-matching using maximum-a-posteriori Markov Random Fields (MAP-MRF) algorithms and (b) Active Appearance Model (AAM) fitting of both images of the stereo pair and using the fitted AAM mesh as the sparse correspondences. Experiments are performed regarding the use of different features extracted from these vertices for face recognition. A comparison between the two approaches (a) and (b) are carried out in this paper. The cumulative rank curves (CMC), which are generated using the proposed framework, confirms the feasibility of the proposed work for long distance recognition of human faces.

1 Introduction

Automatic face recognition is a challenging task that has been an attractive research area in the past three decades (for more details see [1]). At the outset, most efforts were directed towards 2D facial recognition which utilizes the projection of the 3D human face onto the 2D image plane acquired by digital cameras. The face recognition problem is then formulated as follows: given a still image, identify or verify one or more persons in the scene using a stored database of face images. The main theme of the solutions provided by different researchers involves detecting one or more faces from the given image, followed by facial feature extraction which can be used for recognition. Challenges involving 2D face recognition are well-documented in the literature. Intra-subject variations such as illumination, expression, pose, makeup, and aging can severely affect a face recognition system.

To address pose and illumination, researchers recently are focusing on 3D face recognition [2]. 3D face geometry can either be acquired using 3D sensing devices such as laser scanners [3] or reconstructed from one or more images [4-6]. Although 3D sensing devices have been proven to be effective in 3D face recognition [7], their high cost, limited availability and controlled environment settings have created the need for methods that extract 3D information from acquired 2D face images.

Recently, there has been interest in face recognition at-a-distance. Yao, et al. [8] created a face video database, acquired from long distances, high magnifications, and

G. Bebis et al. (Eds.): ISVC 2009, Part I, LNCS 5875, pp. 774–783, 2009.

both indoor and outdoor under uncontrolled surveillance conditions. Medioni, et al. [9] presented an approach to identify non-cooperative individuals at a distance by inferring 3D shape from a sequence of images.

To realize our objectives and the current lack of existing facial stereo databases, we constructed our own passive stereo acquisition setup [10]. The setup consists of a stereo pair of high resolution cameras (and telephoto lenses) with adjustable baseline. It is designed such that user can remotely pan, tilt, zoom and focus the cameras to converge to the center of the cameras' field of views on the subject's nose tip. This system is used to capture stereo pairs of 30 subjects at various distances (3-, 15-, and 33-meter ranges).

The paper is organized as follows: Section 2 discusses stereo reconstruction methods (dense and sparse), Section 3 shows the experimental results, Section 4 validates the best method in Sec. 3 using the FRGC database, and later sections deal with discussions and limitations of the proposed approaches, conclusions and future work.

2   Stereo Matching-Based Reconstruction

Dense, Global Stereo Matching: The objective of the classical stereo problem is to find the pair of corresponding points p and q that result from the projection of the same scene point (X, Y, Z) to the two images of the stereo-pair. Currently, the state-of-the-art in stereo matching is achieved by global optimization algorithms [11], where the problem is formulated as a maximum-a-posteriori Markov Random Field (MAP-MRF) scenario. Given the left and right images, the goal is to find the disparity map D, where at each pixel p, the disparity is $d_p = p_x - q_x$. To correctly solve this problem, the constraints of the visual correspondence should be satisfied: (a) uniqueness, where each pixel in the left image corresponds to at most one pixel in the right image and (b) occlusion, where some pixels do not have correspondences. To achieve these constraints, similar to Kolmogorov's approach [12], we treat the two images symmetrically by computing the disparity maps for both images simultaneously. The disparity map D is computed by minimizing the energy function $E(D) = E_{data}(D) + E_{smooth}(D) + E_{vis}(D)$. The terms refer to the penalty, smoothness, and visibility constraint terms [13][14]. To fill the occluded regions, we propose to interpolate between the correctly reconstructed pixels of each scan line using the cubic splines [15] interpolation model. Finally, after getting a dense disparity map from which we get a set of correspondence points, we reconstruct the 3D points of the face [10]. To remove some artifacts of the reconstruction, an additional surface fitting step is done [16].

Sparse-Stereo Reconstruction: The independent AAM version of [17] is used to find sparse correspondences of the left and right images of the stereo pair. The shape s can be expressed as the sum of a base shape s_0 and a linear combination of n shape vectors s_i, $s = s_0 + \sum_i p_i s_i$, where p_i are the shape parameters. Similarly, the appearance $A(x)$ can be expressed as the sum of the base appearance $A_0(x)$ and a linear combination of basis images $A_i(x)$, $A(x) = A_0(x) + \sum_i \lambda_i A_i(x)$, where the pixels x lie on the base mesh s_0. Fitting the AAM to an input image involves minimizing the error image between the input image warped to the base mesh and the appearance $A(x) = A_0(x) + \sum_i \lambda_i A_i(x)$, that is, $\sum_{x \in s_0} [A_0(x) + \sum_i \lambda_i A_i(x) - I(W(x; p))]^2$. For this work, the error image is minimized using the *project-out* version of the inverse compositional image alignment (ICIA) algorithm [17].

To facilitate a successful fitting process, the AAM mesh is initialized according to detected face landmarks (eyes, mouth center, and nose tip). After detecting these facial features, the AAM base mesh is warped to these points.

The detection of facial features starts with identifying the possible facial regions in the input image, using a combination of the Viola-Jones detector [18] and the skin detector of [19]. The face is then divided into four equal parts to establish a geometrical constraint of the face. The face landmarks are then identified using variants of the Viola-Jones detector, i.e., the face detector is replaced with the corresponding face landmark (e.g., eye detector) detector [20]. False detections are then removed by taking into account the geometrical structure of the face (i.e., expected facial feature locations).

3   Experimental Results

The 3D acquisition system in [10] is used to build a human face database of 30 subjects at different ranges in controlled environments. The database consists of a gallery at 3 meters and three different probe sets at the 3-, 15-, and 33-meter ranges. Table 1 shows the system parameters at different ranges.

Table 1. Stereo-based acquisition system parameters

Range (m)	Baseline B (m)	Zoom f(mm)	Focus $\approx$	Pan ϕ (degree)	Tilt (degree)
3	0.6	150	Range	$\approx 5.6°$	$\approx 0°$
15	1.76	400	Range	$\approx 3.3°$	$\approx 0°$
33	1.76	400	Range	$\approx 1.5°$	$\approx 0°$

Dense, Global Stereo 3D Face Reconstructions: The gallery is constructed by capturing stereo pairs for the 30 subjects at the 3-meter range. We reconstruct the 3D face of each subject using the approach that is described in Section 2. Fig. 1(a) illustrates a sample from this gallery for different subjects. This figure shows the left image of each subject and two different views for the 3D reconstruction with and without the textures.

For the dense, global 3D reconstruction approach, only the images from the probe sets 3-meter and 15-meter ranges are considered. The reason behind this is that the methodology from Sec. 2.1 fails to determine acceptable correspondences of the stereo-pair images of the 33-meter range, leading to unacceptable 3D reconstructions. *This result has led the authors to propose the second method (sparse-stereo) to deal with stereo-pairs that are difficult to extract dense correspondences.* Fig. 1(b) and 1(c) illustrate the samples from these probe sets.

Sparse-Stereo 3D Face Reconstructions: The gallery and probe sets are similar to above, except that the 33-meter images are now included in the probe set. The training of the AAM model involves images from the gallery. The vertices of the final

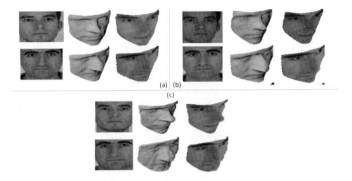

Fig. 1. Dense 3D reconstructions: (a) 3-meter gallery, (b) 3-meter, and (c) 15-meter probes

AAM mesh on both left and right images can be considered as a set of corresponding pair of points, which can be used for stereo reconstruction. To further refine the correspondences, a local NCC search around each point is performed, using the left image as the reference, to get better correspondences. Fig. 2 shows stereo reconstruction results of three subjects, visualized with the x-y, x-z, and y-z projections, after rigid alignment to one of the subjects. Notice that in the x-y projections, the similarity (or difference) of 2D shapes coming from the same (or different) subject is enhanced. This is the main reason behind the use of x-y projections as features in Sec. 3 (**Recognition**).

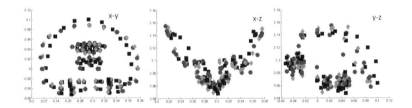

Fig. 2. Reconstruction results. The 3D points are visualized as projections in the x-y, x-z, and y-z planes. Circle and diamond markers belong to the same subject, while the square markers are that of a different subject.

Recognition: For face recognition, we use five approaches for using the 3D face vertices derived from dense- and sparse-stereo (3D-AAM) to identify probe images against the gallery: (a) moment-based approach for dense 3D reconstructions, (b) feature vectors derived from Principal Component Analysis (PCA) of 3D-AAM vertices, (c) goodness-of-fit criterion (*Procrustes*) after rigidly registering the 3D-AAM vertices of a probe with that of a gallery subject, (d) feature vectors from PCA of x-y plane projections of the 3D-AAM vertices, and (e) the same procedure as (c) but using the x-y plane projections of the 3D-AAM vertices of both probe and gallery, after frontal pose normalization.

Moment-based Recognition: For the dense 3D reconstructions, to compare between gallery and probe sets, feature vectors are derived from moments [21] derived from the 3D vertex coordinates. The moments are computed as $\eta_{rst} = \sum_X \sum_Y \sum_Z X^r Y^s Z^t$.

Principal Component Analysis (PCA): To apply PCA [22] for feature classification, the primary step is to solve for the matrix P of principal components from a training database, using a number of matrix operations. The feature vectors Y can then be determined as follows: $Y = P^T X$, where X is a centered input data. The similarity measure used for recognition is the L_2 norm.

Goodness-of-fit (Procrustes): The Procrustes distance [23] between two shapes is a least-squares type of metric that requires one-to-one correspondence between shapes. After some preprocessing steps involving the computation of centroids, rescaling each shape to have equal size and aligning with respect to translation and rotation, the squared Procrustes distance between two shapes x_1 and x_2 is the sum of squared point distances, $P_d^2 = \|x_1 - x_2\|^2$.

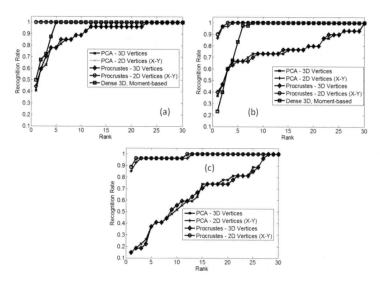

Fig. 3. Cumulative match characteristic (CMC) curve of the: (a) 3-meter probe set, (b) 15-meter probe set, and (c) 33-meter probe set. Note that only the 3-m and 15-m probe sets use moment-based recognition (see Fig. 3).

Discussion of Results: Fig. 3 shows the cumulative rank curves (CMC) curves for the five types of feature extraction methods in the previous section (Sec. 3), using 3-, 15-, and 33-meter probes. We can draw four conclusions: (a) both 2D Procrustes (i.e., x-y projection of 3D-AAM) and 2D PCA outperform both 3D Procrustes and 3D PCA, (b) goodness-of-fit criterion (Procrustes) slightly outperforms PCA in both 2D and 3D, (c) degradation of recognition at increased distances, and (d) the moment-based methods perform poorly at lower ranks but shoots up quickly to 100% at rank-5 and rank-8 for the 3-m and 15-m probe sets, respectively.

The conclusion in (a) can be explained with the help of Fig. 4. The diagram shows the top view of a simple stereo system. O_l and O_r are centers of projection, and p_l, p_r, q_l, q_r are points on the left and right images.

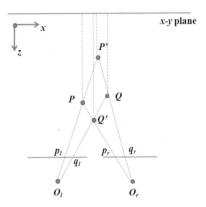

Fig. 4. Simple stereo illustration. 3D reconstruction is sensitive to the correspondence problem but projection to the x-y plane minimizes the error.

Assume that the y-coordinates of the four image points are equal. p_l and p_r will reconstruct P, q_l and q_r will reconstruct Q, and so on. Notice that a small change in the correspondence affects the xyz reconstructions hugely, i.e., their Euclidean distance between each other is huge. But when they are projected to the x-y plane, their 2D Euclidean distances with each other are considerably lesser. This scenario is possibly happening with the correspondence of the AAM vertices between the left and right image.

The conclusion in (b) is related to the primary purpose of PCA, which is optimal reconstruction error, in the mean-square-error (MSE) sense. Projection to a low-dimensional space may remove the classification potential of the vectors. *There is no dimensional change with rigid alignment using Procrustes; similar shapes are expected to have less Procrustes distance after rigid alignment and geometric information of faces (e.g., distance ratios between face parts) are maintained.*

Results are expected to degrade with distance since the captured images are at less ideal conditions (although recognition using 2D x-y projections remain stable). The work in Medioni [9] deals with identification at a distance using recovered dense 3D shape from a sequence of captured images. The results at the 15- and 33-meter range (Figs 5(b,c)) are comparable (and slightly better) than their 9-meter results; however, their experimental setting may be less controlled than ours.

The authors cannot find any concrete reason behind conclusion (d), except the fact that the dense reconstructed shapes may be overfitted by the *gridfit* procedure of [16]. This part of the framework is still a work-in-progress and more elaborate 3D recognition methods will be incorporated as future work (see Conclusions).

Sensitivity Analysis: This section performs a sensitivity analysis of the recognition performance with respect to errors in the AAM fitting of the stereo pair of images. To simulate errors introduced to the system, the fitted AAM vertices of the stereo pair are randomly perturbed with additive white Gaussian noise of a certain variance. Fig. 5 shows the plot of *rank-1* recognition rates versus point sigmas, for the 3-, 15-, and 33-meter ranges. The recognition method used is the 2D *Procrustes* approach of Fig. 3. These findings reinforce the known fact that acceptable AAM fitting is necessary to get satisfactory correspondence, which leads to suitable recognition.

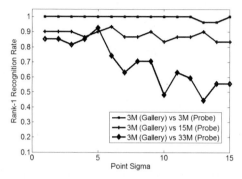

Fig. 5. Sensitivity analysis of recognition with respect to errors in the AAM fitting of the stereo pair of images. Notice that the recognition rates are fairly stable across various values of σ for the 3- and 15-meter probe sets. However, for the 33-meter probe set, the recognition performance severely deteriorates after $\sigma = 5$.

4   Validation with FRGC Database

The main purpose of this section is to test if the 2D (x-y projection of 3D-AAM) *Procrustes results* of Fig. 5 carry over to the much larger FRGC database. Since the FRGC database contains *close-to-perfect* dense 3D information, the advantage of 2D *Procrustes* over its 3D equivalent would also be investigated. The section of the FRGC database [24] with range (3D) data is used. Each range is accompanied by a corresponding texture image. 115 subjects (with three images each, for a total of 345) were chosen out of the total number of subjects, the number being restricted by the manual annotation of AAM training data.

Fig. 6 may provide some insights regarding the better performance of 2D *Procrustes* over 3D *Procrustes*. Note that the 2D+3D (range, depth) partition of the FRGC database contains 2D video images with corresponding range values for each pixel. After the manual/automatic fitting of AAM vertices, the corresponding range (depth) value is extracted for each vertex. In Fig. 6, the red dots represent the extracted depth values using the 2D coordinates of the fitted vertices. Notice that some depth values are undesirable; they do not contain the intended depth of the facial feature points (e.g., nose area of Fig. 6(a) and face boundary of Fig. 6(b)). The 2D coordinates of the AAM vertices (except for the face boundary) are adjusted according to the COSMOS framework [25]; specifically, the vertices are adjusted along a local neighborhood in the horizontal (x) direction, according to some extremum values defined by [25]. The green dots represent the adjusted 3D vertices. The face boundaries are adjusted using ad-hoc methods that investigate the most acceptable face vertex depth of the whole face boundary. Fig. 7 shows both the 2D and 3D Procrustes methods using the FRGC database. Similar to Fig. 3, the 2D approach outperforms the 3D version.

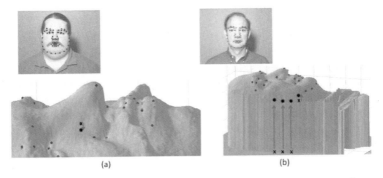

(a)   (b)

Fig. 6. Using the original 2D coordinates of the fitted AAM vertices in the range image do not give the desired feature locations; hence, the adjustment of the vertices: movement of the (**x**) dot to the (●) dot at the (a) nose location and (b) face boundary, indicated by the arrow. The vertex movement is not noticeable in the corresponding 2D video image.

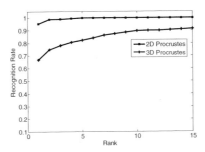

Fig. 7. Cumulative match characteristic (CMC) curve of the FRGC experiment. Notice the superiority of the 2D Procrustes approach over the 3D version. Classification is performed using the leaving-one-out methodology of [22]. The reason behind this is related to Fig. 6; the movement from the red to green dots represents a large variation in the 3D coordinate system, but when projected to the 2D x-y plane, the planar movement is much smaller, e.g., there is no noticeable pixel difference in the 2D image after the vertex movement in the nose of Fig. 6(a).

5   Conclusions and Future Work

This paper described a framework for identifying faces remotely, specifically at distances 3-, 15- and 33-meters. The best approach used AAM to get correspondences between the left and right images of the stereo pair. This study found out that recognition using the point-to-point distance between 2D x-y projected shapes (after rigid registration, Procrustes) outperforms the others. The advantage of the 2D Procrustes approach over its 3D version carried over to the much larger FRGC database (with chosen 115 subjects). Using our database, we have illustrated the potential of using these few vertices, as opposed to the whole set of points of the human face (dense reconstruction).

The authors are aware of more elaborate methods of 3D shape classification related to face recognition (even with the presence of face expression), such as [7]. However, for this application (of identifying faces at far distances), a *close-to-perfect* dense 3D scan of the face is difficult to obtain; therefore, this study currently deals with sparse 3D points related to the AAM vertices. The next step of this work is to densify the AAM mesh to have a reconstruction that has a close semblance to a dense 3D scan but still contain lesser vertices than conventional 3D scans. This study does not currently consider face expression (since the 3D sparse reconstruction can only do so much) but will be considered as future work once the densification of the AAM vertices is taken care of. Additionally, the authors plan to increase the database size (for better statistical significance) and capture images at further distances (with the help of state-of-the art equipment).

References

1. Zhao, W., Chellappa, R., Rosenfeld, A.: Face recognition: a literature survey. ACM Computing Surveys 35, 399–458 (2003)
2. Kittler, J., Hilton, A., Hamouz, M., Illingworth, J.: 3D assisted face recognition: A survey of 3D imaging, modelling and recognition approaches. In: Proc. of CVPR (2005)
3. Pan, G., Han, S., Wu, Z., Wang, Y.: 3D face recognition using mapped depth images. In: Proc. of CVPR- Workshop on Face Recognition Grand Challenge Experiments (2005)
4. Blanz, V., Vetter, T.: Face recognition based on fitting a 3D morphable model. IEEE Trans. on PAMI 25, 1063–1074 (2003)
5. Hu, Y., Jiang, D., Yan, S., Zhang, L., Zhang, H.: Automatic 3D reconstruction for face recognition. In: Proc. of Sixth IEEE International Conference on Face and Gesture Recognition, pp. 843–848 (2004)
6. Chowdhury, A.R., Chellappa, R., Krishnamurthy, S., Vo, T.: 3D face reconstruction from video using a generic model. In: Proc. of IEEE International Conference on Multimedia and Expo., pp. 449–452 (2002)
7. Bronstein, A.M., Bronstein, M.M., Kimmel, R.: Three-dimensional face recognition. Intl. Journal of Computer Vision (IJCV), 5–30 (2005)
8. Yao, et al.: Improving long range and high magnification face recognition: Database acquisition, evaluation, and enhancement. Computer Vision and Image Understanding, CVIU (2008)
9. Medioni, G., Jongmoo, C., Cheng-Hao, K., Choudhury, A., Li, Z., Fidaleo, D.: Noncooperative persons identification at a distance with 3D face modeling. In: Proc. of IEEE International Conference on Biometrics: Theory, Applications, and Systems, BTAS 2007 (2007)
10. Rara, H., Elhabian, S., Ali, A., Miller, M., Starr, T., Farag, A.: Face recognition at-a-distance based on sparse-stereo reconstruction. In: Proc. of CVPR- Biometrics Workshop (2009)
11. Szeliski, R., Zabih, R., Scharstein, D., Veksler, O., Kolmogorov, V., Agarwala, A., Tappen, M.F., Rother, C.: A comparative study of energy minimization methods for markov random fields. In: Leonardis, A., Bischof, H., Pinz, A. (eds.) ECCV 2006. LNCS, vol. 3952, pp. 16–29. Springer, Heidelberg (2006)

12. Kolmogorov, V., Zabih, R.: Multi-camera scene reconstruction via graph cuts. In: Heyden, A., Sparr, G., Nielsen, M., Johansen, P. (eds.) ECCV 2002. LNCS, vol. 2352, pp. 82–96. Springer, Heidelberg (2002)
13. Birchfield, S., Tomasi, C.: A pixel dissimilarity measure that is insensitive to image sampling. IEEE Trans. on PAMI 20, 401–406 (1998)
14. Boykov, Y., Kolmogorov, V.: An experimental comparison of min-cut/max-flow algorithms for energy minimization in vision. IEEE Trans. on PAMI 26, 1124–1137 (2004)
15. Knott, G.D.: Interpolating Cubic Splines. Springer, Heidelberg (2000)
16. D'Errico, J.: Surface Fitting using gridfit. In: Matlab Central File Exchange (2006), http://www.mathworks.com/matlabcentral/fileexchange/
17. Matthews, I., Baker, S.: Active Appearance Models Revisited. In: International Conference on Computer Vision, ICCV (2004)
18. Viola, P., Jones, M.J.: Robust real-time face detection. International Journal of Computer Vision (IJCV), 151–173 (2004)
19. Jones, M., Rehg, J.: Statistical color models with application to skin detection. International Journal of Computer Vision (IJCV), 81–96 (2002)
20. Castrillon-Santana, M., Deniz-Suarez, O., Anton-Canalıs, L., Lorenzo-Navarro, J.: Face And Facial Feature Detection Evaluation: Performance Evaluation of Public Domain Haar Detectors for Face and Facial Feature Detection. In: VISAPP (2008)
21. Elad, M., Tal, A., Ar, S.: Content based retrieval of vrml objects: an iterative and interactive approach. In: Proceedings of the sixth Eurographics workshop on Multimedia, pp. 107–118 (2001)
22. Belhumeur, P., Hespanha, J., Kriegman, D.: Eigenfaces vs. Fisherfaces: Recognition using Class Specific Linear Projection. IEEE Trans. PAMI (1998)
23. Cootes, T.F., Taylor, C.J.: Statistical Models of Appearance for Computer Vision. Technical Report, University of Manchester, UK (2004)
24. Chang, K., Bowyer, K., Flynn, P.: An Evaluation of Multimodal 2D+3D Face Biometrics. IEEE Trans. PAMI (2005)
25. Dorai, C., Jain, A.K.: COSMOS - A Representation Scheme for 3D Free-Form Objects. IEEE Trans. PAMI (1997)

Multi-view Reconstruction of Unknown Objects within a Known Environment

Stefan Kuhn and Dominik Henrich

Lehrstuhl für Angewandte Informatik III (Robotik und Eingebettete Systeme)
Universität Bayreuth, D-95440 Bayreuth, Germany
Stefan.Kuhn@uni-bayreuth.de, Dominik.Henrich@uni-bayreuth.de

Abstract. We present a general vision-based method for reconstructing multiple unknown objects (e.g. humans) within a known environment (e.g. tables, racks, robots) which usually has occlusions. These occlusions have to be explicitly considered since parts of the unknown objects might be hidden in some or even all camera views. In order to avoid cluttered reconstructions, plausibility checks are used to eliminate reconstruction artifacts which actually do not contain any unknown object. One application is a supervision/surveillance system for safe human/robot-coexistence and –cooperation. Experiments for a voxel-based implementation are given.

1 Introduction

Geometrical information about objects is required in many applications. In several cases this information is *known*. For example, most industrial robots act in a geometrically completely known environment. It is indispensable, to guarantee the correctness of this information anytime, in order to avoid collisions. Therefore, fences and safety light barriers are set up to guarantee this correctness and to stop the robot in an unexpected situation. In other cases, the geometrical information is *not known* in advance. Thus, vision sensors can be used, to reconstruct this information. In the example, it would become possible, that a human can walk through the robots workspace, since its geometrical information is reconstructed and included in the robot's environment model, such that even in this case no collision occurs.

The reconstruction of objects based on its silhouettes in multiple cameras is known as *surface from silhouette* or *inferred visual hull* ([12], [15]). Many volume-based ([3], [11], [16]) and surface-based ([5], [13]) approaches have been investigated in the past. Most of them have the assumption that the object(s) to reconstruct resides within the common volume which is seen by all cameras. Furthermore, almost all approaches have the assumptions that the object(s) to reconstruct is not occluded by static obstacles – like tables, racks or the robot itself in the example. It may result in an incomplete reconstruction of a human, since conventional background subtraction methods ([4], [6]) are not able to generate the silhouette of the occluded parts of the human. Thus, it is necessary to explicitly consider these occlusions.

G. Bebis et al. (Eds.): ISVC 2009, Part I, LNCS 5875, pp. 784–795, 2009.

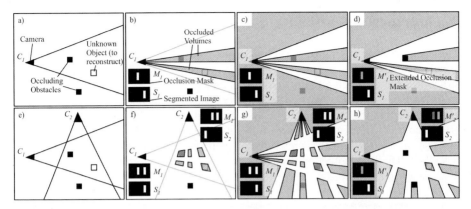

Fig. 1. Illustration of the visibility for a setup with one camera C_1 (first row, a-d) and a setup with multiple cameras C_i (second row, e-h) using different approaches for treating occlusions (b-d and f-h) resulting from the known environment (a, c)

Recently, two concepts to overcome these limitations have been investigated. First, occlusions can be modeled by *occlusion masks* M_i which are binary masks in image space marking regions where occlusions can occur. These marked regions are simply added to the segmented regions S_i detected by the background subtraction method when calculating the inferred visual hull (Fig. 1 b, f). In [2] and [10] the concept of occlusion masks has been used for *masking* dynamic occluding objects (i.e. robots) in order to avoid future collisions with objects, while [7], [8] and [9] automatically *generate* these occlusion masks for static occluding objects by observing active objects in a scene. Unfortunately, using occlusion masks causes more occlusions in the reconstruction than necessary since the volume between the camera and the occluding object is interpreted as occlusion as well even though a background subtraction method could detect objects in front of this occluding object. Second, objects which reside outside the common observed volume can be correctly integrated in the inferred visual hull, if the complement of the fused back-projected free space is calculated [5]. In Fig. 1 c, g this concept is applied in conjunction with the previously described occlusion masks. Note, the object residing outside the common observed volume is now included. But still the reconstruction contains unnecessary occlusions.

Thus, we propose a new approach (Fig.1 d, h), for treating occlusions in a general and more accurate way by additionally utilizing the geometrical information of the known environment (Section 2). Both concepts described above are contained by our approach. Furthermore, the information about how far a pixel can see up to the first occluding object is included (M'_i), resulting in more accurate reconstructions. In most cases, it results in cluttered reconstructions due to the occlusions. Therefore, we propose using plausibility checks, which revise reconstruction artifacts that do not contain an object. Furthermore, a voxel-based algorithm of our approach is provided in Section 3. The memory consumption of look-up tables, which are used for optimization purpose, is analyzed. Experimental results are discussed in Section 4. The paper concludes with Section 5.

2 Occlusions and Visibility

This section comprises a theoretical look at the problem of occlusions and visibility in a multi-camera setup using plausibility checks to eliminate pseudo objects, i.e. reconstruction artifacts which actually do not contain an object.

In the first subsection, the types of objects that are contained by surveyed scenes are described. In the following subsection, visibility and occlusions for a single camera are considered, taking the types of objects into account. Thereafter, the simultaneous use of several cameras with different perspectives onto the surveyed scene is described. The last subsection discusses how reconstruction artifacts can be revised, using plausibility checks.

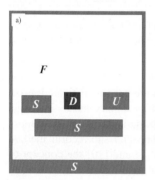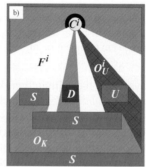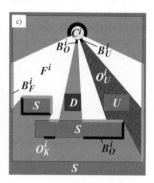

Fig. 2. Illustration of the visibility and the occlusions using color or grayscale cameras with common background subtraction methods. Nomenclature: C^i: Camera i; F^i: Free in camera i; S: Static known object; D: Dynamic known object; U: Unknown object; O^i_K: Known occlusion in i; O^i_U: Unknown occlusion in i; B^i_F: Free boundary in i; B^i_O: Occlusion boundary in i; B^i_U: Unknown boundary in i.

2.1 Object Types

The surveyed scene contains static and dynamic objects. *Static* objects S are racks, tables etc. The geometry, position and the appearance of those objects are *known* and do not change over the time (apart from possibly occurring shadows and from illumination changes caused by the dynamic objects). *Dynamic* objects are robots, conveyor belts, humans etc. This group must be divided into two subgroups. The first subgroup contains *known* dynamic objects D, with changing geometry, position and the appearance but in a known manner. The robots and conveyor belts pertain to this subgroup. The second subgroup contains dynamic objects U with *unknown* changing geometry, position and appearance, e.g. humans. In the majority of cases approximate information about size, volume or similar can be provided. Note that static unknown objects do not exist. The *free space* F of a surveyed scene does not contain any known object but may contain unknown objects. Fig. 2 illustrates the introduced object types. In summary, the following three equations hold true:

$$S \cup D \cup F = \mathbf{E}^n, \text{ with } \mathbf{E}^n: \text{Euclidian Space} \tag{1}$$

$$S \cap D = S \cap F = D \cap F = \varnothing \text{ and } U \subseteq F \tag{2}$$

2.2 Single Camera

A number of N calibrated cameras with focal points $C^i \in \mathbf{E}^n$, $i \in \{1, ..., N\}$ and a frustum $L^i = \{x \in \mathbf{E}^n \mid x \text{ is projected via } C^i \text{ onto the image plane of camera } i\}$, are used to detect and finally reconstruct the unknown objects which reside in-between the known objects as accurate as possible. Here, we only consider color and grayscale cameras, but the approach can easily be extended to depth cameras.

One fundamental characteristic of these vision sensors is that they can only see up to the surface of the nearest opaque object per viewing direction (e.g. pixel center direction). Thus, occlusions always occur at the rear side of an opaque object. Moreover, the visibility of these sensors is limited by the frustum L^i. Outside this frustum, the sensor is not able to see anything. Thus, these parts can be interpreted as occlusions as well.

Now, the terms *visibility* and *occlusions* have to be introduced and detailed (Fig. 2). The *visibility* V^i of a camera i is the region of the free space F where a camera is able to detect unknown objects. The *known occlusion* O_K^i of a camera i is the region of the free space F where a camera can not detect unknown objects due to occlusions caused by known objects. Thus, it can be stated that $O_K^i \cap V^i = \varnothing$ and $O_K^i \cup V^i = F$ for each camera i. The *unknown occlusion* O_U^i of a camera i is the region of the visible space V^i where an unknown object has to be assumed, due to the evaluation of a camera image by a background subtraction method. The free space seen by camera i F^i is defined by $F^i = V^i \backslash O_U^i$. It can be stated, that $O_K^i \cap O_U^i = O_K^i \cap F^i = O_U^i \cap F^i = \varnothing$ and $O_K^i \cup O_U^i \cup F^i = F$ for each camera i. The concrete structures of the sets V^i, O_K^i, O_U^i depend directly on the used camera type with its detection capabilities.

In order to describe our formalism for a specific camera type (here color-/grayscale cameras), we use the concepts of rays and segments in the Euclidean Space. A ray(S, E) and a segm(S, E) are point sets and are defined by

$$\text{ray}(S,E) := \{S + t \cdot (E - S) \mid t \in \mathbf{R}_0^+, S, E \in \mathbf{E}^n\} \tag{3}$$

$$\text{segm}(S,E) := \{S + t \cdot (E - S) \mid t \in \mathbf{R} \wedge 0 \leq t \leq 1, \ S, E \in \mathbf{E}^n\} \tag{4}$$

Furthermore, a distance function dist(P, Q), with $P, Q \in \mathbf{E}^n$ exists, since the Euclidean Space is a metric space. Having the visibility and the occlusions, sets containing the most distant visible points and the nearest occluded points (per viewing direction) can be specified by

$$B_F^i = \{x \in V^i \mid \text{dist}(C^i, x) = \max_{y \in V^i \cap \text{ray}(C^i, x)} \{\text{dist}(C^i, y)\} \tag{5}$$

$$B_O^i = \{x \in O_K^i \mid \text{dist}(C^i, x) = \min_{y \in O_K^i \cap \text{ray}(C^i, x)} \{\text{dist}(C^i, y)\} \tag{6}$$

$$B_U^i = \{x \in O_U^i \mid \text{dist}(C^i, x) = \min_{y \in O_U^i \cap \text{ray}(C^i, x)} \{\text{dist}(C^i, y)\} \tag{7}$$

Using color or grayscale cameras, conventional background subtraction methods can be utilized. These methods segment an image into foreground and background based on the known appearance and a current image of the surveyed scene. If an unknown object resides in the scene and is not occluded by the known environment, it is marked as foreground in the segmented image. But usually the dynamic known objects are also – if not occluded – identified as foreground in the segmented image. Thus, the detection of unknown objects *in front* of a dynamic known object is not possible. Since the change detection method is not able to decide whether the cone between the camera and the dynamic known object is free or contains unknown objects, it must be interpreted as known occlusion. Thus, the visibility is described by

$$V^i = \{x \in L^i \mid \text{segm}(C^i, x) \subseteq F \wedge$$
$$(\text{ray}(C^i, x) \cap D = \varnothing \vee$$
$$(\text{ray}(C^i, x) \cap D \neq \varnothing \wedge \text{ray}(C^i, x) \cap S \neq \varnothing \wedge \tag{8}$$
$$\underset{y \in S \cap \text{ray}(C^i, x)}{\exists} \quad \underset{z \in D \cap \text{ray}(C^i, x)}{\forall} \text{dist}(C^i, y) < \text{dist}(C^i, z)))\}$$

As detailed above, the known occlusions are formulated by $O_K^i = F \setminus V^i$. Since depth values are not available for unknown objects with this sensor type, unknown occlusions start at the camera:

$$O_U^i = \{x \in V^i \mid \text{ray}(C^i, x) \cap U \cap V^i \neq \varnothing\} \tag{9}$$

2.3 Simultaneous Use of Several Cameras

Using several cameras with different perspectives onto the surveyed scene, each camera that is used provides a different occlusion and visibility situation, as discovered in the previous section that now has to be merged.

For every camera i, a collection of sets can be provided describing the known and unknown occlusions as well as the surveyed free space by $Q^i = \{O_K^i, O_U^i, F^i\}$.

The partitioning of the free space via the reconstruction step can be described by

$$R = \{A = q^1 \cap \ldots \cap q^N \mid q^i \in Q^i \wedge$$
$$\forall x, y \in A : \exists f : [0,1] \to A, f(0) = x \wedge f(1) = y\} \tag{10}$$

In words, all different labeled regions of all cameras are intersected among each other. Furthermore, the resulting intersections are grouped into connected components. All these connected components are contained by R (Fig. 3 b). Again, it can be stated that $\cup_{A \in R} A = F$.

Volumes in the free space of the surveyed scene which are actually seen as free (cf. F^i) by at least one camera are not further considered, since no unknown object can reside there. This results in (Fig. 3 c):

$$R' = \{x \in R \mid \forall i = \{0, \ldots, N\} : x \notin F^i \wedge x \neq \varnothing\} \tag{11}$$

To each set of R' a tuple (o, u) can be assigned, containing the number of seen known occlusions and unknown occlusions (Fig. 3 c).

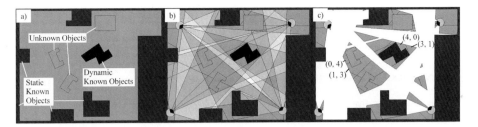

Fig. 3. Illustration of combining several camera views. (a) setup; (b) partitioning of the space; (c) Occluded parts with occlusion-tuples.

2.4 Plausibility Checks

In the majority of applications some information like size, volume, etc. about the unknown objects is available. Several sets of R' actually cannot contain an unknown object. Thus, plausibility checks are used to eliminate those occlusions which do not contain unknown objects. The mentioned plausibility checks aim for a *quasi-static* consideration. Another kind of plausibility checks can utilize *temporal* considerations, like "an unknown object can not suddenly appear in and surrounded by free space".

Note plausibility checks can apply to the whole scene or only to a part of the scene. In the following, a couple of quasi-static plausibility checks are discussed.

Minimum Volume: If only unknown objects like humans with a typical volume of 0.075 m³ should be detected, one might set a maximum volume for occlusions to be eliminated to 0.05 m³. Thus, all connected sets of R' obtained as described in the previous section with a volume smaller than 0.05 m³ can be safely removed. Only unknown objects with a specified minimum volume remain.

Maximum Distance to Ground: Typically, objects do not hover but have contact with the ground. If this can be guaranteed, all connected sets of R' with no contact to the ground S or D can be eliminated. More general, all objects with a distance larger than a specified maximum distance to the ground can be eliminated. Thus, setting the maximum distance to 1 m also a jumping human can be detected and is not removed by this plausibility check.

Surveillance Zones: In most cases, only certain parts of the whole surveyed scene are actually interesting so that unknown objects outside this part can be eliminated.

Occlusion parameter θ: If it can be guaranteed, that an unknown object can be completely occluded by the known environment in a maximum number of θ cameras, all regions where more than θ cameras see a known occlusion can be eliminated if no other region is connected to it with equal or less than θ cameras which see a known occlusion, since no unknown object can reside within this region. For more details about θ, see [10].

3 Reconstruction Algorithm

In this section we provide a voxel-based reconstruction algorithm, which works on the surfaces of the objects and is capable to deal with occlusions. Regarding the camera model, we only assume a two-dimensional field of connected pixels, with their back projected volumes also connected. Furthermore, the position and geometry of the pixels and the back projected volumes have to be known. Thus, we assume neither a pinhole camera model nor undistorted images.

3.1 Surface Voxel Determination

Given several calibrated cameras and images segmented into *free*, *known* and *unknown* and a voxel space, surface voxels can be determined by the following algorithm. Then the result is a voxel space with voxels marked according to the occlusions of all perspectives and a list containing all these voxels. At first, the needed functions are explained.

The classification value (*free*, *known* or *unknown*) of a pixel P is provided by the function *classification(P)*. Assuming that two adjacent pixels are separated by a pixel edge E, a list of voxels that are intersected by the back projection of this pixel edge E down to its visibility depth is provided by the function *voxelList(E)*. The function *neighborClassification(E)* for a pixel edge E provides the value *unknown*, if one of the two pixels is classified as *unknown*. It provides *known*, if one pixel is classified as *known* and the other as *free*. In all other cases, it provides *free*. For each voxel, the pixels it projects to in all cameras are needed. For simplification, here we only use the center of the voxels with the consequence, that objects that are smaller than the half of the voxel diagonal may be reconstructed incorrectly. Thus, it is necessary to choose an appropriate small voxel size. (Another voxel-like but camera centric-representation called conexels [1] could be applied, which avoids this drawback). The pixel of the projection of the voxel center V into a camera image C is provided by the function *projectVoxelCenter(V, C)*. The distance for a voxel center V to a camera C is provided by the function *distance(V, C)*. Per pixel P, the visibility depth (distance to B^i_F) and the occlusion depth (distance to B^i_O) as described in Section 2.2 is provided by the functions *visibleDepth(P)* and *occlusionDepth(P)*. The function *markVoxelAndAddToList(V, O_K, O_U)* marks the voxel V in voxel space by the two counter variables O_K, O_U representing the number of known and unknown occlusions respectively, and adds it to a list containing all surface voxels.

```
foreach camera C do
  foreach silhouette pixel edge E do
    foreach voxel V in voxelList(E) do
      counter O_U = 0, F = 0, O_K = 0
      if neighborClassification(E) == unknown
        O_U = 1
      else if neighborClassification(E) == known
        O_K = 1
      endif
      foreach camera C' != C do
        pixel P = projectVoxelCenter(V, C')
        if classification(P) == known
          or distance(V, C') ≥ occlusionDepth(P)
          O_K++
        else if classification(P) == unknown
          and distance(V, C') ≤ visibleDepth(P)
          O_U++
```

```
         else
            F++
         endif
      done
      if  F == 0
         markVoxelAndAddToList(V, O_K, O_U)
      endif
   done
done
done
```

In summary, each camera provides lists of potential surface voxels due to the segmentation. These voxels are sequentially tested in all other cameras. If no camera marks a voxel as free, it actually is a surface voxel. All actual surface voxels are stored in a list and marked in voxel space by the tuple (O_K, O_U).

Since the surface is not necessarily closed at the known objects, one may use a constrained flood fill algorithm to close it. Furthermore, completely occluded regions exclusively caused by the static environment are not revealed by this algorithm but can be determined in an initialization step by testing each voxel for visibility against the static environment in all cameras.

Having the surface voxels, partitions of related voxels, i.e. voxels with the same known and unknown occlusion counter can be built. Then, the sorted plausibility checks can be applied according to the costs and success probability. Dependent on the plausibility check additional information like volume of a partition, has to be calculated.

Besides pixel discretization, the accuracy of the voxel based algorithm depends on the voxel size and can be described by $e = \pm v / 2 \cdot \sqrt{3}$, with v length of a voxel edge, while the quality of the reconstruction depends on the scene and camera positions, i.e. the visibility and the occlusions.

3.2 Memory Consumption

Some of the used functions can be implemented as look-up tables to enable fast calculations. In order to give a memory consumption estimation M of these look-up tables, the following variables are introduced: A voxel space with dimensions X, Y and Z is used and the resolution of N cameras is provided by W and H.

The visibility depth and occlusion depth per pixel has a memory consumption for all images of:

$$M_1 = 2 \cdot N \cdot W \cdot H \tag{12}$$

The memory consumption for the voxel lists per pixel edges and for all cameras can be estimated by:

$$M_2 \leq N \cdot [((H + W + 2)) \cdot E + ((W + 1) \cdot (H - 1) + (W - 1) \cdot (H + 1)) \cdot G], \tag{13}$$
$$\text{with } G = X + Y + Z \text{ and } E = Z \cdot Y + Z \cdot X, \ X \leq Y \leq Z$$

Furthermore, the distances for each voxel to all cameras results in a memory consumption of:

$$M_3 = 2 \cdot N \cdot X \cdot Y \cdot Z \qquad (14)$$

Thus, the overall memory consumption is bounded by $M \leq M_1 + M_2 + M_3$. As an example the parameters are set to $N = 4$, $W = 320$, $H = 240$ and $X = Y = Z = 100$, with a typical camera placement and voxel-, pixel-addresses and floating point variables of 4 bytes, results in an upper bound of $M \leq 907$ MB and actually of 411 MB.

4 Experiments

In order to evaluate our methods and algorithms, we set up a test environment (Fig. 4 a), with five color cameras mounted around the scene to survey. It is available in a virtual simulation environment, too (Fig 4. b).

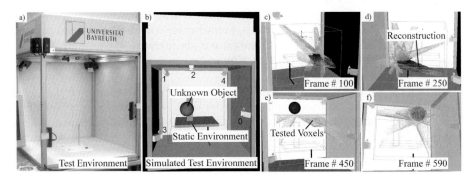

Fig. 4. Our test environment (a), the simulated one (b) and four frames of the experiment (c-f)

4.1 Hardware and Software Configuration

The computer contains an Intel Core™2 Quad CPU, with 2.6 GHz, 6 MB Cache and 4 GB RAM, but currently only one core of the CPU is utilized by our implementation. The graphics card is an NVIDIA GeForce 9600 GT with 512 MB and it is CUDA enabled. The operating system is a SUSE 11.0, with the gcc/g++ compiler suite version 4.3.1. The cubical volume of the test environment is 76 cm × 76 cm × 76 cm. Five Unibrain FireWire Fire-i™ Digital Board Cameras with 15 and 30 fps and a resolution of 640x480 Bayer Pattern are used. The calibration results, obtained by [14] for the images with a resolution of 640x480 have a low 3D position projection error (mean deviation < 1.6 pixels and standard deviation < 1 pixel).

The following performance tests for the reconstruction uses the virtual test environment, based on the real test environment providing a virtual object (here: sphere with a radius of 6 cm) and its segmented camera images. Additionally, a surveillance zone and a static object are included (Fig 4. b).

4.2 Performance Tests

The unknown object in the virtual test environment is moved on a circular path around and through the static known object in the middle of the scene. The virtual object is projected into all camera images simulating a conventional background subtraction. The segmented images are used to reconstruct the unknown object within the predefined surveillance zone and in consideration of the occlusions. Two cycles of this movement with a total of 1200 frames have been recorded. Fig. 4 c-f illustrates four interesting frames of the recorded sequence. The white dots represent voxels which have been tested for being surface voxels. The resulting surface voxels are shown containing the visibility tuples.

Dependent on the position of the unknown object, different numbers of pixels and voxels are marked and thus, different computation times are needed. The diagram in Fig. 5 shows the computation time for reconstructing the unknown object. Further it shows the number of tested voxels and the number of voxels that actually lie on the surface. Obviously, the calculation time corresponds to the number of potential surface voxels which have to be tested for each camera. Furthermore, the number of actual surface voxels must always be smaller or equal to the number of potential surface voxels. The calculation time is high, if the unknown object is seen by all cameras, such that many potential voxels have to be tested (frame# 250). Although the unknown object may be outside the surveillance zone, potential surface voxels have to be tested because of the absent depth information of this unknown object. Only the number of actual surface voxels is zero (frame# 450). In Fig 4. d the lower part of the sphere is only seen by the rightmost camera. Thus, the complete cone of potential surface voxels within the surveillance zone caused by that camera actually results in surface voxels. In this case, the ratio between actual surface voxels and potential surface voxels is relatively high.

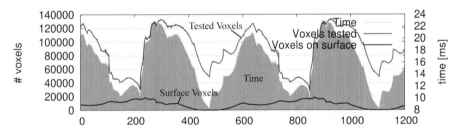

Fig. 5. Diagram of the computation times (gray area) for reconstructing the sphere using five cameras with a resolution of 320x240 and a voxel space of 152x152x138 voxels. Additionally, the number of tested voxels is described by the upper curve and the voxels that actually lie on the surface are described by the lower curve.

Table 2. summarizes the measured computation times by comparing the average values of different configuration pairs. A and B use a camera resolution of 320x240, a voxel space resolution of 152x152x138 and two different number of cameras – 3 and 5. C and D use four cameras, a voxel space of 152x152x138 and two different camera resolutions – 160x120 and 320x240. E and F use four cameras, a camera resolution of 320x240 and two different voxel space resolutions of 76x76x69 and 152x152x138.

The quintessence of this table is that although multiplying the number of pixels or voxels by a factor, the average time increases slower. This behavior is due to the consideration of surfaces and silhouettes instead of volumes and areas, respectively.

Table 2. Comparison of different configuration pairs for reconstructing the sphere

	# cameras	Avg. number of potential surface voxel tested	Avg. number of actual surface voxel tested	Avg. time [ms]
B	5	89021	8281	15.74
A	3	52488	9168	9.88
B/A	1.667	1.696	0.903	1.593

	# pixel			
D	172800	99095	11536	19.06
C	19200	51821	6166	9.01
D/C	9	1.91	1.87	2.11

	# voxel			
F	10760688	134968	15717	24.43
E	398544	31175	3674	5.9
F/E	27	4.33	4.28	4.14

5   Conclusions

For the first time, a general and consistent formalism for describing the visibility and occlusions within a camera surveyed scene with a known environment is provided. To do so, objects are classified as known/unknown and static/dynamic. A voxel-based algorithm constructing the visual hull, which works on surfaces using grayscale/color cameras in combination with a conventional background subtraction method, has been presented. The experimental results show that the computation time for the reconstruction step depends mainly on the number of tested surface voxels. Additionally, the measurements show that the computation time increases slower than the camera resolution and voxel space resolution, due to the surface and silhouette consideration.

In the future, the plausibility checks especially the temporal ones will be considered more intensively, since these promises a valuable enhancement in the reconstruction of unknown objects. The plausibility checks will be integrated into the voxel-based algorithm. Furthermore, the presented algorithm can be parallelized, such that potential surface voxels are tested simultaneously. For this, NVIDIAs CUDA seems to be suited. In addition, non-voxel-based approaches will be investigated.

References

1. Casas, J., Salvador, J.: Image-Based Multi-view Scene Analysis using 'Conexels'. In: ACM Proceedings of the HCSNet workshop on use of vision in human-computer interaction, Canberra, Australia, vol. 56 (2006)
2. Ebert, D., Henrich, D.: Safe Human-Robot-Cooperation: Image-based Collision Detection for Industrial Robots. In: IEEE/RSJ Int. Conf. on Intelligent Robots and Systems, Lausanne (2002)
3. Cailette, F., Howard, T.: Real-Time Markerless Human Body Tracking with Multi-View 3-D Voxel Reconstruction. In: Proc. British Machine Vision Conference (BMVC), Oxford, pp. 597–606 (2004)

4. Elgammal, A., Harwood, D., Davis, L.: Non-parametric Model for Background Subtraction. In: Vernon, D. (ed.) ECCV 2000. LNCS, vol. 1843, pp. 751–767. Springer, Heidelberg (2000)
5. Franco, J.-S., Boyer, E.: Efficient polyhedral modeling from silhouettes. IEEE Transact. on Pattern Analysis and Machine Intelligence 31(3), 414–427 (2009)
6. Fukui, S., Iwahori, Y., Woodham, R.-J.: GPU Based Extraction of Moving Objects without Shadows under Intensity Changes. In: IEEE Congress on Evolutionary Computation (IEEE World Congress on Computational Intelligence), Hong Kong (2008)
7. Guan, L., et al.: Visual Hull Construction in the Presence of Partial Occlusions. In: Proc. of the 3rd International Symposium on 3D Data (2006)
8. Guan, L., Franco, J.-S., Pollefeys, M.: 3D Occlusion Inference from Silhouette Cues. In: Proc. Comp. Vis. and Pattern Rec., CVPR (2007)
9. Keck, M., Davis, J.W.: 3D Occlusion Recovery using Few Cameras. In: Conf. on Computer Vision and Pattern Recognition, CVPR (2008)
10. Kuhn, S., Gecks, T., Henrich, D.: Velocity control for safe robot guidance based on fused vision and force/torque data. In: IEEE Conference on Multisensor Fusion and Integration for Intelligent Systems, Heidelberg, Germany (2006)
11. Ladikos, A., Benhimane, S., Navab, N.: Efficient Visual Hull Computation for Real-Time 3D Reconstruction using CUDA. In: IEEE Computer Society Conference on Computer Vision and Pattern Recognition, Anchorage, Alaska, USA (June 2008)
12. Laurentini, A.: The Visual Hull Concept for Silhouette-Based Image Understanding. IEEE Transactions on Pattern Analysis and Machine Intelligence 16, 150–162 (1994)
13. Lazebnik, S., Furukawa, Y., Ponce, J.: Projective Visual Hulls. International Journal of Computer Vision 74(2) (August 2007)
14. Svoboda, T., Martinec, D., Pajdla, T.: A Convenient Multi-Camera Self-Calibration for Virtual Environments. Presence: Teleoperators and Virtual Environments 14(4) (2005)
15. Slabaugh, G., Culbertson, B., Malzbender, T., Shafer, R.: A survey of methods for volumetric scene reconstruction from photographs. In: Intl. WS on Volume Graphics (2001)
16. Szeliski, R.: Rapid Octree Construction from Image Sequences. CVGIP: Image Understanding 58(1), 23–32 (1993)

Accurate Real-Time Disparity Estimation with Variational Methods

Sergey Kosov, Thorsten Thormählen, and Hans-Peter Seidel

Max-Planck-Institut Informatik (MPII), Saarbrücken, Germany

Abstract. Estimating the disparity field between two stereo images is a common task in computer vision, e.g., to determine a dense depth map. Variational methods currently are among the most accurate techniques for dense disparity map reconstruction. In this paper a multi-level adaptive technique is combined with a multigrid approach that allows the variational method to achieve real-time performance (on a CPU). The multi-level adaptive technique refines the grid only at peculiarities in the solution. Thereby it reduces the computational effort and ensures that the reconstruction quality is kept almost the same. Further, we introduce a technique that adapts the regularizer, used in the variational approach, dependend on the the current state of the optimization. This improves the reconstruction quality. Our real-time approach is evaluated on standard datasets and it is shown to perform better than other real-time disparity estimation approaches.

1 Introduction

A classical correspondence problem in computer vision is the estimation of a disparity field between a stereo image pair. During disparity estimation, for each pixel in one image the corresponding pixel in the other image is sought, so that the corresponding pixels are the projections of the same 3D position. Afterwards, if the camera calibration is known, a depth map can be calculated from the disparity field. If a standard stereo setup is used, the corresponding pixels are constrained to lie on the same row. Thus, the search range for the disparity is 1-dimensional.

Estimation of a 1D disparity field is related to the estimation of a 2D displacement field. A displacement field of corresponding pixels arises, e.g, between consecutive frames in an image sequences. Such a displacement field, represented as a vector field, is called *optic flow*. Variational methods allow to compute a precise and dense estimation of an optic flow field. Moreover, the research by Mémin and Pérez [1] and Brox et al. [2] has proven the variational methods to be among the best techniques for optic flow reconstruction. These techniques minimize an energy functional by solving the corresponding Euler-Lagrange equation. Numerically, the Euler-Lagrange equation is represented as a system of differential equations with finite differences. To optimize the energy functional, iterative solvers, like the Jacobi and Gauss-Seidel methods, are used. The speed of convergence of these methods is quite slow. As a result, processing a single image pair takes several minutes or even up to half an hour on todays CPUs.

G. Bebis et al. (Eds.): ISVC 2009, Part I, LNCS 5875, pp. 796–807, 2009.

As a remedy against this slow convergence, multigrid methods were developed [3], which allow to overcome the rigidity of the single grid approach by using multiple discretization levels. With a single fixed sampling grid, multiple solution components that have different scales may produce conflicting solutions and, thereby, cause slower convergence. For example, the smooth components, which are effectively approximated on coarse grids but slowly converge on fine grids, are often in conflict with high-frequency components, which should be taken into account only on fine grids.

In 1961, Fedorenko [4] formulated the multigrid algorithm for a standard 5-point discretization of the Poisson equation, which allowed to gain a numerical solution in $O(N)$ arithmetical operations (where N is the number of grid nodes). During the 1980s, Brandt [5], Stuben and Trottenberg [6], and Hackbush [7] made important contributions by transfering the multigrid ideas to the area of nonlinear problems, by introducing *multi-level adaptation techniques* (MLAT), and by developing the *full multigrid* (FMG) method. In 2006, Bruhn et al. [8] have demonstrated a real-time variational solver for optic flow reconstruction with discontinuity-preserving techniques. The solver uses a coarse-to-fine strategy in combination with a *full approximation scheme* (FAS).

Because the first-order Taylor expansion to linearize the energy functional is only valid for small disparities, for large disparities, multigrid methods are often combined with so-called *warping steps*. With warping steps [9,2] the original problem is compensated by the already computed solution from all coarser levels before the remaining residual is minimized on the finer level. In this paper, we use linear interpolation to linearize the energy functional. This approach can handle large displacements directly and was shown to be faster and more accurate [10].

In this paper a current variational approach with multigrids is extended by a MLAT in combination with a FAS. In contrast to the current multigrid methods, a grid adaptation technique refines the sampling grid not for the whole image, but locally in regions where interesting structures are located [11]. A similar adaptive mesh algorithm, which is based on a Galerkin finite element method on a triangular mesh for object flow computation [12], is difficult to use with the FAS. It will be shown that with multigrids in combination with the MLAT, heterogeneous adaptive structures can be used with a variational solver for real-time disparity estimation. Thereby, the MLAT allows to quickly perform local and precise adjustment. Furthermore, improved reconstruction quality is achieved by adapting the applied regularizer locally during optimization. A comparison on standard data sets with other real-time disparity estimators shows that our real-time variational approach outperforms the current state-of-the-art.

The paper is organized as follows. The next section gives an introduction to disparity estimation with variational methods. In Section 3 first state-of-the-art multigrid techniques are described and afterwards a different multigrid techniques to improve computation time is suggested. This technique is base on what we call *null-cycles* (O-cycles). Section 4 describes our multi-level adaptive technique for variational solvers. In Section 6 the approach is evaluated and the paper ends with a conclusion.

2   Variational Methods

Let us suppose that we are given a stereo image pair. Each scalar-valued image $I(x, y)$ is stored in a pixel matrix and $(x, y)^\top$ is the coordinate of the pixel within the rectangular image domain Ω. Having two images of a stereo pair $I_1(x, y)$ and $I_2(x, y)$, we try to estimate the position to which every pixel from the first image has moved in the second image. In order to do that, we have to assume that certain image features are still the same in both images. Such features may include the grey value, higher image derivatives (such as the gradient or the Hessian), or scalar-valued expressions (such as the norm of the gradient, the Laplacian, or the determinant of the Hessian) [13]. For simplicity, we will only consider grey value constancy assumption in the remainder of this paper.

If $(x, y)^\top$ is the coordinate of a pixel in the first image and $u(x, y)$ is the disparity, then $(x + u(x, y), y)^\top$ is the new position of the pixel in the second image. By formulating the problem like that, we can state that the computation of the disparity field is actually the computation of the vector field $(u(x, y), 0)^\top$. Now we can write the grey value constancy assumption:

$$I_1(x, y) - I_2(x + u(x, y), y) = 0 \quad . \tag{1}$$

As we are working with continuous real-world data, which is not discrete like the pixel locations in the pixel matrices, the disparities are not necessarily integer values. To perform the linearization, we use a linear interpolation technique [14]. We express the disparity $u(x, y)$ as the sum of two components: integer $A(x, y)$ and floating point $b(x, y)$, such that:

$$u(x, y) = A(x, y) + b(x, y), \quad \text{with} \quad |b(x, y)| < 1 \quad . \tag{2}$$

The linearized form of Eq. (1) is given by:

$$I_1(x, y) - |b(x, y)| \cdot I_2(x + A(x, y), y) - \tag{3}$$
$$(1 - |b(x, y)|) \cdot I_2(x + A(x, y) + \text{sign}(b(x, y)), y) = 0$$

We construct an energy functional, that consists of two terms: a data term that imposes constancy on the grey values, and a smoothness term that regularizes the often non-unique (local) solution of the data term by an additional smoothness assumption.

2.1   Data Term

Due to possible occlusions or unpredictable reflection properties of the object's surfaces, the equality from Eq. (1) can usually not be satisfied perfectly in reality. However, we can fulfill the demand: $\|I_1(x, y) - I_2(x + u(x, y), y)\|^2 \to$ min. The energy functional $E(u(x, y))$, based on the grey value constancy assumptions, can be written as:

$$E(u(x, y)) = \iint_\Omega \|I_1(x, y) - I_2(x + u(x, y), y)\|^2 \, dS \quad . \tag{4}$$

2.2   Smoothness Term

The smoothness term is derived from the assumption that the neighboring regions belong to the same object and thus these regions have similar disparity. The main role of the smoothness term is the redistribution of the computed information and the elimination of local disparity outliers. In case that reliable information from the data term is not available, the smoothness term helps to fill the problematic region with disparities calculated from neighboring regions.

In this paper, we use 3 different regularizers: Tichonovm, Charbonnier, and Perona-Malik regularization. Tichonov regularization assumes overall smoothness and does not adapt to semantically important image or flow structures (Horn and Schunck [15]). Charbonnier's and Perona-Malik's flow-driven regularization assumes piecewise smoothness and respects discontinuities in the flow field (see, e.g., [16,17,18,19]). For all three regularizers, the smoothness term in general form is given by

$$\Psi(|\nabla u(x,y)|^2) \quad . \tag{5}$$

Thus, we can rewrite the energy functional (4) as follows:

$$E(u(x,y)) = \iint_\Omega \|I_1(x,y) - I_2(x + u(x,y), y)\|^2 + \varphi \cdot \Psi(|\nabla u(x,y)|^2)\, dS, \tag{6}$$

where φ is a weighting factor for the smoothness term. In case of the Tichonov regularizer the smoothness term is given by

$$\Psi(s^2) = s^2 \quad , \tag{7}$$

for the Charbonnier regularizer by

$$\Psi(s^2) = 2\lambda^2 \sqrt{1 + \frac{s^2}{\lambda^2}} - 2\lambda^2 \quad , \tag{8}$$

and for the Perona-Malik regularizer by

$$\Psi(s^2) = \lambda^2 ln(\lambda^2 + s^2) - \lambda^2 ln(\lambda^2) \quad . \tag{9}$$

2.3   Euler-Lagrange Equation

The goal of the variational method is to find a function $u(x,y)$, which minimizes the energy functional $E(u(x,y))$. In other words, having constructed the energy functional, we should minimize it in order to find the best solution for the disparity field. Moreover, if the constructed functional is strictly convex, it will have a unique solution that minimizes it.

The Euler-Lagrange equation is an equation satisfied by the unknown function $u(x,y)$ that minimizes the functional $E(u(x,y)) = \iint_\Omega F(x,y,u,u_x,u_y)\, dS$, where $u_x = \frac{\partial u}{\partial x}$, $u_y = \frac{\partial u}{\partial y}$ and F is a given function which has continuous first order partial derivatives. The Euler-Lagrange equation then is the partial differential equation:

$$F_u - \frac{\partial}{\partial x} F_{u_x} - \frac{\partial}{\partial y} F_{u_y} = 0 \quad . \tag{10}$$

For the energy functional (6) the Euler-Lagrange equation for each pixel $(x, y)^{\top}$ is given by

$$I_{2x}(x + u, y)(I_1(x, y) - I_2(x + u, y)) + \varphi \cdot \operatorname{div}(\Psi'(|\nabla u|^2) \cdot \nabla u) = 0 \quad . \quad (11)$$

In order to minimize our energy functional, we solve the resulting system of differential equations with homogeneous Neumann boundary conditions [20]. This step is done via discrete numerical schemes. The Euler-Lagrange equations are discretized, linearized with the Eq. (3), and approximated via finite-differences schemes. In the end, we arrive at a linear (in case of Tichonov regularizer) or non-linear (in case of Charbonnier or Perona-Malik regularizers) system of equations.

3   Multigrid

In general, large equation systems arising from finite difference approximations of elliptic boundary problems are solved with iterations methods, like the Jacobi or Gauss-Seidel method [21,22].

However, such methods converge very slowly for equation systems that are only coupled via a small local neighborhood because numerous iterations are needed to exchange data between unknowns that are coupled indirectly. This leads to efficient computation of high-frequency components, while the lower-frequency components remain almost unchanged. Multigrid methods effectively handle this problem by starting from a fine grid but then perform correction steps that compute the error on a coarser grid and propagate this information back to the finer grid. Thus, lower frequency components of the error reappear as higher ones on the coarser grid and allow an efficient attenuation with basic iterative methods.

To employ this multigrid approach to non-linear problems, the Full Approximation Scheme (FAS) is used. For completeness, in the next subsection a short introduction to the FAS is given.

3.1   Full Approximation Scheme (FAS)

In the following equations the indices H and h indicate entities from a coarser grid and a finer grid, respectively. For the sake of clarity, let us reformulate Eq. (11) as

$$L_h u_h = -f_h \quad . \quad (12)$$

Here L_h is a non-linear operator and f_h stands for the right hand side, which in our particular case is equal to zero. Let u_h^i denotes the approximate solution after i iterations. Then the error is given by

$$e_h^i = u_h - u_h^i \quad . \quad (13)$$

Substituting u_h in equation (12) with u_h from Eq. (13), we obtain

$$L_h(u_h^i + e_h^i) = -f_h \quad . \quad (14)$$

Now we substract $L_h u_h^i$ from the left and right parts of the Eq. (14):

$$L_h(u_h^i + e_h^i) - L_h u_h^i = -f_h - L_h u_h^i \quad . \tag{15}$$

Then we restrict the solution to the coarser grid. Here we have to introduce two operators I_H^h and I_h^H. Let I_h^H denote the *restriction* operator from a fine grid h to a coarse grid H and I_H^h the *interpolation* operator from a coarse grid H to a fine grid h. For the coarser grid we get:

$$L_H(I_h^H u_h^i + e_H^i) - L_H I_h^H u_h^i = -I_h^H f_h - I_h^H L_h u_h^i \quad . \tag{16}$$

If we denote

$$f_H = I_h^H f_h + I_h^H L_h u_h^i - L_H I_h^H u_h^i \quad , \tag{17}$$

then we can rewrite Eq. (16) in a short form: $L_H(I_h^H u_h^i + e_H^i) = -f_H$ (note the similarity with Eq. (12)). Let u_H^i denote the new approximation of the solution on the coarse grid with $u_H^i = I_h^H u_h^i + \tilde{e}_H$, where $\tilde{e}_H$ is the new error approximation after i iterations. From that follows:

$$\tilde{e}_H = u_H^i - I_h^H u_h^i \quad . \tag{18}$$

Now we interpolate the error to the finer grid $\tilde{e}_h = I_H^h \tilde{e}_H$ and after that correct the solution on the finer grid: $\tilde{u}_h = u_h^i + \tilde{e}_h$. The steps of the FAS are summarized in Fig. 1. Note that only the error and the residual are transfered to the finer grid, but not the solution, since only the error and the residual are smooth functions.

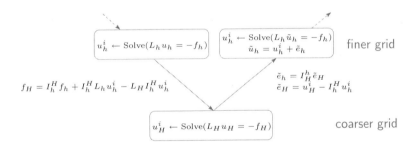

Fig. 1. The steps of the full approximation scheme

3.2 V-Cycle and W-Cycle

The main idea of the multigrid method is that on a coarse grid we are not obliged to solve $L_H u_H = -f_H$ precisely. It is enough to perform a few iterations and achieve an approximate solution u_H^i. On each multigrid level, only a few iterations must be performed to compute the high-frequency components, because the lower-frequency components can be more efficiently computed on the coarser grids. Therefore, in order to increase the computational efficiency, two types of grid cycle are commonly used: V- and W-cycles. While the V-cycles make one recursive call of a two-grid cycle per level, the more reliable W-cycles perform two. The cycles are applied in a hierarchical way as depicted in Fig. 2.

Fig. 2. V-cycles and W-cycles with two, tree and four levels

3.3 Full Multigrid (FMG)

It is possible to significantly improve the convergence of the multigrid methods to the correct solution by applying the full multigrid method (FMG), also known as method of nested iterations. In contrast to the simple multigrid approach, the FMG approach starts from a coarse grid and not from a fine grid. The schematic view of the FMG method with W-cycles is shown in Fig. 3. The full multigrid method combines the solution from coarser grids as initial approximations and than applies V- and W-cycles for calculating the solution at finer grids. Details about the FMG can be found in [5,6].

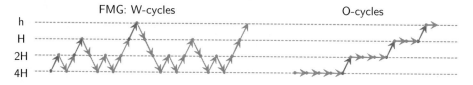

Fig. 3. Left:Full Multigrid (FMG) implementation with W-cycles per resolution level, Right: O-cycles. Refinement steps are marked with red color. Each W- and O-cycle is marked with a blue color.

3.4 O-Cycle

The FMG method assumes that the information from a finer grid is necessary to guide the solver on the coarser grid to the correct solution. This comes at the cost of additional V- and W-cycles. In our experiments we found that these additional V- and W-cycles are redundant for most input images and a similar reconstruction quality can be obtain with less computational effort, which is desirable for real-time applications. To achieve a similar reconstruction quality compared to FMG constructed with V- or W-cycles, we introduce a new approach, based on what we call O-cycle (or null-cycle). The idea is to perform significantly more iterations on the coarse grids, which have a low computational cost since the image resolution is smaller, and thereby increase the likelihood of convergence on the coarse grid. Once convergence on a coarse grid is obtained, the algorithm processes the next finer grid and never returns to the coarser one. As illustrated in Fig. 3, the iteration number m now is dependend on the current multigrid level. In particular, we have obtained good results for $m = m_1 \cdot (n^p)$, where $p = \{1, 2\}$ and $n \in [1, N]$ with $n = 1$ for the finest grid and $n = N$ for the coarsest grid. The user parameter m_1 defines the number of iterations on the finest grid.

4   Multi-Level Adaptive Technique (MLAT)

In this section, a multi-level adaptive technique is described that reduces the computational effort of the variational solver and at the same time retains a high reconstruction quality. From a coarser to a finer grid, we usually only need to refine the reconstruction at some regions of interest in the image. This means that with a static grid we waste resources on grid nodes that do not improve the resulting solution. Therefore, we employ a non-static grid structure.

The whole process starts on the coarsest grid. After finding a solution on the coarsest grid using FAS and O-cycles, we use this solution to identify the grid nodes that need refinement. Therefore, we evaluate the residual error of the energy functional in Eq. (4) as well as the spatial gradient of the solution. These two criteria are used to detect peculiarities of the solution. If either the residual $E(u(x,y)) > \epsilon_E$ or the spatial gradient $||\nabla u(x,y)||^2 > \epsilon_\nabla$, the grid cell is refined. Thereby, the thresholds ϵ_E and ϵ_∇ are user-defined parameters.

As shown in Fig. 4, the number of nodes is always upsampled by to factor of 2 when going from a coarser to a finer grid. Thereby, the area covered by a finer grid cell, can only be a part of the area covered with a coarser grid cell. For example, let us assume that the red arrow in Fig. 4 marks a local image peculiarity. Then the algorithm will employ finer grids at this location.

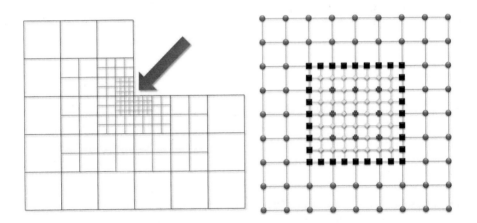

Fig. 4. Irregular grids: **left**: nodes concentration near to the local image peculiarity; **right**: structure of two grid levels G_h and G_H: common nodes of G_h and G_H (red crosses); nodes which belong only to G_h (yellow rhombi); nodes which belong only to G_H (blue circles); boundary nodes of G_h (black squares)

Despite the irregular grid structure, the differential equations can still be solved with a multigrid solver. This is done in a similar way as with regular grids. As shown in Fig. 4, there are different types of nodes:

1. The nodes belonging only to the grid G_H, which will not be refined and no further calculation must be performed for them.
2. The nodes of G_H belonging at the same time to G_h, which will be used for gaining the solution correction with the FAS at nodes on the finer grid G_h.
3. The boundary nodes of the finer grid G_h. In order to connect the solutions from both grids, these nodes are initialized from the coarser grid but are not altered during optimization.

The multi-level grid adaptation can not only be used with O-cycles (which we use for our real-time results), it can also be used with a classical FMG approach. In the areas on the coarse grid G_H, which are not covered with the fine grid G_h, we need not only the error, but the whole solution. Therefore, even if we have a linear problem (e.g., when using Tichonov regularizer in the smoothness term) the FAS must be employed.

Let us consider the error of restriction, when coming from a finer grid to a coarser one. Using Eq. (17), we can write

$$f_H = I_h^H f_h + \tau_h^H \quad , \tag{19}$$

where $\tau_h^H = I_h^H L_h u_h^i - L_H I_h^H u_h^i$ is the error of restriction or the transfer correction from the finer to a coarser grid. This transfer correction for the equation on the coarser level provides a measure for the co-occurrence between the solutions on the coarse and fine grid. When solving the equation $L_H u_H = -F_H$ on the coarse grid G_H, the term F_H is given by Eq. (19), but in the areas where the corresponding node on the fine grid G_h is not available, we assume that τ_h^H is equal to zero.

5   Regularizer Adaptation

We found in our experiments that the reconstruction results can be further improved, if the applied regularizer for each pixel is changed locally within the iteration process. For each pixel we always start with the Tichonov regularizer. Based on thresholds for $E(u(x, y))$ and $||\nabla u(x, y)||^2$ the algorithm decides when to switch to the Charbonnier or Perona-Malik regularizer. Furthermore, we adapt the parameters λ and φ (see Eqs. (6), (8), and (9)). The choice for these parameters depends on the regularizer, the multi-grid level, as well as $||\nabla u(x, y)||^2$ and $E(u(x, y))$. The best parameters for the actual situation are trained off-line by Monte-Carlo simulation on stereo pictures which have a similar characteristic as the test images, e.g., similar range of disparities, comparable image resolution, and equivalent scene illumination conditions. Once these parameters are found they are kept fixed during real-time experiments.

Table 1. Excerpt from the evaluation table generated by the Middlebury stereo evaluation webpage (error threshold = 2 pixels)

Algorithm	Tsukuba			Venus			Teddy			Cones			Average Percent of bad pixels
	nonocc	all	disc	nonocc	all	disc	nonocc	all	disc	nonocc	all	disc	
DoubleBP	0.83	1.24	4.49	0.10	0.35	1.46	1.41	4.13	4.73	1.71	7.02	5.16	2.72
CoopRegion	0.77	1.00	4.14	0.11	0.18	1.53	2.14	3.41	6.61	2.10	5.95	6.24	2.80
. . .													
our method	1.33	3.13	6.94	0.27	1.07	3.16	1.30	2.30	3.87	2.31	3.43	6.90	**3.00**
. . .													
MultiResGC	0.67	1.05	3.64	0.22	0.46	2.97	4.20	7.13	11.6	3.22	8.80	8.07	4.30
RealtimeBP	1.25	3.04	6.66	0.63	1.53	7.68	5.68	8.27	10.2	2.90	9.11	8.27	5.43
RealTimeGPU	1.34	3.27	7.17	1.02	1.90	12.4	3.90	8.65	10.4	4.37	10.8	12.3	6.46
Infection	6.34	7.81	22.8	2.70	3.66	26.0	12.8	18.3	33.5	10.7	16.6	30.1	15.9

Table 2. Computational effort of our algorithm using FMG with V- and W-cycles or O-cycles at the same reconstruction quality, as used in Table. 1

	Speed [fps]		
Scene	FMG:V-cycles	FMG:W-cycles	O-cycles
Tsukuba	0,6	1,3	2,15
Venus	1,15	1,85	3,75
Teddy	1	1,3	3,3
Cones	0,75	2,1	3,5

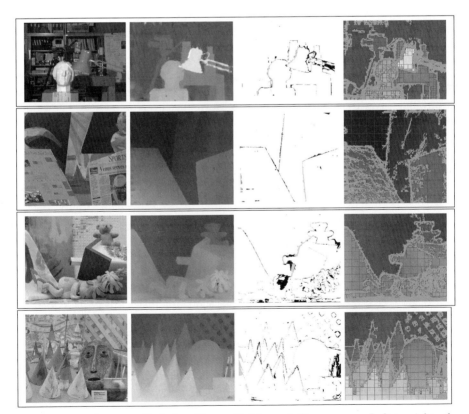

Fig. 5. Top to bottom: Tsukuba, Venus, Teddy, and Cones scene; Left to right: the right image of the stereo pair, solution disparity map, bad pixels (absolute disparity error > 1.0 pixel), the finest MLAT grid

6   Evaluation

To evaluate our novel multigrid variational solver with MLAT, we use the standard stereo data sets provided by the Middlebury Stereo evaluation website [23]. These datasets include stereo images as well as ground-truth disparity maps. The resolution of the provided stereo images is approximately 450×375 pixels and the disparities are in the range $[0; 22]$ pixels.

The Middlebury stereo evaluation website provides a convenient and objective way to evaluate the accuracy of the reconstruction by the percentage of bad pixels (see more details in [24]). As can be seen in Table. 1, our approach is among the most accurate methods. Furthermore, those methods, which have a higher accuracy, are not marked to be real-time approaches like ours.

In Fig. 5 the corresponding results for the Tsukuba, Venus, Teddy, and Cones scene are shown. The timings for our approach are given in Table. 2, where we compare the performance of FMG with V- and W-cycles or O-cycles. All speed measurements are carried out using a standard desktop PC with a 2.83 GHz Intel Pentium CPU executing C++ code.

7   Future Work and Conclusion

In this paper, we have introduced a combination of multigrids and a multi-level adaptive technique for the variational approach to reconstruct a disparity field. Furthermore, a regularizer adaptation technique was proposed. This allows the variational solver to achieve a real-time performance even on a CPU. The gained reconstruction quality is competitive to other state-of-the-art approaches that require more computation time.

In future, we are going to develop and implement a parallel version of our algorithm, which could be capable to run on multiple CPUs or a GPU with stream processing technology. Another direction of future work is to extend our method of 1-dimensional disparity estimation to the problem of 2-dimensional optic flow reconstruction.

References

1. Mémin, E., Pérez, P.: Dense estimation and object-based segmentation of the optical flow with robust techniques. IEEE Transactions on Image Processing 7, 703–719 (1998)
2. Brox, T., Bruhn, A., Papenberg, N., Weickert, J.: High accuracy optical flow estimation based on a theory for warping. In: Pajdla, T., Matas, J(G.) (eds.) ECCV 2004. LNCS, vol. 3024, pp. 25–36. Springer, Heidelberg (2004)
3. Brandt, A.: Multi-level adaptive solutions to boundary-value problems. Mathematics of Computation 31, 333–390 (1977)
4. Fedorenko, R.: Relaxation method for solving elliptic differential equations. Journal of computational mathematics and mathematical phisics 1, 922–927 (1961)
5. Brandt, A.: Multilevel adaptive computations in fluid dynamics. AIAA Journal 18, 1165–1172 (1980)

6. Stüben, K., Trottenberg, U.: Multigrid methods: Fundamental algorithms, model problem analysis and applications. Lecture Notes in Mathematics, vol. 960, pp. 1–176 (1982)
7. Hackbusch, W.: Multigrid Methods and Applications. Springer, New York (1985)
8. Bruhn, A., Weickert, J., Kohlberger, T., Schnörr, C.: A multigrid platform for real-time motion computation with discontinuity-preserving variational methods. Int. J. Comput. Vision 70, 257–277 (2006)
9. Mémin, E., Pérez, P.: A multigrid approach for hierarchical motion estimation. In: IEEE International Conference on Computer Vision, pp. 933–938 (1998)
10. Kosov, S.: 3D map reconstruction with variational methods. Master thesis, Saarland University (2008)
11. Brandt, A.: Multi-level adaptive technique (MLAT) for fast numerical solution to boundary value problems. Lecture Notes in Physics, vol. 18, pp. 82–89 (1973)
12. Condell, J., Scotney, B., Morrow, P.: Adaptive grid refinement procedures for efficient optical flow computation. Int. J. Comput. Vision 61, 31–54 (2005)
13. Papenberg, N., Bruhn, A., Brox, T., Didas, S., Weickert, J.: Highly accurate optic flow computation with theoretically justified warping. Int. J. Comput. Vision 67, 141–158 (2006)
14. Meijering, E.: A chronology of interpolation: From ancient astronomy to modern signal and image processing. In: Proceedings of the IEEE, pp. 319–342 (2002)
15. Horn, B.K.P., Schunck, B.G.: Determining optical flow. Artificial Intelligence 17, 185–203 (1981)
16. Charbonnier, P., Aubert, G., Blanc-Ferraud, M., Barlaud, M.: Two deterministic half-quadratic regularization algorithms for computed imaging. In: International Conference on Image Processing, vol. 2, pp. 168–172 (1994)
17. Cohen, I.: Nonlinear variational method for optical flow computation. In: Eighth Scandinavian Conference on Image Analysis, vol. 1, pp. 523–530 (1993)
18. Schnörr, C.: Segmentation of visual motion by minimizing convex non-quadratic functionals. In: Twelfth International Conference on Pattern Recognition, Jerusalem, Israel, vol. A, pp. 661–663. IEEE Computer Society Press, Los Alamitos (1994)
19. Weickert, J., Schnörr, C.: A theoretical framework for convex regularizers in pde-based computation of image motion. Int. J. Comput. Vision 45, 245–264 (2001)
20. Cheng, A., Cheng, D.T.: Heritage and early history of the boundary element method. Engineering Analysis with Boundary Elements 29, 268–302 (2005)
21. Ortega, J.M., Rheinboldt, W.C.: Iterative solution of nonlinear equations in several variables. Society for Industrial and Applied Mathematics, USA (2000)
22. Young, D.: Iterative Solution of Large Linear Systems. Academic Press, New York (1971)
23. Scharstein, D., Szeliski, R.: High-accuracy stereo depth maps using structured light. In: Proc. Computer Vision and Pattern Recognition, pp. I: 195–202 (2003), datasets: http://vision.middlebury.edu/stereo/
24. Scharstein, D., Szeliski, R.: A taxonomy and evaluation of dense two-frame stereo correspondence algorithms. International Journal of Computer Vision 47, 7–42 (2001)

Real-Time Parallel Implementation of SSD Stereo Vision Algorithm on CSX SIMD Architecture

Fouzhan Hosseini[1], Amir Fijany[1], Saeed Safari[1,2], Ryad Chellali[1], and Jean-Guy Fontaine[1]

[1] Tele Robotics and Applications Department, Italian Institute of Technology,
Via Morego 30, Genova, Italy
[2] School of Electrical and Computer Engineering, University of Tehran,
North Kargar Ave., Tehran 14395-515, Iran
{fouzhan.hosseini,amir.fijany,saeed.safari,ryad.chellali,
jean-guy.fontaine}@iit.it
http://www.iit.it

Abstract. We present a faster than real-time parallel implementation of standard sum of squared differences (SSD) stereo vision algorithm, on an SIMD architecture, the CSX700. To our knowledge, this is the first highly parallel implementation of this algorithm using 192 processing elements. For disparity range of 16 pixels, we have achieved the rate of 160 and 59 stereo pairs per second on 640x480 and 1280x720 images, respectively. Since this implementation is much faster than real time, it leaves enough time for performing other machine vision applications in real time. Our results demonstrate that CSX architecture is a powerful processor for (low level) computer vision applications. Due to the low-power consumption of CSX architecture, it can be a good candidate for mobile computer vision applications.

1 Introduction

Many embedded applications of vision system, such as mobile robots and Humanoid require a supercomputing capability for various complex image processing tasks while being severely limited by the power consumption and size of the computing architecture. Our objective is to develop a flexible low-power lightweight computing architecture for such applications, using new emerging massively parallel low-power computing processors. As starting point, we have focused on digital stereo vision which provides depth information, as the first step of complex and time consuming tasks such as object tracking and obstacle avoidance. To meet the real time processing requirements, digital stereo vision should be performed in such a way to leave enough time for these time consuming tasks.

An extended overview of stereo vision algorithm has been presented in [1]. There are many works in the literature focusing on real time implementation

G. Bebis et al. (Eds.): ISVC 2009, Part I, LNCS 5875, pp. 808–818, 2009.

of stereo vision on various architectures ranging from General Purpose Processors (GPP) to FPGA implementation [2], [3], [4], [5], [6], [7]. Due to the real-time processing requirement, built-in SIMD accelerators, e.g. MMX and SSE in new generation of the general processors, can be employed to become closer to the real-time processing requirement [2], [3]. Sunyoto et al. [3] have presented real time algorithm using the SSE2 instructions available on Intel Pentium 4 and AMD Atlon 64 processors. They have reported 35 frames per second for a 256x256 image and disparity 24. Another way to improve performance is to utilize the processing power of Graphics Processing Units (GPUs) [4], [5]. They typically use texture capability of graphics hardware to compute cost function in stereo vision algorithm. Yang et al. [4] could achieve 289 MPDs (mega-pixel disparities per second) using ATI Radeon 9800 graphics cards. Moreover, increasing density, speed and programmability of Field Programmable Gate Arrays(FPGAs) provide the opportunity to implement stereo vision on these kind of hardwares [6], [7]. Woodfill et al. [6], utilizing 16 Xilinx 4025 FPGAs, have achieved 42 frames per second at 320x240 pixel resolution which is less than 80 MPDs. While utilizing Xilinx Spartan-3 FPGA, Murphy et al. [7] claim that processing 320x240 pixel images at the rate of 150 frames per second is possible.

FPGAs are dedicated hardwares, which should be designed for specific applications. GPU is more flexible and provides some level of programmability, but the main problem is that GPU provides good performance only for some kind of applications. The other issue that should be considered is power consumption. Generally, GPPs and GPUs are high power consuming units, while FPGAs require less power. None of the solutions mentioned above satisfies mobile vision system design goals including low power consumption, flexibility, and real time processing capability.

In this paper, we have used the ClearSpeed CSX [8] user programmable massively parallel SIMD architecture for parallel implementation of the conventional SSD stereo vision algorithm. CSX has 192 Processor Elements (PEs) with a peak of 96 GFOLPS computation power, while its power consumption is less than 9 watts. Implementing SSD on CSX, we have achieved 688 MPDs corresponding to a 3x3 window size, 640x480 image, maximum disparity of 16 , and 160 stereo pairs per second(fps). Experimental results show that CSX architecture is a good candidate to provide low-power supercomputing capability for embedded computer vision applications.

This paper is organized as follows. Sect. 2 explains SSD algorithm. Sect. 3 gives an overview on the CSX architecture and discusses the features used in our implementation. In Sect. 4 parallel implementation of SSD on the CSX architecture is described. Sect. 5 discusses the performance of the architecture. Finally, Sect. 6 concludes the paper and discusses future works.

2   SSD Algorithm

The SSD algorithm is a straightforward window based approach to obtain disparity map on a pair of rectified stereo images.

Assume $I_R(i,j)$ and $I_L(i,j)$ as the intensity of pixel located at row i and column j in the right and left image, respectively. The input parameters of the algorithm are ω, the window size, and β, the maximum disparity. Assuming right image as the reference, the disparity for each pixel (i,j) in the right image is calculated as follow:

– Consider a window centered at (i,j) in the right image
– Consider a window centered at $(i,j+k)$ in the left image where $j \leq k < j+\beta$
– Calculate convolution of the windows in the left and right image using (1).

$$S(i,j,k) = \sum_{l=i-\frac{w-1}{2}}^{i+\frac{w-1}{2}} \sum_{m=j-\frac{w-1}{2}}^{j+\frac{w-1}{2}} \left[I_R(l,m) - I_L(l,m+k) \right]^2 \tag{1}$$

– The pixel which minimize $S(i,j,k)$ is the best match. So,

$$k^* = \arg \min_{j \leq k < j+\beta} S(i,j,k), \tag{2}$$

$$d(i,j) = k^*$$

Briefly, the SSD algorithm consists of the following three steps:

1. Calculating the squared differences of intensity values in a given disparity
2. Summing the squared differences over square windows
3. Finding two matching pixels by minimizing the sum of squared differences.

3   CSX

Our implementation platform, ClearSpeed CSX700 architecture [9], provides both high performance computing capability and low-power consumption. CSX700 has two similar cores, each includes one poly execution unit which is an SIMD architecture, containing 96 PEs. Poly execution unit is in charge of parallel data processing. Each PE consists of a register file, 6KB of SRAM, a high speed I/O channel to adjacent PEs and external I/O, an ALU, integer multiply-accumulate (MAC) unit, and an IEEE 754 compliant floating point unit (FPU) with dual issue pipelined add and multiply (see Fig. 1). The maximum performance, 96 GFLOPs, is reachable if PEs are fully pipelined and vectorized.

Each core is equipped with a DDR2 memory and a 128KB SRAM called external (mono) memory, while PEs' SRAM is called poly memory. PIO unit controls data transfers between mono and poly memory (see Fig. 1). Poly execution unit and PIO unit can operate asynchronously. To utilize the underlying bus bandwidth, the size of the data which transfered between external and poly memory have to be at least 32 bytes, i.e. the time required to transfer 32 byte data and less is almost the same.

Moreover, Each PE is able to communicate with the two neighboring PEs using a dedicated bus called swazzling path. Swazzling path connects the register file of each PE core with the register files of its left and right neighbors (see

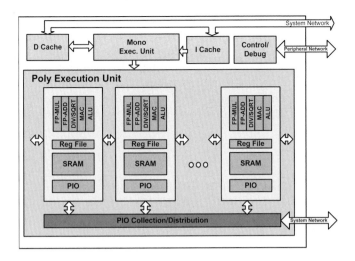

Fig. 1. Simplified CSX Architecture

Fig. 1). Consequently, on each cycle, PEs are able to perform a register-to-register data transfer to either its left or right neighbor, while simultaneously receiving data from the other neighbor. In fact, the swazzling path provides the facility for parallel data communication among PEs. Obviously, data transfer time between two PEs increases linearly with their distance.

Furthermore, to achieve higher floating point operation performance, CSX provides the facility of vector operations which are a set of hardware instructions that perform floating point operation on a block of operands and thus utilize the pipeline nature of the PE's units.

4 Parallel SSD Algorithm Aspects

Like many other low-level image processing tasks, SSD is a highly regular process wherein the same operation is performed on a large set of data. Such a feature enables exploitation of massive data-parallelism and SIMD architecture to obtain a higher processing speed. However, the obtained speedup is affected by data mapping strategy and communication overhead.

4.1 Overlapping Communication and Computation

Each computer application consists of transferring data from memory to the processor, processing data, and transferring data back to the memory (see Fig. 2(a)). So, the total time will be:

$$T = T_r + T_c + T_w \tag{3}$$

where T_r, and T_w are the required time to transfer data to the processor, and back to the memory, respectively, and T_c is the computation time.

To parallelize an application, several processors are employed to run concurrently in order to decrease the processing time. However, reading and writing from/into the memory is still strictly sequential. Even assuming that the processing part can be fully parallelized, according to the Amdahl's law, the communication time could limit the overall speedup available from parallelization. T_s and T_p denote the execution time of serial and parallel applications, respectively. Taking into account the communication time, using p processors, the speedup will be:

$$S = \frac{T_s}{T_p} = \frac{T_r + T_c + T_w}{T_r + \frac{T_c}{p} + T_w} \tag{4}$$

Let $\alpha = \frac{T_r + T_w}{T_c}$, the speedup can be written as follows:

$$S = \frac{\alpha T_c p + T_c p}{\alpha T_c p + T_c} = \frac{\alpha p + p}{\alpha p + 1} = 1 + \frac{p - 1}{\alpha p + 1} \tag{5}$$

This shows that using p processors, to achieve higher speedup, α should be decreased. As $\alpha = \frac{T_r + T_w}{T_c}$ and T_c should be kept as small as possible, decreasing memory communication time is the only way to decrease α. As memory communication has a sequential nature, maximum overlapping of computation and memory communication reduces the effect of this serial part in total time.

Due to the initial and final memory communication overheads, the complete overlap of computation and communication is not possible. In the initial phase, processors wait to receive the first segment of data and in the final phase, processors are ideal and the last segment of data is written back into the memory (see Fig. 2(b)). So, if intermediate memory communication overlaps with computation, and initial and final memory communication times are negligible, the speedup is maximized.

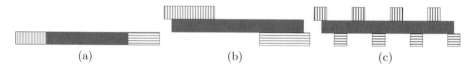

(a)   (b)   (c)

Fig. 2. Computation-Communication Model. Different boxes represent computation (black box), memory read (vertically shaded) and memory write (horizontally shaded). (a). Sequential communication-computation model, (b) Overlapping communication-computation model, (c) Overlapping communication-computation model, in the case that size of PE's memory is limited.

4.2 Data Distribution

In most parallel architectures using distributed memory, data locality is used to minimize communication time. This is due to the fact that global communication cost is much higher than computation cost. Since PE's memory, in CSX architecture, is too limited, it is just able to keep small segment of data. Consequently,

PE obtains data by communicating external memory or other PEs. In CSX architecture, data transfer from external (mono) memory to poly memory is much more expensive than interprocess communication. In fact, interprocess communication cost for neighboring PEs is even less than simple arithmetic operations. Thus, to minimize communication time, it is better to use interprocess communication instead of mono-poly memory communications. Moreover, as mentioned in Sect. 4.1, data mapping strategy should be developed in such a way that overhead of the initial and final memory communications are insignificant.

Considering CSX architecture, we can examine various data distribution strategies to determine the best one. According to the above discussion, comparison between various data distributions can be done in terms of the following parameters.

- initial and final memory communication overheads
- poly memory used for input data
- inter-process communication time

Having an image and a linear array of PEs, several data distributions are possible such as row (column)-stripe distribution, block distribution, and row (column)-cyclic distribution. Selecting any data distribution, to process boundary data, each PE should receives the neighboring data from external memory or other PEs. The former policy means that redundant data should be transferred from memory to the processors. The latter means using interprocess communication.

Assume c and r are the number of columns and rows in image matrix, respectively. Also, w indicates the size of window in SSD algorithm, β is the maximum disparity, and p denotes the number of PEs. Besides, in each memory communication, each PE read or write m bytes of data from/into the external memory, and t_m is the time taken by just one memory communication, i.e. t_m is the time required to transfer $m*p$ bytes of data between external and poly memories.

Block distributions. If the image is divided into $d*s$ blocks, each block has c/d columns and r/s rows. The first block is assigned to the first processor, the second one to the second processor, and so on. As each block is processed in one PE, the number of blocks should be equal to the number of processors. So, each block can be identified by an ordered pair (i, j) where $1 \leq i \leq s$ and $1 \leq j \leq d$. In the same way, each ordered pair can denote the PE which is responsible for processing corresponding block. To process the first segment of arrived data, any PE (i, j) requires last segment of data that will be sent to PE $(i, j-1)$. Here, interprocess communication cannot be used, since PE $(i, j-1)$ has not received data that its neighbor needs. So, this part of data should be sent to both PEs $(i, j-1)$ and (i, j). To process the last segment of data, PE (i, j) needs data which has already been sent to PE $(i, j+1)$. To use interprocess communication, PE $(i, j+1)$ should keep this part of data in its memory. Furthermore, each PE (i, j) requires some boundary data that has been sent to the PE $(i-1, j)$ and vice versa. In this case, also , swazzaling can be used, but the distance between two corresponding PEs is equal to d.

Row-stripe distribution. The rows are divided into several groups, each has r/p rows. Then, the first group is assigned to the first processor, the second one to the second processor, and so on. To process boundary data, PE i receives some data from PE $i - 1$ and vice versa. So, row-stripe distribution can utilize interprocess communication between neighboring PEs.

Row-cyclic distribution. In this scheme, the first row is assigned to the first processor, the second row to the second processor, etc. Each PE needs to communicate with the PEs which are at most at the distance $\frac{w-1}{2}$. Also, each PE can start doing computation, as soon as receiving a small chunk of data.

Table 1. Figure of merit related to the different data distributions. In calculating number of inter-process communication for block and row-stripe distributions, it is assumed that window size is less than number of rows which are assigned to the PEs. The third column shows the memory used for input data in each PE.

Data Dist.	Initial overhead	Inter-Process Comm		PE Memory
Block Dist.	$\frac{r}{s}t\_m$	$\frac{r}{s}\lceil\beta+w-1\rceil + d\frac{c}{d}\beta(w-1)$		$(w-1+2\beta+2m)\frac{r}{s}$
Row-stripe Dist.	$\frac{r}{p}t\_m$	$(w-1)c\beta$		$\frac{r}{p}(2m+\beta)$
Row-cyclic Dist.	t_m	$\frac{r}{p}(w-1)c\beta$		$2m+\beta$

The parameters calculated for each data distributions are summarized in Table 1. Having maximum memory usage and inter-process communication indicates that block distribution does not use CSX interprocess communication effectively. Row-stripe distribution has the minimum interprocess communication and using this strategy needs more memory space. Considering limited memory space of PEs and the minimum initial and final memory communication overhead, row-cyclic distribution is the optimal solution which well matches the CSX architecture features.

4.3 Parallelized SSD

Sequential SSD algorithm has three steps (see Sect. 2): evaluating square of differences, evaluating sum of squared differences over windows, and selecting the minimum. Using row-cyclic data distribution, PE i receives the data of line i and is responsible to calculate the line i of the output. Each PE can perform step 1, evaluating square of difference, just accessing local data, while evaluating sum over windows, it requires the data assigned to the other PEs. Each PE is able to calculate the sum of squared differences over one line of the window. So, performing step 2 is divided into two steps: first, calculating summation over one line using local data, and then receiving the results of other PEs and calculating sum over the window. Finally, during receiving data from other PEs,

Table 2. Algorithm 1- Parallelized SSD

Input: matrix of right & left images, maximum disparity (d), and window size (w)
Output: disparity image

For any PE K, Read the first m $(m + d)$ bytes of the k-th line in the right(left) image
from external memory into WR (WL), the working data buffer of right(left) Image

Main Loop: While there is any unprocessed segments
For any PE K, Read the next m bytes of the k-th line in the right(left) image
from external memory into MR (ML), buffer of right(left) Image, asynchronously

Step 1: Evaluate square of differences
   for i = 1 to m
      for j = 1 to d
         $SD[i, j] = (WR[k, i] - WL[k, i + j])^2$

Step 2: Evaluating sum of squared differences over lines
   for j = 1 to d
      for i = 1 to m
         $SSD[i, j] = SSD[i - 1, j] - SD[i - \frac{w+1}{2}, j] + SD[i + \frac{w-1}{2}, j]$

Step 3: Evaluating SSD over windows and selecting the minimum
   $min\_value = \infty$ for i = 1 to m
      for j = 1 to d
         3.1: Sazzling and Sum over windows
           $SSD[i, j] = SSD[i, j] + swazzle\_down((w - 1)/2) + swazzle\_up((w - 1)/2)$
         3.2: Select the minimum
           if $SSD[i, j] < min\_value$
              $min\_value = SSD[i, j]$, $min\_idx = j$
         $WO[i] = min\_idx$

Write the result buffer, WO, into external memory asynchronously

---

Swap working and memory input, output buffers

---

the minimum value which has been meet by that time can be evaluated. Thus,
the parallelized SSD has three steps as follows: evaluating square of differences,
evaluating sum over one line of the window(local-sum), and finally, evaluating
sum over windows and selecting the minimum.

Due to the memory limitation of PEs in CSX architecture, memory management has an important rule in algorithm implementation. Indeed, PEs' memory
cannot keep large amount of data, and after processing each segment of data, the
input segment should be substituted with unprocessed data and result should be
written back into the external memory as soon as possible. So, double-buffering is
the only way to realize communication-computation overlapping (see Fig. 2(c)).
When PEs are working with one buffer, the other one is reading or writing
data from/into the external memory. The pseudo code of the parallelized SSD
algorithm is shown in Table 2.

5   Implementation and Performance Results

To evaluate the performance, we have developed the proposed parallelized SSD algorithm. In order to utilize both cores of CSX700 processor, the input images should be divided into two nearly equal parts. The first $\lceil r/2 \rceil + (w-1)/2$ rows are assigned to the first core, and the last $\lfloor r/2 \rfloor + (w-1)/2$ lines are assigned to the second core. Sending boundary lines to both cores enables each core to perform all computation locally. Table 3 gives the timing of various steps, running the code on the 640x480 input images with disparity range of 16 pixels.

Table 3. Timing of the parallelized SSD steps on CSX architecture for 640x480 input images and disparity range of 16 pixels. The second and third columns show the timing related to ordinary and vector floating point operations, respectively.

Step	Timing	Timing (vector operation)
Initial & Final Overhead	17.37 μs	No Change
Square-of-differences	3.36 ms	1.176 ms
Sum(over line)	1.974	.888 ms
Sum(over window)& Selecting the Min	4.08 ms	No Change
Total	9.6 ms	6.25 ms

To analyze the performance of different steps of the algorithm and memory communication, we have used the CSX visual profiler tool. Fig. 3 depicts initial part of the log file related to the initial memory read. It shows that reading the first segment of data, the processors are ideal. Receiving the first data, computation starts, and next memory communication are overlapped with computation. The output of the visual profiler is consistent with the model illustrated in Fig. 2(c). It shows that ignoring the initial and final steps, the poly execution unit is working and never becomes ideal to receive data from external memory. According to the Table 3, the initial and final memory communication takes 17.37 μs which is just 0.2% of the whole time and PEs' execution unit are busy 98.3% of the time. The remaining 1.5% is control overhead.

As all PEs are running concurrently all the time, the only way to improve performance is to decrease the PE processing time. Vector operations can be used to decrease the computation time. Vector operations are applicable in steps 1

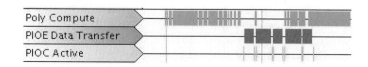

Fig. 3. Running Profile of Parallelized SSD on CSX processor related to initial memory read. Poly Compute shows the time poly execution unit is busy. PIOE Data Transfer indicates data transfer between poly and external memory. Receiving the first data segment, PEs start computation, and computations and communications are overlapped.

and 2, evaluation square of differences and summation over the lines. The third column of Table 3 shows the vectorization effect on timing. Utilizing vector operation in Step 1 and 2 yields an speedup around 2.8 and 2.2, respectively. It is not possible to use vector operations in the third step which takes more than 65% of the total time. As the assembly code generated by the compiler is not optimized, developing assembly code results in better performance in this step.

The number of floating point operation per second (FLOPS) is used as a performance evaluation measurement. Here, the number of operations required to compute disparity map should be divided by running time. Assuming that the algorithm applies 3x3 windows, the FLOPS will be calculated using (6).

$$\frac{\text{Number of Floating Point Operation}}{\text{Total Time}} = \frac{7cr\beta}{\text{Total Time}} \tag{6}$$

Our implementation computes disparity map of 640x480 images with maximum disparity 16 at rate 160 stereo pairs per second (6.25 ms for processing each pair). Increasing the maximum disparity to 32 results in rate of 83 stereo pairs per second (12 ms for processing each pair). Even for the resultion of 1280x720, the disparity map can be calculated in real time. For maximum disparity 16 and 32 the running time is 16.8 and 32 ms, respectively. Table 4 summarizes the running time and GFLOPs for different image sizes. In the last iteration of the algorithm, for 1280x720 and 640x480 images the number of the idle PEs are 16 and 42, respectively. So, due to the more utilization of PEs, we achieve better GFLOPS for 1280X720 images.

Table 4. Perforamnce of parallelized SSD Algorithm. Execution time depends on both image size and maximum disparity, while GFLOPS is just related to the image size.

Image Size	Max Disparity = 16	Max Disparity = 32	GFLOPS
640x480	6.25 ms	12 ms	5.7
1280x720	16.8 ms	32.08 ms	6.4

6   Conclusion and Future Works

This paper focuses on parallel implementation of SSD algorithm on CSX SIMD architecture. To do so, we have investigated different data distributions and PE scheduling. Our analysis shows that row-cyclic data distribution is the optimal solution considering CSX architecture and SSD algorithm. Our approach maximizes the communication-computation overlap to minimize the effect of memory communication which is always sequential on the speedup. Experimental results show that the PEs are computing 98.4% of the time. Our architecture computes 16 stereo disparities on 640x480 images at rate 160 stereo pairs per second, which is 80% faster than real-time. Also, for the 32 stereo disparities on the same image size, the rate is 83 stereo pairs per second. This result through the proposed architecture has two major consequences:

- It allows to free more time to support more advanced and more complex algorithms,
- It needs less power consumption enabling energetically efficient embedded solutions

We can think about more advanced scenes 3D descriptions like objects of interest localization, tracking or more globally, navigation-oriented tasks like obstacle avoidance, path planing or SLAM etc., with minimizing the critical resources for autonomous mobile systems, namely, energy.

References

1. Scharstein, D., Szeliski, R.: A taxonomy and evaluation of dense two-frame stereo correspondence algorithms. International Journal of Computer Vision 47, 7–42 (2001)
2. Hirschmüller, H., Innocent, P.R., Garibaldi, J.: Real-time correlation-based stereo vision with reduced border errors. Int. J. Comput. Vision 47, 229–246 (2002)
3. Sunyoto, H., van der Mark, W., Gavrila, D.M.: A comparative study of fast dense stereo vision algorithms. In: 2004 IEEE Intelligent Vehicles Symposium, pp. 319–324 (2004)
4. Yang, R., Pollefeys, M.: A versatile stereo implementation on commodity graphics hardware. Real-Time Imaging 11, 7–18 (2005)
5. Wang, L., Liao, M., Gong, M., Yang, R., Nister, D.: High-quality real-time stereo using adaptive cost aggregation and dynamic programming. In: 3DPVT 2006: Proceedings of the Third International Symposium on 3D Data Processing, Visualization, and Transmission (3DPVT 2006), Washington, DC, USA, pp. 798–805. IEEE Computer Society, Los Alamitos (2006)
6. Woodfill, J., Herzen, B.V.: Real-time stereo vision on the parts reconfigurable computer. In: IEEE Symposium on FPGAs for Custom Computing Machines, pp. 242–250. IEEE Computer Society Press, Los Alamitos (1997)
7. Murphy, C., Lindquist, D., Rynning, A.M., Cecil, T., Leavitt, S., Chang, M.L.: Low-cost stereo vision on an fpga. In: FCCM 2007: Proceedings of the 15th Annual IEEE Symposium on Field-Programmable Custom Computing Machines, Washington, DC, USA, pp. 333–334. IEEE Computer Society, Los Alamitos (2007)
8. ClearSpeed (2009), http://www.clearspeed.com
9. ClearSpeed Technology: ClearSpeed Whitepaper: CSX Processor Architecture (2007), http://www.clearspeed.com
10. Baldonado, M., Chang, C.C., Gravano, L., Paepcke, A.: The stanford digital library metadata architecture. Int. J. Digit. Libr. 1, 108–121 (1997)

Revisiting the P*n*P Problem with a GPS

Timo Pylvänäinen, Lixin Fan, and Vincent Lepetit*

Nokia Research Center

Abstract. This paper revisits the pose estimation from point correspondences problem to properly exploit data provided by a GPS. In practice, the location given by the GPS is only a noisy estimate, and some point correspondences may be erroneous. Our method therefore starts from the GPS location estimate to progressively refine the full pose estimate by hypothesizing correct correspondences. We show how the GPS location estimate and the choice of a first random correspondence dramatically reduce the possibility for a second correspondence, which in turn constrains even more the remaining possible correspondences. This results in an efficient sampling of the solution space. Experimental results on a large 3D scene show that our method outperforms standard approaches and a recent related method [1] in terms of accuracy and robustness.

1 Introduction

The recent development of mobile devices has made applications such as localization using a simple embedded camera realistic on such devices. Research in this direction, however, has mostly focused on image retrieval techniques to consider large-scale environments [2]. Such approach can only provide a coarse pose, and more accuracy will be needed, for example in Augmented Reality applications.

In this paper, we focus on the estimation of an accurate 3D pose from correspondences between 3D points and their projections in the image. This is most certainly one of the oldest problems in Computer Vision, however in this work, we explicitly target pose estimation of photographs taken with a hand held device. In particular, this implies that we can exploit the other sensors these devices are typically equipped with beside the camera.

These sensors include accelerometers, magnetometers (i.e. electronic compasses), and GPS. Accelerometers, or inertial sensors, have been used for several years in tracking applications [3]. They can measure the camera motion, which is of no use in our application[1], and the camera orientation. In practice, however, it is not uncommon to see errors of over 90 degrees in the obtained orientation estimate, and so we choose not to use orientation measurements. Instead, we only consider the GPS.

---

* Computer Vision Laboratory, École Polytechnique Fédérale de Lausanne (EPFL).
[1] In practice the accuracy of the accelerometers available for mobile devices combined with the erratic motion of such a device together make it virtually impossible to integrate acceleration signal to obtain location.

G. Bebis et al. (Eds.): ISVC 2009, Part I, LNCS 5875, pp. 819–830, 2009.

The standard GPS is typically said to have an accuracy of around few meters. This gives a strong prior on the camera pose, but the uncertainty must still be properly taken into account for an accurate pose estimation. Our method starts from the GPS location estimate to progressively refine the full pose estimate. This is done by sequentially hypothesizing correct correspondences. Because the choices for the previous correspondences dramatically constrain the possibilities of the next correspondence, this allows a particularly efficient sampling of the solution space.

The closest method in the literature is [1], which starts from a prior on the full pose and applies an Extended Kalman filter each time a correspondence is picked to shrink the search space for the next hypotheses. Our experimental results show that our method performs better in terms of robustness and accuracy. This is most likely due to the analytical solution to constrain the search where [1] has to linearize the Kalman filters equations.

In the remainder of the paper, previous work which has used GPS and other sensors is first briefly reviewed. A formalization of the problem is given in Section 3, and Section 4 describes and analyses the sequential sampling procedure. Finally Section 5 compares the proposed method to the standard solution using RANSAC and [1] on a large real 3D scene.

2   Related Work

Carceroni et al. [4] have studied a similar problem of estimating camera orientation from multiple views, given the locations of the viewpoints and a set of point correspondences between views. The uncertainty in the viewpoint locations, however, has not been taken into account. Their method essentially reduces the problem to three degrees of freedom, while we optimize over the full six degrees of freedom of the camera pose.

Some work has been done with inertial sensors and GPS where pose estimation is relevant [3,5,6,7]. Many of these systems do not truly use all the redundancy of the given measurements, but rather use sensors to initialize and help the visual system [3]. When low level sensor fusion is used, it is often done as the prediction part of a Kalman filter such as in [5] and is only useful for tracking applications.

Pollefeys et al. [7] use GPS and inertial tracking for camera pose estimation, and then correspondences for reconstruction. There is therefore no fusion of image and sensor information. GPS and orientation sensor data are used in [6] to compute an initial pose estimate by solving a linear system. This estimate is then refined using a non-linear optimization of an objective function that incorporates a GPS term with an arbitrary weighting factor. Both of these methods require a measurement of the full pose, while we rely only on the GPS and its uncertainty to initialize our method.

We want to be able to deal with outlier correspondences, and RANSAC-like methods have proved their robustness and efficiency. Many variants have been proposed over the years, and we can roughly classify them into four categories:

1. Methods to reduce cost of evaluation to increase the number of iterations that can be done in a given time [8,9].
2. Methods that use prior knowledge to guide sampling [10,11,12].
3. Methods that exploit the connectivity of nearly optimal models [13,14,15,16,17].
4. Methods that modify the evaluation function to get better results [18,19].

Most improvements of RANSAC fall into Categories 1 and 3. In most cases, these improvements can be combined with our method. For instance, our method has been successfully combined with the Hill Climbing strategy [13] to create a combined optimization strategy which performs better than Hill Climbing strategy alone.

The method we propose falls mostly in the second category, taking advantage of very specific prior knowledge. The closest method in this category is perhaps [11] in the sense they use a model specific sampling strategy. They, however, consider a different problem than ours since they focus on homography estimation. They use a different consistency constraint, enforced as a preprocessing step which introduces a fixed overhead.

Like our method, NAPSAC [15] uses the idea of generating sample sets sequentially, but its only assumption was that neighbouring points of an inlier are more likely to be inliers. This method does not use any prior knowledge and falls in Category 3.

Moreno-Noguer [1] also uses sequential sample generation and considers the camera pose estimation problem but requires a prior on the full pose, while we use a prior on the camera center location only. In [1], the correspondences between 3D points and their reprojection are not assumed to be known, and each consecutive point is treated as an observation of the unknown pose in a Kalman filter setting. In contrast, we take a fundamentally different approach as each consecutive point is instead used to reduce degrees of freedom of the pose estimate and the covariances are propagated to represent the uncertainty in the fixed degrees.

However, the algorithm presented in [1] is probably the closest one to ours in that it solves the same problem, uses similar prior knowledge and builds minimal sets sequentially. We therefore compare it against our method in Section 5.

3   Problem Statement

This section gives a formal mathematical definition of the problem we solve. The camera pose estimation is formulated as a minimization problem of a cost function that considers the log likelihood of the pose given the observed correspondence and of the location measurement from the GPS. Because these measurements have meaningful units, the relative weights of the two different kinds of measurements are properly described by the covariance matrices.

We assume we are given a set of world points $\mathbf{X}_i \in \mathbb{R}^3$ and their reprojections $\mathbf{x}_i \in \mathbb{R}^2$ in the image. In practice we use SURF [20] to find these correspondences. The reprojections $\mathbf{x}_i$ are corrupted by noise, and some can even be completely mismatched.

We denote by θ the camera pose and by $\text{proj}_\theta(\mathbf{x}_i)$ the projection of point $\mathbf{X}_i$ by the camera. In other words, for the inliers it is expected that

$$\mathbf{x}_i = \text{proj}_\theta(\mathbf{X}_i) + \epsilon, \tag{1}$$

where ϵ is a Gaussian noise term.

The location provided by the GPS is denoted by $\mathbf{g}$, and to simplify we assume its true value $\bar{g}$ can be computed directly from the true value $\bar{\theta}$ of the camera parameters as the camera center $\bar{g} = c(\bar{\theta})$.

The problem can finally be stated as recovering the pose θ which minimizes the cost function

$$cost(\theta) = E_c(\theta) + \sum_i \rho(e_i^2), \tag{2}$$

where

$$E_c(\theta) = (\mathbf{g} - c(\theta))^\top \Sigma_c^{-1}(\mathbf{g} - c(\theta))$$
$$e_i^2 = (\mathbf{x}_i - \text{proj}_\theta(\mathbf{X}_i))^\top \Sigma_x^{-1}(\mathbf{x}_i - \text{proj}_\theta(\mathbf{X}_i)) \tag{3}$$

and $\rho(.)$ is a robust estimator:

$$\rho(e_i^2) = \begin{cases} e_i^2 & e_i^2 < T^2 \\ T^2 & e_i^2 \geq T^2 \end{cases}. \tag{4}$$

4   Sequential Sampling

For the pose estimation problem, three correspondences define a pose. In the proposed sequential sampling, this minimal set is generated by selecting each consecutive correspondence from a different distribution, starting from the uniform distribution. Each consecutive correspondence reduces the degrees of freedom of the unknown pose. The probability distribution of the location measurement is mapped to a probability distribution of the correspondences, with the assumption that the first correspondence was correct.

This effectively uses the location measurement to guide the generated minimal sets to be consistent with the location measurement. The proposed poses are evaluated against the robust cost function to find a good inlier set.

4.1   Sampling the First Correspondence

Assuming the rotation of the camera is unknown, then even with known location any single correspondence is a priori equally likely. It is always possible to align any point in 3D to any point in the camera image by rotating the camera. So the first correspondence $\mathbf{x}_1 \leftrightarrow \mathbf{X}_1$ is randomly selected from a uniform distribution.

4.2   Sampling the Second Correspondence

Lets first assume that the camera location and the image location $\mathbf{x}_1$ of the first correspondence are known exactly. The only remaining degree of freedom is the rotation about the axis formed by the camera center and the corresponding 3D point $\mathbf{X}_1$. Under this constraint, the projection $\mathbf{x}_2$ of a given world point $\mathbf{X}_2$ lies on an ellipse in the image plane.

Since the value of the angle defined by $\mathbf{X}_1$, the camera center $\mathbf{g}$, and $\mathbf{X}_2$ should remain the same when expressed in the camera coordinate system and in the world coordinate system, the following formula should hold:

$$\frac{\mathbf{x}_1^\top \mathbf{x}_2}{\|\mathbf{x}_1\|\|\mathbf{x}_2\|} - \frac{(\mathbf{X}_1 - \mathbf{g})^\top (\mathbf{X}_2 - \mathbf{g})}{\|\mathbf{X}_1 - \mathbf{g}\|\|\mathbf{X}_2 - \mathbf{g}\|} = 0 \,, \tag{5}$$

where $\mathbf{x}_1$ and $\mathbf{x}_2$ are considered to be in homogeneous coordinates. This equation defines an implicit function constraint of the type:

$$f(\mathbf{x}_2, \mathbf{x}_1, \mathbf{g}) = 0 \,. \tag{6}$$

In practice, however, the location $\mathbf{g}$ provided by the GPS and the projection $\mathbf{x}_1$ are corrupted by noise. For a given $\mathbf{x}_2$, there is then some probability that Equation (6) is satisfied. The exact computation of this probability involves integrating over the sets of $\mathbf{x}_1$ and $\mathbf{g}$ for which the constraint holds. This is not computationally feasible, so instead, we linearize f in the neighbourhood of the observations:

$$f(\mathbf{x}_2, \mathbf{x}_1, \mathbf{g}) \approx J_{\mathbf{x}_2}\mathbf{x}_2 + J_{\mathbf{x}_1}\mathbf{x}_1 + J_{\mathbf{g}}\mathbf{g}, \tag{7}$$

where $J_{\mathbf{x}_1}$, $J_{\mathbf{x}_2}$ and $J_{\mathbf{g}}$ are the Jacobians of (6) with respect to to $\mathbf{x}_1$, $\mathbf{x}_2$ and $\mathbf{g}$ respectively, and evaluated at the measured points.

Since we assumed that $\mathbf{x}_1$, $\mathbf{x}_2$ and $\mathbf{g}$ are normally distributed, the residual $f(\mathbf{x}_2, \mathbf{x}_1, \mathbf{g})$ can now be thought of as a normally distributed random variable:

$$f(\mathbf{x}_2, \mathbf{x}_1, \mathbf{g}) \sim N(0, \sigma_e^2)$$
$$\sigma_e^2 = J_{\mathbf{x}_2}\Sigma_{\mathbf{x}}J_{\mathbf{x}_2}^\top + J_{\mathbf{x}_1}\Sigma_{\mathbf{x}}J_{\mathbf{x}_1}^\top + J_{\mathbf{g}}\Sigma_c J_{\mathbf{g}}^\top \,. \tag{8}$$

Figure 1 illustrates the error introduced by the linear approximation. The ground truth was obtained by randomly sampling the camera center and first correspondence according to the assumed normal distributions. For each sample, the possible exact projections were accumulated.

The likelihood of any point $\mathbf{x}_i$ satisfying the constraint, assuming that $\mathbf{x}_1$ is an inlier corrupted by noise, can now be computed by taking the residual of Equation (6) where $\mathbf{x}_2 = \mathbf{x}_i$ and plugging it into Equation (8). The second point is randomly selected proportional to these likelihoods.

4.3   Sampling the Third Correspondence

Even if a given location and two correspondences constitute an over-constrained pose estimation problem, we found it is still better to consider a third correspondence because of the uncertainty in the location measurement. We give here an

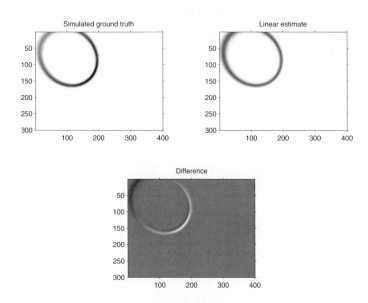

Fig. 1. An example of the linear approximation of the correspondence likelihood. The biggest error happens close to the edges of the image. Light areas of the difference map indicate where the linear approximation under estimates the probability.

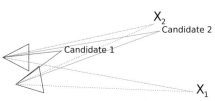

Fig. 2. The scene used in the experiments contains 99 cameras and 10000 world points. The world point visibility in the cameras is based on the original matching information in the original reconstruction.

Fig. 3. The sensitivity of the projection of the third point depends on its location relative to the first two points and the camera. Points such as Candidate 1 that are far from the first points are sensitive to camera location. Candidate 2, on the other hand, is close to one of the first two selected points and is less sensitive to camera location.

approximation of the covariance for the projection $\mathbf{x}_3$ of a selected third point $\mathbf{X}_3$ that works well in practice and keeps the computation tractable.

We first assume that the biggest contributor to the variance comes from the camera location uncertainty and all other sources are neglected. We compute a current estimate for the camera pose by taking for the moment the GPS location measurement $\mathbf{g}$ as camera center, and by using the first two correspondences to estimate the rotation $\mathbf{R}$. Intuitively, as shown in Figure 3, when the third point $\mathbf{X}_3$ is close to the two first points $\mathbf{X}_1$ and $\mathbf{X}_2$, the covariance $\Sigma_{\mathbf{x}_3}$ of $\mathbf{x}_3$ will be small. When it is moved away from these points, the covariance will increase. We therefore use the following approximation for $\Sigma_{\mathbf{x}_3}$:

$$\Sigma_{\mathbf{x}_3} = \kappa_1^2 \kappa_2^2 \mathbf{W}^\top \Sigma_c \mathbf{W} \tag{9}$$

where

$$\kappa_1 = \min\left(\frac{\|\mathbf{X}_3 - \mathbf{X}_1\|}{\|\mathbf{X}_1 - \mathbf{g}\|}, \frac{\|\mathbf{X}_3 - \mathbf{X}_2\|}{\|\mathbf{X}_2 - \mathbf{g}\|}\right)^2$$

$$\kappa_2 = \min\left(\frac{1}{\|\mathbf{X}_1 - \mathbf{g}\|}, \frac{1}{\|\mathbf{X}_2 - \mathbf{g}\|}\right)^2 \tag{10}$$

$$\mathbf{W} = \mathbf{R}^\top \begin{pmatrix} 1 & 0 \\ 0 & 1 \\ 0 & 0 \end{pmatrix}.$$

The matrix $\mathbf{W}$ maps the covariance of the camera location to the image plane.

We still have to explain how we compute the rotation matrix $\mathbf{R}$ that appears in the third row of (10). The camera maps the 3D points $\mathbf{X}_1$ and $\mathbf{X}_2$ to their respective 2D locations $\mathbf{x}_1$ and $\mathbf{x}_2$ and its rotation $\mathbf{R}$ must satisfy

$$\mathbf{R}\hat{\mathbf{X}}_1 = \hat{\mathbf{x}}_1$$

$$\mathbf{R}\hat{\mathbf{X}}_2 = \hat{\mathbf{x}}_2 \tag{11}$$

$$\mathbf{R}(\hat{\mathbf{X}}_1 \times \hat{\mathbf{X}}_2) = \hat{\mathbf{x}}_1 \times \hat{\mathbf{x}}_2 ,$$

where $\hat{\mathbf{X}}_i = \frac{\mathbf{X}_i - \mathbf{g}}{\|\mathbf{X}_i - \mathbf{g}\|}$ and $\hat{\mathbf{x}}_i = \frac{\mathbf{x}_i}{\|\mathbf{x}_i\|}$. The last equation in (11) comes from the properties of a rotation matrix.

This can be written as a linear system, and solved in the least-squares sense to obtain the elements of the rotation matrix. We force the solution $\mathbf{R}'$ to a proper rotation matrix by computing the singular value decomposition $\mathbf{R}' = \mathbf{U}\mathbf{D}\mathbf{V}^\top$ and then taking $\mathbf{R} = \mathbf{U}\mathbf{V}^\top$.

A summary of our method is given in Algorithm 1. This version has no other stopping criterion than a limited time budget. The function *cost* is the cost function defined in (2). Functions *EllipseLikelihood* and *PointLikelihood* will randomly draw a correspondence index according to the likelihoods defined in Sections 4.2 and 4.3 respectively. Function *PossiblePoses* returns the four solutions of the P3P problem.

Input: A set of N possible correspondences and location information $\mathbf{g}$
Output: Camera pose estimate
while *time left* **do**
    $i_1 := Uniform(1..N)$;
    $i_2 := EllipseLikelihood(\mathbf{g}, i_1, \mathbf{x}_1, ..., \mathbf{x}_N)$;
    $i_3 := PointLikelihood(\mathbf{g}, i_1, i_2, \mathbf{x}_1, ..., \mathbf{x}_N)$;
    $\Theta := PossiblePoses(\mathbf{x}_{i_1}, \mathbf{x}_{i_2}, \mathbf{x}_{i_3})$;
    foreach $\theta \in \Theta$ **do**
        if $cost(\theta, \mathbf{x}_1, ..., \mathbf{x}_N, \mathbf{g}) < c_b$ **then**
            $c_b := cost(\theta, \mathbf{x}_1, ..., \mathbf{x}_N, \mathbf{g})$;
            $\theta^* := \theta$;
        end
    end
end

Algorithm 1. Overview of pose estimation

5   Experiments

5.1   Setup

We reconstructed a real 3D scene from 99 images with GPS measurements using our own reconstruction pipeline based on SURF features [20] and the sparse bundle adjustment library described in [21]. The resulting reconstruction contains over 20000 world points. 10000 of the world points were randomly selected for the test set. The resulting scenario is shown in Figure 2.

To analyze the effect of keypoint localization noise and GPS noise, we first corrected the keypoint locations in the images to match the reconstructed world exactly. This created a noise free reconstruction with perfectly known camera poses, which nevertheless represents a real world scenario.

We then added noise to the keypoint locations. Inliers were corrupted with Gaussian noise with a standard deviation of 5 pixels in the original 800×600 pixel images used to capture the scene. With a given probability a keypoint is treated as an outlier, in which case its location is randomly drawn from the uniform distribution over the image. We tests with 4 different outlier ratios from 10% to 70%.

The GPS measurement was generated by adding noise to the camera location in the reconstruction, which after the corrected projections is effectively noiseless. The noise was drawn from the three dimensional Gaussian distribution and normalized to a fixed length. Tests were run with 6 different GPS offsets from 0 to 5 meters. The distribution used for the GPS error in the cost function and in sequential sampling had a standard deviation of 5 meters.

This results in a total of 24 noise scenarios. Three methods were tested under these conditions:

1. Standard RANSAC to optimize the compound cost function.
2. The pose prior method of [1] modified as described below.
3. The proposed sequential sampling method.

The original method described in [1] does not assume known correspondences and performs an exhaustive search. It was adapted to the Random Sampling

Consensus framework as follows. Starting from the given pose prior, the pose estimate and its covariance are updated according to the Kalman filter rules as new points are selected. The selection of the next point is done according to the likelihoods defined by the reprojection error obtained using the current pose estimate. The reprojection error covariance is obtained by propagating the pose estimate covariance using the Jacobian of the projection function.

The pose prior method requires a full pose. To be fair, it was tested with realistic noise which corresponds to the kind of orientation estimate that might be obtained from sensors embedded in a mobile device. The rotation matrix was corrupted by random noise until a rotation matrix was obtained where the mean angle between the axis of the original rotation matrix and the corrupted rotation matrix was between 20 and 25 degrees.

A sample run consists of 10 iterations of the algorithm to estimate the pose of a randomly selected image from the set. A test comprises of 1000 sample runs of an algorithm under specific noise conditions.

5.2   Results

The three histograms of the left of Figure 4 show the results obtained with the three different methods for various GPS errors when there are only 10% outliers. In that case, the three methods perform about the same.

The proposed GPS method, however, is most valuable when the images are difficult to match. In practice, urban scenes have many repetitive structures, for example the windows of a building, and to guarantee the existence of the real correspondence, multiple hypotheses from feature matching should be retained,

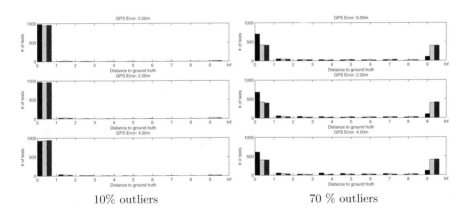

Fig. 4. Comparison with RANSAC and the modified pose prior method of [1] for two different ratios of outliers, and three levels of GPS errors. The bars in the stacks correspond, from left to right, to the proposed method (blue), the modified method of [1] (green), and RANSAC (red). **Left:** For small ratios of outliers, the three methods perform about the same. **Right:** For large ratios of outliers, which correspond to more realistic scenarios, our method is clearly more accurate.

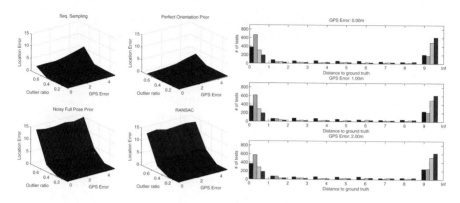

Fig. 5. Left: Median distance to ground truth location for different noise conditions and for all the methods. In this test we also included, for comparison, the modified pose prior method of [1] with no error in the orientation. It can be seen that RANSAC and pose prior with realistic noise in the pose both tend to break down after 50% outliers. Although our method shows degraded performance as the GPS error approaches values unlikely according to the assumed covariance, it still outperforms previous methods. Of course, the full pose prior method with perfect orientation performs very well as with perfect orientation measurement there is little to optimize. **Right:** The histograms of distance to ground truth location for 70% outliers for the proposed method and the full pose prior with different levels of noise in the orientation. The bars in the stacks are from left to right: the proposed method (blue), full pose prior with maximum 5 (cyan), 5-10 (yellow) and 10-15 (red) degrees of error. As is to be expected, with no error in the location and very little error in the orientation, the full pose prior method works very well. When the GPS error is increased, the performance of the proposed method approaches that of the method with nearly perfect orientation. The full pose prior with noisy orientation does not perform as well as the proposed method. It can be observed that if the orientation measurement is available, it must have error less than 5 degrees before the switch to full pose prior method is justified.

resulting in a large number of outliers. As shown by the histograms on the right of Figure 4, our method efficiently takes advantage of the GPS data, and outperforms the other methods.

Figure 5 (left) summarizes the results of more experiments. We also include the results obtained with the modified method of [1] provided by the exact orientation. This version performs remarkably well, unfortunately currently no sensors are able to provide such accuracy on the orientation. It can be seen that RANSAC and pose prior with realistic noise in the pose both tend to break down after 50% outliers.

Finally, we tested our method against the full pose prior method with different levels of noise in the orientation and location measurements. The test case contained 70% outliers and is based on 1000 sample runs. The results are shown in Figure 5 (right). We tested against three different orientation noise cases: where the average angluar error to the axes of the ground truth rotation was less than 5 degrees, when it was between 5 and 10 degrees and between 10 and

15 degrees. Obviously, with nearly perfect prior pose information the full pose prior method performs extremely well. It can be observed, however, that as the GPS error is increased the performance difference becomes smaller.

The results show that if the orientation measurement is available, it has to have an error less than 5 degrees for it to be useful. The proposed method, which uses only GPS, outperforms the full pose prior method with 5-10 degrees of error in the orientation.

6   Conclusion

We showed how GPS information can be used to guide sampling in a RANSAC setting to estimate inliers of the pose estimation problem. This novel sequential sampling method was shown to effectively guide the sampling towards the correct solution.

In the experiments, the method shows clear performance advantage when the number of outliers is high. In real world applications, extremely high outlier ratios commonly occur when multiple hypotheses from feature matching are retained. In the case of repeated patterns, multiple hypotheses can lead to multiple consensus sets only one of which represents the correct pose. The use of GPS effectively resolves this ambiguity and the proposed method does this efficiently.

It should be noted, however, that the evaluation of the likelihoods for each candidate match in steps 2 and 3 is roughly equivalent to one evaluation of the objective function. One iteration of the proposed algorithm in a naive implementation therefore equals to roughly three iterations of standard RANSAC in terms of CPU time. This means, unfortunately, that in practice it is usually faster to not apply the GPS based weighting on the candidate matches unless the outlier ratio is very high. It might be possible to develop more advanced selection strategies which would avoid full evaluation of the likelihoods for each point.

References

1. Moreno-Noguer, F., Lepetit, V., Fua, P.: Pose priors for simultaneously solving alignment and correspondence. In: Forsyth, D., Torr, P., Zisserman, A. (eds.) ECCV 2008, Part II. LNCS, vol. 5303, pp. 405–418. Springer, Heidelberg (2008)
2. Takacs, G., Chandrasekhar, V., Gelfand, N., Xiong, Y., Chen, W.C., Bismpigiannis, T., Grzeszczuk, R., Pulli, K., Girod, B.: Outdoors augmented reality on mobile phone using loxel-based visual feature organization. In: ACM international conference on Multimedia information retrieval, pp. 427–434. ACM, New York (2008)
3. Klein, G., Drummond, T.: Tightly integrated sensor fusion for robust visual tracking. Image and Vision Computing 22 (2004)
4. Carceroni, R., Kumar, A., Daniilidis, K.: Structure from motion with known camera positions. In: CVPR, pp. 477–484 (2006)
5. You, S., Neumann, U.: Fusion of vision and gyro tracking for robust augmented reality registration. In: IEEE Conference on Virtual Reality, pp. 71–78 (2001)

6. Ng, T.K., Kanade, T.: PALM: portable sensor-augmented vision system for large-scene modeling. In: 3-D Digital Imaging and Modeling, pp. 473–482 (1999)
7. Pollefeys, M., Nistèr, D., Frahm, J.M., Akbarzadeh, A., Mordohai, P., Clipp, B., Engels, C., Gallup, D., Kim, S.J., Merrell, P., Sinha, S., Talton, B., Wang, L., Yang, Q., Stewènius, H., Yang, R., Weclh, G., Towles, H.: Detailed real-time urban 3d reconstruction from video. International Journal of Computer Vision (2007)
8. Nister, D.: Preemptive RANSAC for live structure and motion estimation. In: ICCV, vol. 1, pp. 199–206 (2003)
9. Chum, O., Matas, J.: Randomized RANSAC with Td,d test. In: BMVC, pp. 448–457 (2002)
10. Tordoff, B., Murray, D.W.: Guided sampling and consensus for motion estimation. In: Heyden, A., Sparr, G., Nielsen, M., Johansen, P. (eds.) ECCV 2002. LNCS, vol. 2350, pp. 82–96. Springer, Heidelberg (2002)
11. Guo, F., Aggarwal, G., Shafique, K., Cao, X., Rasheed, Z., Haering, N.: An efficient data driven algorithm for multi-sensor alignment. In: Workshop on Multi-camera and Multi-modal Sensor Fusion Algorithms and Applications, ECCV (2008)
12. Chum, O., Matas, J.: Matching with PROSAC - progressive sample consensus. In: CVPR, pp. 220–226 (2005)
13. Pylvänäinen, T., Fan, L.: Hill climbing method for random sample consensus methods. In: International Symposium on Visual Computing (2007)
14. Chum, O., Matas, J., Kittler, J.: Locally optimized RANSAC. In: Porceedings of the 25th DAGM Symposium Pattern Recognition, pp. 236–243 (2003)
15. Myatt, D., Bishop, J., Craddock, R., Nasuto, S., Torr, P.H.: NAPSAC: High noise, high dimensional robust estimation — it's in the bag. In: BMVC, pp. 458–467 (2002)
16. Tordoff, B.J., Murray, D.W.: Guided-MLESAC: Faster image transform estimation by using matching priors. IEEE Transactions on Pattern Analysis and Machine Intelligence 27, 1523–1535 (2005)
17. Raguram, R., Frahm, J.M., Pollefeys, M.: A comparative analysis of ransac techniques leading to adaptive real-time random sample consensus. In: Forsyth, D., Torr, P., Zisserman, A. (eds.) ECCV 2008, Part II. LNCS, vol. 5303, pp. 500–513. Springer, Heidelberg (2008)
18. Torr, P.H., Zisserman, A.: MLESAC: A new robust estimator with application to estimating image geometry. Computer Vision and Image Understanding 78, 138–156 (2000)
19. Torr, P.H., Davidson, C.: IMPSAC: Synthesis of importance sampling and random sample consensus. IEEE Transactions on Pattern Analysis and Machine Intelligence 25, 354–364 (2003)
20. Bay, H., Ess, A., Tuytelaars, T., Van Gool, L.: SURF: Speeded up robust features. In: Computer Vision and Image Understanding (2008)
21. Louarkis, M., Argyros, A.: The design and implementation of a generic sparse bundle adjustment software package based on the Levenberg-Marquardt algorithm. Technical Report 340, Computer Science-FORTH, Heraklion, Crete, Greece (2004)

On Using Projective Relations for Calibration and Estimation in a Structured-Light Scanner

Daljit Singh Dhillon and Venu Madhav Govindu

Department of Electrical Engineering, Indian Institute of Science,
Bangalore-560012, India
djdhillon@gmail.com, venu@ee.iisc.ernet.in

Abstract. For structured-light scanners, the projective geometry between a projector-camera pair is identical to that of a camera-camera pair. Consequently, in conjunction with calibration, a variety of geometric relations are available for three-dimensional Euclidean reconstruction. In this paper, we use projector-camera epipolar properties and the projective invariance of the cross-ratio to solve for 3D geometry. A key contribution of our approach is the use of homographies induced by reference planes, along with a calibrated camera, resulting in a simple parametric representation for projector and system calibration. Compared to existing solutions that require an elaborate calibration process, our method is simple while ensuring geometric consistency. Our formulation using the invariance of the cross-ratio is also extensible to multiple estimates of 3D geometry that can be analysed in a statistical sense. The performance of our system is demonstrated on some cultural artifacts and geometric surfaces.

Keywords: 3D Reconstruction, Structured-Light Scanner, System Calibration, Projector Calibration, Homography, Cross-Ratio.

1 Introduction

While the two-view (epipolar) geometry can be used to solve for the 3D scene geometry via triangulation [1], the need to obtain reliable point correspondences is a significant limitation. In structured-light systems, this problem is overcome by projecting a known series of light patterns on the object of interest that is imaged through the camera. As the projected patterns are known, a dense set of correspondences can be established between the projector pixels and the camera pixels. Typically, structured-light patterns are efficiently projected by codifying the view-space either spatially [2], temporally [3] or by geometric arrangement of cameras [4]. See [5] for a review of various pattern codification strategies.

A projector can be effectively modelled as a 'direction-reversed' camera, implying that in projective-geometric terms it can be treated as a camera. Consequently, the projector-camera pair can be treated as a two-view camera pair. The geometry of such a camera pair is well understood and can be effectively applied to the dense correspondences available in structured-light systems. While the

G. Bebis et al. (Eds.): ISVC 2009, Part I, LNCS 5875, pp. 831–842, 2009.

epipolar relations are sufficient to recover structure by triangulation, such depth estimates are only available in a projective sense. To recover the true Euclidean shape (or depth) of an object, we need to carry out a calibration procedure. A common approach for system calibration involves the projection of coded-light patterns on one or more reference planes that allows for the calibration information to be recovered [6,7,8,9]. In terms of the overall system accuracy, there exists a trade-off between the complexity and precision of the calibration method and the ease and accuracy of the estimation of shape. While some re-construction methods work without any calibration, such approaches suffer from limited accuracy. In [10] the system accuracy is improved by a non-linear mini-mization of an energy functional involving all 3D-points. However, this implies a significant computational complexity. In [9], the complexity of calibration is transferred to precisely controlled mechanical movement of the reference planes that maintain a fixed orientation, i.e. the reference planes have to be parallel. In contrast, systems that calibrate all components explicitly often involve an elab-orate calibration process [8]. In [11] and [7] the projection planes are calibrated resulting in an over-parameterisation of the geometry thus requiring additional consistency constraints. As will be described later, our approach of parameter-ising the projector and system calibration information by using homographies induced between camera and projector by known reference planes reduces the complexity of the calibration process and achieves accurate estimation.

In our approach, the frame of reference for measurements is attached to the camera which is calibrated using a commonly used method [12]. To calibrate the scanner system, one common approach is to project sinusoidal patterns onto reference planes to generate reference phase maps [6,9]. By relating the phase-information between the projector and image pixels, a dense mapping is ob-tained. However, this is an over-parameterisation of the calibration information. Since the reference object is a plane, we can use the projective relationships induced by a plane. In our approach, we estimate the homographies between the projector-camera pair induced by the reference planes to encode the reference phase maps. Given the large number of correspondences between the projector and camera pixels, such homographies can be accurately estimated, resulting in an accurate parameterisation of the reference phase maps. For the estimation of the three-dimensional structure of the imaged object, we utilise the fact that the cross-ratio of four collinear points is invariant to projective transformations. In conjunction with epipolar geometry, we utilise the cross-ratio to solve for the three-dimensional geometry that utilises the calibration information of two reference planes. In contrast with the methods that require precise placement of the reference planes, our method does not impose any such restriction. Our formulation is also extensible to utilise more than two reference planes. Multiple reference planes allow us to use multiple cross-ratio relationships that allow us to compute more than one estimate of the depth of a given point. Such estimates are amenable to statistical analysis which allows us to derive both the shape of an object and associate reliability scores to each estimated point location.

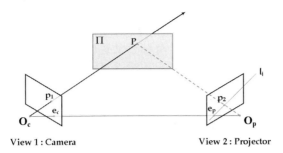

View 1 : Camera   View 2 : Projector

Fig. 1. The correspondences between *camera* and *projector* are related by the epipolar relationship described by the fundamental matrix. In addition, the *3D plane Π* induces a homography that relates the correspondences in the two views. See text for details.

2 Projective Geometric Relations

For two views (projector-camera pair in our case), given a point in the first view $\mathbf{p}_1$ (in homogeneous form) and its corresponding point $\mathbf{p}_2$ in the second view, the epipolar relation is stated as $\mathbf{p}_2^T F \mathbf{p}_1 = 0$, where F is the fundamental matrix. While this is always true for object points in general position, for an imaged plane (i.e. reference plane in our case), there is an additional linear mapping that is satisfied by corresponding points, i.e. matched points satisfy the relationship $\mathbf{p}_2 = H\mathbf{p}_1$ where H is the homography induced by the reference plane, see Fig. 1. In the case of an implicit system calibration process using a reference plane, each reference phase map act as a function that uniquely maps each camera pixel location to the corresponding location in projector's display plane. While such a mapping function exists for any reference object, in the case of a plane this mapping can be fully described by a linear transformation determined by the homography induced by it. In other words, in general each reference phase map has as many degrees of freedom as the number of pixels in image plane. However, in the case of a reference plane, there are only eight degrees of freedom (9 elements for the 3×3 homography minus 1 global scale). As a result, for each reference plane used, the entire mapping between the camera pixels and the projector plane can be easily and robustly estimated in the form of the induced homography H.

Throughout the above formulation using the induced homography, we use a pin-hole model for the projector and the camera. In [6] and [9], the authors use a complete reference phase map so as to account for non-linearities like radial distortion in the projector that deviate from an ideal pin-hole model. However, [6] ignores such non-linearities in their computation of the projector center and 3D point locations. In general, the degree of such non-linear distortion is only of significance at the periphery of the projected set of rays and in an area centered around the optical axis of the projector, the linear pin-hole model is accurate enough. As a result, we choose to model the projector as a pin-hole camera and only use the central region of the projected image for scanning objects. By approximating the non-linear effects we gain significantly in that the entire reference phase map can be described by a 3×3 homography matrix which can be estimated with high accuracy using the high redundancy of correspondences available. A key advantage of this approach is that the parametric form of the homography significantly simplifies both the acquisition of calibration data and its estimation. Using the homographies allows us to place the reference planes in general position. This eliminates the need for any precisely controlled placement of reference planes as well.

To compute homographies, we project a single fixed grid pattern on each reference plane, see Fig. 3(b), and establish correspondences for the grid corners in the respective camera image. These correspondences are then used to solve for the homography induced between the projector and the camera. Since all such correspondences (i.e. for each reference plane) satisfy the epipolar geometry as well, we use all of them to robustly estimate the fundamental matrix [13]. Using a single projected pattern per reference plane simplifies the acquisition process and estimation for system calibration. Further, we can use the same calibration plane for acquiring both the image of the calibration marker pattern, see Fig. 3, and the image of the grid pattern projected onto it for computing homography. This incorporates the system calibration into the process of camera calibration.

3   3D Estimation Using the Cross-Ratio

Once the implicit calibration is carried out using the reference planes we use the following approach for estimate the 3D geometry of an object. Throughout, we use a 'camera-centric' approach, i.e. we attach the frame of reference to the camera, see Fig. 2. For every image pixel $\mathbf{i}$, there is a ray $\mathbf{R}_i$ passing through this pixel and the camera optical centre at $\mathbf{O}_c = (0,0,0)$. Let this ray intersect the reference planes $\mathbf{\Pi_A}$ and $\mathbf{\Pi_B}$ at points $\mathbf{A}_i$ and $\mathbf{B}_i$ respectively. The location and orientation of the reference planes in general positions can be resolved using a calibrated camera [12] [1]. Then for each camera pixel $\mathbf{i}$ we can compute 3D locations of the reference points $\mathbf{A}_i$ and $\mathbf{B}_i$. Let (u_i, v_i) be the co-ordinates of pixel $\mathbf{i}$ and $\mathbf{K}$ be the camera calibration matrix. Then the ray $\mathbf{R}_i$ passes through the origin $\mathbf{O}_c$ and the normalized co-ordinates are $\mathbf{K}^{-1}[u_i, v_i, 1]^T$, i.e. any point on the ray is given by $\lambda \mathbf{K}^{-1}[u_i, v_i, 1]^T$ for some λ. Consequently we can solve for the 3D points $\mathbf{A}_i$ and $\mathbf{B}_i$ as the intersection of the ray $\mathbf{R}_i$ with the planes $\mathbf{\Pi_A}$ and $\mathbf{\Pi_B}$ respectively.

During scanning, the unknown 3D point that is imaged at the pixel $\mathbf{i}$ also lies on $\mathbf{R}_i$ and we denote it as $\mathbf{X}_i$. Thus, for every camera pixel $\mathbf{i}$ we have 4 collinear points on line $\mathbf{R}_i$, i.e. $\{\mathbf{O}_c, \mathbf{A}_i, \mathbf{B}_i, \mathbf{X}_i\}$. The ray $\mathbf{R}_i$ projects onto the projector plane as a line $\mathbf{l}_i$. From epipolar geometry we know that the camera optical centre $\mathbf{O}_c$ projects onto the projector plane at the epipole $\mathbf{e}_p$ that can be computed from the fundamental matrix [1]. Using the estimated homographies H_a and H_b, we can project the points $\mathbf{A}_i$ and $\mathbf{B}_i$ onto the projector plane at

---

[1] We use a calibration plane with known marker locations in lieu of every reference plane. Thus, each reference plane's location & orientation and the global scale for the scan are resolved using Euclidean distances between markers.

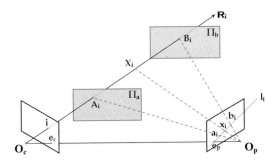

Fig. 2. Camera with optical center O_c and projector with optical center O_p are placed in general positions. Two reference planes Π_a and Π_b are sequentially placed in general positions. Points A_i and B_i lie on planes Π_a and Π_b respectively. Point X_i is at an unknown location. All of points A_i, B_i and X_i project at pixel location i in image plane. Points A_i, B_i and X_i are illuminated by light from pixel locations a_i, b_i and x_i respectively, in projector's display plane. e_c is the epipole in camera's image plane and e_p is the epipole in projector's display plane.

$\mathbf{a}_i$ and $\mathbf{b}_i$. Finally, by decoding the phase relationship for pixel $\mathbf{i}$ during the scanning process, see Section 3.1, we can find the corresponding point for the camera pixel $\mathbf{i}$ as $\mathbf{x}_i$ on the projector plane.

It is well known that for any four collinear points, there exists a cross-ratio that is invariant under projective transformations [1]. For collinear points $\mathbf{p}_1, \mathbf{p}_2, \mathbf{p}_3, \mathbf{p}_4$, the cross-ratio is defined as

$$CR(\mathbf{p}_1, \mathbf{p}_2, \mathbf{p}_3, \mathbf{p}_4) = \frac{||(\mathbf{p}_1 - \mathbf{p}_3)(\mathbf{p}_2 - \mathbf{p}_4)||}{||(\mathbf{p}_2 - \mathbf{p}_3)(\mathbf{p}_1 - \mathbf{p}_4)||}$$

Since the mapping from the ray $\mathbf{R}_i$ to the line $\mathbf{l}_i$ on the projector plane is a projective transformation, we note that the cross-ratio is preserved for the corresponding points, i.e. $CR(\mathbf{O}_c, \mathbf{A}_i, \mathbf{X}_i, \mathbf{B}_i) = CR(\mathbf{e}_p, \mathbf{a}_i, \mathbf{x}_i, \mathbf{b}_i)^2$. In this relationship, the only unknown quantity is the location of the 3D point $\mathbf{X}_i$ and since the cross-ratio on the right-hand side can be estimated, we can solve for the location of the 3D object point $\mathbf{X}_i$.

3.1 Phase-Shifted Patterns

While we are unable to present all details due to space constraints, here we briefly sketch the nature of the sinusoidal projection patterns used in our approach. For a set of N phase-shifted sinusoidal patterns, where N is a power of 2, the projection image S^n, $n = 0 \cdots (N-1)$, is given as $S_c^n = A + B sin(2\pi(c/P +$

---

[2] Since every ray passes through the origin $\mathbf{O}_c$, the origin and the epipole $\mathbf{e}_p$ are present in every cross-ratio that we compute. Thus, we need to use only two reference planes. To avoid limiting conditions for cross-ratios, we ensure that reference planes don't intersect in the field of view. This can be done easily without any precise control.

n/N)). Here A is a dc-offset that ensures that the signal S_c^n is always non-negative, c is the column index of the projector pixel and P is the period of the sinusoidal in pixels units. Since phase value is not unique beyond a single period, usually multiple sets of sinusoidal patterns with different periods (P) are used in conjunction to uniquely encode view-space [5,6]. In our case, instead we project additional graycode patterns such that, for $n = 0$, area corresponding to each sinusoidal period in the projector's display plane has one unique graycode.

For a camera image pixel $\mathbf{i}$, it's phase map value is computed from N images I^k, $k = 0 \cdots (N-1)$, corresponding to N sinusoidal patterns projected on unknown surface using following equation

$$\phi_i = arctan \left(\frac{\sum_k I_i^k cos(2\pi k/N)}{\sum_k I_i^k sin(2\pi k/N)} \right) \qquad (1)$$

By utilising decoded graycodes to unwrap the phase map, we derive a phase map that establishes unique correspondence between every camera pixel and its corresponding projector plane location. In addition, we also suppress shadow areas from which no meaningful structure information can be derived. Further, we note that the cross-ratio of $\{\mathbf{e}_p, \mathbf{a}_i, \mathbf{x}_i, \mathbf{b}_i\}$ can be measured using column co-ordinates alone by projecting the epipolar line $\mathbf{l}_i$ onto the column-axis. While doing this does not change the cross-ratio, it makes use of row co-ordinates redundant. Thus, sinusoidal patterns encoding the projector plane along only one-dimension are sufficient, thereby reducing the number of patterns needed to half.

3.2   Error Statistics

If more than two reference planes are used, we can derive multiple estimates of the location of the 3D point. Since the process of acquiring images for calibrating reference plane homographies is simple and incorporated into camera calibration process, we can easily use more than two reference frames. If we use N reference frames, we can have as many as $\frac{N(N-1)}{2}$ pairs of reference frames, each of which gives us an estimate of the location of the 3D point imaged at a pixel $\mathbf{i}$. As a result we use these estimates of the 3D location of a point to derive both an average 3D location and also its variance along the ray $\mathbf{R}_i$. Thus, in addition to solving for a 3D surface we can also associate a measure of accuracy with each estimated 3D point. This accuracy (or conversely variance) information can be used to improve post-processing algorithms that either smooth the point cloud representation to reduce the effects of noise or to develop a robust estimate of the 3D point location.

4   Experiments

A set of quantitative and qualitative tests were carried out to evaluate the accuracy and efficiency of the proposed system. Throughout, we used a Canon

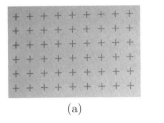
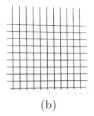

(a)   (b)

Fig. 3. (a) Camera Image of calibration pattern with plus marks (b) Camera Image of the grid pattern projected onto reference plane to compute homographies (with colours inverted)

S5IS digital camera and an NEC NP400 LCD projector set at resolutions of 2048×1536 and 1024×768 pixels respectively. The object was usually placed at around 1.1 meters from the camera. The reference plane was placed between 900 mm to 1300 mm from the camera in different positions. The projector was placed at around 250 mm on right-side of camera at an angle between $5°-20°$. For calibrating the camera and also to solve for the reference plane equations, we used the Caltech Calibration Toolkit [14]. Each induced homography was estimated using 121 grid point correspondences (see Fig. 3(b)). All such correspondences for all reference planes satisfy epipolar geometry and their collective set was used to estimate fundamental matrix using MAPSAC from Torr's library [15]. We used 32 phase-shifted sinusoidal patterns varying along the column axis with an offset($\mathbf{A}$) of 128 and amplitude($\mathbf{B}$) of 96. In addition, 4-bit graycode patterns were used to label the sinusoidal periods ($\mathbf{P} = 64$ pixels) , ($1024/64 = 2^4$), along with an extra bit for accurately resolving phase unwraps at boundary pixels for sinusoidal periods. Finally, after thresholding the amplitude at a level of 25 to identify shadow pixels, we smoothed the phase signal using a Gaussian low-pass filter. For visualising the results, we generated mesh either by using Delaunay triangulation with X and Y co-ordinate planes for the point cloud or by generating grid mesh for adjacent pixels in image plane. Texture information was extracted from an image of the scene and the final results were viewed in MeshLab [16].

4.1   Quantitative Evaluation (Parametric Surfaces)

A plane surface was placed in an unknown position & orientation and scanned. Parametric equation of this plane was estimated by fitting a plane on reconstructed 3D points. Reconstruction error was determined for each individual point as its distance from the estimated parametric plane. Root Mean Square (R.M.S) value for these error distances was computed to analyse the efficiency of the scan. For *Plane 1*, we computed error measures with selective inclusion or exclusion of various steps for proposed approach and results are presented in Table 1. Up to 5 reference planes were used for comparison. In addition, basic cross-ratio scanning approach using three complete reference phase maps was implemented. We used the same set of reference planes and image-dataset for all the scan results, for *Plane 1*, presented in Table 1. As compared to basic method,

Table 1. Error measures for planar surfaces with different methods/conditions : For *Plane 1*, (a) represents basic cross-ratio method using 3 complete reference phase maps (no homographies). (b) is for method with 3 reference maps represented using homographies; (c),(d),(e) & (f) used 2,3,4 & 5 reference map homographies along with epipolar geometry. For *Planes 2,3 & 4* four reference map homographies along with epipolar geometry were used. **All Units are in (mm)**

Absolute Error	Plane 1						Plane 2	Plane 3	Plane 4
	(a)	(b)	(c)	(d)	(e)	(f)			
Median	0.0988	0.0979	0.0974	0.0980	0.0979	0.0977	0.073	0.047	0.085
Max	0.6593	0.4463	0.4444	0.4470	0.4482	0.4468	0.783	0.323	0.592
R.M.S	**0.1426**	**0.1266**	**0.1264**	**0.1262**	**0.1262**	**0.1259**	**0.110**	**0.071**	**0.125**

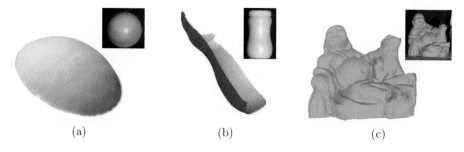

(a)   (b)   (c)

Fig. 4. Scan results for simple surfaces. Insets on top-right corners show object images : (a) A Spherical surface (b) A Bottle that is a surface of revolution. The outer surface is rendered in darker shade (c) Buddha figure with smooth surface variations.

using induced homographies improved scan accuracy by about 12%. Table 1 also shows error measures for the reconstruction of the plane when placed in 3 other unknown positions. Four reference planes were used along with fundamental matrix for each of these reconstructions. R.M.S error for proposed approach is observed to be on the order of one-tenth of a millimeter for each of these scans[3].

4.2   Qualitative Evaluation

For a qualitative evaluation, we scanned objects with varying degrees of surface complexity and the results were visually examined. Fig. 4 shows scan results for simple surfaces that are seen to be accurately recovered. Fig. 5 shows results for an object with surface of medium complexity, a clay figurine of the Hindu god, Ganesh. Some fine details on the trunk, head and limbs are distinctly noticeable in the surface rendering.

---

[3] For evaluations with plane fitting we did not perform low-pass filtering on the phase map.

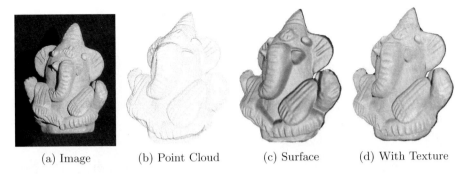

| (a) Image | (b) Point Cloud | (c) Surface | (d) With Texture |

Fig. 5. Scan results for an object (70mm by 90 mm) with medium complexity, a clay figurine of the Hindu god, Ganesh

Table 2. Error Statistics for non-parametric surfaces: Point-wise standard deviation computed from multiple estimates for each surface using 5 reference planes

Error Measure (in mm)	Ganesh idol	Hanuman idol
Minimum σ	0.277	0.084
Maximum σ	1.111	1.334
Median of σ	**0.546**	**0.439**

We also scanned another artifact with a significantly high degree of surface complexity, a clay idol with five heads, i.e. the *panchamukhi* Hanuman. The reconstruction results are shown in Fig. 6. As it is seen, details like that of iris in eyes, nostrils, hand prints and ornamental details on the chest are accurately reconstructed. We also note that even finer details like etching marks on legs are clearly visible. These details are shown in Fig. 6(c) with close-up views of selected regions of the scan. Fig. 7 shows enlarged images for those regions.

4.3   Quantitative Evaluation (Non-parametric Surfaces)

Error statistics proposed in Section 3.2 were computed for Ganesh and Hanuman idols. Results are presented in Table 2. Median value for standard deviation for each scan is about half a millimeter. This implies, multiple estimates using different pairs of reference planes give consistent results.

For the Hanuman idol, we physically measured a set of 5 distances using Vernier Calliper with a precision of 0.05 mm. To account for human error, each distance was measured 10 times and its average value was used for comparison. Measured distances spanned across all 3 Euclidean dimensions. The same set of distances were estimated from the reconstructed model as follows. The end-points for each 'distance segment' were carefully located in the camera image. The reconstructed 3D points corresponding to located pixels were used to estimate the respective distance from the model. Results are shown in Table 3 and absolute errors in distances estimated from reconstructed model are observed to

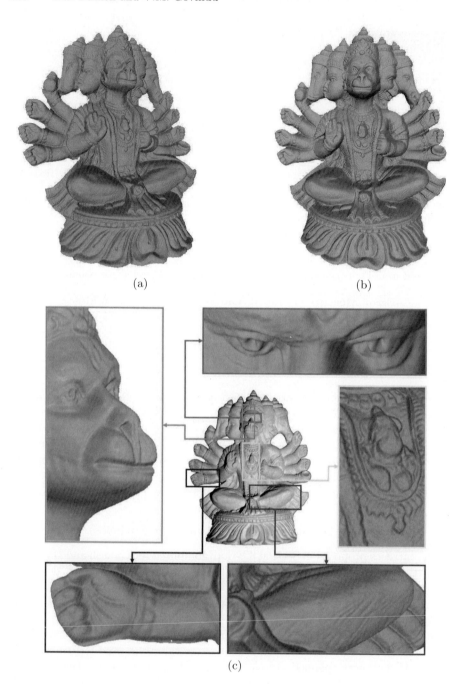

(a)

(b)

(c)

Fig. 6. Scan results for a complex object, a clay idol of the *panchamukhi* Hanuman that is 250 mm wide and 280 mm high : (a) Lateral-view of the reconstruction (b) Front-view (c) Image of the idol with associated close-up views for (i) Head, (ii) Eyes, (iii) Ornament on chest, (iv) Rightmost hand & (v) Etching marks on left leg

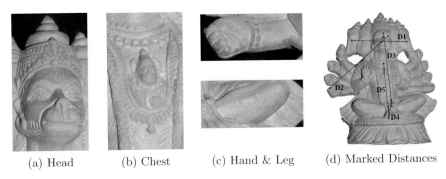

(a) Head   (b) Chest   (c) Hand & Leg   (d) Marked Distances

Fig. 7. Some images of Hanuman idol: (a), (b) and (c) are enlarged images for the regions with close-up view. (d) It indicates the Euclidean distances measured for quantitative evaluation of non-parametric surfaces.

Table 3. Distances measured physically and from reconstructed model for Hanuman idol are shown along with absolute errors: (D_1) From rightmost head's eye to leftmost head's eye, (D_2) From left-eye to the tip of rightmost hand's thumb, (D_3) From tip of the head-mark to center of the flower at base (D_4) From a key-point on abdominal ornament to tip of the rightmost groove at base, (D_5) From center of the necklace to tip of the rightmost groove at base. **All units are in (mm)**

Source	Distance Measured				
	D_1	D_2	D_3	D_4	D_5
Physical Object	128.515	168.415	161.540	69.565	137.290
Reconstructed Model	128.625	168.386	161.421	69.407	137.312
Absolute Error	**0.110**	**0.029**	**0.119**	**0.158**	**0.022**

be on the order of 0.1 mm. Thus, the accuracy of our approach to reconstruct surfaces, in Euclidean sense, is observed to be high.

5   Conclusion

In this paper we have presented a method that uses homographies induced by a reference plane to calibrate a structured-light scanner. By using the projective invariance of the cross-ratio, we solved for the 3D geometry of a scanned surface. We demonstrate the fact that accurate 3D geometric information can be derived through this combination of implicit parameterisation for system calibration and using the cross-ratio to solve for 3D geometry. In addition, by using more than two reference planes we introduce simple statistical measures for the estimated 3D geometry that can be utilised for further intelligent processing. Future work will introduce a maximum likelihood estimator applied to the cross-ratio that will result in a statistically principled estimate of 3D geometry.

References

1. Hartley, R., Zisserman, A.: Multiple View Geometry in Computer Vision, 1st edn. Cambridge University Press, Cambridge (1999)
2. Morano, R.A., Ozturk, C., Conn, R., Dubin, S., Zietz, S., Nissanov, J.: Structured Light Using Pseudorandom Codes. IEEE Transactions on Pattern Analysis and Machine Intelligence 20, 322–327 (1998)
3. Posdamer, J.L., Altschuler, M.D.: Surface Measurement by Space-Encoded Projected Beam System. Computer Graphics and Image Processing 18, 1–17 (1982)
4. Young, M., Beeson, E., Davis, J., Rusinkiewicz, S., Ramamoorthi, R.: Viewpoint-Coded Structured Light. In: IEEE Computer Society Conference on Computer Vision and Pattern Recognition, CVPR (2007)
5. Jordi, J.S., Pages, J., Batlle, J.: Pattern Codification Strategies in Structured Light Systems. Pattern Recognition 37, 827–849 (2004)
6. Peng, T., Gupta, S.K.: Model and Algorithms for Point Cloud Construction Using Digital Projection Patterns. Journal of Computing and Information Science in Engineering 7, 372–381 (2007)
7. Rocchini, C., Cignoni, P., Montani, C., Pingi, P., Scopigno, R.: A low cost 3D scanner based on structured light. Computer Graphics Forum 20 (part 3), 299–308 (2001)
8. Zhang, S., Huang, P.S.: Novel method for structured light system calibration. In: Optical Engineering, vol. 45, SPIE (2006)
9. Ishiyama, R., Okatani, T., Deguchi, K.: Precise 3-D Measurement using Uncalibrated Pattern Projection. In: IEEE International Conference on Image Processing, pp. 225–228 (2007)
10. Furukawa, R., Kawasaki, H.: Dense 3D Reconstruction with an Uncalibrated Stereo System using Coded Structured Light. In: Computer Vision and Pattern Recognition - Workshops, 2005 (2005)
11. Li, Y.F., Lu, R.S.: Uncalibrated Euclidean 3-D Reconstruction Using an Active Vision System. IEEE Transactions on Robotics and Automation 20, 15–25 (2004)
12. Zhang, Z.: A Flexible New Technique for Camera Calibration. IEEE Transactions on Pattern Analysis and Machine Intelligence 22, 1330–1334 (1998)
13. Torr, P.H.S.: Bayesian Model Estimation and Selection for Epipolar Geometry and Generic Manifold Fitting. International Journal on Computer Vision 50, 35–61 (2002)
14. Caltech Calibration Toolbox,
    http://www.vision.caltech.edu/bouguetj/calib_doc/
15. Torr, P.: Structure and Motion Toolkit in MATLAB,
    http://www.mathworks.com/matlabcentral/fileexchange/authors/12514
16. MeshLab Toolkit, http://meshlab.sourceforge.net/

Depth from Encoded Sliding Projections

Chris Hermans, Yannick Francken, Tom Cuypers, and Philippe Bekaert

Hasselt University - tUL - IBBT
Expertise Centre for Digital Media
Belgium

Abstract. We present a novel method for 3D shape acquisition, based on mobile structured light. Unlike classical structured light methods, in which a static projector illuminates the scene with dynamic illumination patterns, mobile structured light employs a moving projector translated at a constant velocity in the direction of the projector's horizontal axis, emitting static or dynamic illumination. For our approach, a time multiplexed mix of two signals is used: (1) a wave pattern, enabling the recovery of point-projector distances for each point observed by the camera, and (2) a 2D De Bruijn pattern, used to uniquely encode a sparse subset of projector pixels. Based on this information, retrieved on a per (camera) pixel basis, we are able to estimate a sparse reconstruction of the scene. As this sparse set of 2D-3D camera-scene correspondences is sufficient to recover the camera location and orientation within the scene, we are able to convert the dense set of point-projector distances into a dense set of camera depths, effectively providing us with a dense reconstruction of the observed scene. We have verified our technique using both synthetic and real-world data. Our experiments display the same level of robustness as previous mobile structured light methods, combined with the ability to accurately estimate dense scene structure and accurate camera/projector motion without the need for prior calibration.

1 Introduction

In the last few decades, there has been a continuous improvement in the quality of synthetic imagery produced by rendering techniques. As a result, the visual quality of present day rendered graphics mostly depends on the quality of the provided 3D models, and less on the choice of a specific method. As physical entities often exhibit small shape imperfections, modeling them into detail can be a tedious task. To this end, automatic shape acquisition systems have been proposed. In this work we will present such a novel system.

One traditional class of shape acquisition methods, which utilizes cameras to obtain a 3D reconstruction, is based on matching distinguishable features observed from multiple viewing locations [1]. However, acquiring these correspondences is non-trivial (*e.g.* disappearing features due to specular highlights, depth discontinuities, . . .) or correspondences are ambiguous as multiple feature candidates match due to the uniform or repetitive nature of the texture around the feature. In this light, alternative techniques have been proposed.

G. Bebis et al. (Eds.): ISVC 2009, Part I, LNCS 5875, pp. 843–854, 2009.

A traditional approach to circumvent the correspondence problem is known as structured light. In this group of methods, one or more cameras are replaced by a projector, which is capable of creating easily distinguishable features through the use of controlled illumination [2,3,4,5,6]. These techniques assume Lambertian reflection properties, an assumption that almost never holds in practice. In real world scenes, the majority of materials exhibit much more complex reflectance properties such as specular reflections off a piece of plastic, or subsurface scattering within a piece of marble. These deviations from the assumed Lambertian reflection model are a common source of mismatches.

In the work of Hermans *et al.* [7] only the presence of a reasonable amount of Lambertian reflection is assumed, while the influence of specular reflections and diffusion due to scattering is regarded as minimal, as these phenomena only have a minimal effect on the observed per pixel principal frequency caused by the translating projected stripe pattern [6,8]. This principal frequency is linearly related to the per pixel depth value, providing an excellent cue for dense depth reconstruction. In a similar fashion, their method is robust against interreflections, as high frequency patterns become sufficiently blurred by second and higher order reflections [9,10]. However, depth is recovered with respect to the projector's principal plane. Hence, if we want the computed depth values to coincide with each pixel's camera depth, the camera's principal plane needs to be perfectly aligned with respect to the projector's principal plane. In our work, no delicate hardware alignment is required and depth values are measured directly in *camera* space instead of *projector* space.

In practice, projectors suffer from a limited depth-of-field range. This is due to the fact that they are most commonly used to project on planar surfaces, more or less aligned with the image plane, and that they are required to provide bright projections, which suggests a large aperture [11]. Although the majority of projector based acquisition methods have a limited scanning volume because of this property, there are existing methods that are insensitive to this problem [7,12], or even exploit it as a depth cue [11,13]. In a similar fashion, most methods are sensitive to camera defocus, while others ignore [7] or exploit it [14,15,16]. Our approach is to a large extent insensitive to camera and projector defocus, allowing for scanning extended volumes compared to the defocus sensitive approaches.

In this work we present a mobile structured light method which obtains depth estimates directly in camera space, combining spatial encoding and frequency analysis using a sliding projector [7] alternating between a sine wave and a De Bruijn pattern. As the per pixel frequency analysis is robust against specular highlights, subsurface scattering, interreflections and camera/projector defocus, projector depths can robustly be found. After calibration, based on the information from the spatial encoding, we obtain a 3D reconstruction of the observed scene. After estimating the position and orientation of the camera in the scene, we can convert the dense set of point-projector depths into a dense set of point-camera depth values. Compared to previous work, this increases the flexibility of the approach considerably, as a delicate setup alignment is no longer required.

2   Related Work

Throughout the years, a large body of work has been developed on camera based depth acquisition methods. For the purpose of this paper, we will focus on two distinct categories: *triangulation* based and *plane sweeping* methods.

2.1   Triangulation Based Methods

The common denominator of all triangulation based methods is their search for corresponding features, visible in multiple camera images taken under multiple viewpoints. The images are recorded by two or more cameras [17,18] placed in a (multi-viewpoint) stereo configuration, or by a single moving camera [19,20,21]. Although this is very common way of obtaining depth information for a scene, this class of algorithms generally has problems with reconstructing uniform regions, depth discontinuities, specular highlights or other more exotic BRDFs [17,18]. Our proposed technique can deal with the previous difficulties, although perfectly mirroring and fully transparant materials are beyond its scope.

An alternative to *detecting* corresponding features is the *creation* of easily distinguishable features using controlled illumination (usually in the form of a projector). This enables a more robust labeling, especially within previously uniform regions. This labeling comes in a variety of forms, from projecting a single spatially encoded pattern [3,4] such as De Bruijn codes [22], to a temporal encoding scheme such as binary/Gray codes [2,5] or viewpoint encoding [23]. Single pattern methods typically yield very fast but lower quality approximations, whereas methods that employ multiple patterns are slower, but higher quality acquisitions. However, both types are generally sensitive to specular outliers, have problems with global light transport (i.e. subsurface scattering and interreflections), or encounter difficulties on depth discontinuities. Problems due to the global light transport can be reduced, employing high frequency illumination modulation [10,24,8], light polarization [25,6] or immersing the scene in a fluorescent dye [26]. Recently, Mohit *et al.* [12] have been able to fully eliminate global illumination. However, their method does not allow for any camera defocus. Our proposed technique is highly insensitive to the typical problems of global light transport, and it does not suffer from camera defocus.

2.2   Plane Sweeping and Related Methods

Plane sweeping methods work fundamentally different, compared to their triangulation based counterparts. The specific details behind each technique aside, plane sweeping and related methods have a set of labels, each corresponding to one depth value, for which they can compute a cost function. Each pixel thus can be assigned a label which minimizes the given cost function. These cost functions come in many forms, *e.g.* by analyzing camera (de)focus [14,15,16], projector (de)focus [11,13], quadratic light attenuation [27] or frequency analysis of a physically linearly moving vertical stripe projection [7].

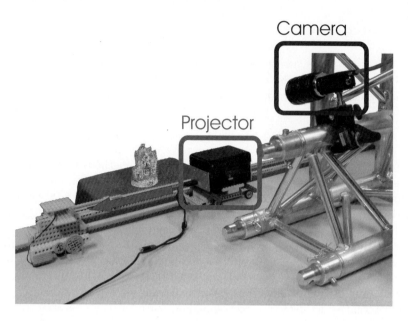

Fig. 1. A picture of our setup: a projector is translated in the direction of its horizontal axis, while a static camera observes the scene

A common advantage to these methods consist of the possibility to create a dense depth map, as they are considerably less prone to occlusion problems. Unfortunately, these methods are either limited to a narrow class of materials, or they come with setups that are rather complex to build or handle. For example, defocus kernels and their corresponding depths have to be calibrated [11,12], camera and projector have to be confocal [11,12] or the linear translation has to be perfectly parallel to the camera's image plane [7].

2.3   A Hybrid Method

The mobile structured light method proposed in this work is a hybrid method, combining ideas from both triangulation and plane sweep techniques. By alternating between a vertical sine wave pattern and a spatially encoded pattern, projected by an *unaligned* sliding projector, we are able to obtain depth values with respect to both the *projector* as well as *camera*. Previously, it was impossible to obtain a depth map in camera space using a sliding projector technique for arbitrary camera positions. Our approach is considerably less delicate since the projector does not have to be perfectly aligned with the camera for a desirable reconstruction in camera space. Additionally, because we have a euclidian reconstruction, we are able to calibrate our system without any knowledge of the projector's speed or display angle, stripe width, or the camera's recording speed, using a sphere with a known radius as a calibration object.

3   Our Approach

Our method produces a depth estimate for each individual pixel, as a result of a three-step process. First, we analyze the temporal intensities of a projected wave pattern on each pixel, providing us with pixel depths from the perspective of the projector. Then, we use a 2D De Bruijn pattern fitted onto this wave pattern to uniquely label a sparse set of projector pixel coordinates. This labeling allows us to perform a *sparse* reconstruction of the scene, based on projector depths only. In order to establish camera depths for a *dense* reconstruction, we compute the camera matrix from this sparse set of 2D-3D correspondences.

3.1   Recovering Point-Projector Distances

For the first step of our algorithm, we assume that the mobile projector illuminates the scene with a sinusoidal stripe pattern. As has been shown by Hermans *et al.* [7], there exists a linear relationship between the projector depth of the points and the period of their observed wave patterns. Converting the observed changes in intensity over time into an equivalent frequency domain representation allows for a search for the dominant frequency f, which can be directly converted into a corresponding depth value μ:

$$\mu = \frac{s}{f} \tag{1}$$

The scaling factor s is uniquely defined by the parameters of the projector and the camera. For more detail on this phase, including its robustness in case of various material properties, we refer to the paper mentioned above.

3.2   Labeling the Projector Pixels

Temporal Encoding. Once we have a projector depth estimate for all our scene points, we want to retrieve the projector's position at each timestep. In order to achieve this, we need to have a set of projector points which we can track throughout the entire sequence. This requires uniquely encoding a sparse set of spatial positions in the illumination pattern in such a way that, based solely on the analysis of the temporal intensities, camera-projector correspondences can be inferred. Thus, we create a mapping from a 1D temporal code to a 2D spatial position.

De Bruijn Sequences. In combinatorial mathematics, a k-ary De Bruijn sequence $B(k, n)$ of order n is a cyclic sequence of a given alphabet A of size k, for which every possible subsequence of length n in A appears as a sequence of consecutive characters exactly once. As such, every position in a $B(2, n)$ De Bruijn sequence is directly linked to a unique binary subsequence of length n, making it an excellent candidate for our 1D pattern.

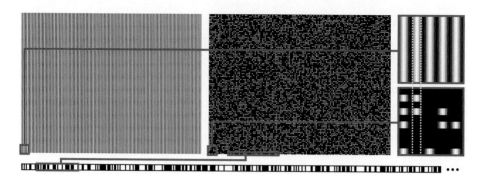

Fig. 2. (left) 1D sine wave pattern used to recover point-projector depth estimates, **(center)** 2D spatially encoded pattern, **(right)** close-up of two patches of both patterns, in which it becomes clear the employed binary code is mapped onto the pre-existing wave pattern, **(bottom)** 1D De Bruijn sequence used to spatially encode a sparse set of 2D positions in the center image. Note that each of the encoded 2D positions is encoded by a unique 16 × 1 window which maps to a unique 16-bit subsequence in the De Bruijn sequence.

Whilest many algorithms exist for the efficient generation of general De Bruijn sequences, we have opted for the use of (16-bit Fibonacci) Linear Feedback Shift Registers [28]. Starting from a random 16-bit pattern, excluding the state in which all bits are equal to zero, each next bit can be generated as a linear function of its previous state. We can continue this process until we have cycled through all states except the all-zero state.

Signal-to-Code Conversion. Assuming our camera has observed a subsequence of a De Bruijn sequence, there still remains the matter of scale that needs to be resolved in order to convert a temporal intensity signal into a binary code. We need to be able to establish the beginning and the end of each single bit in the code. To this end, we have opted to map the De Bruijn sequence onto the same wave pattern we have used to establish point-projector distances (see Figure 2). Each recorded wave has a known period p, directly related to point's distance from the projector's principal plane, which was already established in an earlier step using eq.(1). Also, in order to avoid aliasing effects due to light bleeding from one line of code into another, we have opted to perform horizontal interlacing into the pattern.

Assuming we are able to simultaneously observe both the wave pattern and the De Bruijn pattern, we are able to locate the timesteps at which the indivual bits of the De Bruijn code are visible at the local maxima of each wave period. Comparing the intensity value of the observed De Bruin signal to the local extrema within the observed wave period provides us with a simple and robust method for resolving the De Bruijn code. If the observed De Bruijn signal is sufficiently close to the local maximum, we assign a 1, otherwise we assign a 0.

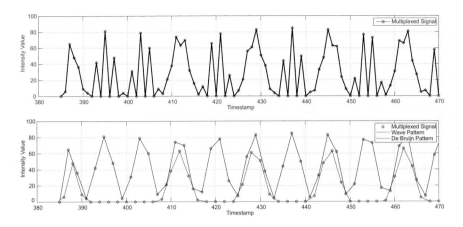

Fig. 3. The observed intensities in the top sequence, contain information of two separate signals: the scene point illuminated by the wave pattern (bottom sequence, blue), and the scene point illuminated by the De Bruijn pattern (bottom sequence, red). Separating these signals, and interpolating the missing information allows for (1) frequency/period estimation on the wave pattern, and (2) recovering the binary code from both signals (*e.g.* the depicted sequence clearly encoded the sequence 1001010101). In the example above, the De Bruijn pattern is projected at even timesteps, while the wave pattern occurs at odd timesteps.

Time Multiplexing. In the previous section, we have assumed the simultaneous observation of both the wave pattern and the De Bruijn pattern. In practice, a projector can only emit one illumination pattern at a time. We simulate this by time multiplexing the patterns displayed by our projector, alternating between the wave and De Bruijn pattern at the same frequency as the camera, and demultiplexing the observed intensities into two separate signals. Missing values can then be interpolated from the available data if needed be (Figure 3).

3.3   Recovering Structure and Projector Motion

Once we have a sparse set of camera-projector pixel correspondences, the next step is to perform a sparse reconstruction of the scene. For the remainder of this section, we will assume the intrinsic parameters of the projector are known.

For each scene point $X^i = [x^i, y^i, \mu^i]^T$, we have a set of projector pixels $[u_t, v_t, 1]^T$:

$$
\begin{aligned}
P_t X^i &= \begin{bmatrix} 1 & 0 & 0 & -\lambda t \\ 0 & 1 & 0 & 0 \\ 0 & 0 & 1 & 0 \end{bmatrix} \begin{bmatrix} x^i \\ y^i \\ \mu^i \\ 1 \end{bmatrix} \\
&= \begin{bmatrix} x^i - \lambda t \\ y^i \\ \mu^i \end{bmatrix} \propto \begin{bmatrix} u_t \\ v_t \\ 1 \end{bmatrix}
\end{aligned}
\tag{2}
$$

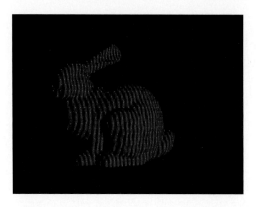 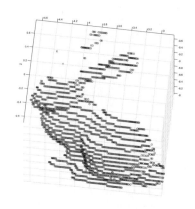

Fig. 4. **(left)** A random image from our Stanford bunny dataset, with the superimposed sparse set of identified projector pixels at that timestep. The color code represents the projector pixel's horizontal position, from left (yellow) to right (red). **(right)** Visualization of the computed sparse reconstruction, based on the information from the projector pixel correspondences.

From eq.(2) one can see the only unknown parameter in this equation is λ, related to the projector's speed. An initial estimate of this parameter is computed by averaging over all λ_{irs} values, r and s being random timesteps at which point X^i registered known projector pixel positions $[u_r, v_r, 1]^T$ and $[u_s, v_s, 1]^T$, where

$$\lambda_{irs} = \mu^i \left(\frac{u_r - u_s}{s - r} \right) \tag{3}$$

Using the initial estimate, we employ Levenberg-Marquardt minimization to find the optimal λ which minimizes the overall reprojection error of all X^i for all P_t. This results in a sparse reconstruction of the scene, *e.g.* as depicted in Figure 4.

3.4   Recovering Camera Parameters and Dense Depth Map

For each point X^i of the sparse dataset, we have the corresponding camera pixel coordinates $[u^i, v^i, 1]^T$. From this set of correspondences, we can estimate the parameters of the camera matrix, using one of many available methods [29].

Once the position and orientation of the camera are known, we convert the dense map of point-projector distances into a corresponding camera depth map, by intersecting each camera pixel's back-projected ray with the plane $\pi(0, 0, 1, -\mu^i)$.

3.5   Calibrating the Setup

The initial per pixel depth computation, as briefly described in section 3.1, assumes that the scaling factor s in eq.(1) is known. However, in practice it is not always easy to accurately recover this constant, as it is dependent on many parameters of both the projector and the camera [7]. However, using a single sphere of a known radius, we can accurately recover this scale factor.

Table 1. Projector pixel reprojection errors for the displayed datasets. Note that these errors are for superpixels consisting of 8 × 8 pixels, and should be interpreted accordingly.

Dataset	Bunny	4 Objects	Sphere	Flag	House
Mean Error	0.997067	0.992447	1.092911	0.858623	1.042102
Standard Deviation	0.012627	0.016163	0.014703	0.014192	0.014237

Each choice of the parameter s gives rise to a sparse set of 3D points $X^i = [x^i, y^i, \frac{s}{f^i}]^T$. By fitting a sphere through these points, we can compute the overall error for each instance. Applying iterative minimization of this error allows us to quickly find the optimal scale factor.

4   Discussion

There are several aspects of to our method that require verification: (1) the robustness of the depth estimation component for different material properties; (2) the influence of time multiplexing to these depth estimation results; (3) the robustness of the signal-to-code conversion algorithm; and (4) the quality of the estimated reconstruction. As the first of these four items has already been the subject of discussion in the work of Hermans *et al.* [7], we will focus in this work on the other three items. Results from our method are shown in Figure 5.

Time Multiplexing. Compared to previous work, time multiplexing effectively cuts our amount of potential depth planes in half. In order to combat this issue, we could add color multiplexing to our method, alternating between the R/GB and GB/R channels for the wave/spatial code patterns, but we have yet to further explore this option.

Signal-to-Code Conversion. Our experiments have indicated that. In order to avoid mismatches, we need to use tri-state codes instead of binary codes, assigning 'unknown' bits in the code when there is no clear inclination towards a local minimum or maximum. A simple backtracking algorithm suffices to uniquely identify the observed code. However, as mismatches are impossible to avoid, outlier rejection is required at the end of this step.

Estimated Reconstruction. The error metric we discuss in order to validate our estimated reconstruction is the (projector pixel) reprojection error. Using a projector stripe width of 8 pixels, we are able to encode $\frac{width \times height}{64 \times 4}$ different projector pixel positions. After estimating the sparse reconstruction, these positions are reprojected back to the projector's image plane at each timestep, with an average reprojection error of approximately a pixel for each block of 8 × 8 pixels per position (see table 1).

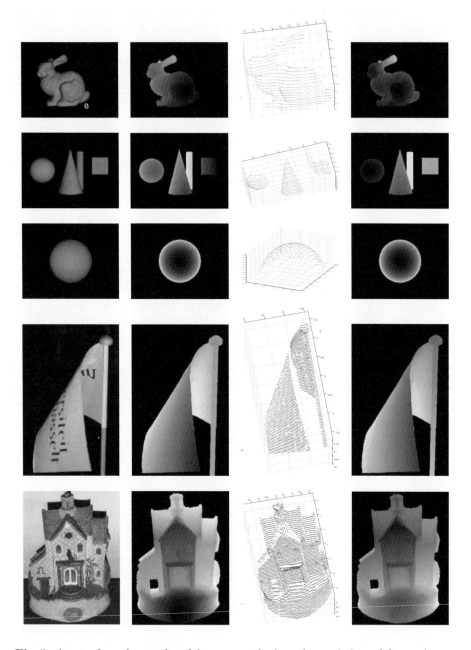

Fig. 5. A set of results produced by our method, each consisting of four columns: (1) the fully illuminated scene; (2) the estimated dense depth map, computed from the projector's perspective; (3) the computed sparse reconstruction; (4) the converted depth map, with respect to the camera instead of the projector. All results were generated with 1500 frames or less, using patterns with a wave period of 8 projector pixels.

Fig. 6. Rendered reconstructions of the 'house' dataset. **(left)** A reconstruction with the superimposed camera pixel depth. For reference, we display a plane parallel to the camera's image plane. **(right)** A reconstruction with the superimposed color image. Note that we are able to capture very fine detail such as the relief on the door and windows, or the tiles on the roof.

5   Conclusions

In this paper we have extended the scope of mobile structured light for dense depth estimation, by relieving the need for finely tuned camera-projector configurations. Our method combines the ability of previous mobile structured light methods to cope with a wide variety of materials, with the ability to estimate both structure and motion for the captured scene. Additionally, we provide a way to calibrate the proposed setup by automatically determining the scale factor, without prior knowledge of the related camera and projector parameters.

Acknowledgements

The authors acknowlege financial support by the ERDF (European Regional Development Fund), the European Commission (FP7 IP "2020 3D media") and the Flemish government. Furthermore, we would like to thank our colleagues and reviewers for their useful comments and suggestions.

References

1. Faugeras, O.: Three-dimensional computer vision: a geometric viewpoint. MIT Press, Cambridge (1993)
2. Posdamer, J., Altschuler, M.: Surface measurement by space-encoded projected beam systems. Computer Graphics and Image Processing 18, 1–17 (1982)
3. Boyer, K.L., Kak, A.C.: Color-encoded structured light for rapid active ranging. PAMI 9, 14–28 (1987)
4. Batlle, J., Mouaddib, E., Salvi, J.: Recent progress in coded structured light as a technique to solve the correspondence problem: a survey. Pattern Recognition 31, 963–982 (1998)

5. Scharstein, D., Szeliski, R.: High-accuracy stereo depth maps using structured light. In: Proceedings of CVPR, p. 195 (2003)
6. Chen, T., Lensch, H.P.A., Fuchs, C., Seidel, H.: Polarization and phase-shifting for 3d scanning of translucent objects. In: Proceedings of CVPR, pp. 1829–1836 (2007)
7. Hermans, C., Francken, Y., Cuypers, T., Bekaert, P.: Depth from sliding projections. In: Proceedings of CVPR, pp. 1865–1872 (2009)
8. Chen, T., Seidel, H., Lensch, H.: Modulated phase-shifting for 3D scanning. In: Proceedings of CVPR, pp. 1–8 (2008)
9. Ramamoorthi, R., Hanrahan, P.: A signal-processing framework for inverse rendering. In: Proceedings of SIGGRAPH, New York, NY, USA, pp. 117–128 (2001)
10. Nayar, S., Krishnan, G., Grossberg, M.D., Raskar, R.: Fast Separation of Direct and Global Components of a Scene using High Frequency Illumination. In: Proceedings in SIGGRAPH (2006)
11. Zhang, L., Nayar, S.K.: Projection Defocus Analysis for Scene Capture and Image Display. In: Proceedings of SIGGRAPH (2006)
12. Gupta, M., Tian, Y., Narasimhan, S., Zhang, L.: (de)focusing on global light transport for active scene recovery. In: Proceedings of CVPR, pp. 2969–2976 (2009)
13. Moreno-Noguer, F., Belhumeur, P., Nayar, S.: Active Refocusing of Images and Videos. In: Proceedings of SIGGRAPH (2007)
14. Nayar, S.K., Nakagawa, Y.: Shape from focus. PAMI 16, 824–831 (1994)
15. Watanabe, M., Nayar, S.K.: Rational filters for passive depth from defocus. IJCV 27, 203–225 (1998)
16. Hasinoff, S.W., Kutulakos, K.N.: Confocal stereo. IJCV 81, 82–104 (2009)
17. Scharstein, D., Szeliski, R.: A taxonomy and evaluation of dense two-frame stereo correspondence algorithms. IJCV 47, 7–42 (2002)
18. Steven, M.S., Curless, B., Diebel, J., Scharstein, D., Szeliski, R.: A comparison and evaluation of multi-view stereo reconstruction algorithms. In: Proceedings of CVPR, pp. 519–528 (2006)
19. Bolles, R., Baker, H., Marimont, D.: Epipolar-plane image analysis: An approach to determining structure from motion. IJCV 1, 7–55 (1987)
20. Sturm, P.F., Triggs, B.: A factorization based algorithm for multi-image projective structure and motion. In: Proceedings of ECCV, pp. 709–720 (1996)
21. Pollefeys, M., Gool, L.V., Vergauwen, M., Verbiest, F., Cornelis, K., Tops, J., Koch, R.: Visual modeling with a hand-held camera. IJCV 59, 207–232 (2004)
22. Vuylsteke, P., Oosterlinck, A.: Range image acquisition with a single binary-encoded light pattern. PAMI 12, 148–164 (1990)
23. Young, M., Beeson, E., Davis, J., Rusinkiewicz, S., Ramamoorthi, R.: Viewpoint-coded structured light. In: Proceedings of CVPR (2007)
24. Zhang, S., Yau, S.: High-resolution, real-time 3D absolute coordinate measurement based on a phase-shifting method. Optics Express 14, 2644–2649 (2006)
25. Wolff, L.: Using polarization to separate reflection components. In: Proceedings of CVPR, pp. 363–369 (1989)
26. Hullin, M.B., Fuchs, M., Ihrke, I., Seidel, H., Lensch, H.P.A.: Fluorescent immersion range scanning. In: Proceedings of SIGGRAPH, pp. 1–10 (2008)
27. Liao, M., Wang, L., Yang, R., Gong, M.: Light fall-off stereo. In: Proceedings of CVPR, pp. 1–8 (2007)
28. Golomb, S.: Shift register sequences. Aegean Park Press, Laguna Hills (1981)
29. Hartley, R.I., Zisserman, A.: Multiple View Geometry in Computer Vision, 2nd edn. Cambridge University Press, Cambridge (2004)

A New Algorithm for Inverse Consistent Image Registration

Xiaojing Ye and Yunmei Chen

Department of Mathematics, University of Florida, Gainesville, FL 32611, USA

Abstract. This paper presents a novel variational model for inverse consistent deformable image registration. This model deforms the source and target image simultaneously, and aligns the deformed source and deformed target images in the way that the both transformations are inverse consistent. The model does not computes the inverse transforms explicitly, alternatively it finds two more deformation fields satisfying the invertibility constraints. Moreover, to improve the robustness of the model to noises and the choice of parameters, the dissimilarity measure is derived from the likelihood estimation of the residue image. The proposed model is formulated as an energy minimization problem, which involves the regularization for four deformation fields, the dissimilarity measure of the deformed source and deformed target images, and the invertibility constraints. The experimental results on clinical data indicate the efficiency of the proposed method, and improvements in robustness, accuracy and inverse consistency.

1 Introduction

Image registration is a very important subject and has been widely applied in medical research and clinic applications. The task of image registration is to find a transformation field that relates points in the source image to their corresponding points in the target image. Deformable image registration allows localized transformations, and is able to account for internal organ deformations. Therefore, it has been increasingly used in health care to assist diagnosis and treatments. In particular, deformable image registration has become a critical technique for image guided radiation therapy. It allows more precise tumor targeting and normal tissue preservation. A comprehensive review for image registration in radiation therapy can be found in [1].

A deformable image registration is called inverse consistent, if the correspondence between two images is invariant to the order of choice of source and target. More precisely, let S and T be the source and target images, and h and g be the forward and backward transformations, respectively, i.e. $S \circ h = T$ and $T \circ g = S$, then an inverse consistent registration satisfies $h \circ g = id$ and $g \circ h = id$, where id is the identity map. By applying an inverse consistent registration, measurements or segmentations from one image can be precisely transferred to the other. In imaging guided radiation therapy, the inverse consistent deformable registration technique provides the voxel-to-voxel mapping between the reference phase

G. Bebis et al. (Eds.): ISVC 2009, Part I, LNCS 5875, pp. 855–864, 2009.

and the test phase in four-dimensional (4D) radiotherapy [2]. This technique is referred to as "automatic re-contouring".

Inverse consistent deformable image registration has been an active subject of study in the literature. There have been a group of work that developed various of models in the context of large deformation by diffeomorphic metric mapping, e.g. [3,4,5,6]. The main idea of this method is modeling the forward and backward transformations as a one-parameter diffeomorphism group. Then, a geodesic path connecting two images is obtained by minimizing an energy functional symmetric to the forward and backward transformations.

Variational method is one of the popular approaches. This method minimizes the energy functional(s) symmetric to the forward and backward transforms, and in general, consists of three parts: regularization of deformation fields, dissimilarity measure on the target and deformed source image, and penalty of inverse inconsistency [7,8,9,10]. In [7], Christensen and Johnson proposed to minimize the following coupled energy functionals with respect to h and g alternatively:

$$\begin{cases} E(h) = & E_1 + \rho \|h - g^{-1}\|^2_{L^2(\Omega)} \\ E(g) = & E_2 + \rho \|g - h^{-1}\|^2_{L^2(\Omega)} \end{cases} \tag{1}$$

where $E_1 = \lambda E_s(S \circ h, T) + E_r(u)$, $E_2 = \lambda E_s(T \circ g, S) + E_r(v)$, $h(x) = x + u(x)$ and $g(x) = x + v(x)$. The dissimilarity E_s is defined as $E_s(S \circ h, T) = \|S \circ h - T\|^2_{L^2(\Omega)}$, and the regularity of deformation field u is defined as $E_r(u) = \|a \Delta u + b \nabla (\text{div } u) - cu\|^2_{L^2(\Omega)}$ where $a, b, c > 0$ are constants. The last term in both energy functionals in (1) enforces the transforms h and g to be inverse to each other. This model solves a system of two evolution equations associated with their Euler-Lagrange (EL) equations iteratively, and gives considerably good results with parameters chosen carefully. However, it needs to compute the inverse mappings g^{-1} and h^{-1} explicitly in each iteration. This may cause accumulative numerical error in the estimation of inverses and is computationally intensive.

The variational models developed in [8] and [10] have the same framework as in [7] with different representations of E_s, E_r, and inverse consistent constraints. In both of work the terms $\|h \circ g(x) - x\|^2_{L^2(\Omega)}$ and $\|g \circ h(x) - x\|^2_{L^2(\Omega)}$ are used in the energy functional to enforce the inverse consistency. By using these terms the explicit computation of the inverse transforms of h and g can be avoided during the process of finding optimal forward and backward transformations. The similarity measure in [10] is the mutual information for multi-modal image registration. The $E_s(S \circ h, T)$ in [8] is $\|S \circ h - T\|^2_{L^2(\Omega)} / \max |DT|$. The regularization term $E_r(u)$ in [10] is a function of Du, and that in [8] is a tensor based smoothing which is designed to prevent the transformation fields from being smoothed across the boundaries of features. In [11,12] the proposed models incorporated stochastic errors in the inverse consistent constraints for both forward and backward transformations.

In [13], Leow *et al.* proposed an approach that updates the forward and backward transformations simultaneously by a force that reduces the energies E_1 and E_2 in (1) and preserves the inverse consistency. However, it only takes linear order terms in the Taylor expression to approximate the inverse consistency and

hence the truncating errors can be accumulated and exaggerated during iterations. This results in large inverse consistent error, despite that it can produce
a good matching quickly [14].

In this paper we propose a novel variational model to improve the accuracy, robustness and efficiency of inverse consistent deformable registration. The current
framework of variational method finds the forward and backward transformations that deform a source image S to match a target image T and vice versa.
In this work, we propose to deform S and T simultaneously, and the registration
matches the deformed source and deformed target images. Since the disparity
between deformed S and deformed T is smaller than that between deformed S
and fixed T or deformed T and fixed S. Therefore, the deformation by the bidirectional simultaneous deformations is in general smaller than the deformation
by unidirectional deformation that deforms S full way to T or T full way to S.
Therefore, as shown in our experimental results deforming S and T simultaneously leads to a faster and better alignment than deforming S to the fixed T
or vice versa. Let u and $\tilde{u}$ represent the deformation fields such that $S(x + u)$
matches $T(x + \tilde{u})$. It is not difficult to verify that if u and $\tilde{u}$ are invertible, then
the registrations from S to T, and T to S are inverse consistent. To avoid the
direct computation of the inverse transformations of $x + u(x)$ and $x + \tilde{u}(x)$, our
model seeks for two additional deformation fields v, $\tilde{v}$ such that $x + u(x)$ and
$x + v(x)$ are inverse to each other, and the same for $x + \tilde{u}(x)$ and $x + \tilde{v}(x)$.
Moreover, the registration process enforces certain regularity for these four deformation fields, and aligns the deformed S and deformed T. Then, the optimal
inverse consistent transformations from S to T, and T to S can be obtained
simply by appropriate compositions of these four transformations. Although the
idea of deforming S and T simultaneously has been applied in the models where
the forward and backward transformations as a one-parameter diffeomorphism
group [5], our approach is different from the aspect that our model finds regularized invertible deformation fields rather than a one-parameter diffeomorphism
group in [5], which brings expensive computational cost, and hence limits its
application in clinic. Moreover, our model allows parallel computations for all
the deformation fields to significantly reduce the computational time.

Furthermore, to improve the robustness of the model to noises and the choice
of the parameter λ (1) in that balances the goodness of matching and smoothness
of the deformation fields, we adopt the maximum likelihood estimation (MLE).
This results in a self-adjustable weighting factor that makes the choice of λ more
flexible, and also speeds up the convergence to the optimal deformation fields.

2   Proposed Method

For simplicity we suppose both the source image S and target image T are
defined on $\Omega \subset \mathbb{R}^2$, and let $\| \cdot \|$ denote the L^2 norm hereafter.

In this paper, we propose to deform S and T simultaneously, and align the
deformed S and deformed T. This means that ideally we pursuit for a pair of
half-way transformations $\phi, \tilde{\phi} : \Omega \to \Omega$ such that $S \circ \phi = T \circ \tilde{\phi}$. To ensure

the transformations from S to T and T to S are inverse consistent, ϕ and $\tilde{\phi}$ are required to be invertible. To avoid direct computation of the inverses of ϕ and $\tilde{\phi}$ during iterations, we enforce the invertibility of ϕ and $\tilde{\phi}$ by seeking for $\psi, \tilde{\psi} : \Omega \to \Omega$, such that

$$\psi \circ \phi = id, \ \phi \circ \psi = id, \ \text{and} \ \tilde{\psi} \circ \tilde{\phi} = id, \ \tilde{\phi} \circ \tilde{\psi} = id. \tag{2}$$

Then, once ψ and $\tilde{\psi}$ are obtained, we can construct the objective full-way transformations h and g that deform S to T and T to S using $h = \phi \circ \tilde{\psi}$ and $g = \tilde{\phi} \circ \psi$. From (2), it is clear that h and g also satisfy the inverse consistent constraints $h \circ g = g \circ h = id$. Let the functions u, $\tilde{u}$, v and $\tilde{v}$ represent the corresponding deformation fields of the transformations ϕ, $\tilde{\phi}$, ψ and $\tilde{\psi}$, respectively, i.e.

$$\phi(x) = x + u(x), \ \tilde{\phi}(x) = x + \tilde{u}(x), \ \psi(x) = x + v(x), \ \tilde{\psi}(x) = x + \tilde{v}(x). \tag{3}$$

Then, the constraints (2) can be rewritten as

$$u + v(x + u) = v + u(x + v) = 0, \ \tilde{u} + \tilde{v}(x + \tilde{u}) = \tilde{v} + \tilde{u}(x + \tilde{v}) = 0. \tag{4}$$

Moreover, to improve the robustness of the algorithm, we use the negative log-likelihood of the residual image as a measure of mismatching. Consider pixel intensities of the residue image $W := S \circ \phi - T \circ \tilde{\phi}$ as independent samples drawn from a Gaussian distribution of zero mean and to be optimized variance σ^2. Then we define the negative log-likelihood as the fitting term F:

$$F(u, \tilde{u}, \sigma) := \|S(x + u) - T(x + \tilde{u})\|^2 / 2\sigma^2 + |\Omega| \log \sigma. \tag{5}$$

The advantages of using MLE over the Sum of Squared Distance (SSD) as the fitting term will be shown later in this section and in the experimental results.

Now we propose our model as the following minimization problem:

$$\min_{u, \tilde{u}, v, \tilde{v}; \sigma} R(u, \tilde{u}, v, \tilde{v}) + \lambda F(u, \tilde{u}, \sigma), \quad \text{s.t. condition (4) holds}, \tag{6}$$

where R is the regularization of the deformation fields $(u, \tilde{u}, v, \tilde{v})$ defined by

$$R(u, \tilde{u}, v, \tilde{v}) := \|Du\|^2 + \|D\tilde{u}\|^2 + \|Dv\|^2 + \|D\tilde{v}\|^2. \tag{7}$$

By using (3), it is easy to see that the final full-way forward and backward deformation fields $\bar{u}$ and $\bar{v}$ can be obtained by

$$\bar{u} = \tilde{v} + u(x + \tilde{v}) \quad \text{and} \quad \bar{v} = v + \tilde{u}(x + v), \tag{8}$$

respectively. Also, the inverse consistency of h and g can be preserved and $\bar{u} + \bar{v}(x + \bar{u}) = \bar{v} + \bar{u}(x + \bar{v}) = 0$, which can be derived from (4).

The term $F(u, \tilde{u}, \sigma)$ in (6) is from the negative log-likelihood of the residue image as in (5). Minimizing this term forces the mean intensity of the residue image to be zero, and allows it having a variance to accommodate certain variability. This makes the model more robust to noise and artifacts, and less sensitive to the choice of the parameter λ than the model with the SSD. The parameter λ balances the smoothness of deformation fields and goodness of alignments, and affects the registration result greatly. In the proposed model, the ratio of the SSD of the residue image over the smoothing terms is $\lambda/2\sigma^2$ rather than a prescribed λ. Since σ is to be optimized, and from its EL equation σ is the standard deviation of the residue image., it updates the weight on the matching term during iterations. When the alignment gets better, the σ, i.e. the standard deviation of the residue as shown in (11), decreases, and hence the weight on the matching term automatically increases. This self-adjustable feature of the weight not only enhances the accuracy of alignment, but also makes the choice of λ flexible, and results in a fast convergence.

To solve problem (6), we convert the constraints to classical quadratic penalties in the energy functional, and obtain an unconstrained minimization problem

$$\min_{u,\tilde{u},v,\tilde{v};\sigma} R(u,\tilde{u},v,\tilde{v}) + \lambda F(u,\tilde{u},\sigma) + \mu\left(\mathcal{I}(u,v) + \mathcal{I}(\tilde{u},\tilde{v})\right), \qquad (9)$$

where $\mathcal{I}(u,v) = \mathcal{I}_v(u) + \mathcal{I}_u(v)$, $\mathcal{I}_v(u) = \|u + v(x+u)\|^2$ and $\mathcal{I}_u(v) = \|v + u(x + v)\|^2$. Similarly, we have $\mathcal{I}(\tilde{u},\tilde{v})$. With sufficiently large μ, solving (9) gives an approximation to the solution of (6).

3   Numerical Scheme and Experimental Results

Since the compositions in the invertible constraints $\mathcal{I}_u(v)$ and $\mathcal{I}_v(u)$ bring in a difficulty to get explicit form of the EL equations for the deformation fields and their inverses, we solve the following two coupled minimization problems alternately instead of solving (9) directly:

$$\begin{cases} \min_{u,\tilde{u}} E_{v,\tilde{v}}(u,\tilde{u}) = \|Du\|^2 + \|D\tilde{u}\|^2 + \lambda\,F(u,\tilde{u},\sigma) + \mu\left(\mathcal{I}_v(u) + \mathcal{I}_{\tilde{v}}(\tilde{u})\right) \\ \min_{v,\tilde{v}} E_{u,\tilde{u}}(v,\tilde{v}) = \|Dv\|^2 + \|D\tilde{v}\|^2 + \mu\left(\mathcal{I}_u(v) + \mathcal{I}_{\tilde{u}}(\tilde{v})\right). \end{cases} \qquad (10)$$

whose EL equations can be computed in a straight forward manner. Also, the first variation of σ gives

$$\sigma = \|S(x+u) - T(x+\tilde{u})\|/|\Omega|^{1/2}. \qquad (11)$$

It is important to point out that, in each iteration, the computations of $u,\tilde{u},v,\tilde{v}$ can be carried out in parallel. To measure the accuracy of matching, we use the correlation coefficients (CC) between the target and deformed source images. Then, for fixed μ, our iterative process is terminated when the mean of $CC(S(x+\bar{u}),T)$ and $CC(T(x+\bar{v}),S)$ converges, and we can obtain $(u,\tilde{u},v,\tilde{v})$ and hence $\bar{u}$ and $\bar{v}$ by (8). If the maximum inverse consistency error (ICE) δ_c, defined by

$$\delta_c = \max\left\{|\bar{u} + \bar{v}(x+\bar{u})|, |\bar{v} + \bar{u}(x+\bar{v})| \,|\, x \in \Omega\right\}, \qquad (12)$$

is large, one can gradually increase the parameter μ in (9) and use the previous $(u,\tilde{u},v,\tilde{v})$ as a warm start. Then the whole computation can be stopped once

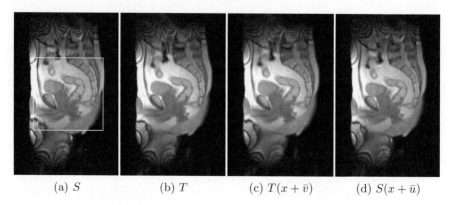

(a) S (b) T (c) $T(x + \bar{v})$ (d) $S(x + \bar{u})$

Fig. 1. Inverse consistent registration result by proposed model (9) on the prostate data, (c) and (d) are the deformed T and deformed S respectively

δ_c is less than a prescribed tolerance ϵ. In our experiments, we set $\epsilon = .5$, in which case the maximum ICE δ_c would be less than half of the grid size between two concatenate pixels/voxels and hence the invertibility is exactly satisfied with respect to the original resolution of the images.

We first test the accuracy of registration and auto re-contouring of the proposed algorithm on a clinical data set of 100 2D-prostate MR images. Each image, called a phase, is a 2D image of dimension 288×192 that focuses on prostate area. The first phase is used as a source image S, as shown in Fig 1a. The boundaries of the regions of interests (ROI) in S are delineated by contours and superimposed by medical experts, as enlarged and shown in Fig. 2d. The rest 99 phases are considered as target images. In this experiment we applied the proposed model (9) with parameters (λ, μ) set as $(.05, .2)$. For demonstration we only show the results using the 21st phase as T, as shown in Fig. 1b. The deformed T and deformed S are shown in Fig. 1c and 1d. The errors after the alignment $|S(x + \bar{u}) - T|$ and $|S - T(x + \bar{v})|$ in the squared area (as shown in Fig. 1a) are displayed in Fig. 2a and 2c, respectively. With comparison to the original error $|S - T|$ in Fig. 2b, we can see the errors after alignment are significantly reduced. This indicates that the proposed registration model (9) is highly accurate in matching two images. The final optimal forward and backward deformation fields $\bar{u}$ and $\bar{v}$ are displayed by applying them to a domain of regular grids, shown as Fig. 3a and 3b, respectively. Furthermore, to validate the well-preserved inverse consistency by model (9) we applied $\bar{u} + \bar{v}(x + \bar{u})$ on a domain of regular grids, and the resulting grids are plotted in Fig. 3c. The resulting grids by $\bar{v} + \bar{u}(x + \bar{v})$ has the same pattern so we omitted it here. From Fig. 3c, we can see that the resulting grids are the same as the original regular grids. This indicates that the inverse consistent constraints $\bar{u} + \bar{v}(x + \bar{u}) = \bar{v} + \bar{u}(x + \bar{v}) = 0$ are almost exactly satisfied. We also computed the maximum ICE δ_c using $\bar{u}, \bar{v}$ and (12), which is .46 of a pixel. The mean ICE $(\|\bar{u} + \bar{v}(x + \bar{u})\| + \|\bar{v} + \bar{u}(x + \bar{v})\|)/2|\Omega|$

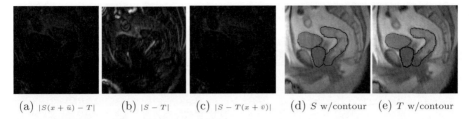

(a) $|S(x+\bar{u})-T|$ (b) $|S-T|$ (c) $|S-T(x+\bar{v})|$ (d) S w/contour   (e) T w/contour

Fig. 2. Results of model (9) applied to S and T in the squared area shown in Fig. 1a. Contours in (e) are obtained by deforming the contours in (d) using $\bar{u}$.

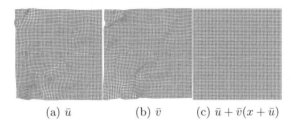

(a) $\bar{u}$ (b) $\bar{v}$ (c) $\bar{u}+\bar{v}(x+\bar{u})$

Fig. 3. Deformation fields applied to regular grids

versus the number of iterations are plotted on the right of Fig. 3, which shows the inverse consistency is preserved in the registration.

One of the applications of this algorithm is auto re-contouring, that deforms the expert's contours from a planning image to new images during the course of radiation therapy. In this experiment, we had expert's contours superimposed in the square area of the source image S as shown in Fig. 2d. Then by applying the deformation field $\bar{u}$ on these contours we get the deformed contours on the target image T as shown in Fig. 2e. The accuracy in auto re-contouring is evident.

The second experiment was aimed to test the effectiveness of the proposed model (9) in registering 3D images. We applied this model to a pair of 3D brain MR images of dimension $128 \times 128 \times 73$ taken from two different subjects. The parameters (λ, μ) were set to be $(.05, .1)$. The registration is performed in 3D, but for demonstration, we only show the corresponding sagittal (xy plane with $z = 53$), coronal (yz plane with $x = 47$) and axial (zx plane with $y = 97$) slices as in the top, middle and bottom rows of Fig. 4, where the columns in Fig. 4a and 4b show the corresponding slices of S and T, respectively, and columns in Fig. 4c and 4d deformed T and deformed S, i.e. $\bar{T} := T(x+\bar{v})$ and $\bar{S} := S(x+\bar{u})$, respectively. Columns in Fig. 4e, 4f and 4g are $|S-T|$, $|S(x+\bar{u})-T|$ and $|S-T(x+\bar{v})|$, respectively. The initial $CC(S,T) = .736$, and $CC(\bar{S},T)$ and $CC(\bar{T},S)$ reach .946 and .941 after 200 iterations, and the mean of inverse consistency errors is .027. The results show the high accuracy and well preserved inverse consistency obtained by using proposed model (9).

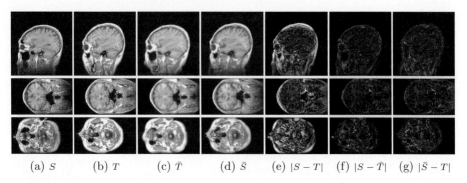

(a) S (b) T (c) $\bar{T}$ (d) $\bar{S}$ (e) $|S - T|$ (f) $|S - \bar{T}|$ (g) $|\bar{S} - T|$

Fig. 4. Registration result of proposed model (9) applied to 3D brain MR image. The results of slices $z = 53$, $x = 47$ and $y = 97$ are plotted on the top, middle and bottom rows, respectively. Here $\bar{T} := T(x + \bar{v})$ and $\bar{S} := S(x + \bar{u})$.

The third experiment was aimed to compare the efficiency of model (9) with the conventional full-way inverse consistent deformable registration model:

$$\min_{u,v,\sigma_u,\sigma_v} \|Du\|^2 + \|Dv\|^2 + \lambda J(u, v, \sigma_u, \sigma_v) + \mu \left(\mathcal{I}_v(u) + \mathcal{I}_u(v) \right) \qquad (13)$$

where u and v are forward and backward deformation fields, respectively, and the term J is defined by $J(u, v, \sigma_u, \sigma_v) = \|S(x + u) - T\|^2 / 2\sigma_u^2 + \|T(x + v) - S\|^2 / 2\sigma_v^2 + |\Omega| \log \sigma_v \sigma_u$. The comparison was made on the efficiency and accuracy of matching, as well as the ability of preserving inverse consistency. In this experiment we applied models (13) and (9) to the images in the first experiment shown in Fig. 1 with parameters (λ, μ) in both models set to be $(.05, .2)$. On the left Fig. 5, we plotted the CC obtained by model (13) and proposed model (9) at each iteration. It can be observed that the CC obtained by model (9) is higher and increases faster than that by model (13). This demonstrates that proposed model (9) is more efficient than the conventional full-way model. The reason is that the disparity between both deformed S and deformed T must be smaller than that between deformed S and fixed T or deformed T and fixed S. When S and T are deformed simultaneously, the two deformation fields u and $\tilde{u}$ are not necessarily to be large even if the underlying deformation field is large, which usually makes it difficult for the full-way based registration model to reach a satisfactory alignment in reasonable time.

The last experiment was aimed to test the robustness of the model to noises and the choice of the parameter λ with the use of MLE based disparity measure (5). The images S and T in Fig. 1 with additive Gaussian noises (using Matlab function `imnoise` with standard deviation being 3% of largest intensity value of S) were used in this experiment. The CC between S and T before registration was $CC(S, T) = .901$. We applied model (9) to the noisy data with σ updated using its EL equation (11), as well as σ fixed to be 1, which is equivalent to using SSD as similarity measure. We proceeded the registration with various values of λ, but kept other parameters unchanged. Then the numbers of iterations (Iter) for convergence and the final CC were recorded and shown in Table 1. One can

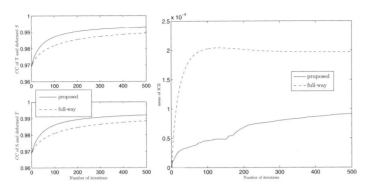

Fig. 5. Left: CC in each iteration obtained by full-way model (13) and proposed model (9). Right: Mean of inverse consistent errors (ICE) of the final deformation fields obtained by using full-way model (13) and proposed model (9).

see that while λ decreases, the accuracy of model (9) using fixed σ reduces as the final CC become much smaller, and it also takes significantly longer time for the algorithm to converge. On the other hand, with σ being updated (whose computational cost is extremely cheap) model (9) can obtain good matching in much less iterations for a large range of λ. This shows that model with MLE fitting is much less sensitive to noise and the choice of λ, and can achieve fast and accurate results compared with the model using SSD as the disparity measure.

Table 1. Number of iterations used for convergence and the final CC obtained by proposed model (9) with σ updated/fixed. For a large range of λ, updating σ in each iteration consistently leads to faster convergence and higher accuracy.

	Update σ		Fix σ	
λ	CC	Iter	CC	Iter
1e2	.962	48	.955	89
1e1	.962	97	.946	420
1e0	.960	356	.933	1762

4   Conclusion

In this work, we proposed a novel registration model that is featured at finding two invertible transformations that deform the source and target image simultaneously, and the registration process aligns both of the deformed source and deformed target images. From the theoretical analysis and experimental results, one can see the proposed model produces a fast convergence of the objective deformation field as well as provides a better correspondence for the ROIs in source and target images. This makes the auto re-contouring results with data involved in the course of radiation therapy much more accurate and efficient.

References

1. Kessler, M.L.: Image registration and data fusion in radiation therapy. Br. J. Radiol. 79, 99–108 (2006)
2. Lu, W., Olivera, G.H., Chen, Q., Chen, M., Ruchala, K.: Automatic re-contouring in 4d radiotherapy. Phys. Med. Biol. 51, 1077–1099 (2006)
3. He, J.C., Christensen, G.E.: Large deformation inverse consistent elastic image registration. In: Taylor, C.J., Noble, J.A. (eds.) IPMI 2003. LNCS, vol. 2732, pp. 438–449. Springer, Heidelberg (2003)
4. Joshiand, S., Davis, B., Jomier, M., Gerig, G.: Unbiased diffeomorphic atlas construction for computational anatomy. Neuroimag. Suppl. 23, 151–160 (2004)
5. Avants, B.B., Grossman, M., Gee, J.C.: Symmetric diffeomorphic image registration: Evaluating automated labeling of elderly and neurodegenerative cortex and frontal lobe. In: Pluim, J.P.W., Likar, B., Gerritsen, F.A. (eds.) WBIR 2006. LNCS, vol. 4057, pp. 50–57. Springer, Heidelberg (2006)
6. Beg, M.F., Khan, A.: Symmetric data attachment terms for large deformation image registration. IEEE Trans. Med. Imag. 26, 1179–1189 (2007)
7. Christensen, G.E., Johnson, H.J.: Consistent image registration. IEEE Trans. Med. Imag. 20, 721–735 (2001)
8. Alvarez, L., Deriche, R., Papadopoulo, T., Sanchez, J.: Symmetrical dense optical flow estimation with occlusions detection. In: Euro. Conf. Comput. Vision, pp. 721–735 (2002)
9. Rogelj, P., Kovacic, S.: Symmetric image registration. Med. Imag. Anal. 10, 484–494 (2006)
10. Zhang, Z., Jiang, Y., Tsui, H.: Consistent multi-modal non-rigid registration based on a variational approach. Pattern Recognit. Lett. 27, 715–725 (2006)
11. Yeung, S., Shi, P.: Stochastic inverse consistency in medical image registration. In: International Conference on Medical Image Computing and Computer-Assisted Intervention, vol. 8, pp. 188–196 (2005)
12. Yeung, S.K., Tang, C.K., Shi, P., Pluim, J.P., Viergever, M.A., Chung, A.C., Shen, H.C.: Enforcing stochastic inverse consistency in non-rigid image registration and matching. In: IEEE Conf. CVPR, pp. 1–8 (2008)
13. Leow, A., Huang, S., Geng, A., Becker, J., Davis, S., Toga, A., Thompson, P.: Inverse consistent mapping in 3d deformable image registration: its construction and statistical properties. In: Proc. Inf. Process. Med. Imag., pp. 493–503 (2005)
14. Zeng, Q., Chen, Y.: Accurate inverse consistent non-rigid image registration and its application on automatic re-contouring. In: Int. Symp. Bioinfo. Res. App., pp. 293–304 (2008)

A 3D Active Surface Model for the Accurate Segmentation of *Drosophila* Schneider Cell Nuclei and Nucleoli

Margret Keuper[1,3], Jan Padeken[2,3], Patrick Heun[2,3],
Hans Burkhardt[1,3], and Olaf Ronneberger[1,3]

[1] Chair of Pattern Recognition and Image Processing,
University of Freiburg, Germany
[2] Max-Planck Institute of Immunbiology, Freiburg, Germany
[3] Centre of Biological Signalling Studies (bioss), University of Freiburg, Germany
keuper@informatik.uni-freiburg.de

Abstract. We present an active surface model designed for the segmentation of *Drosophila* Schneider cell nuclei and nucleoli from wide-field microscopic data. The imaging technique as well as the biological application impose some major challenges to the segmentation. On the one hand, we have to deal with strong blurring of the 3D data, especially in z-direction. On the other hand, concerning the biological application, we have to deal with non-closed object boundaries and touching objects. To cope with these problems, we have designed a fully 3D active surface model. Our model prefers roundish object shapes and especially imposes roughly spherical surfaces where there is little gradient information. We have adapted an external force field for this model, which is based on gradient vector flow (*GVF*) and has a much larger capture range than standard *GVF* force fields.

1 Introduction

For the analysis of living cells, wide-field fluorescence microscopy still plays an important role, because it is prevalently available and, compared to confocal microscopy, has some advantages concerning temporal resolution and phototoxicity. The major disadvantage is the recorded defocused light - volume datasets recorded in wide-field microscopy suffer from strong blurring.

For the analysis of cellular mechanisms, exact knowledge about the subcellular anatomy is necessary. We are looking for a method to accurately detect and segment *Drosophila* cell nuclei and a subcellular structure, the nucleolus, from three dimensional recordings of cell cultures from a Schneider cell line. Nucleus and nucleolus have been recorded in two separate channels. Channel 1 shows the cell nuclei stained with the fluorescent stain 4',6-diamidino-2-phenylindole (DAPI), channel 2 shows the green fluorescent protein (GFP) stained fibrillarin inside the nucleolus. The voxel size in xy-direction is $0.064\mu m$ in z-direction $0.2\mu m$. For the segmentation of this data, we are dealing with a variety of problems inherent in the biological setting as well as with problems originating from

G. Bebis et al. (Eds.): ISVC 2009, Part I, LNCS 5875, pp. 865–874, 2009.
© Springer-Verlag Berlin Heidelberg 2009

the microscopy technique. In figure 1, an example slice from each channel is displayed, as well as orthogonal slices of one of the nuclei in channel 1. In channel 1, the brighter spots of dense chromatin and the low intensity regions, where there is no chromatin, lead to non-closed boundaries. This and the fact, that there are typically touching cells in the dataset make the segmentation of the nuclei challenging. Additionally, due to artifacts caused by the point spread function, the nucleolus in channel 2 often seems to range outside the nucleus, which, in a biological sense, cannot be the case.

For the detection and segmentation of the nucleoli, the use of all available information in both recorded channels is therefore necessary. We thus designed a preprocessing step, in which we combine the two channels to a Channel Differential Structure. This procedure is inspired by the color differential structure described in [4]. The description of this preprocessing step constitutes the first section of this paper. After this preprocessing step, the segmentation of nuclei and nucleoli are both addressed with a two-step procedure. First, the nuclei are detected by the generalized Hough-Transform [1] for the detection of spherical objects, as it has been used e.g. in [3] for the 3D detection of *Arabidopsis thaliana* root nuclei in confocal microscopic data and in [2] for the 2D detection of pollen grains in brightfield data. The detection method will be presented in section 3. The second step, which constitutes the main part of this paper, consists in specifying the objects boundary using a three dimensional active surface model. In the field of biomedical image analysis, active contour methods are widely used for the segmentation and modeling of anatomical structures. The mathematical foundamentals can be found e.g. in [5], a review of deformable models is given e.g. in [6] and [7]. Here, we design a 3D active surface model especially adapted for the segmentation and representation of smooth, 2-sphere like shapes. The grid representation as well as the external forces have been adapted, providing a robust and accurate segmentation with a large capture range even under strong blurring. The detection of the nucleoli is done as for the nuclei with the generalized Hough transform, the segmentation is achieved by the presented active surface model as well - thus proving its generalization ability. Finally, we present

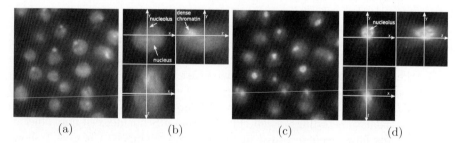

    (a)   (b)   (c)   (d)

Fig. 1. Example slices from the two channels of the *Drosophila* cell raw data and orthogonal slice views of one of the nuclei. (a) and (b): channel 1 shows the cell nuclei stained in DAPI. Brighter regions are caused by denser chromatin. In the location of the nucleoli, no signal is recorded in this channel, thus causing a hole. (c) and (d): in channel 2, the GFP-stained fibrillarin inside the nucleoli has been recorded.

the results of our segmentation. These results will be compared to the results that can be achieved with an active surface implementation using a classical gradient vector flow force field and to a standard state of the art method: the user-guided level set implementation ITK-SNAP [8]. For the detection as well as for the segmentation, we decided to use undeconvolved data. Deconvolution makes the images appear clearer in xy-direction but the deconvolution artifacts in z-direction made a good segmentation nearly impossible.

2   Channel Differential Structure

For the detection of the nuclei, only channel 1 is used, since here, the boundaries can be seen best (see figure 2(a)). For the nucleoli, the situation is different. In channel 2 (see figure 2(b)), only the fibrillarin inside the nucleoli has been stained,which results in a bright region inside the nucleolus - but also the whole nucleus can be seen in this channel at a much lower intensity. In channel 1, one can see a hole at the nucleolus' position, which is larger than the stained region in channel 2 - the correct nucleolus boundaries are between the borders of this hole and the fibrillarin in channel 2. Therefore, it makes sense to use the information of both channels for the detection and segmentation of the nucleoli. To do so, the arctangent of the ratio of the intensities in channel 2 I_{ch2} and channel 1 I_{ch1} is computed in each position, yielding some kind of channel diffenrential structure CDS. This was inspired by the color differential structure defined in [4]. In [4], this color differential structure is computed as the convolution of the spectral color information with a gaussian derivative to detect gradients between complementary colors.

For our task, not the perceptual color difference is important, which heavily depends on human color perception, but the relation of the channels. The arctangent of the ratio can thus be considered a good distance measure. The two-channel volume data is considered as function $\mathbf{I} : \mathbb{R}^3 \to \mathbb{R}_{\geq 0}^2$. The $CDS(\mathbf{I})$ can then be computed as

$$CDS(\mathbf{I}) = \nabla \left(\frac{2}{\pi} \arctan \left(\frac{I_{ch2}}{I_{ch1}} \right) \right), \tag{1}$$

where $\nabla = \left(\frac{\partial}{\partial x_1}, \frac{\partial}{\partial x_2}, \frac{\partial}{\partial x_3} \right)$ is the the grad operator and $\arctan \left(\frac{I_{ch2}}{I_{ch1}} \right)$ is scalar valued between 0 and $\frac{\pi}{2}$. Positions, where the ratio of the channels has large

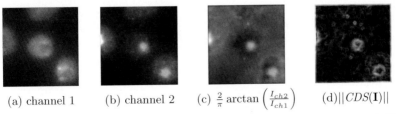

(a) channel 1   (b) channel 2   (c) $\frac{2}{\pi} \arctan \left(\frac{I_{ch2}}{I_{ch1}} \right)$ (d)$\|CDS(\mathbf{I})\|$

Fig. 2. xy-slices from the raw data channels, their ratio and the magnitude of $CDS(\mathbf{I})$

values are likely to correspond to the nucleolus. The ratio of the intensities inside the nucleus are small, such that the boundaries of nucleus and nucleolus cannot be confounded. Compare figure 2(c). The $CDS(\mathbf{I})$ is used instead of ∇I_{ch2} for the further segmentation of the nucleoli.

3   Detection

The detection of nucleus and nucleolus is done with the generalized Hough transform for spherical objects [1] as presented in [3]. The main idea is to let each voxel vote for possible positions $\mathbf{c}$ of sphere centers at specific radii r.

For this, the dataset has to be smoothed first. We use a Gaussian smoothing with $\sigma_1 = 2\mu$m in all directions. Given the recorded resolution, the resulting estimation for the radii r can therefore not be very precise, but as these radii are used only for the initialization of the active surfaces, these rough estimations are sufficient. Then, one has to select the voting voxels. We choose to use all voxels for the voting and weight the votes with the respective gradient magnitude values. A threshold would not make sense, because the intensity variations of the nuclei within one dataset are too strong. The gradient magnitude and direction of the voxels is used to determine the position of the votes. Finally, the votes are combined by integration. Formally, the four dimensional voting field P of a function $I : \mathbb{R}^3 \to \mathbb{R}$ is computed as follows.

$$P(\mathbf{c}, r) = \int_{\mathbb{R}^3} G_{\sigma_2} \left(\mathbf{c} - r \frac{(\nabla(G_{\sigma_1} * I))(\mathbf{x})}{||(\nabla(G_{\sigma_1} * I))(\mathbf{x})||} \right) ||(\nabla(G_{\sigma_1} * I))(\mathbf{x})||d\mathbf{x}, \qquad (2)$$

where G_σ is the 3D Gaussian distribution with standard deviation σ in all directions. G_{σ_2} is used as an indicator function giving contribution only if the argument is nearby zero.

The detection is then done by determining the local maxima of the voting field P.

4   Active Surfaces

After the detection step, we have an estimated center $\mathbf{c}$ and radius r for each nucleus and nucleolus, but in reality, both structures are not spherical. Thus, the best radius can only give a very rough estimation of the object's size. An exact segmentation has been done based on this estimation by employing active surface models.

The three dimensional active surface can be described as a function $\mathbf{X} : [0, 1] \times [0, 1] \to \mathbb{R}^3$ which is placed on a dataset $I : \mathbb{R}^3 \to \mathbb{R}$. These active surfaces have internal energies, depending only on the shape of the model itself, and are subjected to external energies coming from the underlying dataset to which the model shall be adapted. The total energy of an active surface is thus

$$E = E_{int}(\mathbf{X}) + E_{ext}(\mathbf{X}). \qquad (3)$$

The adaption takes place in minimizing this energy.

4.1   Implementation

Mesh Design. For the realization of the active surface model in three dimensional space, a good choice of the grid structure is crucial. As we suppose a roughly spherical shape of the objects to detect, it is intuitive to initialize the active surface with a spherical grid. To provide an equidistant sampling of the sphere, an icosahedron can be used as initial structure. We have used subdivisions of the icosahedron, which still provide a nearly equidistant sampling, with a higher resolution, leading to a more accurate segmentation.

As initialization, we used the parameters found in the detection step, i.e. we initialized the grid as a subdivided icosahedron with 162 vertices $v_i \in V$ located at positions $\mathbf{x}_{v_i}$ around the center $\mathbf{c}$ of the nucleus or nucleolus. All vertices v_i have distance r from the center.

Fig. 3. The sampling of the spherical surface can be done equidistantly by using an icosahedron. Icosahedron subdivisions yield a nearly equidistant sampling.

Mesh Operations. To ensure that, during the evolution of the active surface, the resolution of the grid is always high enough, we implement splitting and merging operations. After each iteration, the length l of every edge is checked. Edges longer than a threshold t_u are split and a new vertex is inserted. This new vertex has to be connected to all its neighboring vertices. If edges are shorter than threshold t_l, they are deleted and the corresponding vertices are merged. See figure 4 for an example.

Fig. 4. (left) If the length $l > t_u$, the edge is split. (right) Edges for which $l < t_l$ are merged.

4.2   Internal Forces

As internal energy, the weighted first and second derivative of the surface are used, corresponding to the classical continuity and curvature energies:

$$E_{int} = \frac{1}{2} \int_s \alpha \left\| \frac{\partial \mathbf{X}}{\partial \mathbf{s}} \right\|^2 + \beta \left\| \frac{\partial^2 \mathbf{X}}{\partial \mathbf{s}^2} \right\|^2 d\mathbf{s} \qquad (4)$$

where $\mathbf{s} \in [0,1] \times [0,1]$ and α and β are the weighting coefficients. The minimization of E thus leads to the Euler-Lagrange equation

$$\frac{\partial}{\partial \mathbf{s}}\left(\alpha \frac{\partial \mathbf{X}}{\partial \mathbf{s}}\right) - \frac{\partial^2}{\partial \mathbf{s}^2}\left(\beta \frac{\partial^2 \mathbf{X}}{\partial \mathbf{s}^2}\right) - \nabla E_{ext} = 0, \tag{5}$$

where the internal forces $\mathbf{F}_{ela} := \frac{\partial}{\partial \mathbf{s}}\left(\alpha \frac{\partial \mathbf{X}}{\partial \mathbf{s}}\right)$ and $\mathbf{F}_{rig} := \frac{\partial^2}{\partial \mathbf{s}^2}\left(\beta \frac{\partial^2 \mathbf{X}}{\partial \mathbf{s}^2}\right)$ prevent the surface from stretching and bending too much. These energies are minimal for planar surfaces. Therefore, on spherical structures, they act as shrinking forces.

The internal forces of the active surface have to be adapted for the three dimensional grid implementation. Since we want to preserve the equidistant sampling, the elasticity force can be approximated with

$$\mathbf{F}_{ela}(v_i) = \frac{1}{c^2}\left(\sum_{j,v_j \in N(v_i)} \frac{\mathbf{x}_{v_j} - \mathbf{x}_{v_i}}{|N(vi)|}\right), \tag{6}$$

where $N(v_i)$ is the set of all neighbors of vertex v_i and c is the average distance between two neighboring vertices

$$c = |V| \sum_i \sum_{j,v_j \in N(v_i)} \frac{|N(v_i)|}{||\mathbf{x}_{v_j} - \mathbf{x}_{v_i}||} \tag{7}$$

$|V|$ is the cardinality of the set of vertices. The rigidity force corresponding to the forth derivative can by analogy be approximated as

$$\mathbf{F}_{rig}(v_i) = \frac{1}{c^4}\frac{1}{|N(v_i)|(|N(v_j)|-1)} \sum_{j,v_j \in N(v_i)} \sum_{\substack{k,v_k \in N(v_j) \\ k \neq i}} (4\mathbf{x}_{v_j} - \mathbf{x}_{v_k} - 3\mathbf{x}_{v_i}). \tag{8}$$

4.3   External Forces

Since the external forces are responsible for the attraction of the active surface to the underlying data, these forces have to be adapted very carefully to the specific task. A classical choice would be $\mathbf{F}_{ext} = -\nabla E_{ext}$ with $E_{ext} = -||\nabla I||^2$, but for the segmentation of the nuclei, some application specific challenges are given. As the chromatin displayed in channel 1 is not homogeneous, some blob structures and holes can be seen in the nucleus. Thus, neither the intensity values nor the pure gradient information can be used as an external energy for the segmentation of the nuclei. Especially nucleoli lying near the nucleus boundary cause the image gradients to pull the contour inwards into the nucleus. To address this problem, we use external forces based on the gradient information of the data coupled with prior knowledge from the detection step. We assume the nuclei to have a star-shaped surface, i.e. every surface point can be reached from the detected center $\mathbf{c}$. As proposed in [2] for the 2D case in pollen segmentation, we then used a projection of the dataset gradients onto radial vectors pointing away from the detected center $(\nabla I_{radial})(\mathbf{x}) = \left\langle (\nabla I)(\mathbf{x}), \frac{\mathbf{x}-\mathbf{c}}{||\mathbf{x}-\mathbf{c}||} \right\rangle$, thus reducing the influence of vectors pointing in other directions. Additionally, as done in [2], those vectors originating from darker inner structures and thus pointing outwards were set to zero length. The resulting gradient image contains by far less gradients corresponding to structures other than the nucleus, but the vectors set to zero length still cause problems in the next step.

Instead of applying the Canny edge detector as it was done in [2], we directly use the resulting gradient magnitude as edge information. We compute the gradients of this edge image and, to get rid of the gradients now caused by the zero-magnitude regions, we use the radial projection of these gradients. This results in a vector valued function $\mathbf{A} : \mathbb{R}^3 \rightarrow \mathbb{R}^3$ with

$$\mathbf{A}(\mathbf{x}) = \left\langle (\nabla(s(\nabla I_{radial})))(\mathbf{x}), \frac{\mathbf{x}-\mathbf{c}}{||\mathbf{x}-\mathbf{c}||} \right\rangle \cdot \frac{\mathbf{x}-\mathbf{c}}{||\mathbf{x}-\mathbf{c}||}, \tag{9}$$

where $\langle ., . \rangle$ is the scalar product and $s(x)$ is defined as

$$s(x) = \begin{cases} x, & \text{if } x > 0 \\ 0, & \text{otherwise.} \end{cases} \tag{10}$$

The external force was finally found as the weighted sum of the gradient vector flow $GVF(\mathbf{A})$ (e.g. in [5]) and the radially projected gradients, pulling the surface outside the object. ∇I_{radial} counteracts the shrinking effect of the internal forces.

$$\mathbf{F}_{ext} = \gamma GVF(\mathbf{A}) - \delta \nabla I_{radial} \tag{11}$$

γ and δ are the weighting coefficients. The in this way defined external force field has some major advantages compared to standard gradient based force fields. On the one hand, the projection onto the radial vectors promotes 2-spherical shapes, on the other hand, these projections and the deletion of gradient vectors pointing in the wrong direction has the effect that the capture range of the resulting force field is much larger. This is important, because of the touching cells in the dataset.

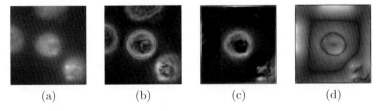

(a)   (b)   (c)   (d)

Fig. 5. (a) xy-slice from the original dataset (b) xy-slice from the gradient magnitude of the data with the estimated center and radius (c) xy-slice from the magnitude of the projected gradients $||\nabla I_{radial}||$, (d) xy-slice from the magnitudes of the gradient vector flow $||GVF(\mathbf{A})||$

5   Experiments and Results

The method was tested on 45 datasets containing 440 cells. For the detection
of the nuclei, we searched for spheres with diameters between 3.6 and 6.4 μm.
Although more or less strong deformations of the nuclei can be observed, a quite
reliable detection of the nuclei was possible. Nuclei clearly lying on the border of
the captured dataset were not detected and left out of the evaluation. Out of 440
nuclei, we have correctly detected 437 nuclei, 3 nuclei have not been detected.
There were 16 false positives: 8 nuclei were detected where there was no data
and 8 defect cells have been detected as nuclei. For some of the nuclei, the estimated radii were too small (compare figure 6), which results from the relatively
strong smoothing. For the detection of the nucleoli, we searched for spheres with
diameters between 0.6 and 1.6 μm. The result of the nuclei segmentation was
used as a mask for the detection of the nucleoli, i.e. every nucleolus has to lie
inside a nucleus. To ensure this, only the Hough-votes within the nucleus were
evaluated and exactly one nucleolus was searched inside each nucleus, since for
healthy cells, there should only be one nucleolus. Despite this fact, there are cells
in the datasets containing more than one clearly defined nucleolus. Detection results for an example dataset can be seen in figure 6. Correct nucleoli positions
were found in all of the analyzed datasets - only where there was more than one
nucleolus inside the same nucleus, one of the nucleoli was missed.

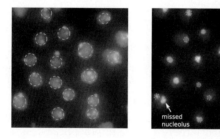

Fig. 6. (left) Detection results of nuclei displayed in the maximum intensity projection
(MIP) of channel 1. (right) Detection results of nucleoli in the MIP of channel 2.

For the accurate segmentation, we manually tested some parameter sets for
three example nuclei and then used the best parameters for the segmentation
of all the nuclei. Finally, we used $\alpha = \beta = 0.2$, $\gamma = 0.9$, and $\delta = 0.7$, but the
method turned out to be very robust against smaller parameter variations. For
the nucleoli, we picked $\alpha = \beta = 0.1$, $\gamma = 0.9$, and $\delta = 0.5$. With those parameters,
satisfying results could be achieved for the segmentation of all nuclei and nucleoli
of the dataset. Some randomly chosen example results can be seen in figure 7.

To evaluate the results of the segmentation, we compared the segmentation
carried out with the presented method with the results that could be achieved
with an active surface implementation using a standard GVF force field as well
as with results from the ITK segmentation tool ITK-SNAP [8], which is based on

a level set implementation. The standard GVF force field was computed directly from the image gradients as $GVF(\nabla||\nabla I||)$. The ITK-SNAP segmentation implements two algorithms: 3D geodesic active contours, where the internal forces are based on the gradient magnitude in the dataset, and a region competition method, based on voxel probability maps, which are estimated by manually adjusted intensity thresholds. In all cases, it was not possible to find parameters that worked for the whole dataset. For our four example cells, we manually adapted the parameters for each nucleus as good as possible, but even though we could not get good results for all of the cells. The 3D geodesic active contours even did not work at all, because of the blurring in z-direction. It was not possible to find parameters, that prevented the contour from running out of the object in upper and lower dataset regions before filling the nucleus' volume in the center, where there are in fact stronger gradients. For a comparison of the other two methods to our presented method, see figure 7. Although the region competition method from ITK-SNAP in most cases yields acceptable results if the manual threshold is carefully adjusted, our method worked best for all of the nuclei.

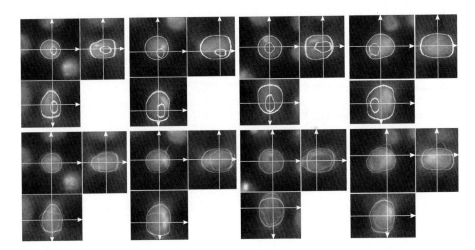

Fig. 7. (above) Orthogonal views of the segmentation results of four nuclei and nucleoli displayed in channel 1. The overall result is satisfying for nuclei as for nucleoli. (below) Segmentation results for the same nuclei segmented with our method (green), active surfaces with standard force field (red) and the region competition method from ITK-SNAP (blue). The red contours are attracted by inner structures as well as neighboring cells and thus yield bad segmentation results, the blue contours are quite good, but elongated in z-direction, which is caused by the blurring in the dataset.

6   Conclusion

We have presented an active surface model on an icosahedron subdivision grid structure, which is specially adapted to the segmentation of 2-sphere like objects.

We have designed an external force field that is able to address the problems caused by wide-field microscopy imaging as well as the specific challenges of the segmentation of *Drosophila* Schneider cell nuclei. The segmentation results with our method were not only better and more reliable than the results found with standard methods, it was also possible to segment all correctly detected nuclei with the same parameter set, such that no further tedious manual adjustments were necessary. The model has shown its generalization ability in yielding very good results for the segmentation of the nucleoli inside these nuclei.

Acknowledgments

This study was supported by the Excellence Initiative of the German Federal and State Governments (EXC 294).

References

1. Ballard, D.H.: Generalizing the hough transform to detect arbitrary shapes. Pattern Recognition 13(2), 111–122 (1981)
2. Ronneberger, O., Wang, Q., Burkhardt, H.: Fast and robust segmentation of spherical particles in volumetric data sets from brightfield microscopy. In: Proc. of the ISBI, pp. 372–375 (2008)
3. Schulz, J., Schmidt, T., Ronneberger, O., Burkhardt, H., Pasternak, T., Dovzhenko, A., Palme, K.: Fast scalar and vectorial grayscale based invariant features for 3d cell nuclei localization and classification. In: Proc. of the DAGM, Berlin (2006)
4. Geusebroek, J.-M., Ter Haar Romeny, B., Koenderink, J., van den Boomgaard, R., van Osta, P.: Color differential structure. In: Front-End Vision and Multi-Scale Image Analysis, Computational Imaging and Vision, vol. 27. Springer, Netherlands (2003)
5. Xu, C., Prince, J.L.: Snakes, shapes, and gradient vector flow. IEEE Trans. Imag. Proc. 7(3), 321–345 (1998)
6. Montagnat, J., Delingette, H., Ayache, N.: A review of deformable surfaces: topology, geometry and deformation. Image and Vision Computing 19/14, 1023–1040 (2001)
7. He, L., Peng, Z., Everding, B., Wang, X., Han, C.Y., Weiss, K.L., Wee, W.G.: A comparative study of deformable contour methods on medical image segmentation. Image and Vision Computing 26/2, 141–163 (2008)
8. Yushkevich, P.A., Piven, J., Hazlett, C., Smith, H., Smith, G., Ho, R., Ho, S., Gee, J.C., Gerig, G.: User-Guided 3D Active Contour Segmentation of Anatomical Structures: Significantly Improved Efficiency and Reliability. Neuroimage 31/3, 1116–1128 (2006)

Weight, Sex, and Facial Expressions: On the Manipulation of Attributes in Generative 3D Face Models

Brian Amberg, Pascal Paysan, and Thomas Vetter

University of Basel

Abstract. Generative 3D Face Models are expressive models with applications in modelling and editing. They are learned from example faces, and offer a compact representation of the continuous space of faces. While they have proven to be useful as strong priors in face reconstruction they remain to be difficult to use in artistic editing tasks. We describe a way to navigate face space by changing meaningful parameters learned from the training data. This makes it possible to fix attributes such as height, weight, age, expression or 'lack of sleep' while letting the infinity of unfixed other attributes vary in a statistically meaningful way.

We propose an inverse approach based on learning the distribution of faces in attribute space. Given a set of target attributes we then find the face which has the target attributes with high probability, and is as similar as possible to the input face.

1 Introduction

When producing movies or computer games it is common to use low dimensional generative 3D face models, which encode the space of possible faces with a few hundred parameters. These models encode the variability of the 3D shape, albedo, and reflectance properties. By varying the parameters of the model new 3D faces are created, which can then be rendered in the movie or game. The parameters of face models like 3D-MM [1] are a compact description of faces, but they are generally not meaningful. And even in manually constructed face models one can only change characteristics of the mesh like the size of the chin, but what the artist really wants is not to change the chin but to create a face which looks older, more male, or even more trustworthy. We propose a system which allows such manipulations directly in attribute space. To this end we learn an association between face and attribute space. We demonstrate our system with a 3D-Morphable Model (3D-MM) of Shape and Texture, but it is equally applicable to other models, which might be better at capturing statistics of wrinkles, hair, or reflectance properties of the skin. 3D-MMs are interesting because the starting face can be easily initialised from a real person using a reconstruction method as proposed for images in [1], videos in [2] or 3D scans in [3].

In this paper we use a relatively wide definition of *attribute*, ranging from physical attributes like crookedness of the nose or testosterone level over latent attributes like lack of sleep to cultural attributes like trustworthiness.

Our method is not only useful as a tool for character generation, but also for the investigation and visualisation of the dependencies between attributes and faces. We can

G. Bebis et al. (Eds.): ISVC 2009, Part I, LNCS 5875, pp. 875–885, 2009.

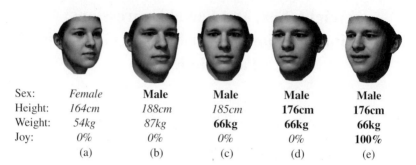

Sex:	Female	Male	Male	Male	Male
Height:	164cm	188cm	185cm	176cm	176cm
Weight:	54kg	87kg	66kg	66kg	66kg
Joy:	0%	0%	0%	0%	100%
	(a)	(b)	(c)	(d)	(e)

Fig. 1. Generating new faces by changing some attributes of a face while keeping other attributes fixed. Starting from an initial face we can fix some attributes (typeset in bold) and let other attributes (typeset in italic) vary according to the statistical distribution of attributes in our training set. When an attribute is fixed we determine the face which simultaneously 1) is most similar to the starting face, 2) has the requested attributes with high probability and 3) is a likely face.

visualise the change of face features associated with attributes, or generate an infinity of psychological stimuly with precisely defined attributes, where it is possible to vary only a single attribute and keep others constant. We can for example generate images which vary in handsomeness while keeping perceived sex, height and weight constant.

2   Facial Attribute Manipulation

We state the problem as follows. For a given starting face and a new set of attributes, find the face which has the chosen attribute vector while being as close as possible to the input face in the natural distance in face space. With our method the artist generates in an intuitive way different distance functions in face space by specifying which attributes can vary and which should be fixed. This is explained in more detail in the next section.

We want to find the distribution $p(\bar{x} \mid x, a)$ of faces $\bar{x}$ which are similar to a starting face x and have the attributes a. Our training data consists of a collection of N example faces, described by their model coefficient $x_i \in \mathbb{R}^k = \mathcal{F}$, and corresponding attribute vectors $a_i \in \mathbb{R}^l = \mathcal{A}$. Attributes can be physical measurements like age, height, weight and sex which can be determined by measurement, subjective attributes like trustworthiness which are gauged with questionnaires, latent attributes like lack of sleep, and rapidly changing attributes like facial expressions, which were classified by asking the subjects to perform a specific expression. We model the likelihood as

$$p(\bar{x} \mid x, a) \propto p(x, a \mid \bar{x})p(\bar{x}) \tag{1}$$
$$= p(x \mid \bar{x})p(a \mid \bar{x})p(\bar{x}), \tag{2}$$

where the probability $p(a \mid \bar{x})$ is learned from labelled training data. If enough data is available it is sensible to use a Gaussian Process for the regression, to get varying variances, or one can assume homoscedastic uncertainty and learn only a single σ_a and model

$$p(a \mid \bar{x}) = \mathcal{N}(a \mid M(\bar{x}), \sigma_a I) \quad . \tag{3}$$

where $M(\bar{x})$ is a regressed mapping from face space to attribute space. A normal distribution is used for the similarity measure, which assumes that face space is smooth in face appearance. This corresponds to the Mahalanobis distance in face space, which has shown good performance in face recognition experiments (e.g. [3]).

$$p(x \mid \bar{x}) = \mathcal{N}(x \mid \bar{x}, \sigma_d I) \quad . \tag{4}$$

The choice of normal distributions is motivated by the ease of handling and the small number of parameters, but when enough data is available more sophisticated probability densities can be used. With the proposed distributions one arrives at the likelihood

$$p(\bar{x} \mid x, a) \propto p(a \mid \bar{x})p(x \mid \bar{x})p(\bar{x}) = \tag{5}$$

$$\frac{1}{2\pi\sqrt{\sigma_a \sigma_d}} \exp\left(-\frac{\|M(\bar{x}) - a\|^2}{\sigma_a^2} - \frac{\|\bar{x} - x\|^2}{\sigma_d^2} - \|\bar{x}\|^2 \right)$$

We could now sample from this distribution, but our purpose is to generate a single predictable answer. Therefore we calculate the maximum likelihood solution.

$$T_{\mathrm{ML}}(x, a) = \arg\min_{\bar{x}} \|M(\bar{x}) - a\|^2 + \gamma_1 \|x - \bar{x}\|^2 + \gamma_2 \|\bar{x}\|^2 \tag{6}$$

where γ_1, γ_2 are trade-off parameters derived from σ_d, σ_a, which determine how closely the attributes should be matched.

Note that even though we used simple distributions the resulting model can be arbitrarily expressive – and the optimisation problem arbitrarily complicated – depending on the choice of the regression function. By taking the limit case of modelling $p(a \mid \bar{x})$ as a Dirac function we can also get a "hard" formulation of our problem, which when combined with linear regression leads to a very efficient optimisation problem. More details are given in section 2.2. We found that while the "hard" formulation works well for some attributes, it is for many attributes better to use the probabilistic formulation. The probabilistic version is harder to optimize but is still fast enough for interactive use. Our method is illustrated in figure 2.

2.1 Regression Methods

We evaluated a number of linear and nonlinear regressors and will describe them in this section. We found, that for the attributes attractiveness and, trustworthiness a *linear regression* worked best. For age, weight, height we got good results with Support Vector Regression (SVR) and the binary attribute sex and the expressions were best modelled by using the probability of one of the classes as the attribute. This is explained in section 2.1.

Measured Attributes. In addition to the attributes established with questionaires we included nonlinear measures on the shape model, namely distance between two points and angle between three points to allow changing the nose size and shape and to open and close the eyes and to fix the inter eye distance. These measures are (nonlinear) functions of the shape parameters and can be directly used instead of regressed attributes in our framework.

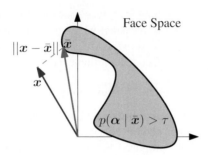

Fig. 2. Given a current face x and a target attribute vector a, we find the face $\bar{x}$ which has the target attributes (with a high probability) as measured by the regression function $M(\bar{x}) = a$, while having the least change in identity as measured by the distance function $||x - \bar{x}||$. Additionally we incorporate the face-prior, which is not depicted here.

Linear Classifier. The simplest classifier is a linear function of the input coefficients

$$M(x) := Mx + g \quad . \tag{7}$$

Each row in M is essentially what has been used as an *attribute vector* in [1], it encodes how much a change in the attribute changes the coefficients of the face parameter.

Binary Attributes. Some of the attributes we are using are of a discrete nature. Most prominently sex takes on only two distinct values, male and female.[1] So while for height or attractiveness we can acquire continuous values on which a regression can be performed, this is impossible for discretely labeled attributes. Typically this was handled by determining the direction of main variance by linear regression where the classes c_1, c_2 are identified with the labels -1, 1. We found, that for this types of attributes a more intuitive and accurate method is to use the probability $p(c_1 \mid x)$ as the attribute. The probability is determined with Bayes theorem as

$$\begin{aligned} p(c_1 \mid x) &= 1 - p(c_2 \mid x) \\ &= p(x \mid c_1)p(c_1)/p(x) \\ &= p(x \mid c_1)p(c_1)/(p(x \mid c_1)p(c_1) + p(x \mid c_2)p(c_2)) \quad . \end{aligned} \tag{8}$$

The priors $p(c_i)$ are determined either from the samples, or in the case of sex we fixed them at $1/2$. The probabilities $p(x \mid c_i)$ are modelled with parametric distributions, where we found that in the case of sex these probabilities were well modelled by normal distributions. A comparison between using linear regressor to model the change of sex and using equation 8 is shown in figure 3. Though the effect is subtle, it is visible that a change of sex does not correspond to a movement in the same direction in feature space for each face. Males and females are nonlinearly distributed, which is captured nicely by the mixture model. This concurs with the fact that when interpreted as a classifier, the probability $p(c_1 \mid x)$ does better in a cross-validation than the linear classifier. This method can be extended to n-ary attributes by splitting them into $n - 1$ two valued attributes.

---

[1] Obviously there are other taxonomies, but we regard only this simple system.

2.2 Solving Attribute Manipulation

We solve equation 6 with a Levenberg-Marquardt method. The optimisation was fast enough to allow us to build an interactive application.

An efficient algorithm for the linear regression. A very efficient algorithm can be derived when using (1) the Mahalanobis Distance for face similarity and (2) a linear classifier with an associated dirac distribution for the attribute distribution and (3) a uniform face prior. In this case we are searching for the minimum of $\|\bar{x} - x\|^2$ under the constraint that $M(\bar{x}) = a$. For the linear classifier the solutions lie in a displaced subspace of face-space, such that the closest face fullfilling the attributes can be found by a projection. We can derive the solution from the Lagrange equation

$$L(\bar{x}, \lambda) = \|\bar{x} - x\|^2 + \lambda^T(M\bar{x} + g - a) \quad , \tag{9}$$

which takes on its extremum at

$$2\bar{x} - 2x - M^T\lambda = 0 \tag{10}$$
$$M\bar{x} + g - a = 0 \quad .$$

This linear system can be written in matrix notation as

$$\begin{bmatrix} I & -2^{-1}M^T \\ M & 0 \end{bmatrix} \begin{bmatrix} \bar{x} \\ \lambda \end{bmatrix} = \begin{bmatrix} x \\ a - g \end{bmatrix} \quad . \tag{11}$$

The solution is a linear function of input face and target parameters

$$\bar{x} = \underbrace{[I\ 0] \begin{bmatrix} I & -2^{-1}M^T \\ M & 0 \end{bmatrix}^+}_{\text{constant}} \begin{bmatrix} x \\ a - g \end{bmatrix} \quad , \tag{12}$$

where we use $\cdot^+$ to denote the pseudo inverse. Note that this is different from the method in [1], as it allows the simultaneous handling of multiple parameters, where attributes can be fixed or left free to vary at the discretion of the user. Also, we can prescribe absolute values, such as 30 years and 83 kilos, instead of only offsets from the current shape.

3 Evaluation

As this method is targeted at humans, it is inevitable to assess the performance of the system based on human judgement – *it has to look good*. But additionally, as the basis of this method are regression methods, it is possible to evaluate the regressions on a test set, which we used to determine the kind of classifier to use for each attribute. If the classifier is unable to correctly classify the test set, then the method must also fail human judgement. We will now present some results, more are shown in the accompanying online material.

3.1   Model

We constructed the underlying 3D model in the same way as [3]. We acquired 800 facial scans from over 300 IDs, which were brought into correspondence with a nonrigid ICP method similar to [4] but with a different regularisation term. To increase the database we added mirrored versions of all scans. Every ID had at least one neutral scan. From the neutral scans we build separate PCA models of shape and texture, which we call the identity models. Two more PCA models where built from the offsets between the expression scans and the corresponding neutral scans. We call this the expression model. The two models are concatenated, such that we have two sets of coefficients, the identity coefficients and the expression coefficients. For the experiments in this paper we used 80 neutral shape, 40 expression shape and 20 neutral texture components, resulting in a 140 dimensional face space.

3.2   Speed

The runtime of the optimisation depends on the evaluation of the regression functions and their derivatives. For the experiments used in this paper we have real-time results when using only linear constraints and the method proposed in section 2.2. When using the nonlinear classifiers presented here, we have typically a runtime of less than half a second for the iterative methods, allowing interactive exploration of attribute space. To speed up the iterative methods we first fit 25 steps from the reference face a and from the current estimate from the previous fitting, and continue then a full fitting from the position with the lower residual, which speeds up the search time in interactive use.

3.3   Binary Attributes

We show in figure 3 that changing the sex when using the nonlinear probability $p(\text{male} \mid x)$ as the attribute results in a face which has convincingly changed sex but is closer to the starting face than that resulting from linear regression. While fixing the target probability $p(\text{male} \mid x)$ to 0 or 1 results in convincing faces, it is more difficult to determine a suitable value in the linear scale, where -1 or +1 often correspond to too pronounced or not enough pronounced changes in sex. For this comparison we fixed the value of the linearly regressed sex to that which we determined for the face generated from the nonlinear regression. This should result in a fair comparisons of the methods.

   The second set of attributes that we applied this method to were expressions. Our dataset includes labeled examples of five different expressions, and we trained one classifier per expression, in a one against all scheme. In figure 5 we show a sequence, where the starting face has a sad expression, which we then remove by setting all expressions to neutral. By changing then the joy value to one and keeping the other expression attributes fixed we changed the initial sad face into a happy face. As we have relatively little data in this high dimensional face-space, we had to regularize the estimation of the Gaussian for each class. This was done by setting the small eigenvalues of the covariance to a constant.

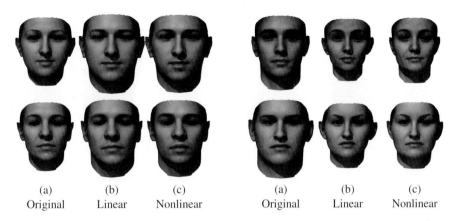

| (a) | (b) | (c) | (a) | (b) | (c) |
| Original | Linear | Nonlinear | Original | Linear | Nonlinear |

Fig. 3. Modelling the distribution of sex in face space as a Gaussian mixture is superior to determining just the direction of main variance. Using the probability of 'male' as an attribute as in equation 8 results in faces (c) which retain more of the characteristics of the source face (a) while undergoing a perceived change of sex than the result achievable with linear regression (a).

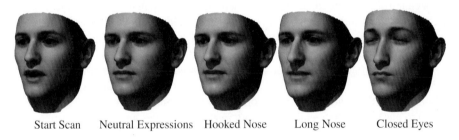

| Start Scan | Neutral Expressions | Hooked Nose | Long Nose | Closed Eyes |

Fig. 4. Face Space Measures allow the manipulation of length and angles in the face. They integrate naturally into our framework, such that we determine the most likely face for the given length and angle constraints. To demonstrate the integration of expressions and measures we first remove the expressions from the starting face using the expression attribute, then make its nose more hooked, enlarge the nose and finally close the eyes.

3.4   Measured Attributes

Distance and angular measures as introduced in section 2.1 are another powerful editing method, which is closer to traditional mesh editing. The difference is, that as we determine the maximum likelihood solution of the model, a local change always has global influence. So changing the length of the nose will also change the size of the nose, and if that should not happen, then the size of the nose has to be constrained by another measure and set to fixed. And the strength of our approach is that both applications are possible, depending on the use case at hand. We demonstrate in figure 4 a number of changes to the size and shape of the nose of the input face, and show that this can be combined with other classifiers by finally changing the expression to sad.

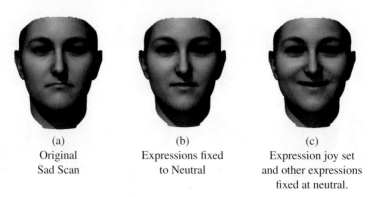

(a)	(b)	(c)
Original	Expressions fixed	Expression joy set
Sad Scan	to Neutral	and other expressions
		fixed at neutral.

Fig. 5. Our method allows the removal (b) of an expression from the input scan (a) and subsequent addition of other expressions (c)

4   Prior Work

Manipulation of facial attributes with the help of statistics learned from example faces has been introduced already by [1] in the seminal paper on 3D-MMs. They learned the direction of maximum variance of an attribute from a labelled set of faces and changed the face parameters of a given identity according to that direction. In addition to not being able to manipulate "pure" attributes, i.e. changing only the weight but not the height, this approach can neither prescribe attributes in an absolute scale, i.e. to age a person to 43 years, nor does the change of attributes depend on the starting identity – every person ages in the same way. Later work [5] of the same group which is most closely related to our approach extended this work to absolute scales and multiple attributes by constraining the generated face to be in the solution space of a linear regression from faces to attributes. Their proposal is a special case of our method.

Other authors propose to learn a function from attributes to shapes [6,7,8,9,10,11], which addresses a different problem. Only the mapping from shapes to attributes is surjective, i.e. to a single face there is always only a single value of each attribute, but not the mapping from attributes to shapes, many faces share the same attributes. Therefore, these approaches can only generate an "average" face for each attribute. The mentioned papers explore combinations of anthropometric or subjective attributes, and linear or nonlinear mappings and linear, multilinear or nonlinear models, but all of them do not constrain the solution by similarity to a starting face, as we do. To generate faces conforming to anthropometric measurements [7] fits a linear function from measurements to faces, which is correct for distance measurements, but not for angle measurements. Similarly, [10] interpolate the example faces based on their distance in measurement space with RBF interpolation, which also does not guarantee correct distance. Our method works with correct measurements, and, as stated, overcomes the injectivity problem by further constraining the mapping by distance to a starting face.

Note that while multilinear models as in [12,11] separate meaningful groups of latent factors, they do not offer meaningful parameters inside of a group (tensor mode). Because multilinear models do not have meaningful parameters [11] suggested to drive face manipulation by motion capture data of actors, and combine identity and expression tensor parameters from different measurements for expression transfer. This is also an interesting application, but it is very different from our goal of attribute and expression editing in a meaningful parameters space. The same applies to [13] where a generative body model is fitted to motion capture data.

We expect, that the change in face parameters for a change of attributes would depend on the starting face. This was addressed in [14] They learned a nonlinear age regression, and directions of change which are parallel to the gradient of the regression function were used to determine a trajectory through face space. This addresses the problem that different individuals should change differently and makes it possible to set an absolute value for an attribute. The disadvantage of that method is, that it only handles a single attribute, and does not model the covariance of multiple attributes. Changing first a persons weight also changes the height, so it is impossible to change height and weight simultaneously to a fixed value. [15] recently proposed a method to manipulate attractiveness in 2D frontal photographs. The method is similar to [14]. They also perform a regression from faces described by the distance between feature points to an attractiveness rating. The input face is then morphed such that its attractiveness rating increases. Our approach, which is also based on a regression, allows simultaneous manipulation of many attributes, enables the artist to specify a precise set of target attributes, and makes it possible to explicitly choose the attributes which should covary and which should stay fixed.

While [14] assume that the direction of the gradient of the nonlinear age regression will leave the attributes which make up the identity the same, as this corresponds to the smallest Mahalanobis distance in face space which achieves the desired change of attribute. Correspondingly [15] assume that closer faces in a euclidean distance in their parametrization of face space corresponds to the smallest change in identity. In reality, when the age of a person changes it is for example the case that the inter-eye distance stays fixed and the sex of the person does not change, measures like this are not incorporated in the distance function used in the mentioned papers. We propose to learn

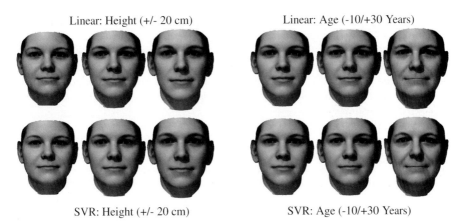

Linear: Height (+/- 20 cm)   Linear: Age (-10/+30 Years)

SVR: Height (+/- 20 cm)   SVR: Age (-10/+30 Years)

Fig. 6. For some attributes we can achieve better control with support vector regression than with linear regression. The effect is relatively subtle in this presentation, but notice e.g. the mouth shape which stays more constant and also the more similar global shape.

many attributes and decide *depending on the task at hand* which attributes should stay the same, and which are allowed to vary. So with our method it is possible to specify the understanding of what makes two faces similar by deciding on attributes which should not vary, while previous methods only assumed that a short distance in face space leaves the identity unchanged.

5 Conclusion

We presented a method to manipulate attributes of faces in generative face models, where the attribute space is learned from labelled examples. We demonstrated the method using a 3D Morphable Model. There exist an infinity of possible attributes ranging from physical values like the curvature of the ears to cultural valuations like attractiveness. Attributes are distributed nonlinearly in face space, which we addressed by learning a nonlinear regression from face space to attribute space. And attributes co-vary, when navigating attribute space it is desirable to be able to choose which should vary, and which should be fixed. For example a change of attractiveness should not change the sex of the modelled face, but might be allowed to change the curvature of the nose. We addressed this in a probabilistic framework.

Our method has three applications. It enables psychologists to generate stimuli with systematically varying attributes, it gives artists a powerful and intuitive new tool for character design based on learned statistics of real faces, and additionally, it is fun to play with.

References

1. Blanz, V., Vetter, T.: A morphable model for the synthesis of 3d faces. In: SIGGRAPH 1999, pp. 187–194. ACM Press/Addison-Wesley Publishing Co., New York (1999)
2. Ilić, S., Salzmann, M., Fua, P.: Implicit meshes for effective silhouette handling. Int. J. Comput. Vision 72, 159–178 (2007)
3. Amberg, B., Knothe, R., Vetter, T.: Expression invariant face recognition with a morphable model. In: Face and Gesture 2008 (2008)
4. Amberg, B., Romdhani, S., Vetter, T.: Optimal step nonrigid ICP algorithms for surface registration. In: IEEE Conference on Computer Vision and Pattern Recognition, 2007. CVPR 2007, pp. 1–8 (2007)
5. Blanz, V., Albrecht, I., Haber, J., Seidel, H.P.: Creating face models from vague mental images. In: Gröller, E., Szirmay-Kalos, L. (eds.) EUROGRAPHICS 2006 (EG 2006), Vienna, Austria. Computer Graphics Forum, vol. 25(3), pp. 645–654. Blackwell, Malden (2006)
6. Allen, B., Curless, B., Popović, Z.: The space of human body shapes: reconstruction and parameterization from range scans. ACM Trans. on Graphics, SIGGRAPH 2003 22, 587–594 (2003)
7. Allen, B., Curless, B., Popović, Z.: Exploring the space of human body shapes: Data-driven synthesis under anthropometric control. In: Proc. Digital Human Modeling for Design and Engineering Conference, Rochester, MI, vol. 113(1), pp. 245–248. SAE International (2004)
8. Skorupski, J., Yee, J., McCoy, J., Davis, J.: Facial type, expression, and viseme generation. In: SIGGRAPH 2007: ACM SIGGRAPH 2007 posters, vol. 34 (2007)

9. Allen, B., Curless, B., Popović, Z., Hertzmann, A.: Learning a correlated model of identity and pose-dependent body shape variation for real-time synthesis. In: SCA 2006: Proceedings of the 2006 ACM SIGGRAPH/Eurographics symposium on Computer animation, Aire-la-Ville, Switzerland, pp. 147–156. Eurographics Association, Switzerland (2006)

10. Zhang, Y., Badler, N.I.: Face modeling and editing with statistical local feature control models. International Journal of Imaging Systems and Technology 17, 341–358 (2007)

11. Vlasic, D., Brand, M., Pfister, H., Popovic, J.: Face transfer with multilinear models. ACM Trans. Graph. 24, 426–433 (2005)

12. Vasilescu, M.A.O., Terzopoulos, D.: Multilinear analysis of image ensembles: Tensorfaces. In: Proceedings of the European Conference on Computer Vision, vol. 1, pp. 447–460 (2002)

13. Anguelov, D., Srinivasan, P., Koller, D., Thrun, S., Rodgers, J., Davis, J.: Scape: shape completion and animation of people. ACM Trans. Graph. 24, 408–416 (2005)

14. Scherbaum, K., Sunkel, M., Seidel, H.P., Blanz, V.: Prediction of individual non-linear aging trajectories of faces. In: Eurographics 2007, Prague, Czech Republic. Computer Graphics Forum, vol. 26(3), pp. 285–294. Blackwell, Malden (2007)

15. Leyvand, T., Cohen-Or, D., Dror, G., Lischinski, D.: Data-driven enhancement of facial attractiveness. ACM Trans. on Graphics, SIGGRAPH 2008 27 (2008) (to appear)

Contrast Constrained Local Binary Fitting for Image Segmentation

Xiaojing Bai[1], Chunming Li[2,*], Quansen Sun[1], and Deshen Xia[1]

[1] School of Computer Science and Technology, Nanjing University of Science and Technology, Nanjing, China

[2] Vanderbilt University Institute of Imaging Science, USA
lchunming@gmail.com

Abstract. This paper presents a more robust and efficient level set method than the original Local binary fitting (LBF) model in [6] for image segmentation under a constrained energy minimization framework. Image segmentation is formulated as a problem of seeking an optimal contour and two fitting functions that best approximate local intensities on the two sides of the contour. The contribution in this paper is twofold. First, we introduce a contrast constraint on the fitting functions to effectively prevent the contour from being stuck in spurious local minima, which thereby makes our model more robust to the initialization of contour. Second, we provide an efficient narrow band implementation to greatly reduce the computational cost of the original LBF algorithm. The proposed algorithm is validated on synthetic and real images with desirable performance in the presence of intensity inhomogeneities and weak object boundaries. Comparisons with the LBF model and the piecewise smooth (PS) model demonstrate the superior performance of our model in terms of robustness, accuracy, and efficiency.

1 Introduction

Image segmentation, an important and fundamental task in computer vision and image analysis, has been extensively studied in the past decades. However, it is still a challenging problem in applications to real-world images. Difficulties in image segmentation typically arise from image noise, low contrast, and artifacts in the images. In particular, the widely used region-based methods [3,11,12,14,10] can be misled by intensity inhomogeneities, which often occur in real-world images due to spatial illumination variations or imperfections of imaging devices. Standard region-based methods aim to identify each region of interest by using a certain region descriptor, such as the mean of the intensity of a region, for image segmentation. However, it is rather difficult to define an appropriate region descriptor for images with intensity inhomogeneities.

Most of region-based models are based on the assumption of intensity homogeneity [3,9,11,14]. A typical example is the *piecewise constant (PC)* model proposed in [3]. These models are not applicable to images with intensity inhomogeneities. Sophisticated methods [4,14], such as the *PS* models [1,2,10,13,12], are able to address intensity inhomogeneities. Based on the well-know Mumford-Shah model [8], Vese and

---

* Corresponding author.

G. Bebis et al. (Eds.): ISVC 2009, Part I, LNCS 5875, pp. 886–895, 2009.

Chan [13] introduced an energy functional on a level set function ϕ and two smooth fitting functions u^+ and u^- that approximate the intensities outside and inside the zero level contour of ϕ, respectively. PS models overcome the limitation of the PC models in segmenting images with intensity inhomogeneities. However, the smoothness condition of the functions is ensured by solving a set of PDEs during the level set evolution. Therefore, the computation in such method is rather expensive. Moreover, due to the non-convexity of the underlying energy functionals, the corresponding energy minimization algorithms may converge to local minima. This renders the algorithm quite sensitive to the initialization of the contour, and the contour may be stuck in the background or foreground.

Recently, Li *et al.*[6] proposed a region-based level set method, called *local binary fitting (LBF)* model, which is able to deal with intensity inhomogeneities. The LBF model formulates image segmentation as a problem of seeking an optical contour and two spatially varying fitting functions that locally approximate the intensities on the two sides of the contour. This method is significantly faster than the PS model [5,6]. Moreover, the LBF model has desirable performance for images with weak object boundaries, and its implementation is simple and straight-forward. However, because there is no constraint on the fitting functions in the LBF model as the PS model, the corresponding energy minimization algorithms may still converge to the local minima. Therefore, the LBF model is still sensitive to the initialization of contour, and the contour may be stuck in the background or foreground. In addition, the LBF model is simply implemented by the full domain algorithm in [5,6], so that there is still plenty of room for improvement in computational efficiency.

In this paper, we propose a robust and efficient level set method for image segmentation under a constrained energy minimization framework, aiming to overcome the sensitivity to initialization and computational inefficiency of the original LBF model. Image segmentation is formulated as a problem of seeking an optimal contour and two fitting functions that best approximate local intensities on the two sides of the contour. We introduce a contrast constraint on the fitting functions to effectively prevent the contour from being stuck in spurious local minima. As a result, our proposed model, which we call a *contrast constrained local binary fitting (CCLBF)* model, is more robust to the choice of initialization than the original LBF model. This presents a great improvement over the original LBF model, which is rather sensitive to initialization. Furthermore, we provide an efficient narrow band algorithm to implement the level set evolution in the proposed model, which greatly improves the computational efficiency over the original LBF model.

2   Contrast Constrained Local Binary Fitting Model

2.1   Local Binary Fitting Model

The LBF model is based on the assumption that the intensities on the two sides of the object boundary can be approximated by two constants [5,6]. Therefore, image segmentation can be formulated as a problem of seeking an optimal contour C and two spatially varying fitting functions f_1 and f_2 that locally approximate the intensities on the two sides of the contour. This optimization problem can be formally defined to minimize the local binary fitting energy

$$\mathcal{E}_{\mathbf{x}}^{Fit}(C, f_1(\mathbf{x}), f_2(\mathbf{x})) = \lambda_1 \int_{outside(C)} K(\mathbf{x} - \mathbf{y})|I(\mathbf{y}) - f_1(\mathbf{x})|^2 d\mathbf{y} \qquad (1)$$
$$+ \lambda_2 \int_{inside(C)} K(\mathbf{x} - \mathbf{y})|I(\mathbf{y}) - f_2(\mathbf{x})|^2 d\mathbf{y},$$

where $K(\mathbf{x} - \mathbf{y})$ is a weighting function, which decreases to zero as $\mathbf{y}$ goes away from $\mathbf{x}$. The weighting function can be chosen as a truncated Gaussian kernel

$$K(\mathbf{u}) = \begin{cases} \frac{1}{a}e^{-|\mathbf{u}|^2/2\sigma^2} & \text{for } |\mathbf{u}| \le \rho \\ 0 & \text{else,} \end{cases}$$

where a is a normalization constant such that $\int K(\mathbf{u}) = 1$, σ is the standard deviation of the Gaussian function, and ρ is the radius of the neighborhood. λ_1 and λ_2 are positive constants, and $f_1(\mathbf{x})$ and $f_2(\mathbf{x})$ are two values that approximate image intensities in $outside(C)$ and $inside(C)$ respectively. The two fitting values approximate the image intensities in a region centered at the point $\mathbf{x}$, whose size can be controlled by the scale parameter σ. In this sense, the fitting energy $\mathcal{E}_{\mathbf{x}}^{Fit}$ is localized around the point $\mathbf{x}$. This local binary fitting energy should be minimized for all $\mathbf{x}$ in the image domain Ω, which can be achieved by minimizing the integral of $\mathcal{E}_{\mathbf{x}}^{Fit}$, namely $\int_{\Omega} \mathcal{E}_{\mathbf{x}}^{Fit}(C, f_1(\mathbf{x}), f_2(\mathbf{x}))d\mathbf{x}$.

By embedding the contour C as the zero level set of a function ϕ, which is referred to as a level set function, the regions $outside(C)$ and $inside(C)$ can be represented by $\{x : \phi(x) > 0\}$ and $\{x : \phi(x) < 0\}$. Therefore, the above local binary fitting energy for all $\mathbf{x}$ in image domain Ω can be expressed in a level set formulation

$$\mathcal{E}^{Fit}(\phi, f_1, f_2) = \int_{\Omega} \mathcal{E}_{\mathbf{x}}^{Fit}(\phi, f_1(\mathbf{x}), f_2(\mathbf{x}))d\mathbf{x}$$
$$= \lambda_1 \int \left(\int K_{\sigma}(\mathbf{x} - \mathbf{y})|I(\mathbf{y}) - f_1(\mathbf{x})|^2 H(\phi(\mathbf{y}))d\mathbf{y} \right)d\mathbf{x} \qquad (2)$$
$$+ \lambda_2 \int \left(\int K_{\sigma}(\mathbf{x} - \mathbf{y})|I(\mathbf{y}) - f_2(\mathbf{x})|^2 (1 - H(\phi(\mathbf{y})))d\mathbf{y} \right)d\mathbf{x},$$

where H is the Heaviside function. In addition, it is necessary to smooth the contour C by penalizing its arc length $\mathcal{L}(\phi) = \int_{\Omega} |\nabla H(\phi(\mathbf{x}))|d\mathbf{x}$, and preserve the regularity of the level set function ϕ by imposing the regularization term $\mathcal{P}(\phi) = \int \frac{1}{2}(|\nabla \phi(\mathbf{x})| - 1)^2 d\mathbf{x}$. Thus, the entire energy functional of LBF model [6] is defined as

$$\mathcal{F}(\phi, f_1, f_2) = \mathcal{E}^{Fit}(\phi, f_1, f_2) + \mu \mathcal{P}(\phi) + \nu \mathcal{L}(\phi). \qquad (3)$$

where, μ and ν are positive constants. This energy functional will be minimized to find the object boundaries.

The LBF model is able to segment images with intensity inhomogeneities as demonstrated in an experiment on a synthetic image. Given an appropriate initial contour (e.g. the dashed rectangle in Fig. 1(a)), the LBF model produces a desirable segmentation result as shown in Fig. 1(a). However, a different initial contour (e.g. the dashed rectangle in Fig. 1(b)) may lead to a spurious segmentation result as shown in Fig. 1(b), from which we can see that two segments of the contour are stuck within the background and foreground. In fact, the undesirable result in Fig. 1(b) is a local minimum of the energy $\mathcal{F}(\phi, f_1, f_2)$. This spurious result is due to the localization property of the LBF model, which allows for the fitting functions f_1 and f_2 to take different values at different locations, as long as they approximate the local intensities. For instance, at parts of the contour that are stuck in the background and foreground in Fig. 1(b), the final fitting functions f_1 and f_2 are close to each other, both approximating the local intensities there. The local binary fitting energy $\mathcal{E}_{\mathbf{x}}^{Fit}(C, f_1(\mathbf{x}), f_2(\mathbf{x}))$ in Eq. (1) is indeed minimized by such f_1 and f_2, even for a contour that is partly stuck at locations far away from the true object boundary.

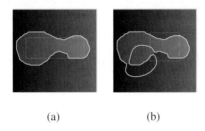

(a)   (b)

Fig. 1. Sensitivity of original LBF model to initialization. Figures (a) and (b) show different segmentation results (solid contours) with two different initial contours (dashed contours).

2.2   Contrast Constrained Local Binary Fitting Model

To overcome the above drawback of the LBF model, we impose a constraint on the fitting functions f_1 and f_2 to prevent the contour from being stuck in the background or foreground. We assume that there is a difference between the local intensities on the two sides of the object boundary. Therefore, we propose to impose a constraint on the fitting functions f_1 and f_2 based on a property of the local intensities around object boundaries — there is a contrast between the background and foreground. To reflect this property, we impose a contrast constraint $|f_2(\mathbf{x}) - f_1(\mathbf{x})| \geq c$, i.e.

$$f_2(\mathbf{x}) - f_1(\mathbf{x}) \geq c \text{ or } f_2(\mathbf{x}) - f_1(\mathbf{x}) \leq -c, \qquad (4)$$

where $c \geq 0$ is a constant, and usually set to a small number (e.g. $c = 1$). To simplify the problem without loss of generality, we consider the first case in Eq. (4), namely $f_2(\mathbf{x}) - f_1(\mathbf{x}) \geq c$, as the contrast constraint on f_1 and f_2. Therefore, we define

$$\mathcal{A} = \{(f_1, f_2) : f_2(\mathbf{x}) - f_1(\mathbf{x}) \geq c, \text{ for all } \mathbf{x} \in \Omega\},$$

Thus, we propose to solve the following constrained energy minimization problem:

$$\text{Minimize } \mathcal{F}(\phi, f_1, f_2) \text{ subject to } (f_1, f_2) \in \mathcal{A}. \qquad (5)$$

2.3   Energy Minimization

We minimize the energy functional $\mathcal{F}(\phi, f_1, f_2)$ in Eq. (3) with respect to ϕ using the standard gradient descent method by solving the gradient flow equation as follows

$$\frac{\partial \phi}{\partial t} = -\delta(\phi)(\lambda_1 e_1 - \lambda_2 e_2) + \nu\delta(\phi)\text{div}\left(\frac{\nabla\phi}{|\nabla\phi|}\right) + \mu\left(\nabla^2\phi - \text{div}\left(\frac{\nabla\phi}{|\nabla\phi|}\right)\right), \quad (6)$$

where δ is the Dirac function, and e_1 and e_2 are the functions

$$e_i(\mathbf{x}) = \int K_\sigma(\mathbf{y} - \mathbf{x})|I(\mathbf{x}) - f_i(\mathbf{y})|^2 d\mathbf{y}, \quad i = 1, 2.$$

The functions f_1 and f_2 are updated after every iteration of the level set function ϕ. The update of f_1 and f_2 is performed by solving the above contrast constrained energy minimization problem in Eq. (5) given the updated level set function ϕ.

3   Narrow Band Algorithm for Contrast Constrained Local Binary Fitting Model

In this section, we present a narrow band level set evolution algorithm for the above proposed CCLBF model. A pixel $P = (i, j)$ is a *zero crossing pixel* of a function ϕ on the grid, if either $\phi_{i-1,j}$ and $\phi_{i+1,j}$, or $\phi_{i,j-1}$ and $\phi_{i,j+1}$ are of opposite signs. Given a set of zero crossing pixels $\mathscr{Z}$ of the function ϕ, we define the corresponding narrow band as a neighborhood of $\mathscr{Z}$ by

$$\mathscr{B} = \bigcup_{\mathbf{x}\in\mathscr{Z}} \mathscr{N}_r(\mathbf{x}),$$

where $\mathscr{N}_r(\mathbf{x})$ is a $(2r + 1) \times (2r + 1)$ square block centered at the pixel $\mathbf{x}$. This parameter r can be chosen as the minimum value $r = 1$ to reduce the computation. The level set function is updated only at the points on the narrow band $\mathscr{B}$ during the level set evolution.

Our narrow band algorithm for the CCLBF model is described below:

Step 1.   **Initialization.** Set $k = 0$ and initialize ϕ to an initial function ϕ_0. Then, build the initial narrow band $\mathscr{B}^0 = \bigcup_{\mathbf{x}\in\mathscr{Z}^0} \mathscr{N}_r(\mathbf{x})$ where $\mathscr{Z}^0$ is the set of the zero crossing pixels of ϕ^0. Go to Step 4.

Step 2.   **Update narrow band.** Determine the set of all the zero crossing pixels of ϕ^k on $\mathscr{B}^{k-1}$, denoted by $\mathscr{Z}^k$. Then, update the narrow band by setting $\mathscr{B}^k = \bigcup_{\mathbf{x}\in\mathscr{Z}^k} \mathscr{N}_r(\mathbf{x})$.

Step 3.   **Assign values to new pixels on narrow band.** For every pixel $\mathbf{x}$ in $\mathscr{B}^k$ but was not in $\mathscr{B}^{k-1}$, set $\phi^k(\mathbf{x})$ to s if $\phi^k(\mathbf{x}) > 0$, or else set $\phi^k(\mathbf{x})$ to $-s$, where $s = r + 1$.

Step 4.   **Update level set function at all the points on narrow band.** Update ϕ^{k+1} for all pixels $\mathbf{x} \in \mathscr{B}^k$ according to Eq. (6).

Step 5.   Determine the termination of iteration. If either the L^2 distance between ϕ^{k+1} and ϕ^k is less than a prescribed threshold $\epsilon > 0$ or k exceeds a prescribed maximum number of iteration, then stop the iteration, otherwise, go to Step 2.

To illustrate the above described narrow band algorithm, we show an application to a synthetic image in Fig. 2. The evolution of the zero level contour is depicted in the first row, and the corresponding narrow band is shown as white band in the second row. For $r = 1$, there are four pixels across the narrow band according to the above definition of narrow band. The computation on the narrow band with such a small width is much more efficient than the computation on the full domain.

Fig. 2. Illustration of curve evolution and the corresponding narrow band. Row 1: Original image with the initial contour and the contours at iterations 100, 250, 440; Row 2: Narrow band (the white band) with zero level contour (the black contour) corresponding to Row 1.

The level set regularization is important in the CCLBF model. In traditional level set methods, level set functions can be degraded during the evolution by developing steep and flat shapes and noisy features [7]. To show the necessity of level set regularization in the proposed model, we remove the level set regularization term in our model (by setting $\mu = 0$), and apply the model to the synthetic image in Fig. 2. The corresponding results are shown in Fig. 3(a), from which we can clearly see the severe irregularity of the zero level set, with many small contours densely scattered in the image as shown in its close-up view of Fig. 3(b). Such irregularity is a result of the level set evolution without regularization. The densely scattered small contours significantly increase the size of narrow band as shown in Fig. 3(c) and its close-up view of Fig. 3(d), which introduces additional computation in each iteration. In the CCLBF model, the zero level contour and the narrow band are maintained in a regular shape throughout the level set evolution as shown in Fig. 2. The above experiments show the necessity of the level set regularization in the proposed CCLBF model.

4   Experimental Results

We implement the proposed CCLBF model using the above described narrow band algorithm. Our model has been validated on both synthetic and real images from different

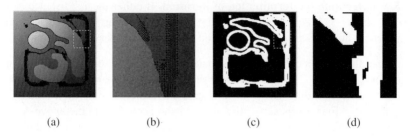

(a)   (b)   (c)   (d)

Fig. 3. Zero level set and the corresponding narrow band of the CCLBF model without regularization. (a) Zero level contour at 440 iteration; (b) Close-up view of the squared region in (a); (c) Narrow band corresponding to (a); (d) Close-up view of the squared region in (c).

modalities. To demonstrate the performance of our model in the presence of intensity inhomogeneities, we tested it on images with intensity inhomogeneities in all the experiments in this paper. We use the following parameters for all images in this paper: $\lambda_1 = \lambda_2 = 1.0$, $\mu = 1.0$, $r = 1$, and time step $\Delta t = 0.1$. We note that, the segmentation results are mainly influenced by the scale parameter σ and the coefficient ν of the arc length term. The value of σ should be chosen appropriately according to the degree of the intensity inhomogeneity. For images with severe intensity inhomogeneity, a relatively smaller σ provides more localized computation, so that the object boundaries can be located more precisely. Parameter ν should be tuned according to the level of image noise. For images with high level noise, a relatively larger ν can be used to avoid creating small scattered contours caused by noise. However, the CCLBF model is in general not sensitive to the variation of parameters.

4.1 Results of Contrast Constrained Local Binary Model

We first test the proposed CCLBF model on five real images shown in Fig. 4. These images exhibit significant intensity inhomogeneities caused by spatial illumination variations. We use $\nu = 0.01 \times 255 \times 255$ and $\sigma = 3$ for all these images. Due to the intensity inhomogeneities, there are overlaps between the distribution of the intensities in the backgrounds and foregrounds. In addition, the contrast between parts of the backgrounds and foregrounds are quite low, which introduces additional difficulty in the segmentation of these images. Despite the above difficulties, our method is able to provide desirable results for these challenging images as shown in the second row of Fig. 4.

The contrast constraint imposed on the fitting functions of our CCLBF model ameliorates the problem of the solutions falling into spurious local minima, which makes our model robust to the initialization of the contour. For example, we applied the CCLBF model to a image of a T-shaped object and an X-ray vessel image shown in Fig. 5. Both images exhibit obvious intensity inhomogeneities, and part of the vessel boundaries are quite weak in the X-ray image. We use the parameters $\nu = 0.006 \times 255 \times 255$ and $\sigma = 3$ in this experiment. For each image, we apply our model with five different initializations of contour (the dashed circles), which are shown together with the resulted

Fig. 4. Segmentation results of the CCLBF model on five real images. Row 1: Original images with initializations of contour. Row 2: Segmentation results of the CCLBF model.

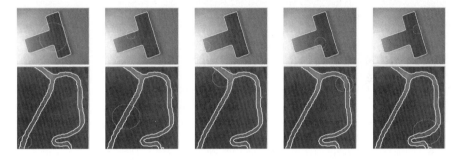

Fig. 5. Results of the CCLBF model (the solid contours) for real images with different initializations (the dashed circles)

contours (the solid contours) in Fig. 5. Obviously, despite significant intensity inhomogeneities and weak object boundaries, the results are all in good agreement with the true object boundaries. This demonstrates the robustness of our model to the choice of initial conditions.

4.2   Comparisons with Local Binary Fitting Model and Piecewise Smooth Model

To demonstrate the advantage of our proposed CCLBF model in terms of accuracy evaluation and computational efficiency over the LBF and PS models [6,13], we apply the three models on a group of synthetic images to extract the object boundaries. For these synthetic images, we have the ground truth that can be used to quantitatively evaluate the accuracy of the results. These 15 images are generated by smoothing an ideal binary image of size 128×128, and then adding different intensity inhomogeneities and different levels of noise. Fig. 6 shows five of these images as examples, with the corresponding results of the CCLBF, LBF and PS models in Rows 2, 3 and 4 respectively. We use the same initial contour (the circle in Row 1), and parameters $\nu = 0.006 \times 255 \times 255$ and $\sigma = 4$ for the three models. It is obvious that the CCLBF model produces more reliable segmentation results than the LBF and PS models. For both the LBF and PS models, part of contours are stuck in the background and the foreground as shown in the first column of Rows 3 and 4. By contrast, the results of the CCLBF model are all

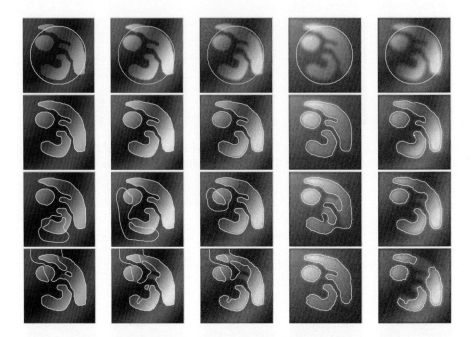

Fig. 6. Comparisons of the CCLBF, LBF and PS models. Row 1: Original images with initial contours; Rows 2-4: Results of the CCLBF, LBF, and PS models.

in good agreement with the true object boundaries. Particularly, no contour is stuck in the background or foreground.

Meanwhile, the narrow band implementation of the CCLBF model is much more efficient than the LBF and PS models, which is demonstrated by the CPU times consumed by the three models for the 15 images. Our model is about 10 times faster than the LBF model, and 200 to 470 times faster than the PS model. Take the 12th image shown in the middle of Fig. 6 as example, the CPU times of the CCLBF, LBF, and PS models are 0.42, 5.13, 174.18 seconds, respectively. The CPU times in these experiments are obtained by running Matlab programs on a Lenovo ThinkPad notebook with Intel (R) Core (TM)2 Duo CPU, 2.4 GHz, 2GB RAM, with Matlab 7.4 on Windows Vista.

The above comparisons of the CCLBF, LBF, and PS models demonstrate that the CLBF model possesses much better performance than the other two models for images under difficult situations, such as in the presence of intensity inhomogeneities and weak object boundaries. Furthermore, the narrow band implementation of the CCLBF model is much computational efficient than the other two models.

5   Conclusion

We have presented a *contrast constrained local binary fitting (CCLBF)* model for image segmentation, which is able to deal with intensity inhomogeneities. This model is more robust to initialization due to the introduced contrast constraint in our model. Furthermore, we provide an efficient narrow band implementation to greatly reduce the computational cost. Our model has been validated on synthetic and real images with desirable performance in the presence of intensity inhomogeneities and weak object boundaries. Comparisons with the LBF model and the PS model demonstrate the superior performance of our model in terms of robustness, accuracy, and efficiency.

References

1. An, J., Rousson, M., Xu, C.: Γ-convergence approximation to piecewise smooth medical image segmentation. In: Ayache, N., Ourselin, S., Maeder, A. (eds.) MICCAI 2007, Part II. LNCS, vol. 4792, pp. 495–502. Springer, Heidelberg (2007)
2. Brox, T., Cremers, D.: On the statistical interpretation of the piecewise smooth Mumford-Shah functional. In: Sgallari, F., Murli, A., Paragios, N. (eds.) SSVM 2007. LNCS, vol. 4485, pp. 203–213. Springer, Heidelberg (2007)
3. Chan, T., Vese, L.: Active contours without edges. IEEE Trans. Imag. Proc. 10(2), 266–277 (2001)
4. Chen, Y., Tagare, H., Thiruvenkadam, S., Huang, F., Wilson, D., Gopinath, K., Briggs, R., Geiser, E.: Using prior shapes in geometric active contours in a variational framework. Int'l J. Comp. Vis. 50(3), 315–328 (2002)
5. Li, C., Kao, C., Gore, J., Ding, Z.: Implicit active contours driven by local binary fitting energy. In: Proceedings of IEEE Conference on Computer Vision and Pattern Recognition (CVPR), Washington, DC, USA, pp. 1–7. IEEE Computer Society, Los Alamitos (2007)
6. Li, C., Kao, C., Gore, J.C., Ding, Z.: Minimization of region-scalable fitting energy for image segmentation. IEEE Trans. Imag. Proc. 17(10), 1940–1949 (2008)
7. Li, C., Xu, C., Gui, C., Fox, M.D.: Distance regularized level set evolution and applications to image segmentation. IEEE Trans. Imag. Proc. (to appear)
8. Mumford, D., Shah, J.: Optimal approximations by piecewise smooth functions and associated variational problems. Commun. Pure Appl. Math. 42(5), 577–685 (1989)
9. Paragios, N., Mellina-Gottardo, O., Ramesh, V.: Gradient vector flow fast geometric active contours. IEEE Trans. Patt. Anal. Mach. Intell. 26(3), 402–407 (2004)
10. Piovano, J., Rousson, M., Papadopoulo, T.: Efficient segmentation of piecewise smooth images. In: Sgallari, F., Murli, A., Paragios, N. (eds.) SSVM 2007. LNCS, vol. 4485, pp. 709–720. Springer, Heidelberg (2007)
11. Ronfard, R.: Region-based strategies for active contour models. Int'l J. Comp. Vis. 13(2), 229–251 (1994)
12. Tsai, A., Yezzi, A., Willsky, A.S.: Curve evolution implementation of the Mumford-Shah functional for image segmentation, denoising, interpolation, and magnification. IEEE Trans. Imag. Proc. 10(8), 1169–1186 (2001)
13. Vese, L., Chan, T.: A multiphase level set framework for image segmentation using the Mumford and Shah model. Int'l J. Comp. Vis. 50(3), 271–293 (2002)
14. Zhu, S.-C., Yuille, A.: Region competition: Unifying snakes, region growing, and Bayes/MDL for multiband image segmentation. IEEE Trans. Patt. Anal. Mach. Intell. 18(9), 884–900 (1996)

Modeling and Rendering
Physically-Based Wood Combustion

Roderick M. Riensche[1] and Robert R. Lewis[2]

[1] Pacific Northwest National Laboratory, National Security Directorate
rmr@pnl.gov
[2] Washington State University, School of EECS
bobl@tricity.wsu.edu

Abstract. Rendering of wood combustion has received some attention recently, but prior work has not incorporated effects of internal wood properties such as density variation (i.e. "grain") and pre-combustion processes such as drying. In this paper we present the status and results of our extensions to prior work of others in the graphics community, leveraging insights and modeling results from physical science fields to simulate the influence of wood grain and moisture on the speed and pattern of decomposition of a burning wooden object.

1 Introduction

Natural phenomena provide some of the most compelling challenges in computer graphics, producing distinct visuals that can elicit strong reactions in human observers. In computer graphics applications, visually accurate representations of natural phenomena can help to suspend disbelief for the viewer, making a rendered scene appear real and helping the viewer understand and relate to the scene being depicted. Ultimately, graphic artists producing animation sequences (e.g., for movies and video games) seek to produce the same emotional reactions that a person might experience were they to witness an image of a rendered scene in real life.

Among these natural phenomena, fire is a particularly interesting case. The process of combustion is extremely complicated, making it a challenging subject for modeling and simulation. At the same time, the visual effects are immediately recognizable and can provide powerful reactions, as humans readily recognize the inherent power of one of nature's most destructive forces.

There are other applications in which simulation of the effects of combustion is desirable. Fire safety researchers work to understand how building materials will degrade when burned, in order to design appropriate safety margins in structures and benefit from visualizing that process.

While there has been much research into rendering flames, showing the underlying degradation of the material being burned has received less attention. Here, we first provide an overview of the physical processes at work during the combustion process and some of the fundamental simulation and graphics advances

G. Bebis et al. (Eds.): ISVC 2009, Part I, LNCS 5875, pp. 896–905, 2009.

that may be used to model these processes. We then review previous work on simulation and visualization of combustion. Finally, we present the early status of our ongoing efforts in extending work that has been done to model combustion effects, with an ultimate goal of improving realism in rendering of the physical degradation of wooden and other fibrous objects as they undergo combustion.

2   Background

A good first step in rendering any natural phenomenon is to understand the physical processes at work. We may then produce simulation and visualization algorithms that correspond to reality.

2.1   The Physics of Combustion

Combustion of wood (and similar fuels) is a particularly complex series of physical reactions and processes. While we naturally understand that a piece of wood will burn when subjected to sufficient heat, in reality the wood is undergoing a breakdown into component materials, which provide the necessary catalysts for combustion. There are two distinct precursors to combustion: *drying* and *pyrolysis*.

Drying is commonly understood. Water entrained in a solid fuel such as wood will begin to evaporate as the temperature of the wood exceeds the boiling point of water (100 °C under normal conditions). The resulting water vapor may then migrate through, and escape from, the porous fuel.

The second, pyrolysis, is perhaps less familiar and is the key process that enables combustion to occur. When wood reaches a sufficiently high temperature (e.g., 300 °C), chemical reactions cause the solid material to decompose into a combination of gases (volatile and nonvolatile), tar, and char (the leftover solids made primarily of carbon)[1]. When the released volatile gases encounter oxygen, the resulting oxidation can lead to flaming combustion[2].

Accompanying these physical changes to the structure of wood are corresponding changes to the physical environment within the wood. The chemical reactions involved generate additional heat, raising the interior temperature. In addition, the release of gaseous products causes an increase in internal pressure, which in turn causes migration of the gases through the porous solid[3]. Some of the gaseous products will migrate into cooler regions of the solid, where they may recondense[4].

2.2   Level Sets

The temperature-driven nature of the drying and pyrolysis phenomena produce moving fronts within the solid. These processes in turn produce physical changes to the topology of the wood object. Level set methods[5] provide widely used techniques for tracking moving geometric interfaces. In level set methods, an implicit surface is represented by a grid of values, with a particular isovalue (typically zero) defining the points at which the surface lies.

2.3   Physical Characteristics of Wood

Within a solid piece of wood, there are variations in physical characteristics that result from the seasonal growth of the wood. In particular, concentric rings of higher density are formed due to the differing growth processes in "early wood" and "late wood" [6].

These variations affect the physical processes described above, for example, by causing variation in the rates at which gaseous products migrate through the solid. In the densely packed late wood regions (the tree rings), the reduction in air space through which gas may migrate has a significant effect on the overall combustion process. Empirical studies of burning rates show that thermal conductivity is considerably greater "across" the grain of the wood (e.g., in a direction parallel to the axis about which the wood grew) than "along" the grain [3].

3   Previous Work

Previous work relevant to the problem at hand falls into two distinct categories: mathematical models of combustion and rendering of combustion effects. In both cases, advances in computing capabilities have enabled progressively greater levels of detail to be considered.

3.1   Mathematical Models of Combustion

Mathematical models of wood combustion are prevalent in fields related to structural engineering[3,4] and energy production[7]. These models simulate the underlying physical processes of combustion in varying (but generally high) levels of detail. The mathematical formulations of these models are used to predict how fast wooden bodies will burn, including features such as char depth and its rate of progression, and the associated structural weakening of the wooden body. In the case of wood combustion for energy production, models generally deal with small particles of wood heated in furnace conditions.

Typically these models are expressed as systems of equations, and the only visual communications of results are graphs showing the evolution of particular characteristics (e.g., temperature profile, pyrolysis depth, etc.) over time under varying heating conditions.

Summary review papers dealing with mathematical combustion models provide a good source of information for the interested reader to learn more about the history and state-of-the-art in pyrolysis models.[1]

3.2   Rendering Combustion

In the field of computer graphics, the emphasis is on translating the combustion process into a visual representation. Models employed in the rendering of combustion are less detailed when considering the internal physical processes, as the desired end result is a graphic or animation that looks real, but need not predict the decomposition of a burning solid with the same degree of accuracy as the aforementioned structural and energy models.

As the most immediate visual indication of the burning process is the presence of flames, it is not surprising that early work in rendering combustion focused largely on depictions of flames and their interactions with their encompassing environments. More recently, there has been pioneering work by Melek and Keyser on modeling the underlying decomposition of a burning solid in addition to the rendering of flames[8,9]. Melek and Keyser also worked on providing visual representations of deformations that result from combustion[10]. In this previous work on rendering of combustion, the focus is largely on techniques that may be used by animators, requiring some degree of control by an animator to guide or shape the processes before allowing the simulation to generate animation frames. Losasso et al. rendered burning of solids but with a focus on thin shells rather than volumetric solids[11].

4   Synthesis

In our work to extend the state-of-the-art in rendering of combustion, a primary goal is to increase realism. While previous work has incorporated topological changes to the overall shape of a wood object undergoing combustion[8], internal physical characteristics of the wood were not used to influence the combustion process (although the idea of incorporating variations of thermal conductivity in nonuniform material was posited). Subsequently, follow-on work to simulate deformation of objects under combustion[10] has required simulation-specific definition of proxy objects that respond to environmental changes according to rules set up by an animator.

Our motivation is to provide graphical simulations in which internal physical characteristics of the wood influence combustion effects with a minimum of simulation-specific user- or programmer-driven setup. Ultimately, we envision a system in which complicated secondary processes—such as the weakening and eventual breakage of a wooden object—may be simulated and represented based only on the initial geometry of the wooden object and initial ignition conditions.

As noted in previous sections, while the effects of internal wood characteristics on wood combustion have not been considered in computer graphics, they have received considerable attention in fire safety and engineering fields. Our approach is therefore to synthesize results from these disparate yet overlapping fields.

4.1   Reconstructing Prior Work

We have chosen to use the work of Melek and Keyser[8] as a starting point. For the sake of familiarity, we provide the following summary of the portions of the method which we have used (with some slight modifications), and refer the reader to that work for a complete description.

- **Volumetric grid:** Object properties (amount V and temperature T of solid fuel) and a signed distance field ϕ representing the surface level set are stored on voxel grids.
- **Heat transfer:** Heat is transferred to the solid object from the surrounding environment. Here we should note that while [8] couples the solid simulation with a surrounding fluid simulation and transfers heat between the two, we use a simpler method (as we are only concerning ourselves with the solid behavior at this point). In our case, we allow the introduction of heat by specification of any of the solid grid cells as ignition points, and define a constant value H_i which is the amount by which we increase the temperature of a solid cell at each time step if the cell is in an ignition state. Similarly, we define a value H_p as a per-time-step incremental heat increase to be introduced when a cell is in a pyrolysis state. As these are gross simplifications of the true underlying processes, we also institute a cap on the solid temperature to prevent runaway temperature increases. It should be noted that, since we do not yet track heat outside the solid in our current implementation, the resultant effects on flame spread at the surface of the model are not treated.
- **Heat diffusion:** Heat is diffused through the solid object at a rate based on a constant k which encapsulates a number of physical properties (e.g., density, specific heat) into a thermal conductivity constant, using a diffusion solver defined by Stam.[12]
- **Decomposition:** When T surpasses a pyrolysis threshold, the cell begins to decompose, resulting in a reduction in solid fuel amount V. Constant parameters are used to control the rate and strength (or completeness) of decomposition. The change in V drives a change in ϕ (according to one of two pyrolysis rate formulae; in our work we have used Constant Rate Pyrolysis as described in [8]).
- **Polygonization:** Tetrahedral decomposition[13] of the signed distance field is used to generate an isosurface for rendering of the solid object.

4.2   Extension 1: Wood Grain

The first extension we introduce is incorporation of the appearance and effects of wood grain. We model wood grain as a Boolean value G stored on a voxel grid, such that a value of true indicates that a given cell within the grid is part of the grain.

Our wood grain generation routine is fairly simple. We assume that the object in question has been cut from a solid block of wood that came from a cylindrical log and provide control over the position and orientation of the resulting wood grain relative to the carved object. We allow specification of two defining parameters: a point P_c that lies on the central axis of the log and a unit vector $\widehat{V_a}$, the axial direction of the log. This axis need not lie within the simulation space. To assign G for each point P on the grid, we find the shortest distance from P to the axis defined by P_c and $\widehat{V_a}$. We then add a scaled Perlin noise function [14], (as Perlin himself did to avoid creating unnaturally perfectly concentric rings) and call the result the radial function $R(P)$:

$$R(P) = \left| (P - P_c) - \left((P - P_c) \cdot \widehat{V_a} \right) \widehat{V_a} \right| + f_n N(P) \tag{1}$$

where $N(P)$ is a Perlin noise function and f_n is a scale factor, typically 0.1, to indicate the degree of modulation.

Given $R(P)$, we use a simple test such as

$$G = \left\{ \frac{R(P)}{r_g} \right\} < f_g \tag{2}$$

where $\{x\}$ is the fractional part of x, r_g is the "grain spacing" an (approximate) spacing between growth rings, and f_g is a "grain fraction" that controls how much grain is in the wood. If we take $f_g = \frac{1}{4}$, we specify that the grain rings will be one quarter of the thickness of the non-grain rings that lie between them.

We apply the values of G in two ways. First, we define a 3D texture as a color modulator, with a darker color if G is false and a lighter one if it is true. This will provide the appearance of wood grain in our solid, but does not in itself affect the underlying physical characteristics of the simulation.

The second application of G is used to modify the behavior of the underlying simulation. Since the darker areas of wood in the rings are the result of higher density, they tend to inhibit diffusion processes within the wood. We model this by selecting an alternate value, k_g, for the diffusion rate constant k. While we do not yet calculate this alternate value based on all possible physical properties— which would vary based on a number of factors such as wood species and age— we alter the thermal conductivity of the grain material based on results from combustion literature. Consistent with [3], we choose k_g to be equal to $k/4$.

A limiting factor that must be considered when applying this method is the coupling of visual and physical scales. The slower heat diffusion within the grain material causes the appearance of ridges in the burned wood. In order for these to align with the visual representation of the grain (the 3D texture), the solid properties V and T must be defined at the same resolution as G, which needs to be comparatively higher in order to display detailed wood grain.

4.3   Extension 2: Moisture

The second extension of physical characteristics is the consideration of a moisture level. We track as a fractional moisture content M defined over the same grid as the other solid properties. We also define a water vapor content variable v on the same grid for use in tracking evaporated water within the solid wood. Wood will absorb water up to a saturation point. For our work we use a saturation point of M_{sat} equal to 0.3 [15].

For the visual representation of moisture, we simply darken the model by applying an ad-hoc scalar multiplier d_m to each of the color components:

$$d_m = 1 - 0.4 \frac{M}{M_{sat}} \tag{3}$$

This entrained moisture has significant effects on precombustion and combustion processes. de Souza Costa and Sandberg describe formulas for thermal conductivity showing higher conductivity for moist wood [16]. From these formulas we derive an approximate scaling factor

$$k_{mod} = 4.03 + 6.15M \tag{4}$$

which scales k when $M > 0$.

As T exceeds $100\,^{\circ}\text{C}$, water begins to evaporate. We model this process by defining an arbitrary constant per-time-step evaporation fraction R_e. When T in a cell exceeds the evaporation threshold, we define a conversion of a portion of the water to vapor as

$$\Delta v_e = VMR_e \tag{5}$$

This value is added to the vapor amount v, and decreases M by

$$M \to M - \frac{\Delta v_e}{V} \tag{6}$$

To account for the heat of vaporization, we apply a rough estimation by imposing an energy sink such that T for any cell that is actively undergoing evaporation is limited to a maximum temperature T_m. We note that beyond this approximated energy sink, we have not yet attempted to model secondary temperature diffusion effects that would result from migration of heated steam through the solid material.

This evaporated water vapor is then able to diffuse through the porous solid. While the actual driving factors are derived from saturation pressure and partial pressure of water vapor [7], in our case we are not modeling internal pressure. We therefore use a similar diffusion process as for heat diffusion, defining a diffusion rate k_v. As with heat diffusion, this rate is modified by existing moisture content in a cell, although in the opposite fashion: whereas the heat diffusion rate is increased by moisture content, diffusion of gaseous products through the wood should be slowed by moisture, as adsorbed moisture will occupy some of what would otherwise be air space. We model this as a simple linear relationship, and assume that there is no modification to k_v in dry wood, and a complete blockage of vapor diffusion in saturated wood, by

$$k_v = \begin{cases} 0 & \text{if } M > M_{sat} \\ (1 - \frac{M}{M_{sat}})k_v & \text{if } M \leq M_{sat} \end{cases} \tag{7}$$

As water vapor migrates through the solid, some will recondense as it cools [4,7]. We account for this condensation in much the same way as we handled evaporation, except that this process occurs in cells where T is less than a condensation threshold, the change in vapor is negative and is driven by an arbitrary condensation fraction R_c, and the change in cell moisture is positive, e.g.:

$$\Delta v_c = -vR_c \tag{8}$$

$$M \to M - \frac{\Delta v_c}{V} \tag{9}$$

4.4   Extension 3: Arbitrary Distance Field

In order to facilitate use of a variety of models, we handle three distinct model import cases:

1. Models may be specified as a set of polygons (e.g., in a PLY file [17]). We find the minimum distance to any polygon in the set from each point in our voxel grid, assigning a positive sign if the point lies outside the object and negative if it lies inside.

   Note that these polygons are used only to initialize the distance field, and not for rendering. Polygons for rendering are generated at each time step by tetrahedral decomposition.

2. Models may be specified by direct population of the signed distance field on the grid. For example, shapes that can be described by an implicit formula can be represented directly.

3. Models may be loaded from our cached signed distance field files.

5   Results

Figure 1 shows the decomposition of a pyramid under two cases: The three images on the left show decomposition with constant k, and were generated using our implementation of the baseline method [8]. The images on the right show decomposition at the same time steps, applying our extension of variable k_g for grain cells.

Figure 2 shows two views of a sphere, half of which is soaked (but not saturated) with $M = 0.2$. The image on the left shows an ignition point that begins in the dry portion, while the right shows an ignition point on the other side of the sphere beginning in a wet portion after the same number of time steps. The most readily apparent effect of moisture is an overall slowing of the burning process, as shown by the larger volume of decomposition in the dry portion. A secondary effect can be observed by looking at the edge of the burning front, and in particular the small region of material surrounding the decomposed area. In this region, the temperature of the material has increased (indicated by lighter color) but the wood has not yet begun to decompose: The presence of moisture in a cell causes a delay in the start of pyrolysis while the moisture is evaporated.

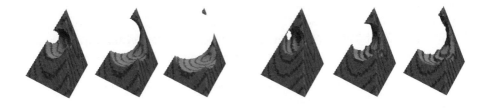

Fig. 1. Burning of a pyramid, ignoring grain (left), with grain (right)

Fig. 2. Half-soaked sphere, combustion started at two points (one dry, one wet)

Fig. 3. Burning of an arbitrary model

Figure 3 demonstrates burning of an arbitrary object, in this case loaded from a PLY file[18].

6   Conclusions and Future Work

In this paper we have demonstrated extensions to prior combustion simulation and rendering work to improve visual realism. Through these extensions, we can simulate the effects of wood grain orientation and moisture on an arbitrary burning wooden object, with no simulation-specific tuning of control parameters. There are a number of potential areas that we may explore in future work:

- Replacing arbitrary constants with physically derived values
- Modeling weakening of wood as it burns
- Handling of deep pyrolysis (in solid cells not exposed to air)
- Decoupling of grain and solid grid resolutions
- Defining alternative solid representations to replace the voxel grid.

References

1. Moghtaderi, B.: The state-of-the-art in pyrolysis modelling of lignocellulosic solid fuels. Fire and Materials 30, 1–34 (2006)
2. Di Blasi, C.: Modeling and simulation of combustion processes of charring and non-charring solid fuels. Progress in Energy and Combustion Science 19, 71–104 (1993)
3. Spearpoint, M., Quintiere, J.: Predicting the burning of wood using an integral model. Combustion and Flame 123, 308–325 (2000)
4. Janssens, M.L.: Modeling of the thermal degradation of structural wood members exposed to fire. Fire and Materials 28, 199–207 (2004)
5. Osher, S., Fedkiw, R., Piechor, K.: Level set methods and dynamic implicit surfaces. Applied Mechanics Reviews 57, B15–B15 (2004)
6. Plomion, C., Leprovost, G., Stokes, A.: Wood formation in trees. Plant Physiology 127, 1513–1523 (2001)
7. Bryden, K.M., Ragland, K.W., Rutland, C.J.: Modeling thermally thick pyrolysis of wood. Biomass and Bioenergy 22, 41–53 (2002)
8. Melek, Z., Keyser, J.: Multi-representation interaction for physically based modeling. In: SPM 2005: Proceedings of the 2005 ACM symposium on Solid and physical modeling, pp. 187–196. ACM, New York (2005)
9. Melek, Z., Keyser, J.: Interactive simulation of burning objects. In: PG 2003: Proceedings of the 11th Pacific Conference on Computer Graphics and Applications, Washington, DC, USA, p. 462. IEEE Computer Society, Los Alamitos (2003)
10. Melek, Z., Keyser, J.: Driving object deformations from internal physical processes. In: SPM 2007: Proceedings of the 2007 ACM symposium on Solid and physical modeling, pp. 51–59. ACM, New York (2007)
11. Losasso, F., Irving, G., Guendelman, E., Fedkiw, R.: Melting and burning solids into liquids and gases. IEEE Transactions on Visualization and Computer Graphics 12, 343–352 (2006)
12. Stam, J.: Real-time fluid dynamics for games. In: Proceedings of the Game Developer Conference, vol. 18 (2003)
13. Bloomenthal, J.: An implicit surface polygonizer. In: Graphics Gems IV, pp. 324–349. Academic Press, London (1994)
14. Perlin, K., Hoffert, E.: Hypertexture. ACM SIGGRAPH Computer Graphics 23, 253–262 (1989)
15. Bridgwater, A.V., Boocock, D.G.B.: Developments in thermochemical biomass conversion. Springer, Heidelberg (1997)
16. de Souza Costa, F., Sandberg, D.: Mathematical model of a smoldering log. Combustion and Flame 139, 227–238 (2004)
17. Anonymous: The Stanford 3D Scanning Repository (2009),
    http://graphics.stanford.edu/data/3Dscanrep
18. Anonymous: Aim@Shape Project (2004),
    http://shapes.aimatshape.net/view.php?id=22

A Unifying View of Contour Length Bias Correction

Christina Pavlopoulou and Stella X. Yu

Computer Science Department
Boston College, Chestnut Hill, MA 02469, USA
{pavlo,syu}@cs.bc.edu

Abstract. Original active contour formulations may become ill-posed especially for boundaries characterized by prominent features. Attempts to yield well-posed formulations lead to bias towards short contours. We provide a framework to unify existing bias correcting energy methods and propose a novel local bias correcting scheme similar to non-maximum suppression. Our method can be seen as an approximation to a well-known algorithm that transforms a graph with positive and negative weights to a graph with only positive weights while preserving the shortest paths among the nodes.

1 Introduction

One of the most well-known energy criteria for modeling and extracting object boundaries is that of Snakes, initially proposed in [1]:

$$E[C(s)] = \int_{C(s)} \frac{1}{2}(\alpha|C'(s)|^2 + \beta|C''(s)|^2)ds - \lambda \int_{C(s)} \|\nabla I\|ds \qquad (1)$$

$C(s)$ denotes the contour parametrized by s. The first two terms favor smooth contours, whereas the third favors contours adhering to prominent image features like strong discontinuities. The above energy has no intrinsic preference towards short boundaries, however it may become ill-posed. Good boundary segments receive negative cost and the minimum of the objective may become $-\infty$. Past approaches that attempted to correct the formulation ([2,3,4]) led to criteria strongly biased towards short segments. An example is shown in Figure 1. Given two points on the object boundary, the criterion in [2] will extract the shortest possible curve instead of the actual boundary.

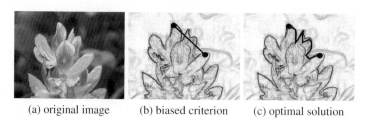

| (a) original image | (b) biased criterion | (c) optimal solution |

Fig. 1. Traditional energy criteria suffer from bias towards segments of short length. Given two points on the object boundary, the criterion of [2] will produce a straight line (shown in (b)). The desired boundary is shown in (c).

G. Bebis et al. (Eds.): ISVC 2009, Part I, LNCS 5875, pp. 906–913, 2009.

We offer a novel interpretation of the bias problem and introduce a framework for correcting it. Our framework unifies existing approaches like min ratio cycles [5], piecewise extension of the contour [6], non maximum-suppression [7], and our probabilistic formulation in [8].

The length bias is a result of converting Criterion 1 to positive by adding a large constant. Such a transformation leads to a well-posed functional, however the minima are not preserved. The new optimum solution is strongly biased towards short boundaries.

To remove the bias we turn to a discrete representation of Criterion 1. We represent the image with a graph where each node corresponds to a pixel and neighboring pixels are connected. The edge weights are derived from the biased criterion that is, Equation 1 plus constant, and are positive. The goal is to find the quantity α to remove from the weights so that they remain positive and the bias is eliminated. We show that earlier bias elimination approaches follow this framework and provide different choices regarding α. We additionally propose a local bias correction scheme, which is an approximation of a well-known algorithm of converting a graph with positive and negative weights to a graph with only positive weights while preserving the shortest paths among nodes.

The problematic nature of Functional 1 has been recognized early on and some of the problems consistently appearing in the literature include: The contour fails to latch to prominent image discontinuities and shrinks to a point. The contour produced is often too smooth and can not model geometrically complex boundaries. Self-intersecting contours are allowed and cannot be easily avoided. These problems have been mostly attributed to either the suboptimal nature of the optimization method, or the parametric form of the energy functional. Rarely have they been linked to the bias introduced when the energy criterion is converted to positive.

Earlier approaches required initialization of the contour very close to the actual boundary [1,9]. The intelligent scissors method described in [2,3,4] provided a novel way for the user to guide the delineation process. Usually, these approaches require a lot of user interaction to delineate the boundary. Level set methods [10] employ an intrinsic representation of the curve and thus are not prone to problems related to parametrization. However, it is difficult to impose topological constraints, for example extraction of a single region. Methods like the ones in [11,6,7] have incorporated heuristics in the optimization process; they essentially extract the boundary in a piecewise manner. Additional image features ([12,13,14]) and stronger contour priors ([15,16,17]) have also been explored. Such methods impose additional constraints but do not correct the built-in bias of the original criterion. The most direct attempt to address the bias problem has been to normalize the quality score of the contour by the length of the contour [5,18].

2   The Boundary Length Bias Problem

To better understand the nature of the bias we will employ a discrete version of Functional 1 and we will omit the second-order derivative. We assume that a curve C is discretized into n points. Let c_i be the i-th point. Then, the energy 1 is given by:

$$E[C] = \sum_{i=1}^{n} \{d(c_{i+1}, c_i) - \lambda \|\nabla I\|_{c_i}\} \tag{2}$$

where $d(c_{i+1}, c_i)$ is an approximation of the first derivative of the curve and $\|\nabla I\|_{c_i}$ is the gradient intensity at point c_i. $d(c_{i+1}, c_i)$ can be defined as the Euclidean length of the linear segment connecting neighboring points c_{i+1} and c_i.

Criterion 2 can be globally optimized with dynamic programming. To this end, the image is represented with a graph. Each arc (u, v) is weighted according to Eq. 2:

$$w(u, v) = d(u, v) - \lambda f(u, v) \tag{3}$$

where $f(u, v)$ refers to the image-derived features term.

The weights of Eq. 3 become negative at image locations with prominent image features. In the case where negatively weighted cycles are formed, the minimum of Eq. 2 is $-\infty$ and the problem becomes ill-posed. A negative cost cycle acts as a black hole in the energy landscape and forces all candidate boundary segments to include that cycle. Such an example is illustrated in Fig. 2. When the weights are positive, the shortest paths from S to all the other nodes include the bold edges. However, when negatively weighted cycles are introduced (Fig. 2(c)), the shortest paths are altered entirely so that they include negative cycles.

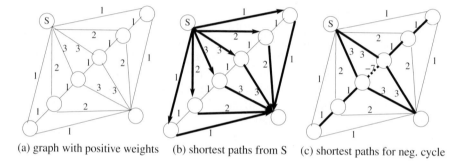

(a) graph with positive weights   (b) shortest paths from S   (c) shortest paths for neg. cycle

Fig. 2. Negatively weighted cycles act as black holes in the energy landscape. (a) Original graph with positive weights. (b) Shortest paths from S to all the other nodes for the weights of (a) (shown with bold arrowed lines). (c) The edge indicated with dashed line has obtained negative weight -7 and negative cycles have been created. The shortest path from S (shown with bold lines) are forced to include such cycles.

Removing negatively weighted cycles is computationally very difficult. An image will typically consist of many prominent features which will cause the creation of an exponential number of such cycles. Further, because the optima of the solution change drastically when such a cycle is created, it is difficult to impose simple constraints that will ensure the extraction of contours adhering to the object boundaries.

In practice, algorithms like [2,3,4] assume that the weights are positive. This is equivalent to adding a large positive constant M to the original weights such that:

$$M = \{ \max_{w(u,v)<0} |w(u, v)| \} + c \tag{4}$$

where $c > 0$. The weights obtained will be

$$w_M(u, v) = w(u, v) + M \tag{5}$$

Such a transformation does not preserve the optima of the objective criterion since the length of the contour is implicitly part of the optimization process. The objective criterion optimized instead, is:

$$E_M(\mathbf{C}) = \sum_{(u,v) \in C} \{d(u,v) - \lambda f(u,v)\} + nM \qquad (6)$$

This difference between Equations 1 and 6 is the term nM which is an additional smoothing term proportional to the length of the contour. Its introduction is arbitrary and its effect can be significant when long and geometrically complex contours are to be extracted. When such criteria are used for interactive contour extraction a large amount of human input is required, as has been observed in [19].

3   Removing the Bias

To remove the bias introduced by adding a constant (Eq. 5), we seek $\hat{w}$ of the form:

$$\hat{w}(u,v) = w_M(u,v) - \alpha(u,v) \qquad (7)$$

Our goal is to estimate $\alpha(u,v)$ so that $\hat{w}(\cdot) > 0$ and we will do so in a local fashion. Previously proposed bias correction methods provide different choices for $\alpha(u,v)$.

3.1   Local Bias Correction

The role of negative weights is to encourage the inclusion of boundary segments in the final solution. Thus, we need to assign very low positive weights to good boundary segments. The quality of a segment can be assessed based on the quality of its neighbors: a segment should receive low value if it is significantly better than nearby segments. The simplest segment is the edge between two nodes and we define:

$$w^+(u,v) = w_M(u,v) - \max_v w_M(u,v) \qquad (8)$$

where u and v are adjacent.

Non-maximum suppression and piecewise extension of the boundary are very similar to this transformation. Non-maximum suppression assigns high values to locally best pixels. Piecewise boundary extension, extracts a boundary in an incremental fashion so that it is composed from high-score segments.

Converting Negative Weights to Positive. *Provided there are no negatively weighted cycles*, a graph with negative and positive weights can be converted to a graph with positive weights so that the shortest paths among the nodes are preserved. Such a transformation is part of Johnson's all pairs shortest paths algorithm ([20]) and defines a new weighting function $w^+(u,v)$ as:

$$w^+(u,v) = w(u,v) + h(u) - h(v) \qquad (9)$$

The function $h(\cdot)$ is computed as follows. We create a new graph G' consisting of all the nodes of the original graph G and an additional dummy node s. Node s is connected to all the other nodes with weights equal to 0. Then, $h(u)$ is defined as the cost of the

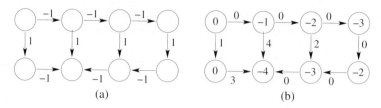

Fig. 3. (a) Original graph with positive and negative weights. (b) Graph with transformed weights. The numbers inside the nodes indicate the shortest path costs from a dummy node s.

shortest path from s to u. The weights thus defined are positive. Figure 3 shows an example of such a transformation.

Since a criterion with both positive and negative weights (but without negatively weighted cycles) does not suffer from the length bias, it follows that there exist criteria with positive terms which do not have an implicit bias.

In practice however, this algorithm cannot be applied since the weights induced by the image features will lead to negative cycles. Our local correction method can be seen as an approximation to the optimal algorithm. Instead of a single dummy node s, we use as many dummy nodes as the nodes of the graph, and find the shortest paths in a small neighborhood of each node. If there is a single best contour, our method yields the same contour as the optimal method.

3.2 Ratio Weight Cycles

The ratio weight criterion in [5] minimizes a normalized version of the original energy functional given by:

$$w(C) = \frac{\sum_e w(e)}{\sum_e n(e)} \qquad (10)$$

where $w(C)$ is the weight of a contour C.

Finding the minimum of Eq. 10 is equivalent to converting the original graph weights $w(e)$ to $w(e) - \lambda n(e)$ and finding zero cost cycles, i.e.:

$$\hat{w}(C) = w(e) - \lambda n(e) = 0 \qquad (11)$$

This is equivalent to *finding the largest λ such that no negatively weighted cycles are created*. The approach as presented in [5] does not model open curves and does not admit user interaction. It provides a way of estimating λ given a fixed $n(e)$.

The authors explore two types of $n(e)$. When $n(e) = 1$, the shortest mean cycle is found and the bias of Eq. 6 is reduced to M. The data term is also altered so that the contour extracted has on average good features. When $n(e) = 1/|\nabla I|(e)$, the criterion minimized is very similar to the original snakes criterion. The stronger the gradient intensity, the shorter the contour extracted.

3.3 Probabilistic Formulation

In [8], we proposed a probabilistic formulation which is capable of extracting geometrically complex boundaries. The weights defined by this method are of the form:

$$w(u, v) = x_j - \log \sum_i e^{-x_i} \qquad (12)$$

where x_i, x_j are the scores of neighboring segments, in our case edges emanating from the same node. The approximation scheme used to compute the log of the summation of exponentials results in weights similar to the ones of the local correcting method. That is, locally best segments receive very low costs.

4 Evaluation

4.1 Toy Example

Figure 4 demonstrates how the methods presented transform the weights and alter the optima of the criteria. Bold lines indicate the desired boundary. In the optimal case, (Fig. 4(a)) the graph contains both positive and negative weights and the desired boundary is the petal-shaped one. Inside the parentheses, the weights obtained from the optimal transformation are indicated. If the weights are translated by a constant, as is shown

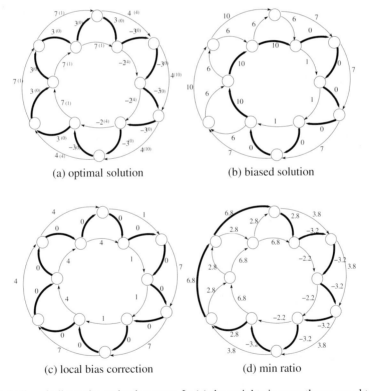

(a) optimal solution

(b) biased solution

(c) local bias correction

(d) min ratio

Fig. 4. Bold lines indicate the optimal contour. In (a) the weights in parentheses are obtained by optimally converting the negative weights to positive. Graph (b) is obtained by adding 3 to the original weights of graph (a). In (c) the local bias correction is applied to the weights of (b). In (d) the weights of (b) are transformed according to the min ratio cycle method.

(a) original image   (b) seed points   (c) biased   (d) locally corrected

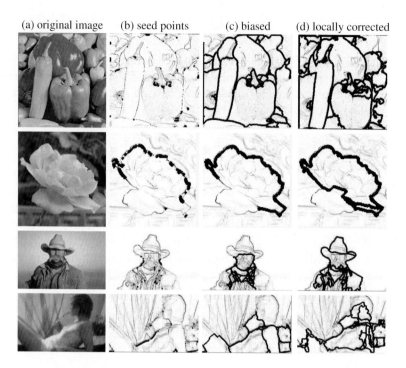

Fig. 5. Seed points were selected as strong edge points on the object boundary (b). Shortest paths between all seed points were found and displayed for the biased method (c) and the locally corrected method (d).

in Figure 4(b), then the optima of the criterion are not preserved. As a result, the optimum contour has been smoothed out (part of the inner cycle is included). The local bias transformation is shown in Figure 4(c); for this example the weights calculated are not similar with the optimal ones but the optimal contour is the desired boundary. Finally, in Figure 4(d) the weights obtained from the mean ration are shown. In this case, the optimal contour includes part of the outer cycle.

4.2   Contour Completion

Figure 5 shows results obtained for some real images for the task of contour completion. The gradient of the image was only used to guide the process. Seed points were selected by identifying consecutive strong gradient intensity points. The shortest paths among all seed points were found using biased weights and the locally corrected weights. In the case of biased weights one can see the tendency towards simpler contours. Further, boundaries which are not characterized with high intensity gradient are not always followed, as for example in the black and white flower. On the other hand, the locally corrected weights produce more detailed edge maps and oftentimes complete the contours in a more conceptually compatible fashion. On the downside, they may lead to irregular boundaries as in the case of the woman example.

Acknowledgments. This research is funded by NSF CAREER IIS-0644204 and a Clare Boothe Luce Professorship to Yu.

References

1. Kass, M., Witkin, A., Terzopoulos, D.: Snakes: Active Contour Models. In: Int'l Conference on Computer Vision. IEEE, Los Alamitos (1987)
2. Mortensen, E., Barrett, W.: Intelligent Scissors for Image Composition. In: SIGGRAPH (1995)
3. Mortensen, E., Barrett, W.: Interactive Segmentation with Intelligent Scissors. Graphical Models and Image Processing 60 (1998)
4. Falcao, A., Udupa, J., Samarasekera, S., Sharma, S.: User-Steered Image Segmentation Paradigms: Live Wire and Live Lane. Graphical Models and Image Processing 60, 233–260 (1998)
5. Jermyn, I., Ishikawa, H.: Globally optimal regions and boundaries as minimum ratio weight cycles. IEEE Transactions on Pattern Analysis and Machine Intelligence 23, 1075–1088 (2001)
6. Mortensen, E., Barrett, W.: A Confidence Measure for Boundary Detection and Object Selection. In: Proc. Computer Vision and Pattern Recognition. IEEE, Los Alamitos (2001)
7. Mortensen, E., Jia, J.: A Bayesian Network Framework for RealTime Object Selection. In: Proc. Workshop on Perceptual Organization on Computer Vision. IEEE, Los Alamitos (2004)
8. Pavlopoulou, C., Yu, S.X.: Boundaries as Contours with Optimal Appearance and Area of Support. In: 7th Int'l Conf. on Energy Minimization Methods in Computer Vision and Pattern Recognition (2009)
9. Geiger, D., Gupta, A., Costa, L., Vlotzos, J.: Dynamic Programming for Detecting, Tracking, and Matching Deformable Contours. IEEE Transactions on Pattern Analysis and Machine Intelligence 17, 294–302 (1995)
10. Malladi, R., Sethian, J.A., Vemuri, B.C.: Shape Modeling with Front Propagation: A Level Set Approach. IEEE Transactions on Pattern Analysis and Machine Intelligence 17, 158–175 (1995)
11. Neuenschwander, W., Fua, P., Iverson, L., Szekely, G., Kubler, O.: Ziplock Snakes. Int'l Journal of Computer Vision 25, 191–201 (1997)
12. Cohen, L.: On Active Contour Models and Balloons. Computer Vision Graphics and Image Processing: Image Understanding 52, 211–218 (1991)
13. Paragios, N., Deriche, R.: Geodesic Active Contours and Level Set Methods for Supervised Texture Segmentation. International Journal on Computer Vision 46, 223–247 (2002)
14. Gérard, O., Deschamps, T., Greff, M., Cohen, L.D.: Real-time Interactive Path Extraction with on-the-fly Adaptation of the External Forces. In: European Conference in Computer Vision (2002)
15. Sullivan, J., Blake, A., Isard, M., MacCormick, J.: Bayesian Object Localisation in Images. Int. J. Comput. Vision 44, 111–135 (2001)
16. Allili, M., Ziou, D.: Active contours for video object tracking using region, boundary and shape information. Signal, Image and Video Processing 1, 101–117 (2007)
17. Joshi, S.H., Srivastava, A.: Intrinsic bayesian active contours for extraction of object boundaries in images. Int. J. Comput. Vision 81, 331–355 (2009)
18. Schoenemann, T., Cremers, D.: Globally optimal image segmentation with an elastic shape prior. In: IEEE International Conference on Computer Vision, ICCV (2007)
19. Rother, C., Kolmogorov, V., Blake, A.: Grab-cut Interactive Foreground Extraction Using Iterated Graph Cuts. ACM Trans. Graph (SIGGRAPH) 23, 309–314 (2004)
20. Cormen, T., Leiserson, C., Rivest, R.: Introduction to Algorithms. McGraw-Hill, New York (1990)

A Novel Method for Enhanced Needle Localization Using Ultrasound-Guidance[*]

Bin Dong[1], Eric Savitsky[2], and Stanley Osher[3]

[1] Department of Mathematics, University of California, San Diego,
9500 Gilman Drive, La Jolla, CA, 92093-0112
[2] Department of Emergency Medicine, University of California Los Angeles,
BOX 951777, 924 Westwood Blvd, Ste 300, Los Angeles, CA 90095-1777
[3] Department of Mathematics, University of California, Los Angeles, CA, 90095-1555

Abstract. In this paper, we propose a novel and fast method to localize and track needles during image-guided interventions. Our proposed method is comprised of framework of needle detection and tracking in highly noisy ultrasound images via level set and PDE (partial differential equation) based methods. Major advantages of the method are: (1) efficiency, the entire numerical procedure can be finished in real-time: (2) robustness, insensitive to noise in the ultrasound images and: (3) flexibility, the motion of the needle can be arbitrary. Our method will enhance the ability of medical care-providers to track and localize needles in relation to objects of interest during image-guided interventions.

1   Medical Background

Image guided interventions have become the standard of care for many surgical procedures. Optimal visualization of the object of interest and biopsy needle in ultrasound images requires the use of specialized biopsy needles and high cost, cart-based ultrasound units. The success of image guided interventions is dependent on anatomic knowledge, visualization, and precise tracking and control of the biopsy needle. A majority of medical care-providers utilize low resolution ultrasound units. In addition, many office-based or emergency department procedures are performed using generic (non-specialized) needles. Unfortunately, the quality of the imagery obtained by most ultrasound units does not allow for clear and concise visualization of a regular needle during many needle-based procedures. The inability to clearly see the tip of a needle in relation to the object of interest (e.g., a vein, artery, or mass) makes such image guided interventions less accurate.

In view of the inadequacy of ultrasound technology identifying inserted needles with desired resolution, a new and improved system for tracking such needles needs to be developed. A more accurate method for localizing the distal tip of

---

[*] The research is supported by SN-30014, Center for Computational Biology NIH Toga; and the Telemedicine and Advanced Technology Research Center (TATRC) of the US Army Medical Research and Material Command (MRMC).

G. Bebis et al. (Eds.): ISVC 2009, Part I, LNCS 5875, pp. 914–923, 2009.
© Springer-Verlag Berlin Heidelberg 2009

inserted needles will greatly improve the efficacy and safety of ultrasound image-guided interventions. In this paper, we shall employ modern level set and PDE methods and fast numerical algorithms to solve the needle tracking problem for ultrasound images.

The rest of the paper is organized as follows. In Section 2, we shall lay down the fundamental mathematical model which is the core of solving our problem. In Section 3 we shall describe the complete schematic procedure of needle localization. Numerical experiments on ultrasound image frames will be given in Section 4 and concluding remarks will be given in Section 5.

2 Mathematical Model

We denote the video frames of ultrasound images as $I(x,t)$ with $0 \le I(x,t) \le 1$, and define the integrated difference of frames as

$$f(x,t) := \int_{t-\delta}^{t} \big|G_\sigma(x) * \partial_\tau I(x,\tau)\big| d\tau, \quad \delta > 0, \tag{1}$$

where G_σ is Gaussian with standard deviation σ. We note that the parameter σ is not essential for our method, we will fix it throughout our experiments.

If the motions of the needle, e.g. jiggling or insertion, are different from the motions of the tissues and organs, which is usually the case, then in $f(x,t)$ we can see regions with such motions highlighted. However these regions in $f(x,t)$ are usually not very clear and have noisy boundaries. Therefore, a robust and efficient segmentation on $f(x,t)$ for each t is needed. Since we will focus on the segmentation of $f(x,t)$ for each fixed t, we now omit the variable τ and denote $f(x,t)$ as $f(x)$ for simplicity.

There are numerous image segmentation methods in the literature [1,2,3,4, 5,6,7]. In this paper, we shall consider the following energy introduced in [1]

$$E(u) = \int g(x)|\nabla u(x)|dx + \lambda \int |u(x) - f(x)|dx. \tag{2}$$

Here $g(x)$ is some edge indicator function defined as $g(x) = \frac{1}{1+\beta|\nabla(G_{\tilde{\sigma}}*f)|}$ (see e.g. [1,3]). It is shown in [1] that for any minimizer u of (2) and for almost all threshold $\mu \in [0,1]$, the characteristic function

$$\mathbf{1}_{\Omega(\mu)=\{x:u(x)>\mu\}}(x)$$

is a *global minimizer* of the corresponding geometric active contour model (see [1] for more details). Therefore, a segmentation of $f(x)$ can be obtained by first computing a minimizer of (2) and then letting $\Omega := \{x : u(x) > 0.5\}$. Now the key issue here is to minimize (2) efficiently.

To minimize the energy (2) efficiently, we adopt the idea of the split Bregman method introduced in [8]. Define

$$|d|_* := g(x)\sqrt{d_1^2 + d_2^2} + \lambda|d_3| \quad \text{and} \quad Fu := (\nabla u^T, u - f)^T,$$

then minimizing energy (2) is equivalent to

$$\text{Minimize} \quad \int |d|_*$$
$$\text{s.t.} \quad d = Fu. \tag{3}$$

After "Bregmanizing" the constrained optimization problem (3), we obtain the following algorithm which minimizes the original energy (2) rather efficiently (the derivation is similar to that in [8]),

$$(u^{k+1}, d^{k+1}) = \text{argmin}_{u,d} \int |d|_* + \frac{\mu}{2} \|d - Fu - b^k\|_2^2$$
$$b^{k+1} = b^k + \left(Fu^{k+1} - d^{k+1}\right). \tag{4}$$

For convenience, we denote $\bar{d} = (d_1, d_2)^T$ and hence $d = (\bar{d}, d_3)^T$. Similarly, we can define $\bar{b}$ and b. Then we introduce the following algorithm to solve (4):

Algorithm 1. *We start with $d^0 = \mathbf{0}$ and $b^0 = \mathbf{0}$.*

1. First update u by solving

$$(-\Delta + I)u^{k+1} = \nabla \cdot (\bar{b}^k - \bar{d}^k) + d_3^k + f - b_3^k;$$

2. Then update d by

$$d_1^{k+1} = \max(s^k - \frac{g(x)}{\mu}, 0) \cdot \frac{u_x^k + b_1^k}{s^k},$$
$$d_2^{k+1} = \max(s^k - \frac{g(x)}{\mu}, 0) \cdot \frac{u_y^k + b_2^k}{s^k},$$
$$d_3^{k+1} = shrink(u^k - f + b_3^k, \frac{\lambda}{\mu}),$$

where $s^k = |\nabla u^k + \bar{b}^k|$.
3. Finally update b^{k+1} by

$$b^{k+1} = b^k + \left(F(u^{k+1}) - d^{k+1}\right);$$

4. If $\frac{\|u^{k+1} - u^k\|}{\|u^k\|} > tol$, go back to step 1 and repeat.

The Algorithm 1 is very efficient in terms of total number of iterations and the cost for each iteration. According to our experiments, it usually only takes about 30 iterations until $\frac{\|u^{k+1} - u^k\|}{\|u^k\|} \approx 10^{-3}$. For each iteration in Algorithm 1, the major calculation is in step 1, where the PDE can be solved rather efficiently by either FFT, for periodic boundary condition, or multigrid method, for Neumann and Dirichlet boundary conditions. An example is given in the following Figure 1 where noise was added to the original image. We note that the image is provided by Laboratory of Neural Imaging, Center for Computational Biology, UCLA.

For the special image $f(x)$ obtained from frames of ultrasound images by (1), the object of interest in $f(x)$ is either a needle or the tip of the needle, which are both simple geometric objects. Therefore, we can stop our iteration at an even earlier stage (e.g. in our experiments, we only perform two iterations) and the segmentation results would not change much if more iterations were carried out. The efficiency of Algorithm 1 ensures that the entire needle localization procedure can be finished in real-time. To be precise, by "real-time" we mean that the total time spent by the entire numerical procedure is no greater than that spent by the ultrasound machine in acquiring each image frame. A detailed description of the needle localization procedure will be given in next section.

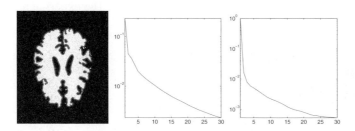

Fig. 1. The left figure shows segmentation result using Algorithm 1; the middle one is the decay of $\frac{\|d - Fu^k\|}{\|d\|}$; and the right one is the decay of $\frac{\|u^{k+1} - u^k\|}{\|u^k\|}$

3   Schematic Descriptions of Needle Detection and Tracking Procedure

The entire needle localization procedure can be decomposed into two phases. The first phase is to locate the needle in the images at the very beginning, based on a few seconds' image frames. During this phase, one can jiggle the needle or gently poke the tissues to help our algorithm locate the needle fast and accurately. The second phase is to track the motion of the tip of the needle when it moves.

3.1   Phase I

To locate the needle when it is first inserted into the tissue, we perform the following operations:

1. Obtain $f(x)$ using (1) based on the previous 1-2 seconds' frames, denoted as $I(x, t)$;
2. Segment the region that indicates needle movements using (4) via the Algorithm 1 (with 2 iterations);
3. Regularize the region obtained by step 2 via the fast algorithm of area-preserving mean curvature motion in [9];
4. Obtain the skeleton of the regularized region to represent the needle, and then the tip of the needle can be located from the skeleton.

To help localize the needle based on $f(x)$, one could gently jiggle the needle, in order to differentiate its motion from that of the tissues or organs. The following Figure 2 illustrates the four steps described above. We first note that it is obviously crucial to consider $f(x)$ instead of any single frame in order to rule out other regions with comparable intensities as the needle (e.g. some tissues or organs). The left two figures in Figure 3 show that if we perform segmentation directly on a single frame, we will capture several regions besides the needle. We also note that the third step above is important because otherwise, we may not get a single line representing the needle, but several branches (see the right figure in Figure 3). In step 4, there is always an ambiguity of the tip (it could be the alternative end of the line). However the ambiguity can be easily removed whenever the needle starts moving. Therefore, here and in the experiments below, we assume the tip is picked up correctly.

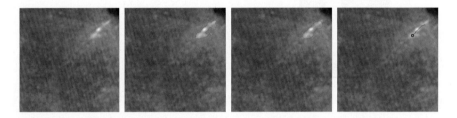

Fig. 2. The four figures from left to right describes the four steps, and the four images are the same one $f(x)$ obtained by (1)

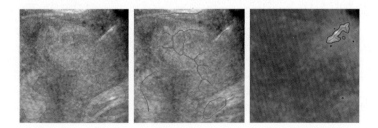

Fig. 3. Left figure shows direct segmentation of one single frame; middle one shows the skeletons extracted from the segmented regions; right one shows the importance of step 3 in Phase I, where the blue curve is represented by the solution u obtained form step 2, and the red one is the skeleton by step 4

3.2   Phase II

The second phase is to track the movements of the tip of the needle starting from the location we obtained from Phase I. We perform the following operations:

1. Obtain $f(x)$ using (1) based on the current and the previous 1-2 frames;
2. Segment the region indicating needle movements using (4) via the Algorithm 1 (with 2 iterations);

3. Regularize the region obtained by step 2 via the fast algorithm of area-preserving mean curvature motion in [9];
4. Shrink the (possibly disconnected) region to points, and then choose one point from them that is closest to the previously tracked location.

The following Figure 4 illustrates the four steps described above. We note that when the noise level is high or some irregular motions exist in tissues or organs, multiple locations may be captured in step 3, most of which are false detections. Therefore, step 4 affects the smoothness of the overall tracking. Evidently, there are more sophisticated ways to regularize the trajectory of the tracking. For example, if we know a priori that the needle moves in a smooth fashion, then we can estimate the current location of the tip based on the approximated locations in previous frames such that the overall motion curve is smooth. For our experiments in Section 4, we only use the simple operation described in step 4 because the needle moves in an irregular fashion. However, the result of the overall tracking is still quite satisfactory. We also note that in step 1, instead of considering the entire image $f(x)$, we can just consider a patch of $f(x)$ that centered at the previously located point (location of the tip in the previous frame). In this way, we can save some computations and also increase the smoothness of the overall tracking. Again, this only works when the motion of the needle is not too fast (which is usually the case in practice). In our experiments in Section 4, we will still use the entire image $f(x)$.

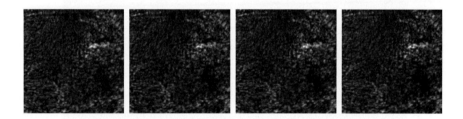

Fig. 4. The four figures from left to right describes the four steps

4   Numerical Result

All of the frames of ultrasound images are obtained by a Sonosite (Titan) ultrasound machine. The ultrasound machine captures 20 frames per second. In our following experiments, 120 frames are used, including 20 frames in Phase I and 100 frames in Phase II. Each image is of size 251×251. In Figure 5 we present 5 of the 20 frames in Phase I, and in Figure 7 we present 12 of the 100 frames in Phase II.

The numerical results for Phase I are given in Figure 6, and those for Phase II are given in Figure 8. We note that the PDE in (1) of Algorithm 1 is solved by FFT. Here we also provide a ground truth in Figure 9 as validation of our results, where we manually selected the positions of the needle based on neighboring

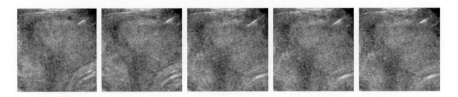

Fig. 5. Images from left to right are 5 sample frames among total 20 frames of ultra-sound images during Phase I

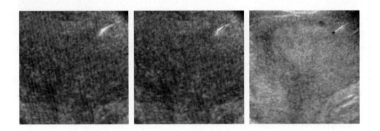

Fig. 6. Left figure is $f(x)$ obtained from the 20 frames; middle one shows the result of localization of the body of the needle; right one shows the result of localization on the first image frame in Figure 5, where the blue dot indicates the tip of the needle

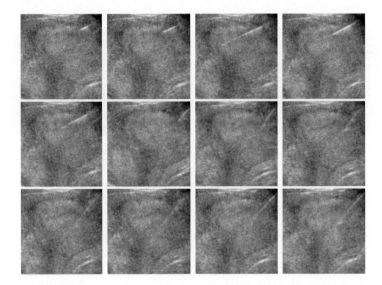

Fig. 7. Images above are 12 sample frames among total 100 frames of ultrasound images during Phase II

frames. We note that for almost all of the frames during Phase II, the tracking is rather accurate. However for some of the frames, the localization is not very accurate, for example the fourth figure in the first row of Figure 8. The reason

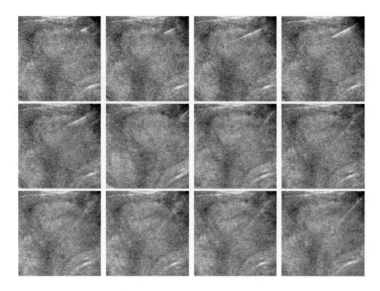

Fig. 8. Tracking results of the 12 sample frames in Phase II shown in Figure 7

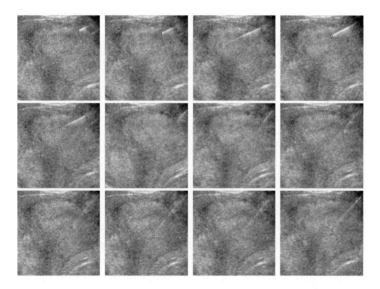

Fig. 9. Manual segmentation results of the 12 sample frames in Phase II shown in Figure 7

is because of acoustic shadows in some image frames, which appear in $f(x)$ with high intensities and conceal the movement of the tip of the needle (see the middle figure of Figure 10). However, an acoustic shadow only seems to appear in $f(x)$ occasionally when we extract the needle, instead of inserting the needle, and an

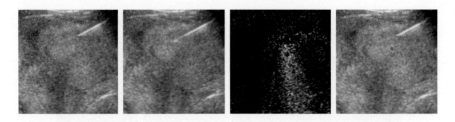

Fig. 10. First figure is the current frame as shown in the fourth figure in first row of Figure 7; second figure is the previous frame of the first figure; third figure shows the corresponding $f(x)$ obtained from the first two figures and the red dot is the tracking result; the last one shows the tracking result on the current frame which is the same figure as in the upper fight figure of Figure 8

accurate tracking of the needle is only required during insertion. Therefore in practice, this error is not an issue and will not affect the safety concerns during image guided surgical operations.

5   Conclusion

Image guided interventions have become the standard of care for many surgical procedures. One of the most important problems in image guided interventions for ultrasound images is the precise tracking and control of biopsy needles. In this paper, we introduced a novel and efficient method for needle localization in highly noisy ultrasound images. Our numerical experiments showed that our proposed method tracks the tip of needle efficiently with satisfactory accuracy.

There are also improvements of the current method that can be done. Firstly, the regularity of the tracking results can be improved. For the current version of the method, we are only segmenting $f(x,t)$ for each t independently, while ideally speaking the segmentation should depend on both x and t so that the approximated needle locations will lie on a smooth path. However, taking t into account during segmentation will increase computational complexity. Therefore, a very efficient algorithm is required.

Another possible improvement is to detect and remove some of the known artifacts, e.g. acoustic shadow, in the images before performing segmentation. This will improve the accuracy of tracking and also helps to improve regularity. Again, the challenge is that the process of artifact removal needs to be done rather efficiently.

Acknowledgements

We would like to thank Jyotsna Vitale from the Radiology Department of UCLA for providing the ultrasound data.

References

1. Bresson, X., Esedoglu, S., Vandergheynst, P., Thiran, J., Osher, S.: Fast global minimization of the active contour/snake model. Journal of Mathematical Imaging and Vision 28, 151–167 (2007)
2. Kass, M., Witkin, A., Terzopoulos, D.: Snakes: Active contour models. International journal of computer vision 1, 321–331 (1988)
3. Caselles, V., Kimmel, R., Sapiro, G.: Geodesic active contours. International journal of computer vision 22, 61–79 (1997)
4. Osher, S., Fedkiw, R.: Level set methods and dynamic implicit surfaces (2003)
5. Chan, T., Vese, L.: Active contours without edges. IEEE Transactions on image processing 10, 266–277 (2001)
6. Chan, T., Glu, S., Nikolova, M.: Algorithms for finding global minimizers of image segmentation and denoising models. Algorithms 66, 1632–1648
7. Goldstein, T., Bresson, X., Osher, S.: Geometric Applications of the Split Bregman Method: Segmentation and Surface Reconstruction. UCLA CAM Report, 09–06 (2009)
8. Gilboa, G., Osher, S.: Nonlocal linear image regularization and supervised segmentation. Multiscale Modeling and Simulation 6, 595–630 (2008)
9. Ruuth, S., Wetton, B.: A simple scheme for volume-preserving motion by mean curvature. Journal of Scientific Computing 19, 373–384 (2003)

Woolz IIP: A Tiled On-the-Fly Sectioning Server for 3D Volumetric Atlases*

Zsolt L. Husz[1], Thomas P. Perry[2], Bill Hill[1], and Richard A. Baldock[1]

[1] MRC Human Genetics Unit, Institute of Genetic and Molecular Medicine, Western General Hospital, Edinburgh EH4 2XU, UK
{Zsolt.Husz,Bill.Hill,Richard.Baldock}@hgu.mrc.ac.uk
[2] Institute for System Level Integration, Alba Centre, Alba Campus, Livingston, EH54 7EG, UK
Thomas.Perry@sli-institute.ac.uk

Abstract. We present a novel method to provide fast access to large 3D volumetric data sets from biological or medical imaging atlases. We extend the Internet Imaging Protocol with an open specification for requesting tiled sections of 3D objects. We evaluate the performance of the protocol and demonstrate it with a platform independent web viewer that allows on-the-fly browsing of section views of multi-gigabyte 3D objects.

The method uses *Woolz*, an efficient image processing library, to provide very fast access to section views of the volumetric data. The server has been implemented to run on standard Linux systems and it avoids the requirement for high-performance parallel processing or expensive software. We have tested the system on data volumes up to 13.4 GB and demonstrated no loss of responsiveness for the user.

1 Introduction

In the field of biomedical science, the ability to access 3D image objects over a wide-area network such as the internet is often imperative. Previous solutions involve the Internet Imaging Protocol (IIP)[1], which is an open protocol that provides fast tiled delivery of large images through a multi-resolution image representation, but which may only be used with 2D images. Since a similar presentation method for 3D objects does not exist, we have developed extensions to IIP which we call *Woolz Internet Imaging Protocol* (WlzIIP), implemented a server to provide this service and built a web application to use it.

Recent work has demonstrated the clear advantages of tile-based image transmission and many *zoom-viewers* have been developed for example by Google and Zoomify. The IIP server has been developed as an open-source resource and is used in tele-pathology and educational archives [2]. It allows a user to select a region of interest at a desired zoom level and provides efficient image transmission.

The importance of virtual slicing systems for remote access of images was previously noted [3] and the IIP protocol was identified as a suitable interface

---

* We acknowledge support from the Medical Research Council, UK and NIH support under grant #1R01MH070370-01A2.

G. Bebis et al. (Eds.): ISVC 2009, Part I, LNCS 5875, pp. 924–933, 2009.

for independent client-server applications. However the availability and flexibility of these imaging systems was limited by the proprietary (and costly) nature of existing implementations. Some existing image servers are able to deliver 3D image data, such as in BrainMaps.org, they provide only predefined 2D sections. Glatz-Krieger et al. [3] consider virtual slices only in the original focal planes of the biological material in the context of a 2D microscope slide. In this paper, we cut an *arbitrary* virtual section from the digitised 3D model. Such sectioning software exists, either as standalone (e.g. Amira [4]) or as online Java applications (e.g. NeuroTerrain [5]). The latter aims for compatibility and platform independence, but sometimes falls short of this in practice: for example, Iowa Slidebox[6] suffers from its binding to an obsolete Java runtime environment.

Tile based image delivery, that transmits the target as smaller image blocks, is known from commercial web applications such as Google maps. This runs in any web browser and does not require additional software or an applet.

The Visible Human project has generated several internet based image servers and clients. The EPFL server [7] is the most similar to ours. It is a high through-put parallelised sectioning server using a FastCGI (FCGI) web interface. However, compared to WlzIIP, it does not allow tiled requests and has a proprietary protocol. To deliver section data the EPFL server requires a high performance cluster. In contrast, the WlzIIP server will run on standard Linux-based servers with the only requirement that the installed memory is larger than the image volume.

Sections 2 and 3 present the internal image representation and 3D sectioning that provide fast image generation. Sect. 4 explains our WlzIIP extension of IIP. Then, a visualisation interface using the WlzIIP is presented in Sect. 5, and in Sect. 6 we compare WlzIIP to the NeuroTerrain image server. The paper ends with a discussion and conclusions.

2 3D Object Representation

As part of our work, we use the image processing library known as *Woolz* [8]. Internally *Woolz* uses an interval coded representation for objects which is efficient with regard to both sparse data storage and image processing operations.

In 2D, an image is defined over an arbitrary region of a discrete 2D space with coordinates (k, l) where k is the column coordinate and l the line coordinate. For each line in the image there is a list of intervals which gives the start and end points of the image along that line. There is a list (possibly empty) of intervals for each line and it is clear that an arbitrarily complex region of the discrete space can be defined in this way. It is assumed that the discretisation in the x and y directions is at fixed regular intervals, constant in both directions but not necessarily equal. The 3D structure is simply a stack of 2D images. The plane coordinate is defined to be p, where the planes are evenly spaced, each with a 2D image, or possibly an empty structure.

The advantage of the Woolz encoding is that only grey-level information within the domain of the image is stored rather than for the whole rectangular box defined by the column, row and plane bounds. For a biological atlas this reduces storage and memory requirements with lossless compression.

3D object reconstructions are built with specialised Woolz tools for section data registration. In this paper the reconstruction technique is immaterial and not discussed and in fact the WlzIIP can be applied to any volumetric image.

Each object has its own internal discrete coordinate system with an associated affine transform which will provide the link between internal coordinates and external, biologically relevant coordinates.

2.1 Coordinate Transformation

There are many ways to define an arbitrary rotation, scaling and translation of one coordinate frame into another. For the purposes of sectioning we use a set of parameters that are chosen to correspond to those used in the MAPaint Woolz viewing tool to select arbitrary planes through reconstructions. The underlying coordinate transformation methods have been extensively used for developing the e-MouseAtlas models and gene-expression database.

We define a viewing plane with a new set of coordinate axes such that the new z-axis is along the *line-of-sight*. The viewing plane is defined to be perpendicular to the viewing direction given by angles θ and ϕ which are yaw and pitch respectively. The actual plane is distance d from the fixed point $\mathbf{f}$. Internally the full rotation transformation is defined in terms of the Euler angles [9, p. 107] with British definition [10, p. 9]. The third degree of freedom (d.o.f.) is called *roll* and for the user corresponds to rotating the section image as viewed on the screen. In many cases the user will want a *standard* view of the data without the requirement of an additional control to set the viewing angle, so we have implemented a number of viewing modes which automatically determine this angle.

3   Viewing Modes

With views that are perpendicular to the line of sight, we present four options to determine the orientation of the section image on the screen.

Statue mode: The viewing plane is *flat* but the image is oriented as if the viewer were *walking around* the object. The actual displayed image is then obtained by rotating the viewed plane about an axis parallel to the line of intersection of the view plane and the *horizontal* which is defined to be a plane of constant z. This has the merit of providing clear feedback of the position of the plane within the whole but is not ideal because for some angles the projection will introduce perspective distortion of the image.

Up-is-Up mode: The projection of a predefined direction *up* will always be displayed as the vertical in the section view. If the viewing direction is parallel to this vector then the angle of rotation around the viewing direction is not defined and an arbitrary choice can be made. As a consequence, small changes in viewing direction around the *up* vector may give rise to arbitrarily large changes in the display orientation.

Fixed point mode: Navigation through a 3D volume can give rise to confusion if unfamiliar views are presented. However it may often be possible to identify one or more points within the image volume that the user wishes to be visible. If one point is fixed then there are two d.o.f. left to set the view and if there are two fixed points then there is only one d.o.f.

The transformation is defined so that by setting one fixed point, $\mathbf{f}$, the orientation parameters, θ and ϕ, will rotate the view plane about this point.

Fixed line mode: If two points are fixed then θ and ϕ are dependent and can be represented in parametric form using a third angle parameter, ψ, which corresponds to the angle around the line joining the two fixed points.

The two fixed points $\mathbf{f_1}$ and $\mathbf{f_2}$ give direction vector $\mathbf{n_1} = \frac{\mathbf{f_2} - \mathbf{f_1}}{|\mathbf{f_2} - \mathbf{f_1}|}$ which must remain in the view plane. The values of pitch and yaw of the original plane in which the fixed line was established define a direction perpendicular to this vector $\mathbf{n_1}$ and can be used to establish the formula linking ψ to new viewing angles.

This technique has proved very powerful and is widely used in MAPaint.

4 Tile Based Imaging with WlzIIP

In this section we present the extension of WlzIIP to the IIP protocol.

4.1 Image Tiles

In WlzIIP, we keep the tile based imaging capability of IIP and so each image is divided into fixed sized tiles (except right-most or bottom-most tiles which might be smaller).

Displaying 3D objects involves multiple coordinate systems, and WlzIIP will automatically perform transformations between these. For clarity, the conventions used are explained in Fig. 1.

4.2 Protocol Extension

The added commands and feature queries are shown in Tables 1 and 2.

The commands specify an object, set the viewing section parameters and request image data or metadata, similar to existing IIP parameters. The commands for image requests are the same as in the original IIP specification[1]: **CVT** for full frame; **JTL** and **TIL** for jpeg-compressed and uncompressed tile answers. For a Woolz object, **SCL** specifies an arbitrary scaling factor, so resolution number is ignored in **JTL** or **TIL** commands.

For 2D images, pyramidal tiled TIFF images are specified by the **FIF** command, while for WlzIIP the **WLZ** command sets the 3D Woolz object. These are cached in the server's memory for efficiency.

MOD specifies the projection mode being STATUE, UP_IS_UP, FIXED_LINE or ZETA (fixed point) as described in section 3.

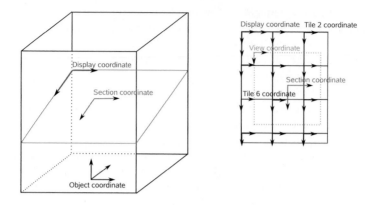

Fig. 1. Coordinate systems. The object coordinate is defined by the object, and the section coordinates result from the sectioning process. The origin of the sectioning coordinates is in an arbitrary position with respect to the visible pixels of the sectioning plane. However, images are normally represented using pixels with positive coordinates. Hence, the display coordinates translate section coordinates such that the lower bound of the bounding box of the visible pixels is at $(0,0)$. Further, on the right, the section is divided into non-overlapping tiles covering the whole section. Tiles are numbered with 0, 1, 2, etc., with the coordinates of the 0th tile (i.e. top-left corner) matching the display coordinates. The 1st, 2nd, etc. tiles continue from left to right and top to bottom in raster fashion. Finally, the view coordinates are defined for the region of the reassembled tiles displayed in the viewer application.

Table 1. WlzIIP command extension summary

Command	Purpose	Syntax
WLZ	Specify the Woolz object	WLZ=$path$
MOD	Specify the projection mode	MOD=$mode$
DST	Specify the distance of the sectioning plane	DST=dis
PIT	Specify the pitch angle of the sectioning rotation	PIT=$angle$
ROL	Specify the roll angle of the sectioning rotation	ROL=$angle$
YAW	Specify the yaw angle of the sectioning rotation	YAW=$angle$
SCL	Specify the scale used in the sectioning transformation	SCL=$scale$
FXP	Specify the fixed point of the viewing section rotation	FXP=X,Y,Z
FXT	Specify the second fixed point of the viewing section rotation	FXT=X,Y,Z
PAB	Specify the 3D query point absolute in the object coordinate	PAB=X,Y,Z
PRL	Specify the 2D query point relative in tile or display or tile coordinate	PRL=T,X,Y
UPV	Specify the up vector for the up is up mode	UPV=X,Y,Z

DST sets the viewing plane distance, while **PIT**, **ROL** and **YAW** set the plane angles. Other commands set parameters specific to the viewing mode.

The retrieval of a tile is an HTTP request that includes a combination of the above commands. Such an example is shown in Fig. 2.

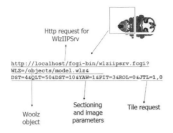

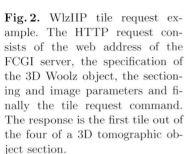

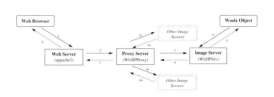

Fig. 2. WlzIIP tile request example. The HTTP request consists of the web address of the FCGI server, the specification of the 3D Woolz object, the sectioning and image parameters and finally the tile request command. The response is the first tile out of the four of a 3D tomographic object section.

Fig. 3. Architecture of WlzIIP server using a proxy server. The web server passes the user requests to the proxy, which forwards them to individual IIP servers. These servers have direct access to the Woolz Object and return the requested data. The numbered lines show the order of the requests (continuous lines) and the replies (dotted lines).

Table 2. WlzIIP object request extension summary

Object	Purpose
IIP-server	Identify if WlzIIP server is running
Max-size	The size of the section
Tile-size	The size of a tile
Wlz-true-voxel-size	The voxel size of the object
Wlz-volume	The volume of the object
Wlz-distance-range	The range of the sectioning plane distance
Wlz-sectioning-angles	The pitch, yaw and roll angles of of the sectioning plane
Wlz-3d-bounding-box	The first and last plane, line and column number of the object
Wlz-coordinate-3D	The 3D coordinates defined in 2D by the **PRL** command
Wlz-grey-value	The grey or RGB value of a point specified either the **PRL** or the **PAB** commands

Features of a given section can be obtained with **OBJ** queries listed in Table 2. We provide also coordinate translations and voxel queries. An example query for a sectioning plane distance range is

```
http://localhost/fcgi-bin/iipsrv.fcgi?YAW=61&PIT=3&ROL=0&MOD=ZETA
&WLZ=/objects/small.wlz&OBJ=Wlz-distance-range
```

which results in the reply

```
Wlz-distance-range:0 171
```

Our C++ software is based on a GPL implementation of the IIP server by Pillay and Pitzalis[11]. This is a FastCGI (FCGI) web server module that is called by the web server (e.g. Apache).

4.3   WlzIIP Proxy

To handle multiple requests, large objects and to provide a single access point to image servers separated from the Internet by a firewall, we have developed a tool called WlzIIPProxy that filters FCGI requests and forwards them to different WlzIIP servers. The communication conforms to the FCGI protocol. Though it was designated to work for IIP and Woolz requests, it is generic and can route any FCGI request, hence it is also possible to chain multiple proxies.

The multiple WlzIIP server architecture is shown in Fig. 3. WlzIIPProxy is an independent program running on the proxy server. The web server (e.g. Apache 2) forwards the FCGI request to this server on a configurable port, then the HTML request string is checked by WlzIIPProxy and if the definition string of any remote WlzIIP server is a substring of the request parameters then this query is forwarded to the matching server. If no correspondence was found then the request is passed to the default server.

5   Web Viewer Prototype with WlzIIP

For testing WlzIIP, we have developed a JavaScript application that runs in a web browser, based on the viewer of Pillay [12]. The WlzIIP viewer allows browsing through the objects with four controls. These change the pitch and yaw angles of the sectioning plane in the fixed point mode, alter the sectioning plane distance and the zoom level, and pick the current viewing region in a thumbnail view.

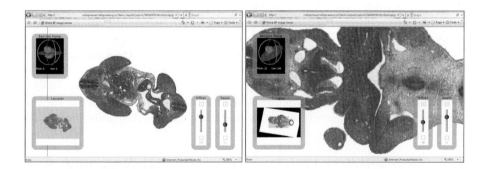

Fig. 4. Web interface using WlzIIP showing two views of a 3D object. The *Section Plane* selects pitch and yaw angles, the *Locator* control provides a thumbnail and allows viewing zone selection, the *Offset* control sets the sectioning plane distance, and the fourth control provides a zoom in the range of 0.25x–4x.

The browser requests tiles only for the currently viewed regions at the nominal screen resolution, so no transmission bandwidth is lost for non-visible regions or for unrepresentable details. When a user pans the object, new tiles are requested and displayed. For an $N \times N$ image, scrolling in one direction requires at most $N + 1$ new tiles. By reducing the size of the transmitted data in this way, the transmission throughput is increased by $\frac{N^2}{N+1}$. For $N = 10$, plausible for biological images, this provides a performance increase factor of 9.09.

The caching mechanisms of the web browser and of the WlzIIP server reduce the response time for tiles, sections and objects.

Tests were performed from a university network, a home ISP and free low-bandwidth wireless access on a train, and all tests show fast response times and great capability for interaction. Currently our application is compatible with Firefox, Internet Explorer, Safari and Opera browsers on Microsoft Windows, Linux and Mac OS X operating systems.

The WlzIIP project webpage[1] provides demos and further information about both WlzIIP server and viewer.

6   Evaluation

We have evaluated the WlzIIP server and compared it to NeuroTerrain [5]. Our test requested 1056 consecutive sections of a 3D object imported from NeuroTerrain. Each grey level section of the default resolution image consists of 3×5 tiles of 128×128 pixels. The average retrieval time of 10 repeated tests provided the results from the bottom five lines of Table 3.

The table includes NeuroTerrain high and low bandwidth results, full section requests using the **CVT** command and four tiled retrievals on a local client (1 Gbps LAN), on two remote clients on the JANET[2] network (1 Gbps backbone) and on a client with 2 Mbps home broadband.

In NeuroTerrain, the full sized, uncompressed grey image is transmitted, therefore real and browsing frame rates are equal. Also, the pixel throughput equals the data throughput.

First, to compare WlzIIP with NeuroTerrain, full frame requests with **CVT** were tested. This has a throughput 6.4 times lower than the highest throughput of NeuroTerrain. However, for WlzIIP the pixel transfer rate is higher due to the compressed image data, which results in a higher frame rate.

Tiled **JTL** requests have a lower real frame rate. However, the tiled approach has the advantage of browser caching. Therefore, the browsing frame rate takes into account an increase of 9.09 times (as estimated in Sect.5). Note that the non-zoomed NeuroTerrain frames are too small to benefit from this directly. However, the speed of the magnified image and of larger datasets improves considerably. The browsing frame rate is superior to the full image based transmission even using the home broadband connection, i.e. equivalent to the DSL tests on NeuroTerrain[5].

---

[1] http://www.EMouseAtlas.org/Software/WlzIIP
[2] JANET is the UK educational and research network.

Table 3. WlzIIP server evaluation. The throughput is the speed of the (compressed) data transmission; the pixel throughput is the pixel transmission rate; the real frame rate is the speed of transmission of full images; the browsing frame rate is the estimated frame rate that users experience.

	Through-put [KB/s]	Pixel through-put [Kpixel/s]	Real frame rate [fps]	Browsing frame rate [fps]
NeuroTerrain[5]				
LAN	4060.00	4060.00	5.47	5.47
DSL	141.00	141.00	0.20	0.20
WlzIIP full (CVT)				
LAN	634.10	6993.91	37.31	37.31
WlzIIP tiled (JTL)				
LAN	107.49	929.61	4.96	45.08
JANET metropolitan	80.26	694.09	3.70	33.66
JANET remote	54.68	472.83	2.52	22.93
Home broadband	17.50	151.31	0.81	7.34

7   Discussion and Conclusions

The main contribution of this paper is the extension of the IIP protocol for 3D objects that allows fast sectional data browsing over the Internet.

Compared to local image and object viewers, the WlzIIP server offers central management of the image content and storage of the object at the provider, thus it allows simple update and deployment of new content. Specification of the details (e.g. zoom level and spatial localisation) of the region the user is interested in permits a reduction in the size of the transmitted information, and hence allows fast interactive access to large data objects.

The main disadvantage of our current WlzIIP server is the requirement of sufficient memory to load 3D the object. However, this drawback is limited: using WlzIIPProxy, large objects can be distributed over multiple servers for which memory has become a low cost resource. Also, up to our largest dataset of 18.5 GB we have not observed performance degradation after the initial object disk read. With the underlying Woolz architecture, the extension to read partial object is straightforward, although currently object sizes don't require this.

Tiled images increase browsing frame rate whilst reducing the size of the transmitted data. Compatibility with HTTP allows portability and simple integration into client applications. Being browser based, the WlzIIP client does not need local installation for viewing nor does it need an additional IP port for communication.

Other web based delivery systems, such as the original IIP specification and Google Maps, are restricted to 2D. However, WlzIIP delivers sections from 3D objects. Similar to IIP, WlzIIP is an open protocol and the server code is freely available from us with a GPL licence.

To summarise, our main achievement was to extend the standard IIP protocol to allow fast sectioning of 3D objects using different sectioning modes and parameters, metadata and section-coordinate queries of the object. This has had the effect of increasing the frame rate experienced by the user. Using WlzIIP-Proxy, the method is scalable. The portable JavaScript browser does not require special software or applets to run locally and provides a highly interactive browsing environment. The WlzIIP server is currently being integrated in biological atlases such as EMAP, EurExpress and EuReGene and we are looking forward to further applications.

References

1. I3A: Internet imaging protocol, version 1.0.5 (1997)
2. Mea, V.D., Roberto, V., Beltrami, C.A.: Visualization issues in telepathology: The role of the internet imaging protocol. In: 5th Int'l Conf. on Information Visualization, pp. 717–722 (2001)
3. Glatz-Krieger, K., Glatz, D., Mihatsch, M.J.: Virtual slides: high-quality demand, physical limitations, and affordability. Human Pathology 34(10), 968–974 (2003)
4. Stalling, D., Westerhoff, M., Christian Hege, H.: Amira: A highly interactive system for visual data analysis. In: The Visualization Handbook, pp. 749–767. Elsevier, Amsterdam (2005)
5. Gustafson, C., Bug, W.J., Nissanov, J.: Neuroterrain – a client-server system for browsing 3D biomedical image data sets. BMC Bioinformatics 2007 8(40) (2007)
6. Heidger Jr., P.M., Dee, F., Consoer, D., Leaven, T., Duncan, J., Kreiter, C.: Integrated approach to teaching and testing in histology with real and virtual imaging. The Anatomical Record Part B: The New Anatomist 269(2), 107–112 (2002)
7. Bessaud, J.C., Hersch, R.D.: The visible human slice sequence animation web server. In: 3rd Visible Human Project Conf. Proc. (2000)
8. Piper, J., Rutovitz, D.: Data structures for image processing in a c language and unix environment. Pattern Recognition Letters 3, 119–129 (1985)
9. Goldstein, H.: Classical Mechanics, 2nd edn. Addison-Wesley, Reading (1950)
10. Whittaker, E.T.: A Treatise on the Analytical Dynamics of Particles and Rigid Bodies, 3rd edn. Cambridge University Press, London (1927)
11. Pillay, R., Pitzalis, D.: IIPSrv, v. 0.9.7 (1997),
    http://prdownloads.sourceforge.net/iipimage/iipsrv-0.9.7.tar.bz2
12. Pillay, R.: IIPMooViewer, v. 1.0 (2007), http://iipimage.sourceforge.net

New Scalar Measures for Diffusion-Weighted MRI Visualization

Tim McGraw, Takamitsu Kawai, Inas Yassine, and Lierong Zhu

Department of Computer Science and Electrical Engineering,
West Virginia University

Abstract. We present new scalar measures for diffusion-weighted MRI visualization which are based on operations of tensor calculus and have a connection to topological visualization. These operators are generalizations of the familiar divergence and curl operations in vector calculus. We also present a method for computing the Helmholtz decomposition of tensor fields which can make the new scalar measures more robust. The methods we present are general with respect to tensor order, so they apply to traditional 2nd order diffusion tensor MRI, as well as 4th and high order models used in high angular resolution diffusion imaging. Results are shown for synthetic tensor fields of orders 2 and 4 and also real diffusion tensor MRI data of orders 2 and 4.

1 Introduction

Random molecular motion (Brownian motion) causes transport of water at a microscopic scale within biological systems. The properties of the surrounding tissue can affect the magnitude of diffusion, as well as direction. Tissue can form a barrier to diffusion, restricting molecular motion. Within an oriented structure, such as a bundle of axonal fibers within white matter of the brain, diffusion can be highly anisotropic. MRI measurements have been developed which are sensitive to diffusion. Diffusion-weighted MRI provides a characterization of the restricted motion of water through tissue that can be used to infer tissue structure. This behavior can be concisely characterized using a tensor.

Basser and others [1] have presented general methods of acquiring and processing the diffusion tensor. In early studies the diffusion tensor was assumed to be a 2nd-order tensor which can be represented by a 3×3 matrix. Inferring the integrity and trajectory of white-matter pathways in the central nervous system has long been a goal of diffusion MRI analysis. The tensor model makes it possible to compute, in vivo, many useful quantities, including estimates of structural connectivity within neural tissue [2].

1.1 Related Work

In diffusion tensor MRI (DT-MRI) indices of anisotropy include Fractional anisotropy (FA) [3], relative anisotropy [3], volume ratio [4] and lattice anisotropy

G. Bebis et al. (Eds.): ISVC 2009, Part I, LNCS 5875, pp. 934–943, 2009.

[4]. These measures have found success in clinical applications because they are useful because many neurological disorders are characterized by changes in brain white-matter anisotropy, for example stroke, trauma, and multiple sclerosis. Recently higher order tensors have been proposed as a model for diffusion in the context of diffusion-weighted MRI [5]. New measures of anisotropy based on variance and entropy have been propose [6]. In this paper we propose new scalar measures based on differential quantities computed from tensor fields of arbitrary order. The quantities we study are generalizations of those which have proven useful in vector field analysis - namely the divergence and curl.

The Helmholtz decomposition separates a flow field into divergence-free (solenoidal) and curl-free (irrotational) components. These parts may be analyzed separately to robustly identify different types of critical points in the field. To date, however, there has been no previous work employing the Helmholtz decomposition of tensor fields. The Helmholtz decomposition has recently proved to be useful in the topological analysis of vector fields. Polthier and Preuss [7] use a discrete Helmholtz decomposition to robustly locate singularities in vector fields. Li et al. [8] used the Helmholtz decomposition to segment 2D discrete vector fields. Tong et al. [9] describe vector fields in a multiscale framework by defining a vector field scale space in terms of the separate scale spaces of the solenoidal and irrotational parts of the field. We apply a similar principal, and decompose the high order tensor field into multiple components, and visualize each separately.

Topological methods for analyzing and visualizing tensor fields hold promise for simplifying these rich and complex datasets. Some developments in topological tensor field visualization have proceeded by generalizing the concepts of vector field topology. Degenerate points (in 2D tensor fields) and degenerate lines (in 3D tensor fields) have commonly been defined in terms of eigenvectors of the tensors. Separatrices in the tensor case are hyperstreamlines, or integral curves of the eigenvector field. Local maxima of our scalar measures can be interpreted as topological features since they serve to identify generalized sources, sinks and vortices of the field. Several approaches to topological tensor field visualization have been described in previous literature. Many consider the topology of the dominant eigenvector field [10,11] and define degenerate points as locations where two or more eigenvalues are equal to each other. Zheng et al. [12] described categories of feature points and numerically stable methods for extracting them and then joining them to form feature lines.

Approaches specific to diffusion tensor MRI have considered the topology of scalar fields of tensor invariants as defined by crease lines. Tricoche et al. [13] use this framework applied to tensor mode (which is related to the skewness of eigenvalues), and Kindlmann et al. [14] used fractional anisotropy (which is related to the variance of eigenvalues). Another approach based on degenerate lines derived from probabilistic tractography has been described by Schultz et al. [15]. We hope to mitigate some of the concerns expressed in their work by not relying on eigenvectors or streamlines at all.

2   Background

In this section we will describe the divergence, curl and gradient of Cartesian tensors as given by Heinbockel [16], and then define the generalized Helmholtz decomposition for tensor fields.

2.1   Tensor Notation

The *order* of the tensor (referred to as *rank* in some literature) is the number of indices into it. Tensors of order 0, 1, and 2 are represented by scalars, vectors and matrices respectively. If d is the *dimension* of the tensor, then each index can take one of d different values. In 3 dimensions a rank-ℓ tensor then has 3^{ℓ} components. A tensor may also have two different types of indices, covariant and contravariant, usually denoted using subscripts and superscripts. For Cartesian tensors these are equivalent, so will denote indices using only subscripts.

In writing an expression containing tensors, we will use the Einstein summation convention. This means that repeated indices are to be multiplied pairwise, and summed over all possible values,

$$A_{i_1 i_2 \ldots i_\ell} B_{i_1 i_2 \ldots i_\ell} = \sum_{i_1=1}^{d} \sum_{i_2=1}^{d} \ldots \sum_{i_\ell=1}^{d} A_{i_1 i_2 \ldots i_\ell} B_{i_1 i_2 \ldots i_\ell}. \tag{1}$$

Partial derivatives will be denoted by the ∂ symbol where $\partial_i \equiv \frac{\partial}{\partial x_i}$. This notation permits tensor equations to be expressed in a very compact manner.

2.2   Tensor Field Differential Operators

Tensor field divergence. In general, the divergence of an order n tensor field is an order $(n-1)$ tensor field. For $n = 2$ the divergence is given in Einstein notation as $\operatorname{div}(D_{ij}) = \partial_i D_{ij}$. This notation indicates that for all possible values of index i, the tensor components are differentiated with respect to that index and summed over. Expressions for the divergence of higher order tensors can be obtained by placing free indices on the right, for example, $\operatorname{div}(D_{ijkl}) = \partial_i D_{ijkl}$.

The divergence often appears in conservation laws (such as Fick's second law governing diffusion) which impose conservation of mass. In the context of diffusion-weighted MRI, we expect divergence to predict the presence of fanning, bifurcating and crossing fiber tracts.

Tensor field curl. The curl of an order n tensor field is an order $(n + d - 3)$ tensor field in d dimensions. For the order 4 case in 3 dimensions it is defined as

$$\operatorname{curl} D = \varepsilon_{ijk}(\partial_j D_{klmn}) \tag{2}$$

where ε_{ijk} is the permutation tensor

$$\varepsilon_{ijk} = \begin{cases} +1 & (i, j, k) \text{ is an even permutation of indices} \\ -1 & (i, j, k) \text{ is an odd permutation of indices} \\ 0 & \text{otherwise.} \end{cases} \tag{3}$$

The permutation tensor is often used to define the vector cross product $u \times v = \varepsilon_{ijk} u_j v_k$. The curl is commonly used to characterize vortices and shear in flows. In the context of diffusion-weighted MRI, we expect curl to predict the presence of bending fiber tracts.

Tensor field gradient. The gradient of an order n tensor field is an order $(n + 1)$ tensor field. For $n = 4$ the gradient is given by

$$\text{grad}\, D = \partial_i D_{jklm}. \tag{4}$$

2.3   Helmholtz Decomposition of Tensor Fields

The Helmholtz decomposition [17] of a vector field, v, is given by

$$v = \nabla\phi + \nabla \times \psi + h \tag{5}$$

where $\nabla\phi$ is the gradient of a scalar potential field ϕ, $\nabla \times \psi$ is the curl of a vector stream field ψ and h is a harmonic vector field. Note that $\nabla\phi$ is irrotational, so it is useful for identifying features such as local maxima and minima of divergence (foci of sources and sinks) in v without interference from curl-based features. Likewise, $\nabla \times \psi$ is solenoidal, and is useful for isolating centers of vortices in v. The harmonic vector field, h, is both solenoidal and irrotational and typically is of small magnitude.

Using the previously defined operators we extend the Helmholtz decomposition to 2nd and 4th order tensor fields as

$$D_{ij} = \partial_i\phi_j + \varepsilon_{imn}(\partial_m\psi_{nj}) + H_{ij} \tag{6}$$

$$D_{ijkl} = \partial_i\phi_{jkl} + \varepsilon_{imn}(\partial_m\psi_{njkl}) + H_{ijkl} \tag{7}$$

Just as in the vector field case, we have $\text{div}(\text{curl}\,\psi) = 0$ and $\text{curl}(\text{grad}\,\phi) = 0$. The formulation can be made for tensors of arbitrary order, but we present the order 2 and 4 decompositions since those are the basis for the experiments in Section 4.

3   Methods

For computational purposes we will be reshaping tensor fields into column vectors. For each tensor component, the elements of the field are vectorized in lexical order of the spatial coordinates (x, y, z). The components are ordered within the vector according to lexical order of indices. An input 2nd-order tensor field, D, with spatial dimensions $n \times m \times p$ is then vectorized as $[D_{xx} D_{xy} D_{xz} D_{yx} D_{yy} D_{yz} D_{zx} D_{zy} D_{zz}]^T$ which has $9mnp$ rows.

We will represent the discretized operators as block matrices where the blocks correspond to finite difference operators applied to a single tensor component. For 3D fields the multidimensional difference matrices are given by

$$\Delta^x = I_{p\times p} \otimes I_{m\times m} \otimes \Delta_{n\times n}$$
$$\Delta^y = I_{p\times p} \otimes \Delta_{m\times m} \otimes I_{n\times n}$$
$$\Delta^z = \Delta_{p\times p} \otimes I_{m\times m} \otimes I_{n\times n} \tag{8}$$

where $I_{n \times n}$ is an $n \times n$ identity matrix, $\otimes$ is the Kronecker product and $\Delta_{n \times n}$ is an $n \times n$ finite difference matrix. We use central differences for approximating derivatives, in which case Δ is given by

$$\Delta = \frac{1}{2} \begin{bmatrix} 0 & +1 & 0 & \cdots & 0 \\ -1 & 0 & +1 & \ddots & \vdots \\ 0 & -1 & 0 & \ddots & 0 \\ \vdots & \ddots & \ddots & \ddots & +1 \\ 0 & \cdots & 0 & -1 & 0 \end{bmatrix}. \tag{9}$$

This definition of this matrix may be modified as needed to impose boundary conditions on the tensor field.

We can approximate the curl of the second order tensor field ψ_{ij} as $\mathbf{C}\psi$, where

$$\mathbf{C} = \begin{bmatrix} 0 & 0 & 0 & -\Delta^z & 0 & 0 & \Delta^y & 0 & 0 \\ 0 & 0 & 0 & 0 & -\Delta^z & 0 & 0 & \Delta^y & 0 \\ 0 & 0 & 0 & 0 & 0 & -\Delta^z & 0 & 0 & \Delta^y \\ \Delta^z & 0 & 0 & 0 & 0 & 0 & -\Delta^x & 0 & 0 \\ 0 & \Delta^z & 0 & 0 & 0 & 0 & 0 & -\Delta^x & 0 \\ 0 & 0 & \Delta^z & 0 & 0 & 0 & 0 & 0 & -\Delta^x \\ -\Delta^y & 0 & 0 & \Delta^x & 0 & 0 & 0 & 0 & 0 \\ 0 & -\Delta^y & 0 & 0 & \Delta^x & 0 & 0 & 0 & 0 \\ 0 & 0 & -\Delta^y & 0 & 0 & \Delta^x & 0 & 0 & 0 \end{bmatrix}, \psi = \begin{bmatrix} \psi_{xx} \\ \psi_{xy} \\ \psi_{xz} \\ \psi_{yx} \\ \psi_{yy} \\ \psi_{yz} \\ \psi_{zx} \\ \psi_{zy} \\ \psi_{zz} \end{bmatrix}. \tag{10}$$

Similarly, the gradient of the first order tensor field ϕ_i is given by $\mathbf{G}\phi$ where

$$\mathbf{G} = \begin{bmatrix} \Delta^x & 0 & 0 \\ 0 & \Delta^x & 0 \\ 0 & 0 & \Delta^x \\ \Delta^y & 0 & 0 \\ 0 & \Delta^y & 0 \\ 0 & 0 & \Delta^y \\ \Delta^z & 0 & 0 \\ 0 & \Delta^z & 0 \\ 0 & 0 & \Delta^z \end{bmatrix}, \phi = \begin{bmatrix} \phi_x \\ \phi_y \\ \phi_z \end{bmatrix}. \tag{11}$$

The discretized operators for 4th order tensors are not given here since they contain 81 rows each, but they are easily generated from the equations in the previous sections.

To perform the generalized Helmholtz decomposition we solve the least squares problem

$$\min_{\psi, \phi} ||D - \mathbf{C}\psi - \mathbf{G}\phi||_F^2 \tag{12}$$

where $||\cdot||_F$ denotes the Frobenius norm of the tensor $||X_{ik}||_F = \text{trace}(X_{ij}X_{jk})$.

Using the fact that $\mathbf{C}^T\mathbf{G} = \mathbf{G}^T\mathbf{C} = 0$ we implement this numerically by alternately solving the normal equations

$$\mathbf{C}^T\mathbf{C}\psi = \mathbf{C}^T D \qquad (13)$$
$$\mathbf{G}^T\mathbf{G}\phi = \mathbf{G}^T D$$

using a stabilized biconjugate gradients method until convergence is reached. Although the matrices on the left-hand sides of Equation (13) are symmetric, they are not positive-definite, so the standard conjugate gradients method cannot be used. The derivatives of all tensor components are constrained to be zero across each boundary. We do not explicitly solve for H, the harmonic part of the field, but instead let $H = D - \operatorname{grad}\phi - \operatorname{curl}\psi$.

4   Results

The generalized Helmholtz decomposition was implemented in Matlab and run on a system with Intel Quad Core QX6700 2.66 GHz CPU and 4 GB RAM. The algorithm was applied to the synthetic and real datasets as described below.

A synthetic second order tensor field was generated from the sources and vortices shown in Figure (1) by computing $D = (D_1 + D_2 + D_3 + D_4)^2$. The tensor fields in Figures (1,2) are visualized by plotting the radial surfaces $r(x) = D_{ij}x_ix_j$ for unit vectors x. The surface is colored blue when r is positive and red when r is negative. The results of the generalized Helmholtz decomposition are shown in Figure (2).

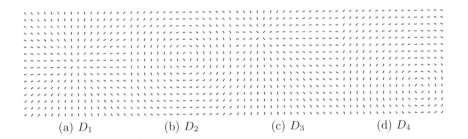

(a) D_1 (b) D_2 (c) D_3 (d) D_4

Fig. 1. Vortices and sources used to construct the synthetic field

Another synthetic tensor field was generated from sources and vortices similar to those shown in Figure (1), but modeled as fourth order tensors. The results of the generalized Helmholtz decomposition of this field are shown in Figure(3). The tensor fields are visualized by plotting the radial surfaces $r(x) = D_{ijkl}x_ix_jx_kx_l$ for unit vectors x. The surface is colored blue when r is positive and red when r is negative. Several interesting observations can be made from these results. The critical points in the original field Figure (2a) are not clearly visible, but in the decomposed fields they are quite evident. In the decomposed fields there

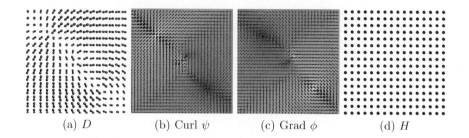

(a) D (b) Curl ψ (c) Grad ϕ (d) H

Fig. 2. Helmholtz decomposition results for 2nd order synthetic tensor field

seems to be a correspondence between sources of positive-definite tensors and vortices of negative-definite tensors. The harmonic field, which is typically of small magnitude for vector field decompositions, can be substantial in terms of the tensor trace, but it is extremely smooth - nearly constant in all of our synthetic field experiments.

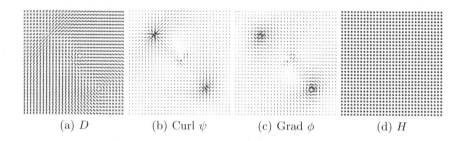

(a) D (b) Curl ψ (c) Grad ϕ (d) H

Fig. 3. Helmholtz decomposition results for 4th order synthetic tensor field

The decomposition was also applied to diffusion tensor MRI of the human brain. The data were acquired at the Center for Advanced Imaging at WVU on a 3.0 Tesla General Electric Medical Systems Horizon LX imaging system with a diffusion weighted spin echo pulse sequence. Imaging parameters were : effective TR = 9000 ms, TE = 78 ms, NEX = 1. Diffusion-weighted images were acquired with 25 different gradient directions with $b = 1000 \frac{s}{mm^2}$ and a single image was acquired with $b \approx 0$. The image field of view was 24×24cm and the acquisition matrix was $256 \times 256 \times 30$. Order 2 and 4 tensors were computed from the diffusion weighted images by performing a least squares fit to the logarithm of the signal attenuation.

Denoting the irrotational part of the field as $D_\phi = \text{grad}\,\phi$ and the solenoidal part as $D_\psi = \text{curl}\,\psi$, we show images of $\|\,\text{div}\,D_\phi\|$ and $\|\,\text{curl}\,D_\psi\|$ in Figure (5). Images of fractional anisotropy [18] are also presented for comparison. Note, however, that our new scalar measures are based on differential operators applied to tensor fields generated by a global optimization procedure, unlike FA which is simply computed on a voxel-by-voxel basis. As such, these new measures are

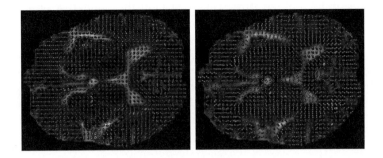

Fig. 4. Real data, order 2. Curl ψ (left), Grad ϕ (right).

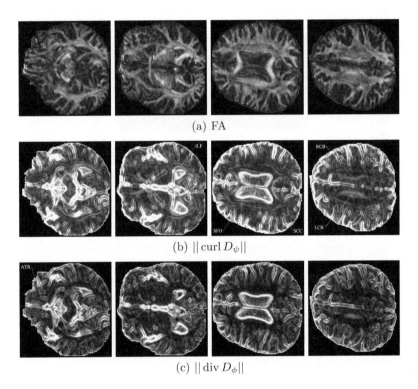

(a) FA

(b) $\|\operatorname{curl} D_\psi\|$

(c) $\|\operatorname{div} D_\phi\|$

Fig. 5. A comparison of scalar measures of white-matter structure. ILF: interior longitudinal fasciculus, SFO: superior fronto-occipital fasciculus, SCC: splenium of corpus callosum, RCB/LCB: right/left cingulum bundle, ATR: Anterior thalamic radiation.

sensitive to the large changes in diffusivity which occur at the cortical surface and the boundaries of the ventricles. Away from these boundaries it is clear that the critical points do form coherent linear and curved regions in the field, as predicted by previous work.

Compared to FA, our measures seem to be more discriminative, often revealing thinned structures. This can be understood in relation to vector field topological visualization technique which use divergence and curl to locate centers of features such as vortices, sources and sinks. These form line structures in 3D flows, and we see analogous behavior from the generalized measures in 3D tensor fields. In the bottom image of Figure (5b) the left and right cingulum bundles are visible as a pair of bright horizontal regions. We note that the curl image seems to convey much more structural information than the divergence image. This may be due to the incompressibility of water resulting in smaller fluctuations in divergence.

Results for 4th order tensors computed from the same diffusion weighted data as above are presented in Figure (6). The displayed slice is the same as the second column from the left in Figure (5).

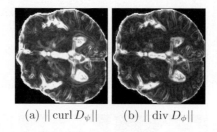

(a) $\|\operatorname{curl} D_\psi\|$ (b) $\|\operatorname{div} D_\phi\|$

Fig. 6. Scalar measures of the 4th order tensor field

5   Conclusions and Future Work

The generalized Helmholtz decomposition can provide intuitive and useful information about the structure of tensor fields. Based on this decomposition, new scalar measures for DT-MRI can be formulated which convey topological information. Specifically, local peaks in magnitude of divergence and curl correspond to critical lines in the tensor field. The formulations we presented are general with respect to tensor order and do not require eigenvalues to be computed. The decomposition and the new scalar measures are easy to compute. In the future we wish to develop a more complete topological characterization of high order tensor fields, including more types of critical points and separatrices. We also wish to explore the potential field ϕ and stream field ψ to see if useful information can be extracted directly from them.

References

1. Basser, P.J., Mattiello, J., LeBihan, D.: MR diffusion tensor spectroscopy and imaging. Biophys. J. 66, 259–267 (1994)
2. Basser, P.J.: Inferring microstructural features and the physiological state of tissues from diffusion weighted images. NMR in Biomed. 8, 333–344 (1995)

3. Basser, P.J.: New histological and physiological stains derived from diffusion-tensor MR images. Ann. N. Y. Acad. Sci. 820, 123–138 (1997)
4. Pierpaoli, C., Basser, P.J.: Toward a quantitative assessment of diffusion anisotropy. Magn. Reson. Med. 36, 893–906 (1996)
5. Özarslan, E., Mareci, T.: Generalized diffusion tensor imaging and analytical relationships between diffusion tensor imaging and high angular resolution diffusion imaging. Magnetic Resonance in Medicine 50, 955–965 (2003)
6. Ozarslan, E., Vemuri, B., Mareci, T.: Generalized scalar measures for diffusion MRI using trace, variance, and entropy. Magnetic Resonance in Medicine 53, 866–876 (2005)
7. Polthier, K., Preuss, E.: Identifying vector fields singularities using a discrete hodge decomposition. Visualization and Mathematics 3, 113–134 (2003)
8. Li, H., Chen, W., Shen, I.: Segmentation of Discrete Vector Fields. IEEE Transactions on Visualization and Computer Graphics, 289–300 (2006)
9. Tong, Y., Lombeyda, S., Hirani, A., Desbrun, M.: Discrete multiscale vector field decomposition. ACM Transactions on Graphics 22, 445–452 (2003)
10. Delmarcelle, T., Hesselink, L.: The topology of symmetric, second-order tensor fields. In: Proceedings of the conference on Visualization 1994, pp. 140–147. IEEE Computer Society Press, Los Alamitos (1994)
11. Hesselink, L., Levy, Y., Lavin, Y.: The topology of symmetric, second-order 3D tensor fields. IEEE Transactions on Visualization and Computer Graphics 3, 1–11 (1997)
12. Zheng, X., Parlett, B., Pang, A.: Topological lines in 3D tensor fields and discriminant Hessian factorization. IEEE Transactions on Visualization and Computer Graphics 11, 395–407 (2005)
13. Tricoche, X., Kindlmann, G., Westin, C.: Invariant Crease Lines for Topological and Structural Analysis of Tensor Fields. IEEE Transactions on Visualization and Computer Graphics 14, 1627–1634 (2008)
14. Kindlmann, G., Tricoche, X., Westin, C.F.: Delineating white matter structure in diffusion tensor MRI with anisotropy creases. Medical Image Analysis 11, 492–502 (2007)
15. Schultz, T., Theisel, H., Seidel, H.P.: Topological visualization of brain diffusion MRI data. IEEE Transactions on Visualization and Computer Graphics 13, 1496–1503 (2007)
16. Heinbockel, J.: Introduction to tensor calculus and continuum mechanics. Trafford Publishing (2001)
17. Arfken, G., Weber, H.: Mathematical Methods For Physicists. Academic Press, London (2005)
18. Basser, P., Pierpaoli, C.: Microstructural and physiological features of tissues elucidated by quantitative-diffusion-tensor MRI. Journal of Magnetic Resonance, Series B 111, 209–219 (1996)

Automatic Data-Driven Parameterization for Phase-Based Bone Localization in US Using Log-Gabor Filters

Ilker Hacihaliloglu[1], Rafeef Abugharbieh[1], Antony Hodgson[2], and Robert Rohling[1,2]

[1] Department of Electrical and Computer Engineering
[2] Department of Mechanical Engineering, University of British Columbia, Vancouver, BC, Canada
ilkerh@ece.ubc.ca, rafeef@ece.ubc.ca, ahodgson@mech.ubc.ca, rohling@ece.ubc.ca

Abstract. Intensity-invariant local phase-based feature extraction techniques have been previously proposed for both soft tissue and bone surface localization in ultrasound. A key challenge with such techniques is optimizing the selection of appropriate filter parameters whose values are typically chosen empirically and kept fixed for a given image. In this paper we present a novel method for contextual parameter selection that is adaptive to image content. Our technique automatically selects the scale, bandwidth and orientation parameters of Log-Gabor filters for optimizing the local phase symmetry in ultrasound images. The proposed approach incorporates principle curvature computed from the Hessian matrix and directional filter banks in a phase scale-space framework. Evaluations performed on in vivo and in vitro data demonstrate the improvement in accuracy of bone surface localization compared to empirically set parameterization results.

Keywords: Ultrasound, local phase features, principle curvature, automatic parameter selection, phase symmetry, bone localization, Log Gabor filters.

1 Introduction

Extraction of tissue and bone boundaries from ultrasound (US) images is particularly challenging due to the typically low signal-to-noise ratio and the presence of artifacts which significantly complicate image interpretation and automatic processing. Image intensity and gradient based methods have shown some promise but still remain highly influenced by the image intensity variations and imaging artifacts [1,2]. Intensity invariant local phase based feature extraction has attracted some attention and has been shown to be very promising for processing US images of soft tissue [3] and, more recently, bone surfaces [4].

Although local phase measures can be quite successful in extracting important image features, they remain somewhat sensitive to the underlying filter parameters used. Previous approaches using local phase relied on empirical selection of appropriate filter parameters, which was typically performed by trial and error and ad hoc investigation of filter outputs on samples of US images depicting a certain anatomical area

G. Bebis et al. (Eds.): ISVC 2009, Part I, LNCS 5875, pp. 944–954, 2009.

of interest [3-5]. Once acceptable filter parameters are found, they are typically fixed for subsequent application to new data. The difficulty in relating correct parameter choices to the properties of the image and image-processing task has inhibited more widespread use of phase-based techniques.

In this work, we present a novel method for automatically selecting the scale, bandwidth, orientation and angular bandwidth parameters of the Log-Gabor filter for calculating phase symmetry (PS) measures in US images, specifically in the context of bone surface localization. The proposed approach incorporates the use of principal curvature computed from the Hessian matrix and directional filter banks in a phase scale-space framework. To the best of our knowledge, this is the first study that investigates the automatic selection of these different parameters for ultrasound images. Our technique relies on contextual information obtained solely from the image content. Qualitative and quantitative evaluations performed on *in vivo* and *in vitro* scans demonstrate the utility of the our parameter selection approach, its insensitivity to artifacts when detecting bone boundaries, and the improvements achieved in terms of surface localization accuracy.

2   Methods

We address the problem of filter parameter selection within the context of localization of bone surfaces in US images, which is one of the emerging imaging modalities in computer assisted orthopaedic surgery (CAOS) applications [1,2,4,5]. Hacihaliloglu *et al* recently presented a local phase-based method for extracting ridge-like features similar to those that occur at soft tissue/bone interfaces using a PS measure [4]. We propose an improvement to such an approach by proposing a complete automation of the parameter selection process. The current paper focuses on extraction of ridge-like features but could be extended to other feature types.

Local phase information is computed by convolving the image with a quadrature pair of band pass filters. In this work, we use the Log-Gabor filter [3-5]. This orientation-dependent 2D Log-Gabor filter is defined in the frequency domain (ω) by multiplying a one dimensional Log Gabor function that controls the frequencies to which the filter responds with an angular Gaussian function that controls the orientation selectivity of the filter (1):

$$G(\omega,\phi) = \exp(\frac{(\log(\omega/\omega_0))^2}{2(\log(\kappa/\omega_0))}) \times \exp(\frac{(\phi-\phi_0)^2}{2\sigma_\phi}) \tag{1}$$

where κ is the standard deviation of the filter in the radial direction and ω_0 is the filter's center spatial frequency. The term κ/ω_0 is related to the bandwidth (β) of the filter with $\beta = -2 \, (2/\ln2)^{(-0.5)}\ln(\kappa/\omega_0)$. The scaling of the radial Log Gabor function is achieved by using different wavelengths which are based on multiples of a minimum wavelength, λ_{min}. Angular bandwidth σ_ϕ is the standard deviation of the Gaussian spreading function in the angular direction that describes the filter's angular selectivity. To obtain higher orientation selectivity, the angular function must become narrower. Steering of the filter is achieved by changing its angle (ϕ_0). In the following sections, we analyze the Log-Gabor filter response in detail and present a data-driven approach for contextual selection of the main filter parameters: bandwidth, orientation, scale and angular bandwidth. We first demonstrate how to select the optimal filter bandwidth based on image acquisition parameters, and then proceed to select the filter orientations, scale and angular bandwidth.

2.1 Filter Bandwidth Selection

The proper filter bandwidth (β= -2 $(2/\ln2)^{(-0.5)}\ln(\kappa/\omega_0)$) in the radial direction is related to both the spatial extents of the speckle and boundary responses in the image. Therefore, we first estimate the image speckle size by selecting a set of images covering a range of depths and acquired by the same US transducer (center ultrasound frequency = 7.5 MHz, image sizes ranged between 1.9cm-7.2cm), isolating a region with fully developed speckle from each image, computing the autocorrelation of each region, and extracting the full-width at half-maximum (FWHM) of these autocorrelations as a measure of the speckle size [6]. We compute the ratio, κ/ω_0, for each image using (2):

$$\kappa/\omega_0 = \exp(-\frac{1}{4}\sqrt{2 \times ln(2)} \times FWHM \times r) \qquad (2)$$

where r is the pixel size in mm, and average the κ/ω_0 ratio over the set of 25 different B-mode US test images; the resulting average is chosen as the filter bandwidth. Selecting a bandwidth significantly greater than this (ie, selecting a smaller value for κ/ω_0) will result in a filter that fails to separate small scale speckle features from larger scale boundary responses. Selecting a significantly lower bandwidth will reduce the accuracy of the boundary detection and cause blurring of the detected bone boundary (Fig.1).

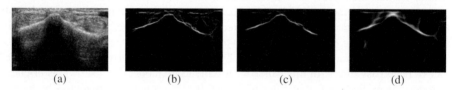

(a)   (b)   (c)   (d)

Fig. 1. Effect of filter bandwidth on local phase based bone detection. (a) *in vivo* B-mode US image of human distal radius, (b) – (d) PS images obtained using κ/ω_0 values of 0.05, 0.24, and 0.55 respectively. (b) illustrates unintended speckle detection at high bandwidths and (d) illustrates bone boundary blurring at low bandwidths.

2.2 Filter Scale Selection

Local image PS is computed by convolving the image with a number of scaled Log-Gabor filters. Each scaling is designed to pick out particular features of the image being analyzed with results typically integrated over multiple scales (in addition to multiple orientations) [4]. Since boundaries are extracted by analyzing the PS measure over a range of scales, correct scale selection is of major importance.

When using very small scales, the filters become highly sensitive to speckle. Selecting larger scales blurs the extracted bone features. Simply integrating different filter scales for PS calculations is insufficient resulting in PS images that extract speckle or blurring the detected features (in our case bone boundaries), as demonstrated in Fig. 2.

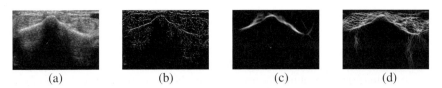

(a)   (b)   (c)   (d)

Fig. 2. Effects of filter scale selection. (a) Original B-mode US image of *in vivo* distal radius, (b) PS obtained using a scale value of λ_{min} =2, (c) PS obtained using a scale value of λ_{min} =88, (d) PS obtained by combining the results of both scales (2 and 88).

Line enhancing filters based on multiscale eigenvalue analysis of the Hessian matrix have been commonly used to extract vessel-like structures in 2D and 3D medical images [7]. The scale selection approach we present in this paper is inspired by these studies where we use the Log-Gabor filter response as the input to the Hessian matrix defined as in (2):

$$ H = \begin{bmatrix} L_{xx} & L_{xy} \\ L_{yx} & L_{yy} \end{bmatrix}, \text{with } L_{ab} = \frac{\partial^2 L}{\partial a \partial b} \tag{2} $$

L is an image obtained convolving the US image with a Log-Gabor filter at a particular scale. Here, the subscripts x and y represent spatial derivatives in the x and y directions. At this stage, the orientation of the Log-Gabor filter during the scale setting step is set to the initial filter angle calculated from the B-mode US image as explained below in the filter orientation selection step We calculate a ridge strength measure as $A_y = t^{2\gamma}((L_{xx} - L_{yy})^2 + 4L_{xy}^2)$, which is the square of the γ normalized eigenvalue difference, and t is the scale of the filter ($t = \lambda_{min}$) [8]; see Fig. 3. This metric in our context measures the 'ridgeness content' of an image, since our main interest here is in localizing bone contours, which generally appear as ridges in US images. The optimal scale is then defined as the one corresponding to the maximal ridgeness content in the Gabor filtered image. In order to define the optimum global filter scale, which gives

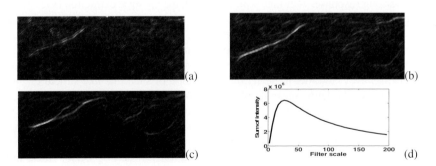

Fig. 3. Effects of filter bandwidth selection. A_y ridge strength obtained from B-mode US image in Fig.1 (a) for a fixed filter orientation (140°) and scale (a) λ_{min}=10, (b) λ_{min}=35, (c) λ_{min}=140. Investigating (a)-(c) we see that the bone ridge content in (b) is the strongest and the most continuous. (d) Filter scale versus sum of intensity values of A_y.

the most significant ridge content in the image, we analyze the intensity distribution of A_γ over all possible scales (e.g. ranging from 2-150) and select the scale where the sum of the intensities reaches a maximum value as our optimal filter scale (Fig.3 (d)). This is based on the observation that at the optimal scale the response of the filter will produce a sharp ridge feature aligned with the bone surface, whereas significantly different scales will result either in detection of speckle or blurred bone surfaces which will give a smaller intensity sum (Fig. 3 (a)–(c)). This analysis is repeated for each orientation separately.

2.3   Filter Orientation Selection

The orientation of the Log Gabor filter is controlled by the angular Gaussian function used in (1). During the calculation of the PS metric, the filter is directed at a number of orientations. Commonly, six orientations are employed to cover the entire angular range (0°-180° with 30° increments) with the responses subsequently averaged [3-5].

| (a) | (b) | (c) | (d) |

Fig. 4. Effects of filter orientation selection. (a) B-mode US of *in vivo* distal radius, (b) filter response at ϕ=60°, (c) filter response at ϕ=120°, (d) filter response at ϕ=0°. All images were produced at a fixed filter scale of λ_{min} =25 and κ/ω_0 =0.25.

   However, given the highly directional nature of ultrasound bone image data, integration of the responses at all of these different filter orientations in fact largely degrades the PS response due to the inclusion of many non-relevant filter orientations. Noting that the strongest ridge features appear when the filter orientation is perpendicular to the bone surface (Fig. 4), identifying and combining filter angles which produce strong responses will therefore likely enhance feature extraction (Fig. 5).

   Bone surfaces in US images typically appear as elongated line-like objects with a higher intensity compared to the other features. The same analysis is true for the PS images. Therefore, integration along a bony feature produces a higher intensity value than doing the integration along a non-bony feature. We thus make use of a simple radon transform (RT) in order to detect the orientation of such line-like structures. In order to automatically define meaningful starting angles for our filter, we initially cluster the RT (obtained from the B-mode US image) image using k-means clustering (Fig. 6).

   The projection angles corresponding to the peak values of the RT reflect the angles that are perpendicular to the high intensity features, such as bone surfaces. Those angles are therefore used for initializing the orientations of the Log-Gabor filter. In order to obtain three initial filter angles, we chose the cluster that corresponds to the peak values of the RT (Fig. 6. (c)) and calculated the mean value of the projection angles corresponding to the RT values in that chosen cluster. This calculated mean value is used as the initial angle, and we add two additional angles set at ±1 standard deviation within the thresholded region. These three initial angles are used as the filter angles during the calculation of the filter scale as explained in section 2.2.

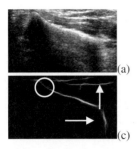
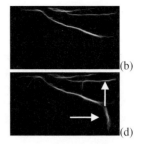

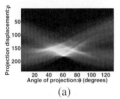
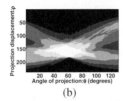
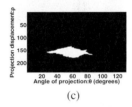

Fig. 5. Effect of varying the number of orientations used. (a) B-mode US image of *in vivo* human distal radius. (b) PS image using 3 orientations (58° 74° 89°), (b) PS image using 6 orientations (0°-150° with 30° increments), (c) PS image using 10 orientations (0°-270° with 30° increments). White arrows point out the extracted phase features which are not bone surfaces obtained by combining orientations which are not perpendicular to the bone surface during the calculation of PS. The white circle points to a location of a degraded bone surface due to the inclusion of less informative orientations with weaker bone responses.

Fig. 6. Filter orientation selection. (a) RT of B-mode US image in Fig.1 (a), (b) clustered RT of (a), (c) the cluster corresponding to the highest RT values. The three initial angles deduced from this cluster are 66°, 88° and 106° (refer to text for details about the calculation of these three angles).

In order to obtain the final filter orientations, the RT is re-calculated for the ridge strength image A_γ as obtained using the scale calculated in section 2.2. Figure 7 shows the calculated RT of the A_γ for the initial angles of 66° and 106°. Investigating the figures, we note that the RT has high intensity locations indicating the presence of line-like structures in the image. The maximum value of the RT indicates the main orientation of the bone, since it has the strongest filter response, and is thus used to set the filter orientation. Figure 7 (c) and (d) show an example where the angles corresponding to the peak occur at 62° and 115°, hence the initial angles will be corrected based on these new calculated angles.

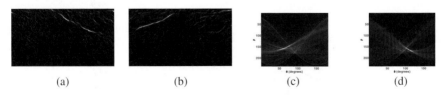

Fig. 7. Filter orientation selection. (a) & (b): A_γ obtained using the initial filter angle ϕ=66° (a) and ϕ=106° (b) which are calculated from the RT of the B-mode image, (c) & (d) RT of (a) & (b) showing new peaks at 62° and 115°, respectively.

2.4 Filter Angular Bandwidth Selection

The angular bandwidth σ_ϕ parameter corresponds to the standard deviation of the Gaussian spreading function in the angular direction and describes the filter's angular selectivity. Investigating the example in Fig. 7, we can see that at large angular bandwidths the Log-Gabor filter acts as a smoothing filter without being sensitive to any orientation. On the other hand, for small angular bandwidths, the filter acts like a line detector degrading the curvature of the bone surface as it becomes less sensitive to curvature making the extracted features look like short line segments. Therefore, using the same analysis we used in our filter scale selection process would not be suitable to set σ_ϕ since the intensity distribution of A_γ over all possible angular bandwidths will give a peak for very large angular bandwidths (Fig. 9 (a)) which would correspond to a filter response shown in Fig.8 (a).

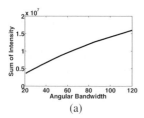
(a)

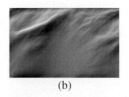
(b)

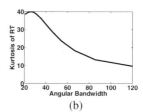
(c)

Fig. 8. Effect of varying angular bandwidth on the Log-Gabor filter output for filter orientation 115°. (a) $\sigma_\phi=120°$, (b) $\sigma_\phi=30°$, (c) $\sigma_\phi=7.5°$.

Based on the above observations, we analyze the kurtosis of the RT of A_γ over different angular bandwidth values. We select the bandwidth corresponding to the peak kurtosis value (Fig.8 (b)). During this stage, the A_γ images used are obtained using the optimum filter scale as calculated in section 2.2.

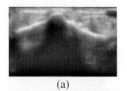
(a)

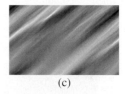
(b)

Fig. 9. Angular bandwidth selection. (a) Filter angular bandwidth versus sum of intensity values of A_γ. (b) filter angular bandwidth versus kurtosis of RT obtained from calculating the RT of A_γ.

2.5 Experimental Setup for Quantitative Validation

We constructed a phantom comprised of an *ex vivo* bovine femur specimen inside an open-topped plexiglass cylindrical tube (Fig. 10). Twenty-eight markers (1mm diameter steel balls) were added to the construct with fourteen beads placed on each side of the bone (longitudinally) and spaced at equal axial intervals over a distance of 75 mm. We obtained US scans of this specimen where the volumes containing 16 fiducials (8 on each side) spanning a region of 37.8mm. To hold the specimen and fiducials in place during the scanning procedures, the tube was filled with a firm gel (Super Soft Plastic, M-F Manufacturing, Texas, USA). We note that we did compensate for the difference in the speed of sound in soft tissue and gel during the image reconstruction process.

The constructed phantom was scanned in an Xtreme CT machine with isometric 0.25 mm voxels. US scanning was performed using a 3D GE Voluson 730 Expert system (GE Healthcare, Waukesha, WI) with a 3D RSP5-12 probe. The US data was subsequently rescaled to match the resolution of the CT image. Also, a fiducial-based rigid-body registration was applied to align the CT and US volumes. Following registration, bone surfaces were extracted from CT scan using a simple thresholding operation to establish a gold standard bone surface. Bone surfaces were then extracted from the US data using the discussed PS features, both with and without the parameters optimized.

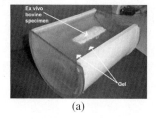
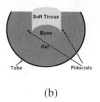

| (a) | (b) |

Fig. 10. Quantitative validation experiment. (a) Picture of the constructed phantom, (b) Axial view schematic sketch of the phantom.

The accuracy of bone localization in US (PS filtered scans) was quantitatively assessed by averaging the PS response at voxels corresponding to varying distances from the real bone surface (as obtained from the CT gold standard). To achieve this, we studied a set of (intensity, distance) pairs using the signed distance transform. High intensity values confined to a zone near zero distance would indicate an accurately located surface. Our surface matching error was hence defined by the average signed distance values corresponding to the maximum phase intensity value along each vertical column of the 2D PS images.

3   Results

Implementation Details: The proposed method was implemented in MATLAB (The Mathworks Inc., Natick, MA, USA). The filter bandwidth was calculated as 0.24 by using the proposed method in section 2.1. During the scale selection process the γ value was set to 0.75 since this was reported to be optimal value for ridge feature detection [8]. For filter orientation we chose to work with three angles since choosing greater than three orientations had an insignificant effect on the results.

Quantitative Results: The surface matching mean error was 0.64 mm (std 1.34) with the best empirically-set parameters compared to 0.51mm (std 0.64) for our proposed automatically-set parameters. Choosing two scales for the empirical method decreases the surface matching mean error to 0.54mm (std: 1.82) but introduces more outlier points away from the zero signed distance indicating an increase in the detection of US artifacts.

Qualitative Results: Figure 11 shows a qualitative comparison of PS images of a *in vivo* human distal radius and pelvis obtained with the proposed optimized Gabor filter paramters and contrasted to the best values that could be empirically set. Note how the local phase images obtained empirically using 2 scales extracted more US artifacts and resulted in a thicker bone boundary due the unsuitable scale combination. Moreover, integrating the zero angle as one of the filter orientations caused the detection of unwanted features on the sides of the bone surface (Fig. 6.white arrows). Decreasing the filter scale to 1 in the empirical case caused gaps in the extracted bone surfaces (Fig. 6.white circles). Our surface results on the other hand, which used optimized filter parameters, were consistently sharper with reduced unwanted features on the

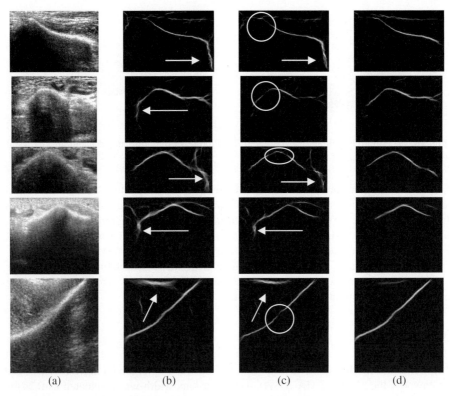

(a)   (b)   (c)   (d)

Fig. 11. Qualitative results. (a) B-mode US image of in vivo distal radius (rows1-4) and pelvis (row 5) areas. Imaging depth of the US machine was 3.5, 3.5, 1.9, 4.5 and 4.9cm, respectively. (b) Non-optimized PS images using two different filter scales (25 and 75), (c) Non-optimized PS images using one scale only (25)(d) PS images obtained using the proposed optimized parameters. White arrows point out the extracted phase features which are not bone surfaces obtained by combining orientations which are not perpendicular to the bone surface during the calculation of PS. The white circles points to a location of a degraded bone surface due to the inclusion of less informative orientations with weaker bone responses.

<table>
<tr><td>(a)</td><td>(b)</td><td>(c)</td><td>(d)</td></tr>
</table>

Fig. 12. Qualitative results on fractured *ex vivo* porcine tibia fibula specimen. (a) B-mode US image, (b) Non-optimized PS images using two filter scales (25 and 75), (c) Non-optimized PS images using one scale only (25), (d) PS images obtained using the proposed optimized parameters. The white circles point out the extracted phase features which are not bone surfaces obtained by combining orientations which are not perpendicular to the bone surface during the calculation of PS.

bone sides and with no gaps in the detected surfaces. Figure 12 shows other qualitative examples where scans of a fractured *ex-vivo* porcine tibia fibula specimen were acquired. Note how the proposed method produced a cleaner identification of the bone fracture.

4   Discussion and Conclusions

Though local image phase information has been successfully applied for extracting US image features, none of the prior studies has investigated the effects of parameter selection on the extracted features nor provided guidelines on how this could be achieved. In this paper, we proposed a novel approach for automatic selection of the scale, bandwidth and orientation of Log-Gabor filters for calculating phase symmetry responses in bone US. For scale selection, we used a ridgeness content measure obtained from the Hessian matrix eigenvalues to investigate the information content extracted at different scales. For orientation selection, the appearance of bone surfaces was incorporated within our framework where the RT obtained from the image ridgeness content measure was used to deduce the optimal angles of the directional filter. We analyzed US images with fully developed speckle and measured the image speckle size by calculating the autocorrelation function to determine the filter bandwidth. Finally the angular bandwidth was calculated by investigating the kurtosis of the RT obtained from the image ridgness content measure. We presented qualitative results obtained from *in vivo* and *ex vivo* scans and demonstrated the critical importance of selecting the correct scales and orientations in local phase based US processing. Quantitative results were also presented on a specially constructed bone phantom where the gold standard surface of the bone was established through CT imaging. An improvement of close to 0.2mm in bone localization accuracy was observed. Future work will include the extension of this automatic parameter selection method to 3D and a clinical study where the proposed method will be tested on scans obtained from patients with distal radius and pelvis fractures.

References

1. Barratt, D.C., Penney, P.G., Chan, S.K., Slomczykowski, M., Carter, T.J., Edwards, P.J., Hawkes, D.J.: Self calibrating 3D-ultrasound-based bone registration for minimally invasive orthopaedic surgery. IEEE Transactions on Medical Imaging 25, 312–323 (2006)

2. Kowal, J., Amstutz, C., Langlotz, F., Talib, H., Ballester, M.: Automated bone contour detection in ultrasound B-mode images for minimally invasive registration in computer assisted surgery an *in vitro* evaluation. The International Journal of Medical Robotics and Computer Assisted Surgery, 341–348 (2007)
3. Mulet-Parada, M., Noble, J.A.: 2D+T boundary detection in echocardiography. Medical Image Analysis 4(1), 21–30 (2000)
4. Hacihaliloglu, I., Abugharbieh, R., Hodgson, A.J., Rohling, R.: Bone segmentation and fracture detection in ultrasound using 3D local phase features. In: Metaxas, D., Axel, L., Fichtinger, G., Székely, G. (eds.) MICCAI 2008, Part I. LNCS, vol. 5241, pp. 287–295. Springer, Heidelberg (2008)
5. Hacihaliloglu, I., Abugharbieh, R., Hodgson, A.J., Rohling, R.N.: Enhancement of bone surface visualization from 3D ultrasound based on local phase information. In: Proc. IEEE Ultrasonics Symposium, pp. 21–24 (2006)
6. Wagner, R.F., Smith, S.W., Sandrik, J.M., Lopez, H.: Statistics of Speckle in Ultrasound B-Scans. IEEE Transactions on Sonics and Ultrasonics, 156–163 (1983)
7. Frangi, A., Niessen, W., Vincken, K., Viergever, M.: Multiscale vessel enhancement filtering. In: Wells, W.M., Colchester, A.C.F., Delp, S.L. (eds.) MICCAI 1998. LNCS, vol. 1496, pp. 130–137. Springer, Heidelberg (1998)
8. Lindeberg, T.: Edge detection and ridge detection with automatic scale selection. International Journal of Computer Vision 30, 117–154 (1998)
9. Coifman, R.R., Wickerhouser, M.V.: Entropy based algorithms for best basis selection. IEEE Transaction on Information Theory (1992)

Wavelet-Based Representation of Biological Shapes

Bin Dong[1], Yu Mao[2], Ivo D. Dinov[3], Zhuowen Tu[3], Yonggang Shi[3], Yalin Wang[3], and Arthur W. Toga[3]

[1] Department of Mathematics, University of California, San Diego,
9500 Gilman Drive, La Jolla, CA, 92093-0112
[2] Department of Mathematics, University of California,
Los Angeles, CA, 90095-1555
[3] Center for Computational Biology, Laboratory of Neuro Imaging,
635 S. Charles Young Dr., #225, University of California,
Los Angeles, CA, USA, 90055

Abstract. Modeling, characterization and analysis of biological shapes and forms are important in many computational biology studies. Shape representation challenges span the spectrum from small scales (e.g., microarray imaging and protein structure) to the macro scale (e.g., neuroimaging of human brains). In this paper, we present a new approach to represent and analyze biological shapes using wavelets. We apply the new technique to multi-spectral shape decomposition and study shape variability between populations using brain cortical and subcortical surfaces. The wavelet-space-induced shape representation allows us to study the multi-spectral nature of the shape's geometry, topology and features. Our results are very promising and, comparing to the spherical-wavelets method, our approach is more compact and allows utilization of diverse wavelet bases.

1 Literature Reviews

Imaging, representation, geometric modelling and topological characterization of shape and form are important components of Computational Biology. They apply across the vast length scales between genotypes to phenotypes, from the small scale of microarray imaging for genomic, to the larger scale of neuroimaging of human brains. Here we review the existing techniques and algorithms and present a new approach for representation and analysis of biological shapes using wavelets. We apply the new method to multi-spectral shape decomposition and study shape variability between populations using brain cortical and subcortical surfaces.

Recently, N. Hacker et al. used conformal mapping and spherical wavelets to analyze biological shapes (see [1,2,3]). Their idea is first mapping the original shape onto a unit 2-sphere using a certain conformal mapping so that one obtains a $\mathbb{R}^3$-valued function f defined on the sphere; and then interpolate the function onto the regular triangular mesh on the sphere (which is generated by recursively

G. Bebis et al. (Eds.): ISVC 2009, Part I, LNCS 5875, pp. 955–964, 2009.

subdividing an icosahedron); and then finally, apply spherical wavelet transform to the interpolated function. The spherical wavelets they used were introduced by P. Schröder and W. Sweldens in [4], which were constructed using lifting scheme (see W. Sweldens [5] and F. Arandiga et. al. [6]).

In this section, we also start from a $\mathbb{R}^3$-valued function f determined by a certain mapping from $\mathbb{R}^3$ to S^2. After that, we linearly interpolate the function onto a triangular mesh, which is generated by recursively subdividing an octahedron in $\mathbb{R}^3$ (not restricted on the sphere) and then transforming the mesh onto the sphere. This method was first introduced by E. Praun and H. Hoppe (see [7]) in the context of computer graphics. The major advantage of it is that we can transform the subdivided octahedron to a unit square so that we obtain an image with $\mathbb{R}^3$-valued entries (which were called geometric image in [7]), and then we can apply traditional X-lets (e.g. wavelets, framelets, curvelets etc.) decomposition. In this way we have plenty of good bases and frames (redundant systems) to choose according the application we have.

Understanding the relationship between the structure and function of the human brain in vivo has been the driving motivation for many neurosciences research for centuries. The research efforts not only focus on studying normal development but also understanding alterations in various clinical populations including schizophrenia, Huntington's disease, Alzheimer's disease, Williams Syndrome, autism, stroke, chronic drug abuse, as well as pharmacological interventions. For instance, there are multiple studies underway to quantify the differences between the brain structure of schizophrenic patients and healthy individuals in different stages of this disease. Detection of these significant differences via neuroimaging studies is not only useful to elucidate the link between change in cognitive profile and change in brain structure, but also to improve diagnosis particularly in early stages of the disease. With the increasing interest in carrying out such studies with large numbers of subjects, there is a need for a unified framework for image segmentation to identify the structure of interest (e.g. caudate, ventricles, cerebral cortex, sulcal regions) (see Z. Tu et. al. [8]), and morphometric analysis which requires methods for shape representation, shape comparison, and change in shape measurement (see P. Thompson and A. Toga [9]).

2   Method

In this section we describe our test data and the specific approach we took to represent shape using wavelets, as well as the statistical analysis we carry on the wavelet-based shape decomposition to identify group, population, time or variation differences. Throughout this paper, all shapes are assumed to be close surfaces in $\mathbb{R}^3$ with genus zero.

2.1   Data

Cortical Models: Surface objects of normal subjects and Williams syndrome patients were used to explore the power of our method to synthesize the energy of the shape content in a few wavelet coefficients. The demographics of the population included age (29.2 9.0), genders (approximately 50/50) and IQ scores, P. Thompson et. al. [10]. Non-brain tissue (i.e., scalp, orbits) was removed from the images, and each image volume was re-sliced into a standard orientation who "tagged" 20 standardized anatomical landmarks in each subject's image data set that corresponded to the same 20 anatomical landmarks defined on the ICBM53 average brain (see Mazziotta et al. [11], Thompson et. al. [12]). Automated tissue segmentation was conducted for each volume data set to classify voxels as most representative of gray matter, white matter, CSF, or a background class (representing extra cerebral voxels in the image) on the basis of signal intensity. The procedure fits a mixture of Gaussian distributions to the intensities in each image before assigning each voxel to the class with the highest probability, Shattuck et. al. [13]. Then each individual's cortical surface was extracted and three-dimensionally rendered using automated software, MacDonald [14]. Each resulting cortical surface was represented as a high-resolution mesh of 131,072 surface triangles spanning 65,536 surface points.

Hippocampal surfaces: High-resolution MRI scans were acquired from 12 AD patients ages 68.4 1.9 and 14 matched controls 71.4 0.9, each scanned twice, 2.1 0.4 years apart. 3D parametric mesh models of the left and right hippocampi and temporal horns were manually, Thompson et. al. [15]. For each scan, a radio frequency bias field correction algorithm eliminated intensity drifts due to scanner field inhomogeneity, using a histogram spline sharpening method, Sled et. al. [16]. Images were then normalized by transforming them to ICBM53 stereotaxic space, Evens et. al. [17], with automated image registration software, Collins et. al. [18]. To equalize image intensities across subjects, registered scans were histogram-equalized.

Each 3D surface is mapped to a unit sphere in $\mathbb{R}^3$ with 1-to-1 correspondence. For cortical surfaces, the conformal mapping method of Shi et. al. [19,20], is used for hippocampal surfaces and that of Gu et. al. [21] for cortical surfaces. For hippocampal surfaces, harmonic maps to the sphere are computed under the constraints of a set of automatically detected landmark curves (see [19,20]). Figures (a) and (b) in Figure 1 shows a hippocampal shape and the mapping of it on a unit sphere respectively, where in the latter, (x, y, z) values on each vertex of the sphere are color-coded by RGB.

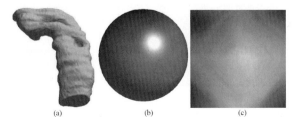

(a)   (b)   (c)

Fig. 1. Figure (a) is the given shape; (b) is the spherical function obtained by surface mapping; (c) is the geometric image

2.2 Wavelet-Based Representation for Shapes

From the previous section, we got a function f defined on the unit sphere S^2, which has vector values in $\mathbb{R}^3$. However, the values of the function were only given on an irregular grid on the sphere. To apply the wavelet transform, we need to get the value of the function on a much more regular spherical grid. There are many approaches to get such kind of grids. The construction of the spherical mesh grid, sometimes called spherical triangular map, is an interesting subject itself (see e.g. Buss and Fillmore [22], and Praun and Hoppe [7]). The basic idea is to start from a polyhedral base, which gives a simple but perfect grid on sphere, and then use some appropriate scheme to subdivide the mesh. A comparison of such techniques can be found in [7].

In our approach, we start from a recursive subdivision of the octahedral base. By mapping the subdivision grid onto the unit sphere, we get a regular grid structure on the sphere, and the function values on such a spherical grid can be obtained by linear interpolation. The reason we choose octahedron is that it can be unfolded to a plain image easily. Therefore, we can build a 1-1 map between a sphere and an image without too much distortion, and the data of a shape is transformed to a $\mathbb{R}^3$-valued function defined on a plain image, which gives a geometric image (as shown in (c) Figure 1). Since the mesh on the plain image is nothing but a Cartesian grid, a huge family of X-lets can be used to analyze properties of the geometric image. The wavelets that we shall use in the following experiments are Daubechies' Biorthogonal Wavelets [23]. We note that the boundary condition is a little complicated in this case. Topological saying, the two halves of each side of the image must be identified with each other (see [7] for more details). Thus, we need to setup corresponding boundary rules for the wavelet filters.

We now summarize the entire multiscale representation process in the following Algorithm 1. Figure 2 shows how the decomposition is carried out to the

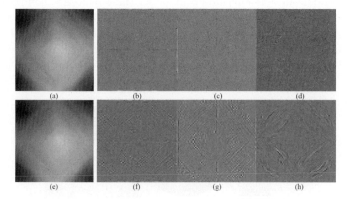

Fig. 2. Two-levels wavelet decomposition: (a) and (e) are the low frequency coefficients of level one and two; (b)-(d) are wavelet coefficients of level 1 and band 1-3; (f)-(h) are wavelet coefficients of level 2 and band 1-3

geometric image we have. Since all low frequency and wavelet coefficients has x, y, and z three components, all coefficients are visualized as color images.

Algorithm 1. Wavelet-Based Representation for Shapes

Given some triangulated biological shape (V, T), where $V \in \mathbb{R}^3$ is the vertex set and T is triangulation.

1. Find the mapping $\mathcal{M} : V \mapsto S^2$, which also induces a triangulation on S^2 denoted as $(V_S, T_S) =: M_S$. Define $f_0 = \mathcal{M}^{-1} : S^2 \mapsto V$ (Figure 1(b)).

2. Recursively subdivide an octahedron uniformly in $\mathbb{R}^3$ up to certain level N. Then project the mesh on to S^2 and obtain a mesh on S^2 denoted as $(V_N, T_N) =: M_N$.

3. Interpolate f_0 from M_S to M_N and obtain a new spherical function f, which, by construction, can be easily transformed to a geometric image (Figure 1(c)).

4. Perform regular X-let decomposition and reconstruction (with proper boundary conditions).

2.3   Multiscale Curvature-Like Characterization

As shown in Figure 2 above, for each level and band of the wavelet coefficients, we have x, y, and z three components. The coefficient vectors reflect details of the shape at each position and scale. Indeed, the wavelet vectors can be treated as the displacement between the observed position and the predicted position calculated from the convolution of the wavelet filter and the scale coefficients of the neighboring vertices. Therefore, the direction of the wavelet vector gives us some information of the local geometric properties. For example, if we consider the wavelet vector at a local sunken area, then the approximated position interpolated from the neighboring vertices should be outer than the observed vertex, which means that the wavelet vector is pointing outwards.

However, we cannot tell the geometric property of the shape from the wavelet coefficients directly, and a single wavelet coefficient itself is geometrically meaningless, so we combine the three components by calculating the inner product of the wavelet vector and the normal vector of the shape. As what we explained above, these inner products reflect geometric property at the corresponding position. In this way, a so called multiscale curvature-like characterization of the shape is obtained: For given scale (or level), we compute normal of the shape

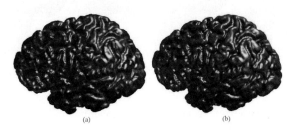

(a)   (b)

Fig. 3. Red regions in figure (a) and (b) are the sulcal and gyral regions respectively

at all positions under that scale and take the inner product of the normal with the wavelet coefficient vector. We then obtain a set of curvature-like coefficients within each level and band. The statistical analysis given in the following section is based on this representation. Figure 3 shows how this representation can be used to find cortical sulci and gyri.

3   Numerical Experiments

3.1   Sparsity of the Representation

One of the most important properties of traditional wavelet transform is that it gives a MSR of the underlying function and the representation is sparse. We now show that our method as discussed in the previous section also gives a sparse MSR for the biological shapes we have. Figure 4 and 5 shows the MSR provided by the wavelet transform, and Figure 6 and 7 shows the sparsity of the representation, where one can see that even with only 2500 coefficients, the reconstructed shapes preserve most of the features of the original shapes.

Fig. 4. Figures left to right present a MSR of the hippocampus from courser approximation to finer approximation. The last figure is the original hippocampus.

Fig. 5. Figures from left to right present a MSR of the cortex from courser approximation to finer approximation. The last figure is the original cortex.

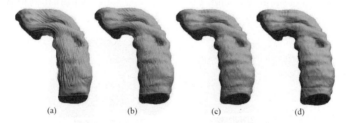

(a)   (b)   (c)   (d)

Fig. 6. Figures (a)-(c) are the reconstructed hippocampus using 1000, 2500 and 5000 coefficients, and the relative errors of them from the original shape are 1.171871e-004, 6.172233e-005 and 3.915038e-005 respectively. Figure (d) is the original hippocampus.

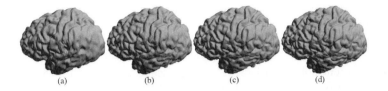

(a)   (b)   (c)   (d)

Fig. 7. Figures (a)-(c) are the reconstructed cortices using 1000, 2500 and 5000 coefficients, and the relative errors of them from the original shape are 6.449351e-004, 3.154665e-004 and 1.664914e-004 respectively. Figure (d) is the original cortex.

One advantage of our method over spherical wavelet transform in analyzing biological shapes is that we have a much more flexible choice of wavelets. In particular, we can choose one wavelet with very high vanishing moments so that the representation is very sparse. Figure 8 below shows a comparison of our method to spherical wavelets as used in [1,2,3].

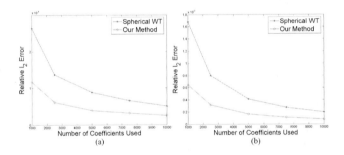

(a)   (b)

Fig. 8. Figures (a) and (b) are the decay of relative ℓ_2 error verse number of coefficients used, where the underline shapes are the hippocampus and cortex as shown in Figure 6 and 7

The multiscale sparse representation provided by the wavelet transform has many applications. For example, one can do shape compression, or in other words, feature dimension reduction for shapes. One can also do shape denoising via thresholding or shrinkage of wavelet coefficients. Since these kinds of applications are not of our main interest, we shall not explore them in further details.

3.2   Non-parametric Tests

For given two groups of hippocampus, one from healthy population, the other from the one with Alzheimer's disease, we apply Wilcoxon's rank sum test (see e.g. [24]) to the multiscale curvature-like representation of shapes to find regions on the shape where the two groups are different. The -value we choose in the results below is 0.05. We note that the tests are more reliable in higher levels than those in lower levels. This is because the shape corresponding to a higher level is a smoothed version of the original shape, which means we have more statistical inference of the object. As one can see from below that the higher the level is, the larger each area of significance will be. In Figure 9 below, we used the mean hippocampus as the reference shape, which is calculated by simply taking an average of all the vector values of hippocampus at every position.

(a)   (b)

Fig. 9. Figures (a) and (b) show regions of significance of level 5 and 4 respectively. Here, each pair of hippocampus is viewed from the bottom.

4   Discussion

The wavelet-based shape representation technique proposed here allows one to study the geometry, topology and features of general biological shapes using any of the standard wavelet-bases on real-valued Euclidean spaces. The results we obtained are robust and consistent across individuals and populations. In addition to direct representation and shape characterization, this technique allows us to compute mean shapes and improve the shape-analysis statistical power by concentrating the energy of the shapecharacteristics in few significant wavelet coefficients, Dinov et. al. [5]. We are in the process of validating the new methodology using larger number of subjects, different types of applications (e.g., studying the population-specific differences in the proportion of gyri to sulcal area) and quantitative comparison with spherical-harmonics, spherical wavelets and tensor-based morphometry techniques.

The computational complexity of the algorithm is $O(N \log N)$ relative to the volume size N. We have a Matlab implementation that we are in the process of converting to stand-alone C++ code. We tested the actual computation time of the wavelet decomposition and reconstruction on a PC with Inter(R) Core(TM) 2, 2.13 GHz and 1G physical memory. For a given shape with 65,536 surface points, the computation time is 2-20 seconds, depending on the choice of basis and level of decomposition.

Acknowledgement

This work is funded by the National Institutes of Health through the NIH Roadmap for Medical Research, Grant U54 RR021813. Information on the National Centers for Biomedical Computing can be obtained from http://nihroadmap.nih.gov/bioinformatics

References

1. Nain, D., Haker, S., Bobick, A., Tannenbaum, A.: Multiscale 3d shape analysis using spherical wavelets. In: Duncan, J.S., Gerig, G. (eds.) MICCAI 2005. LNCS, vol. 3750, pp. 459–467. Springer, Heidelberg (2005)
2. Nain, D., Haker, S., Bobick, A., Tannenbaum, A.: Shape-driven 3D segmentation using spherical wavelets. In: Larsen, R., Nielsen, M., Sporring, J. (eds.) MICCAI 2006. LNCS, vol. 4190, pp. 66–74. Springer, Heidelberg (2006)
3. Nain, D., Haker, S., Bobick, A., Tannenbaum, A.: Multiscale 3-d shape representation and segmentation using spherical wavelets. IEEE Transactions on Medical Imaging 26, 598 (2007)
4. Schröder, P., Sweldens, W.: Spherical wavelets: Efficiently representing functions on the sphere, pp. 161–172 (1995)
5. Sweldens, W.: The lifting scheme: A construction of second generation wavelets. SIAM Journal on Mathematical Analysis 29, 511 (1998)
6. Aràndiga, F., Donat, R., Harten, A.: Multiresolution based on weighted averages of the hat function I: Linear reconstruction techniques. SIAM Journal on Numerical Analysis 36, 160–203 (1999)
7. Praun, E., Hoppe, H.: Spherical parametrization and remeshing. ACM Transactions on Graphics 22, 340 (2003)
8. Tu, Z., Zheng, S., Yuille, A., Reiss, A., Dutton, R., Lee, A., Galaburda, A., Dinov, I., Thompson, P., Toga, A.: Automated extraction of the cortical sulci based on a supervised learning approach. IEEE Transactions on Medical Imaging 26, 541 (2007)
9. Thompson, P., Toga, A.: A framework for computational anatomy. Computing and Visualization in Science 5, 13–34 (2002)
10. Thompson, P., Lee, A., Dutton, R., Geaga, J., Hayashi, K., Eckert, M., Bellugi, U., Galaburda, A., Korenberg, J., Mills, D., et al.: Abnormal cortical complexity and thickness profiles mapped in Williams syndrome. Journal of Neuroscience 25, 4146–4158 (2005)
11. Mazziotta, J., Toga, A., Evans, A., Fox, P., Lancaster, J., Zilles, K., Woods, R., Paus, T., Simpson, G., Pike, B., et al.: A probabilistic atlas and reference system for the human brain: International Consortium for Brain Mapping (ICBM). Philosophical Transactions of the Royal Society of London Series B 356, 1293 (2001)
12. Thompson, P., Hayashi, K., De Zubicaray, G., Janke, A., Rose, S., Semple, J., Herman, D., Hong, M., Dittmer, S., Doddrell, D., et al.: Dynamics of gray matter loss in Alzheimer's disease. Journal of Neuroscience 23, 994 (2003)
13. Shattuck, D., Sandor-Leahy, S., Schaper, K., Rottenberg, D., Leahy, R.: Magnetic resonance image tissue classification using a partial volume model. NeuroImage 13, 856–876 (2001)
14. MacDonald, J.: A method for identifying geometrically simple surfaces from three-dimensional images. PhD Thesis (1998)

15. Thompson, P., Hayashi, K., de Zubicaray, G., Janke, A., Rose, S., Semple, J., Hong, M., Herman, D., Gravano, D., Doddrell, D., et al.: Mapping hippocampal and ventricular change in Alzheimer disease. Neuroimage 22, 1754–1766 (2004)
16. Sled, J., Zijdenbos, A., Evans, A.: A nonparametric method for automatic correction of intensity nonuniformity in MRI data. IEEE Transactions on Medical Imaging 17, 87–97 (1998)
17. Evans, A., Collins, D., Neelin, P., MacDonald, D., Kamber, M., Marrett, T.: Three-dimensional correlative imaging: applications in human brain mapping. In: Thatcher, R.W., Hallett, M., Zeffiro, T., John, E.R., Huerta, M. (eds.) Functional Neuroimaging: Technical Foundations, pp. 145–162 (1994)
18. Collins, D., Neelin, P., Peters, T., Evans, A.: Automatic 3D intersubject registration of MR volumetric data in standardized Talairach space. Journal of computer assisted tomography 18, 192 (1994)
19. Shi, Y., Thompson, P., de Zubicaray, G., Rose, S., Tu, Z., Dinov, I., Toga, A.: Direct mapping of hippocampal surfaces with intrinsic shape context. Neuroimage 37, 792–807 (2007)
20. Shi, Y., Thompson, P., Dinov, I., Osher, S., Toga, A.: Direct cortical mapping via solving partial differential equations on implicit surfaces. Medical image analysis 11, 207–223 (2007)
21. Gu, X., Wang, Y., Chan, T., Thompson, P., Yau, S.: Genus zero surface conformal mapping and its application to brain surface mapping. IEEE Transactions on Medical Imaging 23, 949–958 (2004)
22. Buss, S., Fillmore, J.: Spherical averages and applications to spherical splines and interpolation. ACM Transactions on Graphics (TOG) 20, 95–126 (2001)
23. Daubechies, I.: Ten lectures on wavelets. CBMS-NSF Lecture Notes, SIAM, nr. 61 (1992)
24. Gibbons, J., Chakraborti, S.: Nonparametric statistical inference (2003)

Detection of Unusual Objects and Temporal Patterns in EEG Video Recordings

Kostadin Koroutchev[1],

Elka Korutcheva[2], Kamen Kanev[3], Apolinar Rodríguez Albariño[4],
Jose Luis Muñiz Gutierrez[5], and Fernando Fariñaz Balseiro[6]

[1] Universidad Autónoma de Madrid, Spain
[2] UNED, Madrid, Spain and ISSP, BAS, Sofia, Bulgaria
[3] Research Inst. Electronics, Shizuoka University, Tokio, Japan
[4] Hospital Universitario La Paz, Madrid, Spain
[5] CIEMAT, Madrid, Spain
[6] Hospital de Talavera de la Reina, Spain

Abstract. In this paper we show that by using a modification of our previously developed probabilistic method for finding the most unusual part of a 3D digital image, we can detect the temporal intervals and areas of interest in the signals/video and mark the corresponding objects that behave in an unusual way.

Due to the different dynamics along the temporal and the spatial axes, namely the prevalence of the cylinder-like objects in the video and the pseudo-periodic slowly changing spectral characteristics of the bio-electrical signals, an additional step is needed to treat the temporal axis.

One of the possible practical applications of the method can be in Intensive Care hospital Units (ICU), where EEG video recording is a standard practice to ensure that a potentially life-threatening event can be detected even if its indications are present only in a fraction of the observed signals.

Keywords: EEG, video, novelty detection, epilepsy.

1 Introduction

Increased availability of simultaneous bio-electric and video recordings in hospitals prompts for their automatic processing. There are various algorithms for extracting information from single modality recordings although integral systems that simultaneously observe sets of different measurements are quite rare. Recordings are often monitored and processed by highly trained professionals but such experts are difficult to secure on 24-hours shift teams. Automated monitoring of recordings and the reliable detection of seizures with no obvious manifestations in the ICU is therefore very important and could save many lives.

The goal of the current method is to detect all temporal interval and spatial segments of the EEG video (Fig. 1) that are not similar to the rest of the recording and eventually make them available to skilled personnel and/or raise automatic alarm.

G. Bebis et al. (Eds.): ISVC 2009, Part I, LNCS 5875, pp. 965–974, 2009.

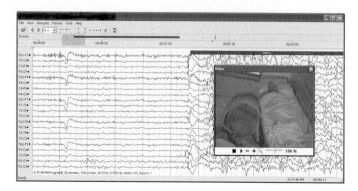

Fig. 1. Example of EEG Video. The EEG as well as another bio-electrical signals are registered synchronously with a video.

As a final practical goal we are interested in detection of epileptic and non-epileptic seizures, particularly the silent ones without apparent manifestations.

Let us first construct a 3D geometric figure from the video by stacking the frames one on top of the other. The goal of the algorithm is to find 3D objects $V_S(t, x, y)$ with shape $x, y \in S$ that are rare in the sense that the distance between $V_S(t, x, y)$ and any other translated object $V_S(t + t', x + x', y + y')$ is the largest or close to the largest possible. The exact solution of this problem is computationally hard, because it requires comparison between the translated objects. Therefore we follow our previous method designed to solve probabilistically that problem. We give a summary of the methods in the following paragraphs.

As we have previously stated [1,2], we need first of all a mathematical definition of the term "most unusual part". For doing this, we chose some shape S within the image A, that could contain that part and we denote the cut of the figure A with shape S and origin $\boldsymbol{r}$ by $B_S(\boldsymbol{\rho}; \boldsymbol{r})$, i.e.

$$B_S(\boldsymbol{\rho}; \boldsymbol{r}) \equiv S(\boldsymbol{\rho})A(\boldsymbol{\rho} + \boldsymbol{r}),$$

where $\boldsymbol{\rho}$ is the in-shape coordinate vector, $\boldsymbol{r}$ is the origin of the cut B_S and $S(.)$ is the characteristic function of the shape S. Further in this paper we will omit the arguments of B_S when not necessary. We can suppose that the most unusual part is the one that has the largest distance with the rest of the cuts with the same shape. Strictly mathematically, we can suppose that the most unusual part is located at the point $\boldsymbol{r}$, defined by:

$$\boldsymbol{r} = \arg\max_{\boldsymbol{r}} \min_{\boldsymbol{r}':|\boldsymbol{r}'-\boldsymbol{r}|>\mathrm{diam}(S)} ||B_S(\boldsymbol{r}) - B_S(\boldsymbol{r}')||. \tag{1}$$

Here we assume that the shifts do not cross the border of the image. We use L_2 norm. As the parts of an image that intersect significantly are similar, we do not allow the shapes located at r' and r to intersect, avoiding this by the restriction on $r' : |\boldsymbol{r}' - \boldsymbol{r}| > \mathrm{diam}(S)$.

2   The Method

The minima estimation of Eq. (1) is complicated because the blocks are multi-dimensional. However one can simplify it by projecting the blocks $B \equiv B_S(\boldsymbol{r})$ and $B_1 \equiv B_S(\boldsymbol{r_1})$ in one dimension using some projection operator X. For this aim, we consider the following quantity:

$$b = |X.B_1 - X.B| = |X.(B_1 - B)|, \quad |X| = 1. \tag{2}$$

The dot product in the above equation is the sum over all ρ-s:

$$X.B \equiv \sum_{\rho} X(\boldsymbol{\rho}) B(\boldsymbol{\rho}; \boldsymbol{r}). \tag{3}$$

If X is random, and uniformly distributed on the sphere of the corresponding dimension, then the mean value of b is proportional to $|B_1 - B|$; $\langle b \rangle = c|B_1 - B|$ and the coefficient c depends only on the number of points of the block, that can be treated as being its dimensionality, considering the projection operator. However, when the size of the block, e.g. its dimensionality increases, the two random vectors $(B_1 - B$ and $X)$ are almost orthogonal and the typical projection is small. If some block is far away from all the other blocks, then with some probability, the projection will be large.

We will regard only projections orthogonal to the vector with components proportional to $X_0(\rho) = 1, \forall \rho$. The projection on the direction of X_0 is proportional to the mean brightness of the area and thus can be considered as not so important characteristics of the image.

Mathematically the projections orthogonal to X_0 have the property $\sum_{\rho} X(\boldsymbol{\rho}) = 0$. The distribution of the values of the projections satisfying this property is well known and universal [3] for the 2D natural images and video cuts. The same distribution seems to be valid for a vast majority of the images.

Further we quantize the projections. If $B(r)$ and $B(r')$ have similar projections, then they will belong to one and the same or to adjacent bins. As a first approximation, we can just consider the projections and score the points according to the bin they belong to. The distribution can be described by a single parameter that, for convenience, can be chosen to be the standard deviation σ_X of the distribution of $X.B$.

The notion of "large value of the projection" will be different for different projections but will always be proportional to the standard deviation. Therefore we can define a parameter a and score the blocks with $|X.B| > a\sigma_X$.

Based on the above scheme, in order to find the most unusual blocks of shape S in an image A, we propose the following algorithm:

0. Initialize: Construct a figure B with the same shape as A and with all pixels equal to zero. The result of the algorithm will be saved in B.

1. Generate a random projection operator X, with carrier with shape S, zero mean and norm one.

2. Project all blocks, convoluting the image.

3. Calculate the standard derivation σ_X of the result of the convolution.

4. For all points of C with absolute values greater than $a\sigma_X$, increment the corresponding pixel in B.

Repeat steps 1 to 4 M times.

5. Select the maximal values of B as the most singular part of the image.

2.1   Video Signal

Let us first consider the video signal, which is a video recording of the patient in two plans and is different from medical imaging such as X-Ray, magnetic resonance, etc.

We are trying to extract the most unusual parts of the video and our first attempt would be to regard the video as a stack of two dimensional images and use the algorithm described in [2] as it is. Unfortunately this approach does not work and there are at least three problems with it.

The first one is that the objects in the video are more or less constant and therefore they form in the so constructed 3D image elongated cylinders with finite dimensions along the spatial axes and very huge extension along the time axis (Fig.2). The size of the area we choose ought to be larger than the size of the objects we detect and this requirement cannot be fulfilled in the temporal direction.

The second problem is rather specific for the type of video we use. Namely, the infrared camera is very noisy in absence of artificial light source. This gives a speckle noise comparable with the signal. The high frequency temporal component will actually mask off the signal.

The third problem is the scan frequency of the images. Having some 24-30 frames per second we need to process a huge amount of information in near real time mode. Therefore even if the method could be theoretically feasible, in practice it will not work because of the lack of productivity – in a typical ICU we have about 40 patients each one with his own video. One will need almost the full power of a modern PC in order to process even few of the video signals.

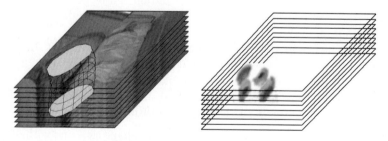

Fig. 2. Time-axes elongated cylinders as predominant elements in the video (left). The filtered video with non-zero low-cut frequency limit (right). The extend of the volumes is finite. The zero frequency component is just 2D still image.

These three problems make the method of random projections not useful in its original form. We need to modify the method in order to: (1) limit objects in the temporal dimension, (2) decrease high frequency noise and (3) reduce temporal data flow. As will be shown below, we can actually achieve the three objectives using a single technique.

In the video records we consider the objects that move to be the parts of the patient body. They do not move constantly. There are some limited time intervals in which the objects move. We can extract only the parts of the frames that are not constant and consider only these parts. Representing this as a 3D stack of frames, the duration of the movement is limited and therefore the non-constant parts have a limited size in the time dimension. However, the analysis of the scene, the segmentation and the extraction of moving areas cannot be performed very fast. A much easier technique that can be applied is to filter each pixel value using some high-pass filter along the time axis.

Analogously the high frequency speckle noise, due to the poor light conditions, can be filtered by low pass filter, which combined with the previous one can be implemented as band-pass filter. In this way we have the signal filtered with a frequency spectrum between say F_0 and F_1. According to the Nyquist theorem, if the signal is limited into the band $[F_0, F_1]$, we can sub-sample it with no loss of information for any frequency above $2|F_1 - F_0|$. Thus for 3 Hz bands we can sub-sample the signal with 6 Hz, which drops the frame rate of 24 fps video four times.

If we are particularly interested in epileptic and non-epileptic seizures, we can use the fact that most of the movements during the ictus are temblor movements with typical frequency of 3-5 Hz.

In a typical EEG video the rapid movements are rare. Most frequently in EEG video rapid temporal changes can be considered as a noise. The movements of the objects, mainly limbs, are usually smooth in the video and can be decomposed as few sinusoidal waves. Therefore among the time axes we can assume that the speed of change is relatively slow and smooth.

An additional argument in this direction is the performance of the video compression algorithms. Namely for still images the wavelets provide one of the best known compression methods, but concerning video, the digital cosine transform seems to give better results. Using this observation we introduce a temporal filter that restricts the video frequencies to the 2-5 Hz range. The filter effectively removes most of the artifacts from the video, for example the snow like noise from the infra-red (IR) videos. We found that no further enhancement of the video is required.

2.2   One Dimensional Signals

Something similar occurs with the one-dimensional bio-electrical signals. In this case it is the temporal development of the signal what we are interested partic-ularly. However, because of the nature of the bio-electrical activity we consider (EEG, Electrocardiogram (ECG), Electromyogram (EMG)), the spectrum of the signals is slowly changing in time, with exception to the artifacts and the temporal patterns (spikes, spindles, K-waves) in the signals.

Surprisingly the video and the one-dimensional signal processing seem very similar. At the first stage we filter all signals and the video in 3 Hz bands, then sub-sample the signal above the Nyquist frequency and finally process the results using the algorithm described in [2].

There are many methods that are suitable for analyzing EEG signals, (see [4] and the citations therein). Practically all procedures described in the literature concern the detection of specific events in the signal, building a model for these events. The method we describe in this paper, on the contrary, searches for the parts of the signal that are unusual compared to the rest of the signal. A similar approach but applicable only to 1D signals is presented in [5]. The random projections are used as a tool to find nearest neighbors in [6].

3   Results

3.1   Data Collection

We use records of the sleep unit of the La Paz University Hospital, Madrid, Spain as well as EEG recordings from epileptic patients at pre-operation observation from the same Hospital. We use 33 channels recoding of bio-monitoring signals. 20 channels of them have been used for EEG and the rest of the channels have been used to collect other biometric monitoring data. Most of the measurements are performed with a scan frequency 256Hz. The video signal is an infrared video with resolution 352 x 288 pixels and 24 fps. We only use the intensity component of the image. In low-light conditions and with an infrared camera, the color components carry little information. The recording time varies from 20 minutes to 8 hours. A typical screen capture of an observation is given in Fig.1. The EEG recording device is the Oxford One$^{(TM)}$. As far as we know there are no EEG video records of sufficient lengths that are public domain. To simplify the interpretation and the reproducibility of the results in this article we have used only the video, 10 EEG differential recordings and the EMG recordings.

In the case of EEG data recording without a video, we have also used the public domain data in the collection of epilepsy studies [7].

3.2   Data Processing

Historically 20 seconds windows were used to analyze the EEG records. However, the main reason for the 20 seconds intervals is the fact that the paper EEG recordings were folded in pages of 20 seconds. We found that the use of 15 second windows, starting each 5-th second of the signal, is more suitable for the analysis. Thus we can give the result of the analysis every 5 seconds. We used the first 5 seconds of each window only for padding in order to eliminate the transitional period of the filters. The maximal delay of the result of 15 seconds seems reasonable from the medical point of view.

The records were processed using the methods described earlier. It was observed that there were two types of bio-electrical signals. On the one side, the periodic signals as ECG, SaO2 and similar, that are produced as a result of the activity of some autonomous system. The histogram of the projection density of a typical signal of that type is shown in Fig. 3 (left). One can observe a typical Gaussian density. Because of a rapid fall of the tail in that case, it is relatively difficult to use these signals with the method described.

The second types of signals are produced by non-periodic complex system. As an example, a single channel EEG projection is shown in Fig. 3(right). In this article we used only this type of signals.

We de-composed all the inputs to narrow frequency bands (0.5-2.5Hz, 2.5-4.5Hz, 4-7 Hz, 7-11Hz) and subsequently sub-sampled the resulting signals at 6.4 Hz. Thus we obtained more signals with lower scan frequency, which as a set are equivalent to the original ones. Regarding seizures, the most informative band is 2.5-4.5 Hz. Using only this band gives practically the same results as using all four frequency bands. The filter we used was the Chebyshev Type I Infinite Impulse Response (IIR) filter. The pass/stop frequency band was assumed to belong to the interval from -0.5 to +0.5 Hz of the all frequency band limits with an attenuation of -40dB. The allowed pass band ripple was set to 1dB. The original scan frequency for the video was 24 fps and for the EEG and EMG signals it was fixed to 256 Hz. The filters were designed using Matlab fdatool. The number of filter stages given the bands was from 3 to 4 so the computational effort was very modest.

We tried also to use a 154 tap finite impulse response filter and a non causal IIR filter (re-passing the filter data with time axes inverted). These filters gave the same results obtained in all the previous cases. That is why we used only Chebyshev Type I causal filters.

We used 10 random projection matrices. The use of 30 random projection matrices, as in the case of 3D medical images, did not give significant advantages in that case. This is probably due to the rejection of the very low frequencies

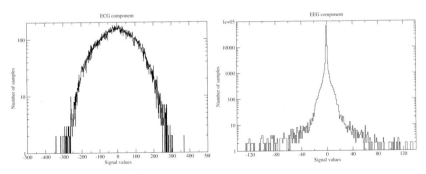

Fig. 3. Statistically we observe two types of signals - with Gaussian (left) and Exponential (right) distribution of the values

and the fact that we have used one dimensional band limited signals. For these signals the standard deviation is smaller than for the case of video.

3.3 Experimental Results and Interpretation

We analyzed the following cases:

A. Extracting the intervals with unusual events only from the one dimensional signals (principally EEG).

In this case we have also used the data from the collection of epilepsy studies [7], and the epileptic records were detected with 99% probability (only one case not detected) with one false alarm (artifacts in the record). Having in mind that the detector was not specifically designed to detect epilepsy, the results seem very good. However, although the data set [7] is good to compare different methods for epilepsy detection, we do not consider the results representative, because the records are pre-selected not to contain artifact and more importantly, they are post-selected regarding the results of the surgery – if the results were poor, the corresponding records were not included in the collection. Both conditions are artificial and not realistic in the medical practice.

The result is shown in Fig. 4. After an inspection by human specialist and annotating the EEG using different EEG records we can observe that the intervals marked as "unusual" by the algorithm are:

1) Epileptic seizures. Detected virtually 100% of the events.
2) Short arousals and artifact produced by electrode problems.
3) Snoring. Detected more than 70% of the events.
4) REM. Detected more than 50% of the events.
5) Phase II morphological elements (K-complex, spindles) when they occupy more than 30% of the 10 seconds interval. Detected about 50% of the events.

Changing the parameter a of the algorithm we can select the level of sensitivity of the system. With $a = 4$ only the first two categories are detected. With $a = 3$ all 5 categories are detected.

B. Extracting intervals and areas that correspond to unusual events from the video recording only.

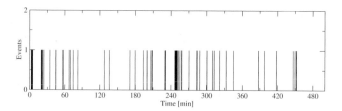

Fig. 4. Events extracted only from EEG data ($a = 3$, $M = 1$). The main features are well detected even with only one projection.

Evidently, the unusual events during the sleep or unconscious state are the movements of the person. In this sense if one is really interested of the events, the technique detects all significant movements of the parts of the body (hands, legs, head) with virtually 100% precision. The best detected is the motor reaction of the epileptic seizure, detectable at level $a = 4$. For $a = 3$ all significant limb and corporal movement can be detected. Note that by using different size of the shape of the area of interest, we can selectively detect the movement of the parts of the body without detecting the global movements. In some of the cases snoring is detectable too. We are aware that there exist many methods for video novelty interpretation (for review see [8,9]), but the computational cost of most of them exceeds the practical limits. Most of them require significant image processing, that is not plausible in the clinical setting where we intend to use our approach.

C. Mixed use of the EEG and video records.

The main problem is to give a relative weight to the projections corresponding to the video and to the bio-electrical records respectively. We choose equal weight of all EEG signals and the video. The resulting system can be tuned to capture any epileptoform motion and brain activity even in presense of electrode problems and when there is no apparent movement (Fig.5). The system has absolutely no a priori knowledge of what could be regarded as "normal" and what is "abnormal" activity. Therefore detecting seizure periods, REM activity and snoring can be considered as very good result.

Performance: Using Double Core Intel Pentium D 2.8 GHz processor, e.g. normal PC, we can process 6.4 s record in less than 5 sec. By optimizing the algorithm it is possible to achieve better performance. The method does not require memory for more than 2 frames (the result of filtering and the current frame). The time required to find the unusual parts is insignificant compared to the rest of the processing. The memory and the memory requirements for the one-dimensional signals are significantly less than that corresponding to the video. This gives us confidence that the algorithm can be used in real time.

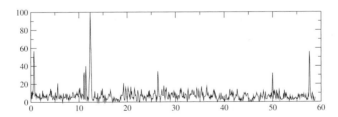

Fig. 5. Events extracted from EEG and Video ($a = 3$ $M = 20$). At level 30 (30% of the proections) only seizures and electrode problems are observed. At lower level one can also see the sleep patterns (REM etc).

4   Discussion

We presented a method that finds unusual intervals and areas in mixed one dimensional signals and video recording. The method uses random projections in order to represent the signal and it is probabilistic. The temporal evolution of the signal must be treated differently than the spatial one, due to the different dynamics involved. Namely, all signals ought to be filtered in narrow frequency bands. This is a significant change in respect to the algorithm for 3D images, because it is not previously clear that the band-limited projections can extract the unusual parts of the signal. Without having any knowledge about the different EEG events and human sleep development, the method extracts most of the important events deserving consideration. The analysis approximates the human specialist performance and therefore, although the results are preliminary, they seem rather promising. The method can be used to present to the neurologist solely the parts detected as unusual. Another possible application of the method is to add an automatic analysis on the extracted parts and to set off an alarm in the case of life treating conditions in ICU.

Acknowledgments

The work is financially supported by Spanish Grants TIN 2004–07676-G01-01, TIN 2007–66862 (K.K.) and Ministerio de Ciencia e Innovacion project FIS2009-9870 (E.K.). The work is partially supported by the Cooperative Research Project of RIE, Shizuoka University, Japan. The authors thank for the kind hospitality of the ICTP, Trieste, Italy, where this paper was completed.

References

1. Koroutchev, K., Korutcheva, E.: Detecting the Most Unusual Part of a Digital Image. In: Brimkov, V.E., Barneva, R.P., Hauptman, H.A. (eds.) IWCIA 2008. LNCS, vol. 4958, pp. 286–294. Springer, Heidelberg (2008)
2. Koroutchev, K., Korutcheva, E.: Detecting the Most Unusual Part of Two and Three-Dimensional Digital Images. Pattern Recognition
3. Ruderman, D.: The Statistics of Natural Images. Network: Computation in Neural Systems 5, 517–548 (1999)
4. Sanei, S., Chambers, J.: EEG Signal Processing. J. Wiley, Chichester (2007)
5. Keogh, E., Lin, J., Fu, A.: HOT SAX: Efficiently Finding the Most Unusual Time Series Subsequence. In: The Proceedings of 39th ACM Symposium on Theory of Computing (2007)
6. Indyk, P.: Uncertainty Principles, Extractors, and Explicit Embedding of L2 into L1. In: The Proceedings of 39th ACM Symposium on Theory of Computing (2007)
7. Polat, K., Gunes, S.: Artificial Immune Recognition System with Fuzzy Resource Allocation Mechanism Classifier, Principal Component Analysis and FFT Method Based New Hybrid Automated Identification System for Classification of EEG Signals. Expert Systems with Applications 34, 2039–2048 (2008)
8. Markou, M., Singh, S.: Novelty Detection: a Review: Statistical Approaches. Signal Processing 83(12), 2481–2521 (2003)
9. Markou, M., Singh, S., Gordiou Desmou, L.: A Neural Network-Based Novelty Detector for Image Sequence Analysis. IEEE PAMI 28(10), 1664–1677 (2006)

DRONE: A Flexible Framework for Distributed Rendering and Display

Michael Repplinger[1,2], Alexander Löffler[1],
Dmitri Rubinstein[1,2], and Philipp Slusallek[1,2]

[1] Computer Graphics Lab, Saarland University, Saarbrücken, Germany
[2] German Research Center for Artificial Intelligence (DFKI),
Agents & Simulated Reality, Saarbrücken, Germany
michael.repplinger@dfki.de, loeffler@cs.uni-sb.de,
dmitri.rubinstein@dfki.de, philipp.slusallek@dfki.de

Abstract. The available rendering performance on current computers increases constantly, primarily by employing parallel algorithms using the newest many-core hardware, as for example multi-core CPUs or GPUs. This development enables faster rasterization, as well as conspicuously faster software-based real-time ray tracing. Despite the tremendous progress in rendering power, there are and always will be applications in classical computer graphics and Virtual Reality, which require distributed configurations employing multiple machines for both rendering and display.

In this paper we address this problem and use NMM, a distributed multimedia middleware, to build a powerful and flexible rendering framework. Our framework is highly modular, and can be easily reconfigured – even at runtime – to meet the changing demands of applications built on top of it. We show that the flexibility of our approach comes at a negligible cost in comparison to a specialized and highly-optimized implementation of distributed rendering.

1 Introduction

Even though the performance available for rendering on today's hardware increases continuously, there will always be demanding applications for which a single computer is not enough, and the workload has to be distributed in a network to accomplish the desired tasks. This demand for distribution is not restricted to the rendering end, but also the display of rendered images frequently requires distribution: be it for a multi-wall projection-based Virtual Reality setup, or server-based rendering where rendering cluster and thin display client are connected across the Internet.

All these scenarios have something in common: they require a foundation that is able to provide access to a maximum of available hardware resources for their particular rendering implementation, be that in the form of processing on a single machine, or by distribution onto several machines. There are strong requirements for timing and synchronization as well, since the distribution of

G. Bebis et al. (Eds.): ISVC 2009, Part I, LNCS 5875, pp. 975–986, 2009.

rendering tasks and the display of their results are highly dependent on each other and have to be done in due time and often synchronized across different physical devices. In this paper, we present the DRONE (Distributed Rendering Object Network) architecture as a framework solution that addresses all these requirements. DRONE is based on NMM, a distributed multimedia middleware. Together, DRONE supports all of the above scenarios to meet the requirements of current and future applications.

This paper is structured as follows: in Section 2, we present related work and derive requirements for our framework. Section 3 then provides an overview of the basic technology of DRONE, before Section 4 explains and discusses its architecture in more detail. Section 5 describes the versatile command language we use for an easy setup of render graphs. Section 6 evaluates the performance of the framework and shows that the overhead of the framework is negligible in comparison to a highly optimized implementation. We conclude our paper and highlight future work in Section 7.

2   Related Work

Molnar et al. [1] presented a classification scheme for distributed rendering. The authors subdivide techniques that distribute geometry according to screen-space tiles (*sort-first*), distribute geometry arbitrarily while doing a final z-compositing (*sort-last*), or distribute primitives arbitrarily, but do per-fragment processing in screen-space after sorting them during rasterization (*sort-middle*). This separation of techniques is based on rasterization, and where the rasterization pipeline distributes the workload across multiple processors. It is difficult to apply the scheme for a generic rendering and visualization architecture supporting other techniques besides rasterization, as for example ray tracing. Here, Molnar's classification approach is no longer applicable, as geometry processing and screen-space projection are combined in the single operation of sampling the scene with rays. Instead, we will discuss several typical application scenarios that our framework should support and discuss available solution strategies.

The first application scenario (AS1), we call it *single-screen rendering*, comprises presenting rendered images on a single screen while using multiple systems for rendering. The major demand of flexibility for (AS1) is the possibility of using available systems in the network both for rendering and displaying a scene, while being independent of the network infrastructure connecting them. A distributed middleware like NMM provides network transparency, which in turn allows transparent access to distributed objects, and aids in achieving this high degree of flexibility. Another desired aspect of (AS1) is the possibility to use different rendering techniques such as ray tracing or rasterization, all working on the same scenes.

Previously presented frameworks for distributed rendering like WireGL [2], and Chromium [3] are limited to the rasterization approach. The Real-Time Scene Graph (RTSG) [4], on the other hand, provides a strict separation of the scene graph and a specific implementation of a renderer, thus making it applicable to use both rasterization and ray tracing. Equalizer [5] concentrates on rasterization as well, but also supports ray tracing as shown in RTT's Scale software [6]. However, a drawback of Equalizer as well as the other rendering frameworks is that they have fixed pipelines and do not allow flexible post-processing of rendered images.

The second application scenario (AS2), *multi-screen rendering*, extends (AS1) by splitting the resulting frame and presenting it on multiple displays simultaneously and fully synchronized. This is required for display walls, for example for large-scale terrain or industrial visualization. The major desired aspect of (AS2) is the flexibility of combining multiple displays as if they were a single one. This also includes presenting of multiple views of the same scene at the same point in time. For example, this is required for stereo imagery required for Virtual Reality installations like the CAVE.

For (AS2), we need distributed synchronization to present an image simultaneously on all displays: hardware-based solutions, e.g., using the genlock signal of the video output of special graphics cards [7] allow for exact frame synchronization, while software-based solutions like NTP, are able to synchronize PCs over the Internet with a few milliseconds of variance. We believe a flexible rendering framework should be able to support arbitrary synchronization mechanisms.

The third application scenario (AS3) is *remote rendering*. It covers situations where rendered images have to be transmitted through a network connection with limited bandwidth, often because the original data sets have to stay at a controlled and secure location. The main demand for of (AS3) is the ability to add different post processing steps, e.g., the encoding of rendered images before a network transmission. Here, the application of a *distributed flow graph* within our rendering shows its full potential by providing the means to transparently insert new processing elements in the data processing pipeline. FlowVR [8] is also based on a flow graph but neither provides the same capability of video post-processing nor the possibility to select transmission protocols (e.g., RTP) that are more suitable for sending multimedia data through an Internet connection.

The last application scenario (AS4) is *collaborative rendering*, an arbitrary number of combinations of the previously described scenarios. An ideal system scenario should allow, for example, a large control center with tiled display walls, and simultaneously thin clients only receiving some important aspects of large rendered images. This is especially interesting for collaborative work where people on different locations have to work with the same view of a scene. This requires that the application is able to share the same rendered images between multiple users while each user may be able to interact with the scene. A similar solution to a collaborative rendering scenario is provided by the CO-VISE [9] system. It does not transfer images but renderer-specific data making it renderer-dependent. Also, multiplexed views for displaying are not possible.

In summary, to our knowledge, there exists no rendering framework that is able to support all the described application scenarios.

3   Overview

The DRONE framework is build on the *Network-Integrated Multimedia Middleware* (NMM) [10]. NMM uses the concept of a *distributed flow graph* for distributed media processing, which perfectly fits the requirement of flexibility we defined for the framework. This approach provides a strict separation between media processing and media transmission as well as a transparent access to local and remote components. The *nodes* of a distributed flow graph represent specific operations (e.g., rendering, or compressing images), whereas *edges* represent the transmission between those nodes (e.g., pointer forwarding for local connections, or TCP for a network connection). Nodes can be connected to each other via their *input jacks* and *output jacks*; depending on the type of operation a node implements, their numbers may vary. Nodes and edges allow the application to configure and control media processing and transmission transparently, for instance by choosing a certain transport protocol from the application layer. Prerequisite for the successful connection of two nodes is a common *format*, which must be identical for the output jack of the predecessor node and the input jack of the successor node to be connected. NMM incorporates a unified messaging system, which allows to send control events together with multimedia data from sources to sinks, being processed by each node in-between.

3.1   The DRONE Flow Graph

The DRONE framework builds on top of an NMM flow graph consisting of custom processing nodes supplemented by existing nodes of core NMM. The ability of NMM to distribute nodes arbitrarily in the network, but still access them transparently from within an application allows the placement of application sub-tasks on arbitrary hosts, enabling high flexibility and efficient use of a cluster.

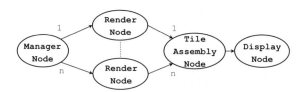

Fig. 1. This flow graph shows the general idea of distributed and parallel rendering in DRONE using n rendering nodes to render to a single output device

RenderNode: A *render node* performs the actual rendering of a scene description to a 2D image. Key principle of DRONE is rendering a single frame distributed on multiple render nodes in the flow graph. All render nodes have access to an identical copy of the scene graph. This node renders the assigned scene to a memory buffer for further processing. Since NMM transparently supports many-core architectures, e.g., GPUs and Cell [11], rendering engines using many-core architectures can also be integrated into DRONE. Requirements for integrating a rendering engine into DRONE are (1) the possibility of rendering tile frames and (2) extending a buffer into the rendering engine that is used for rendering.

ManagerNode: The single source node of the DRONE flow graph is called *manager node*; its job is to distribute the workload of rendering an image to the available render nodes. The manager node distributes the workload between render nodes by splitting the frame to be rendered into many *frame tiles* and assigning them dynamically to render nodes.

DisplayNode: A *display node* constitutes a sink of the flow graph, and simply presents any incoming image buffer synchronized according to its timestamp. Display nodes are part of core NMM and usually platform-dependent: for example, an XDisplayNode would be used on a Unix platform running the X window system.

TileAssemblyNode: A *tile assembly node* in general can receive frame tiles from all rendering nodes, and assembles them to a composite image buffer. As there is one dedicated tile assembly node for each downstream display node, the nodes receive only those tiles of the rendered image stream that are relevant for the particular display node they precede.

4   Architecture

Based on the NMM flow graph components presented in Section 3.1, DRONE provides its functionality to the application in the form of *processing blocks*, which bundle their underlying modules and provide high-level access to an application developer. Furthermore, *composite blocks* allow the application to group different processing blocks that can be treated in the same way as a single processing block afterwards. Below, we will use the application scenarios presented in Section 2 as a guide through the specific design decisions of the framework, and explain how the different processing and composite blocks fit together.

4.1   Single-Screen Rendering (AS1)

The primary processing block, which occurs in every DRONE application, is the *rendering block*. It contains those NMM components that are responsible for rendering a two-dimensional image from a 3D scene: In particular, it consists of a single manager node and at least one render node. Together, all render nodes take care of rendering a frame. The distribution among nodes is done by tiling the frame, and assigning single frame tiles to separate render nodes.

The rendering block is connected to at least one *presentation block*, which combines the tiles and displays the frame on an actual physical display device. A presentation block is a composite block that can be extended by additional nodes for post processing but contains at least two NMM nodes: a tile assembly node, and a display node. All those tiles of the rendered frame that are sent from the render nodes to the corresponding assembly node have to be displayed by the succeeding display node. Since information about the specific view is part of the connection format between a render and tile assembly node, each render node knows which frame tiles have to be sent to which tile assembly node. In the trivial case of having just one display (depicted in Figure 2), the tile assembly node receives all tiles of each frame.

DRONE also allows *user interaction* with the rendered scene. Because all render nodes render the scene, each one has to be informed about user events, such as changes to the viewpoint. In our setup, interaction events, key presses or mouse movements, first are sent as events from the application to the manager node. The manager node in turn forwards all incoming events to all connected render nodes. The key point in doing so is that input events are propagated only between tiles of *different* successive frames to avoid changes of viewpoint before the processing of a single frame is fully completed.

Load Balancing. The manager node is responsible for load balancing because it sends information about the next tile to be rendered as so called *tile events* to its successive render nodes. If the manager node and render nodes run in the same address space, the manager node is directly informed by a render node about processed tile events and the corresponding render node receives a new one as soon as its current tile event was processed. In case of a TCP connection between manager and a render node, DRONE configures the underlying TCP connection such that it stores exactly one tile event in the network stack on the side of the manager and renderer, respectively. This is possible because NMM provides access to the underlying network connection between two nodes. As soon as the render node starts rendering a tile event, NMM reads the next tile event from the network stack on the side of the render node and forwards it to the node for processing. The tile event stored in the network stack on the side of the manager node is automatically requested by the flow control mechanism of TCP and transmitted to the network stack on the render side. Furthermore, the manager node is informed by NMM that this connection is no longer blocked

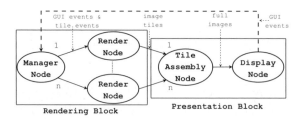

Fig. 2. DRONE encapsulates the flow graph in different basic processing and composite blocks. The rendering block includes the manager node as well as all rendering nodes. The presentation block includes all remaining nodes required to present rendered images. GUI events received from a display node are forwarded to the render nodes through the manager node.

and new data can be sent through this connection. The manager node in turn sends a new tile event to the corresponding render node such that a network connection acts as queue of exactly three tile events. This allows DRONE to reuse the flow control mechanism of TCP for load balancing without any additional communication between manager node and distributed render nodes.

This simple scheduling approach leads to an efficient dynamic load balancing between the render nodes, because render nodes that finish rendering tiles earlier, do receive new rendering event earlier as well. This approach automatically considers differences in rendering time that can be caused by different scene complexity, or different processing power of different rendering machines. Moreover, NMM informs the manager node about a failed network connection to a render node, so that the manager node no longer tries to send tile events to this node. All this is only possible due to the scalable transparency approach of NMM.

4.2   Multi-screen Rendering (AS2)

The general idea to support applications that need to present rendered images on multiple screens can be seen in Figure 3. The application specifies multiple presentation blocks as well as the partial frame configuration to be displayed by each block. All these presentation blocks are then connected to the same rendering block by the framework. To support rendering multi-view images for stereo or Virtual Reality scenarios, each eye is conceptually represented as a separate presentation block. Those independent images are then treated as a single frame and have to be presented at the identical point in time.

Synchronized presentation of rendered images is achieved by adding presentation blocks to a specialized composite block, called *synchronization block*. This block connects a *synchronizer* component to all display nodes of child presentation blocks. The synchronization block is then connected to the rendering block, while the framework automatically connects all presentation blocks to the rendering block, and in doing so adds the information about the partial frame to be presented as part of the connection format between render nodes and tile

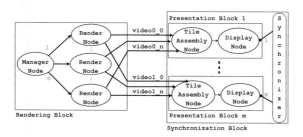

Fig. 3. DRONE allows combining multiple independent presentation blocks (e.g., for realizing video walls). The synchronized presentation of rendered images is achieved by adding these presentation blocks into a synchronization block which then connects a synchronizer to these presentation blocks.

assembly nodes. In summary, any rendered frame can be presented on any of the screens simultaneously; either in full or in part for realizing a video wall setup. For scenarios where blending between adjacent projectors is required, the overlap between presentation blocks can be freely adjusted. Since our specific synchronization component is encapsulated into a composite block, it is not coupled with the framework itself, so that arbitrary synchronization techniques can be integrated into DRONE by implementing a new synchronization block. The synchronizer realized in the DRONE framework is described in [12] and allows for synchronizing the presentation of partial or full frames in multiple display configurations.

4.3   Remote Rendering (AS3)

To enable sending a stream of rendered images across a high-latency network like the Internet and still enable an interactive manipulation of the rendered scene as described in Section 4.1, the bandwidth of the rendered raw video stream has to be reduced drastically. The necessary reduction of the data rate is typically done by means of encoding the image stream before sending; for example using an MPEG-4 or H.263 video codec. Besides encoding of the stream, one can imagine many more potential operations to be performed on the rendered images. For instance, a color correction of neighboring projections of a video wall setup [13], tone mapping or arbitrary other operations in pixel space.

   To enable all these scenarios, we allow the insertion of one or more *post-processing blocks* into a presentation block. This is automatically supported by the framework because the presentation block is a composite block itself. Figure 4 shows a presentation block enhanced by two postprocessing blocks: one for brightness adjustment, and one for encoding and decoding of the stream. A postprocessing block with all its internal nodes is either inserted in front of the tile assembly node or between the tile assembly and the display node of any presentation block. Here, the application of a multimedia middleware like NMM, shows its full potential by providing the means to transparently insert new processing elements in the data processing pipeline.

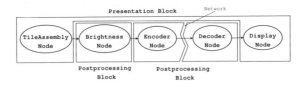

Fig. 4. Extended presentation block: DRONE allows to add an arbitrary number of postprocessing blocks to a presentation block. In this example, we first adjust brightness and then encode rendered images before sending them through an Internet connection.

Fig. 5. Real-time ray tracing simultaneously displayed on three presentation hosts, all of which are fully interactive and synchronized. The rendered images are displayed in a single window on one computer (left) and a display-wall via two split video streams on two additional machines (right).

4.4   Collaborative Rendering (AS4)

The final application scenario to be covered by DRONE is the situation of multiple parties working on and interacting with one and the same rendering block, realizing a collaborative environment, as for example industrial collaborations in which 3D models are synchronously displayed to engineers in distinct offices around the globe. In terms of the DRONE framework, this scenario represents an arbitrary combination of (AS1) to (AS3) as presented above.

As before, the framework configuration for (AS4) includes a single rendering block with potentially multiple presentation blocks attached. The flexible architecture of DRONE allows, for example, to realize different encoded streams for each one of the presentation blocks, and arbitrary display setups for the participating parties. Moreover, different applications can access and share the same rendering block while adding their specific presentation blocks. For example, this could be used for application scenarios where users permanently enter or leave a collaborative virtual 3D environment. Since this scenario may incorporate arbitrary rules of interaction with the scene viewpoint for different applications written against the framework, DRONE can not define access control scenarios that are appropriate purposes. Instead, we only provide access to exclusive manipulation of the virtual camera of a single viewpoint for one presentation block at a time. The access right to manipulate the scene is requested from the corresponding presentation block itself. As soon as an application has requested this access right, it can change the viewpoint through the interface of the presentation block.

The possibility to realize this application scenario by combining and grouping previously presented results again shows the high degree of flexibility of our framework as well as the benefit for applications build on top of it.

5   A Simple Command Language

To be able to easily specify and manipulate the components of a DRONE render graph, we defined the command line application *renderclic*, able to play back *render graph descriptions* (RGDs) defined in respective RGD files. Both application and descriptions are inspired by the graph description format used to specify NMM flow graphs. The RGD syntax is built upon the following context-free grammar:

```
<render_graph>        ::= <rendering_block> "|" <composite_block>
<rendering_block>     ::= <identifier> [ <method>+ ]
<composite_block>     ::= <identifier> [ <method>+ ] "{" <composite_block>+ "}"
                        | <presentation_block>
<presentation_block>  ::= <identifier> [ <method>+ ] [ <presentation_body> ]
<presentation_body>   ::= "{" <postprocessing_graph> "}"
<postproc_graph>      ::= <postprocessing_block> [ "|" <postprocessing_block> ]
<postproc_block>      ::= <identifier> [ <method>+ ] [ "[" <nmm_graph> "]" ]
<method>              ::= "$" <identifier> "(" <arguments> ")" <state>
<state>               ::= "CONSTRUCTED" | "INITIALIZED" | "STARTED"
```

It defines the basic DRONE render graph (`render_graph`) components, namely a rendering block (`rendering_block`), a composite block and its specialization presentation block (`composite_block`, `presentation_block`). All blocks have interfaces enabling a definition of methods (`method`) in an interface definition language, as well as of the internal state (`state`) the block should be in upon their execution. The RGD command language also features a direct specification of post-processing blocks, which may contain inline NMM flow graphs (`nmm_graph`). Here, we omit their specification and the further resolution of identifier and argument symbols (`identifier`, `arguments`) for brevity, though.

With the RGD language, we can define the example depicted in Figure 5, which is real-time ray tracing rendered on two hosts and synchronously displayed on three hosts, two of which configured in a video-wall setup. We can directly run it with the renderclic application afterwards:

```
RenderingBlock $addHost("render1") INITIALIZED  # more render hosts optional
              $setSceneURL("~/box.wrl") INITIALIZED |
SyncBlock $setResolution(1200, 768) INITIALIZED  # used for all children
{
  PresentationBlock $setHost("display1") INITIALIZED  # full frame
  PresentationBlock $setHost("display2") INITIALIZED  # half frame 1
                    $setViewport(600, 768) INITIALIZED  # no offset here
  PresentationBlock $setHost("display3") INITIALIZED  # half frame 2
                    $setViewport(600, 768) INITIALIZED
                    $setOffset(600, 0) INITIALIZED  # viewport offset
}
```

6   Performance Measurements

In order to measure the overhead of our framework, we developed a rendering node on top of RTSG and integrated OpenRT as ray tracer into RTSG which also provides support for distributed ray tracing. As OpenRT is able to distribute rendering in itself, our test environment allows for measuring the overhead when DRONE is used for local or distributed rendering compared to the highly specialized implementation OpenRT.

The test scene we use contains more than 1.3 million triangles and uses reflective and refractive surfaces. Each frame is rendered in a resolution of 1024x512 pixels using a fixed tile size of 64x64 pixel. Our hardware setup consists of 4 rendering PCs, each equipped with two quad core Intel(R) Xeon(R) CPU 3GHz,

64GB RAM and are connected over Infiniband. In Test (1), we measure the overhead of our flexible render graph in comparison to the monolithic rendering application like OpenRT by rendering on a single core without any DRONE-specific distribution. As can be seen in Table 1, DRONE achieves a frame rate that is 0.9 % lower than the frame rate of standalone OpenRT. In order to measure the overhead of the DRONE network communication, we gradually increased in Test (2)-(5) the used cores by eight while presenting images on a different PC. In this case, DRONE achieves a frame rate that is 1.3 % lower than the frame rate of OpenRT. Since an overhead of 0.9 % is caused by using a flow graph, the overhead caused by the network communication has an influence of 0.4 % on the frame rate. We then perform the same tests but with a second presentation block in DRONE in order to show the overhead of the synchronization mechanism. However, when using two presentation blocks, presenting half of each frame, no additional overhead of the synchronization is introduced.

From our point of view, both performance and memory overhead introduced by DRONE are negligible, because applications greatly benefit when using DRONE due to the flexibility of the framework.

Table 1. Performance results using standalone OpenRT vs. OpenRT integrated in DRONE. Frame rate is measured when presenting images on a single display as well as on two displays, each presenting half of the frame.

Test	Cores	OpenRT 1 Display	DRONE 1 Display	DRONE 2 Displays
(1)	1	0.434 fps	0.43 fps	0.43 fps
(2)	8	3.14 fps	3.10 fps	3.10 fps
(3)	16	6.05 fps	5.93 fps	5.93 fps
(4)	24	9.08 fps	8.97 fps	8.97 fps
(5)	32	12.12 fps	12.00 fps	12.00 fps

7 Conclusion and Future Work

In this paper we presented the DRONE architecture, an application development framework for distributed rendering and display. Using NMM as an underlying communication architecture provides an unprecedented flexibility in parallelizing and distributing all aspects of a rendering system: user input, load-balancing, rendering, post-processing, display, and synchronization. By designing a small set of modules that can be combined easily, an application can flexibly configure distributed rendering and display – even dynamically during runtime. As shown in Section 6, this flexibility comes at a negligible cost over specialized and highly optimized implementations of the same functionality.

In the future, we want to explore ways to even further make use of all hardware resources available for rendering in the network. We plan to integrate next-generation multi-core technologies such as the CUDA and Cell architectures in our rendering pipelines.

References

1. Molnar, S., Cox, M., Ellsworth, D., Fuchs, H.: A Sorting Classification of Parallel Rendering. IEEE Computer Graphics & Applications 14, 23–32 (1994)
2. Humphreys, G., Eldridge, M., Buck, I., Stoll, G., Everett, M., Hanrahan, P.: WireGL: a Scalable Graphics System for Clusters. In: SIGGRAPH 2001: Proceedings of the 28th Annual Conference on Computer Graphics and Interactive Techniques, pp. 129–140 (2001)
3. Humphreys, G., Houston, M., Ng, R., Frank, R., Ahern, S., Kirchner, P.D., Klosowski, J.T.: Chromium: a Stream-Processing Framework for Interactive Rendering on Clusters. In: SIGGRAPH 2002: Proceedings of the 29th Annual Conference on Computer Graphics and Interactive Techniques, pp. 693–702 (2002)
4. Rubinstein, D., Georgiev, I., Schug, B., Slusallek, P.: RTSG: Ray Tracing for X3D via a Flexible Rendering Framework. In: Proceedings of the 14th International Conference on Web3D Technology 2009 (Web3D Symposium 2009), pp. 43–50. ACM, New York (2009)
5. Eilemann, S., Pajarola, R.: The Equalizer Parallel Rendering Framework. Technical Report IFI 2007.06, Department of Informatics, University of Zürich (2007)
6. Technology, R.: Rtt scale homepage (2009),
   http://www.realtime-technology.com/
7. NVIDIA Corporation: NVIDIA Quadro G-Sync.,
   http://www.nvidia.com/page/quadrofx_gsync.html
8. Arcila, T., Allard, J., Ménier, C., Boyer, E., Raffin, B.: FlowVR: A Framework For Distributed Virtual Reality Applications. In: 1^{iere} journées de l'Association Française de Réalité Virtuelle, Augmentée, Mixte et d'Interaction 3D, Rocquencourt, France (2006)
9. Rantzau, D., Lang, U., Lang, R., Nebel, H., Wierse, A., Ruehle, R.: Collaborative and Interactive Visualization in a Distributed High Performance Software Environment. In: Chen, M., Townsend, P., Vince, J.A. (eds.) Proceedings of the International Workshop on High Performance Computing for Graphics and Visualization, pp. 207–216. Springer, Heidelberg (1996)
10. Lohse, M., Winter, F., Repplinger, M., Slusallek, P.: Network-Integrated Multimedia Middleware (NMM). In: MM 2008: Proceedings of the 16th ACM international conference on Multimedia, pp. 1081–1084 (2008)
11. Repplinger, M., Beyer, M., Slusallek, P.: Multimedia Processing on Many-Core Technologies Using Distributed Multimedia Middleware. In: Proceedings of The 13th IASTED International Conference on Internet and Multimedia Systems and Applications, IMSA 2009 (to appear, 2009)
12. Repplinger, M., Löffler, A., Rubinstein, D., Slusallek, P.: URay: A Flexible Framework for Distibuted Rendering and Display. Technical Report 2008-01, Department of Computer Science, Saarland University, Germany (2008)
13. Kresse, W., Reiners, D., Knöpfle, C.: Color Consistency for Digital Multi-Projector Stereo Display Systems: the HEyeWall and the Digital CAVE. In: EGVE 2003: Proceedings of the workshop on Virtual environments 2003, pp. 271–279 (2003)

Efficient Strategies for Acceleration Structure Updates in Interactive Ray Tracing Applications on the Cell Processor

Martin Weier, Thorsten Roth, and André Hinkenjann

Computer Graphics Lab
Bonn-Rhein-Sieg University of Applied Sciences
Sankt Augustin, Germany
{martin.weier,thorsten.roth}@smail.inf.h-brs.de,
andre.hinkenjann@h-brs.de

Abstract. We present fast complete rebuild strategies, as well as adapted intelligent local update strategies for acceleration data structures for interactive ray tracing environments. Both approaches can be combined. Although the proposed strategies could be used with other data structures and architectures as well, they are currently tailored to the Bounding Interval Hierarchy on the Cell chip.

1 Introduction

Recent hardware and software developments in the field of fast ray tracing allow for the use of these renderers in interactive environments. While the focus of research of the last two decades was mainly on efficient ray tracing of static scenes, current research focuses on dynamic, interactive scenes. Recently, many approaches came up that deal with ray tracing of dynamic deformable scenes. Current approaches use kd-trees [1,2], grids [3] or Bounding Volume Hierarchies (BVHs) on commodity CPUs or GPUs [4,5,6]. These either do a complete rebuild of the scene's acceleration structure or provide methods to perform an intelligent dynamic update. However, there is always a trade-off between the tree's quality and the time that is needed to perform the rebuild or update operations.

Recent publications [7,8] on using kd-trees for deformable scenes always propose trying to perform a complete rebuild from scratch. However, these approaches do not seem to scale well in parallel [7]. Another approach is [9], a GPU based construction of kd-trees in breadth first manner. In order to use the fine-grained parallelism of GPUs, a novel strategy for processing large nodes and schemes for fast evaluation of the nodes' split costs is introduced. A disadvantage of this approach is that its memory overhead is very high. Grids, on the other hand can be constructed very fast and in parallel [3]. Although grids are usually not as efficient as adaptive structures when it comes to complex scenes, coherence can be exploited here as well [10].

In recent years, BVHs seem to have become the first choice for ray tracing of deformable scenes. BVHs are well suited for dynamic updates. However,

G. Bebis et al. (Eds.): ISVC 2009, Part I, LNCS 5875, pp. 987–998, 2009.

this often leads to increasing render times since the trees degenerate over time. To decide whether it makes sense to perform a complete rebuild, Lauterbach et al. [4] developed a metric to determine the tree's quality. Another approach was proposed by Wald et al. [11]. They use an asynchronous construction that runs in parallel during vertex update and rendering. A novel approach is [12], performing a BVH construction on GPUs. There, the upper tree levels are parallelized using Morton curves. This idea is based on [13] and [14], where surface reconstruction on GPUs using space filling curves is performed. We have chosen the Bounding Interval Hierarchy (BIH) [15] for interactive ray tracing of dynamic scenes on the Cell Broadband Engine Architecture [16]. The Bounding Interval Hierarchy has some advantages compared to kd-trees and BVHs. A BIH node describes one axis aligned split plane like a kd-tree node. However, the split planes in the BIH represent axis aligned bounding boxes (AABBs). Starting from the scene's BB the intervals are always fitted to the right's leftmost and the left's rightmost values of the resulting partitions. Thus, the BIH can be seen as a hybrid of kd-trees and BVHs. One advantage of BIHs compared to kd-trees and BVHs is their easy and fast construction. In addition the BIH's representation of the nodes is very compact since no complete AABBs need to be stored. Due to the growing parallelism in modern architectures, effectively utilizing many cores and vector units is crucial. In fact the BIH construction has many similarities with quicksort which is a well studied algorithm in the field of parallel computing. Modern architectures like the Cell processor or GPUs all have a memory access model (memory coherence model) where data from main memory must be explicitly loaded from or distributed to the threads. For most of these memory models bandwidth is high, but latency as well. This makes it especially important to place memory calls wisely. In the following section, we describe fast methods for completely rebuilding the acceleration data structure each frame. In section 3 we present intelligent local updates of the data structure. Note that both methods can be combined: The intelligent update method finds the subtree that has to be rebuilt after scene changes. This update can then be done by using the methods from the following section. After that we show the results of some benchmarks and evaluate both approaches.

2   Fast Complete Rebuild of Acceleration Data Structures

The most important step in BIH construction is the sorting routine which sorts primitives to the right or left of the split plane. The Cell's SPEs only have a limited amount of local storage. To do sorting on all primitives, which usually exceed the 256kB local store on the SPE, data has to be explicitly requested and written back to main memory. Peak performance for these DMA calls is reached for blocks of size 16kB, which is the maximum that can be obtained by one DMA call. Since all of these DMA operations are executed non-blocking (asynchronously), it is important to do as much work as possible in between consecutive calls. Common approaches for doing an in-place quicksort like in [17] usually distribute the array in blocks of fixed size to the threads. The threads then do a local in-place sort and write their split values to a common new allocated array. Then prefix sums over these values are calculated and the results are distributed back to the threads. By doing so, each thread knows the position where it can write its values back to. Using this algorithm on the Cell, each SPE would have to load one 16kB block, sort it and write the split values and the sorted array back to main memory. After that, the prefix sum operation on the split values of the 16kB blocks would need to be performed. Since this is a relatively cheap operation, parallelizing it using the SPEs leads to new overhead. Additionally, each SPE needs to read the 16kB blocks again to write them back to main memory in correct order. For this reason this method should be avoided. In our algorithm we only use one SPE to perform the sorting of one region. Each region refers to one interval of arbitrary size. To avoid any kind of recombination and reordering after one block is sorted, the values need to be written back to their correct location right away.

If the size of the interval to be sorted is smaller than 16kB, sorting is easy. The SPE can load it, sort it in-place and write it back to where it was read from. Sorting intervals larger than 16kB is a bit more complicated. Values of a block smaller than the split plane value are written to the beginning of the interval to be sorted and the ones larger to the end. To make sure that no values are overwritten that were not already in the SPE's local store, the values from the end of the interval also need to be loaded by the SPE. In order to maximize the computation during the MFC calls five buffers are needed. Figure 1 shows the use of the five buffers.

In the beginning, four buffers are filled. Buffers A and C load values from the end of the triangle array, B and D load values from the beginning (Step 1). After that, values in D are sorted (Step 2). Then the pointer from buffer D and the OUT buffer are swapped and the OUT buffer is written back to main

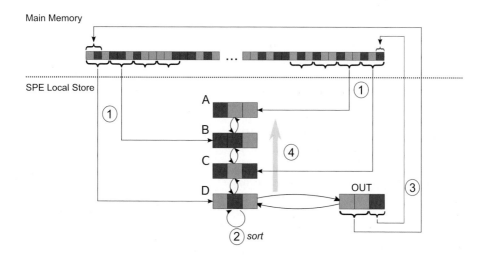

Fig. 1. SPE in-place sorting. Elements larger than the pivot in red (dark grey) and elements smaller in green (light grey).

memory (Step 3). The elements in the OUT buffer smaller than the split plane are written to the beginning of the interval. The elements larger are written back to the end of the interval.After this the buffers need to be swapped again. This can be accomplished by swapping the pointers, i.e. buffer D becomes buffer C, buffer C becomes buffer B and buffer B becomes buffer A respectively (Step 4). Finally it is determined if more values that have not already been processed were read from the beginning or from the end. This information is then used to decide whether the next block needs to be loaded from the beginning or the end. This ensures that no values are overwritten that were not already loaded in the SPE's local store. The newly loaded block is then again stored in buffer A which was the former and thus already processed buffer OUT. One possible extension that is used for blocks smaller than 16kB is job agglomeration. Since the peak performance is achieved for 16kB blocks, jobs can be agglomerated so that the interval of triangles to be sorted fits into one 16kB DMA call (see breadth first construction).

Depth first construction. One naive approach to perform BIH construction is depth first. In this approach only sorting is done on the SPE, the recursion and creation of nodes is still entirely done on the PPE. The algorithm starts on the PPE by choosing a global pivot, i.e. the split plane to divide the current BB. The split plane and the intervals are then submitted to the SPE. By signaling, the SPE now begins with the in-place sorting as described above. When the SPE has finished, it writes the split index and the right's leftmost and left's rightmost values back to main memory and signals this to the PPE. With these values the PPE is now able to create the new node and to make a recursive call on the new intervals.

Breadth first construction. The breadth first construction of data structures on architectures like the Cell B. E. is advantageous for two main reasons:

1. The primitive array is traversed per tree level entirely from the beginning to the end ⇒ memory access is very efficient
2. Many small jobs with primitive counts smaller 16kB can be agglomerated ⇒ reduction of DMA calls

We propose a method to do a breadth first construction on the Cell B. E. where the SPE can run construction of the BIH almost entirely independent from the PPE. The PPE is only responsible for allocating new memory since these operations cannot be performed from the SPEs. Efficient memory management is a particularly important aspect. The naive construction of the BIH allocates memory for each created node. These memory allocation calls are costly. Even though there are various publications like [18] and [19] stating that own implementations of memory management usually do not increase the application performance, memory management on today's architectures still remains a bottleneck. One way of avoiding a large number of memory allocation calls is using an `ObjectArena`, as proposed in PBRT [20]. There, memory is allocated as multiple arrays of fixed size where each pointer to the beginning of an array is stored in a list. Freeing the allocated memory can easily be done by iterating over the list and freeing the referenced arrays. One advantage is, that many threads can be assigned to different lists. Thus, it is possible for them to compute the address where to write back to without the need of any synchronization between the different SPEs. This is particularly important for an implementation to do the BIH construction independently of the PPE. Being able to compute the new node's address, build-up can run more independent of the PPE since the frequency of new allocations is reduced. Unfortunately, such an implementation also has disadvantages as memory can be wasted because the last arrays can only be partially filled. In addition, intelligent updates and the need of deleting single nodes leads to fragmentation of the arrays. To implement a construction in BFS manner, jobs, i.e. the intervals to be sorted on the next tree level, need to be managed by an additional job queue. At the beginning the PPE allocates two additional buffers of sizes as the primitives count. These buffers are denoted in the following as `WorkAPPE` and `WorkBPPE`. This can be seen as a definite upper bound of jobs that could be created while processing one tree level. The actual construction is divided into two phases. In each phase, the `WorkAPPE` and `WorkBPPE` switch roles, i.e. in the first phase jobs are read from `WorkAPPE` and newly created ones are stored in `WorkBPPE`, while this is done vice versa in the second phase. Figure 2 clarifies the two-phase execution.

To do so, two additional buffers are allocated in the SPE's local store as well. We denote these buffers as `WorkA` and `WorkB`. In `WorkA` the jobs from the current tree level, in the beginning that is the job for the root node, are stored. From each job in `WorkA` at most two new jobs are created which are then stored in

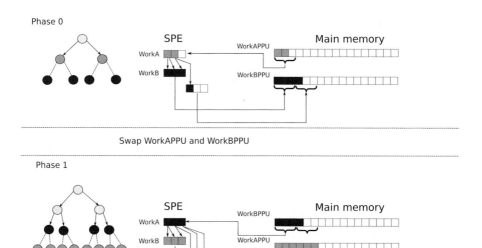

Fig. 2. Two-phase management of the job queue

WorkB. If there are no more jobs in WorkA and the current job is not the last one and if there are no more jobs in main memory, the current tree level is entirely processed.

Should there be no more space left in WorkB, it is written back to the PPE. To keep track of how many blocks were written back in the last phase, an additional counter is used. This counter can also be used to determine if there are jobs left in main memory that have not already been processed. In addition, this approach can be further optimized. While processing the first tree levels, the jobs can be entirely stored on the SPE, so there is no need to write them back to main memory after the tree level was processed. Therefore, to avoid these unnecessary DMA calls, the buffers on the SPE can be swapped as well if there is no need to provide further storage for the newly created jobs. Our method also has some disadvantages. While requesting new jobs from main memory can be handled asynchronously this is not the case when WorkB needs to be written back to main memory. Therefore a double buffering scheme for WorkA and WorkB could be used.

3   Intelligent Local Update

To accelerate object transformations, an algorithm for performing dynamic updates of the data structure is needed. The basic idea is to search the BIH region affected by given transformations and avoid the complete rebuild if possible. This is adapted from [6], where a similar approach for Bounding Volume Hierarchies based on [21] was presented. Results of our algorithm as well as information on potential future optimizations are also stated in sections 4.2 and 5. The algorithm uses the PPE (a traditional CPU core on the Cell chip), so it could be easily ported to other architectures. The traversal method used during the update procedure is shown in algorithm 1.

Finding the affected interval of the underlying triangle array is of importance. To achieve this, states of geometry before and after transformation have to be considered, as transformations may obviously result in an object being moved to another region of the scene. Geometry is always represented as complete objects in the current implementation. For being able to search for affected regions, a bounding box is constructed which encloses the old and new geometry states. We call this the *Enclosing Bounding Box* (EBB). To avoid quality degradation or degeneration of the data structure, a complete rebuild is performed if the scene bounding box changes due to the transformation. Using the EBB, the algorithm is able to find the subtree which contains all modified geometry. Thus, it also encloses the according array interval when traversing down to the leftmost and rightmost leafs of this subtree, as this yields the starting and ending indices of the corresponding interval. As shown in the algorithm, traversal is stopped as soon as both intervals overlap the EBB. This is necessary because there is no information whether the geometry belongs to the lower or upper interval. An alternative approach is explained in section 5. Finding the primitives to be overwritten in the triangle array is now done by iterating over the array and replacing all triangles with the corresponding geometry ID. This is stored in each triangle for material and texturing purposes. Subsequently, all nodes below the enclosing node are deleted recursively and the usual construction algorithm is performed for the subset of triangles given by the array interval. Note that it is necessary to keep track of the bounding box for the tree level reached, which is at this point used as if it was the actual scene bounding box. This is the case because we make use of the global heuristic in [15]. The resulting BIH can then simply be added to the existing tree. The new sub-tree may need adjustment of the pointer to its root, as the split axis may have changed. Results of this approach as well as some advantages and shortcomings are presented in section 4.2.

Algorithm 1. BIH traversal algorithm used to find the smallest enclosing interval for EBB

1	**while** *traverse* **do**
2	**if** *leaf reached* ∨ *both intervals overlap EBB* **then**
3	traverse ← false;
4	**end**
5	**else if** *Only one interval overlaps EBB* **then**
6	**if** *Interval does not contain EBB completely* **then**
7	**if** *EBB overlaps splitting plane* **then**
8	Abort traversal;
9	**end**
10	**else**
11	Set clipping plane for interval to corresponding value of EBB;
12	**end**
13	**end**
14	Set traversal data to successive node;
15	**if** *Child is not a leaf* **then**
16	Set split axis;
17	**end**
18	**end**
19	**else**
20	traverse ← false;
21	**end**
22	**end**

---

4   Results and Evaluation

All measurements of this section were obtained by running the reported algorithms on a Sony Playstation 3.

4.1   Complete Rebuild

Table 1 gives an overview of the different construction times and the resulting speedups. We tested four different versions. Two were PPE based, i.e. the naive and the approximate sorting approach proposed in [15]. The other two were SPE

Table 1. Construction times for four different models in ms averaged from 10 runs

		PPE		
Model	#Triangles	Naive	Approximate sorting	Speedup
ISS	17,633	106 ms	63 ms	1.68
Bunny	69,451	340 ms	201 ms	1.69
Fairy Forest	174,117	1,478 ms	781 ms	1.89
Dragon	871,414	5,956 ms	Out of Memory	
		SPE		
Model	#Triangles	Depth first	Speedup	Breadth first	Speedup
ISS	17,633	47 ms	2.26	24 ms	4.42
Bunny	69,451	133 ms	2.56	68 ms	5.00
Fairy Forest	174,117	385 ms	3.84	195 ms	7.58
Dragon	871,414	1,828 ms	3.26	897 ms	6.64

based, one in DFS and the other one in BFS manner. All speedups relate to the naive approach on the PPE.

The naive approach was the fastest available algorithm on the PPE. This approach already includes vectorization. However, construction times even for the small models like the ISS or the Stanford Bunny are far from interactive. Even though the speedups from the approximate sorting are stated by [15] to be about 3-4, in our implementation speedups of about 1.8 can be achieved. This is due to the maximum primitives per node constraint in our ray tracing system. Nodes with more than 12 triangles need to be further subdivided. For this the naive approach is used.

By looking at the two SPE based approaches it is apparent that the construction times for the breadth first approach are always better than the depth first approach. This is because the breadth first approach uses fewer synchronization calls between the SPE and the PPE. Table 1 shows that speedups of about 5 to 7 can be expected. Increasing speedups with different larger models cannot be concluded from the table, because they are not only dependent on the model's size but also on layout of the overall scene. Since all of the benchmarks are made using only one SPE utilizing more SPEs should lead to another significant performance gain. This is the topic of future work.

4.2   Partial Rebuild

When multiple, arbitrarily distributed objects are transformed, the result might be a very large EBB. This will in turn result in rebuilding large parts of the scene, gaining almost no performance boost (or none at all). A similar effect is caused by large triangles, as they induce an overlapping of intervals. As shown before, this will lead to an early termination of the traversal algorithm. Having large empty parts in a scene and otherwise small geometry, a huge performance boost can be expected. Though, performance may suffer due to counterintuitive reasons, e.g. moving an object through a huge empty region of the scene, but overlapping the root split axis. This would virtually result in a complete rebuild of the data structure. Nevertheless, good performance gains can be obtained, as is shown in the following.

Figure 3 illustrates results for the scene *rtPerf*, which consists of spheres pseudorandomly distributed across the scene. Tests were run for triangle counts from 10k to 100k and moving a number of spheres in a predefined scheme. This has been performed several times and results were averaged. Figure 3a shows the correlation between the total number of triangles and triangles used for rebuilds. While the number of transformed triangles stays constant, there is in most cases an increase in triangle numbers used for rebuilds. This does not simply increase by the number of triangles added to the scene, but just by a fraction of that (which may vary depending on geometry distribution). It can also be seen that only a fraction of data structure updates results in a complete rebuild, thus yielding a huge difference between average and maximum update time, as shown in figure 3b. Here, the correspondence between times needed for complete rebuilds and times achieved with the implemented update strategies is shown. While having an almost linear increase for both, the performance is vastly better than with just using brute force rebuild.

Table 2 shows results for other testing scenes. Note that these scenes have fixed triangle numbers. *carVis* is a car model which can be decomposed into its

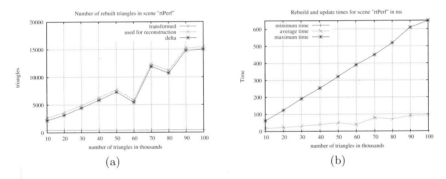

(a)   (b)

Fig. 3. Measurements for the scene "rtPerf" with triangle counts from 10k to 100k

Table 2. Results for the scenes *carVis* and *cupVis*; from left to right: Number of triangles; number of updates; number of complete rebuilds; ratio updates/rebuilds; min, max and averaged update/rebuild times; average number of transformed triangles; average number of triangles used for rebuild; average difference of these values; average traversal depth

scene	tris	updates	rebuilds	ratio	t_{min}	t_{max}	$\bar{t}$	$trans$	rec	Δ	$trav$
carVis	22,106	383	112	3.42	0.48	154.09	69.71	468.47	7,549.25	7,080.78	2.22
cupVis	7,040	6	148	0.04	26.89	35.43	29.1	7,040	7,040	2,181.66	0

parts, while *cupVis* is just a simple model of a glass without any surrounding geometry. A result from that is that in *cupVis* a complete rebuild is performed each time the geometry is moved. The small fraction of rebuilds in the table results from the object being marked for transformation, but not performing a transformation at all. Thus the scene bounding box as well as all the triangles remain the same and an update (without anything happening at all) can be performed.

5   Conclusion and Future Work

In the current implementation of the ray tracer there was only the possibility to utilize one SPE for tree construction since the others were used for rendering. However, since rendering only takes place in between the data structure updates in principle all SPEs could be utilized. Even though an implementation for more SPEs has not been done yet, we want to introduce some ideas for a reasonable parallel construction.

One way of utilizing more threads could be realized in the depth first based construction. Here, the recursive call could be delegated to different SPEs. However, this is not very efficient because on the upper tree levels many SPEs would be idle. The same problem arises when such a parallelization would be done using the breadth first construction and allowing all threads to have access to one common job queue. Additionally, this would make further synchronization between the SPEs and the PPEs necessary since the access to the job queue must be synchronized. In order to get the best performance out of the SPEs, synchronization must be reduced and the execution should be entirely independent from the PPE.

One method that could be used in the context of BIH parallelization was proposed in [12] for BVH construction on GPUs. There a pre-processing step using Morton curves is performed to find regions of primitives that then could be built-up totally independent. The construction of the Morton curve can be efficiently parallelized since Morton code construction of a primitive can be done totally independent from the others. After the Morton code construction a radix sort needs to be performed. Radix sort is advantageous because it does not involve comparisons. Fortunately there are methods for Radix Sort on the Cell B. E., like [22] or [23]. By doing such a pre-computation step, better load balancing while using the SPEs in parallel can be achieved. Other possible improvements regard the job queue. One improvement could be made using a double buffering scheme for the two buffers on the SPE to avoid synchronous memory access. The necessary DMA calls might benefit from the usage of DMA-lists. Another possible way of dealing with the management of the job queue is the use of Cell's software managed cache whereas some improvements have been made recently [24].

To achieve a better performance concerning the update strategy, e.g. an approach for further traversal of the data structure could be beneficial. This way, much less triangles could be involved in the update step by keeping subtrees which are not affected by transformations. Though, it has to be analyzed how deep the tree should be traversed, as keeping subtrees will often lead to index problems with the underlying triangle array. A metric for estimation of needed time for reorganization steps is needed to cope with that, as further traversal potentially leads to more reorganization overhead due tue memcopy operations. Also, more leafs' indices might have to be adjusted accordingly.

Acknowledgements

This work was sponsored by the German Federal Ministry of Education and Research (BMBF) under grant no 1762X07.

References

1. Popov, S., Gunther, J., Seidel, H.P., Slusallek, P.: Stackless kd-Tree Traversal for High Performance GPU Ray Tracing. In: Computer Graphics Forum (Proc. Eurographics) 26, vol. 3, pp. 415–424 (2007)
2. Wald, I., Havran, V.: On Building Fast kd-trees for Ray Tracing, and on doing that in O(NlogN). In: Proc. of IEEE Symp. on Interactive Ray Tracing, pp. 61–69 (2006)
3. Ize, T., Wald, I., Robertson, C., Parker, S.: An Evaluation of Parallel Grid Construction for Ray Tracing Dynamic Scenes. In: Proceedings of the 2006 IEEE Symposium on Interactive Ray Tracing, Salt Lake City, Utah, pp. 47–55 (2006)
4. Lauterbach, C., Yoon, S., Tuft, D., Manocha, D.: RT-DEFORM: Interactive Ray Tracing of Dynamic Scenes using BVHs. In: IEEE Symposium on Interactive Ray Tracing, Salt Lake City, Utah (2006)
5. Wald, I., Boulos, S., Shirley, P.: Ray Tracing Deformable Scenes using Dynamic Bounding Volume Hierarchies. ACM Transactions on Graphics 26(1) (2007)
6. Eisemann, M., Grosch, T., Magnor, M., Müller, S.: Automatic Creation of Object Hierarchies for Ray Tracing Dynamic Scenes. In: Skala, V., ed.: WSCG Short Papers Post-Conference Proceedings, WSCG (2007)
7. Popov, S., Günther, J., Seidel, H.P., Slusallek, P.: Experiences with Streaming Construction of SAH kd-Trees. In: Proceedings of the 2006 IEEE Symposium on Interactive Ray Tracing, Utah, pp. 89–94 (2006)
8. Hunt, W., Mark, W.R., Stoll, G.: Fast kd-tree Construction with an Adaptive Error-Bounded Heuristic. In: 2006 IEEE Symposium on Interactive Ray Tracing (2006)
9. Zhou, K., Hou, Q., Wang, R., Guo, B.: Real-time kd-tree Construction on Graphics Hardware. ACM Trans. Graph. 27, 1–11 (2008)
10. Wald, I., Ize, T., Kensler, A., Knoll, A., Parker, S.G.: Ray Tracing Animated Scenes using Coherent Grid Traversal. In: ACM Transactions on Graphics 25: Proceedings of ACM SIGGRAPH 2006, Boston, MA, vol. 3, pp. 485–493 (2006)
11. Ize, T., Wald, I., Parker, S.G.: Asynchronous BVH Construction for Ray Tracing Dynamic Scenes on Parallel Multi-Core Architectures. In: Favre, J.M., dos Santos, L.P., Reiners, D. (eds.) Eurographics Symposium on Parallel Graphics and Visualization (2007)

12. Lauterbach, C., Garland, M., Sengupta, S., Luebke, D., Manocha, D.: Fast BVH Construction on GPUs. In: Proc. Eurographics 2009, München, Germany, vol. 28 (2009)
13. Zhou, K., Gong, M., Huang, X., Guo, B.: Highly Parallel Surface Reconstruction. Technical Report MSR-TR-2008-53, Microsoft Technical Report (2008)
14. Ajmera, P., Goradia, R., Chandran, S., Aluru, S.: Fast, Parallel, GPU-based Construction of Space Filling Curves and Octrees. In: SI3D 2008: Proceedings of the 2008 Symposium on Interactive 3D Graphics and Games, Electronic Arts Campus. ACM, New York (2008)
15. Wächter, C., Keller, E.: Instant Ray Tracing: The Bounding Interval Hierarchy. In: Rendering Techniques 2006: Proceedings of the 17th Eurographics Symposium on Rendering, Nicosia, Cyprus, pp. 139–149 (2006)
16. Kahle, J.A., Day, M.N., Hofstee, H.P., Johns, C.R., Maeurer, T.R., Shippy, D.: Introduction to the Cell Multiprocessor. IBM J. Res. Dev. 49, 589–604 (2005)
17. Grama, A., Gupta, A., Karypis, G., Kumar, V.: Introduction to Parallel Computing, vol. 2. Addison Wesley Pub. Co. Inc., Reading (2003)
18. Johnstone, M.S., Wilson, P.R.: The Memory Fragmentation Problem: Solved? In: ISMM 1998: Proceedings of the 1st International Symposium on Memory Management, pp. 26–36. ACM, New York (1998)
19. Berger, E.D., Zorn, B.G., McKinley, K.S.: Composing High-Performance Memory Allocators. In: PLDI 2001: Proceedings of the ACM SIGPLAN 2001 conference on Programming language design and implementation, pp. 114–124. ACM, New York (2001)
20. Pharr, M., Humphreys, G.: Physically Based Rendering: From Theory to Implementation (The Interactive 3d Technology Series). Morgan Kaufmann, San Francisco (2004)
21. Goldsmith, J., Salmon, J.: Automatic Creation of Object Hierarchies for Ray Tracing. IEEE Comput. Graph. Appl. 7, 14–20 (1987)
22. No author given: Parallel radix sort on cell (2007), http://sourceforge.net/projects/sorting-on-cell/ (last viewed, 28.02.09)
23. Ramprasad, N., Baruah, P.K.: Radix sort on the Cell Broadband Engine. Department of Mathematics and Computer Science, Sri Sathya Sai University (2007), www.hipc.org/hipc2007/posters/radix-sort.pdf (last viewed, 10.02.09)
24. Guofu, F., Xiaoshe, D., Xuhao, W., Ying, C., Xingjun, Z.: An Efficient Software-Managed Cache Based on Cell Broadband Engine Architecture. Int. J. Distrib. Sen. Netw. 5, 16–16 (2009)

Interactive Assembly Guide Using Augmented Reality

M. Andersen[1], R. Andersen[1], C. Larsen[1], T.B. Moeslund[1], and O. Madsen[2]

[1] Laboratory of Computer Vision and Media Technology
[2] Department of Production
Aalborg University, Denmark

Abstract. This paper presents an Augmented Reality system for aiding a pump assembling process at Grundfos, one of the leading pump producers. Stable pose estimation of the pump is required in order to augment the graphics correctly. This is achieved by matching image edges with synthesized edges from CAD models. To ensure a system which operates at interactive-time the CAD models are pruned off-line and a two-step matching strategy is introduced. On-line the visual edges of the current synthesized model are extracted and compared with the image edges using chamfer matching together with a truncated L2 norm. A dynamic visualization of the augmented graphics provides the user with guidance. Usability tests show that the accuracy of the system is sufficient for assembling the pump.

1   Introduction

With more than 16 million pumps sold annually Grundfos is one of the major players within its field. Most of the pumps produced are standard pumps where most of production and assembling are done automatically. This is however not the case for the so-called CR pumps, which are larger pumps made individually to fit the needs of the customers. In 2008 around 160,000 CR pumps were produced at Grundfos. Of these pumps only an average of three pumps with identical configurations were produced at a time. Due to this diversity the assembly process is done manually and can pose difficulties for especially new employees. Currently the assembly process is guided by a sequence of parts appearing on a screen. In [1] it has been shown that such an assembly process can be optimized by enhancing the guidance with relevant information. In this paper we follow this idea and present the technical aspects of an augmented reality system for doing just this.

1.1   Augmented Reality

Augmented Reality (AR) is somewhere on the continuum spanned by a real environment and a virtual environment [2]. In the last decade AR has been seen in more and more applications, where the most well known is perhaps during TV news transmission, where graphics is added to provide some type of

G. Bebis et al. (Eds.): ISVC 2009, Part I, LNCS 5875, pp. 999–1008, 2009.

information, e.g., who the interviewed person is. In TV sports transmissions AR is also becoming more and more popular, e.g., in football/soccer to illustrate different aspects of the game like distance from the ball to the goal during a free kick [3]. Inspired by the head-up displays used by fighter pilots, the automotive industry recently included AR in vehicles to provide the driver with additional (or important) information projected on the front window, e.g., night vision. AR is also seen in for example computer games [4] and mobile social events [5]. Lastly AR is also seen in assembly, inspection, and maintenance, where the overlayed graphics ease the task of the operator by providing supporting information [6].

The pose of the object to be augmented needs to be known. This is often estimated by locating and clustering a number of features extracted for the image of the object [7]. Due to perspective transform of the features, non-unique features, a non-rigid object, and in general a non-perfect feature extraction process, the estimated pose (and hence the overlayed graphics) is likely to be contaminated with both jitter and off-sets. This is especially problematic when AR is used on see-though displays and the result ranges from mere irritation to actual motion sickness [8]. To avoid these problems different systems use alternative ways to estimate the pose of the object: GPS [9], known transformation between the camera and the real environment as seen in AR sport applications [6], or pre-made patterns are attached to the object to be pose estimated, like seen in ARToolKit [10].

1.2   The Approach

In our case we cannot place markers on the different parts of the pump to be assembled and are therefore faced with the general pose estimation problem. However, we only have rigid parts and have very detailed models of each part (CAD models) and know how they are connected. We therefore apply an approach where the models are synthesized into the image domain and compared with the image data directly. For this purpose different features can be utilized, e.g., appearance, circular features, depth data, or edges [11,12,13,14,15]. We choose edges since they are relatively simple to obtain and less variant with respect to changes in the lighting. In figure 1 the block diagram of our approach is illustrated.

The different blocks correspond to the remaining sections in this paper. In section 2 and 3 the image features and synthesized features are extracted, respectively. In section 4 the matching of the two different feature sets are described. In section 5 it is explained how the synthesized poses are chosen. In section 6 the actual pose is estimated and in section 7 this pose is used to augment graphics. Finally section 8 evaluates the system and section 9 concludes the work.

It should be mentioned that AR systems are not only affected negatively by jitter and offsets, but also by delays when superimposing the graphics. Such delays can render a totally unusable AR system and the different methods applied in the rest of the paper are chosen with the computational complexity in mind.

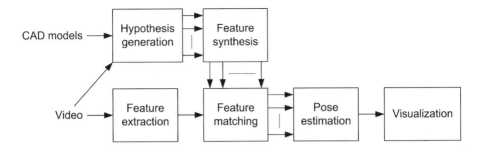

Fig. 1. A block diagram of the approach

2   Feature Extraction

Before the synthesized model can be compared with the image data, the relevant information needs to be extracted from each incoming image. To this end we first find a ROI containing the partly assembled pump and then extract the relevant type of information, i.e., edges.

Since the system is to be operating in a controlled environment, the characteristics of the surface on which the pumps are assembled can be decided before hand. This allows for the opportunity of creating a good contrast between the object (pump) and the surroundings. From this follows that a simple background subtraction can segment the pump in the image. Other objects present in the field-of-view of the camera are sometimes also detected. These, together with pixel noise, are removed using blob-analysis. A small margin is added to the extracted ROI to increase flexibility in the forthcoming matching process, see figure 2.a.

The system uses edge data to represent the synthesized model, so edges need to be extracted in the image. Since the synthesized edges are 1-pixel thin, the

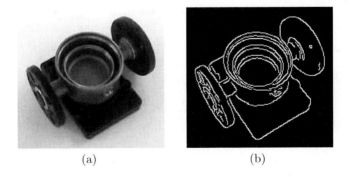

(a)   (b)

Fig. 2. a) The ROI estimated from background subtracting and blob-analysis. b) The detected edges.

same should be the case for the image edges. We therefore apply the Canny edge detector [16], see figure 2.b.

3   Feature Synthesis

In order to extract the edges of a hypothesized model we must consider two issues. First, the 3D model can not be too complicated, due to the inherent computational complexity, and second the extracted edges must be similar to the ones extracted in the image. Both issues lead to the conclusion that the highly detailed 3D models provided by Grundfos need to be simplified. We therefore prune the models to only include the major components. In figure 3 the original high resolution model and the simplified low resolution model are shown. The edges in the low resolution model are now similar to those that can be found in the image. Furthermore, by significantly lowering the amount of polygons we have reduced the computational complexity.

Two types of edges need to be extracted from the polygons and evaluated whether they will be visible or not when synthesized into the image. These are the *sharp edges* and the *contour edges*. A sharp edge is defined as an edge between two polygons having significantly different normal vectors, see figure 3.d. These edges are independent of the viewpoint and can therefore be found off-line. A contour edge is defined as an edge between a polygon (the object) and the background, see figure 3.e. Contour edges clearly depend on the viewpoint and can therefore not be determined off-line. However, what we can do off-line is to divide all (non-sharp) edges into either inner-edges or outer-edges using the expressions below, see also figure 4. If $a \leq b$ then it is an inner edge (or coplanar), which can never become a contour edge no matter the viewpoint. If on the other hand $a > b$ then the edge is an outer edge and a candidate for a contour edge.

$$a = \left|(v_2 + \boldsymbol{n}_2) - (v_1 + \boldsymbol{n}_1)\right| \qquad b = \left|(v_2 + \boldsymbol{n}_1) - (v_1 + \boldsymbol{n}_2)\right| \qquad (1)$$

The potential contour edges are further pruned using the current viewpoint. Since the camera applied in the setup is calibrated we know the viewpoint, i.e.,

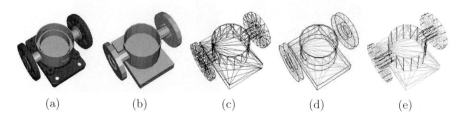

| (a) | (b) | (c) | (d) | (e) |

Fig. 3. a) High resolution model. b) Low resolution model. c) Polygons in b). d) Sharp edges. e) Outer edges.

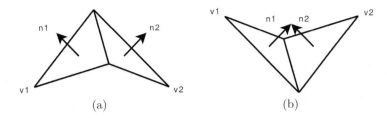

Fig. 4. a) Outer edge. b) Inner edge. v_1 and v_2 are 3D vertices and $\boldsymbol{n}_1$ and $\boldsymbol{n}_2$ are normal vectors.

the direction in which the model will be projected. If the angle between this directional vector and the normal of a polygon is bigger than 90 degrees then the polygon is not facing the camera and opposite for angles below 90 degrees. From this follows a definition of a polygon either being *front-face* or *back-face* with respect to the camera. A contour edge is now characterized as being shared by a front-face *and* a back-face.

In order to assess which of the candidate edges in the model that are actually visible and should be compared with the image data we do the following. First all the sharp edges and contour edges (or rather their vertices) are projected into the camera frame using the projective matrix from the camera calibration. Next we apply standard methods from computer graphics to determine which edges are occluded and which are visible (intersection in the projected image plane and depth ordering). Edges may be partially occluded by faces in the model, making it necessary to split edges into multiple segments. This operation requires more computation, but it is not necessary because missing edges in the edge map do not result in a penalty in the matching cost - as explained in the next section.

4   Feature Matching

For each hypothesis an edge image is generated as explained above. In order to measure the fitness of this hypothesis the edges are compared with the edges extracted in the image. In figure 5 an example of the two different edge images are shown.

A fitness function is required, which can measure the difference between the two edge images. The straight forward approach is to either AND or XOR the two images. If, however, the best hypothesis is just one pixel off, the result of the logic operation will be erroneous. We therefore apply a Chamfer matching approach, where the edge image is replaced by a distance image where each pixel holds a value depending on its distance to the nearest edge pixel in the edge image. The distance image is found using the Distance transform [17], see figure 5.

The fitness function is now a matter of comparing each non-zero pixel in the synthesized edge image with the corresponding pixel in the edge image and

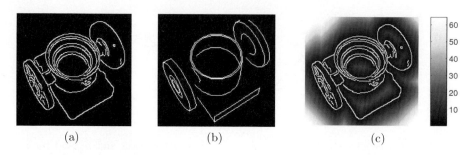

(a)   (b)   (c)

Fig. 5. a) Detected edges. b) Synthesized edges. c). Distance transform with the detected edges superimposed.

summing the result. We experimented with different measures for the actual comparison and found that the truncated L2 norm performs best.

5   Pose Hypothesis Generation

The object model has six degrees of freedom leading to a very high number of hypotheses - depending on the resolution, of course. To lower the computational complexity the number of possible hypotheses needs to be pruned. First of all we know the 2D ROI of the object, which significantly prunes the translational parameters. Second, since the pump is being assembled on a heavy base (see figure 2.a) we can reduce the three rotational degrees of freedom to just one, namely the rotation around the normal of the table. Furthermore, to speed up the system we follow a two-step (coarser-to-finer) hypothesis testing strategy.

At the first level the position of the object is approximated by projecting the 2D center of the ROI onto the table (known from calibration) as a 3D point. This approximation of the 3D position of the object is rather good in the horizontal direction, but more uncertain in the vertical direction. A small improvement is introduced by correcting the vertical position with one third of the ROI's height measured from the bottom. This 3D position may not be correct, but the perspective of the hypothesized model will be nearly correct. The errors in the position and the size of the pump can after the feature synthesis of the pump be compensated by centering and scaling the synthesized edge image according to the ROI. This gives an approximation of a synthesized edge map from the "correct" position.

To determine both if the rough position is good enough, and which rotation step size is required in stage one, a plot of the cost of matching poses with the rough position is seen in figure 6.a together with the cost of the poses based on a refined ("true") position. This plot suggests that it is possible to find the approximate global minimum using the position approximated from the ROI. Furthermore, the step should be less than 20 degrees to avoid local minima. Other plots with different configurations and poses give similar results, and thus the roughly estimated position and a rotation with 10 degree steps is concluded to be sufficient for finding the global minimum. This gives a total of 180/10=18 generated hypotheses in stage one. The best matching pose from the plot is seen in figure 6.B.

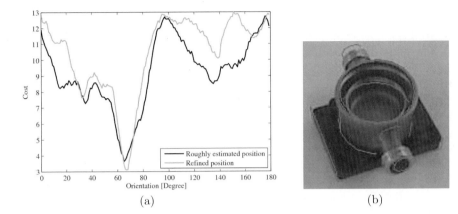

(a)   (b)

Fig. 6. a) Cost of matching hypothesized poses with varying orientations with the image to the right. The plot is made with two different positions (a rough estimate and a refined estimate); both with an angle resolution of 1 degree. Note that since the object is self-symmetric only 180 degrees are required. b) The best fit from the plot superimposed.

In step two the "correct" pose is located somewhere around the best pose of stage one, but to be sure we use the two best poses from the first step. A more precise position of the origin is found by examining the 2D location of the origin in the synthesized edge image. Hereafter the rotation is refined to a resolution of two degrees within a range of ± 5 degrees. In total $18 + 2 \times 5 = 28$ different poses are synthesized and compared with the edge image.

6 Pose Estimation

The synthesized pose which best matches the current edge image defines the current pose. In the following frames there is no need to repeat the above process since the result will be more or less the same unless the object is moved. We therefore define a band around the ROI and watch for changes in this band using image differencing. If no difference is detected, no further processing is preformed. If on the other hand motion *is* detected, then a new pose estimation is performed. This strategy both saves processing time, but also stabilizes the overlayed graphics by ignoring slight changes.

7 Visualization

Recall that the purpose of estimating the pose of the object in the first place is to be able to overlay assembly information (to aid the user) no matter the pose of the object. The layout and type of information to be visualized is determined through a couple of design and user-test iterations where both lo-fi and hi-fi prototypes are applied. In figure 7 the final result is shown. We can identify three regions:

AR-region. In this region live video from the camera is shown together with graphics illustrating the next part to be mounted. In order to emphasize, where and how the next part is to be mounted a small animation is shown, see figure 7 and [18].

Info-region. In this region the next part to be mounted is shown together with the name of the part. To draw attention and to make the user familiar with the part, it slowly rotates in 3D. Furthermore, this region also holds useful information for the user, e.g., "Force may be required for mounting". Such information is also accommodated by a suitable icon. An example is shown in figure 7.

State-region. To provide an overview, this region contains a number of images each illustrating a certain state (past, current, future) of the object. By "object state" is meant how many parts are assembled. The next state is enhanced by a surrounding square and in each illustration the new part is highlighted, see figure 7.

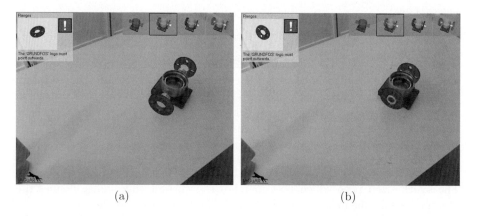

(a)   (b)

Fig. 7. Two screen shots of the AR interface shown to the user during assembly. The screen shots are separated by approximately 0.5 second. Note the movement of the virtual object in both the AR-region and Info-region.

8   Evaluation

The system is implemented on a 2 GHz dual core computer. One thread runs the graphics in order to ensure smooth visualization. Another thread runs the pose estimation at around 5 fps for the largest models. User tests have confirmed that this is a sufficient response time. Regarding the accuracy of the system, qualitative results are presented below. Usability tests showed that the accuracy of the system is sufficient for assembling the pump. A video example can be found here [18].

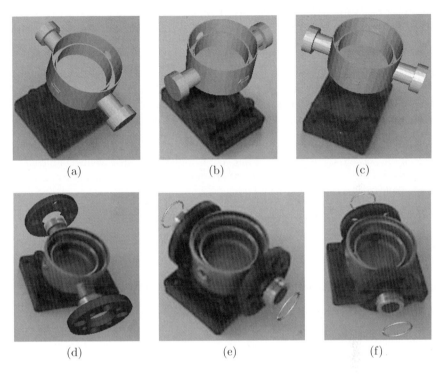

Fig. 8. Different poses of the partly assembled pump together with graphics overlay with the next part. a,b,c) Main champer (the grey object). d) Flanges (two round black objects). e,f) Snap rings (two silver rings).

9   Conclusion

We have presented a robust system for estimating the pose of a pump at different stages in an assembly process. The pose estimation is based on a two-step matching strategy, where different CAD model poses are synthesized into the image and compared with the extracted edges in the image. To this end chamfer matching together with a truncated L2 norm is applied. Significant work has been put into synthesizing visual edges in a less computational manner. This is indeed needed in AR systems, where a smooth visualization is a must. The layout of the user interface and the augmented graphics is developed though a number of usability tests and the final result presents the relevant information to the user. Especially the introduction of small animations to visualize where and how to mount a part has resulted in positive feedback.

Future work will include a fine tuning of the system followed by a large field test where the effects of the system need to be compared with current practise regarding assembly time and errors made. If this is successful and the system is to be including in the assembly process, three issues need to be dealt with: i) integration the system with the existing production planning systems, ii) automatic generation of simplified CAD models. The simplified model used in this work, see for example figure 3.b has been handcrafted, and iii) automatic generation of the guiding animations based on the above data.

Acknowledgement

We would like to thank Grundfos A/S for providing material for this work.

References

1. Tang, A., Owen, C., Biocca, F., Mou, W.: Comparative effectiveness of augmented reality in object assembly. In: Conference on Human factors in computing systems, Ft. Lauderdale, Florida, USA (2003)
2. Milgram, P., Kishino, A.F.: Taxonomy of mixed reality visual displays. IEICE Transactions on Information and Systems E77-D, 1321–1329 (1994)
3. Chandaria, J.: Real-time camera tracking in the matris project. In: International Broadcasting Convention, Amsterdam, NL (2006)
4. Broll, W., Lindt, I., Herbst, I., Ohlenburg, J., Braun, A.K., Wetzel, R.: Toward next-gen mobile ar games. Computer Graphics and Applications 28, 40–48 (2008)
5. IPCity, EU-IST Project (2009), www.ipcity.eu
6. Hakkarainen, M., Woodward, C., Billinghurst, M.: Augmented assembly using a mobile phone. In: Symposium on Mixed and Augmented Reality, Cambridge, UK (2008)
7. Trucco, E., Verri, A.: Introductory Techniques for 3D Computer Vision. Prentice Hall, Englewood Cliffs (1998)
8. Azuma, R.T.: A survey of augmented reality. Presence 6, 355–385 (1997)
9. SPRXMOBILE (2009), www.sprxmobile.com
10. ARToolKit (2009), www.hitl.washington.edu/artoolkit
11. Moeslund, T.B., Kirkegaard, J.: Pose estimation of randomly organised stator housings with circular features. In: Kalviainen, H., Parkkinen, J., Kaarna, A. (eds.) SCIA 2005. LNCS, vol. 3540, pp. 679–688. Springer, Heidelberg (2005)
12. Balslev, I., Eriksen, R.D.: From belt picking to bin picking. International Society for Optical Engineering, vol. 4902 (2002)
13. Salvi, J., Pags, J., Battle, J.: Pattern codification strategies in structured light systems. Pattern Recognition 37, 827–849 (2004)
14. Schraft, R.D., Ledermann, T.: Intelligent picking of chaotically stored objects. Assembly Automation 23, 38–42 (2003)
15. Kirkegaard, J., Moeslund, T.: Bin-picking based on harmonic shape contexts and graph-based matching. In: International Conference on Pattern Recognition, Hong Kong, China (2006)
16. Canny, J.: A computational approach to edge detection. Pattern Analysis and Machine Intelligence 8, 679–698 (1986)
17. Borgefors, G.: Hierarchical chamfer matching: A parametric edge matching algorithm. Pattern Analysis and Machine Intelligence 10, 849–865 (1988)
18. Video download (2009), http://www.cvmt.dk/AR

V-Volcano: Addressing Students' Misconceptions in Earth Sciences Learning through Virtual Reality Simulations

Hollie Boudreaux[1], Paul Bible[1], Carolina Cruz-Neira[1], Thomas Parham[2], Cinzia Cervato[2], William Gallus[2], and Pete Stelling[3]

[1] University of Louisiana at Lafayette, Lafayette LA 70501
[2] Iowa State University
[3] Western Washington University

Abstract. Research in teaching and learning about Earth Sciences indicates that first year geology students not only lack knowledge about basic concepts, but that they may also have developed their own potentially incorrect explanations of those phenomena. Understanding volcanic concepts is one of the areas in which noticeable misconceptions occur, as a significant number of students seem to acquire their knowledge from non-traditional sources such as sensationalist media and catastrophic films. This paper presents V-Volcano, a virtual reality volcano activity learning environment that immerses students in a scientifically-accurate simulation of volcanic systems. Students are able to generate and manipulate volcanic eruptions in real-time with data monitoring to explore the effects of changing conditions. The goal is to provide a geoscience tool that can be used to correct student misunderstandings about volcanic phenomena.

1 Introduction

In recent years, movies, such as Dante's Peak or Volcano, and popular media have glamorized volcanoes. The general public has been charmed to the point that many incoming college students take introductory geology classes wanting to learn about volcanic phenomena. However, these students bring with them many misconceptions and misunderstandings about where and how volcanoes are formed, the underlying mechanisms that cause eruptions, and the effects of volcanoes on human endeavors in the area.

Prior to V-Volcano, a group from Iowa State University [1] [2], developed the Interactive Virtual Earth Science Teaching (InVEST) Volcanic Concept Survey to look into the prevalence and nature of the misunderstandings and alternate conceptions about volcanoes among first year geoscience students. Five universities were included in the study: Iowa State University, University of Texas El Paso, University of Georgia, Western Washington University, and Fort Valley State University.

The results from these surveys were analyzed, with the goal of identifying areas where introductory geology students were lacking or mistaken in their

G. Bebis et al. (Eds.): ISVC 2009, Part I, LNCS 5875, pp. 1009–1018, 2009.

knowledge of volcanoes [1] [2]. The following four main misconceptions about volcanic formation were identified:

1. Volcanoes only form near bodies of water
2. Volcanoes are common only in areas near the equator or other warm areas
3. Volcanoes appear in areas of rocky terrain
4. There is no pattern to volcano formation

In addition, several concepts were very poorly understood. These include the inner workings of a volcano, what controls explosive activity, the role of silica in explosive activity, and the effects of volcanic activity on the surrounding environment and human civilization.

These results led us to the design of V-Volcano, a virtual reality volcano activity learning environment that immerses students in a scientifically-accurate simulation of volcanic systems. Using V-Volcano, students are able to generate and manipulate volcanic eruptions in real-time with data monitoring to explore the effects of changing conditions. Furthermore, teachers can define specific assignments for the students to perform in V-Volcano and review the assignment results with the class in a large-scale virtual reality space, such as a CAVE or large tiled walls. The goal of V-Volcano is to provide a geoscience tool that can be used to correct student misunderstandings about volcanic phenomena.

1.1 Previous Work

Many simulations and visualizations exist that explain various aspects of volcanic activity. First, many animations exist that are Flash-based and offer a simplistic overview of volcanic activity. These include interactive Flash programs such as one provided on the Discovery Channel website [3] and one provided by the Alaska Museum of Natural History [4]. These provide simple controls of silica (SiO_2) and water vapor which effect an animation of a volcanic eruption. Some visualizations exist to supplement the geological research of an institute. For example, the New Millennium Observatory (NeMO) [5] offers a few panoramic views of an underwater volcano and simple animations of an underwater exploration craft used by the geological researchers. Another application called Eruption! is a role playing, low-graphics, scenario simulator that puts students in the role of volcanologists, political leaders, journalists, and village farmers [6].

High fidelity mathematical models of the geological phenomena are another category of visualization. These approaches seek to find approximations to the actual chemical and physical processes of a certain feature such as a lava flow. Stora et al [7] conducted such a lava flow simulation. High fidelity to the natural process in a simulation inevitably requires more computational complexity and thus makes these simulations far too slow to provide students with an interactive experience. Stora et al reported computation time between 20 sec. to 2 minutes per frame for varying numbers of particles used to simulate the lava.

All these tools are designed for informal science education, presenting basic but critical concepts about volcanoes using traditional multimedia techniques, such as videos and Flash animations. Interaction with the volcano is minimal, limited to setting one or two initial constraints and viewing the resulting 2D animation. However, none of these directly addresses the misconceptions people have about volcanic activity, nor do they provide a three-dimensional space to explore the effects of volcanic eruptions in a realistic setting.

Prior to V-Volcano, our group carried out a feasibility study for the development of a virtual reality tool to assist in atmospheric science education. Specifically, the project focused on the exploration of a tornadic supercell thunderstorm, allowing students to perform weather measurements while the storm was taking place as shown in Fig. 1. The Virtual Storm was deployed in several entry-level meteorology classes and a three-year study [8] showed improved learning through its use. It is currently being used in multiple universities as a companion teaching tool in earth sciences curricula. These results encouraged us to move forward with the work presented in this paper.

(a) Immersive Environment   (b) Desktop Environment

Fig. 1. The Virtual Storm running in an immersive environment 1(a) and on the desktop 1(b)

2   Design of Interactive Educational Simulation

The results of the Volcanic Concept Survey combined with the experience gained through the deployment of the Virtual Storm have guided the design of the V-Volcano educational tool. For the Virtual Storm activity, the most interesting features that kept the students engaged were the extensive range of parameter manipulation and the ability to perform 3D exploration of the entire storm space. These two features made the application similar to a video game. The player's choices determine the experience and rewards gained.

Taking the approach of an educational game, the design of V-Volcano focuses on keeping the student engaged in the learning activity by providing exciting and rewarding interactions with the different types of volcanic eruptions. We focused on scientific accuracy and high quality physical representations.

3   Design Approach

Here is a typical scenario for V-Volcano: Students using V-Volcano manipulate the formation of their own volcano in a broad tectonic setting, ideally minimizing any reinforcement of hydrological and climatic misconceptions about volcano formation. Once the volcano is formed, students can travel (by flying and by driving at ground level) through the terrain. In addition, they can explore the interior structure of their volcano and probe the mechanisms controlling its eruption over time. There are three aspects that are key to the design of V-Volcano to create scenarios as described above: interaction flexibility, monitoring, and multiple display settings.

3.1   Interaction Flexibility

The diagram and table shown in Fig. 2 below summarizes the different conditions that that students can manipulate through their interactions with V-Volcano. The boxes with bold borders in the diagram indicate the parameters that define the eruption style. The lighter columns in the table are parameters the students can set and the darker columns indicate the resulting effect of those parameters. Based on this table, we have designed a set of visual simulations representing the different effects, such as ash plumes, fire fountains, volcanic bombs, and pyroclastic flows. We have also created three-dimensional models of representative volcanoes for the different geological types: strato, shield, and dome. V-Volcano can interactively match the visual effects with the type of volcano based on user manipulation of the parameters. This design approach gives us a great deal of flexibility as nothing in the experience is preset. It is truly an explorative experience. Furthermore, as part of the interaction we incorporate the ability to fly

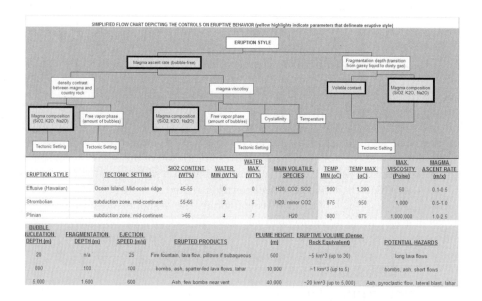

Fig. 2. Diagram of parameters controlling eruption style

and travel through the terrain, even into the volcano and down to the magma chamber. This way, students also learn about the structural details of volcanoes and can grasp the size and spread of their eruption activities.

3.2 Monitoring

As students are interacting with V-Volcano, additional information about the phenomena being explored is presented to the student. For example, if the student interactions generate an Hawaiian eruption in a shield volcano, most likely they will apply their misconception and immediately think of the volcanoes in Hawaii and those along Japan. To address this, V-Volcano will present the student with several other volcanoes, such as the Erta Ale volcano in Ethiopia, which is of the same type but does not conform to the misconception that shield volcanoes are found only in islands or by the ocean. We anticipate that this approach will help to clear some misconceptions and provide a stronger scientific understanding of the geology and dynamics of volcanoes.

3.3 Multiple Display Settings

Multiple displays and interaction modalities are a key feature of V-Volcano. At the basic level, V-Volcano operates like any other conventional desktop graphics application. It presents the user with a graphics window to visualize the volcano environment and a traditional Graphics User Interface (GUI) to input the different parameters, as shown in Fig. 3(a). For game-oriented students, V-Volcano can incorporate a game pad for interaction, adding an extra "thrill" to the experience of exploring the space (see Fig. 3(b)).

Part of the previous work with the Virtual Storm showed that group discussions and teacher involvement in the interactive exploration are important to understanding the more complex concepts. To address this finding, V-Volcano can also be operated in high-end virtual environments such as tiled walls and CAVEs, as shown in Fig. 4. In these settings a group of people can participate in the activity, sharing their knowledge and asking questions as they explore a volcano.

(a) Keyboard Control   (b) Controller Control

Fig. 3. V-Volcano in use on a desktop. Both keyboard 3(a) and controller 3(b) controls are provided.

Fig. 4. V-Volcano running on a rear-projection wall

V-Volcano also incorporates information about subsurface features, allowing for exploration of the magma chamber and the magma conduit to the surface. This component of the project presented a challenge, as the subsurface is solid and it is hard to find a way to peel off subsurface components to see the interesting features. The solution to this problem was inspired by the traditional cut-away diagram shown in most geology textbooks. We developed a three-dimensional model of this, which, as seen in Fig. 5 has worked out very nicely to show the subsurface features of volcanoes.

Fig. 5. A 3D representation of a traditional subsurface cutout showing the magma chamber with shader effect

4   Technical Design

Because V-Volcano was designed to improve pedagogy on large scales and widespread dissemination to academic institutions was desired, we decided to develop it utilizing Open Source tools. Specifically, the simulation uses the virtual environment development framework VR Juggler [9], the scene graph manager OpenSG [10] for the three-dimensional interactive environment, and Java for the GUI. These three very popular tools give us an easy-to-distribute framework to develop V-Volcano, as well as a very powerful environment to integrate all the functionality described in the previous sections.

We used an object-oriented approach to design the activity features for the volcanoes. Physical simulations of fire fountains, ash plumes, volcanic bombs, pyroclastic flows, and other features are built using real-time computer graphics techniques such as particle systems and shaders, as seen in Fig. 5 above. The models were designed to be representative of volcanoes in general, and not specific to any volcano used in the application. Fig. 6 shows a fire fountain and an ash plume. Each one of these features is encapsulated in a software module and then, through the external run-time configuration available through VR Juggler, the valid combinations of volcano types and eruption features can be specified. This software design approach was used to facilitate future expansion of the project, which may require additional features such as impacts on urban areas, air traffic, etc.

(a) Fire Fountain   (b) Ash Plume

Fig. 6. The fire fountain 6(a) and ash plume 6(b)

5   Interaction with the Simulation

The GUI, implemented in Java, allows students to interact with the simulation in the traditional manner of interactive desktop applications. However, through VR Juggler, the same GUI can be launched on a tablet PC or PDA without loss of functionality and gaining mobility in an immersive space. Fig. 7 shows the user on the left manipulating parameters via the tablet PC in front of a high resolution stereoscopic wall.

The interface, as seen in Fig. 8, provides three sliders that control silica, water vapor, and temperature. These are the main eruptive controls that dictate the type and caliber of eruption. Students can modify these values and observe the changes in the visual environment. In addition to these controls, a point-and-click compass allows modification of the wind direction at runtime. The GUI also

Fig. 7. Using a Tablet PC to control eruption parameters

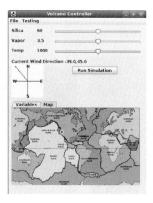

Fig. 8. Depiction of the Graphical User Interface

provides dependent variable information such as eruptive volume and nucleation depth. A full list of these variables can be found in Fig. 2. Perhaps the most important non-control feature of the GUI is a map that indicates the location of the volcano with respect to tectonic plates. This feature imparts to the students the role of plate tectonics in inciting volcanic activity, which, as identified in the Volcanic Concept Survey, is one of the concepts not well understood.

6   Discussion

The main goal of the application is to clear up the misunderstandings and misconceptions students have about volcanoes. The InVEST survey highlights a serious lack of understanding of many basic concepts related to volcanoes and the fact that students are making their own, often incorrect, assumptions about those basic concepts. Though popular films have generated interest about volcanoes, the survey indicates that these same films could be responsible for fostering some of the misconceptions identified by the survey. The students' inability to identify plate tectonics as dictating the placement of volcanoes, for example, illustrates a significant gap in science literacy. Plate tectonics is an overarching geological theory that explains volcanoes and earthquakes, two of the most destructive natural disasters. Knowledge of volcanoes is becoming increasingly important due to the role they can play in global climate change.

The V-Volcano application addresses these misconceptions in a number of ways. First, it provides a good platform to demonstrate the role of plate tectonics in the formation and activity of volcanoes by use of the GUI's map. Another conceptual gap exists where eruptive controls are concerned. According to the Volcanic Concept Survey, no student demonstrated mastery of this concept. To address this deficiency, we designed the GUI's interaction around silica, water vapor, and temperature. Thus, these variables are the first that the students must understand to interact with the simulation. Students will have to explore exactly what combination of values will cause an eruption to occur. Integrating these variables with the tectonic setting of a particular volcano will add to the students' understanding because some combinations of values will work for one setting but not for another. Additionally, V-Volcano also includes monitoring capabilities to present users with additional information developed by the teacher and targeted towards the clarification of misunderstood concepts and assumptions.

The first volcano chosen to be implemented was Mount St. Helens. Our geoscience collaborators believed this volcano addresses the most basic misconceptions as discovered by the Volcanic Concept Survey. Students will more easily relate to this volcano because it is still active, located in North America, and an abundance of eruption footage is available.

7   Conclusion

The V-Volcano application provides many opportunities for students and teachers to engage in teaching and learning activities. The application can be used in the classroom along with other information on plate tectonics and volcanoes in the curriculum. The ability to move freely and manipulate variables to change the simulation will encourage students to ask important questions such as "What will happen if I max-out a variable?" Cross-platform compatibility in the design will allow students to answer these questions outside of the classroom on their own computers. V-Volcano will be an effective teaching tool because of its entertaining nature and conceptual depth. Use of the application in a virtual reality environment has more potential to be memorable and capable of holding students' interest.

8   Future Work

After undergoing a semester of testing by introductory level geology students and their professors, the suggestions made by the users will be considered when making improvements to the application. Thanks to the modularized nature of the environment's design, adding functionality for additional volcanoes will proceed rapidly. Eventually a scene annotation program could provide teachers with the ability to place volcanic features in the desired place on new volcano models developed in the future.

Furthermore, the technical flexibility of the software framework makes it possible for V-Volcano to be played in informal science locations, such as science museums, planetariums, and digital cinemas, opening a new form of science dissemination to the general public.

References

1. Parham, T., Cervato, C., Gallus, W., Larsen, M., Stelling, P., Hobbs, J., Greenbowe, T., Gupta, T., Knox, J., Thomas, E.: The InVEST Volcanic Concept Survey: Exploring Student Understanding About Volcanoes
2. Gallus, W., Cervato, C., Parham, T., Larsen, M., Cruz-Neira, C., Boudreaux, H.: The Interactive Virtual Earth Science Teaching (InVEST) project: preliminary results. In: European Geosciences Union General Assembly (May 2009)
3. Riggs, N.: Virtual Volcano: Discovery Channel. US Geological Survey. Discovery Communications Inc. copyright 2004 (2004),
   http://dsc.discovery.com/convergence/pompeii/interactive/interactive.html
4. Stage 2 Studios LLC. Volcano Simulation. CG Science copyright 2006 (2006),
   http://www.alaskamuseum.org/features/volcano/
5. New Millennium Observatory. Oregon State University,
   http://www.pmel.noaa.gov/vents/nemo/explorer.html
6. Bursik, M.: Eruption! University at Buffalo, The State University of New York Department of Geology, Interactive Models for Geological Education Online (IM-GEO), http://www.glyfac.buffalo.edu/gerp/eruption.html
7. Stora, D., Agliati, P., Cani, M., Neyret, F., Gascuel, J.: Animating Lava Flows. iMAGIS-GRAVIR/ IMAG BP 53, F-38041 Grenoble codex 09, France
8. Gallus Jr., W.A., Cervato, C., Cruz-Neira, C., Faidley, G.: A virtual tornadic thunderstorm enabling students to construct knowledge about storm dynamics through data collection and analysis. Advances in Geoscience, 27–32, SRef-ID: 1680-7359/adgeo/2006-8-27
9. VR Juggler, http://www.vrjuggler.org
10. OpenSG, http://www.opensg.org

A Framework for Object-Oriented Shader Design

Roland Kuck and Gerold Wesche

Fraunhofer IAIS, Competence Center Virtual Environments

Abstract. Shaders offer a wide range of possibilities, but at the same time limit the flexibility of an application as combining shader components is difficult. We present a novel object model for writing shaders for modern graphics hardware. These objects are defined and instantiated within an application. They are then propagated to the different programmable pipeline stages using a well-defined concept of copy-construction. Objects can reference each other and thus offer a flexible way to configure the shading set-up at run-time.

Our framework is built on top of the object model for the standard illumination situation of surfaces and light sources. We show that many modern shading models can be expressed in this framework in a uniform and integrated way. Both, the object model and the framework, make the reuse of components practical and allow object-oriented design to be applied to the development of shaders.

1 Introduction

Modern graphics hardware replaced the fixed-function implementation of graphics algorithms with programmable processors. While this does offer a wide range of new possibilities, it also removes the orthogonal design of the fixed-function units. Light sources, materials, and transformations could be modified separately from each other. Shaders replace all functionality and have to perform all required tasks.

Applications however require the orthogonal behavior. The number of possible combinations of different settings leads either to a complex *super shader* or to a large number of individual shaders that are difficult to maintain.

Reusable components of shader code are a solution to this problem, but they are difficult to design. One major problem is the pipelined structure of graphics hardware. It consists of multiple stages, each responsible for different kinds of primitives. Components often require information in later stages that has to be properly transferred from earlier stages while at the same time has to remain properly encapsulated.

Object-oriented design provides solutions to this problem but this methodology cannot be directly applied to writing shader code. Where should objects be stored and defined? And how does the graphics hardware with its massively parallel pipeline stages access objects concurrently? This paper gives the answers.

G. Bebis et al. (Eds.): ISVC 2009, Part I, LNCS 5875, pp. 1019–1030, 2009.

Fig. 1. A teapot with different combinations of materials. From left to right: a simple plastic material, a bump-mapped plastic, and a geometry shader that tessellates and displaces the teapot.

The main contributions of this paper are:

- An object model designed to be used with modern and future graphics pipelines addressing the combination problem of shader components.
- A framework built on top of the object model for the standard illumination situation of surfaces and light sources. This framework allows different shading and rendering techniques to be combined.

Both contributions together enable a flexible way to configure the rendering state. Objects can be combined to define materials or lighting set-ups. These objects are completely and directly exposed to the application and the connection of objects can be changed at run-time.

We discuss related work in section 2. We then explain the concepts and definitions of our object model in section 3 and the shading framework in section 4. After describing the implementation aspects in section 5, we evaluate several examples in section 6 demonstrating the feasibility of our approach and we review its performance.

2   Related Work

Our object model is based on [1]. We extend it to remove two major shortcomings: the support for only a subset of available pipeline stages and the need for explicit set-up methods, requiring the root object to act as a controller object that knows all other objects. We describe the new mechanisms in section 3.1. Furthermore, we extend the collection of patterns described in [1] to build a flexible framework that allows us to model a wide range of rendering and shading techniques, as shown in section 6.1.

Cg [2] is a low-level shading language. It allows structures to contain functions and it uses interfaces to provide a limited way of polymorphism. Structures cannot contain types with different qualifiers. As shown in section 4, this is often required due to the multi-stage design of the graphics pipeline. A light source usually performs some internal computations in the vertex shader from uniform values, like e.g. the position of the light source, and it uses the result in the fragment shader. This is not possible to encapsulate using structures in *Cg*.

Sh [3] uses meta-programming techniques to generate shader code. Due to its design, it can directly use objects from the *C++* language. While a *C++* class can store read-only attributes for all pipeline stages (called *parameters* in *Sh*), other storage types called *attributes* in Sh are only available as input and output data from functions and are not object data. No implicit connections between them exist. In [4] a complete algebra is defined that uses these functions in *Sh* as components to build an execution pipeline. Different operators in the algebra allow inputs and outputs of functions to be connected. Complete knowledge of data passed between objects and stages has to be available to an application to build the connection between components. Therefore changes to the internal implementation of a shader require a modification of the application.

The *SuperShader* [5] is a complex shader that allows different predefined materials and effects to be selected at run-time. A more extensible system is described in [6], but it is still limited to a simple concatenation of shader code. Our object model allows complex interactions of components. This is used extensively in our shading framework.

The shading framework in section 4 is based on the Renderman Shading Language [7] with its clear separation of light, displacement and surface shaders as first class objects. A recent addition was the introduction of messages between the light and the surface shader [8]. Instead of making this a special case, we can take advantage of our object model (see section 4.2).

The *Stanford shading language* [9] provides a similar high-level abstraction using light and surface shaders. Pipeline stages are not differentiated in this language. Instead, qualifiers are used to denote the primitive that should be associated with the variable, like a vertex or a fragment. The compiler then automatically splits the compiled and linked shaders into different portions of code to run on the pipeline stages. While it provides a framework similar to ours, we provide a clear separation between the object model and a possible framework implemented on top. Our framework can simply be extended by providing new classes.

Brook [10] and *CUDA* [11] are extensions to the *C* programming language that allow writing *kernel* functions to run on the GPU. However, relevant parts of the graphics pipeline, like the rasterizer, are not accessible. Future GPUs will likely support these interfaces more directly, while at the same time processors will continue to contain more cores and can be programmed similarly. These flexible new capabilities might eventually lead to new implementations and variations of the graphics pipeline for real-time rendering tasks since this abstraction provides a good performance compared to other algorithms [12], potentially integrating fixed-function pipeline stages where these provide better performance [13]. Our object model and shading framework relies on a graphics pipeline abstraction and not on a particular hardware model. Therefore it can be layered on top of these new interfaces. In particular it can be directly mapped to the example pipelines from [13].

3   Object Model

We summarize the key ideas of our object model, which is an extension of the work described in [1]. The object model forms a thin layer on top of *OpenGL* and uses the *OpenGL Shading Language* (*GLSL*; see e.g. [14]). As usual we define objects as instances of classes that itself define methods and attributes. The pipeline structure of *OpenGL* is fully exposed by requiring each method and attribute to have a qualifier indicating in which stage it can be accessed. For each instanced object its attributes are declared as global variables using the given type and qualifier.

The following code declares a *C++* class. It declares the *color* attribute and the *illuminance()* method. Note that *void_* is a special class used to transport information about which *GLSL* function implements the method:

```
class Diffuse : public ShaderBase<Diffuse, IlluminatedMaterial>
{
 public:
  void_<> illuminance(vec3<>, vec3<>)
    { return invoke< void_<> >("Diffuse_illum_impl"); }

  ValueReference<vec4, uniform> color;
 private:
  DERIVED_DECL(Diffuse, IlluminatedMaterial)
};

CLASS_INIT(Diffuse, "Diffuse.glsl", NONE, DEFS((color)) )
```

The class derives from *IlluminatedMaterial*. This base class is used by materials that receive light from light sources. It requires an implementation of the *illuminance()* method. The *C++* implementation only provides a name that references the implementation in *GLSL* as follows. It uses methods of the base class to retrieve the normal and to return the final color:

```
void Diffuse_illum_impl
   (Diffuse_SELF, vec3 light_color, vec3 light_direction)
{
   vec3 normal = IlluminatedMaterial_get_normal(self);
   float intensity = max(0., dot(light_direction, normal));
   vec4 color = intensity * vec4(light_color, 1.)
                          * Diffuse_get_color(self);
   IlluminatedMaterial_accum_color(self, color);
}
```

Objects are referenced using a type called *OBJREF*. Objects are instanced only in the application to simplify resource allocation on the GPU. One *root object* provides the entry point for all stages and invokes methods on other referenced objects. The entry points are defined in the *Enterable* interface, which a root object has to implement.

3.1   Extension of the Object Model

We introduce a conceptual *deep copy* of objects between each stage, i.e. all objects including objects they reference are copied. This behavior is well known and directly maps to the copying of values between stages in *GLSL*. Note that sharing object data between multiple stages would require synchronization mechanisms that have high runtime costs and are not available in *GLSL* (opposed to e.g. CUDA).

Extending the idea of a copy operation between the stages we also introduce an initialization phase after each copy operation. This allows an object to perform operations in a stage where it is not called by another object. Consider e.g. an object wrapping the access to texture coordinates for later access in the fragment stage by a material object. This wrapper will be only called in the fragment stage but has to copy the texture coordinates in the vertex stage.

Figure 2 depicts the current *OpenGL* pipeline as used in our object model. We assume data are only copied between stages and that data amplification, like rasterization, is happening within one stage. Our qualifiers can be classified by the stage where they can be read and written. The varying attribute e.g. copies data from the vertex stage to the fragment stage, i.e. between two stages that are not directly connected.

Amplifying stages thus need to define how these values are interpolated. As different values have different requirements, qualifiers can additionally be used to define behavior, again mirroring *GLSL*. If a framework allows using the geometry stage, it has to define its interpolation behavior. This is realized using framework-defined types on top of raw qualifiers that provide only *GLSL* semantics. See section 4.4 for an example.

We also introduce a new stage after a primitive has been emitted in the geometry stage with the same copy semantic as every other stage. This allows dependent values in other objects to be calculated before the rasterization starts.

Another difference of the geometry stage is that it can access multiple primitives, e.g. the vertices of a triangle. In our object model a separate copy of all objects exists for each accessible primitive and is accessible via a unique reference. Utility functions are provided that take an index n and a reference as input and return a reference to the n-th copy. This design allows access to all available

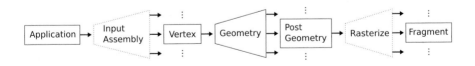

Fig. 2. The graphics pipeline in our object model. The *Application* stage is the application running on the CPU; the *Vertex, Geometry,* and *Fragment* stages are running on the GPU. The *Post Geometry* stage is invoked when a vertex is emitted in the geometry shader. The *Input Assembly* and *Rasterize* stages are not (yet) programmable. The root object is deep-copied between all stages.

objects, while at the same time it offers a consistent way to access objects and their attributes across all stages.

4   Framework and Patterns

The object model can be used to formulate a framework for a normal shading set-up, i.e. the shading of surfaces that are illuminated by light sources. The framework is designed to ease the reuse of individual components. Note that all objects and types from the framework, like e.g. *Surface* and *Varying*, are implemented on top of the object model. This allows the framework to be extended or replaced if required without changing the underlying object model.

4.1   Value References

One important design pattern used in the following is *composition* [15]. We use it in a special scenario where we will refer to it as *value reference*. The idea is simple: instead of storing an attribute directly in a class (i.e. an aggregation), we only store a reference to an object holding this value (i.e. an association). This object implements a simple interface (called *Gettable*) that allows the value to be retrieved.

The simplest class that implements this interface is the *Value* class, which stores a single value with a given qualifier. The *Texture* class provides access to a texture map using itself the *Gettable* interface for the texture coordinates. Other possible implementations include procedurally generated values.

4.2   Surfaces and Lights

The root object in our framework is a *Surface*. A *Surface* references objects of the following classes:

- *Material* is the base class for all materials. It provides a *shade()* method. Most materials derive from the sub-class *IlluminatedMaterial*, which provides the typical code for interactions with light sources. Materials then implement the *illuminance()* method to receive light.
- *Light* is the base class for light sources. A light source should provide an implementation for *illuminate()* to emit light and *transform_light()* in case it needs to transform positional data into the used coordinate system.
- *CoordinateSystem*: this class knows about the coordinate system the lighting calculation is performed in and provides a *transform()* method to transform positional data from world space into the used coordinate system. It also transforms and stores the position and the normal of the point being shaded.

4.3   Calling Sequence

The interaction of these objects is shown in figure 3. In the vertex stage the *CoordinateSystem* is initialized and it transforms the viewing direction into the coordinate system where the lighting calculations are performed. It is required to store this information, which can be used in later stages to retrieve the transformed value.

The material then invokes the *transform_lights()* method on the *Coordinate-System*, providing the list of light sources of the surface as parameters. The *CoordinateSystem* invokes the *transform_light()* method on the light sources, again passing a reference to itself. Each light source then uses its reference to transform its position and stores the result in its attributes to be used later in the fragment stage.

In the fragment stage, the surface object invokes the *shade()* method of the material object. If the material wants to receive light, it calls the *illuminate()* method for each light source, passing along the position of the fragment to be shaded and a reference to itself. Each light source then calculates the amount of energy the fragment receives and the direction of light and calls the *illuminance()*

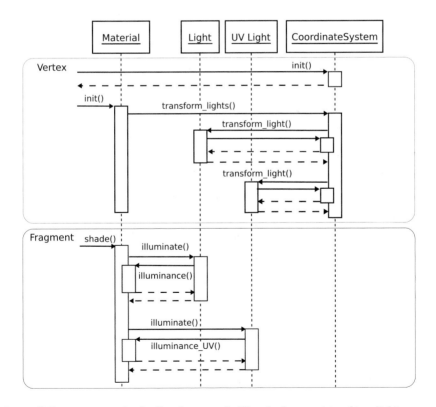

Fig. 3. Calling sequence in shading framework. The shader consists of two light sources, one material, and a coordinate system. The surface holding the material is not shown. One light source is emitting ultraviolet (UV) light.

method of the given material object. This indirection of calls is called *double dispatch* and allows us to use the visitor pattern [15] to implement different shading behavior for special combinations of light sources and materials.

4.4 Subdivision and Displacements

The shading framework supports subdivision of triangles in the geometry shader. It requires that all positions of the new primitives are linear combinations of the positions of the input primitives followed by an arbitrary transformation. Subdivision surfaces and displacement-mapping (see e.g. [16] for a combination of both) fulfill this requirement. An example is shown in figure 1.

A concrete subclass of the *Displacement* class specifies how geometry is generated. It has to provide the framework with the factors of the combination and the transformation. These values are used to correctly interpolate and propagate values stored in the *Varying* type. This type has to be used with all values that always need to be copied from the vertex stage to the fragment stage. It is given an *Interpolator* class as a template parameter to specify the semantics of the value.

We have currently implemented three different interpolators: the *PositionInterpolator* is used for values that represent positions or directions. It interpolates its input values using the given factor and then applies the transformation. The *NormalInterpolator* is similar, but uses the appropriate transformation for normal vectors. The *DefaultInterpolator* is used e.g. for color values, and only interpolates its input values.

5   Implementation

In other non object-oriented languages objects are usually represented by a data structure that is passed as a reference to functions. As *GLSL* lacks the support for references, we enumerate all objects and identify the objects with their number. To access an attribute a *dispatch* function is used. These dispatch functions are also used for method invocations in general to allow for polymorphic behavior. If a dispatch function is called with a reference that is not matched by any branch, a default value is returned instead. A simple introspection layer is used to retrieve the method signature from the supplied *C++* definition.

A dispatch function in *GLSL* follows (taken from the example above). Attributes are stored in global variables. The *color* attribute is a value reference and does not store the value itself but defers to another object:

```
vec4 obj_0x5_value;
vec4 obj_0x7_value;

vec4 Gettable_vec4_get(OBJREF self) {
  if (self == 5) return obj_0x5_value;
  else if (self == 7) return obj_0x7_value;
  return vec4(0, 0, 0, 0);
}
```

```
vec4 Diffuse_get_color(OBJREF self) {
  if (self == 1) return Gettable_vec4_get(5);
  else if (self == 2) return Gettable_vec4_get(7);
  return vec4(0, 0, 0, 0);
}
```

The generation is split into two steps: for each known object we iterate its attributes and ask each for further references. Special types like *Varying* use this for state dependent behavior: if no *Displacement* is active the qualifier instances a *Value* object to use direct hardware support to copy data between the vertex and the fragment stage. Otherwise, an *Interpolator* is instanced that connects two *Value* objects which copy data from the vertex to the geometry stage and from the geometry to the fragment stage. These *Value* objects are accessed using the getter and setter function of the *Varying* type.

For each given unique class and each of its methods and attributes a dispatcher function is generated. Together with the implementation of the methods these form the complete (and valid *GLSL*) shader code.

The use of generic interfaces can lead to static recursion of the dispatch functions. As an example consider a pattern generator implementing the *Gettable* interface which uses a texture object via the same interface. We therefore have to allow recursion for the same method called on different objects. We follow the call graph of the shader in a second processing step and replace all calls to dispatch functions with a constant *self* reference with a simple forwarding function call. This also ensures that no overhead is introduced by the object model.

If the *self* reference is non-constant, we have to leave the dispatcher as is, but call renamed copies of the functions. This ensures that all recursions are removed from the resulting code. We also remember which dispatch functions were called with which *self* reference and will never call this pair in the subgraph again. The number of objects is bounded and thus also the number of copied functions.

6   Evaluation

Given the above framework, we will now evaluate how it can be used to implement different shading techniques. We will also analyze the costs of the object model and the shading framework.

6.1   Applications

Simple materials, like the Phong material [17], only require the re-implementation of the *illuminance()* method, which contains the shading calculation. Simple light sources have to provide two methods to ensure that their own position and directions are properly converted in the vertex stage and to provide the amount of energy the light receives in the fragment stage.

Light sources can cast shadows using shadow mapping [18]. The point being shaded has to be transformed into the coordinate system of the light source to perform the required perspective projection and texture lookup. This transformation can be performed in the *transform_light()* method, using the provided *CoordinateSystem*.

Bump mapping is a simple but effective method to add details to an object. In [19] an efficient implementation is described that uses the tangent space for lighting. To use a tangent space normal map in our framework one simply implements a *CoordinateSystem*, which builds the tangent space transformation from attribute data associated with each vertex.

The geometry shader in modern graphics hardware can be used to generate a complete **cubic environment-map** in a single-pass. Each triangle of the input mesh is output six times. Each face is reprojected to fit onto one of the faces of the cube. The faces should still be rendered as usual and thus all information from the vertex stage has to be correctly propagated to the fragment shader. As described above our framework handles this transparently so that this technique can be used without modifications to any components.

Environment maps are normally used for mirror reflections but they can also be used together with other BRDFs [20] or for *pre-computed radiance transfer* (PRT) [21]. The framework supports **environment maps as light sources** when used with specific materials.

An environment map light invokes *illuminance_envmap()* on the material providing a reference to the environment map. This reference can be used by the material to query the energy received from a given area of the illumination sphere. Similarly a PRT light invokes *illuminate_PRT()* and provides the illumination sphere directly encoded in an appropriate base.

Each new *illuminate()* method is defined in an interface. Only the special materials need to provide an implementation, as the default handler of the dispatch functions properly handles unimplemented methods (see section 5).

Using **deferred shading** (see e.g. [22]) is straightforward: in the geometry pass, we use materials with no light interaction that output data such as position, normal, color, specularity etc. The following light passes render geometry for each light source that covers its area of influence. A special material is used that retrieves the information from the geometry pass and is illuminated by the current light source.

6.2   Run-Time Costs

Our framework results in additional costs. We provide measurements for the examples of figure 1 in table 1. Note that the application controls the compilation and that a re-compilation is only required if objects are removed or added or if other constant values are changed. Also note that we did not optimize for compilation speed. All processing steps run in linear time in relation to the number of classes and the number of method invocations.

As can be seen, the rendering costs that occur due to the generated shader source code, i.e. the number of shader instructions, are negligible. They are also bounded and predictable:

Table 1. Ratio of measurements of our framework compared to hand-written code. The time to *Assemble, Traverse,* and *Compile* are each compared to the compilation time of the hand-written code. *Source Code* uses the number of source code lines and *Instructions* uses the number of instructions in the optimized assembler code.

Example	Assemble	Traverse	Compile	Source Code	Instructions
Plastic	41.4%	609.9%	187.5%	1180.3%	100%
Bump-Mapping	49.5%	777.7%	210.1%	1081.6%	100%
Displacement	25.5%	669.5%	1598.9%	1238.8%	103.8%

– The *GLSL* code provided for the implementation of methods is not modified.
– Method invocations with a *constant* reference have no costs compared to a normal function invocation (see section 5).
– Method invocations with *non-constant* references have bounded costs. The dispatch function will remain in the code and thus conditional code is run in the shader. Note that code that performs the same dispatching without objects will probably be implemented the same way and thus incur the same costs. As the dispatcher uses a constant *self* reference to invoke the delegated function, all further calls to methods on itself have no additional costs.

7   Conclusion and Future Work

We presented an object model for current and future graphics hardware that uses well-known object-oriented mechanisms. A framework is layered on top of the object model using well-known design patterns. The framework allows libraries of light sources and materials to be written independently and combined with different shading or rendering models. The components can be connected at run-time. This makes it possible to dynamically import data models with complex materials.

Other scenarios with different rendering models can be easily integrated as shown above. Some models require direct support for different combinations, i.e. special code handling the exact combination of a material and a light source. Based on our framework, these cases can be handled in an easy and transparent way.

We want to continue to integrate different shading and rendering techniques in our framework. Multi-pass techniques are of special interest as they could require extensions to the object model.

References

1. Kuck, R.: Object-oriented shader design. In: Eurographics Short Papers, pp. 65–68. Eurographics Association (2007)
2. Mark, W.R., Glanville, R.S., Akeley, K., Kilgard, M.J.: Cg: A system for programming graphics hardware in a C-like language. In: Proc. SIGGRAPH 2003, pp. 896–907. ACM, New York (2003)

3. McCool, M.D., Qin, Z., Popa, T.S.: Shader metaprogramming. In: Proc. SIG-GRAPH/Eurographics Graphics Hardware Workshop 2002, pp. 57–68. Eurographics Association (2002)

4. McCool, M., Toit, S.D., Popa, T., Chan, B., Moule, K.: Shader algebra. In: Proc. SIGGRAPH 2004, pp. 787–795. ACM, New York (2004)

5. McGuire, M.: The supershader. In: ShaderX4, pp. 485–498 (2005)

6. Trapp, M., Döllner, J.: Automated combination of real-time shader programs. In: Eurographics Short Papers, pp. 53–56. Eurographics Association (2007)

7. Hanrahan, P., Lawson, J.: A language for shading and lighting calculations. In: Proc. SIGGRAPH 1990, pp. 289–298. ACM, New York (1990)

8. Apodaca, A.A., Gritz, L.: Advanced RenderMan: Creating CGI for Motion Picture, pp. 222–224. Morgan Kaufmann Publishers Inc., San Francisco (1999)

9. Proudfoot, K., Mark, W.R., Tzvetkov, S., Hanrahan, P.: A real-time procedural shading system for programmable graphics hardware. In: Proc. SIGGRAPH 2001, pp. 159–170. ACM, New York (2001)

10. Buck, I., Foley, T., Horn, D., Sugerman, J., Fatahalian, K., Houston, M., Hanrahan, P.: Brook for GPUs: stream computing on graphics hardware. In: Proc. SIGGRAPH 2004, pp. 777–786. ACM, New York (2004)

11. NVIDIA: CUDA Programming Guide (2007)

12. Owens, J.D., Khailany, B., Towles, B., Dally, W.J.: Comparing Reyes and OpenGL on a stream architecture. In: Proc. SIGGRAPH/EUROGRAPHICS conference on Graphics Hardware, pp. 47–56. Eurographics Association (2002)

13. Sugerman, J., Fatahalian, K., Boulos, S., Akeley, K., Hanrahan, P.: GRAMPS: A programming model for graphics pipelines. ACM Trans. Graph. 28, 1–11 (2009)

14. Rost, R.J.: OpenGL Shading Language. Addison Wesley Longman Publishing Co., Inc., Amsterdam (2006)

15. Gamma, E., Helm, R., Johnson, R., Vlissides, J.: Design Patterns: Abstraction and Reuse of Object - Oriented Design. Addison Wesley Longman Publishing Co., Inc., Amsterdam (1993)

16. Lee, A., Moreton, H., Hoppe, H.: Displaced subdivision surfaces. In: Proc. SIG-GRAPH 2000, pp. 85–94. ACM, New York (2000)

17. Phong, B.T.: Illumination for computer generated pictures. Communications of the ACM 18, 311–317 (1975)

18. Williams, L.: Casting curved shadows on curved surfaces. In: Proc. SIGGRAPH 1978, pp. 270–274. ACM, New York (1978)

19. Peercy, M., Airey, J., Cabral, B.: Efficient bump mapping hardware. In: Proc. SIGGRAPH 1997, pp. 303–306. ACM, New York (1997)

20. Heidrich, W., Seidel, H.P.: Realistic, hardware-accelerated shading and lighting. In: Proc. SIGGRAPH 1999, pp. 171–178. ACM, New York (1999)

21. Sloan, P.P., Kautz, J., Snyder, J.: Precomputed radiance transfer for real-time rendering in dynamic, low-frequency lighting environments. In: Proc. SIGGRAPH 2002. ACM, New York (2002)

22. Akenini-Möller, T., Haines, E., Hoffman, N.: Real-Time Rendering, 3rd edn., pp. 279–283. Addison Wesley Longman Publishing Co., Inc., Amsterdam (2008)

Ray Traced Virtual Reality

Christian N.S. Odom, Nikhil J. Shetty, and Dirk Reiners

Center for Advanced Computer Studies
University of Louisiana at Lafayette

Abstract. One of the goals of Virtual Reality (VR) has always been the pursuit of more realistic display of virtual worlds. Many algorithms have been developed to add effects to rasterization-based rendering systems, but for many effects ray tracing is still the only method that can accurately simulate them. Algorithmic advances and the power of computer systems in recent years has made ray tracing a feasible alternative to rasterization.

This work describes the integration of the real-time ray tracing system Manta into the VR framework system VR Juggler. The results show that it is possible to create ray traced Virtual Environments (VE) running at interactive rates with latencies around 100 msec. This is made possible by using compute clusters, parallelization, high-speed image compression and dynamic scaling of image quality to the desired update rates.

1 Introduction

Rendering quality is an important component of realistic, high-quality, immersive VE. High-quality rendering methods cannot deliver images at the frame rates needed in a VE. To achieve interactivity and immersion, image quality is sacrificed for speed.

High-end VE's now routinely use lighting and shading models using shader-based technologies as well as global effects like shadows and reflections. But all those high-level effects have limitations stemming from the fact that they are driven by a rasterization engine: shadows have limited resolution and exhibit discretization artifacts, reflections assume an infinitely large environment, not to talk about self-reflections and reflections on concave surfaces. While hard to impossible to implement using a rasterizer approach, all these problems are trivially solved by ray tracing.

Ray tracing has been the method of choice for high-quality rendering and effects since its inception in 1980 [18]. The basic algorithm is simple and easy to understand. Although it is computationally expensive, there are no alternatives to ray tracing that enable the creation of images at the same level of quality and geometric flexibility. These qualities and the ability of modern day ray tracing algorithms to reach interactive frame rates is the motivation for the work described in this paper.

The paper is structured as follows. After describing previous work in VE's and real-time ray tracing, two systems that form the basis of this work, Manta

G. Bebis et al. (Eds.): ISVC 2009, Part I, LNCS 5875, pp. 1031–1042, 2009.

and VRJuggler are introduced. The next section describes the architecture and development details necessary to integrate them into MantaJuggler, followed by a description of the approaches taken to enhance performance and interactivity. The final chapters present the results and open areas for future work.

2   Previous Work

2.1   High-Quality VE's and Real-Time Ray Tracing

Many applications depend on high-quality images for VE's to be a viable alternative to real life. Design applications have a very high demand for realistic display and high quality in all respects. [4] employs global illumination approximations to an interior design problem in a CAVE system, that is rendered using standard rasterization renderer. [10] uses a fast approximative radiosity algorithm for interactive interreflection updates in a VE in interior design-style applications.

[11] tries to use ray tracing for displaying a VE. The work focuses on using optical flow methods to approximate the actual ray traced image, whith varying degrees of success.

Using a very large supercomputer, the authors of [8] were the first to reach interactive frame rates for large-scale models. Interactive ray tracing got a large push by Wald et al in [14] and again in [17] when they distributed the work across a cluster, eventually producing comfortably interactive frame rates. An overview of the basic concepts and current developments in ray tracing is given in [9] and its use in the review of large CAD models in [3][16].

2.2   Limitations of Previous Work

Most of the efforts in the cited work have been focused on conventional interaction, which is focused on interactive rates for the desktop (or a large screen projection [16]). When applying ray tracing to VE's, the complexity of the interactive problem is much higher. First we need to generate multiple stereo image pairs, increasing the number of needed pixels.

Second, the camera parameters in VE's are defined through 6-degree of freedom trackers typically sending those parameters at speeds of 30 Hz or higher. Finally, the resolution required for the stereo pairs is fairly high compared to typical lower resolution windows used in most real-time ray tracing research.

3   Framework Components

Our approach is to leverage existing Open Source systems that have most of the base capabilities for VR and ray tracing. After evaluating the different available systems the decision was to use Manta [1],the only available real-time ray tracer that would run on linux IA64 machines , as the ray tracing engine and VRJuggler [6] for the VR framework.

3.1 Manta

Manta is an Open Source ray tracing system developed at the Scientific Computing and Imaging (SCI) Institute's Center for Interactive Ray Tracing and Photorealistic Visualization at the University of Utah.

It is based on earlier work [8] and has been in active development since 2004. It is used in a variety of projects for large scale interactive ray tracing [15,13] and features methods for asynchronous updates and overlapping calculation/display operation to reduce latency. See [1] for further details.

One shortcoming of freely available ray tracers (including Manta) is that it does not support distributed operation on clusters, instead it depends on a shared memory system.

3.2 VR Juggler

VRJuggler [5] is an Open Source VR framework developed by a team from the Virtual Reality Applications Center at Iowa State University. VRJuggler takes care of opening windows, setting up OpenGL contexts and viewing parameters, and it will create and synchronize multiple threads to drive multiple graphics cards in a system.

4 Integrating Manta and VRJuggler

4.1 Basic Design and Networked Operation

MantaJuggler follows a Client-Server architecture. The client, VRJuggler, is the user interface of the system. It connects to multiple render servers, Manta instances, which provide the images over the network.

VRJuggler subdivides the whole of the displayed surface into separate viewports. Each viewport is rendered individually, i.e. the application's ::draw() method is called for each one of them. In addition to that it might be called twice, once for left and right eye in stereo setups.

These VRJuggler viewports form a natural subdivision of the workload into separate pieces. Each viewport is connected and serviced by a separate Manta server. This allows us to overcome Manta's shortcoming, which is the inability to distribute the workload to multiple computers.

The principal operation sequence is as follows: whenever VRJuggler asks the application to redraw a viewport the current view information (detailed in section 4.3) is sent over the network to the corresponding Manta server, which generates the ray traced image and sends it back to the MantaJuggler application, which then takes it and displays it on the screen.

4.2 Asynchronous Operation and Forced Synchronicity

The basic pipeline as described in sec. 4.1 is fully functional, but highly inefficient. The main reason for this inefficiency is its sequential and synchronized nature of the receiver. To improve performance, we decouple the the receiving side by creating a separate receiver thread for each connection to a Manta server.

VRJuggler streams the latest camera information to the Manta on every frame. The multi-threaded pipeline model in Manta sends the rendered images to VRJuggler receiver threads, all while the next image is simultaneously being calculated. This avoids adding the transfer time to the latency of the next frame.

One shortcoming of the described system is that it is totally asynchronous. Each viewport is updated and displayed as quickly as possible and totally independently. This can lead to very noticeable differences in update rate for each viewport. Furthermore, the left and right eye images are calculated independently. Given that the load imbalance between the left and right eye images is fairly small, as that they are normally very close together, this is not as unacceptable as it may sound. Nonetheless, in critical applications this asynchronicity might be unacceptable.

Adding synchronization to the described system could be solved by using traditional swaplock mechanisms, but this would not allow us to process user interaction at an acceptable refresh rate. We solved the problem through a relatively simple modification. Instead of constantly sending new camera information, only one camera set is sent to the Manta servers. Then, instead of displaying incoming images as soon as they arrive, they are uploaded to a double-buffered texture and kept until all viewports have received their respective image for the same camera settings. At this point, the buffers get swapped, the new images are displayed, and new camera information is sent to Manta.

This change enforces full synchronicity, at the cost of increased communications overhead and added latency. As described in sec. 5.3 the main impact of this change is reduced image quality, the actual system latency stays at interactive rates. Given that the basic system remains unchanged, it is possible to switch between synchronized and asynchronous mode at runtime easily to gain the benefit of both modes if needed.

4.3   Arbitrary Viewing Parameters

Most standard rendering engines are designed for standard viewing and projection transformations. VR systems have very different requirements. The viewing frustums are often asymmetric and fairly arbitrary compared to standard rendering engines.

Ultimately, they all provide enough information to render the virtual world with OpenGL (or Direct3D). OpenGL specifies the viewing/projection information in the form of two matrices, the GL_MODELVIEW and GL_PROJECTION matrices. Given this commonality, we could use these matrices to calculate the necessary rays.

The basic task is to calculate a ray from the viewer position into the scene, based on a parametrization of the image, usually a normalized system with x and y coordinates from 0 to 1 or from -1 to 1. This is the exact inverse of what OpenGL does using the GL_MODELVIEW and GL_PROJECTION matrices.

Based on this observation it is simple to calculate a ray for a given -1 to 1 x and y position on the screen. Calculate a matrix M as the inverse of the product of the GL_PROJECTION and GL_MODELVIEW matrices. Given this, the viewer position can be calculated by multiplying a constant $(0, 0, 0, 1)$ vector with this M, while a point on the ray through pixel is given my multiplying $(x, y, 1, 1)$ by the same matrix.

Thus the network protocol between MantaJuggler and Manta itself consists of the $(\text{GL\_PROJECTION} * \text{GL\_MODELVIEW})^{-1}$ matrix. The image created by those rays is a perfect match to the OpenGL rendering, making it possible to replace OpenGL objects with ray traced counterparts and vice versa (see also sec. 7.1).

5   Performance

VR systems are always performance critical, and our system is no exception. On the contrary, using a ray tracing engine running on a cluster adds a number of performance considerations beyond those of a "normal" VR setup.

5.1   Bottlenecks

Compared to a standard VR system the potential bottlenecks are somewhat shifted. The core rendering loop is very tight as all it has to do is render a single, screen-filling polygon with the current image textured on it. This will run at full refresh rate at all times. The obvious bottleneck is the calculation time of the ray traced images on the Manta servers. Because of the scaling characteristics of ray tracing this time will generally linearly depend on the image resolution. So for a given image resolution, this time is fixed. The only way to change the time spent in here is putting more processors in the servers or reducing the number of pixels to be calculated (see sec. 5.3).

The final potential bottleneck is the network between the Manta servers and the display machine(s). A single image can range from 2.3 to 6.2 MB, given stereo and multi-screen systems like CAVEs, the maximum required bandwidth can even exceed the capacity of high-speed commercially available network infrastructure like Infiniband.

5.2   Compression

A fairly obvious approach to reduce the network bandwidth for image transfer is compression. There are a large number of image compression algorithms that can reduce image sizes by orders of magnitude. However, even if the algorithms can reach great compression ratios on the images, it might not make sense to use them, as they have to be faster than just transfering the image (at 30 Mpixels/sec for GBit Ethernet). Only very fast and simple algorithms can reach this level of performance (see sec 6.3 for results).

5.3   Dynamic Image Scaling

An alternative approach to help reduce the needed network bandwidth as well as the needed calculation time is scaling the calculated image. Both ray tracing and network transfer are impacted linearly by the number of pixels in an image, so reducing the size of the image will have a positive impact on both ends.

To achieve this, a Dynamic Image Scaling approach was implemented. The approach is similar to Dynamic Video Resolution (DVR) developed by SGI for the infiniteReality systems. The resolution, for each frame, is adjusted depending on the difference between the targeted and reached latency. As an exact match cannot be expected, a tolerance band was added to avoid oscillation.

However in practice the tolerance band alone was not enough to prevent oscillation. While both tracing and transfer depend linearly on the number of pixels, the assumption of linear scaling with respect to time assumes the same image, as the calculation is based on last frame's timing. In general the image content will change from frame to frame, which will offset the real time requirements. To avoid oscillation the controller was dampened (the impact of the adjustment was scaled down), which resulted in smoother frame rates.

6   Results

All described approaches have been implemented based on Manta and VRJuggler. The implementation was tested on a common test object from Stanford 3D Scanning Repository and the results analyzed.

6.1   System Environment

Display Systems. The tests were done on a 3 channel head-tracked active stereo system driven by a SGI Prism with 3 graphics pipes (see fig. 1). Each screen is run using active stereo (left eye/right eye on one pipe) and has a resolution of 1400x1050 pixels. The system was connected to the house network that connects to the compute engines by a single GBit Ethernet connection.

Fig. 1. Raytraced VE on a 3-pipe Curved Stereo Screen System

Compute Systems. The ray tracer was run on a cluster of 6 SGI Altix 350 machines. This choice was more a matter of availability than preference. Each one of them has 32 Itanium 2 processors (31 of which were used for ray calculation) running at 1.5 GHz, and is connected to the house network by GBit Ethernet. Measuring the bandwidth between compute and display systems shows a maximum data transfer rate of $\sim$ 100 MB/sec.

Since SSE compiler does not work on our architecture, we were forced to use Manta without it. As a result, the Itanium processors reached a performance that is only $\sim \frac{1}{16}$th of a current 4-core AMD or Intel Core Duo processor, making the 32 processor Itanium system comparable to a current 2 CPU system. This however does not impact our results because of the inclusion of dynamic image scaling.

Test Scenario. The test scenario contains a single object (the Stanford dragon containing 202,520 triangles) using a refractive glass material, to show ray tracing-specific capabilities, and lit by two point light sources while being situated on top of a marble-textured plane. This object does stress the capabilities of the compute machines, at the full resolution of 1400x1050 with a dragon covering somewhat less than half the image the pure ray tracing engine without any display only achieves about .7 fps.

To achieve repeatable test results the tests were done using a predefined animation path that simulates a user moving left to right in front of the test object. All tests are time-controlled to last 45 seconds. The more efficient scenarios render more frames, achieving lower average frame latencies.

Note that the measurements here are based primarily on latency. Due to the pipelined nature of the system the achieved framerate is higher than $\frac{1}{latency}$, on average $\sim$ 30% higher.

6.2   Latency

The latency is measured starting at the time when the camera information was put on the network until the calculated image was sent to OpenGL for display. For a total system latency the time from the tracker to the VRJuggler application and the time from finishing the OpenGL commands until the image is shown on the screen should be added. These additional times apply to all VR applications and therefore are not characteristic of the ray traced display approach described here, the focus lies on the additional latency introduced by the distributed ray tracing.

Figure 2 shows the results of running a test with full resolution and without synchronization, compression or scaling. The graph represents the average latency for the 6 different viewports.

The average latency over all viewports and frames is 488.60 msec (standard deviation: 330.64 msec). The average latency difference between viewports is 1595.53 msec (standard deviation: 441.40 msec), i.e. on average the views of the three screens and the left and right eyes are 1595.53 msec apart! The results

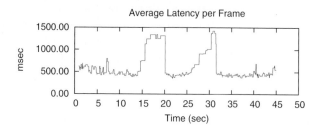

Fig. 2. Average frame latency without synchronization, compression or scaling

are clearly much too slow and much too unsynchronized to be usable for an immersive or even interactive experience.

The two main bottlenecks are the network transfer and ray tracing times. The average time spent in network transfer is 105.41 msec (standard deviation: 66.80 msec), the average time spent in tracing rays is 222.82 msec (standard deviation: 367.46 msec), with the respective maxima significantly higher.

Two approaches were realized to alleviate these bottlenecks: image compression and dynamic image scaling.

6.3   Image Compression

Image compression was implemented based on the Lempel-Ziv-Oberhumer (LZO) [7] library. This library achieves an average compression speed of 103.67 MByte/sec and an average decompression speed of 153.50 MByte/sec. The achieved compression ratio was 8.35 on average, with the compressed images having an average size of 527285 bytes compared the to 4410000 bytes without. Compression reduces the average network transfer time from 105.41 to 36.82 msec, resulting in an average frame latency of 335.08 msec compared to 399.20 msec without.

The result is slightly better than without compression, but unable to reach $\leq$ 100 ms needed for immersion. The main bottleneck is the core ray tracing calculations. This bottleneck is alleviated by using dynamic image scaling.

6.4   Dynamic Image Scaling

Dynamic image scaling tries to adjust the image size (i.e. number of pixels) to reach a given target framerate or latency, by adjusting a factor between 0 and 1 that is used to the scale the image. The basic reasoning for the control logic is that the needed time is linearly dependent on the number of pixels that need to be calculated and sent over the network. Given that the control is reactive and therefore always at least one frame behind the actual computational load a tolerance zone was added to reduce oscillation. Figure 3 shows the results for a 100 msec target latency with a 2 fps tolerance band.

The latency curve is much smoother than fig. 2 (note the different scales), with an average latency of 98.79 msec (standard deviation: 80.07 msec), resulting in

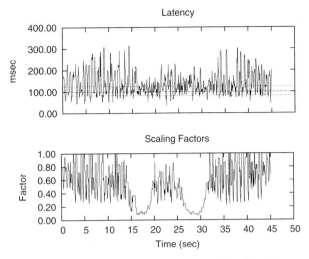

Fig. 3. Latency & Scale factors with dynamic image scaling for 100 msec target latency

an average frame rate of 13.20 fps. This is achieved by scaling the images by an average factor of 0.50 (standard deviation: 0.30), i.e. about half of the screen resolution is used on average.

The result is a faster, lower latency experience but at the cost of having larger variations in latency due to strong over/undershoots. This is especially critical for the variations between viewports/eyes, which can lead to visual tearing between screens. The average maximum difference between viewports/eyes is 180.45 msec (standard deviation: 98.50 msec).

This wide spread of latencies and the fairly significant overshoot and undershoot times are a problem that needs to be solved. One way to reduce these over/undershoots is adding a dampening step in the control.

Dampening. The basic dampening approach used is based on not using the corrected scale factor completely, but instead using a weighted average between the old and the new scale factor. This slows down the adoption of the new factors, but at the same time reduces over- and undershoots and leads to a more stable latency (see fig. 4).

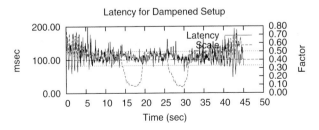

Fig. 4. Latency and scaling needed with dampening

The most effective dampening weights were determined through running a large number of tests to be 1 for the old and 1 for the new scale factor, i.e use the average of both. These values change the average latency from 98.79 msec to 103.32 msec, which is just a small difference. The main effect is reducing the average maximum difference between viewports/eyes from 180.45 msec to 64.99 msec, resulting in significantly more consistent image update rates. 15.03 fps.

6.5   Synchronization

Even with all the above mentioned mechanisms, there is still a noticeable latency difference between different viewports and even between different eyes. The impact of this difference depends on the tolerance of the user, but for a generally acceptable perfect experience it needs to be eliminated. As the synchronized mode tends to oscillate more, 2:1 dampening was used here. The latency difference between viewports is nearly eliminated (average difference between viewports 1.40 msec). As the system is still run at a target latency of 100 msec, the latency itself changes little (Avg: 102.83 msec, standard deviation: 33.35 msec), but the scale factors have to go down to an average of 0.06 (from 0.50 for non-synced operation) to counter the increased time requirements and lost parallelism due to synchronization, resulting in significantly lower quality images.

The results are in sync and consistent, but at the cost of image quality.

7   Conclusions

The presented work demonstrates that it is possible to use real-time ray tracing to drive large-scale immersive projection displays. This can be done by integrating available Open Source components like VRJuggler and Manta using simple message passing and a specialized camera that can bridge the OpenGL-based viewing and projection specification and the ray-based ray tracing one.

Typically available computing power is not capable of generating and transmitting full-resolution low-latency images generated using ray tracing. By using fast compression and decompression in conjunction with dynamically scaling the number of pixels that need to be computed and transmitted it is possible to reach interactive latencies in the 100 msec or less range with somewhat reduced image quality.

The described results were achieved based on a single GBit Ethernet network between compute and display systems. Given the availability of high-speed networks like LambdaRail, the described work opens the door for wide-area distributed interactive ray tracing.

7.1   Future Work

The availability of a cluster based realtime raytracer along with a non-outdated cluster should provide better perfomance measurements. This change will allow us to compare N-1 to N-N communication schemes and the associated load balancing effects of each.

An interesting alternative to scaling the image is using a frameless rendering approach [19,2], replacing the reduced resolution with a noisier image, which might be a more acceptable compromise for certain situations.

Currently interaction is not possible with the rendered scene. Two approaches are possible to alleviate this shortcoming. One would be to extend the communication protocol to allow sending interaction rays to identify objects and to allow manipulation of the ray tracing scene. The second approach is more focused on optical integration and is based in transferring a depth image in addition to the visual image. This depth image can then be used to preset the OpenGL depth buffer to match the ray traced visuals. This would allow hybrid display systems that use ray tracing for the high-quality, relatively slowly updated parts while using OpenGL to display e.g. interaction components or selected objects, using a similar approach to the Multi-Frame Rate Display introduced by [12].

Acknowledgements

The authors wish to thank the Manta user list for their help, especially Abe Stephens, Tiago Ize and Solomon Boulos. This work was supported by the Louisiana Immersive Technologies Enterprise (LITE). Thanks to the Stanford Computer Graphics Laboratory for the test model.

References

1. Bigler, J., Stephens, A., Parker, S.: Design for parallel interactive ray tracing systems. In: IEEE Symposium on Interactive Ray Tracing (2006)
2. Dayal, A., Woolley, C., Watson, B., Luebke, D.: Adaptive frameless rendering. In: SIGGRAPH 2005: ACM SIGGRAPH 2005 Courses, p. 24. ACM, New York (2005), http://doi.acm.org/10.1145/1198555.1198763
3. Dietrich, A., Wald, I., Schmidt, H., Sons, K., Slusallek, P.: Realtime ray tracing for advanced visualization in the aerospace industry. In: Proceedings of the 5th Paderborner Workshop Augmented und Virtual Reality in der Produktentstehung (2006)
4. Dmitriev, K., Annen, T., Krawczyk, G., Myszkowski, K., Seidel, H.P.: A cave system for interactive modeling of global illumination in car interior. In: VRST 2004: Proceedings of the ACM symposium on Virtual reality software and technology, pp. 137–145. ACM Press, New York (2004), http://doi.acm.org/10.1145/1077534.1077560
5. Hartling, P., Bierbaum, A., Cruz-Neira, C.: Virtual reality interfaces using tweek. In: SIGGRAPH 2002: ACM SIGGRAPH 2002 conference abstracts and applications, p. 278. ACM Press, New York (2002), http://doi.acm.org/10.1145/1242073.1242291
6. Just, C., Bierbaum, A., Baker, A., Cruz-Neira, C.: Vr juggler: A framework for virtual reality development. In: 2nd Immersive Projection Technology Workshop, IPT 1998 (1998)
7. Oberhumer, M.F.: Lempel-Ziff-Oberhumer Compression (2007), http://www.oberhumer.com/opensource/lzo/

8. Parker, S., Shirley, P., Livnat, Y., Hansen, C., Sloan, P.P.: Interactive ray tracing for isosurface rendering pp. 233–238 (1998)
9. Recheis, M.: Realtime ray tracing (2007),
   `http://www.cg.tuwien.ac.at/courses/Seminar/SS2007/`
   `RealtimeRayTracing-Recheis.pdf`
10. Schöffel, F.: Online radiosity in interactive virtual reality applications. In: VRST 1997: Proceedings of the ACM symposium on Virtual reality software and technology, pp. 201–208. ACM Press, New York (1997),
    `http://doi.acm.org/10.1145/261135.261172`
11. Siddiqi, S.: Real-time interactive raytracing in distributed environments for virtual reality systems. Honors thesis at Mt. Holyoke College (2003)
12. Springer, J.P., Beck, S., Weiszig, F., Reiners, D., Fröehlich, B.: Multi-frame rate rendering and display. In: Proceedings IEEE Virtual Reality 2007 Conference, pp. 195–202 (2007)
13. Stephens, A., Boulos, S., Bigler, J., Wald, I., Parker, S.: An application of scalable massive model interaction using shared-memory systems. p. (accepted) (2006),
    `http://www.sci.utah.edu/publications/abe06/manta-egpgv06-final.pdf`
14. Wald, I., Benthin, C., Wagner, M., Slusallek, P.: Interactive rendering with coherent ray tracing. In: Chalmers, A., Rhyne, T.M. (eds.) Computer Graphics Forum (Proceedings of EUROGRAPHICS 2001), vol. 20. Blackwell Publishers, Oxford (2001), `http://graphics.cs.uni-sb.de/~wald/Publications`
15. Wald, I., Boulos, S., Shirley, P.: Ray tracing deformable scenes using dynamic bounding volume hierarchies. ACM Transactions on Graphics 26(1) (2007),
    `http://www.sci.utah.edu/publications/wald07/togbvh.pdf`
16. Wald, I., Dietrich, A., Benthin, C., Efremov, A., Dahmen, T., Günther, J., Havran, V., Seidel, H.P., Slusallek, P.: Applying ray tracing for virtual reality and industrial design. In: Proceedings of the 2006 IEEE Symposium on Interactive Ray Tracing, pp. 177–185 (2006)
17. Wald, I., Slusallek, P., Benthin, C., Wagner, M.: Interactive distributed ray tracing of highly complex models. In: Proceedings of the 12th Eurographics Workshop on Rendering Techniques, pp. 277–288. Springer, London (2001)
18. Whitted, T.: An improved illumination model for shaded display. Commun. ACM 23(6), 343–349 (1980), `http://doi.acm.org/10.1145/358876.358882`
19. Wloka, M.M., Zeleznik, R.C., Miller, T.: Practically frameless rendering. Tech. Rep. CS-95-06 (1995), `http://citeseer.ist.psu.edu/wloka95practically.html`

Stochastic Optimization for Rigid Point Set Registration

Chavdar Papazov and Darius Burschka

Machine Vision and Perception Group (MVP)
Department of Computer Science
Technische Universität München, Germany
{papazov,burschka}@in.tum.de

Abstract. In this paper we propose a new method for pairwise rigid point set registration. We pay special attention to noise robustness, outlier resistance and global optimal alignment. The problem of registering two point clouds in space is converted to a minimization of a nonlinear cost function. We propose a cost function that aims to reduce the impact of noise and outliers. Its definition is based on the input point sets and is directly related to the quality of a concrete rigid transform between them. In order to achieve a global optimal registration, without the need of a good initial alignment, we develop a new stochastic approach for global minimization. Tests on a variety of point sets show that the proposed registration algorithm performs very well on noisy, outlier corrupted and incomplete data.

1   Introduction and Related Work

Point set registration is a fundamental problem in computational geometry with applications in the fields of computer vision, computer graphics, image processing and many others. The problem can be formulated as follows. Given two finite point sets $\mathbf{M} = \{\mathbf{x}_1, \ldots, \mathbf{x}_m\} \subset \mathbb{R}^3$ and $\mathbf{D} = \{\mathbf{y}_1, \ldots, \mathbf{y}_n\} \subset \mathbb{R}^3$ find a mapping $T : \mathbb{R}^3 \to \mathbb{R}^3$ such that the point set $T(\mathbf{D}) = \{T(\mathbf{y}_1), \ldots, T(\mathbf{y}_n)\}$ is optimally aligned in some sense to $\mathbf{M}$. $\mathbf{M}$ is referred to as the model point set (or just the model) and $\mathbf{D}$ is termed the data point set. Points from $\mathbf{M}$ and $\mathbf{D}$ are called model points and data points respectively. If T is a rigid transform, i.e., $T(\mathbf{x}) = R(\mathbf{x}) + \mathbf{t}$ for a rotation R and a translation $\mathbf{t}$, we have the problem of rigid point set registration. The problem is especially hard when no initial pose estimation is available and the data point set is noisy, outlier corrupted and incomplete.

Point Set Registration. Algorithms for the rigid registration problem belong to two general classes. One class consists of methods designed to solve the initial pose estimation problem. These methods compute a (more or less) coarse alignment between the point sets without making any assumptions about their initial position and orientation in space. Johnson and Hebert introduce in their work [1] local geometric descriptors, called spin images, and use them for pose estimation and object recognition. The presented results are impressive, but no tests

G. Bebis et al. (Eds.): ISVC 2009, Part I, LNCS 5875, pp. 1043–1054, 2009.

with noisy or outlier corrupted data are performed. Gelfand *et al.* [2] develop a local descriptor which performs well under noisy conditions, but still, defining robust local descriptors in the presence of significant noise and a great amount of outliers remains a difficult task. A more recent approach to the initial pose estimation problem is the 4PCS algorithm introduced by Aiger *et al.* [3]. It is an efficient randomized generate-and-test approach: For an appropriate quadruple $\mathbf{B}$ (called a basis) of nearly coplanar points from the model set $\mathbf{M}$, compute the optimal rigid transform between $\mathbf{B}$ and each of the potential bases in the data set $\mathbf{D}$ and choose the optimal one. In order to achieve high probability for finding the global optimal transform, the procedure is repeated several times for different bases $\mathbf{B} \subset \mathbf{M}$. Note, however, that the rigid transform, found by the algorithm, is optimal only for the two bases (i.e., for eight points). In contrast to this, the rigid transform we compute is optimal for *all* points of the input sets, and thus we expect to achieve higher accuracy than the 4PCS algorithm. This is further supported by the experimental results in Section 4 of our paper.

Since the accuracy of the pose computed by the above mentioned methods is insufficient for many applications, an additional pose refinement step needs to be performed. The pose refining algorithms build the second class of registration approaches. The most popular one is the Iterative Closest Point (ICP) algorithm. Since its introduction by Chen and Medioni [4], and Besl and McKay [5], a variety of improvements have been proposed in the literature. A good summary as well as new results on acceleration of ICP algorithms has been given by Rusinkiewicz and Levoy [6]. A major drawback of these ICP variants is that they assume a good initial guess for the orientation of the data point set (with respect to the model point set). This orientation is improved in an iterative fashion until an optimal rigid transform is found. The quality of the solution depends heavily on the initial guess. Another disadvantage of the methods compared by Rusinkiewicz and Levoy [6] is that they use local surface features like surface normals which cannot be computed very reliably in the presence of noise.

The approach we develop is most related to the ones proposed by Mitra *et al.* [7] and Pottmann *et al.* [8]. They also express the registration problem as a minimization of a cost function. Its definition is based on the distance of the data points to the surface defined by the model points. For its minimization, however, a local optimization method is used. This results in the already mentioned strong dependence on a good initial transform estimation.

Stochastic Optimization. Stochastic optimization has received considerable attention in the literature over the last three decades. Much of the work has been devoted to the theory and applications of simulated annealing (SA) as a minimization technique [9], [10], [11]. A comprehensive overview of this field is given in [12]. A major property of SA algorithms is their "willingness" to explore regions around points in search space at which the objective function takes values greater than the current minimum. This is what makes SA algorithms able to escape from local minima and makes them suitable for the task of global minimization. A known drawback of SA algorithms is the fact that they waste a lot of iterations generating candidate points, evaluating the objective function at these points, and finally rejecting them [12]. In order to reduce the number of rejections, Bilbro and Snyder [13] select candidate points from "promising" regions of the search space, i.e., from regions in which the objective function is likely to have low values. They achieve this by adapting a spatial data structure (an n-dimensional binary tree) to the objective function each time a new candidate point is accepted. If, however, the current point is not accepted, the tree remains unchanged. This is—in the case of candidate rejection—a considerable waste of computation time, since the information gained by the (expensive) evaluation of the objective function is not used at all. In contrast to that, our algorithm adapts the n-dimensional tree at every iteration and thus uses all the information collected during the minimization.

Contributions and Overview. Our registration algorithm aims to solve the initial pose estimation problem with a sufficient accuracy, so that no additional refinement is necessary. Our main contributions are (i) the introduction of a new *noise and outlier resistant* cost function and (ii) a new stochastic approach for its *global* minimization.

The rest of the paper is organized as follows. In Section 2, we define the task of aligning two point sets as a nonlinear minimization problem and define our cost function. In Section 3, we introduce a stochastic approach for global minimization. Section 4 presents experimental results obtained by our registration algorithm. Conclusions are drawn in the final Section 5 of this paper.

2   Registration as a Minimization Problem

Consider we are given a model point set $\mathbf{M} = \{\mathbf{x}_1, \ldots, \mathbf{x}_m\} \subset \mathbb{R}^3$ and a data point set $\mathbf{D} = \{\mathbf{y}_1, \ldots, \mathbf{y}_n\} \subset \mathbb{R}^3$. Suppose we have a continuous function $S : \mathbb{R}^3 \to \mathbb{R}$, called the model scalar field, which takes small values when evaluated at (or near) the model points $\mathbf{x}_j$, $j \in \{1, \ldots, m\}$ and increases with increasing distance between the evaluation point and the closest model point. The model scalar field S will be precisely defined in Section 2.1. Consider for now it is given and it has the above mentioned property. Our aim is to find a rigid transform $T : \mathbb{R}^3 \to \mathbb{R}^3$ of the form $T(\mathbf{x}) = R \cdot \mathbf{x} + \mathbf{t}$ for a rotation matrix $R \in \mathbb{R}^{3 \times 3}$ and a translation vector $\mathbf{t} \in \mathbb{R}^3$ such that the functional

$$\mathcal{F}(T) = \sum_{i=1}^{n} S(T(\mathbf{y}_i)), \quad \mathbf{y}_i \in \mathbf{D}. \tag{1}$$

gets minimized. This definition of $\mathcal{F}$ is based on the following quite natural idea common for most registration algorithms: We seek a rigid transform that brings the data points as close as possible to the model points.

2.1   Definition of the Model Scalar Field

Given the model point set $\mathbf{M} = \{\mathbf{x}_1, \ldots, \mathbf{x}_m\}$, we want to have a function $S : \mathbb{R}^3 \to \mathbb{R}$ which takes its minimal value at the model points, i.e.,

$$S(\mathbf{x}_j) = s_{\min} \in \mathbb{R}, \quad \forall j \in \{1, \ldots, m\}, \tag{2}$$

and takes greater values for all other points in $\mathbb{R}^3$, i.e.,

$$S(\mathbf{x}) > s_{\min}, \quad \forall \mathbf{x} \in \mathbb{R}^3 \setminus \{\mathbf{x}_1, \ldots, \mathbf{x}_m\}. \tag{3}$$

Define

$$d_{\mathbf{M}}(\mathbf{x}) := \min_{\mathbf{x}_j \in \mathbf{M}} \|\mathbf{x} - \mathbf{x}_j\| \tag{4}$$

to be the distance between a point $\mathbf{x} \in \mathbb{R}^3$ and the set $\mathbf{M}$, where $\| \cdot \|$ is the Euclidean norm in $\mathbb{R}^n$. If we set

$$S(\mathbf{x}) := d_{\mathbf{M}}(\mathbf{x}), \tag{5}$$

we get an unsigned distance field which is implicitly used by ICP. It is obvious that this choice for S fulfills both criteria (2) and (3).

Mitra *et al.* [7] and Pottmann *et al.* [8] consider in their work more sophisticated scalar fields. They assume that the model point set $\mathbf{M}$ consists of points sampled from some underlying surface Φ. The scalar field S at a point $\mathbf{x} \in \mathbb{R}^3$ is defined to be the squared distance from $\mathbf{x}$ to Φ. In this context, S is called the squared distance function to the surface Φ. We refer to [7] for details on computing the squared distance function and its approximation for point sets.

The version of S given in (5) and the one used by Mitra *et al.* [7] are both essentially distance fields. This means that $\lim_{\|\mathbf{x}\| \to \infty} S(\mathbf{x}) = \infty$, i.e., $S(\mathbf{x})$ approaches to infinity as the point $\mathbf{x}$ gets infinitely far from the point set. This has the practical consequence that a registration technique based on an unbounded scalar field S will be sensitive to outliers in the data set, because data points lying far away from the model point set will have great impact on the functional value in Eq. (1) and thus will prevent the minimization algorithm from converging towards the global optimal alignment.

To avoid this problem we propose to use a bounded scalar field satisfying (2) and (3) and having the additional property

$$\lim_{\|\mathbf{x}\| \to \infty} S(\mathbf{x}) = 0. \tag{6}$$

We set

$$S(\mathbf{x}) := -\varphi(d_{\mathbf{M}}(\mathbf{x})), \tag{7}$$

where $\varphi : \mathbb{R}^+ \to \mathbb{R}^+$, for $\mathbb{R}^+ := \{x \in \mathbb{R} : x \geq 0\}$, is a strictly monotonically decreasing continuous function with

$$\max_{x \in \mathbb{R}^+} \varphi(x) = \varphi(0) \quad \text{and} \tag{8}$$

$$\lim_{x \to \infty} \varphi(x) = 0. \tag{9}$$

In our implementation we use a rational function of the form $1/(1+\alpha x^2)$ because it is computationally efficient to evaluate and can be controlled by a single parameter α. This results in the following scalar field:

$$S_\alpha^{\mathbf{M}}(\mathbf{x}) = -\frac{1}{1 + \alpha(d_{\mathbf{M}}(\mathbf{x}))^2}, \quad \alpha > 0. \tag{10}$$

It is easy to see that (2), (3) and (6) hold. Different α's in Eq. (10) lead to different scalar fields. The greater the value the faster $S_\alpha^M(\mathbf{x})$ convergences to zero as $\|\mathbf{x}\| \to \infty$. In the next Section, we will discuss how to choose a suitable value for α.

2.2 Cost Function Definition

At the beginning of Section 2, we formulated the rigid point set registration problem as a functional minimization problem: Minimize $\mathcal{F}$ (see Eq. (1)) over the set of rigid transforms. We convert $\mathcal{F}$ to a real-valued scalar field $F : \mathbb{R}^6 \to \mathbb{R}$ of the form

$$F(\theta, \phi, \psi, x, y, z) = \sum_{i=1}^{n} S_\alpha^M(R_{\theta,\phi,\psi} \cdot \mathbf{y}_i + (x, y, z)), \tag{11}$$

for the data points $\mathbf{y}_1, \ldots, \mathbf{y}_n$ and for S_α^M defined in Eq. (10). $R_{\theta,\phi,\psi}$ is a rotation matrix describing a rotation by θ about the x-axis, followed by a rotation by ϕ about the y-axis and a rotation by ψ about the z-axis. A global minimizer $\mathbf{x}^* \in \mathbb{R}^6$ of F defines a rigid transform that brings the data points as close as possible to the model points.

What makes the proposed cost function robust to outliers is the fact, that outlier data points have a marginal contribution to the sum in Eq. (11). More precisely, given a positive real number d, we can compute a value for α, such that $|S_\alpha^M(\mathbf{x})|$ is less than an arbitrary $\delta > 0$, if $d_M(\mathbf{x}) > d$ holds. In this way the contribution of an outlier point to the sum in Eq. (11) can be made arbitrary close to zero, hence F behaves like an outlier rejector. Too large values for α, however, will lead to the rejection of data points which do not have exact counterparts in the model set, but still are not outliers. In our implementation we set $d = \frac{1}{5} diag(BB(\mathbf{M}))$ and $\delta = 0.1$, where $diag(BB(\mathbf{M}))$ denotes the diagonal length of the axis-aligned minimum bounding box of the model point set. Using the absolute value of the right side of Eq. (10) and solving for α yields

$$\alpha = \frac{1-\delta}{\delta d^2}. \tag{12}$$

The cost function given in (11) is nonlinear and nonconvex. This results in a great number of local minima of F over the search space. Using a local optimization procedure—common for the most registration methods in the literature—will lead in most cases to a local minimizer of F and thus will not give the best alignment between model and data.

We employ a new stochastic approach for global minimization, described in the next Section of this paper. We seek the global minimum of F over the search space

$$\mathbf{X} := [-\pi/2, \pi/2] \times [-\pi, \pi] \times [-\pi, \pi] \times BB(\mathbf{M}), \tag{13}$$

where $BB(\mathbf{M})$ denotes the axis-aligned minimum bounding box of the model point set. The first three intervals in (13) build the search space for the rotational part and the bounding box for the translational part of the rigid transform.

3   Adaptive Search for Global Minimization

Our stochastic minimization approach is inspired by the work of Bilbro and Snyder [13]. The algorithm shares two properties with the one presented in [13]: (i) we use the same data structure (an n-dimensional binary tree) to represent the search space and (ii) we adapt the tree during the minimization process to the objective function. In contrast to [13], where the tree is updated only when a new candidate point is accepted, we update it at every iteration, so we use *all* the information gained by the evaluation of the objective function. This apparently minor modification leads to a rather different algorithm (than [13]) and enables a faster rejection of regions in which the objective function is likely to have high (i.e., poor) values and thus speeds up the convergence.

3.1   Problem Definition

We call a set $\mathbf{X} \subset \mathbb{R}^n$ an n-dimensional (or n-d) box if there are n intervals $[a_i, b_i] \subset \mathbb{R}$ such that

$$\mathbf{X} = [a_0, b_0] \times \ldots \times [a_{n-1}, b_{n-1}]. \tag{14}$$

Given an n-dimensional box $\mathbf{X}$ and a bounded continuous function $f : \mathbf{X} \to \mathbb{R}$ our aim is to find an $\mathbf{x}^* \in \mathbf{X}$ with $f(\mathbf{x}^*) \leq f(\mathbf{x})$ for all $\mathbf{x} \in \mathbf{X}$.

3.2   Overall Algorithm Description

We use an n-dimensional binary tree to represent the search space $\mathbf{X}$. The root η_0^0 is at the 0th level of the tree and represents the whole box $\mathbf{X}_0 := \mathbf{X}$. η_0^0 has two children η_{00}^1 and η_{01}^1, which are at the next level of the tree. They represent the n-d boxes $\mathbf{X}_{00}$ respectively $\mathbf{X}_{01}$ resulting from bisecting the 0th interval (this is $[a_0, b_0]$ in (14)) of $\mathbf{X}_0$ and assigning the first half to $\mathbf{X}_{01}$ and the second half to $\mathbf{X}_{11}$. In general, a node η_s^k (where $k \geq 0$ and s is a binary string of length $k+1$) is at the kth level of the tree and has two children η_{s0}^{k+1} and η_{s1}^{k+1} which are at the next, $(k+1)$th, level. The child nodes represent the same n-d box as the one represented by η_s^k (this is $\mathbf{X}_s$) except for that the $(k \bmod n)$th interval of $\mathbf{X}_s$ is bisected and the first and second half is assigned to η_{s0}^{k+1} and η_{s1}^{k+1} respectively.

During the minimization the tree is built in an iterative fashion beginning with the root. The algorithm adds more resolution to promising regions in the search space, i.e., the tree is built with greater detail in the vicinity of points in $\mathbf{X}$ at which the objective function has low values. The overall procedure can be outlined as follows:

1. Initialize the tree (see Section 3.3) and set an iteration counter $j := 0$.
2. Select a "promising" leaf according to a probabilistic selection scheme (see Section 3.4).
3. Expand the tree by bisecting the selected leaf. This results in the creation of two new child nodes. Evaluate the objective function at a point which is uniformly sampled within the n-d box of one of the two children (see Section 3.5).
4. If a stopping criterion is not met, increment the iteration counter j and go to step 2, otherwise terminate the algorithm (see Section 3.6).

3.3   Initializing the Tree

For every tree node η_s^k the following items are stored: (i) an n-d box $\mathbf{X}_s \subset \mathbf{X}$ and (ii) a pair $(\mathbf{x}_s, f(\mathbf{x}_s))$ consisting of a point $\mathbf{x}_s$, randomly selected from $\mathbf{X}_s$, and the corresponding function value $f(\mathbf{x}_s)$. The tree is initialized by storing the bounds of the whole search space $\mathbf{X}$ and a pair $(\mathbf{x}_0, f(\mathbf{x}_0))$ in the root.

3.4   Selecting a Leaf

At every iteration the search for a global minimum begins at the root and proceeds down the tree until a leaf (node without children) is reached. In order to reach a leaf, we have to choose a concrete path from the root down to this leaf. At each node, we have to decide, whether to take its left or right child as the next station. This decision is made probabilistically. For every node two numbers $p_0, p_1 \in (0, 1)$ are computed in a way that $p_0 + p_1 = 1$. Arriving at a node, we choose to descend via either its left or right child with probability p_0 respectively p_1. We make these left/right decisions until we encounter a leaf.

Computing the Probabilities.   The idea is to compute the probabilities in a way, that the "better" child, i.e., the one with the lower function value, has greater chance to be selected. We compute p_0 and p_1 for each node η_s^k based on the function values associated with its children η_{s0}^{k+1} and η_{s1}^{k+1}. Let f_{s0} and f_{s1} be the function values associated with η_{s0}^{k+1} respectively η_{s1}^{k+1}. The following criterion should be fulfilled:

$$f_{s0} < f_{s1} \quad \Leftrightarrow \quad p_0 > p_1. \tag{15}$$

For $f_{s0} < f_{s1}$ we set

$$p_0 = (t + 1)/(1 + 2t), \quad p_1 = t/(1 + 2t), \tag{16}$$

for a parameter $t \geq 0$. For $t \to \infty$ we get $p_0 = p_1 = \frac{1}{2}$ and our minimization algorithm becomes a pure random search. Setting $t = 0$ results in $p_0 = 1$ and $p_1 = 0$ and makes the algorithm deterministically choosing the "better" child of every node, which leads to the exclusion of a great portion of the search space and in general prevents the algorithm from finding a global minimum. For $f_{s1} < f_{s0}$ we set

$$p_0 = t/(1 + 2t), \quad p_1 = (t + 1)/(1 + 2t). \tag{17}$$

Updating the Probabilities.   From the discussion above it becomes evident that t should be chosen from the interval $(0, \infty)$. For our algorithm the parameter t plays a similar role as the temperature parameter for a simulated annealing algorithm [9], so we will refer to t as temperature as well. Like in simulated annealing, the search begins on a high temperature level (large t), so the algorithm samples the cost function quite uniformly. The temperature is decreased gradually during the search process, so that promising regions of the search space are explored in greater detail. More precisely, we update t according to the following cooling schedule:

$$t = t_{\max} \exp(-vj). \tag{18}$$

$j \in \mathbb{N}$ is the current iteration number, $t_{\max} > 0$ is the temperature at the beginning of the search (for $j = 0$) and $v > 0$ is the cooling speed which determines how fast the temperature decreases.

3.5   Expanding the Tree

After reaching a leaf η_s^k, the n-d box $\mathbf{X}_s$ associated with it gets bisected in the way described at the beginning of Section 3.2. This results in the creation of two n-d boxes $\mathbf{X}_{s0}$ and $\mathbf{X}_{s1}$ associated with two new children η_{s0}^{k+1} and η_{s1}^{k+1} respectively. In this way, we add more resolution in this region of the search space. Next, we evaluate the new children, i.e., we assign to the left and right one a pair $(\mathbf{x}_{s0}, f(\mathbf{x}_{s0}))$ and $(\mathbf{x}_{s1}, f(\mathbf{x}_{s1}))$ respectively.

Note that the parent node η_s^k stores a pair $(\mathbf{x}_s, f(\mathbf{x}_s))$. Since we have $\mathbf{X}_s = \mathbf{X}_{s0} \cup \mathbf{X}_{s1}$ and $\mathbf{X}_{s0} \cap \mathbf{X}_{s1} = \emptyset$ it follows that $\mathbf{x}_s$ is contained either in $\mathbf{X}_{s0}$ or in $\mathbf{X}_{s1}$. Thus we set

$$(\mathbf{x}_{s0}, f(\mathbf{x}_{s0})) := (\mathbf{x}_s, f(\mathbf{x}_s)) \text{ if } \mathbf{x}_s \in \mathbf{X}_{s0} \text{ or} \tag{19}$$

$$(\mathbf{x}_{s1}, f(\mathbf{x}_{s1})) := (\mathbf{x}_s, f(\mathbf{x}_s)) \text{ if } \mathbf{x}_s \in \mathbf{X}_{s1}. \tag{20}$$

To compute the other pair we sample a point uniformly over the appropriate n-d box ($\mathbf{X}_{s0}$ or $\mathbf{X}_{s1}$) and evaluate the function at this point.

Updating the Tree. During the search we want to compute the random paths from the root down to a certain leaf such that promising regions—leafs with low function values—are visited more often than non-promising ones. Thus, after evaluating a new created leaf, we propagate its (possibly very low) function value as close as possible to the root. This is done by the following updating procedure. Suppose that the parent point $\mathbf{x}_s$ is contained in the set $\mathbf{X}_{s1}$ belonging to the new created child η_{s1}^{k+1}. Therefore, we randomly generate $\mathbf{x}_{s0} \in \mathbf{X}_{s0}$, compute $f(\mathbf{x}_{s0})$ and assign the pair $(\mathbf{x}_{s0}, f(\mathbf{x}_{s0}))$ to the other child η_{s0}^{k+1}. Updating the tree consists of ascending from η_{s0}^{k+1} (via its ancestors) to the root and comparing at every parent node η_u^j the function value $f(\mathbf{x}_{s0})$ with the function value of η_u^j, i.e., with $f(\mathbf{x}_u)$. If $f(\mathbf{x}_{s0}) < f(\mathbf{x}_u)$ we update the current node by setting $(\mathbf{x}_u, f(\mathbf{x}_u)) := (\mathbf{x}_{s0}, f(\mathbf{x}_{s0}))$ and proceed to the parent of η_u^j. The updating procedure terminates if we reach the root or no improvement for the current node is possible, i.e., if $f(\mathbf{x}_{s0}) \geq f(\mathbf{x}_u)$.

Note that if $f(\mathbf{x}_{s0})$ is the lowest function value found so far, it will be propagated to the root, otherwise it will be propagated only to a certain level $l \in \{1, \ldots, k+1\}$. This means, that every node contains the minimum function value (and the point at which f takes this value) found in the n-d box associated with this node. Since the root represents the whole search space, it contains the point we are interested in, namely the point at which f takes the lowest value found up to the current iteration.

3.6   Stopping Rule

We break the search, if for the last N iterations the absolute difference between the last sample of the objective function and the sample before is less than a predefined $\epsilon > 0$.

4   Experimental Results

In this Section, we test our registration method on several point sets. Since the algorithm is a probabilistic one, it computes each time a (slightly) different result. In order to make a statistical meaningful statement about its performance, we run 100 registration trials for every pair of inputs. We measure the success rate and the accuracy of the algorithm under varying amount of noise and outliers in the data point sets. The success rate gives the percentage of registration trials in which a transform which is close to the global optimal one is found. The accuracy is measured using the RMS error between the point sets after alignment [2]. The type of noise added to some of the data sets is Gaussian and the outliers are simulated by drawing points from a uniform distribution within the bounding box of the corresponding data set. We also measure the number of cost function evaluations and the computation time for varying cooling speed v (defined in (18)). In the following, we describe each test scenario in detail.

First, we use our algorithm to register four data point sets to a noiseless model of the Stanford bunny. The data sets are at a lower level of detail (compared to the model), contain only parts of the bunny and three of them are contaminated by a significant amount of outliers (see Fig. 1). We examine each of the 100 registration results. The upper row in Fig. 3 shows exemplary one result for each data point set.

In the second test case, we register several versions of the Stanford dragon under varying noisy conditions. We use a noiseless point set as the model. The

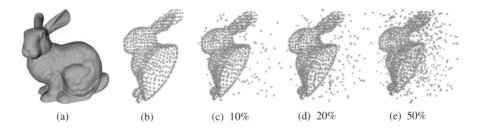

| (a) | (b) | (c) 10% | (d) 20% | (e) 50% |

Fig. 1. (a) The bunny model point set. Although it is shown as a mesh, no surface information is used for registration. (b) The data point set without outliers. Note that the data set is incomplete and very sparsely sampled (compared to the model). (c)–(e) Contaminated data point sets. The number of outliers as percentage of the original number of points are shown below each figure. Note that local descriptors, like spin images [1] or integral invariants [2], are very difficult to compute for such sparsely sampled and outlier corrupted point sets.

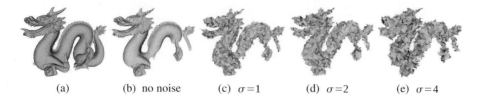

(a)   (b) no noise   (c) $\sigma=1$ (d) $\sigma=2$ (e) $\sigma=4$

Fig. 2. The dragon model and data sets. Although all of them are shown as meshes, only points are used for registration. (a) The dragon model. (b) Noiseless data set. Note that it has a lower level of detail compared to the model and parts of the dragon are missing. (c)–(e) Data point sets corrupted by zero-mean additive Gaussian noise with variance σ which is expressed in percentage of the bounding box diagonal length of the noiseless data set. Again, computing reliable local descriptors for the point sets (d) and (e) is a very challenging task.

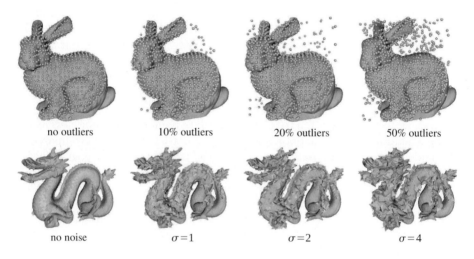

no outliers   10% outliers   20% outliers   50% outliers

no noise $\sigma=1$ $\sigma=2$ $\sigma=4$

Fig. 3. (Upper row) Typical registration results for the incomplete and outlier corrupted data point sets shown in Fig. 1. The amount of outliers is indicated below the corresponding figure. (Lower row) Typical registration results for the incomplete and noise contaminated data point sets shown in Fig. 2. The value for σ of the Gaussian noise added to the data point sets is shown below each figure.

data sets have lower resolution, do not contain all parts of the dragon and three of them are corrupted by Gaussian noise (see Fig. 2). As in the bunny test case, we inspect all registration results. Four of them are shown in the lower row in Fig. 3.

We compute the success rate and the mean RMS error based on all 800 registration results in the bunny and in the dragon test cases. For comparison, we show how the newly proposed 4PCS registration algorithm [3] performs under similar conditions. Note that 4PCS has been tested on different point sets, so an exact comparison is not possible. In Fig. 4, we plot our results together with the

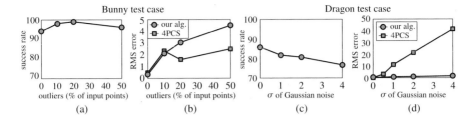

Fig. 4. Success rate and mean RMS error computed from the registration results in the bunny test case (a), (b) and in the dragon test case (c), (d). In (a) and (b) the success rate and the mean RMS error are shown as a function of the number of outliers, whereas in (c) and (d) they are a function of σ of Gaussian noise. In (b) and (d), we compare the accuracy of our method with the accuracy of the 4PCS algorithm [3]. One RMS error unit equals 1% of the bounding box diagonal length of the data point set.

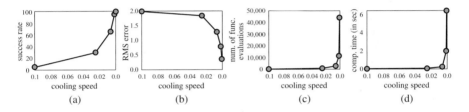

Fig. 5. Success rate (a), mean RMS error (b), mean number of cost function evaluations (c) and mean computation time (d) of our registration algorithm as a function of the cooling speed v. All tests presented in this paper run on a low-cost computer with a 2.2 GHz CPU. For all registration trials we set $t_{\max} = 40$ (see Eq. (18)). Model and data set are copies of the point set shown in Fig. 1(b). One RMS error unit equals 1% of the bounding box diagonal length of the point set.

ones reported in [3]. Observe that the success rate of our algorithm is immune against outliers and shows low sensitivity to noise. For outlier corrupted point sets, the 4PCS algorithm is apparently more accurate than ours (see Fig. 4(b)). Note, however, that Aiger *et al.* [3] add outliers to *both* data and model set. Thus, outliers from both point sets are close to each other and contribute little to the RMS error. In contrast to this, we corrupt only the data point set, so its outliers do not have close counterparts in the model point set, hence we get a greater RMS error. According to Fig. 4(d), our method is far more accurate on noisy point sets than the 4PCS algorithm.

Finally, we measure the performance of our algorithm for varying cooling speed v. We report the results in Fig. 5. Our algorithm achieves a success rate of 100% and a mean RMS error less than 0.5 for 6.5 seconds. For comparison, the best success rate (for similar point sets) achieved by the registration algorithms studied in [8] is 15.951% (see last column of Table 3 in [8]).

5   Conclusions

In this paper we introduced a new technique for pairwise rigid registration of point sets. Our method is based on a noise robust and outlier resistant cost function and on a new stochastic approach for global minimization. Characteristic to the proposed algorithm is (i) that it does not rely on an initial estimation of the globally optimal rigid transform and (ii) that it has low sensitivity to outliers, noise and missing data. Both claims were further supported by a variety of experiments on noisy, outlier corrupted and incomplete point sets.

Acknowledgement. The research leading to these results has received funding from the European Community's Seventh Framework Programme FP7/2007-2013 under grant agreement 215821 (GRASP project).

References

1. Johnson, A., Hebert, M.: Using Spin Images for Efficient Object Recognition in Cluttered 3D Scenes. IEEE Trans. PAMI 21, 433–449 (1999)
2. Gelfand, N., Mitra, N., Guibas, L., Pottmann, H.: Robust Global Registration. In: Eurographics Symposium on Geometry Processing, pp. 197–206 (2005)
3. Aiger, D., Mitra, N., Cohen-Or, D.: 4-Points Congruent Sets for Robust Pairwise Surface Registration. ACM Trans. Graph. 27 (2008)
4. Chen, Y., Medioni, G.: Object Modeling by Registration of Multiple Range Images. In: Proceedings of IEEE International Conference on Robotics and Automation, vol. 3, pp. 2724–2729 (1991)
5. Besl, P., McKay, N.: A Method for Registration of 3-D Shapes. IEEE Trans. PAMI 14 (1992)
6. Rusinkiewicz, S., Levoy, M.: Efficient Variants of the ICP Algorithm. In: 3DIM, pp. 145–152 (2001)
7. Mitra, N., Gelfand, N., Pottmann, H., Guibas, L.: Registration of Point Cloud Data from a Geometric Optimization Perspective. In: Symposium on Geometry Processing, pp. 23–32 (2004)
8. Pottmann, H., Huang, Q.X., Yang, Y.L., Hu, S.M.: Geometry and Convergence Analysis of Algorithms for Registration of 3D Shapes. International Journal of Computer Vision 67, 277–296 (2006)
9. Metropolis, N., Rosenbluth, A., Rosenbluth, M., Teller, A., Teller, E.: Equation of State Calculations by Fast Computing Machines. The Journal of Chemical Physics 21, 1087–1092 (1953)
10. Cerny, V.: Thermodynamical Approach to the Traveling Salesman Problem: An Efficient Simulation Algorithm. Journal of Optimization Theory and Applications 45, 41–51 (1985)
11. Kirkpatrick, S., Gelatt, C., Vecchi, M.: Optimization by Simmulated Annealing. Science 220, 671–680 (1983)
12. Pardalos, P., Romeijn, E. (eds.): Handbook of Global Optimization 2. Nonconvex Optimization and Its Applications. Kluwer Academic Publishers, Dordrecht (2002)
13. Bilbro, G., Snyder, W.: Optimization of Functions with Many Minima. IEEE Trans. on Systems, Man, and Cybernetics 21, 840–849 (1991)

Multi-label MRF Optimization
via a Least Squares $s - t$ Cut

Ghassan Hamarneh

School of Computing Science, Simon Fraser University, Canada

Abstract. We approximate the k-label Markov random field optimization by a single binary $(s−t)$ graph cut. Each vertex in the original graph is replaced by only $ceil(log_2(k))$ new vertices and the new edge weights are obtained via a novel least squares solution approximating the original data and label interaction penalties. The $s − t$ cut produces a binary "Gray" encoding that is unambiguously decoded into any of the original k labels. We analyze the properties of the approximation and present quantitative and qualitative image segmentation results, one of the several computer vision applications of multi label-MRF optimization.

1 Introduction

Many visual computing tasks can be formulated as graph labeling problems, *e.g.* segmentation and stereo-reconstruction [1], in which one out of k labels is assigned to each graph vertex. This may be formulated as a k-way cut problem: Given graph $G(V, E)$ with $|V|$ vertices $v_j \in V$ and $|E|$ edges $e_{v_i,v_j} = e_{ij} \in E \subseteq V \times V$ with weights $w(e_{ij}) = w_{ij} > 0$, find an optimal k-cut $C^* \subset E$ with minimal cost $|C^*| = argmin_C|C|$, where $|C| = \sum_{e_{ij} \in C} w_{ij}$, such that $E \backslash C$ breaks the graph into k groups of labelled vertices. This k-cut formulation encodes the semantics of the problem at hand (e.g. segmentation) into w_{ij}. However, if the optimal label assigned to a vertex depends on the labels assigned to other vertices (e.g. to regularize the label field), setting w_{ij} $\forall i, j$ becomes less straightforward. The Markov random field (MRF) formulation captures this desired label interaction via an energy $\xi(l)$ to be minimized with respect to the vertex labels l.

$$\xi(l) = \sum_{v_i \in V} D_i(l_i) + \lambda \sum_{(v_i,v_j) \in E} V_{ij}(l_i, l_j, d_i, d_j) \tag{1}$$

where $D_i(l_i)$ penalizes labeling v_i with l_i, and V_{ij}, aka prior, penalizes assigning labels (l_i, l_j) to neighboring vertices[1]. V_{ij} may be influenced by the data value d_i at v_i (*e.g.* image intensity). λ controls the relative importance of D_i and V_{ij}.

For labeling a P-pixel image, typically a graph G is constructed with $|V| = P$. To encode $D_i(l_i)$, G may be augmented with k new terminal vertices $\{t_j\}_{j=1}^k$; each representing one of the k labels (Figure 2(a)) and w_{v_i,t_j} set inversely proportional to $D_i(l_j)$. When $V_{ij} = V_{ij}(d_i, d_j)$, *i.e.* independent of l_i and l_j, V_{ij} may be encoded by $w_{v_i,v_j} \propto V_{ij}(d_i, d_j)$. The random walker [2] globally solves a

---

[1] Higher order priors, e.g. 3rd order $V_{ijk}(l_i, l_j, l_k)$, are also possible.

G. Bebis et al. (Eds.): ISVC 2009, Part I, LNCS 5875, pp. 1055–1066, 2009.

labeling problem of this type, *i.e.* disregarding label interaction. Solving multi-label MRF optimization for any interaction penalty remains an active research area. In [3], the globally optimal binary ($k=2$) labeling is found using min-cut max-flow. For $k > 2$ with convex prior, the global minimizer is attained by replacing each single k-label variable with k [4] or by using $k - 1$ [5] boolean variables. However, convex priors tend to over-smooth the label field. For $k > 2$ with metric or semi-metric priors, Boykov *et al.* performed range moves using binary cuts to expand or swap labels [1]. Other range moves were proposed in [6,7]. More recent approaches to multi-label MRF optimization were proposed based on linear programming relaxation using primal-dual [8], message passing or belief propagation [9], and partial optimality [10] (see [11] for a recent survey).

In this paper, we focus on optimal encoding of the k-label MRF energy solely into the edge weights of a graph. We impose no restrictions on k, or on the order (2nd or higher) or type (*e.g.* non-convex, non-metric, or spatially varying) of the label interaction penalty. The calculated edge weights are optimal in the sense that they minimize the least squares (LS) error when solving a linear system of equations capturing the original MRF penalties. Further, we transform the multi-labelling problem to a binary $s-t$ cut, in which each vertex in the original graph is replaced by the most compact boolean representation; only $ceil(log_2(k))$ vertices represent each k-label variable. In [12], a general framework for converting multi-label problems to binary ones is presented. In contrast to our work, [12] solved a system of equations *to find the boolean encoding function* (not the edge weights), they did not use LS, and their resulting binary problem can still include label interaction. We perform a single (non-iterative and initialization-independent) $s - t$ cut to obtain a "Gray" binary encoding, which is then unambiguously decoded into the k labels. Besides its optimality features, LS enables offline *pre-computation* of pseudoinverse matrices that can be re-used for different graphs.

2   Method

2.1   Reformulating the Multi-label MRF as an $s - t$ Cut

Given a graph $G(V, E)$, the objective is to label each vertex $v_i \in V$ with a label $l_i \in \mathcal{L}_k = \{l_0, l_1, ..., l_{k-1}\}$. Rather than labeling v_i with $l_i \in \mathcal{L}_k$, we replace v_i with b vertices $(v_{ij})_{j=1}^b$, and binary-label them with $(l_{ij})_{j=1}^b$, *i.e.* $l_{ij} \in \mathcal{L}_2 = \{l_0, l_1\}$. b is chosen such that $2^b \geq k$ or $b = ceil(log_2(k))$, *i.e.* a long enough sequence of *bits* to be decoded into $l_i \in \mathcal{L}_k^2$. To this end, we transform $G(V, E)$ into a new graph $G_2(V_2, E_2)$ with additional source s and sink t nodes, *i.e.* $|V_2| = b|V| + 2$. E_2 includes terminal links $E_2^{tlinks} = E_2^t \cup E_2^s$ where $|E_2^t| = |E_2^s| = |V_2|$; neighborhood links $E_2^{nlinks} = E_2^{ns} \cup E_2^{nf}$ where $|E_2^{nlinks}| = b^2|E|$, $|E_2^{ns}| = b|E|$, and $|E_2^{nf}| = (b^2 - b)|E|$; and intra-links E_2^{intra} where $|E_2^{intra}| = \binom{b}{2}|V|$. Figure 1 shows these different types of edges. Following an $s - t$ cut on G_2, vertices v_{ij} that remain connected to s are assigned label 0, and the remaining are connected

---

2 We distinguish between the decimal (base 10) and binary (base 2) encoding of the labels using the notation $(l_i)_{10}$ and $(l_i)_2 = (l_{i1}, l_{i2}, \cdots, l_{ib})_2$, respectively.

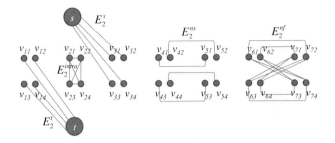

Fig. 1. Edge types in the $s - t$ graph. Shown are seven groups of vertex quadruplets, $b=4$, and only sample edges from $E_2^t, E_2^s, E_2^{ns}, E_2^{nf}$, and E_2^{intra}.

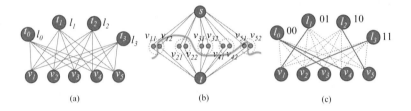

(a)   (b)   (c)

Fig. 2. Reformulating the multi-label problem as an $s - t$ cut. (a) Labeling vertices $\{v_i\}_{i=1}^{5}$ with labels $\{l_j\}_{j=0}^{3}$ (only t-links are shown). (b) New graph with 2 terminal nodes $\{s, t\}$, $b = 2$ new vertices (v_{i1} and v_{i2} inside the dashed circles) replacing each v_i in (a), and 2 terminal edges for each v_{ij}. An $s - t$ cut on (b) is depicted as the green curve. (c) Labeling v_i in (a) is based on the $s - t$ cut in (b): Pairs of (v_{i1}, v_{i2}) assigned to (s, s) are labeled with binary string 00, (s, t) with 01, (t, s) with 10, and (t, t) with 11. The binary encodings $\{00, 01, 10, 11\}$ in turn reflect the original 4 labels $\{l_j\}_{j=0}^{3}$.

to t and assigned label 1. The string of b binary labels $l_{ij} \in \mathcal{L}_2$ assigned to v_{ij} are then decoded back into a decimal number indicating the label $l_i \in \mathcal{L}_k$ assigned to v_i (Figure 2).

It is important to set the edge weights of E_2 in such a way that decoding the binary labels resulting from the $s - t$ cut of G_2 results in optimal (or close to optimal) labels for the original multi-label problem. To achieve this, we derive a system of linear equations capturing the relation between the original multi-label MRF penalties and the $s - t$ cut cost incurred when generating different label configurations. We then calculate the weights of E_2 as the LS error solution to these equations. The next sections expose the details.

2.2 Data Term Penalty: Severing T-Links and Intra-Links

The 1st order penalty $D_i(l_i)$ in (1) is the cost of assigning l_i to v_i in G, which entails assigning a corresponding sequence of binary labels $(l_{ij})_{j=1}^{b}$ to $(v_{ij})_{j=1}^{b}$ in G_2. To assign $(l_i)_2$ to a string of b vertices, appropriate terminal links must be cut. To assign a 0 (resp. 1) label to v_{ij}, the edge connecting v_{ij} to t (resp.

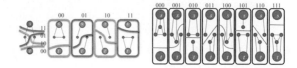

Fig. 3. The 2^b ways of cutting through $\{v_{ij}\}_{j=1}^b$ for $b = 2$ (left) and $b = 3$ (right) with the resulting binary codes $\{00, 01, 10, 11\}$ and $\{000, 001, \cdots, 111\}$

s) must be severed (Figure 3). Therefore, the cost of severing t-links in G_2 to assign l_i to vertex v_i in G is calculated as

$$D_i^{tlinks}(l_i) = \sum_{j=1}^{b} l_{ij} w_{v_{ij},s} + \bar{l}_{ij} w_{v_{ij},t} \tag{2}$$

where $\bar{l}_{ij}$ denotes the unary complement (NOT) of l_{ij}. The G_2 $s-t$ cut severing the t-links, as per (2), will also result in severing edges in E_2^{intra} (Figure 1). In particular, $e_{im,in} \in E_2^{intra}$ will be severed iff the $s-t$ cut leaves v_{im} connected to one terminal, say s (resp. t), while v_{in} remains connected to the other terminal t (resp. s). If this condition holds, then $w_{v_{im},v_{in}}$ will contribute to the cost. Therefore, the cost of severing intra-links in G_2 to assign l_i to vertex v_i in G is

$$D_i^{intra}(l_i) = \sum_{m=1}^{b} \sum_{n=m+1}^{b} (l_{im} \oplus l_{in}) w_{v_{im},v_{in}} \tag{3}$$

where $\oplus$ denotes binary XOR. The total data penalty is the sum of (2) and (3),

$$D_i(l_i) = D_i^{tlinks}(l_i) + D_i^{intra}(l_i). \tag{4}$$

2.3   Prior Term Penalty: Severing N-Links

The interaction penalty $V_{ij}(l_i, l_j, d_i, d_j)$ for assigning l_i to v_i and l_j to neighboring v_j in G must equal the cost of assigning a sequence of binary labels $(l_{im})_{m=1}^b$ to $(v_{im})_{m=1}^b$ and $(l_{jn})_{n=1}^b$ to $(v_{in})_{n=1}^b$ in G_2. The cost of this cut can be calculated as (Figure 4)

$$V_{ij}(l_i, l_j, d_i, d_j) = \sum_{m=1}^{b} \sum_{n=1}^{b} (l_{im} \oplus l_{jn}) w_{v_{im},v_{jn}}. \tag{5}$$

This effectively adds the edge weight between v_{im} and v_{jn} to the cut cost iff the cut results in one vertex of the edge connected to one terminal (s or t) while the other vertex connected to the other terminal (t or s). Note that we impose no restrictions on the left hand side of (5), *e.g.* it could reflect non-convex or non-metric priors, spatially-varying, or even higher order label interaction.

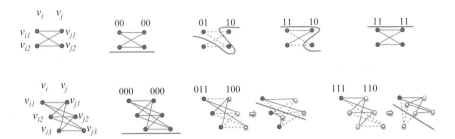

Fig. 4. Severing n-links between neighboring vertices v_i and v_j for $b = 2$ (four examples are shown in the top row) and $b = 3$ (three examples in the bottom row). The cut is depicted as a red curve. In the last two examples for $b = 3$, the colored vertices are translated while maintaining the n-links in order to clearly show that the severed n-links for each case follow (5).

2.4   Edge Weight Approximation with Least Squares

Equations (4) and (5) dictate the relationship between the penalty terms (D_i and V_{ij}) of the original multi-label problem and the severed edge weights $w_{ij,mn}$; $\forall e_{ij,mn} \in E_2$ of the $s - t$ graph G_2. What remains missing before applying the $s - t$ cut, however, is to find these edge weights.

Edge weights of t-links and intra-links. For $b = 1$ (*i.e.* binary labelling), (3) simplifies to $D_i^{intra}(l_i) = 0$ and (4) simplifies to $D_i(l_i) = l_{i1}w_{v_{i1},s} + \bar{l}_{i1}w_{v_{i1},t}$. With $l_i = l_{i1}$ for $b = 1$, substituting the two possible values for $l_i = \{l_0, l_1\}$, we obtain

$$l_i = l_0 \Rightarrow D_i(l_0) = l_0 w_{v_{i1},s} + \bar{l}_0 w_{v_{i1},t} = 0w_{v_{i1},s} + 1w_{v_{i1},t}$$
$$l_i = l_1 \Rightarrow D_i(l_1) = l_1 w_{v_{i1},s} + \bar{l}_1 w_{v_{i1},t} = 1w_{v_{i1},s} + 0w_{v_{i1},t} \tag{6}$$

which can be written in matrix form $A_1 X_1^i = B_1^i$ as $\begin{pmatrix} 0 & 1 \\ 1 & 0 \end{pmatrix} \begin{pmatrix} w_{v_{i1},s} \\ w_{v_{i1},t} \end{pmatrix} = \begin{pmatrix} D_i(l_0) \\ D_i(l_1) \end{pmatrix}$ where X_1^i is the vector of unknown edge weights connecting vertex v_{i1} to s and t, B_1^i is the data penalty for v_i, and A_1 is the matrix of coefficients. The subscript 1 in A_1, X_1^i, and B_1^i indicates that this matrix equation is for $b = 1$. Clearly, the solution is trivial and expected: $w_{v_{i1},s} = D_i(l_1)$ and $w_{v_{i1},t} = D_i(l_0)$

For $b = 2$, we address multi-label problems of $k = \{3, 4\}$, or $2^{b-1} = 2 < k \le 2^b = 4$ labels. Substituting the $2^b = 4$ possible label values, $((0,0),(0,1),(1,0),$ and $(1,1))$, of $(l_i)_2 = (l_{i1}, l_{i2})$ in (4) we obtain

$$(0,0) \Rightarrow D_i(l_0) = 0w_{v_{i1},s} + 1w_{v_{i1},t} + 0w_{v_{i2},s} + 1w_{v_{i2},t} + 0w_{v_{i1},v_{i2}}$$
$$(0,1) \Rightarrow D_i(l_1) = 0w_{v_{i1},s} + 1w_{v_{i1},t} + 1w_{v_{i2},s} + 0w_{v_{i2},t} + 1w_{v_{i1},v_{i2}}$$
$$(1,0) \Rightarrow D_i(l_2) = 1w_{v_{i1},s} + 0w_{v_{i1},t} + 0w_{v_{i2},s} + 1w_{v_{i2},t} + 1w_{v_{i1},v_{i2}} \tag{7}$$
$$(1,1) \Rightarrow D_i(l_3) = 1w_{v_{i1},s} + 0w_{v_{i1},t} + 1w_{v_{i2},s} + 0w_{v_{i2},t} + 0w_{v_{i1},v_{i2}}$$

which can be written in matrix form $A_2 X_2^i = B_2^i$ as

$$
\begin{pmatrix} 0\ 1\ 0\ 1\ 0 \\ 0\ 1\ 1\ 0\ 1 \\ 1\ 0\ 0\ 1\ 1 \\ 1\ 0\ 1\ 0\ 0 \end{pmatrix}
\begin{pmatrix} w_{v_{i1},s} \\ w_{v_{i1},t} \\ w_{v_{i2},s} \\ w_{v_{i2},t} \\ w_{v_{i1},v_{i2}} \end{pmatrix}
=
\begin{pmatrix} D_i(l_0) \\ D_i(l_1) \\ D_i(l_2) \\ D_i(l_3) \end{pmatrix}.
\tag{8}
$$

In general, for any b, we have

$$
A_b X_b^i = B_b^i
\tag{9}
$$

where X_b^i is a column vector of length $2b + \binom{b}{2}$ (superscript t denotes transpose),

$$
\begin{aligned}
X_b^i = (&w_{v_{i1},s}, w_{v_{i1},t}, w_{v_{i2},s}, w_{v_{i2},t}, \cdots, w_{v_{ib},s}, w_{v_{ib},t}, \\
&w_{v_{i1},v_{i2}}, w_{v_{i1},v_{i3}}, \cdots, w_{v_{i1},v_{ib}}, w_{v_{i2},v_{i3}}, \cdots, w_{v_{i2},v_{ib}}, \cdots, w_{v_{i,b-1},v_{ib}})^t
\end{aligned}
\tag{10}
$$

A_b is a $2^b \times (2b + \binom{b}{2})$ matrix whose jth row $A_b(j,:)$ is given by

$$
\begin{aligned}
A_b(dec(l_{i1}l_{i2}\cdots l_{ib}),:) = (&l_{i1}, \bar{l}_{i1}, l_{i2}, \bar{l}_{i2}, \cdots, l_{ib}, \bar{l}_{ib}, \\
&l_{i1} \oplus l_{i2}, l_{i1} \oplus l_{i3}, \cdots, l_{i1} \oplus l_{ib}, l_{i2} \oplus l_{i3}, l_{i2} \oplus l_{ib}, \cdots, l_{i,b-1} \oplus l_{ib})
\end{aligned}
\tag{11}
$$

where $dec(.)$ is the decimal equivalent of its binary argument. B_b^i is a 2^b-long column vector given by $B_b^i = (D_i(l_0), D_i(l_1), D_i(l_2), \cdots, D_i(l_{2^b-1}))^t$.

We now solve the linear system of equations in (9) to find the optimal, in a LS sense, t-links and intra-links edge weights $\hat{X}_b^i$ related to every vertex v_i using

$$
\hat{X}_b^i = A_b^+ B_b^i
\tag{12}
$$

where A^+ is the (Moore-Penrose) pseudoinverse of A calculated using singular value decomposition (SVD)[13].

Edge weights of n-links. For $b = 1$ (i.e. binary labelling), (5) simplifies to $(l_i \oplus l_j)w_{ij} = V_{ij}(l_i, l_j, d_i, d_j)$, where $w_{v_{i1},v_{j1}}$ has been replaced by $w_{i,j}$ and l_{i1} and l_{j1} have been replaced by l_i and l_j, since they are equivalent for $b = 1$. If $V_{ij}(l_i, l_j, d_i, d_j) = V_{ij}(d_i, d_j)$, i.e. label-independent, we can simply ignore the outcome of $l_i \oplus l_j$ by setting it to a constant. Then, the solution is trivial and as expected (Section 1): $w_{i,j} \propto V_{ij}(d_i, d_j)$. However, in the general case when V_{ij} depends on the labels l_i and l_j of the neighboring vertices v_i and v_j, a single edge weight is insufficient to capture such elaborate label interactions, intuitively, because $w_{i,j}$ needs to take on a different value for every pair of labels. To address this problem, we substitute in (5) each of the $2^b 2^b = 2^{2b} = 2^2 = 4$ possible combinations of pairs of labels $(l_i, l_j) \in \{l_0, l_1\} \times \{l_0, l_1\} = \{0, 1\} \times \{0, 1\}$, and obtain the following system of linear equations:

$$
\begin{aligned}
(l_0, l_0) = (0, 0) &\Rightarrow V_{ij}(l_0, l_0, d_i, d_j) = (0 \oplus 0)\, w_{i,j} = 0 \\
(l_0, l_1) = (0, 1) &\Rightarrow V_{ij}(l_0, l_1, d_i, d_j) = (0 \oplus 1)\, w_{i,j} = w_{i,j} \\
(l_1, l_0) = (1, 0) &\Rightarrow V_{ij}(l_1, l_0, d_i, d_j) = (1 \oplus 0)\, w_{i,j} = w_{i,j} \\
(l_1, l_1) = (1, 1) &\Rightarrow V_{ij}(l_1, l_1, d_i, d_j) = (1 \oplus 1)\, w_{i,j} = 0
\end{aligned}
\tag{13}
$$

which is written in matrix form $S_1 Y_1^{ij} = T_1^{ij}$ as

$$
\begin{pmatrix} 0 \\ 1 \\ 1 \\ 0 \end{pmatrix} (w_{i,j}) =
\begin{pmatrix} V_{ij}(l_0, l_0, d_i, d_j) \\ V_{ij}(l_0, l_1, d_i, d_j) \\ V_{ij}(l_1, l_0, d_i, d_j) \\ V_{ij}(l_1, l_1, d_i, d_j) \end{pmatrix}
\tag{14}
$$

where Y_1^{ij} is the unknown n-link weight $w_{i,j}$ connecting v_i to neighboring v_j. The 1st and 4th equations capture the condition that in order to guarantee the same label for neighboring vertices then the edge weight connecting them should be infinite $(0/V_{ij})$ and, hence, never severed. Solving for w_{ij} using pseudoinverse gives $w_{ij} = S_1^+ T_1^i = \frac{1}{2}(V_{ij}(l_0, l_1, d_i, d_j) + V_{ij}(l_1, l_0, d_i, d_j))$ since $S_1^+ = (0, 0.5, 0.5, 0)$, i.e. w_{ij} is equal to the average between the interaction penalties of the two cases when the labels are different.

For $b = 2$, (5) simplifies to

$$V_{ij}(l_i, l_j, d_i, d_j) = (l_{i1} \oplus l_{j1}) w_{v_{i1}, v_{j1}} + (l_{i1} \oplus l_{j2}) w_{v_{i1}, v_{j2}} + \qquad (15)$$
$$(l_{i2} \oplus l_{j1}) w_{v_{i2}, v_{j1}} + (l_{i2} \oplus l_{j2}) w_{v_{i2}, v_{j2}}$$

We can now substitute all possible $2^b 2^b = 2^{2b} = 16$ combinations of the pairs of interacting labels $(l_i, l_j) \in \{l_0, l_1, l_2, l_3\} \times \{l_0, l_1, l_2, l_3\}$, or equivalently, $((l_i)_2, (l_j)_2) \in \{00, 01, 10, 11\} \times \{00, 01, 10, 11\}$. Here are a few examples,

$$
\begin{aligned}
(l_0, l_0) = (00, 00) &\Rightarrow V_{ij}(l_0, l_0, d_i, d_j) = 0 w_{v_{i1}, v_{j1}} + 0 w_{v_{i1}, v_{j2}} + 0 w_{v_{i2}, v_{j1}} + 0 w_{v_{i2}, v_{j2}} \\
(l_1, l_2) = (01, 10) &\Rightarrow V_{ij}(l_1, l_2, d_i, d_j) = 1 w_{v_{i1}, v_{j1}} + 0 w_{v_{i1}, v_{j2}} + 0 w_{v_{i2}, v_{j1}} + 1 w_{v_{i2}, v_{j2}} \quad (16) \\
(l_3, l_3) = (11, 11) &\Rightarrow V_{ij}(l_3, l_3, d_i, d_j) = 0 w_{v_{i1}, v_{j1}} + 0 w_{v_{i1}, v_{j2}} + 0 w_{v_{i2}, v_{j1}} + 0 w_{v_{i2}, v_{j2}}
\end{aligned}
$$

Writing all the 16 equations, we obtain the system of linear equations in matrix format as $S_2 Y_2^{ij} = T_2^{ij}$, where $Y_2^{ij} = (w_{v_{i1}, v_{j1}}, w_{v_{i1}, v_{j2}}, w_{v_{i2}, v_{j1}}, w_{v_{i2}, v_{j2}})^t$ is the 4×1 vector of unknown n-link edge weights, T_2^{ij} is a 16×1 vector whose entries are the different possible interaction penalties $((V_{ij}(l_i, l_j, d_i, d_j))_{i=0}^3)_{j=0}^3$, and S_2 is a 16×4 matrix with 0 or 1 entires resulting from $\oplus$.

In general, for any b, we obtain the following linear system of equations

$$S_b Y_b^{ij} = T_b^{ij} \qquad (17)$$

where Y_b^{ij} is the $b^2 \times 1$ vector of unknown n-link edge weights, S_b is $2^{2b} \times b^2$ matrix of 0s and 1s, and T_b^{ij} is a $2^{2b} \times 1$ vector of interaction penalties.

We now solve the linear system of equations in (17) to find the optimal, in a LS sense, n-links edge weights $\hat{Y}_b^{ij}$ related to a pair of vertices v_i and v_j using

$$\hat{Y}_b^{ij} = S_b^+ T_b^{ij}. \qquad (18)$$

Solving (18) for every pair of neighboring vertices v_i and v_j, we obtain the weights of all edges in E_2^{inter}, and solving (12) for every vertex v_i, we obtain the weights of all edges in $E_2^{tlinks} \cup E_2^{intra}$, i.e. $w_{ij,mn}, \forall e_{ij,mn} \in E_2$ are now known. It is important to note that some of these resulting edge weights may turn out negative. In order to guarantee positive weights and hence guarantee a globally optimal cut of G_2 in polynomial time, we simply add the same constant to all the edge weights in G_2 to translate all the values to become larger than zero. We now calculate the minimal $s - t$ cut of G_2 to obtain the binary labeling of every vertex in $V_2 = \{\{v_{ij}\}_{i=1}^{|V|}\}_{j=1}^b$. Finally, every sequence of b binary labels $(v_{ij})_{j=1}^b$ is decoded to a decimal label $l_i \in \mathcal{L}_k = \{l_0, l_1, ..., l_{k-1}\}, \forall v_i \in V$, i.e. the solution to the original multi-label MRF problem.

2.5   Gray Encoding of Extra Labels

When $k < 2^b$, an $s - t$ cut may generate *extra labels* (the nth label l_{n-1} is extra iff $k < n \leq 2^b$), which must be replaced or merged with a *non-extra* label (the mth label l_{m-1} is non-extra iff $2^{b-1} < m \leq k$). To replace l_n with l_m, we replace $D_i(l_n)$ with $D_i(l_m)$ in (4), $V_{ij}(l_n, l_j, d_i, d_j)$ with $V_{ij}(l_m, l_j, d_i, d_j)$, and $V_{ij}(l_i, l_n, d_i, d_j)$ with $V_{ij}(l_i, l_m, d_i, d_j)$ in (5). Rather than merging arbitrary labels, we adopt a Gray encoding scheme that minimizes the Hamming distance between the binary codes of merged labels. We first note that the most significant bit of any extra label l_n will always be 1. Then, l_n is merged with the non-extra label whose binary code is identical to l_n *except* for having 0 as its most significant bit: *i.e.* $(l_n)_2 = (1, l_2, \cdots, l_b)_2$ is merged with $(l_m)_2 = (0, l_2, \cdots, l_b)_2$.

3   Results

3.1   LS Error and Rank Deficiency Analysis

LS approximation error is a well studied topic (e.g. [13]). Table 1 summarizes the main properties (number of equations, unknowns, and rank) when solving for the weights of t-links and intra-links (A_b in (9)) and for inter-links (S_b in (17)) for increasing bits b. We note that, not surprisingly, the only full-rank case is A_1 (*i.e.* binary segmentation). A_b is underdetermined for $b = 2, 3$ and overdetermined for $b \geq 4$. All cases of S_b are rank deficient and overdetermined.

In Figure 5, we present empirical results of LS error $e_b = |B_b^i - \hat{B}_b^i|/|B_b^i| = |(I - A_b A_b^+) B_b^i|/|B_b^i|$ (when solving for t-links and intra-links) and $e_t = |T_b^{ij} - \hat{T}_b^{ij}|/|T_b^{ij}| = |(I - S_b S_b^+) T_b^{ij}|/|T_b^{ij}|$ (for n-links), for increasing number of labels k, where I is the identity matrix and $|.|$ is the l^2-norm. The plots are the result of a Monte Carlo simulation of 500 random realizations of B_b^i and T_b^{ij} for every k. Note how e_b starts at exactly zero for binary segmentation ($b = 1$), as expected. As k increases, the average of e_b increases with an (empirical) upper bound of 0.5, while its variance decreases. e_t is non-zero even for $b = 1$ (Section 2.4) and converges to 0.5 with increasing k.

Table 1. Properties of the systems of linear equations (9) and (17). For increasing b, the number of equations e, number of unknowns u, and ranks r of A_b and S_b are shown. For A_b, u_0 is when intra-links are not used ($E_2^{intra} = \emptyset$) and, for S_b, u_0 is when only sparse n-links are used ($E_2^{nf} = \emptyset$).

b	A_b in (9)					S_b in (17)				
bits	$e = 2^b$	$u = 2b + \binom{b}{2}$	r	$u_0 = 2b$	r_0	$e = 2^{2b}$	$u = b^2$	r	$u_0 = b$	r_0
1	2	2	2	2	2	4	1	1	1	1
2	4	5	4	4	3	16	4	4	2	2
3	8	9	7	6	4	64	9	9	3	3
4	16	14	11	8	5	256	16	16	4	4
5	32	20	16	10	6	1024	25	25	5	5
6	64	27	22	12	7	4096	36	36	6	6
7	128	35	29	14	8	16384	49	49	7	7
8	256	44	37	16	9	65536	64	64	8	8

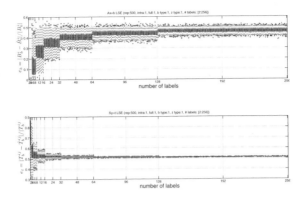

Fig. 5. LS error for increasing number of labels. (top) Error e_b in estimating the weights of $E_2^{tlinks} \cup E_2^{intra}$. (bottom) Error e_t in estimating the weights of E_2^{nlinks}. The general behavior of the error is clear from the figure. The reader may refer to the electronic copy of this paper for color and scalable graphics.

3.2 Effect of LS Error in Edge Weights on the $s - t$ Cut

The LS error in edge weights induces error in the $s - t$ cut or binary labeling, which is decoded into a suboptimal solution to the multi-label problem. To quantify the cut cost error $\triangle|C|$ and the labeling accuracy ACC due to edge weight error, we create a graph G with a proper topology (*i.e.* reflecting the 4-connectedness of 2D image pixels) and edge weights sampled from a uniform probability distribution function (PDF) with support $[0, 1]$. We then construct G_{LSE}, a noisy version of G, by adding uniformly distributed noise with support $[0, \text{noise level}]$ to the edge weights. Figure 6 shows the results of $\triangle|C| = ||C| - |C_{LSE}||/|C|$ and $ACC = (TP + TN)/|V|$, where $|C| = \sum_{e_{ij} \in C} w_{ij}$ is the cut cost of G, $|C_{LSE}|$ is the cut cost of G_{LSE}, and $TP + TN$ is the number of correctly labelled vertices (*i.e.* true positive and true negatives), and $|V|$ is the number of vertices in G. The plots are the results of a Monte Carlo simulation of 20 realizations of G and G_{LSE} each with 10,000 vertices.

3.3 Image Segmentation Results

We evaluated our method's segmentation accuracy by calculating the average (over all labels) Dice similarity coefficient $\overline{DSC}$ [14](Figure 7(left)) on synthetic (with known ground truth) images: $I(x, y) : R^2 \rightarrow [0, 1]$, containing ellipses with random major and minor axes and varying pixel intensities (Figure 7(right)). We tested increasing levels of Gaussian noise $\sim \mathcal{N}(0, \sigma \in \{0, 0.05, 0.10, \cdots, 0.40\})$, labels $k = \{2, 3, \cdots, 16\}$, and with non-convex Pott's label interaction weighted by a spatially varying Gaussian image intensity penalty [3]. We ran 10 realization for each test case. For pixel i with intensity d_i, $D_i(l_i) = (p_l(\mu) - p_l(x_i))/p_l(\mu)$, where $p_l(d) \sim \mathcal{N}(\mu_l, \sigma_l)$ is a Gaussian PDF learned from 50% of the pixels of

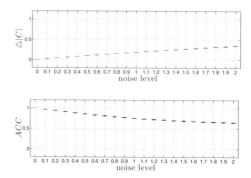

Fig. 6. Cut cost error $\triangle|C|$ and labeling accuracy ACC as we corrupt the edge weights with increasing levels of noise

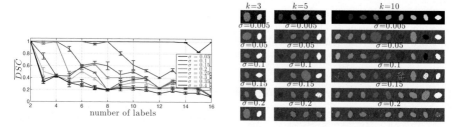

Fig. 7. Segmentation results on images of ellipses. (left) $\overline{DSC}$ between the ground truth and our method's segmentation with increasing number of labels and noise levels (different colors). (right) Sample qualitative results with k labels ($k-1$ ellipses plus background) and noise level σ. (top row) sample intensity images; (remaining rows) labeling results.

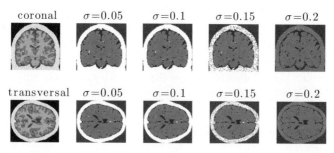

Fig. 8. Brain MRI segmentation on coronal (top) and transversal (bottom) slices for increasing noise σ

each region (or label) l of the noisy image (mimicking seeding). Note that $\overline{DSC}$ gradually decreases from unity with increasing σ or k, e.g. the topmost curve (blue, $\sigma = 0.05$) shows almost perfect segmentation; $\overline{DSC} = 1 \ \forall k$, whereas $\overline{DSC}$ drops below 1 for $k \geq 9$ for the second-from-top curve (green, $\sigma = 0.1$), and for $k \geq 5$ for third curve (red, $\sigma = 0.15$).

We present qualitative segmentation results on synthetic data (Figure 7(right)) and on magnetic resonance brain images (Figure 8) from BrainWeb [15].

4   Conclusions

Multi-label MRF optimization with non-trivial priors is a challenging problem with several computer vision applications. In our proposed approach, rather than labeling a vertex with one of k labels, the vertex is replaced by $b = ceil(log_2(k))$ new vertices that are binary-labelled to encode the original k labels; effectively approximating the multi-label problem with a globally and non-iteratively solvable $s - t$ cut. The new $s - t$ graph is optimal in a least squares sense because its edge weights are the LS error solution of a system of linear equations capturing the original multi-label MRF energy, without any restrictions on the interaction priors. To the best of our knowledge, this is the first work to use LS to approximate the multi-label MRF with any order of label interaction solely via the edge wights of a graph (with no label interaction). Offline pre-computation of A_b^+ and S_b^+ in (12) and (18) is performed only once for each b value then re-used for different vertices and graphs. We quantitatively evaluated different properties of the proposed approximation and demonstrated its application to image segmentation (with qualitative and quantitative results on synthetic and brain images). More elaborate analysis of the algorithm (*e.g.* error bounds, value of the minimized energy, computational complexity, running times) and comparison with state-of-the-art approaches on standard benchmarks is left for future work. Further, we are exploring the use of non-negative least squares (*e.g.* Chapter 23 in [16]) to guarantee non-negative edge weights as well as quantifying the benefits of the Gray encoding.

Acknowledgements

We thank Yuri Boykov, Vladimir Kolmogorov, and Olga Veksler for the MRF optimization code and Shai Bagon for its MATLAB wrapper; Marc Koppert and Ben Smith for discussions and assistance in conducting some initial experiments; and the reviewers for their valuable feedback.

References

1. Boykov, Y., Veksler, O., Zabih, R.: Fast approximate energy minimization via graph cuts. IEEE PAMI 23, 1222–1239 (2001)
2. Grady, L.: Random walks for image segmentation. IEEE PAMI 28, 1768–1783 (2006)
3. Boykov, Y., Funka-Lea, G.: Graph cuts and efficient N-D image segmentation. International Journal of Computer Vision 70, 109–131 (2006)
4. Ishikawa, H.: Exact optimization for markov random fields with convex priors. IEEE PAMI 25, 1333–1336 (2003)

5. Schlesinger, D., Flach, B.: Transforming an arbitrary min-sum problem into a binary one. Tech. Report TUD-FI06-01, Dresden University (2006)
6. Veksler, O.: Graph cut based optimization for MRFs with truncated convex priors. In: IEEE CVPR, pp. 1–8 (2007)
7. Lempitsky, V., Rother, C., Blake, A.: Logcut - efficient graph cut optimization for markov random fields. In: IEEE ICCV, pp. 1–8 (2007)
8. Komodakis, N., Tziritas, G., Paragios, N.: Performance vs computational efficiency for optimizing single and dynamic MRFs: Setting the state of the art with primal-dual strategies. Comput. Vis. Image Underst. 112, 14–29 (2008)
9. Kolmogorov, V.: Convergent tree-reweighted message passing for energy minimization. IEEE PAMI 28, 1568–1583 (2006)
10. Kohli, P., Shekhovtsov, A., Rother, C., Kolmogorov, V., Torr, P.: On partial optimality in multi-label MRFs. In: International Conference on Machine learning (ICML), pp. 480–487. ACM, New York (2008)
11. Szeliski, R., Zabih, R., Scharstein, D., Veksler, O., Kolmogorov, V., Agarwala, A., Tappen, M., Rother, C.: A comparative study of energy minimization methods for markov random fields with smoothness-based priors. IEEE PAMI 30, 1068–1080 (2008)
12. Ramalingam, S., Kohli, P., Alahari, K., Torr, P.: Exact inference in multi-label CRFs with higher order cliques. In: IEEE Conference on Computer Vision and Pattern Recognition, 2008. CVPR 2008, pp. 1–8 (2008)
13. Björk, A.: Numerical Methods for Least Squares Problem. SIAM, Philadelphia (1996)
14. Dice, L.R.: Measures of the amount of ecologic association between species. Ecology 26, 279–302 (1945)
15. Cocosco, C.A., Kollokian, V., Kwan, R.K.-s., Pike, G.B., Evans, A.C.: Brainweb: Online interface to a 3D mri simulated brain database. NeuroImage 5, 425 (1997)
16. Lawson, C., Hanson, R.: Solving Least Squares Problems. Society for Industrial Mathematics (1974)

Combinatorial Preconditioners and Multilevel Solvers for Problems in Computer Vision and Image Processing*

Ioannis Koutis, Gary L. Miller, and David Tolliver

Computer Science Department
Carnegie Mellon University

Abstract. Linear systems and eigen-calculations on symmetric diagonally dominant matrices (SDDs) occur ubiquitously in computer vision, computer graphics, and machine learning. In the past decade a multitude of specialized solvers have been developed to tackle restricted instances of SDD systems for a diverse collection of problems including segmentation, gradient inpainting and total variation. In this paper we explain and apply the support theory of graphs, a set of of techniques developed by the computer science theory community, to construct SDD solvers with provable properties. To demonstrate the power of these techniques, we describe an efficient multigrid-like solver which is based on support theory principles. The solver tackles problems in fairly general and arbitrarily weighted topologies not supported by prior solvers. It achieves state of the art empirical results while providing robust guarantees on the speed of convergence. The method is evaluated on a variety of vision applications.

1 Introduction

The Laplacian operator ∇^2 has played a central role in computer vision for nearly 40 years. In Horn's early work, he employed finite element methods for elliptical operators in shape from shading [1], to produce albedo maps [2], and flow estimates [3]. In Witkin's seminal work [4] he studied the diffusion properties of matrix equations derived from ∇^2 for linear filtering, later generalized by Perona and Malik [5] to the anisotropic case.

In recent years, combinatorial Laplacians of graphs have formed the algorithmic core of spectral methods [6,7,8,9,10,11,12,13], random walks segmentation [11], inpainting [14,15,16], and matting methods [12]. Given the power of modern iterative solvers, we believe that reducing traditional image processing problems, such as Grady *et al.*'s work [17] on Mumford-Shah segmentation, to SDD systems at the inner loop is a critical endeavor. To this end, we note that non-linear filtering operations such as ℓ_2, ℓ_1 Total Variation [18,19] and Non-Local Means [20,21] can also be formulated as optimizations with these linear systems at their core. Further, we provide timing and modern complexity bounds for computer vision methods in §4 that require the solutions to SDD systems at their core.

---

* This work was partially supported by the National Science Foundation under grant number CCF-0635257 and the University of Pittsburgh Medical Center under award number A-006461.

From a practical standpoint, modern photos and videos, and medical images derived NMR and CT scanners provide enormously detailed portraits of a scene. As the resolution of imaging hardware has pushed at the limits of computational feasibility, researchers inevitably arrived at the study of iterative and **hybrid** solvers. Recently, vision and graphics researchers have developed specialized solvers [14,22,16,15], and heuristic solvers with impressive empirical performance [23,15,24,25]. In either case, the methods place strict requirements on the system, such as unit weight edges or 4-connectivity. For the methods that handle general weights, including Algebraic Multigrid (AMG) [26,27], the solvers are based on heuristics and offer no guarantees on the speed of convergence. Indeed many applications, such as the spectral segmentation and convex programming, require wildly varying weights and often employ randomly sampled and loosely localized topologies. Furthermore, the heuristic nature of the solvers is generally undesirable in certain commercial applications, e.g. medical, where robust and timely behavior is a critical issue.

Given modern data volumes and reliability requirements it is clear that a SDD solver with provable convergence properties and sound theoretical machinery is important for the advancement and real-world success of methods based on linear system solutions. In this work we introduce the Combinatorial Multigrid Solver (CMG), a state of the art solver with provable properties. The CMG solver is based on principles of *support theory* for graphs, a set of techniques developed for the construction of *combinatorial preconditioners*, i.e. graphs that are simpler than a given graph and approximate it well in a precisely defined sense. An ancillary goal of this paper is to review certain useful fragments of support theory and apply them to analyze solvers.

2   Support Theory for Graphs

Support Theory was developed for the study of Combinatorial subgraph preconditioners, introduced by Vaidya [28,29]. It has been at the heart of impressive theoretical results which culminated in the work of Spielman and Teng [30] who demonstrated that SDD systems can be solved in nearly-optimal $\tilde{O}(n \log^{O(1)} n)$ time and later in the work of Koutis and Miller [31] who formally proved that SDD matrices with planar connection topologies (*e.g.* 4-connectivity in the image plane) can be solved asymptotically optimally, in $O(n)$ time. We dub these solvers **hybrid solvers** since they combine algorithms and ideas from direct solvers, preconditioned Conjugate Gradient, and recursion.

2.1   Reduction of SDDs to Laplacians

A matrix A is SDD if it is real symmetric and $A_{ii} \geq \sum_{j \neq i} |A_{ij}|$ for $1 \leq i \leq n$. The *Laplacian* A of a graph $G = (V, E, w)$, where w is a non-negative weight function on the edges, is defined by $A_{i,j} = A(j, i) = -w_{i,j}$ and $A_{i,i} = \sum_{i \neq j} w_{i,j}$. Thus Laplacians are SDD matrices having non-positive off diagonals and zero row sums. We briefly describe how any SDD system can be reduced to a Laplacian. SDD systems with positive off-diagonals can be reduced to the case of non-positive off diagonals using a very light-weight reduction known as the double-cover construction [32]. Assuming now negative off-diagonals, nodes with positive row sums can be viewed as nodes that have an implicit edge to a new "grounded node". In general they do not cause any significant changes in the Laplacian solver [33,34].

2.2   Preconditioners — Motivating Support Theory

Iterative algorithms, such as the Chebyshev iteration or the Conjugate Gradient, converge to a solution using only matrix-vector products with A. It is well known that iterative algorithms suffer from slow convergence properties when the conditioning of A, $\kappa(A)$, - defined as the ratio of the largest over the minimum eigenvalue of A - is large [35].

Preconditioned iterative methods attempt to remedy the problem by changing the linear system to $B^{-1}Ax = B^{-1}b$. In this case, the algorithms use matrix-vector products with A, and solve linear systems of the form $By = z$. The speed of convergence now depends on the **condition number** $\kappa(A, B)$, defined as

$$\kappa(A, B) = \max_x \frac{x^T A x}{x^T B x} \cdot \max_x \frac{x^T B x}{x^T A x} \tag{1}$$

where x is taken to be outside the null space of A. In constructing a preconditioner B, one has to deal with two contradictory goals: (i) Linear systems in B must be easier than those in A to solve, (ii) The condition number must be small to minimize the number of iterations.

Historically, preconditioners were natural parts of the matrix A. For example, if B is taken as the diagonal of A we get the Jacobi Iteration, and when B is the upper triangular part of A, we get the Gauss-Seidel iteration.

The cornerstone of combinatorial preconditioners is the following intuitive yet paradigm-shifting idea explicitly proposed by Vaidya: *A preconditioner for the Laplacian of a graph A should be the Laplacian of a simpler graph B, derived in a principled fashion from A.*

2.3   Graphs as Electric Networks – Support Basics

There is a fairly well known analogy between graph Laplacians and resistive networks [36]. If G is seen as an electrical network with the resistance between nodes i and j being $1/w_{i,j}$, then in the equation $Av = i$, if v is the vector of voltages at the node, i is the vector of currents. Also, the quadratic form $v^T A v = \sum_{i,j} w_{i,j}(v_i - v_j)^2$ expresses the *power dissipation* on G, given the node voltages v. In view of this, the construction of a good preconditioner B amounts to the construction of a simpler resistive network (for example by deleting some resistances) with an energy profile close to that of A.

The **support** of A by B, defined as $\sigma(A/B) = \max_v v^T A v / v^T B v$ is the number of copies of B that are needed to support the power dissipation in A, for all settings of voltages. The principal reason behind the introduction of the notion of support, is to express its local nature, captured by the Splitting Lemma.

Lemma 1 (Splitting Lemma). *If $A = \sum_{i=1}^{m} A_i$ and $B = \sum_{i=1}^{m} B_i$, where A_i, B_i are Laplacians, then $\sigma(A, B) \le max_i \sigma(A_i, B_i)$.*

The Splitting Lemma allows us to bound the support of A by B, by splitting the power dissipation in A into small local pieces, and "supporting" them by also local pieces in B.

For example, in his work Vaidya proposed to take B as the maximal weight spanning tree of A. Then, it is easy to show that $\sigma(B, A) \leq 1$, intuitively because more resistances always dissipate more power. In order to bound $\sigma(A, B)$, the basic idea to let the A_i be edges on A (the ones not existing in B), and let B_i be the unique path in the tree that connects the two end-points of A_i. Then one can bound separately each $\sigma(A_i, B_i)$. In fact, it can be shown that any edge in A that doesn't exist in B, can be supported *only* by the path B_i.

As a toy example, consider the example in Figure 1(a) of the two (dashed) edges A_1, A_2 and their two paths in the spanning tree (solid) that share one edge e.

In this example, the **dilation** of the mapping is equal to 3, i.e. the length of the longest of two paths. Also, as e is uses two times, we say that the **congestion** of the mapping is equal to 2. A core Lemma in Support Theory [37,33] is that the support can be upper bounded by the product **congestion∗dilation**.

2.4   Steiner Preconditioners

Steiner preconditioners, introduced in [32] and extended in [38] introduce external nodes into preconditioners. The proposed preconditioner is based on a partitioning of the n vertices in V into m vertex-disjoint clusters V_i. For each V_i, the preconditioner contains a star graph S_i with leaves corresponding to the vertices in V_i rooted at a vertex r_i. The roots r_i are connected and form the **quotient** graph Q. This general setting is illustrated in Figure 1(b), consisting of good clusters.

Let D' be the total degree of the leaves in the Steiner preconditioner S. Let the **restriction** R be an $n \times m$ matrix, where $R(i, j) = 1$ if vertex i is in cluster j and 0 otherwise. Then, the Laplacian of S has $n + m$ vertices, and the algebraic form

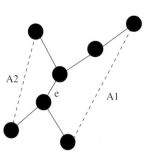
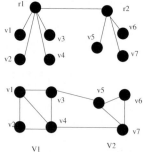
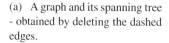

(a)   A graph and its spanning tree - obtained by deleting the dashed edges.

(b)   A graph and its Steiner preconditioner

Fig. 1.

$$S = \begin{pmatrix} D' & -D'R \\ -R^T D' & Q + R^T D'R \end{pmatrix}. \tag{2}$$

A worrying feature of the Steiner preconditioner S is the extra number of vertices. So how do we even use it? Gremban and Miller [32] proposed that every time a system of the form $Bz = y$ is solved in an usual preconditioned method, the system $S \begin{pmatrix} z \\ z' \end{pmatrix} = \begin{pmatrix} y \\ 0 \end{pmatrix}$ should be solved instead, for a set of *don't care* variables z'. They also showed that the operation is equivalent to preconditioning with the dense matrix

$$B = D' - V(Q + D_Q)^{-1}V^T \tag{3}$$

where $V = D'R$, and $D_Q = R^T D'R$. The matrix B is called the Schur complement of S with respect to the elimination of the roots r_i, further it is a well known that B is also a Laplacian.

The analysis of the support $\sigma(A/S)$, is identical to that for the case of subgraph preconditioners. For example, going back to Figure 1(b), the edge (v_1, v_4) can only be supported by the path (v_1, r_1, v_4), and the edge (v_4, v_7) only by the path (v_4, r_1, r_2, v_7). Similarly we can see the mappings from edges in A to paths in S for every edge in A. In the example, the **dilation** of the mapping is 3, and it can be seen that to minimize the **congestion** on every edge of S (i.e. make it equal to 1), we need to take $D' = D$, where D are the total degrees of the nodes in A, and $w(r_1, r_2) = w(v_3, v_5) + w(v_4, v_7)$. More generally, for two roots r_i, r_j we should have $w(r_i, r_j) = \sum_{i' \in V_i, j' \in V_j} w_{i,j}$. Under this construction, the algebraic form of the quotient Q can be seen to be $Q = R^T AR$.

So far no special properties of the clustering have been used. Those come into play in bounding the support of S by A, $\sigma(S/A)$. In [38] it was shown that the support $\sigma(S/A)$ reduces to bounding the support $\sigma(S_i, A[V_i])$, for all i, where $A[V_i]$ denotes the graph induced in A by the vertices V_i. When are these bounded? Before we answer this question, let us recall the definition of **conductance**.

Definition 1. *The conductance $\phi(A)$ of a graph $A = (V, E, w)$ is defined as*

$$\phi(A) = \min_{S \subseteq V} \frac{w(S, V - S)}{\min(w(S), w(V - S))}$$

where $w(S, V - S)$ denotes the total weight connecting the sets S and $V - S$, and where $w(S)$ denotes the total weight incident to the vertices in S.

The main result of [38] is captured by the following Theorem.

Theorem 1. *The support $\sigma(S/A)$ is bounded by a constant c independent from n, if and only for all i the conductance of the graph $A^\circ[V_i]$ induced by the nodes in V_i augmented by the edges leaving V_i is bounded by a constant c'.*

Although Theorem 1 doesn't give a way to pick to clusters, it does provide a way to *avoid* bad clusterings.

2.5   Support Theory and Grady's Clusterings

In recent work [25], Grady proposed a multigrid method where the construction of the "coarse" grid follows exactly the construction of the **quotient** graph in the previous section. Specifically, Grady proposes a clustering such that every cluster contains exactly one of certain pre-specified "coarse" nodes. He then defines the restriction matrix R and he lets the coarse grid be $Q = R^T A R$, identically to the construction of the previous Section. The question then is whether the proposed clustering provides the guarantees that by Theorem 1 are necessary to construct a good Steiner preconditioner. In the following Figure, we replicate Figure 2 of [25], with a choice of weights that force the depicted clustering.

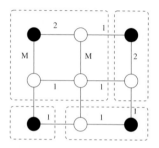

Fig. 2. A bad clustering

Every cluster in Figure 2 contains exactly one black/coarse node. The problem with the clustering is that the top left cluster, has a very low conductance when $M >> 1$. In general, in order to satisfy the requirement of Theorem 1, there are cases where the clustering has to contain clusters with *no* coarse nodes in them.

It is interesting that the Maximally Connected Neighbor (MCN) algorithm proposed in [25] comes very close to the clustering algorithm proposed in [38]. Of course, it is imaginable that there are instances where MCN may not induce bad clusterings. On such instances, Grady's clustering has provable properties. Grady's solver is a multigrid solver, but as we will see in Section 3, multigrid solvers and Steiner preconditioners are closely related.

3   The Combinatorial Multigrid Solver

In this section we describe the Combinatorial Multigrid Solver. As we will see, the CMG solver matches the simple form of AMG, but with two distinguishing features: (i) The "coarsening" strategy is markedly different; it is in fact easier to implement and faster than the various AMG coarsening strategies. (ii) The algorithm is truly "black-box", in stark contrast which AMG which employs an extensive list of algorithmic knobs.

3.1   A Graph Decomposition Algorithm

According to the discussion of §2.4, the crucial step for the construction of a good Steiner preconditioner is the computation of a group decomposition that satisfies, as best as possible, the requirements of Theorem 1. Before the presentation of the **Decompose-Graph** algorithm, that extends the ideas of [38], we need to introduce a couple of definitions. Let $vol_G(v)$ denote the total weight incident to node v in graph G. The *weighted degree* of a vertex v is defined as the ratio

$$wd(v) = \frac{vol(v)}{\max_{u \in N(v)} w(u, v)}.$$

The *average weighted degree* of the graph is defined as $awd(G) = (1/n) \sum_{v \in V} wd(v)$.

Algorithm **Decompose-Graph**

Input: Graph $A = (V, E, w)$
Output: Disjoint Clusters V_i with $V = \bigcup_i V_i$

1. Let $W \subseteq V$ be the set of nodes satisfying $wd(v) > \kappa \cdot awd(A)$, for some constant $\kappa > 4$.
2. Form a forest graph F, by keeping the heaviest incident edge of v for each vertex $v \in V$ in A.
3. For every vertex $w \in W$ such that $vol_T(w) < vol_G(w)/awd(A)$ remove from F the edge contributed by w in Step 2.
4. Decompose each tree T in F into vertex-disjoint trees of constant conductance.

It is not very difficult to prove that the algorithm **Decompose-Graph** produces a partitioning where the conductance of each cluster depends only on $awd(A)$ and the constant κ. In fairly general topologies that allow high degree nodes, $awd(A)$ is constant and the number of clusters m returned by the algorithm is such that $n/m > 2$ (and in practice larger than 3 or 4). There are many easy ways to implement Step 3. Our current implementation makes about three passes of A. Of course, one can imagine variations of the algorithm (i.e. a correction step, etc) that may make the clustering phase a little more expensive with the goal of getting a better conductance and an improved condition number, if the application at hand requires many iterations of the solver.

3.2   From Steiner Preconditioners to Multigrid

Multigrid algorithms have been a very active research area for nearly three decades. There are many expository article and books, among which [39]. In order to describe the reasoning that leads to our Combinatorial Multigrid Algorithm, we will need to shortly review the basic principles behind the generic two-level iteration.

Algebraically, any of the classic preconditioned iterative methods, such as the Jacobi and Gauss-Seidel iteration, is nothing but a matrix $\mathcal{S}$, which gets applied implicitly to the current error vector e, to produce a new error vector $e' = \mathcal{S}e$. For example, in the Jacobi iteration we have $\mathcal{S} = (I - D^{-1}A)$. This has the effect that it reduces effectively only part of the error in a given iterate, namely the components that lie in the low eigenspaces of S (usually referred to as high frequencies of A). The main idea behind a two-level multigrid is that the current *smooth* residual error $r = b - Ax$, can be used to calculate a correction $P^T Q^{-1} Pr$, where Q is a smaller graph and P is an $m \times n$ restriction operator. The correction is then added to the iterate x. The hope here is that for smooth residuals, the low-rank matrix $P^T Q^{-1} P$ is a good approximation of A^{-1}. Algebraically, this correction is the application of the operator $T = (I - P^T Q^{-1} PA)$ to the error vector e. The choice of P and Q is such that T is a projection operator with respect to the A-inner product, a construction known as the *Galerkin* condition. Two-level convergece proofs are then based on bounds on the angle between the subspace $Null(P)$ and the high frequency subspace of S.

At a high level, the key idea behind CMG is that the provably small condition number $\kappa(A, B)$ where B is given in expression 3, is equal to the condition number $\kappa(\hat{A}, \hat{B})$ where $\hat{A} = D^{-1/2} A D^{-1/2}$ and $\hat{B} = D^{-1/2} B D^{-1/2}$. This in turn implies a bound on the angle between the low frequency of $\hat{A}$ and the high frequency of $\hat{B}$ [38]. The latter subspace includes $Null(R^T D^{1/2})$. This fact suggests to choose $R^T D^{1/2}$ as the projection operator while performing relaxation with $(I - \hat{A})$ on the system $\hat{A}y = D^{-1/2}b$, with $y = D^{1/2}x$. Combining everything, we get the following two-level algorithm.

Two-level Combinatorial Multigrid

Input: Laplacian $A = (V, E, w)$, vector b, approximate solution x, $n \times m$ restriction matrix R

Output: Updated solution x for $Ax = b$

1. $D := diag(A)$; $\hat{A} := D^{-1/2} A D^{-1/2}$;
2. $z := (I - \hat{A})D^{1/2}x + D^{-1/2}b$;
3. $r := D^{-1/2}b - \hat{A}z$; $w := R^T D^{1/2}r$;
4. $Q := R^T A R$; Solve $Qy = w$;
5. $z := z + D^{1/2}Ry$
5. $x := D^{-1/2}((I - \hat{A})z + D^{-1/2}b)z$

The two-level algorithm can naturally be extended into a full multigrid algorithm, by recursively calling the algorithm when the solution to the system with Q is requested.

4   Experiments

Many computer vision problems naturally suggest a graph structure - for example the vertices often correspond to samples (*e.g.* pixels, patches, images), the edge set establishes pairwise comparisons or constraints encoded in the graph and the weights are either data driven (for clustering) or the result of an ongoing optimization procedure (weights in the tth iteration of Newton's method).

In this section we demonstrate our general case combinatorial multigrid preconditioner (CMG) solver on a suite of applications taken from computer vision. The data is presented relative to the timing of a direct methods, found in MATLAB and LAPACK, for reference. We emphasize that our goal is not to perform numerical comparisons with any of the previous solvers, but rather to demonstrate that fast solvers with provable properties (such as CMG) are within the realm of practical implementation. The presented solver was written in a combination of C and MATLAB with modest attention paid to code optimization.

Notation. The weighted laplacian matrix A can be factored as $A = \Gamma^T W \Gamma$ where Γ is a edge-node incidence matrix, and W is a nonnegative diagonal weight matrix over the edges. The normalized Laplacian is defined $\hat{A} = D^{-1/2} A D^{-1/2} = I - D^{-1/2} G D^{-1/2}$ where D is the weighted degree matrix and G the weighted adjacency matrix. Recall that quadratic form $x^T A x$ can be written in several forms $x^T D x - x^T G x = \sum_i d_i x_i^2 - 2 \sum_{ij} w_{ij} x_i x_j = \sum_{ij} w_{ij} (x_i - x_j)^2$.

4.1 SDD Linear Systems

Sparse, unit weight, SDD linear systems arise in non-local means[20,21], gradient in-painting [14,15,16], segmentation [40], regression and classification [41] and related data interpolation optimizations. For example, the in-painting functional $f(x) = min_x :$ $(\Gamma x - \Delta)^T W (\Gamma x - \Delta)$, where Δ is a vector of target gradient values to be exhibited in the image x and Γ is a generalized gradient operator, requires the solution to $\Gamma^T W \Gamma x = \Gamma^T W \Delta$. When W is the identity and Γ embodies a 4-connected topology these systems can be efficiently solved (provably) by geometric multigrid methods including the method described herein. For unit and weighted general planar systems the asymptotic complexity of solving $Ax = b$ is $O(n)$ [38].

When W is a more general nonnegative diagonal matrix and Γ encodes non-planar connectivity many multigrid methods will fail. Figure 3 shows such a class of problems in a log-log plot for weighted 3D graphs derived from a 3D CT Study of an oncology phantom – with the edge weighting: $w_{uv} = exp\left(-(I_u - I_v)^2/\sigma^2\right)$ between neighboring voxels u and v, where I_u denotes the intensity of voxel u. Solving such 3D lattice SDDs, and general topologies, requires only $O(n log^{o(1)} n)$ [30] work, however we observe linear work empirically.

Eigencalculations: Calculating a minimal, say $k-$dimensional, eigenspace of an SDD matrix forms the computational core of the spectral relaxation for NCuts [6], spectral clustering [7], Laplacian eigenmaps [8], diffusion maps [10], and the typical case for Levin *et. al.*'s image matting algorithm [12]. Recall that eigensystems satisfy the following equations for a Laplacian $A z_i = \phi_i z_i$ and generalize to $A x_i = \lambda_i D x_i$ where we assume D is positive definite diagonal in general, and typically D is the weighted degree matrix (see §2.4).

For the generalized problem, efficient computation of the k eigenpairs (x_i, λ_i), such that $\lambda_1 \le \lambda_i \le \lambda_{k+1}$, depends upon the relationship to a normalized Laplacian, $\hat{A}$. $\hat{A}$ is possessed of the same eigenvalues $\{\lambda_i\}$ as the generalized problem, with eigenvectors y_i that map to generalized vectors under the operator $D^{1/2}$: $x_i = D^{1/2} y_i$.

We find the set of eigenvalues and vectors by inverse powering, *i.e.* repeated solves of the problem $\hat{A} q^{t+1} = q^t$, coupled with a Krylov space method such as the Lanczos algorithm. In [30] the number of inverse powers required was shown to be $O(\log n)$ to calculate a vector with a Rayleigh quotient arbitrarily close to that of the minimal eigenspace.

Thus the general case complexity of an eigencalculation remains $O(n \log^{o(1)} n)$. In essence by employing the CMG solver we achieve approximate NCuts solutions in time roughly proportional to sorting the vertices (pixels) by intensity. Timing results for estimating (x_2, λ_2) on three panoramic landscape image derived graphs is shown in column 4 of Table 1 (edge weightings as in above).

Convex Programming and Reweighted Problems: The optimization of convex functionals such as "$\ell_2 - \ell_1$ Total Variation"[18], given by: $f(x) = min_x : ||x - s||_2 + \lambda |\nabla x|_1$ for an input signal s, and related problems can be accomplished using Newton's method with log-barriers in $O(n^{1.5} \log^{o(1)}(n))$ time with modern solvers. For TV, the computational crux of each iteration is the solution of a SDD system with iteration dependent edge weights for the Laplacian modeling the ℓ_1 penalty on spatial gradients $|\nabla x|_1$. As the program iterates, the weights on edges between regions of different intensity approach zero; such weightings radically violate the conditions required by most

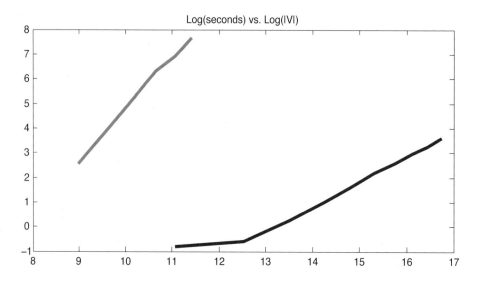

Fig. 3. Relative speeds of the CMG solver (black) and MATLAB's "\" operator (grey) on weighted graphs with 6 connected 3D lattices derived from a CT study. The X-axis is shown in log scale over $|V|$, the number of variables in the system. For reference, the execution time for the CMG solver on a problem with 27 million variables is $\approx$ 50 seconds.

Table 1. Two dimensional comparison with MATLAB for solutions on weighted 2D problems (with $O(|V|)$ super-lattice topology) at 2, 10, and 50 megapixels

Size in Mega-pixels	"\"	CMG	"\"+eigs	CMG+eigs	ℓ_2, ℓ_1 w "\"	ℓ_2, ℓ_1 w CMG
2M	45s	8s	1.6m	59s	5.1m	31s
10M	4.9m	22s	7.6m	2.7m	49.4m	3.7m
50M	NA	1.1m	NA	7.1m	NA	14.6m

multi-grid methods. CMG is suitable for such weightings, and further, the required clusterings (see §3) are linearly updated in the preconditioner, unlike direct methods that must recompute a factorization from scratch with each weighted instance (incurring a large computational cost). This property is also useful for other iterative methods such as robust least squares. Timing results on the panoramic image problems can be found in columns 5 and 6 of Table 1.

5   Discussion

Finally, by segregating the code for solvers (and eigen-calculations) from that of the applications we harvest improved modularity, reliability and factorizations of the system. Thus, 1) errors can be isolated to either the solver or the application, 2) as new solvers become available they can be easily adopted and perhaps most importantly 3) it relieves the application designer of the burden of implementing the state-of-the-art in solver technology.

We feel that significant improvements are still to be made in the solver and eigensolver technology. We envision major new applications for these solvers in scientific computing, image processing, data-mining, and machine learning. Finally, to ease adoption of **hybrid** solvers, an implementation of the CMG solver will be made available in the near future.

References

1. Horn, B.: Shape from shading: A method for obtaining the shape of a smooth opaque object from one view. Technical Report 232, MIT AI Laboratory (1970)
2. Horn, B.: Determining lightness from an image. Computer Graphics and Image Processing 3, 277–299 (1974)
3. Horn, B.: Determining optical flow. MIT AI Laboratory 17, 185–203 (1981)
4. Witkin, A.: Scale-space filtering. IJCAI, 1019–1022 (1983)
5. Perona, P., Malik, J.: Scale-space and edge detection using anisotropic diffusion. PAMI 7, 629–639 (1990)
6. Shi, J., Malik, J.: Normalized cuts and image segmentation. PAMI (2000)
7. Ng, A., Jordan, M., Weiss, Y.: On spectral clustering: Analysis and an algorithm. In: NIPS (2002)
8. Belkin, M., Niyogi, P.: Laplacian eigenmaps for dimensionality reduction and data representation. Neural Computation (15), 1373–1396
9. Yu, S., Shi, J.: Segmentation given partial grouping constraints. PAMI 26, 173–183 (2004)
10. Coifman, R.R., Lafon, S., Lee, A.B., Maggioni, M., Nadler, B., Warner, F., Zucker, S.W.: Diffusion maps geometric diffusions as a tool for harmonic analysis and structure definition of data. PNAS 102, 7426–7431 (2005)
11. Grady, L.: Random walks for image segmentation. IEEE Trans. on Pattern Analysis and Machine Intelligence 2, 1768–1783 (2006)
12. Levin, A., Rav-Acha, A., Lischinski, D.: Spectral matting. In: CVPR (2007)
13. Cour, T., Shi, J.: Solving markov random fields with spectral relaxation. AISTATS (2007)
14. Szeliski, R.: Locally adapted hierarchical basis preconditioning. SIGGRAPH 25, 1135–1143 (2006)
15. McCann, J., Pollard, N.S.: Real-time gradient-domain painting. SIGGRAPH 27 (2008)

16. Bhat, P., Curless, B., Cohen, M., Zitnick, C.L.: Fourier analysis of the 2d screened poisson equation for gradient domain problems. In: Forsyth, D., Torr, P., Zisserman, A. (eds.) ECCV 2008, Part II. LNCS, vol. 5303, pp. 114–128. Springer, Heidelberg (2008)
17. Grady, L., Alvino, C.: Reformulating and optimizing the mumford-shah functional on a graph - a faster, lower energy solution. In: Forsyth, D., Torr, P., Zisserman, A. (eds.) ECCV 2008, Part I. LNCS, vol. 5302, pp. 248–261. Springer, Heidelberg (2008)
18. Rudin, L., Osher, S., Fatemi, E.: Nonlinear total variation based noise removal algorithm. Physica D, 259–268 (1992)
19. Chan, T., Shen, J.: Image Processing And Analysis: Variational, Pde, Wavelet, And Stochastic Methods. SIAM, Philadelphia (2005)
20. Buades, A., Coll, B., Morel, J.: Nonlocal image and movie denoising. IJCV 76, 123–139 (2008)
21. Brox, T., Kleinschmidt, O., Cremers, D.: Efficient nonlocal means for denoising of textural patterns. Trans. on Image Processing
22. Gilboa, G., Osher, S.: Nonlocal operators with applications to image processing (2007)
23. Sharon, E., Brandt, A., Basri, R.: Segmentation and boundary detection using multiscale intensity measurements. In: CVPR (2001)
24. Grady, L., Sinop, A.K.: Fast approximate random walker segmentation using eigenvector precomputation. In: CVPR (2008)
25. Grady, L.: A lattice-preserving multigrid method for solving the inhomogeneous poisson equations used in image analysis. In: Forsyth, D., Torr, P., Zisserman, A. (eds.) ECCV 2008, Part II. LNCS, vol. 5303, pp. 252–264. Springer, Heidelberg (2008)
26. Ruge, J.W., Stüben, K.: Algebraic multigrid (AMG). In: McCormick, S.F. (ed.) Multigrid Methods. Frontiers in Applied Mathematics, vol. 3, pp. 73–130. SIAM, Philadelphia (1987)
27. Brandt, A.: General highly accurate algebraic coarsening (2000)
28. Vaidya, P.M.: Solving linear equations with symmetric diagonally dominant matrices by constructing good preconditioners. A talk based on this manuscript (1991)
29. Joshi, A.: Topics in Optimization and Sparse Linear Systems. PhD thesis, University of Illinois at Urbana Champaing (1997)
30. Spielman, D.A., Teng, S.H.: Nearly-linear time algorithms for preconditioning and solving symmetric, diagonally dominant linear systems (2006)
31. Koutis, I., Miller, G.L.: A linear work, $O(n^{1/6})$ time, parallel algorithm for solving planar Laplacians. In: SODA 2007 (2007)
32. Gremban, K.: Combinatorial Preconditioners for Sparse, Symmetric, Diagonally Dominant Linear Systems. PhD thesis, Carnegie Mellon University, Pittsburgh (1996); CMU CS Tech Report CMU-CS-96-123
33. Boman, E.G., Hendrickson, B.: Support theory for preconditioning. SIAM J. Matrix Anal. Appl. 25, 694–717 (2003)
34. Koutis, I.: Combinatorial and algebraic algorithms for optimal multilevel algorithms. PhD thesis, Carnegie Mellon University, Pittsburgh (2007); CMU CS Tech Report CMU-CS-07-131
35. Axelsson, O.: Iterative Solution Methods. Cambridge University Press, New York (1994)
36. Doyle, P.G., Snell, J.L.: Random walks and electric networks (2000)
37. Bern, M., Gilbert, J.R., Hendrickson, B., Nguyen, N., Toledo, S.: Support-graph preconditioners. SIAM J. Matrix Anal. Appl. 4, 930–951 (2006)
38. Koutis, I., Miller, G.L.: Graph partitioning into isolated, high conductance clusters: Theory, computation and applications to preconditioning. In: SPAA (2008)
39. Trottenberg, U., Schuller, A., Oosterlee, C.: Multigrid, 1st edn. Academic Press, London (2000)
40. Grady, L.: Multilabel random walker image segmentation using prior models. In: CVPR, San Diego, IEEE. CVPR, vol. 1, pp. 763–770. IEEE, Los Alamitos (2005)
41. Zhu, X., Ghahramani, Z., Lafferty, J.D.: Semi-supervised learning using gaussian fields and harmonic functions. In: ICML (2003)

Optimal Weights for Convex Functionals in Medical Image Segmentation

Chris McIntosh and Ghassan Hamarneh

Medical Image Analysis Lab
School of Computing Science, Simon Fraser University, Canada
{cmcintos,hamarneh}@cs.sfu.ca
http://mial.cs.sfu.ca

Abstract. Energy functional minimization is a popular technique for medical image segmentation. The segmentation must be initialized, weights for competing terms of an energy functional must be tuned, and the functional minimized. There is a substantial amount of guesswork involved. We reduce this guesswork by analytically determining the optimal weights and minimizing a convex energy functional independent of the initialization. We demonstrate improved results over state of the art on a set of 470 clinical examples.

1   Introduction

Image segmentation is a key task in visual computing. For medical image analysis, segmentation is important for quantifying the progression of diseases and quantifying anatomical variation. In such applications, a high degree of accuracy is sought, as the anatomical variation itself can be small and thus consumed by segmentation error. For this and other reasons, the automatic segmentation of medical images remains a daunting task. Many segmentation approaches rely on the minimization of objective functions, including several landmark papers: from the seminal paper of Snakes for 2D segmentation [1] and other explicit models [2] to implicit models [3,4], graph approaches [5,6], and variants thereof.

Objective function-based methods are commonly built using five essential building blocks: (i) an objective function whose minima provide good segmentations; (ii) an appropriate shape representation; (iii) a set of parameters including weights to balance the competing terms of the energy functional; (iv) an initialization; and (v) a method for minimization, whether it be local or global, continuous or combinatorial.

Each of these common blocks is known to have certain challenges. In particular, the parameter setting, initialization, and minimization phases are well known to be problematic. Often there are unanswered questions: what if a different initialization was used, what about a different minimizer, what if different weights were used between competing terms of the energy functional? Hence, erroneous segmentations cannot be directly attributed to the energy functional or one of the many unknowns of the segmentation process. The goal is to reduce and ultimately remove these points of uncertainty.

G. Bebis et al. (Eds.): ISVC 2009, Part I, LNCS 5875, pp. 1079–1088, 2009.

Uncertainty with initializations and minimizers can be addressed by formulating problems as convex energy functionals over convex domains, or submodular in the case of combinatorial approaches. Papers dealing with this issue are now common ground for both continuous [7,8] and discrete [5,6] optimization, but they all have uncertainty stemming from the free weights in their energy functionals.

Uncertainty with energy functional weights can be addressed by determining the optimal weights for each image to be segmented, else another set of weights may exist that provides better results. Recently, we developed an analytical expression describing the optimal functional weights [9]. Our method solves for the optimal functional weights for a training set of image-segmentation pairs, and then infers the optimal parameters for a novel image via geodesic interpolation over the training set. Our results demonstrated the importance of not only using the optimal weights for a functional, but how those weights vary from image to image. However, the method was not without its drawbacks.

We build upon and extend our earlier work by addressing two key issues. First, our previous work was done using non-convex functionals. There is uncertainty that a different initialization or optimization process may have yielded improved results. Instead, in this work, we focus on convex functionals, ensuring global optima and thus removing uncertainty related to local minima and initializations. Second, our analytical expression for optimal parameters included an implicit weighting between its two competing terms; a weighting which we seek to address here. We remove the implicit weighting using a convex quadratic formulation under a linear constraint, and thus remove the uncertainty implied by it. Our lastest results show significantly improved accuracy.

Though we are focusing on continuous functionals, a related field of approaches has come up in the study of combinatorial problems. The first set is based on recent advances in maximum margin estimation, wherein the parameters of the objective function are sought such that the highest scoring structures (in our case segmentations) are as close as possible to the ground truth [10,11,12,13]. However, in addition to being limited to combinatorial objective functions rather than continuous ones, these methods propose a fixed set of parameters for novel samples (in our case images), whereas we follow the direction of [9] using geodesic interpolation to infer the optimal parameters on a per-image basis. In other words, these works assume that a single set of parameters works for an entire test set. As shown in [9], this is often not the case in image segmentation, and greatly improved results can be obtained by adapting the parameters to the individual images (as we do). The second related direction, was introduced recently in [14]. Though this work is also restricted to combinatorial objective functions, an optimal parameter is indeed sought on a per-image basis. Given a parameter range, the method simultaneously solves the objective function for a set of parameters that bound how the parameters influence the solution. Each solution is then treated as a potential segmentation. They propose a number of heuristics, including user intervention, to select the best segmentation from a set of potential ones.

In what follows we describe the theory behind our method (Sec. 2), how we build convex functionals (Sec. refsec:energy), and how to analyticaly describe the optimal weights of a functional (Sec. 2.2). We then detail how to apply our method to novel images (Sec. refsec:methods) and validate our improvements over [9] (Sec. 4). Finally, we discuss our results and future work (Sec. 4).

2 Theory: Notations and Uncertainty in Segmentation

In order to more formally explain where the uncertainty lies, and how it needs to be addressed, we first give a more detailed view on the energy minimization based segmentation process. We define a gray-level image I, and its corresponding segmentation S. Then $\mathbf{I} = \{I_1, I_2, ..., I_\mathbb{N}\}$ and $\mathbf{S} = \{S_1, S_2, ..., S_\mathbb{N}\}$ are training sets of images and their corresponding, correct segmentations.

The first step is the identification of the form of the energy functional. It may be convex or non-convex, as can the shape space over which it is minimized. A common general form is $E(S, I, \mathbf{w}) = w_1 \times internal(S) + w_2 \times external(S, I)$. Notice the free parameter $\mathbf{w} = [w_1, w_2]$. Depending on its value, minima of E favor the internal energy, or the external energy.

The segmentation problem is to solve $S^* = \arg\min_S E(S|I, \mathbf{w})$, which involves choosing a $\mathbf{w}$ and, depending on the nature of the energy functional, may also require training appearance and/or shape priors, and setting an initialization. A gradient descent-based solver is typically used but combinatorial approaches have also been explored for discretized versions of the problem [5]. Here we focus on continuous problems, and thus assume a gradient descent solver.

When using gradient descent, non-convexity can be quite problematic. There is no guarantee that another solution does not exist which better minimizes the energy, and thus is potentially a better segmentation. Ideally both functional and shape space are convex; guaranteeing globally optimal solutions.

Simply obtaining a global optima does not, however, guarantee a correct segmentation in the general case. If not appropriately set, the weights $\mathbf{w}$ can cause significant error. Optimizing the weights has been shown to have dramatic effects; reducing error in large data sets by as much as 30% [9]. However, optimizing the weights by hand for even a single image can be a long and tedious task, with no real guarantee of obtaining the correct segmentation.

Instead of *guessing* the optimal weights, suppose we write a function $\gamma(\mathbf{w}|I_j, S_j)$ evaluating how well weight $\mathbf{w}$ works for a given image-segmentation pair (I_j, S_j); such that a parameter is deemed better when it causes S^* to approach S_j, i.e. the minimum of E to be the correct segmentation. Given S_j, we could then calculate the ideal weights for a particular image I_j by solving $\mathbf{w}^* = \arg\min_\mathbf{w} \gamma(\mathbf{w}|I_j, S_j)$. It is important that γ itself be convex or globally solvable in $\mathbf{w}$. If γ was not globally solvable, uncertainty would remain in that another $\mathbf{w}^*$ may better minimize γ, and thus better segment the image. Similarly, γ can not contain free parameters, else those parameters would themselves introduce uncertainty; as was the case in [9].

2.1   Convex Energy Functionals

We make use of recent research into convex functionals for image segmentation, specifically that of Cremers et al. where a convex energy functional E is minimized over a convex shape space represented as probability maps, i.e. $S(x) \in [0,1]$ for all points x in the image domain Ω, to yield a convex segmentation problem [15]. A shape model is then constructed via principal component analysis (PCA) on a set of training shapes forming a k-dimensional approximation to the shape space, with $\alpha_1, ..., \alpha_k$ eigen coefficients, a mean shape $\bar{S}$, and eigenvectors $\psi_1, ..., \psi_k$. Shapes can now be reconstructed as $S = \bar{S} + \sum_{i=1}^{k} \alpha_i \psi_i$. Writing S in terms of the vector of shape parameters $\alpha = \{\alpha_i\}_{i=1}^k$, convex E can be written as a sum of convex energy terms:

$$E(\alpha | I = I_j, \mathbf{w} = \widehat{\mathbf{w}}) = \int_\Omega \widehat{w}_1 J_1(\alpha | I_j) + ... + \widehat{w}_n J_n(\alpha | I_j) d\mathbf{x} \qquad (1)$$

for a fixed image I_j and arbitrary, fixed weights $\widehat{\mathbf{w}}$, where J_i is a convex energy term and $\mathbf{w} = [w_1, ..., w_n]$ with $w_i \in [0,1]$ are weights. Consequently, E is a convex functional since the positively weighted sum of a set of convex terms is itself convex. For proofs of convexity and more details see [15].

Minimizing E optimally can then be performed via gradient descent on α using derivative: $E_\alpha(\alpha | I = I_j, \mathbf{w} = \widehat{\mathbf{w}}) = \widehat{w}_1 T_1(\alpha | I_j) + ... + \widehat{w}_n T_n(\alpha | I_j)$ where E_α denotes the derivative of E with respect to α, and T_i is the derivative of J_ith term. However, since $\widehat{\mathbf{w}}$ is arbitrary nothing can be said about its optimality for the particular image I_j.

2.2   Optimal Energy Functional Weights

For each (I_j, S_j), $I_j \in \mathbf{I}$ and $S_j \in \mathbf{S}$, the task is to find the optimal values for the free weights $\mathbf{w}(I_j)$. This section explores the notion of 'optimal'.

One computationally intractable approach for finding $\mathbf{w}^*$ is to try all possible weight combinations and run the segmentation method then select the weights with the least segmentation error. A better approach, as outlined in [9], is to find the weights $\mathbf{w}^*$ that minimize the magnitude of the derivative, in our case E_α, of the energy functional at the correct segmentation α^j (i.e. $\alpha^j = [\psi_1 \psi_2 ... \psi_k]^+ (S_j - \bar{S}))^1$. Doing so encourages α^j to be a minimum of E (i.e. $E_\alpha(\alpha^j | I_j, \mathbf{w}^*) = 0$). Since E_α is in our case a vector of length k and $w_i(I_j)$ a scalar function, we measure its magnitude as $\left| E_\alpha(\mathbf{w} | \alpha^j, I_j) \right|^2$. McIntosh and Hamarneh go further to minimize $\left| E_\alpha(\mathbf{w} | \alpha, I_j) \right|^2$ for $\alpha = \alpha^j$ while maximizing it for all other possible shapes (in a direction toward the optimal solution). Adopting their approach, for the time being, but with the new convex setup, we proceed as follows.

For a given shape α^i, a vector $(\alpha^i - \alpha^j)$ in $\mathbb{R}^k$ represents the direction towards α^j. Since $E_\alpha(\mathbf{w} | \alpha^i, I_j)$ is the vector in $\mathbb{R}^k$ dictating in what direction, and

---

[1] We assume the chosen eigenvectors explain 99% of the variance and thus the error incured by representing S as α is negligible.

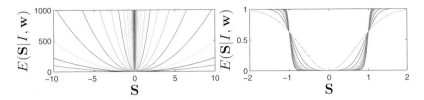

Fig. 1. Varying the shape of energy functionals. Left: Various functionals that show an increasing extent to which the gradient at neighboring shapes points towards the correct segmentation, represented by $S = 0$. Right: Various functionals that show how a set of neighboring shapes can also become minima, a degenerate case.

by what amount, the solution will change at the point $\boldsymbol{\alpha}^i$, a normalized dot-product (projection-like approach) will measure how much in the right direction $E_{\boldsymbol{\alpha}}(\mathbf{w}|\boldsymbol{\alpha}^i, I_j)$ points.

So for an energy functional with a form like those in (1), and following [9], for now, we define $\gamma(\mathbf{w}|I_j, S_j)$ as

$$\gamma(\mathbf{w}) = \left(\left| E_{\boldsymbol{\alpha}}(\mathbf{w}|\boldsymbol{\alpha}^j, I_j) \right|^2 - \lambda \sum_{i \in \mathcal{N}_{\mathbf{S}}} F_{\mathcal{N}_{\mathbf{S}}}(\boldsymbol{\alpha}^j, \boldsymbol{\alpha}^i) \frac{E_{\boldsymbol{\alpha}}(\mathbf{w}|\boldsymbol{\alpha}^i, I_j) \cdot (\boldsymbol{\alpha}^i - \boldsymbol{\alpha}^j)}{|\boldsymbol{\alpha}^i - \boldsymbol{\alpha}^j|} \right) \quad (2)$$

where $\mathcal{N}_{\mathbf{S}}$ denotes a set of nearby (or similar) shapes in the domain of E, and $F_{\mathcal{N}_{\mathbf{S}}}$ is used to weight closer segmentations according to their proximity. The neighborhood $\mathcal{N}_{\mathbf{S}}$ is used instead of the entire shape space to reduce computational complexity. The second term, dubbed the neighborhood term, is negative and $|E_{\boldsymbol{\alpha}}(\mathbf{w}|\boldsymbol{\alpha}^i, I_j)|$ is omitted from the normalized dot-product to reward large steps in the correct direction. Solving $\mathbf{w}^*(I_j) = \arg\min_{\mathbf{w}} \gamma(\mathbf{w}|I_j, S_j)$, yields the optimal weights for image I_j.

However, notice that there is a weighting λ between competing terms of (2), which was implicit in [9], i.e. was assumed equal to unity and not addressed. With two competing terms, a balance must be struck between: (i) the degree by which $\boldsymbol{\alpha}^j$ is a minimum of E; and (ii) the degree by which the derivative at neighboring points in the shape space points towards $\boldsymbol{\alpha}^j$ (Fig. 1-left). Make λ too small and $\boldsymbol{\alpha}^j$ might be a minimum, but so might the entire neighborhood (Fig. 1-right). Make λ too large and the neighborhood will point in the right direction, but $\boldsymbol{\alpha}^j$ might no longer be a minimum.

To rectify this problem, we make the following observation: when the energy functional E is convex our only concern is making $\boldsymbol{\alpha}^j$ as much a minimum as possible, while avoiding the degenerate case that the neighboring points are minima (Fig. 1-right). As a result we can replace the neighborhood term by a *constraint rather than a cost term* since the degree to which the neighbors point towards $\boldsymbol{\alpha}^j$ does not change whether or not $E_{\boldsymbol{\alpha}}(\mathbf{w}|\boldsymbol{\alpha}^j, I_j) = 0$ (i.e. we must avoid forcing the gradient in the neighborhood to point at $\boldsymbol{\alpha}^j$ at the cost of making $E_{\boldsymbol{\alpha}}(\mathbf{w}|\boldsymbol{\alpha}^j, I_j) \neq 0$). Thus instead of (2) we re-define $\mathbf{w}^*(I_j)$ as

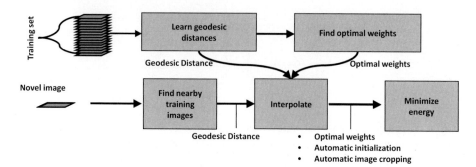

Fig. 2. Overview of our proposed method

$$
\mathbf{w}^* = \arg\min_{\mathbf{w}} \gamma(\mathbf{w}) = \arg\min_{\mathbf{w}} \left| E_{\boldsymbol{\alpha}}(\mathbf{w}|\boldsymbol{\alpha}^j, I_j) \right|^2
$$

$$
\text{s.t.} \sum_{i \in \mathcal{N}_{\mathbf{S}}} F_{\mathcal{N}_{\mathbf{S}}}(\boldsymbol{\alpha}^j, \boldsymbol{\alpha}^i) E_{\boldsymbol{\alpha}}(\mathbf{w}|\boldsymbol{\alpha}^i, I_j) \cdot \frac{(\boldsymbol{\alpha}^j - \boldsymbol{\alpha}^i)}{|\boldsymbol{\alpha}^j - \boldsymbol{\alpha}^i|} \geq 0 \tag{3}
$$

The result is a convex function in $\mathbf{w}$ under a linear constraint, since the first term in (2) has been shown to be a convex quadratic [9] and the second term linear. A convex function under a linear constraint can be solved via convex optimization, and thus the optimal $\mathbf{w}(I_j)$ is guaranteed.

3   Method: Segmenting Novel Images

Given our set of training image-segmentation pairs we will have $\mathbb{N}$ samples of $\mathbf{w}^*(I)$, from which we can interpolate to find values at new points (i.e. novel images). In order to interpolate, we need a metric for measuring distances between images[2]. The set of images with the shortest distances constitutes the neighboring images, $\mathcal{N}_{\mathbf{I}}$, and $\mathcal{N}_{\mathbf{S}}$ are their corresponding correct segmentations.

We assume $\mathbf{I}$ is smooth over its domain, the space of a particular class of images (e.g. MRI brain scans of normal adults), and that the mapping from images to segmentations is smooth. In other words, we assume that similar images have similar parameters, and similar segmentations. As such, we use a normalized Gaussian kernel, defined over the image distances, to interpolate both the parameters and initializations. For shape and appearance priors, we limit the training data to $\mathcal{N}_{\mathbf{I}}$ and $\mathcal{N}_{\mathbf{S}}$, since we are more confident that the correct shape and appearance information is similar to the training data lying in those neighborhoods. The process is summarized in figure 2.

Manifold learning methods are a special class of nonlinear dimensionality reduction techniques that enable the calculation of geodesic distances between data points. We make use of these techniques to calculate distances between both images and segmentations. Distances between neighboring segmentations allow us to define

---

[2] The choice of metric is beyond the scope of this paper (Sec. 5).

$$F_{\mathcal{N}_\mathbf{S}}(S_j, S_i) = 1 - \frac{g(S_i, S_j)}{\sum\limits_{t \in \mathcal{N}_\mathbf{S}} g(S_t, S_j)} \qquad (4)$$

where $g(S_i, S_j)$ is the learned geodesic distance between shape S_i and shape S_j. The function $F_{\mathcal{N}_\mathbf{S}}(S_j, S_i)$ then acts as a weighting for the given neighborhood $\mathcal{N}_\mathbf{S}$, normalized to sum to one (i.e. the weight decreases as a function of distance from the center of the neighborhood). Here we use the geodesic distance between shapes, as opposed to their linear distance in the PCA subspace, for greater accuracy. The linear shape space is well suited for our shape representation because it forms the basis for a convex optimization problem (as previously noted). In essence, we assume an underlying non-linear shape space exists, but use a higher dimensional linear space to represent it. This of course, allows non-valid shapes to be represented, but it brings the benefit a convex energy functional, and with a good energy functional those non-valid shapes will not be minima anyway (as our results show).

Though numerous points of uncertainty have been addressed by our method, there are a few remaining free parameters: (i) k, the dimensionality of the PCA shape space used for our shape representation; (ii) the dimensionality of the shape manifold; (iii) the dimensionality of the image manifold; (iv) the input image-distance metric used as input to the manifold learning algorithm; (v) the manifold learning algorithm to be used; and finally (vi) the interpolation function used to determine parameters for novel images, as a function of their distance to similar images. As these choices are somewhat application dependent, we include a specification of their values in the experiment section. We also include a discussion of the implications of these parameters in section 5.

4   Experiments

We validate our method on a set of 470 256×256 affine registered mid-sagittal MR images, with corresponding expert-segmented corpora callosa (CC). Our energy functional takes the form:

$$\begin{aligned} E(\alpha) = \int_\Omega &\left(w_1(I) f(x) S(x) + w_2(I) g(x)(1 - S(x)) \right. \\ &\left. + w_3(I) h(x) \left| \nabla S(x) \right| \right) dx + w_4(I) \alpha^T \Sigma^{-1} \alpha \end{aligned} \qquad (5)$$

where $f = -log(P_{obj}(I))$, $g = -log(P_{bk}(I))$, for object and background histograms P_{obj}, P_{bk}, $h = \frac{1}{1+|\nabla I|}$, and Σ^{-1} characterizes the allowable shape distribution (see [15] for details).

To learn the distances, we used a MATLAB implementation of K-ISOMAP [16] from http://isomap.stanford.edu/, with Euclidean distance between images as the input distance matrix. As this paper is about the application of optimal parameters to segmentation, issues related to learning the manifold will not be addressed in this work. For K-ISOMAP we set $K = 10$, and reduce the image space to a 5-manifold; chosen as the elbow of the scree plot. One caveat with using ISOMAP is that it does not directly extend to novel samples. Though

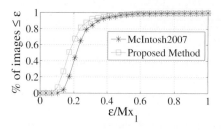
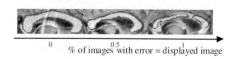

Fig. 3. CC segmentation results. (Left) Error plot where Mx_1 is the maximum measured value for ε. (Right) Segmentations demonstrating the full range of error.

out-of-sample extensions for ISOMAP have been published, for simplicity we choose to simply re-run ISOMAP to include the novel image, as this only takes a few seconds. For PCA on the shape space we set $k = 10$; the elbow of the corresponding scree plot.

Firstly, to show that eq. (3) can balance an ideal energy functional (one for which the minimum can be positioned exactly at the correct segmentation using the optimal weights) we calculate the error of energy functional (5) plus a fifth term whose unique global minimum is always the correct segmentation for the given image (i.e. we gave the functional weights **w** the power to achieve 100% accuracy by adding a rigged, strictly convex term). We measure the error using a modified Dice metric: $\varepsilon = Area(A \cup G - A \cap G)/Area(G)$, where A and G are the binary automatic segmentation and the ground truth, respectively. Using this rigged term, we obtained $\varepsilon = 0$ for all 470 images, validating that our method can achieve the full potential of a given functional.

To compare to the original weight optimization equation (eq. (2)), as originally presented in [9], we performed validation on the set of 470 images using energy functional (5). For each image, the optimal parameters are learned directly, using the ground truth segmentation, rather than interpolation from the manifold. Doing so isolates the error induced by the weights, not the interpolation method or the ability to locate the position of novel images on the manifold. Using eq. (2) we obtained an error of 0.1201 vs 0.1099 with eq. (3), a clear improvement.

To validate our segmentation method as it pertains to novel data, we perform leave-one-out validation on the set of 470 images using energy functional (5). Under error metric ε, we found the average error to be 0.13, improving over the average error of 0.16 reported in [9]. Our results are summarized in Fig. 3. The figure shows the percentage of images with $\varepsilon \leq$ the specified value on the x-axis. So, for example, with our proposed method approximately 65% of the images have $\varepsilon \leq 0.2$ as opposed to only 44% using the method proposed in [9]. If an error of 0.2 was the cut off for the segmentation method to be clincially useful, our method would have succesfully segmented an additional 21% of the data, or about 100 images more than [9].

5   Discussion

Our results demonstrate our method's ability to optimize the weights for convex energy functionals. In doing so, we have addressed a few key areas of uncertainty typically found in objective function based segmentation methods. Specifically, our method does not suffer from uncertainty with local minima, initializations, or hand-tuned parameters. However, new questions remain: Was the optimal manifold learned? Is this the best way to describe the optimal weights? And is this the best interpolation function for novel images? As already shown in [9], even with the inherit manifold uncertainty, this technique of analytically describing the optimal weights is better than the alternative (hand-tuning the weights, and/or fixing the weights as constant values over the set of application images).

The difference $\delta\varepsilon = 0.02$ between what our method achieved in practice ($\varepsilon = 0.13$) and the error using weights calculated directly from the ground truth segmentations, ($\varepsilon = 0.1099$), is due to our localization of novel images on the manifold and the interpolation over $\mathbf{w}^*(I)$. An important area of future work is thus to lower this difference by using better manifold learning techniques. Improved image distance metrics [17] may also work, as they can simplify the learning problem.

Finally, we have provided two improvements over [9]: we removed the uncertainty with local minima and initializations; and we improved their weight optimization equation by removing its implicit weight. We demonstrated how our proposed weight optimization equation yields improved weights, and that our method has a significantly lower error overall. In the end, we were able to segment an additional 100 images under a reasonable cutoff, which is an important improvement in a clinical setting.

References

1. Kass, M., Witkin, A., Terzopoulos, D.: Snakes: Active contour models. IJCV 1(4), 321–331 (1987)
2. Lobregt, S., Viergever, M.: A discrete dynamic contour model. IEEE TMI 14, 12–24 (1995)
3. Caselles, V., Kimmel, R., Sapiro, G.: Geodesic active contours. IJCV 22(1), 61–79 (1997)
4. Chan, T., Vese, L.: Active contours without edges. IEEE TIP, 266–277 (2001)
5. Boykov, Y., Kolmogorov, V.: Computing geodesics and minimal surfaces via graph cuts. In: ICCV (2003)
6. Grady, L.: Random walks for image segmentation. IEEE TPAMI 28(11), 1768–1783 (2006)
7. Nikolova, M., Esedoglu, S., Chan, T.F.: Algorithms for finding global minimizers of image segmentation and denoising models. SIAM Journal on Applied Mathematics 66(5), 1632–1648 (2006)
8. Bresson, X., Esedoglu, S., Vandergheynst, P., Thiran, J.P., Osher, S.: Fast global minimization of the active contour/snake model. Journal of Mathematical Imaging and Vision 28(2), 151–167 (2007)

9. McIntosh, C., Hamarneh, G.: Is a single energy functional sufficient? adaptive energy functionals and automatic initialization. In: Ayache, N., Ourselin, S., Maeder, A. (eds.) MICCAI 2007, Part II. LNCS, vol. 4792, pp. 503–510. Springer, Heidelberg (2007)

10. Taskar, B., Chatalbashev, V., Koller, D., Guestrin, C.: Learning structured prediction models: a large margin approach. In: ICML 2005: Proceedings of the 22nd international conference on Machine learning, pp. 896–903. ACM, New York (2005)

11. Anguelov, D., Taskar, B., Chatalbashev, V., Koller, D., Gupta, D., Heitz, G., Ng, A.: Discriminative learning of markov random fields for segmentation of 3d scan data. In: CVPR 2005: Proceedings of the 2005 IEEE Computer Society Conference on Computer Vision and Pattern Recognition (CVPR 2005), Washington, DC, USA, vol. 2, pp. 169–176. IEEE Computer Society, Los Alamitos (2005)

12. Szummer, M., Kohli, P., Hoiem, D.: Learning crfs using graph cuts. In: Forsyth, D., Torr, P., Zisserman, A. (eds.) ECCV 2008, Part II. LNCS, vol. 5303, pp. 582–595. Springer, Heidelberg (2008)

13. Finley, T., Joachims, T.: Training structural svms when exact inference is intractable. In: ICML 2008: Proceedings of the 25th international conference on Machine learning, pp. 304–311. ACM, New York (2008)

14. Kolmogorov, V., Boykov, Y., Rother, C.: Applications of parametric maxflow in computer vision. In: IEEE 11th International Conference on Computer Vision, 2007. ICCV 2007, October 2007, pp. 1–8 (2007)

15. Cremers, D., Schmidt, F.R., Barthel, F.: Shape priors in variational image segmentation: Convexity, lipschitz continuity and globally optimal solutions. In: IEEE CVPR (2008)

16. Tenenbaum, J.B., de Silva, V., Langford, J.C.: A global geometric framework for nonlinear dimensionality reduction. Science 290, 2319–2323 (2000)

17. Souvenir, R., Pless, R.: Image distance functions for manifold learning. Image and Vision Computing 25(3), 365–373 (2007); Articulated and Non-rigid motion

Adaptive Contextual Energy Parameterization for Automated Image Segmentation

Josna Rao[1], Ghassan Hamarneh[2], and Rafeef Abugharbieh[1]

[1] Biomedical Signal and Image Computing Lab,
University of British Columbia, Canada
[2] Medical Image Analysis Lab, Simon Fraser University, Canada
josnar@ece.ubc.ca, hamarneh@cs.sfu.ca, rafeef@ece.ubc.ca

Abstract. Image segmentation techniques are predominately based on parameter-laden optimization processes. The segmentation objective function traditionally involves parameters (i.e. weights) that need to be tuned in order to balance the underlying competing cost terms of image data fidelity and contour regularization. In this paper, we propose a novel approach for automatic adaptive energy parameterization. In particular, our contributions are three-fold; 1) We spatially adapt fidelity and regularization weights to local image content in an autonomous manner. 2) We modulate the weight using a novel contextual measure of image quality based on the concept of spectral flatness. 3) We incorporate our proposed parameterization into a general segmentation framework and demonstrate its superiority to two alternative approaches: the best possible spatially-fixed parameterization and the globally optimal spatially-varying, but non-contextual, parameters. Our segmentation results are evaluated on real and synthetic data and produce a reduction in mean segmentation error when compared to alternative approaches.

Keywords: Adaptive regularization, contextual weights, image segmentation, energy minimization, adapting energy functional, spectral flatness, noise estimation.

1 Introduction

Robust automated image segmentation is a highly sought after goal that continues to defy solution. In medical images, for example, natural and pathological variability as well as noise often result in unpredictable image and shape features that significantly complicate segmentation tasks. Furthermore, spatially nonuniform noise can result from numerous reconstruction and postprocessing techniques on MR images to correct for intensity inhomogeneity effects and from images obtained with decreased acquisition times and high speedup factors [1]. Current state-of-the-art segmentation methods are predominantly based on optimization procedures that produce so called 'optimal' segmentations at their minimum. The optimization methods typically incorporate a tradeoff between two classes of cost terms: data fidelity and contour regularization. This is the case not only in image segmentation, but also in image registration, shape matching,

G. Bebis et al. (Eds.): ISVC 2009, Part I, LNCS 5875, pp. 1089–1100, 2009.

and other computer vision tasks. This basic tradeoff scheme is ubiquitous, relating to Occam's razor and Akaike/Bayesian information criteria [2], and is seen in many forms, such as likelihood versus prior in Bayesian methods [3] and loss versus penalty in machine learning [4]. Making progress toward determining how to best control such balancing between competing cost terms within the optimization process is therefore of great importance to many related algorithmic formulations in computer vision as well as numerous applications most notably in medical image analysis. Examples of optimization-based segmentation methods that are fragile and highly sensitive to the aforementioned tradeoffs are plentiful, including active contours techniques [5][6][7][8], graph cut methods [9], and optimal path approaches [10].

For simplifying the exposition of the ideas in this paper, we will adopt the simplified but general form of the cost or energy function:

$$E(S|I, \alpha, \beta) = \alpha E_{int}(S) + \beta E_{ext}(S|I) \qquad (1)$$

where S is the segmentation and I is the image. E_{int} is the internal cost term contributing to the regularization of the segmentation, most often by enforcing some smoothness constraints, in order to counteract the effects of imaging artifacts. E_{ext} is the external cost term contributing to the contour's conformity to desired image features, e.g. edges. The weights α and β are typically set empirically by the users based on their judgment of how to best balance the requirements for regularization and adherence to image content. In most cases, this is a very difficult task and the parameters may be unintuitive for a typical non-technical end user, e.g. a clinician, who lacks knowledge of the underlying algorithm's inner working. Also the resultant segmentations can vary drastically based on how this balance is set. Avoiding the practice of ad-hoc setting of such weights is the driving motivation for our work here.

To the best of our knowledge, regularization weights have traditionally been determined empirically and are fixed across the image domain (i.e. do not vary spatially). In Pluempitiwiriyawej *et al.* [8], the weights are changed *as the optimization progresses*, albeit in an ad-hoc predetermined manner. McIntosh and Hamarneh [11] demonstrated that adapting the regularization weights *across a set of images* is necessary in addressing the variability in real clinical image data. However, neither approach varies the weights spatially across the image and hence are not responsive or adaptive to local features within a single image.

Image regions with noise, weak or missing boundaries, and/or occlusions are commonly encountered in real image data. For example, degradation in medical images can occur due to tissue heterogeneity ("graded decomposition" [12]), patient motion, or imaging artifacts, e.g. echo dropouts in ultrasound or non-uniformity in magnetic resonance. In such cases and in order to increase segmentation robustness and accuracy, *more regularization is needed in less reliable image regions* which suffer from greater deterioration. Although an optimal regularization weight can be found for a single image in a set [11], the same weight may not be optimal for all regions of that image. Spatially adapting the regularization weights provides greater control over the segmentation result, allowing it to adapt not only to images with spatially varying noise levels and edge strength, but also to objects with spatially-varying shape characteristics, e.g. smooth in some parts and jagged in others.

Some form of spatially adaptive regularization over a single image appeared in a recent work by Dong *et al.* [13]. For segmenting an aneurysm, they varied the amount of regularization based on the surface curvature of a *pre-segmented* vessel. The results demonstrated improvements due to adaptive regularization. However, the regularization weights did not rely on the properties of the image itself, which limited the generality of the method. Kokkinos *et al.* [14] investigated the use of adaptive weights for the task of separating edge areas from textured regions using a probabilistic framework, where the posterior probabilities of edge, texture, and smoothness cues were used as weights for curve evolution. Similarly, Malik *et al.* [15] and, very recently, Erdem and Tari [16] tackled the problem of texture separation and selected weights based on data cues. However, while these methods focused on curve evolution frameworks, our current work focuses on graph-based segmentation. Additionally, we emphasize balancing the cost terms by adapting the regularization for images plagued by nonuniform noise and weak or diffused edges rather than textured patterns in natural images.

In this paper, we advocate the strong need for spatially-adaptive balancing of cost terms in an automated, robust, data-driven manner to relax the requirement on the user to painstakingly tweak these parameters. We also demonstrate how existing fixed-weight approaches (even if globally optimized) are often inadequate for achieving accurate segmentation. To address the problem, we propose a novel data-driven method for spatial adaptation of optimization weights. We develop a new spectral flatness measure of local image noise to balance the energy cost terms at every pixel, without any prior knowledge or fine-tuning.

We validate our method on synthetic, medical, and natural images and compare its performance against two alternative approaches for regularization: using the best possible spatially-fixed weight, and using the globally optimal set of spatially-varying weights as found automatically through dynamic programming.

2   Methods

2.1   Energy-Minimizing Segmentation

Our formulation employs energy-minimizing boundary-based segmentation, where the objective is to find a contour that correctly separates an object from background. We embed a parametric contour $C(q) = C(x(q), y(q)) : [0, 1] \to \Omega \subset \mathbf{R}^2$ in image $I : \Omega \to \mathbf{R}$. We use a single adaptive weight $w(q) \in [0, 1]$ that varies over the length of the contour and re-write (1) as:

$$E(C(q), w(q)) = \int_0^1 (w(q)E_{int}(C(q)) + (1 - w(q))E_{ext}(C(q))) \, dq \qquad (2)$$

where

$$E_{ext}(C(q)) = 1 - |\nabla I(C(q))| \, / \max_{\Omega} |\nabla I(C(q))| \qquad (3)$$

penalizes weak boundaries and

$$E_{int}(C(q)) = |dC(q)/dq| \tag{4}$$

penalizes longer and more jagged contours. Sec. 2.4 discusses the discrete formulation to minimize E with respect to $C(q)$ in (2).

2.2   Spatially Adaptive Energy Parameterization

Our approach for balancing the cost terms is to gauge the levels of signal vs. noise in local image regions. We estimate the edge evidence $G(x,y)$ and noise level $N(x,y)$ in each region of the image and set $w(x,y)$ in (2) such that regions with high noise and low boundary evidence (i.e. low reliability) have greater regularization, and vice versa. Hence, image reliability $R(x,y) \in [0,1]$ is mapped to $w(x,y)$ as

$$w(x,y) = 1 - R(x,y) \tag{5}$$

where

$$R(x,y) = (1 - N(x,y))\, G(x,y). \tag{6}$$

Assuming additive white noise, uncorrelated between pixels, we propose to estimate spatially-varying noise levels $N(x,y)$ using a local image spectral flatness (SF) measure. SF is a well-known Fourier-domain measure that has been employed in audio signal processing and compression applications [17][18]. SF exploits the property that white noise exhibits similar power levels in all spectral bands and thus results in a flat power spectrum, whereas uncorrupted signals have power concentrated in certain spectral bands and thus result in a more impulse-like power spectrum. In this paper, we extend the SF measure to 2D and measure $N(x,y)$ as

$$N(x,y) = \frac{\exp\left(\frac{1}{4\pi^2}\int_{-\pi}^{\pi}\int_{-\pi}^{\pi} \ln S\left(\omega_x,\omega_y\right) d\omega_x d\omega_y\right)}{\frac{1}{4\pi^2}\int_{-\pi}^{\pi}\int_{-\pi}^{\pi} S\left(\omega_x,\omega_y\right) d\omega_x d\omega_y} \tag{7}$$

where $S(\omega_x,\omega_y) = |F(\omega_x,\omega_y)|^2$ is the 2D power spectrum of the image. $F(\omega_x,\omega_y)$ is the Fourier spectrum of the image and (ω_x,ω_y) are spatial radian frequencies. Note that (7) can be easily extended to 3D images via a triple integral. We use $G(x,y) = \max\left(|\nabla I_x(x,y)|, |\nabla I_y(x,y)|\right)$, where $\nabla I_x(x,y)$ and $\nabla I_y(x,y)$ represent the x and y components of the image gradient. We chose this measure rather than the standard gradient magnitude for its rotational invariance in the discrete domain.

2.3   Non-contextual Globally Optimal Weights

A theoretically appealing and intuitive approach for setting the regularization weight is to optimize E in (2) for the *weight* $w(q)$ *itself* in addition to optimizing the contour. In our discrete setting, this involves a 'three dimensional' graph search that computes the globally optimal, spatially-adaptive regularization weight $w(q)$, in conjunction with the contour's spatial coordinates, i.e. we optimize[1] $\tilde{C}(q) = (x(q), y(q), w(q))$. Sec. 2.4 discusses the method for implementing the graph search.

As we demonstrate later in Sec. 3, there are three main drawbacks to this globally optimum (in (x, y, w)) method: (i) it does not explicitly encode image reliability, even though regularization is essential in regions with low reliability; (ii) it encourages a bimodal behavior of the regularization weight (this is easy to observe as the weight will be allocated to whichever cost is smaller):

$$w(q) = \begin{cases} 0, & E_{int}(q) > E_{ext}(q) \\ 1, & \text{otherwise} \end{cases}, \tag{8}$$

and (iii) it combines the weight and segmentation optimization into one process, thus reducing the generality of the method as finding globally-optimal weights for other segmentation frameworks would require significant changes to the energy minimization process. In short, even though *optimal* with respect to E in (2), the solution is *incorrect* and, as we later demonstrate, inferior to the spatially adaptive balancing of energy cost terms proposed in Sec. 2.2.

2.4   Implementation Details

To minimize E with respect to $C(q)$ in (2), we model the image as a graph where each pixel is represented by a vertex v_i in the graph, and graph edges $e_{ij} = \langle v_i, v_j \rangle$ capture the pixel's connectedness (e.g. 8-connectedness in 2D images). A local cost $c_{ij} = w E_{int}(v_i, v_j) + (1 - w) E_{ext}(v_i)$ is assigned to each edge e_{ij}, where $E_{int}(v_i, v_j)$ is the Euclidean distances between v_i and v_j (e.g. 1 for 4-connected neighbors and $\sqrt{2}$ for diagonal neighbors in 2D). The contour that minimizes the total energy $E = \sum_{e_{ij} \in C} c_{ij}$ represents the optimal solution for the segmentation and is found by solving a minimal path problem, e.g. the global snake minimization method between two end points presented in [21] which does not require any initialization other than specifying the end points.

The non-contextual globally optimal weight (Sec. 2.3) is determined through dynamic programming. In our formulation, each vertex in the original graph is now replaced by K vertices representing the different choices of the weight value at each pixel. In addition, graph edges now connect vertices corresponding to neighboring image pixels for all possible weights. Note that the optimal path $\tilde{C}(q) = (x(q), y(q), w(q))$ cannot pass through the same $(x(q), y(q))$ for different w, i.e. only a single weight can be assigned per pixel. Our graph search abides by this simple and logical constraint. The optimal $C(q)$ and $w(q)$ that globally minimize (2) are again calculated using dynamic programming but now on this new (x, y, w) graph.

---

[1] This is similar in spirit to [19] and [20] where they also optimize for a non-spatial variable: vessel radius or scale, in addition to the spatial coordinates of the segmentation contour.

To further demonstrate the global utility of our proposed contextual parameterization approach, we tested incorporating our adaptive weights into a traditional graph cuts (GC) segmentation framework [22][23], where the segmentation energy is

$$E(f) = \sum_{p,q \in N} V_{p,q}(f_p, f_q) + \sum_{p \in P} D_p(f_p). \tag{9}$$

$f \in \mathcal{L}$ is the labeling for all pixels $p \in P$ where $\mathcal{L}$ is the label space and P is the set of pixels in image I. $V_{p,q}$ is the pairwise interaction penalty between pixel pairs (i.e. the penalty of assigning labels f_p and f_q to pixels p and q), N is the set of interacting pairs of pixels, and D_p measures how well label f_p fits pixel p given the observed data. D_p is calculated as the difference in intensity between pixel p and the mean intensity of seeds within label f_p, and the interaction penalty is calculated as $V_{p,q} = e^{-|\nabla I(x,y)|^2}$ such that low gradient magnitude regions have a high interaction penalty. We modified this standard GC optimization process by replacing $V_{p,q}$ by our proposed spatially adaptive weight in (5).

3   Results and Discussion

We first performed quantitative tests on 16 synthetic images carefully designed to cover extreme shape and appearance variations (two examples are shown in Fig. 1). We created the test data by modeling an object boundary as a sinusoidal function with spatially-varying frequency to simulate varying contour smoothness conditions, and we added spatially-varying (non-stationary) additive white Gaussian noise patterns of increasing variance. We also spatially varied the gradient magnitude of the object boundary across each image by applying Gaussian blurring kernels at different scales in different locations. Computationally, the proposed method required less than 5 minutes for a 768×576 image when run on a Pentium 4 (3.6GHz) machine using MATLAB code.

Our resulting image reliability measure is exemplified in Fig. 1 for two synthetic images with the resulting segmentations shown in Fig. 2. The contour obtained using the globally-optimal weights method (Sec. 2.3) is also shown, along with the contour obtained using a *spatially-fixed* regularization weight, set to the value producing *the smallest* (via brute force search) segmentation error.

We quantitatively examined our method's performance using ANOVA testing on 25 noise realizations of each image in the dataset, where the error was determined by the Hausdorff distance to the ground truth contour. Our method resulted in a mean error (in pixels) of 6.33 ± 1.36, whereas the best fixed-weight method had a mean error of 12.05 ± 1.61, and the globally-optimum weight method had a mean error of 33.06 ± 3.66. Furthermore, for each image, we found our method to be significantly more accurate with all p-values $<< 0.05$.

We also tested our method on clinical MR images of the corpus callosum (CC), which exhibits the known problem of a weak boundary where the CC meets the fornix (Fig. 3(a)). Note how the contour obtained using globally optimal weights exhibits an optimal, yet undesirable, bimodal behavior (either blue

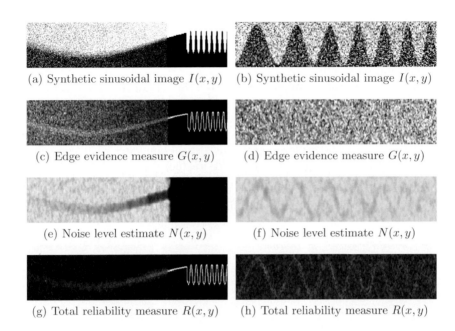

(a) Synthetic sinusoidal image $I(x,y)$ (b) Synthetic sinusoidal image $I(x,y)$

(c) Edge evidence measure $G(x,y)$ (d) Edge evidence measure $G(x,y)$

(e) Noise level estimate $N(x,y)$ (f) Noise level estimate $N(x,y)$

(g) Total reliability measure $R(x,y)$ (h) Total reliability measure $R(x,y)$

Fig. 1. Two sample synthetic images used in our validation tests. The left column image has spatially varying noise and blurring (increasing from right to left) and with changing boundary smoothness (smooth on the left and jagged on the right). The right column image has higher curvature and noise levels. Black intensities corresponds to 0 and white to 1. The result confirms the desired behavior of the reliability measure.

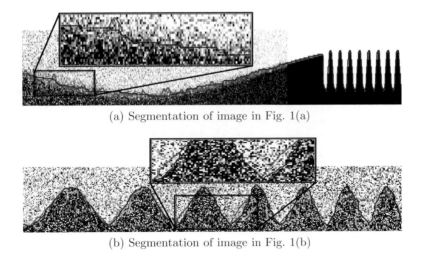

(a) Segmentation of image in Fig. 1(a)

(b) Segmentation of image in Fig. 1(b)

Fig. 2. *Color is essential for proper viewing, please refer to the e-copy.* Contours obtained from: (*blue*) proposed adaptive weights, (*red*) best fixed weight, and (*cyan*) globally optimum weight.

or red in Fig. 3(a)) completely favoring only one of the terms at a time. In comparison, our method automatically boosts up the regularization (stronger red in Fig. 3(b)) at the CC-fornix boundary producing a better delineation, as seen in the segmentation results (Fig. 3(c)).

In addition, we tested our method on MR data from BrainWeb [24]. Fig. 4 shows the segmentation of the cortical surface in a proton density (PD) image with a noise level of 5%. This example is a difficult scenario due to the high level of noise and low resolution of the image. The proposed method provided a smoother contour while conforming to the cortical boundary when compared to the other methods (although the difference was not too large).

To demonstrate the general applicability of our method, we also used natural images in our testing, such as the tree leaf on a complicated background shown in Fig. 5(a). The resulting reliability measure (Fig. 5(b)) has lower image reliability

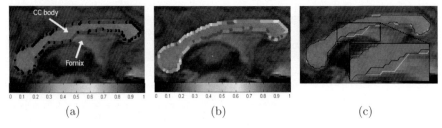

(a)   (b)   (c)

Fig. 3. (*Color figure, refer to e-copy*). Results of (a) globally-optimum weight method and (b) proposed adaptive-weight method for a corpus callosum MR image. The coloring of the contours reflects the value of the spatially-adaptive weight. The same color map is used for both figures, with pure blue corresponding to $w = 0$ and pure red to $w = 1$. The proposed method results in greater regularization in the difficult fornix region and has smoother transition between weights. (c) Contours produced by using the proposed adaptive weight (*blue*), best fixed-weight (*red*), and the globally-optimum weight (*cyan*).

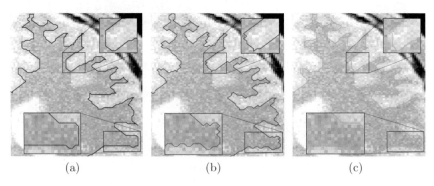

(a)   (b)   (c)

Fig. 4. (*Color figure, refer to e-copy*). Segmentation results on BrainWeb data of cortical surface in a proton density image with noise level of 5%. Contours produced by using (a) the proposed adaptive weight (*blue*), (b) best fixed-weight (*red*), and (c) the globally-optimum weight (*cyan*). Improved regularization resulted from our method.

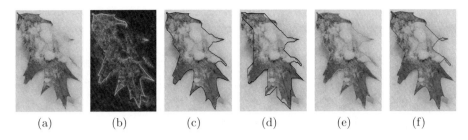

(a)   (b)   (c)   (d)   (e)   (f)

Fig. 5. (*Color figure, refer to e-copy*). Segmenting a natural image. (a) Original leaf image. (b) Reliability calculated by our proposed method. Contours produced by using (c) our method (*blue*), (d) fixed-weight of 1 (*black*), (e) 0.5 (*green*), and (f) 0 (*red*).

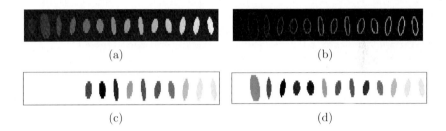

(a)   (b)

(c)   (d)

Fig. 6. (*Color figure, refer to e-copy*). Segmentation of a synthetic image using GC with adaptive regularization. (a) Synthetic image of 14 ellipses with image contrast increasing from left to right. (b) Reliability calculated by our proposed method. (c) Segmentation using standard GC, where each color represents a separate label. (d) Segmentation using adaptive regularization GC.

and, hence, higher regularization at the regions of the leaf obscured by snow, whereas reliable boundaries light up (bright white boundary segments). The resulting segmentations are shown in Figs. 5(c) to 5(f) (note that no best-fixed weight was determined since a true segmentation of the image is not known).

We validated Graph Cuts with our proposed method on simulated noisy images of variably-sized ellipses with complicated background patterns, e.g. with image contrast decreasing from right to left, as in Fig. 6(a). The leftmost ellipses with lower contrast have a lower SNR than the rightmost ellipses and thus require greater regularization. Note how our resulting reliability measure (Fig. 6(b)) indicates lower image reliability for low contrast ellipses. When comparing our segmentation results to those of standard GC (Sec. 2.4), as shown in Fig. 6(c), 6 ellipses out of 14 were mislabeled, whereas GC with adaptive regularization correctly labeled 12 ellipses (Fig. 6(d)). To quantify the advantage of our approach, we tested a synthetic dataset of images containing 2 to 40 ellipses at various noise levels. We calculated the Dice similarity coefficient (DSC) of the segmentation to the ground truth for each ellipse and averaged over all the ellipses in the image. Fig. 7 plots the difference in average DSC between adaptive regularization GC and standard GC for images of increasing ellipse numbers.

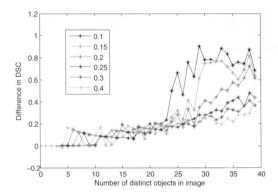

Fig. 7. (*Color figure, refer to e-copy*). Difference in average DSC between adaptive regularization GC and standard GC for images with increasing numbers of ellipses. Different curves represent different noise standard deviations as shown in legend. Positive DSC difference indicates adaptive regularization GC is more successful than standard GC at labeling ellipses with low image quality.

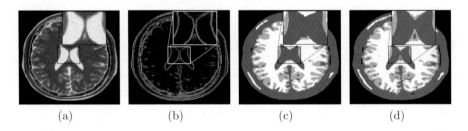

(a)   (b)   (c)   (d)

Fig. 8. Segmentation of MR data from BrainWeb using GC with adaptive regularization. (a) Original T2 slice with 3% noise level and 40% intensity non-uniformity. (b) Reliability calculated by our proposed method. Note increased reliability along the ventricular boundary. (c) Segmentation from standard GC. (d) Segmentation from GC with adaptive regularization.

The same images were also tested at various noise levels. Note that a positive difference in the DSC indicates that our proposed regularization method with GC had greater success detecting low contrast ellipses.

Finally, we tested Graph Cuts with our proposed method on MR data from BrainWeb [24] using a low number of seeds (0.1% of image pixels for each label). Fig. 8(a) shows a T2 image with a noise level of 3% and an intensity non-uniformity of 40%. Increased reliability along the ventricle boundary (Fig. 8(b)) results in lower regularization and greater accuracy in the adaptive regularization GC result (Fig. 8(d)) when compared to the standard GC result (Fig. 8(c)) for a segmentation of the ventricles and cortical surface.

4   Conclusion

We proposed a novel approach for addressing a ubiquitous problem that plagues most energy minimization based segmentation techniques; how to properly balance the weights of competing data fidelity and regularization energy terms. Our technique spatially adapts the regularization weight based on a novel measure of image data reliability. The proposed spectral flatness metric reflects the spatially-varying evidence of signal versus noise and is automatically derived without any tuning. We incorporated our proposed contextual parameterization technique into a general segmentation framework and demonstrated its superiority to non-contextual parameterization even when the latter employed globally optimized values of the objective function parameters. Using quantitative and qualitative tests, we demonstrated that regularization needs to vary spatially and must increase where image evidence is less reliable. Our simple approach is powerful and was shown to be capable of handling varying image data ranging from MRI scans to natural images. Additionally, we demonstrated our approach's general applicability in that it can be extended to state-of-the-art energy-minimization segmentation frameworks such as Graph Cuts.

We are currently extending our approach to other variational and graph-based segmentation approaches such as [5][25]. Additionally, we intend to expand our technique to handle energy functionals where multiple weights balance the different energy terms and to explore alternative image reliability metrics. Another important conclusion of our findings is that globally optimal weights do not necessarily reflect correct segmentations. We intend to further explore this issue and its implications in more detail.

References

1. Samsonov, A.A., Johnson, C.R.: Noise-adaptive nonlinear diffusion filtering of MR images with spatially varying noise levels. Magnetic Resonance in Medicine 52(4), 798–806 (2004)
2. Burnham, K.P., Anderson, D.R.: Multimodel inference: Understanding AIC and BIC in model selection. Sociological Methods & Research 33(2), 291–304 (2004)
3. Akselrod-Ballin, A., Galun, M., Gomori, M.J., Brandt, A., Basri, R.: Prior knowledge driven multiscale segmentation of brain MRI. In: Ayache, N., Ourselin, S., Maeder, A. (eds.) MICCAI 2007, Part II. LNCS, vol. 4792, pp. 118–126. Springer, Heidelberg (2007)
4. Zhao, P., Yu, B.: Stagewise lasso. Journal of Machine Learning Research 8, 2701–2726 (2007)
5. Kass, M., Witkin, A., Terzopoulos, D.: Snakes: Active contour models. International Journal of Computer Vision 1(4), 321–331 (1988)
6. Caselles, V., Kimmel, R., Sapiro, G.: Geodesic active contours. International Journal of Computer Vision 22(1), 61–79 (1997)
7. Osher, S.J., Paragios, N.: Geometric Level Set Methods in Imaging, Vision, and Graphics. Springer, Heidelberg (2003)
8. Pluempitiwiriyawej, C., Moura, J.M.F., Wu, Y.J.L., Ho, C.: STACS: New active contour scheme for cardiac MR image segmentation. IEEE Transactions on Medical Imaging 24(5), 593–603 (2005)

9. Boykov, Y., Funka-Lea, G.: Graph cuts and efficient N-D image segmentation. Int. J. Comput. Vision 70(2), 109–131 (2006)
10. Barrett, W.A., Mortensen, E.N.: Interactive live-wire boundary extraction. Medical Image Analysis 1, 331–341 (1997)
11. McIntosh, C., Hamarneh, G.: Is a single energy functional sufficient? Adaptive energy functionals and automatic initialization. In: Ayache, N., Ourselin, S., Maeder, A. (eds.) MICCAI 2007, Part II. LNCS, vol. 4792, pp. 503–510. Springer, Heidelberg (2007)
12. Udupa, J.K., Grevera, G.J.: Go digital, go fuzzy. Pattern Recognition Letters 23(6), 743–754 (2002)
13. Dong, B., Chien, A., Mao, Y., Ye, J., Osher, S.: Level set based surface capturing in 3D medical images. In: Metaxas, D., Axel, L., Fichtinger, G., Székely, G. (eds.) MICCAI 2008, Part I. LNCS, vol. 5241, pp. 162–169. Springer, Heidelberg (2008)
14. Kokkinos, I., Evangelopoulos, G., Maragos, P.: Texture analysis and segmentation using modulation features, generative models, and weighted curve evolution. IEEE Trans. Pattern Analysis and Machine Intelligence 31(1), 142–157 (2009)
15. Malik, J., Belongie, S., Leung, T.K., Shi, J.: Contour and texture analysis for image segmentation. International Journal of Computer Vision 43(1), 7–27 (2001)
16. Erdem, E., Tari, S.: Mumford-Shah regularizer with contextual feedback. Journal of Mathematical Imaging and Vision 33(1), 67–84 (2009)
17. Jayant, N.S., Noll, P.: Digital Coding of Waveforms. Prentice-Hall, Englewood Cliffs (1984)
18. Taubman, D.S., Marcellin, M.W.: JPEG 2000: Image Compression Fundamentals, Standards and Practice. Kluwer Academic Publishers, Norwell (2001)
19. Poon, K., Hamarneh, G., Abugharbieh, R.: Live-vessel: Extending livewire for simultaneous extraction of optimal medial and boundary paths in vascular images. In: Ayache, N., Ourselin, S., Maeder, A. (eds.) MICCAI 2007, Part II. LNCS, vol. 4792, pp. 444–451. Springer, Heidelberg (2007)
20. Li, H., Yezzi, A.: Vessels as 4-D curves: Global minimal 4-D paths to extract 3-D tubular surfaces and centerlines. IEEE Transactions on Medical Imaging 26(9), 1213–1223 (2007)
21. Cohen, L.D., Kimmel, R.: Global minimum for active contour models: A minimal path approach. International Journal of Computer Vision 24(1), 57–78 (1997)
22. Bagon, S.: Matlab wrapper for graph cut (December 2006)
23. Boykov, Y., Veksler, O., Zabih, R.: Fast approximate energy minimization via graph cuts. IEEE transactions on Pattern Analysis and Machine Intelligence 20(12), 1222–1239 (2001)
24. Cocosco, C.A., Kollokian, V., Kwan, R.K.S., Evans, A.C.: BrainWeb: Online interface to a 3D MRI simulated brain database. In: Friberg, L., Gjedde, A., Holm, S., Lassen, N.A., Nowak, M. (eds.) Third International Conference on Functional Mapping of the Human Brain, NeuroImage, vol. 5. Academic Press, London (1997)
25. Chan, T.F., Vese, L.A.: Active contours without edges. IEEE Transactions on Image Processing 10(2), 266–277 (2001)

Approximated Curvature Penalty in Non-rigid Registration Using Pairwise MRFs

Ben Glocker[1,2,*], Nikos Komodakis[3], Nikos Paragios[2,4], and Nassir Navab[1]

[1] Computer Aided Medical Procedures (CAMP), TU München, Germany
[2] Laboratoire MAS, Ecole Centrale Paris, Chatenay-Malabry, France
[3] Computer Science Department, University of Crete, Greece
[4] Equipe GALEN, INRIA Saclay - Ile-de-France, Orsay, France
glocker@in.tum.de

Abstract. Labeling of discrete Markov Random Fields (MRFs) has become an attractive approach for solving the problem of non-rigid image registration. Here, regularization plays an important role in order to obtain smooth deformations for the inherent ill-posed problem. Smoothness is achieved by penalizing the derivatives of the displacement field. However, efficient optimization strategies (based on iterative graph-cuts) are only available for second-order MRFs which contain cliques of size up to two. Higher-order cliques require graph modifications and insertion of auxiliary nodes, while pairwise interactions actually allow only regularization based on the first-order derivatives. In this paper, we propose an approximated curvature penalty using second-order derivatives defined on the MRF pairwise potentials. In our experiments, we demonstrate that our approximated term has similar properties as higher-order approaches (invariance to linear transformations), while the computational efficiency of pairwise models is preserved.

1 Introduction

Non-rigid image registration is an important problem in computer vision and medical imaging. Given two images I and J, one seeks a transformation T which aligns the corresponding objects visible in the images. This is commonly solved by posing an energy minimization problem where the objective function is a sum of a matching criteria S and a regularization term R,

$$\hat{T} = \arg\min_{T} S(I, J \circ T) + \alpha R(T) \ . \tag{1}$$

Here, α is a weighting factor controlling the influence of the regularization term. In the case of non-rigid registration, the transformation is often defined as the identity transformation plus a dense displacement field D. The new location of an image point $\mathbf{x}$ is then computed by

$$T(\mathbf{x}) = \mathbf{x} + D(\mathbf{x}) \ . \tag{2}$$

---

* This work is partially supported by Siemens Healthcare, Erlangen, Germany.

G. Bebis et al. (Eds.): ISVC 2009, Part I, LNCS 5875, pp. 1101–1109, 2009.

Regularization plays an important role due to the inherent ill-posedness of the problem [1]. A natural approach for regularization is to penalize the derivatives of the displacement field. Smoothness terms based on the first-order derivatives penalize high gradients and thus, piecewise constant deformations are favored. Such smoothness models require a proper pre-alignment by linear registration prior to the non-linear one, since penalizing the gradients is only invariant to global translation. If still some linear transformation (such as rotation or scaling) is present, penalizing the gradients might prohibit a proper non-rigid alignment. Since a perfect linear alignment is not trivial to achieve when deformations are present, one can consider to define a penalty term based on the second-order derivatives [2,3]. Such a term penalizes high curvature in the displacement field, is invariant to linear transformations, and thus, favors deformations which are piecewise linear.

Recently, labeling of discrete Markov Random Fields has become an attractive approach for solving the problem of non-rigid image registration [4,5,6]. We will give a short introduction into the general framework in Section 2. Most of the methods share a similar model for the registration which is based on a pairwise MRF, i.e. is an MRF with cliques of size up to two. Then, the unary terms[1] (cliques of size one) play the role of the matching criteria, while pairwise terms are used to encode the regularization of the displacement field. In [6], the regularization is based on the norm of the displacement vector differences between neighboring control points which is an approximation of penalizing the gradients of the displacement field. The assumption is that neighboring nodes should follow a similar motion. A more robust measure is used in [4,5] which allows more freedom on the deformation. However, this measure is still based on the gradient approximation. As remarked in [7], all these approaches penalize linear transformations such as rotation and scaling which in practice is often not desired. To this end, in [7] a regularization term based on the second-order derivatives is introduced by adding triple cliques of collinear neighboring control points to the MRF model. Each triple clique is in charge of penalizing the local curvature of the displacement field. The main problem of this approach are in fact the triple cliques, which require complex graph modifications in order to use efficient optimization techniques based on message-passing. In [7], the third-order MRF is converted to a pairwise one and then the TRW-S algorithm [8] is utilized to infer the MRF variables. Unfortunately, no running time is provided, but it is assumed to be much higher [9] than for the method proposed in [6] which uses the FastPD algorithm [10,11] (based on iterative graph-cuts).

One may ask if it is possible to define a regularization term which has similar properties as the curvature penalty based on triple cliques while keeping the efficiency of a pairwise model.

In this paper, we investigate the use of an approximated curvature penalty term in a pairwise MRF. Our experiments demonstrate the practicability of such a regularization for non-rigid registration when the optimal transformation

---

[1] Please note, that in [4,5] a decomposition of the unary terms is used but the general model is similar to [6].

contains a linear part. Compared to prior work, no higher-order cliques have to be employed for our curvature term and thus, our approach is efficient in terms of computational speed.

The remainder of the paper is organized as follows: the general framework for non-rigid registration using MRFs is described in Section 2. The proposed approximated curvature penalty is introduced in Section 3. Section 4 demonstrates the practicability of our regularization through a set of experiments, while Section 5 concludes our paper.

2   Non-rigid Registration Using MRFs

Markov Random Field inference is a popular approach for parameter estimation. Given a set of parameters, one can define a graph $\mathcal{G} = (\mathcal{V}, \mathcal{C})$ consisting of a set of nodes $\mathcal{V}$ (one node per parameter) and a set of cliques $\mathcal{C}$ (where each clique is a subset of $\mathcal{V}$). Assuming that each node i takes a label l_i from a discrete set $\mathcal{L}$, the task becomes to find the optimal *labeling* $\mathbf{l}$ which minimizes

$$E_{\mathrm{mrf}}(\mathbf{l}) = \sum_{c \in \mathcal{C}} \psi_c(\mathbf{l}_c) \ , \tag{3}$$

which is a sum of *clique potentials* ψ_c determining the costs of certain label assignments and $\mathbf{l}_c$ is the vector of labels assigned to the parameter subset c.

The most common MRF model used in computer vision tasks (e.g. segmentation) is the second-order (pairwise) model containing at most cliques of size two. Many efficient algorithms have been proposed [12,8,10,13] to solve the inference problem for this special case. For the second-order MRF the energy becomes the sum of unary and pairwise potentials

$$E_{\mathrm{mrf}}(\mathbf{l}) = \sum_{i \in \mathcal{G}} \psi_i(l_i) + \sum_{i \in \mathcal{G}} \sum_{j \in \mathcal{N}_i} \psi_{ij}(l_i, l_j) \ , \tag{4}$$

where $\mathcal{N}_i \subset \mathcal{G}$ defines the neighborhood system of the graph.

In case of non-rigid registration, the MRF variables correspond to locations in the image domain at which we want to estimate the motion (i.e. a displacement vector). Each discrete label is mapped to a displacement from a discretized version of the search space. For simplicity, we will denote the displacement vector associated with label l as $\mathbf{d}^l$. A simple approach would be to introduce an MRF variable for each pixel [14]. Then, the unary terms play the role of the data term or matching cost. Exemplary, we can define the costs for the sum of absolute differences (SAD) criteria based on image intensities as

$$\psi_i(l_i) = |(I(\mathbf{x}_i) - J(\mathbf{x}_i + D^{t-1}(\mathbf{x}_i) + \mathbf{d}^{l_i})| \ , \tag{5}$$

where D^{t-1} is the dense field from the previous iteration and $\mathbf{d}^{l_i}$ the potential displacement corresponding to label l_i. The pairwise terms encode the regularization. A simple smoothness term penalizing high gradients can be defined as

$$\psi_{ij}(l_i, l_j) = \alpha \, \|(D^{t-1}(\mathbf{x}_i) + \mathbf{d}^{l_i}) - (D^{t-1}(\mathbf{x}_j) + \mathbf{d}^{l_j})\| \ . \tag{6}$$

The optimization problem for dense registration in (1) is now completely defined as a *discrete labeling* of an MRF. The main problem for such an approach is the number of variables. One variable per pixel becomes computationally very expensive and is not feasible for large volumes in case of 3D registration. To this end, we can reduce the dimensionality of the problem by introducing a transformation model based on a sparse set of control points and an interpolation strategy. The dense displacement field in (2) is then defined as

$$D(\mathbf{x}) = \sum_{i}^{M} \eta_i(\mathbf{x})\,\mathbf{d}_i \quad , \tag{7}$$

where M is the number of control points and η_i is a weighting function (e.g. based on cubic B-splines) determining the contribution of the control point displacement $\mathbf{d}_i$ to the displacement of an image point $\mathbf{x}$. In this paper, we consider free form deformations (FFDs) [15] as the transformation model, where the control points are defined on a regular lattice and each control point has only local influence on the deformation. Let us now reformulate the matching cost (5) w.r.t. to the control points

$$\psi_i(l_i) = \sum_{\mathbf{x}\in\Omega_i} |I(\mathbf{x}) - J(\mathbf{x} + D^{t-1}(\mathbf{x}) + \mathbf{d}^{l_i})| \quad , \tag{8}$$

where Ω_i is a local image patch centered at the control point i. Intuitively, (8) can be understood as a block matching cost where the whole block Ω_i is potentially moved by $\mathbf{d}^{l_i}$. The size of the blocks is automatically defined by the distance between control points of the deformation grid. Additionally, [6] proposes a *weighted block matching* by incorporating the weighting functions η_i into the matching cost. The idea is that the influence of an image point to the matching criteria of a control point should be proportional to the contribution of that control point to the displacement of the image point. In other words, image points far away from a control point should have less influence on its cost than points in the immediate vicinity. Besides the reduction of the number of MRF variables, the block matching has additional advantages. For instance, it is straightforward to encode more sophisticated matching criteria such as correlation or mutual information which often provide more reliable matches than intensity differences. A comparison of different measures can be found in [6]. The regularization term (6) is similar as before, but now evaluated only on the control points instead of all image points

$$\psi_{ij}(l_i, l_j) = \alpha \, \|(\mathbf{d}_i^{t-1} + \mathbf{d}^{l_i}) - (\mathbf{d}_j^{t-1} + \mathbf{d}^{l_j})\| \quad , \tag{9}$$

where $\mathbf{d}_i^{t-1}$ is the displacement of control point i from the previous iteration. The final pairwise MRF energy for the non-rigid registration in (1) based on a deformation grid is then defined as

$$E_{\mathrm{mrf}}(\mathbf{l}) = \sum_i^M \psi_i(l_i) + \sum_i^M \sum_{j \in \mathcal{N}_i} \psi_{ij}$$

$$= \underbrace{\sum_i^M \sum_{\mathbf{x} \in \Omega_i} |I(\mathbf{x}) - J(\mathbf{x} + D^{t-1}(\mathbf{x}) + \mathbf{d}^{l_i})|}_{\approx S(I, J \circ T)} \tag{10}$$

$$+ \underbrace{\sum_i^M \sum_{j \in \mathcal{N}_i} \alpha \, \|(\mathbf{d}_i^{t-1} + \mathbf{d}^{l_i}) - (\mathbf{d}_j^{t-1} + \mathbf{d}^{l_j})\|}_{\approx \alpha \, R(T)}$$

3   Approximated Curvature Penalty

The main limitation of the registration framework based on pairwise MRFs are the constraints for regularization. The second-order cliques can only model interactions between two variables. The smoothness terms proposed so far, which all penalize high gradients on the displacement field, have the disadvantage of not being invariant to linear transformations such as rotation and scaling. Therefore, in [7] regularization is employed by introducing triple cliques which are able to encode a smoothness prior based on the discrete approximation of the second-order derivatives. The potential functions can be defined as

$$\psi_{ijk}(l_i, l_j, l_k) = c(\mathbf{d}_i^{t-1} + \mathbf{d}^{l_i}, \, \mathbf{d}_j^{t-1} + \mathbf{d}^{l_j}, \, \mathbf{d}_k^{t-1} + \mathbf{d}^{l_k}) \;, \tag{11}$$

$$c(\mathbf{a}, \mathbf{b}, \mathbf{c}) = \frac{1}{\delta^2} \sum_d^n (-a_d + 2b_d - c_d)^2 \;, \tag{12}$$

where c approximates the local curvature at location $\mathbf{b}$, b_d denotes the d-th component of the n-dimensional vector space, and δ is the control point distance. Such a smoothness term is invariant to linear transformations. The drawback of this approach is the complex handling of triple cliques. Graph modifications and insertion of auxiliary nodes are necessary in order to use efficient message-passing optimization techniques [16,7]. The performance of message-passing algorithms in terms of computational speed is much lower than methods based on iterative graph-cuts [9,11].

Therefore, we propose a regularization term based on second-order derivatives which works on pairwise potential functions and which we call approximated curvature penalty (ACP). Since in pairwise terms only the potential label assignment of two variables is known, we approximate the local curvature by assuming the other variables to stay fixed. In detail, for two neighboring variables i and j, we compute the approximated curvature at both locations and average them. To this end, we define different pairwise potentials depending on the axis on which the two variables are neighboring. In 2D, we have a set of potential functions for the horizontal and vertical axis

$$\psi_{i,j}^H(l_i, l_j) = \frac{1}{2} \left(c(\mathbf{d}_{i-1}^{t-1}, \mathbf{d}_i^{t-1} + \mathbf{d}^{l_i}, \mathbf{d}_j^{t-1} + \mathbf{d}^{l_j}) \right.$$
$$\left. + c(\mathbf{d}_i^{t-1} + \mathbf{d}^{l_i}, \mathbf{d}_j^{t-1} + \mathbf{d}^{l_j}, \mathbf{d}_{j+1}^{t-1}) \right) \;,$$
$$\psi_{i,j}^V(l_i, l_j) = \frac{1}{2} \left(c(\mathbf{d}_{i-M_x}^{t-1}, \mathbf{d}_i^{t-1} + \mathbf{d}^{l_i}, \mathbf{d}_j^{t-1} + \mathbf{d}^{l_j}) \right.$$
$$\left. + c(\mathbf{d}_i^{t-1} + \mathbf{d}^{l_i}, \mathbf{d}_j^{t-1} + \mathbf{d}^{l_j}, \mathbf{d}_{j+M_x}^{t-1}) \right) \;,$$

(13)

where M_x is the number of control points on the deformation grid in horizontal direction. The definition of an additional term for the third axis in 3D registration is straightforward. In each evaluation of the ACP we determine the average of the two local curvatures by considering the displacements of four variables: i and j which are the variables with potential movement $\mathbf{d}^{l_i}$ and $\mathbf{d}^{l_j}$ and the two surrounding variables with the displacements from the previous iteration.

Considering the properties of such a smoothness prior, we claim that it allows much more flexibility on the deformation w.r.t. to linear transformations compared to other terms based on pairwise potentials. In the beginning of every registration process, the matching criteria is usually the driving force towards the correct alignment, while the regularization increases its importance on the global energy in later iterations. The incremental deformation at the end of the process is getting smaller and the ACP will favor deformations which are piecewise linear instead of piecewise constant as for the gradient penalty. The practicability of our proposed regularization is demonstrated in the following experiments.

4   Experiments

We perform several experiments which hopefully illustrate the advantages of the proposed ACP as regularization. Throughout the tests we use the FastPD algorithm for MRF inference. In the first two experiments, we generate synthetic target images by warping a source image with different linear transformations. The first one is a 60° rotation and the second one an anisotropic scaling (see Fig. 1(a) and 1(e)). For each target, four point correspondences are distributed around the image center which exactly define these linear transformations. Please note, that in these two experiments our aim is to investigate the properties of the regularization only. Therefore, we choose a geometric matching term based on perfect point correspondences. Thus, the data term is reliable and will guide the registration towards an optimal alignment of the correspondences which allows us to study the behavior of the regularization. We use a simple Euclidean distance measure as the matching criteria w.r.t the correspondences. The unary potentials are therefore defined as

$$\psi_i(l_i) = \sum_k^K \| \mathbf{p}_k - (\mathbf{q}_k + D^{t-1}(\mathbf{q}_k) + \mathbf{d}^{l_i}) \| \;,$$

(14)

where $\mathbf{p}_k$ and $\mathbf{q}_k$ are the corresponding points in the target and source image, respectively. Now we compare the behavior of three different regularization terms,

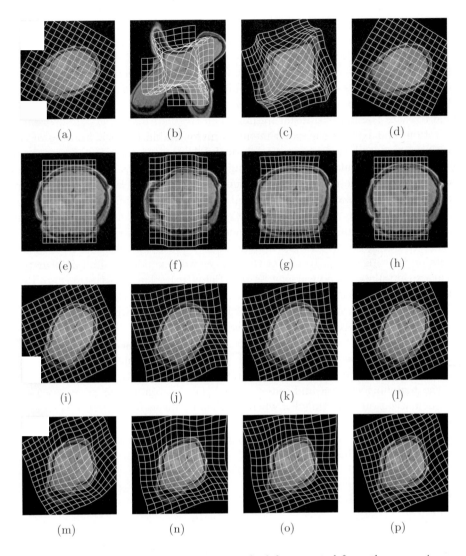

Fig. 1. First two rows: Target images on the left generated from the source image by a 60° rotation (a) and anisotropic scaling (e). The registration results are shown for the absolute difference (b,f), quadratic difference (c,g), and the approximated curvature penalty (d,h). The Euclidean distance on four point correspondences is used to define the registration data term. **Last two rows:** Target images on the left generated by a 25° rotation (i). For the last row, additional random deformation is added (m). From left to right the registration results (j-l,n-p) in the same order as shown in the upper rows. This time, no information about point correspondences is used and the registration is purely based on image intensities. Please note, that we use backward warping why the actual transformations visualized as grids appear to point in the opposite direction as the warped images.

namely the absolute vector difference (cp. (9)), the quadratic vector difference (i.e. the squared version of (9)), and our ACP defined in (13). For all registrations, we use a 7×7 deformation grid. The results for the registration with the different regularization terms are shown in Fig. 1(b-d) and (f-h). We should note, that in all cases the final mean distance for the correspondences is less than one pixel, which indicates a very good minimization of the matching costs. However, the ACP is the only method which is able to correctly regularize the deformation field towards the linear transformations.

The second part of the experiments is investigating the performance of the ACP in intensity-based registration. In fact, large rotations such as $60°$ are very unlikely to be present if a proper pre-alignment via rigid registration has been performed prior to the non-rigid one. Additionally, a block matching strategy which is mainly based on a translational search most likely fails to recover large rotations or scaling. However, in practice it is likely that a certain amount of linear transformation is still present when starting the non-rigid alignment [17]. Again, we generate two target images from a source image, both with a $25°$ rotation (cp. Fig. 1(i)). To one of the images we also add random deformation using a thin-plate spline warping [2] (cp. Fig. 1(m)). The registrations are then performed purely based on intensities using the matching criteria defined in (8). The results for the different regularization terms are shown in Fig. 1(j-l) and (n-p). Again, the ACP outperforms the gradient penalty terms in the ability of regularization towards linear transformations. In the last case of random deformation combined with rotation, the resulting transformation using ACP is very close to the ground truth. This is remarkable since only intensities are used in the matching criteria and outer control points obtain their positions solely by regularization. Additionally, when we visually inspect the warped images after registration the results for the ACP are almost perfect, while the gradient penalty terms prohibit a proper alignment due to the increasing costs for linear transformations. This is consistent with the observations in [7].

5   Conclusion

We propose a novel regularization term based on an approximated curvature penalty for pairwise MRFs. Our results demonstrate the superior performance of this approach compared to previous smoothness terms based on gradient penalties. Our regularization can successfully recover linear transformations and thus, has similar properties as a curvature penalty using triple cliques, while the computational efficiency of a pairwise MRFs is preserved. In fact, the running time using ACP increases only very little compared to the gradient penalty terms. All shown registrations are performed within a few seconds. We believe that the proposed regularization is an important extension to the MRF registration framework. Furthermore, we could show that introducing approximated terms in pairwise MRFs can lead to very promising results. Future work includes the comparison to recent advances in optimization of higher-order MRFs [18].

References

1. Tikhonov, A.: Ill-posed problems in natural sciences (1992); Coronet
2. Bookstein, F.L.: Principal warps: thin-plate splines and the decomposition of deformations. In: Pattern Analysis and Machine Intelligence (1989)
3. Modersitzki, J.: Numerical Methods for Image Registration. Oxford University Press, Oxford (2004)
4. Lee, K.J., Kwon, D., Yun, I.D., Lee, S.U.: Deformable 3d volume registration using efficient MRFS model with decomposed nodes. In: British Machine Vision Conference (2008)
5. Shekhovtsov, A., Kovtun, I., Hlavac, V.: Efficient mrf deformation model for nonrigid image matching. Comput. Vis. Image Underst. 112, 91–99 (2008)
6. Glocker, B., Komodakis, N., Tziritas, G., Navab, N., Paragios, N.: Dense image registration through MRFS and efficient linear programming. Medical Image Analysis 12, 731–741 (2008)
7. Kwon, D., Lee, K.J., Yun, I.D., Lee, S.U.: Nonrigid image registration using dynamic higher-order MRF model. In: European Conference on Computer Vision (2008)
8. Kolmogorov, V.: Convergent tree-reweighted message passing for energy minimization. Pattern Analysis and Machine Intelligence 28, 1568–1583 (2006)
9. Szeliski, R., Zabih, R., Scharstein, D., Veksler, O., Kolmogorov, V., Agarwala, A., Tappen, M., Rother, C.: A comparative study of energy minimization methods for markov random fields with smoothness-based priors. Pattern Analysis and Machine Intelligence 30, 1068–1080 (2008)
10. Komodakis, N., Tziritas, G., Paragios, N.: Fast, approximately optimal solutions for single and dynamic MRFS. Computer Vision and Pattern Recognition (2007)
11. Komodakis, N., Tziritas, G., Paragios, N.: Performance vs computational efficiency for optimizing single and dynamic MRFS: Setting the state of the art with primal-dual strategies. Comput. Vis. Image Underst. 112, 14–29 (2008)
12. Boykov, Y., Veksler, O., Zabih, R.: Fast approximate energy minimization via graph cuts. Pattern Analysis and Machine Intelligence 23, 1222–1239 (2001)
13. Lempitsky, V., Rother, C., Blake, A.: Logcut - efficient graph cut optimization for markov random fields. In: International Conference on Computer Vision (2007)
14. Boykov, Y., Veksler, O.: Graph Cuts in Vision and Graphics: Theories and Applications. In: Handbook of Mathematical Models in Computer Vision, pp. 79–96. Springer, Heidelberg (2005)
15. Sederberg, T.W., Parry, S.R.: Free-form deformation of solid geometric models. In: SIGGRAPH 1986: Proceedings of the 13th annual conference on Computer graphics and interactive techniques. ACM Press, New York (1986)
16. Potetz, B., Lee, T.S.: Efficient belief propagation for higher-order cliques using linear constraint nodes. Comput. Vis. Image Underst. 112, 39–54 (2008)
17. Zikic, D., Hansen, M.S., Glocker, B., Khamene, A., Larsen, R., Navab, N.: Computing minimal deformations: Application to construction of statistical shape models. In: Computer Vision and Pattern Recognition (2008)
18. Komodakis, N., Paragios, N.: Beyond pairwise energies: Efficient optimization for higher-order mrfs. In: Computer Vision and Pattern Recognition (2009)

Author Index

Printing: Mercedes-Druck, Berlin
Binding: Stein+Lehmann, Berlin